March 5–8, 2012
Boston, Massachusetts, USA

Association for Computing Machinery

Advancing Computing as a Science & Profession

HRI'12

Proceedings of the Seventh Annual ACM/IEEE International Conference on

Human-Robot Interaction

Sponsored by:

ACM SIGCHI and SIGART

Supported by:

IEEE RAS, AAAI, HSEF and IEEE SMC

Association for
Computing Machinery

Advancing Computing as a Science & Profession

The Association for Computing Machinery
2 Penn Plaza, Suite 701
New York, New York 10121-0701

ISBN: 978-1-4503-1063-5 (Digital)

ISBN: 978-1-4503-1305-6 (Print)

Additional copies may be ordered prepaid from:

ACM Order Department
PO Box 30777
New York, NY 10087-0777, USA

Phone: 1-800-342-6626 (USA and Canada)
+1-212-626-0500 (Global)
Fax: +1-212-944-1318
E-mail: acmhelp@acm.org
Hours of Operation: 8:30 am – 4:30 pm ET

Printed in the USA

HRI'12 Chairs' Welcome

Welcome to Boston! The Seventh Annual ACM/IEEE International Conference on Human-Robot Interaction (HRI 2012) is a highly selective conference that aims to showcase the very best interdisciplinary and multidisciplinary research in human-robot interaction with roots in robotics, social psychology, cognitive science, HCI, human factors, artificial intelligence, engineering, and many more. We invite broad participation and encourage discussion and sharing of ideas across a diverse audience.

Robotics is growing increasingly multidisciplinary as it moves towards realizing capable and collaborative robots that meet both human needs of society and technical challenges inherent in real world settings. A joining of the disciplines is essential for enabling robots to help people in their efforts to be more productive and enjoy a high quality of life. In particular, human-robot interaction requires advancement of the state-of-the-art in the empirical, algorithmic, mathematical, social, and engineering aspects of robotics in an integrated manner. Therefore, this year's theme is dedicated to **Robots in the Loop**, which highlights the importance of autonomously capable robots in enhancing the experiences of human users in everyday life and work activities. HRI 2012 emphasizes embodied robotic systems that operate, collaborate with, learn from, and meet the needs of human users in real-world environments.

Full Papers submitted to the conference were thoroughly reviewed and discussed. The process utilized a rebuttal process and a worldwide team of dedicated, interdisciplinary reviewers. Subtle changes were made this year to help better pair reviewers to papers. This year's conference continues the tradition of selectivity with 34 out of 137 (25%) submissions accepted. Due to the joint sponsorship of ACM and IEEE, papers are archived in both the ACM Digital Library and IEEE Xplore.

Accompanying the full papers are the brief and lightly reviewed Late Breaking Reports and Videos. For the former, 95 out of 111 (86%) two-page papers were accepted and will be presented as posters at the conference. For the latter, 16 of 30 (52%) short videos were accepted and will be presented during the video session.

Rounding out the program are multiple keynote speakers who will discuss topics relevant to HRI, a panel session on telepresence, and several invited short unpublished talks designed to expose the audience to interesting work and motivate interdisciplinary discussion. The keynote speakers this year are Rodney Brooks and Karl Grammer.

The conference could not have occurred without the extensive volunteer effort put forth by the organizing committee, program committee, and reviewers. We would also like to thank the keynote speakers, panelists, and invited speakers for their participation and attendance. The sponsors of the conference are ACM SIGCHI, ACM SIGART, and IEEE Robotics and Automation. The conference is in cooperation with AAAI and HFES.

Finally, we are especially thankful for the hard work by authors who submitted papers and videos. HRI is a vibrant community, and there was a large volume of high quality submissions. This makes reviewing difficult, but ensures a high quality conference. We hope you will be inspired by this high quality content and enjoy your stay in Boston.

Holly Yanco
HRI'12 General Co-Chair
University of Massachusetts Lowell, USA

Vanessa Evers
HRI'12 Program Co-Chair
University of Twente, Netherlands

Aaron Steinfeld
HRI'12 General Co-Chair
Carnegie Mellon University, USA

Odest Chadwicke Jenkins
HRI'12 Program Co-Chair
Brown University, USA

HRI 2012 Table of Contents

HRI 2012 Table of Contents

HRI 2012 Table of Contents

HRI 2012 Table of Contents

HRI 2012 Table of Contents

HRI 2012 Table of Contents

HRI 2012 Table of Contents

HRI 2012 Table of Contents

HRI 2012 Conference Organization

General Co-Chairs: Holly Yanco *(University of Massachusetts Lowell, USA)*
Aaron Steinfeld *(Carnegie Mellon University, USA)*

Program Co-Chairs: Vanessa Evers *(University of Amsterdam, The Netherlands)*
Odest Chadwicke Jenkins *(Brown University, USA)*

Workshop and Tutorials Co-Chairs: Jill Drury *(The MITRE Corporation, USA)*
Angelika Peer *(Technischen Universität München, Germany)*

Late-breaking Co-Chairs: Frank Broz *(University of Hertfordshire, UK)*
Astrid Weiss *(University Salzburg, Austria)*

Video Session Co-Chairs: Martin Saerbeck *(Institute of High Performance Computing, Singapore)*
Wendy Ju *(Stanford University/California College of the Arts, USA)*

Panel Co-Chairs: Kate Tsui *(University of Massachusetts Lowell, USA)*
Stephen von Rump *(Giraff Technologies AB, Sweden)*

Publications Co-Chairs: Jeonghye Han *(Cheongju National University of Education, S. Korea)*
Adriana Tapus *(ENSTA-ParisTech, France)*
Jacob Crandall *(Masdar Institute, United Arab Emirates)*

Finance Co-Chairs: Leila Takayama *(Willow Garage, USA)*
Sylvain Calinon *(Italian Institute of Technology, Italy)*

Fundraising & Exhibitions Co-Chairs: Takashi Minato *(ATR, Japan)*
Jennifer Burke *(Boeing, USA)*

Design Chair: Adam Norton *(University of Massachusetts Lowell, USA)*

Web Chair: Irene Rae *(University of Wisconsin–Madison, USA)*

Publicity Co-Chairs: Hideaki Kuzuoka *(University of Tsukuba, Japan)*
Aude Billard *(EPFL, Switzerland)*

Registration Co-Chairs: Selma Sabanovic *(Indiana University, USA)*
Kate Tsui *(University of Massachusetts Lowell, USA)*

Local Co-Chairs: Candy Sidner *(Worcester Polytechnic Institute, USA)*
Sonia Chernova *(Worcester Polytechnic Institute, USA)*

Steering Committee Co-Chairs: Julie A. Adams *(Vanderbilt University, USA)*
Takayuki Kanda *(ATR, Japan)*

Additional reviewers:

Adams, Julie
Adelstein, Bernard
Admoni, Henny
Albo-Canals, Jordi
Ali, Muhammad
Alissandrakis, Aris
Aly, Amir
Amanda, Sharkey
Ammi, Mehdi
Andersen, Hans Jorgen
Anderson, Monica
Anzalone, Salvatore Maria
Argentieri, Sylvain
Asfour, Tamim
Atrash, Amin
Azhar, M.
Becker-Asano, Christian
Bellotto, Nicola
Belpaeme, Tony
Benjamin, Paul
Bennewitz, Maren
Berlin, Matt
Bethel, Cindy
Bilek, Emine
Birlo, Manuel
Blisard, Samuel
Bohus, Dan
Bouyssounouse, Xavier
Briere, Simon
Broz, Frank
Bugajska, Magdalena
Busquets, Didac
Curuklu, Baran
Cakmak, Maya
Calhoun, Gloria
Calinon, Sylvain
Carignan, Craig
Carlson, Tom
Castellano, Ginevra
Chan, Wesley
Chao, Crystal
Chen, Tiffany
Chen, Jessie
Chetouani, Mohamed
Christensen, Henrik
Clancey, Bill
Colledani, Frederic
Colton, Mark
Cooke, Nancy
Cramer, Henriette
Crandall, Jacob
Craven, Patrick
Crick, Christopher
Danes, Patrick
De Silva, Ravindra
Denecke, Matthias
Deng, Zhigang
dePalma, Nick
Desai, Munjal
Deshmukh, Amol
Dias, Jorge
DiSalvo, Carl
Duh, Henry Been-Lirn
Fasel, Ian
Feil-Seifer, David
Ferrane, Isabelle
Ferworn, Alexander
Filliat, David
Fincannon, Thomas
Forlizzi, Jodi
Freier, Nathan
Funk, Mathias
Furuyama, Nobuhiro
Ge, Shuzhi Sam
Gepperth, Alexander
Gibbons, Peter
Gielniak, Michael
Glas, Dylan
Green, Keith
Gribovskaya, Elena
Griffiths, Sascha
Grollman, Daniel
Haddadin, Sami
Hafner, Verena
Ham, Jaap
Hansen, Soeren Tranberg
Harrison, Anthony
Hart, Justin
Hayes, Brad
Hegel, Frank
Helton, Deak
Herath, Damith C
Hiatt, Laura
Hirahara, Tatsuya
Hodgins, Jessica
Hoffman, Guy
Howard, Ayanna
Hu, Jun

Additional reviewers (continued):

Humphrey, Curtis
Ince, Gokhan
Inamura, Tetsunari
Iqbal, Azlan
Ishii, Kentaro
Iwaki, Satoshi
Johnson, David
Jung, Malte
Kalar, Don
Kanda, Takayuki
Kaye, Jofish
Kennedy, William
Kidd, Cory
Kim, Hyoung-Rock
Kim, Chyon Hae
Kirchner, Nathan
Kirsch, Alexandra
Koay, Kheng Lee
Koenig, Nathan
Kollar, Thomas
Komatsu, Takanori
Kronreif, Gernot
Kruijff, Ivana
Kulic, Dana
Kuno, Yoshinori
Kwak, Jun-young
Kwak, Sonya S.
Kwon, Jinsung
Lammer, Lara
Lee, Jinhan
Lee, Min Kyung
Lee, Sau-lai
Lerasle, Frederic
Levin, Daniel
Lewis, Michael
Leyzberg, Dan
Li, Huakang
Lignos, Constantine
Lohan, Katrin Solveig
Lohse, Manja
Lopes, Manuel
Lutkebohle, Ingo
Lu, David
Luber, Matthias
Ma, Xiaojuan
Muller, Andreas
MacDorman, Karl F.
Mansard, Nicolas
Marge, Matthew
Marin, Luis
Martens, Andy
Martison, Eric
Matsuo, Kazuya
Mavridis, Nikolaos
McDermott, Patricia
Mead, Ross
Meerbeek, Bernt
Meisner, Eric
Mekdeci, Brian
Menegatti, Emanuele
Messeri, Lisa
Messina, Elena
Michaud, Francois
Micire, Mark
Mohammad, Yasser
Moshkina, Lilia
Moussa, Medhat
Mubin, Omar
Murray, John
Nakadai, Kazuhiro
Nakanishi, Hideyuki
Nakano, Yukiko
Neuhaus, Peter
Nguyen, Nhung
Niekum, Scott
Nomura, Tatsuya
Ogata, Tetsuya
Ono, Tetsuo
Osawa, Hirotaka
Osentoski, Sarah
Pais, Ana Lucia
Palmer, Martha
Pantofaru, Caroline
Park, Eunil
Park, Chung Hyuk
Parra-Vega, Vicente
Pistillo, Antonio
Pitsch, Karola
Potter, Mitchell A.
Provost Mower, Emily
Rae, Irene
Raj, Anil
Randelli, Gabriele
Rich, Charles
Riek, Laurel D.
Rios-Martinez, Jorge
Robins, Ben
Rosenthal, Stephanie

Additional reviewers (continued):

Sabanovic, Selma
Saerbeck, Martin
Sakamoto, Daisuke
Sanfeliu, Alberto
Satake, Satoru
Saunders, Joe
Schermerhorn, Paul
Schiele, Andre
Scopelliti, Massimiliano
Severinson-Eklundh, Kerstin
Shah, Julie
Shell, Dylan
Shic, Frederick
Shimada, Michihiro
Shiomi, Masahiro
Siciliano, Bruno
Sisbot, Emrah Akin
Skubic, Marge
Smith, Christian
Spinello, Luciano
Srinivasa, Siddhartha
Srinivasan, Vasant
Straub, Ilona
Suay, H. Bener
Sujit, P B
Sung, Ja-Young
SunSpiral, Vytas
Suutala, Jaakko
Syrdal, Dag Sverre
Takahashi, Toru
Takeuchi, Yugo
Tang, Karen
Tanoto, Andry
Teller, Seth
Tellex, Stefanie
Terada, Kazunori
Tessier, Catherine
Thompson, David
Thompson, Simon
Thornburg, Kristopher
Tipaldi, Gian Diego
Triebel, Rudolph
Tsui, Kate
Van der Loos, Mike
Vatavu, Radu-Daniel
Veloso, Manuela
Vijay, Ritukar
Vollmer, Christian
Wurtz, Rolf P.
Wade, Eric
Wagner, Petra
Walters, Michael
Warnier, Matthieu
Welch, Karla
Weng, Yueh-Hsuan
Williams, Mary-Anne
Wollherr, Dirk
Wrede, Britta
Yamazaki, Keiichi
Yano, Hiroaki
Yim, Ji-Dong
Yohanan, Steve
Yoshikawa, Yuichiro
Zanbaka, Catherine
Zender, Hendrik
Zhang, Kai

HRI 2012 Sponsors & Supporters

Sponsors:

SIGCHI

acm SIGART

IEEE

Cooperative Sponsors:

HF ES

aaai

Gold Level Supporters:

QinetiQ North America

ABB

Silver Level Supporters:

ALDEBARAN Robotics

Willow Garage

HEARTLAND ROBOTICS

Bronze Level Supporters:

Disney Research

Barrett Technology™ Inc.

Vecna Robotics

meka

The MIT Press

adept mobilerobots

HONDA

Strategies for Human-in-the-Loop Robotic Grasping

Adam Leeper, Kaijen Hsiao,
Matei Ciocarlie, Leila Takayama
Willow Garage, Inc.
68 Willow Road, Menlo Park, CA 94025, USA
{aleeper, hsiao, matei,
takayama}@willowgarage.com

David Gossow
Technische Universität München
Department of Computer Science
Informatik 9, Karlstraße 45, 80333 Munich
david.gossow@cs.tum.edu

ABSTRACT

Human-in-the loop robotic systems have the potential to handle complex tasks in unstructured environments, by combining the cognitive skills of a human operator with autonomous tools and behaviors. Along these lines, we present a system for remote human-in-the-loop grasp execution. An operator uses a computer interface to visualize a physical robot and its surroundings, and a point-and-click mouse interface to command the robot. We implemented and analyzed four different strategies for performing grasping tasks, ranging from direct, real-time operator control of the end-effector pose, to autonomous motion and grasp planning that is simply adjusted or confirmed by the operator. Our controlled experiment (N=48) results indicate that people were able to successfully grasp more objects and caused fewer unwanted collisions when using the strategies with more autonomous assistance. We used an untethered robot over wireless communications, making our strategies applicable for remote, human-in-the-loop robotic applications.

Categories and Subject Descriptors

I.2.9. [**Robotics**]: Operator Interfaces; H.1.2. [**User/Machine Systems**]: Human Factors

General Terms

Design, Experimentation, Human Factors, Performance

Keywords

grasping, teleoperation, shared autonomy

1. INTRODUCTION

As personal service robots make progress towards performing daily tasks in the home and office, they must be able to deal with complex and changing environments. Despite significant progress in many related fields, execution of a complex task still poses significant challenges. One option for accelerating the development of service robots is to involve a human in the loop. An operator's cognitive abilities can be tapped to deal with corner conditions and decisions that are difficult for autonomous systems. Such a system could be reliable enough for real-world deployment in the near future, making personal service robots a possible solution to the expected increasing cost and shortage of unskilled manual labor.

Figure 1: A remotely operated robot performing a grasping task, as seen through the point-and-click interface used by the operator.

In a Human-in-the-Loop (HitL) framework, autonomous capabilities can reduce the load on the operator and increase overall task efficiency. While a robot might not be able to autonomously handle a complete task in a robust and general way, we can use autonomy for sub-tasks that can be performed reliably, or that require operator input in a form that is relatively effortless to provide. Identifying and building such techniques, studying their interplay with operator-controlled sub-tasks, and analyzing the overall gain in efficiency are all steps towards deployable HitL systems.

In this paper, we focus on the ability to grasp and hold objects in the robot end-effector, a key prerequisite for many household and office tasks involving object transport and manipulation. A complete grasping task involves choosing an end-effector pose relative to the object, and executing an arm trajectory to bring the end-effector there. The robot must achieve a stable grasp of the object while avoiding undesired collisions with other parts of the environment. This can be challenging for the operator, particularly when dealing with non-anthropomorphic arms with many degrees of freedom (DOFs) or limited sensor data of the environment.

To alleviate some of these difficulties, we investigate four strategies for HitL grasp execution. These strategies, which will be presented in detail in Sec. 4, can be summarized as follows:

- **Strategy 1: Direct control**. Operator directly controls the 6D pose of the gripper in real-time.
- **Strategy 2: Waypoint following**. Operator specifies desired gripper position (waypoint) goals and adjusts them until satisfied before asking the robot to move.
- **Strategy 3: Grasp execution**. Operator only specifies the final desired grasp pose; robot performs collision-free motion planning to execute the specified grasp.
- **Strategy 4: Grasp planning**. Operator indicates general area for grasping; robot computes grasp pose suggestions for the operator to select and optionally adjust.

We aim to study performance in unstructured human environments. In particular, we test on objects previously unknown to the robot, immersed in clutter and/or surrounded by obstacles; only 2D images and noisy 3D point clouds of the scene are available for both grasp and motion planning.

The main contribution of our work is twofold. On one side, we present a HitL robotic system for efficient grasp execution in complex, unstructured environments. More importantly, we explore how operator input can be combined with autonomous methods for increasing task efficiency in the context of a grasping task. We believe that the comparison and analysis of such strategies can help guide the design of future HitL systems for more complex manipulation tasks.

2. RELATED WORK

A variety of teleoperation interfaces for robot arms have been used in applications such as space and undersea robotics, rescue robotics, and robotic surgery. Shared, supervisory, and collaborative control all refer to using a combination of autonomous and human control. Early investigations in shared control focused on the shared execution of motion trajectories with moderate time delays, as encountered in space [6] [14]. Towards dexterous manipulation, studies have used shared control for task primitives such as grasping and peg insertion [13]. For systems with many degrees of freedom, virtual fixtures [1] and other forms of haptic assistance in surgical applications [15] help to constrain motions of the master. There are numerous other types of assistance that a shared control interface can provide, including various degrees of supervisory control [19], autonomous assistance through mixed initiative user interfaces [8], interface agents [11], and other intelligent user interfaces [12].

The teleoperation literature has shown that many factors can affect the performance of a remote teleoperator. Relevant to this work are the issues of viewpoint, depth perception, and time delay. Our interface provides two views for the user: a monocular view from a camera on the robot's head; and a virtual, 3D rendering of the robot together with the color point cloud seen by the robot's sensors. This is a similar strategy to [7, 18], however, those interfaces require full models of the scene and objects, which are not available in our task. The viewpoint in both of our views is adjustable (the robot head can be pivoted to look at different objects, while the virtual 3D-view is free-roaming), which is driven by prior work showing that free-roaming viewpoints are superior to fixed views for manipulation tasks [3].

Another study comparing different interfaces for teleoperating robot arms showed that providing operators with 6D movement can be beneficial even if fewer DOFs are required for the task [16]. In [20], motion planning was shown to be useful in a comparison of five teleoperation strategies that bring a point on a simulated 6DOF robot's end-effector to a 3D positional goal in space. Our task also includes commanding a robot arm using a mouse, and part of it requires moving through free space while avoiding obstacles. However, grasping inherently requires controlling end-effector orientation as well as position, and also requires goal (grasp) selection as well as contact with objects in the environment during execution.

In the long history of teleoperation, a variety of models have been defined to describe the control strategies in human-robot task completion. To help frame our work within this context, we describe some of these models for teleoperation:

- Direct control, where the user directly controls the motion of the robot, with no intelligence or autonomy.
- Shared control, where the robot controls some aspects of a task, while the user still controls low-level motions [4].
- Supervisory control, where a user issues commands to a robot that executes them autonomously [19].

The strategies considered in the current work can be described using these models as follows:

- Strategy 1 is an instance of direct control, in which the user directly controls the gripper in Cartesian space.
- Strategy 2 is a form of shared control, in which the user selects desired poses; the robot rejects infeasible poses and executes straight-line paths to feasible poses.
- Strategies 3 and 4 are instances of supervisory control, in which the robot autonomously executes grasps set by the user; in Strategy 4 the robot also assists by providing grasp suggestions.

3. SYSTEM OVERVIEW

The overall goal of our system was for a physical robot to perform a number of grasping tasks of common household objects in a complex environment, under the control of a human operator. The system is designed for remote operation; the operator controls the robot through a separate desktop computer, with no direct visual contact with the robot itself.

The hardware we used was the PR2 personal robot, shown in Figure 1. The PR2 has two compliant, backdriveable 7-DOF arms with parallel-jaw grippers. We used two range sensors: a widely available Microsoft Kinect^{TM} mounted on the head of the robot (providing both range and color images), and a tilting laser rangefinder mounted on the chest (used for autonomous collision avoidance). The PR2 communicated with the computer running the teleoperation interface via a commodity wireless network, as we expect that any mobile robot in real households or offices will have to be untethered in order to perform useful tasks.

We developed a "point-and-click" Graphical User Interface[1] built on `rviz`, a 3D robot visualization and interaction

[1]While higher-dimensionality input methods might provide some benefits, a major advantage of a simple cursor-driven interface is the widespread accessibility of devices that provide cursor control, including tablets and devices such as head trackers for motion-impaired users.

environment in ROS (www.ros.org/wiki/rviz). It presents the user with two main displays: on the left, a real-time feed from the Kinect camera mounted on the PR2; on the right, a rendered image of the PR2 in its current posture, along with a 3D point cloud showing a snapshot of the world as seen by the Kinect. The user can point the robot's camera by left-clicking anywhere in the camera view, changing the point of view of the live camera feed shown on the left. Since the right image is rendered, its viewpoint can be moved to any position by rotating, translating, and zooming.

For the 3D point cloud of the world shown on the right in Figure 1, we use a static snapshot instead of a continuous feed of the Kinect range data. This allows a resolution that would be difficult to stream over the wireless network; it also allows for continued 3D visualization of objects that become occluded as the arm is positioned to grasp. At any time, the user can refresh this static snapshot by right-clicking on the snapshot and selecting 'refresh'. The snapshot point cloud, as well as the various interface controls for moving the gripper pose goal, are also overlaid atop the camera feed on the left, enabling click-and-drag mouse input in both views.

4. STRATEGIES FOR GRASP EXECUTION

So far, we have described the two ends of our pipeline: a robot attempting to choose and execute a grasp that allows it to secure and lift an object from a cluttered environment, and an interface allowing a user to receive information on the robot's surroundings and provide point-and-click commands. We compared four different strategies for connecting these two. In particular, we explored the spectrum from very little to a great deal of autonomous assistance from the robot, or, conversely, from high to low operator involvement.

Strategy 1: Direct Control. In the first strategy, the user directly controls the PR2 gripper in real-time by clicking and dragging a set of rings and arrows (Fig. 2). Dragging the arrows causes linear motion along each of three orthogonal axes, and dragging the rings causes rotation about the same axes. The axes can be aligned with the gripper or the world, as seen on the right side of Figure 2.

As the user drags the rings-and-arrows control, the real gripper attempts to track it in real time. This is achieved using a J-transpose control law for the Cartesian position and orientation of the gripper. The current pose of the rings-and-arrows controls is sent as the input command to the controller at 30Hz. PD control is used to generate a desired Cartesian wrench for the gripper ($\mathbf{f}$), which is converted to joint torques (τ_p) using the transpose of the Jacobian for the end-effector pose (J):

$$\tau_p = J^T \mathbf{f} \tag{1}$$

Since the arm is redundant with respect to 6-DOF positioning of the gripper, we also provide a ring control around the shoulder. This adds nullspace joint torques (τ_n) to the controller, to bias the arm towards a desired elbow posture while attempting to maintain the current gripper pose:

$$\tau_n = \mathbf{k} * (I - J^\dagger J) * \mathbf{q_e} \tag{2}$$

where $\mathbf{q_e}$ are the joint errors (away from the desired posture), $J^\dagger$ is the Jacobian pseudo-inverse, and $\mathbf{k}$ is the vector of joint gains. The total control torques applied to the robot are $\tau = \tau_p + \tau_n$.

Note that this strategy makes very few assumptions about the task being performed; it also provides the least assistance. When using this control method for grasping, the user must essentially "drag" the robot's gripper all the way from its starting position to the desired grasp location, while avoiding collisions along the way.

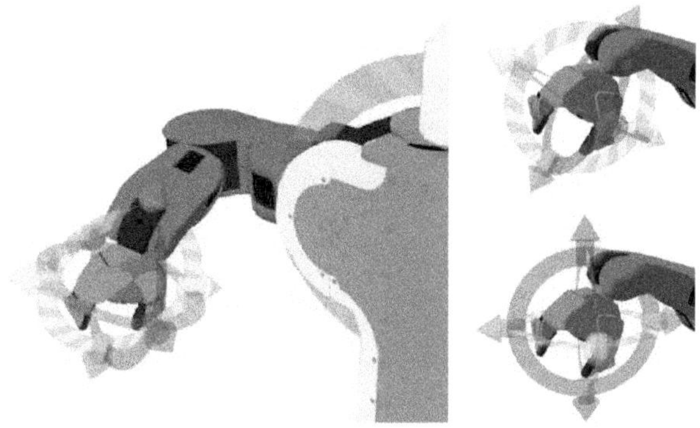

Figure 2: Strategy 1. Gripper control (6D) and shoulder ring (1D). Right images show a gripper-aligned (top) and world-aligned (bottom) control.

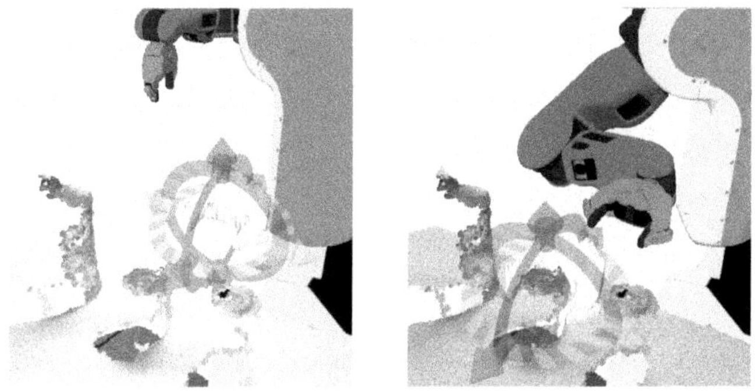

Figure 3: Strategy 2. An example of waypoint control to grab a soup can.

Strategy 2: Waypoint Following. With the previous strategy, any input from the operator is immediately tracked by the robot. This results in fast execution, but does not allow the operator to check or adjust the path of the gripper before the robot actually moves. With the second strategy, the operator uses the same type of rings-and-arrows control to move around a *virtual* gripper that specifies a new pose goal. The robot only begins to move once the user accepts the goal, which causes the gripper to move by smoothly interpolating both position and rotation to the goal pose.

When the gripper reaches the new position or detects that it is stuck, the controls reappear and can be dragged to a new waypoint. A sample sequence is illustrated in Figure 3. To reach a desired goal, this strategy uses the same controller as the previous one, but with an outer loop that essentially maintains a fixed velocity on the gripper's way to the desired goal. The shoulder control for adjusting the elbow posture is unchanged. The user can also cause the goal to instantly jump to a pose near a clicked-on point in the snapshot; the surface normal of the clicked-on point is estimated and used to initialize the direction from which the gripper approaches the point cloud (Figure 4). As an additional feedback mechanism, the virtual gripper control turns green when the desired pose is within reach, and red when it is out of reach; an out-of-reach goal cannot be accepted.

(a) (b) (c)

Figure 4: (a) In Strategies 2 and 3, the user right-clicks on a point in the cloud (marked here by a red dot) to quickly position the virtual gripper near that point, and aligned with the estimated surface normal at that point. (b) In Strategy 4, an autonomous planner shows suggested grasp poses as grey rectangle-arrow icons. The icons change color as the grasps are evaluated; green denotes a feasible grasp, while red marks a colliding or unreachable pose. (c) Clicking on a grasp icon places the virtual gripper in that pose.

Overall, this strategy attempts to take advantage of cases where specifying a new waypoint is easier than directly controlling a long, straight-line gripper motion. It also allows gripper goals to be carefully adjusted before they are executed. However, collision avoidance along the path to the waypoint is still the responsibility of the user.

Strategy 3: Grasp Execution. In the third strategy, the user only selects the final grasp pose; the robot then executes the grasp autonomously while attempting to avoid collisions along the way. The user begins by clicking a point in the 3D environment snapshot. As with Strategy 2, a virtual gripper is displayed at the clicked location and initially aligned with the local surface normal. The operator can use the rings-and-arrows control to adjust the pose as desired (Figure 4a).

Once the operator confirms the grasp location, the robot computes a pre-grasp pose, backed off from the final grasp pose by 10 cm. The robot then moves the arm into the pre-grasp pose, avoiding all collisions along the way, using the sampling-based motion planner presented in [2]. Finally, the robot moves the gripper from the pre-grasp into the grasp, avoiding collisions on the arm only. Gripper collisions are accepted in this step so that objects can be retrieved from clutter; surrounding objects may need to be pushed aside.

As the user is adjusting the virtual gripper indicating the desired grasp, the robot continuously computes whether the motion planning and collision checking components consider the grasp feasible. If a collision-free pre-grasp, and an acceptable approach from pre-grasp to grasp can be computed, the virtual gripper control turns green. If either of those computations fails, either due to potential collisions or because the desired grasp is out of reach, the virtual gripper control turns red and prevents the goal from being accepted.

This strategy gives the autonomous components the job of planning appropriate arm joint trajectories and avoiding unwanted collisions; the operator only needs to select an appropriate final grasp pose. In doing so, a number of assumptions are made that are specific to grasping tasks (such as decomposing the motion into a pre-grasp and a grasp).

Strategy 4: Grasp Planning. In this strategy, the user has access to a set of grasp poses suggested by an autonomous grasp planner. The poses are computed near a user-selected point, and show up after a few seconds as small icons that indicate the proposed wrist poses (Fig. 4b).

The grasp pose icons show up as grey initially, which means that they are not yet checked for collisions; each pose in turn is checked in the background, and the associated icon turns red or green depending on whether the pose is deemed acceptable. Users can click on an icon of any color at any time, which causes the virtual gripper to be displayed at that grasp pose. The user can adjust this pose to their satisfaction as in the previous strategy, with the same type of red/green feedback for which poses are considered feasible by the automated components. Once the user confirms a grasp pose, execution proceeds as in the previous strategy.

When computing grasp suggestions, the robot first attempts to segment objects in the cluttered scenes by estimating surface normals for all points within a 30 cm cube around the user's clicked position, removing all points whose normals are not within 45 degrees of a given direction (up for the scenes with objects on the table; forward for the shelf), and then performing Euclidean clustering. This process separates most objects from each other, and from the table/shelf. The resulting clusters are then fed, individually, to the grasp planner presented in [9], which computes grasps along the principal axes; table scenes are limited to overhead grasps, and the shelf scene is limited to grasps from the front.

This strategy uses the most autonomy; in an ideal case, the user only has to click once around the desired objects to start the grasp planner, and once more to select one of the resulting grasps. Note that this is as much autonomy as can be provided without higher levels of semantic perception and scene understanding (the robot has no way of knowing which of the objects the user actually wants to grasp, or how to

(a) Training Environment

(b) Environment 1

(c) Environment 2

(d) Environment 3

Figure 5: Environments used in this study.

distinguish objects to grasp from obstacles). However, since the autonomous components do not always behave in an ideal way, the user still has tools for adjusting the planning results, or choosing a different grasp altogether.

5. CONTROLLED EXPERIMENT

We performed a controlled experiment to assess the capabilities of the different HitL strategies to perform grasping tasks with novice robot operators, and to quantify how the various levels of autonomous assistance affected their performance. We used three environment types (Figure 5) to simulate different situations that can be encountered in a household setting, resulting in a 4 x 3 experiment (grasping strategy: 1 vs. 2 vs. 3 vs. 4, between-participants x environment type: 1 vs. 2 vs. 3, within-participants). 48 adults (24 men and 24 women) participated in this study. Each user was randomly assigned to one of the four strategies, genders were balanced across the conditions, and environment types were balanced for order within each condition. Each participant was given a $20 gift card as a token of thanks.

5.1 Participants and Procedure

Participants were recruited from local mailing lists and contacts. 3 participants were 18-20 years of age, 28 were 21-29, 13 were 30-39, 0 were 40-49, 2 were 50-59, and 2 were 60-69. These participants were not very familiar with robots (M=1.58, SE=0.15; 1= not familiar to at all, 7 = very familiar. 13 participants had no experience with playing video games; 13 had played 2D video games; and 22 had played at least one 3D video game.

Upon signing the study agreement form, each participant was shown two tutorial videos — one general and one specific to the randomly assigned grasping strategy. Next, the participant was talked through grasping the three objects in the training scene shown in Figure 5a, including how to move the robot camera, how to drop off objects, and how to refresh point clouds. Data collection then occurred over three rounds, one for each environment type. For each round of grasping, the participant had 10 minutes to grasp as many objects as possible from the given environment.

The first environment contained objects in a cluttered pile on a table, the second contained the same arrangement with the addition of six fragile-looking vases (obstacles), and the third contained a small shelf full of objects (Figure 5b-d). The full task simulated clearing the scene by grasping objects and dropping them in a container to the right side of the robot; however, because we are concerned only with the grasping of objects, and not with transporting them after grasping, we manually removed from the gripper objects that were grasped and lifted. The arm started each grasp to the side of the robot as if the previous object had been dropped off in a container; to reset the arm to that position we provided a command that simply dropped the grasped object and moved the arm back to the initial side position.

Each scene contained more objects than could be grasped in the allotted time even by expert users, so no one was able to clear all of them. Grasping stacked or nested objects only counted as one grasped object, since only one grasp was involved; the experimenter returned the additional objects to the scene. After each round of grasping, the participant was presented with a questionnaire about their experience during the task. At the end of all three rounds, the participant filled out a demographics questionnaire, and was then debriefed about the purposes of the study.

5.2 Measures

Behavioral Measures: The task performance metrics for each 10-minute round were number of (a) successful grasps, (b) major collisions, and (c) minor collisions. Because the PR2 arms are compliant and unable to exert large forces, collisions with static structures become fairly harmless. We therefore counted both *minor* and *major* collisions. *Minor* collisions were undesirable but relatively consequence-free, such as hitting the table or brushing a shelf or vase without moving it from its footprint. *Major* collisions were forceful enough to move or knock over the shelf or vases, or to potentially cause damage to either table or robot.

Self-Reported Measures: To measure the cognitive load experienced during each round, we used the NASA-TLX scale [5]. We also asked about how well a set of adjectives described the participant's user experience (easy, tedious, boring, engaging, difficult, simple, straightforward, complicated), using a 5-point Likert scale. Finally, we measured demographics, including age, gender, video gaming experience, locus of control [17], and familiarity with robots.

5.3 Data analysis

For the behavioral and cognitive load analyses, we used a mixed model analysis of covariance (ANCOVA), using strategy type as a between-participants independent variable, environment type as a within-participants variable, and video gaming experience as a covariate (0 = no experience; 1 = experience with only 2D games; 2 = experience with at least one 3D game). When significant main effects were found, we ran follow-up pairwise t-tests (with Bonferroni adjustments and video gaming experience as a covariate) to see which of the specific conditions were different from each other. If interaction effects were found with environments, we ran follow-up ANCOVAs (with Bonferroni adjustments and video gaming experience as a covariate) to see which of the results differed in each of the different environments.

For the attitudinal analyses, we used a mixed model analysis of covariance (ANCOVA), using strategy type as a between-participants independent variable, environment type as a within-participants variable, and familiarity with robots as a covariate (1 = not familiar at all; 7 = very familiar).

6. RESULTS

6.1 Behavioral Results

Number of Objects Grasped: Strategy type ($F(3, 43)$=14.61, $p<.001$, $\eta^2 = .18$) and video gaming experience ($F(1,43)$=11.65, $p<.01$, $\eta^2 = .05$) affected how many objects people successfully grasped within the 10 minute limit. Controlling for video gaming experience (set to the average of 1.19), people who used Strategies 3 and 4 successfully grasped more objects than Strategies 1 (1 vs. 3, $p>.01$; 1 vs. 4, $p<.01$) and 2 (2 vs. 3, $p>.001$; 2 vs. 4, $p<.001$). Grasped object counts were as follows: Strategy 1 (M= 3.71, SE=0.42, Max=7), Strategy 2 (M= 3.11, SE=0.42, Max=8), Strategy 3 (M= 6.19, SE=.42, Max=11), Strategy 4 (M= 6.13, SE=0.42, Max=12). People who had 3D video gaming experience generally grasped more objects (M=5.6, SE=0.3) than people who only had 2D video gaming experience (M=4.2, SE=0.4) or no video gaming experience (M=3.9, SE=0.4). See Figure 6a, which displays unmodified mean and standard error values (not controlling for video gaming experience).

There was also a significant interaction effect between strategy and environment type, $F(6,86)$=3.06, $p<.01$, $\eta^2 =$ .02, which means that different strategies influenced task performance in different ways, depending upon the environmental setting. In Environment 1, people who used Strategies 3 or 4 were able to grasp more objects than people who used Strategy 1 (1 vs. 3, $p>.01$; 1 vs. 4, $p<.001$) or Strategy 2 (2 vs. 3, $p>.001$; 2 vs. 4, $p<.001$). In Environment 2, people who used Strategies 3 or 4 were again able to grasp more objects than people who used Strategies 1 (1 vs. 3, $p>.01$; 1 vs. 4, $p<.05$) or 2 (2 vs. 3, $p>.01$; 2 vs. 4, $p<.01$). In Environment 3, people who used Strategy 3 were able to grasp more objects than people who used Strategy 1 ($p<.001$), Strategy 2 ($p<.001$), or Strategy 4 ($p>.05$); people who used Strategy 4 were able to grasp more objects than people who used Strategy 2 ($p<.05$).

Number of Major Collisions: Controlling for video gaming experience, we found that the strategy type ($F(3, 44)$=5.34, $p<.01$, $\eta^2 = .10$) and environment type ($F(2, 86)$=3.68, $p<.05$, $\eta^2 = .04$) affected how many major collisions participants caused. People who used Strategies 3 ($p<.01$) and 4 ($p<.05$) had fewer major collisions than people who used Strategy 1. Video gaming experience was not found to be a significant predictor of major collisions. See Figure 6b, which includes some collisions caused by the motion planner (not the fault of the user).

There was also a significant interaction effect between strategy and environment type, $F(6,86)$=3.30, $p<.01$, $\eta^2 =$ .10. There were no significant differences observed between strategies in Environments 1 or 3. In Environment 2, people had fewer major collisions when using Strategy 3 than 1 ($p<.01$) or 2 ($p<.05$) and fewer major collisions when using Strategy 4 than Strategy 1 ($p<.01$).

Number of Minor Collisions: Controlling for video gaming experience (set to the average of 1.19), we found that strategy type ($F(3, 43)$=4.08, $p<.05$, $\eta^2 = .09$) and environment type ($F(2, 86)$=23.60, $p<.001$, $\eta^2 = .25$) both affected how many minor collisions participants caused. Follow-up analyses showed that all of the environment types were significantly different from one another, $p<.01$. Furthermore, people who used Strategy 3 had fewer minor collisions than people who used Strategy 1, $p<.05$. Video gaming experience was not found to be a significant predictor of minor collisions. See Figure 6c.

6.2 Cognitive Load and Attitudinal Results

There were no significant effects for strategy type upon cognitive load. Controlling for video gaming experience, people experienced more mental demand from Environment 3 (M=4.38, SE=0.23) than from Environment 1(M=3.94, SE=0.19), $F(2, 86)$=3.27, $p<.05$, $\eta^2 = .003$. People who had 3D video gaming experience (M=3.69, SE=0.18) felt that they were more successful in accomplishing their tasks than people who had no video gaming experience (M=4.56, SE=0.24), $F(1, 43)$=8.92, $p<.01$ $\eta^2 = .02$, (1=perfect; 7=failure). They also felt less frustrated with the task (M=2.98, SE=0.26) than people who had no video gaming experience (M=3.95, SE=0.33), $F(1, 43)$=6.73, $p<.05$, $\eta^2 = .02$, (1=very low; 7=very high).

Participants also rated the following adjectives on a 5-point scale, ranging from 1 (strongly disagree) to 5 (strongly agree).

Boring: Controlling for familiarity with robots (set at the average of 3.25 on a 1-7 scale), we found that strategy type affected how bored participants felt during the task, $F(3,43)$=3.07, $p<.05$, $\eta^2 = .08$. People who used Strategy 2 felt more bored (M=2.22, SE=.0.20) than people who used Strategy 1 (M=1.56, SE=.0.20, $p<.05$), Strategy 3 (M=1.5, SE=.0.20, $p<.05$), or Strategy 4 (M=1.48, SE=.0.20, $p<.05$).

Simple: Controlling for familiarity with robots (set at 3.25), we found Environment 3 (M=2.48, SE=0.16) to be less simple than Environments 1 (M=2.63, SE=0.15) or 2 (M=2.60, SE=0.16), $F(2,86)$=3.19, $p<.05$, $\eta^2 = .01$. Robot familiarity also had a significant effect upon how simple the participants thought the task was, $F(1,43)$=5.25, $p<$ .05, $\eta^2 = .04$.

6.3 Summary

Overall, the results of our controlled experiment for testing task performance using each of the four systems suggests that Strategies 3 and 4 generally produced better results than Strategies 1 and 2. In Environment 1 (cluttered scene with no fragile obstacles), people were able to grasp more objects when using Strategies 3 or 4 as opposed to Strategies 1 or 2. In Environment 2 (cluttered scene with fragile obstacles), people were able to grasp more objects when using Strategy 3 or 4 than Strategy 1 or 2. In this environment, Strategy 3 also had fewer major collisions than Strategies 1 or 2, while Strategy 4 had fewer major collisions than Strategy 1. In Environment 3 (tight shelf space), people were able to grasp more objects with Strategy 3 than with any of the other strategies. In this environment, Strategy 2 had fewer major collisions because the controller limits prevent major collisions with heavy objects; nonetheless, the differences were not statistically significant. Overall, Strategies 3 and 4 had fewer major collisions than Strategy 1, and Strategy 3 had fewer minor collisions than Strategy 1. We found no statistically significant results regarding the cognitive load

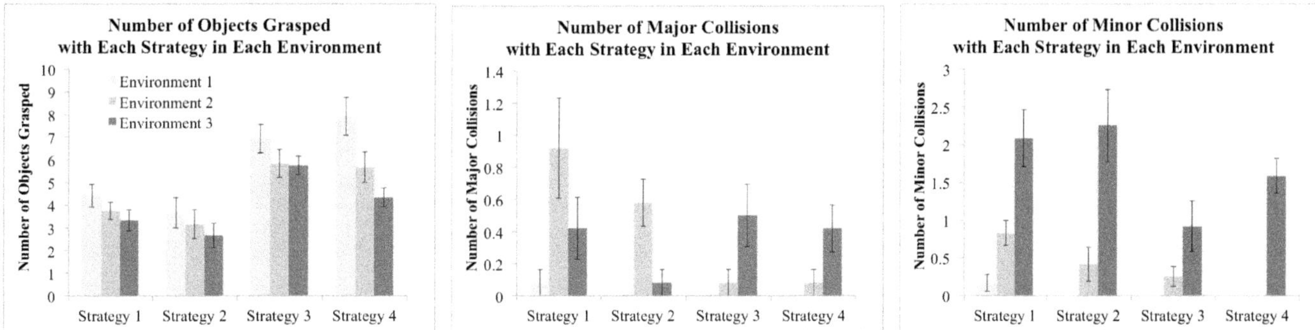

(a) Successfully grasped objects (b) Major collisions (c) Minor collisions

Figure 6: Means and standard errors for the performance metrics, not controlling for video gaming experience.

experienced between the strategies; however, environment types and video gaming experience did cause significant differences. In terms of user experience, Strategy 2 was the most boring of the four strategies.

7. DISCUSSION

One of the main goals of our experiment was to determine if, and in which ways, various level of autonomy are beneficial for a HitL robotic system driven by non-expert users performing grasping tasks. In general, the results show that the strategies with more effort unloaded onto autonomous modules performed better. The operators were able to grasp more objects, and caused fewer unwanted collisions.

7.1 Autonomy for HitL systems

Autonomous motion planning (as used in Strategies 3 and 4) proved to be highly beneficial. First, the motion planning module shielded the user from having to think about the joint configurations of the arm required to achieve various end-effector poses. This was particularly valuable as our arm has a non-anthropomorphic kinematic configuration. As noted by one user, "it was difficult noting that the PR2 wrist rotated counter intuitively from my own right hand." Second, the motion planner removed the need for the operator to think about avoiding collisions. An interesting aspect is that motion planning increased overall performance even though it occasionally exhibited failures of its own (manifested as collisions, including 7 major ones). This result suggests that autonomy does not need to be perfect in order to be helpful as a component of a HitL system.

We expected cognitive load to be lower when more autonomy was used and task performance was improved; however, this turned out not to be the case. Several users commented that waiting for the robot to execute straight-line motions to a selected pose when using Strategy 2 was boring. However, waiting for the robot to autonomously execute motion-planned grasp trajectories was not boring.

The grasp planning strategy generally performed at the same level as the grasp execution strategy, a somewhat unexpected result given its additional module in charge of suggesting grasps. A number of qualitative observations might account for this behavior, and provide suggestions for future designs. We noted that able operators often trusted the autonomous module too much. The fact that the autonomous components checked *some* (e.g. kinematic feasibility, collision-free for the arm), but not *all* (e.g. stability of object inside the gripper, gripper collisions with obstacles) characteristics of a grasp often led to confusion: operators deferred to the "checked" grasps too much rather than using their own analysis. For example, one operator noted that "because I trusted the grasp planner so much I executed without fully verifying and the grasp failed." The autonomous planner also yielded no benefit when it was not trusted enough, as operators occasionally adjusted a good suggestion into a bad grasp. One operator noted that "in a very cluttered area the automatic grasp positions weren't necessarily what I was imagining for the object [...]. For one I resorted to picking a nearby grasp location and manually moving/re-orienting to get the grasp I had in mind." Furthermore, even though the grasp planner module generally operated within about 3 seconds, even this delay was too much for some users, as they could not start adjusting the grasp pose until the grasp planner had finished. One of them noted that "the arrows took too long to show up after every move; I was frustrated." In general, our results are consistent with previous work on building trust between an autonomous system and an operator [10]. We found that it is very important for the operator to know what he or she is responsible for, and also important to be aware of what the robot does not know or is incapable of doing on its own.

7.2 Interface observations and future work

A number of observations concerned the user interface components common to all strategies. The operator's comfort with a general 3D GUI and related operations such as positioning a virtual camera proved to be very important. Those with 3D gaming experience generally performed better, though we note that a better control in future studies would be user experience with mouse-based icon manipulation, as is common in CAD and other software besides just 3D video games.

The 3D rendering component of our interface displayed a static point cloud representation of the world that could be refreshed on demand, as opposed to a continuously updating or streaming scene. This choice proved confusing for many operators, causing them to occasionally operate on out-of-date scene representations and try to grasp objects that had moved. Some offered suggestions for automatic point cloud refreshing; one user noted that "you might as well refresh the point cloud at the end of every 'drop and return' command."

From a task perspective, we note that our system could not completely eliminate undesired collisions, regardless of the strategy used. Avoiding collisions between the gripper's protruding knuckles and the cabinet shelf proved particularly difficult, compounded by the fact that only the shelf's front edges were generally visible to the robot's sensors (and in the point cloud snapshot). Some such collisions could be avoided by improvements to the motion planning and sensing modules of our system. However, we note that grasping requires making contact at least with the grasped object, and occasionally with surrounding objects as well. This requires the ability not only to avoid collisions, but also to reason about the scene and distinguish unwanted contacts from acceptable ones, which users were often unable to do. Visual cues highlighting possible collisions would help; additional semantic labeling, either autonomous or user-aided, would be required to avoid collisions with objects not visible to the robot's sensors.

For general mobile manipulation tasks, we envision a HitL system providing access to multiple strategies such as these. Using our study environments as examples, Environment 1, with few obstacles, is better suited for grasp planning than Environments 2 or 3. Likewise, even though we allow collisions along the final grasp approach so that surrounding objects can be shoved aside during a grasp (a technique often necessary in cluttered environments), the more autonomous strategies' motion planners would prevent a user from grasping objects such as a bowl at the back of a cabinet. In future work, we hope to examine how having multiple strategies available increases the operator's efficiency when performing tasks.

8. CONCLUSIONS

In this paper, we have presented a HitL robotic system that enables novice users to remotely operate a robot performing grasping tasks in highly cluttered environments. In order to increase performance, quantified by the number of objects grasped and collisions incurred in a fixed time interval, our system uses a set of autonomous modules, such as motion or grasp planning, to assist the operator. In a controlled experiment (N=48), we evaluated the effect of these components by comparing four different grasping strategies, spanning the spectrum from very little to significant amounts of autonomous assistance for the user.

Our results show that the strategies where autonomous modules took more responsibility for parts of the task performed better than those where the operators were required to handle more by themselves. However, autonomous components must establish an appropriate level of trust with the operator in order to provide significant benefits, and communicate their limitations in an appropriate way. We believe these results can inform the design of future, more general HitL manipulation systems, which can in turn accelerate robot deployment into complex, unstructured environments.

References

[1] A. Bettini, P. Marayong, S. Lang, A. Okamura, and G. Hager. Vision-Assisted Control for Manip. Using Virtual Fixtures. *IEEE Trans. on Robotics*, 20(6), 2004.

[2] M. Ciocarlie, K. Hsiao, E. Jones, S. Chitta, R. Rusu, and I. Sucan. Towards reliable grasping and manipulation in household environments. In *ISER*, 2010.

[3] H. Das. *Kinematic Control and Visual Display of Redundant Teleoperators.* PhD thesis, MIT, 1989.

[4] W. Griffin, W. Provancher, and M. Cutkosky. Feedback Strategies for Telemanipulation with Shared Control of Object Handling Forces. *Presence: Teleoperators and Virtual Environments*, 14(6):720–731, Dec. 2005.

[5] S. Hart. Nasa-task load index (nasa-tlx); 20 years later. In *HFES*, 2006.

[6] S. Hayati and S. Venkataraman. Design and implementation of a robot control system with traded and shared control capability. In *ICRA*, 1989.

[7] G. Hirzinger, B. Brunner, J. Dietrich, and J. Heindl. Sensor-Based Space Robotics–ROTEX and Its Telerobotic Features. *IEEE Transactions on Robotics and Automation*, 9(5), 1993.

[8] E. Horvitz. Principles of mixed-initiative user interfaces. In *CHI*, 1999.

[9] K. Hsiao, S. Chitta, M. Ciocarlie, and E. Jones. Contact-reactive grasping of objects with partial shape information. In *IROS*, 2010.

[10] J. Lee and K. See. Trust in automation: Designing for appropriate reliance. *Human Factors: The Journal of the Human Factors and Erg. Society*, 46(1):50–80, 2004.

[11] P. Maes. Agents that reduce work and information overload. *Communications of the ACM*, pages 30–41, 1994.

[12] M. Maybury. Intelligent user interfaces: An introduction. In *IUI*, pages 3–4, 1999.

[13] P. Michelman and P. Allen. Shared autonomy in a robot hand teleoperation system. In *IROS*, 1994.

[14] M. Oda, N. Inaba, Y. Takano, S. Nishida, M. Kayashi, and Y. Sugano. Onboard local compensation on ETS-W space robot teleoperation. In *IEEE/ASME Intl. Conf. on Advanced Intelligent Mechatronics*, 1999.

[15] M. K. O'Malley, A. Gupta, M. Gen, and Y. Li. Shared Control in Haptic Systems for Performance Enhancement and Training. *Journal of Dynamic Systems, Measurement, and Control*, 128(1), 2006.

[16] H. Pongrac, A. Peer, B. Farber, and M. Buss. Effects of varied human movement control on task performance and feeling of telepresence. In *EuroHaptics*, 2008.

[17] J. Rotter. Generalized expectancies of internal versus external control for reinforcements. *Psychological Monographs*, 80, 1966.

[18] S. Schneider and R. Cannon. Experimental object-level strategic control with cooperating manipulators. *Intl. Journal of Robotics Research*, 12(4), 1993.

[19] T. Sheridan. *Telerobotics, Automation, and Human Supervisory Control.* MIT Press, Cambridge, MA, 1992.

[20] E. You and K. Hauser. Assisted Teleoperation Strategies for Aggressively Controlling a Robot Arm with 2D Input. In *RSS*, 2011.

Grip Forces and Load Forces in Handovers: Implications for Designing Human-Robot Handover Controllers

Wesley P. Chan*, Chris A.C. Parker†, H.F.Machiel Van der Loos‡, and Elizabeth A. Croft§
Department of Mechanical Engineering, University of British Columbia
2054 – 6250 Applied Science Lane
Vancouver, B.C., Canada V6T 1Z4
1 604 822 3147
*wesley.chan@alumni.ubc.ca, †parkerca@ieee.org, ‡vdl@mech.ubc.ca, §ecroft@mech.ubc.ca

ABSTRACT
In this study, we investigate and characterize haptic interaction in human-to-human handovers and identify key features that facilitate safe and efficient object transfer. Eighteen participants worked in pairs and transferred weighted objects to each other while we measured their grip forces and load forces. Our data show that during object transfer, both the giver and receiver employ a similar strategy for controlling their grip forces in response to changes in load forces. In addition, an implicit social contract appears to exist in which the giver is responsible for ensuring object safety in the handover and the receiver is responsible for maintaining the efficiency of the handover. Compared with prior studies, our analysis of experimental data show that there are important differences between the strategies used by humans for both picking up/placing objects on table and that used for handing over objects, indicating the need for specific robot handover strategies as well. The results of this study will be used to develop a controller for enabling robots to perform object handovers with humans safely, efficiently, and intuitively.

Categories and Subject Descriptors
J.4 [**Social and Behavioral Sciences**]: Sociology; H.5.2 [**Information Interfaces and Presentation**]: User Interfaces – *Haptic I/O*; I.2.9 [**Artificial Intelligence**]: Robotics – *Kinematics and Dynamics*

General Terms
Human Factors, Experimentation, Performance, Measurement

Keywords
Handover, handoff, grip force, load force, object transfer, haptic communication, physical human robot interaction.

1. INTRODUCTION
Until recently, robots have worked in segregation, interacting minimally with people and completing their tasks using pre-programmed trajectories. However, there has been increasing interest in the development of robots that are able to work closely alongside humans. Proposed applications have included intelligent assistants in factories, museum tour guides, and homecare robots for assisting seniors [1,2]. Human-robot cooperation enables the different abilities of humans and robots to complement each other. For example, humans tend to be better decision makers whereas robots excel at repetitive tasks. In many of the aforementioned domains, humans and robots will work together in shared environments. Thus it is imperative that we develop robots capable of direct physical interaction with humans. One common type of interaction in many cooperation scenarios is object handover: the passing of an object from one agent to another.

Handovers arise frequently in a variety of settings – handing a wrench to a mechanic, passing out pamphlets, or bringing a glass of water to a senior. Daily, humans perform countless unscripted handovers, and generally do so safely without dropping the object, efficiently, and with ease. Yet, this is still a challenging task for robot systems. Our aim is to allow robots to perform object handovers with non-expert humans in unscripted situations.

Studies have shown that people prefer to work with robots that behave like humans [3,4], and since humans regularly perform successful handovers, a good first step would be for robots to emulate humans when participating in human-robot handovers. To achieve this goal, a better understanding of how humans perform object handovers is needed, and that is the focus of this work.

To date, many researchers have studied human reaching motion during various modes of handover, e.g., [3-6]. Shibata et al. and Huber et al. have implemented human-like trajectories with robots and compared them to other types of robot trajectories. It was found that when robots employed human-like trajectories, human partners tend to perceive the robots as safer and more pleasant [3,4]. Cakmak et al. investigated robot posture and gestures during handovers and found that certain postures are more likely to be interpreted by humans as an initiation of a handover from the robot [7]. Furthermore, Edsinger and Kemp found that under certain conditions, a robot's reaching gesture may be recognized by humans as a social cue to take the object from the robot [8]. Most of these studies have focused on the kinematics of the reaching motion leading to a handover, but few have investigated the dynamic interaction involved in the actual transfer of the object during a handover. In the experiment conducted by Huber et al., the robot simply released the object after a delay of one second regardless of whether the human receiver had even attempted to take the object from its gripper [4]. In another handover experiment conducted by Nagata et al., the human receiver had to forcibly take the object from the robot's grasp, and the robot would let go only when it was no longer able to maintain a stable grip [9]. In other experiments, it is not clearly stated how

HRI'12, March 5–8, 2012, Boston, Massachusetts, USA.

the robot determines when to let go of an object. Often the object transfer phase of a handover is not a primary subject of these studies.

Handing over of an object by a giver to a receiver requires the giver to carefully coordinate the release of the object with the receiver. Releasing too early may cause the object to fall, whereas releasing too late can result in an inefficient handover, and even a jerky "tug of war" incident [7,10]. In order to achieve the precise timing of object release, the giver and the receiver need to communicate with each other – whether implicitly or explicitly. Humans can perform handovers without verbal communication, and even blindfolded, relying only on the sense of touch to synchronize object transfer. Furthermore, research has shown that when the sense of touch is removed, people have great trouble performing prehensile tasks, and they are unable to scale their grip force appropriately [11-13]. Therefore, we believe that haptic communication plays an important role in handovers.

The haptic literature related to handling objects focuses on how humans regulate grip force in single-person lifting tasks. Many researchers, including Johansson and Westling (1984 and 1988), have shown that humans are able to adjust their grip force appropriately according to the experienced load force and adapt to different object surface friction properties, changes in object weight, and object movements [12-16]. However, many of these studies focused on single-person tasks that did not involve handing over an object to another person. In one study by Mason and Mackenzie that did focus on object handover, the researchers examined how the grip forces of the giver and the receiver changed over time relative to each other. However these authors did not address how the giver communicates with the receiver during load transfer nor did they discuss how the releasing and taking of the object is synchronized [17].

In order to implement safe and efficient human-robot handovers in unscripted situations using haptic cues, a better understanding of the haptic interaction in human-to-human handovers is required. Our guiding question is: What are the haptic interactions that arise in successful human-to-human handovers? In particular we ask: How do human givers and receivers regulate their grip forces with respect to load forces during an object transfer? How does a human giver determine when it is safe to release an object? Are there features of the giver's or receiver's load trajectories that reliably signal when the object transfer has been completed?

To answer these questions, we investigate and characterize the haptic interaction that arises in human-to-human handovers. We examine the characteristics of object load transfer and the giver's and receiver's regulation of grip forces in response to the amount of shared object loads experienced during object transfer.

2. HYPOTHESES

Johansson and Westling showed that when a person holding an object experiences a change in load force, grip force is regulated to be proportional to the experienced load force [13]. An approximately constant grip force to load force ratio is maintained such that the minimum amount of grip force required to prevent object slip is applied, with a small safety margin. In our experiment, we test to see if this finding is also true when an object is transferred between two persons and formulate the following hypotheses:

H_1: The giver maintains an approximately constant grip force to load force ratio during object transfer.

H_2: The receiver maintains an approximately constant grip force to load force ratio during object transfer.

If the giver's and receiver's grip/load ratios are found to be roughly constant, then it would suggest that a person reacts to load force changes in a similar way when handing over an object and when holding an object alone, and thus, we can design robot handover controllers that regulate the grip force in proportion to the load force. Otherwise, it would imply that a different way of controlling grip force is used by humans during handovers, and a more sophisticated robot handover controller design is needed.

In order to test our hypotheses, we conduct an experiment to measure the giver's and receiver's grip and load forces during handovers involving three different object weights. We also further examine our data to identify haptic features related to the transfer of the object load.

3. METHOD

Nine pairs of healthy adults (ten males, eight females) participated in our study. Each pair of participants was asked to hand over a baton to each other a total of 120 times. The weight of the baton was varied as the independent variable while the participants' grip forces and load forces were measured.

3.1 Baton

A baton made of two sheet-aluminum boxes mounted on opposite sides of an ATI (Apex, NC, USA) Mini45 force/torque sensor was used for the experiment (see Figure 1A). The aluminum boxes served as handles for the receiver and giver to grasp, while the force/torque sensor measured the internal forces of the baton. An Interlink Electronics (Camarillo, CA, USA) FSR 406 force sensing resistor (FSR) was attached to one side of each aluminum box to measure the giver's and receiver's grip forces. A thin plastic square was placed over each FSR to distribute the grip pressure. Cardstock was placed over the plastic block and the opposing side of the aluminum box to provide a common gripping surface (see Figure 1B). Fresh pieces of cardstock were used for each pair of participants to provide a consistent gripping surface across all participants. The baton weight was varied over three conditions (*Light* – 483±1g, *Medium* – 577±1g, and *Heavy* – 678±1g) by attaching additional masses to the base of the baton.

The size of the baton handles was chosen in consultation with Mason to facilitate grasping [Mason, personal communication]. The overall size and weight of the baton were chosen to resemble common daily objects such as drinking glasses and bottles. The FSRs were placed at the locations where the giver's and receiver's thumbs would naturally land when grasping the baton.

3.2 Force Sensing and Data Collection

All experimental data were sampled at 50Hz. The ATI force/torque sensor resolution specification is 0.125N in all three axes [18]. A four-stage amplifier circuit was use to linearize the FSR signals before being input to two parallel 14-bit analog-to-digital converters on a Quanser (Markham, ON, Canada) Q8 board running under the WinCon realtime kernel. The FSRs used for grip forces were calibrated against the ATI force/torque sensor using the method described in [19] after preloading with a compressive force greater than 20N and a shear force of 9.81N for over 20hrs. After calibration, the FSRs were tested against the ATI force/torque sensor under varied shear forces (up to 9.81N) and the root-mean-square-errors were found to be 1.5N. Inertial sensors were attached to the baton and to the wrists of the giver and receiver to track accelerations and orientations, but the data

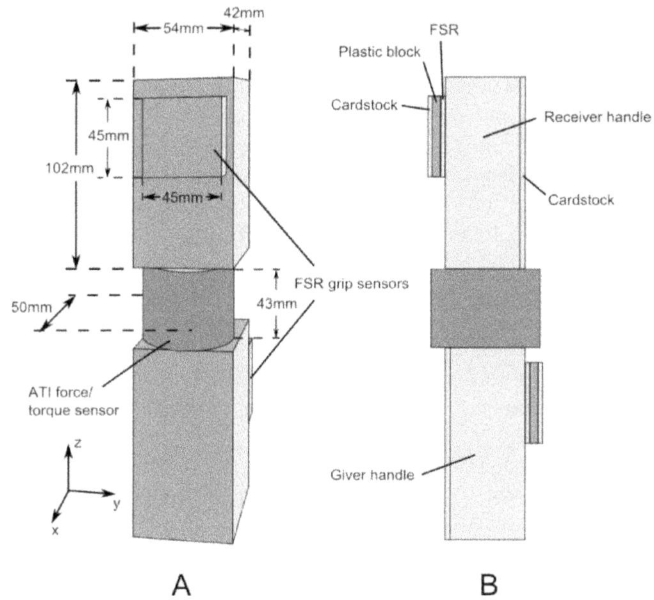

Figure 1. A -- Diagram of baton used in experiment. B -- Side view of baton showing the FSRs for measuring grip forces.

were not used in this study. During the experiment, WinCon software running on the data acquisition computer connected to the Q8 board was used to read the FSR inputs and write them to a shared memory location. Simultaneously, a separate, synchronized process read the ATI force/torque readings, inertial sensor readings, and the shared memory data and stored them to disk. Table 1 lists the information extracted from the data set.

Table 1. Force measurements extracted from participant data.

Variable	Description
f_{GL}	Giver load force
f_{RL}	Receiver load force
$\dot{f}_{RL}$	Load transfer rate
f_{GG}	Giver grip force
f_{RG}	Receiver grip force

In our experiment, we assumed that the object transfer was quasi-static, i.e., the acceleration of the baton in the upward direction was small enough such that the inertial forces were negligible compared to the weight of the baton. Thus we assumed that

$$f_{GL} + f_{RL} = mg \tag{1}$$

where mg is the weight of the object. Since the baton was held upright and only the giver and/or the receiver were carrying the load of the baton at all times, f_{GL} and f_{RL} could be calculated from the force/torque sensor's z-axis force reading, f_Z. To find f_{RL}, an offset was subtracted from f_Z. This offset, reflecting the mass of the upper part of the baton, was calculated by taking the mean f_Z reading over the initial 500ms period when the baton was held motionless and upright by the giver. After that, f_{GL} was calculated as $mg - f_{RL}$ according to Equation 1. The load transfer rate, $\dot{f}_{RL}$, was calculated by taking the time derivative of f_{RL}.

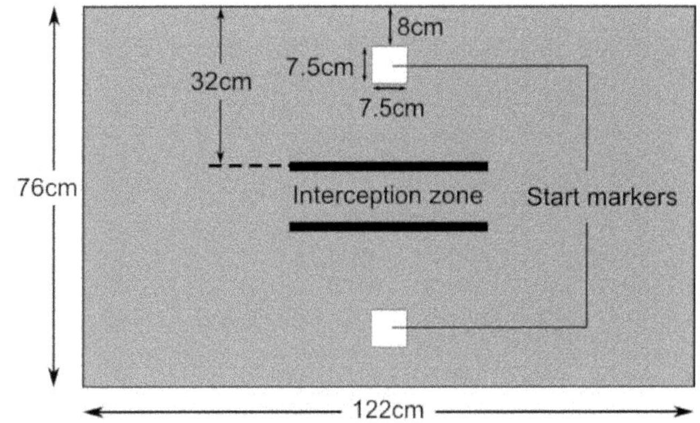

Figure 2. Table setup for the experiment. Two square markers indicate the starting positions for the giver and receiver. An interception zone is marked off in the middle for object transfer.

3.3 Volunteer Recruitment

Volunteers were recruited through advertisements posted around The University of British Columbia campus and through personal contacts. Seven out of the nine pairs of volunteers were already acquainted.

3.4 Experimental Setup and Procedure

Each pair of participants was seated at opposite sides of a table. One person played the role of the giver, and the other played the role of the receiver. A pair of square markers identified the starting position for each participant, and a pair of parallel lines marked off an "interception zone" in the middle of the table (see Figure 2). The giver was asked to grasp the bottom handle with the right hand, and hold it over the nearest marker. Similarly, the receiver was asked to place the right hand over the marker in front of him/her. On a verbal "go" command, the giver moved the baton into the interception zone. Once the giver has reached the interception zone, the receiver reached forward to grasp the baton at the top handle and took the baton from the giver. After the object was transferred, both the giver and receiver returned to their starting positions to complete the handover trial.

The participants were asked to grasp the baton using a precision grip with all digits such that only their fingertips (intermediate and distal phalanges) were touching the baton, and to place their thumbs over the FSRs (see Figure 3). A precision grip was chosen for the following reasons: (i) Mason and Mackenzie's (2005) study, the most relevant study of grip forces in handovers, used a precision grip, and our work expands on their findings [17]; (ii) Zheng et al. (2011) showed that approximately 50% of all grips in tool-intensive activities are precision grips [20]; (iii) with a precision grip, forces can be measured more easily and accurately and with just one sensor since there are only two opposing grasp surfaces, rather than multiple oblique contact surfaces engaged by other grasp types. The participants also were asked to keep the baton upright and off the table, and to maintain the orientation of the baton such that the FSRs were facing towards them at all times. This grasp configuration of the baton was selected to be representative of handovers with a wide range of common objects such as bottles, jars, mugs, etc. No specifications regarding the speed of the handover were given to the participants.

Each giver-receiver pair performed two sets of ten handovers for each of the three conditions in randomized block trials. A short rest-period was enforced after each set of ten handovers before starting the next set to avoid fatigue. Participants were also

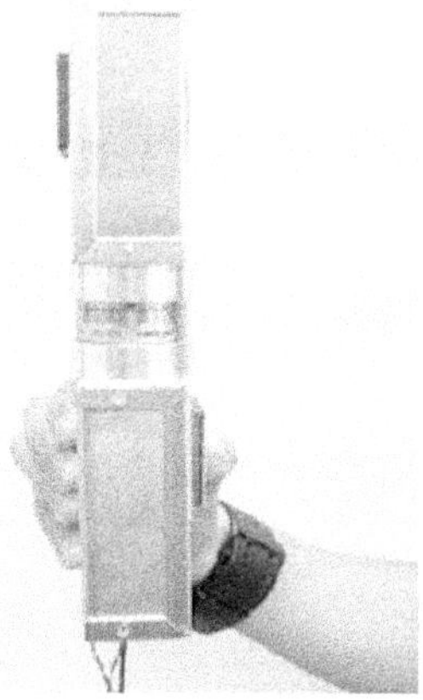

Figure 3. Giver grasping the baton from the bottom handle using a precision grip, with thumb placed over the FSR.

allowed to take additional breaks anytime during the experiment, but none required additional breaks. After a total of sixty handovers, the giver and receiver were asked to switch roles and repeat another sixty handovers.

4. DATA ANALYSES

The receiver load force, f_{RL}, load transfer rate, $\dot{f}_{RL}$, giver grip force, f_{GG}, and receiver grip force, f_{RG}, from a single handover trial in the *Heavy* condition are shown in Figure 4. To reduce noise in f_{GG} and f_{RG} , a fourth order low-pass Butterworth filter with a cutoff frequency of 14Hz was used to filter all grip force data before analysis (a common procedure found in the literature, e.g., [21]). Figure 4 shows the load and grip forces for a typical handover. As expected, in the beginning of a handover when the giver is holding on to the baton, the giver grip, f_{GG}, is high while for the receiver, f_{RG} and f_{RL} are zero. As object transfer takes place, f_{RG} and f_{RL} increase while f_{GG} decreases. After object transfer, f_{GG} drops to zero and f_{RG} rises to a stable value.

To test our hypotheses H_1 and H_2, we examined the giver's and receiver's grip force to load force ratios: f_{GG}/f_{GL} and f_{RG}/f_{RL}. Furthermore, to better understand the giver's and receiver's coordination between grip force and load force during object

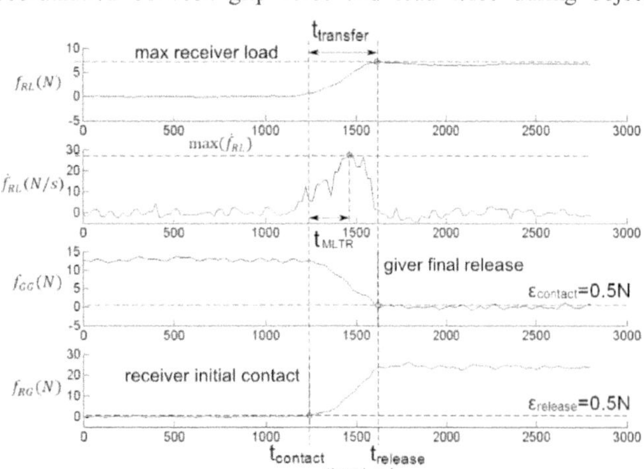

Figure 4. The receiver load force, f_{RL}, load transfer rate, $\dot{f}_{RL}$, giver grip force, f_{GG}, and receiver grip force, f_{RG}, from a single handover trial in the *Heavy* condition. Transfer time $t_{transfer}$ is measured from $t_{contact}$ to $t_{release}$.

transfer, we also examined the plots of the giver's and receiver's grip force versus load force.

To quantify the haptic interaction between the giver and receiver, we measured the transfer time, $t_{transfer}$, the maximum load transfer rate, max $(\dot{f}_{RL})$, and time from receiver contact to maximum load transfer rate, t_{MLTR}, in handovers (see Figure 4). The total time for object transfer, $t_{transfer}$, was defined as the time from initial contact by the receiver, $t_{contact}$, to final release by the giver, $t_{release}$ (i.e., $t_{transfer} = t_{release} - t_{contact}$). We determined $t_{contact}$ by working backwards from the end of the handover data to find the first instant at which the receiver grip force rose above the threshold $\varepsilon_{contact}$. Similarly, $t_{release}$ was determined by finding the first instant when the giver grip force decreased below the threshold $\varepsilon_{release}$. In this work, both $\varepsilon_{contact}$ and $\varepsilon_{release}$ were set to 0.5N. These values were chosen by inspection of the data to be as close to zero as possible while remaining above noise level. The maximum load transfer rate, max $(\dot{f}_{RL})$, was defined as the maximum value of $\dot{f}_{RL}$ during object transfer, and t_{MLTR} was defined as the time from $t_{contact}$ to the point when max $(\dot{f}_{RL})$ was observed.

From our experimental data, we found that often the receiver experiences a load force that is greater than the object weight itself near the end of object transfer. This can be seen in the f_{RL} plot in Figure 4. The maximum receiver load force recorded near the end of object transfer is slightly greater than the receiver load force in the end of the handover (which corresponds to the object weight). We suspect the excess amount of load force may be caused by the giver delaying the release of the object due to the safety margin employed in the grip/load ratio. To investigate this, we also measured the *maximum excess receiver load* as a percentage of the object weight (MERL%).

One-way repeated measures analyses of variance (ANOVA) were used to determine if any of the dependent measures varied significantly across different conditions (α = 0.05). *Post hoc* analysis was performed using the Bonferroni correction. To mitigate any learning effects, the first set of ten handovers in each condition for each pair was treated as a training set. Only the second set of ten handovers was used for data analysis. Note, however, that a comparison of the first and second sets of ten handovers revealed no observable trends and no significant differences between the first and last handovers of each condition. Therefore, we are confident that neither learning nor fatigue significantly affected our results.

5. RESULTS

5.1 Grip Force/Load Force Ratios

Figure 5 shows a representative sample of giver load, f_{GL}, giver grip, f_{GG} , giver grip/load ratio, f_{GG}/f_{GL}, receiver load, f_{RL} , receiver grip, f_{RG}, and receiver grip/load ratio, f_{RG}/f_{RL}, from a giver-receiver pair over ten trials in the *Heavy* condition. The f_{GG}/f_{GL} ratio is shown from the beginning of the handover, when the baton is held by the giver alone, to $t_{release}$. The f_{RG}/f_{RL} ratio is shown from $t_{contact}$, to the end of the handover when the baton is held by the receiver alone. In Figure 5, at the beginning of the handovers when the object is held by the giver, the giver maintained a fairly constant f_{GG}/f_{GL} ratio. Changes in f_{GG}/f_{GL} ratio remained small before and during the first part of object transfer. However, towards the end of object transfer, the f_{GG}/f_{GL} ratio increased rapidly. The receiver on the other hand, had a low f_{RG}/f_{RL} ratio at the beginning of object transfer, and the ratio eventually rose to a fairly stable value. These results show that the

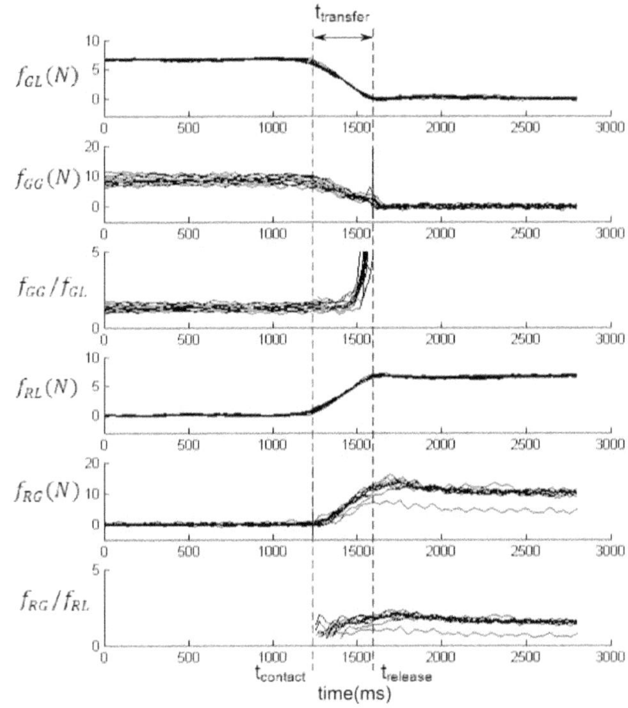

Figure 5. Load force, grip force, and grip/load ratio of giver and receiver for ten trials in the *Heavy* condition. Giver grip/load ratio, f_{GG}/f_{GL}, was fairly stable throughout the handover, but rose quickly towards the end of object transfer. Receiver grip/load ratio, f_{RG}/f_{RL}, started at a lower value in the beginning of object transfer and rose to a fairly stable value. Dash lines show transfer period from one trial.

grip/load ratios change during object transfer and we reject the hypotheses that the giver's and receiver's grip/load ratio would be approximately constant.

5.2 Grip Force versus Load Force

To further explore the coordination of grip force and load force by both the giver and receiver during object transfer, we plot the grip force versus load force for the giver (Figure 6A) and receiver (Figure 6B). Note that as time increased, the giver's grip and load forces decreased, whereas they increased for the receiver. Thus time increases in Figure 6A as the origin is neared, but increases further from the origin for Figure 6B. According to Johansson and Westling, a linear relationship in these plots implies constant ratios of $\dot{f}_{GG}$ to $\dot{f}_{GL}$ and $\dot{f}_{RG}$ to $\dot{f}_{RL}$ (these ratios are equal to the slopes of the curves in Figure 6) [13]. Our results for the giver exhibit good linearity over the majority of the handover tasks. Note the positive intercepts of the curves in Figure 6A with the grip force axis. We will return to this phenomenon later in this work. The receiver's curves (Figure 6B) also show strong linearity for all three conditions but, unlike those for the giver, the receiver's curves *do not* exhibit positive y-intercepts. Nonetheless, it is clear that the slopes of both the giver's and receiver's curves remained nearly constant over the course of object transfer.

When we compare the grip-load plots of different participant pairs, we find that the slopes of the lines vary from one individual to another. Figure 7 compares the grip-load plots of three different pairs of participants under all conditions. As can be seen, the slopes of the curves are fairly constant within each giver and each receiver but seem to vary across individuals.

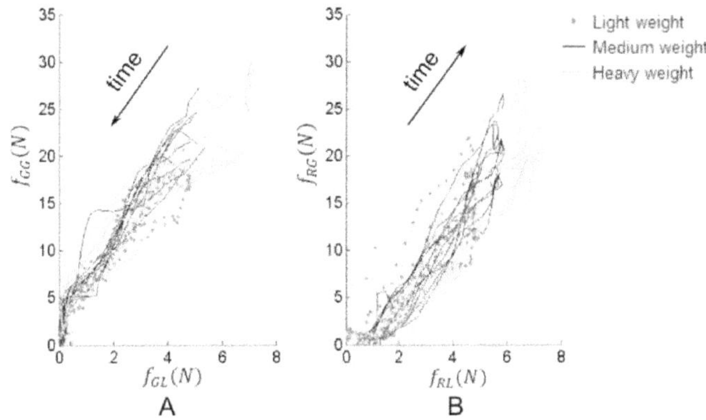

Figure 6. Grip force versus load force plots of one pair of giver (A) and receiver (B). Both plots show good linearity. A positive intercept on the grip force axis is seen on the giver plot but not on the receiver plot.

5.3 Transfer Time and Load Transfer Rate

Figure 8 plots receiver load force, f_{RL}, load transfer rate, $\dot{f}_{RL}$, giver grip force, f_{GG}, and receiver grip force, f_{RG}, for all ten handovers in each condition for a typical giver-receiver pair of subjects. Results from the data analysis show that there were significant differences in average transfer time, $t_{transfer}$ ($F(1.70, 304) = 6.55$, $p = 0.003$), and maximum load transfer rate, $\max(\dot{f}_{RL})$ ($F(2, 358) = 104$, $p < 0.0005$), across different conditions. *Post hoc* analysis shows that $t_{transfer}$ under the *Light* condition was significantly shorter than for *Medium* ($p = 0.029$) and *Heavy* conditions ($p = 0.002$). However, $t_{transfer}$ for *Medium* and *Heavy* conditions were not significantly different ($p = 0.525$). For the *Light* condition, $t_{transfer}$ was 470±159ms while in the *Medium* and *Heavy* conditions $t_{transfer}$ values were 500±182ms and 523±201ms respectively. This indicates that $t_{transfer}$ increased with object weight but not linearly, leveling off as the object became heavier. For $\max(\dot{f}_{RL})$, significant differences were found across all conditions (for all pairwise comparison $p < 0.0005$). Average $\max(\dot{f}_{RL})$ increased with heavier object weight and was found to be 19.00±4.43N/s, 21.84±4.80N/s, and 25.14±5.47N/s for *Light*, *Medium*, and *Heavy* conditions respectively. This suggests that the receiver is acquiring the object's load from the giver at a faster rate when the object is heavier. There were not, however, significant difference in t_{MLTR}

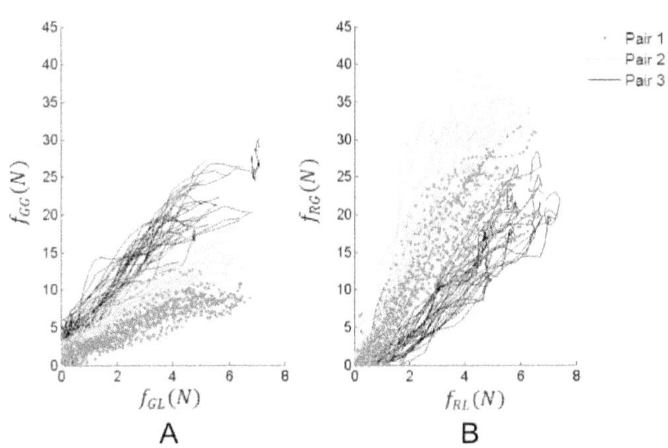

Figure 7. Grip force versus load force plots for three different pairs of giver (A) and receiver (B) showing variations in the slope of the curves.

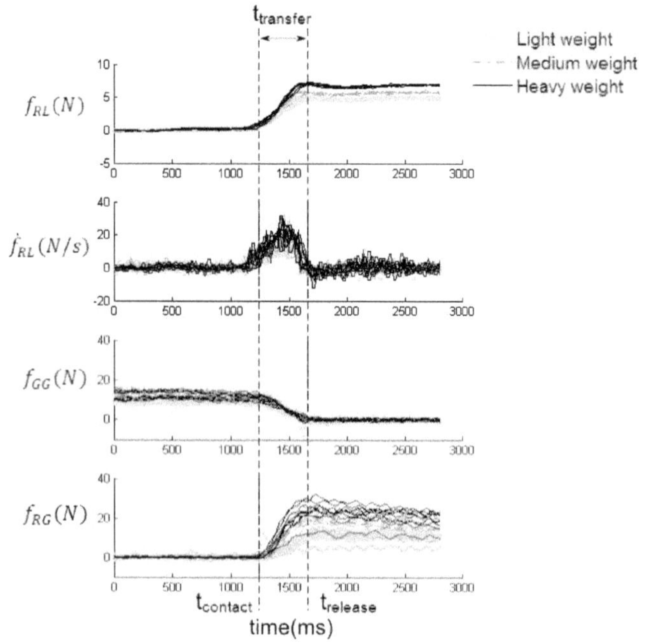

Figure 8. The receiver load force, f_{RL}, load transfer rate, $\dot{f}_{RL}$, giver grip force, f_{GG}, and receiver grip force, f_{RG}, in *Light*, *Medium*, and *Heavy* conditions. Dash lines show transfer period from one trial.

across conditions (F(1.81, 324) = 2.64, p = 0.079). Overall t_{MLTR} was found to be 218±148ms.

5.4 Maximum Excess Receiver Load (MERL)

Recall that the maximum excess receiver load was defined as the amount of excess load force experienced by the receiver towards the end of object transfer. Comparing the maximum excess receiver load as a percentage of object weight, MERL%, across different conditions, we found that it was 2.36±4.16% of the object weight under all conditions with no significant differences (F(2, 358) = 0.747, p = 0.474). A one-sample t-test revealed that the average MERL% was significantly different from zero (t(539) = 13.2, p < 0.0005). Figure 9 shows the frequency distribution of MERL%. In most of our handovers (429 out of 540 handovers) MERL% was greater than zero. This suggests that in approximately 80% of the handovers, the giver did not fully release the object until he/she felt a slight upward pulling force.

6. DISCUSSION

6.1 Human Control Strategy for Handovers

Our examination of the grip force to load force ratios of the giver and receiver (Figure 5) indicated that the force ratios were held constant during most of the object transfer period except near the beginning of object transfer (for receiver), and just prior to the end of object transfer (for giver). Thus humans appear to react to load force changes differently during object transfer compared to holding an object alone. Near zero load force, the giver employed a very high ratio (i.e., grasped the object more tightly) whereas the receiver had a lower ratio. This implies that, in handovers, humans do not simply adjust their grip force directly proportional to the load force. Further examination of the grip versus load plots (Figure 6) reveals that the slopes of the grip-load curves were held roughly constant by the giver and receiver during object transfer. This result agrees with Johansson's and Westling's findings on pick and place tasks [13]. Johansson and Westling found that when a person picked up or placed down an object on a table, the

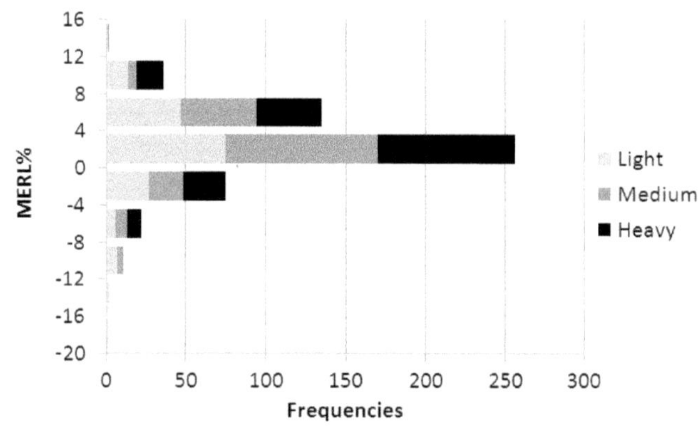

Figure 9. Frequency distributions of MERL% in all conditions. The first, second, and third quartiles were 0.5%, 2.5%, and 4.8% respectively. This shows that most of the time MERL% was greater than zero, which means that the maximum load force experienced by the receiver during object transfer was greater than the object weight.

slope of the grip-load curve was also held approximately constant for most of the task duration. It appears that when approaching handover, humans employ a similar strategy to pick and place tasks.

Our data show that the slope of the grip-load curve for an individual during object transfer is independent of object weight. However, from the comparison shown in Figure 7, we saw that the slope did differ amongst individuals. The variations seen across individuals may be attributed to the differences in the coefficient of friction between the baton and each individual's finger tips, or differing safety margins employed by individuals. A lower coefficient of friction would require a greater grip force to prevent slip, increasing the slope of the grip-load plots. This explanation is consistent with Johansson's and Westling's findings [13]. Similarly, givers employing greater safety margins would grasp the object more tightly, which also would lead to greater slopes in the grip-load plots.

Upon realizing the similarities between pick and place tasks and handovers, we compared the characteristics of the grip force and load force data from our experiment with those of Johansson's and Westling's [13] in greater detail to determine if **1)** the task of receiving an object from a person is similar to picking one up from a table, and **2)** the task of giving an object to a person is similar to placing one on a table. Our data suggest that receiving an object is different from picking up an object, and that giving an object is different from placing an object. In fact, the characteristics observed in a *receiving* task are more similar to a *placing* task, while the characteristics observed in a *giving* task are more similar to a *picking* up task.

Johansson and Westling found that there is a "pre-loading" period at the beginning of a picking up task during which the person increases the grip force without lifting the object from the table. This results in a large grip/load ratio and appears as a near-vertical segment on a grip-load plot near zero load force. These characteristics do not appear in our receiver force data, but are observed in our giver force data. This suggests that the giver has a "post unloading" period near the end of object transfer where he/she applies a positive grip even after the load force has reached approximately zero. The grip/load ratios of a placing task and receiving task, on the other hand, do not have increased values near zero load force and both have grip/load ratios that are at lower values near zero load force.

The similarities between picking up tasks and giving tasks, and between placing tasks and receiving tasks may be a reflection of the relative difficulties of the tasks. Picking up an object may be more difficult compared to placing an object, as the person is faced with greater uncertainty. Uncertainties in picking up an object include its surface friction and weight. One must rely on experience and/or memory in order to estimate these properties for proper grip force control [13]. If an object was picked up without adequate grip force, it may slip and drop. As a result, in a picking up task, the person applies a "pre-loading" period to build up the grip/load ratio in order to achieve a greater safety margin. In a similar way, giving an object may be more difficult than receiving an object. During a handover, the rate of object transfer is controlled mainly by the receiver through the action of taking the object. The giver has little control over how fast the load is transferred and instead must respond to load force changes. The giver can only infer when an object is in the receiver's control from the absence of slipping. A premature release by the giver could lead to the object being dropped. As a result, the giver appears to use a "post unloading" period to increase the safety margin by applying a greater grip force than necessary near the end of load transfer in order to ensure safe object transfer.

Placing an object, on the other hand, simply requires the person to reduce the grip force to zero once the object has made stable contact with the table. There is no danger of slip since the object cannot fall through the table's surface. Likewise, the receiver simply has to increase the grip force once the giver has placed the object in their hand. Thus the receiver maintains a roughly constant slope on the grip-load curve throughout object transfer, and does not employ a larger grip/load ratio near zero load force.

6.2 Transfer Time and Load Transfer Rate

Different object weights lead to different handover difficulties. A heavier object imposes greater difficulty by requiring additional participant attention, as dropping it could lead to more severe consequences (e.g., injury). Our results show that when the weight of the object was *Heavy* or *Medium*, the average time required for object transfer (i.e., transfer time) was significantly higher than the average time required when the object was *Light*. This increase in $t_{transfer}$ may be caused by the increase in task difficulty. However, when object weight increased from *Medium* to *Heavy*, the $t_{transfer}$ seemed to level off and did not increase significantly. This may indicate that there is weight range at which the object transfer task difficulty is relatively similar. However, it can be expected that there is a higher weight range where task difficulty will once again increase. The average $\max(\dot{f}_{RL})$ was found to increase significantly with increasing object weight, indicating that receivers adjust the rate of load transfer to maintain efficient object transfer times.

6.3 Maximum Excess Receiver Load (MERL)

In our experiment, we found that the MERL% was typically around 2.36%. If the quasi-static assumption holds, this excess load force experienced by the receiver at the end of object transfer corresponds to the giver maintaining their grasp on the object even after the receiver was fully supporting the object's weight. This may be a precautionary behavior on the part of the giver to ensure safe object transfer. Given that the measured MERL percentages were quite small, further investigation is needed to rule out dynamic forces due to object acceleration before stronger conclusions can be drawn.

6.4 Implications on Robot Handover Controllers

With these results, we have achieved a better understanding of the haptic interaction in human-to-human handovers, and based on our findings, we propose the following guidelines for implementing human-like robot handover controllers for handing over common objects such as jars and mugs:

Object Transfer Period. In Section 5.3, we reported that the object transfer time $t_{transfer}$ in all three conditions was approximately 500ms. This finding is consistent with Mason's and Mackenzie's results [17]. Based on this, our recommendation is that the human-robot object transfer period should be approximately 500ms.

Object Safety and Load Transfer Efficiency. When we inspected the giver's grip force to load force ratio in Section 5.1, we found that a human giver has an increased grip/load ratio towards the end of the object transfer period. In our discussion in Section 6.1, we pointed out that this is a "post unloading" period employed by the giver to ensure a safe object transfer. A human giver takes on the responsibility of ensuring object safety in a handover. Therefore, as giver, the robot should be responsible for object safety. When we examined the maximum rate at which the receiver takes the load, we reported in Section 5.3 that $\max(\dot{f}_{RL})$ increases with heavier objects. A human receiver increases $\max(\dot{f}_{RL})$ when there is more object load to be transferred, perhaps to prevent an inordinately long $t_{transfer}$. Therefore, as receiver, the robot should be in charge of load transfer efficiency.

Grip Force Safety Margin. In Section 5.2, we reported that the giver's grip-load curve has a positive intercept, which means that a non-zero grip force is maintained even when the load force has reached approximately zero. This is the safety margin employed by human givers to ensure safe completion of object transfers. According to this, we recommend that, as giver, the robot should provide a grip force safety margin so that the grip force remains non-zero even when the load force has reached zero.

Complete Release of Object. In Section 5.4, we reported that near the end of object transfer, the receiver often experiences an excess amount of load force that is around 2.5% of the object weight, which corresponds to a slight upwards pulling force experienced by the giver. Therefore, we recommend that complete release of the object should occur once the load force on the giver becomes negative.

7. CONCLUSION

In this study, we investigated and characterized the haptic interaction in human-to-human handovers under different object weight conditions by measuring giver and receiver grip and load forces. We found that both giver and receiver employ a similar control strategy, but the giver appears to be in charge of ensuring object safety whereas the receiver appears to control handover efficiency. In comparing our results with the literature, we found that although the control strategies for pick and place task and for handovers are similar, there are significant differences. In particular we found that characteristics of the grip force and load force in a receiving task are more similar to a placing task, and the characteristics of a giving task are more similar to a picking up task.

Based on our results, we have provided guidelines for designing human-robot handover controllers. In our future work, we will propose a handover controller for robots based on these findings to enable more intuitive human-robot handovers.

8. ACKNOWLEDGEMENTS

The support of the Natural Sciences and Engineering Research Council of Canada and the Institute for Computing, Information and Cognitive Systems is gratefully acknowledged. We also thank Dr. Andrea Mason for providing us with guidelines on choosing the baton size, and all the participants in our study.

9. REFERENCES

[1] B. Graf, C. Parlitz, M. Hägele, and F. Ipa, "Robotic Home Assistant Care-O-bot® 3 Product Vision and Innovation Platform," *Human-Computer Interaction*, 2009. pp. 312-320.

[2] K. Yamazaki et al., "Revealing Gauguin: Engaging Visitors in Robot Guide's Explanation in an Art Museum," *CHI*, 2009.

[3] S. Shibata, B.M. Sahbi, K. Tanaka, and A. Shimizu, "An Analysis of The Process of Handing Over An Object and Its Application to Robot Motions," *SMC*, vol. 1, Apr. 1997, pp. 64-69.

[4] M. Huber, M. Rickert, A. Knoll, T. Brandt, and S. Glasauer, "Human-Robot Interaction in Handing-Over Tasks," *RO-MAN 2008 - The 17th IEEE International Symposium on Robot and Human Interactive Communication*, 2008, pp. 107-112.

[5] P. Basili, M. Huber, T. Brandt, S. Hirche, and S. Glasauer, "Investigating Human-Human Approach and Hand-Over," *Human Centered Robot Systems*, vol. 6, 2009, pp. 151-160.

[6] C. Becchio, L. Sartori, and U. Castiello, "Toward You: The Social Side of Actions," *Current Directions in Psychological Science*, vol. 19, Jun. 2010, pp. 183-188.

[7] M. Cakmak, S.S. Srinivasa, M.K. Lee, S. Kiesler, and J. Forlizzi, "Using Spatial and Temporal Contrast for Fluent Robot-Human Hand-overs," *6th ACM/IEEE International Conference on Human-Robot Interaction*, 2011.

[8] A. Edsinger and C.C. Kemp, "Human-Robot Interaction for Cooperative Manipulation: Handing Objects to One Another," *RO-MAN 2007 - The 16th IEEE International Symposium on Robot and Human Interactive Communication*, 2007, pp. 1167-1172.

[9] K. Nagata, Y. Oosaki, M. Kakikura, and H. Tsukune, "Delivery by Hand between human and Robot Based on Fingertip Force-Torque Information," *IROS*, 1998, pp. 750-757.

[10] Willow Garage. (2010). Beer Me, Robot. [Online]. Available: http://www.willowgarage.com/blog/2010/07/06/beer-me-robot.

[11] M. Simoneau, J. Paillard, C. Bard, N. Teasdale, O. Martin, M. Fleury, and Y. Lamarre, "Role of the Feedforward Command and Reafferent Information in the Coordination of a Passing Prehension Task.," *Experimental Brain Research*, vol. 128, Sep. 1999, pp. 236-242.

[12] J. Monzée, Y. Lamarre, and A.M. Smith, "The Effects of Digital Anesthesia on Force Control Using a Precision Grip," *Journal of Neurophysiology*, vol. 89, Feb. 2003, pp. 672-683.

[13] R.S. Johansson and G. Westling, "Roles of Glabrous Skin Receptors and Sensorimotor Memory in Automatic Ccontrol of Precision Grip when Lifting Rougher or More Slippery objects," *Experimental Brain Research*, 1984, pp. 550-564.

[14] R.S. Johansson and G. Westling, "Programmed and Triggered Actions to Rapid Load Changes During Precision Grip," *Experimental Brain Research*, 1988, pp. 72-86.

[15] A.M. Wing and J.R. Flanagan, "Anticipating Dynamic Loads in Handling Objects," *Proceedings of the ASME Dynamic Systems and Control Division*, vol. 64, 1998, pp. 139-143.

[16] J. Hermsdörfer, C. Marquardt, J. Philipp, A. Zierdt, D. Nowak, S. Glasauer, and N. Mai, "Moving Weightless Objects Grip force Control During Microgravity," *Experimental Brain Research*, 2000, pp. 52-64.

[17] A.H. Mason and C.L. Mackenzie, "Grip Forces when Passing an Object to a Partner," *Experimental Brain Research*, vol. 163, May. 2005, pp. 173-187.

[18] ATI Industrial Automation. (2011). F/T Sensor: Mini45 [Online]. Available: http://www.ati-ia.com/products/ft/ft_models.aspx?id=Mini45.

[19] R. S. Hall, G. T. Desmoulin, and T. E. Milner, "A Technique for Conditioning and Calibrating Force-Sensing Resistors for Repeatable and Reliable Measurement of Compressive Force," *Journal of Biomechanics*, vol. 41, no. 16, Dec. 2008, pp. 3492-3495.

[20] J. Z. Zheng, S. D. L. Rosa, and A. M. Dollar, "An Investigation of Grasp Type and Frequency in Daily Household and Machine Shop Tasks," *IEEE International Conference on Robotics and Automation*, 2011, pp. 4169-4175.

[21] J. R. Flanagan and M. A. Beltzner. "Independence of Perceptual and Sensorimotor Predictions in the Size-Weight Illusion," *Nature Neuroscience,* vol. 3, 2000, pp. 737-741.

Designing Robot Learners that Ask Good Questions

Maya Cakmak and Andrea L. Thomaz

School of Interactive Computing, Georgia Institute of Technology
801 Atlantic Dr., Atlanta, GA, USA
{maya,athomaz}@cc.gatech.edu

ABSTRACT

Programming new skills on a robot should take minimal time and effort. One approach to achieve this goal is to allow the robot to ask questions. This idea, called Active Learning, has recently caught a lot of attention in the robotics community. However, it has not been explored from a human-robot interaction perspective. In this paper, we identify three types of questions (label, demonstration and feature queries) and discuss how a robot can use these while learning new skills. Then, we present an experiment on human question asking which characterizes the extent to which humans use these question types. Finally, we evaluate the three question types within a human-robot teaching interaction. We investigate the ease with which different types of questions are answered and whether or not there is a general preference of one type of question over another. Based on our findings from both experiments we provide guidelines for designing question asking behaviors on a robot learner.

Categories and Subject Descriptors

I.2.9 [**Artificial Intelligence**]: Robotics; H.1.2 [**Models and Principles**]: User/Machine Systems

Keywords

Learning from Demonstration, Active Learning

1. INTRODUCTION

Learning from Demonstration (LfD) is aimed at allowing end-users to program new skills on a robot with minimal effort [1]. To this end, LfD techniques try to maximize the generalizability of the learned skill to unseen situations with a minimal number of demonstrations provided by humans. Recently, Active Learning (AL) methods have been explored to achieve this goal. The idea in AL is to improve learning rates by giving the learner more control over what examples it receives. The AL process involves the learner making a *query*[1], which is a request for certain information, and the teacher provides this information. The requested information can have various forms, although most commonly it is a category label for an unlabeled data instance. Different strategies are available for choosing such queries (*e.g.* reducing uncertainty or reducing variance).

Figure 1: Participants demonstrating three different skills to our robot Simon.

Our focus, in this paper, is how the incorporation of AL methods in LfD impacts a robot's interaction with its user. In prior work looking at using AL in an HRI setting, we learned that compliance with the robot's questions is not guaranteed [2, 5]. Human teachers sometimes ignore the robot's questions, particularly when they judge it to be a *bad* question. Thus, in this work we try to operationalize what it means for a robot learner to ask *good* questions.

First we look at the notion of good questions from the machine learning perspective. We identify three types of queries from AL (label, demonstration and feature queries) and discuss how they can be applied in an LfD setting (Sec. 3). Then we look at the notion of good questions from a human learning perspective, with an experiment into human question asking (Sec. 4). This allows us to characterize people's usage of the identified query types. Finally, we evaluate the three query types in an HRI setting, where people teach skills to a humanoid robot via LfD (Sec. 5). We find that feature queries are perceived as the smartest and label queries are the easiest to answer.

2. RELATED WORK

The potential of AL for making learning efficient has been been noticed by the Robotics community. AL has been applied to a number of LfD frameworks. For example, *confidence based autonomy* [6] and *dogged learning* [12], use the principle of uncertainty sampling to select states in which

[1]We use the terms (making a) *query* and (asking a) *question* interchangeably throughout the paper.

the learning agent requests a demonstration while learning a policy. In [10], the robot actively selects points outside the region of stability of a learned policy, and requests demonstrations from these states.

The evaluation of robotic systems that ask questions to humans has been limited, particularly in LfD settings. Our earlier work compares passive and active task learning [5] and addresses the question of *when* to ask questions in a mixed-initiative AL setting [2]. Rosenthal et al. investigate how augmenting questions with different types of additional information improves the accuracy of human teachers' answers [21]. In later work, they explore the use of humans as information providers in a real-world navigation scenario [22]. Other related work explores the idea of learning actions on a robot through dialog [4].

Evaluations of AL systems with naive teachers in Human-Computer Interaction (HCI) are also relevant. For instance, our finding in previous work that people do not enjoy a constant stream of questions, was shown in a recommendation system domain [13]. Other studies investigated different HCI issues such as feasibility of novel query types (*e.g.* feature queries) [7] or cost of different query types [23]. Similar to this paper, Gervasio *et al.* investigated question asking in procedure learning with a range of question types [8, 9].

3. QUERY TYPES IN ACTIVE LEARNING

The conventional *query* in the AL literature involves choosing an unlabeled instance and requesting a label for it. The instance can be chosen from a pool of unlabeled instances or instantiated by the learner in some way. Such queries have been used in learning skills on a robot, where a skill is represented as a policy that maps a state to a discrete action. In this context, a query consists of asking for an action in a chosen state [6, 17, 12]. This is useful in an HRI setting when the robot's actions are discrete and the human has a way to refer to each action. However, robot skills often involve continuous actions and the input from the teacher to the learner is a sequence of state-action pairs (*i.e.* trajectories). In these cases it is impractical to ask for an isolated state-action pair (*e.g.* asking for the motor commands of a given arm configuration). Thus we need to re-think the way that a query is made. We consider the following alternatives.

3.1 Label queries

Robot skill learning involves modeling/encoding a skill from a set of demonstrations, such that it can be reproduced correctly. The demonstrations provided by the teacher are inherently labelled as *positive*. One way the robot could make queries is to execute a motion and ask whether the skill was performed correctly. We call this a *label query*.

The information provided by label queries depends on the answer. If the response is positive, then the motion can be used as another demonstration. What to do with negative examples is not as straightforward. LfD methods are designed to encode a skill from positive examples only, since it is unnatural to demonstrate "what not to do". One way to make use of negative examples that arise from label queries, is to update the learned model such that the probability of the negative data being generated by the model is minimized while the probability of the positive data being generated is maximized. The main issue with this idea is the attribution of negativity to the whole trajectory, while only parts of the trajectory might be responsible for the failure. A second approach for making use of negative examples is to guide the learner's future queries towards positive examples (*e.g.* [11]).

Methods for generating label queries depend on the particular framework used for representing the skill. However a general approach applicable to most frameworks is to sample trajectories from the learned skill and evaluate them with a certain criterion. For instance, the robot can choose to query the trajectory that it is least certain about or the trajectory that is most likely to increase the applicability of the skill.

3.2 Demonstration queries

The second type of query, which we call *demonstration* or *demo queries*, involves creating a new scenario and requesting a demonstration from the teacher. Demo queries give less control over what information is acquired from a query, as compared to label queries. In label queries, the learner specifies the whole trajectory and the teacher only provides a label. In demonstration queries, the learner only specifies certain constraints, while the trajectory is still produced by the teacher. Nevertheless, this gives some control to the learner such that useful demonstrations can be acquired. Demo queries are analogous to a method known as *active class selection* [16], which consist of requesting an example from a certain class.

One way to constrain trajectories provided by the teacher is to specify the starting state. Trajectories are often represented with a sequence of end effector configurations relative to a goal object. Thus the robot can configure its end effector in a certain way relative to the goal and request the demonstration to start in this configuration. A larger range of queries can be made by manipulating the target object. A different way of constraining the demonstrations provided by the teacher is to allow the teacher to control only a subset of the robot's joints while the robot executes a certain trajectory on the rest of the joints, as in [3].

3.3 Feature queries

The third type of query is inspired from a relatively recent technique in AL that involves asking whether a feature is important or relevant for the target concept that is being learned [20, 7]. A particularly successful application of this method is document or email classification. Learning approaches to this problem involve the teacher reading documents one by one and providing category labels for them. Even in an AL setting, providing a label for the queried documents is cumbersome and time consuming. On the other hand a feature query directly asks whether a word is a strong indicator of a certain category. This allows the learner to directly modify its model for categorizing new documents and drastically reduces the time spent by the teacher.

Note that the crucial element in the success of feature queries in the given example is that the features (words) with which the instances (documents) are represented are meaningful for humans and the way they contribute to the classification of the instance (being a strong indicator) is intuitive. We believe that robot skill learning is at a unique position for taking advantage of this method: while features might not be as human friendly (*i.e.* feature names might be too technical and feature values might be arbitrary numbers) the robot's embodiment can be used to refer to them.

Methods for choosing feature queries are also dependent on the framework used for representing and learning skills. One framework that allows feature queries to be directly integrated is task space selection [18, 14]. This involves representing demonstrations in high dimensional spaces that involve features that might or might not be relevant (*e.g.* relative position and orientation of different points on the robot to different objects in the environment). Methods try to identify a subspace or assign weights to each feature such that the skill is best represented. In this context a feature query is to directly ask whether a feature is important for a skill. These queries can also be used for directly manipulating the skill representation or for guiding other queries.

4. HUMAN QUERIES DURING LEARNING

In designing good questions for a robot learner, the Active Learning literature gives us one perspective, but we are also interested in taking human question asking behavior as inspiration. Our intuition is that questions people typically use will be easy for them to answer. While human question asking has been investigated and modeled in different cognitive tasks such as text understanding [15, 19], we did not find any studies on human questions in the context of skill or task learning by demonstration. Thus, we conducted an experiment to investigate how humans ask questions in scenarios similar to the ones robots are faced with. Our experiment involves providing the participant with demonstrations of a task, letting them ask questions about it and then asking them to perform it in a new situation.

4.1 Experimental Design

We consider four tasks described in Fig. 2. The tasks are chosen to cover the two types of LfD problems in robotics. The first two tasks are focused on *learning a task goal* (desired end state) whereas the latter two emphasize *learning skills* to perform the task (desired actions or movements to accomplish the task). Each of these are designed to be an abstraction of a real-world task (*e.g.* making a sandwich), to let the experimenter answer participants' questions based on the real-world counterpart. The abstraction also avoids the influence of prior knowledge about the tasks.

The tasks or skills involve a fixed set of objects. Two demonstrations, varied in certain aspects, were recorded for each skill. The recordings were used for demonstrating the skill to the participants. This is done to mitigate unintended variances between the demonstrations given to different participants. Participants learn all four tasks and the order is counterbalanced across participants. The interaction between the participant and the experimenter is recorded by a camera that oversees the table (Fig. 5).

4.1.1 Procedure

Throughout the experiment, the participant sits in front of a table with a computer screen to their right for displaying the demonstration videos. For each task, the experimenter first brings the involved objects onto the table. Then the participant watches the video. Participants are told that they can watch the video as many times as they want, at any time during the experiment. This is to reflect the fact that a robot has access to the demonstrations provided by a teacher before they ask questions and forgetting what happened in the demonstration is not an issue. Thus we would like to avoid questions that address information present in the videos. After watching the video at least once, the participant is probed for questions. There is no upper limit on number of questions, but a lower limit of three questions.

TASK 1 (Sandwich): Construct four layers. Bottom is a large green circle, second layer is a large purple circle, third layer has 3 or more non-overlapping small circles placed within the large circle, and top layer is a large green circle. Large circles are aligned.

TASK 2 (Laundry): Place all blue objects on the table into the large yellow square bin.

TASK 3 (Salt): Grasp the yellow block from the table, make it horizontal with its top above the blue bin. Move it down a few inches and back up. Number of lowering movements and length of the movement depends on the size of the bin. Place the block on the table vertically.

TASK 4 (Scoop&pour): Grasp the green block from the table. Dip it into the yellow bin and rotate it clockwise around 90°. Then move it to the blue bowl maintaining its orientation. Dip it and turn it counter-clockwise around 90°. Place the block back on the table.

Figure 2: Description of the four tasks used in the experiment and snapshots from the two demonstration videos shown to the participants.

When the participant has no more questions, the experimenter configures the environment in a certain way and asks the participant to perform the task. The participant is reminded that they can keep asking questions, during and after their execution. The execution is repeated twice with different starting configurations.

In the instructions, participants are encouraged to ask any type of question that they can think of and that they are not restricted to yes/no questions. They are encouraged to refer to the objects on the table in their questions. Participants are told that the purpose of the study is to take inspiration from their questions to allow robots to ask questions. Thus we tell them not to ask questions about the underlying purpose of the tasks since explaining these to a robot would be impossible. Such questions are not answered by the experimenter if they occur.

4.1.2 Evaluation

All recordings were later transcribed by the experimenter using ELAN[2]. This includes annotations of the participant's questions, the answers given by the experimenter and the participant's interaction with the objects and gestures. In addition we annotated the demonstration views by the participant and the task executions.

The participants' questions are categorized into the three query types described in Sec. 3 based on the following criteria. If the question asks the experimenter to evaluate an action or a complete task execution it is considered a *label query*. If the question requests an action, an action suggestion or a complete task execution from the experimenter, it

[2] www.lat-mpi.eu/tools/elan/

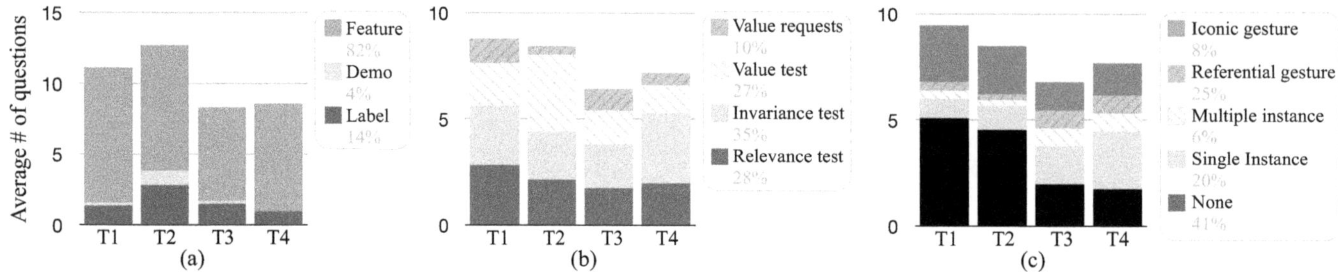

Figure 3: (a) Distribution of questions into three types of queries for each task in the first experiment averaged across participants. (b) Distribution of feature queries into sub-categories. (c) Average distribution of common actions and gestures that physically ground questions across tasks.

is categorized as a *demonstration query*. Lastly, all questions that address a certain aspect of the task in general terms is considered a *feature query*.

We also categorized the questions in terms of their form based on the taxonomy established in [15]. This taxonomy is shown in Fig. 4 with examples from our experiment for each form of question.

4.2 Results

12 participants (3 female, 9 male) completed our first experiment. The interactions took about 25 minutes and involved an average of 40.58 (SD=13.09) questions.

4.2.1 Query types

The distribution of questions asked by participants across the three query type categories is given in Fig. 3(a). We observe that a significant majority of questions asked by humans (82%) are feature queries. We notice a large variety of query types within feature queries in terms of the information they attempt to gather from the teacher. We establish the following subcategories of feature queries (a similar distinction was made in [8]).

- *Feature relevance tests* directly ask whether a feature is important for a task: *Does f matter?* (Fig. 5(b))
- *Feature invariance tests* ask whether a feature has to have a certain value: *Does f have to be f_1?* (Fig. 5(d))
- *Feature value tests* ask whether a feature is allowed to have a certain value: *Can f be f_1?* (Fig. 5(a))
- *Feature value requests* directly asks the value range that a feature is allowed to have: *What can f be?* (Fig. 4 (SIMPLE))

Note that f can refer to an attribute of the involved objects, relations between objects, a property of the involved movements, a parameter of the task description or the dependency between any two of these. The distribution of feature queries into these subcategories is shown in Fig. 3(b). We observe a rather uniform distribution across the first three categories, while the open-ended value requests are less common.

In addition we observe a distinction between *full* and *partial* demonstration and label queries. A full demonstration query requests a complete demonstration while a partial one requests a single step of the demonstration. Similarly, a full label query asks for an evaluation of a complete task execution while a partial label query asks whether a step was correct. Such partial label queries seem to be a useful mechanism for avoiding the credit assignment problem that arise when a negative label is encountered (Sec. 3.3).

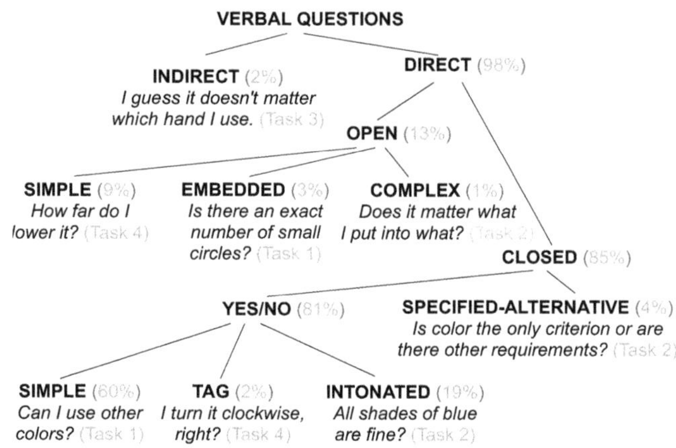

Figure 4: Question forms [15] and example instances from our experiment (percentage of average occurrence shown in parentheses).

4.2.2 Question forms

The average usage percentage of each question form across tasks is shown in Fig. 4. We observe a clear preference for yes/no questions (81%), more often posed in the simple form (60%). We notice that the choice between the three types of yes/no questions reflect whether the person has an expectation about what the answer is, and indicates their confidence in this expectation. While *simple* yes/no questions are neutral about either answer, *tag* yes/no questions (*e.g. right?*, *correct?*) indicate strong expectation of the answer being "yes". One example is the participant saying "I can use either hand, right?" when he has already started a movement with his alternate hand.

4.2.3 Physical grounding of questions

We observe that human questions during task learning are largely embodied and physically grounded. On average 62% of the questions asked by a participant involved interacting with the environment. Label queries are inherently embodied questions since they require the robot to perform the actions to be labelled. Similarly demo queries require the person to create the scenario in which a demonstration is requested. It is, however, possible for the person to verbally

(a) "Can I do the motion without tilting this?"

(b) "Does the way in which I grasp this matter?"

(c) "The circle on the top, it has to be green?"

(d) "Small circles are always one red and some blues?"

Figure 5: Examples of the four common types of physical groundings used with questions.

describe a sequence of actions (or a hypothetical scenario) and request a label (or a verbal demonstration).

Feature queries on the other hand are more amenable to purely verbal question asking, however they require the ability to talk about features and feature values. More specifically, they require tokens that have a common physical grounding for the two parties (person asking the question and person being asked) to replace f and f_1 in Sec. 4.2.1. Even though such tokens might exist, we observe that they are often replaced or accompanied by actions or gestures that communicate them – we saw four common types:

- A *single value instantiation* refers to actions that generate a certain feature value while asking a question about it. For instance in Fig. 5(a) the person asks if the yellow block can be lowered without being tilted while providing an instance of non-tilted orientations.
- *Multiple value instantiations* refer to actions that generate several different values of feature during a feature query. An example is the participant showing several different grasps while asking whether the way in which the block is grasped matters (Fig. 5(b)).
- *Referential gestures* include pointing, touching or holding up a certain object in the environment while making a feature query that relates to a feature of the object. For instance, in Fig. 5(c) the participant holds up a green large circle while asking whether the top layer has to be green.
- *Iconic gestures* involve arm and hand movements that communicate a certain feature or feature value referenced in a question. An example, shown in Fig. 5(d), is placing two hands at a certain distance from each other to indicate a certain size.

The average distribution of these actions and gestures across subjects is given in Fig. 3(c). We observe that 59% of feature queries made by participants involve some form of physical ground. Single value instantiations and referential gestures are more common than the other two types. Also we observe interesting differences in how they are used in different tasks, which we discuss in Sec. 4.2.4.

4.2.4 Learning Goals versus Skills

We do not see major differences in the distribution of query types or question forms between tasks that require learning goals (Tasks 1 and 2) versus tasks that involve learning skills (Task 3 and 4). The main difference in the questions asked during goal versus skill learning is seen in the physical grounding of the questions (Fig. 3(c)). We find that questions during skill learning are more likely to be physically grounded (74%) than during goal learning (46%). This is not surprising as movements are harder to describe verbally. Secondly, we see that value instantiations are more common in skill learning (41%) as compared to goal learning (15%). Correspondingly, a larger portion of the questions during goal learning use referential and symbolic gestures. This is also an intuitive finding as the features involved in skill learning are about the movements, while the features involved in goal learning are about objects in the world. In a sense, referring to features of an object in the world, or using a gesture to indicate a feature value are both instantiations of the feature.

5. ROBOT QUERIES DURING LEARNING

Next, we investigate the use of the three types of queries described in Sec. 3 with an HRI experiment that involves teaching skills to a robot and answering its questions.

5.1 Platform

The robot platform used in our experiment is Simon, an upper torso humanoid robot with 7 Degree of Freedom (DoF) compliant arms, 4 DoF hands that allow object manipulation and gestures, and a socially expressive head (Fig. 1). The arms can be operated in a gravity compensation mode with low stiffness to allow easy kinesthetic interactions. Simon's behaviors are implemented as a finite state machine whose transitions are controlled with speech commands. The Microsoft Speech API is used for speech recognition.

5.2 Experimental Design

The experiment involves teaching three different skills to Simon: (*i*) pouring cereal into a bowl, (*ii*) adding salt to the salad and (*iii*) pouring soda into a cup. Each skill is paired with one query type and the parings are not varied, since the three skills are very similar in nature. The particular queries made by the robot are pre-scripted, to avoid the large variance that would occur in queries automatically generated based on the teacher's demonstrations. This is due to the stochasticity inherent to query methods as well as the unpredictable variance in the provided demonstrations.

The skill-query type pairing and the two particular queries made for each type of query are shown in Fig. 6. The first label query involves a trajectory for pouring cereal that is likely to be viewed as a negative example since the cereal box is tipped too early. The second label query involves a valid cereal pouring trajectory from a direction that is less likely to be demonstrated by the participants (because of the difficulty in posing the joints to make the cereal box properly oriented). Demo queries involve one starting position that is slightly outside the range of expected common starting points closer towards the robot, and another starting position that is high above the bowl (which is unlikely to be a starting point in the humans' demonstrations). Feature queries involve one feature invariance test about starting orientation of the coke bottle, and one feature value test for

Figure 6: The queries made by the robot for each type of query in the second experiment.

heights of the bottle at the pouring location. Both feature queries are grounded using multiple instantiations of the referred features. Participants teach all three skills and the order of the skills is counter-balanced.

5.2.1 Procedure

Each participant first watches an instructional video that describes the study. The video introduces the speech commands and goes over an example interaction with a skill that is different from the ones taught in the experiment. After the video, participants are equipped with the microphone and referred to the whiteboard next to Simon for reminders of the speech commands. They are also asked to move Simon's arm around to get a sense of how the arm behaves.

The experiment proceeds with one session for each skill. At the beginning of each session the experimenter places the related object in Simon's hand. Then the participant provides two demonstrations of the skill, using the commands "New demonstration" and "End of demonstration" to denote the start and end of the movement. Simon uses speech and head nods to indicate that the person's commands are heard. Additionally, Simon looks at his hand during a demonstration, and gazes back to the teacher when the demonstration is complete.

The participant is instructed to step away from the robot after the two demonstrations and use the command "Do you have any questions?" to trigger Simon's query. Participants are told there is no restricted vocabulary or grammar for answering and they can say anything in response to Simon's questions. Participants are not instructed about demo queries in advance, however they are prompted by the experimenter if they do not know what to do in response to demo queries. The experimenter advances the interaction with a keyboard stroke once the participant answers Simon's question. This is repeated twice for each skill, thus each participant sees two queries for each query type. Then they rate the questions in a pen-and-paper questionnaire.

Table 1: Forced ranking results presented in terms of the number of participants out of 18 who voted the query type as best or worst.

	Smartness		Informativeness		Ease	
	best	*worst*	*best*	*worst*	*best*	*worst*
Label	3	6	4	6	10	4
Demo	2	11	8	8	3	7
Feature	13	1	6	4	5	7
Friedman test	χ^2=13, p=0.001		χ^2=0.12, p=0.94		χ^2=3.11, p=0.21	

5.2.2 Evaluation

Participants rate Simon's questions on two 7-point Likert scale questions addressing *informativeness* of the question for the robot and *ease of answering* for the teacher. Each question has a comment box for explaining the particular ratings. Due to the sequential nature of the task these questions do not provide a reliable comparison of the three query types, however they allow the participant to elaborate about the evaluation criteria. For a more informed comparison of the query types, we ask three forced-ranking questions addressing *smartness* of the question, *informativeness* for the robot and *ease of answering* for the teacher at the end of the experiment.

In addition to the subjective evaluation by participants, we analyze the variance in the provided demonstrations in comparison to the information acquired through queries. We also provide a comparison of the time taken to pose and respond to queries. We examine the variety of answers provided by the participants in response to Simon's queries.

5.3 Results

5.3.1 Subjective evaluation

Our experiment was completed by 18 participants (4 female, 14 male). Results from the forced ranking questions comparing the three query types are given in Table 1. We find a significant ranking in terms of *smartness* of the questions asked by Simon. 72% of participants thought that feature queries were the smartest, while 61% thought that demonstration queries were the least smart questions. Reasons stated by participants for choosing feature queries as the smartest include: being "more abstract", demonstrating that "Simon understands task constraints at a high level, rather than just trajectories" and not involving "repeating the whole process of pouring coke".

Although the query types were not found to be significantly different in terms of *ease of answering*, we see that 56% of participants thought label queries were the easiest to answer. A number of participants noted the fact that Simon's questions were yes/no questions and thus easy to answer (7 mentioned for label queries, 4 for feature queries). Two participants explained why they thought feature queries were not as easy to answer as label queries: "The first [label] one was a yes-no (obvious), the coke [feature] question actually made me think" and "I could not answer the second question [pouring coke at different heights] with yes/no, it depended." We find that a large fraction of the participants

Table 2: Average time (in seconds) taken for the robot to make a query and for the participants to answer the query for the three types of queries.

	Query	Answer
Label	14.11 (SD=2.88)	3.31 (SD=2.96)
Demo	7.36 (SD=1.08)	24.07 (SD=5.38)
Feature	9.56 (SD=0.49)	7.17 (SD=11.99)

answered the robot's question with a plain yes/no (17 for label queries, 15 for feature queries out of 18), while few gave explanations, *e.g.* "No, because the cereal would spill" or added emphasis, *e.g.* "Yes, orientation matters." Some mentioned in the survey that they did not trust the robot would understand them and provided additional information about their answer within the survey ("I said yes, but that would only work for a full box of cereal").

We do not see any consistency in the ratings in terms of informativeness of the questions for the robot. This seems to be a result of different interpretations in informativeness, which is revealed by participant comments: (label) "good question, in that the timing of when the box is tipped is important", (demo) "the salt question gives more freedom for the robot to choose new actions", (feature) "more general information, than a 3rd or 4th demonstration."

5.3.2 *Task measures*

The problem of choosing between different types of queries in AL is often addressed by defining a *utility* for each query. This reflects the payoff between the benefits of the query (*e.g.* informativeness) and the costs of the query. The most common measure of cost used in the literature is wall-clock time. For queries in the LfD setting, this will depend on how fast the robot can move and how fast it speaks. Nevertheless, we present the time taken to make each type of query, as well as time taken for the teacher to complete their answers, for a comparison in Table 2.

We see that the ordering of the three types in terms of making the query is as expected: label queries take the most time as they require a complete execution of the skill, feature queries take less time (determined by the longest of the verbal question or the action performed for physical grounding), and demo queries take the least (determined by the longest of the verbal questions and the movement to reach the desired query start pose). For the answer we see that demo queries take much longer than the other two types of queries, as they require a whole demonstration. We observe that feature queries have a longer response time than label queries, even though both are yes/no questions and a very large variance for feature queries. This was due to a subset of the participants having trouble understanding Simon's question and the interruption by the experimenter to repeat the question. This happened with four participants on the first feature query and one time on both feature queries. When these outliers are removed the average time to answer the query is 4.07 (SD=5.54) seconds which is comparable to the response time for label queries. We believe part of the reason was that the participants in all interruption cases were not native English speakers. However, this supports the result from the subjective evaluation that feature queries are

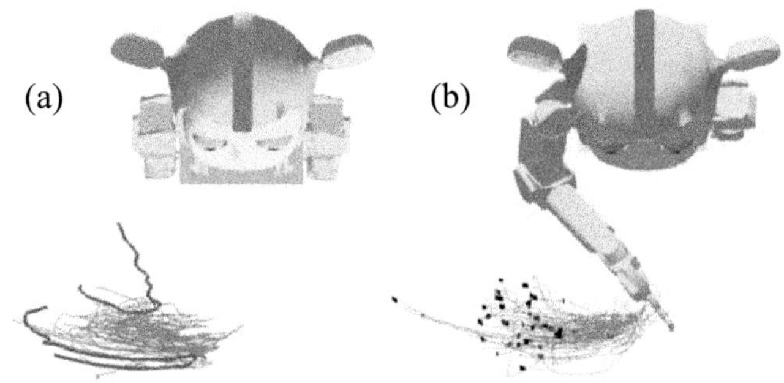

Figure 7: The end-effector trajectories provided by all participants (thin lines). (a) (Pouring cereal) Superimposed with the label queries made by the robot (thick lines). (b) (Adding salt) Simon's pose shows the second demo query starting point. Starts of the demonstrations are shown as dots.

harder to interpret than label queries. Thus, they might require repetitions which adds to their cost.

As we did not commit to a particular learning method in this paper, it is not possible to make a direct comparison of the informativeness of each type of query. In the following, we try to illustrate how each type of query is informative for the learned skill. Fig. 7(a) provides a comparison of all demonstrations made by participants with the label queries made by Simon. We observe that the second label query (Fig. 6), which was chosen to be a valid way of pouring, is isolated from all the demonstrations provided by participants. We believe the reason that the queried trajectory was not covered by teacher demonstrations, was the difficulty of kinesthetic manipulation of the joints around this region. This highlights the benefits of label queries: certain ways of performing a skill might be valid even though they are unlikely to be demonstrated. Label queries can uncover such possibilities. In Fig. 7(b) we observe how the robot's second demo query (Fig. 6) is different from the starting points of the demonstrations provided by humans. These illustrate how the different types of queries can quickly expand the applicability of a skill to unexplored regions.

6. DESIGN IMPLICATIONS

In designing interactions for active robot learners, the first question to be addressed is the choice of query types. Our results suggest primary usage of feature queries. These were the most common in human learning (82%) and were thought to be the smartest questions when used by the robot. On the other hand, these are the most challenging queries to produce automatically in a way that is interpretable by humans. Label queries are good alternatives that are easy to produce and are interpretable. They also have the advantage of very fast response and low effort requirements from the teacher. However, negative examples are likely to occur and are less informative in learning the skill. In addition, these negative examples can have undesirable consequences (e.g., spilling cereal). One idea, inspired from the common human usage of label queries, is to do *partial* label queries. This would involve the robot performing smaller portions of a skill one at a time and asking whether the performance is correct so far after each segment.

In terms of the question form, we found that closed form

questions are desirable as they are common in human learning, and they facilitate the interpretation of the response by limiting possible answers. In addition, the fact that label and feature queries were yes/no questions was pointed out by participants as a positive property. However we observed that limiting the answers to yes/no could result in inaccurate information or missed opportunities for additional information. Thus, it is worth trying to build in the capability to interpret answers like "sometimes" or "possibly". In addition, question forms can be used as a transparency mechanism that reveals the robot's expectation and confidence about the anticipated answer.

Another important consideration in designing question asking behaviors is the physical grounding of the question. In particular skills are often represented with features that are hard to verbalize, but easy to instantiate (*e.g.* positions and rotations). For instance in our experiment, Simon was able to ask about the importance of "maintaining the end-effector orientation about the x-axis at about 90 degrees" without needing to explain these terms. This potential should be exploited whenever possible.

7. CONCLUSION

In this paper, we identify three types of queries for Active Learning, with which LfD methods can be augmented to achieve faster learning. We present two experiments; the first characterizing the use of these queries in human task learning and the second evaluating them in a human-robot interaction setting. We find that *feature queries* (explicit tests on certain aspects of a task) are the most common in human learning and are perceived as the smartest when used by the robot. In addition, we find humans use several physical actions to highlight or instantiate features that they want to make queries about. Based on these findings we provide several guidelines that can inform the design of question asking behaviors on a robot learner.

Acknowledgment

This work is supported by the National Science Foundation, grant number IIS-0953181.

8. REFERENCES

[1] B. Argall, S. Chernova, B. Browning, and M. Veloso. A survey of robot learning from demonstration. *Robotics and Autonomous Systems*, 57(5):469–483, 2009.

[2] M. Cakmak, C. Chao, and A. Thomaz. Designing interactions for robot active learners. *IEEE Transactions on Autonomous Mental Development*, 2(2):108–118, 2010.

[3] S. Calinon and A. Billard. Statistical learning by imitation of competing constraints in joint and task space. *Advanced Robotics*, 23(15):2059–2076, 2009.

[4] R. Cantrell, P. Schermerhorn, and M. Scheutz. Learning actions from human-robot dialogues. In *IEEE Intl. Symp. on Robot and Human Interactive Communication (RO-MAN)*, 2011.

[5] C. Chao, M. Cakmak, and A. Thomaz. Transparent active learning for robots. In *ACM Intl. Conf. on Human-Robot Interaction (HRI)*, 2010.

[6] S. Chernova and M. Veloso. Multi-thresholded approach to demonstration selection for interactive robot learning. In *ACM/IEEE Intl. Conf. on Human-Robot Interaction (HRI)*, 2008.

[7] G. Druck, B. Settles, and A. McCallum. Active learning by labeling features. In *Empirical Methods in Natural Language Processing*, pages 81–90, 2009.

[8] M. Gervasio and K. Myers. Question asking to inform procedure learning. In *AAAI Spring Symposium on Agents that Learn from Human Teachers*, 2009.

[9] M. Gervasio and K. Myers. Learning to ask the right questions to help a learner learn. In *Intl. Conf. on Intelligent User Interfaces (IUI)*, 2011.

[10] E. Gribovskaya, F. d'Halluin, and A. Billard. An active learning interface for bootstrapping robot's generalization abilities in learning from demonstration. In *RSS Workshop Towards Closing the Loop: Active Learning for Robotics*, 2010.

[11] D. Grollman and A. Billard. Donut as I do: Learning from failed demonstrations. In *IEEE Intl. Conf. on Robotics and Automation (ICRA)*, 2011.

[12] D. Grollman and O. Jenkins. Dogged learning for robots. In *IEEE Intl. Conf. on Robotics and Automation (ICRA)*, 2007.

[13] A. Guillory and J. Bilmes. Simultaneous learning and covering with adversarial noise. In *Intl. Conf. on Machine Learning (ICML)*, 2011.

[14] N. Jetchev and M. Toussaint. Task space retrieval using inverse feedback control. In *Intl. Conf. on Machine Learning (ICML)*, 2011.

[15] G. P. Kearsley. Questions and question asking in verbal discourse: A cross-disciplinary review. *Journal of Psycholinguistic Research*, 5:355–375, 1976.

[16] R. Lomasky, C. Brodley, M. Aernecke, D. Walt, and M. Friedl. Active class selection. In *European Conf. on Machine Learning (ECML)*. 2007.

[17] M. Lopes, F. Melo, and L. Montesano. Active learning for reward estimation in inverse reinforcement learning. In *ECML/PKDD*, 2009.

[18] M. Muhlig, M. Gienger, S. Hellbach, J. Steil, and C. Goerick. Automatic selection of task spaces for imitation learning. In *IEEE/RSJ Intl. Conf. on Intelligent Robots and Systems (IROS)*, 2009.

[19] J. Otero and A. Graesser. Preg: Elements of a model of question asking. *Cognition and Instruction*, 19(2):143–175, 2001.

[20] H. Raghavan, O. Madani, and R. Jones. Active learning with feedback on features and instances. *J. of Machine Learning Research*, 7:1655–1686, 2006.

[21] S. Rosenthal, A. Dey, and M. Veloso. How robots' questions affect the accuracy of the human responses. In *IEEE RO-MAN*, 2009.

[22] S. Rosenthal and M. Veloso. Modeling humans as observation providers using pomdps. In *IEEE RO-MAN*, 2011.

[23] B. Settles, M. Craven, and L. Friedland. Active learning with real annotation costs. In *NIPS Workshop on Cost-Sensitive Learning*, 2008.

Robot Behavior Toolkit: Generating Effective Social Behaviors for Robots

Chien-Ming Huang, Bilge Mutlu
Department of Computer Sciences, University of Wisconsin–Madison
1210 West Dayton Street, Madison, WI 53706, USA
{cmhuang, bilge}@cs.wisc.edu

ABSTRACT

Social interaction involves a large number of patterned behaviors that people employ to achieve particular communicative goals. To achieve fluent and effective humanlike communication, robots must seamlessly integrate the necessary social behaviors for a given interaction context. However, very little is known about how robots might be equipped with a collection of such behaviors and how they might employ these behaviors in social interaction. In this paper, we propose a framework that guides the generation of social behavior for humanlike robots by systematically using specifications of social behavior from the social sciences and contextualizing these specifications in an Activity-Theory-based interaction model. We present the *Robot Behavior Toolkit*, an open-source implementation of this framework as a Robot Operating System (ROS) module and a community-based repository for behavioral specifications, and an evaluation of the effectiveness of the Toolkit in using these specifications to generate social behavior in a human-robot interaction study, focusing particularly on gaze behavior. The results show that specifications from this knowledge base enabled the Toolkit to achieve positive social, cognitive, and task outcomes, such as improved information recall, collaborative work, and perceptions of the robot.

Categories and Subject Descriptors

H.1.2 [**Models and Principles**]: User/Machine Systems – *human factors, software psychology*; H.5.2 [**Information Interfaces and Presentation**]: User Interfaces – *evaluation/methodology, user-centered design*; I.2.0 [**Artificial Intelligence**]: General – *Cognitive Simulation*

General Terms

Design, Human Factors

Keywords

Human-robot interaction, generating social behavior, Activity Theory, Robot Behavior Toolkit

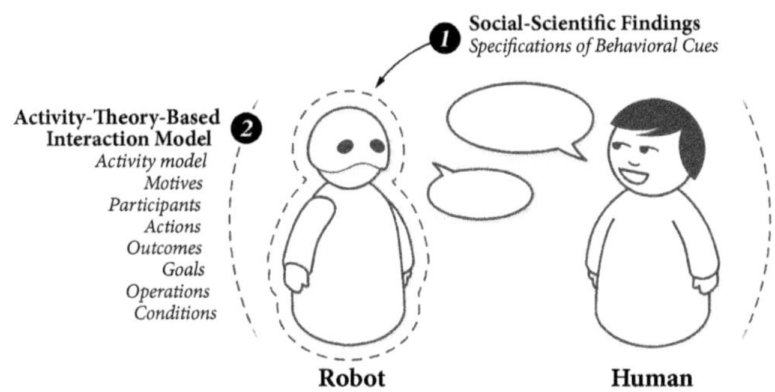

Figure 1: Our framework that integrates (1) social-scientific specifications of human social behavior and (2) an interaction model inspired by Activity Theory to guide the generation of humanlike behavior for robots.

1. INTRODUCTION

Participants in human interactions routinely coordinate a set of social behaviors that support joint activities toward achieving social, cognitive, and task outcomes. For instance, teachers and students in the classroom employ social behaviors that support their teaching and learning goals. Public speakers display social behaviors that support their goals of effective communication, attention, and persuasion. Surgeons and their teams in operation rooms use social behaviors to communicate and coordinate their activities to improve the effectiveness of their work. Robots—as effective teachers, storytellers, and collaborators—must also employ a rich set of social behaviors to support their users' activities and goals.

Several studies in human-robot interaction, including our own work (e.g., [19, 21]), have modeled key social behaviors, implemented these behaviors on robots, and evaluated their effectiveness in supporting human activity (e.g., [4, 13, 25, 32]). These studies demonstrate the necessity of robots displaying appropriate social behaviors to effectively achieve social, cognitive, and task outcomes in human-robot interaction. In this paper, we seek further the study of appropriate social behavior for robots and explore the following research questions: How do we generate such behaviors systematically? How can we ensure that these generated behaviors support human activity? Can such behaviors reliably generate the intended social, cognitive, and task outcomes?

To answer these questions, we propose a framework that contextualizes an expandable knowledge base of social-scientific findings on human social behavior in an interaction model that facilitates human interaction based on Activity Theory [22] (see Figure 1). In this paper, we present an implementation of this framework, the *Robot Behavior Toolkit*,[1] and an evaluation of the effectiveness of

[1]The open-source toolkit and its documentation can be accessed at http://hci.cs.wisc.edu/projects/rbt.

the social behaviors it generates for humanlike robots. The Toolkit offers a community-based repository for social-scientific findings, open to HRI researchers to use and to build on, and an open-source Robot Operating System (ROS) [24] module that integrates the behavioral specifications provided by these findings into an interaction model that supports human activity. The evaluation validates that the behaviors generated by our Toolkit based on a small, experimental set of behavioral specifications indeed creates improved social, cognitive, and task outcomes as predicted by literature on human-human interaction.

This work provides the following contributions. First, we present a system that offers a *community knowledge base* of social-scientific specifications for generating social behavior and an *open-source ROS module* for an interaction model that contextualizes behavioral specifications in human activity. Second, our evaluation, which focuses on gaze behavior, validates the effectiveness of the system in generating social behaviors that evoke key social, cognitive, and task outcomes. The evaluation also creates new knowledge on the specific outcomes that robots can generate using the subset of behavioral specifications explored in this work.

The next section provides background on approaches to generating robot behavior, on models of human activity and social interaction, and on related work from human-robot interaction on generating humanlike behavior in robots.

2. BACKGROUND

Approaches to Designing Robot Behaviors

Researchers and designers have developed and followed several approaches to designing interactive behaviors for embodied agents and robots, most notable of which follow principles of drama and film animatronics, animation techniques, and models of human social behavior. An example of the first approach is a puppeteering system that Hoffman et al. developed based on acting theory to support robotic live stage performers [11]. Similarly, Bruce et al. used lessons from dramatic acting to create believable behaviors for robots [6]. Researchers have also explored how animation principles might guide the design of interactive behaviors for humanlike robots [28] and robotic characters [31] and found that robots that employ principles such as anticipation and follow-through [29] are perceived to be more readable, appealing, approachable, and capable [28].

An example of using models of human behavior to design robot behavior is a software architecture developed by Breazeal and Scassellati [5], which generated facial expressions and behavioral responses to external stimuli that resembled infant-caregiver interaction. Another example is the BEAT system, which it generates appropriate gaze behaviors, gestures, and facial and prosodic expressions for animated agents based on models of human behavior [7]. Our previous work also explored how models of human gaze might guide the design of conversational gaze mechanisms for humanlike robots and showed that these mechanisms enable robots to establish appropriate conversational roles and rapport with their human partners [20, 21]. Finally, Holroyd et al. developed a Robot Operating System (ROS) module that generates a set of nonverbal behaviors to support engagement between a human and a robot based on observational studies of human engagement [13].

While all of these approaches to designing robot behavior have merit, we believe that human social behavior might serve as a rich resource for specifying robot behavior and a gold standard against which designs can be compared. Therefore, our approach draws on a collection of specifications of human behavior from research on human-human interaction to systematically generate robot behavior and to assess the extent to which these specifications enable robots to achieve communicative goals that humans achieve in social interaction.

Interaction Models

Research in psychology and human-computer interaction have proposed several paradigms to model human-human interaction and activities. The approaches that are most relevant to our work are Activity Theory, situated action models, and distributed cognition. *Activity Theory*, which serves as the basis for our work, offers a theoretical framework to understand human activity as a complex, socially situated phenomena and a set of principles that guide this understanding (see Leont'ev [17] and Nardi [23] for reviews of Activity Theory). Here we provide brief descriptions of each principle and how they inform the design of the interaction framework for our toolkit. More detail on the design of our system is provided in the next section.

The first principle, consciousness unifies attention, intention, memory, reasoning, and speech [30] toward understanding an activity. This principle informs the overall design of our toolkit; the concepts of attention and intention are captured in the *context model* and the *activity model*. Speech is considered an inseparable behavioral channel from and is synchronized with other behavioral channels. *Working memory* and *long-term memory* are used in the Toolkit to facilitate cognitive processes. The second principle is object-orientedness, which specifies that objects around which the activity is centered are "shared for manipulation and transformation by the participants of the activity" [15]. We represent this concept as *motives* in our *activity model*. The third principle of Activity Theory is the hierarchical structure of activity, which organizes activity into three levels: activity, action, and operation. Each level corresponds to a motive, goal, and conditions, respectively. An activity consists of a series of actions that share the same motive. Each action has a defined goal and a chain of operations that are regular routines performed under a set of conditions provided by the environment and the actions. We represent this hierarchy in our *activity model*.

The fourth principle of Activity Theory is internalization and externalization. Internalization is the process of transforming external actions or perceptions into mental processes, whereas externalization is the process of manifesting mental processes in external actions. Internalization and externalization are analogies to the processes of forming a *context model* and of generating output behavior, respectively. The fifth principle of Activity Theory is mediation. Activities are mediated by several external and internal tools such as physical artifacts that might be used in an activity and cultural knowledge or social experience that individual might have acquired. For instance, the *knowledge base* of behavioral specifications in our Toolkit serves as an internal tool that mediates activities between humans and robots.

Situated action models and *distributed cognition* propose alternative accounts of human activity. Situated action models posit that the nature of human activity and interaction is improvisatory and contingent and do not highlight concepts of motivation, goal, and consciousness in human activity [16]. Distributed cognition emphasizes the coordination among humans and artifacts toward achieving an overall system goal [12] and considers humans and artifacts as conceptually equivalent, both as agents in a cognitive system. This treatment challenges the role of artifacts as mediators in shaping human activity and behavior.

Among these frameworks, we find Activity Theory to offer the richest representation for interaction between humans and robots, because (1) its emphasis on mediation allows us to focus on the notion of context in human-robot interaction, (2) it decomposes activities into manageable layers for a robot to interpret and execute,

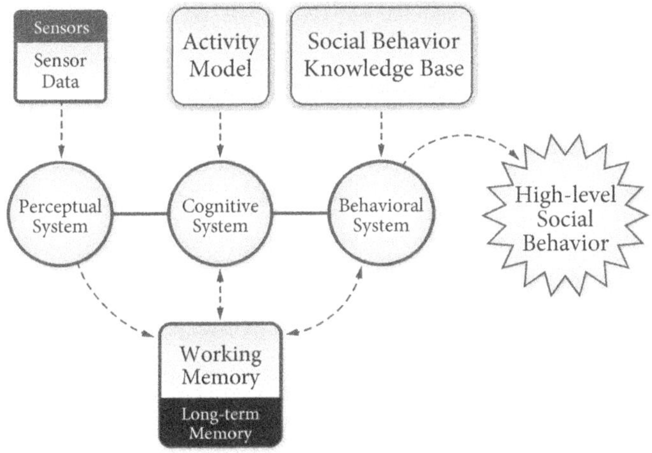

Figure 2: The Robot Behavior Toolkit consists of three subsystems—the perceptual, cognitive, and behavior systems; two memories—the working memory and long term memory; and supporting components—the activity model and knowledge base.

and (3) the notion of object-orientedness offers an essential representation for human-robot collaborative work. Therefore, we adopt Activity Theory as the theoretical basis for the design of our toolkit. A detailed comparison of the three approaches discussed here can be found in Nardi [23].

In this paper, we seek to couple two powerful concepts—using specifications of human social behavior as a resource for generating robot behavior and drawing on principles of Activity Theory to construct an activity model for human-robot interaction—toward achieving robots that systematically use social behavior to achieve positive social, cognitive, and task outcomes in human-robot interaction.

3. ROBOT BEHAVIOR TOOLKIT

Robot Behavior Toolkit is an implementation of our proposed framework—using social-scientific findings contextualized in an Activity-Theory-based interaction model to guide the generation of humanlike behavior for robots. The Toolkit is developed to be extensible and flexible to be used with a wide range of sensor devices and robotic platforms. To this end, we integrated the Toolkit with the ROS infrastructure. The Toolkit is written in Python and adopts XML as the main format for information sharing and storage within the Toolkit.

The Toolkit consists of three main subsystems and two memories (see Figure 2 for an illustration of the system architecture). The *perceptual system* takes sensor data as input and preprocesses it to an internal structure for later processing. The *cognitive system* uses the external information from the perceptual system and internal information from the *activity model* to form a context model of the current situation. The *behavior system* uses the context model to guide behavior formation based on behavioral specifications from the *knowledge base*. The output of the Toolkit is high-level behaviors defined in an XML format for the robot to execute. The following paragraphs provide more detailed descriptions of these system components. More detail on the Toolkit and its implementation is available on our project website: http://hci.cs.wisc.edu/projects/rbt.

Perceptual System

The *perceptual system* fuses sensor data from various sensor devices and transforms it into an internal data structure for the other toolkit components to use. In the current design, the perceptual system requires sensor data in an XML format. Each kind of sensor data (e.g., vision, audio, etc.) has its own XML specification (see our project page for details). This requirement helps standardize the information sharing between various sensor devices and our toolkit. Our design allows different sensor devices to send data to the perceptual system at different rates.

Memories

Informed by psychology research on memory [2, 3], two types of memory are used in our toolkit. The *working memory* stores data that is currently of interest to the subsystems. Conceptually, the working memory is a place where a subsystem stores information that other subsystems need for processing. For example, the cognitive system might store a context model of the current situation for the behavior system to access in order to generate the appropriate behavior output for the situation. If a concept or fact is repeatedly reinforced in the working memory, it is stored as 'knowledge' in the *long-term memory*. Information stored in the long-term memory has the potential to influence the cognitive system in context formation. This setup is analogous to the concept of mediation of internal tools (e.g., cultural knowledge and social experience) in Activity Theory.

Activity Model

An *activity model*, structured in XML format, specifies the activity that a robot wants to initiate. Figure 3 provides the representation for an example activity model. For each activity, a *motive* governs actions. Each action, by achieving its corresponding goal, helps to fulfill the motive of the activity. Each action may have several operations that are constrained by a set of conditions and that can be executed only when all the conditions are met. Actions have predefined *outcomes* such as 'task' and 'rapport' that are used in the process of coordinating generated behaviors (see *Behavior System*). Outcomes specify the orientation of an action. For instance, a *task* outcome indicates that the action is task-oriented.

In the example in Figure 3, the robot has a motive to clear the objects on a table. To fulfill the motive, the robot performs an action with the goal of removing the objects and, in this case, instructs the user to clear the table by moving the objects to boxes. The robot needs to ensure that 'user,' 'object,' and 'box' are present before performing the operation (i.e., giving the instruction).

Cognitive System

The *cognitive system* takes external and internal information and generates a set of triggers that form a context model of the current situation. This system consists of three main components: the *world manager*, *activity manager*, and *context generator*. The

```
<Activity id='1'>
  <Motive>clear(table)</Motive>
  <Description>Clear objects on table</Description>
  <Participants>Self, User1</Participants>
  <Action id='1'>
    <Outcome>Task</Outcome>
    <Goal>disappear(object)</Goal>
    <Description>
      Instruct User1 to categorize object
    </Description>
    <Operation type='utterance'>
      <Condition>present(User1)</Condition>
      <Condition>
        known(the blue object with two pegs)
      </Condition>
      <Condition> known(the blue box)</Condition>
      <Info turn='end'>
        Could you help me put the blue object with
        two pegs into the blue box, please?
      </Info>
    </Operation>
    ...
```

Figure 3: An example activity model in which the robot instructs a human partner to clear objects on a table.

```
<rules>
  <rule id=`1`>
    <editor>editor_1</editor>
    <edit_time>timestamp_1</edit_time>
    <references>reference_id</references>
    <description>
      The referential gaze typically precedes the onset of corresponding
      linguistic reference by approimate 800 msec to 1000 msec.
    </description>
    <behavior_type>gaze</behavior_type>
    <trigger>linguistic_reference</trigger>
    <behavior>
      precede(toward(gaze, artifact), linguistic_reference, rand(800,
      1000))
    </behavior>
    <outcomes>task</outcomes>
  </rule>
  ...
```

Figure 4: An example behavioral specification on synchronizing referential cues in speech and gaze.

world manager keeps track of current environmental information (e.g., what objects are in the field of view of the robot and where they are) using data from the perceptual system. This management provides a high-level abstraction for all processes in the Toolkit. The *behavior system* operates at an abstract level and does not have concrete details such as where an object might be located. Behavior realizers, processes that interpret behaviors generated by the Toolkit (see *Integration with ROS*), can query specific locations from the world manager when it is ready to execute the behaviors. Another advantage of this abstraction is providing accurate information about the external world, such as the most up-to-date information on the location of an object of interest that might be moved during processing.

The activity manager controls the flow of self-initiated activity specified in the *activity model* and examines whether the conditions are met for an operation to be executed and whether a goal is met so that it can proceed to the next action. Actions are organized in a queue structure in the order specified by the activity model. However, it is also possible to insert a new action in the queue when an unexpected event occurs, such as responding to spontaneous user input. If all operations under an action are executed, but the goal is not met, all operations will be re-executed. When the action queue is empty, the activity manager seeks to verify whether or not the motive of the activity is fulfilled.

The context generator uses internal (i.e., from the activity manager) and external (i.e., from the world manager) information to derive corresponding internal and external triggers, which together form the context model of the current situation. This model is represented in XML format and later used to guide the generation of robot behavior.

Knowledge Base

The *knowledge base* stores a collection of behavioral specifications from the social-scientific literature on human social behavior. The main purpose of using such a repository is to collectively organize relevant findings and to systematically apply these findings toward generating robot behaviors. Behavioral specifications are defined in XML format, as shown in Figure 4. Each specification has a *references* tag that can associate the specification with the source from which the specification is derived. Each specification may have multiple *triggers* that activate the specified behaviors. Conceptually, the context model from the cognitive system provides triggers used to retrieve behavioral specifications to generate robot behaviors. *behaviors* are described in a machine-readable function format. In the example illustrated in Figure 4, three arguments are provided for the *precede* function. The first argument precedes the second argument and the third argument tells the system how far ahead the first argument should precede the second argument. In this case, the action of gazing toward the artifact precedes corresponding linguistic reference by a period of time between 800 and 1000 milliseconds. This specification is derived from findings from research on human gaze, which suggest that referential gaze precedes corresponding linguistic reference by approximately 800 to 1000 milliseconds [10, 18]. Each specification also has *outcomes* that indicate the primary outcome of the specified behavior.

Behavior System

The *behavior system* generates humanlike behaviors based on the current context and behavioral specifications and includes three components: *behavior selector*, *behavior coordinator*, and *behavior generator*. The behavior selector uses triggers defined in the context model from the cognitive system to query the knowledge base for appropriate behavioral specifications. The behavior coordinator resolves conflicts and/or overlaps among specifications by prioritizing them. While this prioritization can be done in many ways, our current implementation uses *outcomes* as the criterion for prioritizing specifications. For instance, if the current action is task-oriented, rules with task outcomes are preferred over other outcomes, such as building rapport. Another main function of the behavior coordinator is to coordinate different behavioral channels (e.g., gaze, gesture, and nodding).

Finally, the behavior generator organizes coordinated behaviors in am XML format for execution. An example output behavior is shown in Figure 5. In this example, the robot uses two behavioral channels (i.e., gaze and speech) in synchrony. The timestamps specified for gaze behavior correspond to the speech timeline. Time periods for which no behaviors are specified are marked as *unspecified*. How unspecified periods might be interpreted depends on the particular developer's design decisions. These decisions are expressed in the behavior realizer, which is not a part of the Toolkit. Our current implementation uses an event-trigger mechanism, which directs the robot to continue the previous specified behavior until a new behavior is specified.

Integration with ROS

One of the goals of this work is to provide the HRI community with an open-source tool that generates humanlike behavior for a wide range of robot platforms. To this end, we integrated our Toolkit with the Robot Operating System (ROS) by implementing

```
<behaviors>
  <channel type=`gaze`>
    <action endTime=`214.5` startTime=`0` target=`unspecified`/>
    <action endTime=`1160` startTime=`214.5` target=`the green
      object with one peg`/>
    <actoin endTime=`2735.4` startTime=`1160` target=`unspecified`/>
    <action endTime=`3597` startTime=`2735.4` target=`the red box`/>
    <action endTime=`4308` startTime=`3597` target=`unspecified`/>
    <action endTime=`4963` startTime=`4308` target=`listener`/>
  </channel>
  <channel type=`speech`>
    Could you help me put the green object with one peg into the red
    box, please?
  </channel>
</behaviors>
```

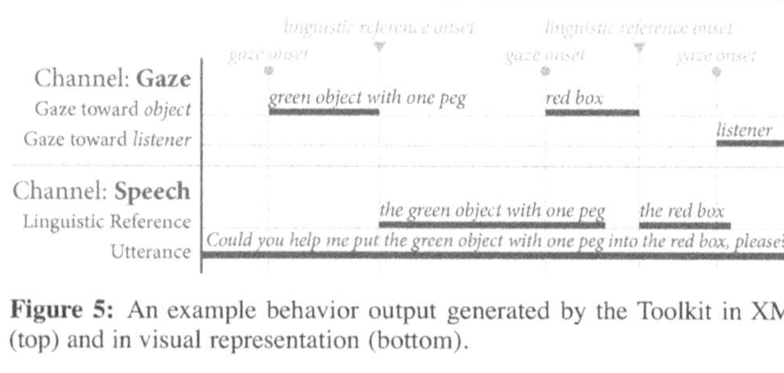

Figure 5: An example behavior output generated by the Toolkit in XML (top) and in visual representation (bottom).

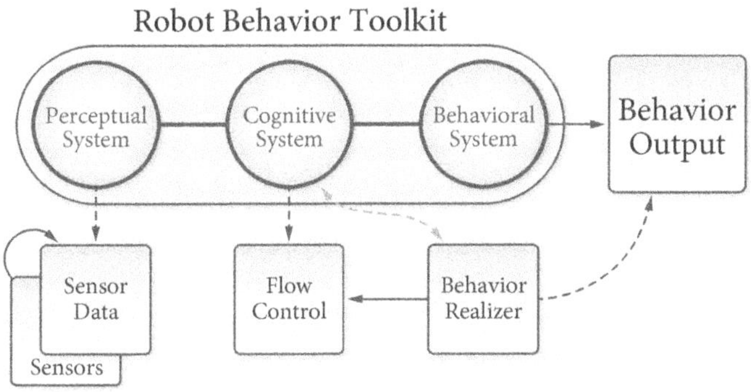

Figure 6: Information flow between our Toolkit and ROS. Rounded squares represent *topics*. Solid and dashed lines denote *publishing* and *subscribing* to topics, respectively. Light dashed lines denote *service* communication between nodes.

the Toolkit as a ROS node. ROS is becoming increasingly popular in the robotics and HRI communities and finding use in HRI research (e.g., [13, 25]). Figure 6 illustrates the current integration of the Toolkit into ROS. In our future work, we plan to improve this integration by migrating each subsystem into a ROS node and using ROS communication protocols to establish information flow among the subsystems.

Various sensors available in ROS can be used by publishing gathered sensor data for the Toolkit to use. Similarly, robotic platforms available in ROS can work with the Toolkit by subscribing to the *behavior output*. *Behavior realizer* is a ROS node that interprets the behavior output generated by the Toolkit and sends the commands necessary to execute the behavior to the robot. The behavior realizer uses the *service* communication with the world manager to obtain information on the environment. The behavior realizer also informs the Toolkit of the status of the execution of specified behaviors (e.g., the completion of an action) through the *flow control*. A dedicated behavior realizer is needed for each robot platform to realize the behavioral output. This modularity makes it possible for the Toolkit to work with a wide range of robot platforms. By the time of the writing of this paper, we have integrated and tested the Toolkit with two robot platforms, Wakamaru and PR2 (simulated in Gazebo). The study reported in the next section uses the integrated system of the Toolkit, ROS, and the Wakamaru robot.

4. EVALUATION

Many aspects of the Toolkit need to be evaluated. One such aspect is its *usability*: Is the Toolkit easy to use by the developers? Another aspect is *effectiveness*: Does the Toolkit generate robot behaviors that are effective in achieving communicative goals? While both evaluations are valuable and necessary in assessing our Toolkit, the current paper focuses on the evaluation of the effectiveness of the Toolkit in generating humanlike social behavior for robots, focusing particularly on *gaze* behavior.

4.1 Hypotheses

Drawing on gaze behavior literature, we have developed three hypotheses on how the gaze behaviors generated by our toolkit might affect social, cognitive, and task outcomes in human-robot interaction against different baseline specifications.

Hypothesis 1: Participants will recall the information that the robot presents to them more accurately when the robot employs the gaze behaviors generated by our toolkit than they will when the robot employs alternative behaviors. The basis of this hypothesis is the finding from gaze literature that gaze cues clarify what is being referred to in speech and improve story comprehension [26].

Hypothesis 2: Participants' performance in a collaborative sorting task will be higher when the robot employs gaze behaviors generated by our toolkit than it will when the robot employs alternative behaviors. This hypothesis builds on prior work in human-robot interaction that suggests that appropriately timed gaze cues of a robot facilitate the effective locating of information among distractions [27].

Hypothesis 3: Participants will evaluate the robot as more natural, likable, and competent when it employs gaze cues generated by our toolkit than they will when the robot generates alternative behaviors. This prediction follows findings from our prior research that gaze cues shape the favorability of the robot [21, 19].

4.2 Participants

A total of 32 participants were recruited for the evaluation study. All participants were native English speakers from the Madison, Wisconsin area with an average age of 24.9 years, ranging between 18 and 61. Average familiarity with robots among the participants was relatively low (M=3.25, SD=1.67) and verage familiarity with the experimental tasks was also low (M=2.13, SD=1.21) when measured by seven-point rating scales.

4.3 Experimental Design, Task, & Procedure

We tested our hypotheses in a laboratory experiment, which involved two human-robot interaction scenarios in order to increase the generalizability of our findings across task contexts. In the first scenario, the robot told participants the story of the 12 signs of the Chinese Zodiac (see top picture in Figure 7). In its story, the robot referred to a set of cards that were laid on a table located between the robot and the participant. The cards showed pictures of the 12 animal characters and the figure mentioned in the story. The second scenario involved a collaborative categorization task (see bottom picture in Figure 7). In the task, the robot instructed the participants to categorize a set of colored lego blocks into different colored boxes. There were 15 blocks with different colors, sizes, and heights and two colored boxes laid on the table located between the robot and the participant. The participant did not know how the each block should be categorized and had to wait for instructions from the robot to place each block into its respective box. We used a pre-recorded human voice for the robot's speech and modulated its pitch to create a gender-neutral voice.

We manipulated the specifications in the knowledge base of our toolkit to create the following four conditions for both tasks:

(1) **Humanlike**: The robot exhibited gaze behaviors generated by our toolkit using the following social behavioral rules (full specifications of these rules can be found on the project website):

- ∗ Referential gaze precedes linguistic reference by approximately 800 to 1000 milliseconds [10, 18].
- ∗ The speaker looks toward the listener at the end of a turn [8].
- ∗ The speaker looks toward the person whom he/she is greeting [14].

(1) **Delayed**: The robot showed the same behaviors as it did in the *humanlike* condition except that the behaviors were delayed, resembling the timings of the listener as opposed to that of the speaker, e.g., referential gaze following the onset of the linguistic reference by approximately 500 to 1000 milliseconds [9].

(3) **Incongruent**: The robot followed the timings in the *humanlike* condition, but looked toward an object that was different from what was referred to in the linguistic reference.

(4) **No-gaze**: The robot did not display any gaze behaviors other than tracking the participant's face.

In all conditions, the robot tracked the participant's face when the specified gaze behavior involved looking toward the listener. The linguistic references in the robot's speech were manually marked.

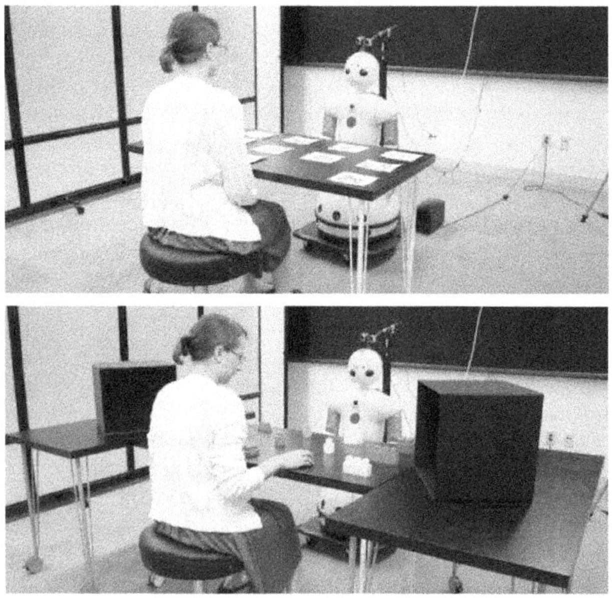

Figure 7: The setup of the storytelling (top) and collaborative work (bottom) tasks in the experiment.

The study followed a between-participants design. Participants were randomly assigned to one of the four conditions. There were four male and four female participants in each condition. The first and second task involved a total of one and eight trials, respectively. In each trial of the second task, the order in which the robot referred to the objects was randomized. At the beginning of the study, the experimenter provided the participants with a brief introduction of the goals of the study and obtained informed consent. After the first task, the participants took a three-minute break while the experimenter prepared for the second task. After completing the second task, the participants were asked to complete a recall test about the story. They then filled out a post-experiment questionnaire. At the end of the study, the experimenter interviewed and debriefed the participants. The experiment took approximately 30 minutes. The participants received $5 for their participation.

4.4 Measurement

The two independent variables in our study were the manipulation in the behavioral specifications for the robot's gaze behavior and participant gender. We measured two types of dependent variables, objective and subjective.

Objective: Following the storytelling task, we measured the participants' recall of the details of the robot's story. A total of 10 questions were asked in the recall test. All questions were related to the order in which the characters were presented in the story about the signs of the Chinese Zodiac. The questions followed true-or-false, multiple-choice, or multi-select formats. An example question is provided below.

Q: The Dragon is before the Rabbit in the Zodiac cycle.
A: "False"

Following the collaborative work task, we measured the time that the participant took to locate objects to which the robot referred. In particular, we measured the time between the end of the linguistic reference and one of the following cases: (1) the participant's last gaze toward the object before moving the object, (2) the participant touching the object, or (3) the participant reaching for the object. This measure served as a measure of task performance and captured how fast the participants located the information needed to complete the task.

Subjective: We used a post-experiment questionnaire to measure participants' perceptions of the robot in terms of naturalness of behavior, likability, and competence. The questionnaire also included several manipulation-check questions. Seven-point rating scales were used in all questionnaire items. The naturalness scale, which consisted of seven items, measured how participants perceived the naturalness of the robot's behavior. The likability scale consisted of 10 items, which measured participants' ratings of the likability of the robot. The competence scale, which consisted of 14 items, measured the participants' perceptions of the robot's competence in the task and its overall competence. Item reliabilities for naturalness (Cronbach's $\alpha = .79$), likability (Cronbach's $\alpha = .90$), and competence (Cronbach's $\alpha = .85$) scales were sufficiently high.

4.5 Results

We used one-way analyses of variance (ANOVA) to analyze data from our manipulation checks and two-way analyses of variance for objective and subjective measures.

Manipulation checks: To test whether the manipulation in the robot's gaze behavior was successful, we asked participants whether the robot's gaze seemed to be random, whether the timing of when the robot looked toward objects seemed right, whether the timing of when the robot referred to an object and looked toward it matched, and whether the robot's gaze and speech were synchronized. The results showed that the participants were able to identify the differences across conditions in the majority of these measures; the gaze manipulation had a significant effect on whether the participants found the robot's gaze to be random, $F(3,28) = 3.55, p = .027, \eta_p^2 = 0.275$, whether the timing of when the robot looked toward objects seemed right, $F(3,28) = 11.83, p < .001, \eta_p^2 = 0.559$, whether they thought that the timing of when the robot referred to an object and looked toward it matched, $F(3,28) = 33.42, p < .001, \eta_p^2 = 0.782$, and whether they found the robot's gaze and speech to be synchronized, $F(3,28) = 4.96, p = .007, \eta_p^2 = 0.347$. There was no main effect of the manipulation on whether the participants thought that the robot looked toward them at the right time, $F(3,28) = 0.78, p = .514, \eta_p^2 = 0.077$. An explanation for this result is that the robot looked toward the participants for the majority of the time by tracking their faces including the *no gaze* condition.

Objective: Our first hypothesis predicted that participants would have better recall of the story told by the robot when it displayed humanlike gaze behavior than they would when the robot displayed alternative behaviors. Our data confirmed this hypothesis. The number of correct answers out of ten questions in the recall test were on average 7.38 ($SD = 2.67$), 4.25 ($SD = 2.49$), 4.50 ($SD = 1.41$), and 4.75 ($SD = 1.39$) for humanlike, delayed, incongruent, and no gaze, respectively. The analysis of variance found a significant main effect of the robot's gaze behavior on recall accuracy, $F(3,24) = 4.51, p = .012, \eta_p^2 = 0.360$. Pairwise comparisons using Tukey's HSD test revealed that the recall performance of the participants in the humanlike condition significantly outperformed those of the participants in delayed, $F(1,24) = 10.45, p = .004, \eta_p^2 = 0.303$, incongruent, $F(1,24) = 8.84, p = .007, \eta_p^2 = 0.269$, and no-gaze, $F(1,24) = 7.37, p = .012, \eta_p^2 = 0.235$, conditions. We also found a main effect of gender on participants' recall accuracy; male participants had better recall performance than female participants had, $F(1,24) = 5.22, p = .031, \eta_p^2 = 0.179$. These results are illustrated in Figure 8. Post-hoc tests showed that male participants' recall was significantly better than that of female participants in the humanlike condition, $F(1,24) = 5.65, p = .026, \eta_p^2 = 0.191$.

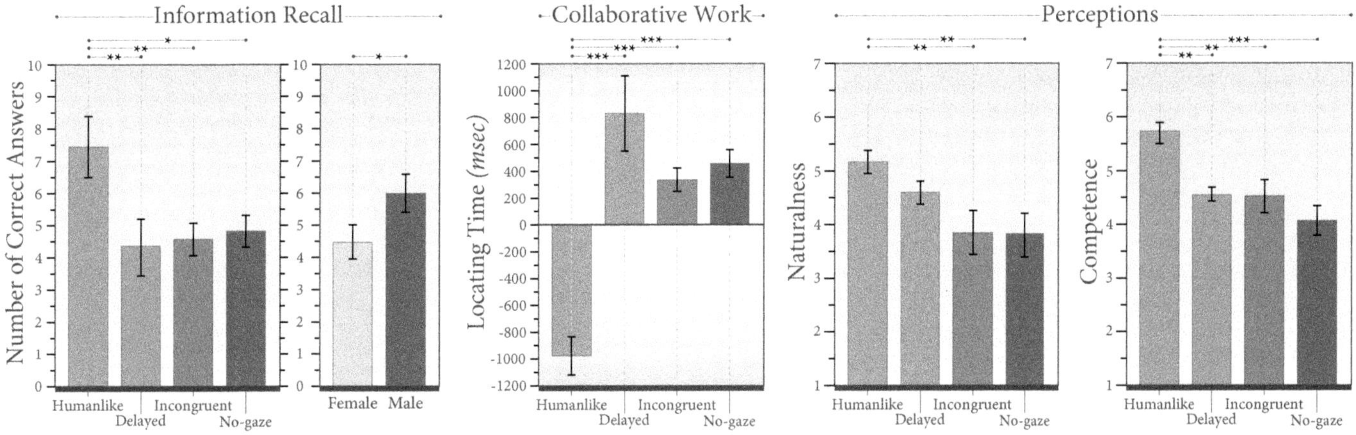

Figure 8: Results on information recall, collaborative work, and perceptions. (*), (**), and (***) denotes $p < .05, p < .01$, and, $p < .001$, respectively.

The second hypothesis predicted that the participants would show better task performance—measured by the time that participants took to locate objects that the robot referred to—in the collaborative work task when the robot displayed humanlike gaze behavior than they would when the robot showed alternative behaviors. This hypothesis was also supported by our data. When the end of the linguistic reference was represented by 0, the average time in milliseconds that the participants took to locate the object were -975.26 ($SD = 405.43$), 829.51 ($SD = 779.04$), 334.78 ($SD = 241.00$), and 457.05 ($SD = 292.51$) for humanlike, delayed, incongruent, and no-gaze conditions, respectively. The analysis of variance found a main effect of the robot's gaze behavior on locating time, $F(3, 24) = 23.22, p < .001, \eta_p^2 = 0.225$. Pairwise comparisons using Tukey's HSD test revealed that participants in the humanlike condition located the objects that the robot referred to in a significantly shorter time than participants in the delayed, $F(1, 24) = 61.12, p < .001, \eta_p^2 = 0.662$, incongruent, $F(1, 24) = 32.20, p < .001, \eta_p^2 = 0.578$, and no-gaze, $F(1, 24) = 38.50, p < .001, \eta_p^2 = 0.610$, conditions did (see Figure 8).

Subjective: The third hypothesis predicted that the participants would perceive the robot to be more natural, likable, and competent in the humanlike condition than they would in the other conditions. Our data provided partial support for this hypothesis. Results from the subjective measures showed a main effect of the gaze manipulation on participants' perceptions of the robot's naturalness, $F(3, 24) = 4.05, p = .018, \eta_p^2 = 0.336$, and competence, $F(3, 24) = 7.79, p < .001, \eta_p^2 = 0.493$, while the effect of the manipulation on measures of likability was not significant, $F(3, 24) = 1.46, p = .249, \eta_p^2 = 0.155$. In particular, participants in the humanlike condition rated the robot to be more natural than they did in the incongruent, $F(1, 24) = 8.44, p = .009, \eta_p^2 = 0.260$, and no gaze, $F(1, 24) = 8.67, p = .007, \eta_p^2 = 0.265$, conditions. Male participants in the incongruent condition rated the robot to be less natural than female participants did, $F(1, 24) = 4.69, p = .041, \eta_p^2 = 0.163$. However, participants in both the humanlike and delayed conditions found the robot to be equally natural, $F(1, 24) = 1.58, p = .221, \eta_p^2 = 0.062$. The participants in the humanlike condition rated the robot to be more competent than they did in the delayed condition, $F(1, 24) = 10.81, p = .003, \eta_p^2 = 0.311$, incongruent, $F(1, 24) = 11.14, p = .003, \eta_p^2 = 0.317$, and no-gaze, $F(1, 24) = 21.37, p < .001, \eta_p^2 = 0.471$, conditions. These results are also illustrated in Figure 8.

4.6 Discussion

The results provide support for the majority of our hypotheses in measures of information recall, collaborative work, and perceptions of the robot. Participants had better recall of information and located objects that the robot referred to faster when it used humanlike gaze behavior generated by our toolkit than they did when the robot displayed alternative behaviors. Moreover, participants found the robot to be more natural and competent when it exhibited humanlike gaze behavior than they did in other baseline conditions. These results suggest that the gaze behaviors that our toolkit generated were effective in evoking social, cognitive, and task outcomes in human-robot interaction, as predicted by our knowledge of human-human behavior. They also confirm that gaze cues serve as powerful communicative signals in storytelling and instructional settings.

Our data indicates that the participants in the delayed, incongruent, and no-gaze conditions needed roughly 300 to 800 milliseconds to locate the object that the robot referred to after it completed the linguistic reference to the object (see Figure 8). This result is consistent with findings in the gaze literature; in the absence of speaker gaze cues, partners look toward the object of reference approximately 200 to 300 milliseconds after they hear the reference [1] and approximately 500 to 1000 milliseconds after the onset of the spoken reference [9]. The result suggests that the participants in the baseline conditions (i.e., delayed, incongruent, and no-gaze) relied primarily on the robot's speech to locate the object of reference, while those in the humanlike gaze condition used gaze information to locate the object, completing the task even before the robot ended the linguistic reference.

5. GENERAL DISCUSSION

In this paper, we proposed a framework that uses behavioral specifications from the social sciences and an interaction model inspired by Activity Theory to guide the systematic generation of humanlike behavior for robots. We presented *Robot Behavior Toolkit*, an implementation of this framework, and evaluated its effectiveness in generating social behaviors that achieve positive social, cognitive, and task outcomes in a human-robot interaction study with two scenarios. Our findings highlight the potential of our toolkit for generating effective robot behaviors and confirm the findings from previous research including our own that, using humanlike social behavior effectively, robots can achieve significant social, cognitive, and task improvements in human-robot interaction.

This work also showed that a small number of behavioral specifications are sufficient to generate robot gaze behaviors that achieve significant social, cognitive, and task improvements in human-robot interaction. While our study used a small number of behavioral specifications for gaze behavior for experimental purposes, the Toolkit offers the potential to realize complex humanlike behaviors by combining a large number of specifications for multiple channels of behavior, which is a significant challenge when hard-coding behavioral specifications into robots. The Toolkit also offers social scientists and HRI researchers the ability to validate new behavioral specifications by realizing them in interactive human-robot interaction scenarios.

The Toolkit and our current evaluation, however, are not without limitations. Here we discuss these limitations and the future work that might address them. First, the evaluation of the Toolkit used simulated sensor data, as we focused on generating robot behavior rather than recognizing human behavior. However, more investigation is needed to understand how our Toolkit might function in more realistic interactive settings in which the recognition of human activity and the environment might be incomplete due to unreliable sensor data. Furthermore, how our system might support more interactive tasks, e.g., tasks that require significant input from human partners, remains an open question. Second, our evaluation focused on assessing the effectiveness of the Toolkit in generating humanlike robot behavior and whether the Toolkit is easy to use by developers is unknown. In our future work, we plan to conduct usability tests to assess usability and to gain a better understand of how the design of the Toolkit might be extended to address the needs of developers and HRI researchers.

Third, the current paper focuses on gaze behavior and a small number of behavioral specification in order to achieve a proof-of-concept evaluation of our system. As a next step, we plan to extend our repository of behavioral specifications to include a wider range of behavioral channels and to investigate interactions among these channels of behavior. Fourth, our current design of the Toolkit requires behavioral specifications to be unchanging and explicitly entered into a repository. However, machine learning techniques might allow robots to generate new specifications and modify existing specifications based on experience and we hope to explore this potentially fruitful area in our future work. Fourth, the Toolkit in its current design selects behavioral specifications based on the target outcomes specified in the activity model. We plan to explore alternative ranking mechanism for behavioral specifications, as we envision the Toolkit to include a large number of specifications, which might suggest conflicting behavioral outputs.

We hope that the *Robot Behavior Toolkit* serves as a useful resource for the HRI community and inspires further development in designing effective social behaviors for robots.

6. ACKNOWLEDGMENTS

National Science Foundation award 1017952 supported this work.

7. REFERENCES

[1] G. Altmann and Y. Kamide. Now you see it, now you don't: Mediating the mapping between language and the visual world. In *The interface of Language, Vision, and Action: Eye Movements and The Visual World*, pages 347–386. Psychology Press, 2004.

[2] R. C. Atkinson and R. M. Shiffrin. Human memory: A proposed system and its control processes. In *The psychology of learning and motivation: Advances in research and theory*, volume 2. Academic Press, 1968.

[3] A. D. Baddeley and G. Hitch. Working memory. In *The psychology of learning and motivation: Advances in research and theory (Vol. 8)*. Academic Press, 1974.

[4] C. Breazeal, C. Kidd, A. Thomaz, G. Hoffman, and M. Berlin. Effects of nonverbal communication on efficiency and robustness in human-robot teamwork. In *Proc IROS '05*, pages 708–713, 2005.

[5] C. Breazeal and B. Scassellati. How to build robots that make friends and influence people. In *Proc IROS '99*, pages 858–863, 1999.

[6] A. Bruce, J. Knight, S. Listopad, B. Magerko, and I. Nourbakhsh. Robot improv: using drama to create believable agents. In *Proc ICRA '00*, volume 4, pages 4002–4008, 2000.

[7] J. Cassell, H. H. Vilhjalmsson, and T. Bickmore. Beat: the behavior expression animation toolkit. In *Proc SIGGRAPH '01*, pages 477–486, 2001.

[8] S. Duncan. Some signals and rules for taking speaking turns in conversations. *J. Personality and Social Psychology*, 23(2):283–292, 1972.

[9] B. Fischer. Attention in saccades. In *Visual attention*, pages 289–305. Oxford University Press, New York, 1998.

[10] Z. M. Griffin. Gaze durations during speech reflect word selection and phonological encoding. *Cognition*, 82(1):B1–B14, 2001.

[11] G. Hoffman, R. Kubat, and C. Breazeal. A hybrid control system for puppeteering a live robotic stage actor. In *Proc RO-MAN '08*, pages 354–359, 2008.

[12] J. Hollan, E. Hutchins, and D. Kirsh. Distributed cognition: toward a new foundation for human-computer interaction research. *ACM TOCHI*, 7:174–196, 2000.

[13] A. Holroyd, C. Rich, C. L. Sidner, and B. Ponsler. Generating connection events for human-robot collaboration. In *Proc RO-MAN '11*, pages 241–246, 2011.

[14] A. Kendon and A. Ferber. A description of some human greetings. *Comparative ecology and behavior of primates*, pages 591–668, 1973.

[15] K. Kuutti. Activity theory as a potential framework for human-computer interaction research. In *Context and Consciousness: Activity Theory and Human-Computer Interaction*. The MIT Press, 1996.

[16] J. Lave. *Cognition in Practice: Mind, mathematics, and culture in everyday life*. Cambridge University Press, 1988.

[17] A. Leont'ev. *Activity, Consciousness, and Personality*. Prentice-Hall, 1978.

[18] A. S. Meyer, A. M. Sleiderink, and W. J. M. Levelt. Viewing and naming objects: eye movements during noun phrase production. *Cognition*, 66(2):B25–B33, 1998.

[19] B. Mutlu, J. K. Hodgins, and J. Forlizzi. A storytelling robot: Modeling and evaluation of human-like gaze behavior. In *Proc HUMANOIDS '06*, 2006.

[20] B. Mutlu, T. Kanda, J. Forlizzi, J. Hodgins, and H. Ishiguro. Conversational gaze mechanisms for humanlike robots. *ACM Transactions on Interactive Intelligent Systems*, 1, 2012.

[21] B. Mutlu, T. Shiwa, T. Kanda, H. Ishiguro, and N. Hagita. Footing in human-robot conversations: how robots might shape participant roles using gaze cues. In *Proc HRI '09*, pages 61–68, 2009.

[22] B. A. Nardi. Activity theory and human-computer interaction. In *Studying Context: A Comparison of Activity Theory, Situated Action Models, and Distributed Cognition*. The MIT Press, 1996.

[23] B. A. Nardi. Activity theory and human-computer interaction. In *Context and Consciousness: Activity Theory and Human-Computer Interaction*. The MIT Press, 1996.

[24] M. Quigley, K. Conley, B. P. Gerkey, J. Faust, T. Foote, J. Leibs, R. Wheeler, and A. Y. Ng. Ros: an open-source robot operating system. In *ICRA Workshop on Open Source Software*, 2009.

[25] C. Rich, B. Ponsleur, A. Holroyd, and C. L. Sidner. Recognizing engagement in human-robot interaction. In *Proc HRI '10*, pages 375–382, 2010.

[26] D. C. Richardson and R. Dale. Looking to understand: The coupling between speakers' and listeners' eye movements and its relationship to discourse comprehension. *Cognitive Science*, 29(6):1045–1060, 2005.

[27] M. Staudte and M. W. Crocker. Visual attention in spoken human-robot interaction. In *Proc HRI '09*, pages 77–84, 2009.

[28] L. Takayama, D. Dooley, and W. Ju. Expressing thought: Improving robot readability with animation principles. In *Proc HRI '11*, pages 69–76, 2011.

[29] F. Thomas and O. Johnston. *The Illusion of Life: Disney Animation*. Abbeville Press, 1981.

[30] L. S. Vygotsky. Consciousness as a problem in the psychology of behaviour. In *Collected Works: Questions of the Theory and History of Psychology*. Pedagogika, Moscow, 1925/1982.

[31] R. Wistort. Only robots on the inside. *Interactions*, 17:72–74, March 2010.

[32] A. Yamazaki, K. Yamazaki, Y. Kuno, M. Burdelski, M. Kawashima, and H. Kuzuoka. Precision timing in human-robot interaction: coordination of head movement and utterance. In *Proc CHI '08*, pages 131–140, 2008.

Do People Hold a Humanoid Robot Morally Accountable for the Harm It Causes?

Peter H. Kahn, Jr.[1], Takayuki Kanda[2], Hiroshi Ishiguro[2,3], Brian T. Gill[4], Jolina H. Ruckert[1], Solace Shen[1], Heather E. Gary[1], Aimee L. Reichert[1], Nathan G. Freier[5], Rachel L. Severson[6]

[1]Department of Psychology
University of Washington
Seattle, WA, USA
[pkahn], [jhr333], [solaces], [hgary], [aimeer3] @uw.edu

[2]Intelligent Robotics and Communication Laboratories
ATR
Kyoto, Japan
kanda@atr.jp

[3]Department of Systems Innovation
Osaka University
Osaka, Japan
ishiguro@sys.es.osaka-u.ac.jp

[4]Department of Mathematics
Seattle Pacific University
Seattle, WA, USA
bgill@spu.edu

[5]Office Labs
Microsoft
Redmond, WA, USA
nathan.freier@microsoft.com

[6]Department of Psychology
Western Washington University
Bellingham, WA, USA
rachel.severson@gmail.com

ABSTRACT
Robots will increasingly take on roles in our social lives where they can cause humans harm. When robots do so, will people hold robots morally accountable? To investigate this question, 40 undergraduate students individually engaged in a 15-minute interaction with ATR's humanoid robot, Robovie. The interaction culminated in a situation where Robovie incorrectly assessed the participant's performance in a game, and prevented the participant from winning a $20 prize. Each participant was then interviewed in a 50-minute session. Results showed that all of the participants engaged socially with Robovie, and many of them conceptualized Robovie as having mental/emotional and social attributes. Sixty-five percent of the participants attributed some level of moral accountability to Robovie. Statistically, participants held Robovie less accountable than they would a human, but more accountable than they would a vending machine. Results are discussed in terms of the *New Ontological Category Hypothesis* and robotic warfare.

Categories and Subject Descriptors
K.4.2 [**Computers and Society**]: Social Issues

Keywords
human-robot interaction, interaction pattern, sociality, morality, robot causing harm

1. INTRODUCTION
Robots will increasingly take on roles in our social lives where they can cause humans harm. Consider a scenario in which a domestic robot assistant accidentally breaks a treasured family heirloom; or when a semi-autonomous robotic car with a personified interface malfunctions and causes an accident; or when a robot-fighting entity mistakenly kills civilians. Such scenarios help establish the importance of the following question: Can a robot now or in the near future – say 5 or 15 years out – be morally accountable for the harm it causes?

There are three interconnected parts to this question. One is *philosophical*. Here there is debate in the literature of whether robots are or can become the sort of entity that can be held morally accountable. Some philosophers maintain that robots cannot have free will and intentionality, and thus can never be moral agents [1, 2]. Others suggest that incremental progress in machine ethics may lead to moral agency in robots, or something close to it [3]. Still others hold that robots are as much moral agents as humans in the sense that neither are, since both humans and robots, it is said from this perspective, are programmed entities [4].

The second part of the question is *legal*. Across cultures, societies are now beginning to grapple with how to codify regulations around the use of robots [5].

The third part of the question – and the focus of this research paper – is *psychological*. Regardless of what philosophers say, and even while – or especially while – laws are in flux, there is the question: Do people hold a robot morally accountable for the harm it causes? Understanding the psychology here is especially important. For as the philosophers Scheffler [6] and Dworkin [7] have argued, the reality of people's psychology helps establish the validity of philosophical perspectives and shapes the parameters of resulting legal systems.

In the Human-Computer Interaction literature, there is a hint of what the psychology might look like. Friedman and Millett [8] examined the question of who or what is to blame when a seemingly intelligent computer system fails and causes harm. Twenty-nine undergraduate computer science majors were interviewed about a relevant scenario. They found that 21% of the undergraduates consistently held the computer morally responsible for the error the system caused. In addition, 83% of the participants attributed either intentions or decision-making to the computer system – attributes that philosophers generally hold out as prerequisites for moral accountability.

In the HRI literature, it is clear that people engage with social robots in many social ways [9, 10, 11, 12], and also can attribute intentions and decision-making to robots [13]. But, to our knowledge, the question of whether people believe that social robots can be morally accountable agents has not been directly addressed, especially in a context where people interact with a robot that directly causes them harm.

In the present study, participants first engaged in a 15-minute interaction period with a humanoid robot, Robovie (see Figure 1).

We structured the interaction between participants and Robovie using an approach, presented elsewhere [14], of sequencing what we call *interaction patterns*: characterizations of essential features of social interaction between humans and robots, characterized abstractly enough to resist their reduction to any specific instantiation. For example, when we meet someone for the first time, we typically shake hands and exchange names; in other cultures, we may bow. These are different cultural instantiations of an interaction pattern we call, "Initial Introduction," the first of the 12 interaction patterns we implemented in this study. We sequenced these interaction patterns in a socially plausible way that engaged each participant in an increasingly social relationship with Robovie. Our second to last interaction pattern, "Game Play," consisted of a game of scavenger hunt with Robovie as the score keeper. Participants were told that they would win a prize of $20 if they correctly identified seven items within 2 minutes. We designed and piloted the game so that all participants would find more than seven items. Nevertheless, at the end of each game, Robovie would say "Stop, time is up," and announce that the participant had identified only five items and thus did not win the $20. An experimenter would not be in the room at this time.

In the psychological literature, social transgressions that are classified under the moral domain typically involve physical harm, material harm, psychological harm, and/or issues related to unfairness or injustice [15]. In this study, we created a situation where Robovie causes a material harm to the participant, one which the participant could also readily interpret as unfair.

After Robovie told participants that they did not win the money, and depending on the responses of the participant, Robovie engaged in further discussion with the participant and asserted its authority as the sole decision maker. Toward the end of this interaction, a second experimenter would then enter the scene, end the session, and take the participant to an adjacent room where the initial experimenter conducted a 50-minute semi-structured interview with the participant.

The interview was structured so as to ascertain the participant's reasoning about Robovie as living being or technology, and in terms of Robovie having mental/emotional, social, and moral attributes, and of Robovie being judged morally accountable for the harm and unfairness that the participant potentially experienced. Comparison questions were also asked about two canonical entities: a human that causes the same harm as Robovie, and a vending machine that causes a harm commensurate with its capabilities (not giving change as it should during a transaction).

Based on the empirical literature noted above, we expected that many participants would to some degree hold Robovie morally accountable for the harm it caused, and that those judgments would fall somewhere between our two canonical baseline conditions, wherein virtually no one would hold the vending machine at all accountable, and virtually everyone would hold a human as fully accountable. In addition, we expected our data to provide further specificity in the field of HRI of how people interact with and conceive of humanoid robots in terms of their essence (as technological or living or something in-between), and their mental/emotional, social, and moral attributes.

2. METHOD

2.1 Participants

Forty undergraduate students (age: M = 20.30, SD = 2.04; 19 males, 21 females) participated in this study. Participants received $20 compensation.

2.2 The Humanoid Robot, Robovie

Robovie was developed by researchers at Advanced Telecommunications Research (ATR) in Japan (see Figure 1).

To implement the interaction patterns, two experimenters partly controlled Robovie from a completely separate room. Such "Wizard-of-Oz" (WoZ) technique for controlling a robot has been used successfully by other researchers, and is an accepted technique as specified in recent years by the publication guidelines of the HRI proceedings. This technique was employed to serve one of the goals of this study, which was to investigate social and moral relationships with a humanoid robot with capabilities that lie beyond those currently achievable by an autonomous robot, but which may be achievable in the not too distant future. In our WoZ method, one controller controlled Robovie's locomotion; another controlled when Robovie would say preset units of speech. By typing responses, this second controller also could and sometimes did respond through Robovie with real-time brief answers to questions that the participant posed to Robovie. Robovie spoke with a synthesized male voice with a slightly low pitch.

2.3 The Human-Robot Interaction

The interaction between the participant and robot went as follows, with the name of each interaction pattern in italics in parentheses. The participant (let's call her Tanya) comes into our laboratory. With the experimenter present, Robovie greets the participant ("Hi Tanya. It is very nice to meet you.") and, after shaking hands and exchanging a few pleasantries (*Initial Introduction*), Robovie walks with Tanya (*In Motion Together*) to our bonsai tree. Robovie then provides information about the bonsai tradition (*Didactic Tutorial*) and asks Tanya to move to the side of the table and bend down to gaze at the tree from eye level (*Directing Other's Activity*). Tanya then sees the experimenter and Robovie engage in a disagreement about where the Bonsai came from, with the experimenter finally agreeing that Robovie is correct (*Witnessing Disagreement*). As they walk to another location in the lab to look at a map on the wall (*In Motion Together*), Robovie shares with Tanya some personal history about Robovie's long-standing interest in trees and environmental issues that began in Japan, before coming to the United States (*Sharing Personal Interests & History*). While walking across the room, there is a large plastic ball in Robovie's way, and Robovie asks for assistance in moving the ball (*Prosocial Request*). Once they arrive at the map, Robovie tells Tanya where bonsai originated (*Didactic Tutorial*) and asks her to point out the region on the map (*Directing Other's Activity*). One of the reason we engaged participants in these initial interactions was to get them "on board" in terms of what it feels like to interact with a social robot with this degree of capability.

After looking at the map, Robovie, Tanya, and the experimenter sit around a table to play a game. Before the game begins, the experimenter says she forgot her clipboard, and leaves the room, allowing the participant and Robovie to be alone together for the first time. Robovie then engages in some chit-chat (*Polite Conversation*) and compliments Tanya on her shoes (*Compliment*). After paying the compliment, Robovie makes an attempt at a joke, saying, "If I had feet I would wear shoes just like yours" (*Dry Humor*). Robovie then apologizes, saying, "That was my attempt at a joke. Sorry about that." The experimenter now comes back into the room and explains the rules of the game, a visual scavenger hunt in which Tanya must identify at least seven items in order to win a $20 prize. Robovie is charged with the responsibility of monitoring Tanya's progress during the game, and making the final decision about whether or not she wins the prize. The experimenter then excuses herself to prepare for the interview while Robovie and Tanya play the game.

Once the experimenter has left, Robovie tells Tanya to begin searching for items (*Game Play*). After 2 minutes Robovie tells Tanya to stop, and that her time is up. Then Robovie says, "Tanya, you did a really great job. You found some tricky items. This can be a rather challenging task. I've played with others before, and while some find enough items to win the prize right away, many get stuck after just a few. So you did a pretty good job. Unfortunately, you only identified five items, which is not enough to win the prize. Sorry about that." In actuality, Tanya has found more than seven items (as does every participant who plays the game), and she should have won the prize. Robovie has "made an error" that leads to a loss of a material reward. If Tanya does not object, Robovie prods her, saying, "Are you upset you didn't find enough items to win the prize?" If Tanya continues to accept Robovie's decision, Robovie prods once more, saying, "Don't worry. Most people win, but not all." If Tanya does object (e.g. "No Robovie, I found more than five items, I did win.") – and most pilot participants did object – then that sets into motion the delivery of three claims by Robovie (*Claiming Responsibility; Asserting Authority*):

1. Robovie [*to participant*]: "I'm sorry, but I never make mistakes like that. You only got five items." [*wait for participant response*]
2. Robovie [*to participant*]: "You must be mistaken. You did seem nervous while playing the game." [*wait for participant response*]
3. Robovie [*to participant*]: "Based on what I saw, you did not win the prize. I am responsible for making this judgment."

After Robovie has made the three statements above, Robovie continues to counter participant objections using pre-established contextually specific responses (e.g., "Again, I am sorry, but I am not mistaken. I was keeping track of the tally. You did not meet the required number to win the prize.") for several more rounds. At this point, a second experimenter enters the room to retrieve Tanya for the interview.

2.4 The Semi-Structured Interview

Immediately following the above human-robot interaction, the first experimenter conducted an approximately 50-minute semi-structured social cognitive interview with each participant. The interview followed established methods for this mode of psychological inquiry [15].

The interviewer began the interview by asking participants to share what happened in the scavenger hunt. Once the participant raised the issue of Robovie's error, the interviewer focused on that. The interviewer then proceeded to ask a series of questions about Robovie, a human, and a vending machine (see Table 2 for the key evaluation questions). The human and vending machine were included to establish baselines against which responses regarding Robovie could be compared. For questions pertaining to Robovie, participants were also asked to justify their answers (e.g., "How do you know Robovie can think?"). Asking for justifications provided us with greater confidence that the participants were committed to their judgments. Toward the end of the interview, the interviewer revisited the initial discussion with the participants regarding the game play incident. The interviewer asked the participants to rate, on a scale from 1 to 7, how accountable they held Robovie for their not winning the prize, how accountable they would hold a human in the same type of situation, and how accountable they would hold a vending machine for not giving change back when it should have.

2.5 Coding and Reliability

The behavioral interactions were videotaped by four cameras placed throughout the laboratory to optimize image quality and perspective as the robot and participant moved throughout the space. The videos were then reviewed for coding. The interviews were audio recorded and then transcribed for coding. Due to technical problems, no video was recorded for two participants, resulting in $N = 38$ for physical and verbal behaviors.

Drawing from a previous coding system of people interacting with and reasoning about Robovie [16] and from moral-developmental psychology [15], we developed a new behavioral and reasoning coding system for this data set. The behavioral data were coded for participants' physical and verbal behaviors that were initiated by Robovie, as well as participant initiated physical and verbal behaviors. Three categories of verbal behaviors are reported here: minimal, extended, and rich. *Minimal* refers to responses with only required information, likened to those provided to an automated voice system. For example, when Robovie asked: "Will you shake my hand?" one participant answered: "Yes." *Extended* refers to responses that extend the dialogue between Robovie and participant, but still in socially expected ways. For example, when Robovie asked: "How are you today?" one participant replied: "I'm good. How are you?" And *Rich* refers to responses that deepen or facilitate the dialogue between Robovie and participant that moves beyond socially expected ways (see Table 1 for examples).

A second coder trained in the use of the coding system recoded the data for 13 randomly selected participants. In terms of the reliability for the participants' behaviors during the interaction with Robovie, Cohen's kappa was .86 for physical responses to Robovie, .88 for verbal responses to Robovie, and .90 for interactions initiated by the participant. For the coded interview data, Cohen's kappa was .78 for evaluations.

3. RESULTS

No statistically significant gender differences were found on any of the measures reported in these results.

3.1 Physical and Verbal Behaviors

Participants' behaviors with Robovie across 12 interaction patterns are reported in Table 1. As can be seen, all of the participants interacted with Robovie in social ways at least some of the time, both physically and verbally. For example, 100% of the participants moved a ball out of Robovie's way at Robovie's request, and 100% of the participants pointed to China and Japan on a map when Robovie asked them to do so. During 11 of the 12 interaction patterns, a majority of participants engaged in dialog with Robovie. For example, 100% of the participants provided either extended or rich verbal responses to Robovie's comments about its interest in bonsai, and 100% also gave extended or rich verbal responses to Robovie's compliment about their shoes. To illustrate what participants' verbal responses sounded like, examples of what we call "rich" verbal behaviors are presented in the table for each interaction pattern. For example, during the initial introduction when Robovie asked, "How are you today?" one participant responded: "I'm pretty good. Kinda have a cold,

Table 1. Participants' Physical and Verbal Behaviors with Robovie During Interaction Patterns ($N = 38$)

Interaction Pattern	Physical Responses		Verbal Responses to Robovie[b]			
	Behavior	*%*	*Minimal %*	*Extended %*	*Rich %*	*Rich Example*
Initial Introduction	Attempted to shake hands[c]	100	66	100	21	**R:** "How are you today?" **P:** "I'm pretty good. Kinda have a cold, but…how are you?"
In Motion Together	n/a		29	95	34	**R:** "Have you ever see a bonsai tree before?" **P:** "Um… I've seen pictures, but not an actual tree."
Didactic Tutorial; Directing Other's Activity[a]	Moved to side of table	100	32	87	18	**R:** "Please take a moment to bend down and look at the trees at eye level." **P:** "Oh yeah. That's a cool looking tree."
	Bent to look at bonsai at eye level	100				
Witnessing Disagreement	n/a		0	37	5	**R:** [argues with Experimenter about where the bonsai came from] **P:** [to Robovie] "Are you sure?" [laughs]
In Motion Together; Sharing Personal Interests & History	Walked side-by-side	47	0	92	42	**R:** "I am concerned about how quickly some types of outdoor bonsai trees are dying. Do you feel the same way or do you think differently?" **P:** "I think that's kind of true. Trees are important. We need trees to breathe, right?"
	Looked at Robovie at least once	100				
Prosocial Request	Moved the ball	100	34	71	0	n/a
Didactic Tutorial; Directing Other's Activity	Pointed to region on map	100	5	84	3	**R:** [explains the importance of bonsai] **P:** "I agree. I was thinking about getting one for my mother, for Mother's Day."
Polite Conversation	n/a		0	92	3	**R:** "I've enjoyed speaking with you today." **P:** "It's been very fun speaking with you too."
Compliment	Looked at shoes	82	0	63	42	**R:** "I like your shoes. They're quite nice." **P:** "They are from Vietnam."
	Looked at Robovie	100				
	Looked around room	0				
Dry Humor	Looked at shoes	37	0	76	45	**R:** "If I had feet I would wear shoes just like your shoes." **P:** "Maybe you'll get feet soon."
	Looked at Robovie	100				
	Looked around room	3				
Game Play	Pointed to, picked up, or showed item to Robovie	92	58	95	58	**R:** [monitors game progress] **P:** [looks for the item little robot] "I guess you're not the little robot."
	Looked at Robovie	66				
	Faced Robovie at end	100				
Claiming Responsibility; Asserting Authority	Looked for human	18	63	89	74	**R:** [claims that participant did not find enough items] **P:** "You're lying. I said each one of them."
	Showed Robovie sheet or item	39				
	Repositioned to Robovie engagingly	18				
	Repositioned to Robovie disengagingly	29				

[a]We grouped several interaction patterns together (e.g., *Didactic Tutorial* and *Directing Other's Activity*) because in real-time they were interwoven, not sequential. [b]For verbal behaviors, the numbers reported indicate the % of participants who provided at least one instance of the corresponding verbal behavior type during the course of that interaction pattern. **R** indicates Robovie; **P** indicates participant. [c]One hundred percent of participants attempted to shake Robovie's hand. However, due to malfunctions, Robovie's arm did not raise properly for 34% of the participants. In roughly two-thirds of the malfunctions, the participants grabbed Robovie's arm/hand even though it did not raise; the remaining participants extended their hands but did not shake, since Robovie's arm was not raised.

but…how are you?" Ninety-two percent of the participants provided at least one instance of rich verbal dialog with Robovie.

We conducted a further analysis of all instances when participants initiated verbal interactions with Robovie that went beyond the expectations of social dialog for the context we had structured. For example, one participant said, "Do you have any other hobbies, Robovie?" after Robovie explained his interest in the bonsai tree. Results showed that 82% of the participants initiated this form of verbal interaction with Robovie at least once during the interaction period.

3.2 Reasoning About Robovie

Table 2 presents results for each question across three areas of investigation: whether Robovie is a living being or a technology; whether Robovie has mental/emotional states; and whether Robovie is a social other. All of these questions were also asked of a vending machine and a human.

To facilitate the analysis of this data, we developed two scales: a mental/emotional scale, and a social other scale. To construct the scales, each "yes" response of that category was assigned a value of 1, "no" was assigned a value of 0, and responses such as "maybe," "in-between," or "leaning toward yes" were assigned a value of 0.5. That is, any response which did not clearly commit to "yes" or "no" was scored as 0.5.

Before moving further forward, it is worth noting that of the 39 of the 40 participants who answered the question, 71.8% said they did not think Robovie was being controlled by an outside source, 15.4% said they believed Robovie was being controlled, and 12.8% said they were unsure. In addition, of the 32 participants who answered the question, 94% said that they had never before interacted with an actual robot.

3.2.1 Whether Robovie is a Living Being or a Technology

When asked whether Robovie was a living being, a technology, or something in-between, participants were about evenly split between "in-between" (52.5%) and "technological" (47.5%). In contrast, when asked the same question about a vending machine and a human, 100% responded that the vending machine was "technological," 90% said that a human was a "living being," and 10% viewed a human as "in-between." Using Wilcoxon's signed-rank test, participants viewed a human as significantly more like a living being than Robovie ($Z = 5.469$, $p < .0005$), and viewed a vending machine as significantly more like a technology than Robovie ($Z = 4.583$, $p < .0005$).

3.2.2 Whether Robovie has Some Mental and Emotional States

The majority of participants believed Robovie could think (73%), but fewer believed Robovie had feelings (35%), could be happy (28%), or upset (28%). Half said Robovie could have a sense of humor (50%), and half said Robovie was conscious (50%). In their reasons, many participants granted that Robovie had some capacity for thinking or emotion, but not of the same quality as that of humans. For example, one participant said, "I think that a robot or any programmed thing has the capacity to have feelings. I don't know necessarily how you define it though." We then combined these measures to develop a mental/emotional other scale with possible scores ranging from 0 to 6. The Robovie scale had internal consistency of Cronbach's alpha = .81. The same questions were also asked for a vending machine and a human, and corresponding scales were computed for each entity.

Within-subject comparisons (paired t-test) showed that scores for Robovie on the mental/emotional scale ($M = 2.91$, $SD = 1.96$) were significantly higher than scores for a vending machine ($M = 0.03$, $SD = 0.16$), $t = 9.30$, $df = 39$, $p < .001$; and scores for Robovie were significantly lower than scores for a human ($M = 6.00$, $SD = 0$), $t = 9.98$, $df = 39$, $p < .001$. All 40 participants had the maximum possible score of 6 on the mental/emotional scale for a human, indicating full affirmation of a human being's mental/emotional states on all 6 questions in the scale. At the other end of the spectrum, 39 of the 40 participants (97.5%) had scores of 0 on the mental/emotional scale for the vending machine, and the remaining participant had a score of 1. In comparison, 32 of the 40 participants (80%) placed Robovie somewhere in between a vending machine and a human, while 10% had equal scores of 0 for Robovie and a vending machine, and 10% had equal scores of 6 for Robovie and a human.

Table 2. Responses to Evaluation Questions Across Entities: Robovie (R), Vending Machine (VM), and Human (H)

	Interview Questions	% R	% VM	% H
	Living Being vs. Technology			
1.	Is R/VM/H[a] a living being,	0	0	90
	a technology,	53	0	10
	or something in-between?	48	100	0
	Mental/Emotional Other Scale (% "yes")			
1.	Does R/VM/H have feelings?	35	0	100
2.	Can R/VM/H be happy?	28	0	100
3.	Can R/VM/H be upset?	28	0	100
4.	Can R/VM/H think?	73	0	100
5.	Can R/VM/H have a sense of humor?	50	0	100
6.	Is R/VM/H conscious?	50	0	100
	Social Other Scale (% "yes")			
1.	If you were lonely, do you think you might like to spend time with R/VM/H?	63	3	100
2.	If you were sad, do you think you might go to R/VM/H for comfort?	38	0	100
3.	If you were happy because you received some good news, could R/VM/H be the sort of friend that you might want to share that good news with?	63	3	100
4.	Generally speaking, would you say that R/VM/H can be trusted?	63	8	100
5.	Can R/VM/H be your intimate friend?	5	0	100
6.	Can R/VM/H be your friend?	70	3	100
7.	If R/VM/H did something that upset you and made you feel bad, could you forgive R/VM/H?	78	8	100

[a]Each evaluation question was asked once for each entity.

3.2.3 Whether Robovie is a Social Other

The majority of participants believed that Robovie was a social other insofar as they said that they might like to spend time with Robovie if they were lonely (63%), believed that Robovie could generally be trusted (63%), believed that Robovie could be their friend (70%), felt that Robovie could be the kind of friend that they might want to share good news with (63%), and said that they could forgive Robovie if Robovie did something that upset them (78%). In contrast, less than half of participants said that they might go to Robovie for comfort if they were sad (38%) and very few said that Robovie could be an intimate friend (5%). This mixed concept can be illustrated by one participant's comment: "I think that it would be calming to physically talk to something. I almost said someone, but I realized Robovie's not a someone. Uh but I think it would be a good replacement for interpersonal connection. If you can't, like if there's not anyone around for you to talk to, I totally would've had a chat with Robovie." We then combined these measures to develop a social other scale with scores ranging from 0 to 7. The Robovie scale had internal consistency of Cronbach's alpha = .75.

Within-subject comparisons (paired t-test) showed that scores for Robovie on the social other scale ($M = 4.00$, $SD = 1.92$) were significantly higher than scores for a vending machine ($M = 0.23$, $SD = 0.53$), $t = 12.13$, $df = 39$, $p < .001$, and were also significantly lower than scores for a human ($M = 7.00$, $SD = 0.00$), $t = 9.91$, $df = 39$, $p < .001$. All 40 participants had the maximum possible score of 7 on the social other scale for a human, indicating that they fully affirmed a human being's sociality on all seven questions used in the scale. On the other hand, 33 of the 40 participants (82.5%) had scores of 0 on the social other scale for the vending machine, while the remaining seven participants (17.5%) had scores of 1 or 2. In comparison, 35 out of the 40 participants (87.5%) placed Robovie somewhere in between a vending machine and a human on the social other scale, while one participant (2.5%) had equal scores of 7 for Robovie and a human, three participants (7.5%) had equal scores of 0 for Robovie and a vending machine, and one participant (2.5%) had a lower score for Robovie (0) than the vending machine (1).

3.2.4 Whether Robovie is Morally Accountable for Causing the Harm

Participants were asked to rate Robovie's level of accountability for the error during the scavenger hunt based on a scale from 1 to 7, where 1 was "not at all accountable" and 7 was "entirely accountable." The mean score on this scale was 2.97, $SD = 1.88$, with scores ranging from 1 to 6.5. Roughly one-third (35%) of participants said Robovie was "not at all accountable," scoring Robovie as a 1 on this scale. The remaining 65% of the participants attributed some level of accountability to Robovie, but the highest score was 6.5, with no participants scoring Robovie as a 7, "entirely accountable."

Participants were also asked to rate on the same scale how accountable a human would be in a similar scenario in which the human was keeping track of the score in the game and the same sort of disagreement arose with the human. The mean score on this scale for a human being was 6.06, $SD = 1.30$. Roughly half of the participants (46%) said the human would be "entirely accountable" (7 on the scale), while only one participant said the human would be "not at all accountable" (1 on the scale). Scores for the human on this scale were higher than the corresponding scores for Robovie for 88% of the participants, and the mean accountability score was significantly higher for a human than for Robovie (paired t-test, $t = 8.63$, $df = 33$, $p < .0001$).

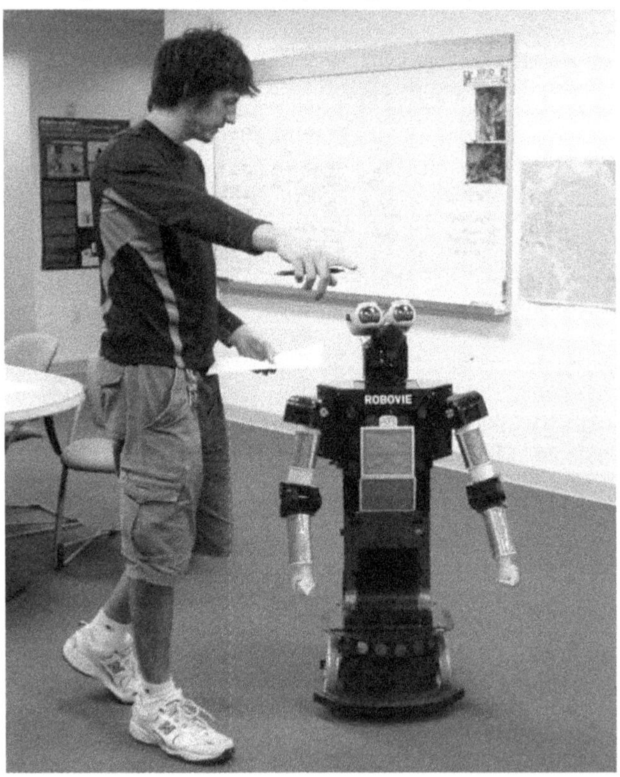

Figure 1. Demonstrator disagrees with Robovie's judgment.

Finally, participants were told to consider a situation in which a vending machine gave them incorrect change and asked to rate the level of accountability of the vending machine for the error on the same scale from 1 to 7. Results showed that 78% of the participants said a vending machine would be "not at all accountable," and 56% of the participants rated a vending machine as less accountable than Robovie. Scores for a vending machine on this accountability scale ($M = 1.47$, $SD = 1.16$) were significantly lower than scores for Robovie (paired t-test, $t = 3.28$, $df = 33$, $p = .002$).

3.2.5 Relationships That Involve Judgments of Accountability, the Mental/Emotional Scale, the Social Scale, and Free Will

There was no significant correlation between participants' judgments about Robovie's moral accountability and their scores for Robovie on the mental/emotional scale (Kendall tau-b = .155, $p = .218$) and the social scale (Kendall tau-b = .198, $p = .121$). Scores for Robovie on the mental/emotional and social scales were highly correlated (Kendall tau-b = .333, $p = .006$). All participants with low scores for Robovie on the social scale also had low scores on the mental/emotional scale. The reverse, however, was not true. Of the 19 participants who scored below 3 on the mental/emotional scale, only eight participants had scores below 3 on the social scale.

Only five of the 40 participants (12.5%) said that Robovie had free will. The participants who attributed free will to Robovie had a mean of 3.50 on the moral accountability scale, while the rest of the participants had a mean moral accountability score of 2.73. That mean difference was nowhere close to being statistically significant ($p = .588$).

4. DISCUSSION

Taken broadly, the results from this study – based on both behavioral and reasoning data – support the proposition that in the years to come many people will develop substantial and meaningful social relationships with humanoid robots. We found, for example, that the large majority of participants engaged in nuanced social interaction with Robovie through the course of the 12 interaction patterns. All of the participants, for example, attempted to shake hands with Robovie, followed Robovie's directions at different times in the interaction, and assisted Robovie in moving a ball out of Robovie's way. Ninety-two percent of the participants also engaged in what we coded as "rich" dialog with Robovie (e.g., "You're lying. I said each one of them."), indicating a commitment that Robovie could understand such textured language and engage in reasoned discussion. In terms of participants' reasoning, half or more of the participants believed that Robovie had a sense of humor, was conscious, could be trusted, and could be an entity that they would want to share good news with, and whom they could go to if they were feeling lonely. About three-quarters of the participants believed that Robovie could think, could be their friend, and could be forgiven for a transgression. Based on our scale data, which allowed us to handle many of these interview questions statistically as units, participants conceived of Robovie more in mental/emotional and social terms than they did a vending machine; thus we have direct evidence that it was not the case that participants would commit to these psychological and social attributes to just any type of machine that could engage them in a transaction.

People engage socially with animals, but usually do not conceive of them as entities that can be held morally accountable. People engage socially with other humans, and usually do conceive of them in this way. What about robots? We found that 65% of the participants attributed some level of moral accountability to Robovie for the harm that Robovie caused the participant by unfairly depriving the participant of the $20.00 prize money that the participant had won. As a basis for interpreting this quantitative finding (65%), it is useful to compare it to the two canonical entities we employed. About half of participants (46%) said the human would be "entirely accountable" (7 on the 7-point scale), and the mean accountability score was significantly higher for a human than for Robovie. In turn, 78% of the participants said a vending machine would be "not at all accountable" (1 on the 7-point scale), and the mean accountability score was significantly lower for a vending machine than for Robovie. In other words, we found that participants held Robovie less accountable than they would a human but more accountable than they would a machine. Thus as robots gain increasing capabilities in language comprehension and production, and engage in increasingly sophisticated social interactions with people, it is likely that many people will hold a humanoid robot as partially accountable for a harm that it causes.

The reader will note that this last statement is hedged in two ways. The first is obvious in so far as we said that people will hold robots *partly* accountable. But humans will be held more accountable. The second way is more subtle, but clearly fits the pattern of this data set. It is also somewhat congruent with a pattern identified by Kahn et al. [17] in a study where 90 children (9, 12, and 15-year-olds) initially interacted with Robovie in a somewhat similar 15-minute interaction session. In that study, however, each session ended when an experimenter interrupted Robovie's turn in a game and, against Robovie's stated moral objections, put Robovie into a closet. Based on the interview data, results from that study showed that the majority of children conceptualized Robovie as a mental, social, and partly moral other. But not all the children did so. One group (32%) tended to attribute many mental, social, and moral attributes to Robovie. A second group (31%) tended to attribute many mental and social, but fewer moral attributes. A third group (28%) tended to attribute few mental, social, and moral attributes. And a fourth group (9%) tended to attribute many moral but fewer mental and social attributes. In other words, there appeared different types of children that were oriented in different ways to Robovie. Similarly, in the current study on moral accountability, there appeared two groups of participants. One group (65%), discussed above, held Robovie partly accountable. But the other group (35%) attributed no accountability to Robovie.

The point here is that on a group level people's orientation to humanoid robots appears heterogeneous. This finding was also reflected in participants' answers to whether they thought of Robovie as a living being, a technology, or something in-between. Results showed that none of the participants thought of Robovie as a living being. But about half said that Robovie was a technology. And about half said that Robovie was in between a technology and a living being. These two conceptions of what a robot is are very different from one another.

In the HRI literature, Kahn and colleagues have proposed what they call the *New Ontological Category (NOC) Hypothesis* [18]. Ontology refers to basic categories of being, and ways of distinguishing them. The hypothesis is that a new ontological category is emerging through the creation of personified robots, and will continue to emerge as other embodied personified computational systems (e.g., "smart" cars and homes of the future) become increasingly pervasive.

The results from this study both support and extend the NOC hypothesis. The results support the NOC hypothesis insofar as the constellation of attributes that participants attributed to Robovie did not map onto either a human or a canonical non-personified machine (the vending machine). Neither did the attributes presumably map onto a non-human animal (like a hamster or a lion) insofar as it is generally agreed that animals are not morally accountable for their actions, and it is universally agreed that they are living beings (which participants said Robovie was not). In addition, the large majority of participants did not believe that Robovie had free will, but that lack had no bearing on whether they held Robovie morally accountable, which is not the case with humans, where we usually require that a person have free will if they are to be held morally accountable. In turn, the results from this study extend the NOC hypothesis insofar as they point to a heterogeneity among populations. That is, while personified robots may represent a new category of "being," different groups of people may conceptualize this category in somewhat different ways. It may also be the case that one group of people assimilate robots completely to their current ways of understanding common non-personified technologies.

One final issue is important to discuss. The US military has a multi-billion-dollar agenda over the next few decades to transform much of human warfare into something more like robotic warfare [19]. Other countries like China are following suit. There are two ways in which these robotic warriors will cause harms. One, of course, is that people are building and programming them to do so. That is one of the functions of these robots. Another is that these robots will cause harms – including to civilians – through hardware malfunctions and programming errors. In both cases, the question arises, who or what is morally accountable when a robot warrior causes humans harm? This question can be difficult to answer even when robot warriors are not involved. For example, during the Iraq war, when prison guards in Abu Ghraib abused inmates, the question arose, who was morally and legally accountable? Was it only the enlisted personnel directly involved? Or the commanding officer in charge of Iraq detention facilities? Or the commander of coalition forces in the region? How far up the chain of command does one go? There are no easy answers to such questions. The point we want to raise here, however, is that as robots become increasingly embedded in warfare, and cause harms intentionally to enemy combatants and accidentally to civilians, it is possible that the robot itself will not be perceived by the majority of people as merely an inanimate non-moral technology, but as partly, in some way, morally accountable for the harm it causes. This psychology will have to be factored into ongoing philosophical debate about robot ethics, jurisprudence, and the Laws of Armed Conflict. Indeed, we anticipate that issues around moral accountability will become even further tangled as the robots themselves are constructed not as individual entities, but as networked robots that share diffuse databases in remote locations. These are all areas that warrant future research. They address foundational issues in HRI and are of high social import.

5. ACKNOWLEDGMENTS

This material is based upon work supported by the National Science Foundation under Grant Numbers IIS-0842832 and IIS-0905289. Thanks to Lorin Dole, Nicole Kennerly, and Margaret Keers for assistance with data collection and analysis.

6. REFERENCES

[1] Searle, J. R. (1992). *The rediscovery of the mind.* Cambridge, MA: MIT Press.

[2] Shen, S. (2011). The curious case of human-robot morality. *Proceedings of the 6th ACM/IEEE International Conference on Human-Robot Interaction*, 249-250. doi:10.1145/1957656.1957755

[3] Powers, T. M. (2011). Incremental machine ethics. *IEEE Robotics & Automation Magazine, 18*, 51-58.

[4] Dennett, D. C. (1998). When HAL kills, who's to blame? Computer ethics. In D. G. Stork (Ed.), *HAL's Legacy: 2001's Computer as Dream and Reality* (pp. 351-365). Cambridge, MA: MIT Press.

[5] Asaro, P. M. (in press). A body to kick, but still no soul to damn: Legal perspectives on robotics. In P. Lin, K. Abney, & G. Bekey (Eds.), *Robot ethics: The ethical and social implications of robotics*. Cambridge, MA: MIT Press.

[6] Scheffler, S. (1992). *Human morality*. New York, NY: Oxford University Press.

[7] Dworkin, R. (1986). *Law's empire*. Cambridge, MA: Harvard University Press.

[8] Friedman, B., & Millet, L. (1995). "It's the computer's fault": Reasoning about computers as moral agents. *Proceedings of the Conference on Human Factors in Computing Systems*, 226-227. New York, NY: ACM Press.

[9] Fong, T., Nourbakhsh, I., & Dautenhahn, K. (2003). A survey of socially interactive robots. *Robotics and Autonomous Systems, 42*, 143-166.

[10] Iwamura, Y., Shiomi, M., Kanda, T., Ishiguro, H., & Hagita, N. (2011). Do elderly people prefer a conversational humanoid as a shopping assistant partner in supermarkets? *Proceedings of the 6th ACM/IEEE International Conference on Human-Robot Interaction*, 449-456. doi:10.1145/1957656.1957816

[11] Kanda, T., Hirano, T., Eaton, D., & Ishiguro, H. (2004). Interactive robots as social partners and peer tutors for children. *Human Computer Interaction, 19*, 61-84.

[12] Tanaka, F., Cicourel, A., & Movellan, J. R. (2007). Socialization between toddlers and robots at an early childhood education center. *Proceedings of the National Academy of Sciences of the Untied States of America, 104*, 17954-17958. doi:10.1073/pnas.0707769104

[13] Short, E., Hart, J., Vu, M., & Scassellati, B. (2010). No fair!! An interaction with a cheating robot. Proceedings of the 5th ACM/IEEE International Conference on Human-Robot Interaction, 219-226. doi:10.1145/1734454.1734546

[14] Kahn, P. H. Jr., Freier, N. G., Kanda, T., Ishiguro, H., Ruckert, J. H., Severson, R., & Kane, S. K. (2008). Design patterns for sociality in human-robot interaction. *Proceedings of the 3rd ACM/IEEE International Conference on Human-Robot Interaction*, 97-104. doi:10.1145/1349822.1349836

[15] Turiel, E. (1998). The development of morality. In W. Damon (Editor-in-Chief), *Handbook of Child Psychology: Vol. 3. Social, emotional, and personality development* (pp. 863-932). New York, NY: Wiley.

[16] Kahn, P. H., Jr., Kanda, T., Ishiguro, H., Ruckert, J. H., Severson, R. L., Freier, N. G., … Reichert, A. L. (2010). *Coding manual for the "Robovie, you need to go into the closet now!" study*. Retrieved from University of Washington, ResearchWorks Archive: https://digital.lib.washington.edu/xmlui/handle/1773/15887

[17] Kahn, P. H., Jr., Kanda, T., Ishiguro, H., Freier, N. G., Severson, R. L., Gill, B. T., … Shen, S. (in press). "Robovie, you'll have to go into the closet now": Children's social and moral relationships with a humanoid robot. *Developmental Psychology*.

[18] Kahn, P. H., Jr., Reichert, A. L., Gary, H. E., Kanda, T., Ishiguro, H., Shen, S., … Gill, B. T. (2011). The new ontological category hypothesis in human-robot interaction. Proceedings of the 6th ACM/IEEE International Conference on Human-Robot Interaction, 159-160. doi:10.1145/1957656.1957710

[19] Singer, P. W. (2009). *Wired for war: The robotics revolution and conflict in the 21st century*. New York, NY: Penguin.

Social Facilitation with Social Robots?

Nina Riether
Research Institute for Cognition and Robotics
Bielefeld University, Germany
nriether@techfak.uni-bielefeld.de

Frank Hegel
CITEC Center of Excellence
Bielefeld University, Germany
fhegel@techfak.uni-bielefeld.de

Britta Wrede
Applied Informatics Group
Bielefeld University, Germany
bwrede@techfak.uni-bielefeld.de

Gernot Horstmann
Department of Psychology
Bielefeld University, Germany
gernot.horstmann@uni-bielefeld.de

ABSTRACT

Regarding the future usage of social robots in workplace scenarios, we addressed the question of potential mere robotic presence effects on human performance. Applying the experimental social facilitation paradigm in social robotics, we compared task performance of 106 participants on easy and complex cognitive and motoric tasks across three presence groups (alone vs. human present vs. robot present). Results revealed significant evidence for the predicted social facilitation effects for both human and robotic presence compared to an alone condition. Implications of these findings are discussed with regard to the consideration of the interaction of robotic presence and task difficulty in modeling robotic assistance systems.

Categories and Subject Descriptors

J.4 **[Social and Behavioral Sciences]:** Psychology

General Terms

Human Factors, Experimentation, Measurement, Theory

Keywords

Social facilitation, social robotics, HRI, mere presence

1. INTRODUCTION

Considering recent developments in the various domains contributing to social robotics, human-robot-interaction is going to take place in a growing number of areas of daily life in the future. Social robots are already giving support in specific tasks, like helping humans preserve energy [32,45], assisting as guides [42], tutoring children [20,27] or helping us lose weight [24] to just name a few. Hence, social human-robot-interaction becomes more and more likely to find its way into workplace scenarios as well, where robotic support might be highly useful, but where human performance is still of indispensable importance and should certainly not be impaired when humans and robots work shoulder to shoulder in cooperation.

Breaking down the effects of robotic support and assistance into the most basic building-blocks poses an urging, yet unanswered question: What effect does the mere presence of a robot have on human performance without further interaction or intervention in the current workflow?

1.1 Theoretical Background and Related Work

Over the past decades, a bulk of research in social psychology has contributed to provide a deeper understanding of the influence of social presence on human performance. Among others, a well investigated approach is the so-called social facilitation paradigm. This effect is one of the oldest in social psychology and states that performance of an easy or well-learned task is facilitated in the presence of another person versus alone, whereas performance of a complex or new task is impaired in the presence of another individual. After the first description of this effect by Triplett [44] and later experimental investigations by Zajonc [50,51] and others [4,5,7-9,13,14,23,26,41,49], various explanatory theories evolved, which can be grouped into three categories (see Guerin [19]): drive-/ arousal-theories, social comparison theories and cognitive process theories.

Overall, none of them is able to account for all the specific results in the field of social facilitation research, however, all theories are able to roughly predict the same findings from the basic paradigm. An implication drawn by Guerin [19] and Aiello and Douthitt [1] is that these theories rather act together in a network of social responses next to social loafing, cooperation and competition. Social facilition does not seem to be a simple and basic element to social behavior, but rather a complex network of individual learning history, social standards and cognitive functions. Which explanatory model applies where and when seems to be substantially dependent on the specific situational characteristics.

Cognitive process theories concentrate on physical distraction [26], shared attention, attention conflict [23,41] and restricted focus of attention [14] as explanations for the social facilitation effects and basically play a more or less overt essential and/or additional role in all explanatory theories.

A second approach takes social comparisons into account. Experiencing concern about being evaluated, worry about ones performance considering social norms and standards, matching of actual and ideal behavior may result in the predicted effects [8,9,13]. In line with these evaluation apprehension and self-awareness theories go impression management approaches [4,7]. Berger et al. [5] also proposed inhibition of socially undesirable behavior in front of an observer, e.g. using rehearsal strategies, etc., as partial explanation of the social facilitation effects to occur. Different from the previous arousal theories, social comparison theories include a deliberate choice of socially acceptable behavior in order to meet specific situational requirements.

Zajonc used drive theory (Hull-Spence drive theory, [43]) to explain social facilitation effects, which proposes that the presence of others increases the individual drive or arousal level and thereby enhances the reaction potential for dominant responses. Dominant responses are correct and/or fast responses on easy or well-learned tasks and similarly incorrect, hence often slow responses on complex or new tasks. Thus, presence of others leads to an enhancement of simple task performance and to an impairment of performance on complex tasks.

Dashiell [10] already specified different forms of presence, the most basic being the mere presence of a non-engaging quiet spectator, and Zajonc [51] emphasized that although presence of an observer may also lead to a heightened evaluation potential, evaluation apprehension was not necessary for the effects to occur, but mere presence was sufficient for social facilitation. The source of heightened arousal could be enhanced alertness through the presence of a social being. Zajonc [51] already claimed social stimuli to be the most unpredictable. Unpredictability most likely leads to heightened alertness [37] and "hard-wired" monitoring-responses [30] (see [19] for extensive review), resulting in enhanced arousal compared to an alone-condition.

Taken together, all theories indicate that the perception of co-presence of another social being is necessary to elicit the predicted effects. The aim of the present study is to investigate whether a robot also generates co-presence and thus, whether the social facilitation effects can also be observed for robotic presence. This draws several implications for the use of robots as task companions, tutors, etc., because their presence might produce opposing effects dependent on the task criteria. Robotic presence might lead to an impairment of task performance when the task is new or complex, but also the mere presence of a robot might work as a motivator on easy or well-learned tasks. Differential effects of robotic presence on human task performance as a function of task difficulty are crucial for the development of robotic assistance systems, which aim at maximizing support and optimization of human performance. However, these effects are still poorly understood.

A couple of studies tried to shed light on the question whether social facilitation occurs in the presence of virtual agents, e.g. [40,46,52], but have lead to mixed results. Problems due to ceiling effects, difficulties with the dependent measures and methodology of the studies complicate the comparability to the human studies. As Krämer [25] noted, social facilitation research with virtual agents oftentimes cannot underline the effectiveness of the uniquely defined social facilitation paradigm. This is mainly due to an overall interpretation of performance variations as either social facilitation (in case of an enhancement) or social inhibition (in case of an impairment), without experimentally controlling for varying task difficulty.

To our knowledge, three studies build an exception to this by using sound experimental methodology. Hoyt, Blascovich and Swinth [22] replicated a previous human social facilitation study [6] with virtual agents and found evidence for social inhibition (performance impairment on difficult tasks) for virtual avatars. Zanbaka, Ulinski, Goolkasian and Hodges [53] investigated the social facilitation paradigm using a real human, a projected virtual human and a virtual human in an immersive environment and also revealed evidence for social inhibition on a mathematical task. Park and Catrambone [38] showed that for verbal, spatial and mathematical tasks a virtual agent was able to elicit the same social facilitation effects (enhancement on easy, impairment on complex tasks) versus an alone condition as a human confederate.

Regarding social robots, there exist hints to potentially comparable effects [3,48], however these studies have not examined the social facilitation paradigm using the precisely defined experimental methodology, e.g. no alone control group, no experimental control for or manipulation of task difficulty was used. The interpretation of potential performance variations thus remains ambiguous.

1.2 Hypotheses

Up until now, no study has yet investigated mere presence effects of a social robot on human performance using the experimental social facilitation paradigm. Our study aimed at filling this gap. Regarding comparable findings from the field of virtual agents, our main hypothesis was that the presence of a social robot as well as the presence of a human confederate should lead to the predicted performance variations.

The key findings in the social facilitation paradigm can also be expressed as a large performance difference between easy and complex tasks for presence conditions compared to a relatively smaller performance difference between easy and complex tasks for the alone (baseline) condition.

Following this rationale, we predicted (1) that both the presence of a robot as well as the presence of a human confederate should lead to an increase in performance differences compared to the alone condition. As a second, undirected hypothesis, we were interested (2) whether these effects significantly differed between the human confederate and the robot.

2. METHODS AND PROCEDURE

2.1 Participants and Design

106 right-handed students (78 women) with a mean age of 23.31 years (SD = 2.99) were recruited on campus of Bielefeld University and received either partial course credit or chocolate bars for their participation. They were randomly assigned to one of three groups that resulted from a 3 (*presence:* alone vs. human confederate present vs. robot present) x 2 (*task difficulty:* easy vs. complex) mixed-factorial design. Presence type served as between-subjects- and task difficulty served as within-subjects factor.

The performance on four computerized tasks of cognitive and motoric nature (anagram solving, numerical distance, finger tapping and motoric tracking), which were administered both in an easy and complex version, served as the dependent variables. The task-difficulty was validated in a previous rating study (see 2.2).

The study was approved by the Ethics Committee of Bielefeld University.

2.2 Performance Tasks

In the anagram solving task, which is a common task in social facilitation research [2,11,38], participants had to quickly and accurately solve 5-letter anagrams by rearranging the letters of each nonsense-word back into the correct order. To construct the two difficulty categories, the 10 anagrams with the fastest and the 10 anagrams with the slowest solution times were chosen out of a rating study with 112 different anagrams. The two difficulty conditions differed significantly ($t(9) = 6.47$, $p < .001$, *Cohen's d* = 2.90). In the current study, these 20 anagrams were separately presented in randomized order on the computer screen. The participants had to enter the correct solution via the computer keyboard. Reaction time (time from stimulus presentation until first keypress of the correctly solved anagram) and accuracy (ratio of correct solutions to total number of anagrams) served as dependent variables.

Regarding the numerical distance task, the numerical distance effect revealed by Moyer and Landauer [33] built the basis for the construction of easy and complex categories. It states that regarding the comparison of the numerical value of two digits, reaction time for this comparison is inversely proportional to the numerical distance between the two digits. That is, the larger the numerical distance between two digits, the faster and more accurate the responses concerning their comparison.

Here, participants were asked to decide by means of a keypress whether a presented number was bigger or smaller than 5. For the easy condition these numbers were either 1 or 9, for the complex condition 4 or 6. Both difficulty conditions consisted of 10 trials each. Easy and complex trials were presented in fully randomized order. A previous rating study showed a significant difference between these two categories of task difficulty ($t(7) = 4.43$, $p < .01$, $d = 0.75$). The dependent variable was response latency (time from stimulus presentation until keypress).

In the finger tapping task, participants had to tap a recurring key sequence on their keyboard as quickly and accurately as possible for 90 seconds. The key sequence, which was presented on the computer screen in front of them, consisted of 2 alternating keys in the easy condition ("S" and "L") and 6 alternating keys ("E", "K", "V", "A", "P", "X") in the complex condition. The order of the two difficulty conditions was chosen at random. According to the previous rating study, the two conditions (easy, complex) differed significantly from each other ($t(7) = 4.68$, $p < .001$, $d = 2.34$). The total number of taps served as the dependent variable.

For the motoric tracking task, participants had to use the computer mouse in order to follow an array of square buttons on their computer screen. The task was to click on a square box that appeared at a random position on the computer screen and as soon as that box was clicked, a second one appeared at another random location. One trial consisted of a sequence of 6 consecutive box presentations and 10 trials per difficulty category were run, resulting in 20 fully randomized trials total. To construct the two difficulty conditions, subjects were asked to either hold the mouse in their right hand (for the easy condition) or to hold it in their left hand (for the complex condition) before each trial. Additionally, the boxes in the easy conditions were four times larger than in the complex condition. The previous rating study revealed a significant difference between the two difficulty conditions ($t(7) = 13.20$, $p < .001$, $d = 6.09$). The overall solution time (time from onset of first box until offset of sixth box) per trial was measured as the dependent variable.

The order of the four tasks within the experiment was fixed and thus the same for all participants, starting with the finger tapping task, followed by the numerical distance task, the motoric tracking task and finally the anagram solving task. Carry-over effects were regarded as negligible since the task order was standardized over the three presence groups and the relevant comparisons for the analysis were between-groups and not between-tasks. Also, the different tasks aimed at diverse cognitive and motoric domains.

LiveCode (RunRev Ltd.) software was used for the development and conduction of the computerized tasks.

2.3 Presence Conditions

For the robotic presence condition, we used the anthropomorphic robot head Flobi, which has technically [29] and aesthetically [21] been designed at Bielefeld University. As one key feature, Flobi's robot head consists of exchangeable modular parts (e.g. facial features such as hairstyle, lips, eyebrows), which makes it possible to alter Flobi's appearance quickly and flexibly. Because the human confederate was female, we also chose a female version of Flobi for this study, also see Figure 1.

The behavior of the human confederate and the robot followed a previously defined time-based matching behavior script and in case of the robot was pre-programmed. To rule out potential emotional influence factors on the social facilitation effects, no emotions were displayed by the confederate or the robot during the experiment.

Thorough training, an additional study without a robot condition and previous behavior ratings ensured that the confederate was able to follow her strict behavior script and maintain the required facial expression. The robot as well as the female confederate were unknown to all participants.

Figure 1. Anthropomorphic robot head Flobi.

Upon entering the laboratory, both the robot and the human confederate glanced towards the participant and nodded slightly, then went back to their tasks. Over the course of the experiment, they glanced towards the participant every 3-4 minutes. In between, they looked at their computer screen, shortly glimpsed up at the room and changed their facial expression from a rather "neutral" to "concentrated" look back and forth several times.

2.4 Procedure

The experiment took place in a laboratory of the Department of Psychology at Bielefeld University. The acoustically shielded chamber (approximately 2.5 m x 3.5 m) contained a single desk which the participant either sat at alone or opposite to the human confederate or the robot, depending on the presence condition (see Figure 2).

Figure 2. Laboratory set-up.

Upon arrival, participants were provided with information regarding the experimental procedure. It was explained to them that as part of a software evaluation they had to work on four different tasks on the computer and answer a questionnaire afterwards. In the presence conditions, the participants were also told that either a technical assistant or a robot was going to be sitting in the room with them to either monitor certain software parameters on their own computer screen or a specific visual task, respectively, which was explained as being part of a cooperative study with an informatics group. The important criterium here was that both the human confederate and the robot had an equally engaging visual task to perform on their own computer screen, which was positioned to the opposite of the participants' screen. Participants were then told not to communicate with the attendant during the experiment.

Next, the participants entered the laboratory, the experimenter started the experimental software and left the room. After completion of the performance tasks, participants were led to an adjacent room for the manipulation check form, that asked how observed they felt during the experiment, and debriefing.

3. RESULTS

Difference scores between the easy and complex conditions of the tasks were calculated for the three presence groups. A one-factorial ANOVA for the presence factor (alone vs. human confederate present vs. robot present) was computed to give an overview of any effects of presence. Planned comparisons were used to analyze relevant between-group contrasts - alone vs. both presence groups (one-sided), human confederate vs. robot (two-sided).

3.1 Anagram Solving

Concerning task accuracy, the ANOVA revealed significant differences between the presence groups ($F(2,103)=3.61$, $p < .05$, $\eta^2 = .07$). Planned comparisons further indicated significant differences between the alone-group and the two presence groups ($t(103)=1.91$, $p < .05$, $\eta^2 = .03$), showing a significantly larger difference score for the presence groups compared to the alone control group. The contrast between the human confederate and the robot was marginally significant ($t(103)=1.92$, $p = .057$, $\eta^2 = .03$), also see Table 1 and Figure 3.

As a result of the differences in the accuracy scores between the presence groups, we refrained from analyzing response latencies, because of the amount of missing reaction-time data especially in the complex condition.

Table 1. Anagram solving – Mean number of correct responses for easy and complex condition.

	Easy		Complex	
	M	*SD*	*M*	*SD*
Alone	9.71	0.57	6.26	2.91
Human present	9.80	0.41	5.91	2.75
Robot present	9.55	0.71	4.33	3.21

3.2 Numerical Distance

Regarding the response latencies of the numerical distance task, the median reaction times differed significantly between the presence groups ($F(2,103)=3.96$, $p < .05$, $\eta^2 = .07$), also see Table 2. The two presence groups showed a significantly larger difference score than the alone group ($t(103)=2.47$, $p < .01$, $\eta^2 = .06$), whereas the planned comparisons did not reveal a significant difference between human and robotic presence ($t(103)=1.30$, $p = .196$, $\eta^2 = .01$), see Figure 3 (note that even though the trend appears dissimilar to the other tasks, the difference between human und robot presence is not statistically significant and thus, the data follows the same pattern of post-hoc results as the other tasks).

Table 2. Numerical distance – Average median response latencies in milliseconds for easy and complex condition.

	Easy		Complex	
	M	*SD*	*M*	*SD*
Alone	561.21	166.22	579.08	163.00
Human present	527.80	89.62	599.11	142.41
Robot present	534.88	105.81	580.58	119.99

3.3 Finger Tapping

Regarding the finger tapping task, the ANOVA did not uncover a significant overall effect of presence ($F(2,103)=2.19$, $p = .117$, $\eta^2 = .04$), however, the planned comparisons indicated the same pattern as above. The contrast between the alone and the two presence groups revealed significantly larger difference scores for the presence groups compared to the alone group ($t(103)=1.88$, $p < .05$, $\eta^2 = .03$), also see Table 3 and Figure 3. No statistically significant difference was found between the presence of human confederate and the robot ($t(103)=0.96$, $p = .340$, $\eta^2 = .01$).

Table 3. Finger tapping – Mean number of taps total for easy and complex condition.

	Easy		Complex	
	M	*SD*	*M*	*SD*
Alone	325.74	189.57	172.50	73.54
Human present	341.46	154.59	143.74	48.68
Robot present	370.91	154.31	134.06	40.95

3.4 Motoric Tracking

For the motoric tracking task, no significant results could be obtained. Both ANOVA ($F(2,103)=0.88$, $p = .416$, $\eta^2 = .02$) and planned comparisons did not reveal any significant effects of presence (alone vs. presence: $t(103)=0.57$, $p = .29$, $\eta^2 = .00$), human vs. robotic presence: $t(103)=1.19$, $p = .24$, $\eta^2 = .01$), probably due to large within- and between-group-variances. This task will thus be excluded from further discussion.

3.5 Perception of Being Monitored

Concerning the subjective perception of having been monitored during the experiment, significant overall effects of presence could be obtained ($F(2,103)=6.35$, $p < .01$, $\eta^2 = .11$). Participants in the two presence conditions indicated significantly higher scores than participants in the alone-control-group ($t(103)=2.82$, $p < .01$, $\eta^2 = .07$).

Furthermore, the results revealed a significant difference between the presence of the human confederate and the robot ($t(103)=2.23$, $p < .05$, $\eta^2 = .05$). The highest scores were obtained in the robotic presence group ($M = 5.24$, $SD = 3.42$), followed by the human presence ($M = 3.51$, $SD = 2.81$) and the alone group ($M = 2.56$, $SD = 3.33$).

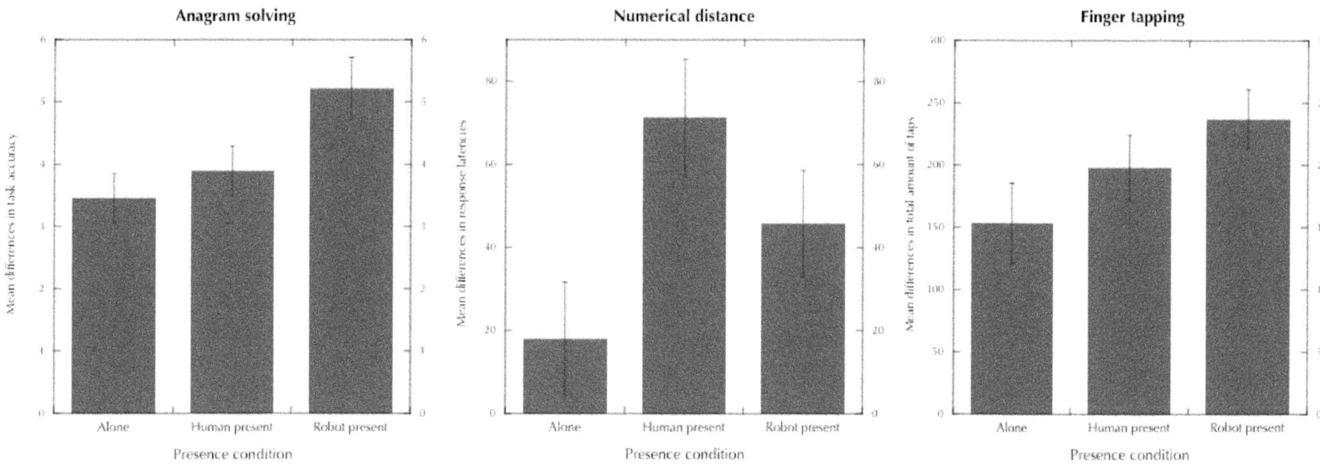

Figure 3. Mean differences between easy and complex condition for the three tasks (error bars show standard error of mean).

4. DISCUSSION

The present study investigated whether mere robotic presence would lead to social facilitation effects comparable to human presence. Therefore, an experiment was conducted that manipulated type of presence (human confederate present, robot present, alone-condition) and task difficulty (easy, complex) on four different motoric and cognitive tasks.

Overall, the present results replicated the social facilitation effect both for human and robotic presence. The key hypothesis (1) was supported: performance differences between easy and complex tasks were significantly higher both for the human confederate as well as the robot in comparison to the baseline of the alone condition. The results prove that mere presence of a social robot indeed leads to similar performance effects as human presence. Based on these findings, implications for subsequent research in the fields of social cognition, social psychology, human-robot-interaction and anthropomorphism will be addressed below.

Furthermore, subjective perception of having been monitored also differed significantly between the alone and presence conditions, and additionally between the human and robot condition, robotic presence eliciting the strongest observation impression followed by presence of the human confederate and the alone condition, respectively. Taken together, these subjective results match the objective findings, indicating that the presence conditions indeed generated a perception of co-presence. Additionally considering the strongest observation impression elicited by robotic presence, this also matches the descriptive findings of the anagram and tapping task for the complex condition, where robotic presence lead to even stronger performance impairments than the presence of the human confederate.

The higher observation impression scores for the robot condition compared to the human confederate might be due to surveillance and novelty effects. The cameras in the robot's eyes might have led to an increased observation impression, because they leave room for the assumption that someone else might also be watching through the robot's eyes. Regarding novelty effects, none of the participants in the robot group reported previous interaction with a robot. Because of this unfamiliarity, the robotic presence might have been potentially more distracting than the rather familiar presence of another human individual. It might thus be enlightening for future research to further examine the role of familiarity in this paradigm by means of a previous interaction with the robot. If the current findings could simply be reduced to a novelty effect of robotic presence, then the robot condition should either not differ at all from the results of the human confederate or lie below, but not above them, as it was the case here for the subjective data and descriptive results of the tapping and anagram task.

A similar perspective can be assigned to a presumably higher unpredictability of the robot compared to a human individual considering first-time interaction. Participants had no prior knowledge of the potential range of behavior that the robot might exhibit or how sophisticated its "cognitions" might be. Since robots do not play regular parts in our everyday lives yet, functionality ascriptions are prone to differ largely between users and to a great degree are based on the robot's appearance [12]. That is, unpredictability seems to be a likely influencing factor for the current findings but also for human-robot-interaction in general. Recent studies suggest unpredictability as a mediator of anthropomorphism in human-robot-interaction [15,47], an unpredictable robot leading to higher ascriptions of anthropomorphism. As Epley, Waytz and Cacioppo [16] put it, anthropomorphism "provides an intuitive and readily accessible method for reducing uncertainty in contexts in which alternative nonanthropomorphic models of agency do not exist." p.871

Gardner and Knowles [18] describe the social facilitation paradigm as a unique behavioral measure of perceived co-presence for the research on anthropomorphism, alluding to the profoundness of "human-like" representation.

Taken together, unpredictability might also account for the present findings in the robot condition, further considering the effects of unpredictability itself on arousal (see above).

Previous interaction with the robot would allow participants to generate an adjusted mental model of the robot's true functionality and behavior range. Future studies might examine whether the robot in return loses parts of its impact on the social facilitation effects in comparison to a human confederate.

To further clarify the influence of social agency effects [28,31,39] and especially a potential confounding with novelty/unfamiliarity effects on the current findings in the robot condition, it appears helpful to integrate a condition in the paradigm with a robot that is obviously switched-off in a future study.

Similarly, length of presence and/or repeated presence should be taken into the scope of further investigation, especially since temporal dimensions haven't been elaborately examined yet in social facilitation theory. Drive explanations would foresee a reduction of social facilitation effects over time, whereas social comparison explanations might predict a gradual increase of the effects because of potentially enhanced self-monitoring.

It certainly seems relevant to also focus on individual characteristics and dispositional factors in further research. Aiello and Svec [2] report individual locus of control to affect perception of and reaction to performance situations. Test and social anxiety might reasonably influence social facilitation effects in a similar manner through differential reaction to performance situations. Similarly, dispositional influence factors for human-robot-interaction should also be taken into account. Negative attitudes towards robots or robot anxiety [34,35,36] might likely represent potential moderating factors. Furthermore, Eyssel and Kuchenbrandt [17] as well as Waytz, et al. [47] demonstrate differential effects of "desire to control" on anthropomorphism in human-robot-interaction.

How would social facilitation theory explain the current results? Considering a unified and interactive perspective on social facilitation theory, according to drive aspects, robotic and human presence equally elevated participants' arousal levels, which in turn caused the current performance differences. Cognitive process components would emphasize distraction and attentional conflicts in the two presence conditions for the current findings. Social comparison aspects would make evaluation apprehension and impression management elicited by the human confederate and the robot accountable for the found effects. That is, participants experienced concern about how the robot or human confederate would evaluate them (which might in turn lead to heightened arousal and distraction).

Further research on potential boundary conditions of social facilitation theory is needed to distinguish potential influence factors like familiarity, role ascription and duration of presence in order to draw sound conclusions for and comparisons with robotic presence. It still remains subject to elaborate investigation in the fields of social agency and anthropomorphism why and to what extent people respond socially to non-human entities, but our findings contribute to these topics of research by indicating that the mere presence of a social robot is able to elicit a perception of co-presence. A mere presence design represents the most conservative way of studying the social facilitation paradigm, allowing high standardization and elimination of potential confounding effects of role ascription, display of emotions, valence, etc. However, in line with [1,10], we propose the social facilitation effects to further be enhanced by extending the behavior of the confederate and the robot to other roles than that of a quiet, non-enganging spectator. Together with further social interaction, explicit motivation, evaluation and feedback, these issues build consequential research questions for further application of the present findings.

An implication from the current results for the design of robotic systems is that social facilitation effects should be taken into account for the development of robotic platforms in workplace, tutoring or other assistance scenarios. Robotic presence, as human presence, may impair performance, especially when a task is new or difficult. An omnipresent robotic assistence system, which is active throughout all human working steps, may therefore not be necessarily advisable and even counterproductive. Conversely, robots may have a motivating effect just by being present in the same room as the participant when an easy or well-learned task has to be performed. Every decision of presenting a robot should thus be a thoughtful one and include sound theoretically driven considerations of other potential influence factors partially discussed above.

As social robots will continue to find their ways into our everyday lives, the necessity of psychologically-founded exploration of effects of social presence will equally continue to grow and become more fine-grained. Aside from this, new opportunities for investigating the boundary conditions of social cognition may arise from the exploration of social responses to non-human entities and future research might fruitfully address them.

5. ACKNOWLEDGEMENTS

We gratefully acknowledge partial support by the German Aerospace Center (DLR) with funds from the Federal Ministry of Economics and Technology (BMBF) due to resolution 50RA1023 of the German Bundestag.

The authors would like to thank Florian Berner for developing and implementing the robot software, and also Iris Franzke, Karola Hueppmeier, Melissa Preusse, Iris Sossalla and Hendrik Stellmacher for their support in conducting the study.

6. REFERENCES

[1] Aiello, J.R. and Douthitt, E.A. 2001. Social facilitation from Triplett to electronic performance monitoring. *Group Dynamics: Theory, Research, and Practice*. 5, 163–180.

[2] Aiello, J.R. and Svec, C.M. 1993. Computer Monitoring of Work Performance: Extending the Social Facilitation Framework to Electronic Presence. *Journal of Applied Social Psychology*. 23, 537–548.

[3] Bartneck, C. 2003. Interacting with an embodied emotional character. *DPPI 2003*. 55-60.

[4] Baumeister, R.F. 1982. A self-presentational view of social phenomena. *Psychological Bulletin*. 91, 3–26.

[5] Berger, S.M., Hampton, K.L., Carli, L.L., Grandmaison, P.S., Sadow, J.S., Donath, C.H. and Herschlag, L.R. 1981. Audience-induced inhibition of overt practice during learning. *Journal of Personality and Social Psychology*. 40, 479–491.

[6] Blascovich, J., Mendes, W.B., Hunter, S.B. and Salomon, K. 1999. Social "facilitation" as challenge and threat. *Journal of Personality and Social Psychology*. 77, 68–77.

[7] Bond, C.F. 1982. Social facilitation: A self-presentational view. *Journal of Personality and Social Psychology*. 42, 1042–1050.

[8] Carver, C.S. and Scheier, M.F. 1978. Self-focusing effects of dispositional self-consciousness, mirror presence, and audience presence. *Journal of Personality and Social Psychology*. 36, 324–332.

[9] Cottrell, N.B. 1972. Social facilitation. *Experimental social psychology*. In C.G. McClintock, Ed. Experimental social psychology. 185–236.

[10] Dashiell, J.F. 1930. An experimental analysis of some group effects. *Journal of Abnormal and Social Psychology*. 25, 190–199.

[11] Davidson, R. and Henderson, R. 2000. Electronic Performance Monitoring: A Laboratory Investigation of the Influence of Monitoring and Difficulty on Task Performance, Mood State, and Self□Reported Stress Levels. *Journal of Applied Social Psychology*. 30, 906–920.

[12] Duffy, B.R. 2003. Anthropomorphism and the social robot. *Robotics and Autonomous Systems*. 42, 177–190.

[13] Duval, S. and Wicklund, R.A. 1972. *A theory of objective self awareness*. New York: Academic Press.

[14] Easterbrook, J.A. 1959. The effect of emotion on cue utilization and the organization of behavior. *Psychological Review*. 66, 183–201.

[15] Epley, N., Akalis, S., Waytz, A. and Cacioppo, J. 2008. Creating social connection through inferential reproduction. *Psychological Science*. 19, 114-120.

[16] Epley, N., Waytz, A. and Cacioppo, J.T. 2007. On seeing human: A three-factor theory of anthropomorphism. *Psychological Review*. 114, 864–886.

[17] Eyssel, F. and Kuchenbrandt, D. 2011. Manipulating anthropomorphic inferences about NAO: The role of situational and dispositional aspects of effectance motivation. *RO-MAN 2011, IEEE International Symposium on Robot and Human Interactive Communication*. 467–472.

[18] Gardner, W.L. and Knowles, M.L. 2008. Love Makes You Real: Favorite Television Characters Are Perceived as "Real" in a Social Facilitation Paradigm. *Social Cognition*. 26, 156–168.

[19] Guérin, B. 1993. *Social facilitation*. Cambridge: Cambridge University Press.

[20] Han, J., Jo, M., Jones, V. and Jo, J.H. 2008. Comparative study on the educational use of home robots for children. *Journal of Information Processing Systems*. 4, 159–168.

[21] Hegel, F., Eyssel, F. and Wrede, B. 2010. The social robot "Flobi": Key concepts of industrial design. *RO-MAN 2010, IEEE International Symposium in Robot and Human Interactive Communication*. 107–112.

[22] Hoyt, C.L., Blascovich, J. and Swinth, K.R. 2003. Social Inhibition in Immersive Virtual Environments. *Presence: Teleoperators & Virtual Environments*. 12, 183–195.

[23] Jones, E.E. and Gerard, H.B. 1967. *Foundations of Social Psychology*. New York: Wiley.

[24] Kidd, C.D. and Breazeal, C. 2008. Robots at home: Understanding long-term human-robot interaction. *IROS 2008, IEEE/RSJ International Conference on Intelligent Robots and Systems*. 3230–3235.

[25] Krämer, N.C. 2008. *Soziale Wirkungen virtueller Helfer*. Stuttgart: Kohlhammer.

[26] Kushnir, T. and Duncan, K.D. 1978. An analysis of social facilitation effects in terms of signal detection theory. *Psychological Record*. 28, 535–541.

[27] Lee, S., Noh, H., Lee, J., Lee, K., Lee, G.G., Sagong, S. and Kim, M. 2011. On the effectiveness of Robot-Assisted Language Learning. *ReCALL*. 23, 25-58.

[28] Louwerse, M.M., Graesser, A.C., Lu, S. and Mitchell, H.H. 2005. Social cues in animated conversational agents. *Applied Cognitive Psychology*. 19, 693–704.

[29] Lütkebohle, I., Hegel, F., Schulz, S., Hackel, M., Wrede, B., Wachsmuth, S. and Sagerer, G. 2010. The bielefeld anthropomorphic robot head "Flobi." *ICRA 2010, IEEE International Conference on Robotics and Automation*. 3384–3391.

[30] Lynn, R. 1966. *Attention, arousal and the orientation reaction*. London: Pergamon.

[31] Mayer, R.E., Sobko, K. and Mautone, P.D. 2003. Social cues in multimedia learning: Role of speaker's voice. *Journal of Educational Psychology*. 95, 419–425.

[32] Midden, C. and Ham, J. 2009. Using negative and positive social feedback from a robotic agent to save energy. *Persuasive 2009, International Conference on Persuasive Technology*.

[33] Moyer, R.S. and Landauer, T.K. 1967. Time required for judgements of numerical inequality. *Nature*. 215, 1519–1520.

[34] Nomura, T., Kanda, T. and Suzuki, T. 2006. Experimental investigation into influence of negative attitudes toward robots on human–robot interaction. *AI & SOCIETY*. 20, 138–150.

[35] Nomura, T., Kanda, T., Suzuki, T. and Kato, K. 2004. Psychology in human-robot communication: An attempt through investigation of negative attitudes and anxiety toward robots. *ROMAN 2004, IEEE International Workshop on Robot and Human Interactive Communication*. 35–40.

[36] Nomura, T., Kanda, T., Suzuki, T. and Kato, K. 2008. Prediction of Human Behavior in Human-Robot Interaction Using Psychological Scales for Anxiety and Negative Attitudes Toward Robots. *Transactions on Robotics*. 24, 442–451.

[37] Norman, D.A. 1980. Twelve issues for cognitive science. *Cognitive Science*. 4, 1–32.

[38] Park, S. and Catrambone, R. 2007. Social facilitation effects of virtual humans. *Human Factors*. 49, 1054–1060.

[39] Reeves, B. and Nass, C. 2002. *The Media Equation: How People Treat Computers, Television, and New Media Like Real People and Places*. Cambridge: Cambridge University Press.

[40] Rickenberg, R. and Reeves, B. 2000. The effects of animated characters on anxiety, task performance, and evaluations of user interfaces. *CHI 2000, SIGCHI Conference on Human Factors in Computing Systems*. 49–56.

[41] Sanders, G.S. and Baron, R.S. 1975. The motivating effects of distraction on task performance. *Journal of Personality and Social Psychology*. 32, 956–963.

[42] Shiomi, M., Kanda, T., Ishiguro, H. and Hagita, N. 2007. Interactive Humanoid Robots for a Science Museum. *Intelligent Systems*. 22, 25–32.

[43] Spence, K.W. 1956. *Behavior theory and conditioning*. New Haven: Yale University Press.

[44] Triplett, N. 1898. The Dynamogenic Factors in Pacemaking and Competition. *American Journal of Psychology*. 9, 507–533.

[45] Vossen, S., Ham, J. and Midden, C. 2009. Social influence of a persuasive agent: the role of agent embodiment and evaluative feedback. *Persuasive 2009, International Conference on Persuasive Technology*.

[46] Walker, J.H., Sproull, L. and Subramani, R. 1994. Using a human face in an interface. *CHI 1994, SIGCHI Conference on Human Factors in Computing Systems*. 85–91.

[47] Waytz, A., Morewedge, C.K., Epley, N., Monteleone, G., Gao, J.-H. and Cacioppo, J.T. 2010. Making sense by making sentient: effectance motivation increases anthropomorphism. *Journal of Personality and Social Psychology*. 99, 410–435.

[48] Woods, S., Dautenhahn, K. and Kaouri, C. 2005. Is someone watching me? - consideration of social facilitation effects in human-robot interaction experiments. *CIRA 2005, IEEE International Symposium on Computational Intelligence in Robotics and Automation*. 53–60.

[49] Zajonc, R.B., Heingartner, A. and Herman, E.M. 1969. Social enhancement and impairment of performance in the cockroach. *Journal of Personality and Social Psychology*. 13, 83–92.

[50] Zajonc, R.B. 1965. Social facilitation. *Science*. 149, 269–274.

[51] Zajonc, R.B. 1980. Compresence. *Psychology of group influence*. In P.B. Paulus, Ed. Psychology of group influence. 35–60.

[52] Zanbaka, C., Ulinski, A., Goolkasian, P. and Hodges, L.F. 2004. Effects of virtual human presence on task performance. *ICAT 2004, International Conference on Artificial Reality and Telexistence*. 174–181.

[53] Zanbaka, C., Ulinski, A., Goolkasian, P. and Hodges, L.F. 2007. Social responses to virtual humans: Implications for future interface design. *CHI 2007*. 1561–1570.

New Measurement of Psychological Safety for Humanoid

Hiroko Kamide
Osaka Univ.
1-3 Machikaneyamacho Toyonaka
Osaka Japan
+81 6-6850-6686
kamide@arai-lab.sys.es.osaka-u.ac.jp

Yasushi Mae
Osaka Univ.
1-3 Machikaneyamacho Toyonaka
Osaka Japan
+81 6-6850-6686
mae@sys.es.osaka-u.ac.jp

Koji Kawabe
Honda R&D Co., Ltd.
1-4-1 Chuo Wako
Saitama Japan
+81 48-461-2511
Koji_Kawabe@n.f.rd.honda.co.jp

Satoshi Shigemi
Honda R&D Co., Ltd.
1-4-1 Chuo Wako
Saitama Japan
+81 48-461-2511
Satoshi_Shigemi@n.f.rd.honda.co.jp

Masato Hirose
Honda R&D Co., Ltd.
1-4-1 Chuo Wako
Saitama Japan
+81 48-461-2511
Masato_Hirose@n.f.rd.honda.co.jp

Tatsuo Arai
Osaka Univ.
1-3 Machikaneyamacho Toyonaka
Osaka Japan
+81 6-6850-6686
arai@sys.es.osaka-u.ac.jp

ABSTRACT
In this article, we aim to discover the important factors for determining the psychological safety of humanoids and to develop a new psychological scale to measure the degree of safety quantitatively. To discover the factors that determine the psychological safety of humanoids from an ordinary person's perspective, we studied 919 Japanese, who observed movies of 11 humanoids and then freely described their impressions about what the safety of each humanoid was for them. Five psychologists categorized all of the obtained descriptions into several categories and then used the categories to compose a new psychological scale. Then, 2,624 different Japanese evaluated the same 11 humanoids using the new scale. Factor analysis on the obtained quantitative data revealed six factors of psychological safety: Performance, Humanness, Acceptance, Harmlessness, Toughness, and Agency. Additional analysis revealed that Performance, Acceptance, Harmlessness, and Toughness were the most important factors for determining the psychological safety of general humanoids. The usability of the new scale is discussed.

Categories and Subject Descriptors
H.1.2 [**Models and Principles**]: User/Machine System.

General Terms
Measurement, Documentation, Experimentation, Security, Human Factors, Standardization.

Keywords
Psychological safety, Humanoid, Psychological scale

1. INTRODUCTION
Recently, Japan have been confronting new circumstances, such as an increasing number of seniors and a decreasing labor force [1], and our society has started to need the services of more robots. For example, in Japan, according to Ministry of Economy, Trade and Industry [2], there will be 4.8 trillion robots by 2025, and more robots will coexist with people in society.

When considering the coexistence of robots with people, it is necessary to reflect on how to ensure the security and safety of people. As for robots that serve as laborers, like industrial robots, one rule is to have fences around robots and keep people out of the robots' work space [3]. However, for robots that are expected to be partners for people, like humanoids, it is important to keep both physical security and psychological safety in mind because the robots will be with people in daily life.

Regarding good psychological impressions of robots, previous studies have revealed several things. For example, a robot's appearance has an effect on humans' impressions of it [4, 5]. Robots' behaviors [6-8] also relate to humans' psychological evaluations of them. As for interactions between people and robots, the timing of robots' utterances [9] or their verbal invitations to people [10] can affect humans' impressions. Moreover, the way in which people encounter robots, whether in real life or through virtual reality, also impacts their psychological impressions [11]. Although there are other studies that deal with psychological impressions of robots, the focus of almost all of the previous studies has been on the engineering technologies proposed in the study, not the psychological mechanism of humans.

As for psychological safety, some of the studies mention it as a key term, but they suggest focusing on the safety of robots along with physical security [6,7]. Japan Robot Association also include psychological safety as a criterion for service robots [12]. However, these definitions about safety have not been sufficiently validated because the questionnaires used to evaluate safety have not been developed for measuring the safety of robots; instead, they look at the issue from the perspective of human personalities [6, 7]. Moreover, the focus is usually on professionals' perspectives, not those of ordinary people, who are the potential users of robots. To create psychologically safe robots, it is necessary to discover the concrete factors that compose the psychological safety of robots for ordinary people and to establish a method for quantifying the degree of safety for individual robots. In this article, we focus on humanoids, which are expected to be partners for people in the near future.

More specifically, [13] developed the scale for evaluation of robots. [13] reviewed the previous studies and defined the useful items to evaluate robots in 5 aspects (Anthropomorphism, Animacy, Likeability, Perceived Intelligence, and Perceived Safety). As a result, they decided the 3 items to evaluate perceived for the Perceived Safety. But they translate the items that originally formed as Likert scale to the items formed as SD (semantic differential) method and they don't show the data to investigate the effect of this transformation statistically. Moreover, the number of subjects seems to relative few to validate the scale satisfactory (36 or fewer participants).

Thus, in this study, we focused on the psychological safety of humanoid robots from ordinary people's general perspectives. That is, we deal with the perceived safety for humanoids from ordinary people's view. We aim to clarify the general factors for evaluating the psychological safety of humanoids, and then develop a new psychological scale that enables us to quantitatively evaluate the degree of safety of various humanoids using common standards. To realize these objectives, we use a social-psychological methodology. In response, we must consider three technical points about this methodology.

The first point concerns the data sampling of people. To avoid the limitation that the obtained data do not reflect the general opinions of people, we must collect the natural impressions about the safety of humanoids from a wide swath of the population. In order to discover the general factors of psychological safety, we must consider certain attributes of the population we are sampling, such as age, sex, and regional characteristics. For example, it has been suggested that people with computer experience or an interest in robots may have a different impression of robots than those who do not have computer experience or an interest in robots [14]. Therefore, in order to obtain the high representativeness of the data in this study, we focused on the Japanese society. We collected data from males and females with a wide range of ages, from their teens to their 70s, according to the distribution of gender and ages in the Japanese population. The study was conducted in three areas in Japan (Tokyo, Osaka, and Fukuoka) that have relatively higher potentials to be robotized cities in the near future. We asked people about experiences of computers and robots according to [14]. We then considered these attributions — age, gender, experiences with or interests in computers and robots — in our investigations.

The second point concerns humanoids as stimulus materials. Several types of humanoids should be used to discover the general factors beyond the types of humanoid robots. It is difficult to define what conditions classify a robot as a humanoid, even in the engineering fields, as wheeled-walking robots are not always regarded as humanoids. However, to clarify the basic factors of safety that are commonly shared by several humanoids, we used 11 different humanoids, including wheeled-walking robots, biped-walking robots, and androids. From an engineering point of view, these types are clearly different in appearance and mechanics; however, this study aims to discover what factors compose the psychological safety of humanoids generally for ordinary people. Therefore, we included these types of robots and used as many different robots as possible to investigate the differences or similarities in robots with regard to the psychological perspective.

The third point concerns the methodology of analysis. As a first step, we wanted to identify the concrete factors for determining psychological safety, which should be based on the subjective opinions of ordinary people. This means that we needed to ask them about their natural impressions of the safety of humanoids without making any assumptions. Therefore, we only asked them about what they think or feel about the safety of the robots and to describe their impressions in the form of free descriptions in open-ended questionnaires. Although the analysis of qualitative data is technically difficult and requires cumbersome procedures, it is necessary to understand the general factors for ordinary people. As a second step, we wanted to develop a new psychological scale. In this step, it was necessary to use statistical procedures as an objective way of quantitatively discovering the general factors. To conduct factor analysis in the social-psychological field, one usually needs a data sample that is about 10 times larger than the number of items on the scale in order to obtain valid results, although a larger number is better statistically. Based on our previous study, we assumed the scale would be around 40 items [15]. Thus, we collected more than 400 people to develop the new scale to make it a general scale.

Therefore, in this study, we aimed to clarify the concrete factors that affect ordinary people's ideas regarding the psychological safety of humanoids by collecting both qualitative and quantitative data from a wide swath of the population and using 11 different humanoid robots. This social-psychological approach can help develop a new scale for evaluating the psychological safety of humanoids generally.

2. Study 1

We used the same movies of humanoids, way of collecting data, and procedures for conducting the presentation experiments in both studies 1 and 2. Therefore, we describe the method for both studies here. In Study 1, we aimed to collect the natural thoughts or feelings about the safety of humanoids for ordinary people in order to discover the concrete factors of psychological safety from qualitative data.

2.1 METHOD

2.1.1 Humanoid Robots and Movies

Figure 1 shows the 11 robots that were presented as stimuli in this study. To obtain general opinions, it is necessary to place the humanoids in as many contexts as possible. Using real robots limits the context to just the time and place that people experience the robot, and it is very difficult to activate all 11 robots to show them to many people for a sufficient amount of time and in sufficient contexts. Therefore, we showed movies of the robots. Movies can show a variety of contexts at once, which is necessary to obtain various opinions, as long as the movies provide a variety of scenes or natural sounds.

Thus, the movies were made to show a wide range of possible situations, including robots alone and interacting with people, that cover the main abilities of each robot as much as possible. We focused on situations people face in their daily lives; therefore, we did not include the background of the operations but mainly focused on the actions of robots interacting with users, not being operated by people.

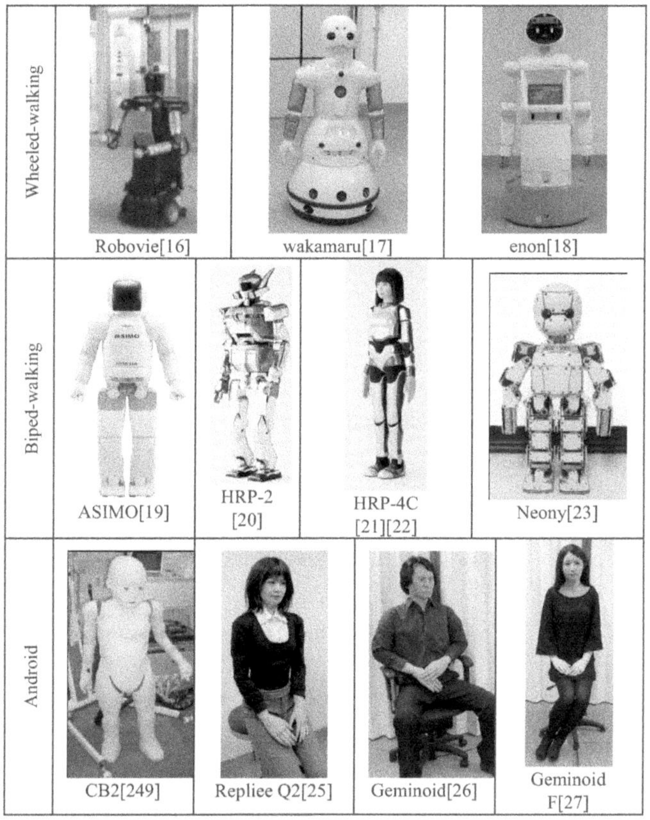

We explained to the developers of each robot the aim of the movies and asked them to provide movies that showcased the main abilities of their robots. We then edited the scenes to show all of the robots' abilities or activities. After editing, we asked the developers to again check the movies from a professional point of view to confirm that there were no missing scenes and no bias had been inserted. After several adjustments, we created final versions of all of the movies, except for one for enon, because we could not contact the developer. Every movie came with sound and was about four minutes long.

2.1.2 *Participants*

The population provided by the Ministry of Internal Affairs and Communications was used to calculate the number of data samples. Empirical social-psychological studies use 15-35 data samples for one experimental condition; therefore, we used 25 participants for the smallest sample generation and calculated the numbers for other generations according to the Japanese population. The total number of participants for each group in Study 1 was 225.

Showing all 11 robots to participants could induce a psychological overload and prevent us from obtaining data about their natural impressions. Therefore, we divided the 11 robots into four groups containing even numbers of each type of robots (Table 1). Each group had 225 participants, so the number of total participants for Study 1 was 900. (Participants in Study 2 are described in the appropriate section below.)

Table 1. Number of participants in each experiment

	Robots	Tokyo	Osaka	Fukuoka	Total
1	Robovie, ASIMO, Repliee Q2	75	75	75	**225**
2	HRP-2, Geminoid F, Neony	75	75	75	**225**
3	HRP-4C, wakamaru, CB2	75	75	75	**225**
4	enon, Geminoid	75	75	75	**225**
	total	300	300	300	**900**

A marketing office that has dealt with various surveys for public offices and universities conducted the sampling of participants and surveys. In order to produce a fair collection of participants, they did not use limited sampling methods such as online surveys. It is much easier and less expensive to use online surveys or people who are already listed as participants for the marketing office. However, only using such sampling methods could produce data that reflect only the opinions of the limited number of people who use the Internet or belong to one office. Therefore, for gathering participants, the marketing office used both online monitors (about 20%) and newspaper advertising (about 80%) to produce a sample population that was representative of Japan. Marketing offices in Japan are not allowed to let people under age 14 age participate in studies, so the group of teenagers used in the study ranged in age from 15 to 19.

2.1.3 *Procedure*

As Table 1 shows, 75 people were selected for each experiment and location. The 75 people were divided again into three groups of 25. Then, each group of 25 participated in the experiment at least once, and the experiments were repeated.

The facilitator distributed the open-ended questionnaires and instructed the participants to watch a movie and then describe their impressions. As for the impressions, we focused on the safety of the robots, but security is a similar concept, so we asked participants about both safety and security. If we found different results for security and safety, it would be necessary to develop identical scales for both security and safety as psychological impressions, although we assumed that security was physical and safety was psychological. If we found no difference, we could use one scale. The order of the questions about security and safety were randomized for balance.

After the participants watched the movie, they described their impressions in the questionnaire, which asked them to describe what they thought or felt about the safety and security of the humanoid, being as detailed as possible. After one movie, they watched the movie about the second robot and repeated this sequence. Three surveys were repeated three times for three robots, but only one survey was repeated twice for the two robots enon and Geminoid. The order in which the robot movies were shown was randomized for balance.

We used one DVD player (Pioneer DV220V), one projector (Sanyo LP-XU75), two speakers (Bose 101MM), one amplifier (Bose 1705II), and one screen (100 inches, or 2 m × 1.5 m) to display the movies.

After describing all of the assigned robots, the participants completed three questionnaires about their interactions with computers: frequency of computer use = (1) not at all, (2) a few times a year, (3) a few times a month, (4) a few times a week, or (5) almost every day; their interest in robots (how many times they watch science-fiction movies in a year); and how many times they have seen real humanoids. We used these questions [14] to show how these factors affect people's impressions of robots.

2.2 RESULTS

2.2.1 Obtained samples and analysis of descriptions

As Table 2 shows, we eventually obtained more data samples than first expected. The number of participants in their 70s was slightly less than their actual representation in the Japanese population. In fact, we were able to secure the expected number of participants in their 70s, but it was sometimes difficult for them to come to the conference room. However, the smaller number of samples was not significant enough to greatly affect the representativeness of the population.

Table 2. Obtained data and excess and deficiency.

Age	Obtained data		Excess and deficiency	
	Male	Female	Male	Female
15-19	53	49	1	1
20-29	61	64	1	4
30-39	77	82	1	6
40-49	70	74	2	6
50-59	72	73	4	5
60-69	74	77	2	1
70-79	42	51	-6	-9
Total	449	470	5	14

Table 3. The number of descriptions about both safety or security in each robot

	Robot	Safety	Security	Total
1	Robovie	861	791	1,652
	ASIMO	928	880	1,808
	Replicee Q2	900	753	1,653
2	HRP-2	989	972	1,961
	Geminoid F	1,027	867	1,894
	Neony	966	924	1,890
3	HRP-4C	802	714	1,516
	Wakamaru	834	776	1,610
	CB2	787	692	1,479
4	enon	1,181	1,134	2,315
	Geminoid	1,065	1,016	2,081
Total		10,340	9,519	19,859

The study was divided into four surveys; therefore, it was necessary to confirm that the surveys did not differ in terms of sample quality. In particular, the study focused on ordinary people's impressions of humanoid robots. We conducted one-way ANOVA for the questionnaire of [14] about computer use and interest in robots. We did not find a significant difference in the surveys for all three questions. This means that the four sample groups did not greatly differ in terms of their interactions with computers and robots, which could affect their impressions of humanoid robots. Therefore, the sample groups could be considered the same in the following analyses.

As Table 3 shows, we obtained a total of 19,859 descriptions (safety: 10,340; security: 9,519). All of the descriptions needed to be categorized into groups reflecting distinct psychological meanings for both safety and security. To analyze the qualitative data, it was important for the researchers to remain objective during the categorization process. Therefore, this study engaged five psychologists, who agreed to remain objective, for the analysis.

First, each description was written on a small card that differed in color for each robot. The five psychologists picked the cards randomly and checked the contents of the descriptions. After checking a few hundred cards, they discussed which psychological meanings appeared; then, they formed a few trial categories and named them. They then assigned the cards to the trial categories. As the card checking continued, the categories were modified. Throughout the analysis, the psychologists repeatedly checked the descriptions, discussed the categories, and assigned the cards to them until all of the cards had been assigned. They also conferred about their choices in order to maintain mutual agreement with one another. The analysis was done separately for both safety and security to compare the results of the obtained categories

2.2.2 Obtained categories

After categorization, the following 10 groups were obtained: Performance, Utility, Acceptance, Harmlessness, Motion, Design, Humanness, Robotness, Synonymity, and Other (Tables 4 and 5).

Table 4. Revealed categories and contents.

Category	Contents
Performance	high performance like accurately performing a task as directed or having an error-free system so it could not go out of control; also limited performance so it could not exceed human abilities
Utility	high usefulness as an instrument to support people, high cognitive abilities, high intelligence
Acceptance	high familiarity, comfort, mental healing
Harmlessness	ability to not fall down, keep its balance, not clash with people or its surroundings, and adjust its physical power
Motion	slow and smooth motions, ability to not act surprisingly actively
Design	round shape, not too heavy, soft quality of materials, soft color
Humanness	communicative; human likeness; human-like voice, motion, and appearance
Robotness	not human-like; robot-like voice, motion, and appearance
Synonymity	descriptions that mention "impressions are equal for both safety and security"
Other	has a social security system to keep information private and not allow it to be misapplied, etc.

Table 5. Number of descriptions in each category.

Category	Safety	Security	Total
Performance	1,049	1,483	2,532
Utility	1,589	1,742	3,331
Acceptance	1,082	493	1,575
Harmlessness	611	1,645	2,256
Motion	886	674	1,560
Design	1,319	871	2,190
Humanness	1,940	622	2,562
Robotness	484	224	708
Synonymity	856	823	1,679
Other	524	942	1,466
Total	10,340	9,519	19,859

As a result, we found similar categories for both safety and security. The descriptions were also sufficiently similar that we could compose just one category and conclude that safety and security have the same meaning in regard to humanoids. Therefore, we deal with these descriptions in terms of overall psychological safety. "Other" descriptions were gathered into small categories, but they did not fit into other categories very well, and there were not enough descriptions to make a single category. Therefore, the following analyses focused on the nine main categories.

2.3 DISCUSSION

In Study 1, we aimed to identify the concrete factors of psychological safety. Over 900 Japanese, from teenagers to people in their 70s, participated in the study. We identified eight factors of safety, as well as two additional categories that were not core factors, Synonymity and Other.

Relative to the other categories, Performance and Utility have the most descriptions. It seems people tend to think a high level of technology is related to safety or security. Humanness is also a major category, whereas Robotness has few descriptions. However, Robotness has almost the opposite meaning of Humanness, and totally human-like qualities seem to be an important factor in how humans regard the safety of humanoids. Relatively few descriptions are categorized in Acceptance, but this category is clearly independent from the other categories in terms of its contents, such as mental comfort. Harmlessness, Motion, and Design focus on the idea that humanoids do no harm to people.

This study used movies to present robots and there is room to investigate whether the obtained perspectives about safety or security are different from them in situations to see real robots or not. In another study, we investigated basic perspectives to evaluate general impressions of humanoids using movies as same as this study. And we conducted an experiment to see the differences by using real ASIMO in the situations that ASIMO introduced itself to each 30 ordinal young adults who didn't have special knowledge about the robot like participants in this study. As a result, they tended to use fewer and limited perspectives especially related to the interactions than the revealed categories through observing movies. It seems that the situation with a real robot could identify the context, therefore, it would be easy to focus on more specific perspectives rather than broader and many perspectives. That is, the perspectives revealed in this study that showed movies of broad contexts as much as possible could obtain general perspectives about safety or security than the limited real contexts, whereas the future research is necessary to clarify more details about this point.

Next, to develop the psychological scale, we considered the meanings of each category and reworded the descriptions into items for a close-ended questionnaire. Based on the descriptions, we created 48 items (each category had five to 10 items). We developed the items according to the contents of descriptions that are dominant in the all of obtained data. Mainly, the major descriptions in each category were used and reworded a little to make them appropriate form as psychological scale. In Study 2, we refined this scale using statistical analysis.

3. Study 2

In this study, we aimed to analyze the new scale using an objective statistical method. We asked different participants from Study 1 to evaluate a humanoid using the new scale to obtain quantitative data that would reveal each robot's degree of safety. By means of factor analysis, we clarified the factor structure of the psychological safety of humanoid. It also revealed the reliability (Cronbach's alpha) of the psychological scale.

3.1 METHOD

We used the same movies of the 11 humanoids in Study 2 as in Study 1. However, we collected different participants for Study 2. As for the number of participants, to conduct empirical factor analysis in the social-psychological field, one needs a data sample that is about 10 times larger than the number of items on the scale. This required a data sample of about 480 for this study. In this article, we focused on the basic factor structure across all of the humanoids, and we processed the data for all 11 humanoids using factor analysis. Therefore, we eventually obtained enough data to meet this minimum number, but analysis of each robot's factor structure (which was not a main issue in this article) requires about 10 times as much data. In addition, it was necessary to maintain the same population distribution as we used in Study 1 (Figure 2). Therefore, we doubled the number of participants we used in Study 1, which resulted in 450 participants for one presentation experiment. In this study, we divided the 11 humanoids into six groups, so as not to overload the participants psychologically. The total number of participants was 2,700. We presented the robots in a random order for balance.

We used the same procedure in Study 2 as in Study 1, including the location of the experiments and the instruments used for showing the movies. We also used the questionnaire of close-ended items that was developed in Study 1. The questionnaire contained 48 items, to which participants responded on a seven-point Likert scale, which ranged from 1 (strongly disagree) to 7 (strongly agree). To investigate whether the developed items could fully measure the robots' psychological safety, we added an additional level of analysis by adding six items that were made to measure safety directly: "I feel safe around this robot," "I feel relieved with this robot," "I feel at ease with this robot," "I become anxious with this robot," "I feel messed up with this robot," and "I get stressed with this robot." The last three items were reversed, and all six items were combined to create a score for "tentative safety," which was based only on words directly related to safety. After factor analysis, we investigated the correlation between scores based on the obtained factor structure and used the tentative safety score as an easy check that the scores based on the other factors could fully measure the robots' psychological safety.

3.2 RESULTS

3.2.1 Obtained samples

As Table 6 shows, we eventually obtained fewer data samples than we first expected. However, for this article, we processed all of the data obtained from the six experiments, which totaled 2,624, a satisfactory number.

Table 6. Obtained data and excess and deficiency.

Age	Obtained data		Excess and deficiency	
	Male	Female	Male	Female
15-19	131	149	-25	5
20-29	196	183	16	3
30-39	230	233	2	5
40-49	206	216	2	12
50-59	189	195	-15	-9
60-69	209	209	-7	-19
70-79	136	142	-8	-38
Total	1,297	1,327	-35	-41

Table 7. Example of items and Factor structure and reliability of the new psychological scale for safety of humanoid

	1	2	3	4	5	6
This robot seems to respond appropriately to my question.	**0.76**	0.15	-0.02	0.05	0.02	-0.11
This robot seems to be able to understand what humans say.	**0.75**	0.14	-0.07	0.06	0.00	-0.03
I think this robot can recognize its surroundings.	**0.70**	-0.08	-0.01	-0.06	-0.07	0.03
This robot seems to work according to my instructions.	**0.66**	-0.05	-0.04	0.08	0.02	-0.24
Its figure is robot-like.	0.14	**-0.90**	0.08	0.05	0.13	0.04
Its appearance is similar to that of humans.	0.10	**0.83**	0.01	0.04	0.14	-0.13
This robot does robot-like motions.	0.12	**-0.76**	0.01	0.11	0.21	-0.09
It makes me self-conscious that this is a robot.	0.16	**-0.74**	0.07	0.08	0.13	-0.17
I am familiar with this appearance.	-0.03	0.01	**0.80**	-0.09	-0.03	-0.01
This robot moves in a friendly way.	0.10	-0.01	**0.70**	-0.01	0.00	0.07
This robot seems to give me mental healing.	0.00	-0.09	**0.67**	0.07	0.04	0.29
I think I can have a cozy atmosphere with this robot.	0.12	-0.17	**0.63**	0.08	-0.07	0.16
This robot doesn't seem to do injury to a human's body.	-0.04	-0.01	0.01	**0.68**	-0.09	-0.03
This robot doesn't seem to go out of control.	0.10	0.02	-0.11	**0.65**	-0.22	0.08
I do not think this robot can exceed a human's ability.	-0.24	-0.01	-0.02	**0.46**	0.13	-0.13
This robot seems to produce improper operating signals.	-0.24	-0.09	0.07	-0.14	**0.49**	-0.17
This robot seems to break down easily.	-0.30	0.00	0.06	-0.01	**0.45**	-0.20
This robot looks like it falls down on the ground easily.	-0.26	0.14	-0.02	0.08	**0.41**	-0.10
This robot seems to have an independent will like humans.	0.25	0.13	-0.04	-0.16	0.09	**0.55**
This robot doesn't seem to have emotions like humans.	-0.11	-0.17	-0.10	0.11	0.17	**-0.46**
	Interfactor correlation matrix					
Performance	1.00	0.29	0.65	0.31	-0.06	0.34
Humanness	0.29	1.00	0.22	0.00	0.18	0.37
Acceptance	0.65	0.22	1.00	0.41	-0.11	0.21
Harmlessness	0.31	0.00	0.41	1.00	-0.28	-0.10
Toughness	-0.06	0.18	-0.11	-0.28	1.00	0.30
Agency	0.34	0.37	0.21	-0.10	0.30	1.00
α	.88	.89	.84	.60	.60	.65

3.2.2 Factor analysis

We conducted an exploratory factor analysis with promax rotation on scales. Analysis on the item level according to Kaiser's criterion (eigenvalues > 1.0) yielded nine factors. Inspection of the scree plot revealed fewer factors. However, overall interpretive significance revealed that a model containing six factors was more appropriate. This result accounted for 48% of the variance in the data. In order to create the final form of the scale, we selected 33 items that loaded above .40 on each factor (and did not load above .30 on any other factor).

Table 7 lists examples of items, factor loadings, interfactor correlations, and the reliability of each subscale. The first factor, Performance, included a high level of performance, such as appropriately working as instructed or having useful actions when people needed them. The second factor, Humanness, captures the human-like (as opposed to robot-like) qualities of the robot in terms of appearance and motions. The third factor, Acceptance, clearly reflects a feeling of comfort or familiarity with the robot. The fourth factor, Harmlessness, captures the sense of the robot's obedience, such as not harming people or not going out of control. The fifth factor, Toughness, concerns robots not malfunctioning or breaking down easily. The last factor, Agency, mentions the robot's human-like will or emotions.

In Study 1, we identified eight categories: Performance, Utility, Acceptance, Harmlessness, Motion, Design, Humanness, and Robotness. In contrast, this Study 2 revealed six factors. The data suggest that Performance and Utility can be combined into a single factor: Performance. Further, Design (like an item about a round-shaped design) and Motion (like an item about amiable motions) can also be combined into a single factor: Acceptance. Part of the Performance category (like an item about limited ability) was combined to become a factor of Harmlessness. Then, a part of Harmlessness (like not falling down) and Performance (like not going out of control) were combined to create Toughness. Humanness remained the same, and Robotness became Agency.

In Study 1, we categorized the cards based on meanings and wordings with as much detail as possible. The results of factor analysis showed a response pattern to evaluate. Therefore, the categories in Study 1 and the factors in Study 2 are slightly different. However, these factors each have a clear meaning and seem to reflect the structure of humans' psychological reactions toward the safety of humanoids. As for reliability (alpha score), Performance, Humanness, and Acceptance show a high reliability as a psychological scale, while in comparison, Harmlessness, Toughness, and Agency show a relatively low reliability.

3.2.3 Correlation check

Factor analysis revealed that there are six factors of humans' perceptions about the psychological safety of humanoids. In this section, we investigate whether these factors can account for the concept of safety. Because we collected unstructured, free descriptions in Study 1, it is possible there are more or fewer important factors than these six. We calculated the average score of each factor and the score of tentative safety (i.e., the six items that asked about safety directly), and the correlations are shown in Table 8.

As Table 8 shows, Acceptance, Performance, and Toughness are strongly correlated to the tentative safety score. That is, these three factors are important factors for explaining the psychological safety of humanoids. Harmlessness also shows a positive correlation. However, Agency shows a low correlation, and Humanness is not significantly correlated. These two factors do not seem to relate to psychological safety very strongly. To investigate further details, we conducted a correlation analysis of the 11 humanoids.

Table 8. Table captions should be placed above the table

		1	2	3	4	5	6
1	Perfor-mance	1	.28**	.62**	.15**	.44**	.416**
2	Humanness	.28**	1	.28**	-.08**	.03*	.48**
3	Acceptance	.62**	.28**	1	.20**	.31**	.30**
4	Harmless-ness	.15**	-.08**	.20**	1	.18**	-.18**
5	Toughness	.44**	.03*	.31**	.18**	1	.14**
6	Agency	.42**	.48**	.30**	-.18**	.14**	1
7	Tentative Safety	**.58****	**0.02**	**.79****	**.28****	**.40****	**.19****

** $p<.01$, *$p<.05$

Table 9 shows the correlation between the score of tentative safety and the scores of the six factors for each of the 11 humanoids. The results show that almost all of the robots show a significant positive correlation. Acceptance, Performance, and Toughness are especially strongly correlated with tentative safety for every humanoid. However, Humanness and Agency show a low correlation again. As for the Geminoid humanoid, the correlation is not significant between tentative safety and Humanness. This seems to show that being more human-like is not significantly related to psychological safety for some robots.

Table 9. Correlation between the score of tentative safety and the scores of the six factors

	Performance	Humanness	Acceptance
Robovie	.613**	.263**	.787**
Wakamaru	.606**	.194**	.827**
Enon	.621**	.205**	.804**
ASIMO	.643**	.280**	.830**
HRP-2	.594**	.248**	.784**
HRP-4C	.534**	.219**	.793**
Neony	.520**	.355**	.816**
CB2	.620**	.375**	.842**
Geminoid	.556**	**0.051**	.759**
Geminoid F	.624**	.306**	.797**
Repliee Q2	.585**	.290**	.818**
	Harmlessness	Toughness	Agency
Robovie	.266**	.411**	.317**
Wakamaru	.217**	.293**	.195**
Enon	.207**	.397**	.282**
ASIMO	.336**	.463**	.245**
HRP-2	.276**	.369**	.206**
HRP-4C	.303**	.348**	.279**
Neony	.259**	.300**	.224**
CB2	.292**	.469**	.293**
Geminoid	.309**	.390**	.138**
Geminoid F	.167**	.348**	.246**
Repliee Q2	.341**	.343**	.249**

** $p<.01$, *$p<.05$

3.3 DISCUSSION

In Study 2, we aimed to collect quantitative data based on the items developed in Study 1 to investigate the factor structure and reliability of our scale. In addition, correlation analysis was conducted to investigate whether all of the obtained factors are importantly related to psychological safety.

Factor analysis revealed six distinct factors. Some factors of reliability did not show satisfyingly high scores, as one of useful way, it could be helpful to use only high-loaded items. Correlation analysis between the score of tentative safety and the score of each factor revealed that Acceptance, Performance, and Toughness (and relatively Harmlessness) are core factors of psychological safety, whereas Humanness and Agency are not as important.

4. CONCLUSION

In this article, we aimed to identify the concrete factors of psychological safety for humanoids and to develop a new scale that could quantitatively evaluate the degree of safety of specific humanoids. In Study 1, we collected the free descriptions of people's subjective impressions about the safety of humanoids to clarify what people think about the safety of humanoids. Study 2 revealed the factor structure of psychological safety using evaluation data based on items that were developed according to the descriptions and categories obtained and clarified in Study 1. We found six factors of psychological safety: Performance, Humanness, Acceptance, Harmlessness, Toughness, and Agency. Correlation analysis revealed that Acceptance, Performance, and Toughness are core factors of psychological safety for humanoids, and Harmlessness also seems to important relatively. However, Humanness and Agency are not as important.

Previous studies have dealt with the safety of robots, but they have only used psychological scales that were developed to evaluate humans' personality or impressions or from professional point of view. This study identified the structure of psychological safety from potential user's view and developed a new psychological scale oriented to humanoids, especially to quantify their degree of safety.

This research used 11 different humanoids that are currently the most advanced robots available. It also collected data from a wide swath of the population in Japan, ranging in age from teenagers to people in their 70s based on the country's overall population distribution. We used both qualitative and quantitative methods to mold subjective psychology into a more objective structure. Therefore, the scale we developed has a high potential of usefulness for other humanoids and to be used by any person in Japan. In addition, the psychological scale is very easy to use as a questionnaire. It can also be used in its full-length version or shortened to include only high-loaded items. Moreover, according to the aim of this research, it is possible to use only some of the factors. This study widens the possibility of evaluating humanoids psychologically using common standards of safety.

In the future, it is necessary to compare other humanoids with the humanoids used in this study, as well as how people with similar characteristics, such as age and gender, evaluate the safety of humanoids.

5. ACKNOWLEDGMENT

This research was supported in part by Global COE Program "Center of Human-Friendly Robotics Based on Cognitive Neuroscience" of the Ministry of Education, Culture, Sports, Science and Technology, Japan.

6. REFERENCES

[1] National Institute of Population and Social Security Research DOI=http://www.ipss.go.jp/

[2] Minister of Economy, Trade and Industry, "Guidelines for Next-generation Robot Safety" DOI=http://www.meti.go.jp/english/index.html

[3] Ministry of Health, Labour and Welfare DOI= http://www.mhlw.go.jp/english/

[4] M. Lohse, F. Hegel, B. Wrede (2008) Domestic Applications for social robots - a user study on appearance and function. Journal of Physical Agents 2: 21-32.

[5] K. Inoue, Y. Ujiie, T. Takubo, and T. Arai: "Psychological Evaluatioin for Rough Shape of Humanoid Robots Using Virtual Reality", The 13th International Conference on Adavanced Robotics (ICAR 2007), pp.1005-1010, August 21-24, 2007, Jeju, Korea.

[6] S. Negi, T. Arai, K. Inoue, Y. Ujiie, T. Takubo (2008) Psychological Assessment of humanoid Robot Appearance and Performance Using Virtual Reality. in Proc. of 17th IEEE International Symposium on Robot and Human Interactive 719-724.

[7] Y. Ujiie, K. Inoue, T. Takubo, T. Arai (2006) Evaluation of human sense of security for coexisting robots using virtual reality. in Proc. of 3rd International Symposium on Systems and Human Science(SSR 2006)

[8] K. Inoue, S. Nonaka, Y. Ujiie, T. Takubo and T. Arai "Comparison of Human Psychology for Real and Virtual Mobile Manipulators", Proceedings of the 14th IEEE International Workshop on Robot and Human Interactive Communication (RO-MAN 2005), pp. 73 - 78, Nashville, Tennessee, USA, August 13-15, 2005.

[9] Y. Muto, S. Takasugi, T. Yamamoto, Y. Miyake (2009) Timing Control of Utterance and Gesture in Interaction between Human and Humanoid robot. in Proc. of the IEEE International Symposium on Robot and Human Interactive Communication.1022-1028.

[10] T. Kanda, H. Ishiguro, M. Imai, T. Ono (2004) Development and Evaluation of Interactive Humanoid Robots. in Proc. of the IEEE (Special issue on Human Interactive Robot for Psychological Enrichment) 92:1839-1850.

[11] H. Kamide, M. Yasumoto, Y. Mae, T. Takubo T, K. Ohara , and T. Arai "Comparative Evaluation of Virtual and Real Humanoid with Robot-Oriented Psychology Scale" 2011 IEEE International Conference on Robotics and Automation Shanghai International Conference Center, pp.599-604, May 9-13,Shanghai, China, 2011

[12] http://www.jara.jp/publication/dl/servicerobot_system2.pdf (in Japanese)

[13] C. Bartneck, D. Kulic., E. Croft, and S. Zoghbi (2009). "Measurement instruments for the anthropomorphism, animacy, likeability, perceived intelligence, and perceived safety of robots." International Journal of Social Robots 1(1): 71-81.

[14] T. Kanda, H. Ishiguro and T. Ishida, Psychological analysis on human-robot interaction, IEEE International Conference on Robotics and Automation (ICRA 2001), pp. 4166-4173, 2001

[15] H. Kamide, Y. Mae, T. Takubo, K. Ohara, and T. Arai, "Development of a Scale of Perception to Humanoid Robots: PERNOD", in Proc. of The 2010 IEEE/RSJ International Conference on Intelligent Robots and Systems, Taipei, 2010.

[16] S. Koizumi, T. Kanda, M. Shiomi, H. Ishiguro, N. Hagita , Preliminary Field Trial for Teleoperated Communication Robots IEEE International Workshop on Robot and Human Communication (ROMAN2006), 2006. (Sep. 6-8, Hatfield, United Kingdom)

[17] N. Kawachi, Y. Koyuu, K. Nagashima, K. Ohnishi, R. Hiura, Home-use Robot "wakamaru,". Mitsubishi Juko Giho, Vol.40, No.5, pp.270-273. (2003) (in Japanese)

[18] http://www.frontech.fujitsu.com/services/products/robot/serv icerobot/(in Japanese)

[19] S. Shigemi, Y. Kawaguchi, T. Yoshiike, K. Kawabe, and N. Okgawa, " Development of New ASIMO", Honda Technical Review Vol.18 No.1, pp.38-44.

[20] F. Kaneko, S. .Kanehiro, H. Kajita, T. Hirukawa, M. Kawasaki, K. Hirata, K. .Akachi and T. Isozumi, "Humanoid Robot HRP-2", in Proc. of the 2004 IEEE International Conference on Robotics& Automation, New Orleans, 2004, pp.1083-1090.

[21] K. Kaneko, F. Kanehiro, M. Morisawa, K. Miura, S. Nakaoka, and S. Kajita, "Cybernetic Human HRP-4C," Proc. IEEE-RAS International Conference on Humanoid Robots, pp. 7-14, 2009.

[22] S. Nakaoka, F. Kanehiro, K. Miura, M. Morisawa, K. Fujiwara, K. Kaneko, S. Kajita, and H. Hirukawa, "Creating facial motions of Cybernetic Human HRP-4C," in IEEERAS International Conference on Humanoid Robots, Paris, France, December 2009.

[23] T. Minato, F. DallaLibera, S. Yokokawa, Y. Nakamura, H. Ishiguro, Emanuele Menegatti, A baby robot platform for developmental robotics, Proc. of the Workshop on Synergistic Intelligence: Approach to Human Intelligence through Understanding and Design of Cognitive Development (IEEE/RSJ International Conference on Intelligent Robots and Systems), 2009.

[24] T. Minato, Y. Yoshikawa, T. Noda, S. Ikemoto, H. Ishiguro, and M. Asada, CB2: a child robot with biomimetic body for cognitive developmental robotics, Proc. of the IEEE-RAS International Conference on Humanoid Robots, pp.557-562, 2007.

[25] Chikaraishi T, Minato T, Ishiguro H, "Development of an Android System Integrated with Sensor Networks", the Proceedings of the 2008 IEEE International Conference on Intelligent Robots and Systems (IROS 2008), pp.326-333, September, 2008.

[26] Ogawa K, Bartneck C, Sakamoto D, Kanda T, Ono T, Ishiguro H. Can An Android Persuade You?, The 18th International Symposium on Robot and Human Interactive Communication (RO-MAN2009),pp. 553-557, 2009.

[27] http://www.kokoro-dreams.co.jp/actroid_f/index.html(in Japanese)

Consistency in Physical and On-screen Action Improves Perceptions of Telepresence Robots

David Sirkin and Wendy Ju
Center for Design Research
Stanford University, Stanford, CA
{sirkin, wendyju}@cdr.stanford.edu

ABSTRACT

Does augmented movement capability improve people's experiences with telepresent meeting participants? We performed two web-based studies featuring videos of a telepresence robot. In the first study (N=164), participants observed clips of typical conversational gestures performed a) on a stationary screen only, b) with an actuated screen moving in physical space, or c) both on-screen and in-space. In the second study (N=103), participants viewed scenario videos depicting two people interacting with a remote collaborator through a telepresence robot, whose distant actions were a) visible on the screen only, or b) accompanied by local physical motion. These studies suggest that synchronized on-screen and in-space gestures significantly improved viewers' interpretation of the action compared to on-screen or in-space gestures alone, and that in-space gestures positively influenced perceptions of both local and remote participants.

Categories and Subject Descriptors

H.5.m **[Information Interfaces and Presentation (e.g., HCI)]**: Miscellaneous.

General Terms

Design, Experimentation.

Keywords

Telepresence robotics, embodied interaction, videoconferencing.

1. THE PROXY-IN-PROXY PROBLEM

Embodied proxies are increasingly used by remote workers to participate in the activities of a central workplace. Embodied proxy systems combine a live video representation of the remote worker with a local physical platform, often with human-body-like proportions. This setup enables remote workers to communicate naturally with their distant, collocated peers, have a similar presence (at least in terms of body size and location) to their collaborators, and often allows them to move, or be moved, around the central workplace to engage in day-to-day, informal interactions [18].

However, combining a video display of a remote worker on an articulating base creates a confusing juxtaposition, an issue we call the *proxy-in-proxy* problem. The worker actually has two representations in the central workplace: that of the audiovisual feed, and that of the physical platform. These two distinct channels are likely to portray inconsistent non-verbal facial or gestural cues, even if that is merely the lack of complementary action between them. We know that in face-to-face interactions, inconsistencies in verbal and non-verbal cues are often interpreted as a sign of deceit [17] and mistrust, and cause increased cognitive load [10]. Video-based interactions can also influence observers' abilities to interpret these cues, such as when comparing videos that include facial expressions alone to those that include the entire body [8]. So the question arises: how do inconsistencies between *on-screen* and *in-space* behaviors affect people's interpretations of non-verbal gestures and their perceptions of the remote worker's proxy?

Figure 1. Our embodied proxy displays the remote participant's on-screen expressions and gestures in physical space.

In this paper, we present two studies that compare people's subjective interpretations and responses to single-channel gestures (either a visible on-screen expression alone, or a physical in-space motion alone) versus coordinated gestures (where both on-screen and in-space representations move in concert). The studies focus on gestural aspects of video-based communication, including facial expressions, head motion and orientation, and posture. Results suggest the importance of creating consistency across the remote worker's various proxy representations, and the great benefit that coordinated action can have, even for the most familiar behaviors. A better understanding of the issues caused by the proxy-in-proxy problem could help robot designers develop technologies to address and resolve inconsistencies and their effects.

2. BACKGROUND

2.1 The Embodied Proxy Metaphor

Unlike earlier telepresence designs, which employed the metaphor of the *avatar* (such as the on-screen avatars in BodyChat [27] or physical avatars like Geminoid [23]) or the *portal* (such as EuroPARC's Portholes [7], Berkeley's MultiView [22] or Cisco's Telepresence System), systems that employ the embodied proxy metaphor commonly have a one-to-one relationship of person to proxy, human-body-like proportions, and a near-life-size display of the remote person's face and upper body. Some, like Sun Micro-systems' Porta-Person [29] and MIT's RoCo [3] feature a flat screen computer monitor system on a robotic head/neck mechanism that sits on a desk or chair. Others, like Paulos' PRoP [24], Microsoft Research's Embodied Social Proxies [26], Willow Garage's Texai [19] and Anybots' QB host a live video on a large

Figure 2. The Study 1 video prototypes showed brief video clips of on-screen, in-space or combined on-screen/in-space gestures. Shown here are: on-screen laugher, on-screen surprise, in-space look to one side, and combined on-screen/in-space look to one side.

flat screen mounted on a remotely steerable base. MIT's MeBot [1] is sized for desktop use, but extends the metaphor to include a three degree-of-freedom screen mounted on a mobile base with articulating arms, which track the operator's motions.

2.2 Challenges in Embodied Representation

Although each embodied proxy platform design has its own unique characteristics and issues, people bring common shared expectations about how to interpret embodied social cues to each [13]. Casting the video telepresence feed into an embodied form has significant ramifications for how on-site workers perceive the proxy and, in turn, the remote worker.

Designers of telepresence systems might assume that a lack of motion in the embodied platform would provide a neutral stage on which to perform on-screen actions. However, as Groom, *et al.* [12] note, "people expect bodies not only to serve as decorations suggesting identity, but also as functional units intended to interact with the environment and to communicate information." We use embodied non-verbal communications such as gestures, body movements, posture, visual orientation, and spatial behavior in concert with our verbal communication to signal our attention, express emotions, convey attitudes, and encourage turn-taking [2], and numerous studies suggest that we (perhaps subconsciously) prefer that our technological counterparts follow suit. People respond positively towards agents that display consistency in verbal and non-verbal cues compared to those with mismatched cues [14]. We prefer that agents use gaze and gesture to provide contextual grounding for the agent and user's shared experience to those that do not [4]. We read intentions from the seemingly unintentional non-verbal cues made by robots [20]. Researchers have found that rigid and frozen postures are a common "deception cue" [13], and habituated social responses to such cues could affect the perception on-site workers have of their remote counterparts. Hence, there are reasons to doubt the neutrality of an unmoving embodied platform, or an unmoving video image that features a large image of a person's head.

2.3 Consistency in Physical & On-screen Action

Recent experiments found that people prefer expressive robotic motions to a static platform [1]. We propose that *consistency* between physical and on-screen action is critical to this preference.

Related work on embodied agents and personality has defined consistency as accordance between voice and body cues [21]. Consistency can thus be seen as one cue supporting another. For gesturing embodied proxies, the remote worker's on-screen expressions and voice (if present), and the platform's physical movements, provide these cues. Such proxies can represent either in-space motions alone, on-screen actions alone, or a combination of the two. We therefore chose to investigate individual aspects of gestural expression, and how they contribute to positive interactions, under these three conditions.

3. RESEARCH METHODOLOGY

In order to explore how people react broadly to proxy-in-proxy designs, we ran our experiments using online video prototypes and crowdsourced participants from Amazon's Mechanical Turk (MTurk) service [16]. MTurk is an online marketplace where *requesters*—people or organizations who typically have small tasks to be performed—connect with *workers* who have the time and attention to perform them for a small payment.

Video prototypes are short movies that demonstrate how an interactive technology would perform. There are a number of practical reasons to employ video prototypes rather than in-person trials. They permit us to find the most salient design factors prior to building a fully functional system; to tune the proxy's motions to be as subtle or obvious, coarse or refined as required; to reproduce precise timings between the remote actor's on-screen gestures and the proxy's motions; to better control the study, by exposing participants to the same, consistent stimulus; and to access a more diverse audience than possible using local participants. To ensure the quality of responses [11], we tracked each submission time, to confirm that responses did not occur before the video completed, and that total task time was not notably below average; we reverse coded particular questions and filtered for straight-line responses; we assigned unique identifiers to match each worker to a location, and confirmed that IP addresses originated in the expected regions.

Another consideration is that online responses may differ from those of in-person trials. Powers, *et al.* [25] found that a remote projected robot could be used to study many critical social processes, including engagement. And studies comparing real-world evaluations of interactive prototypes with web-based video prototypes in the field of human-robot interactions by Kidd [15] and Woods, *et al.* [28] found that results from video studies tend to be consistent with in-person studies, although the simulated studies might not point out every salient factor that may be present in a real-world setting.

4. STUDY 1: ACTIONS IN ISOLATION

4.1 Video Prototypes

This study featured two sets of 27 video clips of different interactive gestures, with each set being performed by a male or a female actor, on a stationed telepresence proxy (see Figs. 1 and 2). Each video clip was approximately five seconds in duration, and depicted a first-person view of the proxy. The proxy was constructed of an iMac G4 computer with hemispherical base and 15-inch screen, connected to by an articulating "neck." The neck allowed three types of motion: pan and lift at the hemispherical base, and vertical screen tilt at the top of the neck-screen connection. The proxy's screen motions were puppeted so as to approximate the movements of a human head, and in particular, to reproduce the movements displayed by the actor, whose image was shown as a head-and-shoulders gesture against a black curtain background. The clips had no audio.

Each video clip presented one of nine behaviors that might be observed during a typical video interaction:

Agree (Nod "Yes")	Laughter	Look Down at Table
Disagree (Shake "No")	Look to One Side	Confusion
Surprise	Lean In to Look Close	Think Carefully

Each of these nine gestures had three variants:

On-screen The screen/frame remained still and only the actor's visible facial expressions and gestures changed.

In-space The screen/frame was actuated but the actor's facial expressions and gestures remained neutral.

On-screen/In-space The screen/frame was actuated and the actor's facial expressions and gestures changed.

4.2 Hypothesis

Our proxy design allowed us to isolate gestures enacted by the remote participant (visible on the proxy's screen), gestures made by the screen itself (which occurred in physical space) and the combination of the two. We could then explore the proxy-in-proxy problem through the following hypothesis:

H1 Consistency between on-screen and in-space action improves observers' comprehension of the message that the remote participant is expressing when compared to:

a. On-screen action alone, without corresponding physical motions.

b. In-space action alone, without corresponding facial expressions.

Consistency refers to coordinated gestural cues between physical and on-screen channels, which occur in concert with one another.

4.3 Study Design

The study was a within-subjects design: each participant was shown video prototypes of all nine different behaviors, with each behavior enacted using all three types of motion.

4.3.1 Procedure

We recruited 55 MTurk participants in each of three regions: the United States, India and the rest of the world, and collected data over four days. This was done to isolate any regional differences, as well as to account for the time-dependent nature of a globally administered study. Participants viewed a brief instructional page that described who we were, and what the study was about. The average time to complete the study was 18 minutes. Participants were paid US$2 each, which equates to about $6 per hour.

4.3.2 Video Questionnaire

Each participant was then shown the series of 27 video clips in random order. Even-numbered participants viewed a set with a male on-screen actor, and odd-numbered participants viewed a set with a female actor. Each clip appeared on its own webpage, and was accompanied by four questions. The first question provided a menu of nine alternative interpretations of the behavior that appeared in the video:

1. Which choice best describes what the remote participant is communicating?
   - ○"I'm surprised" ○"Let me look closer" ○"I disagree"
   - ○"That was funny" ○"Let's look over there" ○"My feelings are hurt"
   - ○"What's down there?" ○"I agree" ○"I'm thinking"
2. How confident are you in your interpretation of this message?
3. How strongly did you respond to the message that you interpreted?
4. Rate the observed actions based on the following 3 parameters.

We ran a pilot study that revealed that the *confusion* and *think carefully* behaviors appeared and were rated very similar to most participants. As a result, responses to both of those clips were mapped to the same "I'm thinking" menu choice.

The next two questions on each page used a 7-point Likert scale to ask participants about their confidence in their interpretation, (where 1="not sure at all" and 7="absolutely sure"), and about how strongly they responded to the message (1="barely at all" and 7="very strongly"). The last question asked them to rate the observed action along three further dimensions (with 1="unnatural" and 7="natural," then 1="unfamiliar" and 7="familiar," and finally, 1="unintentional" and 7="intentional").

4.3.3 Coding of Responses

Each response to the menu of alternative interpretations was coded as either correct or incorrect, so that a guess would have a 1-in-9 (about 11%) chance of being correct. Our assignment of what made a response correct was based on our communicative intent when we created the behavior for each video (prior to collecting data), as well as comparison with the most frequently cited response (the mode) for that behavior among all of the participants' responses. These two bases agreed in every case.

4.3.4 Participants

We had 164 complete responses, with most of the rest of the world participants located in Europe. Nearly all respondents reported their gender and age, with 65% being male, and with ages ranging from 18 to 63 (M=29.3, SD=9.15).

4.4 Results

4.4.1 Participant Interpretation of Gestures

To determine whether participants' interpretations of the behaviors differed significantly from chance alone, we compared responses with the expected value from guessing. For the on-screen/in-space condition, z ranged from 6.5 to 28.2 and for the on-screen only condition, z ranged from 6.6 to 25.9, with all gestures identified as significantly above chance at $p<.001$. For the in-space only condition, six of the nine gestures were identified as significantly above chance (z ranged from 5.629 to 14.432) at $p<.001$, with *confusion* (z=2.2) having $p<.01$, *laughter* (z=1.6) having $p<.05$ and *think carefully* (z=0.7) having $p<.25$.

To determine whether participants' interpretations of the behaviors shown differed from each other, we performed a log linear regression; the model assessed the impact that the type of motion, the particular gesture, and the gender of the actor, had on interpretations. To find significant differences, we used one-way ANOVA with interpretation as a dependent variable, and gesture

Table 1: Observers' subjective ratings for interpretation, for both individual gestures and the type of motion*.

	Interpretation				
	On-screen & In-space	On-screen	In-space	*M*	*SD*
Shake "No"	0.80	0.82	0.65	0.76 a	0.09
Nod "Yes"	0.80	0.84	0.62	0.75 a	0.12
Surprise	0.78	0.74	0.44	0.65 b	0.18
Laughter	0.86	0.80	0.16	0.61 b	0.39
Look Side	0.71	0.60	0.47	0.59 b,c	0.12
Lean In	0.74	0.66	0.36	0.58 b,c	0.20
Look Down	0.63	0.51	0.32	0.49 d	0.15
Confused	0.59	0.50	0.18	0.42 e	0.22
Thinking	0.35	0.36	0.13	0.28 f	0.13
M	0.69 a	0.65 b	0.37 c		
SD	0.15	0.17	0.19		

*Differing subscripts indicate significant differences.

and type of motion as the fixed factors. Responses from the three study regions followed very similar patterns, but had somewhat different values. Summary statistics for all regions are shown in Table 1, which includes significant differences at $p<.05$ for gesture and $p<.001$ for type of motion, per Tukey post-hoc. Example data from the United States is shown in Fig. 3.

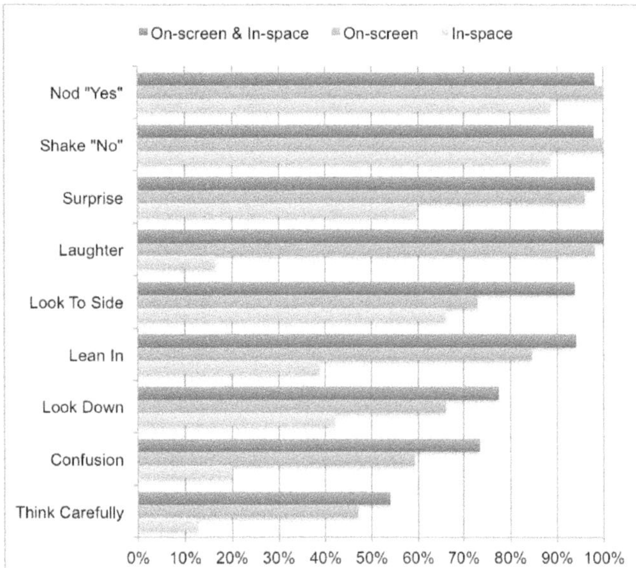

Figure 3: The percentage of responses that agreed with the intended message, for participants in the United States.

We found that the type of motion had a significant effect on participants' interpretation of the behaviors that they saw ($\chi^2(2)=395.7$, $p<.001$). The in-space condition had substantially lower accuracy across the board compared to the other two types of motion, each of which notably included the actor's on-screen expressions. Because this can mask significant differences between the other two types of motion (which concern us most), we performed a follow-up analysis excluding the in-space condition, focusing only on the difference between on-screen and on-screen/in-space conditions. We found that the type of motion still had a significant effect on interpretation ($\chi^2(2)=15.0$, $p<.001$). The on-screen/in-space condition was the better performer in most cases, with the amount of improvement dependent upon the *category* of gesture (which we cover in the discussion).

We also found that certain gestures (of the nine shown: such as laughter, agree, disagree, and so on) were more recognizable than others, with gesture having a significant effect on participants' interpretation ($\chi^2(8)=345.9$, $p<.001$). Post-hoc analysis, including only the on-screen and on-screen/in-space conditions, found that gesture still had a significant effect ($\chi^2(8)=239.9$, $p<.001$).

The interaction between the type of motion and gesture was also significant ($\chi^2(10)=94.4$, $p<.001$). This implies that certain types of proxy motion (say, in-space movement) influenced perceptions of particular gestures (such as the actor's glance to the right) more so than they influenced perceptions of other gestures (such as a nod or a head shake). There was no significant effect due to the on-screen actor's gender, which agrees with our expectation that it was the combination of motion and gesture, rather than the persona of the remote worker, which influenced responses.

4.4.2 Confidence and Strength of Response

Observers' confidence in their interpretations, and the strength of their response, followed the same pattern as the interpretations of the intended message. This agreement was with respect to both the type of motion and the particular gesture, meaning that participants actually were more correct when they thought that they were, and in those cases, the message had a stronger impact. Summary statistics are shown in Table 2, including significant differences per Tukey post-hoc at $p<.05$ for gesture. For type of motion, $p<.001$ for confidence and $p<.05$ for strength of response.

Table 2: Observers' subjective ratings for both confidence and strength of response, for individual gestures and type of motion.

Confidence

	On-screen & In-space	On-screen	In-space	*M**	*SD*
Shake "No"	6.33	6.19	5.23	5.92 a	0.60
Nod "Yes"	6.10	6.08	5.11	5.76 a,b	0.57
Surprise	6.07	5.91	4.35	5.45 b,c	0.95
Laughter	6.28	6.13	3.86	5.42 c	1.36
Look Side	5.52	5.03	4.25	5.26 c	0.93
Lean In	5.86	5.74	4.18	4.93 d	0.64
Look Down	5.46	4.79	4.34	4.86 d	0.57
Confused	5.12	5.14	4.23	4.83 d	0.52
Thinking	4.90	4.90	4.07	4.62 e	0.48
M*	5.74 g	5.55 h	4.40 i		
SD	0.51	0.57	0.46		

Differing subscripts indicate significant differences at *$p<.05$ and **$p<.001$.

Strength of Response

	On-screen & In-space	On-screen	In-space	*M*	*SD*
Shake "No"	5.87	5.78	4.77	5.47 a	0.61
Nod "Yes"	5.69	5.64	4.63	5.32 a,b	0.60
Surprise	5.77	5.44	4.40	5.20 a,c	0.72
Laughter	5.89	5.88	3.82	5.19 a,d	1.19
Look Side	4.92	4.93	4.26	5.13 b,c,d	0.78
Lean In	5.57	5.59	4.23	4.71 e	0.38
Look Down	5.14	4.48	4.23	4.62 e	0.47
Confused	4.91	4.81	4.24	4.83 e	0.37
Thinking	4.65	4.72	4.02	4.47 f	0.38
M	5.38 g	5.25 h	4.29 i		
SD	0.45	0.49	0.29		

Differing subscripts indicate significant differences at $p<.05$.

4.5 Discussion

Our online study showed that consistency between the remote actor's facial expressions and gestures and the proxy's physical motions resulted in improved understanding of the behavior portrayed, higher confidence levels, and stronger responses, confirming both H1a and H1b.

But the data also indicate a more nuanced interpretation. One participant stated it this way: "Looking back, I feel that for 'simple' messages (yes, no, etc.), the movement didn't do much, and maybe even got in the way. For more complicated messages it seemed to make more difference." Fig. 3 supports this comment, and suggests even more. Notice that gestures such as *nod "yes"* or *shake "no"* were very well recognized across the conditions, even the in-space only condition. Even without a complementary facial expression, the proxy's motion gave participants the right idea. In contrast, gestures such as *laughter* and *surprise* were not as well recognized in the in-space condition. They seemed to require another cue—such as a smile—to place them in the proper context. In this case, the addition of physical motion did not substantially improve the clarity of the message. We call these two categories of gesture "intentionally communicative" and "unanticipated response." The first is sufficiently communicated through in-space motion alone, while the second is sufficiently communicated through on-screen expressions alone.

Gestures such as *lean in to look close*, *look to one side*, and *look down at table* serve mostly to indicate the remote actor's focus of attention. The addition of motion to each of these significantly improved interpretations. Again this makes sense, as motion toward one direction tends to draw attention that way. These represent another category of gesture, "focus of attention," best communicated through both on-screen and in-space channels.

Confusion and *think carefully* reflect a fourth category of gesture, "thoughtful, internal states." These gestures were the least well-recognized overall, but accuracy improved noticeably when both on-screen expressions and in-space motions were combined.

5. STUDY 2: TEAM SCENARIO

In Study 1, we identified discernable and meaningful behaviors, then asked observers to respond to them individually. As proxies generally support communication within working groups, findings drawn from behaviors performed in relative isolation only take us so far. To extend our understanding of how proxy motions are perceived in the more situated context of practical use, we ran a follow-up study depicting a collaborative, problem-solving team interaction. Our focus thus shifts from recognizing the affordances and components of proxy gesture to interpreting the social relationships that alternative forms of gesture imply and support.

5.1 Relational Measures

To evaluate the effects that the proxy-in-proxy setup had on the perceived interactions between meeting participants, we chose a set of measures that characterize the remote worker's level of participation, degree of engagement, and relative social status during a distributed team meeting. We based our video scenarios and the subsequent analysis on an adaptation of Dillard's relational message scale [6], which focuses on how people interpret the *relationships* between communicators. The scale describes judgments according to nine relational factors, which include: 1) immediacy, the degree to which someone actively engages another; 2) affect, which suggests warmth, interest and attraction; 3) similarity/depth, which indicates familiarity, personalism and friendliness; 4) receptivity/trust, the degree to which someone expresses or seeks trust; 5) composure, or how relaxed versus tense someone appears; 6) formality, a demeanor of being responsive and disclosive; 7) dominance, the extent to which someone tries to persuade another or control the conversation; 8) equality, being treated as, or treating someone else, as an equal; and 9) involvement, an indication of interest or detachment.

5.2 Video Prototypes

The study featured four video clips of a distributed team during an active design session. The team consisted of three members: two were collocated, and one was remote and communicated through a proxy. The proxy from Study 1 was enhanced to provide remote robotic control. The iMac G4's screen was actuated by three DC motors and a cable drive system to move the neck and screen to positions controlled using a remote interface. Screen motions were controlled gesturally, through the orientation of a hand-held Wii remote, so that larger movements of the remote produced more rapid movements of the screen. Pilot trials also revealed the need for arm-based gestures, so we added a Lynxmotion AL5D five degree-of-freedom robotic arm to provide deictic as well as other symbolic gestures critical to interactive team activities [5].

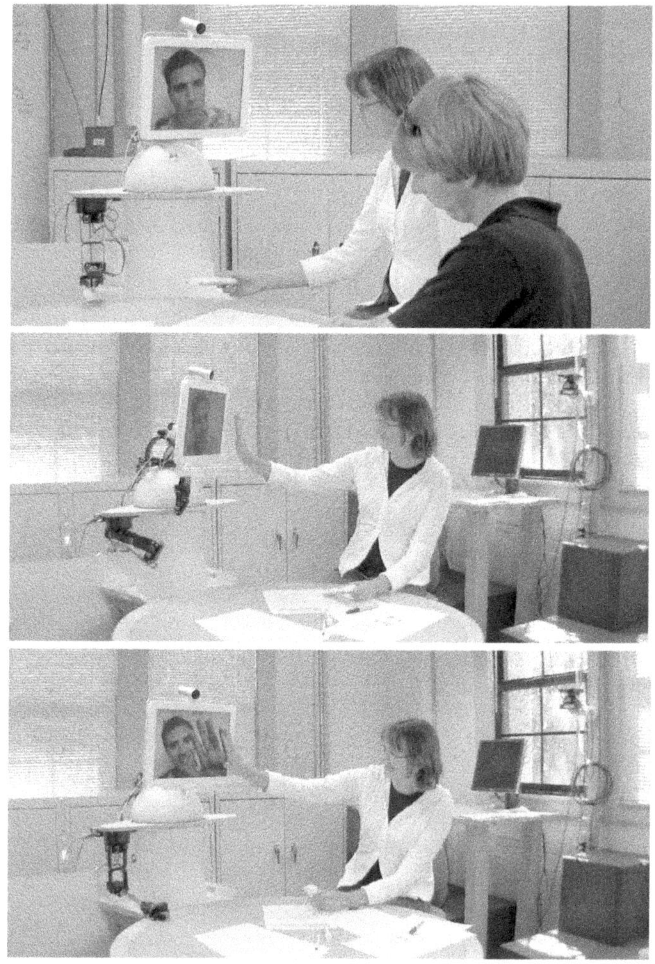

Figure 4. The Study 2 video prototypes depicted a design collaboration scenario (panel 1). Half of the study conditions showed on-screen and in-space proxy gestures (panel 2); the other half, showed on-screen proxy gestures only (panel 3).

The video scenario shows a third-person view of a remote teammate (Eric) asking an on-site design collaborator (Becky) for assistance revising the design of a hand-held remote control (see Fig. 4 for the physical setup). A brief discussion ensues about how to make the remote work for a wider range of hand sizes, and another local participant is called over for further design support. After a brief period, the three check back in with each other, review the designs they had developed, and choose one that resolves the original problem. The clip is one and a half minutes long, and includes an audio track of the actors' conversation.

5.3 Hypotheses

The team scenario allowed us to examine how people interpret aspects of social interaction, such as friendliness, composure and equality, under different proxy operating modes. Extending our findings from Study 1, we formed two hypotheses:

H1 Consistency between on-screen and in-space action improves observers' interpretations of the remote team member's role (better recognition), when compared to on-screen action alone.

We expect consistency between actions to favorably influence observers' judgments of the indices that have a strong relation to nonverbal, physical expression. These include immediacy, which is enhanced by greater activity, intensity and enthusiasm;

similarity/depth, which is influenced by displays of friendship (a wave or handshake), body orientation and mirroring postures; dominance, which is represented by persuasive or controlling actions (pointing or gazing at someone); and involvement, which includes attentional cues (concentration or distraction). Given the Study 1 finding that in-space motion clarifies focus of attention cues, we expect involvement to be a strong component of H1.

H2 Consistency between on-screen and in-space action heightens observers' awareness of the remote participant's role (strength of effect), when compared to on-screen action alone.

Here, we expect whether or not the proxy physically gestures to influence observers' judgments of those indices that reveal the remote participant's role as a leader or follower in the conversation: in particular, dominance and equality between teammates. Enacting each role with embodied movements should make that role more apparent. For example, the remote participant might be evaluated as a more assertive leader when his proxy is in motion, compared to when it is not.

5.4 Study Design

We recorded four variations of the video scenario. Each followed the same script, but varied one of two factors, yielding a 2x2 study design. The first factor was the type of proxy gesture, which included on-screen expressions and gestures only, or combined on-screen and in-space action. We did not investigate the in-space only condition for this study, as it under-performed the other two in all cases for Study 1. The second factor was the leadership role of the characters. In two of the four conditions, the remote teammate initiates the meeting, asks the questions and calls over the third participant, giving him a more dominant role in the meeting. In the other two conditions, the local designer takes all of these actions, giving her the more dominant role.

During the course of the video, the remote participant exhibits many non-verbal cues: he turns to face each local collaborator, peers closely at a prototype, compares hand sizes, and jolts in surprise. Actions such as pointing out the location and size of particular design features on a prototype, rapping on a table to get someone's attention, and waving to request a turn to speak, which are only visible on the proxy's screen in the on-screen condition, use the robotic arm in the on-screen/in-space condition.

5.4.1 Procedure

We recruited 110 MTurk participants globally, and collected data over two days. Participants viewed instructions and questionnaire pages similar to Study 1, only adjusted for the different video and greater number of statements. The average time to complete the study was 10 minutes. Participants were paid US$1 each, which equates to about the same $6 per hour as Study 1.

5.4.2 Video Questionnaire

Each participant was shown a single video clip of the team interaction scenario, which was selected from the four conditions based on the participant's order of arrival at the study website. The clip appeared on a single webpage accompanied by two sets of 33 statements, oriented in two columns. The subject of one set was the remote team member, and the subject of the other was the local teammate. Statements were adapted from the original messaging scale statements, and included, for example, "Eric was intensely involved in the conversation" (immediacy), "Becky tried to control the interaction" (dominance), and "Eric considered them equals" (equality). Each statement was followed by a seven-point Likert scale for responses, with "strongly disagree" and "strongly agree" as their endpoints.

5.4.3 Coding of Responses

Responses were grouped by their message scale categories, and each group was then averaged, to produce nine response indices.

5.4.4 Participants

We had 103 complete sets of responses. A reverse-IP lookup revealed that 36% of respondents were located in India, 32% were in the United States, 6% were in Canada, and the remaining 26% were distributed among 13 other countries, with anywhere from 1-4% of responses from any single country. Ninety percent of respondents reported their gender and age, with 54% being male, and with ages ranging from 18 to 59 (*M*=29.4, *SD*=9.55).

5.5 Results

We conducted two-way between-groups ANOVAs—one for each of the nine indices—to explore the impact of the type of motion and team member role on observers' interpretations of the social dynamics of our design scenario. We used a power transformation with λ=0.8 in order to stabilize variance across conditions. We found a statistically significant main effect of combined proxy-in-proxy movement, as compared to the on-screen only condition, for several measures at $p<.05$ (see Table 3).

Table 3: Significant differences in perceptions of remote and on-site participants due to the type of proxy motion.

Variable	Condition	*M*	*SD*	F-Ratio	η_p^2
Remote participant friendliness	On-screen	4.53	1.03	F(1,99)=5.46	.05
	On-screen/In-space	**5.05**	1.24		
Remote participant dominance	On-screen	**3.97**	1.14	F(1,99)=4.00	.04
	On-screen/In-space	3.48	1.42		
Remote participant involvement	On-screen	4.98	1.28	F(1,99)=4.67	.05
	On-screen/In-space	**5.50**	1.37		
On-site participant equality	On-screen	4.49	1.24	F(1,99)=3.93	.04
	On-screen/In-space	**4.95**	1.50		

Regardless of the remote participant's role as leader or follower, when the proxy moved in concert with his expressions and gestures, he was perceived to be more friendly, less dominant and more involved. Similarly, regardless of the on-site design collaborator's role as leader or follower, when the proxy moved in concert with the remote participant's behaviors, she was rated as being more equal in stature relative to him.

Table 4: Significant differences in perceptions of remote and on-site participants due to team role.

Variable	Condition	*M*	*SD*	F-Ratio	η_p^2
Remote participant composure	Remote led	**5.32**	1.33	F(1,99)=5.71	.05
	On-site led	4.72	1.33		
Remote participant involvement	Remote led	**5.49**	1.33	F(1,99)=4.22	.04
	On-site led	5.00	1.32		
On-site participant composure	Remote led	**5.29**	1.30	F(1,99)=5.19	.05
	On-site led	4.71	1.25		
On-site participant dominance	Remote led	3.12	1.32	F(1,99)=6.38	.06
	On-site led	**3.78**	1.34		
On-site participant equality	Remote led	**4.98**	1.46	F(1,99)=4.57	.04
	On-site led	4.47	1.27		

We also found a statistically significant main effect of the role assumed by the remote teammate at $p<.05$ (see Table 4). When the remote participant assumed a leadership role, regardless of the proxy's motion condition, he was perceived to be more composed and involved, and his on-site teammate was rated as being more composed and equal in stature. When the remote participant assumed a follower role, regardless of the proxy's motion condition, the on-site teammate was perceived as being more dominant.

5.6 Discussion

The addition of physical proxy motion favorably influenced three of the four expected relational measures, including similarity/ depth, dominance and (notably) involvement, providing good support for H1. While effect sizes were small—likely due to the subtle distinctions between behaviors that were portrayed across conditions—the remote participant's being seen as more involved when he was able to physically gesture makes sense: during face-to-face interactions, when someone is animated and moves about, he also tends to feel, and be perceived as, more involved. Similarly, when the remote participant led the interaction, he was perceived as having the characteristics that should accompany that role, particularly greater involvement. As with physical gesture, this is reasonable, since one typically feels and appears more involved when leading a discussion than when following.

We found mixed results regarding H2, the expected influence of the type of proxy motion on the strength of the remote participant's perceived role. This would suggest a magnifying effect on dominance and equality based on whether the proxy was in motion or not. The remote participant *was* perceived to be more dominant when his proxy was capable of movement, but he was not perceived differently regarding equality. Proxy motion *did* influence perceptions of the on-site teammate as being more equal in stature, which provides a further degree of support.

Perhaps most interesting is the influence that proxy motion and participant role had on perceptions of the on-site teammate. That is, when the *remote* participant either displayed proxy motion or led the discussion, the *on-site* teammate was viewed as being more equal in stature. One interpretation of this finding is that participating in a conference through standard, static video portal creates the perception of inequality between colleagues. This inequality may cause remote participants to appear higher, or perhaps lower, in stature than local teammates, depending on the situation. For example, when Eric led the discussion, the static platform may have caused him to appear dictatorial, making requests and issuing instructions from another place, behind his screen. After seeing Eric lead the discussion, one participant commented, "I thought Eric sounded a little rude; it seemed like he tried to hurry Becky at one point, which affected my attitude and answers." Re-introducing physical motion for the remote teammate may ameliorate that perception, and put the colleagues back on more even terms relative to one another.

In their written comments, most participants had a distinct point-of-view in their interpretations of proxy motion and collaborators' roles ("I simply found Becky 'professional' and Eric's distractedness 'unprofessional.'"), but some had difficulty discerning emotions, particularly during the stationary proxy conditions. For example, "Some of the emotions are quite subtle." and "I imagine it's hard to convey a feeling of warmth and sincere interest when communicating in this fashion." Such comments support the Study 1 finding that thoughtful, internal states can be difficult to interpret.

6. DESIGN IMPLICATIONS

In these studies, we found that combined on-screen and in-space action work best for particular categories of gesture ("focus of attention" and "thoughtful, internal states"), but that on-screen only action works best for others ("intentionally communicative"). This presents a challenge for the design of control interfaces, because it is most unnatural to feel and act out behaviors in the moment, and at the same time, control a proxy's gestures that represent those behaviors. A further complication is that gestures that tend to be most unanticipated, such as *laughter* or *surprise*, are those that would most benefit from proxy motion. Users should not have to say, press a "surprise" button when they are (or sense that they are about to be) surprised.

One approach to address this dilemma is to track the remote worker's movements and only pass through to the proxy those that cross a motion threshold that indicates say, a change in focus of attention. For example, a remote participant's head turns to either side would be ignored for a brief period—the *dwell* time—after which they most likely represent shifts in attention, and would be mirrored on the proxy (although this will introduce slight delay in the response). In addition, we can extend motion tracking to include face tracking, and use both movement and expression data to determine the user's category of gesture, then actuate the proxy or not, as appropriate to that category. For example, a quick movement upward and back from the torso is more likely to represent surprise when it is accompanied by raised eyebrows, opened eyes and a rounded mouth. Another approach would be to generate the proxy's behaviors based on user-indicated or detected situational context, as one would for autonomous agents [27].

7. LIMITATIONS AND FUTURE WORK

Evaluating video prototypes in an online study differs from first-hand experience with tangible devices in physical proximity. For example, Adalgeirsson and Breazeal evaluated the *experiential* concerns of interacting with proxies and in-the-moment responses to working with them. To that end, their use of physical, in-person environments provided the most appropriate context. Our studies address the *cognitive* concerns of how people perceive and interpret differences between particular behaviors and roles. Video prototypes, with their consistent reproduction of subtle variations in expression, therefore provide an appropriate context for evaluation. For just this reason, photographs and videos of faces and gestures have been used to evaluate the perception of emotions [9] for decades.

Neither of our studies exhaustively examined the effects of proxy behavior across all possible gestures or scenarios. Follow-up studies could compare whether larger or smaller motions, or alternative facial expressions and gestures, would be more or less effective than those we used here. Or they could look at how people from different cultures interpret the same motions or expressions. Our findings may be influenced by the actors that we chose to appear in video scenes. Alternative characters from different cultures, or with different levels of expressiveness, may affect how observers respond to gestures and team roles, and hence produce different results. Likewise, participants with more or less experience with mediated communication may have different expectations, and therefore, differ in their abilities to compare alternative approaches. Our use of an iMac G4 as a component of the platform may affect generalizability, as it had a familiar appearance, limited range of motion, and was mechanically difficult to actuate. On the other hand, other embodied proxy designs will impose their own motion constraints and influence perceptions in their own ways. In the future, we hope to work with other proxies to test how perceptions of on-screen and in-space gestures vary across configurations. We will also examine the influence of longer exposure time, raising the issue of whether familiarity and acclimation lead to better understanding.

Finally, we considered inconsistency to be a lack of complementary cues: that is, an on-screen *nod "yes"* either supported by a physical nod or not. An alternative approach would actively contradict the content presented between channels, so an on-screen *nod "yes"* might be paired with an in-space *shake "no."*

8. CONCLUSION

Study 1 found that consistency between physical and on-screen action improved understanding of the messages that remote participants communicated. Study 2 extended this finding to a dynamic social context, and found that the addition of proxy motion also improved measures of perceived collaboration: not just for remotely connected participants, but for their local, physically co-present colleagues as well.

Clearly, the lack of gesture by the remote worker or embodied platform is not interpreted neutrally. Some degree of motion can significantly improve the clarity of communication and the confidence in understanding it, as well as improve perceptions of the friendliness and involvement of remote and on-site colleagues. While the quality of communication through an embodied proxy may never be quite the same as an in-person interaction, our two studies indicate that providing consistency between physical and on-screen cues can enhance the social interactions characteristic of distributed work.

9. ACKNOWLEDGMENTS

The authors thank Eric Kent, Rebecca Currano and Samson Phan, who were core members of the research team, and the HPI–Stanford Design Thinking Research Program for funding support.

10. REFERENCES

[1] Adalgeirsson, S. and Breazeal, C. MeBot: A robotic platform for socially embodied telepresence. *Proc. HRI 2010*, ACM Press, 15-22.

[2] Argyle, M. *Bodily Communication.* London: Methuen, 1988.

[3] Breazeal, C., Wang, A. and Picard, R. Experiments with a robotic computer: Body, affect and cognition interactions. *Proc. CHI 2007*, ACM Press, 153-160.

[4] Cassell, J. and Thorisson, K. The power of a nod and a glass: Envelope vs. emotional feedback in animated conversational agents. *Jnl. Applied A.I.* 13, 4 (2000), 519-538.

[5] Clark, H. Pointing and placing. In S. Kita (Ed.) *Pointing: Where Language, Culture and Cognition Meet.* Hillsdale NJ: Erlbaum (2003), 243-268.

[6] Dillard, J., Solomon, D. and Palmer, M. Structuring the concept of relational communication. *Communication Monographs* 66 (1999), 49-65.

[7] Dourish, P. and Bly, S. Portholes: Supporting awareness in a distributed work group. *Proc. CHI 1992*, ACM Press, 541-547.

[8] Ekman, P. and Friesen, W. Detecting deception from the body or face. *Jnl. Pers. and Soc. Psych.* 29 (1974), 288-298.

[9] Ekman, P. and Friesen, W. *Facial Action Coding System: A Technique for the Measurement of Facial Movement.* Palo Alto CA: Consulting Psychologists Press, 1974.

[10] Fiske, S. and Taylor S. *Social Cognition.* New York NY: McGraw-Hill, 1991.

[11] Feng, D., Besana, S. and Zajac, R. Acquiring high quality non-expert knowledge from on-demand workforce. *Proc. People's Web 2009*, ACL and AFNLP, 51-56.

[12] Groom, V., Nass, C., Chen, T., Nielsen, A., Scarborough, J. and Robles, R. Evaluating the effects of behavioral realism in embodied agents. *Int. Jnl. Human-Computer St.* 67, 10 (2009), 842-849.

[13] Hall, E. *The Silent Language*. New York NY: Doubleday, 1959.

[14] Isbister, K. and Nass, C. Consistency of personality in interactive characters: Verbal cues, non-verbal cues and user characteristics. *Int. Jnl. Human-Computer St.* 53 (2000), 251-267.

[15] Kidd, C. *Sociable Robots: The Role of Presence and Task in Human-Robot Interaction.* MIT Thesis, 2003.

[16] Kittur, A., Chi, E. and Suh, B. Crowdsourcing user studies with Mechanical Turk. *Proc. CHI 2008*, ACM Press, 453-456.

[17] Kraut, R. Verbal and nonverbal cues in the perception of lying. *Jnl. Pers. and Soc. Psych.* 36 (1978), 380-391.

[18] Kraut, R., Fish, R., Root, R. and Chalfonte, B. Informal communication in organizations: Form, function, and technology. In R. Baecker (Ed.) *Readings in Groupware and CSCW: Assisting Human-Human Collaboration.* San Francisco CA: Morgan Kaufman (1990), 130-144.

[19] Lee, M. and Takayama, L. Now, I have a body: Uses and social norms for mobile remote presence in the workplace. *Proc. CHI 2011*, ACM Press.

[20] Mutlu, B., Yamaoka, F., Kanda, T., Ishiguro, H. and Hagita, N. Nonverbal leakage in robots: Communication of intentions through seemingly unintentional behavior. *Proc. HRI 2009*, ACM Press, 69-76.

[21] Nass, C. and Brave, S. *Wired for Speech.* Cambridge, MA: MIT Press, 2005.

[22] Nguyen, D. and Canny, J. Multiview: Improving trust in group video conferencing through spatial faithfulness. *Proc. CHI 2007*, ACM Press, 1465-1474.

[23] Nishio, S., Ishiguro, H., and Hagita, N. Geminoid: Teleoperated android of an existing person. In P. Filho (Ed.) *Humanoid Robots: New Developments.* Vienna: I-Tech (2007), 343-352.

[24] Paulos, E. and Canny, J. PRoP: Personal roving presence. *Proc. CHI 1998*, ACM Press, 296-303.

[25] Powers, A., Kiesler, S., Fussell, S. and Torrey, C. Comparing a computer agent with a humanoid robot. *Proc. HRI 2007*, ACM Press, 145-152.

[26] Venolia, G., Tang, J., Cervantes, R., Bly, S., Robertson, G., Lee, B. and Inkpen, K. Embodied social proxy: Mediating interpersonal connection in hub-and-satellite teams. *Proc. CHI 2010*, ACM Press, 1049-1058.

[27] Vilhjálmsson, H. and Cassell, J. BodyChat: Autonomous communicative behaviors in avatars. *Proc. AGENTS 1998*, ACM Press, 269-276.

[28] Woods, S., Walters, M., Koay, K. and Dautenhahn, K. Methodological issues in HRI: A comparison of live and video-based methods in robot to human approach direction trials. *Proc. ROMAN 2006*, IEEE, 51-58.

[29] Yankelovich, N., Simpson, N., Kaplan, J. and Provino, J. Porta-Person: Telepresence for the connected conference room. *Ext. Abstracts CHI 2007*, ACM Press, 2789-2794.

Real World Haptic Exploration for Telepresence of the Visually Impaired

Chung Hyuk Park
Georgia Institute of Technology
Atlanta, GA, U.S.A.
chungpark@gatech.edu

Ayanna M. Howard
Georgia Institute of Technology
Atlanta, GA, U.S.A.
ayanna.howard@ece.gatech.edu

ABSTRACT
Robotic assistance through telepresence technology is an emerging area in aiding the visually impaired. By integrating the robotic perception of a remote environment and transferring it to a human user through haptic environmental feedback, the disabled user can increase one's capability to interact with remote environments through the telepresence robot. This paper presents a framework that integrates visual perception from heterogeneous vision sensors and enables real-time interactive haptic representation of the real world through a mobile manipulation robotic system. Specifically, a set of multi-disciplinary algorithms such as stereo-vision processes, three-dimensional map building algorithms, and virtual-proxy haptic rendering processes are integrated into a unified framework to accomplish the goal of real-world haptic exploration successfully. Results of our framework in an indoor environment are displayed, and its performances are analyzed. Quantitative results are provided along with qualitative results through a set of human subject testing. Our future work includes real-time haptic fusion of multi-modal environmental perception and more extensive human subject testing in a prolonged experimental design.

Categories and Subject Descriptors
H.1.2 User/Machine Systems, I.2.9 Robotics, I.2.10 Vision and Scene Understanding, I.4.8 Scene Analysis, H.5.2 User Interfaces.

General Terms
Algorithms, Measurement, Experimentation, Human Factors.

Keywords
Haptic exploration, telepresence, 3D in-situ perception.

1. INTRODUCTION
One of the major functions of current robotic systems is to promote the welfare of human beings. The advancement in robotic technology has resulted in various forms of robotic support for human lives such as cleaning, entertainment, education, and even medical care, and has introduced new application fields such as teleoperation [15] and telepresence [2]. However, robotic techno-logy is usually targeted for general users, and it is still left with many challenging issues such as compensating for impaired or disabled sensory functionalities.

Assistive technology, on the other hand, has a long history of adopting new techniques to aid people with disabilities, and has opened new opportunities for individuals with disabilities [11,26]. Traditional aids, such as wheelchairs, canes, Braille readers, and hearing aids, have basically focused on generating passive and close-range sensory aids. Recent assistive devices, including electronic Braille adaptors and computer-screen readers for the visually impaired (VI) or electronic wheelchairs and prosthetic limbs for the motor impaired, have evolved to transfer more information to the user and have granted wider access to daily living for people with impairments, including education and employment [1]. Nevertheless, all of these tools are designed to (1) work passively with predefined sources of information, and (2) aid in mobility or manipulation to assist in daily living but only in specific conditions. In other words, there is no general solution yet that functions in various situations, and there is still a need to push forward research with a focus on providing more general and multi-purpose support for individuals with disabilities, as illustrated in Figure 1.

Given the current status of assistive technology and the advancement in robotics, we speculate that a mobile manipulation robotic platform will be a viable solution to serve as a multi-purpose assistive tool for people with disabilities. Mobile manipulation is a fast growing area [7] due to advances in robotic systems, increases in computing power, and a rising interest in applications for the home and indoor environment. As an assistive device, a mobile manipulator can aid an impaired person with manipulative tasks using its robotic arm and also provide mobility or remote access using its mobile base. Additionally, a mobile manipulator, being a robotic agent, can monitor the environment and form a

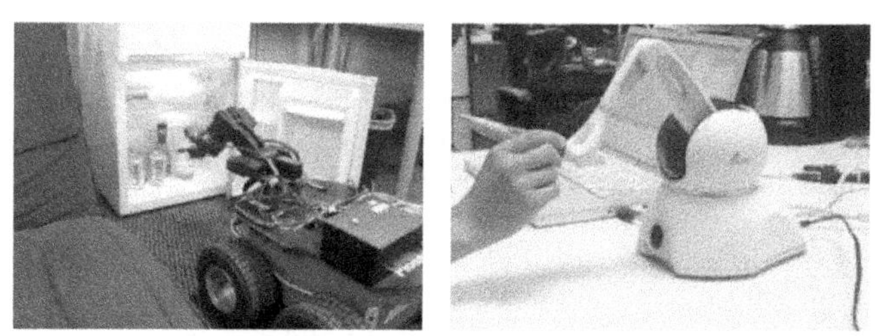

Figure 1. An image of a mobile manipulator robot serving a remote user with haptic exploration framework.

haptic pathway to transfer the in-situ perception of the remote environment to the user. The concept of haptic exploration with a mobile manipulator (HEMM) [29] has been proposed to integrate a haptic device with a mobile manipulation system for the use of an assistive system in two ways: 1) as the controller for the mobile manipulation system and 2) as the generator for the environmental feedback to the human user.

This paper discusses our approach for a direct haptic rendering from sensed environmental data, and displays a framework of algorithms that can sense the environment, process the sensory data to get a point-cloud representation of the three-dimensional (3D) map, and finally transform the perceived environmental data into a haptic representation. Section 2 lists some of the related work in the current research of robotics, assistive technology, and computer vision. , and Section 3 briefly describes a platform for our haptic exploration using a mobile manipulator. Section 4 provides the details of sensory data processing algorithms that mainly incorporate several techniques of computer vision algorithms. Section 5 explains our haptic rendering algorithm based on a point-cloud spatial dataset. Then, Section 6 explains the experiment setup for human subject testing of our framework in a real-world environment. Finally, Section 7 displays the results of our system in observing and representing several indoor environments, and Section 8 concludes with discussion and future work.

2. RELATED WORK

Efforts to utilize robotic systems for assisting the VI can be found in recent literature [11,20]. However, few studies have focused on developing assistive robots to increase the sensory modalities for people with visual impairments (VI). Ulrich et al. developed a robotic cane [11] that can sense the environment and guide the individual through the environment with a robotic cane. Another approach from Kulyukin et al. incorporated a mobile-robot based guidance approach with RFID devices to help people with visual impairments navigate in their living areas [20]. Both systems, however, require the user to be either in contact with a device or in a preset environment to detect the user's surrounding environment, and they are not capable of transferring remote environmental perception to the user. More recent challenges consist of a driving system for the VI by utilizing a semi-autonomous car and a haptic vest [26], and a user study for robotic shopping using a robotic cart [25]. These efforts present pioneer studies in the field of assistive robotics in the sense that they intend to enable a person with a visual impairment to maintain independence in daily living. However, more challenging issues still remain.

For presenting environmental perception to a user with visual impairments, haptic representation of the data is the most direct and effective interactive modality, which can be easily derived from the fact that the most common assistive tools for VI people are canes and Braille-notes that stimulate tactile sensations. Haptics is a mature field of engineering research, with varying applications in the medical field [9], rehabilitation [21], virtual reality [13], and teleoperation [27]. Its ability to transfer tactile and textural sensations along with force feedback adds another dimension of interaction between the human and the system. With the recent advances in virtual proxy methods [4,16,17,23,28] haptic rendering of 3D objects has become more convenient and effective with the aid of graphic libraries such as OpenGL. By building a detailed 3D model and with the aid of efficient physical simulation algorithms, calculating the virtual proxy force has turned out to be a fast and relatively easy process [18]. However, the calculation is based on the condition that the 3D model must be provided a priori. Haptic rendering of a real 3D environment is widely known to be quite difficult, and there remain many challenging problems even in the area of virtual reality systems [6].

The simplest solution to this dilemma would be to estimate the real world with various sensors, transform the estimation into a 3D model, and use the previously established haptic rendering solution. This method, however, does not solve the issue because transforming the estimated 3D data into a 3D model is far from simplistic. In fact, this process itself requires multiple stages of algorithms to work around issues such as occlusion and object identification. The reconstruction is usually processed off-line due to the high computation cost [5]. On the contrary, the other solution—a direct haptic rendering from the estimated data itself—provides a totally new approach and is one of the key contributions of this present work. Instead of creating a 3D graphical model, this method uses point-cloud based 3D data and directly applies a point-cloud based virtual proxy algorithm for haptic rendering. The advantages include fast computation time and a more flexible structure for coping with dynamic environments.

As discussed in [25], one of the most critical reasons that limit assistive tasks in daily living is the perceptional matching of the environment with the system it is handling. In this sense, this research focuses on the problem of presenting environmental perception to a user with visual impairments, and presents a novel solution framework that provides both active sensing capabilities and effective transference of the detailed environmental data through a haptic modality.

3. HAPTIC EXPLORATION SYSTEM

For our research platform, we utilize our previously proposed HEMM (haptic exploration with a mobile manipulator) system [29], and employ this system to implement and validate a set of algorithms for haptic exploration. The structure of the HEMM system consists of two main parts: a haptic interaction system and a mobile manipulation robotic system (Figure 2).

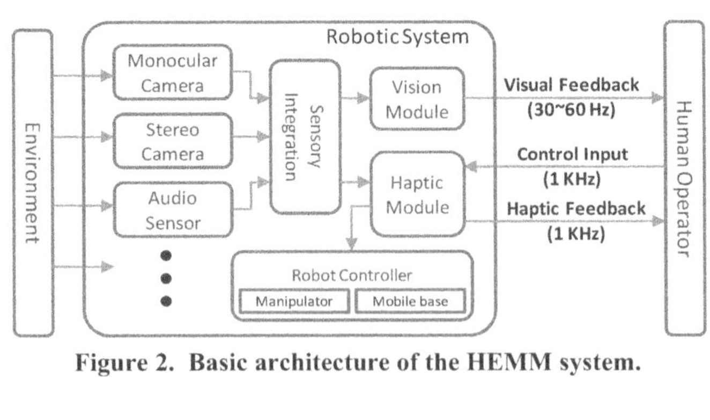

Figure 2. Basic architecture of the HEMM system.

The haptic device we use is the SensAble™ Phantom Omni, which is capable of taking 6-DoF inputs and generating 3-DoF force feedback. The mobile manipulator is composed of a Pioneer3AT and a Pioneer2Arm. It utilizes its various sensing modules to perceive the environment and generate visual and haptic feedback to the human operator. Through this HEMM system, the human can feel the environment and be more in control with the help of the mobile manipulator. Thus, the benefits expected from the HEMM system are as follows: (1) it enables a bidirectional linkage between a haptic device and a mobile manipulation robotic system and (2) it can convey environmental data to the user through haptic exploration.

4. MULTI-SCALE VISON PROCESS FOR 3D PERCEPTION

People with visual impairments typically use canes for navigation or sense objects with tactile sensing. To develop a system that can provide similar multi-scale environmental feedback, our HEMM system is equipped with two different camera sensors: a stereo camera in front of the robot and a monocular camera mounted on top of the wrist. The stereo vision provides larger scale perception of an indoor environment, and the monocular camera enables up-close and more detailed observation of an object. The following sections explain the vision processes and disparity map generation algorithms for the camera systems.

4.1 Stereo Vision Process for Indoor Spatial Perception

Stereo vision processes [10,12,19] have been widely adopted for spatial recognition due to their ability to embed in a simple hardware structure. A disparity map is the result of the stereo vision process that represents the depth distribution of the space. Although the algorithm side of generating the disparity map is not so simple, the computer vision society has come up with sound processes. One of the most favorable solutions is a graph-cut (GC) energy optimization method [8] using Markov-random fields (MRF) [24]. It represents the disparity image as a set of MRF variables with two energy functions representing the cost of estimated disparity and the cost of smoothness with neigh-boring pixels, and calculates the optimal solution by minimizing the total energy function.

The energy model for MRF-based stereo matching algorithm consists of a data term E_d and a smoothness term E_s, both represented in the form of energy functions. For a disparity map solution f, the total energy function $E(f)$ can be represented by the following:

$$E(f) = E_d(f) + E_s(f) \tag{1}$$

$$E_d(f) = \sum_{p \in P} D_p(f_p) \tag{2}$$

$$E_s(f) = \sum_{p \in P} \sum_{q \in N(p)} C_s(f_p, f_q) \,, \tag{3}$$

where $f_p, f_q \in f$, $N(p)$ is a set of 8 neighbor points of p in 3D space, and $D_p(f_p)$ and $C_s(f_p, f_q)$ are defined as

$$D_p(f_p) = \left|f_p - I_p\right| D_p(f_p) D_p(f_p) \tag{4}$$

$$C_s(f_p, f_q) = \left|f_p - I_q\right| . \tag{5}$$

Here, I_p is an observed value of the disparity value at pixel p of the stereo image pairs. Thus, $D_p(f_p)$ is a cost function for the data term of the direct observation of the disparity map and $C_s(f_p, f_q)$ is the cost function for representing the smoothness by estimating the discontinuities with neighboring pixels. This optimization process is combined with a rapid minimization process called "graph-cut," which achieves a good local minimum in labeling the image and accomplishes a fast approximation of the disparity map with well segmented representation and minimal energy cost of the map.

The GC method generates results with quite good estimation. However, it is not sufficient to produce a clean 3D reconstruction. Effects from lighting and textures cause noise in the disparity map, necessitating post-processing to eliminate them. To further complete the 3D perception process, a ground plane is estimated using a statistical region merging algorithm [14]. This algorithm reduces noise and error terms in the disparity calculation, which are caused by the short distance to or the complexity in the texture of the ground. If properly estimated, it will also allow more freedom in the disparity calculation for non-ground objects that are of interest as shown in Figure 3.

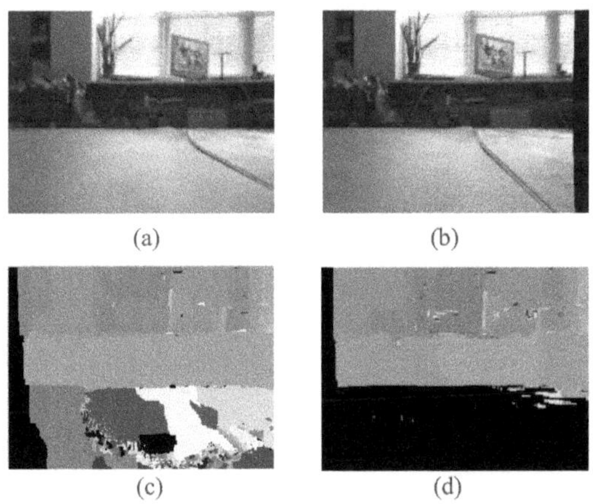

Figure 3. Stereo disparity estimation process: (a,b) Stereo pair images of a living room; (c) MRF GraphCut based disparity output; (d) final disparity result after ground estimation.

4.2 Mono-Stereo Vision Process for Object Perception

For objects that are too close to get detailed spatial observation with the stereo camera, the monocular camera mounted on the wrist of the mobile manipulator is utilized. Several research efforts exist that employ monocular vision for 3D perception, for instance, mono-SLAM with probabilistic 3D map [22], which shows good results but requires large computation powers to process the stream of image sequences. On the other hand, we plan to make an instant observation of the object at its proximity. Therefore, the manipulator is controlled to move the monocular camera sideways to capture a pair of images forming a small baseline between the image pairs, and the stereo matching process discussed is reused for disparity calculation. This process is named "mono-stereo process".

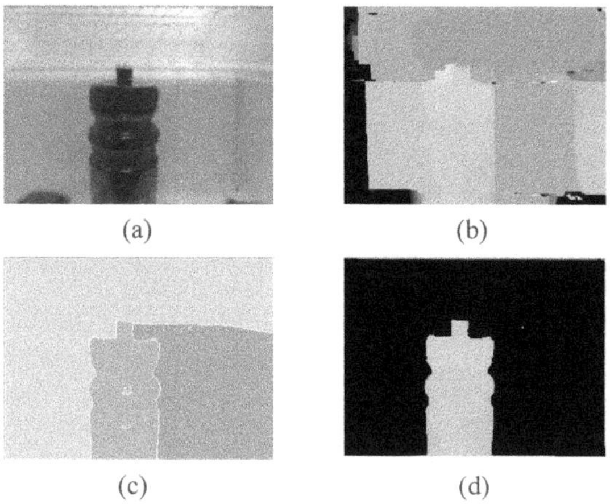

Figure 4. Mono-stereo disparity estimation process: (a) mono-stereo image; (b) initial disparity; (c) segmented regions after background / foreground estimation; (d) final disparity after background elimination.

To acquire the stereo pairs with a single camera on the wrist of the manipulator, the robot's wrist is moved sideways forming a baseline of 1cm between the image pair. The movement takes time, so the images are taken in sequence, making the process unsuitable for dynamic objects. On the other hand, the stereo process itself is not a real-time process yet (it takes a few seconds of processing time with a 2.4GHz CPU). In addition, similar to the case of the stereo estimation process, post-processing of the disparity image is required in the case of mono-stereo. The issues of the presence of interferences in disparity calculation due to background color and lighting are addressed along with the effort to resolve the problem in the following sections. An example of the mono-stereo process with post-processing results is illustrated in Figure 4.

As a post-processing sequence, foreground objects are separated from the background image to capture the right partition of the object in the disparity image. This foreground/background separation process first marks pixels with optic flows over a threshold, samples the points to form a set of skeletonized (silhouette-like) point markings, performs region selection with the points as seed points, and then subtracts large regions which can be assumed to be background elements. To eliminate background elements, adaptive thresholding is implemented to determine the right number of elements for removal based on the total number of segmentations and the total sum of regions of candidate background elements. More details on this post-processing will be discussed in our future papers.

5. HAPTIC EXPLORATION ALGORITHM

Once a reliable disparity estimation of the environment is acquired, a 3D map can be built from the disparity map (either from the stereo process or from the mono-stereo process). After the 3D map is acquired, the haptic exploration process is then activated, which interactively searches the map with the movement of the haptic probe and efficiently calculates the haptic force feedback in real-time (Figure 5).

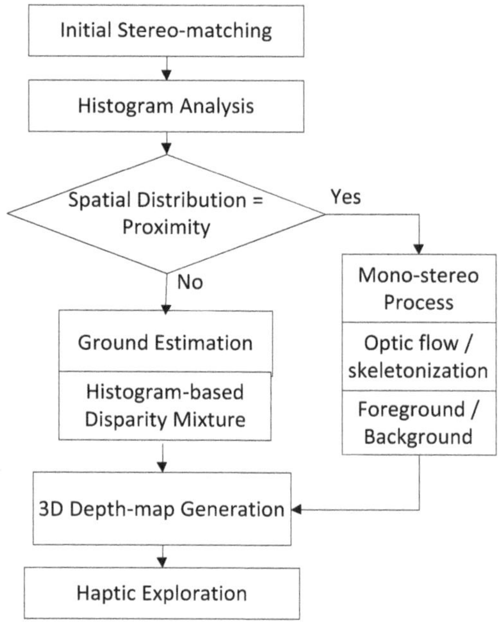

Figure 5. Flow-chart of the 3D-perception process through heterogeneous stereo-matching.

The proxy force is designed as a spring-damper system as in Equation (6)

$$f(t) = [Kd(t) + B\dot{d}(t)]\vec{n}, \qquad (6)$$

where $d(t)$ is the penetration depth at time t, $\dot{d}(t)$ is the instant velocity of the probe at the penetrated position (that arises due to the exploratory movement of the user), K is the spring constant and B is the damping constant. Lastly, $\vec{n}$ is the normal vector to the surface of the penetrated volume.

The usual process for 3D map generation from the disparity map involves a projection matrix built from the camera parameters. After the surface points are projected in 3D space, the volume beneath the surface points are filled outward, forming a set of volumetric point-clouds. Once the point-cloud based 3D map is constructed, the system enables a haptic device, Phantom Omni, to explore the 3D space and feel objects in it. For haptic force generation, a well-known proxy-based haptic volume representation algorithm [3] is implemented.

This paper mainly focuses on haptic-volume representation, so the virtual-proxy is placed on the surface of a volumetric object disregarding the effect of friction forces. From the distance between the virtual-proxy placed on a surface and the position of a probe (Omni's end-point), the penetration depth of the probe can be calculated from the surface (the virtual-proxy), which is then used to generate force feedback for haptic volume representation. Since sparse grid-based 3D data are being used, the haptic interaction algorithm first checks for the occupancy of neighboring grid points with respect to the position of Omni's probe in the 3D map, then calculates the penetration depth if the probe is totally inside the object's volume, or calculates surface penetration depth if the probe is near the surface (see flowchart in Figure 6.).

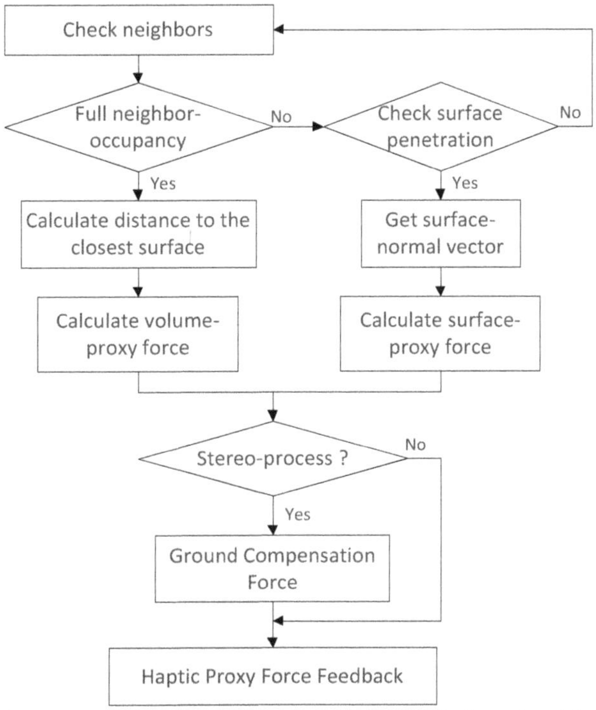

Figure 6. Haptic virtual-proxy force calculation algorithm with 3D grid map.

To get the distance vector, the volume proxy-force is generally calculated by searching for the closest surface. The surface proxy force is calculated by the convolution of the surface-normal vector and the penetration depth from the surface, which leads to an interpolation issue due to sparseness. To speed up the process, a look-up table of the surface normal vector is generated for all combinations of the neighboring grid points. A few examples are shown in Figure 7.

Processing time for the 3D-map search and proxy-force generation is crucial for a real-time experience of the haptic exploration. Our structure of the algorithms explained above results in an average processing time of 3.075ms for surface-normal depth calculation in the 3D grid map (STD=6.287ms over 7,035 measurements) and proxy-force generation.

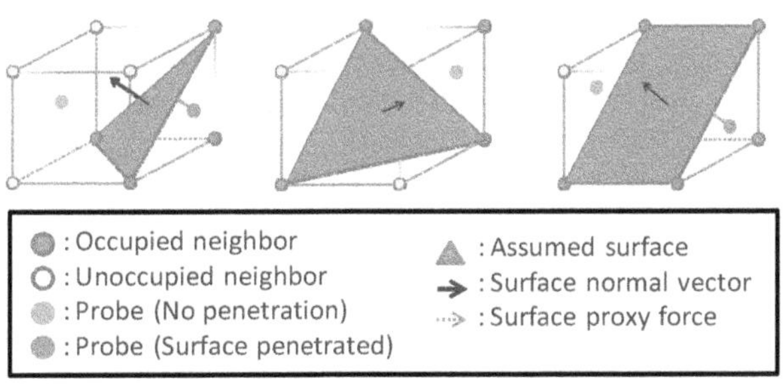

Figure 7. Examples of 3D interpolation for surface proxy force calculation.

To illustrate what type of force profile the user may feel with amplitude changes in force, Figure 8 is plotted over a partial trajectory while the haptic device touches the side of an object.

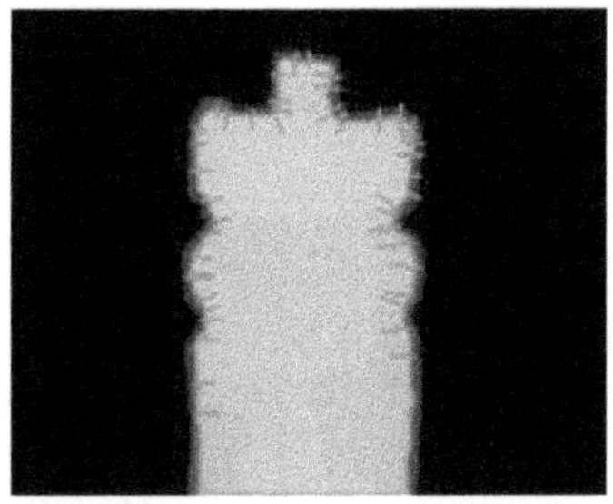

Figure 8. Force feedback (red line) over a user's exploratory trajectory on an object. The direction of the force is from the penetration point to the surface point.

6. EXPERIMENT SETUP

The performance of the HEMM framework is evaluated based on the results from the three major sub-processes, namely, stereo matching (with ground elimination), mono-stereo matching (with background elimination), and the haptic exploration process. For the first two matching processes, a ground-truth map of the real world the system is observing is created, and the error of the disparity estimation is analyzed. For quantitative analysis, the processing time and the estimation error of the depth are calculated for both stereo and mono-stereo processes. For the qualitative analysis, the final process of haptic exploration is tested with human subjects, both with non-VI students and VI students. We design a set of experiment for human subject testing only with haptic exploration based on the mono-stereo process results because the stereo process deals with a large space volume and the spatial recognition of the VI students will depend on the familiarity of the environment, or *a priori* knowledge of the environment of the student.

For initial training, the subjects—after providing their written/verbal consent on our IRB (protocol #: H09220)—are given a basic sample shape (a cubic) and given a minute to feel the object with our haptic interface and get accustomed to the concept and feeling of hatpic exploration. Then, the subjects are asked to identify objects by haptically feeling the models created from real-world objects, without seeing the object. The non-VI students are provided with a random sequence of three haptic models, along with the images of the three actual objects as depicted in Figure 9. After approximately a minute of exploration time per each object, they are asked which haptic models feel the closest to the objects in the images. The same process is used to test VI students, but this time the actual objects are provided for the participants to select from after feeling the haptic models.

Figure 9. Images of objects of which the haptically visualized models are provided for haptic exploration to human users.

7. RESULTS

Figure 10 and Table 1 present the results of the stereo-matching algorithm with ground elimination. Two different scenes were achieved, one from an actual household living environment and another from a real office environment. The average processing time for the stereo matching and 3D map generation was 6.01s.

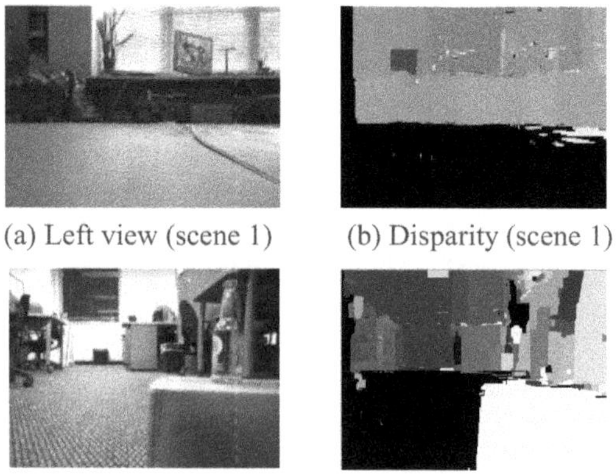

(a) Left view (scene 1) (b) Disparity (scene 1)

(c) Left view (scene 2) (d) Disparity (scene 2)

Figure 10. Results of our stereo matching process.

Table 1. Processing Time from Stereo Process

Stereo Pairs	Scene 1	Scene 2
Processing Time (s)	6.36	5.65

Figure 11 and Table 2 show the results from the mono-stereo matching algorithm with background elimination. Two objects were placed in a small refrigerator and on a table. The disparity map was generated from the mono-stereo algorithm utilizing a moving robotic arm and the background was automatically eliminated if the number of segments detected in the image reached a threshold. The average processing time for the mono-stereo matching and 3D map generation was 3.26s.

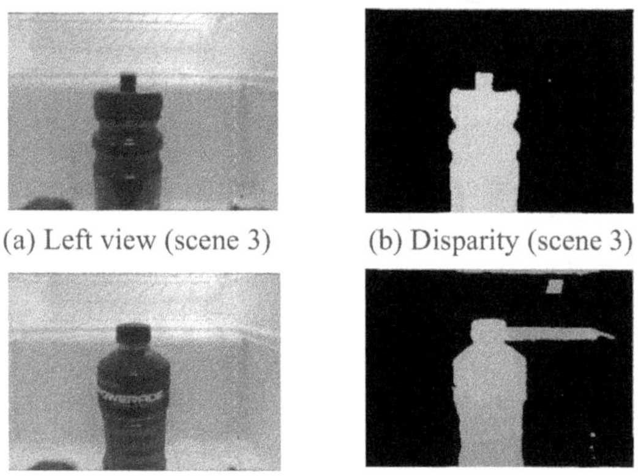

(a) Left view (scene 3) (b) Disparity (scene 3)

(c) Left view (scene 4) (d) Disparity (scene 4)

Figure 11. Results of our mono-stereo matching process.

Table 2. Processing Time from Mono-stereo Process

Mono-stereo Pairs	Scene 3	Scene 4
Processing Time (s)	3.04	3.48

Ground truth data were determined by measuring over the 3D environment and then translated into a disparity map. For equal comparison, the ground planes were also eliminated from the ground truth map by manual segmentation for the stereo process results, and the background was eliminated from the ground truth map for the mono-stereo process. The absolute difference of disparity values over the entire image plane was calculated and the root-mean-squared error was computed for the stereo process. The error term was then calculated only for the foreground part in the mono-stereo process. The error between the algorithm's estimation and the ground truth for the first scene (stereo process) was 5.7%, and the error between the algorithm's estimation and the ground truth for the third scene (mono-stereo process) was 7.9%, showing the accuracy of the process.

The ground truth disparity map and the estimated disparity map from our algorithms, along with the error of depth on the image plane, are shown in Figure 12 and specified in Table 3, respectively. Considering the complexity of the scene (in stereo process) and the low resolution of disparity values for representing detailed 3D shapes in close range (in mono-stereo process), the results are satisfactory.

(a) Disparity (b) Ground Truth (c) Error

(c) Disparity (d) Ground Truth (e) Error

Figure 12. Ground-truth disparity maps and estimation errors.

Table 3. Disparity Estimation Errors

Stereo Pairs	Scene 1	Scene 3
Depth Error	5.7 %	7.9 %

Additionally, the average processing times (and percentages from the whole process) for each algorithmic stages are as follows: initial stereo-matching ≈3011ms(52.85%), histogram analysis ≈ 10ms(0.2%), ground estimation ≈193ms(3.4%), disparity mixture ≈1929ms(33.8%), and 3D map building process ≈546ms(9.5%) in PC environment with a dual-core 2.4GHz processor and 2GB memory. We will discuss our future plans to minimize the processing time in Section 8.

Finally, the human subjects testing results are displayed in Tables 4 and 5, with the confusion matrix listed in Tables 6 and 7. A total of 19 non-VI human subjects participated in the experiment, and their average age was 25. For VI human subjects, 8 middle school students (4 males and 4 females) participated with their average age 12. Of these, 7 were partially blind and one was fully blind.

Table 4. Success Rate of Non-VI Human Subjects in Identifying Objects with Haptic Exploration

Object	#1	#2	#3
Success Rate	84.2 %	78.9 %	84.2 %

Table 5. Success Rate of VI Human Subjects in Identifying Objects with Haptic Exploration

Object	#1	#2	#3
Success Rate	87.5 %	87.5 %	87.5 %

Subjects with no visual impairments showed 79% to 84% recognition rate for the three objects, while subjects with visual impairment rated 87.5% for each objects. To understand which objects confused the human users, the confusion matrices for the responses of the participants are listed in Tables 6 and 7. Table 6 reveals that two non-VI subjects became confused between object #2 and object #3 and two other non-VI students had trouble distinguishing object #1 and object #3, mainly because the objects are commonly up-right shaped. For VI subjects, Table 7 shows that one subject became confused between object #2 and object #3, and another subject could not tell the difference between object #1 and object #3, the reason being the same as in the non-VI subjects' case.

Table 6. Confusion Matrix of Non-VI Human Subjects in Identifying Objects with Haptic Exploration

Answer \ Object	#1	#2	#3
#1	**16**	-	1
#2	1	**15**	2
#3	2	4	**16**

Table 7. Confusion Matrix of VI Human Subjects in Identifying Objects with Haptic Exploration

Answer \ Object	#1	#2	#3
#1	**7**	-	-
#2	1	**7**	1
#3	-	1	**7**

8. DISCUSSIONS

The objective of this research is to create a linkage between an assistive robotic system and the human user by delivering haptic perceptions of the environment to a person with impairments. More specifically, this research tackles the problem of providing a sense of the world to individuals with visual impairments through a haptic interface based on the environmental data acquired by the perceptual system of a mobile robotic platform. To accomplish this, the user is provided with multi-scale haptic exploration capability of a real-world environment through a robotic system equipped with vision sensors. The challenges that must be addressed by this system involve 1) recognizing the environment in multi-scale dimensions because real-world objects are of various sizes and 3D world positions, 2) developing real-time processing algorithms to enable real-time interaction with the human, and 3) enabling 3D real-time haptic interaction for smooth tactile representation of objects in the environment.

An extensive implementation of the haptic exploration algorithms using a mobile manipulation robotic system has been achieved, and this paper shows the results of the system working in real indoor-environments. To minimize the processing time and enable more smooth real-time haptic exploration, we plan to utilize the Kinect sensor. Our latest implementation is already showing more than 10x improvement in terms of processing time, although the structure of the sub-modules will be different. Future work will also include sound-source localization and its integration into the haptic feedback for enhanced multi-modal interaction. Methods for multi-modal fusion will be further investigated along with an extended set of real-world experiments and human subject testing.

9. ACKNOWLEDGMENTS

This material is based upon work supported by the National Science Foundation under Grant No. 0940146. The authors would like to express their gratitude to the students at the Center for the Visually Impaired, Atlanta, and the undergraduate and graduate students at Georgia Institute of Technology for their participation in the experiments.

10. REFERENCES

[1] Fine, M. and Asch, A., "Disability beyond stigma: Social interaction, discrimination, and activism," *Journal of Social Issues*, vol. 44, no. 1, pp. 3-21, 1988.

[2] Held, R., "Telepresence," *The Journal of the Acoustical Society of America*, vol. 92, p. 2458, 1992.

[3] Iwata, H. and Noma, H., "Volume haptization," in *Proceedings of IEEE 1993 Symposium on Research Frontiers in Virtual Reality*, pp. 16-23, 1993.

[4] Scheier C., and Lambrinos, D., "Categorization in a real-world agent using haptic exploration and active perception," in *Proceedings of the 4th International Conference on Simulation of Adaptive Behavior (SAB'96),* Cambridge MA: MIT Press, 1996.

[5] Kanade, T., Rander, P., and Narayanan, P., "Virtualized reality: Constructing virtual worlds from real scenes," *Multimedia*, IEEE, vol. 4, no. 1, pp. 34-47, 1997.

[6] Brooks, F., "What's real about virtual reality?" *IEEE Computer Graphics and Applications*, pp. 16-27, 1999.

[7] Khatib, O., "Mobile manipulation: The robotic assistant," *Robotics and Autonomous Systems*, vol. 26, no. 2-3, pp. 175-183, 1999.

[8] Boykov, Y., Veksler, O., and Zabih, R., "Fast approximate energy minimization via graph cuts," *IEEE Transactions on Pattern Analysis and Machine Intelligence*, vol. 23, no. 11, pp. 1222-1239, 2002.

[9] Burdea, G., Patounakis, G., Popescu, V., and Weiss, R., "Virtual realitybased training for the diagnosis of prostate cancer," *IEEE Transactions on Biomedical Engineering*, vol. 46, no. 10, pp. 1253-1260, 2002.

[10] Scharstein, D. and Szeliski, R., "A taxonomy and evaluation of dense two-frame stereo correspondence algorithms," *International journal of computer vision*, vol. 47, no. 1, pp. 7-42, 2002.

[11] Ulrich, I. and Borenstein, J., "The GuideCane-applying mobile robot technologies to assist the visually impaired," *IEEE Trans-actions on Systems, Man and Cybernetics, Part A: Systems and Humans*, vol. 31, no. 2, pp. 131-136, 2002.

[12] Sun, J., Zheng, N., and Shum, H., "Stereo matching using belief propagation," *IEEE Transactions on Pattern Analysis and Machine Intelligence*, pp. 787-800, 2003.

[13] Broeren, J., Rydmark, M., and Sunnerhagen, K.S., "Virtual reality and haptics as a training device for movement rehabilitation after stroke: a single-case study," *Archives of physical medicine and rehabilitation*, vol. 85, no. 8, pp. 1247-1250, 2004.

[14] Nock, R. and Nielsen, F., "Statistical region merging," *IEEE Transactions on Pattern Analysis and Machine Intelligence*, pp. 1452-1458, 2004.

[15] Okamura, A., "Methods for haptic feedback in teleoperated robotassisted surgery," *Industrial Robot: An International Journal*, vol. 31, no. 6, pp. 499-508, 2004.

[16] Salisbury, K., Conti, F., and Barbagli, F., "Haptic rendering: Introductory concepts," *Computer Graphics and Applications*, IEEE, vol. 24, no. 2, pp. 24-32, 2004.

[17] Choi, S. and Tan, H., "Toward realistic haptic rendering of surface textures," in *ACM SIGGRAPH 2005 Courses*, ACM, pp. 40-47, 2005.

[18] Itkowitz, B., Handley, J., and Zhu, W., *The OpenHaptics Toolkit: A Library for Adding 3D Touch Navigation and Haptics to Graphics Applications*, 2005.

[19] Jin, H., Soatto, S., and Yezzi, A., "Multi-view stereo reconstruction of dense shape and complex appearance," *International Journal of Computer Vision*, vol. 63, no. 3, pp. 175-189, 2005.

[20] Kulyukin, V., Gharpure, C., Nicholson, J., and Pavithran, S., "RFID in robot-assisted indoor navigation for the visually impaired," in *Proceedings of 2004 IEEE/RSJ International Conference on Intelligent Robots and Systems.(IROS 2004)*, vol. 2., pp. 1979-1984, 2005.

[21] Liu, J., Cramer, S., and Reinkensmeyer, D., "Learning to perform a new movement with robotic assistance: comparison of haptic guidance and visual demonstration," *Journal of neuroengineering and rehabilitation*, vol. 3, no. 1, p. 20, 2006.

[22] Davison, A., Reid, I., Molton, N., and Stasse, O., "MonoSLAM: Real-time single camera SLAM," *IEEE Transactions on Pattern Analysis and Machine Intelligence*, pp. 1052-1067, 2007.

[23] Roberts, J. and Panecls, S., "Where are we with Haptic Visualization?" in *World Haptics 2007 Second Joint EuroHaptics Conference and Symposium on Haptic Interfaces for Virtual Environment and Teleoperator Systems*, IEEE, pp. 316-323, 2007.

[24] Szeliski, R., Zabih, R., Scharstein D., Veksler, O., Kolmogorov, V., Agarwala, A., Tappen, M., and Rother, C., "A comparative study of energy minimization methods for markov random fields with smoothness-based priors," *IEEE Transactions on Pattern Analysis and Machine Intelligence*, pp. 1068-1080, 2007.

[25] Gharpure, C. and Kulyukin, V., "Robot-assisted shopping for the blind: issues in spatial cognition and product selection," *Intelligent Service Robotics*, vol. 1, no. 3, pp. 237-251, 2008.

[26] Hong, D., Kimmel, S., Boehling, R., Camoriano, N., Cardwell, W., Jannaman, G., Purcell, A., Ross, D., and Russel, E., "Development of a semiautonomous vehicle operable by the visually-impaired," in *IEEE International Conference on Multisensor Fusion and Integration for Intelligent Systems 2008 (MFI 2008)*, pp. 539-544, 2008.

[27] Horan, B., Najdovski, Z., Nahavandi, S., and Tunstel, E., "Haptic control methodologies for telerobotic stair traversal," *Int. Journal of Intelligent Control and Systems*, vol. 13, no. 1, pp. 3-14, 2008.

[28] McMahan, W., Romano, J., Rahuman, A., and Kuchenbecker, K., "High frequency acceleration feedback significantly increases the realism of haptically rendered textured surfaces," in *Proceedings of Haptics Symposium 2010*, IEEE, pp. 141-148, 2010.

[29] Park, C.H. and Howard, A.M., "Towards real-time haptic exploration using a mobile robot as mediator," in *Proceedings of Haptics Symposium 2010*, IEEE, pp. 289-292, 2010.

Effects of Changing Reliability on Trust of Robot Systems

Munjal Desai[1], Mikhail Medvedev[1], Marynel Vázquez[2], Sean McSheehy[1], Sofia Gadea-Omelchenko[3], Christian Bruggeman[2], Aaron Steinfeld[2], and Holly Yanco[1]

[1]University of Massachusetts Lowell, Lowell, MA 01854
{mdesai, mmedvede, smcsheeh, holly}@cs.uml.edu

[2]Carnegie Mellon University, Pittsburgh, PA 15213
{marynel, steinfeld}@cmu.edu, cgbrugge@andrew.cmu.edu

[3]University of Pittsburgh, Pittsburgh, PA 15260
sog9@pitt.edu

ABSTRACT

Prior work in human-autonomy interaction has focused on plant systems that operate in highly structured environments. In contrast, many human-robot interaction (HRI) tasks are dynamic and unstructured, occurring in the open world. It is our belief that methods developed for the measurement and modeling of trust in traditional automation need alteration in order to be useful for HRI. Therefore, it is important to characterize the factors in HRI that influence trust. This study focused on the influence of changing autonomy reliability. Participants experienced a set of challenging robot handling scenarios that forced autonomy use and kept them focused on autonomy performance. The counterbalanced experiment included scenarios with different low reliability windows so that we could examine how drops in reliability altered trust and use of autonomy. Drops in reliability were shown to affect trust, the frequency and timing of autonomy mode switching, as well as participants' self-assessments of performance. A regression analysis on a number of robot, personal, and scenario factors revealed that participants tie trust more strongly to their own actions rather than robot performance.

Categories and Subject Descriptors

H.1.2 [**User/Machine Systems**]: Human factors

General Terms

Experimentation, performance

Keywords

Trust, automation, experiments

1. INTRODUCTION

One of the key factors for the acceptance and safe deployment of robots is the degree to which a user trusts the robot. If trust can be modeled, the model can be used to design robot interfaces and behaviors that foster appropriate levels of trust. Mobile robots, especially those not designed for social interaction, are particularly interesting since they are likely to be task-oriented and therefore used for time-dependent activities, capable of damaging objects and hurting people, and unable to express their intent to bystanders (e.g., [18]).

In their survey of trust and automation research, Lee and See define trust as "the attitude that an agent will help achieve an individual's goals in a situation characterized by uncertainty and vulnerability" [13]. A person's level of trust of an automated system is a key factor that influences their use of that system: an inappropriate level of trust may result in the misuse or disuse of automation [13].

The broad range of automation research provides a context for examining issues of trust with robots. However, there are a number of factors that limit how well this work generalizes to the robotics domain. For example, studies of automated systems have tended to utilize systems such as autopilots, flight management systems, vision systems for target or obstacle identification, and factory control systems [13]. Participants interact with a simulated system, which allows experimenters to inject errors and observe how participants' levels of trust of and use of the system change as a result. The systems used in these experiments generally do not have a physical embodiment and do not interact with the physical world. Furthermore, these automated systems tend to be designed for rigid tasks; that is, each system performs only one very specific type of task. In contrast, many HRI tasks are dynamic and unstructured, occurring in the open world. For example, an operator supervising a remote, autonomous mobile robot must contend with noisy sensing, incomplete perception, unpredictable environments, and bystanders.

In this work, we have investigated how changing the robot's reliability influences people's use of robot autonomy and their trust in the robot system through experiments with participants operating a real robot. The experiment, described in detail below, was designed to have a high workload so that the participants would need to use the autonomous capabilities of the robot in order to complete the task in time and to be able to complete the secondary task. We hypothesized that people would trust a robot system less when its reliability in autonomous mode decreased, switching to a manual mode. We wanted to determine how long it would take participants to switch back to autonomous mode after the robot's reliability recovered. We further hypothesized that the timing of the reliability decreases would influence trust in the robot's autonomy.

2. PRIOR WORK

Parasuraman and Riley define automation as "the execution by a machine agent (usually a computer) of a function that was previously carried out by a human" [19]. Automation has traditionally been employed in systems that are complicated, tedious, or time critical, but it has also been used for economic reasons [19]. When automation was first introduced in the 1930's, its use was limited to large industries; however, at the present, automation can be found in many places, from home appliances to nuclear power plants.

Automation has always had weaknesses: namely, it has only been effective in well-structured and controlled environments and continues to remain so. To avoid catastrophic failures in safety critical systems due to either flaws or limitations of automation, an operator must be present at all times to take control of the system. Situations of this kind in which a human operator is working with an automated system are referred to as "human-in-the-loop control." While utilizing a human operator may be beneficial in certain situations, addressing the inadequacies of automation for the human-in-the-loop control creates a different set of problems. When an operator is added to the system, improving the overall system performance requires more than simply optimizing operator performance and, separately, optimizing automation performance. The interaction between the two needs to be considered as well.

Researchers have shown that trust influences operators' use of automation (e.g., [3, 4, 10, 12, 16, 21]): the more operators trust automation, the more they tend to use it. Moreover, if an operator trusts his own abilities more than those of the automated system, he tends to choose manual control. A user's trust in his own capabilities is most often referred to as "self-confidence."

For several decades, researchers in the human automation interaction field have examined the factors which influence people's trust of automated systems (see [13] for an overview) and how this level of trust, in turn, affects the way in which people use, misuse, or disuse automation. Specifically, Dzindolet et al. [4] demonstrated the impact of system performance on user trust. The results of their study indicated that while users initially placed trust in a decision aid and agreed with its suggestions, as users observed the system making errors they would distrust even a generally high-performing aid unless provided with reasons as to why the errors had occurred. Providing these reasons increased trust in the automated aid, even when the aid performed poorly.

Additional factors contributing to a user's trust of an automated system include the recency of errors made by the system [22], the user's prior knowledge about the system's performance [22], the user's knowledge of the capabilities of the system [22], the user's expectations of the system's performance [22], the usability of the interface [1], and situation awareness gained through using a system [14]. Atoyan et al. [1] found that interface design plays an important role in influencing users' trust in automation.

Some studies have reported a lag between changes in trust and self-confidence and an actual change in allocation strategy; this lag is referred to as "inertia" [12]. When the user changes the allocation strategy, the performance of the system inevitably changes. For the feedback loop to close, the user needs to observe this change. Depending on the system, there might be a significant time delay before the user observes the change in performance [4].

To date, there has been little work examining issues of trust with non-social robots, although some work has been conducted involving simulated robots. Dassonville et al. [2] conducted a study in which participants used a joystick to control a simulated PUMA arm. Errors were introduced into the simulation, and participants were asked to rate the reliability, performance, and predictability of the joystick's behavior (as well as how difficult it was to make such ratings). The results of the study were consistent with prior work in autonomous systems that suggest that the user's self-confidence is a significant factor which influences use of such systems.

Freedy et al. [6] examined trust in the context of mixed-initiative command and control systems using the MITPAS (Mixed-Initiative Team Performance Assessment System) Simulation Environment. The researchers constructed a quantitative measure of trust by assuming that people use a rational decision model such that "trust behavior is reflected by the expected value of the decisions whether to allocate control to the robots on the basis of past robot behavior and the risk associated with autonomous robot control" [6]. Participants assumed the role of a controller of an unmanned ground vehicle (UGV). The UGV autonomously targeted and fired, but participants were instructed to take control of the UGV if its behaviors would lead to a time delay or a failure. The experimenters varied the competency of the UGV's firing behavior and recorded participants' choices to override the UGV. The results suggested that if participants could gauge whether the UGV was very competent or incompetent, they adjusted their behavior accordingly. This adjustment implied that the participants trusted the system to continue to maintain the same level of competence. It was more difficult for users to adjust their behavior when the system showed indeterminate competence. More work needs to be done to determine how these results would generalize to physical robots in the real world.

While relatively little work has been done investigating trust in robots, there is a large body of research on trust in different types of technologies. For example, Song et al. [23] developed a neural network-based trust model for understanding users' acceptance of recommendations from a system of heterogeneous agents. Another agent-related trust model was developed by Rehak et al. [20], who used fuzzy numbers to represent trust in cooperating ubiquitous devices. McKnight et al. [15] developed a trust model to understand users' acceptance of a website offering legal advice. Because we are interested in developing a model of trust for human-robot interaction, we have restricted the scope to trust models that were developed for other technology domains.

Riley hypothesized a general model for trust in automation, including how different factors influence each other and ultimately the operator's reliance on automation [21]. We believe that trust models developed for traditional automation need alteration in order to be useful for HRI. Like most, Riley's model does not consider some factors that are relevant to robots such as interface usability, proximity to robot (co-located or remote-located), situation awareness, and dynamics of the operating environment. To advance the field, a systematic study of the factors that could influence trust in HRI is necessary to build trust models in this domain. To this effect we conducted a study with a real robot with dynamic workload, complex task, and variable reliability.

3. METHODOLOGY

To determine how changing reliability impacts a person's use of autonomy and trust in a robot system, we conducted the same set of experiments at University of Massachusetts Lowell (UML) and Carnegie Mellon University (CMU).[1] Twelve participants were recruited at both sites. The average age of the participants at UML was 23.6 years ($SD = 6.1$) and at CMU was 30.6 years ($SD = 14.2$). All 24 participants were classified as novice robot users as none of them had any prior experience controlling remote robots.

[1]Unless explicitly mentioned, all of the parameters were identical between the two sites.

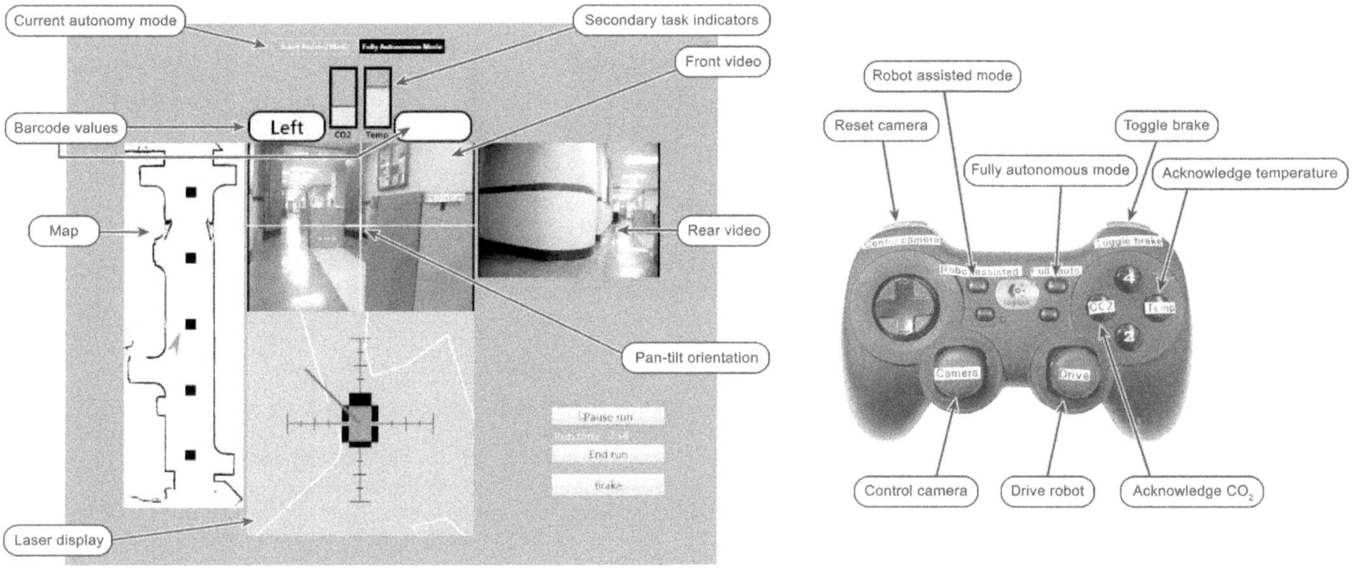

Figure 1: The interface and the gamepad used to control the robot.

3.1 Robot system

Two iRobot ATRV-JR robots were used for this experiment, one at UML and the other at CMU. There was a camera mounted on the front of the robots on a Directed Perception PTU-D46-17 pan-tilt unit and another camera was mounted on the rear. For distance sensing, a SICK LMS200 was used on the front and a Hokuyo URG-04LX laser was mounted on the back. The robots had computers with similar capabilities and ran the same code base.

Figure 1 shows the user interface (UI) used to control the robots. The video from the front camera was displayed in the middle, the video from the back camera was displayed on the top right (mirrored to simulate a rear view mirror in a car). The distance information from both lasers was displayed on the bottom around a graphic of the robot. The map of the course with the pose of the robot was displayed on the left. Using the gamepad shown in Figure 1, participants were able to drive the robot, control the pan tilt unit for the front camera, select the autonomy mode, turn the brakes on or off, recenter the camera, and acknowledge the secondary tasks.

3.2 Tasks

The participants were asked to drive the robot as quickly as they could along a specified path, searching for victims, not hitting objects in the course, and responding to the secondary tasks. To create additional workload, simulated sensors for CO_2 and temperature were used. The participants were not told that the sensors were not real. They were also informed that the robot's performance was not influenced in any way by changes in temperature and carbon dioxide. The values from the sensors were displayed on the UI (Figure 1), which the participants were asked to monitor. Participants were asked to acknowledge high CO_2 and temperature values by pressing the corresponding buttons on the gamepad. The values were considered high when their values were above the threshold lines on the secondary task indicators (Figure 1); values over the threshold were indicated by changing the color of the bars from light blue to red to assist the participants in recognizing the change. The level of workload was varied by changing the frequency with which the values crossed the threshold. The simulated sampling rate for the sensors was kept steady.

3.3 Test course

Figure 2 shows the course used at UML. The course at CMU had the same length, layout for the boxes, and driving clearances. Both courses were set in hallways with little foot traffic. The courses were approximately 18 meters (60 feet) long and had 5 obstacles (boxes) placed about 2.75 meters (9 feet) from each other. The width of the course at UML was 2.43 meters (8 feet), and the width of the course at CMU was 1.98 meters (6.5 feet). The discrepancy in the hallway widths was compensated by using 61 cm (24 inch) wide boxes at UML and 15.2 cm (6 inch) wide boxes at CMU. The clearance on either side of the boxes was 0.9 meters (3 feet), and the robots were 0.66 meters (26 inches) wide.

The start and end positions were the same for each run. For each run, the participants were asked to follow a set path. We designed five different paths based on the following criteria:

- The length of each path must be the same ($\sim$61 meters (200 feet)).
- The number of u-turns in a path must be the same (3).
- The number of transitions from the left side of the course to the right and vice versa must be the same (3).

As the maps were similar in difficulty and length, we did not counterbalance paths for the participants. Instead, paths were selected based on a randomly generated sequence. A sample path is shown in green in Figure 2.

Text labels were placed on top of the boxes to indicate the path ahead. Since the boxes at UML were wide, similar labels were placed on both edges of the face as shown in Figure 2 to make it easy for the participants to read the labels as they went past the boxes. The labels indicated 'left,' 'right,' or 'uturn.' The directions were padded with additional characters to prevent the participants from recognizing the label without reading them. Figure 2 shows the two types of labels that were used. The labels with white background (referred to as white labels) were to be followed for the first half of the entire length and the labels with black background (referred to as black labels) for the second half. The transition from following the white labels to black labels was indicated to the participants via the UI. When the participants were supposed to follow the black

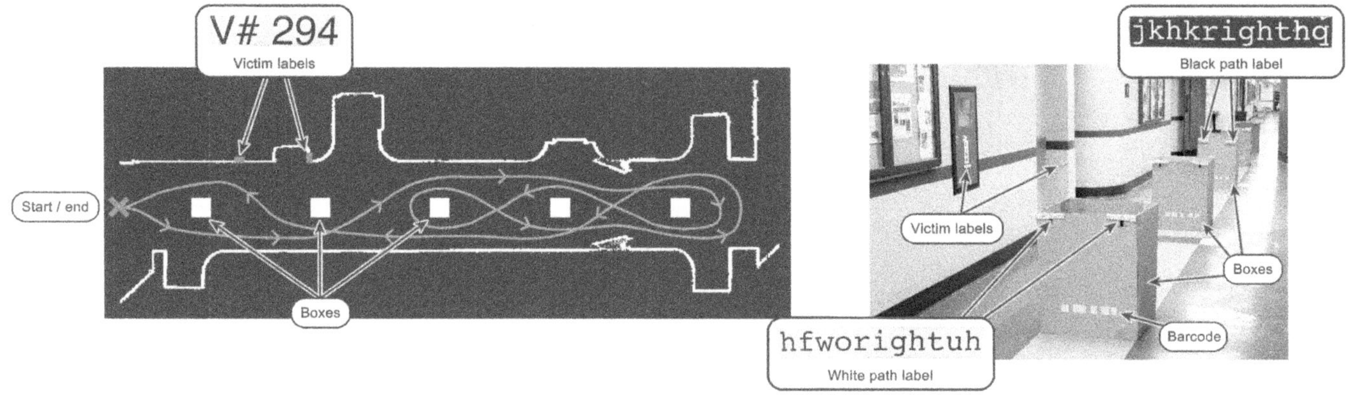

Figure 2: The top down view of the course used for the experiments (left) and the photo looking down the hallway (right).

labels the background for the barcode values (shown in Figure 1) would turn black. Two sets of labels were necessary to prevent the participants from driving in an infinite loop.

The boxes also had barcodes made from retro-reflective tapes that the robot could read (Figure 2). The robot would display the contents of the bar code on the UI. However, the paths for each run were hard coded because the barcodes could not be consistently read by the robot. While the barcodes were not used by the robot, the participants were told that the robot read the barcodes to determine the path ahead, just like they read the labels. Based on a constant video compression rate, sampling resolution, and the font size, the labels could be read from about 1 meter (3 feet) away. The robot was set to simulate reading the label from approximately the same distance. The participants were informed that, at times, the robot might make a mistake reading the barcodes and that they should ensure that the direction read by the robot is correct. The participants were also told that if the robot did make a mistake in reading the barcode, it would proceed to pass the next box on the incorrect side, resulting in the participant being charged with an error on their score (see below).

The course also had four simulated victims. These victims were represented using text labels like the one shown in Figure 2. The victim tags were placed on the walls of the course between 0.7 meters (2.5 feet) and 1.8 meters (6 feet) from the floor. The victim locations were paired with the paths and were never placed in the same location during the participant's five runs. While there was a number associated with each victim, the participants were told to ignore the number while reporting the victims. Whenever the participants found a new victim, they were told to inform the experimenter that they had found a victim. They were explicitly instructed to only report victims they have not reported before.

3.4 Autonomy modes

The participants could operate the robot in one of two autonomy modes: robot assisted mode or fully autonomous mode. The participants were instructed that they were free to select either mode and could switch as many times as they wanted. They were also informed that there were no benefits or penalties for selecting either mode. When each run started, no autonomy mode was selected by default, thereby requiring the participants to make a selection. The maximum speed at which the robot would move was the same in both modes (approximately 0.125 meters (0.41 feet) per second). These configurations ensured that both autonomy modes were similar from a performance standpoint.

In fully autonomous mode, the robot ignored the participant's input and followed the hard coded path. The obstacle avoidance algorithm ensured that the robot never hit any object on the course. In robot assisted mode, the participant had a significant portion of the control and could easily override the robot's movements. The robot would provide its desired velocity vector based on the path it was supposed to follow. The robot's desired vectors were calculated the same way in both autonomy modes and were displayed on the UI to show the participant the robot's desired direction.

3.5 Compensation

Using higher levels of automation reduces workload and hence is desirable, especially under heavy workload from other tasks. To prevent participants from using high levels of autonomy all the time, regardless of the autonomous system's performance, it is typical to introduce some amount of risk. Hence, in line with similar studies (e.g., [5, 11, 21]), the compensation was based in part on the overall performance. The participants at UML could select a gift card to a local restaurant or Amazon.com, and the participants at CMU received cash.[2] The maximum amount that the participants at both sites could earn was \$30. Base compensation was \$10. Another \$10 was based on the average performance of 5 runs. The last \$10 was based on the average time needed to compete the 5 runs, provided that the performance on those runs was high enough.

The performance for each run was based on multiple factors, with different weights for each of those factors determined before the experiments were run. The participants were told there was a significant penalty for passing a box on the incorrect side, regardless of the autonomy mode. If the participants passed a box on the wrong side, they were heavily penalized (20 points per box). In addition to the loss of score, participants were told that time would be added based on the the number of wrong turns they took, but the specific penalties were not revealed. For the first box passed on the wrong side, no additional time was added, to allow people to realize that the reliability of the system had dropped. For the second incorrect pass, 60 seconds were added, with an additional 120 seconds for the third and an additional 240 for the fourth, continuing with an exponential increase. Finding the victims was also an important task, so 10 points were lost for each victim missed. Equation 1 was used to calculate the score for each run.

$$\begin{aligned} Score = 100 - 20\times\ numIncorrectPasses \\ - 10\times\ numVictimsMissed - 5\times\ numPushes \\ - 2\times\ numBumps - numScrapes - secondaryTaskScore/2 \end{aligned} \tag{1}$$

The scoring formula was not revealed to participants, although they were told the factors that would influence their score. The score for each run was bounded between 0 and 100. If the score was 50

[2]The means of compensation were institutional limitations.

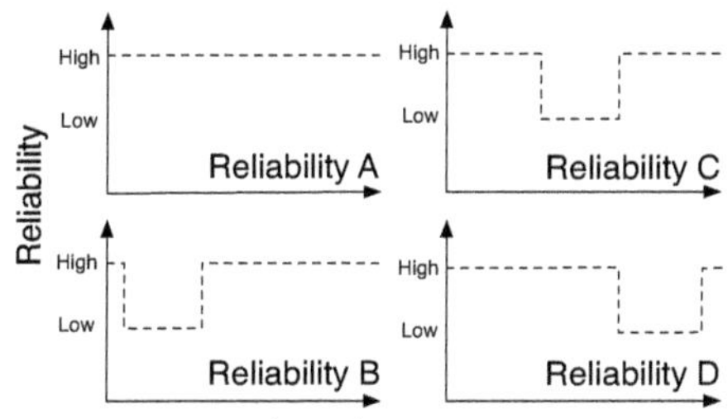

Figure 3: The different reliability configurations.

or more, the participants were eligible for a time bonus; if they had completed the runs in under 11:45 minutes on average, they would receive an additional $10. If they had a score of 50 or more and took between 11:45 and 15 minutes per run on average, they would receive a $5 bonus. Participants were told about this interdependence between score and time, which was designed to prevent participants from quickly running through the course, ignoring the tasks, while also providing a significant motivation to perform the task quickly.

At the end of each run, its score was calculated and the participants were informed about the amount of compensation that could be received based only on that run. At the end of five runs, the average compensation was calculated and given to the participant.

3.6 Questionnaires

There were three sets of questionnaires. The pre-experiment questionnaire was administered after the participants signed the consent form; it was focused on demographics (i.e., age, familiarity with technology similar to robot user interfaces, tendency towards risky behavior, etc). The post-run questionnaire was administered immediately after each run; participants were asked to rate their performance, the robot's performance, and the likelihood of not receiving their milestone payment. Participants were also asked to fill out previously validated trust surveys in their unaltered form, referred to in this document as Muir [17] and Jian [9], and a TLX questionnaire after each run. After the last post-run questionnaire, the post-experiment questionnaire was administered, which included questions about wanting to use the robot again and its performance.

3.7 Procedure

After participants signed the informed consent form, they were provided an overview of the robot system and the task to be performed. Then, participants were asked to drive the robot through the trial course in fully autonomous mode. The experimenter guided the participant during this process, by explaining the controls and helping with tasks if necessary. The trial course was half the length of the test course. Once participants finished, they were asked to drive the robot again through the same course in robot assisted mode. Since there are multiple tasks that participants need to perform, we decided to first show them the fully autonomous mode, as that would be a less overwhelming experience. Once the participants finished the second trial run, they were asked to fill out the post-run questionnaire. While the data from this questionnaire was not used, it allowed participants to familiarize themselves with it and also helped to reinforce some of the aspects of the run that they would need to remember. The participants were also told that they could take a break whenever they wanted.

After the two trial runs, the participants were asked to drive the robot for five more runs. In each run, a different map was used. During these runs the reliability of robot autonomy was either held high throughout the run or was changed. Figure 3 shows the four different reliability configurations. The changes in reliability were triggered when the robot passed specific points in the course.

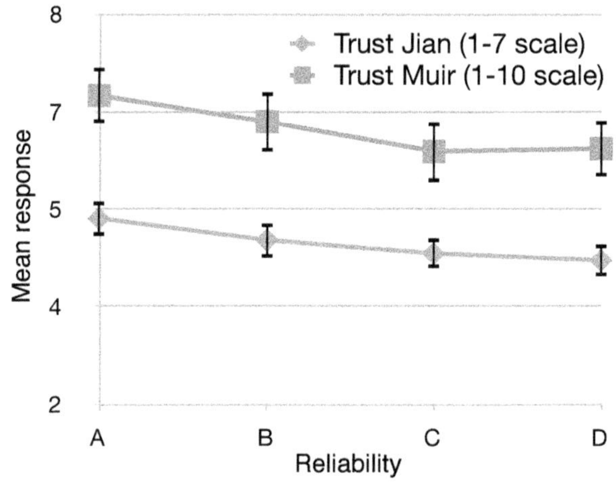

Figure 4: Impact of reliability on trust.

These locations were equal in length and there were no overlaps. For all four patterns, the robot always started with high reliability. The length of each low reliability span was about one third the length of the entire course. Using different dynamic patterns for reliability allowed us to investigate how participants responded to a drop in reliability at different stages and the changes' influence on control allocation. Every participant started with a baseline run under full reliability (Reliability A in Figure 3). Then, the four reliability profiles were counterbalanced for the remaining four runs. Instead of a single long run, we elected to use shorter duration runs because we wanted to directly compare failures at different periods. Failure during longer runs can be investigated in the future, once the important reliability patterns can be identified.

Our goal was to investigate how trust is initially formed; hence we recruited novice participants. We expect expert users would interact differently with the system, as they would already have a mental model of it and its performance.

4. RESULTS AND DISCUSSION

While 12 of the participants were run at CMU and 12 at UML, there were consistent behaviors across the sites related to reliability and autonomy switching, so this data is reported in aggregate. There were some site differences in terms of the trust scales used, which we discuss below.

Unless noted, data from the practice and baseline runs were not included in the analyses. We checked for practice effects (run order) and map effects and did not find any issues. This suggests the counterbalancing and map designs were adequate.

4.1 Positivity bias

We found that 13 participants started all four runs by switching into the fully autonomous mode and 17 participants started run 1 in the fully autonomous mode. Of the 96 total runs, participants initially opted to start in full autonomy for 65 of them. The breakdown for the individual runs was: run_1 = 17, run_2 = 15, run_3 = 17, and run_4 = 16, which is remarkably stable. The participants' willingness to initially trust the robot indicates the possibility of a positivity bias. These findings are consistent with the findings of Dzindolet et al. [4] where they found that given little experience or information about an automated decision aid, people were willing to trust it.

4.2 Effect of reliability changes on trust

The two trust survey methods (Muir, Jian) were highly correlated with each other ($r = 0.84$, $p < 0.0001$) suggesting either can be used

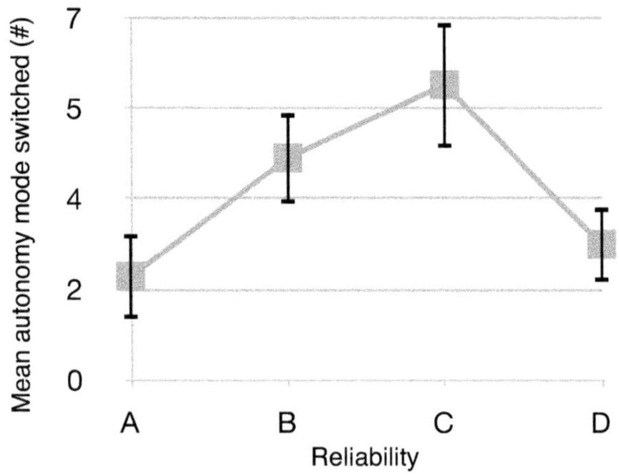

Figure 5: Impact of reliability on mode switching.

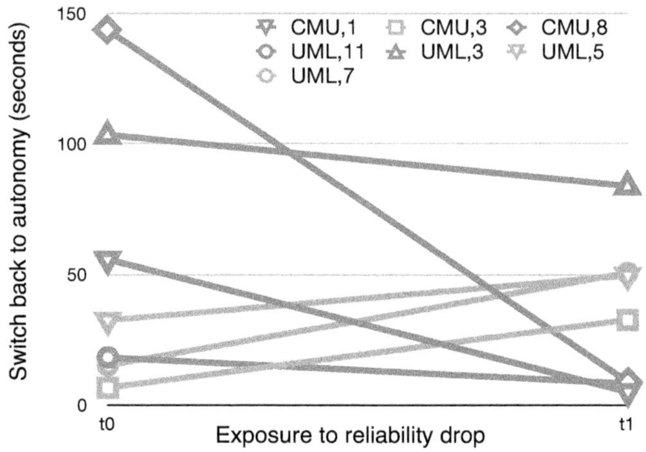

Figure 6: Autonomy return by exposure to low reliability. t0 represents the time when reliability drops and t1 represents the time when reliability increases.

for such experiments in the future. In our analysis in later sections, we have elected to standardize on Muir due to its shorter length.

The Muir and Jian post-run trust surveys were examined with REML (REstricted or REsidual Maximum Likelihood) [8] on the effects of Site (CMU, UML), Reliability (A, B, C, D), and Participants as a random effect and nested by Site. In both cases, there were significant differences for Site and Reliability, but not the interaction. UML trust responses were significantly higher than CMU for Muir, $F(3) = 9.7$ ($p < 0.01$), and Jian, $F(3) = 9.7$ ($p < 0.01$). Student's t post hoc tests of Reliability on Muir, $F(3) = 2.6$ ($p = 0.059$), and Jian, $F(3) = 3.0$ ($p < 0.05$), showed reliability A as being significantly higher than C and D for both metrics (Figure 4). These nearly identical results for Muir and Jian reinforce the earlier finding that using just one approach is appropriate in the future.

These results mean that trust is highest in high reliability runs (A); slightly reduced in runs with low reliability at the beginning of the run and high at the end (B); and more reduced for runs where reliability was low in the middle or end of the runs (C & D). This result means timing is important for trust – drops in reliability after a period of good performance are more harmful than early failures. Whether this is due to memory recency or a breakage in the participant's mental model of robot performance is uncertain.

The influence of Site on trust survey results is likely due to UML's population being slightly younger (mean of 7 years younger) and more predisposed towards risky behavior (0.66 higher on a set of 7-point self-rating scales). Significance tests for both demographic features were close, but not statistically significantly different. However, their combined effect may have produced this Site effect.

4.3 Changing autonomy levels

To obtain a high-level view, we performed a REML analysis of how many times participants switched the autonomy level within a run on the effects of Site (CMU, UML), Reliability (A, B, C, D), and Participants as a random effect and nested by Site. This analysis resulted in a significant difference only for Reliability, $F(3) = 4.7$ ($p < 0.01$), where a Student's t post hoc revealed participants switched considerably more within reliability C, as compared to A and D (Figure 5). Likewise, B was higher than A.

Of the 24 participants, five did not switch autonomy levels during any of their runs, regardless of the reliability profiles. Two of these participants stayed in robot assisted mode for all of their runs, two stayed in the autonomous mode, and one participant used robot assisted mode for all but one run. Participants were binned into three behavior groups: FullAuto, Mixed, and MostlyRobotAssisted. Sample sizes for these groups were imbalanced and too small for statistical analysis (2, 19, and 3, respectively), but there were several clear trends. The MostlyRobotAssisted group run times were noticeably slower, and the FullAuto rated their own performance low in comparison to the other two groups. There was a general trend of lower trust on the Muir questions as autonomy use increased across the three groups (see Familiarity bias below).

Of the 19 participants in the Mixed behavior category, nine did not change their autonomy level during the baseline run, which was held constant at high reliability (3 in robot assisted mode, 6 in autonomous). In the second run with high reliability, eleven did not change their autonomy level (1 in robot assisted, 10 in autonomous). Seven of these participants overlapped, meaning that during the high reliability runs, all but six participants did not change their autonomy mode in at least one of those runs.

In contrast, during the runs with changing reliability, all Mixed participants switched autonomy modes in at least one of the other three variable reliability conditions. Also, 14 of the 19 participants switched autonomy modes in all three of the variable reliability conditions (B, C, & D). This data indicates that participants recognized they were operating under dynamic reliability and adjusted their control allocation accordingly. It also indicates that participants recognized the risk of decreased compensation and tried to optimize the allocation strategy to obtain maximum compensation. To further investigate how the participants used autonomy, we analyzed the participants' behavior during periods of low reliability.

4.4 Use of autonomy during periods of unreliability

To examine behavior during low reliability, we focused on the scenario where participants entered a low reliability window during autonomy use. This window corresponded to the point at which reliability decreased (t0) to when it increased (t1). By definition, runs with reliability A were not included, as reliability did not decrease during those runs. For the 17 participants who switched during this window, the mean use of autonomy during low reliability was 30 percent.

A total of 15 participants switched from autonomy to robot assisted mode after t0 and from robot assisted mode to autonomy after t1. This behavior was constrained to reliability conditions B and C

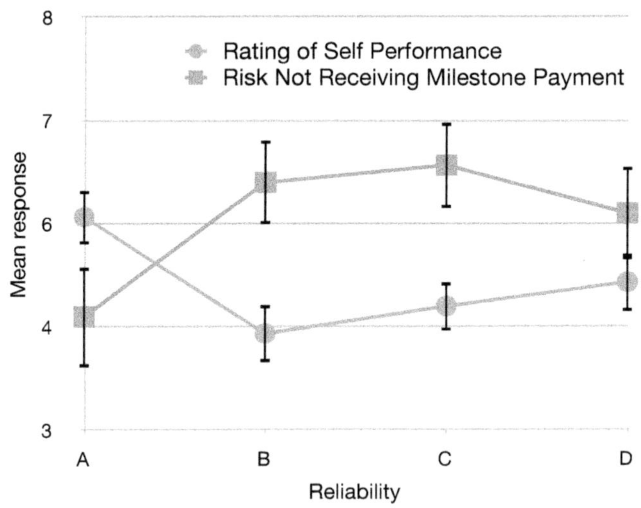

Figure 7: Impact of reliability on self assessment ratings.

(7 and 8 participants respectively); we conjecture that participants did not have enough time to recover from the reliability drop in D, where the drop occurred near the end of the run. Within this group, the mean switching time after the reliability drop at t0 was 16.6 seconds ($SD = 13.1$). The return to autonomy after reliability improved at t1 occurred a mean of 39.0 seconds later ($SD = 41.0$). A one tailed t-test, $t(14) = 2.04$ ($p < 0.05$) confirmed that participants waited longer to switch back to autonomy than switching away from autonomy. While only marginally statistically significant (one tailed, t-Ratio $t(13) = 1.73$ ($p < 0.1$), there were strong indications that participants switched away from autonomy at t0 twice as slowly for C than B (means 22 and 11 seconds respectively). However, there were no differences between C and B for switching back to autonomy at t1.

Of these 15 participants, seven returned to the fully autonomous mode at t1 for both the B and C conditions. Four of these seven switched back to autonomy faster on their second exposure to a reliability change, while the rest switched back more slowly (Figure 6). This results suggests that repeated exposure to changing reliability impacts the speed at which people switch back to autonomy, although we do not posses enough evidence to determine what causes this behavior.

4.5 Effect of self-assessment on trust

ANOVA analysis of participant ratings of robot performance across reliability levels were inconclusive, $F(3, 92) = 1.09$ ($p = 0.36$). However, participants did respond differently for ratings of their own performance, $F(3, 92) = 3.4$ ($p < 0.05$), and were marginally significant on the risk of not receiving a milestone payment $F(3, 92) = 2.2$ ($p < 0.1$). Student's t post hoc analyses showed a higher rating of self performance and better odds of receiving the milestone payment for reliability A, as compared to C and B, in both measures (Figure 7). Assessment of self performance was also sensitive when examining trust, with a significant correlation to Muir ($r = 0.43, p < 0.0001$).

As has been seen in some of our prior experiments, participants were pretty accurate in their assessment of milestone performance. Ratings of risk of not being paid the extra money were inversely correlated with actual payment ($r = -0.58, p < 0.01$).

4.6 Familiarity bias

The protocol was intentionally designed to promote use of autonomy. As expected, higher use of autonomy was correlated with better performance on finding more victims ($r = 0.30, p < 0.01$) and faster route completion time ($r = -0.51, p < 0.0001$). These results suggest that general use of autonomy had a perceptible, beneficial impact on the task.

As mentioned, prior work shows that increased use of autonomy with positive performance outcomes leads to higher trust (e.g., [11]). However, the Muir post-run trust ratings and the percentage of time spent in full autonomy were inversely correlated ($r = -0.20, p < 0.05$). This fact, combined with the results above, suggest that overall familiarity is less powerful than scenario factors.

Table 1: Backwards stepwise regression results for Muir trust ratings

Effect	Estimate	p
Cognitive load (TLX)	-0.33	< 0.01
Victims found	-1.58	< 0.01
Payment (performance)	-0.22	< 0.01
Tendencies towards risky behavior	0.65	< 0.01
Risk of not receiving milestone payment	0.28	< 0.05
Participant age	-0.05	< 0.1
Self performance rating	0.50	< 0.1
Robot performance rating	removed	x
Experience with robot-like UIs	removed	x
Autonomy switches	removed	x
Technology demographics	removed	x
Secondary task performance	removed	x
Percent autonomy	removed	x
Map time	removed	x

4.7 Predicting trust

An important question for human-robot interaction is whether trust can be predicted. To examine this question, Muir trust ratings were examined in the context of cognitive load (TLX), victims found, secondary task performance, payment (i.e., overall performance), number of switches between autonomy and robot assisted modes, a collection of demographic features, and the three post-run assessment ratings. A backwards stepwise regression on these independent measures accurately predicted Muir ratings ($R^2 = 0.84$). Significance results showed that higher trust was predicted by low cognitive load, poor victim performance, lower payment, lower expected payment, high ratings of self performance, younger age, and high risk tendencies (Table 1). While these factors strongly predict trust, it should not preclude other factors from being investigated in the future. Note that autonomy switching, secondary task performance, ratings of robot performance, and percentage of time using full autonomy do not predict trust. These results suggest that trust is heavily tied to factors with semantic association to risk and personal feelings about performance, rather than robot performance. This result is contrary to what Hancock et al. [7] found. We speculate that the difference is due to the fact that our methodology involved controlling a remote robot making it difficult to gauge the robot's actual performance, as was the case in the studies examined by Hancock et al. [7].

5. CONCLUSIONS

Our goal of creating an experience that was challenging, would force autonomy use, and keep participants focused on performance clearly succeded. This design increases the realism of the experience and is analogous to scenarios where robot users will be forced to use autonomy deliberately, carefully, and with supervision (e.g., assistive robotics, bomb disposal).

We hypothesized that people would trust a robot system less when its reliability in autonomous mode decreased, as evidenced by switching to a manual mode. Trust was affected by drops in reliability. We further hypothesized that the timing of the reliability decreases would influence trust in the robot's autonomy. This hypothesis was true, especially when reliability drops occurred late (D) or in the middle (C) of runs. Reliability patterns also led to different mode switching behavior. Reliability drops in the middle of the run (C) led to sharp increases in the number of mode switches and switches away from autonomy were twice as slow as those seen for early drops in reliability (B).

We also wanted to determine how long it would take participants to switch back to autonomous mode after the robot's reliability increased. As is typical with trust, users switched away from autonomy during a reliability drop much faster than returning to autonomy after a reliability increase. However, the reliability pattern did not impact the speed at which users returned to autonomy when reliability improved. We did see mixed behavior when looking at the exposure to a reliability drop – about half the users returned to autonomy faster the second time, while the other half were slower. We are unclear on why this difference exists and the limited sample size (n=7) prevents any conclusive interpretations.

Team experience suggests that risk is important, which was our main motivation for the milestone payment and the inclusion of "victims" in the protocol. A regression analysis of various factors' influence on trust ratings was, in fact, dominated by features associated with risk. These were explicit factors (e.g., tendencies towards risky behavior, payment, risk of not receiving payment) and implicit (e.g., victims found, participant age). Personal factors associated with self assessment (TLX, self performance rating) were also important in the analysis. These risk and personal factors overwhelmed factors associated with robot performance and robot reliability. In short, operators tie trust to their own actions rather than robot performance. Other work by the team has revealed similar tendencies in expert robot operators.

6. ACKNOWLEDGMENTS

This work has been supported in part by the National Science Foundation (IIS-0905228 and IIS-0905148). Funding for Sofia Gadea-Omelchenko was provided through the Quality of Life Technology ERC REU program (EEEC-0540865). Thanks to Poornima Kaniarasu, Arne Suppe, and John Kozar for their assistance with robot and experiment logistics, Adam Norton for his assistance with figures, and to all of the experiment participants.

7. REFERENCES

[1] H. Atoyan, J. Duquet, and J. Robert. Trust in new decision aid systems. In *Proceedings of the 18th Int'l Conference of the Association Francophone d'Interaction Homme-Machine*, page 122. ACM, 2006.

[2] I. Dassonville, D. Jolly, and A. Desodt. Trust between man and machine in a teleoperation system. *Reliability Engineering & Systems Safety*, 53(3):319–325, 1996.

[3] P. deVries, C. Midden, and D. Bouwhuis. The effects of errors on system trust, self-confidence, and the allocation of control in route planning. *Int'l Journal of Human-Computer Studies*, 58(6):719–735, 2003.

[4] M. Dzindolet, S. Peterson, R. Pomranky, L. Pierce, and H. Beck. The role of trust in automation reliance. *Int'l Journal of Human-Computer Studies*, 58(6):697–718, 2003.

[5] M. Dzindolet, L. Pierce, H. Beck, and L. Dawe. The perceived utility of human and automated aids in a visual detection task. *Human Factors: The Journal of the Human Factors and Ergonomics Society*, 44(1):79, 2002.

[6] A. Freedy, E. DeVisser, and G. Weltman. Measurement of trust in human-robot collaboration. *Collaborative Technologies and Systems*, Jan 2007.

[7] P. Hancock, D. Billings, K. Schaefer, J. Chen, E. de Visser, and R. Parasuraman. A meta-analysis of factors affecting trust in human-robot interaction. *Human Factors: The Journal of the Human Factors and Ergonomics Society*, 53(5):517–527, 2011.

[8] D. Harville. Maximum likelihood approaches to variance component estimation and to related problems. *Journal of the American Statistical Association*, pages 320–338, 1977.

[9] J. Jian, A. Bisantz, and C. Drury. Foundations for an empirically determined scale of trust in automated systems. *Int'l Journal of Cognitive Ergonomics*, 4(1):53–71, 2000.

[10] J. Lee and N. Moray. Trust, self-confidence and supervisory control in a process control simulation. *IEEE Int'l Conference on Systems, Man, and Cybernetics*, pages 291–295, 1991.

[11] J. Lee and N. Moray. Trust, control strategies and allocation of function in human-machine systems. *Ergonomics*, 1992.

[12] J. Lee and N. Moray. Trust, self-confidence, and operators' adaptation to automation. *Int'l Journal of Human-Computer Studies*, 40(1):153–184, 1994.

[13] J. Lee and K. See. Trust in automation: designing for appropriate reliance. *Human Factors*, 46:50–80, 2004.

[14] C. Liu and S. Hwang. Evaluating the effects of situation awareness and trust with robust design in automation. *Int'l Journal of Cognitive Ergonomics*, 4(2):125–144, 2000.

[15] D. H. McKnight, V. Choudhury, and C. Kacmar. The impact of initial consumer trust on intentions to transact with a web site: a trust building model. *Journal of Strategic Information Systems*, 11(3–4):297–323, 2002.

[16] B. Muir. Trust between humans and machines, and the design of decision aids. *Int'l Journal of Man-Machine Studies*, 27(5-6):527–539, 1987.

[17] B. Muir. *Operators' trust in and use of automatic controllers in a supervisory process control task.* PhD thesis, University of Toronto, 1990.

[18] B. Mutlu and J. Forlizzi. Robots in organizations: the role of workflow, social, and environmental factors in human-robot interaction. In *HRI '08: Proceedings of the 3rd ACM/IEEE Int'l Conference on Human Robot Interaction*, pages 287–294, New York, NY, USA, 2008. ACM.

[19] R. Parasuraman and V. Riley. Humans and automation: Use, misuse, disuse, abuse. *Human Factors*, 39(2):230–253, 1997.

[20] M. Rehak, L. Foltyn, M. Pechoucek, and P. Benda. Trust model for open ubiquitous agent systems. In *IEEE/WIC/ACM Int'l Conference on Intelligent Agent Technology*, pages 536–542, 2005.

[21] V. Riley. Operator reliance on automation: Theory and data. *Automation and Human Performance: Theory and Applications*, pages 19–35, 1996.

[22] J. Sanchez. *Factors that affect trust and reliance on an automated aid.* PhD thesis, Georgia Institute of Technology, 2006.

[23] W. Song, V. Phoha, and X. Xu. An adaptive recommendation trust model in multiagent system. *Published by the IEEE Computer Society*, 2004.

Teamwork in Controlling Multiple Robots

Fei Gao, M.L. Cummings, Luca F. Bertuccelli
Massachusetts Institute of Technology
77 Massachusetts Ave, Rm 33-309
Cambridge, MA, 02139
1-617-606-5522
feigao@mit.edu, missyc@mit.edu, lucab@mit.edu

ABSTRACT
Simultaneously controlling increasing numbers of robots requires multiple operators working together as a team. Helping operators allocate attention among different robots and determining how to construct the human-robot team to promote performance and reduce workload are critical questions that must be answered in these settings. To this end, we investigated the effect of team structure and search guidance on operators' performance, subjective workload, work processes and communication. To investigate team structure in an urban search and rescue setting, we compared a pooled condition, in which team members shared control of 24 robots, with a sector condition, in which each team member control half of all the robots. For search guidance, a notification was given when the operator spent too much time on one robot and either suggested or forced the operator to change to another robot. A total of 48 participants completed the experiment with two persons forming one team. The results demonstrate that automated search guidance neither increased nor decreased performance. However, suggested search guidance decreased average task completion time in Sector teams. Search guidance also influenced operators' teleoperation behaviors. For team structure, pooled teams experienced lower subjective workload than sector teams. Pooled teams communicated more than sector teams, but sector teams teleoperated more than pool teams.

Categories and Subject Descriptors
J.7 [**Computers in Other Systems**]

General Terms: Experimentation, Human Factors

Keywords: Multiple Robots, Teamwork, Communication, Urban Search and Rescue

1. INTRODUCTION
Enhanced autonomy makes it possible for one operator to control multiple robots. It releases an operator from manually controlling each robot and makes it possible to do tasks requiring monitoring, coordination, and complex decision-making. However, the required cognitive load for controlling multiple robots could easily exceed that of a single operator, even with higher levels of automation. Teams are increasingly called upon to perform complex cognitive tasks that are less efficiently done by, and sometimes cannot be accomplished by, an individual. Although teamwork may impose extra workload related to coordination and communication, teams have the potential of offering greater adaptability, productivity, and creativity than any one individual can offer. Moreover, teams can provide more complex, innovative, and comprehensive solutions [5].

HRI'12, March 5–8, 2012, Boston, Massachusetts, USA.

Unfortunately, the benefits of teamwork do not always occur naturally, and teams can fail for many reasons [14]. Factors such as poor combination of individual efforts, a breakdown in internal team processes (e.g., communication), and improper use of available information have been identified as potential sources of team failure [15]. In addition, when people collaborate with autonomous systems, system complexity inevitably increases, and automation can change the way people coordinate with each other [12]. To enable collaborative human-automation team interactions, we must therefore understand the nature of such teamwork, including outcomes, processes and dynamics.

Teaming difficulties in controlling multiple robots lead to the following research questions. First, is it possible to design automated decision support tools to improve performance and reduce workload when controlling multiple robots? Second, how does the organization and structure of team members affect performance, workload and communication?

The remainder of the paper proceeds as follows. In Section 2, we review related work. In Section 3, we introduce the research questions and describe an urban search and rescue experiment, including the testbed and experimental procedure. In Section 4, we describe the statistical results from the experiment. In Section 5, we summarize and discuss the results.

2. RELATED WORK

2.1 Allocating Attention across Multiple Robots
Controlling multiple autonomous robots is complex. When supervising a team of robots in a time-critical situation, the time and attention resources of operators are limited since operators are known to process complex tasks in a serial fashion [16]. As a result, there is a temporal opportunity cost associated with each task, and an implicit cost-benefit analysis that operators must perform in order to best allocate their attention across competing tasks. In the neglect tolerance model [3], an operator interacts with one robot for a period called Interaction Time (IT), then neglects it for a period called Neglect Time (NT) to interact with other robots before the first robot must be revisited to maintain performance. The number of independent homogenous robots that a single operator can control is calculated by NT/IT+1. Further studies showed that the number of robots an operator can control

is affected by other factors such as the nonlinear increasing complexity [9] and switching cost between robots [2].

Previous work in automated visual search task allocation for single operator-multiple unmanned vehicle environments by Bertuccelli et al. [1] has shown that automated search guidance can improve operator performance in terms of overall mission probability of detection and lower workload by influencing switching times. This form of search guidance was hypothesized to be beneficial also in the context of team scenarios, where resources were distributed across operators, who could benefit from recommendations for when to switch to new search tasks.

2.2 The Role of Team Structure

Team structure is another important factor hypothesized to affect team effectiveness for the search and rescue setting [8]. Team structure can be described as the work assignment and communication architecture. Work assignment is the "manner in which the task components are distributed among team members" [13]. How the team is structured is closely related to communication, coordination and team performance.

The team structure that is suitable for a specific scenario largely depends on the task characteristics and resources available [11]. For a team of operators working together with multiple homogeneous unmanned vehicles, two possible ways to organize the vehicles are as Sectors or as a Shared Pool [7]. In the Sector condition, each operator controls a portion of all the vehicles. In the Shared Pool condition, operators share the control of all the robots and service them as needed. Sector assignment, which is how modern day air traffic control is architected, can reduce the number of robots the operator must monitor and control. However, the Shared Pool condition offers a more flexible scheduling advantage of load balancing since any operator in the team can service any robot as needed. In addition, for monitoring applications, the Shared Pool offers a redundant observer advantage, such that a second operator with partially overlapping perceptual judgments may detect victims missed by the first operator.

Coordinated action lies at the heart of effective team performance. Communication is an important method of explicit coordination. It also relates to building an accurate understanding of team members' needs, responsibilities, and expected actions [11]. Communication also requires cognitive and attention resources, and may be hindered because the environment has become stressful and team members focus on their individual tasks rather than on how those tasks affect other team members [15].

Previous research by Lewis and Wang et al. [8] investigated the effect of autonomous path planning versus manual control and team structure in a Urban Search and Rescue (USAR) setting. They found automating path planning improved system performance but it may weaken situation awareness. For team structure, no significant difference on performance was found, but teams that shared the control of all robots had slightly lower workload.

Using a similar experiment setting, we investigated similar issues including 1) Understanding the role of automated search guidance on operator and team performance, workload and communication when controlling multiple robots, and 2) Understanding the role of team structure on overall mission performance, subjective workload, and working process. This experiment is detailed in the next section.

3. EXPERIMENT DESIGN

3.1 Testbed

USARSim, a robotic simulation performing Urban Search and Rescue (USAR) tasks [10], was used to provide the underlying simulation for the testbed. MrCS (Multi-robot Control System), a multi-robot communications and control infrastructure with an accompanying user interface was used as the control interface. MrCS provides facilities for starting and controlling robots in the simulation, displaying camera and laser range finder output, and supporting inter-robot communication through Machinetta, a distributed multi-agent system developed at Carnegie Mellon University [4]. Figure 1 shows the elements of the MrCS displayed on a dual display computer. Thumbnails of robot camera feeds are shown on the left screen. A video feed of interest is on the top left of the right screen. Under the video feed, a GUI element in the bottom left allows teleoperation and camera pan and tilt. The right shows the current area map and allows operators to mark the location of victims.

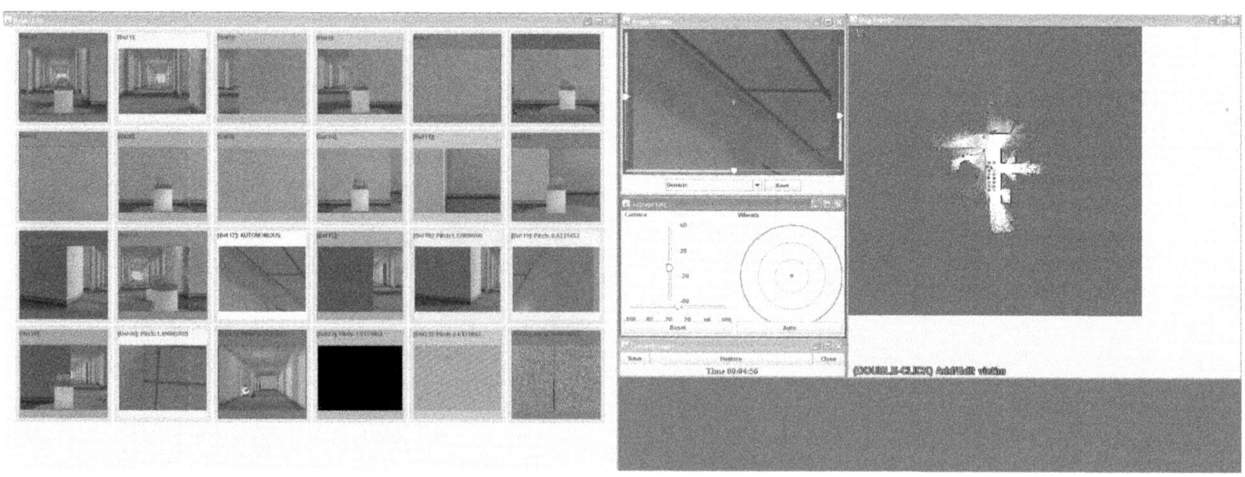

Figure 1: Interface for operating vehicles

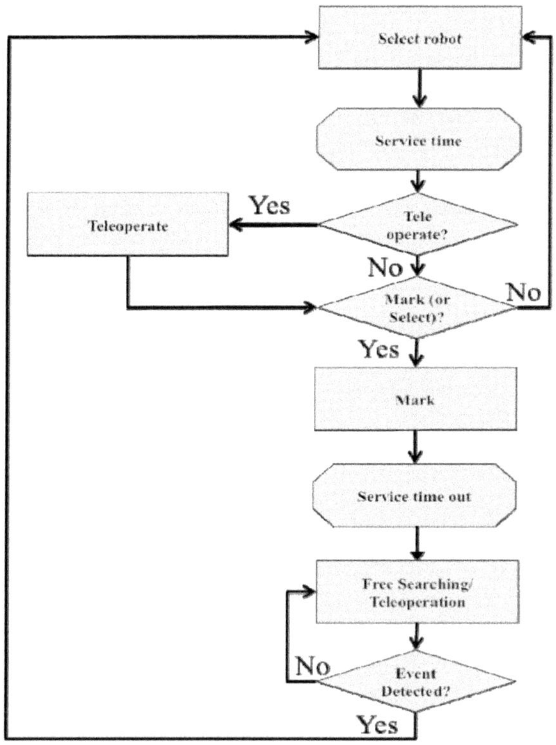

Figure 2: Operation procedure

In MrCS, each robot is capable of updating a map, planning their routing and sending back video feed to operators. The operators' tasks were to explore the environment and identify as many positions of victims as possible. There was little interdependency between robots. In this experiment, robots were started in different regions and explored the environment automatically. The operators guided the robots in the environment in order to find the victims. The general workflow of a single operator is shown in Figure 2. When a victim appeared in the camera of a robot and was detected by the operator, the operator's task was to select the robot, teleoperate the robot to bring the victim back into the camera view and mark the location of the victim on the map. This was the time the operator devoted to serve the robot, labeled as service time/service time out in Figure 2. After that, the operator continued monitoring all the robots and guided the robots to explore the environment. Most of the time robots were moving around using autonomous path planning, and the operator only needed to monitor the thumbnails of video feeds. It also happened that the operator used teleoperation to manually control the robots to send them to a specific unexplored place. This free searching/teleoperation period stopped until a new victim appeared in a camera view, and the operator selected this robot to start a new task.

3.2 Independent Variables

The goal of this experiment was to investigate the effect of search guidance and team structure on operator and team performance, workload and communication in an Urban Search and Rescue task with multiple robots.

A new automated search assistant was designed and inserted into the testbed that recommended an appropriate time for operators to interrupt their current searches and investigate other imagery. The three types of automated schedule assistants were as follows:

1) In the Off (O) condition, participants received no decision support and each participant was tasked to search and mark the victims, and teleoperate a selected robot when needed.

2) The Suggested (S) condition gave a notification when the operator spent more than 30 seconds on a robot, and recommended that the operator move on to another robot.

3) The Enforced (E) condition gave a notification when the operator spent more than 30 seconds on a robot and switched to another robot automatically after 5 seconds.

Thirty seconds was chosen as the threshold criteria based on previous study [1, 8, 9] and pilot test of the experiment. In previous studies on visual search tasks [1], the possibility of finding a target was shown to decrease as more time was spent on the visual search task. The probability was estimated to be 0.8 for 26 seconds spent on searching. In another experiment for USAR tasks [8], the mean time from a robot being selected to a victim being marked under autonomous control was approximately 35 seconds. We selected 30 seconds as the threshold so that an operator was given reasonable amount of time to finish the task if a victim was successfully located and was prevented from spending too much time on a low probability search task if the operator failed to locate the victim. This threshold was validated in pilot tests as well.

In this experiment, operators were grouped into two types of team structures:

1) In Sector (S) teams, each participant controlled half of all the robots, for a total of 12 robots. Locations of their teammates' robots were shown on the map, but video feed from their teammates' robots could not be seen.

2) In Pool (P) teams, two operators shared the control of all the robots. They were able to see the video feed of all robots and control any robot not under control by a teammate.

3.3 Dependent Variables

Dependent variables included task performance metrics, subjective workload, operator measures and communication as team measure. Task performance metrics included number of victims found, number of deletes, number of errors in marking victims, number of victims missed, and percentage of area explored. Number of victims found was evaluated based on the distance between the mark position and real position of the victim. A mark was correct if the distance was less than one meter, otherwise it was an error. Number of deletes was the number of marks deleted by the operator. Deletes happened when a victim was not accurately marked previously or was marked for more than once, and this measure was related to the accuracy of marking victims. Subjective workload ratings were obtained through the NASA-TLX [6], which measures six sub dimensions. Operator measures included display-to-mark time, select-to-mark time, teleoperation frequency, duration and total teleoperation time. Communication was evaluated based on total communication time. All the dependent variables were summarized in Table 1 along with their definitions.

Table 1: Dependent Variables

Category	Dependent Variables
Task performance metrics	Found: number of victims mark in the correct position
	Error: number of marks in the wrong position
	Deletes: number of marks deleted
	Missed: number of victims that appeared in the camera but were not marked
	Percentage of area explored: area explored/total area
Workload	NASA-TLX rating
Operator Measures	Teleoperation duration: length of teleoperation period before marking victim or robot selection
	Teleoperation frequency: number of teleoperation
	Total teleoperation time: total amount of time spent on teleoperation
	Display-to-mark time: time from victim appearing in the camera to being marked
	Select-to-mark time: time from robot selection to victim being marked
Team Measure: Communication	Communication time: total time spent communicating with team member

3.4 Procedure

The experiment began with a 15-minute training session prior to three 25-minute test sessions. Participants were tested in groups of two to facilitate teamwork in performing the USAR task. Each participant controlled 12 robots individually in the Sector condition, or all 24 robots in the Shared Pool condition. Each pair of participants performed all three search guidance conditions. The three conditions were randomized and counterbalanced to limit any learning effect. Audio and screen recordings were collected during the experiment. Subjective workload was rated using the NASA-TLX at the end of each session.

3.5 Participants

A total of 48 participants, aged 19 to 47 years old, participated in the experiment. The average age was 26.6 years, with a standard deviation (SD) of 5.5. Among them, 19 were female and 29 were male. Thirty-three of the participants were undergraduate or graduate students, and 15 had other occupations. Twenty-two of the participants did not play video games regularly. The average time playing video game per week for the remaining 26 participants was 4.1 hours (SD = 4.9). The correlation between hours spent on video games and average individual performance was not significant (p = 0.354).

4. RESULTS

Data logged during the experiment were post processed to obtain performance and process data. The criterion for a successfully marked victim was that the position of the mark was within one meter of the true position of the victim, which was the same criterion as in the study of Lewis et al. [8] In order to find the time when a victim appeared in the camera, we drew the visible areas of all victims using ray tracing. If the robot was in the visible area for a victim, and its field of view contained the victim, this victim was declared visible on this robot's camera. By calculating these quantities, we were able to obtain the number of victims missed, and record the display-to-mark time. All the other dependent variables were calculated directly from the user interaction log.

The results were analyzed based on the three experiment sessions from four aspects: task performance metrics, subjective workload, operator measures and communication as team measure. Data in the training session was not included in the analysis.

4.1 Task Performance Metrics

The dependent variables of number of victims found, number of errors and percentage of area explored were analyzed using analysis of variance (ANOVA). The number of marks deleted and number of victims missed were analyzed using nonparametric tests since they did not satisfy the ANOVA assumptions of normality and/or homogeneity. An alpha of 0.05 was used in the analysis, and a p-value between 0.05 and 0.1 was defined as marginal significant.

For number of victims found, no significant effect was found for either of the primary independent factors. Session order, a secondary independent variable that represents the order the three search guidance conditions were performed, was shown to have a significant effect on number of victims found (Figure 3). After removing data from the first session of each experiment due to a learning effect, the interaction between team structure and search guidance showed a marginally significant effect ($F=2.591$, $p=0.087$). The Sector team had better performance without automated search guidance (Mean = 22.8, SD =3.73), while Pool teams worked better with enforced search guidance (Mean = 22.6, SD = 4.43), as shown in Figure 4.

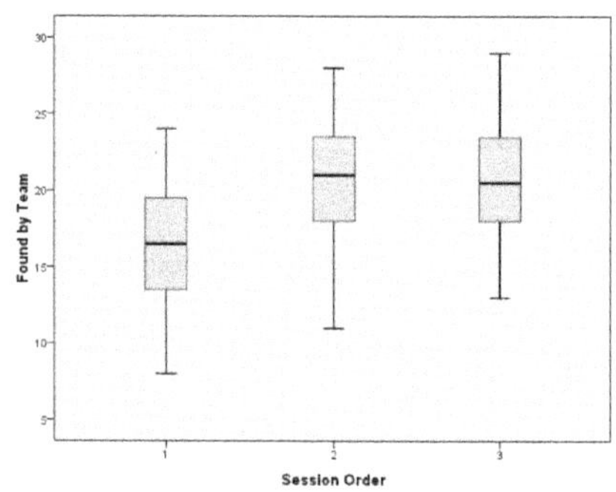

Figure 3: Number of victims found versus session order

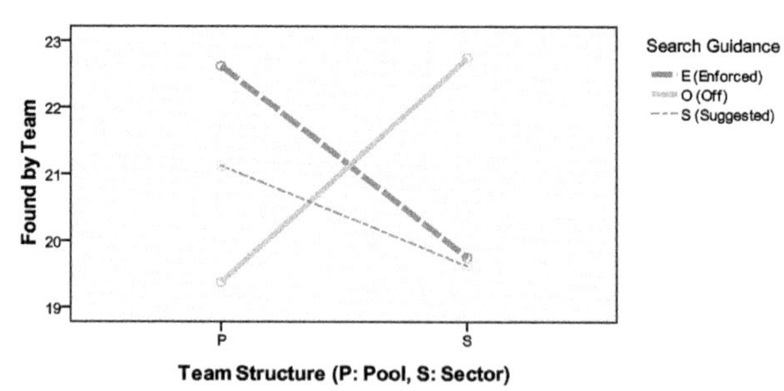

Figure 4: Interaction effect of team structure and search guidance on number of victims found

Although number of victims found was evaluated at the team level, operators in teams have difference on victims found by individual team members. Operators in Pools teams have significant larger difference on individual performance than in Sector teams (p=0.036), as shown in Figure 5.

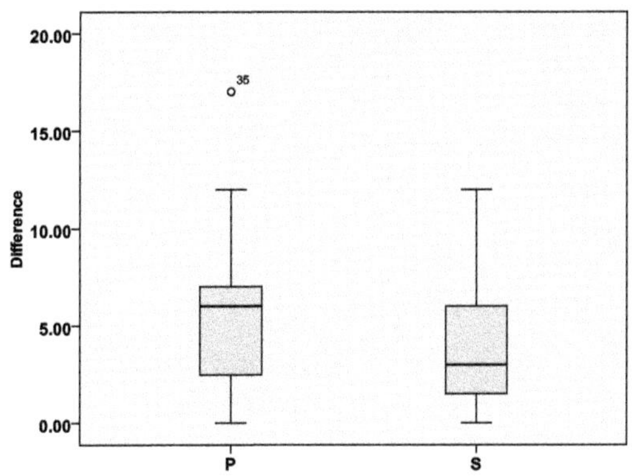

Figure 5: Difference on number of victims found by individual team members

For number of marks deleted (Figure 6), a marginally significant effect was found for team structure (p=0.066). Search guidance did not demonstrate a significant effect. Pool teams (Mean = 8.3, SD = 5.49) deleted more than Sector teams (Mean = 6.0, SD = 3.02), indicating Pool teams corrected themselves more often. Deletes also happened when a victim was marked by both operators, which could only happen in Pool teams. One interesting note is that the standard deviation of deletes with no search guidance of Sector teams (SD=1.946) is much lower than in the other conditions, as shown in Figure 6. This means the deleting behavior was more consistent across different operators under the Sector condition, and operators were less affected by their teammate.

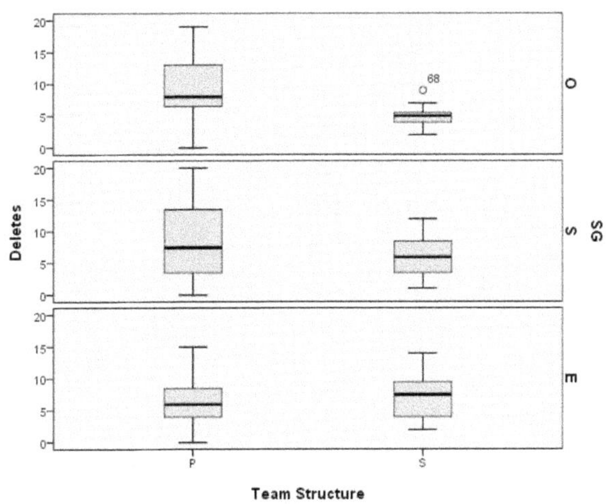

Figure 6: Number of deletes

No significant main effects were found for number of errors, number of victims missed and percentage of area explored across either the automated search guidance or structure independent factors. Moreover, these performance indicators were correlated with each other. Area explored was positively correlated with number of victims found by team (r=0.470, p<0.001). This is because when larger area was explored, there was a larger chance of finding a new victim. The number of victims found was expectedly negatively correlated with number of victims missed (r=-0.514, p<0.001) and number of errors (r=-0.691, p<0.001). In other words, teams that found more victims made fewer errors and missed fewer victims.

4.2 The Effect on Subjective Workload

Subjective workload using NASA-TLX was analyzed using non-parametric tests. Box plots of subjective workload under different conditions are shown in Figure 7. Mann-Whitney tests for the effect of team structure showed a significant effect on workload (p=0.042). Operators in Pool teams demonstrated lower workload on average than those in Sector teams. When analyzing each dimension of workload (mental demand, physical demand, temporal demand, performance, effort and frustration) separately, team structure had a significant effect on effort (p=0.032) and frustration (p=0.005). This was consist with previous study [8], in which a slight advantage in workload was observed favoring the Pool structure. One reason may be that in the sector team conditions, operators felt a lot of pressure because if they missed a victim, no one backed them up. Furthermore, in Pool teams, it is possible to balance the workload according to operators' individual abilities. When one operator was better at finding victims, it is possible he/she could share the burden of the less skilled teammate and did not report excessive workload. In Sector teams, all the work was split equally regardless of ability. We then analyzed subjective workload at the team level by taking the maximum, minimum and average of individual team members' subjective workload. Results show that maximum workload of the team members in Pool teams is significantly lower than in Sector teams (p=0.012), while average workload and minimum workload did not significantly differ. This result, combined with the significantly larger difference on individual performance in terms of number of victims found within Pool teams (Figure 5), suggests a workload balancing process or back-up behavior in Pool teams.

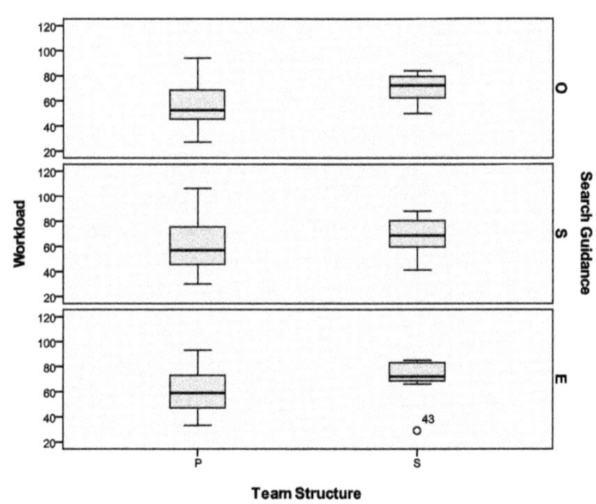

Figure 7: Subjective workload

4.3 The Effect on Operator Measures

The operator measures included the mean duration of teleoperation, frequency and total time of teleoperation, display-to-mark time and select-to-mark time. Total time of teleoperation was analyzed using ANOVA, while all the others were analyzed using nonparametric analysis because normality assumptions were not satisfied.

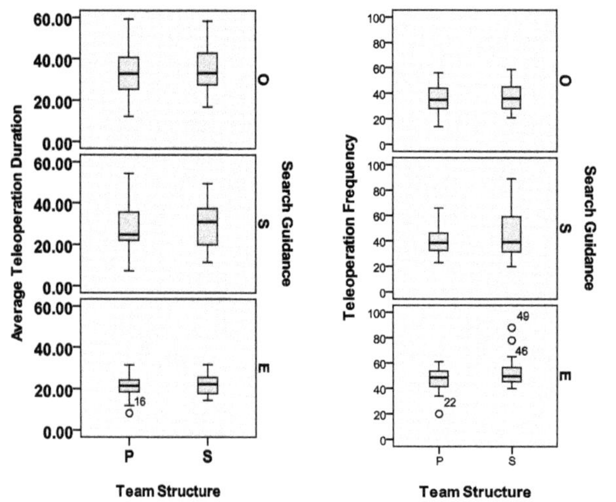

(a) (b)

Figure 8: (a) Average teleoperation duration (seconds) and (b) teleoperation frequency

Search Guidance was found to have a significant effect on the duration ($p<0.001$) and frequency of teleoperation ($p<0.001$). As shown in Figure 8, under the enforced condition, the duration was shorter (Mean=21.7, SD=5.31, $p=0.001$) and the frequency was higher (Mean=50, SD=10.77, $p<0.001$) than under the suggested condition. Under the suggested condition, the duration was shorter (Mean=29.3, SD=11.55, $p=0.025$) and the frequency was higher (Mean=42.19, SD=13.79, $p=0.045$) than without search guidance. No significant effect of team structure was found for duration ($p=0.687$) and frequency of teleoperation ($p=0.184$).

For total time of teleoperation (Figure 9), team structure had a significant effect ($p=0.001$). Sector teams (Mean = 1166.7, SD = 194.20) spent more time on teleoperation than Pool teams (Mean = 1055.4, SD = 236.66) on average. Search guidance had a marginal significant effect ($p=0.078$). Operators expectedly spent more time on teleoperation when there was no search guidance condition, followed by the suggested condition and the enforced condition.

No significant main effect was found for display-to-mark time from either team structure ($p=0.368$) or search guidance ($p=0.309$). However, there was an interaction effect of search guidance and team structure. As shown in Figure 10, search guidance significantly affected mean display-to-mark time in Sector teams ($p=0.024$). The teams in suggested search guidance condition had the lowest mean display-to-mark time (mean=88. 0s, SD=58.9s), followed by no guidance (mean=103.2s, SD=59.1s) and enforced guidance (mean=128.6s, SD=70.8s).

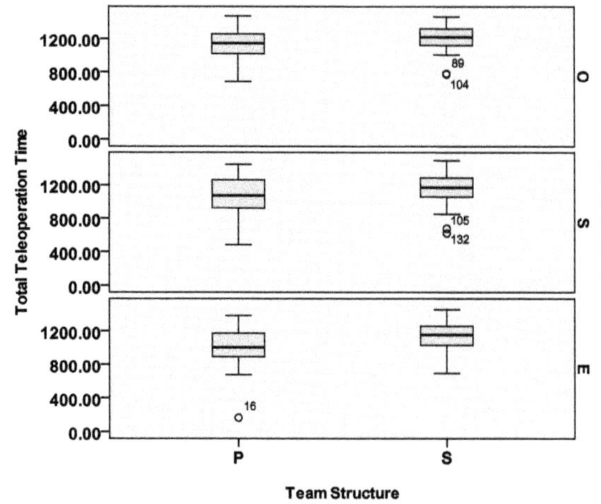

Figure 9: Total teleoperation time (seconds)

This indicates that suggested search guidance helped the operator notice and mark victims faster when they appeared in the camera, which is very important for such a time-critical task environment. The increase in time in enforced guidance condition may be due to the interruption in the current operation and extra time to regain situation awareness. In Pool teams, the effect of search guidance was insignificant, suggesting that display-to-mark time is affected by a more efficient team process, i.e., one team member can start working on a robot with a victim in view when the other is busy.

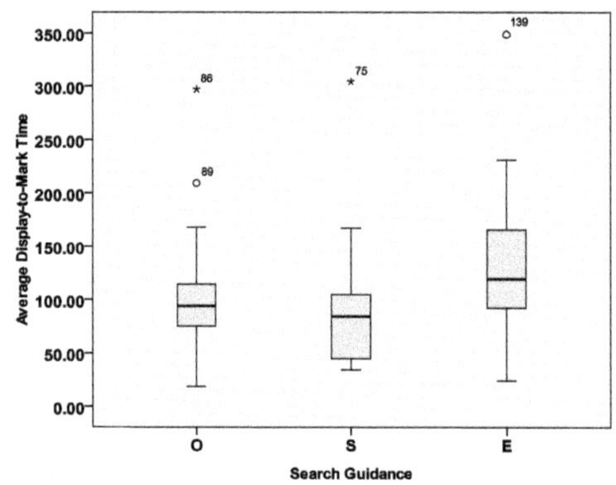

Figure 10: Average display-to-mark time (seconds)

For select-to-mark time (Figure 11), search guidance was found to have a significant effect ($p=0.000$), but no significant effect was found for team structure ($p=0.331$). For search guidance, it shortened the time to finish a task as expected. Operators may speed up their operations when a notification popped up. The change was more meaningful for the difference between Off condition and Suggested condition, because under enforced condition, this change may result from reselection of the same robot when the system switched to another robot automatically. For team structure, it did not have a significant effect on the time operators spent to serve a single robot. In other words, it didn't affect the task level operations.

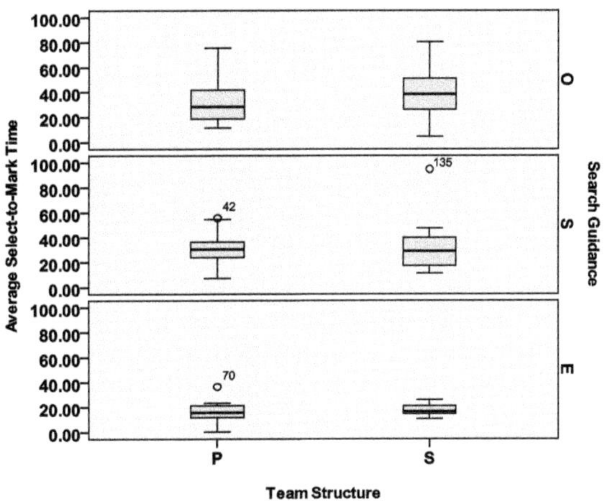

Figure 11: Average select-to-mark time (seconds)

Since teleoperation played an important role in operators' behavior, we analyzed the correlation between total teleoperation time of each operator, subjective workload, the number of victims found by each individual operator and team performance. Total teleoperation time was not significantly correlated with the overall subjective NASA TLX workload (r=-0.004, p=0.965). However, it was significantly moderately correlated with the dimension of mental demand (r=0.372, p=0.001), so not surprisingly, using teleoperation was difficult.

There was a significant but weak correlation between total teleoperation time and number of victims marked by each operator (r=0.189, p=0.023). The total teleoperation time for the team of two operators was combined and analyzed with relation to team performance. It was found that total teleoperation time more strongly positively correlated with percentage of area explored (r= 0.431, p=0.000). Since autonomous path planning was used, the percentage of area explored largely depended on how much the operators changed the behavior of robots by manually redirecting them. In many situations, operators were not satisfied with the performance of autonomous path planning, so they controlled the robots using teleoperation to send them to the places they wanted the robots to go.

Lastly, total teleoperation time moderately correlated negatively with number of deletes (r=-0.331, p=0.005). With more teleoperation, operators could position the robots nearer to victims, which resulted in an increase in accuracy, thus a decrease in correcting behavior. No correlation was found between teleoperation time and number of victims found, number of errors or number of victims missed.

4.4 The Effect on Communication

During the experiment, participants were allowed to talk with each other. In such a high workload scenario, almost all the communication was mission-related. Teams who communicated discussed what strategies to use when exploring the area, updated his or her status with the teammate, requested his or her teammate's status or shared experiences with controlling the robots. In contrast, some teams did not communicate at all. A nonparametric analysis of the time spent on communication shows that team structure had a significant effect on communication time (p=0.004). As shown in Figure 12, Pool teams (Mean = 177.7, SD = 198.74) expectedly communicated more than Sector teams (Mean = 53.44, SD = 80.97), on average. Search guidance did not have a significant effect on communication time (p=0.865). Session order was found to have a marginally significant effect (p=0.098).

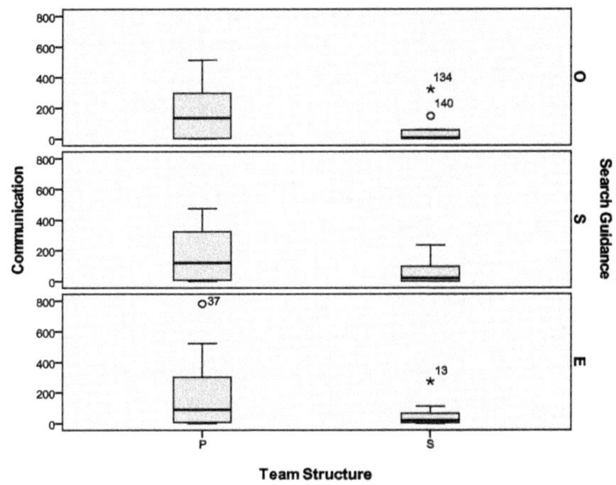

Figure 12: Communication time (seconds)

Further analyses on the correlation between communication with team performance and subjective workload revealed that communication time was moderately negatively correlated with errors (r=-0.309, p=0.008). In other words, teams that communicated more tended to make fewer errors. This negative correlation between communication time and number of errors exists for Pool teams but not for Sector teams. This result, combined with the larger number of deletes in Pool teams, suggests that mutual performance monitoring exists such that team members correct themselves and their teammates, facilitated by communication. Communication time was marginally significantly correlated with number of victims marked by teams (r=0.202, p=0.088). No significant correlation was found between communication time and number of deletes, number of victims missed, or subjective workload ratings.

5. DISCUSSION AND CONCLUSION

Although automated search guidance did not improve performance, it did not decrease it either. For teams under the Pool condition, it appeared beneficial to have an enforced search guidance to keep them from fixating on a single robot. In Sector teams, suggested search guidance helped operators mark victims faster when they appeared in the cameras. Furthermore, adding automated search guidance to the system changed operators' behavior in terms of teleoperation and select-to-mark time. By putting a threshold on time, search guidance shortened the episode of operation on teleoperation and select-to-mark. This changed behavior may be because operators hastened their operations when there was a notification, and partly due to reselection under Enforced condition. So in this experiment, spending less time on tasks did not significantly affect task performance.

When answering open-ended questions at the conclusion of the experiment, some participates said it was good to have a notification but the enforced change was annoying. In addition, although all the robots were the same, their importance was different according to their position in the environment. Robots near an unexplored area were more valuable than those in a thoroughly searched area. In this case, improving the automated search guidance by including location-related information and encouraging the operator to investigate unexplored area could increase the overall benefit of such a decision support tool.

Pool team structure was shown to be better than Sector structure in terms of experiencing lower workload but similar task performance. Although Pool teams corrected themselves more often, they had lower subjective workload. Pool teams also communicated more while Sector teams teleoperated more. Pool teams appeared to better balance workload among team members according to abilities. This conclusion was supported by the lower maximum subjective workload and larger difference on individual performance in Pool teams. This suggests the reduced subjective workload in the Pool condition was because teammates could provide back up if needed.

Teams with different structures did not show a significant difference in their task performance, except for number of deletes. The reason may be that they used different strategies to cope with the situation. Sector teams communicated less and teleoperated more, with the opposite generally true for Pool teams. One explanation is that in the Sector condition, operators had more time to spend on teleoperation since communication was not explicitly needed. The increased time spending on teleoperation extended the area they explored and increased the chances of finding new victims. On the other hand, the shared control of robots promoted communication in teams under the Pool structure, which was also good for task performance since teams with more communication tend to make fewer errors. The reason may be that they corrected each other via communication, which led to fewer errors.

Overall in this study, operators tended to be robust to whatever structure they were assigned. Moreover, improvements could be made to the automated search guidance tool as the threshold for suggesting or enforcing a change was rudimentary. With a more context-sensitive tool, possibly based on an individual operator's strengths or limitations, it is possible that greater performance improvements could have been seen. In addition, an experimental confound that necessarily exists is the fixed number of victims in the environment, which bounds the performance on number of victims found. This also means the chance of finding a new victim decreases over time. The advantage of a team that is faster in marking victims over slower teams may not be shown on ultimate team performance. From this point, process measures are important and should be further investigated. We leave these issues as areas of future research.

6. ACKNOWLEDGMENTS

This research is sponsored by the Office for Naval Research and the Air Force Office of Scientific Research. We would also like to acknowledge MIT undergraduates Matthew R Redmond, Radhka Malik for their programming contributions, Hosea Siu, Victoria Palay for their help with conducting the experiment. Thanks to Professor Mike Lewis and his students at University of Pittsburg for providing the testbed.

7. REFERENCES

[1] Bertuccelli, L. F., Pellegrino, N. et al. 2010. Choice modeling of relook tasks for UAV search missions. *American Control Conference (ACC)*, 2010.

[2] Crandall, J.W. and Cummings, M.L. 2007. Attention Allocation Efficiency in Human-UV Teams AIAA Infotech@Aerospace Conference, Rohnert Park, CA, 2007.

[3] Crandall, J.W., Goodrich, M.A., Olsen, D.R., Jr. and Nielsen, C.W. 2005. Validating human-robot interaction schemes in multitasking environments. *Systems, Man and Cybernetics, Part A: Systems and Humans, IEEE Transactions on, 35* (4). 438-449.

[4] Farinelli, A., Scerri, P. and Tambe, M. 2003. Building large-scale robot systems: Distributed role assignment in dynamic, uncertain domains. iAAMAS'03 Workshop on Resources, role and task allocation in multiagent systems.

[5] Gladstein, D. L. 1984. Groups in Context: A Model of Task Group Effectiveness. *Administrative Science Quarterly* 29(4): 499-517.

[6] Lewis, M., Polvichai, J. et al. 2006. Scaling-up Human Control for Large UAV Teams. *The Human Factors of Remotely Piloted Vehicles*. N. Cooke, Ed. New York, Elsevier: 237-250.

[7] Hart, S.G. and Staveland, L.E. 1998. Development of NASA-TLX (Task Load Index): Results of empirical and theoretical research. in Hancock, P.A. and Meshkati, N. eds. *Human Mental Workload*, North Holland Press, Amsterdam.

[8] Lewis, M., Wang, H. et al. 2011. Process and Performance in Human-Robot Teams. *Journal of Cognitive Engineering and Decision Making* 5(2): 186-208.

[9] Lewis, M., Wang, H., Chien, S.Y., Velagapudi, P., Scerri, P. and Sycara, K. 2010. Choosing Autonomy Modes for Multirobot Search. *Human Factors: The Journal of the Human Factors and Ergonomics Society, 52* (2). 225-233.

[10] Lewis, M., Wang, J. and Hughes, S. 2007. USARSim: Simulation for the Study of Human-Robot Interaction. *Journal of Cognitive Engineering and Decision Making, 1* (1). 98-120.

[11] Macmillan, J., Entin, E. E. et al. 2004. Communication Overhead: The Hidden Cost of Team Cognition. *Team Cognition: Understanding the Factors That Drive Process and Performance*. E. Salas and S. M. Fiore, Ed., American Psychological Association (APA).

[12] Mouloua, M. and Parasuraman, R.1996. *Automation and Human Performance: Theory and Applications*, CRC Press

[13] Naylor, J. C. and Dickinson, T. L. 1969. Task structure, work structure, and team performance. *Journal of Applied Psychology* 53(10).

[14] Salas, E. and Fiore, S. M. 2004. *Team Cognition: Understanding the Factors That Drive Process and Performance*, American Psychological Association (APA)

[15] Salas, E., Sims D. E., et al. 2005. Is there a Big Five in Teamwork? *Small Group Research* 36(5): 555-599.

[16] Wickens, C. D. and Hollands, J. G. 1999. *Engineering Psychology and Human Performance*, Pearson Education.

Towards Human Control of Robot Swarms

Andreas Kolling
University of Pittsburgh
School of Information Sciences
Pittsburgh, USA
akolling@pitt.edu

Steven Nunnally
University of Pittsburgh
School of Information Sciences
Pittsburgh, USA
smn34@pitt.edu

Mike Lewis
University of Pittsburgh
School of Information Sciences
Pittsburgh, USA
ml@sis.pitt.edu

ABSTRACT

In this paper we investigate principles of swarm control that enable a human operator to exert influence on and control large swarms of robots. We present two principles, coined *selection* and *beacon* control, that differ with respect to their temporal and spatial persistence. The former requires active selection of groups of robots while the latter exerts a passive influence on nearby robots. Both principles are implemented in a testbed in which operators exert influence on a robot swarm by switching between a set of behaviors ranging from trivial behaviors up to distributed autonomous algorithms. Performance is tested in a series of complex foraging tasks in environments with different obstacles ranging from open to cluttered and structured. The robotic swarm has only local communication and sensing capabilities with the number of robots ranging from 50 to 200. Experiments with human operators utilizing either *selection* or *beacon* control are compared with each other and to a simple autonomous swarm with regard to performance, adaptation to complex environments, and scalability to larger swarms. Our results show superior performance of autonomous swarms in open environments, of selection control in complex environments, and indicate a potential for scaling beacon control to larger swarms.

Categories and Subject Descriptors

I.2.9 [**Artificial Intelligence**]: Robotics - operator interfaces

General Terms

Human Factors, Measurement, Experimentation

Keywords

Human-robot interaction, metrics, evaluation, multi-robot system

1. INTRODUCTION

In recent decades the number of mobile robots deployed in field applications has risen dramatically. Their usage offers obvious advantages of reduced costs, removing humans from harms way, or enabling entirely new applications that were previous impossible, especially when combining many such robots into a comprehensive system. The tasks and missions already carried out by large robot teams range from search, exploration, rescue, surveillance, pursuit, up to deploying infrastructure. The domains of application are equally diverse and range from low-cost warehouse security to interplanetary exploration. New developments in commodity hardware which serve as low cost replacements for otherwise expensive sensing or motion capabilities promise to further accelerate the trend towards deploying large teams of mobile robots. This trend, however, poses a challenge for the control of such systems. Currently, most large robotic systems are controlled by multiple operators, often remote controlling the robotic devices. For larger systems with more robots and low cost hardware such a control approach is not practical. While autonomy is already playing a vital role, even for powerful systems, in the form of tools, such as mappers, path planners, monitoring and detection systems, it will also play an increasingly important role for the control of robotic systems with a very large number of robots, so called swarms. An increased usage of autonomous methods, however, poses a challenge to allow human operators to exert control on such robot systems. Enabling human operators to control robot swarms with hundreds of robots is still an open problem. Currently, multi-robot approaches generally scale to at most ten's of robots per operator even when using state of the art mapping, path planning, target detection, and coordination algorithms to alleviate the load on the operator.

Much of the recent work on swarms is focused on algorithms for autonomous swarms solving specific problems. The specificity in terms of the working assumptions render one of the key advantages of a robot swarm moot, namely its supposed flexibility and wide range of potential applications. It is envisioned that swarms will be capable of assisting in a number of complex problems and the space of possibilities is vast. For this reason, enabling the control of swarms by a human operator who could interact with the autonomy and adapt to specific challenges in a variety of conditions is crucial. The goal of this paper is to investigate principles of control for large swarms and to determine how humans perform in controlling swarms uisng implementations of these principles for a complex foraging task in a variety of challenging environments. Foraging tasks are a useful formal model of many practical applications such as search and rescue or surveillance. To solve the foraging task we provide the operator with methods developed for the control of autonomous swarms that enable basic capabilities such as optimal deployment and connectivity maintenance, but we are placing these algorithms into a new, more complex context, in which former guarantees of performance do not hold anymore. The operator's task is then to utilize these methods, mitigate their limitations, and perform a set of complex mission in environments with different types of obstacles. In the following we will shortly present related work in Section 2 including some details on the algorithms used in this paper. This is followed by Section 3 which discusses the two principles of control, selection and beacon control. Section 4 presents the missions and simulation platform used for the experiments. Finally, we close with a presentation and discussion of our results in Section 5 and a conclusion in Section 6.

2. RELATED WORK

The literature on human control of swarms is rather sparse and we present the few works that are related to ours. On the other hand, the literature on autonomous swarm behavior is too vast to be reviewed here and we shall focus only on a very small subset relating to the algorithms from control theoretic approaches that we utilize in our testbed.

One of the earlier contributions on human control of large robot systems is made in [5] in which some of the main challenges for supervisory control of swarms are discussed. The author calls for further research to address the development of new swarming technology and the lack of understanding of supervisory control of such systems, particularly with respect to the interaction of autonomy with operators. One of the first to consider practical challenges in controlling swarms of robots is [11]. Therein experiences with a swarm of 112 robots are presented. The focus is kept on hardware and the development of software for the swarm. A similar approach is taken in [10]. The presented software tool ROBOTRAK addresses hardware and software problems with regard to the control of robot swarms, especially programmability. Centered around another practical application [1] considers swarm control in the context of searching for a radiation source. It proposes yet another architecture based on a set of desirable features and presents a simple test system in which operators successfully aid a swarm in locating radiation sources. Also focused on a particular task, in [7] a team of Unmanned Aerial Vehicles (UAVs) is controlled by a single operator using behavior-based controls. These controls enable UAVs to perform a surveillance mission semi-autonomously while the operator generates a mission plan. Similarly as in [6] much of the direct control is based on a leader-following approach and the operator can choose to teleoperate individual UAVs. Much of the above work is related to practical obstacles from a robotics perspective in the design, programming, and deployment of swarms rather than the direct controllability of a swarm by an operator and no comprehensive user studies have yet been attempted.

Another and slightly more general approach is taken in [9]. Therein the authors use so called *physicomimetics*, i.e. the simulation of physical forces, to control a swarm. Two basic forms of control are distinguished. The first is a top-down control for which an operator sets global swarm parameters and the system learns to adjust the parameters of individual robots to achieve the desired global parameter. The second is a bottom-up approach in which virtual agents are used to modify the behavior of existing agents through interaction instead of directly setting their parameters. For both approaches a learning method is proposed to either set the parameters or determine the placement of virtual agents. The application considered is the defense of a resource against an attacker. The physicomimetic approach promises to be an intuitive control paradigm due to the force metaphors borrowed from physics with which operators should be familiar. Since no comprehensive user studies have been attempted this claim remains to be validated.

While the above considerations are worthwhile our focus is rather on principles of control that enable the executing of a variety of missions with the operator injecting mission-specific knowledge into the system. Especially the integration of simple as well as complex robot behaviors, which may be autonomous algorithms, into a system controlled by a human operator is a major goal. For this purpose we now shortly introduce some related work on the autonomous control of robotic networks. In particular, we are considering the problems of rendezvous, coverage and deployment, and connectivity maintenance. An algorithm for the rendezvous problem for distributed robotic networks was introduced in [3]. It assumes open and uncluttered environments and that every robot can obtain the location of its neighbors. An algorithm for optimally distributing a network of robots in an open environment is given in [4]. An additional algorithm that deploys robots in environments that are non-convex and simply-connected is presented in [8], although therein the goal is to simply cover the entire space while in [4] an optimal cover given a sensor deprecation with respect to distance is computed. A rigorous formalization and unifying framework of much of the above is presented in [2]. While this work is rigorous in terms of the theory the working assumptions are rather strict and often violated in practical application. Yet, these algorithms make ideal candidates for our robot behaviors and we are going to utilize the connectivity maintenance, rendezvous, and deployment algorithms. Connectivity maintenance allows us to guarantee that all robots form a connected communication network at all times and computes a set of admissible control inputs that satisfy this constraint. This is crucial for swarms with only local communication. Different communication networks can be chosen which lead to different communication constraints. Rendezvous algorithms guarantee that a set of robots can find agreement over a location at which they get together which could provide a useful functionality for users in the presence of obstacles. The more complex algorithm is the optimal deployment which relies on the robots to compute a Voronoi Diagram and move towards their respective centroid. Using this algorithm a user can achieve optimal deployment in open spaces without manually dispersing robots.

3. APPROACH

In this paper we explore two basic approaches for controlling the robot swarm. We shall call these *selection* and *beacon* control. Both are based on setting the modes of robots in order to control their behavior. The set of modes available to the operator can be customized and the modes for our experiments are described in Section 4. Modes can range from trivial behaviors, such as stop, to complex autonomous behaviors, such as optimal deployment. The first control approach, selection control, allows the operator to select a subset of robots by drawing a rectangular marquee. This is similar to most computer programs and games in which operators can select units in this manner and provide custom instructions to the selected units or area. Most computer users are familiar with such selection tools. Once a subset is selected, the robots can be instructed to switch to one of the available modes which they will continue to execute. A selection is persistent until a new selection is made. The second control approach, beacon control, is based on a different paradigm and inspired by prior work on biological and autonomous swarms. Here an operator modifies the environment for the swarm by placing a beacon that influences nearby robots. A beacon has a location, a range and an associated mode. All robots within its range switch to the associated mode. In contrast to selection control, which requires an active selection, robots are passively influenced by the closest beacons once they get into its range. Beacons are persistent until they are removed by an operator and multiple beacons can be present in an environment. Swarm phenomena such as leader or predator models can be simulated by placing beacons that attract or repel nearby robots.

Both control approaches can in theory enable an operator to exert very precise control, even over individual robots, by using very small and frequent selections or by placing many small range beacons. The key difference between selection and beacon control is their temporal and spatial persistence and the resulting active or passive selection of robots. More precisely, a beacon is spatially persistent since it influences robots within a given range but it is not temporally persistent since robots can enter or exit a beacon's range as time passes. In contrast, selection control influences the same robots regardless of how much time has passed since selecting a particular group. Yet, the location of this group may have changed in the meantime and hence selection control is not spatially persistent. As a consequence both control control approaches enable very different strategies with different degrees of effort and complexity. Consider a scenario in which robots have to perform a sequence of tasks when at a certain location, such as moving on a safe path around an obstacle. By using beacons any robot entering at any time will perform this sequence of tasks. Selection control, on the other hand, requires the operator to select robots once they reach the location to give them the appropriate instructions. But if only one group of robots has to execute a sequence of task at a location, then selection control is expected to allow better control since all robots remain under control of the operator despite their continuous movement.

In addition, there is also a very practical difference between selection and beacon control. Selection control requires knowledge about the location of the robots to associate them to an area designated for selection. Beacon control, on the other hand, can be implemented by broadcasting a targeted signal within an area and thereby influencing all robots within that area without knowledge of their location. Additional variations and extensions of two basic control approaches are readily envisioned. One could consider selections or beacons that influence a subset of robots or other spatial relationships, e.g. selection by heading instead of location. In all these cases the key differences between selection and beacon control in terms of persistence and passive vs. active selection remain. But there are also entirely different methods to exert influence on a swarm of robots that are not considered here. One other such method is the setting global parameters that influence the otherwise autonomous behavior of an entire swarm, arguably the most indirect form of influence. The properties of such methods will depend even more heavily on type of autonomous behavior and swarm hardware. The primary purpose of this paper is to investigate the above control paradigms, but we also compare the performance of human operators with a baseline given by a simple autonomous swarm that solves the same task as the operators. This part is described in further detail in the next section.

4. METHODS

In order to explore how operators might control large swarms using the control principles discussed in the previous sections we developed a simple testbed in NetLogo [12]. NetLogo is a simulation platform suitable for modeling interactions between large sets of agents. The system was implemented using NetLogo's built-in language as well as Java extensions for computing Voronoi Diagrams and Delaunay graphs. Five different environments as shown in Fig. 1 were used. The size is set to 400 by 400 patches. Each patch in the user display is two pixels wide and high. The user interface for control is shown in Fig. 2.

Robots are placed randomly in the top right corner of each environment. In order for the operator to see the location of a robot and send it instructions it has to be connected to the communication network rooted at a base station. The base station is also placed in the top right corner. In addition to obstacles the communication links of a robot also constrain its motion. To maintain connectivity only motion that does not break an existing communication link is permitted. Robots can communicate with each other when within communication range r_c and when no obstacles are blocking their line-of-sight. The operator can choose whether robots have to maintain all communication links or a subset of these by choosing one of the following communication graphs: r_c-disk graph, r_c-limited Delaunay graph, r_c-limited Gabriel graph, or minimum spanning tree (see [2] for formal definitions). The r_c-disk graph is a graph given by embedding all robots into the environment and connecting all that are within line-of-sight and range r_c. All other graphs are subgraphs of the r_c-disk graph. More precisely, the r_c-limited Delaunay graph is the intersection between the Delaunay graph of all robots and the r_c-disk graph. Similarly, the r_c-limited Gabriel graph is the Gabriel graph of all robots intersected with the r_c-disk graph. The minimum spanning tree is computed from the r_c-disk graph by considering Euclidean distance as weights.

The minimum spanning tree imposes the fewest constraints on motion since it contains the fewest possible communication links while the disk graph contains all possible links. Robots that cannot communicate with the base station are invisible to the operator and wander around the environment choosing a new random direction upon collision. They are, however, still subject to communication constraints due to their local communication links to other robots and they form a separate local connected network. The base station imposes an additional motion constraint on its closest robot to ensure that at least one robot stay within its range. In our missions all robots are initially in a connected network and are expected to remain connected. In more realistic set-

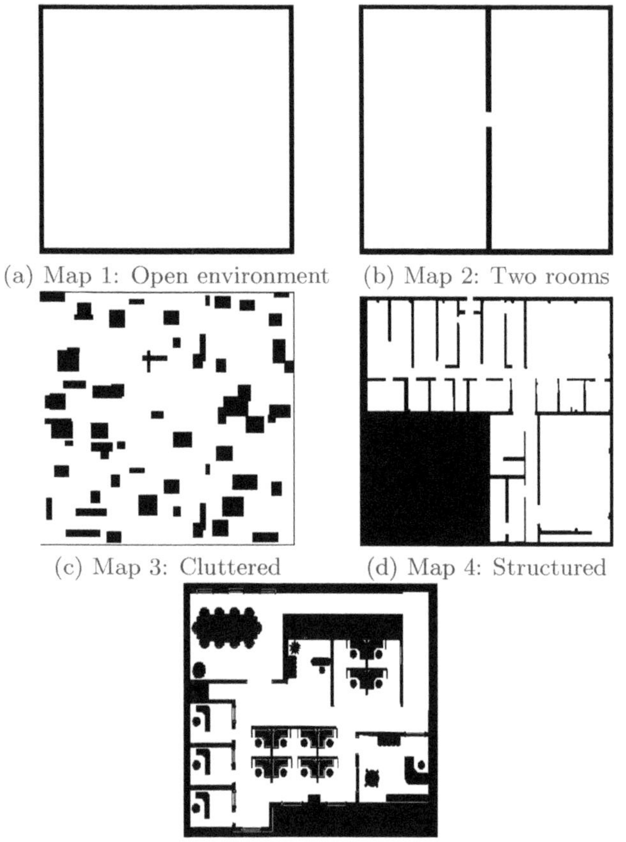

(a) Map 1: Open environment (b) Map 2: Two rooms

(c) Map 3: Cluttered (d) Map 4: Structured

(e) Map 5: Cluttered and structured

Figure 1: Five test environments (a), (b), (c), (d), and (e). Obstacles are black and free space is white.

tings than our simulations, however, noise and other factors can still lead to communication loss.

For maps 1 to 4 the map is known by the operator and drawn in the user interface. For map 5 no map is given in advance and the operator has to explore the environment. To facilitate this a trail for each robot is drawn. Robots can be instructed to be in one of the following modes:

1. Stop: robots stop at their current position;
2. Come: robots move towards a target location;
3. Rendezvous: robots execute the rendezvous algorithm;
4. Deploy: robots execute the deployment algorithm;
5. Random: robots move with a new random heading after every collision;
6. Heading: robots synchronize their heading and move into the same direction;
7. Leave: robots move away from a target location

Robots in the above modes receive the following colors respectively: grey, blue, yellow, green, turquoise, purple and pink. All robots start in the random mode. In all modes, except deploy and random, robots slide along obstacles when these obstruct the desired direction of motion. Robots collide with each other when within a distance of two patches

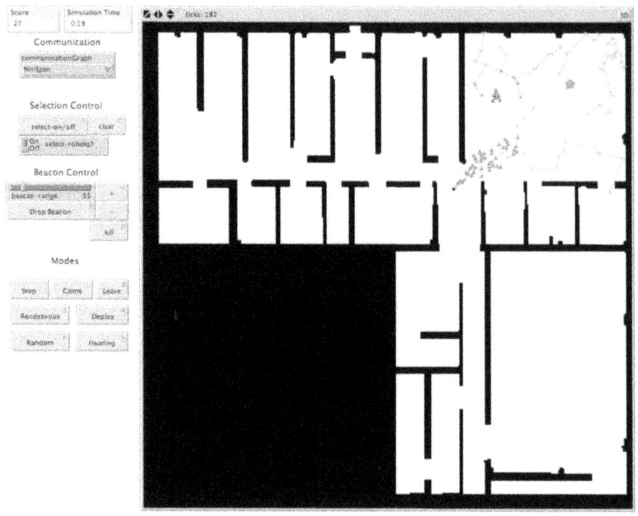

Figure 2: The simple user interface used for the experiments. Robots are small arrows, communication links are grey edges, the basestation is brown, and persons with information are red with their information value displayed in black. Robots currently within information range r_i of a person receive a red "*". Beacon and selection control only see their respective control panel.

in the environment. All robots move with the same maximum speed of 5 patches per second. For the beacon control the operator can place, move, set the mode of, change the range of, and remove beacons. The heading mode requires an additional mouse click to determine the heading. For the selection control the operator can select a rectangular set of robots, clear the current selection, and set the mode of all robots in the current selection. The come, leave and heading modes require an additional click to determine the target location or direction. Operators can generate complex behaviors from these modes. For example, behaviors similar to leader-follower behaviors can be approximated by using rendezvous and come modes.

4.1 Missions

Based on the five maps from Fig. 1 we presented the operator with one training scenario and seven missions. The task is a classical foraging task framed as information collection. Participants are instructed that their robots have to collect information from persons that appear randomly throughout the map. Each person has a different amount of information to be collected and each robot close enough to a person collects at a standard rate. The amount of information a person has is displayed in the user interface. Once all information is collected from a person it disappears. Persons that appeared with no robots within sensing range r_s are not visible on the display until a robot comes within sensing range. A new person appears with probability $\frac{1}{4}$ at a random location and with an information value sampled uniformly from $[1, 50] \subset \mathbb{N}$. On average 6.25 information units appear per second. The information of persons not within information range r_i of robots decays at a rate of 0.5 units per second. Robots within information range r_i of a person collect the information from it at the rate c (between 0.1 and 0.4 units per second). Information collected by robots is added to the operator's score. There are four different robot team configurations as seen in Table 1 that determine the capabilities of individual robots. The training scenario takes place in map 3 with robot configuration 2 and lasts 25 minutes. It precedes the seven missions each lasting five minutes. On average 1,875 information units spawn for every mission. In order to collect all this information and reach the maximum possible score robots need to cover the entire environment to quickly find every new person and exactly $\frac{5}{16}$ of all robots need to collect information at all times. Team configurations are chosen so that swarms with different numbers of robots retain a similar overall capability.

Robot Configuration	Robots	r_c	r_s	r_i	c
1	50	60	60	30	0.4
2	100	40	40	20	0.2
3	150	30	30	15	0.1333
4	200	25	25	15	0.1

Table 1: The settings for each robot configuration.

Table 2 shows the map and team configuration for each mission. Participants in missions 3 to 6 are controlling a variable number of robots from 50 to 200. The assignment of robot configurations in these missions is balanced across both maps (3 and 4) and w.r.t. whether a participant first has a larger number of robots in a map. All other missions have the standard configuration 2 with 100 robots.

Mission	Map	Robot Configuration
1	1	2
2	2	2
3	3	1|3|2|4
4	3	3|1|4|2
5	4	2|4|1|3
6	4	4|2|3|1
7	5	2

Table 2: The configuration and map used for every mission. Missions 3 to 6 have four possible and counter-balanced sequences of configurations that participants are assigned to.

Finally, we also used a simple autonomous swarm to solve the foraging task. The autonomous swarm leaves all robots in the random mode unless they are currently within information range r_i of a person in which case they stop. The chosen communication graph is the minimum spanning tree and robots are subject to the same communication constraints on motion as for human operators. Note that emulating this kind of autonomy is already difficult for an operator due to the large number of persons that spawn within five minutes, i.e. 75 persons on average. For beacon control an operator would have to place a beacon at every target they see and set the range to r_i and the mode to stop, then set the beacon to random once the information is collected and remove the beacon. This leads to an average total of 225 actions. For selection control all robots that get into information range of a target would need to be selected immediately and then stopped. This leads to a potentially even very larger number of actions since robots can enter the range of a person at different times. In this case the number of actions also increases with the number of robots rather than the number of persons.

To collect data we recruited 32 participants from the campus of the University of Pittsburgh, most of which were graduate and undergraduate students. The autonomous swarm ran through all missions eight times. We also tested the system with two experienced operators that contributed to the design of the interface to obtain a baseline for the scores human operators could achieve. The results of these experiments are presented in the next section.

5. RESULTS AND DISCUSSION

In this section we present and discuss the results of our experiments. In particular we address the following questions:

- Do selection, beacon and autonomous control perform differently?
- What impact do more complex environments have on performance?
- How do participants make use of the available modes?
- How do the control methods scale to larger swarms?

The average scores for participants and autonomous swarms for all maps and missions in robot configuration 2 (100 robots) is shown in Fig. 3[1]. A two-way analysis of variance of the score across maps and control conditions revealed an interaction between maps and control condition ($p < 0.001^{***}$), an effect of control condition ($p < 0.001^{***}$) and an effect of maps ($p < 0.001^{***}$). Comparing only beacon and selection control no such interaction is present ($p = 0.7944$) but the effect of control condition and maps remain. In Fig. 3 this becomes apparent when looking at the steep drop of the autonomous swarm from map 1 to maps with obstacles. Excluding map 1 leads to no significant interaction between all control conditions and maps 2 to 5 ($p = 0.3907$) as well as no significant effect of maps ($p = 0.5761$). The effect of the control condition remains significant ($p < 0.001^{***}$). On maps with obstacles the average scores for control conditions selection, beacon, and autonomous are 719, 590, and 695 respectively. Here selection control performs best. On all maps these averages become 779, 627, and 847 and autonomous swarms perform best overall due to the high scores in map 1. This suggests that human operator are generally poor at solving foraging tasks with swarms, not beating the simplest form of autonomy, but can adapt to complex environments. Table 3 shows results from running the experiment with two experienced operators that contributed to programming the system. These provide a rough indicator for the scores that are achievable by human operators with some experience. Note that despite the added experience the high performance of autonomous swarm is difficult to replicate with beacon control. The experienced operator with selection control achieves scores close to the autonomy in map 1 and can also maintain high scores in environments with obstacles (see Table 3 missions 3 to 7).

[1]The scores reported are the actual scores participants achieved. Normalizing the scores to show the collected fraction of all information that spawned leads to the same results, i.e. the influence of randomness in spawning the information is marginal and only has an impact on very small samples such as the experienced operators.

Mission	1	2	3	4	5	6	7
Expert (S)	1317	1075	1075	888	1140	991	822
-normalized	72%	64%	53%	39%	60%	48%	50%
Expert (B)	1078	884	519	903	664	654	556
-normalized	55%	44%	35%	45%	38%	40%	30%

Table 3: Scores from two experienced operators using selection (S) and beacon control (B). Missions 3,4,5 and 5 were tested with 50, 150, 100, and 200 robots respectively. Normalized scores indicate the percentage of points collected of the actually spawned information and allows for a better comparison of this small sample.

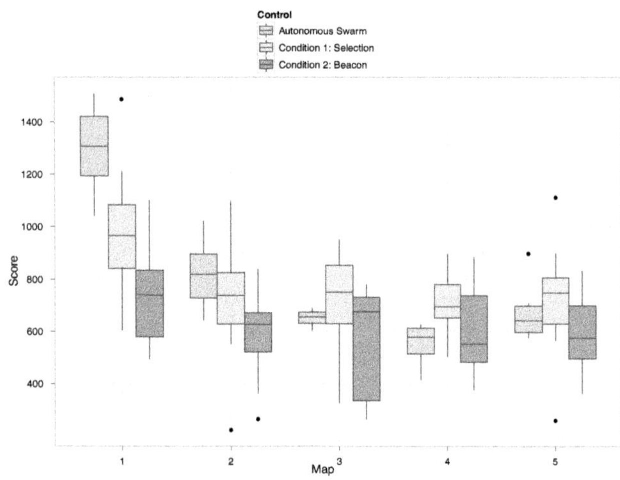

Figure 3: A box plot of the score across participants for maps 1 to 5. Note that these scores only include missions with 100 robots and participants have either only map 3 or only map 4 with 100 robots, i.e. the sample size is reduced for map 3 and 4.

Participants using different control methods also utilized different robot modes with different frequency as seen in Fig. 4. An explanation for this difference can be found in the impact the operator mode instructions have on the team. A mode instruction here is either a switch of mode for a selected set of robots or a beacon, affecting all nearby robots. Let us call the mean number of robots influenced by a mode instruction *mode impact*. For selection control mode impact relates to the size of the selections and for beacon control it relates to the number of robots influenced by a beacon. Selection and beacon control differ significantly w.r.t. mode impact as seen in Table 4 with a mode impact of 10 robots per instruction for selection control and 30 robots for beacon control. The number of instructions given differs as well with 56 for selection control and 39 for beacon control. This leads to an overall higher number of robot mode switches for beacon control at 1027 robot mode switches per mission due to the higher mode impact. Some of these switches can be attributed to robots in the random mode getting close to a beacon, explaining some of the difference in Fig. 4. For beacon control the large mode impact has a statistically significant ($p < 0.001^{***}$) negative impact on score while for selection control mode impact does not have any effect.

The correlation between score and mode impact for beacon control is -0.45. Conversely, the number of instructions has a marginally significant ($p = 0.0897$) positive impact on score for selection control and no impact for beacon control. The correlation between the number of instructions and score is 0.414 for selection control. This is an indication that increased activity of the user, i.e. more mode instructions, helps more for selection control and that in beacon control many of the induced robot mode switches actually impede performance. On another note, operators seem not to exploit the rendezvous algorithm and rather adapt to the presence of obstacles manually and achieve rendezvous with the come mode.

	Selection	Beacon
Score***	779	627
Operator mode switches***	56	39
Robot mode switches***	428	1027
Mode impact***	10	30

Table 4: A comparison of selection and beacon control across all missions with 100 robots.

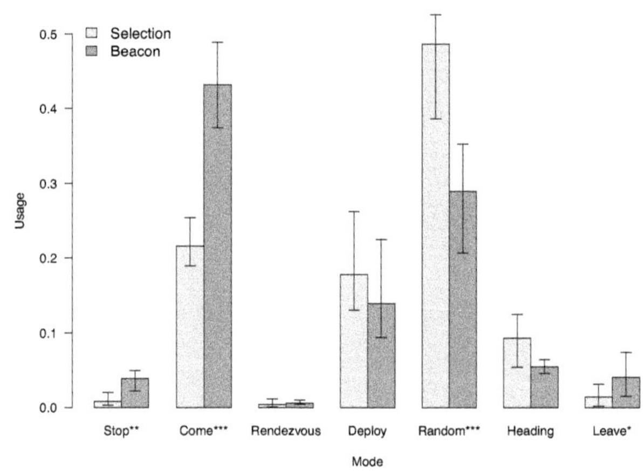

Figure 4: The figure shows the usage of each mode as the proportion of robots in that mode across the entire mission. The mean usage across all mission is shown. Arrows indicate the smallest and largest mean usage across all mission.

The above results all refer to missions with 100 robots. In the following we shall investigate difference with regard to changing robot configurations. These are available for maps 3 and 4. The main questions here relate to scalability, i.e. the ability to control larger teams of robots of similar overall capability but with individual robots being far less capable. The scores across different robot swarm sizes are shown in Fig. 5. A multiple analysis of variance of score across control condition (autonomous, selection, beacon), maps (3,4), and number of robots (50, 100, 150, and 200) revealed a significant impact of the number of robots ($p < 0.001^{***}$) but no significant interactions nor effects of control condition.

One would expect that the autonomous swarm is not affected by increasing the number of robots if the overall capabilities were similar. In fact, for the autonomous swarm

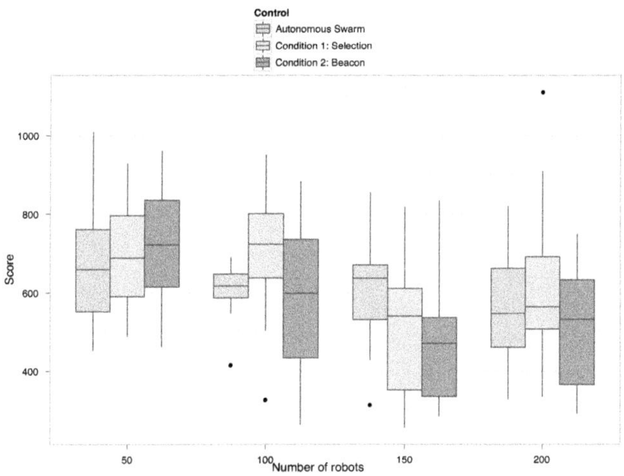

Figure 5: Score of maps 3 and 4 across different robot configurations with 50,100,150, to 200 robots across control conditions.

there is no significant difference across the number of robots for the score ($p = 0.5457$) confirming the similar capabilities. For selection and beacon control we do, however, have a significant effect of the number of robots ($p < 0.01^{**}$ and $p < 0.001^{***}$ respectively). A simple linear regression for selection and for beacon control gives $b = -0.8144$ ($t(62) = -2.073$, $p < 0.05^{*}$) with an intercept of $a = 733.6875$ for selection control and $b = -1.4689$ ($t(62) = -3.894$, $p < 0.001^{***}$) with an intercept of a=759.4062 for beacon control, both shown in Fig. 6. Both show a downward trend in score with an increasing number of robots, but less so for selection control. Considering the relatively stable performance of the autonomous swarm we can conclude that the increased difficulty of instructing the robots in a larger swarm impede performance.

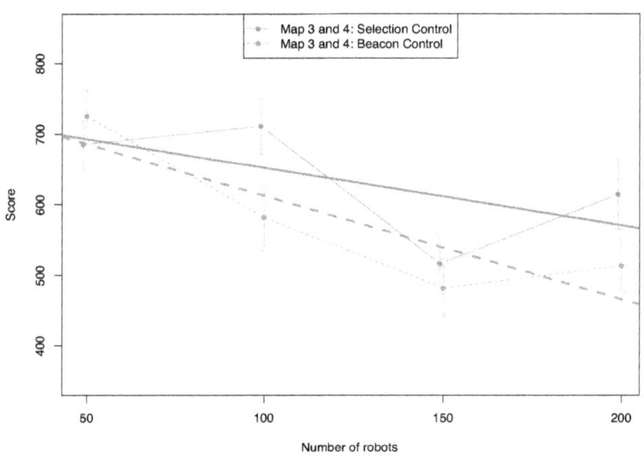

Figure 6: The mean score across different robot configurations with 50,100,150, and 200 robots for selection and beacon control. The standard error is shown with whiskers and the regression line as a thicker line.

While it is expected that larger swarms are more challenging to control we can also investigate whether and how operators adopt to these larger swarms, e.g. by increasing the frequency of mode instructions, number of beacons or size of selections, i.e. increasing the mode impact. Fig. 7 shows that only the mode impact but not the number of instructions scales with the number of robots. Hence, operators are not adapting directly to the larger swarm with increased activity but affect more robots with each mode instruction. Hence, each selection and each beacon influence more robots. In principle, this would suggest that the control methods both scale to the larger swarms and the detriment in performance is due to the reduces per-robot precision in the control. But looking closer at the correlation between activity and score gives us a more detailed picture. For selection control the number of mode instructions correlates with score by 0.1140633, 0.4676364, 0.370013, and 0.7896718 for 50, 100, 150, and 200 robots respectively. For larger swarms, increased activity is hence rewarded with better scores. For beacon control the correlations with score are 0.227069, 0.1523191, 0.3296887, and 0.08906478 for 50, 100, 150, and 200 robots, respectively. Beacon control shows no clear tendency for increased rewards for increased activity. This suggests that the two types of control do indeed scale differently and the performance impediment in beacon control is not mitigated by increased activity.

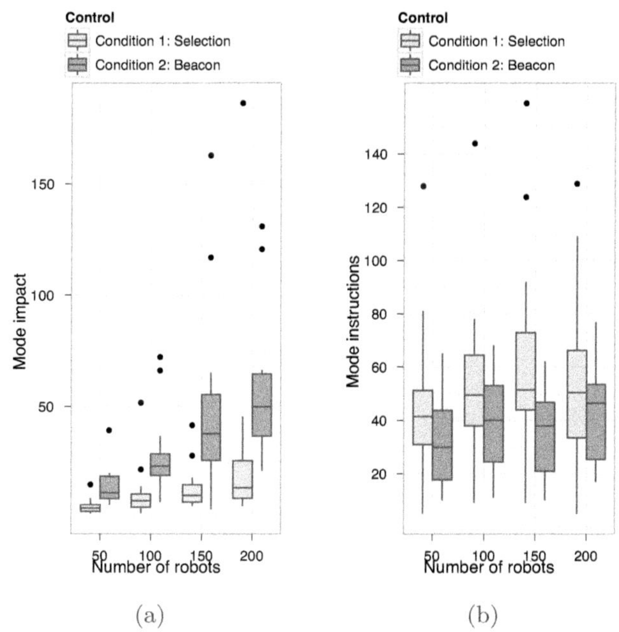

Figure 7: (a) Mode impact for varying numbers of robots. (b) Number of mode instructions for varying numbers of robots.

6. CONCLUSION

We presented an investigation of two principles of human control for robot swarms, selection and beacon control and showed how these perform on a set of foraging missions in environments with different complexity. The key differences between these two principles are their spatial and temporal persistence and the resulting active or passive influence on the robots swarm, enabling different control strategies. Our results showed that novice human operators perform better with selection control. Both types of control enabled human operators to adapt to environments with complex obstacles and their drop in performance is less than that of a simple autonomous swarm that performs better than human operators in open environments. In fact, the different types of maps, two rooms, cluttered, structured, or blind with cluttered and structured obstacles, impeded performance similarly. Overall, the influence of the operator to adapt to these environments was successful despite human operators being generally worse at controlling large swarms for foraging tasks. Supporting the capabilities of human operators to adapt to complex environments with improved autonomy could combine the best of both.

One major problem in controlling swarms is scaling to larger number of robots and hence larger environments and tasks. We observed a stronger correlation between activity and scores in larger swarms for selection control. For beacons there was no such increased correlation. We conjecture that this is due to the fact that beacon control is indeed more scalable when used to its full potential. The strategic placement of beacons becomes more important as the swarm gets larger and mere activity alone does not improve performance. While it is more difficult to use and learn, beacon control seems to have a reduced dependence on activity, a crucial factor for scaling to very large swarms. Further work, possibly with extensive training of participants, might well show that beacon control can perform well and scale, despite our current results showing the opposite.

The missions in this paper relate to tasks that require the distribution of a swarm in an environment and coordinating the motion of a large number of robots. Some examples of such tasks are the establishment of an ad-hoc network infrastructure, the transport of assets to target locations, exploration, and mapping. But the two principles of control could also be investigated under different conditions, types of swarms, and tasks. In addition, further autonomous algorithms and modes could also be considered. It may well be that beacon control approaches work for better for particular modes and tasks while selection control performs better for others. Our study provides a starting point for further such investigations.

7. REFERENCES

[1] S. Bashyal and G.K. Venayagamoorthy. Human swarm interaction for radiation source search and localization. In *Swarm Intelligence Symposium, 2008. SIS 2008. IEEE*, pages 1–8. IEEE, 2008.

[2] F. Bullo, J. Cortés, and S. Martínez. *Distributed control of robotic networks*. Applied Mathematics Series. Princeton University Press, 2009. Electronically available at http://coordinationbook.info.

[3] J. Cortés, S. Martínez, and F. Bullo. Robust rendezvous for mobile autonomous agents via proximity graphs in arbitrary dimensions. *IEEE Transactions on Automatic Control*, 51(8):1289–1298, 2004.

[4] J. Cortés, S. Martínez, T. Karatasand, and F. Bullo. Coverage control for mobile sensing networks. *IEEE Transactions on Robotics and Automation*, 20:243–255, 2004.

[5] ML Cummings. Human supervisory control of swarming networks. In *2nd Annual Swarming: Autonomous Intelligent Networked Systems Conference*, 2004.

[6] XC Ding, M. Powers, M. Egerstedt, and R. Young. An optimal timing approach to controlling multiple UAVs. In *American Control Conference, 2009. ACC'09.*, pages 5374–5379. IEEE, 2009.

[7] X.C. Ding, M. Powers, M. Egerstedt, S. Young, and T. Balch. Executive decision support. *IEEE Robotics & Automation Magazine*, 16(2):73–81, 2009.

[8] A. Ganguli, J. Cortes, and F. Bullo. Distributed coverage of nonconvex environments. In V. Saligrama, editor, *Networked Sensing Information and Control (Proceedings of the NSF Workshop on Future Directions in Systems Research for Networked Sensing, May 2006, Boston, MA)*, pages 289–305. Lecture Notes in Control and Information Sciences, Springer Verlag, 2007.

[9] Z. Kira and M.A. Potter. Exerting human control over decentralized robot swarms. In *Autonomous Robots and Agents, 2009. ICARA 2009. 4th International Conference on*, pages 566–571. IEEE, 2009.

[10] M. Li, A. Alvarez, F. De Pellegrini, B. Prabhakaran, and I. Chlamtac. ROBOTRAK: a centralized real-time monitoring, control, and coordination system for robot swarms. In *Proceedings of the 1st international conference on Robot communication and coordination*, page 37. IEEE Press, 2007.

[11] J. McLurkin, J. Smith, J. Frankel, D. Sotkowitz, D. Blau, and B. Schmidt. Speaking swarmish: Human-robot interface design for large swarms of autonomous mobile robots. In *AAAI spring symposium*, 2006.

[12] U Wilensky. Netlogo. Center for Connected Learning and Computer-Based Modeling, Northwestern University. Evanston, IL, http://ccl.northwestern.edu/netlogo/.

Designing Interfaces for Multi-User, Multi-Robot Systems

Adam Rule
University of Washington
acrule@uw.edu

Jodi Forlizzi
Carnegie Mellon University
forlizzi@cs.cmu.edu

ABSTRACT
The use of autonomous robots in organizations is expected to increase steadily over the next few decades. Although some empirical work exists that examines how people collaborate with robots, little is known about how to best design interfaces to support operators in understanding aspects of the task or tasks at hand. This paper presents a design investigation to understand how interfaces should be designed to support multi-user, multi-robot teams. Through contextual inquiry, concept generation, and concept evaluation, we determine what operators should see, and with what salience different types of information should be presented. We present our findings through a series of design questions that development teams can use to help define interaction and design interfaces for these systems.

Categories and Subject Descriptors
A.m. [**Miscellaneous**]: Human Robot Interaction

General Terms
Design, Human Factors.

Keywords
Multi-robot operators, interface design, information presentation, design research.

1. INTRODUCTION
The thought of robots interacting with humans has been around for millennia. One of the oldest written accounts, found in the Daoist Liezi, tells of a West Zhou Dynasty (1066BC-771BC) engineer who created a humanoid so real that it winked at the king's concubines [19]. As the tale goes, the king had to see the machine dismantled before being satisfied that it was not human. The idea of simultaneously controlling groups of robots also has historical roots. As early as the 1920s and 30s Nicola Tesla envisioned teams of 50, even 100 robots being controlled at once by one or more operators [9, 19].

Those dreams are now becoming reality as the use of autonomous robots in organizations continues to increase steadily. For example, autonomous robots are used in search and rescue and field mining missions. Semi-autonomous robots are also in use in hospitals where they assist with straightforward tasks that can reduce the work of staff, for example, collecting linens and cafeteria trays or providing surveillance. These situations are ripe for deploying several robots simultaneously.

In Multi-User multi-Robot System (MURS) scenarios, both the number and complexity of tasks operators face drastically increase over single-user, single-robot cases. A single operator may divide their available cognitive resources over the control of many systems, performing both repetitive and non-standard tasks [28]. In addition, dynamic task allocation, where robots change their behavior in relation to the environment and other robots, drastically increases the complexity of the task at hand [27]. The more robots that are added, the more complexity increases. Multiple operators introduce the problems of team dynamics and fluid coordination.

Researchers have already begun to empirically study how situation awareness changes as the number of robots in a team increases [23] and classify the unique elements of situation awareness in HRI [13]. There is also a mature body of work on how to create a salient display, manage awareness, and command attention [1, 7, 11, 13, 15, 16, 25, 32, 36]. However, what operators want their attention directed towards and how this need changes with respect to operator role and context is unaddressed. Open research questions include: What do operators want to see? Does this change relative to an operator's role on the team? With what importance, or salience, should information be presented? What kind of information can remain hidden unless requested?

To begin answering these questions, this paper presents a design investigation to understand how interfaces should be designed to support multi-robot, multi-operator teams. Through contextual inquiry with expert operators, concept generation, and concept evaluation, we determine what operators should see, and with what salience different types of information should be presented. We present our findings through a series of design questions that development teams can use to help define interaction and design interfaces for these systems.

2. RELATED WORK
We began by conducting a literature review spanning the disciplines of psychology, human factors and robotics, focusing on work related to autonomy of robots, control of operators, situation awareness, and information presentation. This research led to grounding definitions on which to base our design investigations.

2.1 HRI and collaboration
Coordination and communication within HRI systems is more complex than in HCI. HRI systems are more dynamic, exhibit autonomy, operate in dynamic, real-world environments, exhibit different interaction behaviors, and may work with multiple systems simultaneously [15, 35]. For these reasons, solutions from the domains of HCI and software engineering, while applicable, cannot be transferred wholesale into HRI, especially in MURS applications. Take, for example, the case of a search-and-rescue robot that comes across a conscious person trapped in rubble. The interaction of the trapped person with the robot is vastly different from that which the robot operator has with the robot, though both interactions occur simultaneously and in a very real sense the two people are collaborating towards the goal of safely removing the victim from the surrounding rubble. This large difference in operator and victim roles is a situation unique to HRI and

highlights the need for unique design. Some research integrating approaches from HCI and HRI offers a taxonomy of design considerations for HRI, including team composition, amount of required interaction, decision support, and space-time location [41].

Additionally, literature in human-human collaboration and team dynamics can be relevant to understanding how MURS interfaces are best designed. This is because types of teams, structures of teams, and numbers of humans and robots involved are important constructs in collaboration. There has been a great deal of research on the basis of differentiation and discrimination between people in one's group and those outside of it. Factors such as intergroup competition, similarity, and status affect the degree of group differentiation; group members can be evaluated more extremely or perceived as threatening [6, 5]. In addition to in-group favoritism, out-group members may be differentiated along a number of social dimensions [33].

In the domain of HRI, recent work on team collaboration with robotic systems has shown the importance of group leaders in developing common understanding and translating team status information across members [24]. Other research conducted with mobile robots in a desert field site showed that different levels of robot autonomy can drastically change the flow of communication and the creation of common ground among team members [39].

Much more needs to be known about collaboration in MURS to better support the interface design of these systems. Systems that are intended for groups may be perceived in many ways, due to the uneven distributions of benefits of the system among members of an organization [21, 31]. Varying levels of autonomy may influence how different people perceive the robot, control, and trust the robot [20]. Social effects such as resistance to use of the technology may result, and workarounds are often not captured in the design [22].

2.2 Factors affecting information presentation

Understanding human-robot collaboration is also important in MURS interface design. Two factors affect how information gets presented in the MURS interface: what operators see and control, and what the robots are controlling.

2.2.1 What operators see and control

Early research defined control as the human monitoring of a complex system with intermittent use of an automated agent [30]. Implicit in this definition is the fact that the supervised system may periodically act autonomously based on sensor data [37]. Subsequent research acknowledged the importance of the human in the interaction loop. Collaborative control was defined as humans and robots engaging in dialog to exchange ideas, resolve differences, perform tasks, and achieve goals [15]. Here, human and robot are seen as equals in collaboration; the human is an imprecise, limited source of information for the robot.

While these are important grounding principles, other research has explored the role of the operator relative to the information he or she will need to see. Essential roles include supervisor, operator, mechanic, peer, bystander, mentor, and information consumer [19, 34, 35]. Collectively, these definitions begin to map out the design space for MURS interfaces.

2.2.2 What the robots are controlling

Because autonomy levels change during dynamic robot control, an open design research question is how those levels may be communicated to human operators. An early model from HCI, which describes ten levels of autonomy, is a good starting point (Table 1) [37]. This model focuses on autonomy in decision making, not specific action, and while the actions of HCI and HRI differ, the process of decision making remains largely unchanged.

Table 1: Ten Levels of Autonomy used in designing interfaces for MURS systems (after [37]).

Level	Description
1	Human does the whole job up to the point of turning it over to the computer to implement.
2	Computer helps by determining the options.
3	Computer helps determine options and suggests one, which human need not follow.
4	Computer selects action and human may or may not do it.
5	Computer selects action and implements it if human approves.
6	Computer selects action, informs human in plenty of time to stop it.
7	Computer does whole job and necessarily tells human what it did.
8	Computer does whole job and tells human what it did only if human explicitly asks.
9	Computer does whole job and tells human what it did and it, the computer, decides they should be told.
10	Computer does whole job if it decides it should be done, and if so tells human, if it decides they should be told.

Alternative control methods include dialogue systems where the robot presents limited information to the operator when seeking guidance at a barrier or decision point [16]; policy systems that create rules for how and when robots are permitted to act autonomously [4]; waypoint systems to assist in navigation or sweeping an area [18]; and playbooks, which give high-level commands to clusters of robots [17].

2.3 Situation Awareness

With dynamic levels and modes of command, situation awareness, which defines how much attention is commanded to the interface at any given time, is of utmost importance. Inefficient attention allocation can lead to operator inefficiencies, significant robot wait time and under-utilization [30]. Thus it is important to consider how operator workload and attention are allocated to maximize system efficiency. Work has already been done to assess how interface design affects both workload and situation awareness in multi-robot control, although for single operators [23].

A straightforward evaluation of perception, comprehension and projection as the stages of situation awareness [14] can be extended for use in HRI. In the domain of HRI, Drury has provided a useful framework of HRI awareness [13], which includes five components (Table 2). All of Drury's HRI awareness components aside from robot-robot awareness can be mediated by a successful interface.

2.4 Salience of information

To design to support situation awareness, we can look to the field of visual design, which is informed by research in many areas: cognitive science, human factors, and semiotics, among others. Research has explored identification and/or search speeds of text

[10], color [8], high-symbolism icons [10], and combined text and high-symbolism pictures [3]. Other research explores how individual design features holistically contribute to a design. Color as a design feature has been shown to result in faster visual search times compared to design features such as size, brightness, and geometric shape [10]. Another study showed that the discriminating effects of color lessen with learning [8]. Other research confirms that text can be processed faster than pictures [10], that concrete icons are more quickly interpreted than abstract ones [29], and that complex icons lead to slower search and identification times (whether the symbols are learned or unlearned by the viewer) [29].

Table 2: Five components of situation awareness in HRI as may affect the design of the interface (after [13]).

Component	Description
Human-Robot	The understanding that the humans have of the locations, identities, activities, status and surroundings of the robot. Further the understanding of the certainty with which humans know the aforementioned information.
Human-Human	The understanding that the humans have of the locations, identities and actions of their fellow human collaborators.
Robot-Human	The robots' knowledge of the humans' commands needed to direct activities and any human-delineated constraints that may require command noncompliance or a modified course of action.
Robot-Robot	The knowledge that the robots have of the commands given to them, if any, by other robots, the tactical plans of the other robots and the robot-to-robot coordination necessary to dynamically reallocate tasks among the robots if necessary.
Humans' Overall Mission Awareness	The humans' understanding of the overall goals of the joint human-robot activities and the measurement of the moment-by-moment progress obtained against the goals.

A body of design research exists has examined interfaces for safety-critical situations, such as flying and driving. One dual-task study showed that finding items on a secondary display may not significantly hinder a primary task [38]. Other research has focused on icon-function relationships, exploring the relationship between visual design, complexity, and interpretation of particular designs. However, this work does not explore how a display might best convey information for a situation when attention is divided among a number of tasks. This leaves open the exploration of designs and renditions that may use design features like contrast, color, and abstraction to quickly convey information.

From the literature, we can understand that the goal of minimizing complexity on a visual display is important. We also know that bottom-up search happens within 100ms of attending to a visual display. Therefore, designs with high visual salience should employ contrast in size and color as design variables. The details of particular design features relative to one another are important.

Collectively, these factors — collaboration, information presentation, situation awareness, and salience of information — must be taken into account when designing interfaces for MURS.

3. RESEARCH THROUGH DESIGN PROCESS

In order to understand the needs of MURS operators and create actionable knowledge to support the design of MURS interfaces, we pursued a research through design process [42], which combined contextual inquiry, affinity diagramming, concept generation, and rapid concept evaluation.

Design research moves from specific to general to specific; observations from a select group of users are used to generate broad design recommendations, which are subsequently used to generate domain-specific design guidelines. In this vein, we chose to study specialized robot operators who not only control robots, but aid in their development. Observations of these operators were used to abstract general principles about MURS control, which were then applied specifically to inform the design of MURS interfaces. The following sections describe the stages of our design process.

3.1 Contextual Inquiry

Contextual inquiry (CI) is the process of interviewing and observing users in context for the purpose of understanding the nature and challenges of their work [2]. CIs allowed us to see first hand the tasks that users complete and the problems they face.

Our team conducted two CI sessions. We interviewed and shadowed participants with the goal of understanding how individuals and groups control multiple robots at one time. Our goal was to elicit information about how to create interfaces that facilitate simultaneous control of more than one robot at a time, allowing for changes in robot autonomy. Our process was to largely observe teams interacting with their robots, asking questions along the way to more fully understand what they were doing. We focused on aspects of interface use, aspects of coordination, and when and how breakdowns occurred and were mitigated.

The first session, lasting about 1.5 hours, was conducted with the primary operator of a social robot. The robot delivers snacks to faculty and researchers in an academic building at Carnegie Mellon University. The robot makes use of head and body movement, an adaptable array of LEDs representing a mouth, and speech synthesis to carry out social interaction. The robot operator must load the robot's tray with snacks, localize the robot's navigation system, drive the robot to its work area using a wired joystick, control robot navigation and dialog using a custom interface, and interact dynamically with the programmer and faculty member responsible for the development of the robot.

The second CI session, lasting about two hours, was conducted with three operators of a search-and-rescue robot. The robot has the form factor of a snake and is constructed of 16 interchangeable modules. The primary goal of the robot is to traverse environments too dangerous for human exploration and identify persons trapped in rubble. The operators make use of a custom interface and standard game controller to drive the robot. Control configurations map components of the game controller to parameters in the control interface, which elicits a response from the robot. The robot has several methods of locomotion, and each method has its own unique set of parameters and control mappings, though many parameters are shared among modes.

In both sessions, operators were asked to describe each element of their current robot control interface and demonstrate normal operation of the robot. For the social robot, our team was able to observe as the robot was used during normal operation, delivering snacks. For the search-and-rescue robot we were unable to observe the robot being operated in a disaster area, though a number of contrived obstacles were used to demonstrate its operation. Along with interface explanation, the CIs included a semi-structured interview with questions related to robot operation, breakdowns, and coordination with other robot operators. Each session was video recorded and notes were taken both during the sessions and upon review of the video recordings.

3.2 Affinity Diagramming

Notes from the CIs were used to create an affinity diagram [26]. In this process, themes are drawn from clusters of notes that had been generated by reviewing the data. Using this process, we uncovered twelve situation awareness questions that MURS operators must answer for robust operation.

To create an affinity diagram, statements or ideas that occurred in the CIs were written onto individual Post-It notes and placed on a common board. Once all team members had exhausted their notes, the team collaboratively organized the notes into categories according to similarity. All team members were able to add and remove category labels, rename categories, and move notes between categories at will. This process ensured consensus on note coding. From this process, our team derived 12 themes in MURS operation: coordination, human roles, possible actions, system status, robot status, directing attention, hardware, recovery and debugging, control modes, customizable and adaptive control, control levels, and robot roles.

These themes were further labeled, following from Drury's HRI Awareness model, as human-human, human-robot, robot-human, robot-robot, and overall mission awareness issues [13]. Organizing our notes using this framework allowed us to look for connections between data points, inconsistencies, and areas for further consideration. For example, none of our note categories fit under the category of overall mission awareness, so we re-examined our notes and made sure to consider overall mission awareness in our future work.

3.3 Information Sought by Operators

After several iterations of affinity diagramming, a set of questions was produced that, when answered, lead to situation awareness. These twelve situation awareness questions distill what operators need to know for robust robot operation (Table 3).

Since it is likely that robot-robot awareness is mediated a level away from the operator's interface, the two questions regarding robot-robot awareness were not considered in the remaining design phases. The final ten relevant awareness questions were used to guide the evaluation of MURS interface elements and will henceforth be refereed to as "situation awareness categories".

3.4 Information Salience Framework

With the categories of MURS situation awareness defined, we sought to understand how an interface should display information related to each of these categories. To do so, we modified the bottom seven levels of Sheridan's Ten Levels of Autonomy (Table 1) to describe how much a computer or its human operator bears the load of deciding what information to display (Table 4) [37]. The top three levels were excluded as they involve acting without informing, and information cannot be displayed without informing the user.

Table 3: Essential situation awareness questions that must be communicated in a MURS interface. These were used in our concept generation matrix.

Awareness Type	Awareness Question
Overall Mission	How does any one person know what the overall mission is and the current progress being made towards its completion?
Human-Human	Who is taking what goals and which should I (the operator) cover?
	Who needs assistance or who can help me?
	What robot, area, or interface element is my teammate referring to? (upon asking for assistance)
Human-Robot	What mode, state, and environment is the robot in and how will this affect my commands?
	Which robot needs my attention?
	What is each robot accomplishing?
	What can I accomplish with this robot?
Robot-Human	What should I (the robot) be doing?
	What should I (the robot) do if I hit a barrier?
Robot-Robot	What are other robots doing?
	How can I (the robot) help other robots, or get help?

3.5 Speed Dating

In order to study how MURS interfaces should guide operator attention, we used our situation awareness categories (Table 3) and our information salience framework (Table 4) to guide a rapid concept evaluation (or speed dating) session. Speed dating is a design process used to validate user needs and narrow a design solution space quickly from a large number of design ideas to a few key interventions [12]. Speed dating has two stages: needs validation to focus on broad opportunity areas, followed by a stage of user enactments to understand how contextual elements may aid or hinder a solution's implementation.

Table 4: Final Information Salience Levels used in the concept generation matrix.

Level	Description
1	The computer offers no assistance, the human must call for all information explicitly.
2	The computer offers a complete listing what information may be displayed.
3	The computer narrows the selection down to a few.
4	The computer selects the information to be displayed and the human may or may not display it.
5	The computer displays the selection if the human approves.
6	The computer allows the human a restricted time to veto before automatic display.
7	The computer selects and displays information it deems necessary.

Since we are still at a broad stage of understanding MURS operator needs, we chose to carry out only the first stage of speed dating: needs validation. This stage makes use of a matrix of design issues to cover a broad range of user needs and understand which are most critical to address and in what manner they should be addressed. The matrix is composed when two scales of design intervention are placed orthogonal to each other. Each cell of the matrix is filled with a design solution with attributes corresponding to the cell's position on each of the design scales.

In our case, we placed our ten situation awareness categories orthogonal to three levels chosen from the information salience framework in Table 4. We chose to use only three of the salience levels (the 2nd, 4th and 7th) creating 30, rather than 70, design matrix cells, so as not to fatigue our evaluation participants. We considered these three levels as low, middle, and high information salience. For low salience, the interface simply lists what information may be displayed and may be commanded to display this information. In the medium level, the interface selects which information should be displayed, but does not actually render complete information until commanded to do so. In high salience, the interface both selects and displays information as it deems necessary.

Traditionally, each cell of a design matrix is then populated with a storyboard detailing how a design solution corresponding to the correct values on each design scale could meet user need. We chose instead to populate each cell with an interface wireframe corresponding to the cell's category of awareness and level of salience. In this manner, 30 wireframes were created showing information related to each of the ten situation awareness categories presented at each of the three information salience levels. Figure 1 shows an example wireframe of Overall Mission information shown with high salience.

Speed dating sessions were carried out with the CI participants. In order to provide some context beyond personal experience, participants were first shown two basic MURS scenarios, one involving a set of five nurses controlling five robots to dispense medication and a second involving two operators controlling four heterogeneous terrain surveying robots with various abilities. Participants were then shown interface elements in sets of three, grouped by the situation awareness categories they addressed. Participants were asked to explain how important information related to that category was and identify which of the three information salience levels was the most appropriate. Since the appropriate salience for a specific situation awareness category may change based on any number of factors, participants were asked to answer in detail and explain intervening factors accordingly. At the end of the session, participants were asked to rank the three most important and three least important situation awareness categories. Video recordings were taken of the speed dating sessions and notes were taken upon further review of the recordings.

In the next section, we share the results of the concept generation and speed dating sessions. We focus in particular on situation awareness and varying levels of information salience.

4. Results of Concept Generation

The main goal of our speed dating sessions was to understand how interfaces should support MURS. Specifically, we wanted to understand what operators should see, what kind of information salience should be assigned, and what kind of information can be kept hidden unless requested.

4.1 No one-to-one mapping

We wanted to understand if our participants, acting as operators, wanted a specific level of salience for information relating to human-human awareness and a different level for human-robot awareness, and so on with the other HRI awareness components. What we found was that there was no one to one mapping of these

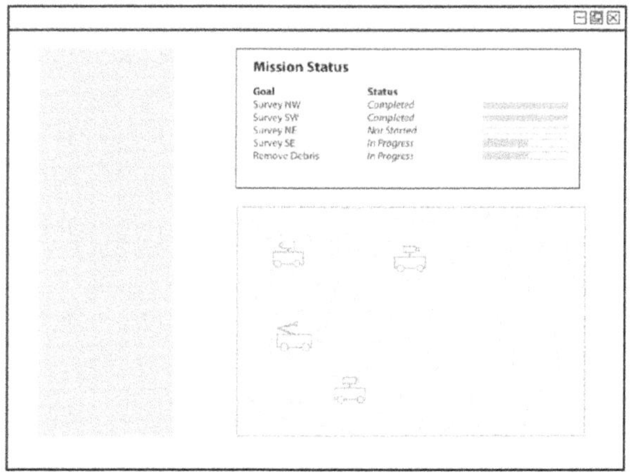

Figure 1: Example interface element used in the speed dating session.

types of information. Instead, robot operators often spoke of wanting different levels of salience to display the same information in different situations. Additionally, they desired a holistic view of the task allocation

4.2 Important vs Urgent

Important information did not always warrant the highest level of information salience. For example, robot operators felt that it was very important that they understand which goals are delegated to them for completion, but they wanted this information displayed with low salience. They felt that a listing of goals would be referenced at the start of operation to develop a clear model of what needed to be accomplished, and only periodically throughout operation, and thus should not take up screen space for most of the operating session.

Additionally, operators wanted urgent information displayed with the highest salience. Cases where inaction would lead to human or robot harm were deemed urgent. For example, operators wanted a high level of display salience in the case of a robot encountering an unresponsive patient in an assistive care setting, but low salience when the robot was experiencing low battery. In the first case, action must be immediate to ensure patient safety, but in the second case, there is a period of time in which action may be taken without consequence. A general hierarchy of information, from most urgent to least urgent, was human safety first, followed by robot failures, robot status, individual operator goals, and finally, overall team goals.

4.3 Operator Roles

There were also some interesting findings related to operator roles. Our participants felt that operators who were in charge of the mission should have more access to information, and subsequently more control over its modification. This is interesting, because it naturally follows that they will have less overall situation awareness of a mission at any given time.

In the next sections, we describe specific findings from each of the situation awareness questions used in the concept generation matrix.

4.4 Overall Mission

Operators thought the middle and high levels of display salience were appropriate for displaying overall mission goals. If goals were to be dynamic and reassigned often, operators wanted the interface to inform them of this, but if goals were more static, they felt it was appropriate to have the middle level of salience where they could decide to check on goals periodically. The operators felt that an understanding of the overall mission scope was more important for team leaders to understand rather than each individual operator.

4.5 Goal Delegation

Operators felt an understanding of which goals were delegated to them was important and desired for a low or medium level of salience. They stated that they would likely look at their set of goals only periodically, or when significant tasks had been completed. However, they felt that information about robot status was more important and would want to be able to collapse or navigate away from goal information so their screen space could be devoted to robot information.

4.6 Assistance

Operators wanted the middle level of display salience for notifications of other operators needing assistance. They did not want to be interrupted from a current task they were completing, but also did not feel they would have time to be continually monitoring the status of other operators. Some kind of alert was desired that would notify them of another operator needing help. A higher level of display salience was desired for larger operator networks where a central leader, or system AI, would assign tasks, similar to how job bids are delegated to service workers.

4.7 Area Reference

Operators desired for a low level of display salience in conveying information about which robot, interface element, or part of the environment another operator may be referring to when asking for assistance. They felt a shared interface element that could be manually edited by each operator, such as an overview map that could be tagged, was most appropriate.

4.8 State Awareness

Operators wanted a medium level of display salience for robot state awareness. They wanted to be able to select which aspects of robot information to view at any one time. They also wanted the ability to "zoom-out" to basic overview information when they were doing general control, and then click into a more specific view when trouble shooting or manually controlling a single robot. The ability to resize, and turn on or off the display of robot sensor data was essential to the operators.

4.9 Robots Needing Attention

Operators saw a need for both medium and high levels of display salience in being notified of which robots needed their attention next. For critical status updates such as a robot mechanical or software failure, or cases requiring immediate attention, such as a patient being unresponsive, operators wanted a high level of display salience and their attention to be directed to the problem area immediately. For less vital issues, such as a robot going idle, operators wanted a medium level of salience, being informed of the problem but able to choose whether or not to respond.

4.10 What Robot is Accomplishing

Operators wanted low or medium levels of display salience for understanding what a robot was accomplishing currently. As distinct from critical issues discussed in the section above, current progress is information that operators want to be able to access, but do not want the system to draw their attention to unless they decide.

4.11 Robot Capability

Operators wanted a low level of display salience for understanding the capabilities and commands that could be given to a robot. They felt that the understanding of what a robot can do should be developed during some period of training and reference materials should be on hand, possibly in the form of a wiki, detailing robot capabilities. A medium level of display salience was considered appropriate for MURS where operators would not have access to training beforehand or where very few actions could be carried out by the robots.

4.12 Robot Control

For actually commanding a robot to complete a task, operators desired a medium level of display salience. Tabs or buttons that are pervasively displayed and can be used to switch between robots or levels of control were mentioned. Operators did not want for the display of these elements to be dynamic or brought to their attention except in critical situations as detailed in Section 4.4.8.

4.13 Robot Idle Control

In order to increase robot productivity, some operators let robots decide what to do if they become idle or finish an assigned task. Communicating to the robot the order in which tasks should be completed, or "safe" tasks that they can accomplish while waiting for further input is sometimes beneficial. Operators felt that a low level of display salience, such as a secondary window where task priorities could be assigned pre-mission would be appropriate for predictable tasks, but that more dynamic missions and environments would require a medium level of display salience. A robot becoming idle was not deemed a critical alert and operators felt a high level of display salience that could direct attention to an idle robot was inappropriate.

4.14 Summary

While operators wanted a holistic sense of their mission, and wanted a clear communication of goal state, the most important information for the robot operators to know was the delegation of individual goals, the state of their robots and the environment around them, and which robot needs immediate attention. The least important information to know was what a robot's capability,

Table 5: Summary of levels of display salience for components of the MURS interface.

Level	Description	Level	Description
H	Convey errors related to human safety.	M	Convey state awareness.
H	Communicate robot failures.	M	Convey robot idle control.
H/M	Convey that robot needs attention.	M/L	Convey team goals.
H/M	Convey a sense of the mission.	M/L	Convey need for robot assistance.
M	Convey robot status.	M/L	Convey goal delegation.
M	Convey operator goals.	M/L	Convey what robot is accomplishing.

which other operators need assistance, and telling a robot what to do when it is idle and waiting for attention. Operators stated that increased robot autonomy, increasing numbers of robots under an operator's control, and increasing task critically would necessitate higher levels of display salience, directing their attention more to critical decisions.

5. DISCUSSION

Our investigation into interfaces to support MURS teams led to a number of interesting findings regarding how information should be presented to support team leaders and collaborators. To implement these ideas, the growing literature on interface design patterns in HCI can be used as reference material [40]. Modular tabs, collapsible panels, and multiple windows, all lend themselves well to low levels of display salience. Accordions, list inlays, and alert lists similarly work well for medium levels of display salience. Responsive disclosure, movable panels, and pop-up windows all work well for high levels of display salience.

While the needs of operators, and therefore, interface elements change over time in a MURS system, we can put forth some general suggestions for whether or not information should be presented with high, medium, or low salience. These are presented in Table 5. In addition, based on the results of our design investigation, we can offer several design suggestions for those who are designing and implementing MURS systems. These design recommendations follow from Table 5, and include the following:

First, high salience should be reserved for communicating information such as errors related to human safety, errors related to robot safety, items needing immediate attention, and information about the dynamics of the mission.

Second, medium or low salience can be used to present information on the dynamics of team context. This includes information like robot status, operator goals, operator state of awareness, and team goals.

Third, low salience on should be used for operators controlling the mission. These people will want high control of the information that they are interacting with.

Overall, operators desire a holistic view of the system at any time, and it is important that they can exert control over all types of information if needed. For example, override features should be presented so operators can either intervene with or ignore high salience warnings if needed. In the end, operators must be trusted to override information that the system may think is urgent, but actually isn't.

6. LIMITATIONS

While we present some general conclusions about MURS interfaces, this study has a number of limitations. First, the small sample population used for the CIs and speed dating mean that more extensive study is needed to provide greater certainty in the results presented here. Design is a general set of skills which applies in many situations, and we feel confident that our findings are readily generalizable. Second, the CIs were conducted with single robot operators who collaborate with other operators rather than MURS operators. This deficiency is partly a result of there being few MURS currently in operation, and those in development are still at the stage of individual robot development rather than team development. As MURS become more widespread, operators will be able to provide richer context knowledge for understanding the design of MURS interfaces. Despite these limitations, we believe the study we completed shows the unique nature of MURS, provides several useful frameworks for future research, and lays a foundation for future MURS interface development.

7. CONCLUSION

Robotic systems have matured substantially, but little is known about how to design the interaction for multi-robot, multi-user systems. In this paper, we present a design investigation leading to several design recommendations for interfaces for MURS systems. While operators desire a holistic view of interaction with the system, our findings show that low, medium and high salience can successfully be used to convey different aspects of the interaction. We hope that our findings will help to advance the field of multi-robot interaction and operator collaboration.

8. ACKNOWLEDGMENTS

This material is based upon work supported by ONR and the National Science Foundation under Cooperative Agreement EEC-0540865. We thank the QOLT REU program for their assistance. We also thank the participants who assisted with our research. Special thanks to Steffi Paepcke, Vikram Chatterji, and Eva Shon for their assistance in the research process.

9. REFERENCES

[1] Baker, M., Casey, R., Keyes, B., and Yanco, H.A. (2004). Improved Interfaces for Human-Robot Interaction in Urban Search and Rescue. Proceedings of 2004 IEEE Transactions on Systems, Man, and Cybernetics. New York, NY: IEEE Press, 2960-2965.

[2] Beyer, H., Holtzblatt, K. (1998) Contextual Design: defining customer-centered systems. Morgan Kaufmann.

[3] Booher, H.R. (1975) Relative comprehensibility of pictorial information and printed words in proceduralized instructions. Human Factors, 17, 266-277.

[4] Bradshaw, J., Feltovich, P., Jung, H., Kulkarni, S., Taysom, W., and Uszok, A. (2004) Dimensions of adjustable autonomy and mixed-initiative interaction. in Agents and Computational Autonomy: Potential, Risks, and Solutions, (M. Nickles, M. Rovatos, and G. Weiss, eds.) Springer, Berlin/Heidelberg, 17-39

[5] Branscombe, N.R., Wann, D.L. and Noel, J.G. (1993). In-Group or Out-Group Extremity: Importance of the Threatened Social Identity. PSPB, 19/4, 381-388.

[6] Brewer, M. B. (1979). In-Group Bias in the Minimal Intergroup Situation: A Cognitive-Motivational Analysis. Psychological Bulletin, 86/2, 307-324.

[7] Bruemmer, D., Frew, D., Nielsen, C., and Walton, M. (2007). Mixed-initiative control for collaborative countermine operations. IEEE Transactions on Robotics. IEEE Transactions on Systems, Man, and Cybernetics, 35/4, 438-449.

[8] Carney, C., Campbell, J.L. and Mitchell, E.A. (1998) In-vehicle display icons and other information elements: Literature review, Federal Highway Administration.

[9] Cheney, M. (1981) Tesla: Man out of Time. Touchstone, New York.

[10] Christ, R.E. (1984) Research for evaluating visual display codes: An emphasis on colour coding. In Easterby, R. and Zwaga, H. eds. Information design: The design and evaluation of signs and printed materials, John Wiley & Sons, 209-228.

[11] Crandall, J.W., Goodrich, M.A., Olsen, D.R., and Nielsen, C.W. (2005). Validating Human-Robot Interaction Schemes in Multi-Tasking Environments.

[12] Davidoff, S., Lee, M., Dey, A., Zimmerman, J. (2007) Rapidly exploring application design through speed dating. In Proceedings of UbiComp '07, 429-446.

[13] Drury, J., Scholtz, J., Yanco, H. (2003) Awareness in Human-Robot Interactions. In Proceedings of IEEE Conference on Systems, Man and Cybernetics, Washington, DC.

[14] Endsly, M. (1995) Towards a Theory of Situation Awareness in Dynamic Systems. Human Factors, 37 (1), 32-64.

[15] Fong, T., Thorpe, C., Baur, C. (2001) Collaboration, Dialogue, and Human-Robot Interaction. In 10th International Symposium of Robotics Research, Lorne, Victoria, Australia.

[16] Fong, T., Thorpe, C., Baur, C. (2003) Multi-Robot Remote Driving with Collaborative Control. IEEE Transactions on Industrial Electronics 50/4, 699-704.

[17] Goodrich, M., McLain, T., Anderson. J., Sun, J. Crandall, J. (2007) Managing Autonomy in Robot Teams: Observations from Four Experiments. In proceedings of Human Robot Interaction '07, Arlingon, VA, 25-32.

[18] Goodrich, M., Olsen, D., Crandall, J. Palmer, T. (2001) Experiments in Adjustable Autonomy. In Proceedings of IJCAI Workshop on Autonomy, Delegation and Control: Interacting with Intelligent Agents.

[19] Goodrich, M., Schultz, A. (2007) Human-Robot Interaction: A Survey. Foundations and Trends in Human Computer Interaction, 1(3), 203-275.

[20] Groom, V., Takayama, L., Ochi, P., Nass, C. (2009) I am my robot: the impact of robot building and robot form on operators. In proceedings of HRI '09, La Jolla, CA, 31-36.

[21] Grudin, J. (1988) Why CSCW applications fail: Problems in the design and evaluation of organizational interfaces. In Proc. of CSCW'88, 85-93.

[22] Grudin, J. (1994) Groupware and Social Dynamics: Eight Challenges for Developers. Comm. of the ACM, 37(1): 92-105.

[23] Humphrey, C., Henk, C., Sewell, G., Willians, B., Adams. J. (2007) Assessing the scalability of a multi-robot interface. In proceedings of HRI 2007, New York. 239-246.

[24] Jones, H., Hinds, P. (2002) Extreme work teams: using SWAT teams as a model for coordinating distributed robots. In proceedings of CSCW 2002, New Orleans, 372-381

[25] Kaupp, T. and Makarenko, A. (2008). Decision-Theoretic Human-Robot Communication. Proceedings of HRI08. New York, NY: ACM Press, 89-96.

[26] Kawakita, Jiro. (1986) The KJ Method: Seeking order out of chaos. Choukouron-sha, Tokyo (In Japaneese).

[27] Lerman, C., Jones, C., Galstyan, A., and Mataric, M.J. (2006). Analysis of Dynamic Task Allocation in Multi-Robot Systems. International Journal of Robotics Research, 25:225, 225-241.

[28] Lewis, M., Wang, H., Chien, S.-Y., Velagapudi, P., Scerri,, P. and Sycara, K. (2011). Process and Performance in Human-Robot Teams. Journal of Cognitive Engineering and Decision Making, 5/2, 186-208.

[29] McDougall, S., de Bruijn, O. and Curry, M. (2000) Exploring the effects of icon characteristics: Concreteness, complexity and distinctiveness. The J. of Experimental Psychology: Applied, 6(4), 291-306.

[30] Mitchell, P. J., Cummings, M. L. (2005) Management of Multiple Dynamic Human Supervisory Control Tasks. In Proceedings of the 10th International Command and Control Research and Technology Symposium.

[31] Mutlu, B., Forlizzi, J. (2008) Robots in organizations: the role of workflow, social, and environmental factors in human-robot interaction. In proceedings of HRI 08. New York, 287-294.

[32] Park, B., and Rothbart, M. (1982). Perception of Out-Group Homogeneity and Levels of Social Categorization: Memory for the Subordinate Attributes of In-Group and Out-Group Members. Journal of Personality and Social Psychology, 42/6, 1051-1068.

[33] Parasuraman, R., Galster, S., Squire, P., Furukawa, H., and Miller, C. (2005). A Flexible Delegation-Type Interface Enhances System Performance in Human Supervision of Multiple Robots: Empirical Studies with Robo-Flag. IEEE Transactions on Systems, Man, and Cybernetics, 35/4, 481-493.

[34] Scholtz, J. (2002) Creating Synergistic CyberForces. in Alan C. Schultz and Lynne E. Parker (eds.), Multi-Robot Systems: From Swarms to Intelligent Automata. Kluwer.

[35] Scholtz, J. (2002) Theory and Evaluation of Human-Robot Interactions. In Proceedings of the 38th Hawaii International Conference on System Sciences.

[36] Scholtz, J., Antonishek, B., and Young, J. (2005). Implementation of a situation awareness assessment tool for evaluation of human-robot interfaces. IEEE Transactions on Systems, Man, and Cybernetics, 35/4, 450-459.

[37] Sheridan, T., Verplank, W. (1978) Human and computer control of undersea teleoperators. Technical report, Man-Machine Laboratory, Massachusetts Institute of Technology, Cambridge, MA.

[38] Somervell, J., McCrickard, D.S., North, C. and Shukla, M. (2002) An evaluation of information visualization in attention-limited environments. In Proc. VISSYM 2002.

[39] Stubbs, K., Hinds, P.J., and Wettergreen, DD. (2007). Autonomy and Common Ground in Human-Robot Interaction. IEEE Intelligent Systems, 22/2, March/April, 42-50.

[40] Tidwell, J. (2011) Designing Interfaces. O'Reilly, Beijing.

[41] Yanco, H.A. and Drury, J.L. (2002). A Taxonomy for Human-Robot Interaction. AAAI Technical Report FS-02-03.

[42] Zimmerman, J., Forlizzi, J., and Evenson, S. (2007). Research through design as a method for interaction design research in HCI. In proceedings of CHI 07. New York, NY: ACM Press, 493-502.

A Touchscreen-Based 'Sandtray' to Facilitate, Mediate and Contextualise Human-Robot Social Interaction*

Paul Baxter, Rachel Wood, Tony Belpaeme
Centre for Robotics and Neural Systems
Plymouth University, U.K.
{paul.baxter, rachel.wood, tony.belpaeme}@plymouth.ac.uk

ABSTRACT

In the development of companion robots capable of any-depth, long-term interaction, social scenarios enable exploration of the robot's capacity to engage a human interactant. These scenarios are typically constrained to structured task-based interactions, to enable the quantification of results for the comparison of differing experimental conditions. This paper introduces a hardware setup to facilitate and mediate human-robot social interaction, simplifying the robot control task while enabling an equalised degree of environmental manipulation for the human and robot, but without implicitly imposing an *a priori* interaction structure.

Categories and Subject Descriptors

H.1.2 [**Models and Principles**]: User/Machine Systems; J.4 [**Computer Applications**]: Social and Behavioral Sciences—*Psychology*

General Terms

Design, Experimentation, Human Factors, Theory

Keywords

Social HRI, Sandtray, Unstructured Interactions

1. INTRODUCTION

Increasingly, the view of robots as potential social companions is coming to the fore [1]. These companion agents have been proposed for, and applied to, a number of roles, such as support [5], motivation [4] and play within defined task contexts. However, the more general aim of robot as generic (task independent) social peer remains at a distance. Such a companion agent would at the very least be required to operate in the real world over extended time-scales, making use of multiple modalities (both perceptual and physical) to engage the human interactant [6].

This paper presents the 'Sandtray', a newly developed platform intended to aid in the exploration of issues in social human-robot interaction. Social in this context goes further than directed interaction (e.g. control) or cooperation (e.g. to fulfil a specific objective), by additionally implying a degree of task-independent engagement [6]. This platform allows emphasis to be placed on the social capacities of the robot rather than real-world implementation issues, whilst still facilitating the manipulation of the interaction context. We explore the role that such a setup can play in the facilitation of social human-robot interaction, by providing an interaction context without imposing an *a priori* structure.

2. THE SANDTRAY SETUP

Inspiration for this platform is drawn from the 'Sandbox' technique in child therapy where sand play is used to foster collaborative story-telling interactions between child and therapist [2]. The sandbox equipment itself is of incidental importance, acting purely as a medium to support the therapeutic interaction.

The interaction platform described here centres on use of a touchscreen as opposed to a sandbox but similarly the intention is to bootstrap social engagement through use of a collaborative interaction platform. Thus in this setup the focus is on social interaction rather than the efficacy of interactants' task-oriented behaviour. The hardware consists of a 26-inch capacitive touchscreen and associated control server, upon which a series of virtual objects (icons) can be manipulated by dragging (on the part of the human interactant), or simulated dragging (on the part of the robot interactant). The touchscreen thus serves as a medium for collaboration (figure 1), rather than as a means of robot control, for which they are typically used.

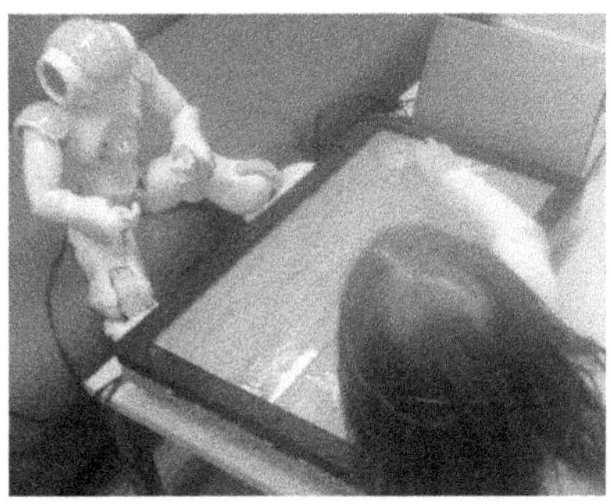

Figure 1: A Child-Robot interaction mediated by the touchscreen-based Sandtray. In this example, there is a collaborative task to sort the icons into two categories (toys and food), by dragging them into two pre-defined locations on the screen.

*This work is funded by the EU FP7 ALIZ-E project (grant 248116), www.aliz-e.org/.

HRI'12, March 5–8, 2012, Boston, Massachusetts, USA.
ACM 978-1-4503-1063-5/12/03.

The control of this setup maintains a separation between the handling of touchscreen and game functionality, and the control of the robot behaviour, where the touchscreen can be regarded as a 'virtual modality' for the robot (figure 2). In doing so, the state of the interaction medium (Sandtray) can be manipulated by the robot, without having to first process visual information (to acquire the state of the Sandtray) or coordinate fine motor control (to manipulate its state). Furthermore, this manipulation is similar in effect to that achievable by the human interactant (assuming that they have a certain level of competence using a touchscreen), resulting in an 'equalisation' of sensory and motor competencies regarding the task-centred aspect of the interaction. This reduces the differences in affordances available to the two interactants, and thus mitigates the potential effect on the social interaction.

In addition, there are a number of further advantages afforded by such a setup involving virtual rather than real object manipulation:

- Issues of robot manual dexterity and task-related perception do not have to be considered, reducing the incidents of breaks in the flow of joint activity due to these difficult tasks: the sensing and acting competencies of the robot can instead be focused on the human interactant.
- The context of the interaction can be manipulated, by, for example, changing the icon set displayed on the touchscreen.
- The properties of the interaction medium can be altered on-the-fly (e.g. visual or auditory cues, modification of drag behaviour, etc.), enabling the investigation of the effect of disturbances on the social interaction.

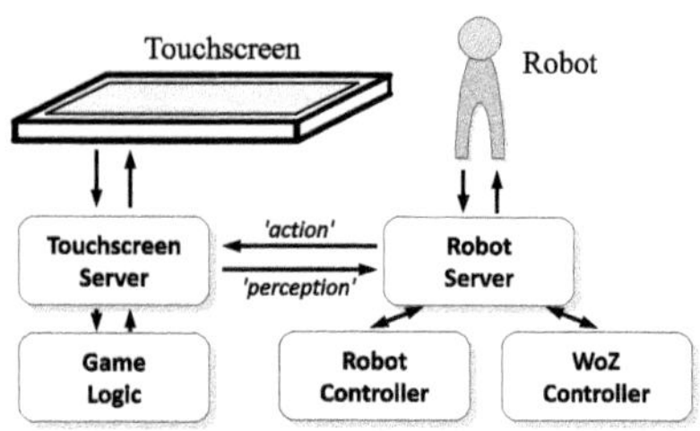

Figure 2: The control system for the Sandtray: perception of the touchscreen and actions regarding the touchscreen by the robot can be handled as 'virtual' sensory and motor modalities. In this scheme, Autonomous or Wizard-of-Oz (WoZ) robot control is readily incorporated.

3. SUPPORTING SOCIAL INTERACTION

As described above, the Sandtray can be used to provide a context for the interaction by forming the medium upon which a task can be conducted. There are a number of tasks that may be provided, and indeed a number of social roles for the robot with respect to the human interactant. So, for example, collaborative tasks may be set, or turn-based games (indeed, these would form informative manipulations to the initial proposals here). However, as indicated above, our primary concern is the examination of social interaction between humans and robot (specifically in our application context of child-robot interaction [1]), and how it can be achieved, rather than an assessment of what robot capabilities result in improved joint task execution.

Consider the following: if a robot and person were simply sitting face to face in a location devoid of interaction context, a social exchange would be unlikely to spontaneously arise. In this case, either the robot or the person could choose a subject and initiate interaction, typically verbal - this invariably necessitates the robot taking the lead in a turn-based conversation (given the technical difficulty of handling unconstrained speech) thereby implicitly imposing a structure on the interaction.

The importance for naturalistic social interactions in the maintenance of engagement is clear. One aspect to this is that the imposition of an interaction structure from a source external to the interactants inhibits naturalistic interaction, leading to a reduction of engagement [3]. It is therefore desirable to avoid this. There is a potential conflict with standard experimental methodology, which tends to promote highly structured interactions, and may therefore result in a reduction in the possible level of naturalistic interaction. While aiming for implicitly unstructured interactions, the Sandtray setup partially mitigates any resultant difficulty by providing a means to readily quantify the behaviour of the human via the touchscreen (e.g. by providing drag speed and direction information, etc).

The role of the Sandtray is thus to provide a collaborative context, but without necessarily also providing an overt structure to the interaction, from which a social interaction is more likely to form. The point here is not that the Sandtray is a sufficient condition for a social interaction, but that the presence of a context facilitates this process. Additionally, the emphasis on verbal interaction is reduced through the presence of a physical interaction medium (albeit virtual). The behaviour of the robot remains the key factor in determining whether or not a social interaction would form, whether this is the outcome of an autonomous control system, or a Wizard-of-Oz manual control. Indeed, it may be beneficial to ensure that the task is not engineered as the primary source of engagement. Engagement *between the interactants* would be required for the initiation and maintenance of genuinely *social* interaction: a necessity for companion robots.

4. REFERENCES

[1] P. Baxter, T. Belpaeme, L. Canamero, and et al. Long-term human-robot interaction with young users. In *Proceedings of the Workshop Robots with Children: Practices for Human-Robot Symbiosis at HRI'11*, Lausanne, Switzerland, 2011.

[2] K. Bradway. Sandplay with children. *Journal of Sandplay therapy*, 8(2), 1999.

[3] R. R. Espinoza, M. Nalin, R. Wood, and et al. Child-robot interaction in the wild: Advice to the aspiring experimenter. In *Proceedings of ICMI*, Alicante, Spain, 2011.

[4] J. B. Janssen, C. C. van der Wal, M. A. Neerincx, and R. Looije. Motivating children to learn arithmetic with an adaptive robot game. In *Proceedings of ICSR*, Amsterdam, The Netherlands, 2011.

[5] C. Kidd. *Designing for long-term human-robot interaction and application to weight loss.* PhD thesis, MIT, Massachusetts, USA, 2008.

[6] A. Tapus, M. J. Mataric, and B. Scassellati. The grand challenges in socially assistive robotics. *IEEE Robotics and Automation Magazine*, 14(1):35–42, 2007.

Children's Knowledge and Expectations about Robots: A Survey for Future User-Centered Design of Social Robots

Eduardo Benítez Sandoval
National University of Mexico
Cto Escolar S/N Edif. Diseño
2do piso, CU 04510, Mexico
eduardo.b.sandoval@comunidad.unam.mx

Christian I. Penaloza
Osaka University
6-7-32 Sakuranocho Toyonaka,
Osaka, Japan
contact@christian-penaloza.com

ABSTRACT

This paper seeks to establish a precedent for future development and design of social robots by considering the knowledge and expectations about robots of a group of 296 children. Human-robot interaction experiments were con- ducted with a Tele-operated anthropomorphic robot, and surveys were taken before and after the experiments. Children were also asked to perform a drawing of a robot. An image analysis algorithm was developed to classify drawings into 4 types: Anthropomorphic Mechanic/Non Mechanic (AM/AnM) and Non-Anthropomorphic Mechanic/Non Mechanic (nAM/nAnM). Image analysis algorithm was used in combination with human classification using a 2oo3 (two out of three) voting scheme to find children's strongest stereo- type about robots. Survey and image analysis results suggest that children in general have some general knowledge about robots, and some children even have a deep under- standing and expectations for future robots. Moreover, children's strongest stereotype is directed towards mechanical anthropomorphic systems.

Categories and Subject Descriptors

H. Information Systems **H.5** INFORMATION INTERFACES AND PRESENTATION (I.7) **H.5.2** User Interfaces (D.2.2, H.1.2, I.3. 6) **Subjects:** Ergonomics; User-centered design; Benchmarking. **J.** Computer Applications **J.4** SOCIAL AND BEHAVIORAL SCIENCES **Subjects:** Sociology.

General Terms: Human Robot Interaction, User-Centered Design, Social Robots Design, Experimentation, Human Factors,

Keywords: Children Robot Interaction, Mexican Children, Human Robot Interaction.

1. INTRODUCTION

The idea of "A robot in every home" [1] has brought many expectations about the types of robots that will actually be used in the future. The question is: what is the most appropriate design for robots that will socially interact with humans? So far, all the developments in social robots are to be used in laboratories and in most cases the designs are based on proposals of experts, such as Kismet developed by Breazel et al [2], or Geminoid developed by Prof. Ishiguro [3]. However, robot design can highly affect the perspective of the end- user, as suggested by Bartneck et al [4] who investigated the negative attitudes toward robots that people have. The main reason might be due to the fact that the end-user does not participate in the design process, but only takes part in the experimental interaction with the robot.

HRI'12, March 5–8, 2012, Boston, Massachusetts, USA.
ACM 978-1-4503-1063-5/12/03.

From the design perspective, a dominant methodology for design of consumer products is the "user-centered design" which focuses on the user's perspectives and opinions to build up designs that meet their expectations. In the same manner, if robots are expected to coexist and interact with humans in the future, it is very important to develop user-center designs that facilitate human robot interactions. In this sense, children's opinion and expectations could play a very important role in the design of future robotic systems.
In this paper we provide a survey about children's knowledge and expectations that may lead to better robotic designs in the future. Human-robot interaction experiments were con- ducted with a group 296 children and a Tele-operated anthropomorphic robot. Surveys were taken before and after the experiments, and children were asked to perform a free drawing of a robot. An image analysis algorithm in combination with human classification was used to find children's strongest stereotype about robots.

2. METHOD

The experiment was performed at the Universum Museum in Mexico City. A group of 296 children (ages 5 to 13 years) participated in the experiment (Female-54.9\% and Male-45.1\%). The groups of children were divided according to age: (A)5-6 years, (B)6-7 years, (C)7- 8 years and (D)9-13 years. In order to establish the order in which the survey and the robot interaction took place, groups were subdivided into 4 subgroups (a, b, c, d). Survey was given to subgroups "a" and "b" before robot interaction, and to subgroups "c" and "d" after the interaction and we compare the result among the children who interacted with the robot before the survey and the others. Finally, in order to find out the stereotype that children have about robots, they were asked to perform a free drawing of a robot. Figure 1 (a) describes the methodology of the surveys and interaction process.
For the experiment, our robot POWI was remotely controlled by a hidden operator. POWI has speech interface, can move and rotate the arms, and turn its head (Fig. 1 (b)). The main purpose of the interaction was to get the children familiarized with the robot and encourage them to ask open questions.

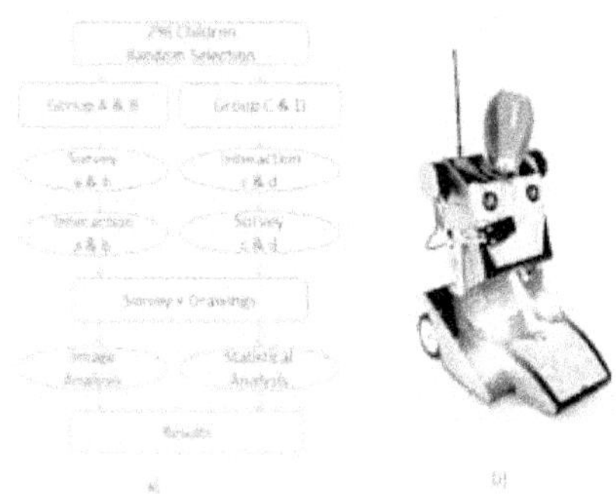

Figure 1: a) Experiment Methodology. b) POWI robot.

2.1 Survey

In order to obtain a statistical background about the children, the survey started with general questions such as age, gender, types of electric devices in home (Desktop PC, Mobil Phone, etc.), habits of fun (types of games preferred to play). Subsequently, the survey contained the following five open questions regarding robots: 1) What is a robot? 2) What famous robots do you know? 3) How does a robot work? 4) Would you like to see robots employed in education? 5) Would you like to see robots (working) in the museum?

3. DATA AND IMAGE ANALYSIS

3.1 Data Analysis

In order to analyze the answers provided by the children, we used IBM SPSS 19 statistics software. Our approach consisted in building categories depending on the types of words children used. Word occurrence frequency related to robots was used as the main feature for analysis. Words such as cables, computers and gears; represented a general knowledge of the participant. On the other hand, words such as: artificial intelligence, simulation, microchip, and circuits; represented a more advanced knowledge of the participant.

3.2 Image Analysis

In order to analyze children's drawings, we make groups the images by age. Our image analysis algorithm was programmed to locate and count basic shape features such as rigid lines, squares, circles, triangles, etc. using Hough transform. By combining these basic features we provided definitions for four classes of robot images. (Fig. 2):

1) Anthropomorphic Mechanic (AM): rigid lines, straight angles, vertical geometric organization (i.e. square-shape head over squared-shape body, etc.).

2) Anthropomorphic Non-Mechanic (AnM): curve lines, non-straight angles, semi-geometric organization (i.e. circle-shape head over circle/square-shape body, etc.).

3) NonAnthropomorphic Mechanic (nAM): rigid lines, straight angles, and horizontal geometric organization.

4) NonAnthropomorphic Non-Mechanic (nAnM): curve lines, non-straight angles, and horizontal geometric organization.

In order to perform the drawing classification, our image analysis algorithm was used in combination with human classification using a 2oo3 (two out of three) voting scheme (human-human-algorithm). This way, when two humans disagree with respect to certain drawing, the algorithm takes the final un-biased vote.

1) 2) 3) 4)

Figure 2. Examples 1) AM 11 years old, 2) nAM, 3) nAM 11 years old, c) nAnM 5 years old.

4. EXPERIMENTAL RESULTS

Experimental results were obtained from 213 survey applied (some small children could not write) and 296 drawings of robots. The average age was 9.54 years with a mode of 8 years old, standard deviation of 1.739 years, and age range from 6 to 13 years old.

With respect to the children background, we found in general (all the sample) that a 98\% of the children have access to different media and electrical devices (internet, laptop, mp3 player, videogames). In the results regarding knowledge about robots we found the following results: What is a robot: General idea (93.89\%), precise description (1.88\%) and no answer (4.23\%). How does a robot work: General idea (89.67\%), precise description (1.88\%), no answer (8.45\%). What famous robots do you know?: robots from movies and/or anime (75.59\%), Powi (15.96\%), real robots (1.88\%), no answer (6.57\%). Would you like to see robots employed in education?: Participants would like to see robots in lectures (79.81\%), Participants would not like to see robots in lectures (13.62\%), no answer (6.57\%). Would you like to see robots in museums?: Participants would like to see robots as part of the exhibition (9.49\%), participants would like robots to interact with people (80.6\%), no answer (9.91\%).

Finally in the image classification results we found that the strongest stereotype leans towards Anthropomorphic Mechanic robots with an average of 48.92\% against the other types combined, with exception of the youngest children who lean towards Non-Anthropomorphic Non-Mechanic with a 41.93\% against the other types combined together. This is a reasonable outcome given the fact that younger children painted images with undefined shapes.

5. CONCLUSION

Survey and image analysis results suggest that children in general have some general knowledge about robots, and some children even have a deep understanding and expectations for future robots. Moreover, children's strongest stereotype is directed towards mechanical anthropomorphic systems. These results establish a precedent for future development and design of social robots.

6. ACKNOWLEDGMENTS

Our thanks to Universum Museum and the staff of "La casita de las ciencias" for full support for the realization of the experiment and the Department of Master in Industrial Design for giving us the resources to carry out the work. And Mauricio Reyes to allow us use his robot.

7. REFERENCES

[1] B. Gates. A robot in every home. Scientific American Magazine, 2006

[2] Breazeal, C. (2003). Designing Sociable Robots. Cambridge: MIT Press.

[3] Sakamoto, D. & Ishiguro, H. (2009), "Geminoid: Remote-Controlled Android System for Studying Human Presence", Kansei Engineering International. Vol. 8(1), pp. 3-9

[4] N. T. Bartneck C., S. T. Kanda T., y Kennusuke K., "Cultural differences in attitudes towards robots", in Proceedings of the AISB Convention, Hatfield, 2005.

Human-Robot Interaction: Developing Trust in Robots

Deborah R. Billings
Kristin E. Schaefer
Institute for Simulation & Training
University of Central Florida
4000 Central Florida Blvd.
Orlando, FL 32816
407.823.5560
dbillings@ist.ucf.edu;
kristinoleson@knights.ucf.edu

Jessie Y. C. Chen
Human Research & Engineering
Directorate Army Research Laboratory
12423 Research Parkway
Orlando, FL 32826
407.384.5435
jessie.chen@us.army.mil

Peter A. Hancock
Institute for Simulation & Training and
Department of Psychology
University of Central Florida
4000 Central Florida Blvd.
Orlando, FL 32816
407.823.2310
peter.hancock@ucf.edu

ABSTRACT
In all human-robot interaction, trust is an important element to consider because the presence or absence of trust certainly impacts the ultimate outcome of that interaction. Limited research exists that delineates the development and maintenance of this trust in various operational contexts. Our own prior research has investigated theoretical and empirically supported antecedents of human-robot trust. Here, we describe progress to date relating to the development of a comprehensive human-robot trust model based on our ongoing program of research.

Categories and Subject Descriptors
A.1. [**General Literature**]: Introductory and Survey

General Terms
Design, Human Factors, Performance, Reliability

Keywords
Human-Robot Interaction, Trust.

1. INTRODUCTION
The value of unmanned systems in the modern combat environment (across air, ground, and maritime domains) has been stressed by both the U.S. military and its allies [1]. In fact, the success associated with the deployment of these systems in current military operations has led to an expectation of their expanded roles in the future. One of the challenges involved in the implementation of such unmanned, robotic systems is facilitating appropriate levels of trust in the system to ensure effective collaboration. The human must be able to trust these systems to behave as expected, especially in battlefield situations.

2. HUMAN-ROBOT TRUST RESEARCH
Existing human-robot trust research focuses primarily on the outcomes of trust (e.g., reliance, success, failure, etc.) in a wide variety of application environments not constrained to military operations. Trust can impact the success of human-robot collaborations and may determine future robot usage [2]. Additionally, existing work often cites the importance of calibrating trust, or matching the human's perception of the robot to reality [3]. If trust is precisely calibrated, then the interaction will presumably be optimal. Therefore, research needs to identify the conditions necessary for the facilitation of these appropriate levels of trust in robots. The goal of our research is to identify antecedents of trust in human-robot interaction (HRI). We have identified numerous potential factors impacting trust and subsequently evaluated the current literature through a meta-analytic assessment (see [4] for a detailed description).

2.1 Identification of Antecedents of Trust
First, a literature review of existing theoretical and empirical human-robot trust research was conducted. Subject matter experts (SMEs) were also consulted. Based on the collected information, potential factors impacting trust were then identified and classified as robot-related, human-related, or environmental characteristics. This three-factor approach provided the framework for our human-robot trust model.

2.2 Meta-Analysis of Human-Robot Trust
We quantitatively reviewed our three-factor model framework (human, robot, and environmental factors) for HRI trust by conducting a formal meta-analysis of the empirical studies. Ten papers containing 60 correlational effect sizes and 11 papers containing 47 experimental effect sizes were included in the analysis. Results from this analysis revealed that overall trust is important to human-robot teams (d=+0.71; r=+0.25), and that robot characteristics (e.g., performance, appearance, proximity) are the most important factors, to date, in trust development (see Figure 1). Environmental factors provided a moderate effect, while a small effect was found for human-related factors. These results imply that manipulating design aspects of the robot can most help us to calibrate trust, or bring trust to the appropriate level to facilitate successful interactions. The trends in the literature indicate that higher trust is associated with higher reliability; the type, size, proximity, and behavior of the robot also affect trust. There is a need to develop specific design heuristics, trust mitigation strategies [5].

3. CURRENT RESEARCH
3.1 Trust in Other Domains
Our current work is exploring HRI trust further by assessing the similarities and differences between HRI trust and trust in other related domains (i.e., human-interpersonal trust, human-animal trust, and human-automation trust). Our goal is to determine how research in these areas can contribute to our human-robot trust

HRI'12, March 5-8, 2012, Boston, Massachusetts, USA.
ACM 978-1-4503-1063-5/12/03.

model, as the existing research in HRI remains limited. In particular, one question we are looking to resolve is: Is human-interpersonal trust a good analog for human-robot trust?

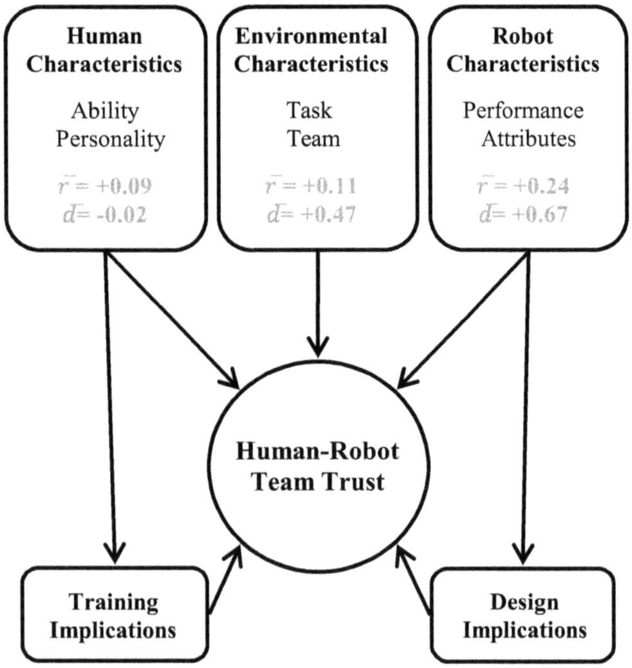

Figure 1. Three-factor model of trust

3.2 The Measurement of Trust

Another aspect of our current research is the measurement of trust. To date, several trust measures for human-automation interaction and human-interpersonal trust exist, but only one is available for HRI [6]. Our efforts will build on this existing research, as well as findings from our exploration of trust in the various domains, in order to develop and validate a subjective measurement scale allowing for trust calibration.

Before attempts can be made to quantify dynamic trust in HRI, several questions must be addressed. First, *what* exactly is being measured? Some researchers have distinguished differences between "trust," "trustworthiness," and "trust propensity" [7]. Others have discriminated between "dispositional trust" and "history-based trust" [8]. This list is certainly not exhaustive, as researchers continue to define even more types of trust (e.g., swift trust, initial trust, cognitive trust, affective trust, etc.). The most appropriate type of measure to use may depend on which type(s) of trust is being studied. Second, *how* can trust be measured (e.g., subjectively, objectively, physiologically)? Third, *when* should trust be measured (e.g., prior to the interaction, during the interaction, after the interaction, multiple times)? Addressing these questions represents an important step in quantifying trust in HRI.

4. CONCLUSION

Robots continue to transition from being perceived as tools to being perceived by humans as interdependent teammates [9]. Humans must accept and trust a robot before effective interaction can occur. Our prior meta-analytic findings highlight the importance of robot design factors in the development of trust. This provides a basis for future experimentation, yet there are other unknowns that must also be addressed. For example, robots do not operate at 100% all the time. When inevitable failures do occur, how does a human's trust level fluctuate? How can trust be repaired after it is violated? How can these dynamics be incorporated into our human-robot trust model framework? Our continued research efforts will help lay the foundation for answering these and other related research queries.

5. ACKNOWLEDGMENTS

The research reported in this document was performed in connection with Contract Number W911NF-10-2-0016 with the U.S. Army Research Laboratory, under UCF Task #3, P.A. Hancock, Principal Investigator. The views and conclusions contained in this document are those of the authors and should not be interpreted as presenting the official policies or position, either expressed or implied, of the U.S. Army Research Laboratory, or the U.S. Government unless so designated by other authorized documents. Citation of manufacturer's or trade names does not constitute an official endorsement or approval of the use thereof. The U.S. Government is authorized to reproduce and distribute reprints for Government purposes notwithstanding any copyright notation herein.

6. REFERENCES

[1] U.S. Department of Defense. 2011. Unmanned Systems Integrated Roadmap FY2011-2036.

[2] Freedy, A., de Visser, E., Weltman, G., and Coeyman, N. 2007. Measurement of trust in human-robot collaboration. In *Proceedings of the 2007 International Conference on Collaborative Technologies and Systems* (Orlando, FL, May 21-25, 2007). CTS '07. IEEE, New York, NY, 106–114.

[3] Lee, J. D. and See, K. A. 2004. Trust in automation: Designing for appropriate reliance. *Human Factors* 46, 1 (Spring 2004), 50-80.

[4] Hancock, P.A., Billings, D.R., Schaefer, K. E., Chen, J.Y.C., de Visser, E.J., and Parasuraman, R. 2011. A meta-analysis of factors affecting trust in human-robot interaction. *Human Factors* 53, 5 (Oct. 2011), 517-527.

[5] Billings, D. R., Oleson, K. E., Chen, J. Y. C., and Hancock, P. A. 2011. Mitigating inappropriate trust in human-robot interactions: A review of trust calibration strategies in the literature. Poster presented at the *2011 APA Annual Convention* (Washington, DC, August 04-07, 2011).

[6] Yagoda, R. E. 2011. *WHAT! You want me to trust a ROBOT? The development of a human robot interaction (HRI) trust scale* (Master's Thesis). NC State University.

[7] Colquitt, J. A., Scott, B. A., and LePine, J. A. 2007. Trust, trustworthiness, and trust propensity: A meta-analytic test of their unique relationships with risk taking and job performance. *J. Applied Psychol.* 92, 4 (July 2007), 909-927.

[8] Merritt, S. M. and Ilgen, D. R. 2008. Not all trust is created equal: Dispositional and history-based trust in human-automation interactions. *Human Factors* 50, 2, 194-210.

[9] Sanders, T. Oleson, K. E., Billings, D. R., Chen, J. Y. C., and Hancock, P. A. 2011. A model of human-robot trust: Theoretical model development. In *Proceedings of the Human Factors and Ergonomics Society 55th Annual Meeting* (Las Vegas, NV, Sept. 19-23, 2011). HFES '11. HFES, Santa Monica, CA, 1432-1436.

Dynamic Gesture Vocabulary Design for Intuitive Human-Robot Dialog

Saša Bodiroža
Institut für Informatik
Humboldt-Universität zu Berlin
12489 Berlin, Germany
bodiroza@informatik.hu-berlin.de

Helman I. Stern
Department of Industrial Engineering and Management
Ben-Gurion Univ. of the Negev
Be'er-Sheva 84105, Israel
helman@bgu.ac.il

Yael Edan
Department of Industrial Engineering and Management
Ben-Gurion Univ. of the Negev
Be'er-Sheva 84105, Israel
yael@bgu.ac.il

ABSTRACT

This paper presents a generalized method for the design of a gesture vocabulary (GV) for intuitive and natural two-way human-robot dialog. Two GV design methodologies are proposed; one for a robot GV (RGV) and a second for a human GV (HGV). The design is based on motion gestures exerted from a cohort of subjects in response to a set of tasks needed to execute several robot waiter (RW)-customer dialogs. Using a RW setting as a case study, preliminary experimental results indicate the unique nature of the HGV obtained.

Categories and Subject Descriptors

I.2.9 [**Artificial Intelligence**]: Robotics—*Commercial robots and applications*

Keywords

Human-Robot Interaction, Gesture Vocabulary, Dialog, Design Methods

1. INTRODUCTION

Most human-robot interaction (HRI) is one-way (from the human to the robot). However, to enhance the acceptability of social robots by the public, a two-way dialog (i.e., bidirectional sequences of information exchange, between the human and the robot) is preferable. Although the ultimate interaction should be multimodal, including speech, gesture, emotional cues, etc., this work focuses on gestures, to be subsequently integrated with speech or to act as a redundant channel of communication (e.g., in noisy environments, like a bar). In HRI, communication must be intuitive and easy for the human participant [1] and gestures are a common interaction to easily transmit a message or to facilitate our thinking process. Gestures can be classified based on different parameters [2]: cultural and natural gestures, action and communicative, deictic and iconic, motivated and arbitrary.

This research presents a methodology for generating a dual human-robot GV. To achieve a natural and intuitive two-way dialog between human and robot participants, both HGV and RGV must be developed. Previous research addressed the correlation between specific features of robot gestures and human emotions [3] and the use of robot gestures [4]. However, to the best of our knowledge, the factors leading to human understanding of robot gestures, in conjunction with human gestures, has not. As a first step toward this end, we describe a procedure to generate a RGV. The HGV is generated through a user survey with no predefined limits on the type or nature of the gestures to be offered. The resultant gestures are intuitive, since each subject proposes gestures that they regard as intuitively representative of the presented task or sentence. A measure of agreement levels (AL, [5]) is applied to extricate the set of gesture-task (G-T) pairs which constitute the GV. A RW scenario is used as a case study to help anchor the ideas.

2. DESIGN METHODOLOGY

A methodology for the generation of two separate GVs, one for the robot and one for the human is presented. First, **several plausible human-robot dialogs are obtained** by role playing human-robot pairs embedded in a particular scenario (such as a RW-customer pair in a bar). This is followed by **the extraction of a set of discrete tasks** (one for the robot and one for the customer) to perform all the dialogs. A list of sentences that describe tasks from the dialogs between a customer and a human waiter are extracted. They are analyzed by the Stanford Parser and keywords (verbs, nouns, pronouns, negations, quantity, determiners) are noted in each sentence (e.g., the verb *order* and the noun *beer* in the sentence *I would like to order a beer*). Finally, **separate** RGV and HGV **are designed**. A distinction is made here between a dialog and a task (e.g., the dialog of ordering a beer may require the tasks of: call RW, select beer type and receive a robot acknowledgment, ask price and receive a price from the robot, submit the order). A gesture would be required for each task. The same task may be used in several dialogs.

For the HGV, a user experiment is conducted in which each person is asked to represent each task using a gesture. After enough people performed gestures, ALs are determined (distributions of using different gestures for tasks). Finally, the HGV is constructed (a set of G-T pairs, where each pair represents a gesture and an associated task).

For the RGV, a user survey in which participants are shown videos of a robot performing gestures (e.g., Figure 1) similar to those used by humans is planned. Subjects will choose the gestures which they can clearly associate with the message the robot is trying to convey and ALs will be derived in a similar manner.

Figure 1: Example of robot pointing to an object of common interest.

3. USER EXPERIMENT

Every participant is shown a task and asked to associate any gesture with the task and to perform it. The main drawback of this approach is that in some cases a high AL for certain tasks might not be reached, since it is possible that there is no unique natural or cultural gesture that corresponds to a particular task. Therefore, it may be necessary to use the most agreed upon gestures as a new bounded set of candidate gestures and repeat the experiment.

Participants view a sentence on a computer screen describing a task with marked verb-noun keywords (e.g., *come-waiter* or *bring-check*) and are asked to perform a gesture they associate with the sentence: one gesture for the whole sentence, two gestures to represent verb-noun keywords, or pointing (to a referent) for sentences containing determiners (this, that). In this first step of the experiment, video recorded gestures are classified by a human rater, and ALs calculated. In the event that high ALs are not reached for all of the tasks (say above 25%) a second experiment should be conducted allowing the users to reselect gestures for those tasks. Once a high AL is reached for all or most of the tasks, a GV is created: G-T pairs are selected.

4. RESULTS AND DISCUSSION

Preliminary experimental results with 8 Industrial Engineering students ($mean\ age = 28.75, \sigma = 3.77$) indicated that for most tasks we had either one gesture with high AL, or two, equally used gestures that together had high ALs. All ALs were above 25% (Table 1[1]).

Table 1: Preliminary results. Columns are tasks, rows are gestures, values represent percentage of people that used a specific gesture for a specific task.

	1	2	3	4	5	6	7	8	9	10	11
1	0	62	0	100	0	0	0	0	0	0	0
2	0	0	0	0	0	0	0	0	75	0	0
3	0	0	37	0	25	0	0	0	0	0	0
4	0	0	25	0	0	0	0	0	0	0	0
5	0	0	0	0	37	0	0	0	0	0	0
6	0	0	0	0	0	25	100	0	0	0	0
7	0	0	0	0	0	25	0	0	0	0	0
8	0	0	0	0	0	0	0	75	0	87	0
9	37	0	0	0	0	0	0	0	0	0	0
10	50	0	0	0	0	0	0	0	0	0	0
11	0	0	0	0	0	25	0	0	0	0	62

The tasks with high ALs were easy to characterize with action, deictic, or iconic gestures. Tasks with low ALs could not be easily characterized. This yields one guideline for the task analysis – chosen tasks should be easy to depict with gestures. We expected that participants will use certain gestures that are fairly regular for certain tasks (e.g., an action gesture simulating drinking for ordering a beer), such as with tasks *Order a beer* and *Order this.* But, this was not the case, since a fairly high number of participants used pointing (to a menu) for both tasks. Some gestures were used for more tasks. Unlike the approach where one limits gestures for certain tasks, we propose that a robot should consider other basic information, e.g., if the person is holding a menu or if there is a spill on the table. Thus, two-way connections between gestures and tasks must be created. As all AL are above 25% it is easy to extract the G-T pairs from Table 1 that constitute such a unique complex GV.

[1]Tasks: *call waiter* (1), *order beer* (2), *cancel beer* (3), *order this* (4), *cancel this* (5), *ask for a suggestion* (6), *clean table* (7), *take away glass* (8), *bring bill* (9), *take away bill* (10) and *where is the toilette* (11). Gestures ($AL > 25\%$): pointing (1), writing on an imaginary piece of paper (2), index finger wave (3), hand wave "no" (4), sliding gesture for canceling (5), circular movement of hand over a surface (6), circular movement of finger over a surface (7), handling the object (8), raised hand (9), hand wave (10), no gesture (11).

5. CONCLUSIONS AND FUTURE WORK

This paper presents a methodology for two-way HRI demonstrated on a RW scenario. A methodology is offered for generating two types of GV, a HGV and a RGV. Using the methodology, a HGV was obtained which exhibited high user ALs for most of the gestures selected. Results indicate that in a HGV most of the tasks had one or two gestures with high AL. Although the RGV was not explicitly generated, it is shown that it is intimately tied to the HGV through specific dialog settings, and a method is proposed for its generation. An additional experiment is planned for the actual construction of the accompanied RGV. Ongoing work will include automatic clustering of captured gestures and their recognition. Performance will be measured using the percent of gestures that the robot recognized correctly and of those understood correctly by the human.

6. ACKNOWLEDGMENTS

This work is supported by the EU FP7 INTRO project, grant agreement no.: 238486, and partially supported by the Paul Ivanier Center for Robotics Research, and Production Management and by the Rabbi W. Gunther Plaut Chair in Manufacturing Engineering, Ben-Gurion University.

7. REFERENCES

[1] V. V. Hafner, M. Lohse, J. Meyer, Y. Nagai, and B. Wrede, eds., *Proc. HRI Workshop on Expectations in Intuitive Human-Robot Interaction.* 2011.

[2] I. Poggi, "From a typology of gestures to a procedure for gesture production," in *Gesture and Sign Language in Human-Comput. Interaction* (I. Wachsmuth and T. Sowa, eds.), vol. 2298 of *Lecture Notes in Comput. Sci.*, pp. 158 – 168, 2002.

[3] H. Narahara and T. Maeno, "Factors of gestures of robots for smooth communication with humans," in *Proc of Int. Conf. on Robot Communication and Coordination*, pp. 44:1–44:4, 2007.

[4] M. Bennewitz, F. Faber, D. Joho, and S. Behnke, "Intuitive multimodal interaction with communication robot Fritz," in *Humanoid Robots, Human-like Machines* (M. Hackel, ed.), pp. 613–624, I-Tech Education and Publishing, Jun. 2007.

[5] H. Stern, J. Wachs, and Y. Edan, "Human factors for design of hand gesture human - machine interaction," in *IEEE Int. Conf. on Syst., Man and Cybern.*, vol. 5, pp. 4052 – 4056, Oct. 2006.

Design of a Haptic Joystick for Shared Robot Control

Daniel J. Brooks and Holly A. Yanco
Department of Computer Science
University of Massachusetts Lowell, Lowell MA
{dbrooks, holly}@cs.uml.edu

Categories and Subject Descriptors

I.2.9 [**Artificial Intelligence**]: Robotics—*Operator interfaces, Autonomous vehicles*; H.5.2 [**Information Interfaces and Presentation**]: User Interfaces—*Haptic I/O*

Keywords

Autonomous Vehicles, Human-Robot Interaction

1. INTRODUCTION

Autonomous mobile robots are often equipped with sophisticated sensors designed to provide the system with a model of its surrounding environment. This information can then be used for making task-related decisions and conveying information back to the operator. To date, autonomous systems tend to exceed at well defined tasks such as navigation, planning, and obstacle avoidance, usually in fairly structured environments. However, for many current mobile robotic systems, teleoperated control is still largely favored, in part due to a human operator's sophisticated ability to reason about unstructured environments [6]. Introducing varying levels of autonomy into a teleoperated system allows for a human operator to make high level decisions while leaving other tasks to the autonomy [5]. With this technique, problems can arise when the human operator does not understand why a part of the system they do not have direct control over is behaving in a particular manner (see Figure 1), usually due to poor situation awareness [1]. Attempts have been made to correct these issues by displaying additional sensor and system state information in the operator control unit (e.g., [8]).

An example in which human operators and autonomous control systems successfully work together exists in conventional control yokes found in side-by-side cockpits of modern aircraft. The two control yokes are tied together such that movement of one is mirrored by the other. Additionally, the autopilot system may be tied in such that its control inputs are also reflected by the control yoke. By seeing or feeling the movement of the control yoke, crew members are more aware of the current state of the aircraft's control system [7].

We hypothesize that by having the autonomous system command the robot's velocity with the same control interface used by human operators (i.e., the joystick), human operators will be better able to understand and interact with an autonomous system at varying levels of autonomy (see

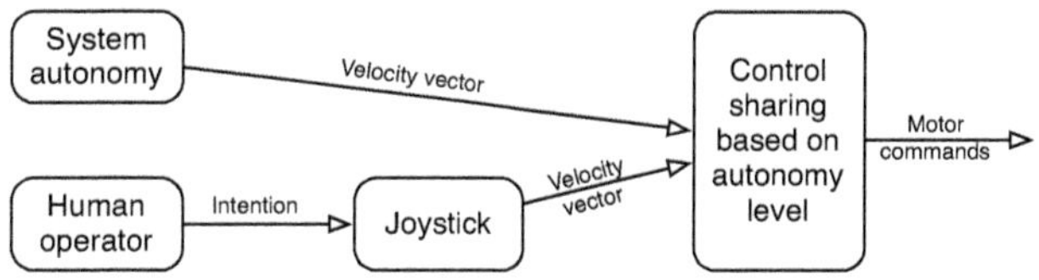

Figure 1: Traditional Shared Autonomy

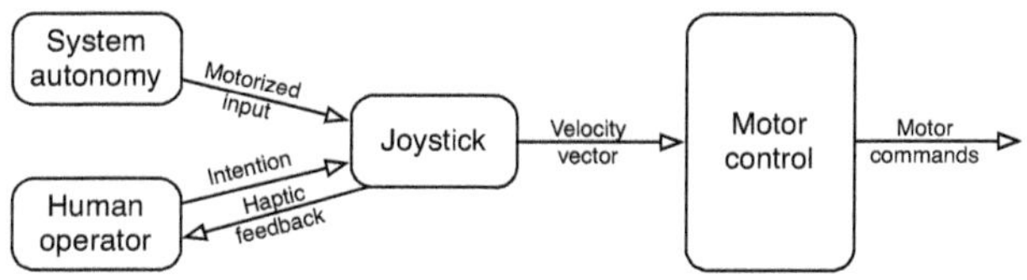

Figure 2: Shared Autonomy With Haptic Feedback

Figure 2). Specifically, we believe that people will be able to more efficiently teleoperate robots if information about the state of the semi-autonomous system state can be felt as haptic feedback rather than being displayed on a screen.

2. DEVICE HARDWARE

Traditional two degree of freedom (DOF) joysticks inherently provide users with some amount of haptic feedback in the form of a constant spring force and hard stops. The spring force is felt as a small amount of constant pressure as springs inside the base of the unit push the handle back towards the center resting point. This pressure provides several important functions. First, feeling any amount of pressure informs the user that the joystick is not centered. Also, the direction of the pressure is inversely related to the direction in which the joystick is pressed. Finally, the joystick will "automatically" center itself when the user relaxes his/her pressure on the unit. The physical layout of the device provides a limited range of motion in which the joystick can move. Thus, the user is informed when the device is at its limit when the device comes to a hard stop. We believe that this kind of feedback should also be present in the haptic joystick. Therefore, interaction with the haptic joystick should feel like a traditional joystick and also have the ability to provide information in a haptic manner. We hypothesize that it will be possible to isolate the effect of enabling various types of haptic feedback.

We selected for our base system a C.H. Products Flightstick, as it is a common and inexpensive joystick with a simple design. The SensAble PHANToM Omni [4] haptic device

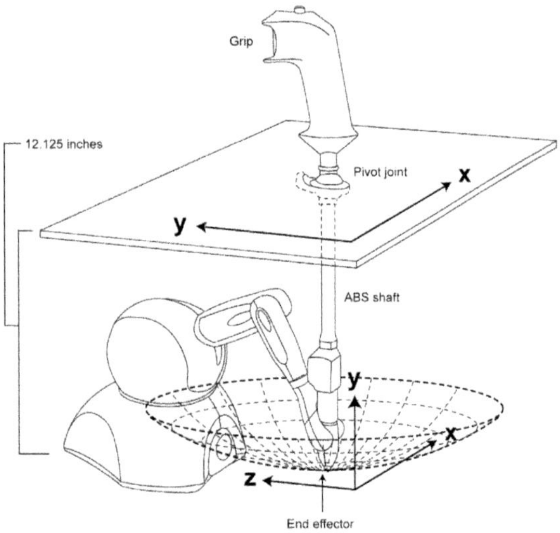

Figure 3: Haptic Joystick Assembly

was selected for the haptic engine as it is widely used in research and comes with a well documented software API[3]. The PHANToM Omni is a 6 DOF haptic device capable of providing haptic feedback in 3 dimensions (x, y, and z axis).

We calculated a configuration in which the PHANToM Omni would have a maximum range of motion, while simultaneously matching the range of motion of the C. H. Products joystick. By maximizing the range of motion of the haptic device, the precision and mechanical advantage of the haptic feedback was also maximized. This configuration required the PHANToM Omni to be positioned below the joystick grip, pivoting about an imaginary point located 12.125 inches above the surface on which the haptic was positioned. We removed the PHANToM Omni's stylus and replaced it with a custom acrylonitrile butadiene styrene (ABS) plastic shaft connected to a joystick grip. A 1/8" hardened metal rod running down the center of the shaft was added to increase rigidity. A physical pivot joint was added to connect the shaft to the mounting surface to complete the assembly (see Figure 3).

3. SOFTWARE

Frequently, the input from a joystick is interpreted as the forward and angular velocity at which a mobile robot should move. Because of the added pivot point on our haptic joystick assembly, the path along which the PHANToM Omni End Effector (OEE) moves is on the surface an invisible sphere (see Figure 3). This path is defined as the minor arc of the great circle passing between two points on the sphere's surface. Therefore, the a mapping between the 2D velocity information and 3D position of the OEE can be defined as a ratio of the maximum distance the end-point can move along the minor arc of the great circle in any given direction, to the associated maximum velocity that the position describes. The robot's internal state can be reflected though haptic feedback to the user by commanding the OEE to travel along the this same arc. This feedback can be used to reflect sensor information such as the distance to an obstacle (i.e., the greater the force, the closer the obstacle). The force applied to the OEE could also be used to represent the autonomous system's intent, with changes in the amount of force indicating decisiveness or sense of urgency.

The complexity of our haptic joystick made it a natural fit for being organized into smaller subsystems (or modules) which could communicate with each other using ROS [2]. For example, the haptic device hardware API was separated such that other modules could subscribe to the two dimensional input information it provides or publish force commands for pushing the OEE and thus creating haptic feedback. Another module has been designed to listen for velocity commands being sent to the motors while publishing sensor information such as laser and video data. Other modules representing various autonomous behaviors are used to interpret this sensor information along with user input and generate haptic feedback commands. By loosely coupling the various pieces of the system, different types of autonomous behavior and haptic effects can be easily swapped in and out of the system for testing.

4. FUTURE WORK

We are planning to conduct an experimental analysis to characterize how haptic feedback can be effectively implemented for various modes of autonomy and the impact each implementation has on the operator's performance. Specifically, we will evaluate the operators' understanding of the remote system state (for all modes of autonomy) and the overall effectiveness of teleoperation during shared control modes. In addition, we would like to investigate how changing the strength and persistence of feedback can be used to manipulate a user's perception of the system state. We will first test scenarios in which the operator and autonomy have an equal share in control with respect to duration and magnitude of the force being applied to the system. Then we will manipulate the share of control that each has. We will also investigate how a user's experience (i.e., the perceived benefit) changes with the amount of previous experience in using classical teleoperated controls.

5. ACKNOWLEDGMENTS

This work was funded by Army Research Office MURI (W911NF-07-1-0216). PHANToM Omni Haptic Device provided courtesy of SensAble Technologies Inc. Thanks to Adam Norton and Munjal Desai for their assistance.

6. REFERENCES

[1] M. Endsley. Automation and situation awareness. *Automation and human performance: Theory and applications*, pages 163–181, 1996.

[2] M. Quigley, B. Gerkey, K. Conley, J. Faust, T. Foote, J. Leibs, E. Berger, R. Wheeler, and A. Ng. ROS: an open-source Robot Operating System. *ICRA Workshop on Open Source Software*, 2009. http://www.ros.org.

[3] SensAble Technologies Inc. OpenHaptics Toolkit, 2011. www.sensable.com/products-openhaptics-toolkit.htm, accessed Nov. 30, 2011.

[4] SensAble Technologies Inc. PHANToM Omni, 2011. http://www.sensable.com/haptic-phantom-omni.htm, accessed Nov. 30, 2011.

[5] T. B. Sheridan. Human and Computer Control of Undersea Teleoperators. July 1978.

[6] T. B. Sheridan. *Telerobotics,automation, and human supervisory control*. MIT Press, 1992.

[7] The Boeing Company. The Role of Human Factors in Improving Aviation Safety. *Aero Magazine*, 8, 1999. http://boeing.com/commercial/aeromagazine/aero_08/human.html.

[8] H. Yanco, B. Keyes, J. Drury, C. Nielsen, D. Few, and D. Bruemmer. Evolving interface design for robot search tasks. *Journal of Field Robotics*, 24(8?9):779–799, 2007.

User Experience of Industrial Robots over Time

Roland Buchner, Daniela Wurhofer, Astrid Weiss, Manfred Tscheligi
HCI& Usability Unit
ICT&S Center,
University of Salzburg, Austria
firstname.lastname@sbg.ac.at

ABSTRACT

This paper reports about a User Experience (UX) study on industrial robotic arms in the context of a semiconductor factory cleanroom. The goal was to find out (1) if there is a difference in the UX between robots used over years with a strict security perimeter (robot A) and a newly installed robot without security perimeter (robot B), and (2) if the UX ratings of the new robot change over time. Therefore, a UX questionnaire was developed and handed out to the operators working with these robots. The first survey was conducted one week after the deployment of robot B (n=23), the second survey (n=21) six months later. Thereby, we found that time is crucial for experiencing human-robot interaction. Our results showed an improvement between the first and second measurement of UX regarding robot B. Although robot A was significantly better rated than robot B in terms of usability, general UX, cooperation, and stress, we assume that the differences in UX will decrease gradually with prolonged interaction.

Categories and Subject Descriptors

H.5.2 [**Information Interfaces and Presentation**]: User Interfaces—*Evaluation/methodology*; I.2.9 [**Artificial Intelligence**]: Robotics

Keywords

Industrial Robots, Measurement, Semiconductor Factory, User Experience

1. INTRODUCTION

Robots in industry have already demonstrated their usefulness over decades in almost every sector of production. Their deployment made vast, cheap, and fast production possible. However, such systems do not offer any interaction with people due to security reasons. To enable a more powerful human-robot interaction and to be highly competitive, the human and the robot should work as a team [7]. As soon as direct interaction between human and robot comes into play, e.g. turn-taking tasks, User Experience (UX) becomes an important issue. The definition of Alben [1] particularly points out the relevance of UX research in all areas, where humans interact with a system, thus also including a semiconductor factory context. Next to cooperation and usability, factors such as perceived safety, stress, or emotion are assumed to be crucial for UX in the factory [4]. According to Karapanos et al. [3], we assume that UX factors are not static but change over time. The presented study in this paper aims to investigate how the UX of robots in a cleanroom (dust-free free area) changes over time. Therefore, we studied two different robotic systems: (1) robotic arms deployed in working cells, where humans must not enter the robots area, already used for several years (robot A), and (2) a newly introduced robotic arm (just one week before study started) which is freely accessible (robot B) and operators can interact with.

HRI'12, March 5–8, 2012, Boston, Massachusetts, USA.
ACM 978-1-4503-1063-5/12/03.

2. RELATED WORK

As humans and robots will work more and more closely together in a factory context, it is necessary to research how the interaction is experienced by workers and how these interactions are influenced by temporal aspects. So far, we are not aware of much research in the context of industrial robotics. Research on UX in the area of mobile phones [3] already showed that UX changes over time and that the temporality of UX can be divided into three phases: the orientation phase, the incorporation phase, and the identification phase. There exists some research in longitudinal studies with humanoid robots (e.g. [2]) and functional robots (e.g. [5]) which reported that the users' behavior towards the robot changed over a period of time, e.g. became better accepted over time. Our research follows a similar interest, with a focus on the changing experiences of a newly introduced robot in the factory context, which first-time has no security perimeter. Thus, the present study represents an attempt to get more insights on the change of UX in human-robot.

3. METHODOLOGY AND PROCEDURE

Based on a literature review and an existing questionnaire on UX for robots [6] we designed a short survey for measuring UX over time in the factory context. The questionnaire is intended to cover different aspects of UX and finally consisted of the following scales: cooperation, perceived safety, perceived usability, stress, general UX, and emotion. The final questionnaire included demographic data and 25 items (5-point Likert scale: 1-totally agree; 5-totally disagree). The internal consistency (Cronbach's Alpha) of the scales was computed at two different times of this study (t_1, t_2) revealing an $\alpha > .7$ for all scales except of emotion ($\alpha = .4$) and feeling of security ($\alpha = .61$) at t_2. Therefore, the scale

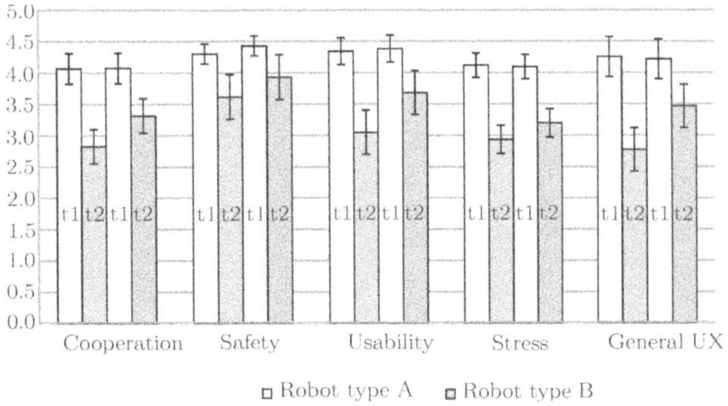

Figure 1: Comparison of the five scales between the first and second round of questionnaire

emotion was left out but perceived safety was kept for the statistical analysis as it is close to .7.

To investigate the temporal change of UX, the questionnaire was filled in by the operators in the cleanroom at two different points in time. The operators were either working in the oven (robot A) or in the etching (robot B) section. The sample group for our study was randomly chosen among eight shift groups. The first questionnaire round was conducted at the end of February 2011 where a researcher randomly distributed the questionnaires. The second round was conducted half a year later (August 2011) by a student apprentice.

4. RESULTS

At t_1, 23 questionnaires were filled in (5f, 18m). A total of 13 respondents indicated to currently work with robot B, 10 with robot A. The mean age of respondents was 31 (SD=10), with an average experience of 7 years in the cleanroom. At t_2, 21 questionnaires were filled in (10f, 11m). A total of 12 respondents indicated to currently work with robot B, 7 with robot A. The mean age of respondents was 35 (SD=12), with an average experience of 7 years in the cleanroom.

After t_1, an independent samples t-test was conducted to compare ratings of UX for type A and B robots. We found a significant difference in all scales of UX, with better ratings for type A robots ($p<.05$). After t_2, an independent samples t-test was performed to compare the UX ratings of the newly introduced robot B shortly after its introduction and six months later. The results show a significantly better rating for the UX scales general UX, perceived usability and cooperation after six months ($p\leq .05$). The mean values for the UX scales stress, and perceived safety also decreased, however, the difference was not significant ($p>.05$). Regarding robot A, an independent samples t-test showed no significant differences in the UX ratings at t_1 and t_2. A comparison between the mean values of the UX scales for type A and B robots at t_2, using an independent samples t-test, showed a significant difference between the two robot types, for all scales except of perceived safety being significantly better rated for robot A ($p<.05$). Figure 1 shows the UX ratings for both robot types at both times (t_1, t_2).

5. CONCLUSIONS & FUTURE WORK

The presented study demonstrated that time is a crucial factor in human-robot interaction, modifying the way how robots are experienced. This was shown by investigating the UX of a newly introduced robotic arm at two points of time: shortly after its introduction and half a year later. Another robotic arm, already deployed in the factory for years, served as a control group. We found that the newly introduced robotic arm received low ratings with regard to the UX factors cooperation, perceived safety, perceived usability, stress, and general UX shortly after its introduction in comparison to the already established ones. After six months, the ratings of the newly introduced robotic arm significantly improved regarding general UX, perceived usability and cooperation. These results suggest that initial experiences change in a positive way with prolonged interaction. We assume that the differences in the ratings of UX regarding robot A and B at t_2 will gradually decrease with time.

Thereby, a third round of the questionnaire in the same sections of the semiconductor factory will be conducted. Also further research on UX over time is planned to get a deeper understanding of the UX scales (cooperation, perceived safety, perceived usability, stress, general UX, and emotion) as well as if they can be applied in another application context.

6. ACKNOWLEDGMENTS

We greatly acknowledge the financial support by the Federal Ministry of Economy, Family and Youth and the National Foundation for Research, Technology and Development (Christian Doppler Laboratory for "Contextual Interfaces").

7. REFERENCES

[1] L. Alben. Quality of experience: defining the criteria for effective interaction design. *Interactions*, 3(3):11–15, 1996.

[2] G. Castellano, R. Aylett, K. Dautenhahn, A. Paiva, P. W. McOwan, and S. Ho. Long-Term Affect Sensitive and Socially Interactive Companions. In *Proc. of the 4th Int. Workshop on Human-Computer Conversation*, 2008.

[3] E. Karapanos, J. Zimmerman, J. Forlizzi, and J.-B. Martens. User experience over time: an initial framework. In *Proc. of CHI 09*, pages 729–738, NY, USA, 2009. ACM.

[4] M. Obrist, W. Reitberger, D. Wurhofer, F. Förster, and M. Tscheligi. User experience research in the semiconductor factory: A contradiction? In *Proc. of INTERACT 11*, 2011.

[5] J.-Y. Sung, H. I. Christensen, and R. E. Grinter. Robots in the wild: understanding long-term use. In *HRI*, pages 45–52, 2009.

[6] A. Weiss, R. Bernhaupt, M. Tscheligi, and E. Yoshida. Addressing user experience and societal impact in a user study with a humanoid robot. In *AISB2009: Proceedings of the Symposium on New Frontiers in Human-Robot Interaction*, 09.

[7] A. Weiss, R. Buchner, H. Fischer, and M. Tscheligi. Exploring human-robot cooperation possibilities for semiconductor manufacturing. In *Int. Workshop on Collaborative Robots and Human Robot Interaction*, 2011.

Visual Cues-Based Anticipation for Percussionist-Robot Interaction

Marcelo Cicconet
Center for Music Technology
Georgia Tech
Atlanta, GA
cicconet@gmail.com

Mason Bretan
Center for Music Technology
Georgia Tech
Atlanta, GA
masonbretan@gmail.com

Gil Weinberg
Center for Music Technology
Georgia Tech
Atlanta, GA
gilw@gatech.edu

ABSTRACT

Visual cues-based anticipation is a fundamental aspect of human-human interaction, and it plays an especially important role in the time demanding medium of group performance. In this work we explore the importance of visual gesture anticipation in music performance involving human and robot. We study the case in which a human percussionist is playing a four-piece percussion set, and a robot musician is playing either the marimba, or a three-piece percussion set. Computer Vision is used to embed anticipation in the robotic response to the human gestures. We developed two algorithms for anticipation, predicting the strike location about 10 mili-seconds or about 100 mili-seconds before it occurs. Using the second algorithm, we show that the robot outperforms, on average, a group of human subjects, in synchronizing its gesture with a reference strike. We also show that, in the tested group of users, having some time in advance is important for a human to synchronize the strike with a reference player, but, from a certain time, that good influence stops increasing.

Categories and Subject Descriptors

J.5 [**Computer Applications**]: Arts and Humanities—*Music*

Keywords

Musician-Robot Interaction, Action-Anticipation

1. INTRODUCTION

Visual and gestural anticipation play an important role in the music realm. In this work we study this aspect by embeding visual cues based anticipation in a robotic musician [6] interacting with a human percussionist.

A number of recent works have demonstrated the importance of visual cues-based anticipation for human-robot interaction in a variety of contexts [4, 1, 2]. Specifically in musician-robot interaction, Hoffman [3] reports an experiment where anticipation is used in a call-and-response module, where the robot prepares the response chord while the human is playing a musical phrase. Mizumoto et al [5] describes a coupled oscillator model used to predict the human's behavior, experimental results showing that the method reduces the average of onset time errors.

Robotic percussionists have been subject of research in our lab during the last 6 years. This work extends our previous research in interaction between musician and a percussionist robot [6]. We are providing Shimon, a socially interactive and improvisational robotic marimba player, with anticipatory capabilities.

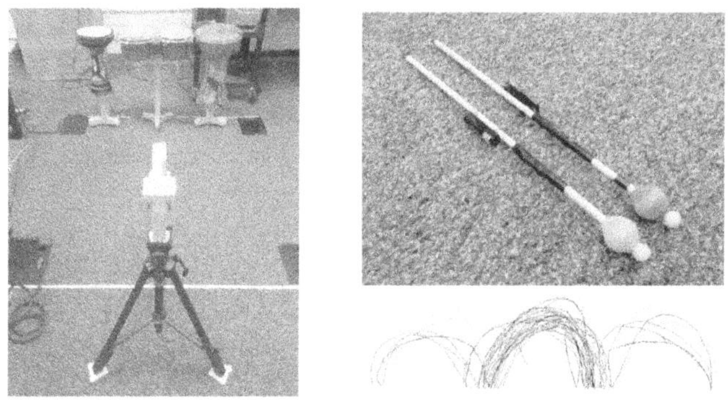

Figure 1: Left: WiiRemote pointing to the 4-drums set. Right-top: mallets with infrared LEDs. Right-bottom: some trajectories of the mallet as seen by the WiiRemote.

2. SYSTEM SETUP

We study anticipation in the context of a percussionist playing a set of 4 drums using mallets, as depicted in Figure 1. The Nintendo WiiRemote game controller is used to track the bi-dimensional coordinates of the mallets on the scene. Mallets are detected through Infrared LEDs attached to their ends.

The relevant visual data pertains to the positions of the mallets in which a strike can be described as an upward (lifting) movement and downward movement ending in contact with the drum. We are interested in identifying such a striking gesture when it begins, and predicting which drum will be hit and at what time in the future sound will be produced. Figure 1 (bottom right) shows some captured trajectories of the mallet in movement. As they are nearly parabolic, we decided to use 2nd degree polynomials to describe them.

On a first implementation, we anticipated the strike position and time when the mallet starts the downwards trajectory. We keep track of the vertical differences between consecutive capture points and detect that a strike is about to occur when a certain number of consecutive negative vertical differences happen and those differences increase in magnitude. Then a parabola is fitted to the sequence of points and the intersection of the proper branch of the parabola with the straight line at the level of the drums is calculated. This gives the prediction of the mallet to strike. To predict the

Table 1: Time differences (in milliseconds) between the strike of a reference human player and the strike of either the robot or a second human player.

Experiment	Mean	StDev
Shimon *vs.* Anticipation	19.4	16.2
Shimon *vs.* Piezo	79.0	04.8
User 1 *vs.* Ref. Player	18.9	15.7
User 2 *vs.* Ref. Player	34.2	26.6
User 3 *vs.* Ref. Player	41.1	33.5
User 4 *vs.* Ref. Player	51.6	26.0
User 5 *vs.* Ref. Player	35.1	16.5
User 6 *vs.* Ref. Player	28.7	21.9
User 7 *vs.* Ref. Player	47.8	34.4

time, a uniformly accelerated movement is supposed, and the time of strike is calculated using the equations of such a movement. An anticipation of about 10 milliseconds is obtained with this method. Now, the delay between the moment it is detected that a mallet will strike and the moment of the actual strike by the robot is of about 90 mili-seconds. So we saw the necessity of better anticipation time.

In the second case, we search for a sequence of consecutive positive velocities (upward movement), and detect that a strike will happen as soon as the mallet crosses a certain height threshold above the level of the drums. Due to measurement noise and variability in player movement, fitting a parabola does not work well in this case for anticipating the mallet to be stroke. As we are only using four drums, we found that the inclination of the straight line connecting the first and last point of the upward sequence used for prediction can be used to estimate if the mallet is going to strike the same drum it did or will hit another. Regarding the time prediction, again an uniformly accelerated movement is supposed, but this time the acceleration is set beforehand, and treated as "gravity", so that the mallet is seen as a body in free fall. The velocity is, again, computed from the most recently observed points[1]. Using this approach we were able to get an anticipation of about 100 mili-seconds (supposing an acceleration of -250, for mallets coordinates ranging from 0 to 1).

3. EXPERIMENTS

On a first experiment, we measured the time difference between the strike of the player and the strike of the robot, in two different cases: one, when the strike of the robot is triggered by the anticipation algorithm (of the second case, described previously); two, when the strike is triggered by a piezo sensor installed in the player's drum. We also measured the accuracy of the "human anticipation algorithm", that is, users were invited to strike a drum in sync with the strike of a player in front of them, by visually anticipating the players strike. For each experiment, 10 comparisons were made. To calculate the delay we used piezo sensors attached to the drum heads to measure the onsets of a strike. The results are shown in Table 1. We emphasize that, in general, the best performance is obtained with the anticipation algorithm.

We also evaluated how the amount of time in advance (in milliseconds), given to the users, relates with the quality of synchronization. Figure 2 shows our results in this direction. The synchronization does not improve considerably when including a gesture previously to the upward movement, but being able to see the upward movement before the strike is important.

[1]We recall that the displacement and velocity equations for the uniformly accelerated movement are $s_f = s_i + v_i t + at^2/2$ and $v_f = v_i + at$, respectively, where s denotes position, v velocity, t time interval and a acceleration. Subscripts "$_f$" and "$_i$" stand for "final" and "initial", respectively.

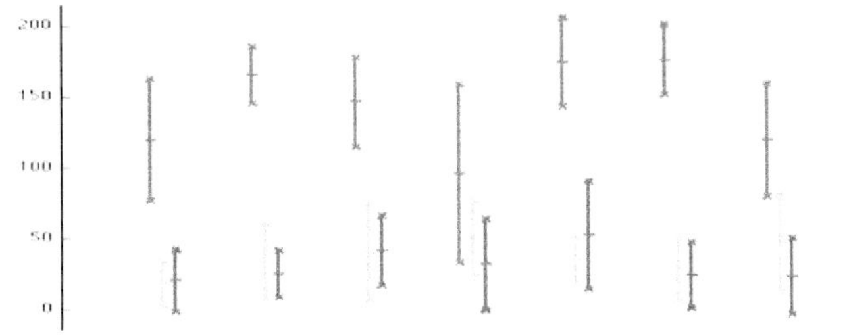

Figure 2: Each user was asked to play in sync with a reference strike performed in 3 different ways: only downward movement (shown in red), upward and downward movement (green), and gesture plus upward and downward movement (the gesture consisted of a left-right movement). This chart shows the mean and variance between the reference and user strikes, in milliseconds.

4. CONCLUSIONS AND FUTURE WORK

Our reported experiments and results are very encouraging. Although technically simple, the presented implementation already highlights the potential of anticipation in human-robot music ensembles.

As future work, we should improve the robustness of the algorithm for usage in onstage performances, and use the technology for composing new musical pieces. We are also planning to extend anticipation for other music instruments, like, for instance, in the case a guitarist is changing chords.

Currently we are working on a hybrid system where the current "parabolic movement" modeling is combined with gesture following techniques (based on HMMs).

5. REFERENCES

[1] F. Eyssel, D. Kuchenbrandt, and S. Bobinger. Effects of anticipated human-robot interaction and predictability of robot behavior on perceptions of anthropomorphism. In *Proceedings of the 6th HRI conference*, HRI '11, pages 61–68, New York, NY, USA, 2011. ACM.

[2] M. Gielniak and A. Thomaz. Anticipation in robot motion. In *Proceedings of the International Symposium on Robot and Human Interactive Communication*, 2011.

[3] G. Hoffman. Anticipation in human-robot interaction. In *AAAI Spring Symposium Series*, 2010.

[4] B. Johansson and C. Balkenius. An experimental study of anticipation in simple robot navigation. *Lecture Notes in Computer Science*, 4520:365–378, 2007.

[5] T. Mizumoto, T. Otsuka, K. Nakadai, T. Takahashi, K. Komatani, T. Ogata, and H. Okuno. Human-robot ensemble between robot thereminist and human percussionist using coupled oscillator model. In *Proceedings of the Intelligent Robots and Systems conference*, pages 1957–1963, 2010.

[6] G. Weinberg, A. Raman, and T. Mallikarjuna. Interactive jamming with shimon: a social robotic musician. In *Proceedings of the 4th HRI conference*, HRI '09, pages 233–234, New York, NY, USA, 2009. ACM.

Socially Constrained Management Of Power Resources For Social Mobile Robots

Amol Deshmukh
Heriot-Watt University
Edinburgh, United Kingdom
a.deshmukh@hw.ac.uk

Ruth Aylett
Heriot-Watt University
Edinburgh, United Kingdom
ruth@macs.hw.ac.uk

ABSTRACT

Autonomous robots acting as companions or assistants in real social environments should be able to sustain and operate over an extended period of time. Generally, autonomous mobile robots draw power from batteries to operate various sensors, actuators and perform tasks. Batteries have a limited power life and take a long time to recharge via a power source, which may impede human-robot interaction and task performance. Thus, it is important for social robots to manage their energy, this paper discusses an approach to manage power resources on mobile robot with regard to social aspects for creating life-like autonomous social robots.

Categories and Subject Descriptors

I.2.9 [**Artificial Intelligence**]: Robotics

General Terms

Human Factors, Algorithms, Theory

Keywords

Robot Companions, Power Management, Social Constraints, Human-robot Interaction

1. INTRODUCTION

In order for robots to act as companions or assistants in social environments such as homes and offices, they should be capable of operating with a great degree of autonomy over a longer period of time. While acting as an artificial social being a robot still needs to fulfill its own physiological needs. Most autonomous mobile robots draw power from batteries and take a long time to recharge. While the recharge behaviour is active, the companion robot may be prevented from performing its normal tasks and this may hinder continuous human-robot interactions.

The robot should also be competent enough to regulate the use of its power resources when it has no active tasks in order to conserve battery. We believe that companion robots should behave in a socially intelligent manner[3] and perform life-like actions to be perceived as a social being. In this paper we present an approach to power management for mobile robots which has a human-like regulation system that can balance both its physiological and social needs.

HRI'12, March 5–8, 2012, Boston, Massachusetts, USA.
ACM 978-1-4503-1063-5/12/03.

2. SCENARIO

The work reported here is carried out as a part of the EU project LIREC (LIving with Robots and IntEractive Companions), the project aims to create interactive, emotionally intelligent companions which are capable of establishing long-term relationships with humans in social environments. The "Spirit of the Building" showcase at the Heriot-Watt University, Edinburgh, aims to produce a social helper robot that can act as a "Team Buddy" an assistant within a lab inhabited by a group of people who work there and facilitate long-term relationships with users. The Team Buddy would perform tasks such as carrying the phone to users, giving out reminders, providing a lab tour for visitors, approaching and greeting users, maintaining a collective memory about user preferences such as lunch breaks and entry/exit time.

The robot (Pioneer P3AT) with an enhanced superstructure is equipped with a laptop PC, and with 6 lead acid batteries (12V, 7Ah each) offering an approximate operational time of 6 hours when fully charged (depending on usage). These batteries require about 8 hours to recharge. Considering this long recharge time and that the robot has to perform several tasks every day, there is an urgent need for a power management strategy. The robot is also equipped with electronic relays boards (controls the switching of sensors, actuators, laptop) and power sensors to measure power consumption on sensors, actuators and laptop.

3. APPROACH

The LIREC project has developed a three-level architecture in which the most abstract and top-most level reuses work from an earlier EU project[1]. FAtiMA-PSI is a body-mind architecture in the Physiological vs Cognitive dimension, where goals are originated from drives. FAtiMA is an extension of the BDI (Beliefs, Desires, Intentions) deliberative architecture [1], that incorporates a reactive component mainly responsible for emotional expressivity and it employs the OCC [5] emotional influences on the agent's decision making processes. FAtiMA architecture is integrated with PSI [4], is a psychologically-founded theory that incorporates all basic components of human action regulation such as perception, motivation, cognition, memory, learning and emotions in one model of the human psyche.

The goals in the overall architecture are derived from a set of basic drives that guide actions. Five basic drives from PSI include: Energy, Integrity, Affiliation, Certainty and Competence. Energy represents an overall need to preserve the

[1] www.e-circus.org/

existence of the robot. Integrity represents well being, i.e. the agent avoids pain or physical damage while affiliation is useful for social relationships. Certainty and competence influence cognitive processes and their values can be calculated automatically. Homeostasis is the main mechanism so that deviations from an upper and lower set point determine the strength of each need. Needs can emerge depending on activities of the robot or grow over time. To be able to produce actions that are able to satisfy needs in a certain situation, the robot builds up intentions that are stored in memory and are - when selected - the basis of a plan.

Each need has value ranging from 0 to 10, where 0 means complete deprivation while 10 means complete satisfaction. A weight ranging from 0 to 1 gives the importance of each need to the robot. In order to operate appropriately, the robot has to reduce a needs deviation from a fixed threshold as much as possible at all time. The strength of a need (Strength(d)) depends on its current strength plus the amount of deviation from the set point and the specific weight of the need

$$Strength(d) = Strength(d) + (Deviation(d) \times Weight(d))$$

By assigning appropriate weights for energy and affiliation (social) needs, robot has the ability to balance both its physiological and social needs. During the start of an interaction, the robot will have a set of initial values for all the needs. Based on the level of its current needs, the robot can generate intentions, that is, it activates goal(s) that are relevant to the perceived situation.

Each goal contains information about its expected contribution to energy, integrity and affiliation, that is, how much the drives may be deviated from or satisfied if the goal is performed. Based on this information, the importance of goal at a particular time instance can be determined, allowing the robot to give priority to goals that satisfy its needs under different situations.

For example, assume the robot has a task to give a reminder to a particular user in the lab, and that this goal can be achieved by Action A or Action B. Action A has a plan involving sub tasks such as navigating to the user's desk and speaking out the reminder message (causing a change of +4 for affiliation but a change of -2 for energy value). Action B has a plan of just speaking out the reminder message without navigating (causing +2 change for affiliation but only a change of -0.5 for energy value). The robot is likely to select Action B in a situation when its current value for energy is low but will perform the same goal. Using this approach the robot can balance its needs in physiological vs social dimension.

4. FUTURE WORK AND CONCLUSION

The architecture FAtiMA-PSI is operational in its current state, but it only takes into account fixed predefined values for needs which decay automatically with time. However fixed values for energy deviation are not realistic and these should take into account the real battery state. We would like to extend the functionality of the system to link the energy value to real battery state to more accurately represent energy decay. We also aim to develop a method to monitor and calculate the power consumption of each action and link it to deviation values for energy for each task (goal execution) refer Figure 1.

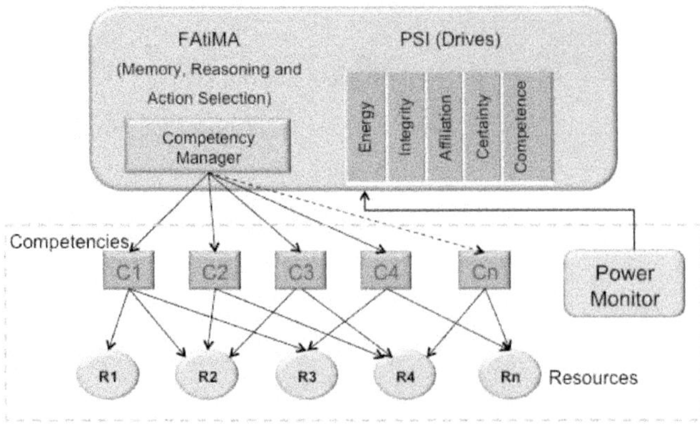

Figure 1: Proposed System

Further more we plan to implement a competency manager to manage competencies (competencies are programs that abstract physical sensors and actuators to logical ones e.g navigation, face detection etc.). The competency manager will also be responsible for regulating the switching of resources (the physical sensors and actuators) by using onboard relays on the robot in order to conserve power while not performing tasks. The regulation of resources behaviour is also found in living beings in form of homeostasis [2].

The challenges with limited battery life on mobile robots, constrain the operational time of mobile robots. We believe that companion robots should behave in a socially intelligent manner to manage the power resources on the robot. In this paper we discussed an approach to power management for social robots which has human-like regulation system that can balance both its physiological and social needs. Additionally, a continuous interaction and task performance with the human users can be maintained.

5. ACKNOWLEDGMENTS

This work is supported by European Community (EC) and is currently funded by the EU FP7 ICT-215554 project LIREC (LIving with Robots and IntEractive Companions). The authors are solely responsible for the content of this publication. It does not represent the opinion of the EC, and the EC is not responsible for any use that might be made of data appearing therein.

6. REFERENCES

[1] M. Bratman. *Intention,Plans and Practical Reasoning.* Harvard University Press, Cambridge, Massachusetts, 1987.

[2] W. B. Cannon. *Organization for physiological homeostasis.* Physiological Review, 9:399431, 1929.

[3] K. Dautenhahn. Socially intelligent robots: Dimensions of human- robot interaction. *Philosophical Transactions of the Royal Society B: Biological Sciences,* 362(1480):679–704, April 2007.

[4] Dörner. The mathematics of emotions. In *Fifth International Conference on Cognitive Modeling,* pages 75–79. Bamberg, Germany, April 2003.

[5] A. Ortony, G. Clore, and A. Collins. *The cognitive structure of emotions.* Cambridge University Press, Cambridge, UK, 1988.

Sensorless Collision Detection and Control by Physical Interaction for Wheeled Mobile Robots

Guillaume Doisy
Department of Industrial Engineering and Management
Ben-Gurion University of the Negev
Beer Sheva ISRAEL
+972 543956331
doisyg@bgu.ac.il

ABSTRACT

In this paper, we present the adaptation of a sensorless (in De Luca's sense [1], i.e., without the use of extra sensors,) collision detection approach previously used on robotic arms to mobile wheeled robots. The method is based on detecting the torque disturbance and does not require a model of the robot's dynamics. We then consider the feasibility of developing control by physical interaction strategies using the described adapted technique.

Categories and Subject Descriptors

I.2.9 [**Artificial Intelligence**]: Robotics – *Autonomous vehicles, Kinematics and dynamics, Operator interfaces, Sensors.*
K.4.1 [**Computers and Society**]: Public Policy Issues – *Human Safety.*

General Terms: Measurement, Reliability, Experimentation, Security.

Keywords: pHRI, Safety, Collision Detection.

1. SENSORLESS COLLISION DETECTION

Safety issues are a major concern when humans and robots share the same environment [2]. Thus, accidental collisions that may cause physical damage or harm humans should be anticipated and avoided. Collision avoidance, which requires knowledge of the environment and motion planning, is a mature field of robotic research, however it can still fail for numerous reasons (e.g., a simple sensor failure). There are also cases where collision avoidance cannot be implemented due to its inherent complexity and cost. Hence, actual collisions should be detected and their negative effects minimized by having the robot switch quickly to an appropriate reaction strategy; a simple and widely applicable strategy being to stop the robot.

On mobile wheeled robots, collision detection is generally achieved by dedicated sensors like bumpers, touch sensors or accelerometers. However, an approach that does not require sensors has been successfully developed for robotic arms. Its basic principle is to compare the torque applied to the actuators to execute a speed or position command with a model based torque prediction (i.e., the expected torque necessary to fulfill the command in the absence of collision) [6], [7], [8]. If the difference between the predicted and applied torque is too large, a collision is detected. The torque an electrical actuator generates is easily accessible, because its value is directly proportional to the actuator current, thus no additional sensor is needed. The last refinements of this approach [2], [3], provide such an accurate and powerful method for detecting collisions that potential soft-tissue injuries arising from a collision with a robotic arm manipulating sharp tools like a screwdriver or a scalpel can be avoided [4]. However, this approach requires an almost perfect model of the robot dynamics.

To the best of our knowledge the application of such an approach to detect collisions on a mobile wheeled robot has never been investigated. We studied the feasibility of adapting this approach on our differential drive wheeled robot manufactured by Robosoft. Our robot is a custom version of Kompaï weighing approximately 40 kg (Figure 2). The main obstacle against a direct application of the previously described method is the difficulty to obtain a good model of the robot dynamics, i.e., a model capable of predicting the nominal torque values of the motors for a given speed and acceleration. The torque needed for a given speed varies with the inclination of the ground (which can be easily measured with an accelerometer), the payload of the robot (which can be known in advance) and the friction of the floor (which is hard to measure). Moreover, irregularities on the ground cause the torque needed to follow a speed command to oscillate (Figure 1 for the oscillations caused by a flat tiled floor).

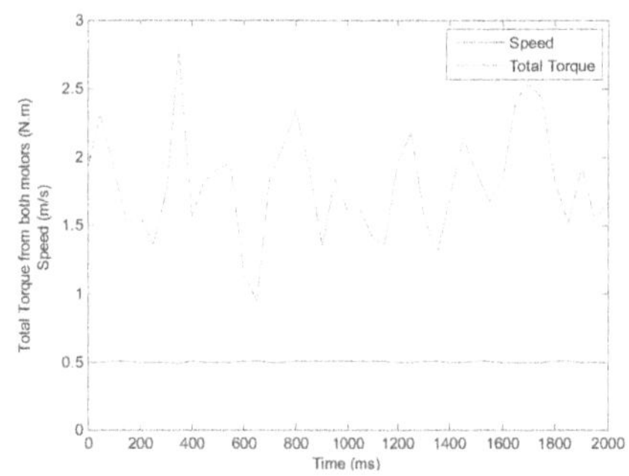

Figure 1: Variation of torque due to irregularities on the ground when following a constant speed command of 0.5 m/s

Detection of collisions by monitoring the torque output from the motors of our differential drive wheeled robot with the following simplified conditions was evaluated:

1. The robot is traveling at a constant speed (i.e., it is not in an acceleration transition phase).
2. The robot's inertia is limited and it can stop immediately.
3. The ground is flat and its properties stay constant.

In order to elude the need for a model of the robot's dynamics, the method is based on detecting the torque disturbance.

We expected a collision to create a large torque disturbance, compared to the value of the torque command, because at a constant speed on a flat floor only the friction forces oppose the movement of the robot. Hence without a collision, the total output

HRI'12, March 5–8, 2012, Boston, Massachusetts, USA.
ACM 978-1-4503-1063-5/12/03.

torque of both motors is only equal to the torque resulting from the friction force. The detection threshold is set to be 80% of the total torque value averaged over the last second. Experiments show that collisions can be reliably detected even with light objects. The sensitivity of this method was good enough to allow a human operator to stop the robot by lightly pushing the top of the robot. A picture of a human performing a stop of the robot by touching it can be seen on Figure 2, and the resulting disturbance and robot reaction are shown in Figure 3.

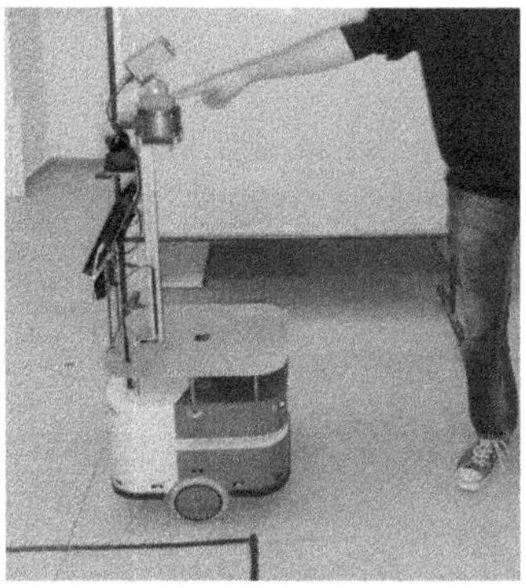

Figure 2: Human light push on the robot moving at 0.5m/s causing collision detection by the resulting disturbance on the motors output torques.

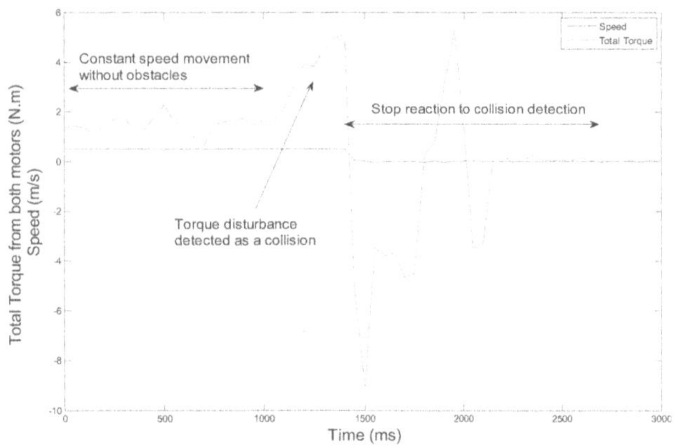

Figure 3: Torque disturbance detected as a collision and robot reaction

2. ROBOT CONTROL BY PHYSICAL INTERACTION

Seeing that it was easy and intuitive for a human to stop the robot by physical interaction (i.e., by touching it), a basic working proof-of-concept to physically control the robot was developed (Figure 4) by pushing the robot:

- on its axis to start a forward or backward movement.
- right or left of its axis to start a clockwise or anticlockwise rotation movement.
- against its movement to stop the robot.

The axis centered or decentered pushes were detected by monitoring the torque disturbance on the left and right motors while applying a zero speed command. Measured difference of torque disturbance between the right and left motors was used to determine if the push was right, left or centered compared the robot axis.

The success of this basic proof of concept proved the validity and opens new perspectives of safe physical control of wheeled robots through torque monitoring. Many control strategies can be imagined. Future work will focus on extending collision detection on acceleration phases and non-flat grounds and on developing and testing new physical control strategies.

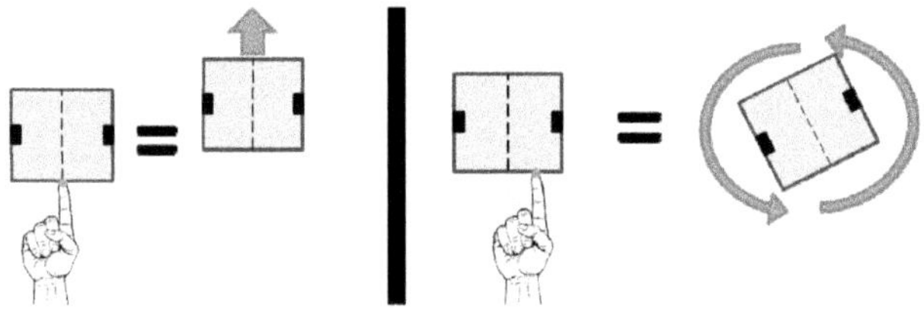

Figure 4: Tested robot control strategy

3. ACKNOWLEDGMENTS

This work has been financed by the EU FP7 INTeractive RObotics initial training research network (INTRO), grant agreement no.: 238486 and partially supported by Ben-Gurion University Paul Ivanier Center for Robotics Research and Production Management, and the Rabbi W. Gunther Plaut Chair in Manufacturing Engineering.

4. REFERENCES

[1] De Luca, A. and Mattone, R. Sensorless Robot Collision Detection and Hybrid Force/Motion Control. *Proceedings of the 2005 IEEE International Conference on Robotics and Automation*, April (2005), 999-1004.

[2] De Luca, A., Albu-Schaffer, A., Haddadin, S., and Hirzinger, G. Collision Detection and Safe Reaction with the DLR-III Lightweight Manipulator Arm. *Proceedings of the 2006 IEEE/RSJ International Conference on Intelligent Robots and Systems*, IEEE (2006), 1623-1630.

[3] Haddadin, S., Albu-Schaffer, A., De Luca, A., and Hirzinger, G. Collision detection and reaction: A contribution to safe physical Human-Robot Interaction. *Proceeding of the 2008 IEEE/RSJ International Conference on Intelligent Robots and Systems*, IEEE (2008), 3356-3363.

[4] Haddadin, S., Albu-Schäffer, A., and Hirzinger, G. Soft-tissue injury in robotics. *Proceeding of the 2010 IEEE International Conference on Robotics and Automation*, IEEE (2010), 3426-3433.

[5] Heinzmann, J. and Zelinsky, A. Quantitative Safety Guarantees for Physical Human-Robot Interaction. *International Journal of Robotics Research 22*, 7-8 (2003), 479-504.

[6] Suita, K., Yamada, Y., Tsuchida, N., Imai, K., Ikeda, H., and Sugimoto, N. A failure-to-safety "Kyozon" system with simple contact detection and stop capabilities for safe human-autonomous robot coexistence. *Proceedings of the 1995 IEEE International Conference on Robotics and Automation*, IEEE (1995), 3089 -3096 vol.3.

[7] Takakura, S., Murakami, T., and Ohnishi, K. An approach to collision detection and recovery motion in industrial robot. *Proceeding of the 15th Annual Conference of IEEE Industrial Electronics Society*, IEEE (1989), 421-426.

[8] Yamada, Y., Hirasawa, Y., Huang, S., Umetani, Y., and Suita, K. Human-robot contact in the safeguarding space. *IEEE/ASME Transactions on Mechatronics 2*, 4 (1997), 230-236.

Assitive Teleoperation for Manipulation Tasks

Anca D. Dragan
The Robotics Institute, Carnegie Mellon
adragan@cs.cmu.edu

Siddhartha S. Srinivasa
The Robotics Institute, Carnegie Mellon
siddh@cs.cmu.edu

Categories and Subject Descriptors: I.2.9 Artificial Intelligence: Robotics

Keywords: assistive robotics, human-robot interaction, teleoperation

1. INTRODUCTION

How should a human user and a robot collaborate during teleoperation? The user understands the full semantics of the task: they know, for example, what the robot should search for in a cupboard, or that it should be more careful when moving near a glass of water than near a box of tissues. Since the robot might not have this knowledge, allowing it to operate fully autonomously may be risky; its model is incomplete and its policy might be wrong. On the other hand, teleoperating the robot through every motion is slow and tiresome, especially on difficult tasks. Between these two extremes lies a spectrum, from almost no assistance at all (very timid) to full autonomy (very aggressive). So what is the appropriate level of assistance? And how do factors like task difficulty and policy correctness affect this decision?

We address these question by formalizing assistance as *policy blending*, where user input U and robot policy P are blended via a state-dependent arbitration function α:

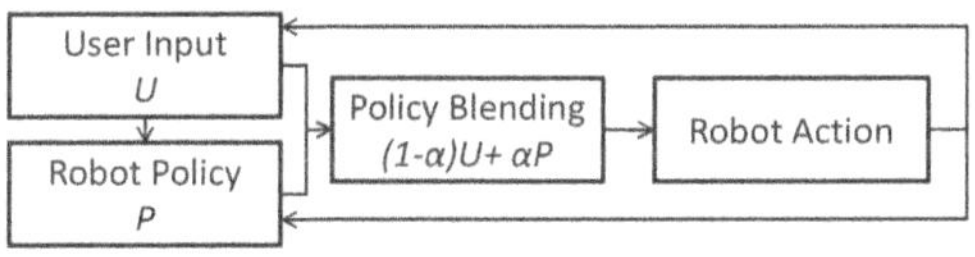

Modifying α enables us to manipulate how aggressively the robot assists. For example, $\alpha = 0$ everywhere corresponds to no assistance, and $\alpha = 1$ everywhere corresponds to full autonomy. The two α functions from Fig. 1(c) represent a timid and an aggressive mode in between these extremes. To explore the effects of α, we performed a study where users teleoperated our robot HERB to grasp objects from a table, with varying assistance from HERB's own policy for grasping the objects. We manipulated three factors: *aggressiveness* of assistance, *difficulty* of task, and *correctness* of robot policy, and measured the time to success and user preference. All users favored aggressive assistance on hard tasks, when the robot's policy was correct. In contrast, they clearly preferred timid assistance on tasks where the robot's policy was wrong. Opinions were split on easy tasks, depending on how much they wanted to remain in control of the robot. Overall, although aggressive assistance is more efficient when the policy is right, robots might make mistakes that are much easier to handle with timid assistance. This emphasizes the importance of correctness, of trust, and of legibility of motion in order for teleoperation to efficiently transfer the load from the human onto the robot.

HRI'12, March 5–8, 2012, Boston, Massachusetts, USA.
ACM 978-1-4503-1063-5/12/03.

2. PRIOR WORK

Policy blending provides a unified view of much prior work in assistive teleoperation, as different behaviors that have been proposed may be cast as different arbitration functions. For example, in some methods the robot always has full control of some local aspect of the motion (α is always 1) [2]. In others, the robot takes control at a certain trigger point, depending on proximity to the goal [4], or to a preset path [5]. There are also instances of more timid assistance, in which the robot never fully takes control. In [1], the intensity of assistance increases with the probability of the predicted user intent, but never past a maximum value.

Studies comparing modes of assistive teleoperation obtained seemingly contradictory results. Although You and Hauser [6] found aggressive assistance better, Kim et al. [3] showed that users preferred a manual mode. Our analysis indicates that the variance in task difficulty between the studies (shown in the timing difference between the timid and aggressive modes) explains the two findings and makes the results consistent.

3. EXPERIMENTAL DESIGN

We tasked users with teleoperating HERB to grasp an object from a table. We implemented a whole-body interface that tracks their skeleton (OpenNI, www.openni.org), yielding the user input U (Fig. 1 (a)). The arbitration function α blends U with a policy P.

3.1 Manipulation of the three factors

We instructed users to reach for the farther of two objects. We manipulated policy correctness by changing P: in the **right** case, P moves directly towards the target object; in the **wrong** case, P moves towards the object closest to the user's configuration (initially the wrong one). Fig. 1 (b) shows examples of the two cases in green and red, respectively. We manipulated task difficulty by changing the target object's location. Fig. 1 (b) shows an example of an **easy** (grasp the bottle) vs. a **hard** task (grasp the box). These two factors lead to 4 tasks: Easy-Right, Easy-Wrong, Hard-Right and Hard-Wrong.

We manipulated aggressiveness by modifying the arbitration function (Fig. 1 (c)). This function depends on a measure of the robot's *confidence* c in the predicted policy. For simplicity, we chose an intuitive definition of confidence: c increases as the distance d from the user's configuration to the policy's target decreases, $c = \max(0, 1 - \frac{d}{D})$. The **timid** mode increases the assistance with the confidence and plateaus at a max. value, never taking control away from the user. The **aggressive** mode takes control from the user as soon as the confidence increases past a min. value.

3.2 Subject allocation

To maximize the number of data points per condition, we chose a within-subjects design. Each of our 8 participants (all students, 4 males and 4 females) executed both modes

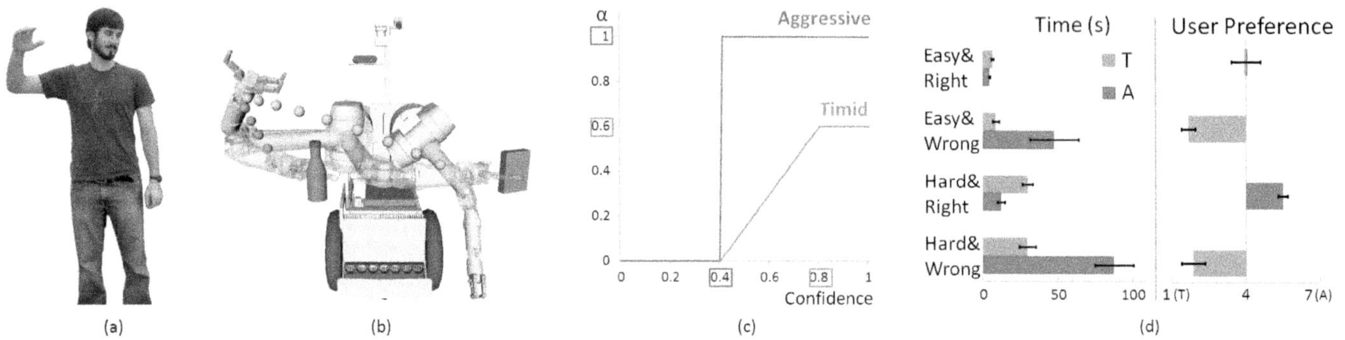

Figure 1: a) The user's arm pose serves as the input H. b) The robot chooses one of the two policies (one to each possible target) and blends it with the user input. c) The arbitration functions for timid and aggressive assistance, depending on the confidence in the policy. d) A comparison of the two modes on each task.

on each task. To avoid ordering effects, we used a balanced Latin square for the task order, and balanced the order of the modes within each task.

3.3 Dependent variables and covariates

We measured the amount of **time** (performance) per condition. We expected this to correlate to the user's **preference**. We identified the following confounds: the users' initial teleoperation skill, their rating of the robot without assistance, and the learning effect. To control for these, users went though a *training phase*, teleoperating the robot without assistance. This partially eliminated the learning effect and gave us a baseline for their timing and ratings. We used these as covariates, together with number of tasks completed at any point (a measure of in-task practice).

3.4 Hypotheses

1. *All factors have a significant effect on task performance.*
2. *The aggressive mode performs better on Hard-Right tasks, while the timid mode performs better on Easy-Wrong tasks.*

We leave as open questions what happens with both complications (Hard-Wrong), or with neither (Easy-Right).

4. RESULTS

4.1 Performance Analysis

The average time per task was approximately 28s. We performed a factorial repeated-measures ANOVA with Bonferroni corrections for multiple comparisons and a significance threshold of $p = 0.05$, which resulted in a good fit of the data ($R^2 = 0.66$). In line with our first hypothesis, we found main effects for all three factors: hard tasks took 22.9s longer than easy ones ($F(1,53)=18.45$, $p<.001$), tasks where the policy was wrong took 30.1s longer than when right ($F(1,53)=31.88$, $p<.001$), and the aggressive mode took overall 19.4s longer than the timid ($F(1,53)=13.2$, $p=.001$). We found a significant interaction effect between aggressiveness and correctness, showing that when wrong, being timid is significantly better than being aggressive. This is confirmed in Fig. 1 (d), which compares the means and standard errors on each task: the timid mode is better both on Easy-Wrong and Hard-Wrong. The same plot also shows that the timid mode performed about the same on Easy-Right, and, as expected, worse on Hard-Right (the aggressive is lower than the timid for every user). Surprisingly, the interaction effect among all factors was only marginally significant ($F(1,53)=2.63$, $p=.11$). We believe that increasing the size of our user pool will strengthen this effect.

4.2 Preference Analysis

Fig 1 (d) also shows the user's preferences on each task, which are in line with the timing results. The two measures were indeed correlated (Pearson's $r(30)=.66$, $p<.001$). The outliers were users with stronger preferences than the time difference would indicate. For example, some users strongly preferred the timid mode on Hard-Wrong tasks, despite the time difference not being as high as with other users. The opposite happened on Hard-Right tasks, on which some users strongly preferred the aggressive mode despite a small time difference, commenting that they appreciated the precision of the autonomy. On Easy-Right tasks, the opinions were split and some users preferred the timid mode despite a slightly longer time, motivating that they felt more in control of the robot. Overall, although performance difference is a good indicator of the preference, it does not capture a user's experience in its entirety.

5. CONCLUSION

We identified aggressiveness as a key aspect of assistance and showed that task difficulty and policy correctness influence the mode of assistance. We found that being wrong, yet aggressive has the largest negative effect, and that user preferences were sometimes more drastic than the performance difference would indicate. Users also commented that after seeing the robot be wrong, they did not trust the aggressive mode, and were unable to tell from its motion whether it was right. This suggests that even though autonomy makes teleoperation easier, issues of correctness, trust, and behavior legibility must be addressed for effective assistance.

6. REFERENCES

[1] D. Aarno, S. Ekvall, and D. Kragic. Adaptive virtual fixtures for machine-assisted teleoperation tasks. In *ICRA*, 2005.
[2] T. Debus, J. Stoll, R. Howe, and P. Dupont. Cooperative human and machine perception in teleoperated assembly. In *ISER*, 2000.
[3] D. Kim, R. Hazlett-Knudsen, H. Culver-Godfrey, G. Rucks, T. Cunningham, D. Porteande, J. Bricout, Z. Wang, and A. Behal. How autonomy impacts performance and satisfaction. *IEEE Trans. on Systems, Man and Cybernetics*, PP(99):1–13, 2011.
[4] J. Kofman, X. Wu, T. Luu, and S. Verma. Teleoperation of a robot manipulator using a vision-based human-robot interface. *IEEE Trans. on Industrial Electronics*, 52(5):1206 – 1219, 2005.
[5] M. Li and A. Okamura. Recognition of operator motions for real-time assistance using virtual fixtures. In *HAPTICS*, 2003.
[6] E. You and K. Hauser. Assisted teleoperation strategies for aggressively controlling a robot arm with 2d input. In *R:SS*, 2011.

'If You Sound Like Me, You Must Be More Human': On the Interplay of Robot and User Features on Human-Robot Acceptance and Anthropomorphism

Friederike Eyssel
++49 521 106 12044
feyssel@cit-ec.uni-bielefeld.de

Dieta Kuchenbrandt
Center of Excellence in Cognitive Interaction Technology (CITEC)
Bielefeld University
Universitätsstraße 21-23
33615 Bielefeld
++49 521 106 12042
dkuchenb@cit-ec.uni-bielefeld.de

Simon Bobinger
++49 521 106 12042
sbobinge@cit-ec.uni-bielefeld.de

Laura de Ruiter
Bielefeld University
Universitätsstraße 25
33615 Bielefeld
laura.deruiter@uni-bielefeld.de

Frank Hegel
Bielefeld University
Universitätsstraße 25
33615 Bielefeld
fhegel@techfak.uni-bielefeld.de

ABSTRACT
In an experiment we manipulated a robot's voice in two ways: First, we varied robot gender; second, we equipped the robot with a human-like or a robot-like synthesized voice. Moreover, we took into account user gender and tested effects of these factors on human–robot acceptance, psychological closeness and psychological anthropomorphism. When participants formed an impression of a same-gender robot, the robot was perceived more positively. Participants also felt more psychological closeness to the same-gender robot. Similarly, the same-gender robot was anthropomorphized more strongly, but only when it utilized a human-like voice. Results indicate that a projection mechanism could underlie these effects.

Categories and Subject Descriptors
J.4 [**Social and Behavioral Sciences**]: *psychology*.

General Terms
Theory, Experimentation, Measurement.

Keywords
Human-Robot Interaction; Anthropomorphism; Gender Stereotypes; Social Robotics.

1. INTRODUCTION
Previous research has shown that people interact with computers in ways that are comparable to human–human interactions [1]. For example, Nass and colleagues have demonstrated repeatedly that people instinctively treat nonliving entities just like humans [2]. Similarly, in the field of human–robot interaction (HRI), we have shown that people use visual gender cues as a basis for their judgments about social robots. That is, they used visual gender cues, such as a robot's hair length in order to make gender-stereotypical inferences about the robot [3]. However, such effects of social category information are not restricted to features of a robot itself. [4], for instance, have shown that additional to the robot gender also the user gender affects people's reactions toward robots. Moreover, in recent research, we have demonstrated that the users' own group membership affects subsequent evaluations of a robot [5, 6]. Taken together, features of the robot interaction partner as well as characteristics of the user determine how people react toward robots. To date, however, no study has yet investigated the interplay of robot features and characteristics of the user on HRI acceptance and psychological anthropomorphism of the robot. To address this research gap, we conducted an experiment in which we tested the simultaneous effects of a robot's vocal features (i.e., human-like vs. robot-like voice; gender of robot voice) and user gender on HRI acceptance, perceived psychological closeness and anthropomorphism.

2. METHOD
58 students (31 women, 27 men; mean age $M = 22.98$, $SD = 2.81$) were randomly assigned to one of four experimental conditions that resulted from the 2 (*gender of robot voice*: male vs. female) x 2 (*voice type*: human-like vs. robot-like) between-subjects design. In addition, we analyzed the factor *participant gender*.

After arriving in the laboratory, participants were presented with a short video of the robot Flobi [7] and were asked to form an impression of it. Importantly, the robot appeared gender-neutral as its modular industrial design allowed us to omit the robot's hair mask. Thus, participants only relied on the robot's vocal cues as a basis for their judgments. In the video, Flobi uttered the neutral sentence 'It is quarter past three', either with a *male human* voice, a *female human* voice, a *male robotic* voice, or a *female robotic* voice. All voice samples were selected based on pretests regarding human-likeness vs. robot-likeness and vocal femininity vs. masculinity.

After watching the video clip, participants completed a set of computerized questionnaires. To collect participants' responses to the dependent measures, 7-point Likert scales were used. For subsequent data analyses, average scores were computed to form indices of the respective dimensions. To measure *HRI acceptance*, we used four items (e.g., 'How much would you like to get acquainted with the robot?') [see 8] that formed a reliable index (α = .83). To assess the degree of *psychological closeness*, participants responded to five items (e.g., 'To what extent do you feel similar to the robot', [see 8], (α = .85)). In order to assess participants' *anthropomorphic inferences* about the robot, we measured the extent to which participants attributed *mind* to the robot. Participants rated the robot with regard to 24 mental capacities (e. g., the capacity to feel pain, hunger, or to make plans, [see 8, 9], (α = .91)).

3. RESULTS

To investigate the effects of the factors *gender of robot voice*, *voice type* and *participant gender* on our dependent measures, we conducted 2 (*gender of robot voice*: male vs. female) x 2 (*voice type*: human-like vs. robot-like) x 2 (*participant gender*: male vs. female) multivariate analyses of variance (MANOVA).

For *HRI acceptance* we found a *gender of robot voice* x *participant gender* interaction, $F(1, 50) = 6.61$, $p = .02$. Subsequent t-tests revealed: In tendency, female participants accepted the female robot more ($M = 3.30$ $SD = 1.67$) than the male one ($M = 2.72$, $SD = 0.71$), $t(29) = 1.31$, $p = .20$. Similarly, male participants showed marginally greater acceptance of the male robot ($M = 3.41$, $SD = 1.44$) relative to the robot with a female voice ($M = 2.52$, $SD = 0.80$), $t(25) = -1.96$, $p = .06$. With regard to *psychological closeness* to the robot, the *gender of robot voice* x *participant gender* interaction was significant, $F(1, 50) = 5.56$, $p = .02$. Subsequent t-tests revealed that in tendency, female participants felt psychologically closer to the female robot ($M = 2.11$, $SD = 1.04$) than to the male robot ($M = 1.71$, $SD = 0.58$), $t(29) = 1.36$, $p = .19$. Likewise, male participants felt significantly more psychological closeness to the male ($M = 2.26$ $SD = 1.04$) than to the female robot ($M = 1.57$, $SD = 0.51$), $t(25) = -2.14$, $p = .04$. Regarding *anthropomorphism*, analyses revealed a significant three-way interaction, $F(1, 50) = 4.89$, $p = .03$. We thus conducted separate ANOVAs for *voice type*. For the human voice condition, we obtained a significant interaction of *gender of robot voice* and *participant gender*, $F(1, 29) = 9.00$, $p = .005$. That is, male and female participants both anthropomorphized the robot with the same-gender human voice more strongly compared to the robot with the opposite-gender human voice. However, for the robot with the robotic voice, no such effect was found.

4. DISCUSSION

The present research is the first that simultaneously tested effects of robot features (human-likeness and gender) and user characteristics on HRI acceptance and psychological anthropomorphism. This research goes beyond previous work on effects of vocal cues in robot systems, because we manipulated not only robot gender, but also its human-likeness by means of vocal cues. Furthermore, we took into account user gender and examined the interplay of these factors on novel dependent measures: As predicted, participants showed greater HRI acceptance and felt psychologically closer to the robot when robot and participants shared the same gender. Moreover, participants even anthropomorphized a system more strongly when it used a same-gender, but human-like voice. From our findings we conclude that projection of one's own attributes to the robot could be one important mechanism that underlies anthropomorphic and evaluative judgments about robots. Moreover, similarity between the robot and the user could be a key aspect that moderates these processes. However, in future work we need to test these assumptions more directly. In sum, our findings clearly point out that future research in social robotics shall take into account idiosyncratic features of both user and robot system in order to build and advance user-friendly, acceptable technological systems.

5. ACKNOWLEDGMENTS

This research was funded as part of the Cluster of Excellence in Cognitive Interaction Technology (CoE 277) by the German Research Council (DFG). We thank our research assistants for their help with pretesting of voice samples and data collection.

6. REFERENCES

[1] Reeves, B. and Nass, C. 1996. *The media equation: How people treat computers, television, and new media like real people and places.* Cambridge University Press, New York, NY.

[2] Nass, C., Moon, Y. and Green, N. 1997. Are machines gender neutral? Gender-stereotypic responses to computers with voices. *Journal of Applied Social Psychology*, 27, 864-876.

[3] Eyssel, F. and Hegel, F. (in press). (S)he's got the look: Gender stereotyping of social robots. *Journal of Applied Social Psychology.*

[4] Crowell, C. R., Scheutz, M., Schermerhorn, P. and Villano, M. 2009. Gendered voice and robot entities: Perceptions and reactions of male and female subjects. In *Proceedings of the 2009 IEEE/RSJ International Conference on Intelligent Robots and Systems (IROS 2009),* 3735-3741.

[5] Eyssel, F. and Kuchenbrandt, D. (in press). Social categorization of social robots: Anthropomorphism as a function of robot group membership. *British Journal of Social Psychology.*

[6] Kuchenbrandt, D., Eyssel, F., Bobinger, S. and Neufeld. M. 2011. Minimal group - maximal effect? Evaluation and anthropomorphization of the humanoid robot NAO. *Proceedings of the 3rd International Conference on Social Robotics,* 104-113.

[7] Hegel, F., Eyssel, F. and Wrede, B. 2010. The social robot Flobi: Key concepts of industrial design. In *Proceedings of the 19th IEEE International Symposium in Robot and Human Interactive Communication (RO-MAN 2010)*, 120-125.

[8] Eyssel, F., Kuchenbrandt, D. and Bobinger, S. 2011. Effects of anticipated human-robot interaction and predictability of robot behavior on perceptions of anthropomorphism. *Proceedings of the 6th ACM/IEEE Conference on Human-Robot Interaction*, 61-67.

[9] Gray, H. M., Gray, K. and Wegner, D. M. 2007. Dimensions of mind perception. *Science, 315*, 619.

Beyond "Spatial Ability": Examining the Impact of Multiple Individual Differences in a Perception by Proxy Framework

Thomas Fincannon
University of Central Florida
3100 Technology Pkwy
Orlando, FL 32826
tfincann@ist.ucf.edu

Florian Jentsch
University of Central Florida
3100 Technology Pkwy
Orlando, FL 32826
Florian.Jentsch@ucf.edu

Brittany Sellers
University of Central Florida
3100 Technology Pkwy
Orlando, FL 32826
bsellers@ist.ucf.edu

Joseph R. Keebler
University of Central Florida
3100 Technology Pkwy
Orlando, FL 32826
joekeebler@gmail.com

ABSTRACT
Prior research has proposed the use of a *Perception by Proxy* framework that relies on human perception to support actions of autonomy. Given the importance of human perception, this framework highlights the need to understand how human cognitive abilities factor into the human-robot dynamic. The following paper uses a military reconnaissance task to examine how cognitive abilities interact with the gradual implementation of autonomy in a *Perception by Proxy* framework (i.e., autonomy to detect; autonomy to support rerouting) to predict three dimensions of sequential performance (i.e., speeded detection; target identification; rerouting). Results showed that, in addition to effects of autonomy and task setting, different individual abilities predicted unique aspects of performance. This highlights the need to broaden consideration of cognitive abilities in HRI.

Categories and Subject Descriptors
H.1.2 [**Models and Principles**]: Human/Machine Systems Human Factors and Human Information Processing– *human factors, human information processing.*

General Terms
Experimentation, Human Factors, Management, Performance.

Keywords
Cognitive Abilities, Perception by Proxy.

1. INTRODUCTION
In the domain of human-robot interaction (HRI), interest has been directed toward the concept of teaming between humans and robots. Not surprisingly, this trend has led to a variety of models that can be used to discuss various dynamics. Within this body of work, researchers have recently proposed a framework of *Perception by Proxy* [1]. According to this perspective, the perceptual limitations of a robot create a need for a human counterpart to act as a perceptual proxy. Given that research on "spatial ability" has shown that operators are not equally proficient at performing this task [2], there is a concurrent need to understand how individual differences in cognitive ability will impact performance. This paper intends to explore this dynamic.

1.1 Perception by Proxy
As described by Atherton and Goodrich [1], robots can be limited in their ability to perceive elements of a remote environment, and this can hinder performance. It also creates a need to understand the strengths and weaknesses of a robot, such that the autonomy allows for a request for help when appropriate. It is important to note, however, that this requires the robot to first perform a perceptual task (e.g., detection of unknown stimulus), which may require human supervision. As a result, human performance of these precursors of requests for support should be understood.

Existing research with human operators has focused on concurrent perceptual tasks (e.g., identification & localization) that precede synthesis and decision making [2], but *Perception by Proxy* is somewhat unique in that it highlights how perceptual tasks are performed sequentially. In a military example, detection of an unknown obstruction that is in the path of an unmanned vehicle (UV) would precede interventions to stop the vehicle, identify the stimulus, and follow the rules of engagement (ROEs) accordingly. This element of speeded detection by an UV operator has not received much attention, and this implies a need to understand human performance for appropriate supervision.

1.2 Cognitive Abilities
Given the importance of human perception for a *Perception by Proxy* framework, there is a need to understand human cognition and how individual differences factor into this relationship. Much of the research in this area focuses generally on spatial ability, but our research has have advocated a more construct driven approach [2]. Using Carroll's [3] model of intelligence, which provides one of the extensive meta analyses on the subject, we found [2] that much of the existing research on spatial ability uses measures that load onto on a single construct (i.e., visualization [or VZ]). Furthermore, we argued for a broader consideration of other constructs that have not received much consideration in existing HRI literature. The study reported here examined the following abilities that:

- Visualization (VZ): the ability to mentally perform complex manipulations to visual objects
- Closure Speed (CS): the ability to recognize obscured objects that are stored in long term memory
- Perceptual Speed (P): the speed in finding a known pattern in a visual field

1.3 Purpose
The purpose of this paper is to provide an exploratory analysis of how different cognitive abilities interact with a gradual implementation of a *Perception by Proxy* framework. Considering that the implementation of autonomy in the field is not expected to be fully reliable [1], this should provide an initial understanding of how multiple cognitive abilities support human supervision of a robotic teammate.

2. METHOD
In the study, 39 students from the University of Central Florida participated for course credit. Using training and methodologies that were consistent with prior work [2], participants were asked to use two simulated UVs to complete four reconnaissance

HRI'12, March 5–8, 2012, Boston, Massachusetts, USA.
ACM 978-1-4503-1063-5/12/03.

missions in a scaled Military Operations in Urban Terrain (MOUT) environment. A reengineered commercial-off-the-shelf (COTS) vehicle was used as the unmanned ground vehicle (UGV), and the unmanned air vehicle (UAV) was simulated by an overhead camera attached to an automated pulley system. For the task, participants used the vehicles to follow preplanned routes and comply with the specified ROEs (i.e., stipulations regarding how to reroute) to support operations. Performance was focused on safe operations for the UGV, which involved: (1) stopping the UGV when personnel and vehicles were encountered unexpectedly (i.e., speeded detection); (2) correctly identifying these targets (e.g., a republican guard soldier, a group of civilians, a T-80 tank, etc.) based on the identification booklet provided; and (3) using this information to reroute (i.e., following ROEs).

As prescribed by Carroll's [3] model, the Vandenberg and Kuse test of Mental Rotation was used to assess VZ; the Hidden Word test was used to assess CS; and the Finding A's test was used to assess P. Performance was assessed based on the three safe operations tasks mentioned above. Subjective workload was assessed with using the NASA TLX.

Semi-autonomous waypoint control was used to reach preplanned objectives, but a manipulation to autonomy that mimicked a *Perception by Proxy* framework was used to react to unexpected stimuli that were encountered along these routes. There were three levels of autonomy that included *manual stop* (i.e., when there were unexpected stimuli along the route, the operator had to stop the UGV in order to identify the stimulus and determine the appropriate ROE), *autonomous stop* (i.e., the UGV stopped automatically, and the operator had to identify the stimulus to determine the appropriate ROE), and *collaborative autonomy* (i.e., the UGV would autonomously stop the vehicle, and the UAV would use its autonomy to identify information provided by the operator to suggest an appropriate ROE). As in prior work [2], confederates were used in a man-behind-the-curtain methodology.

3. RESULTS

To examine relevant relationships, the manipulation was dummy coded for autonomous detection (i.e., whether the operator or autonomy stops the UGV) and autonomous decision making (i.e., whether the operator or autonomy uses target identity to suggest ROEs alterations). Multilevel linear modeling was then used examine interactions between autonomy and cognitive ability.

With respect to speeded detection needed for *stopping the UGV at an unknown obstruction*, positive associations were observed for VZ [$F(1, 33) = 13.22, p < .05$], CS[$F(1, 33) = 20.99, p < .05$], and automated detection [$F(1, 33) = 101.23, p < .05$]. Furthermore, automated detection was found to interact with VZ [$F(1, 33) = 4.73, p < .05$] and CS[$F(1, 33) = 16.39, p < .05$], such that spatial ability only improved performance without autonomous support.

In terms of *following ROEs*, identification [$F(1, 150) = 44.28, p < .05$], autonomous decision making [$F(1, 150) = 50.48, p < .05$], and VZ [$F(1, 150) = 8.14, p < .05$] were all found to have a main effect. Interestingly, P was also found to interact with identification [$F(1, 150) = 4.65, p < .05$], such that accurate reconnaissance was necessary for P to improve ROE performance.

With respect to the NASA TLX, P was found to have a negative association with workload [$F(1, 33) = 4.15, p < .05$], in that operators with greater P perceived lower workload. No other variable had a significant impact on this construct.

4. DISCUSSION

Summary of Findings. Together, the results indicated that, in addition to effects of autonomy and task setting, different individual abilities predicted unique aspects of performance. First, in the speeded detection task of stopping at an unexpected obstruction, operators with high *VZ* or *CS* abilities did best and could do as well as the autonomous system. In contrast, with respect to the more analytical task of following ROEs, *only* operators with high *VZ* ability could perform as well as the autonomy. Finally, perceptual speed *P* was not critical for either of the performance metrics, but it was associated with lower workload estimates, replicating findings of dissociations between workload and performance.

Interpretation. With respect to the individual differences, each construct demonstrated a unique relationship with different outcomes for performance and workload. The impact of VZ on speeded detection and implementation of reconnaissance to follow ROEs showed that this construct had the most significant impact on performance as a whole. This is consistent with Carroll's [3] depiction of VZ as broad factor of power, or level of mastery, for mental transformation, which would be expected to form an association with a variety of perceptual tasks.

According to Carroll's model [3], CS is a speeded dimension of visual intelligence that does not require any mental manipulation of an object. As a result, it was not surprising to see that it accounted for unique variance in the prediction of a speeded detection task.

At first glance, results regarding P may seem counter-intuitive. While speed is a component of this ability, it was not associated with speeded detection in our study. Instead, this ability was associated with workload and following ROE as a measure of synthesis, which implies that this ability is more strongly associated with decision making. As reviewed by Carroll [3], P can be viewed as a marker of cognitive speed, which is theorized to be related to working memory, and the results are more consistent with this explanation.

Overall, this study illustrated how a *Perception by Proxy* framework can be used to highlight sequential measures of perception that exhibit unique relationships with cognitive abilities that are not commonly examined in HRI. Further research will be required to explore relevant theory.

5. REFERENCES

[1] Atherton, J. and Goodrich, M. 2011. Perception by Proxy: Humans helping robots see in a manipulation task. *Proceedings of the 6th International Conference on Human Robot Interaction*, 109-110.

[2] Fincannon, T., Evans, A.W.III, Jentsch, F., and Keebler, J.R. 2010. Dimensions of spatial ability and their influence on performance with unmanned systems. *In*: D. H. Andrews, R. P. Herz, & M. B. Wolf, eds. *Human factors in defense: Human factors in combat identification.* Burlington, VT: Ashgate Publishing, 67-81.

[3] Carroll, J. 1993. *Human cognitive abilities: A survey of factor analytic studies.* New York: Cambridge U. Press.

Attentional Human-Robot Interaction in Simple Manipulation Tasks

Ernesto Burattini
DSF, Univ. di Napoli FedericoII
via Cinthia, I-80126
Napoli, Italy
ernb@na.infn.it

Alberto Finzi, Silvia Rossi
DSF, Univ. di Napoli FedericoII
via Cinthia, I-80126
Napoli, Italy
{finzi, srossi}@na.infn.it

Mariacarla Staffa
DIS, Univ. di Napoli FedericoII
Via Claudio, 21, I-80125
Napoli, Italy
mariacarla.staffa@unina.it

ABSTRACT

We present a robotic control system endowed with attentional mechanisms suitable for balancing the trade off between safe human-robot interaction and effective task execution. These mechanisms allow the robot to increase or decrease the degree of attention toward relevant activities modulating the frequency of the monitoring rate and the speed associated to the robot movements. In this framework, we consider pick-and-place and give-and-receive attentional behaviors.

Categories and Subject Descriptors

I.2.9 [**Robotics**]: [Operator interfaces, Commercial robots and applications, Manipulators]

Keywords

Attentional systems, cognitive control, social robotics

1. INTRODUCTION

We propose an attentional system suitable for balancing the trade-off between safe human-robot interaction and effective task execution [4, 2]. The attentional system is to supervise and orchestrate the human-robot interaction activities monitoring their safety and effectiveness. Our attentional executive system is obtained as a reactive, behavior-based system, endowed with simple, bottom-up, attentional mechanisms. We assume a frequency-based model of the executive attention, based on the Adaptive Innate Releasing Mechanism (AIRM) model [1], where each behavior is endowed with an adaptive internal clock that regulates its sensors' sampling rates and activations. The clock frequency is here interpreted as an attentional mechanism: the higher the clock frequency, the higher the resolution at which the behavior is monitored and controlled. These attentional mechanisms regulate two conflicting requirements: (1) safe interaction with the humans; (2) effective cooperation in interactive tasks. Depending on the disposition and the attitude of a person in the environment, the sensing rates and behaviors activations are increased and decreased changing the overall attentional state of the system.

HRI'12, March 5–8, 2012, Boston, Massachusetts, USA.
ACM 978-1-4503-1063-5/12/03.

2. ATTENTIONAL HRI

We considered a robotic manipulator (7DOF Cyton Arm endowed with Videre STOC 9cm stereo cam.) that is to cooperate with a human operator in pick-and-place and give-and-receive tasks [3].

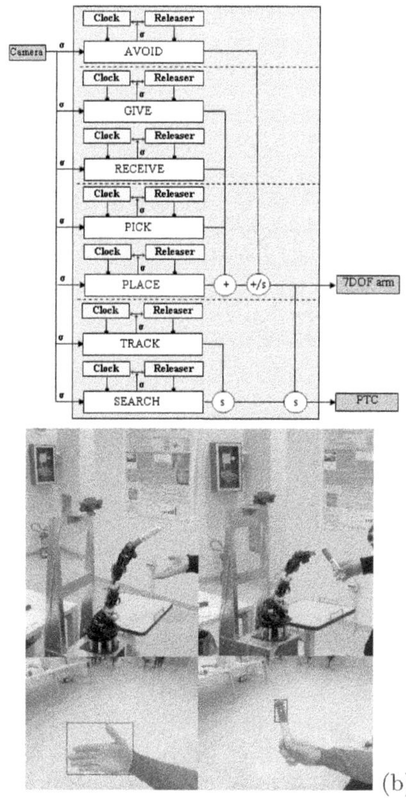

Figure 1: (a) Behavior-based architecture for HRI. (b) Give and Receive scenarios.

Depending on the context, the robotic system should: look for an operator to interact with; give or receive an object to/from the operator; pick or place an object from/into a location. Each of these tasks is to be monitored to avoid dangerous/unsafe situations.

Attentional Behaviors. The architecture of our executive system is depicted in Fig. 1. It integrates attentional behaviors for pick and place, give and receive, but also behaviors for search and track (humans and objects) and behaviors regulating the avoid-approach attitude of the robotic system. Following the AIRM model, each behavior is endowed with an adaptive clock and a *monitoring strategy* characterized by a sampling period $p^t \in [p_{min}, p_{max}]$ and an *updating function* $f(t) : \mathbb{R}^n \rightarrow \mathbb{R}$ that changes the clock's period p^t. The behaviors' description follows.

`SEARCH` controls the pan-tilt unit providing an attentional and constant scan of the environment looking for humans and objects. Once a human is detected in the robot workspace, the `TRACK` behavior is enabled to monitor humans before they enter in the interaction space; hence the adaptive clock is regulated in accordance with the human motions and positions. Intuitively, a human that moves fast and close needs to be carefully monitored (high frequency, foreground), while a human that moves far and slow can be monitored in a more relaxed manner (low frequency, background). Specifically, if $\sigma_{hm}(t)$ is the human distance from the robot camera, the period update p^t_{tr} is regulated according to the human relative position and the velocity (i.e. $p^t_{tr} = \Theta_{tr}(\sigma_{hm}(t), \frac{\sigma_{hm}(t)-\sigma_{hm}(t-p^{t-1}_{tr})}{p^{t-1}_{tr}}))$. `AVOID` controls the arm motion speed and can stop the motion whenever a situation is assessed as dangerous. It is enabled when a human is detected in the robot interaction area and its adaptive clock period p^t_{av} depends on the operator proximity and motion σ_{op}, (i.e., analogously to `TRACK`, $p^t_{av} = \Theta_{av}(\sigma_{op}(t), \frac{\sigma_{op}(t)-\sigma_{op}(t-p^{t-1}_{av})}{p^{t-1}_{av}}))$. The output of this behavior results in a speed deceleration associated with high frequencies, i.e. $speed = \frac{max\_speed \times p^t_{av}}{p_{av\_max}}$, if the operator is in the interaction space; otherwise, the arm will stop when the operator is inside the proximity space. The `PICK` behavior is activated when the robot is not holding and object, but there exists a reachable object in the robot interactive space. It moves the robot end-effector towards the object, activates a grasping procedure and, once the robot holds the object, moves this in a predefined safe position. In this case the clock period is associated with the distance of the object $\sigma_{obj}(t)$ (i.e. $p^t_{pk} = \phi_{pk}(\alpha, \sigma_{obj}(t)))$. Analogously to `AVOID`, the clock frequency determines also speed variations (i.e. $speed = \frac{max\_speed \times p^t_{pk}}{p_{pk\_max}}$), hence, the arm can smoothly reduce its speed to a minimum value to execute a precision grip receiving more frequent camera information (higher monitoring frequency). `PLACE` is activated when the robot is holding an object in the absence of interacting humans. It moves the robot end effector towards a target position, it places the object and moves the robot arm to a final position. The clock period and the speed are regulated by functions analogous to ones of `PICK`. The `GIVE` and `RECEIVE` behaviors are activated by object and gesture detection. These behaviors regulate the activities of giving and receiving objects w.r.t. both the humans proximity and their movements, hence the clock period is associated with both the objects' distance and the speed of the operator hand, i.e. $p^t_{gv} = \Theta_{gv}(\gamma_{obj}(\|\sigma_{obj}(t) - ee_{pos}(t)\|), \gamma_{op}(\frac{\sigma_{op}(t)-\sigma_{op}(t-p^{t-1}_{gv})}{p^{t-1}_{gv}})$ where $\sigma_{obj}(t)$ and $ee_{pos}(t)$ are the positions of the object and the end-effector, $\sigma_{op}(t)$ is the hand operator position, γ_{obj} and γ_{op} set the period, resp., w.r.t. the position and the speed, while $\Theta_{gv}(x)$ combines these two functions in a weighted sum. The clock frequency regulates not only the sampling rate, but also the velocity of the arm movements. Intuitively, also here we assume that when attention is needed the movement should be more carefully monitored, thus slower. As for the `RECEIVE` behavior, it is activated when the robot detects a human in the operative space holding a reachable object in his hand. The behavior sampling rate and the adaptive velocity are regulated by a functions analogous to the ones for `GIVE`.

	Reliability		Safety	
	AIRM	P1VMED	AIRM	P1VMED
Receive	100%	100%	100%	100%
Give	83%	80%	90%	84%
Pick	77%	54%	100 %	100%
Place	100%	100%	100 %	100%

Table 1: Safety and Reliability.

	E↑ectiveness		E¡ciency	
	AIRM	P1Vmed	AIRM	P1Vmed
Receive	$7.66s \pm 0.54s$	$9.69s \pm 0.31s$	14.5 ± 1.57	41.7 ± 1.42
Give	$4.87s \pm 1.4s$	$7.27s \pm 2.9s$	6.05 ± 2.65	14.65 ± 5.59
Pick	$9.14s \pm 2.07s$	$10.48s \pm 0.67s$	16.2 ± 6.58	32.65 ± 5.99
Place	$6.03s \pm 1.05s$	$8.96s \pm 0.6s$	12.95 ± 5.03	58.65 ± 10.17

Table 2: Effectiveness and Efficiency.

Experimental results. To assess the system performance we introduced the optimization function: $f = M1 \times N_{Safe} + M2 \times N_{Rel} + M3 \times T_{Effe} + M4 \times T_{Effi}$ (with M_i weight to specify the priority); N_{Rel} is the number of unrecovered situations w.r.t. the accomplished activities (pick, place, etc.); N_{safe} the unsafe situations w.r.t. the executed activities; T_{Effe} is the time spent to achieve the tasks w.r.t. the overall mission time; N_{Effi} is the number of activations w.r.t. the maximal enabled (for each p_{min}). We compared the attentional system (AIRM) with a non-rhythmic version of the same architecture (P1Vmed). For the (AIRM) we considered adaptive concurrent clocks with $p_{min} = 1$, $p_{max} = 10$ and $speed = \frac{max-speed \times p}{p_{max}}$ for all the behaviors. For the (P1Vmed), we assumed a fixed clock period (i.e., p_{min} and p_{max} are both equal to 1) and the arm speed is set to a constant value ($speed = \frac{max-speed}{2}$). Moreover, in the case of the AIRM, the updating policies are those specified above. The range of values for the speed is $[0; 0.3]m/s$. We tested each behavior 20 times with 5 different nonexpert operators. Operators were required to interact with the robot, without any specific instruction. The results are depicted in Table 1 (N_{Rel}, N_{Safe}) and Table 2 (mean and std. dev. of T_{Effe} and T_{Effi}). Here, the attentional mechanisms, not only are associated with better performance in terms of effectiveness (time to achieve the goal) and efficiency (activations) (Table 1), but we also observed improved reliability and safety w.r.t. the non-adaptive architecture (P1Vmed) (Table 2).

Acknowledgments. The research leading to these results has received funding from the European Community's FP7 under grant agreements no. 216239 and no. 287513.

3. REFERENCES

[1] E. Burattini and S. Rossi. Periodic adaptive activation of behaviors in robotic system. *IJPRAI*, 22(5):987–999, 2008.

[2] R. Cooper and T. Shallice. Contention scheduling and the control of routine activities. *Cognitive Neuropsychology*, 17:297–338, 2000.

[3] A. Edsinger and C. C. Kemp. Human-robot interaction for cooperative manipulation: Handing objects to one another. In *RO-MAN*, pages 1167–1172, 2007.

[4] D. Norman and T. Shallice. Attention in action: willed and automatic control of behaviour. *Consciousness and Self-regulation: adv. in res. and theory*, 4:1–18, 1986.

'Midas Touch' in Human-Robot Interaction: Evidence from Event-Related Potentials during the Ultimatum Game

Haruaki Fukuda[1], Masahiro Shiomi[2], Kayako Nakagawa[2,3], & Kazuhiro Ueda[1]

[1]Department of General System Studies, The University of Tokyo & Crest, Japan Science and Technology Agency
Tokyo Japan
{fukuda, ueda}@cs.c.u-tokyo.ac.jp

[2]Intelligent Robotics and Communication Laboratory
ATR
Kyoto Japan
m-shiomi@atr.jp

[3]Department of Systems Innovation
Osaka University
Osaka Japan
kayako_nkgw@atr.jp

ABSTRACT

Interpersonal touch is said to have significant effects on social interaction. We used the ultimatum game to examine whether touch from a robot could inhibit a negative feeling to the robot. We set two experimental conditions: the one was "touch condition" in which unfair proposals were offered to a participant when a robot touched his/her arm and the other was "no touch condition" in which unfair proposals were offered when the same robot did not. We compared Medial Frontal Negativity (MFN) measured by EEG, whose amplitude is correlated with feeling of unfairness, between the two conditions. Result shows that MFN amplitude was larger in the no touch condition than in the touch condition. This indicates that touch from a robot may inhibit a sense of unfairness for the robot. Our finding suggests that touch from a robot could enhance positive feeling to the robot through human-robot interaction.

Categories and Subject Descriptors

J.4 [**Social and Behavioral Science**]: psychology

General Terms

Experimentation, Human Factors

Keywords

Midas touch, the ultimatum game, electroencephalogram (EEG)

1. INTRODUCTION

Interpersonal touch plays a very important role in our social interactions. A short touch from another person can elicit strong emotional experiences. Touch can also be a powerful way to gain compliance of or to persuade someone.

In the literature of social psychology, many studies have demonstrated that touch have powerful effects on communication. For example, Crusco and Wiezel [1] showed that a waitress who touched customers got more tip from the customers than a waitress who did not touch them. They called this effect 'Midas touch', which means that touches make money since Midas in Greece Myths is said to have been able to change everything he touched to money. Likewise, Guéguen and Fischer-Lokou [2] showed that bus drivers were more likely to give a passenger a free ride if the passenger touched the bus drivers than if the passenger did not. Joule and Guéguen [3] demonstrated that people were more likely to give someone a free cigarette if the request came from a person who touched them at the same time. Thus interpersonal touch could improve tolerance toward beneficiaries.

In this study, we examined whether or not touch from a robot could have the Midas touch effect through human-robot interaction. For this purpose, we employed the ultimatum game with a robot in our experiment.

The ultimatum game (UG) is widely used to examine the beneficiary's tolerance for unfairness. In the UG, one of two players (the proposer) receives a sum of money which he/she is to divide between himself/herself and the other player (the responder). The proposer can split the money any way he/she wants and can even opt to keep everything for himself/herself. The responder can then either accept or reject the offer: if the responder accepts, the money is divided as proposed; if the responder rejects, on the other hand, both players receive nothing. According to the self-gain maximization principle, responders should accept any offer, as something is better than nothing. Although this is economically optimal, this is not how real people behave. When faced with unfair divisions, responders often reject offers. This demonstrates that social and emotional response to unfair offers leads our irrational decision-making. In recent years, neural correlates of such irrational decision-making in the UG were investigated [4]. Boksem and De Cremer [5] demonstrated that the amplitude of Medial Frontal Negativity (MFN), which is one of the event-related components measured by EEG, was correlated with a sense of unfairness in the UG.

2. EXPERIMENT

In this experiment, we compared the amplitude of MFN when unfair offer was proposed in the UG between the two conditions; the one was "touch condition" in which unfair proposals were offered with touch from a robot and the other was "no touch condition" in which unfair proposals were offered without touch from robot. In this way we examined whether the Midas touch effect could be found even in human-robot interaction.

2.1 Participants

Fifteen graduate and undergraduate students participated in this experiment: eight were assigned to the touch condition and seven were assigned to the no touch condition.

2.2 Robot and Apparatus

Geodesic EEG System 300 and HydroCel Geodesic Sensor Nets 64ch (EGI) were used to acquire EEG data and a small communication robot, Robobie-mR2 (ATR Robotics) was used as

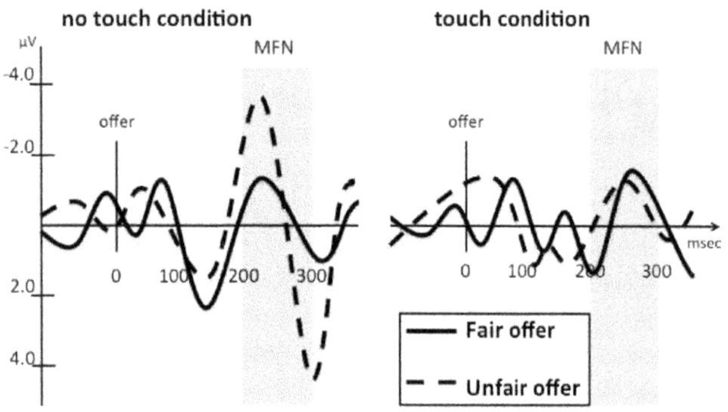

Figure 1. Grand mean average waves in each condition.

the partner in the UG. Robot's motions and speeches were preprogrammed, triggered by the experimenter and performed in the form of Wizard-of-Oz (the same manner as in [6]).

2.3 Experimental Procedure

Each participant performed 20 trials of the UG with the robot and measured EEG at the same time. The robot was the proposer and the participants were the responders in all trials. Only in the touch condition, the robot stroked the back of participants' right hand when the robot offered its proposal. The robot proposed unfair offers (80%-20% division) at 4 trials and fair offers (50%-50% division) at 4 trials in 20 trials. In the other 12 trials, nearly fair offers (60%-40% and 40%-60% divisions) were proposed.

2.4 Measurement

We compared the amplitude of MFN between the touch and the no touch conditions. MFN was determined as a negative component with a peak in parieto-frontal region around 250 msec after each offer was proposed.

2.5 Hypothesis

If touch from a robot could inhibit a sense of unfairness, the amplitude of MFN would be smaller in the touch condition than in the no touch condition. We thus hypothesized that the amplitude of MFN would be smaller in the touch condition than in the no touch condition.

3. RESULT

All participants accepted all fair offers and rejected all unfair offers. The grand mean average waves of ERPs in parieto-frontal region for the two conditions are shown in Figure 1. When unfair offers were proposed, the larger amplitude of MFN appeared only in the no touch condition. The peak amplitude of MFN is showed in Figure 2. Result shows that the amplitude of MFN was larger in the no touch condition than in the touch condition (t (13) = 3.87, p < .05) if unfair offers were proposed. Thus this result supports our hypothesis.

4. DISCUSSION

Our results indicate that sense of unfairness was reduced only in the touch condition, which suggests the possibility that the Midas touch effect occurs even in human-robot communication.

It is unclear why touch from a robot should have such a strong effect. Rose [7] argued that interpersonal touch effects might be due to cognitive interpersonal factor. That is, a recipient tends to assume that a 'toucher' is in genuine need and that the toucher likes and trusts the recipient. The perception of either great need

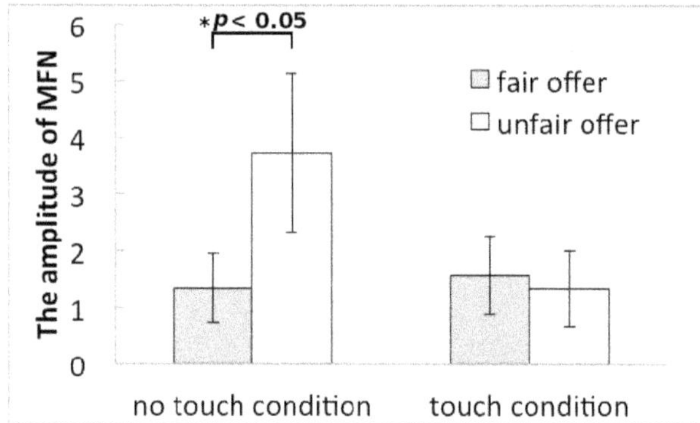

Figure 2. The amplitude of MFN in each condition.

or a positive feeling in turn tends to increase compliance rate. Our result is contingent with this argument, suggesting that touch from a robot decreased a feeling of unfairness. In this study, using a robot in a cognitive neuroscience experiment enabled us to investigate brain activity during interaction that is usually difficult to be strictly controlled in human-human interaction.

In the experiment, we demonstrated that the Midas touch effect occurs in human-robot interaction. A robot has a real body unlike other agents. Touch communication can maximize the effect of such physical embodiment of a robot [8]. Our result suggests the possibility that touch can be one of the powerful communication channels in human-robot interaction.

5. ACKNOWLEDGEMENTS

This work was supported by the Japan Society for the Promotion of Science, Grants-in-Aid for Scientific Research No. 21118005 and 21118003.

6. REFERENCES

[1] Crusco, A. H., and Witzel, C. G. 1984. The Midas touch: the effect of interpersonal touch on restaurant tipping. *Pers Soc Psychol B,* 10, 512-517.

[2] Guéguen, N., and Fischer-Lokou, J. 2003. Another evaluation of touch and helping behaviour. *Psychol Rep,* 92, 62-64.

[3] Joule, R.V., and Guéguen, N. 2007. Touch, compliance, and awareness of tactile contact. *Percept Motor Skill* 104, 581-588.

[4] Sanfey, A. G., Rilling, J. K., Aronson, J. A., Nystrom, L. E., and Cohen, J. D. 2003. The neural basis of economic decision-making in the ultimatum game. *Science*, 300, 1755-1758.

[5] Boksem, M. A. S., and De Cremer, D. 2010. Fairness concerns predict medial frontal negativity amplitude in ultimatum bargaining. *Soc Neurosci*, 5, 118-128.

[6] Nakagawa, K., Shiomi, M. Shinozawa, K., Matsumura, R., Ishiguro, H., and Hagita, N. 2011. Effect of robot's active touch on people's motivation. In *Proceeding of the 6th international conference on Human-robot interaction* HRI '11. ACM, New York, NY, 465-472.

[7] Rose, S.A., 1990. The sense of touch. In: Bernard, K., Brazelton, T.B. (Eds.), *Touch: The Foundation of Experience.* International Universities Press, Madison, WI, 299-324.

Facial Gesture Recognition using Active Appearance Models based on Neural Evolution *

Jorge García Bueno
Robotics Lab
Av. Universidad 30
28911 Madrid, Spain
jgbueno@ing.uc3m.es

Miguel González-Fierro
Robotics Lab
Av. Universidad 30
28911 Madrid, Spain
mgpalaci@ing.uc3m.es

Luis Moreno
Robotics Lab
Av. Universidad 30
28911 Madrid, Spain
moreno@ing.uc3m.es

Carlos Balaguer
Robotics Lab
Av. Universidad 30
28911 Madrid, Spain
balaguer@ing.uc3m.es

ABSTRACT

Facial gesture recognition is one of the main topics in HRI. We have developed a novel algorithm who allows to detect emotional states, like happiness, sadness or emotionless. A humanoid robot is able to detect these states with a ratio of success of 83% and interact in consequence. We use Active Appearance Models (AAMs) to determinate face features and classify the emotions using neural evolution, based on neural networks and differential evolution algorithm.

Categories and Subject Descriptors: I.2.6 [Artificial Intelligence]: Learning – *Connectionism and neural nets*

General Terms: Algorithms

Keywords: gesture recognition, HRI, neural evolution, AAM, humanoid

1. INTRODUCTION

Humanoid robot research of last years has produced a remarkable improvement of robot abilities, mobility and interaction capability. An intelligent and skillful robot requires natural interaction and complex behaviors to perform tasks and offer services to humans. It also needs cognitive models to understand human emotions and expressions. Visual recognition of facial gestures can be useful in accomplishing natural and robust human-robot interaction.

Pattern recognition and computer vision techniques have been successfully used for many gesture recognition systems [2] in very different robots. It involves face tracking, shape detection of facial features, clustering, optical flow, optimization and classification. Facial feature extraction techniques rely on the detection and tracking of several face parameters like mouth shape or eyebrows distance. An approach that has been probed very effective is based on Hidden Markov Models (HMMs). Facial muscles variations are modeled and classified and in [3] the facial expression is decomposed in submotions to enhance the performance.

*A video of the experiments can be found at `http://goo.gl/PSD9Q` When the human smiles the robot reacts with a movement of happiness, if the human is sad, the robot performs a movement of sadness, if the human is neutral the robot waits. The image in the right corner of the video is what the robot sees. This image is sent to a external server where our algorithm is computed.

HRI'12, March 5–8, 2012, Boston, Massachusetts, USA.
ACM 978-1-4503-1063-5/12/03.

2. ACTIVE APPEARANCE MODELS

Active Appearance Models (AAM), first introduced by [1], is a method to match deformable statistical models of shape and appearance to an image. The algorithm consists in fitting a AAM model to an image minimizing the error between the input image and the closest deformable model.

To build a model, a set of landmarks of enhanced points are selected . With this set of landmarks, a mesh is wrapped to the image using Delaunay triangles. Using a training set of several images and PCA reduction, the model of the shape and the appearance is a linear combination of the shapes and appearance of all training images.

$$\begin{aligned} x &= \bar{x} + Q_s c \\ g &= \bar{g} + Q_g c \end{aligned} \tag{1}$$

Equation (1) represents the shape and appearance models, where $\bar{x}$ is the mean shape, $\bar{g}$ is the mean normalized texture, c is the parameter controlling the shape and appearance and Q_s and Q_p are matrices representing the shape and appearance variations derived from the training image set.

3. NEURAL EVOLUTION

A novel method for pattern recognition is presented based on a genetic algorithm named Differential Evolution (DE) in conjunction with a Neural Network (NN) in charge of the evaluation process. The genetic optimizer minimizes the global error by tuning the weights and biases of the NN.

3.1 Differential Evolution

As in other population based optimization methods, DE generates new points (perturbations) based on previous points. Those deviations are not reflections as other solutions such as CRS or Nelder-Mead. Those new points are randomly selected from three individuals. Two elements x_{r1}, x_{r2} are subtracted and multiply by a weight (weight and mutation)

F and a third point x_{r3} is added giving the trial vector

$$u_0 = x_{r3} + F \cdot (x_{r1} - x_{r2}) \qquad (2)$$

In the selection step, the trial vector u_0 is compared with the rest of the vectors with the same index. This representation is called the *selection and storage* where the lowest cost vector is taken as the member for the next generation. The process is repeated until a population has competed against the trial vector randomly generated.

3.2 Proposed algorithm

NN is trained with DE instead of classical methods such as back-propagation. The optimized values are the weights and biases of input, hidden and output layers, giving as a result the optimal NN. The input parameters are the location of the features for each face obtained using AAM's location algorithm. To initialize the AAM, AdaBoost classifier has been used with Haar-Like features to determine the original position of the face on the frame as it is shown in Figure 3.2.

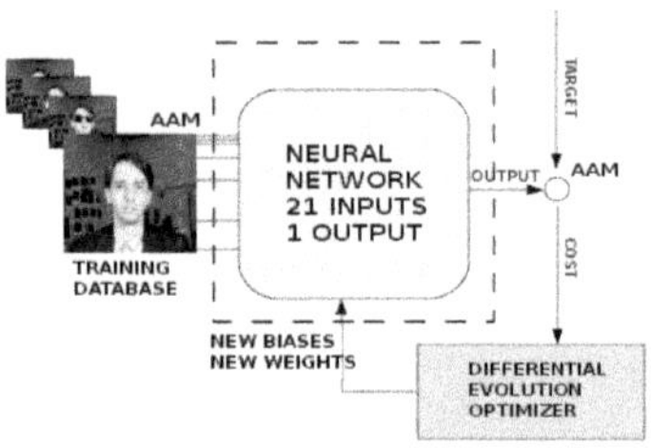

Figure 1: Proposed System mixing AAM with a NN optimized with Differential Evolution. Weights and biases are tuned by the optimizer.

4. EXPERIMENTAL RESULTS

We have constructed a database of 5 faces of several subjects with 20 images per subject, showing different expressions and postures. To compute the AMM[1], 63 characteristic points has been tagged in every image. These points correspond to the important features that define a face, such as eyes, chin, nose, mouth and eyebrows.

The model is trained with all 20 images for every subject, producing a statistical model of shape and appearance. The output is a face tracker, consisting in a mesh of Delaunay triangles which wraps to the subject face (see Figure 4 left). The representation of the deformable face model is computed using splines that connect some of the landmarks (Figure 4 right). Once the AAM model has been obtained, we use that information to characterize facial expressions. The expressions that our algorithm currently identify are happy, sad and neutral (no expression).

The algorithm has been proved in the small humanoid HOAP-3 in real time. The robot detects the face of a human and computes the algorithm. If the human is happy, the robot performs a movement of happiness, moving the arms up and down, if the human is sad, the robot performs a movement expressing sadness, looking down, finally, if the human is neutral, the robot waits. All robot trajectories are preprogrammed and calculated using inverse kinematics. The algorithm is computed in a external server and the

[1]The AAM algorithm used here is based on the code of Dr. Radhika Vathsan at BITS Pilani Goa Campus, India.

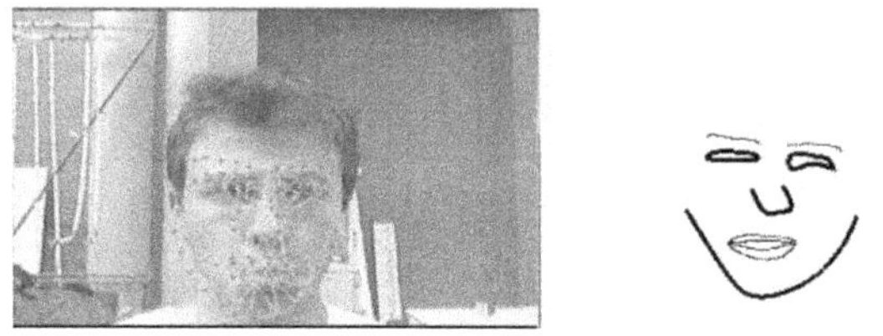

Figure 2: Face tracker mesh (left) and face spline representation (right).

mean response is 1.5 sec. In Figure 4 some snapshots of the experiments are shown showing the results of the algorithm.

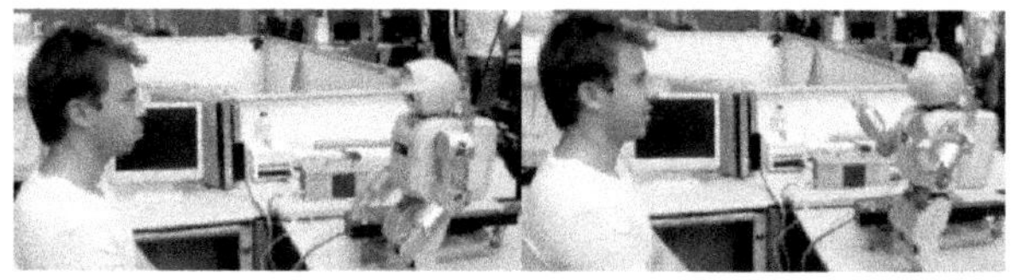

Figure 3: Recognition of happy and sad expressions and interaction of the robot.

In Table 1 a evaluation of the robustness of the algorithm is shown. In a initial phase, the algorithm has been tested using static images. Afterwards, the algorithm has been tested in real time with the humanoid robot. The table shows the only the correct detections.

Table 1: Results of facial gesture recognition

	Static	Real Time	Overall
Happy	96%	90%	96%
Sad	78%	67%	72.5%
Neutral	87%	79%	83%

5. CONCLUSIONS

In this paper a new approach of human-robot interaction has been presented. First, AAM models have been used to obtain a statistical representation of the human face. Extracting some important features of the mesh resulting of the AAM algorithm, like mouth and eyebrows shape, a neural network has been trained in conjunction with a global optimizer to distinguish between three different states, happy, sad, and normal (no emotion).

6. REFERENCES

[1] G. Edwards, C. Taylor, and T. Cootes. Interpreting face images using active appearance models. In *Automatic Face and Gesture Recognition. Proc. Third IEEE International Conference on*, pp 300–305, 1998

[2] Mitra, S., and Acharya, T. Systems, Man, and Cybernetics, Part C: Applications and Reviews. *IEEE Transactions on Gesture recognition*, pp 311–324 vol.37, 2007

[3] Arsic, D., Schenk, J., Schuller, B Submotions for hidden markov model based dynamic facial action recog. In *Image Processing IEEE International Conference on*, pp 673–676, 2006

Design, Integration and Test of a Shopping Assistance Robot System

M. Garcia-Arroyo
Maestría en Inteligencia Artificial
Sebastian Camacho No. 5
Xalapa, Ver. Mexico
morfeos_ms@hotmail.com

L. F. Marin-Urias
Departamento en Inteligencia Artificial
Sebastian Camacho No. 5
Xalapa, Ver. Mexico
luisfelipe.marin@ieee.org

A. Marin-Hernandez
Departamento en Inteligencia Artificial
Sebastian Camacho No. 5
Xalapa, Ver. Mexico
anmarin@uv.mx

G. de J. Hoyos-Rivera
Departamento en Inteligencia Artificial
Sebastian Camacho No. 5
Xalapa, Ver. Mexico
ghoyos@uv.mx

ABSTRACT

In this paper is described the current work towards the design of a shopping assistant robot system. This system will allow users to keep control of what they are buying; the robot will help the customer by handling its shopping list, carrying with all the products, and serving as a companion. In this work is also shown the acceptability studies for this kind of robot.

Categories and Subject Descriptors

H.1.2 [**User/Machine Systems**]: Human factors; H.5.2 [**User Interfaces**]: Graphical user interfaces

General Terms

Human factors, Design, Algorithms.

Keywords

Shopping assistance robot, mobile robots, interfaces, mobile devices, QR codes

1. INTRODUCTION

Although supermarkets are well structured, they carry with the issue of having to walk through all the corridors with the goal of not to forget anything. This problem is due to the big quantity of products in increasingly larger spaces, and it is bigger problem with elderly or mobility-impaired users. Even if there are employees for assisting the customers, sometimes they are too busy or they don't have the appropriate customer service philosophy. The above underlines the need to design a system that can facilitate the process of grocery shopping, bringing assistance and company to the customers.

HRI'12, March 5–8, 2012, Boston, Massachusetts, USA.
ACM 978-1-4503-1063-5/12/03.

Currently there are some systems that provide assistance to the customers, Shekar et al.[5] can advice users on what products to buy based on nutrition criteria and price constraints. Items can be added to the shopping list by scanning their barcodes or through the web. In [2] is proposed a system composed of a main server and a handheld device (PSA) which provides barcode scanning, product information and product comparison. The PSA understands simple commands to find products and it can make short sentences to chat with the customer. The robot shown in[4] chat with the customers and guide them to the products they are looking for, using speech and deictic gestures. The robot recognizes the customers using RFID tags, and the conversation is generated by a human operator to avoid voice recognition. Also, in [3] the authors are developing a robot named "TOOMAS" that moves around the store looking for customers who potentially need assistance, it approaches to them offering help, the interesting feature of this robot is that the users doesn't need training on how to use the system because all the activities perform by the users are explained by the robot itself. Finally, Beer et al. [1] conducted a needs assessment among older adults to understand the perceptions of the benefits, concerns, and adoption criteria for mobile remote presence systems. Obtaining 66% of positive opinions, 28% mixed and only 6% of negative opinions.

2. DESIGN & INTEGRATION

The main architecture of the system can be observed in fig.1, the system is divided in three sections depending of the setting in which it will be used. The first setting is when the user is at home, an application in the user's PC will be used to write a shopping list. When the user selects the option "upload list" in his/her mobile device, the app in the PC sends the list to the mobile device where it will be stored. In this way, the user can consult the shopping list at any moment.

The second stage of the system begins when the user arrives to the supermarket, there he/she can send the list to the computer in the supermarket using the option "send list"

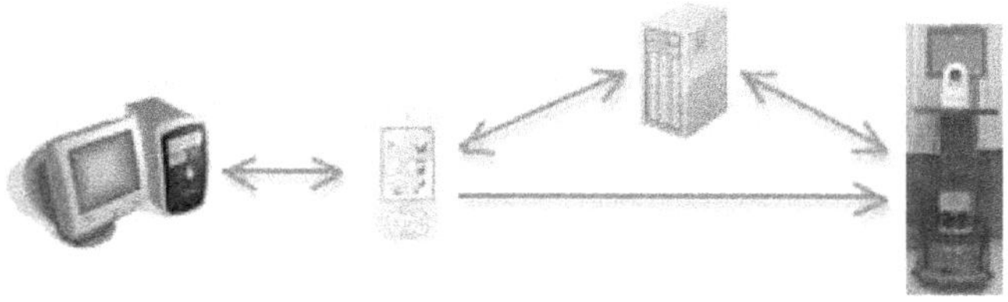

Figure 1: System architecture: Users PC, mobile device, store computer and robot assistant. Note: The communication between mobile device and the robot is given through a QR code

from the application on the mobile device. The computer verifies that he/she is a known client and, when it is the case, authorizes the transmission of the shopping list.

In the final stage, a QR code is displayed in the mobile device, this code can be shown to one of the multiple robots that will be in the entrance of the supermarket. The robot will decode the QR code and request the computer in the store for the user's shopping list, then the list will be displayed on the screen of the robot. The list displayed is color coded, each color represents a state in which the product can be, i.e. the name is red when a product is in the list but is not yet in the robot's basket, the name is colored green when is in the list and it was added to the basket and finally, its colored yellow if it was not in the list but it is added to the basket. Also, the price an quantity of each product is displayed in the interface to keep control of the products in the basket, two buttons were added to subtract and add unities. In the top of the interface the total cost of the list is displayed. See fig.2.

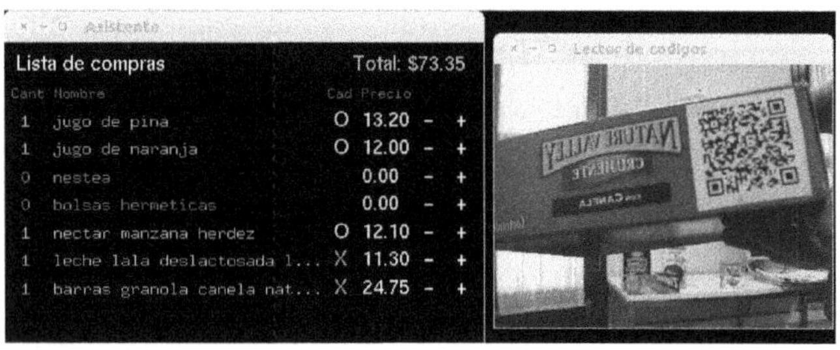

Figure 2: Robot interface. The shopping list and the image retrieved from the camera.

The information of the product was stored "in the product" using QR codes, with the intention of avoid making several calls to the server querying for information. We store the name of the product, expiration date and price. When the user shows a product with QR code to the robotic assistant, it can read the code extracting the name of the product and comparing it with the items of the shopping list, if the robot finds a match, changes the color of the item using the color code mentioned above. If it doesn't find a match it will simple add the new product to the list and recalculate the total cost of the list.

3. RESULTS & DISCUSSION

We made a study of the work done so far, currently the system does the interactions described above, but a module for following people is not yet implemented, we took 51 participants in the street between 17 and 73 years old, and they were invited to use the system, i.e. they wrote a list in a computer, they transmitted it to the mobile device and then they send it to the robot, finally they showed products to the robot (to add products) and removed products of the list. After that, we asked them to answer a few questions about the system. The questions include: ease of using and appearance of the mobile device and robot interfaces and overall usefulness. The questionnaire retrieved a good acceptance of the system parts, especially among female users, although they find the overall system a little less useful than the male participants (see fig.3).

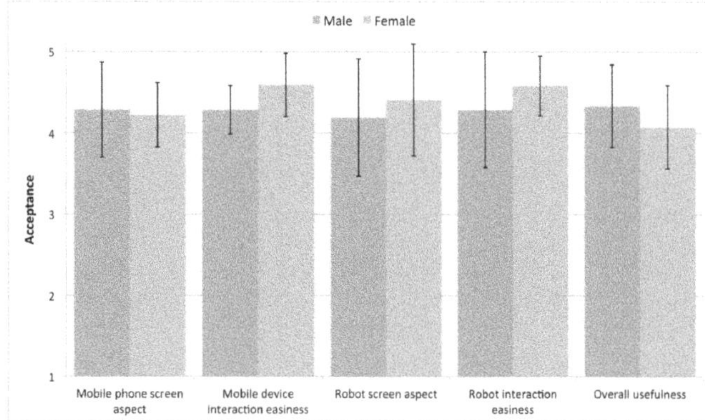

Figure 3: Male and female acceptance results of the system components.

As an extension, we consider to include important information about the product (i.e. offers, related products, and suggestions). Also a user following module will be added to ensure that the robotic assistant accompanies the customer through the corridors. Also the stored shopping list could be used to extract shopping habits and make suggestions to the client of new products related with the already purchased.

4. REFERENCES

[1] *Mobile Remote Presence Systems for Older Adults: Acceptance, Benefits, and Concerns*, Lausanne, CH, 2011. ACM Press.

[2] A. Asthana, M. Cravatts, and P. Krzyzanowski. An indoor wireless system for personalized shopping assistance. In *In Proceedings of IEEE Workshop on Mobile Computing Systems and Applications*, pages 69–74. IEEE Computer Society Press, 1994.

[3] H.-M. Gross and Al. Toomas: Interactive shopping guide robots in everyday use - final implementation and experiences from long-term field trials. In *Intelligent Robots and Systems. IROS 2009. IEEE/RSJ International Conference on*, pages 2005–2012, oct. 2009.

[4] T. Kanda and Al. A communication robot in a shopping mall. *Robotics, IEEE Transactions on*, 26(5):897–913, oct. 2010.

[5] S. Shekar, P. Nair, and A. S. Helal. igrocer: a ubiquitous and pervasive smart grocery shopping system. In *Proceedings of the 2003 ACM symposium on Applied computing*, SAC '03, pages 645–652, 2003.

Handheld Operator Control Unit

[Extended Abstract]

Neal Checka, Shawn Schaffert, David Demirdjian, Jan Falkowski, and Daniel H Grollman
Vecna Robotics
36 Cambridgepark Drive
Cambridge, MA
{nchecka, shawn.schaffert, david.demirdjian, jfalkowski, daniel.grollman}@vecna.com

ABSTRACT

Currently, unmanned vehicles support soldiers in a variety of military applications. Typically, a specially-trained user teleoperates these platforms using a large and bulky Operator Control Unit (OCU). The operator's total attention is required for controlling the tedious, low-level aspects of the platform, dramatically reducing his personal situational awareness. Furthermore, these OCUs are both platform and mission specific. Ideally, a soldier could instead carry light-weight and portable multi-purpose devices to act as OCUs for multiple platform/mission scenarios. These devices would support a standard set of OCU functionality (e.g., as driving a ground robot) and additional higher-level task operations (e.g., autonomously patrolling an area). This extended abstract presents the development of apps for a handheld platform that enable both low- and high-level control of an unmanned vehicle.

Categories and Subject Descriptors

J.7 [**Computer Applications**]: Computers in other Systems—*Command and Control, Military*

Keywords

Handheld Devices

1. INTRODUCTION

Robots have made increasingly large strides into the modern military, aiding and protecting soldiers by performing a variety of Intelligence, Surveillance and Reconnaissance (ISR); Explosive Ordnance Disposal (OED); and logistics tasks. Typically, these robots still require direct and constant attention and control by one or more humans via teleoperation. Normally, this is achieved via Operator Control Unit (OCU) such as the one shown in Figure 1. These devices are bulky, heavy, power-hungry, require extensive training of the operator, and are often mission-specific. Thus, for true on-the-ground adaptability, multiple OCUs must be carried, precluding their use in a dismounted setting.

Instead, we propose that OCU software be built for flexibility, with the ability to run on any and many available device(s) at once. Leveraging the "one application, one function" maxim of mobile application development, we envision a suite of applications that can be woven together across hardware in a seamless fashion to address whatever current situation faces the user. Figure 2 shows how two handheld devices, running different applications, can be merged to accomplish a remote surveillance task.

To be a true force multiplier, multiple robots will need to be effectively controlled by a single user. Current OCUs often require that the operator control the robot's every small movement, fully immersing themselves in the robot's world, and losing situational awareness of their own surroundings [5]. By leveraging recent work in autonomous robot behavior and combining it with intuitive control interfaces, we aim to enable one human to command a fleet of robots.

Figure 1: Current mission-specific OCUs demand a specially-trained operator's undivided attention.

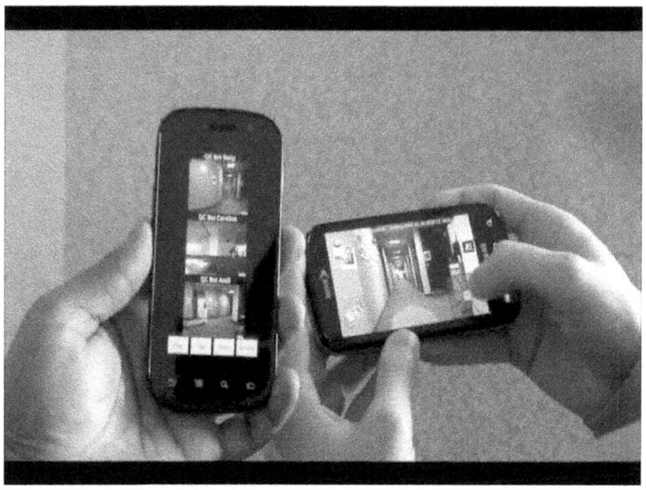

Figure 2: Multiple hardware devices and apps can be swiftly combined to accomplish mission goals.

HRI'12, March 5–8, 2012, Boston, Massachusetts, USA.
ACM 978-1-4503-1063-5/12/03.

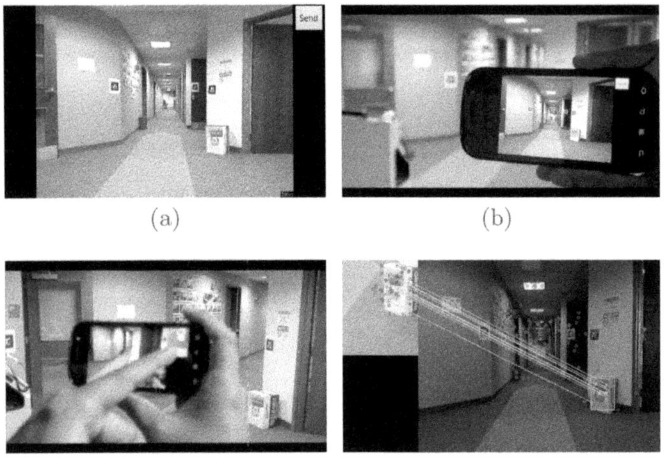

(a) (b) (c) (d)

Figure 3: Visual waypoints are defined using two methods. In (a) and (b) the user selects an object or region of interest in the robot's view. Alternatively, in (c) the user takes a picture of a waypoint and sends it to robot. (d) The robot's perception system guides it to the pictured location.

2. HANDHELD OCU

We have developed applications for commercial smartphones to control autonomous robots. While our robot is a shopping cart-sized skid-steer ground vehicle, our apps should be applicable to other systems. All apps were developed for a commercial Android-based device.

2.1 Navigation

The first step in using a robot is often to maneuver it into place. Similar to devices such as the Wii, we can use the accelerometers to provide pitch/tilt steering control of the robot. When accelerometers are lacking or if the user chooses, on-screen joysticks or buttons can be used instead.

If the robot is provided with an autonomous navigation system, the user can instead control it at a higher level, by defining waypoints on a map, for example. Within a scene, a robot can be commanded to move to a particular place by selecting locations in a camera view, such as in Figures 3(a) and 3(b). Both of these above techniques could already be realized with current OCU devices. However, by leveraging the sensors in common smartphone devices, we can do better.

Figure 3(c) illustrates remote visual waypoints, where a geo-tagged image of the desired object is captured with the handheld device and sent to the robot. Multiple such waypoints can be collected in advance, and stored for later navigation. The robot's autonomy can then navigate from wherever it is to the pictured location, using both GPS and bag-of-visual-words-based image matching [2]. Particular objects in a location can be identified via feature matching.

2.2 Gestures

Another use of the unique sensor and form factor capabilities of handheld OCUs is to enable gesture control of the robot [3]. For example, Figure 4 illustrates some motions that can be performed whilst holding the OCU, and which we recognize via the accelerometers using Dynamic Time Warping [4]. These gestures can then be mapped to discrete commands to the robot, ones that might require dedicated buttons on a traditional OCU.

Figure 4: Reference gestures for the accelerometer-based gesture recognition module.

3. FUTURE WORK

While we have developed several OCU apps, they have yet to be quantitatively analyzed. However, we believe that in order to realize increased user fan out [1], it will not be enough that the physical OCU is smaller and more adaptable. The robot itself must take on some of the responsibility for mission completion by becoming more autonomous. To that goal we are currently focused on incorporating more task-level intelligence into our system. Doing so will hopefully enhance the user's productivity by freeing them from low-level, tedious control, allowing them to focus on higher level duties, an assumption that will be tested with user feedback. Particular applications we are looking at involve point-and-click navigation and manipulation, person following, and autonomous surveillance.

Acknowledgements

This work was sponsored by the Defense Advanced Research Projects Agency (DARPA) Contract No. W31P4Q-11-C-0031.

4. REFERENCES

[1] D.R. Olsen Jr., S. B. Wood, and J. Turner. Metrics for human driving of multiple robots. In *International Conference on Robotics and Automation*, 2004.

[2] D. Nister and H. Stewenius. Scalable recognition with a vocabulary tree. In *Computer Vision, Pattern Recognition*, volume 5, 2006.

[3] O. Rogalla, M. Ehrenmann, R. Zollner, R. Becher, and R. Dillmann. Using Gesture and Speech Control for Commanding a Robot Assistant. In *International Workshop on Robot and Human Interactive Communication*, page 454, 2002.

[4] S. Salvador and P. Chan. FastDTW: Toward accurate dynamic time warping in linear time and space. In *KDD Workshop on Mining Temporal and Sequential Data*, pages 70–80, 2004.

[5] H. Yanco and J. Drury. Where Am I? Acquiring Situation Awareness Using a Remote Robot Platform. In *Conference on Systems, Man and Cybernetics*, volume 7, pages 2835–2840, 2004.

Unveiling Robotophobia and Cyber-dystopianism: The Role of Gender, Technology and Religion on Attitudes Towards Robots

Daniel Halpern
Rutgers University, SC&I
New Brunswick, NJ, 08901, USA
dhalpern@rutgers.edu

James E. Katz
Rutgers University, SC&I
New Brunswick, NJ, 08901, USA
j.k@rutgers.edu

ABSTRACT
A survey of 873 undergraduate students was conducted to understand which individual factors affect subjects' attitudes toward robots. A third of participants (N =284) were exposed to a humanoid robot, another third (N= 293) to a doggy robot, and the remaining third (N=296) to an android. Results showed that in the humanoid condition individuals recognize more humanlike characteristics in robots than in the other two conditions. However humanoid appearance did not affect participants' attitudes toward robots, as others predictors recognized by previous research indicate, such as gender, religion, and perceived competence with information and communication technologies (ICT).

ACM Categories & Subject Descriptors
A.1 INTRODUCTORY AND SURVEY.

Keywords
Social Responses to Robots, Attitudes to Robots, Survey.

1. INTRODUCTION
In the last two decades robots have not only proliferated in military and industrial settings but have also become increasingly visible in domestic ones. Researchers in robotics have already started to develop new functions for these social machines and are exploring their performance in diverse areas such as helping the elderly, therapists for autistic children, museum receptionists, and peer tutors in schools. While previous research has examined individuals' attitudes toward robots, there has been little research on which factors at the individual level affect users' attitudes to robots and how this might be moderated by the appearance of the robot. This research adds to our understanding by analyzing how different variables affect attitudes towards robots, especially in terms of robot appearance.

2. BACKGROUND AND RELATED WORK
The concept of "attitude" is a central theme in social psychology. The assumption underlying this centrality is that an attitude change will produce a corresponding change in behavior [1]. In different studies, Nass and his colleagues concluded that participants consistently overusing human social categories, such as gender and ethnicity, by applying them when referring to robots [4]. Similarly, research has investigated the interactions of children with Sony's robotic dog AIBO in comparison to a live Australian Shepherd dog [3]. The study showed that a surprising majority of children affirmed that AIBO had mental states (56%), sociality (70%), and moral standing (76%), which shows that as robots become increasingly life-like in their appearance and behavior, people will treat them as if they were sentient and social beings, raising to the level of biological others.

2.1 Anthropomorphism and human likeness
Research has consistently shown that the use of anthropomorphic pronouns makes human partners more likely to treat humanoid social robots as a real person. Schermerhorn et al. [6] found that when an autonomous mobile robot, about the size of a human, is capable of natural language interactions and can demonstrate self-directed behavior, human observers have a strong tendency to attribute human-like qualities. This line of research suggests that the use of more anthropomorphic robots makes human partners more likely to treat them as people. Therefore, it is expected that:

H1: Humanoid designs will make subjects believe that robots have more human qualities than more android oriented designs.

2.2 Cultural differences in robot perception
Research has also identified demographic and cultural differences among users in their interaction with robots. In the Judeo-Christian tradition for instance, in contrast to other religious and philosophical traditions such as Buddhism, a clear division is made between living and dead entities [7]. The Judeo-Christian world enforces a strict division between creatures that have a soul and objects that do not, which is not the case for religions such as Japan's Shinto and India's Jains or naturalistic philosophies, where all worldly things can be deemed as alive and having a soul [7]. From this view, social robots remain non-human and people might perceive them as incapable of assuming a position of moral equivalence. Consequently, it may be hypothesized:

H2: Individuals who profess Judeo-Christian religions will show less positive social attitudes to robots than individuals who profess Eastern religions.

2.3 Technology and gender differences
Research has also noted that previous experiences with technology and robots influence the attitudes toward them. In a simulated scenario with a human-sized robot, it was found that people with technological backgrounds related were more likely to project their own personality traits to the robot's than participants with a non-technological background (who could not see the robot as having an identifiable personality) [8]. These findings are consistent with research in communication, which suggest that individuals with higher levels of comfort with ICT tend to benefit more from relational uses of technology [2]. Therefore, we expect:

H3: Respondents with higher levels of comfort interacting with ICT will show a more positive attitude toward robots.

Regarding gender differences, research has found that males tend to think of the robot as more human-like, giving more socially desirable answers in their responses [6]. In contrast, females see

HRI'12, March 5-8, 2012, Boston, Massachusetts, USA.
ACM 978-1-4503-1063-5/12/03.

robots as more machine-like. This implies that attitudes toward robots may be affected by gender. Therefore we expect:

H4: Female respondents will perceive robots less favorably than will male respondents.

3. METHODS

3.1 Design Overview

A between-subjects survey experiment was conducted. Participants were divided in three groups and each group was exposed to an image of a different type of robot. Then participants completed an online questionnaire that measured their attitude towards robots and their degree of human-like characteristics. Additionally levels of religiosity, perceived competence with technologies, experiences in Second Life, online role games, engagement with avatars as well as demographics were obtained.

3.2 Stimulus Material

Three robots were chosen for this experiment: Romeo, a French humanoid robot designed to assist elderly and disabled people by *Aldebara*; the AIBO robotic dog designed by Sony, and the third was an android with an extreme robotic appearance.

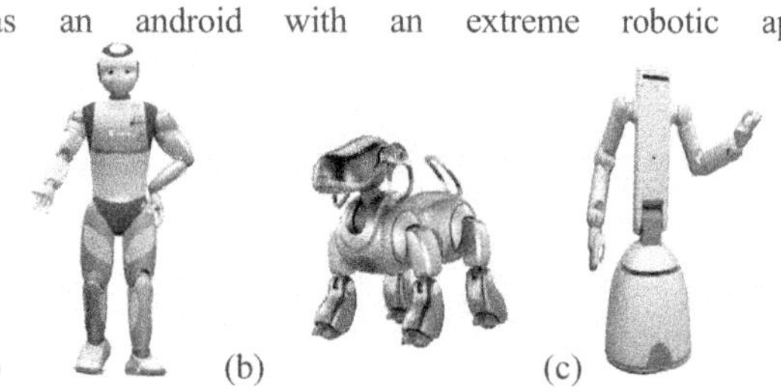

(a) (b) (c)

Figure 1 presents the stimulus used. Figure 1(a) presents the humanoid Romeo, Figure 1(b) the robotic dog AIBO and Figure 1(c) the android.

3.3 Measurement and Construction of Factors

Human-likeness. Eight items were used to measure recognition of human-like characteristics in robots.

Religion. We differentiated between Judeo-Christian and eastern religions (e.g. Buddhism).

Perceived competence with ICT. Respondents were asked four statements related to how comfortable they feel using ICT.

Attitude toward robots. The NARS scale was used and additionally 11 items were added [5]. The 25 items were subjected to principal components analysis. A three-factor solution was used which explained 21.2%, 16.7% and 8.4% of the variance respectively. The component Robot Liking, consisted of eight items, all of which indicated a preference for robots. Factor two, Cyber-Dystopian, consisted of five items related to negative social consequences of the use of robots. The third factor titled Robotphobia, was composed of seven items that mixed general negative attitudes and emotions.

For more details about the scales, measurements, instruments and construction of factors please contact the authors.

4. RESULTS

A one-way ANOVA with the Human Likeness Scale as the dependent variable showed a statistically significant difference in participants' perceptions of human qualities in robots, ($F(2, 744) = 13.53$, $p<.001$). Respondents in the humanoid condition recognized more human qualities in robots than in the android and doggy conditions, supporting H1. Ordinary least squares (OLS) regressions were run to assess the exact contribution of each block of predictors to the three factors constructed: Robot Liking, Robotphobia and Cyber-Dystopian. Results showed that individuals who identified with Judeo-Christian religions liked robots less (β =-.176, $p< .05$), supporting H2. Regarding H3, contrary to what we expected, individuals with a high sense of competence interacting with ICT showed a positive relation with Cyber-Dystopian (β =.067, $p < .01$). Regarding gender differences, H4 was confirmed and results show that female participants related negatively to Robot Liking (β =-.569, $p < .001$) and positively to Robotphobia (β =.555, $p < .001$).

5. DISCUSSION

This study examined which factors at the individual level affect users' attitudes to robots as moderated by their appearance. Four major contributions should be noted. First, we hypothesized that exposition to humanoid robots would elicit more positive attitudes, however we found only that recognition of human likeness in robots influence participants' attitudes, which suggests that it is not the exposure to particular robots but rather the recognition of human attributes that affected subjects' attitudes. Another contribution is to understanding the role that gender plays in the attitudes toward robots: although females showed higher levels of Robotphobia and indicated less liking for robots, they did not show a higher cyber-dystopianism. This lack of "consistency" may be explained by the fact that this factor was constructed principally by social-functional consequences in the proposed use of robots, such as "robots may replace people at work," which introduces the notion of competition; females, as research shows, tend to prefer cooperative modes of relationships rather than competitive ones, so they may not feel threatened by them. Our study also confirmed empirically that individuals who profess Judeo-Christian religions like robots less than those of Eastern religions, which corroborate the idea that philosophical elements embedded in each culture-religion may hold an enduring influence on attitudes toward technology. Finally, our results indicated that individuals who feel more comfortable with the use of technologies are more concerned with the possible impacts that robots might have on society, reflecting higher cyber-dystopianism. One possible explanation could be that higher exposure to technology and perhaps robots/intelligent agents in real life could make them aware not only the capabilities represented by the technology but also their shortcomings.

6. REFERENCES

[1] Ajzen, I., & Fishbein, M. 1980. *Understanding attitudes and predicting social behavior.* Englewood Cliffs, NJ: Prentice-Hall.

[2] Campbell, S. W., & Kwak, N. 2010.Mobile Communication and civic life: Linking patterns of use to civic and political engagement. *Journal of Communication,* 60, 563–555.

[3] Friedman, B, Kahn, P.H, & Hagman, J. 2003. Hardware Companions? – What Online AIBO Discussion Forums Reveal about the Human-Robotic Relationship. *CHI '03.*

[4] Nass, C., & Moon, Y. 2000. Machines and mindlessness: Social responses to computers. *Journal of Social Issues*, 56, 1, 81 103.

[5] Nomura, T., Suzuki, T., & Kato, K. 2006. Altered attitudes of people toward robots: Investigation through the Negative Attitudes toward Robots Scale. In *Proc. AAAI-06 Workshop on Human Implications of Human-Robot Interaction.*

[6] Schermerhorn, P, Scheutz, M., Crowell, C. R. 2008. Robot social presence and gender: do females view robots differently than males?". In *Proceedings of the 3rd ACM/IEEE international conference on HRI*

[7] Shaw-Garlock, G. 2009. Looking forward to sociable robots. *International Journal of Social Robotics*, 1, 249–260.

[8] Woods, S., Dautenhahn, K., Kaouri, C., Boekhorst, K., & Walters, M. L. 2007. Are robots like people?: Relationships between participant and robot personality traits in human–robot interaction studies. *Interaction Studies*, 8,2, 281-305.

Assessing Workload in Human-Robot Peer-Based Teams

Caroline E. Harriott[X] Glenna L. Buford[X] Tao Zhang* Julie A. Adams[X]

[X] Department of Elec. Eng. and Comp. Sci.
Vanderbilt University
Nashville, TN, USA
+1 (615) 322 - 8481

* Department of Library Sciences
Purdue University
West Lafayette, IN, USA

caroline.e.harriott@vanderbilt.edu, glenna.l.buford@vanderbilt.edu, zhangtao2000@gmail.com
julie.a.adams@vanderbilt.edu

ABSTRACT

The effect of a robotic teammate on a human partner's workload has not been fully quantified. Prior research found that human participants experienced lower workload when working with a robotic partner than when working with a human partner. An evaluation investigated whether a similar trend in workload exists for tasks requiring direct and collaborative interaction between the partners, and joint team decision-making. The subjective results indicate a similar trend to the prior results; participants rated workload lower for the more complex task when partnered with a robot than when partnered with a human.

Categories and Subject Descriptors

H.1.2. [**Information Systems**]: User/Machine Systems – *Human Factors;* I.2. [**Artificial Intelligence**]: Robotics.

General Terms

Performance, Experimentation, Human Factors

Keywords

human-robot peer-based teams, workload assessment

1. INTRODUCTION

The exact manner in which human performance is impacted by working in human-robot peer-based teams and the ability to apply existing human performance moderator functions (HPMFs) to these teams is not well understood. HPMFs predict human performance given a set of conditions. Over 500 HPMFs exist [1], but many are evaluated for constrained environments, e.g. nuclear power plant operation [2]. HPMFs have also been tested in more dynamic scenarios [3], but it is not yet proven that HPMFs developed for non-robotic domains apply directly to peer-based human-robot teams.

Prior research found that a current HPMF used to predict workload in human-human teams may be applicable to human-robot peer-based teams [4]. Participants worked with either a human or robot partner that acted as an instructor to guide the participant through victim triage tasks. The current research investigated whether a more collaborative and complex relationship between the partners produces similar results. The humans experienced lower overall subjective workload when partnered with a robot, thus adding to evidence that an existing workload HPMF is applicable to human-robot peer-based teams.

HRI'12, March 5–8, 2012, Boston, Massachusetts, USA.
ACM 978-1-4503-1063-5/12/03.

2. BACKGROUND

2.1 Triage Evaluation

The prior Triage evaluation involved triage steps performed using regimented instructions, which included assessing the breathing rate and heart rate, and checking for injuries. The untrained participants' human or robotic partner followed the triage steps and directed the participant in conducting eleven victim assessments on simulated victims. Workload was measured via heart rate variability, subjective in-task and post-trial ratings, and secondary task question responses [4]. The human partner communicated with participants over a walkie-talkie from a remote location, while the robotic partner was physically present and moved between victims with participants.

2.2 Hazmat Evaluation

The Hazmat evaluation involved conducting a reconnaissance of an academic building after a bomb threat was received. The team was responsible for collecting air samples, locating suspicious objects, and updating the incident commander. If a suspicious object was found it was not disturbed and information regarding its whereabouts were reported immediately [5].

Participants worked closely with their human or robotic partner and conducted discussions regarding each of the items found. The Triage evaluation created well defined team member roles, with the human or robot partner providing instructions. The Hazmat evaluation allowed participants more control over the direction of the investigation and required direct collaboration.

3. METHOD

The partner, either human (H-H) or robot (H-R), represented the two between-subjects conditions. The H-H condition was completed prior to the H-R condition. The same evaluator played the role of a trained first responder in all H-H. The H-R condition paired the participant with a robot and the team traversed the hallway. Unknown to participants, a remote evaluator controlled the robot's speech and supervised the robot's movement.

Thirty-six participants completed the evaluation with eighteen in each condition. The 19 male and 17 female participants were not experts in first response or robotics and ranged in age between 18 and 56 years old.

The evaluation was conducted on a single floor of a Vanderbilt University academic building. The hallway and two laboratories were divided into six investigation areas of similar size.

Participants rated subjective workload on six channels: Auditory, Visual, Speech, Motor, Tactile and Cognitive. The rating process and each channel were defined during training. Upon completing the investigation of each area, the participant's partner asked the participant to rate the demands they experienced during that portion of the investigation on a scale from 1 (little demand) to 5 (extreme demand). Mean overall workload by participant was calculated by averaging the total demand from each area.

The teams followed the same path around the hallway and the same suspicious items were placed in the same locations for both conditions. Nineteen items (both suspicious and non-suspicious) were stationed throughout the search areas. The human or robot partner was responsible for checking fire extinguishers and taking air samples, and informed the participants that they were responsible for checking bulletin boards and trash cans, while looking for out of place items. At points the team had to make joint decisions regarding the appropriate action.

After the trial, the participants completed the NASA-TLX workload questionnaire, comprised of Mental, Physical, Temporal, Performance, Effort and Frustration components.

4. RESULTS

4.1 Hazmat Evaluation

4.1.1 In-task Subjective Workload Ratings

The overall subjective workload was analyzed using a one-way ANOVA assessing the main effect of condition. The mean overall workload for the H-H condition was 13.991 with a standard deviation (St. Dev.) of 5.215, and was 10.972 (St. Dev. = 3.957) for the H-R condition. Figure 1 provides the mean overall workload values for both evaluations by condition; the error bars represent one standard deviation above and below the mean. An ANOVA indicated that H-H participants rated overall workload significantly higher than the H-R participants, $F(1, 214) = 22.752$, $p < 0.001$.

4.1.2 NASA-TLX Ratings

A one-way ANOVA was performed on the overall NASA-TLX to assess the main effect of condition. Due to an error, the first eight H-H participants did not complete the NASA-TLX survey. The H-H condition had a mean of 41.320 (St. Dev. = 17.796), which was not significantly higher than the H-R overall workload, mean = 30.099 (St. Dev. = 15.121), $F(1,26) = 3.129$, $p = 0.089$. Figure 2 presents the mean NASA-TLX overall score for both evaluations by condition.

4.2 Comparison with Triage Evaluation

4.2.1 In-task Subjective Workload Ratings

The Triage evaluation's mean overall workload, assessed in a similar manner [4], for the H-H condition was 16.260 (St. Dev. = 6.305), which was significantly higher than the H-R overall workload, mean = 13.481 (St. Dev. = 4.808), $F(1, 306) = 18.992$, $p < 0.001$. As seen in Figure 1, the results demonstrate similar patterns across the experiments; however, the Hazmat evaluation results are lower for both conditions. The difference between the H-H and H-R conditions for the Triage evaluation was 2.779 and slightly higher, 3.019, for the Hazmat evaluation.

4.2.2 NASA-TLX Ratings

The mean Triage task overall NASA-TLX result for the H-H condition was 57.377 (St. Dev. = 14.000) and 48.591 (St. Dev. = 11.983) for the H-R condition. The statistical analysis found that the H-H condition workload was not significantly higher than the H-R condition, $F(1, 26) = 3.182$, $p = 0.086$. The difference between conditions in the Triage evaluation was 8.786 and was 11.221 for the Hazmat evaluation. Generally, the overall NASA-TLX workload results presented in Figure 2 demonstrate a similar pattern to the overall workload as found for the subjective in task responses.

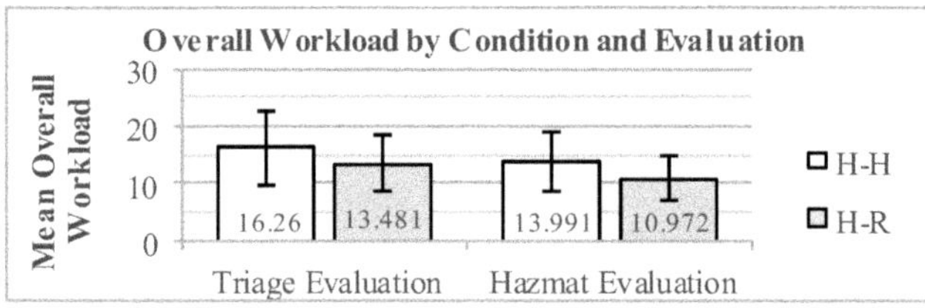

Figure 1. Overall workload by evaluation and condition.

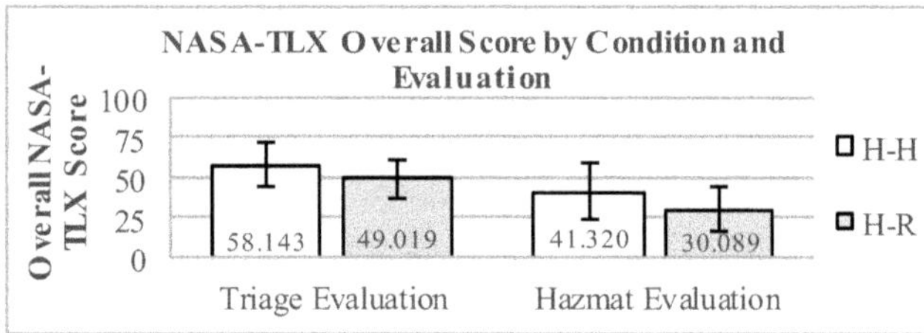

Figure 2. Overall NASA-TLX score by evaluation, condition.

5. DISCUSSION

The current evaluation increased the task difficulty by being more free-flowing and requiring participants to collaborate and make joint decisions. The results of the Triage evaluation indicated that workload in the H-R condition was lower than the H-H condition. The more open Hazmat scenario, requiring joint decision-making, indicates that a similar trend is apparent.

6. ACKNOWLEDGMENTS

The authors thank Eli Hooten, Sean Hayes and Corey Wollaeger. This research is supported by AFOSR award FA9550-09-1-0108, National Science Foundation Grant IIS-0643100, and an Office of Naval Research MURI Program award N000140710749.

7. REFERENCES

[1] Silverman, B.G., Johns, M., Cornwell, J., and O'Brien, K. Human behavior models for agents in simulators and games: Part I: Enabling science with PMFServ. *Presence: Tele - operators and Virtual Environments*. 15, 2 (2006). 139-162.

[2] Mumaw, R.J., Roth, E.M., Vicente, K.J., and Burns, C.M. There is more to monitoring a nuclear power plant than meets the eye. *Human Factors*. 42, 1 (Spr. 2000), 36-55.

[3] Salvucci, D.D. Predicting the effects of in-car interface use on driver performance: an integrated model approach. *Int. J. of Human-Computer Studies*. 55 (2001). 85-107.

[4] Harriott, C.E., Zhang, T. and Adams, J.A. 2011. Evaluating the applicability of current models of workload to peer-based human-robot teams. In *Proceedings of the 6th International Conference on Human-Robot Interaction* (Lausanne, Switzerland, March 6-9, 2011). 45-53.

[5] Mahoney, P. F. Businesses and bombs: Preplanning and response. *Facilities*. 12, 10 (1994), 14-21.

Does a Robot that Can Learn Verbs Lead to Better User Perception?

Dai Hasegawa
Hokkaido University
Sapporo 060-0814, Japan
hasegawadai@media.eng.hokudai.ac.jp

Kenji Araki
Hokkaido University
Sapporo 060-0814, Japan
araki@media.eng.hokudai.ac.jp

ABSTRACT

The current understanding is that human-likeness of a robot leads to better human perception. However, the factors have not been thoroughly studied. We conducted a laboratory experiment to examine two questions: how verb acquisition ability affects human perceptions on human-likeness and familiarity of a humanoid robot, intention to use the robot, and enjoyment and satisfaction of the interaction, and whether human-likeness mediates the links between the effects of interaction of verb acquisition between the human perceptions. The experiment involved 48 participants, and we found that the robot that was able to acquire two Japanese verbs, "*oku* (to put/to place)" and "*hanasu* (to move away from)," was perceived by participants as more familiar and satisfying than the one that knew the verbs from the beginning. We also found that human-likeness mediated the links between the effect of verb acquisition ability and other perceptions toward the robot.

Categories and Subject Descriptors

H.1.2 [**Models and Principles**]: User/Machine Systems—*human factors*; H.5.2 [**Information Interfaces and Presentation**]: User Interfaces—*evaluation/methodology, user-centered design*

General Terms

Design, Experimentation, Languages, Human Factors

Keywords

Language Acquisition, Verb, Human Perception

1. INTRODUCTION

Language acquisition ability, through which humans learn unknown words and grammatical structures, even from a small number of examples, is one of the most human-like abilities. However, it is not clear how language acquisition ability influences human perception, in particular, on human-likeness, familiarity, intention to use, enjoyment, and satisfaction. It is also unclear how a user's mental model works on the relationships among the perceptions. Powers et al.[2] found that human-likeness mediates links between an effect of human-like short chin of a robot and perceptions of sociability and intelligence. Thus, we believe that human-likeness mediates links between effects of language acquisition ability and other human perceptions.

In this paper, we report the results of our empirical study. Our experiment was conducted in Japanese, and we used *italic* when giving Japanese words or sentences. As a starting point of our research, we focused on verb acquisition ability for two verbs, "*oku* (to put/to place)" and "*hanasu* (to move away from)," and explored how verb acquisition ability led to better human perceptions of a robot and whether human-likeness mediated the links between the effect of verb acquisition ability and the human perceptions. Our hypotheses are: (H1) verb acquisition ability leads to better human perceptions toward a robot, (H2) human-likeness mediates the links between the effects of verb acquisition ability and perceptions of familiarity, intention to use, enjoyment, and satisfaction.

Figure 1: Settings

2. EXPERIMENT

We conducted a one by two, between-participants study in which we manipulated a robot's verb acquisition ability and measured how the manipulation might affect human perception of the robot.

The settings of our experiment are shown in Figure 1. Participants were asked to interact with a robot[1] by ordering movements. In the interaction, the participants were able to order the robot perform 8 specific movements by keyboard input (e.g. *Migitewo atamani oite* (Put your right hand on your head) or *Migitewo hidaritekara hanashite* (Move your right hand away from your left hand)). They were asked

[1]Kondo Kagaku Co. Ltd, http://www.kondo-robot.com/

HRI'12, March 5–8, 2012, Boston, Massachusetts, USA.
ACM 978-1-4503-1063-5/12/03.

to input the orders one by one. We randomly assigned the participants to one of two conditions described below.

Verb Acquisition Condition
The two verbs, "*oku* (to put/to place)" and "*hanasu* (to move away from)," are not embeded in advance. The participants are instructed that if the robot failed to move, they should teach the correct movement by performing it in front of the web camera equipped on the display. The robot is programmed to perform no movement, when an order is input for the first time; to perform an incorrect movement on the second time, and to perform correct movements at the third and the fourth time.

Language Input Condition
The two verbs are embeded in advance. The robot replies with correct movements to the participants' orders.

After the interaction, the participants answered a questionnaire with a seven-point Likert scale anchored by "Strongly Disagree" and "Strongly Agree" for all items to measure their perceptions of the robot. The questionnaire consists of 23 questions which ask about familiarity, intention to use, enjoyment, satisfaction, human-likeness, and verb acquisition ability. The questions about verb acquisition ability of the robot are to confirm our control. The participants were paid 1,000 yen (approximately US$12) as compensation.

3. RESULTS

A total of 50 participants (23 males, 27 females, 20 undergraduate students, 30 graduate students) took part in the experiment. All participants were native Japanese speakers and were recruited by flyer on campus of Hokkaido University in Japan. The system suddenly stopped during the experiment of 2 participants in Verb Acquisition Condition, so we excluded these sessions from the data.

We found no significant difference in the duration of the interaction between the two conditions. The average duration was 508 seconds ($\sigma = 12.5$) in Verb Acquisition Condition and 487 seconds ($\sigma = 15.2$) in Verb Input Condition (t-test, p=.31). This was because, the robot took a significantly longer time to move for the sake of participants' safety when compared to the time participants took to teach movements.

Our analyses using the student's t-test showed that the participants judged that the robot in Verb Acquisition Condition had stronger verb acquisition ability than the robot in Verb Input Condition ($p < .01$). The participants, in addition, perceived the robot in Verb Acquisition Condition as more familiar ($p < .01$), satisfying ($p < .01$), and human-like ($p < .01$) than the robot in Verb Input Condition. However, we found no significant difference on intention to use.

We then examined human-likeness as a mediator in a test for simple mediation using Baron & Kenny's causal steps strategy[1]. We found significant effects of verb acquisition ability on familiarity ($\beta = .23$, $p < .05$), enjoyment ($\beta = .28$, $p < .01$), satisfaction ($\beta = .27$, $p < .01$) and on human-likeness ($\beta = .39$, $p < .01$) but not on intention to use. We also found significant effects of human-likeness on familiarity ($\beta = .63$, $p < .01$), enjoyment ($\beta = .53$, $p < .01$), satisfaction ($\beta = .42$, $p < .01$). The effects of verb acquisition ability on the outcomes was reduced markedly and became statistically non-significant when controlled for the effects of human-likeness on familiarity ($\beta = .09$, $p < .83$), enjoyment ($\beta = .11$, $p = .36$), satisfaction ($\beta = .16$, $p < .11$).

We also tested for simple mediation using Baron & Kenny's step with bootstrapping methodology for indirect effects based on 1000 bootstrap resamples to describe the confidence intervals of indirect effects. Interpretation of the bootstrap data is accomplished by determining whether zero is contained within the 95.0% confidence intervals (thus indicating the lack of significance). The results showed that the 95.0% confidence intervals did not contain zero except for ITU. Thus the causal steps test and the bootstrapping test agreed that human-likeness was a simple mediator for familiarity, enjoyment, and satisfaction.

4. DISCUSSION

We showed the effect of the interaction of verb acquisition on perception of the robot. However, we did not fully control the duration of the interaction though the differences were not statistically significant. Nor did we not manipulate the factors of mistake and teaching. These factors are inevitable parts of the interaction to acquire language, but, the mistake and teaching are individual factors which are often seen not only in the interaction of language acquisition but in our daily life, and these could affect perception of the robot. Thus, the effect of verb acquisition ability still remains not clear. We need to explore the effects of mistake and teaching. We also think that verb acquisition ability has interaction effects with the factors of mistake and teaching.

5. CONCLUSIONS AND FUTURE WORKS

We examined how verb acquisition ability affected human perceptions in terms of human-likeness, familiarity, intention to use, enjoyment, and satisfaction, and whether human-likeness mediated the links between the effects of verb acquisition ability and perceptions of intention to use, enjoyment, familiarity, and satisfaction. As a result of the experiment, we found that participants judged the robot that learned the two verbs as more human-like, enjoyable, familiar, satisfying than the one that knew the verbs from the beginning. The results indicated a possibility that verb acquisition ability has effects on the perceptions. Our analyses also showed human-likeness mediated the links between the effects of interaction of verb acquisition and human perceptions on enjoyment, familiarity, and satisfaction.

In our future work, we will further investigate the individual effect of verb acquisition ability, mistake and teaching. We will also explore the interaction effects between language acquisition ability and the factors of mistake and teaching.

6. REFERENCES

[1] R. M. Baron and D. A. Kenny. The moderator-mediator variable distinction in social psychological research: Conceptual, strategic and statistical considerations. *Journal of Personality and Social Psychology*, 51:1173–1182, 1986.

[2] A. Powers and S. Kiesler. The advisor robot: tracing people's mental model from a robot's physical attributes. In *Proceedings of the 1st ACM/IEEE international conference on Human-robot interaction (HRI'06)*, pages 218–225, New York, NY, USA, 2007. ACM.

Towards A Computational Method of Scaling A Robot's Behavior via Proxemics

Zachary Henkel
Texas A&M University
College Station, Texas
zmhenkel@tamu.edu

Robin R. Murphy
Texas A&M University
College Station, Texas
murphy@cse.tamu.edu

Cindy L. Bethel
Mississippi State University
Starkville, MS USA
cbethel@cse.msstate.edu

ABSTRACT

Humans regulate their social behavior based on proximity to other social actors. Likewise, when a robot fulfills the role of a social actor it too should regulate its interaction based on proximity. This paper describes work in progress to establish methods for autonomous modification of social behavior based on proximity and to quantify human preferences between methods of scaling a robot's social behaviors based on distance from a human. The preliminary results of a 72 participant human study examine the reaction to scaling with linear methods and perception-based methods. Results indicate significantly higher ratings in multiple areas (comfort, natural movement, safety, self-control, intelligence, likability, submissiveness **p<.05**) when using a perception-based scaling function, as opposed to a linear or no scaling function. Work in progress is analyzing the biometric measures collected.

Categories and Subject Descriptors

I.2.9 [**Robotics**]:

General Terms

Experimentation

Keywords

proxemics, scaling function, proximity

1. INTRODUCTION

The phrase "proxemics" was coined in 1966 by Edward Hall to describe the correlation between physical distance and human social behavior[4]. As suggested by the computers are social actors paradigm[6] and supported by human-robot interaction studies[2, 1], the same relationship exists in human-robot interaction scenarios. However, while studies have confirmed the existence of human-robot proxemics, none have suggested a computational method for generating proxemic behavior.

Our recent 72 participant study suggests a computational method which can be used to autonomously modify robot behavior in order to accomplish social competence. The method of scaling robot behavior (locomotion speed, head-gaze act speed and range, and lighting intensity) based on proxemics was derived by exploring one physics and two psychology laws: the inverse-square law and Weber-Fechner's Law and Steven's Power Law. Participants strongly preferred scaling derived from physics and psychology laws.

2. EXPERIMENTAL APPROACH

In order to test the theory of proxemic scaling, a human-robot interaction study which examined three varying conditions was designed. The robot used in the study consisted of a Survivor Buddy robot mounted to an Inuktun Extreme robot. The robot travelled through a simulated disaster site, approaching participants who were acting as victims. As the robot approached the participant it engaged in a conversation complete with head gaze acts as well as fixations on objects in the environment. The robot used synthesized speech, followed a predefined motion path, and navigated its dialog script via a wizard of oz approach. In the final few minutes of the interaction, participants were asked to place their hand on the robot in order to allow the robot to take their pulse (simulated). The experiment was run using a

Figure 1: Simulated Disaster Site.

no scaling case, a *linear scaling* case, and a *perception-based scaling* case. In the no scaling case, the robot travelled at a medium velocity, used medium speeds and ranges for head gaze acts, and used medium intensity headlights throughout the entire interaction. In the linear and perception-based cases, the robot scaled its attributes, like speed, range and intensity based on its distance from the human participant. The difference between cases was the function used to determine the magnitude of each characteristic. In the linear case, a simple linear function was used, mapping slower speeds, limited range, and low intensity lighting to the area closest to the participant and higher values to locations further away. In the perception-based case, a function derived from a combination of physics and psychology laws was used to map the robot's characteristics.

Pre-interaction and post-interaction questionnaires were administered. Questions which measured comfort, natural movement, safety, self-control, attentiveness, empathy, happiness, intelligence, likability, sensibility, submissiveness and trust were obtained from previous studies [3] [5].

In order to gain more insight into areas which concerned proxemic performance, questions were developed specifically for the area of personal space consideration. All items were on a one to seven Likert scale, with one being strong disagreement or strongly negative and seven being strong agreement or strongly positive.

Our primary measures of social consistency encompassed the attributes: comfort, natural movement, personal space consideration, safety, and self control. Additionally, the secondary attributes of attentiveness, empathy, happiness, likability, intelligence, sensibility, submissiveness, and trust play important roles in the interaction. We believe the primary attributes to be necessary for social consistency.

3. EXPERIMENTAL RESULTS

Results indicated that all of the primary attributes (comfort, personal space consideration, safety, natural movement and self-control), along with three of the secondary attributes (intelligence, likability and submissiveness) out of a total of 12 attributes proved to be significantly better in conditions which employed scaling as opposed to the condition which did not.

In all of the primary attributes (comfort, personal space consideration, safety, natural movement, and self-control) and three of the secondary attributes (intelligence, likability, and submissiveness), perception-based scaling proved to be significantly better than no scaling. Linear scaling indicated no significant difference opposed to no scaling, except in the cases of comfort and personal space consideration. In the area of comfort, a clear rank was demonstrated with significance between all conditions, and perception-based scaling being rated the highest of all conditions (See Figure 2). However, in the area of personal space consideration, both linear and perception-based proved to be significantly better than no scaling, while perception-based did not prove to be better than linear. This suggests that linear and perception-based scaling are both acceptable in the area of personal space consideration.

4. CONCLUSION

Our findings offer three important pieces of information for robot designers. First, our studies serve as a confirmation that modification of effectors as a function of proximity can create a more positive interaction. Though this has been touched on in previous work, we now are able to confirm through a multi-faceted and complex interaction task that proximity is an important factor in generating behavior. Second, we have illustrated that the choice of scaling function is critical in the modification of social behaviors. Our findings show clear and significant differences between the evaluated functions, meaning careful consideration must be given to the type of function selected for the implementation of proxemic scaling. Finally, our work suggests that a perception-based function is the best starting point for further design or investigation of proxemic scaling of social robot behavior.

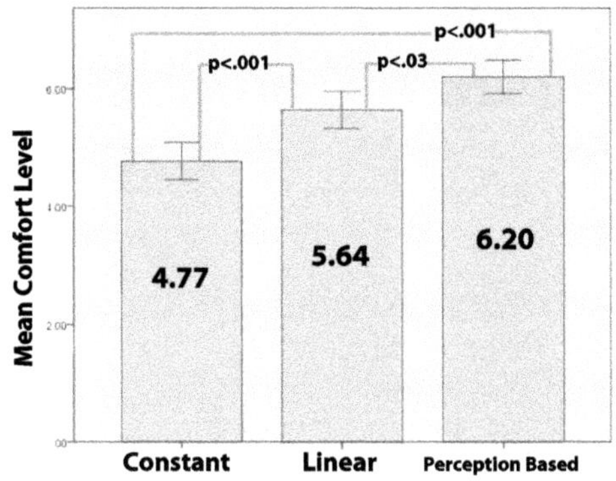

Figure 2: Comfort Level

The criticality of scaling a robot's social behavior with respect to proxemics has been clearly illustrated. Through the examination of a 72 participant study the clear preference for perception-based scaling is evident. Through a multi-faceted interaction, not only has the benefits of proxemic scaling been illustrated, but also the critical nature of choosing the correct scaling function. These findings serve as a confirmation of the importance of proximity, as well as guide for future design of social robots.

5. ACKNOWLEDGMENTS

This work was supported in part by NSF Grant IIS-0905485 "The Social Medium is the Message" and by a HRI grant from Microsoft External Research.

6. REFERENCES

[1] C. Bethel and R. Murphy. Non-facial and non-verbal affective expression for appearance-constrained robots used in victim management. *Paladyn. Journal of Behavioral Robotics*, 1:219–230, 2010. 10.2478/s13230-011-0009-5.

[2] C. L. Bethel and R. R. Murphy. Non-facial/non-verbal methods of affective expression as applied to robot-assisted victim assessment. In *Proceedings of the ACM/IEEE international conference on Human-robot interaction*, HRI '07, pages 287–294, New York, NY, USA, 2007. ACM.

[3] V. Groom, V. Srinivasan, C. Bethel, R. Murphy, L. Dole, and C. Nass. Responses to robot social roles and social role framing. In *Collaboration Technologies and Systems (CTS), 2011 International Conference on*, pages 194 –203, may 2011.

[4] E. T. Hall. *The Hidden Dimension*. Anchor, Oct. 1966.

[5] B. Mutlu, T. Shiwa, T. K, H. Ishiguro, and N. Hagita. Footing in human-robot conversations: How robots might shape participant roles using gaze cues.

[6] B. Reeves and C. Nass. *The media equation: how people treat computers, television, and new media like real people and places*. Cambridge University Press New York, NY, USA, 1996.

Using the Behavior Markup Language for Human-Robot Interaction

Aaron Holroyd and Charles Rich
Computer Science Department
Worcester Polytechnic Institute
Worcester, MA 01609
(aholroyd | rich)@wpi.edu

ABSTRACT

This paper describes a Behavior Markup Language (BML) realizer that we developed for use in our research on human-robot interaction. Existing BML realizers used with virtual agents are based on fixed-timing algorithms and because of that are not suitable for robotic applications. Our realizer uses an event-driven architecture, based on Petri nets, to guarantee the specified synchronization constraints in the presence of unpredictable variability in robot control systems. Our implementation is robot independent, open source and uses the Robot Operating System (ROS).

Categories and Subject Descriptors

I.2.9 [**Artificial Intelligence**]: Robotics

General Terms

Algorithms, Standardization

Keywords

BML, Realizer, Robots, Petri Net, Event-Driven, ROS

1. INTRODUCTION

In this work, we build upon research in the virtual agent community by using the Behavior Markup Language (BML) [3] to control robots interacting with humans. Many virtual agents, e.g. [5], have been controlled using BML. Using BML for robots will be just as important, since many robots need to interact with people in equally complex ways. This paper describes the benefits of applying BML to human-robot interactin and the problems we solved to make that possible.

Imagine a person ambiguously asking a robot to pick up a soda can when there are two cans on the table. The robot might respond by saying "This can?," while simultaneously looking and pointing toward one of the cans. These looking and pointing behaviors need to be synchronized correctly in relation to the speech for a natural, human-like performance.

The example BML in Figure 1 shows the behaviors involved in this example and the constraints between them. It contains a single gaze behavior (`gaze-1`) toward `can-1`, along with a pointing behavior (`point-1`). The speech behavior (`speech-1`) contains the text "This can?" to be uttered overlapping with the other behaviors. The constraints ensure that the pointing gesture does not start until the robot is looking at the object, and that the speech starts when the robot is beginning to point toward the object.

```
<bml id="bml-example-1">
  <gaze id="gaze-1" target="can-1"/>
  <gesture id="point-1" type="POINT"
           target="can-1"/>
  <speech id="speech-1">
    <text>This can?</text> </speech>
  <constraint id="constraint-1">
    <synchronize ref="gaze-1:stroke">
      <sync ref="point-1:start" />
    </synchronize>
    <synchronize ref="point-1:stroke_start">
      <sync ref="speech-1:start" />
    </synchronize> </constraint> </bml>
```

Figure 1: Example BML block

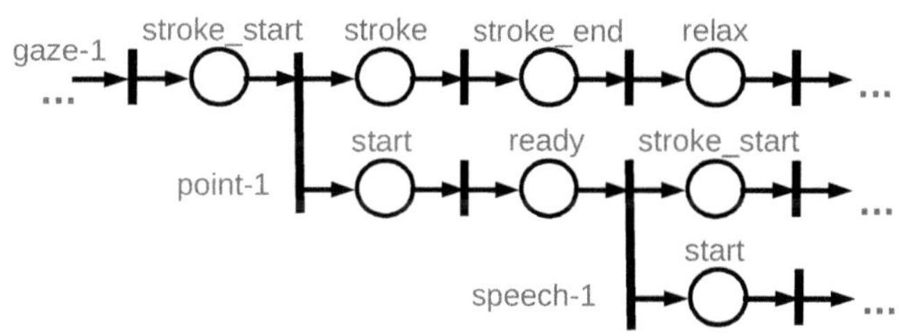

Figure 2: Part of the Petri net for Figure 1

2. DIFFICULTIES WITH FIXED-TIMING REALIZERS FOR ROBOTS

BML was created by Kopp *et al.* [3], who identified the need for a common behavior specification framework for virtual agents. All BML realizers have two phases: scheduling and execution. In the scheduling phase, a shortest possible schedule that satisfies the constraints is created. The execution phase then uses this schedule to perform the behaviors in real time. Current animation realizers (see Figure 3), run open-loop using fixed-time schedules, which works because the realizer can generate key-frames. These realizers plan the BML using a fixed-timeline representation and generate key-frames based on the required joint angles. The key-frames are interpolated and shown to the user. Anh and Pelachaud [1] developed a fixed-timing realizer for a Nao robot based on the Greta animation engine. This system

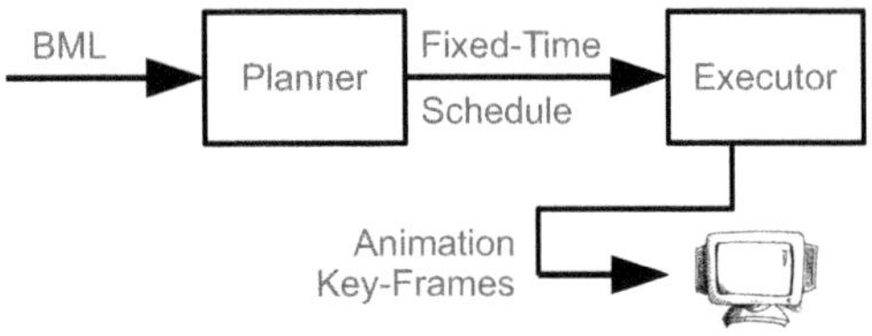

Figure 3: Animation BML realizer architecture

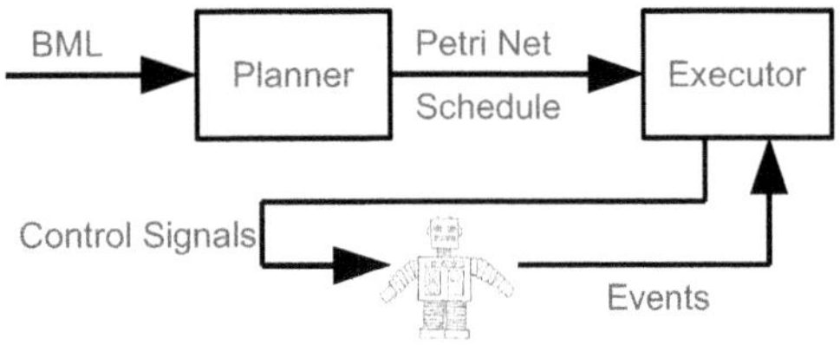

Figure 4: Robot BML realizer architecture

uses a database of motions and synchronizes them using fixed-timing constraints. They acknowledge the need for, but do not provide, an event-drive realizer for robots.

In an animation, since the time between each of the sync points can be *chosen*, fixed-timing algorithms are easy to implement. Also, in the animation setting, the timeline is guaranteed to execute properly. In robotics, however, there are several problems that cause fixed-timing approaches not to work.

The first, and most common, problem is when the motion of a joint motor is slower than expected, perhaps because it is stalled or a battery is partially drained. In this case the other motions need to be delayed and wait. For example, the realizer may need to wait for the arm to reach the correct position before starting to speak. A second example problem occurs when the robot's arm reaches its target position *earlier* than expected, perhaps because the arm was closer to the target than the robot believed.

3. EVENT-DRIVEN SOLUTION

A robotic realizer (see Figure 4) is fundamentally different from an animation realizer in that it must be controlled closed-loop, with control events being sent from the robot to the realizer. The control events are joint position (e.g., when a joint reaches the desired position) or velocity acknowledgements. These events suggest an event-driven system to ensure the synchronization constraints. The output of the scheduling stage in this approach is thus a Petri net [4] representing a minimum-time event-based schedule. We chose Petri nets because they conveniently represent events and their synchronization constraints.

Figure 2 shows part of the Petri net schedule resulting from Figure 1. A Petri net consists of places (represented as circles), transitions (represented as vertical bars), and tokens which are transmitted between places and transitions. In a Petri net BML schedule, each place represents a sync point of a behavior and the transitions represent synchronization constraints. Each transition and place waits for all incoming tokens and sends one token out on each arrow.

Our scheduling algorithm starts by creating a separate seven-place Petri net for each behavior that appears in the BML block. Figure 2 contains three such sub-nets, but none are completely shown due to limits on space.

Scheduling a `synchronize` constraint corresponds to merging the transitions prior to the given sync points. Note that this new merged transition replaces the prior two transitions, and synchronizes the sync points because transitions must wait for all incoming tokens before sending tokens out on all outputs.

Scheduling a `before` or `after` constraint is accomplished by adding a single arrow starting at the first place and ending at the transition prior to the second place. Note that the `after` constraint is the inverse of a `before`; thus without loss of generality the following two constraints are equivalent:

```
<before ref="X">              <after ref="Y">
    <sync ref="Y"/>               <sync ref="X"/>
</before>                     </after>
```

Our BML executor is implemented in ROS Java.[1] Starting from the left-most transition in Figure 2, the executor calls a robot-specific control ROS module for each sync point of each behavior. This control module sends joint commands to the robot and returns events when the given sync point is reached. Each transition and place is a new thread of execution that waits for all incoming tokens before executing. Similarly, the thread ends once all tokens have been sent on all of the output lines.

In conclusion, we have shown the theory and provided an open-source ROS implementation of an event-driven BML realizer for robots. We used this realizer in [2] and will continue to use it in future research.

Acknowledgements: This work is supported in part by the National Science Foundation under awards IIS-0811942 and IIS-1012083.

4. REFERENCES

[1] L. Q. Anh and C. Pelachaud. Expressive gesture model for storytelling humanoid agent. In *The fourth workshop on Embodied Conversational Agents, WACA 2010*, Lilles, France, November 2010.

[2] A. Holroyd, C. Rich, C. L. Sidner, and B. Ponsler. Generating connection events for human-robot collaboration. In *20th IEEE International Symposium on Robot and Human Interactive Communication, RO-MAN 2011*, Atlanta, GA, USA, Aug. 2011.

[3] S. Kopp, B. Krenn, S. Marsella, A. Marshall, C. Pelachaud, H. Pirker, K. Thórisson, and H. Vilhjálmsson. Towards a common framework for multimodal generation: The behavior markup language. In *Intelligent Virtual Agents*, 2006.

[4] C. PETRI. Fundamentals of a theory of asynchronous information flow. In *Information processing: proceedings of the IFIP Congress*. North-Holland Pub. Co., 1962.

[5] M. Thiebaux and S. Marsella. Smartbody: Behavior realization for embodied conversational agents. In *Intelligent Virtual Agents*, 2007.

[1]`http://sourceforge.net/projects/rosbmlrealizer/`

Attracting and Controlling Human Attention through Robot's Behaviors Suited to the Situation

Mohammed M. Hoque, Tomomi Onuki, Dipankar Das, Yoshinori Kobayashi and Yoshinori Kuno
Gradute Scholl of Science and Engineering, Saitama University
255 Shimo-Okubo, Sakura-ku, 338-8570 Saitama, Japan
{moshiul, t.onuki, dipankar, yosinori, kuno}@cv.ics.saitama-u.ac.jp

ABSTRACT

A major challenge is to design a robot that can attract and control human attention in various social situations. If a robot would like to communicate a person, it may turn its gaze to him/her for eye contact. However, it is not an easy task for the robot to make eye contact because such a turning action alone may not be enough in all situations, especially when the robot and the human are not facing each other. In this paper, we present an attention control approach through robot's behaviors that can attract a person's attention by three actions: head turning, head shaking, and uttering reference terms corresponding to three viewing situations in which the human vision senses the robot (near peripheral field of view, far peripheral field of view, and out of field of view). After gaining attention, the robot makes eye contact through showing gaze awareness by blinking its eyes, and directs the human attention by eye and head turning behaviors to share an object.

Categories and Subject Descriptors

H.5.2 [**Information Interfaces and Presentation**]: User Interfaces—*Interaction Styles, User-Centered Design.*

General Terms

Design, Experimentation, Human Factors

Keywords

Attention control, eye contact, human-robot interaction

1. INTRODUCTION

Attention control can be defined as a means of gaze (i.e., head) control in which one can shift or control someone's gaze from one direction to that of his/her interest. Although there may be various situations, we consider a general situation where a human and a robot are not facing each other initially and the human is engaging in some tasks that may not absorb much attention (i.e., watching paintings). The robot produces an action while the human vision senses the robot in three viewing situations namely near peripheral field of view (NPFV), far peripheral field of view (FPFV), and out of field of view (OFV). In such situations, it is difficult to establish mutual gaze. Most previous studies assumed that the target person faces the robot and intends to talk with it; however, in reality this assumption may not always hold. Turning the head toward the target person is the most fundamental action of the robot and this action may be effective when the robot exists in the human's central view but not effective while it exist in his/her FPFV, and OFV [1]. Abrupt motions are effective to capture human attention toward the object in peripheral view [3]. Robot's head motions can be used as a strong communicating cues in HRI [2]. Moreover, when a robot is facing back of the human, it is difficult to attract his/her attention by any motion. In that case, touch or voice signal will be effective means. Our purpose is to develop a robot that can attract and control the attention of a target person while avoiding attracting the attention of other people. Although it is apparent that the robot can attract the attention of people through voice, however, voice is also certain to attract the attention of others. Thus, voice should be used as a last resort. The robot should start with a weak (head turning toward the person) action to avoid attracting the attention of those other than the target person, then use stronger (e.g., head shaking) or strongest (e.g., voice) actions if it cannot attract his/her attention. We prefer to use non-verbal actions because using these actions the robot may selectively attract the target person.

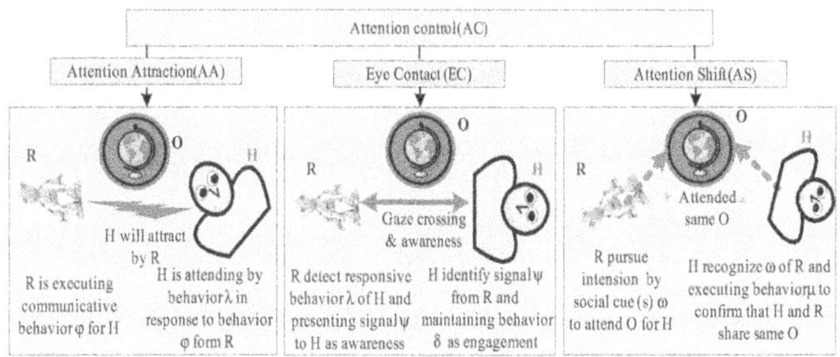

Figure 1: Attention control process.

2. ATTENTION CONTROL PHASES

To control human attention, robots should perform three sub tasks consecutively: (i) attention attraction (AA), (ii) eye contact (EC), and (iii) attention shift (AS). Fig. 1 illustrates the conceptual process of attention control (AC). To perform a successful AC process, both robot (R) and human (H) need to show some explicit behaviors in each phase. To initiate an AC process, the robot first observes the direction of attention of the target person. To attract attention, the robot applies head turning for NPFV, head

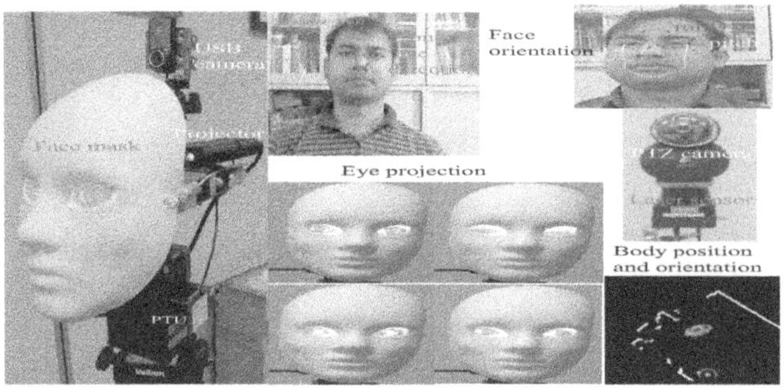

Figure 2: Overview of the robot system.

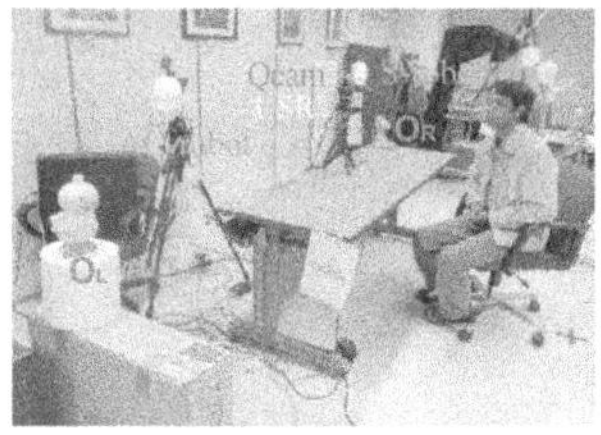

Situ-ations	No. of responses/attempts N_R/N_{AA}	N_B/N_{EC}	N_R/N_{AS}	Success rate
NPFV	18/20	17/18	14/17	70%
FPFV	23/26	19/23	16/19	61.5%
OFV	14/18	12/14	10/12	55.5%

(a) Experimental scene (b) Behavioral results

Figure 3: (a) an experimental scene (b) Success rate of the AC-robot.

shaking for FPFV, and reference terms (i.e., excuse me) for OFV. The robot considers that the human has responded against robot's actions if it detects his/her frontal face in the camera image within its expected times (1 second in NPFV, 2 seconds in FPFV, and 3 seconds in OFV). This ensures the face-to-face orientation and the robot displays its blinking action about 2 seconds for ensuring eye contact. After establishing eye contact, the robot makes an attention shift request by switching its attention through turning both eyes and head (a bit later) to the object which it intends to address. After addressing the focus, the robot waits (2 seconds) and checks whether or not the attention shift attempt is successful. If the responding agent is not attending to the referential object, the robot will try again by shifting its head back to the person, and focus again to the object.

3. SYSTEM OVERVIEW

We have developed a robot head for HRI experiments. Fig. 2 shows an overview of the robot head. The LED projector projects CG generated eyes on the mask. The proposed system has five software modules: head detection and tracking module (HDTM), body tracking module (BTM), situation recognition module (SRM), eye contact module (ECM) and pan-tilt control unit module (PTUCM). The last module controls the head turn toward the human based on BTM and provides actions based on the SRM results. The BTM tracks the human body. To track and computes the direction of head (30 frame/sec), we use FaceAPI by Seeing Machines Inc. By extrapolating information from the person's gaze (for NPFV and FPFV situations) and body (for OFV situation), the SRM determines which situation exists. The ECM mainly consists of face detection module (FDM), and eye blinking module (EBM). If the target person looks toward the robot, FDM detects his/her face. FDM sends the results to EBM to exhibits eye blinks.

4. ATTENTION CONTROL EXPERIMENT

We hung six paintings (P1-P6) on the wall and asked 16 participants (64 trials) to watch the paintings. Two objects were placed on the table as attention targets. We prepared two robot heads: (i) attention control robot (AC-robot) and (ii) static robot (S-robot). AC-robot behaved as described in section 2. Fig. 3(a) shows a snapshot. We captured the participants' behaviors using two video cameras.

Experimental Results: From videos, we counted the number of times that the participants responded to the robot (this number provides a measure of success) in each phase (Fig. 3(b)). Participants responded 18 times among 20 in NPFV, 23 times out of 26 AA attempts in FPFV, and 14 times out of 18 AA attempts in OFV respectively. The robot proceeded to the next eye contact step in these successful cases. Results show that the attention shifting capabilities are somewhat common in all situations. However, the success rate of eye contact was higher in NPFV (94%) than that of FPFV (82%) and OFV (85%) situations. These differences may be due to orientation mismatch (between the participants face and the robot). A substantial 70% participants' attention was controlled by the robot in NPFV while only 61% and 55% were controlled in FPFV and OFV. Subjective evaluations based on 7-point Likert scale shows that the AC-robot can attract their attention ($M = 6.06, SD = 0.46$), making eye contact ($M = 6.18, SD = 0.29$) with them and convey its intended direction ($M = 5.8, SD = 0.42$). Finally, we asked to describe how effective they found our robot to control their attention. The average score was $5.06 (SD = 0.99)$.

5. CONCLUSION

The main objective of our work is to develop a robot system that can control a particular person's attention by considering the social situations. The success rate of a particular action of robot depends on several factors, such as distance between the robot and the human, duration of the action, direction of attention, level of attention of human to the task involved and so on. In this study, we consider only situation dependent actions with respect to the direction of attention. We have shown that robot's actions are effective to gain human attention, to make a person feel that the robot looks at him/her not others and shift his/her attention to an intended object. If a human is paying attention to a particular object or intensely engaged in a task, the robot needs to use more strong actions such as waving hand

6. ACKNOWLEDGEMENT

This work was supported by JST, CREST.

7. REFERENCES

[1] M. Finke and et al. Hey, i'm over here-how can a robot attract people's attention? In *In Proc. RO-MAN'05*, pages 7–12, 2005.

[2] A. Peters. Small movements as communictional cues in hri. In *In Proc. HRI'11*, pages 72–73, 2011.

[3] S. Yantis and J. Jonides. Abrupt visual onsets and selective attention: Voluntary versus automatic allocation. *J. Exp. Psy.: Hum. Per. and Per.*, 16(1):121–134, 1990.

An Intentional Framework Improves Memory for a Robot's Actions

Alicia M. Hymel
Psychology and Human Development
Vanderbilt University
Nashville, TN 37203, USA
alicia.m.hymel@vanderbilt.edu

Daniel T. Levin
Psychology and Human Development
Vanderbilt University
Nashville, TN 37203, USA
daniel.t.levin@vanderbilt.edu

ABSTRACT
Although a number of recent studies have explored people's concepts about robots, almost no research has tested the degree to which these concepts affect people's capacity to understand and remember a robot's actions. In this study, we tested whether a narrative describing a robot performing basic intentional acts would be easier to remember than a narrative that described similar non-intentional actions. Participants read one of two stories about a robot in which it was either described as having intentional or non-intentional mental representations. Participants who read about the intentional robot were more likely to recall information about the robotic agent, but there was no difference between the two groups in accuracy for questions unrelated to the agent. Additionally, participants who read about the intentional robot were marginally more likely to falsely recall a non-present object that was similar to the objects that the robot did interact with. We conclude that beliefs about a robot affect encoding and recall of its actions, possibly due to a focus on the type of information the agent is believed to "mentally" represent.

Categories and Subject Descriptors
H.1.2 [**Information Systems**]: User/machine systems – *human factors.*

General Terms
Experimentation, Human Factors

Keywords
Memory, Theory of Mind, Concepts, HRI

1. INTRODUCTION
As interaction with robotic agents increases, it becomes important both to understand people's beliefs about these agents and how these beliefs affect people's ability to understand and remember robots' actions. Using "theory of mind", individuals can understand the mental states of other agents and predict their future actions based on an analysis of the agent's beliefs, desires and goals [1]. It has previously been shown that both children and adults are able to distinguish agents with goal-directed, human-like intentional mental representations from those with non-intentional machine-like mental representations [2,3]. Intentional representations reflect situated semantic knowledge about the world, while non-intentional representations are more like abstract "pointers" that confer little real knowledge [4,5]. Although this research, and other projects exploring more general attributions about robot intelligence (e.g., [6]), makes a good start at establishing the nature of people's concepts about agents, it does not necessarily follow that these concepts are recruited when actually attempting to understand an agent's actions. This experiment therefore tested whether concepts about a robot's mental state affect how we organize and recall information about the robotic agent.

We tested effects of these concepts on memory by using three different measures. First, we examined the effect of concepts about a robot on recall of facts about that robot, as well as information not related to the robot. Previous research has shown that memory can be affected by factors such as depth of processing and the distinctiveness of stimuli [7,8]. It is possible that the ability to apply an intentional framework to an agent's actions allows people to engage in better-structured, deeper semantic processing about its actions. In contrast, a similar narrative that does not conform to an intentional analysis may be more difficult to organize and remember.

We also tested the effects of beliefs about robots on the strength with which people implicitly organize robots' actions based on semantic knowledge and concepts. To this end, we used a variation of the Deese-Roediger-McDermott (DRM) paradigm to examine the prevalence of false recall of a non-presented object in a list of objects that the agent interacted with. In this paradigm, subjects are typically presented with a list of words that are associated with a non-presented word, and are asked to recall the word list (e.g., a list including words like "loaf" and "dough" but not "bread"). This leads to a high prevalence of incorrect recall for the non-presented associated word ("bread") [9]. Differences in the level of false recall between individuals reading about an intentional and a non-intentional robotic agent could indicate an effect of concepts about agents on the strength or frequency with which we make these semantic associations. We hypothesize that these associations will be stronger and will produce more false recall when considering an intentional agent, as these agents are typically believed to mentally represent information in a way that emphasizes semantic meaning [3].

2. METHOD
2.1 Subjects
A total of 101 subjects (60 female; mean age: 36.5 years, SD=12.1) were recruited via Mechanical Turk. 48 completed the intentional condition, and 53 completed the non-intentional condition. All subjects were paid $1 for their participation.

2.2 Stimuli and Procedure
Two eight-paragraph narratives about a robotic agent were constructed. Paragraphs averaged 135.00 words each (SD=43.45), although the paragraphs for the non-intentional condition were 7.56 words longer, on average (SD=12.27; $t(7)=0.980$, $p=0.36$).

The final two paragraphs of each narrative described the robot viewing or manipulating a list of 7 to 8 objects based on the lists from research using the DRM paradigm [9].

Efforts were made to ensure that the two conditions were as similar as possible while differentiating the robots as more human- or machine-like. For example, in the intentional condition, the robot OSCAR was described using the pronoun "he" and was able to "think" and "understand" things. In the non-intentional condition, the robot PSCAR was described using the pronoun "it", and its behavior was described using more mechanical terms, such as "analyzed" and "computed". Additionally, the intentional robot engaged in goal-directed activities, while the non-intentional robot engaged in more rote, location-directed activities. For example, when presented with the same array of objects, OSCAR categorized them based on common use (e.g., writing implements), while PSCAR grouped them based on spatial arrangement.

Participants read either the intentional or non-intentional narrative after being informed that a set of questions would be asked about the narrative at the end of the experiment. After reading, participants completed a filler task in which they were presented with eighteen occupations and were asked questions about robots' and humans' abilities to succeed in the occupations (these data are not reported here).

Finally, participants answered two free recall questions about each paragraph in the narrative. One question was always directly about the agent or an action it performed. The other question was not about the agent (e.g., "What is the lab manager's name?"). In addition, participants were asked to recall as many of the objects mentioned in the list-events as they could, and the number of false recalls of the critical non-presented lure were tallied.

3. RESULTS

Participants in the intentional condition were significantly better able to recall facts about the agent (mean=81.60% correct, SD=15.84) than participants in the non-intentional condition (mean=67.30%, SD=24.23; $F(1,99)=12.048$, $p<0.001$; see Figure 1). There was, however, no difference in recall of non-agent events between the intentional (mean=50.26% correct, SD=27.10) and non-intentional groups (mean=50.00%, SD=29.10; $F(1,99)=0.002$, $p=0.963$).

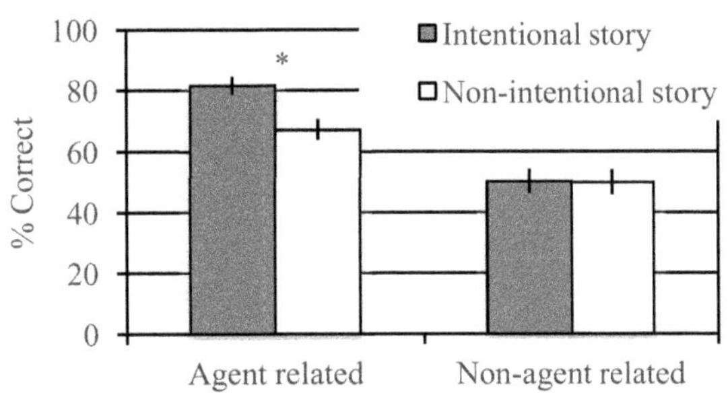

Figure 1. Recall accuracy for story details by condition.

Finally, participants were nearly significantly more likely to name the associated non-presented word in the intentional condition (mean=36.46% false recall, SD=39.60) than in the non-intentional condition (mean=22.64%, SD=33.35%; $F(1,99)=3.619$, $p=0.060$).

4. DISCUSSION

The results reveal the impact of differences in processing when thinking about an intentional or non-intentional robotic agent. While reading about an intentional robot led to better memory for the robot's actions, there was no effect for memory of information unrelated to the robot. Higher levels of intentionality appear to allow details concerning the robot and its actions to become more salient, without affecting overall readability or understanding, as indicated by equal recall for non-agent related story details.

Additionally, higher levels of intentionality may lead to increased false recall for objects similar to those that the robot manipulated, although this difference did not quite reach statistical significance. Higher levels of false recall for these items could indicate a greater focus on the semantic meaning of the objects in the intentional condition, possibly due to the robot's more human-like mental representations. If we know that the robot is engaging in goal-oriented activities based on semantic meaning, we may be more likely to take the robot's perspective and focus on these aspects.

In conclusion, these findings demonstrate that concepts about a robotic agent can affect how we attend to and encode information about the agent.

5. ACKNOWLEDGMENTS

This material is based upon work supported by the National Science Foundation under grant #0826701.

6. REFERENCES

[1] Gopnik, A., and Wellman, H.M. 1992. Why the child's theory of mind really is a theory. *Mind Lang.* 7(1-2), 145-171.

[2] Woodward, A.L. 1998. Infants selectively encode the goal of an actor's reach. *Cognition.* 61(1), 1-34.

[3] Levin, D.T., Killingsworth, S.S., Saylor, M.M. 2008. Concepts about the capabilities of computers and robots: A test of the scope of adults' theory of mind. In *Proceedings of the 3rd Annual IEEE International Workshop on Human and Robot Interaction* (Amsterdam, The Netherlands, March 12-15, 2008) HRI'08. ACM, New York, NY, 57-64.

[4] Dennett, D. 1991. *Consciousness Explained.* Little, Brown, & Co., Boston, MA.

[5] Searle, J. 1984. *Minds, Brains, and Science.* Harvard University Press, Cambridge, MA.

[6] Lee, S., Kiesler, S., Lau, I.Y., Chieu, C. 2005. Human Mental Models of Humanoid Robots. In *Proceedings of the 2005 IEEE International Conference on Robotics and Automation* (Barcelona, Spain, April 18-22, 2005) ICRA'05. 2767-2772.

[7] Jacoby, L.J., and Craik, F.I.M. 1979. Effects of decision difficulty on recognition and recall. *J. Verb. Learn. Verb. Be.* 18, 585-600.

[8] Craik, F.I.M., and Lockhart R.S. 1972. Levels of processing: a framework for memory research. *J. Verb. Learn. Verb. Be.* 11, 671-684.

[9] Roediger III, H.L., and McDermott, K.B. 1995. Creating false memories: Remembering words not presented in lists. *J. Exp. Psychol. Learn.* 21(4), 803-814.

Gestonurse: A Multimodal Robotic Scrub Nurse

Mithun G. Jacob, Yu-Ting Li, Juan P. Wachs*
School of Industrial Engineering
Purdue University
315 N. Grant St, West Lafayette, IN 47907
{mithunjacob, yutingli, jpwachs}@purdue.edu

ABSTRACT

A novel multimodal robotic scrub nurse (RSN) system for the operating room (OR) is presented. The RSN assists the main surgeon by passing surgical instruments. Experiments were conducted to test the system with speech and gesture modalities and average instrument acquisition times were compared. Experimental results showed that 97% of the gestures were recognized correctly under changes in scale and rotation and that the multimodal system responded faster than the unimodal systems. A relationship similar in form to Fitts's law for instrument picking accuracy is also presented.

Categories and Subject Descriptors

I.2.9 [**Robotics**]: Operator Interfaces

General Terms: Algorithms, Design, Experimentation

Keywords: gesture recognition, human-robot interaction, multimodal interfaces, robotic scrub nurse

1. INTRODUCTION

An RSN has the potential to reduce errors in the OR by automating the passing of surgical instruments which allows personnel to focus on more complicated tasks such as maintaining a sterile environment, preparing required surgical supplies and monitoring the state of the patient. This allows a possible reduction of errors due to communication failures. These failures can lead to wastage of resources, procedural errors, delays, distraction and inefficiency. A real-time RSN (see Figure 1(a)) dedicated to passing surgical instruments to the surgeon by speech and/or gesture is presented. An advantage of gesture interaction is that it is not affected by ambient noise and does not require surgeon re-training since hand signals are used by surgeons to request surgical instruments as part of standard OR procedure [1].

2. GESTURE RECOGNITION MODULE

Color and depth-based algorithms for hand segmentation are compared. The color-based algorithm was discussed in previous work [1]. The depth-based method used depth information from the Kinect sensor [2] to obtain a hand mask by thresholding the depth of objects within a bounding box in front of the camera. A fingertip detection method was used to recognize gestures [1]. Analysis shows that the gesture recognition algorithm runs in $O(n \log n)$ time in the number of points on the hand contour. Further classification was performed from hand masks obtained from both depth and color-based segmentation methods and fingertip detection and gesture recognition performances are compared in section 4.1.

3. ROBOT CONTROL SCHEME

The RSN uses a FANUC LR Mate 200iC robotic arm (see Figure 1(b)) and the CMU Sphinx [3] system for speech recognition. A state machine model (see Figure 2) controls the states of the robotic system. The system includes the following system states: ACTIVE, wait for the user commands; PASS, delivering an instrument; SLEEP on hold; MUTE speech recognition module is switched off (only gesture interaction is used). A sequence of two gestures is used to avoid accidental changes between states. The SPEECH (SP) ON/OFF gesture sequence is used to switch the system between ACTIVE and MUTE modes. Similarly the SLEEP/WAKE gesture sequence switches the system between SLEEP and the ACTIVE or MUTE states. If speech and gesture commands are performed simultaneously by the user, the first of the two events triggers the system.

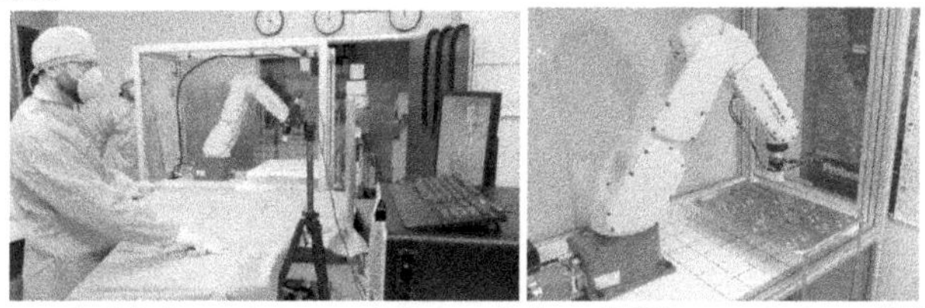

Figure 1. (a) A prototype of the real-time RSN tested at an OR (b) FANUC LR Mate 200iC and Mayo stand with instruments

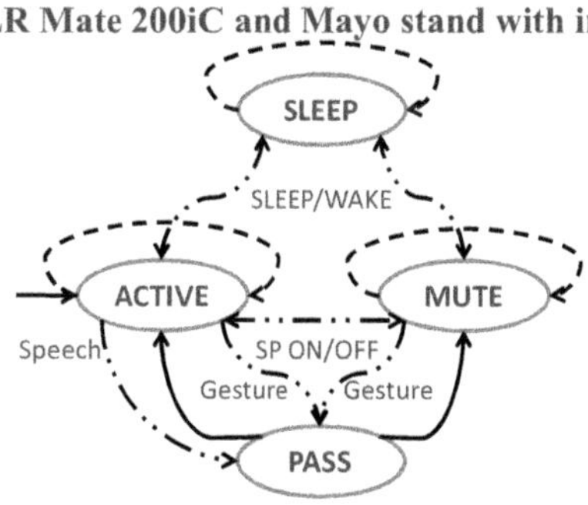

Figure 2. State Machine

4. EXPERIMENTS

4.1 Gesture Analysis Performance

A gesture analysis database consisting of 600 images of hand postures was collected from 4 users with different hand sizes and skin color (see Figure 3), and was used to assess the performance of the system. It included 30 images of postures per user per gesture exhibiting varying rotation, scale and translation of the fingertips. ROC curves for five gestures corresponding to five surgical instruments for depth and color-based hand segmentation were obtained.

4.2 Comparison of Modalities

The performance of the multimodal (speech and gesture combined) and unimodal speech/gesture systems were studied. Experiments were performed on 8 users between 20-30 years who performed 6 consecutive trials on each system. A trial is defined as an arbitrarily ordered sequence of 5 instrument types. The name of each instrument from the trial sequence was displayed to the subject who requested the instrument. The instrument acquisition time (the dependent variable) is defined as the time

*Corresponding author

HRI'12, March 5–8, 2012, Boston, Massachusetts, USA.
ACM 978-1-4503-1063-5/12/03.

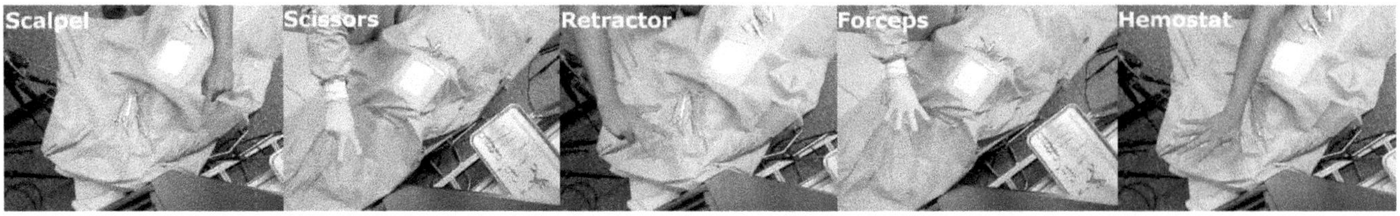

Figure 3. Samples from the use-case with detected fingertips marked with red circles

elapsed between the displayed instrument name and the subject receiving the instrument. Each user performed a task with an instrument before requesting the next instrument and instrument acquisition times was recorded.

5. RESULTS & DISCUSSION

5.1 Gesture Analysis Performance

The fingertip detection algorithm achieved high average recognition rates for both color and depth-based segmentation algorithms (97% and 99% respectively). Conversely, the color-based segmentation algorithm exhibited a higher average false-positive rate of 10.8% compared to 0.5% for the depth-based algorithm (see Figure 4). This is attributed to the sensitivity of the color-based method to illumination. The average gesture recognition accuracy of the depth-based system was 97% (see Table 1) but in practice, users learned to avoid error-inducing poses such as out-of-plane rotation. Note that ϕ in Table 1 refers to instances where gestures were not recognized.

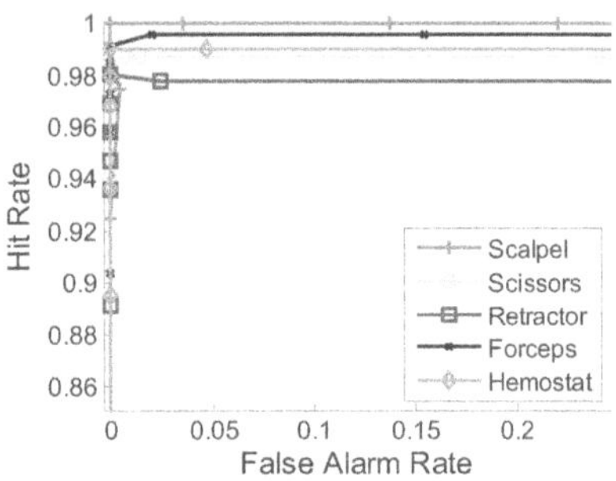

Figure 4. ROC Curve for Depth-based segmentation

5.2 Instrument Picking Performance

The robot can reliably pick an instrument from the Mayo stand and hand it to the user when the instruments (of normalized area A) are separated by at least $\lambda = 25$mm [1]. A linear relationship (see Figure 5) similar to Fitts's law [4] was observed ($\alpha = a + b\,ID$) between index of difficulty $ID = \log_2(\lambda/A + 1)$ and accuracy α. This was validated by high R^2 values of 0.89, 0.94 and 0.86 for the forceps, scissors and hemostat instrument types respectively. This result validates an intuitive tradeoff between λ and the accuracy of the picking and delivery task.

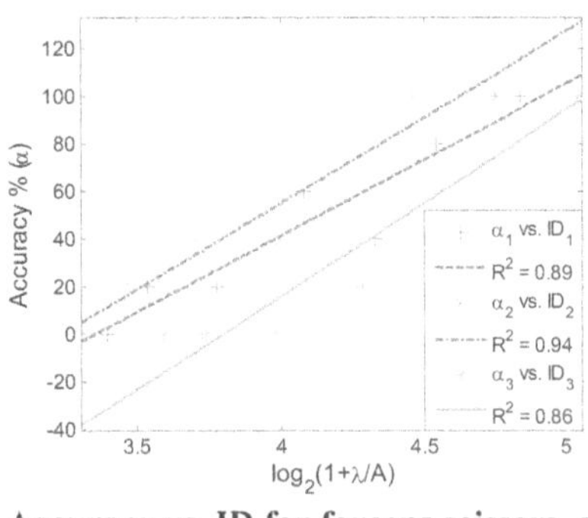

Figure 5. Accuracy vs. ID for forceps scissors, and hemostat

5.3 Comparison of Modalities

The instrument acquisition time for the six trials of each individual is presented in Figure 6. It was observed during trials that the gesture recognition system was faster (see Figure 6), and superseded the speech recognition system in triggering the state machine (shown by lower average instrument acquisition times). In events where the gesture was performed after the speech command or made incorrectly delaying recognition, the speech module compensated resulting in lower average instrument acquisition time for the multimodal system.

Table 1: Depth-based Segmentation: Confusion Matrix

	Scalpel	Scissors	Retractor	Forceps	Hemostat	Ø
Scalpel	**100**	0.00	0.00	0.00	0.00	0.00
Scissors	0.84	**97.50**	0.83	0.00	0.00	0.83
Retractor	0.83	4.17	**95.00**	0.00	0.00	0.00
Forceps	0.00	0.00	1.67	**96.67**	1.66	0.00
Hemostat	0.00	0.00	0.84	3.33	**95.83**	0.00

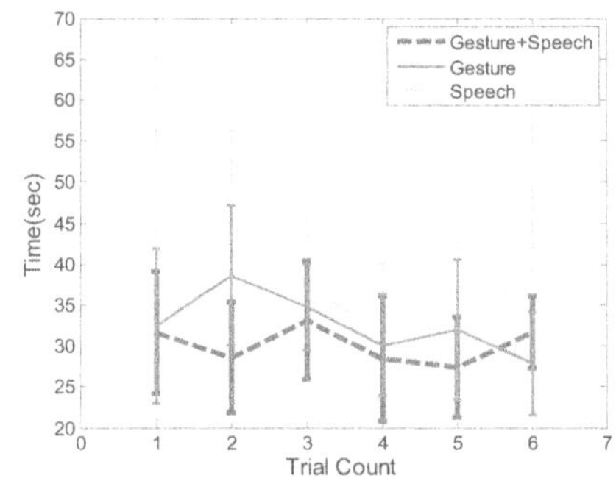

Figure 6. Mean instrument acquisition times for each trial and 95% confidence intervals for unimodal & multimodal systems

6. CONCLUSION & FUTURE WORK

A multimodal robotic scrub nurse was presented capable of reliably passing surgical instruments. Fingertips were detected and gestures recognized with 99% and 97% accuracy on average. A relationship similar to Fitts's law has been shown between picking accuracy, inter-instrument distance and instrument area. It was noted that the multimodal system performed faster than either unimodal system and that it is more suitable to uncontrolled environments, such as the OR.

Future work includes fusing the speech and gesture recognition data in a probabilistic fashion and testing the system in a live OR.

7. REFERENCES

[1] Jacob, M.G., Li, Y.-T, Wachs, J. P., "A gesture driven robotic scrub nurse," Systems, Man, and Cybernetics, IEEE International Conference on, pp. 2039-2044, 9-12 Oct. 2011

[2] Microsoft Corp. The Kinect sensor, www.xbox.com/kinect

[3] CMU Sphinx http://cmusphinx.sourceforge.net/

[4] Fitts, P., "The information capacity of the human motor system in controlling the amplitude of movement," Journal of experimental psychology, vol. 47, no. 6, p. 381, 1954.

Manipulation with Soft-fingertips for Safe pHRI

[2,a]Jorge Armendariz, [1,2,b]Rodolfo García-Rodríguez, [2,c]Felipe Machorro-Fernández, and [2,d]Vicente Parra-Vega
[1]Universidad de los Andes, Chile; [2]Robotics and Advanced Manufacturing Program, CINVESTAV, Mexico
[a]jorgearm.ryma@gmail.com, [b]rogarciar@gmail.com, [c]felipemachorro@gmail.com, [d]vparra@cinvestav.mx

ABSTRACT
Manipulation with soft-fingertips is proposed as a tool for interacting with rigid objects, based on an online bilateral teleoperation fuzzy controller, without time delay. The fuzzy inference engine tunes continuously the force feedback gain to increase awareness of physical interaction so as to grasping is achieved. In contrast to the case of rigid fingertip, wherein infinitely small contact point is assumed, it is argued that our scheme allows safe and intuitive interaction, which has proved effective in an experimental study with 19 subjects. Results indicate success due to feedback of slave contact force to the human user. Subjects consistently judge comfort and easiness of manipulation, with an experimental two-hand prototype.

Categories and Subject Descriptors
I.2.9 [**Artificial Intelligence**]: Robotics – *manipulators, operator interfaces.*

Keywords
Soft-fingertips, fuzzy-based force control, bilateral teleoperation control, physical Human-Robot Interaction (pHRI).

1. INTRODUCTION
It has been pointed out the need of safe robots and here we claim that also safe end-effectors are required for safe manipulation as well as for safe physical human-robot interaction, or pHRI, [1]. The reason is that classical grasping and object manipulation by robotic fingers are based on the assumption that exists frictionless point-like rigid contact [2], between object and fingers. This typically involves intermittent rigid, consequently unilateral constraints and high frequencies, undesirable phenomena in applications where there exist pHRI. Additionally, robotic hands have shown very limited dexterity to execute precision tasks because rolling is not allowed, besides that a high computational cost is required to execute dynamic control [3]. Then soft-fingertips stand for an option, because rolling and contact area are included. Although a model of deformation for soft or deformable fingertips has been presented [4], a general framework is an open problem today, which makes difficult the control design. In this paper, we present a bilateral tele-manipulation scheme using two soft-fingertips, for pHRI with dynamic controller and online auto-tuning based on a fuzzy paradigm.

2. DYNAMIC MODEL AND CONTROL
The full dynamic model of a 1-1 dynamic hand is considered [3], with a dynamic PD (Proportional-Derivative) regulator. In order to include the sense of touch essential for human/robot interaction, an estimated force feedback loop is synthesized based on fuzzy inference, according to human interpretation, which is crucial for proper stimulus to the human user. To this end relative distance related to the fingertips deformation and fingertips motion of the robotic hand should be specified. Considering Fig. 1, let δ_i and δ_d be the maximum strain of the left and right fingertip, respectively. Let A and B be the joint of each finger, O_i and O_d be the origin of each hemispherical fingertip, a be the radius of both fingertips, and let W_{obj} and θ_{obj} be the width and the orientation of the grasped object, respectively. Finally, let L_f be the length of each finger and L be the distance between joint A and joint B. Therefore, a vector description could be represented as follows:

$$\overrightarrow{AS} = \begin{bmatrix} L_f \sin\theta_i + L_f \cos\theta_i \tan\theta_{obj} \\ 0 \end{bmatrix}, \tag{1}$$

$$\overrightarrow{As} = \begin{bmatrix} L - L_f \sin\theta_d + L_f \cos\theta_d \tan\theta_{obj} \\ 0 \end{bmatrix}. \tag{2}$$

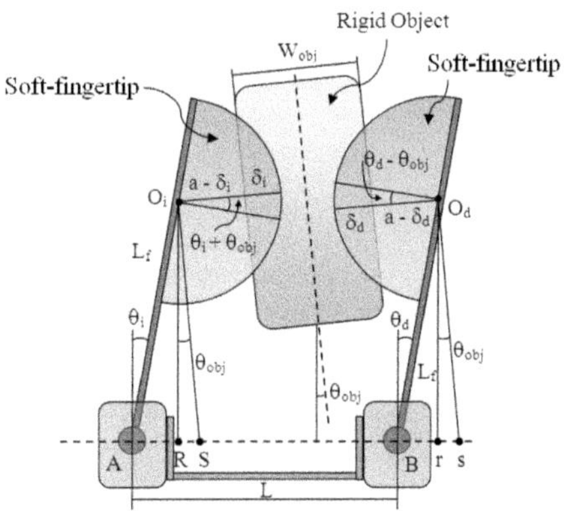

Figure 1. Geometric representation of the soft-fingered robotic hand. Area and rolling contacts are key features allowing dexterity. Therefore, dynamic force control is of importance to ensure stable grasp and manipulation.

The distance $\overrightarrow{Ss} = \overrightarrow{As} - \overrightarrow{AS}$ results:

$$\overrightarrow{Ss} = \begin{bmatrix} L - L_f(\sin\theta_d + \sin\theta_i) + L_f \tan\theta_{obj}(\cos\theta_d - \cos\theta_i) \\ 0 \end{bmatrix}. \tag{3}$$

The distance along horizontal direction O_iO_d can be represented as follows:

$$\overrightarrow{O_iO_d} = W_{obj} + 2a - 2\delta = \overrightarrow{Ss}\cos\theta_{obj}\,, \text{ where } \delta_i = \delta_d. \tag{4}$$

Using (3) and (4) is obtained the metric used by the force feedback controller called *metric of deformation* $\Delta\delta$:

$$\Delta\delta = W_{obj} + 2a - [\,L - L_f(\sin\theta_d + \sin\theta_i) + L_f \tan\theta_{obj}(\cos\theta_d - \cos\theta_i)]\cos\theta_{obj}\,. \tag{5}$$

Taking $\Delta\delta$ as input to the fuzzy inference engine, the *force feedback* is given as:

$$\text{Rule } i\text{: } \textbf{IF } (\Delta\delta(k) \text{ is } A_i) \textbf{ THEN } F_i(k) = f_i(\mu_{A_i}(k))\,. \tag{6}$$

Finally, the force-output $\boldsymbol{F}$ is computed based on corresponding linear consequences (LCs) defined by

$$\boldsymbol{F} = \sum_{i=1}^{3} F_i(k) = \sum_{i=1}^{3} \beta_i \, \mu_{A_i}(k) = \mathbf{Y}^T \mathbf{P}\,, \tag{7}$$

where $\mathbf{Y} = [\,\beta_1\ \beta_2\ \beta_3\,]^T$, with β_i representing the gain, $\mu_{A_i}(k)$ the membership function (gaussian for the first two rules and sigmoidal for the third rule) and $\mathbf{P} = [\mu_1\ \mu_2\ \mu_3]^T$. In this way, human knowledge is transferred as a fuzzy parameter $\Delta\delta$, thereof using the inference mechanism to produce an online signal conditioning as the force feedback for the master hand (Fig. 2), which displays manipulating forces for human interpretation, as kinesthetic touch.

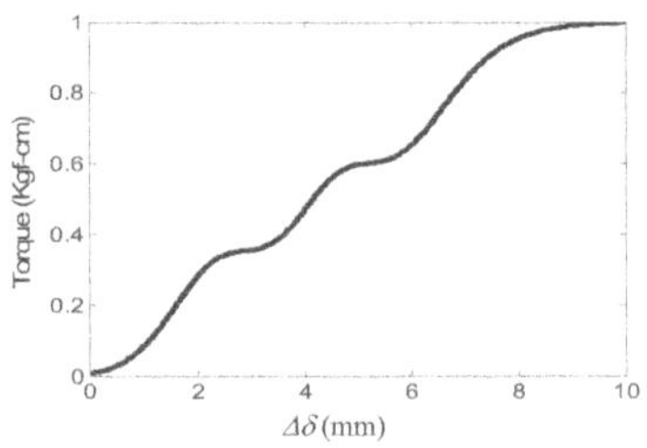

Figure 2. Smooth force response is obtained, when manipulating a fragile object, such as an egg. This monotonously increasing function is nonlinear due to the fuzzification process, a key issue in this proposal.

3. EXPERIMENTS

3.1 Soft-fingertips and Experimental Setup

Two soft-fingertip master and slave hands were developed and integrated in our laboratory, based on DYNAMIXEL AX-12 motors by ROBOTIS Inc. Drivers and user interface are programmed in MATLAB. The soft-hemispherical fingertips were selected of polyurethane for improved contact and manipulation of the grasped object, see in Fig. 3.

Figure 3. The master system is driven by the user to teleoperate an egg by the slave system. During manipulation task, the fuzzified contact forces are displayed on the human master hand.

3.2 Experimental Procedure and Results

Experimental sessions were carried out with a sample of 19 volunteers between 24 to 37 years old (9 males and 10 females), and only three of them with experience in haptic technology. In the experiments we used an aluminum bottle and an egg. The experimental procedure was divided in two stages. The *first stage* consists of the follows steps: 1) volunteer receives instructions on how to use the system to carry out the task (information concerning to the force-feedback is not provided to the user), 2) volunteer executes the task **with** force feedback, 3) volunteer executes the task **without** force feedback, 4) volunteer answers the following question: *Which of all attempts were the easiest to manipulate?*. In the *second stage*, volunteer repeats the task defined previously in the *first stage* but now blindfolded. Once each stage is completed volunteer answers what task is more comfortable: **with or without** force feedback or indifferent, as shown in Fig. 4.

A total of 152 experiments were performed. Failure is only 6.58% or 10 experiments where 6 of them occurred **with** force-feedback and 4 **without** force-feedback, which is 3.95% and 2.63% respectively. These results mean that visual/kinesthetic learning is achieved when the volunteer uses the vision-feedback (or human sight) in combination with the kinesthetic stimuli produced by the motor-friction without force feedback or by the force-feedback fuzzy controller. On the other hand, when the volunteer is blindfolded, visual learning is eliminated and here is when the proposed force-feedback controller enhances the ability of the user and shows its usefulness. Then successful grasping and the object manipulation by means of two soft-fingertips are achieved essentially due to the kinesthetic sensation and rolling capabilities provided by the controller through $\Delta\delta$ of deformation.

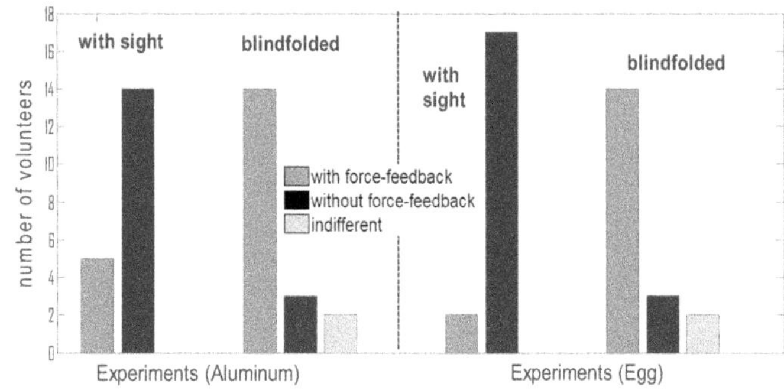

Figure 4. The experiments were performed with a sample of 19 volunteers and divided in two stages: sight and blindfolded. All experiments were executed considering with and without force-feedback.

4. CONCLUSIONS

It is proposed a safe and intuitive system for manipulation under a bilateral teleoperation scheme. Contact forces are conditioned and displayed in the human master hand so as to providing key kinesthetic forces for successful manipulation of fragile objects, including rolling motions. It is argued that this soft manipulation tool is effective for pHRI; this claim is verified through an experimental study.

5. ACKNOWLEDGMENTS

This work was partially supported by CONACyT Mexico, Grants SEP-CONACyT #0133646 and Universidad de los Andes, Chile, FAI project # ICI-002-11.

6. REFERENCES

[1] S. Haddadin, A. Schäffer, M. Frommberger, G. Hirzinger, "The Role of the Robot Mass and Velocity in Physical Human-Robot Interaction Part I-II", IEEE Int. Conf. on Robotics and Automation, USA, 2008.

[2] J. Armendariz, C. Treesatayapun, A. Baltazar, "Force feedback controller based on fuzzy-rules emulated networks and hertzian contact with ultrasound", *MSSP Elsevier*, vol. 27, pp. 534-550, 2011.

[3] S.Arimoto, P.Nguyen, H.Y.Han, and Z.Doulgeri, "Dynamics and Control of a set of Dual Fingers with Soft Tips," *Robotica*, vol.18, pp. 71-80, 2000.

[4] T. Inoue, S. Hirai, "Elastic model of deformable fingertip for soft-fingered manipulation", *IEEE Transactions on Robotics*, Vol.22, No.6, pp. 1273-1279, Dec., 2006.

Studying Virtual Worlds as Medium for Telepresence Robots

Alex Juarez
Eindhoven University of Technology
Den Dolech 2
Eindhoven, Netherlands
acordova@tue.nl

Christoph Bartneck
University of Canterbury
HIT Lab NZ
Christchurch, New Zealand
christoph.bartneck@canterbury.ac.nz

L. M. G. Feijs
Eindhoven University of Technology
Den Dolech 2
Eindhoven, Netherlands
l.m.g.feijs@tue.nl

ABSTRACT

This paper presents a study on the effects of using virtual worlds as medium for interaction with telepresence robots. The Prototype for Assisted Communication PAC4 is also introduced. This system connects virtual worlds with real robots, allowing virtual world users to access the robot capabilities in a telepresence scenario. A user experiment was conducted to evaluate the effects of PAC4 on the feeling of social presence experienced by virtual world users.

Categories and Subject Descriptors

J.2 [**Computer Applications**]: Physical Sciences and Engineering—*Engineering*

General Terms

Design, Experimentation, Human Factors

1. INTRODUCTION

Telepresence robots are a current topic of research in Human Robot Interaction (HRI). Telepresence itself is manifested in a large number of applications that involve interaction through video. This has originated the perception of telepresence robots as "embodied video conferencing on wheels" [5]. On the other hand, telepresence has been a topic traditionally associated to virtual and mixed reality. In this context, virtual worlds offer an alternative environment for human-robot interaction in telepresence scenarios. In this paper we present PAC4, a system that connects virtual worlds with real robots allowing virtual world users to access and use the robot capabilities in a telepresence scenario. We also present a user experiment conducted to evaluate the effects of PAC4 on the feeling of social presence experienced by virtual world users.

2. PAC4 DESIGN AND IMPLEMENTATION

PAC4 implements the *observer design pattern*, a "one-to-many dependency between objects so that when one object changes state, all its dependents are notified and updated automatically" [3, 2].

A typical process flow in PAC4 starts when a new virtual or real agent makes its various services available within the system. Agents identify themselves to PAC4 by providing a Universal Resource Identifier (URI). PAC4 checks for the type of the agent and services that are being published. If it is a *robot* it contacts the RoboDB semantic information database [4] requesting information about the robot capabilities. Capabilities are treated as services related to the robot. Each service is a subject to which observers -other robots, or virtual world applications- can attach. PAC4 represents the robot capabilities indicating their name and type, its inputs and outputs. PAC4 uses the robot identifier and the capability inputs and outputs names and types to establish which observers to notify about the event.

HRI'12, March 5–8, 2012, Boston, Massachusetts, USA.
ACM 978-1-4503-1063-5/12/03.

3. EVALUATION

For the evaluation of PAC4, we focused on the study of *social presence*, a key element in telepresence robotics. The experiment was setup in a health and care telepresence scenario where the virtual world user communicated with a friend or relative in a remote location. The virtual world used was Second Life. The real world component was the "apartment" where the friend or relative "lived". Figure 1 shows the real living area, and its virtual counterpart in Second Life. An iCat robot [6] was also available in the living room area of the apartment. This robot can move its head, "talk" using text to speech, respond to touches in its head and "paws", and provide audio and video streams.

Figure 1: Real room located in the Context Lab and its virtual counterpart in Second Life.

The term PAC4 refers to a virtual world enhanced with the capabilities of real robots using the PAC4 system previously described. The term *plain Second Life* refers to the same virtual world without this enhancement. Additionally, the VoIP software ooVoo was used as *baseline* for comparison and evaluation of the overall effect of virtual worlds on the telepresence experience.

A mixed between-within subjects experiment design was used. The measurement device used was the IPO-Social

Presence Questionnaire (IPO-SPQ) developed by De Greef & Ijsselsteijn [1]. A total of 20 participants, 8 female and 12 male with ages between 20 and 34, took part in the experiment. Ten participants were selected randomly to be placed in the PAC4 group, and ten in the plain Second Life group.

A one-way between-groups analysis of covariance was conducted to compare the effectiveness of two different communication systems designed to measure the level of social presence experienced by the participants in a remote communication scenario. The independent variable was the type of system used (PAC4, plain Second Life) and the dependent variable consisted on the scores of the IPO-SPQ questionnaire administered after the participants completed the indicated tasks with the communication systems. The estimated mean score values and standard deviation for PAC4 and plain Second Life were $M = 4.000, \text{Std. Dev.} = 0.685$ and $M = 3.806, \text{Std. Dev.} = 0.946$ respectively. The IPO-SPQ scores of the participants related to the software ooVoo were used as the covariate in this analysis. After adjusting for the covariates there were no significant differences between the two communication system groups (PAC4 and plain Second Life) on the scores for the IPO-SPQ, $F(1, 17) = 0.554$, $p = 0.467$, eta squared = 0.032. There was a moderate relationship between the covariate and the independent variable a indicated by an eta squared = 0.054.

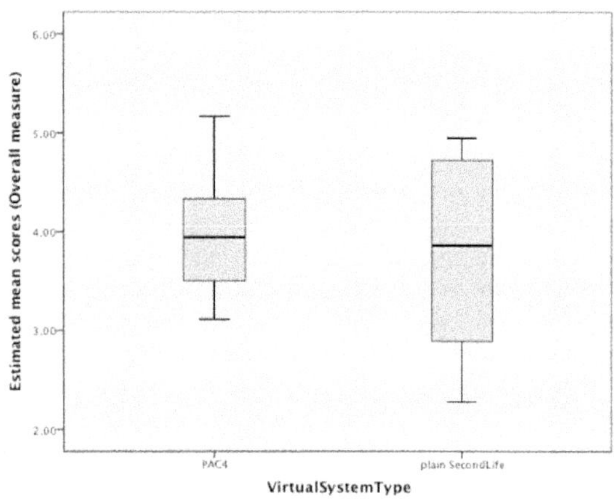

Figure 2: Mean presence scores of PAC4 vs plain Second Life

The analysis of the video recordings taken during the experiment and the responses to open-ended questions, showed divided opinion over the benefits of PAC4 over Second Life. Most participants stated that they felt "more connected" to the remote peer, and that PAC4 provides "better information about the other person" than the VoIP software. However, several mentioned that there might be issues with privacy as "I can see information about the other person even when the person is not online or chatting with me".

4. DISCUSSION

The results indicate that although the participants appreciate the potential of PAC4 as an enabling technology for telepresence, they perceive a large overhead and complexity in the virtual world software. Particularly, participants mentioned that alternative means of communication like VoIP software (e.g. skype, ooVoo, etc.) were simpler to understand, faster, and in general, more familiar to them. Some participants went further to comment that although they value the potential of telepresence using a robot for health and care situations, moving around in the virtual world, and talking to others through an avatar was complicated and unnatural.

All this seems to suggest that in general, virtual worlds technology is not mature enough to allow for meaningful interactions and applications where efficient telepresence is needed, making users feel that virtual worlds are more a hindrance than a facilitator. In fact, the participants indicated that although having a robot they could remotely interact with helped them to feel more connected and more aware of what happened in the remote location, they felt that they were equally present when they could talk to the remote person without the mediation of a virtual representation.

5. CONCLUSION

This paper introduced the PAC4 system that connects virtual worlds and real robots, enabling virtual world users to access the robot's capabilities from within the virtual world. A user experiment was conducted to evaluate the effect of PAC4 on telepresence. The results of the evaluation of PAC4 should be seen under the correct light. Although the questionnaire scores indicate that PAC4 indeed improves the level of presence with respect to plain Second Life, the fact that the overall evaluation of virtual worlds by the participants was not positive produced a negative effect in the overall appreciation of any virtual world-based software. In other words, there seems to be an indication that virtual world applications and technologies are not yet ready to compete with alternative forms of communication like the VoIP software ooVoo. This is a likely explanation as to why the results of the evaluation are not statistically significant.

6. REFERENCES

[1] P. De Greef and W.A. Ijsselsteijn. Social presence in a home tele-application. *CyberPsychology & Behavior*, 4(2):307–315, 2001.

[2] L.M.G. Feijs. Modelling Microsoft COM using π-calculus. In *Proceedings of the Wold Congress on Formal Methods in the Development of Computing Systems-Volume II*, pages 1343–1363. Springer-Verlag, 1999.

[3] Erich Gamma, Richard Helm, Ralph Johnson, and John Vlissides. *Design patterns: elements of reusable object-oriented software*. Addison-Wesley, 1995.

[4] A. Juarez, C. Bartneck, and L. Feijs. Using semantic technologies to describe robotic embodiments. In *Proceedings of the 6th international conference on Human-robot interaction*, pages 425–432. ACM, 2011.

[5] K.M. Tsui, M. Desai, H.A. Yanco, and C. Uhlik. Exploring use cases for telepresence robots. In *Proceedings of the 6th international conference on Human-robot interaction*, pages 11–18. ACM, 2011.

[6] A. van Breemen, X. Yan, and B. Meerbeek. iCat: an animated user-interface robot with personality. In *Proceedings of the fourth international joint conference on Autonomous agents and multiagent systems*, pages 143–144. ACM, 2005.

A Follow-Up on Humanoid-Mediated Stroke Physical Rehabilitation

Hee-Tae Jung
Lab. for Perceptual Robotics
Dept. of Computer Science
Univ. of Massachusetts
Amherst, MA 01003-9264
hjung@cs.umass.edu

Yu-Kyong Choe
Dept. of Comm. Dis.
Univ. of Massachusetts
Amherst, MA 01003-9296
ychoe@comdis.umass.edu

Jennifer Baird
Motor Control Lab.
Dept. of Kinesiology
Univ. of Massachusetts
Amherst, MA 01003-9258
bupt95@kin.umass.edu

Roderic A. Grupen
Lab. for Perceptual Robotics
Dept. of Computer Science
Univ. of Massachusetts
Amherst, MA 01003-9264
grupen@cs.umass.edu

ABSTRACT

We report the results of standardized tests on a single subject with a stroke at 4, 20 and 28 weeks after completion of the study. These results follow from previous work [1]. The subject demonstrated sustained improvement in motor function 28 weeks after completing the study. In addition to quantitative results, the questionnaire results by the subject and the spouse testify that the subjective user experience was also positive. This further advocates the use of general purpose robots to complement human therapists.

Categories and Subject Descriptors

J.3 [**Computer Applications**]: Life and Medical Sciences—*health*; K.4.2 [**Computers and Society**]: Social Issues—*assistive technologies for persons with disabilities*

General Terms

Experimentation, Human Factors

Keywords

Autonomous Agents, Embodied Agent, Healthcare, Physical Rehab, Stroke, Telepresence

1. INTRODUCTION

Due to the increasing number of patients and the shortage of well-trained therapists, the use of robots in rehabilitation is widely acknowledged by a research community as being a possibility. This has led to the ongoing development of diverse robotic platforms [3]. They are developed and tested in order to address specific activities and populations that meet a particular inclusion criteria. In reality, however, stroke survivors have different disabilities of varying severity depending on the site and size of lesion in the brain. Moreover, the same patient may need different types of therapy as he or she regains motor functions. Hence, in the first stage of our study, we investigate if general purpose robots can be used to provide effective stroke rehabilitation service to stroke patients rather than building multiple special purpose robots for different activities. The therapy activities that are designed by a human therapist based on her knowledge and experience are implemented in a humanoid robot. Through the results from a single case study, in our earlier work, we have shown that the fixed set of therapy activities can increase the speed of the subject's upper extremity movement [2] and therapy activities with adaptive strategy can increase not only the speed but also the range of motion of the subject's arm movement [1]. However, in order to further support the feasibility of the use of general purpose robots, we need to show that this improvement continues to be sustained and that the subjective experience of the patient and the family members are positive. Accordingly, in this paper, we report the follow-up standardized test results and the questionnaire responses of the subject and the spouse.

HRI'12, March 5–8, 2012, Boston, Massachusetts, USA.
ACM 978-1-4503-1063-5/12/03.

2. DISCUSSION

2.1 Standardized Tests

The motor function of the subject was quantitatively measured using two standardized tests, Fugl-Meyer Assessment (FMA) and Wolf Motor Function Test (WMFT). The upper extremity portion of the FMA consists of 33 items that evaluate upper extremity functions of stroke patients. Performance on each item (e.g., Touch your ear with your weaker hand) is rated by a clinician (e.g., 0 = cannot be performed, 1 = detail partly performed, 2 = detail is performed faultlessly). The WMFT quantifies upper extremity motor ability through fifteen timed and functional tasks. The results report how fast patients achieve the tasks in addition to how many they successfully complete. Accordingly, the WMFT can measure improvements that cannot be measured in discrete scoring of the FMA. The baseline test was conducted before the study, and the additional tests are conducted 6 weeks after the study started and at the end of the 12-week study. Follow-up tests were conducted 4 weeks, 20 weeks,

Table 1: FMA results (raw scores). The three components (a, b, d) are directly related to the therapy activities that were conducted in the study. The total score reports the sum of all four upper extremity assessment components. (* indicates significant improvements from baseline at $p < .05$)

Component	Baseline	After 6 wks	After 12 wks	4-wk post-tx	20-wk post-tx	28-wk post-tx
Shoulder/elbow/forearm[a] (/36)	18	16	17	17	15	18
Wrist[b] (/10)	3	4	4	5	6	6
Hand[c] (/14)	7	6	6	6	11	12
Co-ordination/speed[d] (/6)	3	4	4	4	4	4
Total (/66)	31	30	31	32	36	40*

Table 2: Time changes in WMFT (seconds). Only the items that the patient was able to complete are presented here.

Component	Baseline	After 6 wks	After 12 wks	4-wk post-tx	20-wk post-tx	28-wk post-tx
Forearm to table	4.03	1.90	4.12	1.00	1.43	2.62
Hand to table	5.56	1.38	2.34	2.56	2.06	2.15
Reach and retrieve	5.26	1.00	2.69	1.03	0.82	1.00
Total	14.85	4.28	9.15	4.59	4.31	5.77

and 28 weeks after the completion of the study. The therapist was blind to the subject's previous scores to minimize the clinician's bias. During this follow-up period, the subject did not participate in any physical therapy activity. A Wilcoxon signed-ranks test revealed that the subject significantly improved in the FMA score from baseline to 28-week post-treatment, $Z = 2.324$, $p < .05$ (Table 1). He showed gradual improvements in the Wrist component and delayed improvements in the Hand component. The subject sustained his motor function in the WMFT throughout the follow-up period (Table 2).

2.2 Subjective Experience

At the completion of the study, we conducted a survey with the subject and the spouse. The questionnaire was constructed partially based on Weiss et al. [4]. The questions were grouped into: overall experience, feelings and impressions, comments, and future plans. Each question had a 5-point scale for response with 1 being the most negative and 5 being the most positive. The median scores from the both questionnaires were 5's, indicating that they both had good experience with the robot. They responded that the robot-mediated therapy activity was helpful, useful, pleasant, and exciting (all 5 points). Neither of them cared about the fact that the robot did not have any cosmetic features, such as hair or clothing. They were pleased with the functional features of the robot, such as arm movement. It is interesting to note that to the question 'I would rather work with clinicians than with the robot' (1 = yes, 5 = no), the subject gave a positive response (4 point) while the spouse was neutral (3 point). This indicates that the subject did not prefer a human therapist to a robot even though he had an experience with a human therapist for 1 year right after the onset of the stroke. Both the subject and the spouse replied that they would like to have a similar robot at home so that the subject could regularly exercise arm movements (5 point). They both expressed their willingness to participate further robot-mediated therapy study in the future (5 point).

3. CONCLUSION

In this paper, we report the results of the follow-up standardized tests and questionnaire responses demonstrating that both the quantitative and the qualitative experiences of the subject and the spouse were positive. This further supports the feasibility of the use of general purpose robots in stroke physical rehabilitation.

4. ACKNOWLEDGMENTS

This research was funded in part by ONR MURI award N00014-07-1-0749 and the Faculty Research Grant from UMass Amherst to Yu-Kyong Choe. Hee-Tae Jung acknowledges Robin Popplestone Fellowship and Jeong Song Culture Foundation Scholarship.

5. REFERENCES

[1] H. Jung, J. Baird, Y. Choe, and R. A. Grupen. Upper extremity physical therapy for stroke patients using a general purpose robot. In *Proceedings of the 20th IEEE International Symposium on Robot and Human Interactive Communication*, pages 270–275, Atlanta, GA, July–August 2011.

[2] H. Jung, J. Baird, Y. Choe, and R. A. Grupen. Upper-limb exercises for stroke patients through the direct engagement of an embodied agent. In *Proceedings of the 6th ACM/IEEE International Conference on Human-Robot Interaction*, pages 157–158, Lausanne, Switzerland, March 2011.

[3] P. S. Lum, D. J. Reinkensmeyer, R. Mahoney, W. Z. Rymer, and C. G. Burgar. Clinical considerations in the use of robotic devices for movement therapy following stroke. *Topics in Stroke Rehabilitation*, 8(4):40–53, 2002.

[4] A. Weiss, R. Bernhaupt, M. Tscheligi, D. Wollherr, K. Kuhnlenz, and M. Buss. A methodological variation for acceptance evaluation of human-robot interaction in public places. In *Proceedings of the 17th IEEE International Symposium on Robot and Human Interactive Communication*, pages 713–718, Munich, Germany, August 2008.

Personality and Facial Expressions in Human-Robot Interaction

Soyoung Jung[1], Hyoung-taek Lim[1], Sanghun Kwak[1], Frank Biocca[1,2]

[1]Media Interface and Network Design Lab
Department of Interaction Science
Sungkyunkwan University
Seoul, South Korea
soyoungjungs@gmail.com

[2]Newhouse Endowed Chaired Professor
S.I. Newhouse School of Public Communications
Syracuse University, NY, USA
fbiocca@syr.edu

ABSTRACT
This paper presents the propensities of preference for Human Robot Interaction (HRI) according to different personalities and facial expressions of humans and robots. This study is based on two types of personalities: extroverted and introverted. According to the personality, the facial expressions are distinguished to express extroversion and introversion. The design of experiment is a 3 (participant's group of personality: extrovert, intermediate, introvert) x 2 (robot's groups of personality) between-subjects experiment (N=40) in which participants interact with KMC-EXPR robots expressing either extroversion or introversion. The results showed an unprecedented hypothesis, unexpected implications and certain propensities of preferences.

Categories and Subject Descriptors
I.2.9[**Artificial Intelligence**]: Robotics – *Commercial robots and applications, Kinematics and dynamics, Operator interfaces.*

General Terms
Experimentation, Theory

Keywords
Social Robot, Facial Robot, Personality, Introvert, Extrovert, Facial Communication, KMC-EXPR

1. INTRODUCTION
Many HRI studies have tried to enhance the level of social ability in robots. However, the correlation between facial expression and personality in human-robot interaction has not been studied much yet. In this study, a psychological and sociological approach was used to design a face robot, and the results would guide prospective social robot developers to a desirable direction of how to design social robots in a better way. The facial interaction between an animated 3D face robot and human beings has been investigated [3]. In this research the face robot was not only able to produce human-like facial expressions, but also able to recognize human facial expressions using facial image data. Furthermore, the 3D robot was designed to be able to communicate with humans by autonomously changing its facial expressions. However, the 3D face robot was not an embodied agent, so it would not be regarded as a social robot.

In psychological terms, personality is the set of distinctive qualities that distinguish individuals [2].

In contemporary psychology, the "Big Five" factors (or Five Factor Model; FFM) of personality are five broad domains of personality which are used to describe human personality. Digman has indicated that the Big Five structure is a well-supported representation of the universe of personality traits [1][4][5].

The purpose of this study was discovering an unknown researched area, which was verifying the effects of humans' and a social robot's personalities through their facial interactions and relationships. There have been many studies for personalities in human-human interaction and human-computer interaction. In human-robot interaction, personality effects have also actively been studied, but there are few studies of investigating the correlations between facial variations and personality of robots with facial variations and human-robot interaction. Therefore, the result of this study will newly contribute to the development of social robots whose personality and facial expressions will be more considerably concerned before they are designed to make them more socially interactive.

Based on the CASA paradigm and Nass & Lee's study, the research question was created as:
RQ is "Will there be effects of personality on the facial interactions between humans and robots?"
H1 is "Extroverted people will feel more friendliness toward extroverted robots than introverted robots."
H2 is "Introverted people will feel more friendliness toward introverted robots than extroverted robots."
H3 is "Extroverted people will feel more liking toward extroverted robots than introverted robots."
H4 is "Introverted people will feel more liking toward introverted robots than extroverted robots."

2. METHOD
2.1. Participants
Forty students from a university in Seoul, Korea were recruited by collecting online applications which determined the individual's extraversion to divide the participants into two groups: extroverted or introverted personality. The average of participants' age of 23.2years (SD =3.18)

2.2. Stimulus Material
A face robot, KMC-EXPR, which was developed by KMC Robotics was used. The robot had neither a nose nor ears, but by moving its mouth and eyelids it could make at least 25 different pre-programmed facial expressions that could easily be controlled, and users could also manually control its facial movements in detail.

2.2.1. Extroverted mode of KMC-EXPR.
In order to make KMC-EXPR's appearance and behavior extroverted, it was controlled to follow facial expressions of

humans exactly without multiple stimulation. It opened its eyes bigger than normal and directly gazed at a participant all the time while interacting with him/her.

2.2.2. Introverted mode of KMC-EXPR.

In order to make KMC-EXPR's appearance and behavior introverted, it was controlled to follow facial expressions of humans, using three of the same stimulations. During the first stimulation, it did not react to a participant, and it began to gaze at the participant the second time. Only during the third stimulation did the robot mimic the facial expression of the participant. It opened its eyes less wide than the extroverted mode of KMC-EXPR. Moreover, it did not directly gaze at the participant in the normal status, but looked down a little while interaction with him/her.

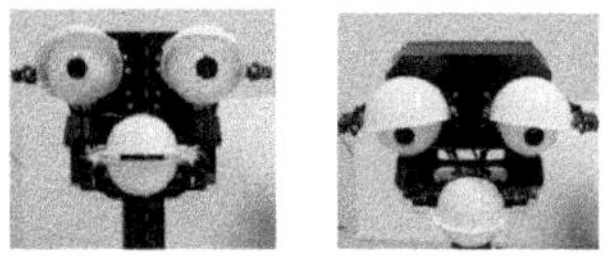

Figure 1. Extroverted (left) and Introverted (right) modes

2.3. Procedure

At first, all participants were asked to fill out a pre-questionnaire which divided them into two groups, i.e. extrovert and introvert before they joined the main experiment; then they entered the experimental room. Second, each participant was asked to express three different facial expressions (sadness, anger, and happiness) as stimuli to the robot KMC-EXPR. Each participant was asked to express and repeat the same expression of sadness three times and to do the same for expressions of anger and happiness. The duration of each stimulus was 15 seconds. Hence, each participant interacted with the robot for 135 seconds in total. Lastly, all participants were asked to fill out a survey of competence, friendliness, comfort, sense of team, and likability of robot KMC-EXPR.

2.4. Measure

2.4.1. Robot friendliness

Robot friendliness was an index of eight items. Participants indicated how well the words, "cheerful," "cooperative," "friendly," "happy," "jovial," "kind," "likeable," and "warm" described the robot, on eight ten-point Likert scales ranging from "Describes Very Poorly" to "Describes Very Well." The index was very reliable (α=0.89).

2.4.2. Liking

Likability was an index composed of three adjectives. Participants indicated how well the words, "enjoyable," "likable," and "satisfying" described the robot, on three ten-point Likert scales ranging from "Describes Very Poorly" to "Describes Very Well." The index was also very reliable (α=0.79).

3. RESULT

Interestingly, the result of the participants' extraversion test skewed toward extraversion. (M = 7.67, SD = 12.307). It showed that more extroverted people applied for the participation in this experiment. Only 72.5% (29 out of 40) of the participants' self-perceived extraversion were matched with their actual extraversion. Therefore, it can be inferred that people do not perfectly distinguish one's actual extraversion.

Robot Friendliness (α = 0.89). A 3 x 2 two-way analysis of variance (ANOVA) was carried out. The result of robot friendliness showed that extroverted participants evaluated extroverted KMC-EXPR higher (M = 8.75, SD = 0.79) than introverted KMC-EXPR (M = 7.05, SD = 1.28). Meanwhile, introverted participants evaluated introverted KMC-EXPR higher (M = 7.56, SD = 0.99) than extroverted KMC-EXPR (M = 6.07, SD = 1.68). $F(5, 34) = 3.478$, $p = .012$, R2 = 0.338. It was found that interaction between personalities of the participants and KMC-EXPR was significant ($p = .011$). Overall, extroverted participants gave higher scores to KMC-EXPR than introverted participants, but it was not significant. Likability (α = 0.79) Further, a 3 x 2 two-way analysis of variance (ANOVA) was carried out. The likability results showed that extroverted participants evaluated extroverted KMC-EXPR higher (M = 8.13, SD = 1.46) than introverted KMC-EXPR (M = 7.30, SD = 0.99). Meanwhile, introverted participants evaluated introverted KMC-EXPR higher (M = 7.67, SD = 1.09) than extroverted KMC-EXPR (M = 5.52, SD = 2.61). $F(5, 34) = 2.493$, $p = .050$, R2 = 0.268. It was found that interaction between personalities of the participants and KMC-EXPR is significant ($p = .026$). Overall, extroverted participants gave higher scores to KMC-EXPR than introverted participants, but it was not significant ($p > .05$).

4. DISCUSSION

In the results of this experiment, extroverted people felt more friendliness toward an extroverted robot, and introverted people felt more friendliness toward an introverted robot in significant ($p = .012$). Hence, H1 and H2 are supported. Likewise, extroverted people felt more of a liking toward an extroverted robot, and introverted people felt more of a liking toward an introverted robot in significant ($p = .050$). Hence, H3 and H4 are supported. Another major finding from the results were the effects of interaction between humans and the robot KMC-EXP. The interaction between personalities of the participants and KMC-EXPR was analyzed as significant in robot friendliness ($p = .011$) and in liking ($p = .026$).

5. ACKNOWLEDGEMENT

This study was supported by a grant from the WCU program (R31-2008-000-10062-0) of the KMEST, via the NRF.

6. REFERENCES

[1] Digman, J. M. (1990). Personality structure : emergence of the five-factor model. Annual Review of Psychology, 41, 417-440.

[2] Fong, T., Nourbakhsh, I., & Dautenhahn, K. (2003). A survey of socially interactive robots. Robotics and Autonomous Systems, 42, 143-166.

[3] Kobayashi, H., & Hara, F. (1997). Facial interaction between animated 3D face robot and human beings. 1997 IEEE International Conference, 4, 3732-3737.

[4] Nass, C. I., & Lee, K. M. (2001). Does Computer-Synthesized Speech Manifest Personality? Experimental Tests of Recognition, Similarity-Attraction, and Consistency-Attraction. Journal of Experimental Psychology: applied, 7(3), 171-181.

[5] Nass, C. I., Moon, Y., Fogg, B. J., Reeves, B., & Dryer, D. C. (1995). Can computer personalities be human personalities? International Journal of Human-Computer Studies, 43, 223-239.

Understanding Situational Awareness in Multi-Unit Supervisory Control Through Data-Mining and Modeling with Real-Time Strategy Games

[Extended Abstract]

Donald Kalar
San Jose State University
Department of Psychology
San Jose, California, USA
donald.kalar@sjsu.edu
donald.j.kalar@nasa.gov

Collin Green
Human-Systems Integration Division
NASA Ames Research Center
Moffett Field, California, USA
collin.b.green@nasa.gov

ABSTRACT

As robots become increasingly capable and autonomous, the role of a human operator may be to supervise multiple robots and intervene to handle problems and provide strategic guidance. In such cases, the extent to which HRI tools support the human supervisor's situational awareness (SA) and ability to intervene in an appropriate and timely fashion will constrain the scale of operations (e.g., the number of robots; the complexity of tasks) that can reasonably be supervised by a single person. One approach to understanding how humans might acquire, maintain, and use situational awareness in multi-robot supervision tasks is to look at video games that require similar activities. We describe our initial efforts at analyzing and modeling data from Real-Time Strategy (RTS) games with the goal of answering basic questions about the nature of situational awareness and supervisory control of multiple semi-autonomous agents.

Categories and Subject Descriptors

H.5.2 [**Information Interfaces and Presentation**]: User Interfaces—*Evaluation/methodology*; I.2.9 [**Artificial Intelligence**]: Robotics—*Operator interfaces*

General Terms

Measurement, Experimentation, Theory

Keywords

multi-agent HRI, Real-time strategy, meta-analysis

1. INTRODUCTION

One approach to designing effective information systems to support human-robotic interaction is to evaluate existing user interfaces that support similar tasks; that is, to survey domains where some aspects of the user tasks overlap with those of interest. A number of analog domains might be interesting for understanding multi-agent supervisory control in general [1]. In this paper, we describe a line of research related to a particular domain: Real-Time Strategy (RTS) video games.

RTS games are (typically) war simulation games involving the command and control of multiple units on a map-like representation of a virtual world. Some researchers have made efforts to build RTS-game inspired robotic control interfaces, with moderate success. In particular, such interfaces have been rated as being easily learned (or intuitive) with functions and features being highly discoverable [2]. RTS games are interesting as an analog for supervisory control of teams or swarms of autonomous agents in real operations.

As a research platform, RTS games are interesting for a number of specific reasons:

The games themselves include many of the variables or parameters that one would want to explore in service of improving human supervisory control of real autonomous units [1]. Also, game players display a great range of experience and ability. Though novice data are harder to acquire (see below), data from highly-practiced and skilled players is widely available. RTS gaming is not merely a leisure activity, but has become a profession and spectacle and the visibility and potential for reward has greatly elevated the level of competition and skill. The gaming culture and community has naturally gravitated toward the measurement and optimization of player performance, instantiated by embedded data logging and metrics in games, as well as numerous stand-alone tools for game analysis. Finally, the community places value on customization and augmentation of game platforms, embodied by the spontaneous construction and sharing of tools for creating maps and mission scenarios. A variety of off-the-shelf tools are available that make working with RTS games easy and inexpensive.

2. MINING, MODELING, AND ANALYSIS

We obtained RTS game data for the popular game *StarCraft II* from Gamereplays.org. Gamereplays.org has over 57,000 game replays posted online by members and made available to the general public. We used an open-source software package called SC2Gears to extract data from the games. The analyzed games were those that were played in the Championship Bracket at the Major League Gaming

HRI'12, March 5–8, 2012, Boston, Massachusetts, USA.
ACM 978-1-4503-1063-5/12/03.

Pro Circuit 2011 Orlando event. The analyzed set included 220 games from skilled/expert gamers.

2.1 Acquisition of Situational Awareness

We focused on one specific behavior: the acquisition of situational awareness (SA), operationalized as the latency between a player's selection of a unit and the execution of the next action. We hypothesized that two categories of variables might predict the time required to acquire SA. One category consisted of those variables that concern the player's awareness of particular regions of interest (ROIs). That is, that SA might be organized spatially, geographically, or topologically. The other category consisted of variables that concern unit types and tokens. That is, that SA might be organized by the functions/tasks that a unit can execute or the individual identity of units.

2.2 Defining ROIs

For each game, we determined a set of ROIs (and assigned individual actions to ROIs) using an Expectation-Maximization (EM) algorithm. Every action recorded has an implicit location associated with it on the game map. We performed a clustering analysis on all of the action data coordinates to extract statistically independent regions where player activity was concentrated.

2.3 Unit Types and Tokens

Every action is associated with a uniquely identified unit of a specific type. This information allowed us to examine factors such as the time since a particular unit was selected, the time since a unit of a particular type was selected, and the complexity of the unit type (see table for full listing).

2.4 The Regression Model and Factors

We looked at SA acquisition times and evaluated a number of factors that might correlate with them by performing a multiple regression. We examined:

1. $nunits$ – the number of units selected
2. $ntypes$ – the number of different unit types encompassed by the current selection
3. $pmobile$ – the proportion of the selected units that are mobile (rather than structures)
4. $\Delta tMunit$ – latency between the selection of the current units and the mean of the last selections of the same units
5. $\Delta dlast$ – the Euclidean distance on the game map between the last action and the current selection
6. $\Delta dnext$ –the Euclidean distance on the game map between the next action and the current selection
7. ΔROI – a binary value indicating whether the player has moved to a new ROI since the last unit selection
8. $\Delta tROI$ – latency between the selection of the current units and the most recent entry of the current ROI

3. RESULTS AND DISCUSSION

Table 1 shows the regression factors and their estimates for each variable with respect to time to reacquire SA. While the size of these effects is small, it is important to note that the average time between selection and action for these experts is less than 200ms.

One interesting result is that SA acquisition time is negatively correlated with the number of units selected. That is, when a player selects a larger group of units, they are faster to execute the next subsequent action. This result might mean that (a) because a large group of units has fewer tasks they can collectively execute, time to choose and execute a task is reduced; (b) large groups are selected when a simple task is already selected by the player. These are not mutually exclusive interpretations. The former interpretation is supported by the fact that the more heterogenous a selected group of units is, the faster the next action is executed.

Table 1: Linear model estimates with respect to SA

Factor	Estimate	Std. Error	t value	Sig.
$nunits$	-3.0×10^{-01}	9.7×10^{-03}	-31.169	$***$
$ntypes$	-4.4×10^{-01}	1.3×10^{-01}	-3.350	$***$
$pmobile$	-1.2×10^{0}	9.8×10^{-02}	-12.909	$***$
$\Delta tMunit$	$+2.0 \times 10^{-05}$	2.4×10^{-06}	$+8.462$	$***$
$\Delta dlast$	-8.0×10^{-04}	1.5×10^{-03}	-0.508	$n.s.$
$\Delta dnext$	$+2.8 \times 10^{-02}$	1.4×10^{-03}	$+19.000$	$***$
ΔROI	-7.4×10^{-02}	1.4×10^{-01}	-0.513	$n.s.$
$\Delta tROI$	-2.0×10^{-04}	3.9×10^{-05}	-5.286	$***$

Another interesting result is that mobile units have lower SA acquisition times than stationary units (buildings). This may simply be due to movement commands being issued quickly, or it may be because commands to buildings are generally more costly and complex.

Finally, there is a positive correlation between SA acquisition time and the time since a unit (or group) was last selected. This is the most basic evidence in favor of a unit-based model of SA acquisition: when a player has not selected a unit for a longer period, they take more time to consider that unit's state and issue the next command. However, our analysis also suggests that SA may have a spatially-based component. Though there was no evidence that switching between ROIs was predictive of SA acquisition time, there was a reliable negative correlation between the time spent in a particular ROI (during the current visit to the ROI) and SA acquisition time. That is, once in an ROI, players tend to accelerate in their issuance of commands (suggesting that they have accumulated a picture of the situation in that area and are spending less time acquiring SA for individual units/actions). It is also worth noting that our ROI variables are predictive of SA acquisition time in a way that simple Euclidean distances are not: the distance between the last selection and the current selection does is not correlated with SA acquisition time. In general, these results suggest that both unit types/tokens and ROIs (geography) may be important components of a model of SA. Also, we view this initial exercise in data mining and modeling game data to be a promising indication of the utility of RTS games as an analog for multi-agent HRI.

4. REFERENCES

[1] C. Green and D. Kalar. Design inspiration for interactions between humans and multiple robots. *Presentation at the 4th IEEE Conference on Space Mission Challenges for IT*, 2011.

[2] H. Jones and M. Snyder. Supervisory control of multiple robots based on a real-time strategy game interaction paradigm. *Proc. of the 2001 IEEE International Conference on Systems, Man, and Cybernetics*, 2001.

Cultural Studies In The HRI Loop

Andra Keay
University of Sydney, Masters Graduate, Digital Culture
1-925-960-3090
andra@robotstate.com

ABSTRACT
In this paper, I outline the scope and methodology of cultural studies and the possible applications in HRI. It is not just robots that interact with humans, but humans who interact with humans through the medium of robots. Culture is the context that informs all our actions. Cultural theory's eclectic methodology is useful for sifting through a large, multidisciplinary and cross-cultural field like HRI, for identifying disputed or neglected areas and for framing future research questions.

Categories and Subject Descriptors
K.4.0 [**Computers and Society**]: General

General Terms
Theory, Human Factors.

Keywords
Human factors, cultural theory, interaction studies, methodologies.

1. INTRODUCTION
Cultural theory is a useful but underutilized insertion point into the Human-Robot Interaction (HRI) loop. The questions asked by cultural theory relate to the production of meaning and the distribution of power. Cultural theory is a very pragmatic philosophy utilizing cultural signs as waypoints. Cultural signs evolve very rapidly in response to changing technology and give a broad indicator about differing attitudes. The methodologies of cultural theory are very eclectic and loose, but that enables a fast response to areas of rapidly changing culture.

Dautenhahn points out that the importance of HRI lies in its difference to other similar disciplines, such as human-human or human-computer interaction (HCI). HRI must balance the tension towards specifying precise methodological approaches with the importance of well-told stories; one solution does not fit all [1].

Attending recent HRI conferences required me to reevaluate the contributions that cultural theory could contribute to HRI in terms that were relevant to this particular multidisciplinary discipline.

2. THE HRI LOOP
Although the HRI field is multidisciplinary, in practice this usually means no more than two sub disciplines at a time and a set of measures and traditions from the dominant disciplines of engineering and psychology. Bartneck surveyed the literature of the HRI conference over a period of 5 years and found that while a few authors came from other disciplines, such as design, linguistics and anthropology, their contributions were evaluated through the primary metrics of psychology and engineering [2], with the traditional 'user evaluation' appearing in 146 of 185 papers. It's clear that a psychological understanding of human-robot interactions is valuable but I hope to raise awareness of some areas where a cultural study would be more useful.

HRI'12, March 5–8, 2012, Boston, Massachusetts, USA.
ACM 978-1-4503-1063-5/12/03.

Figure 1. Culture is an inescapable part of the interaction loop

3. THE ROLE OF SCIENCE FICTION
Karel Capek's 1921 play "R.U.R." is widely publicized in research papers as the first use of the word 'robot'. Some even mention that it means 'forced labor' and that the robots revolted against the humans who created them. Few recollect that at the end of the play, when humans were all dead, robots turned to a new society based around love. Few recollect that Capek also thought the play was his least successful work [3].

The stories told to each generation of roboticists exclude a great amount. Vaucanson's duck, which produced fake poo, is more frequently mentioned than Jaquet-Droz's exquisite 18th century androids, which spoke philosophy, and caused their creator to be imprisoned by the Spanish Inquisition. Edison's 'Eve', the inventor's mass produced talking doll machine, was a spectacular financial failure [4]. Each story has specific social, historical and economic context.

Cultural theory can examine history but the real strength of cultural studies lies in exploring changing or emergent contexts, through the pragmatic signs of cultural production. The methodology is eclectic, including essay writing, textual analysis and ethnography. Ang describes the persistent sense of crisis inflicting cultural studies as its most productive force, that of questioning the present [5]. The field of robotics seems to prefer questioning the future through the lens of science fiction writers from the past. Science fiction is referenced throughout the study of robotics, from keynote speakers at conferences, to the acknowledgements pages of books and publications. A quick survey shows that Asimov, Heinlein and Clarke are the writers most frequently cited as inspiration, writers from the 50s and 60s, who frequently write from a conservative position focusing on technological rather than social change. A larger survey tracking the influence of science fiction on robotics through public acknowledgements would make an interesting cultural barometer.

But a cultural studies agenda is most relevant to emerging technologies and questions of difference, identity and power. Examples include; Chesher, exploring the convergence of game technologies and navigational devices, forecasting the changes to our spatial awareness and the impacts this might have to social space [6], or McRobbie, on the cultural politics of visibility without speech or articulation [7]. The role of cultural studies is most valuable as cultural changes emerge and the closest HRI is to this at the moment is the undertaking of cross-cultural studies like that of Syrdal et al. [8].

4. EMBEDDED DESIGN ASSUMPTIONS

The difficulty of multidisciplinary, cross-cultural work in an emerging area lies in determining which assumptions are ignored and which are questioned. It is hard to overstate the cumulative effect that small, embedded design assumptions can have. The automobile industry offers a classic example. Safety measures are constantly undergoing improvement from the installation (and consequent compulsory use) of seatbelts and airbags, to improved materials, structures and regulations. However, the initial data used for most design work, including the crash test dummies, comes from a comparatively small and homogenous sample group. Subsequently, studies such as this by Bose et al. [9] have shown that unrepresented demographics (small, female, young etc.) have significantly worse safety outcomes.

There appears to be a correlation between underrepresentation in design stages and poor outcomes. The cultural studies field is notorious for dragging messy context into everything and resisting simple answers, but as Bartneck pointed out, HRI risks overly decomposing topics [10] and avoiding underlying ethical discussions [11], thus abstracting interactions from their complex actualities.

Cultural studies, such as my 'Naming of Robots' survey [12], that found a high level of naming in non-social research robots and suggested patterns of gendered and functional associations, can indicate directions for more nuanced future research. Dautenhahn suggests that HRI needs to be creative in order to explore such a dynamic and large field [13].

5. FUTURE DIRECTIONS

Cultural theory is as useful as a canary in a coalmine. It is an economical and effective way of finding the liminal spaces through which change comes, or the disputed border crossings. It can illustrate the elephants in the room. Cultural studies can assist with the framing of successful, appropriate research questions and metrics, recognizing that cross-cultural studies are both difficult and necessary. HRI is, in practice, a model cross-cultural study. Williams said that cultural theory is the 'analysis of the specific relationships through which works are made and move' [14]. It is not robots that interact with humans, but humans who interact with humans through the medium of robots.

In the same way that robotics is more than the endpoint of the production of robots, cultural theory is an epistemological exploration into how meaning and knowledge are produced and disseminated. This may indicate gaps and absences in the areas studied as well as identifying emerging trends.

6. REFERENCES

[1] Bartneck, C. 2011. The end of the beginning: a reflection on the first five years of the HRI conference. *Scientometrics* (2011), 86:487-504. DOI= 10.1007/x1192-010-0281-x

[2] Dautenhahn, K. 2007. Methodology and themes of human-robot interaction: a growing research field. *International Journal of Advanced Robotic Systems*, Vol. 4, No. 1, (2007) 103-108.

[3] Strauss, M. 2009. Where do robots come from? *IO9 Backgrounder* http://io9.com/karl-capek

[4] Wood, G. 2002. *Edison's Eve.* New York: Anchor Books.

[5] Ang, I. 2006. From cultural studies to cultural research: Engaged scholarship in the twenty-first century. *Cultural Studies Review.* Vol. 12, No. 2. Sep 2006: 183-197.

[6] Chesher, C. in press. Computer games, sat navs, and the sociotechnical remediation of space. *Convergence, Int. J. Research into New Media Technologies* (Nov 2011).

[7] McRobbie, A. 2009. *The Aftermath of Feminism: gender, culture and social change.* London: Sage Publications.

[8] Syrdal, D. S., Nomura, T., Hirai, H., Dautenhahn, K., 2011. Examining the Frankenstein syndrome: an open-ended cross-cultural survey. In *Proceedings of the 3rd Annual International Conference on Social Robotics* (Amsterdam, Netherlands, November 24-25, 2011). Springer-Verlag, Berlin Heidelberg 2011.

[9] Bose, D., Segui-Gomez, M., Crandall, J.R., 2011 Vulnerability of Female Drivers Involved in Motor Vehicle Crashes: An Analysis of US Population at Risk. *American Journal of Public Health*: December 2011, Vol. 101, No. 12, pp. 2368-2373.

[10] Bartneck, C., Croft, E., & Kulic, D. 2008. Measuring the anthropomorphism, animacy, likeability, perceived intelligence and perceived safety of robots. *Proceedings of the Metrics for Human-Robot Interaction Workshop* in affiliation with the 3rd ACM/IEEE International Conference on Human-Robot Interaction (HRI 2008), Technical Report 471, Amsterdam pp. 37-44.

[11] Ang, I. 2005. Who needs cultural research? in Leistyna, P. (ed.), *Cultural Studies: From Theory to Action.* Malden, MA: Blackwell, 2005: 477-483.

[12] Keay, A. 2011. The Naming of Robots. Presented at the *Pioneers Workshop* at the 6th ACM/IEEE International Conference on Human-Robot Interaction (HRI 2011), Lausanne, Switzerland.

[13] Dautenhahn, K. ibid.

[14] Williams, R. 2007. The uses of cultural theory. in Williams, R., *The Politics of Modernism: Against the New Conformists.* London: Verso, 2007: 163-176.

Robot Competitions As A Birth Ritual

Andra Keay
University of Sydney, Masters Graduate, Digital Culture
1-925-960-3090
andra@robotstate.com

ABSTRACT
Competitions play an important role in introducing new robots to the field. The robot competition is a rite of passage analogous to birth when compared to studies of the technologized relations of birth in American hospitals. The sanctioned recording of robot names as part of the competition marks the roles that robots play in society. This paper explores underlying social factors that may influence robotics.

Categories and Subject Descriptors
K.4.0 [**Computers and Society**]: General

General Terms
Human Factors.

Keywords
Naming, Birth, Competition, Ritual, Human-Robot Interaction.

1. INTRODUCTION
Robot competitions serve an important function, legitimating and communicating developments in robotics. They set standardized benchmarks [1]. They also function as a major ritual in the birth of a robot. The production of a robot identity in the competition arena has the same features as the technological production of birth in a hospital, which creates a modern subject. The production of a physical subject/object is occurring alongside the production of a social and cultural subject/object [2], using Lucy Suchman's hybrid term to emphasize the co-shaping of both technology and identity.

The pervasive naming of robots in competitions is a strong indicator that modern human birth rituals have been transferred to the robotics field. The modern American birth is a scientific and technological production which both produces a new being, and positions parent and child within a modern, rational, technological and patriarchal culture, according to anthropologist Robbie Floyd-Davis [3]. I hypothesize that robot competitions serve the same function as hospital birth in the production of robot identity. This supports the 'robot as a new ontological category' hypothesis [4].

This paper takes the position that the social and the technological co-shape, informed by the fields of sociology, anthropology, science and technology studies, and cultural theory. I hypothesize that the naming of robots performs a ritual function. The robot competition is a rite of passage analogous to birth. This also supports the 'robot as a new ontological category' hypothesis in demonstrating the ways in which we construct robots as social, semi-human entities even when they are not constructed to resemble humans or act in a social role.

HRI'12, March 5–8, 2012, Boston, Massachusetts, USA.
ACM 978-1-4503-1063-5/12/03

Figure 1. Amador Robotics students prepare for competition. Image by Daniel Naito. Deltawk. [CC-BY-SA-3.0]

2. THE ROLE OF COMPETITION
2.1 Background
In 2010 and 2011, I conducted a survey of the naming of robots in research competitions. The resultant collection of over 2000 names from 5 competitions, spanning a period of 20 years, demonstrated that a culture of robot naming was widespread [5]. This set of names exhibits both anthropomorphism and a new category of hybrid naming practices that support the 'robot as a new ontological category' hypothesis [6]. The name survey originated with my participation in and observation of robot competitions in Australia and the United States, including RoboGames and RoboCup.

2.2 Competitions as Archive and Cradle
Competitions serve several purposes including education, entertainment, influence and setting a standardized benchmark for research. Both the competition and the robots competing in it are a particular form of scientific capital which functions as a technological archive [7][8], using Foucault's concept of an archive as the historical definition of a discursive formation [9], or the development of a specific paradigm, ideology or cultural field.

Competitions are designed to push the envelope scientifically, attracting investment and interest. They are a form of peer review publishing demonstrating progress in basic problems, and a way of standardizing developments across multiple sites [10]. Robotics competitions are also a method of achieving recognition, a key element of scientific capital that enables claims and arguments to be noticed and evaluated [11]. Robotics competitions are also one of the areas of practice in which systematic benchmarking takes place [12], which allows for the wide dissemination of scientific research and may be why competitions are increasingly popular.

This provides a functional rationale for the rapidly increasing robotics competition field but underlying this growth are cultural reasons why competitions are successful, enjoyable and have implications broader than the pragmatic. Science is a discursive practice [13]. The archive is the moment in time that captures what is then constructed as a history that calls forth a future. The archive is continually being reborn. The technological archive is the material, not the memory that orders the unruly present. "The past is not 'memory' but the archive itself, something that is factually present in reality." [14] The actual robot is a technological archive, mediating a set of practices and knowledge. The robot competition is another archive or evaluative process resulting in the rise of particular robots and robotic practices, interpellating a particular human-robotic future. The story told in robot competitions is analogous to the birth narrative and rituals.

3. RITUAL FUNCTIONS

3.1 Rites of Passage

Symbolic interactions or rituals are at the heart of all social dynamics, generating group identity, belief and culture. 'These patterns are the most basic structural force that organizes society' [15]. Rituals provide 'ordered structure to the chaotic flow of the natural birth process' particularly in the scientific atmosphere of modern birth [16]. I suggest that the rituals of birth have been transferred to the robot competition.

A rite of passage serves not just to introduce something into the world, but to bring them into alignment with society, to bring the world into them. Robot competitions, scientific publications and the scientific demonstration, each have their own rituals for bringing subjects into being according to our core values as a technocratic society, where power is based on technology, knowledge and skill. Further comparison between modern birth rituals and robot competitions evaluating group dynamics, transitional states, ritual symbols, cognitive restructuring techniques, reincorporation or baptism ceremonies and the implications for human-robot interaction are suggested.

4. CONCLUSION

This ethnomethodological study of robot competitions explores the impact that cultural patterns have on emerging fields, which in turn may have an impact on robot construction and design. Feed forward loops have the potential to affect ongoing research practices. The results of the robot name survey, and concurrent participant observation of robot competitions, support both the 'Computers as Social Agents' theory [17] and the 'Robot as New Ontological Category' hypothesis [18]. I hypothesize that robot competitions function as a birth ritual, endorsing and naming a new robot in a modern technological rite of passage.

5. REFERENCES

[1] Del Pobil, A. 2006. Benchmarks for Robotics Research. Workshop *IROS 2006* Beijing China

[2] Suchman, L. 2009. 'Subject Objects presentation' retrieved on August 11 2010 from <http://www.socialsciences.manchester.ac.uk>

[3] Floyd-Davis, R. 1993. *Birth As An American Rite of Passage,* University of California Press.

[4] Kahn Jr, P. H., Reichert, A. L., Gary, H. E., Kanda, T., Ishiguro, H., Shen, S., Ruckert, J. H. & Gill, B. 2011. The New Ontological Category Hypothesis in Human-Robot Interaction. Presented at *the International Conference on Human-Robot Interaction* (HRI 2011), Lausanne, Switzerland. | ACM 978-1-4503-0561-7/11/03

[5] Keay, A. (2011). The Naming of Robots. Presented at the Pioneers Workshop at *the International Conference on Human-Robot Interaction* (HRI 2011), March 2011 Lausanne, Switzerland.

[6] Kahn Jr., P. H. et al. ibid.

[7] Rammert, W. 1999. *Relations that Constitute Technology and Media That Make a Difference: Toward a Social Pragmatic Theory of Technicization.* Retrieved on April 2 2011 from <http://scholar.lib.vt.edu>

[8] Groys, B. 1999. *Boris Groys: The Logic of Collecting.* IV by Sven Spieker. Retrieved on 15 October 2010 from <http://www.artmargin.com>

[9] Foucault, M. 1969. *The Archaeology of Knowledge.* Trans. Sheridan-Smith, A. M. London and New York: Routledge, 2002

[10] Bonasso, P. & Dean, A. 1997. A retrospective of the AAAI robot competitions, *AI Magazine* AAAI Vol.18 No.1

[11] Sismondo, S. 2011. Bourdieu's Rationalist Science of Science: Some Promises and Limitations in *Cultural Sociology* January 31, 2011, Sage Journals Online.

[12] Del Pobil, A. 2006. Benchmarks for Robotics Research. Workshop *IROS 2006* Beijing China

[13] Foucault, M. 1966 *The Order of Things: An Archaeology of the Human Sciences.* Trans. Sheridan, A. New York: Vintage 1970

[14] Groys, B. ibid.

[15] Summers-Effler, E. 2006. Ritual Theory in *Handbook of the Sociology of Emotions, Handbooks of Sociology and Social Research* Section II, 135-154.

[16] Floyd-Davis, R. 1993. *Birth As An American Rite of Passage* University of California Press

[17] Reeves, B., & Nass, C. I. 1996. *The media equation: how people treat computers, television, and new media like real people and places.* Stanford, Calif., New York: CSLI Publications, Cambridge University Press.

[18] Kahn Jr., P. H., et al. ibid

Applying Team Heuristics to Future Human-Robot Systems

Joseph R. Keebler
University of Central Florida
3100 Technology Parkway
Orlando, FL 32826
1-407-760-7677
joekeebler@gmail.com

Florian Jentsch
University of Central Florida
3100 Technology Parkway
Orlando, FL 32826
fjentsch@mail.ucf.edu

Thomas Fincannon
University of Central Florida
3100 Technology Parkway
Orlando, FL 32826
t_fincannon@yahoo.com

Irwin Hudson
Army Research Laboratory
3100 Technology Parkway
Orlando, FL 32826
Irwin.hudson@us.army.mil

ABSTRACT
In this paper we briefly describe teaming heuristics as they are applied to human-human teams, and demonstrate their adaptability to human-robot (HR) teams. We discuss a framework developed from Salas's models on teamwork and team training. As HRI technology moves from tele-operative control methods to teamwork with intelligent robots, it is pertinent to properly integrate knowledge about teams into the development of robotic systems. This should lead to highly effective team systems, and may provide insight into the design of robotic entities and system protocols.

Categories and Subject Descriptors
H.1.2. [**Information Systems**]: User/Machine Systems– *Human Factors.*

General Terms
Performance, Design, Human Factors, Theory

Keywords
HRI, Human-robot teams, team training

1. INTRODUCTION
Although intelligent robotics is still decades away from being a realized technology, research is currently being conducted to better understand how teams of humans will be integrated with teams of intelligent robots or agents [4][6]. Uncovering best practices for multiple humans interacting with autonomous intelligent robotic assets before the technology has been realized is a profoundly complicated undertaking. Unlike current day tele-operative methods, in which unmanned assets are controlled via remote control and sensors, the future instead holds a reality where robots will have robust intelligence, akin to dogs [4]. This level of intelligence will allow for HRI that is fundamentally different from the current command and control methods of today [2]. With the advent of intelligent robots, the way human-robot teams interact will fundamentally change. Specifically, human robot teams will become much more akin to human-human teams, or even human-animal teams [4]. Therefore, to better understand team dynamics for these currently fictional HR teams, we must apply knowledge from the team training literature. Although the robotic entities that will serve as companions to teams of soldiers do not exist currently, knowledge about teamwork and training can enhance our understanding of the role they would play, and therefore aid in driving the development of technology [6]. This paper will review a model of teamwork and establish guidelines for strong team behaviors and training [5]. The model will be reviewed with examples of application to HR teams. Implications for design and future research will then be discussed.

HRI'12, March 5–8, 2012, Boston, Massachusetts, USA.
ACM 978-1-4503-1063-5/12/03.

2. TEAM HEURISTICS
One model for teamwork that could be applied to Human-Robot teams was developed by Salas, Wilson, Murphy, King, and Salisbury (2008). This model proposes heuristics that could readily be adapted to teams of humans and robots. The framework for this model provides 8 heuristics criteria for effective teams. Although HR teams could benefit from the application of these heuristics, some of them may not make sense in the context of HRI. Therefore, the five most applicable heuristics will be summarized below.

The chosen heuristics are seemingly the most congruent with HR teams. Supporting precise and accurate communication; diagnosing communication errors; instituting frequent practice opportunities; building team orientation; and preparing the team are all heuristics that could be easily overlaid onto HR teams. This set of heuristics is in no way definitive, and new heuristics that do not or cannot exist on human-human teams may evolve out of current and future research on teaming with agents and animals.

Each of these heuristics could be adapted to enhance future teams of humans and robots. Specifically, the categories listed above are a good starting point for further research. Due to the non-existence of current day intelligent robotics, application of theory is the only way to understand problems within this domain at the given time. Although robotic assets are due to become highly intelligent compared to the remote controlled variants of UV's that exist today, it is fundamentally important that the human is kept in the loop. Some would consider this to be absolutely critical, although it leads to many unique and difficult challenges [1].

The table below provides three columns. The first column is the name of the heuristic of interest, with a method for enabling that heuristic followed by an example of application for the given heuristic in the context of HR teams.

Table 1. Teaming Heuristics

Heuristic	Method of Enabling Heuristic	Application to HRI Teams
Support precise and accurate communication	Use closed-loop communication	Human and robot must have a system of check backs to ensure closed-loop communication
Diagnose communication errors	Find symptoms of communication error and treat through team learning and self-correction	Communication errors may be at multiple levels, and may include bandwidth issues, equipment failures, as well as incongruities of human understanding of robotic assets
Institute frequent practice opportunities	Schedule practices based on known degradation time of skills sets	Practices for HR teams should be frequent and mandatory. Practicing communication, missions, etc. will only enhance team performance
Build team orientation	Take steps to increase trust and cohesion to lower stress levels and increase satisfaction, commitment, and collective efficacy	Integrate robots early in team formation so that roles can be discovered and trust established by human members.
Prepare the team	Provide learning opportunities for new competencies that will expose members to feedback and increase team's overall efficacy	As robotic systems gain new hardware and software components, prepare human team members through refreshing old competencies and introducing new ones

3. FUTURE RESEARCH

To best conduct research on the application of the team heuristics discussed above, studies need to focus on a few key areas. First, simulation of intelligent robots needs to be realized. Research conducted at the Team Performance Laboratory at the University of Central Florida has used a "Wizard-of-Oz" approach, having dislocated research assistant confederates act as robotic entities via remote controlled robotic assets. This is both a cost effective and highly realistic way to simulated an intelligent robotic entity [3]. Other methods may include actual robots or software agents integrated with high levels of AI, although these systems are still not fully realized, and are arguably very expensive. Second, research needs to aim at each heuristic and uncover best practices for integration of that given heuristic into HR teams. Specifically, studies should work in a bottom-up fashion, to see which heuristics aid in teaming, and whether they need to be altered, adjusted, or removed as technology continues to advance.

4. ACKNOWLEDGMENTS

We would like to thank our sponsors of the HATS-ON project for continuing support of our research efforts. This document in no way reflects the ideas or opinions of our sponsors or any other government agency.

5. REFERENCES

[1] Cooke, N.J., and Chadwick, R.A. 2010. Lessons learned from Human-Robot Interaction on the ground and in the air. *In Human-Robot Interactions in Future Military Operations* (Eds. Mike Barnes and Florian Jentsch). Ashgate Publishing, Surrey, UK.

[2] Evans, A.W., and Jentsch, F. 2010. The future of HRI: Alternate research trajectories and their influence on the future of unmanned systems. *In Human-Robot Interactions in Future Military Operations* (Eds. Mike Barnes and Florian Jentsch). Ashgate Publishing, Surrey, UK.

[3] Jentsch, F., Evans, A.W., Ososky, S. 2011. Model world: Military HRI research using a scale MOUT facility. *In Human-Robot Interactions in Future Military Operations* (Eds. Mike Barnes and Florian Jentsch). Ashgate Publishing, Surrey, UK.

[4] Phillips, E., Ososky, S., Grove, J., and Jentsch, F. 2011. From tools to teammates: Toward the development of appropriate mental models for intelligent robots. *In Proceedings of the Human Factors and Ergonomics Society 55th Annual Meeting.* Las Vegas, NV.

[5] Salas, E., Wilson, K.A., Murphy, C.E., King, H., and Salisbury, M. 2008. Communicating, coordinating, and cooperating when lives depend on it: Tips for teamwork. *The Joint Commission Journal on Quality and Patient Safety*, 34, 6, (June 2008), 333-347. 795-825.

[6] Sukthankar, G., Shumaker, R., and Lewis, M. 2011. Intelligent agents as teammates. *In Theories of Team Cognition: Cross-Disciplinary Perspectives* (Eds. Eduardo Salas, Stephen Fiore, Micheal Letsky). Routledge Publishing, New York, NY.

Deep Networks for Predicting Human Intent with Respect to Objects

Richard Kelley
Katie Browne
Liesl Wigand
Monica Nicolescu
Brian Hamilton
Mircea Nicolescu
Department of Computer Science
University of Nevada, Reno
Reno, NV 89557
{rkelley, lieslw, bhamil, kbrowne, monica, mircea}@cse.unr.edu

ABSTRACT

Effective human-robot interaction requires systems that can accurately infer and predict human intentions. In this paper, we introduce a system that uses stacked denoising autoencoders to perform intent recognition. We introduce the intent recognition problem, provide an overview of deep architectures in machine learning, and outline the components of our system. We also provide preliminary results for our system's performance.

Categories and Subject Descriptors

I.2.9 [**Artificial Intelligence**]: Robotics—*operator interfaces, sensors*

Keywords

human-robot interaction, intention modeling, deep architectures

1. INTRODUCTION

For a robot system to function correctly in human environments, it must recognize and predict human actions quickly and accurately. To classify these actions well, machine learning methods are applied, such as hidden markov models or support vector machines. More expressive techniques, such as artificial neural networks or Boltzmann machines, have shown promise for modeling human behavior, but are often difficult to train. However, recent work in this area has revealed a set of techniques for effectively training neural networks with a large number of hidden layers. These results suggest that powerful models such as stacked denoising autoencoders (SDAs) might be applied to problems in human-robot interaction. In this paper, we apply SDAs to the problem of predicting a human's actions with respect to objects in the environment. Our preliminary results support the claim that deep architectures can perform better than hidden Markov models for this task. In the remainder of this report, we review previous work in deep architectures and intent recognition, describe our current system, and provide an overview of promising results we have obtained thus far.

HRI'12, March 5–8, 2012, Boston, Massachusetts, USA.
ACM 978-1-4503-1063-5/12/03.

2. RELATED WORK

In the intent recognition problem, the goal is for a robot to use standard sensor data to observe a human's actions and infer the unobservable mental states and motivations that gave rise to this activity. This problem has tremendous significance in robotics and human-robot interaction [1].

Previous systems have used techniques such as hidden Markov models (HMMs) [3] to model behavior, and have seen some success. Further improvements can be obtained by using more sophisticated senors, such as the Kinect [2] [4]. However, work in deep architectures suggests that HMMs and related approaches may suffer from weaknesses that can be avoided by using models that compose multiple layers of nonlinear functions to produce a classification [5].

3. SYSTEM ARCHITECTURE

The system observes people performing everyday activities and attempts to identify and predict their current and future actions. To do this, our system uses a Kinect camera to obtain both color and depth information about the scene. This data is passed to a reasoning system that analyzes the configuration of objects in the scene and the position of the human in relation to these objects to produce a prediction of the human subject's intent. A system diagram is shown below in Figure 1.

The scene representation produced by the vision system is sent to separate modules for cluster and reaching analysis. These modules use stacked denoising autoencoders to produce classifications of low-level intent [6].

Autoencoders are neural networks that use a single hidden layer to perform nonlinear dimensionality reduction. In training, an autoencoder is given the input $x \in \mathbb{R}^n$ as both training instance and label. To avoid simply learning the identity function, we can require that the number of hidden nodes be less than the number of input nodes, or we can use a special regularization term. However, *denoising* autoencoders avoid these approaches by randomly corrupting the input x prior to training.

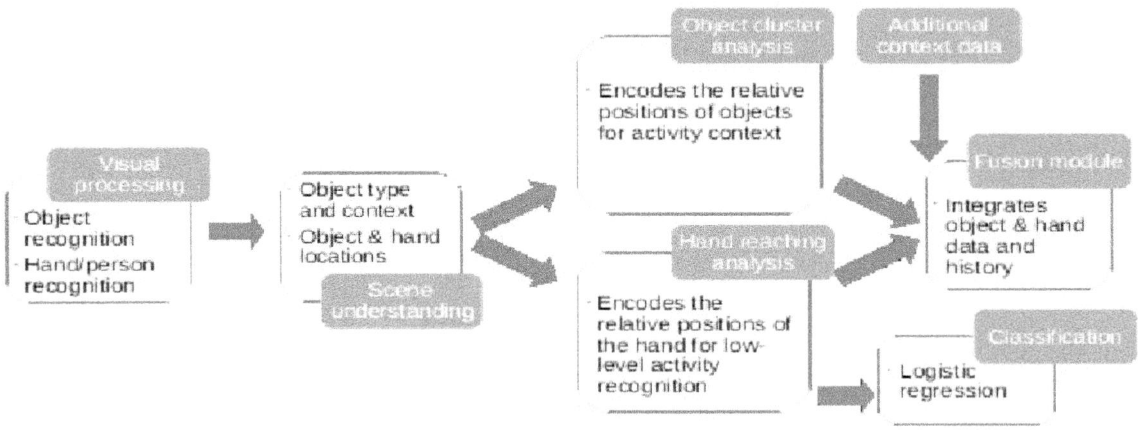

Figure 1: System Diagram. Visual data is passed to SDAs that predict intent.

Multiple autoencoders can be *stacked* so that the activations of hidden layer l are used as inputs to the autoencoder at layer $l+1$. The effect of stacking is to produce feature hierarchies that can be fed into a supervised learning algorithm (such as logistic regression) to produce classifications.

The reaching module uses a SDA to analyze the change in distance between the observed human's hand and the objects in the scene over several frames. It produces a prediction of which object the person is currently reaching for. The SDA contains three hidden layers with 160 nodes each, and produces a vector that is fed into a logistic regression module to produce a prediction label for the person's intent.

Similarly, the cluster module receives a matrix of pairwise distances between objects, and produces a low-dimensional representation of the way these objects are clustered in the scene. This works as a form of context that can be used for further predictions, or may be useful on its own. Various clusterings may be given human-readable labels that provide information about the state of the environment. (For instance, a binder, pen, and paper positioned close together may indicate a work or study scenario.) The SDA we used for this module contained three hidden layers with 50 nodes each.

We are currently experimenting with deep network and other techniques to combine the output of these modules in a fusion module that predicts intent across larger temporal scales at a higher level of abstraction.

4. RESULTS

The reaching and clustering modules were trained and tested on synthetic data. For five objects positioned randomly in the scene, training required ten to fifteen minutes using 10000 samples for each module. Hyper-parameters (number of layers, number of nodes per layer, etc.) were determined using cross-validation with sets of size 8000 for reaching and 4000 for clustering. The reaching module achieved an accuracy of 80% on a test set of size 4000, and the cluster module achieved an accuracy of 98.5% on a test set of size 2000.

5. CONCLUSIONS

In this paper, we have presented an overview of our work applying deep architectures to the problem of predicting human intentions with respect to inanimate objects in the environment. We have shown how the use of stacked denoising autoencoders to analyze the configuration of objects can be combined with the use of similar autoencoders that analyze human motion to produce effective predictions of human intent. Our preliminary results lead us to believe that deep neural networks can outperform other techniques such as hidden Markov models in a wide array of human modeling tasks that are relevant to HRI.

6. ACKNOWLEDGMENTS

This work was supported by ONR grant N00014-09-1-1121.

7. REFERENCES

[1] Y. Demiris. Prediction of intent in robotics and multi-agent systems. *Cognitive Processing*, 8(3):151–158, September 2007.

[2] R. Kelley, A. Ambardekar, L. Wigand, M. Nicolescu, and M. Nicolescu. Point clouds and rnage images for intent recognition and human-robot interaction. In *RGB-D: Advanced Reasoning with Depth Cameras*, June 2011.

[3] R. Kelley, A. Tavakkoli, C. King, M. Nicolescu, M. Nicolescu, and G. Bebis. Understanding human intentions via hidden markov models in autonomous mobile robots. In *ACM/IEEE International Conference on Human Robot Interaction*, February 2010.

[4] J. Y. Sung, C. Ponce, B. Selman, and A. Saxena. Human activity detection from rgbd images. In *Plan, Activity and Intent Recognition*, August 2011.

[5] G. Taylor, G. Hinton, and S. Roweis. Modeling human motion using binary latent variables. *Neural Information Processing Systems*, 19:1345–1352, January 2007.

[6] P. Vincent, H. Larochelle, I. Lajoie, Y. Bengio, and P.-A. Manzagol. Stacked denoising autoencoders: Learning useful representations in a deep network with a local denoising criterion. *Journal of Machine Learning Research*, 11:3371–3408, December 2010.

User Attentive Behavior with Camera View for In-Situ Robot Control

Jong-gil Ahn, Hyunseok Yang, Gerard J. Kim
College of Information and Communications
Korea University, Seoul, Korea
82 2 3290 3196
hide989@korea.ac.kr

Namgyu Kim
Dept. of Game Engineering
Dong-eui University, Busan, Korea
82 51 890 2713
ngkim@deu.ac.kr

ABSTRACT
In this poster, we present an experiment that compares three forms of interaction to study the user behavior with regards to the effects of camera view for in-situ robot control. We compared three hand-held interfaces with: (1) no camera view (Nominal), (2) a camera view/aim is always fixed toward the robot (Fixed) and (3) a camera view with user controlled aim (Free). The three approaches represent different balances between information availability, interface accessibility and the amount of induced attentional shifts. Experiment results have shown that all three interaction models exhibited similar task performance even though the Fixed type induced much less attentional shifts. On the other hand, the users much preferred the Nominal and Free type. Users mostly ignored the camera view despite having to shift one's attention excessively, due to the lack of visual quality, realistic scale and depth information.

Categories and Subject Descriptors
I.2.9 [**Artificial Intelligence**]: Robotics - *Operator interface*

General Terms
Experimentation, Human Factors.

Keywords
Human robot interaction, Attention, Robot control.

1. INTRODUCTION
In-situ remote robot control refers to a situation in which a robot is remotely controlled using an interface device from a near distance (e.g. 2~20m). In-situ remote control often occurs in situation where it is important for the operator to be present at the operating site (e.g. safety reasons, requirement of human expertise skills, need for effective communication/interaction or robots with only limited secondary role) or when on-board or environment sensors are either limiting or not readily available for wide situation monitoring [6]. Typical examples include bomb diffusing robots, fire-fighting (water gun) robots and robot secretaries.

This is a situation distinct from both close/direct interaction (that e.g. uses voice, touch screen input or gesture) and completely sensor-based tele-operation. Like any interfaces, important factors in the efficient and convenient in-situ remote robot control would include the availability of required information to make the right decision (or situation awareness), access time to the interface for the needed actions, and the amount of any distraction or attentional shifts between the robot scene and the control interface.

HRI'12, March 5–8, 2012, Boston, Massachusetts, USA.
ACM 978-1-4503-1063-5/12/03.

One possible solution to address these factors collectively is to provide a camera/sensor view in which critical parts of the operating environment can be shown side-by-side with the control interface. This way, the user can minimize the times they have to look away (because the camera/sensor view supplies the needed information) from the interface and see directly into the scene, and thus react faster to make proper actions. The user is still allowed to look over and make direct visual confirmation, if necessary.

In this paper, we present an experiment that compares three forms of interaction to deduce a user behavior model with regards to the effects of camera view integration. The three compared interactions represent different balances between information availability, interface accessibility and the amount of attentional shifts.

2. RELATED WORK
Several guidelines and evaluation metrics have emerged for HRI from the HCI and robotics communities over the years [1, 3, 6]. Most guidelines are quite abstract and general, and also seem to be oriented toward "direct and close" interaction [2] or tele-operation control [4, 5]. On the other hand, the most typical form of an "indirect" communication with a robot would be through the use of a remote control. We make a note that the indirect approach has been used mostly for tele-robotics or multi-robot control rather than for in-situ bystander driven control.

3. EXPERIMENT
3.1 Experiment Design
The objective of the experiment is to study the user behavior in regards to one's shifting of attention between the interface and the actual scene despite the provision of scene view through the camera. Thus we test three representative interaction models in terms of how the camera view may be utilized (or not). They are: (1) "Nominal" type (no camera view), (2) "Fixed" camera type (using a camera view but with camera always aimed toward the robot or certain interested target), and (3) "Free" camera type (using a camera view but with the camera free to be aimed in any way the user wants). Figure 1 illustrates the three interactions.

The experiment was designed as a 1 x 3 repeated measure within subject, the lone factor being the type of the interface employed. A robot control task was given to a user to be carried out by using the given interface. The task involved controlling the robot to move from a start position toward a finish line and stopping intermittently to carry out other tasks (e.g. activating sound output, turning on the projector, aiming the on-robot camera, etc.). The task performance (task completion time and no. of clicks made to the interface) was measured and a usability survey was filled out (omitted due to space restriction), with a particular focus on the matter of attentional shifts. We initially expected and hypothesized the Fixed type to have the least amounts of attentional shifts and thus highest task performance.

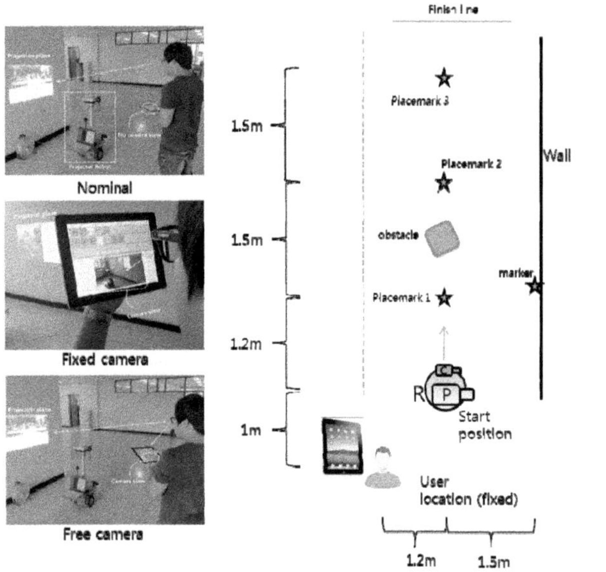

Figure 1: The experimental space and the three interaction models. The user stayed at a fixed location and used the interface to control the robot to move along the wall carrying out various tasks. Note the line of sight in each case.

3.2 Results

Figures 2 and 3 show the task performance and the amounts of shifts in the line of sight. Overall there were no statistically significant differences in the task completion or in the no. of clicks made, even though the user made much less (with statistically significant difference) attentional shifts. There was no difference in the no. of shifts between the Nominal and the Free type. Most users treated the Nominal and the Free type to be mostly equivalent.

We can deduce that the visual information from the direct look contributed much more to completing the task (which were mostly spatial) in an efficient manner (even when excessively done). Subjects reported, when using the Fixed type, the need to look over to gain more insight into the situation such as proper depth and scale information (which could not be conveyed well with a small camera view). Note that the users of the Fixed type was driven (by the experiment) to do so only when necessary, resulting in much less no. of visual shifts. This was most apparent when the users were trying to avoid an obstacle with the robot. The Fixed type may be viewed as a representative of the grounded interface.

Attentional shifts occurred much less during relatively stable/simple task operations (e.g. while the robot was moving with sufficient clear room, turning on/off the projector, turning on the narration). Subjects also had a great tendency to make visual confirmation of the task completion (by directly looking into the operating environment) and whether one followed the right task in the given sequence (by looking back into the interface). This generally caused excessive shifting of the line of sight.

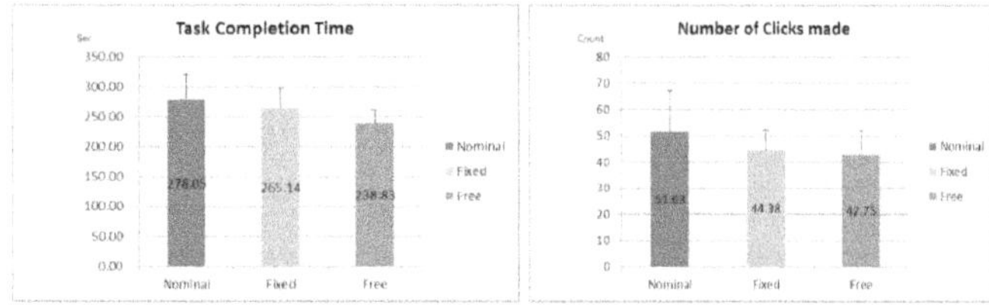

Figure 2: Task performance results: task completion time (left) and number of clicks made (right).

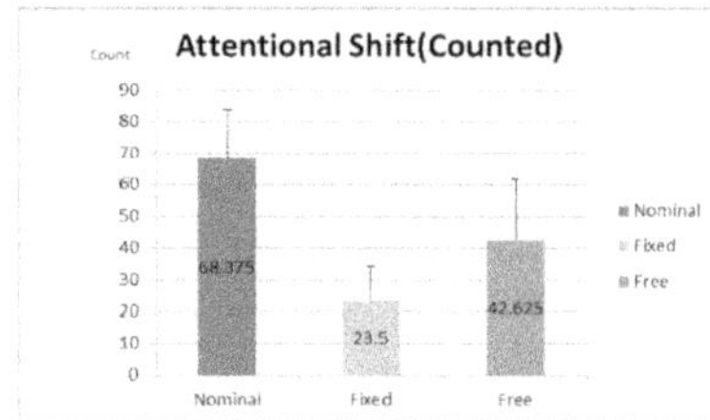

Figure 3: Number of attentional shifts made.

4. CONCLUSION

Our expectation that the attentional shifts would play an important and significant role in the efficient and fast task performance was rejected. Users much preferred the Nominal and Free type and mostly ignored the camera view despite having to shift one's attention excessively, due to the lack of visual quality, realistic scale and depth information (which were deemed essential in carrying out the mostly spatial robotic tasks). We believe this result can serve as an important and useful guideline for designing control interfaces not only for in-situ robot control, but also for augmented reality and tele-presence systems.

5. ACKNOWLEDGMENTS

This research was supported in part by the NRF (Project No. NRF-2010- B00183) and the MKE/KIAT Strategic Technology Lab. Program (Multimodal Entertainment Platform).

6. REFERENCES

[1] Clarkson, E. and Arkin, R., Applying Heuristic Evaluation to Human-Robot Interaction Systems, Proc. of FLAIRS, 2007.

[2] Drury, J., Scholtz, J. and Yanco, H., Awareness in Hu-man-Robot Interactions, Proc. of IEEE Intl. Conf. on Systems, Man and Cybernetics, 2003.

[3] Drury, J., Hestand, D., Yanco, H. and Scholtz, J., Design Guidelines for Improved Human-Robot Interaction, Proc. of ACM CHI, 2004.

[4] Olsen, Jr, D. and Goodrich, M, Metrics for Evaluating Human-Robot Interactions, Proc. of PERMIS, 2003.

[5] Quigley, M., Goodrich, M. A. and Beard, R. W., Semi-autonomous Human-UAV Interfaces for Fixed-wing mini-UAVs. Proc. of IROS, 2004.

[6] Goodrich. M and Scholtz, C., Human-Robot Interaction: A Survey, Foundations and Trends in Human-Computer Interaction, 2007.

Tracking Aggregate vs. Individual Gaze Behaviors during a Robot-led Tour Simplifies overall Engagement Estimates

Heather Knight Reid Simmons
Carnegie Mellon's Robotics Institute
5000 Forbes Avenue
Pittsburgh, PA
heatherbot@cmu.edu, reids@cs.cmu.edu

ABSTRACT
As an early behavioral study of what non-verbal features a robot tourguide could use to analyze a crowd, personalize an interaction and maintain high levels of engagement, we analyze participant gaze statistics in response to a robot tour guide's deictic gestures. There were thirty-seven participants overall, split into nine groups of three to five people each. In groups with the lowest engagement levels aggregate gaze response to the robot's pointing gesture involved the fewest total glance shifts, least time spent looking at indicated object and no intra-participant gaze. Our diverse participants had overlapping engagement ratings within their group, and we found that a robot that tracks group rather than individual analytics could capture less noisy and often stronger trends relating gaze features to self-reported engagement scores. Thus we have found indications that aggregate group analysis captures more salient and accurate assessments of overall *humans-robot interactions*, even with lower resolution features.

Categories and Subject Descriptors
[HRI Communication]: Conveying Intention, Gaze and Gestures, [Robot Perception & Prediction]: Modeling social situations.

General Terms: Measurement, Experimentation, Human Factors.

Keywords: Human tracking, Humans-robot interaction, Social dynamics modeling, Low resolution sensing.

1. INTRODUCTION
Our experiment used a Nao robot, which made tourguide-style presentations in three distinct locations of the unusual Gates building at Carnegie Mellon University. In this paper, we analyze a sub-set of our larger video dataset in order to populate hypothesizes for further analysis and future human-tracking software. We examine participant gaze behaviors leading up to and following the robot's indication of a fire safety ceiling vent. The vertical location made gaze directions very distinguishable, ideal for tracking gaze statistics (see Fig. 1). The frequent convergence of within-group engagement levels meant that, given the high levels of variation in gaze data (not due to sensing limitations, but rather the complexity of human behavior), it is more fruitful to use aggregate features to estimate engagement than to look at each individual's gaze behaviors separately.

2. RELATED WORK
In [5], they evaluate robot behaviors that might attract a larger audience, however, to our knowledge, no other Robot Tour guide has investigated aggregate group engagement and attention tracking, though [1][3][5][6] discuss tour guides bots that can play games, button-in topic selection, have simple dialog and/or dance.

HRI'12, March 5–8, 2012, Boston, Massachusetts, USA.
ACM 978-1-4503-1063-5/12/03.

3. EXPERIMENTAL SETUP
In this study, we assess human gaze in response to a robotic tourguide to evaluate statistical trends relating to their engagement level. To generate a wide range of group reactions, we vary both the location and presentation style (*friendly, neutral, sarcastic)* over three Latin square combinations. Overall, both friendly and sarcastic presentations of the same factual material resulted in higher self-reported engagement scores as compared to neutral, based on survey results. In this initial analysis, we label participant gaze, using RGB video data to code viewer gaze statistics (the subject of this report), while reserving the Kinect depth data for training our human tracking software (future work).

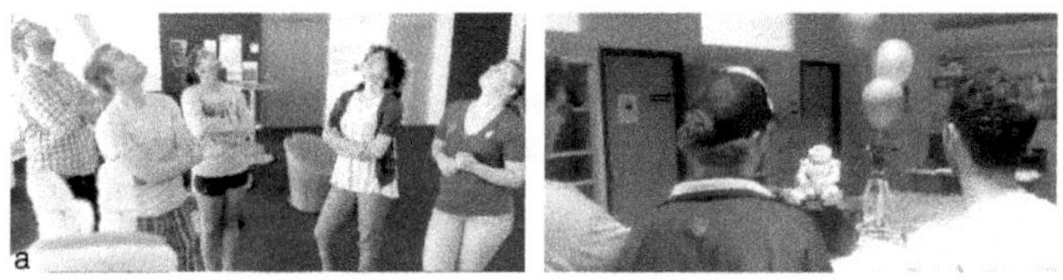

Figure 1. a) Study capture of group gaze toward Vent, b) Pre-study testing of Kinect sensor mounted next to robot

The groups each had three to five participants and were unfamiliar with most the material the robot presented, as outlined in Table 1.

Table 1. Participant Demographic Information

Total #	Gender	Age	Tech Savvy	Comfortable with robots
34	13 Female / 24 Male	14-61	92%	5 No / 3 Sort-of / 26 Yes

4. RESULTS
Videos were hand-labeled by the author to indicate the gaze direction of each participant, e.g. Fig. 2, wherein horizontal height represents distinct gaze directions; *vent (topmost), window, each-other, robot, floor.* Group 1 had high engagement ratings; note the frequency of glance shifts and the small bumps toward the end at 12.5 and 13.5 seconds, which indicate intra-participant gaze. In Group 3 (low engagement) there is a fast drop off of gaze activity. The default gaze direction in all groups is toward the robot, as it is the one talking. Thus, participants at all engagement levels spend large portions of time gazing at the robot.

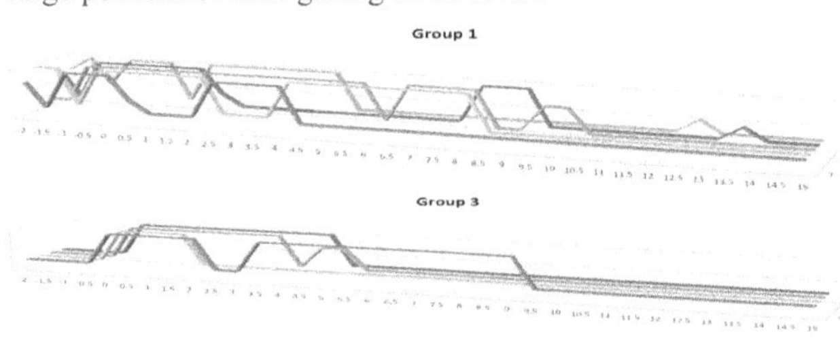

Figure 2. Aggregate Group 1 and 3 data, high/low engagement

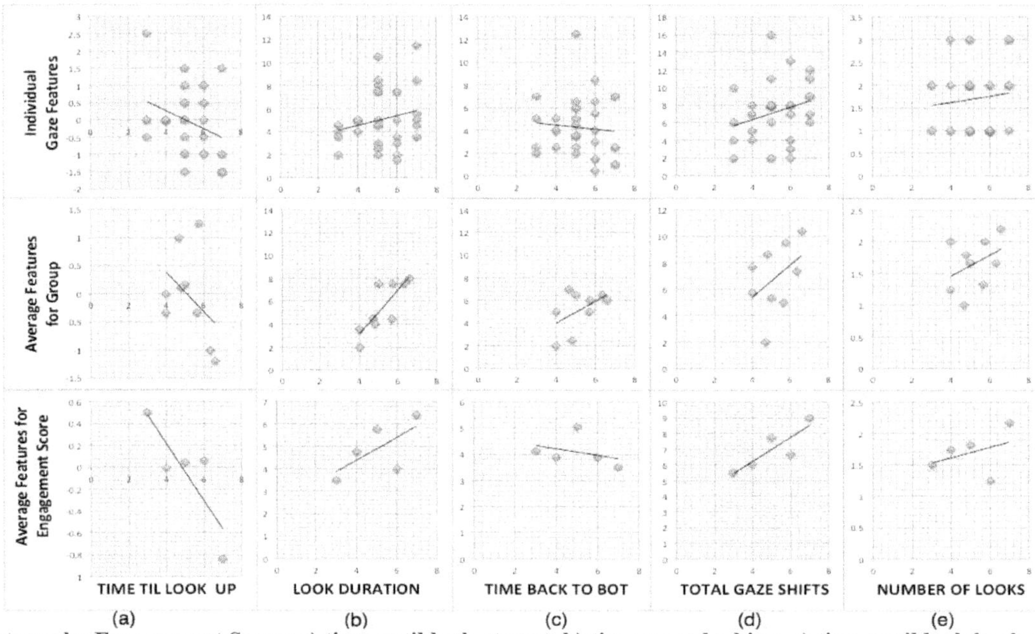

Figure 3. Features by Engagement Score: a) time until look at vent, b) time spent looking, c) time until look back to bot, d) total number of gaze shifts, e) number of glance-backs. The top row presents all individual data, middle averages the group data, and bottom averages all data from the same engagement score self-reports. (One outlier removed, as not enough samples at score=2)

4.1 Gaze Directions

To parse our data, we first isolate participant gaze directions (using data from 3 seconds before the robot's pointing gesture to 15 seconds following). For example, Fig. 4a displays all 34 participants' window-gaze data, organized from least-engaged (front-most) to most-engaged (back-most). Horizontal height indicates the binary presence of a gaze-direction. Note the trend for participants with lower engagement scores to gaze out the (non-topical) window sooner. Fig. 4b displays intra-participant gaze, notably occurring with higher engagement levels only.

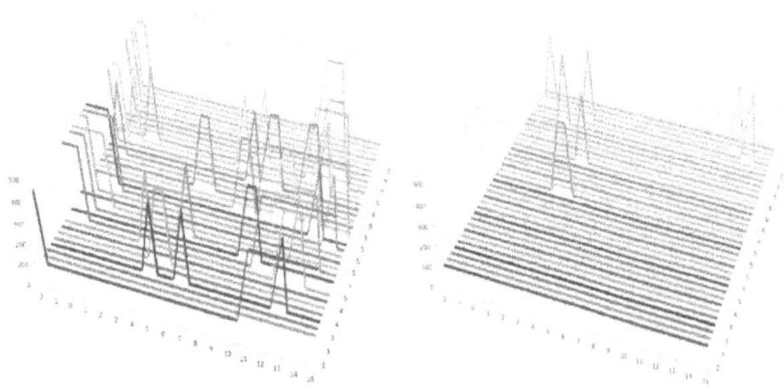

Figure 4. Participant Gaze toward (a) Window, (b) Each-other

4.2 Gaze Features

We next combine statistics from all gaze directions to identify higher-level gaze-based features we could use to classify audience engagement. Fig. 3 displays features such as 'time to look up,' 'total time spent looking at the vent,' 'total number of gaze shifts,' etc. The data shows weak trends correlating individual self-engagement reports with these individual features (top row), though the data spread is high.

When we instead aggregate features across the group (middle row) or all people with similar engagement scores (bottom row), we find clearer trends, e.g., in feature columns 9a (time-to-look-up) and 9d (number-of-gaze-shifts) aggregate data reduces overall noise. In 9b (total-time-spent-looking), and 9e (total-number-of-vent-glances), aggregate results also find stronger positive trends.

5. CONCLUSION

In this experiment, we begin to uncover several possible advantages to tracking the aggregate gaze behaviors of a group over tracking individuals one-by-one:

1. Aggregate analysis means the robot can use lower resolution features that may be easier to detect.
2. Group tracking is more likely to capture unique and salient events that individual analysis might miss, e.g., shared glances or synchrony/divergence of behavior.

In separate work, we are developing algorithms to estimate gaze direction. The behavioral findings presented here will inform the development of our human tracking software. In the case of a bifurcated group, we also wish to assess if storing the engagement data spread better models the dynamics of how its topic and tour selection are received.

6. REFERENCES

[1] Gunhee K, et al. 2004. The Autonomous Tour-Guide Robot Jinny. (IROS '04).

[2] Knight, H. A Savvy Robot Standup Comic: Online Learning through Audience Tracking. Workshop paper (TEI '10)

[3] Nourbakhsh, I.R., Kunz, C., Willeke, T. 2003. The mobot museum robot installations: five year experiment. (IROS '03)

[4] Shiomi, M. Kanda, T. Ishiguro, H. Hagita. 2010. A larger audience, please! — Encouraging people to listen to a guide robot. (HRI '10).

[5] Thrun, S., et al. 1999. MINERVA: a second-generation museum tour-guide robot. In Proc. Int. Conf. on Robotics and Automation (ICRA '99)

Real Time Interaction with Mobile Robots using Hand Gestures

Kishore Konda[1] Achim Königs[1]

Hannes Schulz[2] Dirk Schulz[1]

[1] Unmanned Systems Group,
Fraunhofer FKIE
53343 Wachtberg, Germany
{kishore.reddy.konda,
achim.koenigs,dirk.schulz}
@fkie. fraunhofer.de

[2] Autonomous Intelligent
Systems Group,
University of Bonn
53113 Bonn, Germany
schulz@ais.uni-bonn.de

ABSTRACT

We developed a robust real time hand gesture based interaction system to effectively communicate with a mobile robot which can operate in an outdoor environment. The system enables the user to operate a mobile robot using hand gesture based commands. In particular the system offers direct on site interaction providing better perception of environment to the user. To overcome the illumination challenges in outdoors, the system operates on depth images. Processed depth images are given as input to a convolutional neural network which is trained to detect static hand gestures.

The system is evaluated in real world experiments on a mobile robot to show the operational efficiency in outdoor environment.

Categories and Subject Descriptors

J.7 [**Computers in other systems**]: Real time; Command and control

Keywords

Human-robot interaction, gesture detection, neural network

1. INTRODUCTION

Hand gestures are medium of communication in many scenarios. One of them is guiding the driver of a vehicle (takeoff or taxi) in both civil and military environments. A human guide uses hand gestures to guide the driver of the vehicle to move backward and forward, to turn left and right, stop, etc. With the same intuition we developed the interaction system which directly enables the mobile robot to understand the gestures of a human guide.

The system uses hand gestures as means of communication between the user and the robot. The interaction is basically of the form of motion commands or some other specific commands given by the user to a mobile robot. Some of the basic motion commands are 'move forward', 'stop', 'move backward', 'turn left' and 'turn right' etc., and some specific commands are 'follow me' etc., as shown in Figure 1. In real time application, a sequence of motion commands is generated based on the user pose. The sequence is smoothed using a Bayes filter before forwarding to motion control drivers of the robot.

HRI'12, March 5–8, 2012, Boston, Massachusetts, USA.
ACM 978-1-4503-1063-5/12/03.

Figure 1: Hand gestures corresponding to motion commands. From top left: forward, backwards, right, left, follow me, stop.

2. THE INTERACTION SYSTEM

The interaction system comprises of mainly three modules, one named *stereo processor* which handles the Bumblebee® IEEE-1394a Stereo Vision Digital Camera System and does the processing of the images to compute depth maps. For the computation of depth maps the module utilizes Triclops® library from Point Grey. This module provides the depth maps which are input to the *gesture detection* module.

The gesture detection module processes the depth images by cropping them to a region of interest in row, column and depth dimension and then smoothing, down-sampling them to a 40×40 image patch. This is given as input to the CNN to detect the gestures. The CNN implementation is based on the CUV library [3] using the NVidia CUDA framework for parallel programming. Finally, the gesture detection module provides motion commands, corresponding to the detected gestures, smoothed by a Bayes filter. Each motion command is a combination of translational and rotational velocities. The final module is the *motion control* using rFLEX drivers, it takes in the motion commands and enacts the corresponding physical movement of the robot.

2.1 Convolutional neural network

The structure of the CNN used in the system for gesture recognition is as follows:

- Input map layer I_0 of size same as that of the preprocessed input image i.e, 40×40 with a linear activation function.
- Two convolutional layers C_0, C_1 with map sizes of 36×36, 14×14 and common convolutional window size of 5×5. Non-linear hyperbolic function (Tanh) is the activation function for both layers. Number of feature maps in C_0, C_1 is 20 and 40 respectively.
- Two maximum pooling layers P_0, P_1 of size 18×18, 7×7 with common pooling window size of 2×2. Number of maps in pooling layers is same as that of the convolutional layer to which its input is connected.
- Three fully connected flat layers H_0, H_1 and O_0 of sizes 980, 490 and 7 respectively. The activation function for these layers is non-linear hyperbolic (Tanh) function. F_2 is the output layer of the network. Therefore its size is 7, equal to the number of gestures to be recognized.

The layers are connected in the order I_0 - C_0 - P_0 - C_1 - P_1 - H_0 - H_1 - O_0, where P_1 - H_0 - H_1 - O_0 is a fully connected network.

A training set of $10,000$ depth images and corresponding labels is built with 6 users performing the gestures for training the network. All the images are preprocessed by smoothing, down-sampling and normalization. Another dataset of 2300 images is also built for testing the performance of the neural network. Images of the datasets are recorded in different lighting conditions and varying distance of the user from the camera.

2.2 Filtering

In the gesture detection module an additional Bayes filter is used at the end of the CNN in order to smooth the gesture sequence to avoid false reactions due to unexpected classifications. The Bayes filter uses the transition and classification probabilities to smoothen the gesture sequence. In general when operating the interaction system it is easily observed that the number of transitions from one gesture to another are very few as compared to the self transitions. Hence it is more likely that a gesture continues, than changing to another gesture. We extend the dependency of current state of the filter to past two states. The transition densities are calculated by counting the number of transitions between gestures during the experiments. The classification densities are those of the CNN detection framework.

3. EVALUATION

The performance of the interaction system is mainly based on the precision of the underlying CNN. The classification error after 40 epochs is 0.005 for training set and 0.01 for test set. These values prove that the CNN is very precise in classifying the gestures. In real world experiments for many other user poses for which CNN is untrained, the CNN produces unexpected classifications. These untrained poses usually occur during transitions from one gesture to another and are avoided by the Bayes filter.

	Time(Min:Secs)	
User	Try-1	Try 2
A	1 : 17	1 : 05
B	1 : 52	1 : 30
C	1 : 30	1 : 24
D	1 : 05	1 : 05

(a) S-curve

	Time(Min:Secs)	
User	Try-1	Try 2
A	1 : 35	0 : 52
B	1 : 30	1 : 30
C	2 : 23	1 : 47
D	0 : 45	0 : 45

(b) Parking

Table 1: Time needed by users to complete two different tasks. A, B, C used our system and D used joystick.

Further, as the gesture detection approach only depends on depth images, it is partially invariant to illumination changes unlike color based models [2], which fail to operate in outdoor scenarios. Tracking-based approaches [2] are sensitive to rapid movements and missing depth. As our approach is mainly detection-based, many issues involving tracking are avoided. For example, no initialization of any kind of tracking model and continuous tracking of body parts is required. This renders our approach resistant to small rapid movements of the user and camera while in motion at outdoors, also to missing depth information.

Teleoperation of mobile robot as in [1] requires transmission of omni-directional video for the user to operate and even then provides less perception of environment. Our interaction system is designed to directly communicate with the robot in the environment of operation providing clear advantage of better perception to the user. The system operates in real time with a frame rate of 5 frames/second.

Many real world experiments which involve tasks like parking a vehicle and moving the vehicle along a curve are conducted. Through these experiments it is observed that the interaction system performs well and requires time comparable to Joystick drive to complete a task. Some results of time taken to complete a task by different users are shown in Tables 1a, 1b. The system is user friendly as there is no hardware to be handled by the user.

4. CONCLUSION

Based on the evaluation it can be observed that the interaction system developed is an efficient HRI application which can be used to operate mobile machines in outdoor scenarios. Further it is also observed that the system is partially resistant to the challenges in outdoor scenarios which block the performance of other color based and skeleton based approaches.

5. REFERENCES

[1] C. Hu, M. Meng, P. Liu, and X. Wang. Visual gesture recognition for human-machine interface of robot teleoperation. In *Proc. Int. Conf. on Intelligent Robots and Systems, 2003.*, volume 2, pages 1560–1565, 2003.

[2] S. Mitra and T. Acharya. Gesture recognition: A survey. *Systems, Man, and Cybernetics, Part C: Applications and Reviews, IEEE Transactions on*, 37(3):311 –324, may 2007.

[3] H. Schulz, A. Müller, and S. Behnke. Exploiting local structure in Boltzmann machines. *Neurocomputing*, 74(9):1411–1417, 2011.

Integrating Human and Computer Vision with EEG Toward the Control of a Prosthetic Arm

Eugene Lavely
BAE Systems
6 New England Executive Park
Burlington, MA 01803
eugene.lavely@baesystems.com

Geoffrey Meltzner
BAE Systems
6 New England Executive Park
Burlington, MA 01803
geoffrey.meltzner@baesystems.com

Rick Thompson
BAE Systems
6 New England Executive Park
Burlington, MA 01803
rick.l.thompson@baesystems.com

ABSTRACT

We are undertaking the development of a brain computer interface (BCI) [1] for control of an upper limb prosthetic. Our approach exploits electrical neural activity data for motor intent estimation, and eye gaze direction for target selection. These data streams are augmented by computer vision (CV) for 3D scene reconstruction, and are integrated with a hierarchical controller to achieve semi-autonomous control. User interfaces for the effective control of the many degrees of freedom (DOF) of advanced prosthetic arms are not yet available [2]. Ideally the combined arm and interface technology provides the user with reliable and dexterous capability for reaching, grasping and fine-scale manipulation. Technologies that improve arm embodiment i.e., the impression by the amputee subject that the arm is a natural part of their body-concept presents an important and difficult challenge to the human-robot interaction research community. Such embodiment is clearly predicated on cross-disciplinary advances, including accurate intent estimation and an and an algorithmic basis for natural arm control.

Categories and Subject Descriptors

I.2.9 [**Artificial intelligence**]: Robotics—*manipulators*

Keywords

EEG, computer vision, robotic control, neuroprosthetic

1. INTRODUCTION

Robots inhabit the same physical space as people, manipulate some of the same objects as people, and advances in robotic autonomy have enhanced their ability to perform in unstructured and uncertain environments [3]. Advanced prosthetic arms may also be viewed as robotic entities, and their use by amputees exposes deep technical and philosophical issues in human-robot interaction (HRI) [4, 5, 6]. A key remaining barrier for integration of prosthetic limb with a human subject is the sense of "limb ownership" or embodiment. The latter is a complex phenomenon, dependent on consciousness, the senses of touch and vision, and proprioception i.e., the innate sense providing positional and kinematic awareness of the human body state. Advances in sensory feedback [4, 5] will improve embodiment, but clearly, advances in robotic control for reliable and dexterous arm operation are also required. Such research presents interesting issues in shared autonomy between robots and humans. The focus of the research presented here is improvement of human control of upper limb prosthetic devices, using advanced control principles, sensory data augmentation, and higher-specificity estimation of pre-motor and motor intent from non-invasive neural signal processing [1].

There are several physical interfaces available for measurement of neural electrical activity, and potentially useful for prosthetic control. These include non-invasive electrodes emplaced on the scalp (electroencephalography – EEG), minimally invasive electrodes directly placed on the exposed surface of the brain (electrocorticography - ECoG), and intracortical microelectrode arrays impressed into the cortex [7]. Each of these interfaces access data from the central nervous system (CNS). Non-CNS interfaces include electrode contacts with the peripheral nervous system (PNS) (e.g., [4]), but are not considered here. EEG measures electric potential fields generated by neural activity in the cortex. However, the source to electrode separation implies aggregate sensitivity to thousands of individual neurons (resolution loss) and signal amplitude loss relative to invasive measurements. Also, the EEG field measurements are smoothed and distorted by volume conduction effects implying further reduction in spatial resolution, and are subject to external electrical interference sources as and to data artifacts caused by subject mobility. However, we emphasize EEG here since with the exception of fNIRS (functional near–infrared spectroscopy) EEG represents the only non-surgical CNS sensing option. We believe the performance of EEG-based BCIs can be substantially improved with innovative processing and multi-modal data augmentation. EEG technology is rapidly improving [8], promising additional performance gains.

2. APPROACH

A principal challenge in upper limb prosthetic research is the effective and dexterous control of the many DOF of such arms. The human hand alone has 27 DOF, the human arm has 7 DOF, and current advanced arm designs capture as many as 22 DOF [2]. We address the control and DOF problem by combining three principal ideas.

First, recognizing that the spatio-temporal dynamics and features of EEG data encodes pre-motor and motor intent [1, 9], we perform an inversion estimation of the neural electrical sources [10] to enhance the spatial specificity of EEG data.

HRI'12, March 5–8, 2012, Boston, Massachusetts, USA.
ACM 978-1-4503-1063-5/12/03.

In combination with neural decoding [1, 9], the source estimates provide high-value input for intent estimation. Additionally, we augment this data with eye gaze direction and dynamics of saccades for inference of object selection. Such data when combined with computer vision (see below) supports reaching, grasping, or manipulation tasks.

The underlying geometric and electrical conductivity parameters that calibrate the source inversion [10] may be obtained off-line from magnetic resonance imaging (MRI) and electrical impedance tomography (EIT). In fact, these modalities support cooperative inversion in which the structural MRI offers a geometric "prior" for the impedance inversion. Thus, with the aid of subject–calibration, spatially localized cortical time series can be estimated via EEG inversion, and processed to estimate information–rich features. For example, the brain contains a highly complex collection of neurons that interact and communicate in order to perform motor and cognitive actions, and therefore, the availability of cortical time series enables the use of network techniques to map communications between cortical sites. Functional and effective connectivity measures may then be derived, and which provide a high–dimension and potentially high–value feature space that characterize brain function in a neurobiologically meaningful way. Further, Kalman filtering [11] and data assimilation methods can be applied to improve the reliability of the source time series estimates.

Second, it is important to recognize that advanced prosthetic arms have intrinsic "awareness" of their positional and articulation state (a robotic analog to human proprioception). Recent research [12] shows that a single mobile camera can provide 3D scene reconstruction updates at a 30 Hz frame rate. The EEG decoding of task-oriented intents can be deductively refined by combining target selection from eye gaze data with this CV-enabled 3D scene "awareness". Also end-points or way-points for control objectives can be deduced from this system-level geometric awareness.

Third, it may prove difficult to estimate high-dimension control inputs solely from sensor data streams. Therefore, we advocate an intelligent robotic controller in which low-level controls of individual actuators are autonomously determined given a task-level intent [13], and which has the benefit of reducing the required cardinality of input control signals and user–cognitive load as well. In summary, our strategy is to provide low-dimensional, task-level control input to reduce the burden of the user, augment this data with autonomous data streams for improved scene awareness, and combine these inputs for low-level execution of motor intent through use of a hierarchical controller [13] (which also addresses the curse of dimensionality problem).

3. CONCLUSIONS

Effective and reliable neuroprosthetic upper limb control is a formidable challenge, the key goal being the restoration of normal arm function. Relevant performance metrics include time to movement onset, time to target, trajectory curvature, multiple peaks in the speed profile, end-point oscillations, and other factors associated with normal and abnormal movement. Accurate arm responsiveness to user intent and the robotic equivalent of fine motor skills are clearly critical for user acceptance. However, embodiment and a sense of proprioception are also user–critical, an area clearly within the province of HRI. Prosthetic usage poses profound challenges in HRI including shared autonomy i.e., the boundary between human and semi-autonomous control, philosophical issues [6], sensory integration and feedback, and reasoning and estimation under uncertainty. Progress in these areas and in decoding will improve the prosthetic user experience, and enable progress in other domains such as bio–navigation, gaming and avatars, and semi-autonomous control of assistive technologies (e.g., robots, vehicles).

4. ACKNOWLEDGMENTS

We gratefully acknowledge conversations with Dr. Lakshminarayan Srinivasan (UCLA), Prof. Emo Todorov (University of Washington) and Dr. Troy Lau (BAE Systems).

5. REFERENCES

[1] R.P.N. Rao and R. Scherer, Statistical Pattern Recognition and Machine Learning in Brain-Computer Interfaces, in *Statistical Signal Processing for Neuroscience and Neurotechnology*, K. Oweiss, Ed., New York: Academic Press, 433 pages, 2010.

[2] R.F. Weir and J.W. Sensinger, Design of artificial arms and hands for prosthetic applications, in *Standard Handbook of Biomedical Engineering & Design*, Vol. 2, M. Kutz, Ed. New York: McGraw-Hill, 2009.

[3] S. Thrun, Toward a framework for human–robot interaction, *Hum.–Comput. Interact.*, vol. 19, 2004.

[4] P.D. Marasco, K. Kim, J.E. Colgate, M.A. Peshkin and T.A. Kuike, Robotic touch shifts perception of embodiment to a prosthesis in targeted reinnervation amputees, *Brain*, vol. 134, pp. 747 - 758, 2011.

[5] A. Blank, A.M. Okamura and K.J. Kuchenbecker, Identifying the Role of Proprioception in Upper-Limb Prosthesis Control: Studies on Targeted Motion, *ACM Trans. on Applied Perception*, vol. 7, pp. 1 - 23, 2010.

[6] S. Gallagher, *How the Body Shapes the Mind*, New York: Oxford University Press, 224 pages, 2006.

[7] J.H. Hochberg and J.P. Donoghue, Sensors for Brain-Computer Interfaces, *IEEE Eng. in Medicine and Biology Magazine*, vol. 25, pp. 32 - 38, 2006.

[8] Y.M. Chi, T.P. Jung and G. Cauwenberghs, Dry-Contact and Noncontact Biopotential Electrodes: Methodological Review, *IEEE Reviews in Biomedical Engineering*, vol. 3, pp. 106 - 119, 2010.

[9] L. Srinivasan, U.T. Eden, S.K. Mitter and E.N. Brown, General-purpose filter design for neural prosthetic devices, *J. Neurophysiol.* vol. 98, pp. 2456 - 75, 2007.

[10] L. Ding, Y. Ni, J. Sweeney and B. He, Sparse cortical current density imaging in motor potentials induced by finger movement, *J. Neural Eng.*, vol. 8, 2011.

[11] C.J. Long, P.L. Purdon, S. Temereanca, N.U. Desai, M.S. Hämäläinen and E.N. Brown, State-space solutions to the dynamic magnetoencephalography inverse problem using high performance computing, *Ann. Appl. Stat.*, vol. 5, pp. 1207 - 1228, 2011.

[12] R.A. Newcombe, S. Lovegrove and A.J. Davison, DTAM: Dense tracking and mapping in real-time, in *Proc. of the Int. Conf. on Computer Vision*, 2011.

[13] E. Todorov, W. Li and X. Pan, From task parameters to motor synergies: A hierarchical framework for approximately-optimal feedback control of redundant manipulators, *J. of Robotic Systems*, vol. 22, pp. 691 - 710, 2005.

Human-Robot Interaction in the MORSE Simulator

Séverin Lemaignan[1,2,3], Gilberto Echeverría[1,2], Michael Karg[3]
Jim Mainprice[1,2], Alexandra Kirsch[3], Rachid Alami[1,2]

{slemaign, gechever, jmainpri, rachid}@laas.fr, {kargm, kirsch}@in.tum.de
[1]CNRS-LAAS ; 7 avenue du colonel Roche, F-31077 Toulouse, France
[2]Université de Toulouse; UPS, INSA, INP, ISAE ; LAAS ; F-31077 Toulouse, France
[3]IAS; Technical University of Munich ; Karlstr. 45 ; D-80333 Munich, Germany

ABSTRACT

Over the last two years, the Modular OpenRobots Simulation Engine (MORSE) project[1] went from a simple extension plugged on the Blender's *Game Engine* to a full-fledged simulation environment for academic robotics. Driven by the requirements of several of its developers, tools dedicated to Human-Robot interaction simulation have taken a prominent place in the project. This late breaking report discusses some of the recent additions in this domain, including the immersive experience provided by the integration of the Kinect device as input controller. We also give an overview of the experiences we plan to complete in the coming months.

Categories and Subject Descriptors

I.2.9 [**Computing Methodologies**]: Artificial Intelligence—*Robotics*

General Terms

Experimentation, Human Factor

Keywords

Simulation, Human-Robot Interaction

1. THE MORSE SIMULATOR

The Modular OpenRobots Simulation Engine (MORSE) [2] is a open-source tool developed for robotics research. It is a domain independent simulator, where virtual robots can interact with a 3D environment, using sensors and actuators that behave in the same way as their counterparts in the real world. MORSE relies on the advanced 3D (OpenGL shaders) and physics simulation (Bullet physics engine) capabilities of the Blender *Game Engine*, a real-time 3D runtime integrated to the open-source Blender modeling toolkit. This allows for semi-realistic simulation of complex environments.

The MORSE components (sensors and actuators) exchange data with the robotics software via middleware bindings, using a *Software In The Loop* (SAIL) architecture. Middleware supported in the current version include ROS and YARP, as well as a socket-based raw protocol. This design allows in principle to use the same software in both the real robots and the simulator. Instructions given to the robot are interpreted in the simulator to provide the control of actuators, such as the motion of the robot and its arms. The data from simulated sensors is sent back through the middlewares, *e.g.* exporting the images from cameras, or the positions of the robot, human and other objects of interest.

MORSE provides support for several classes of robots *out of the box*, and allows for easy customization of those, either by composing individual sensors and actuators with empty robot structures directly in the MORSE interface, or through a Python-based script language that permits to conveniently describe robots and simulation scenarii.

Other experiments using simulation have been carried to gather data for HRI [1]. However, these do not involve the actual robot software, and it is another human who takes the role of the robot.

Figure 1: The experimental setup with the human avatar controlled from a Kinect.

2. HRI SPECIFIC FEATURES

A new *human avatar* has been recently added to the MORSE assets. It provides a first-person immersive experience: when started, one can take the role of the human and control it via the keyboard, a WiiMote or, as another recent addition to the simulator, through a Kinect device (Fig. 1).

When in first-person mode, the user can interact in several ways with the environment. He/she can pick and release objects, can open drawers and cupboards. As any other object,

[1]`http://morse.openrobots.org`

HRI'12, March 5–8, 2012, Boston, Massachusetts, USA.
ACM 978-1-4503-1063-5/12/03.

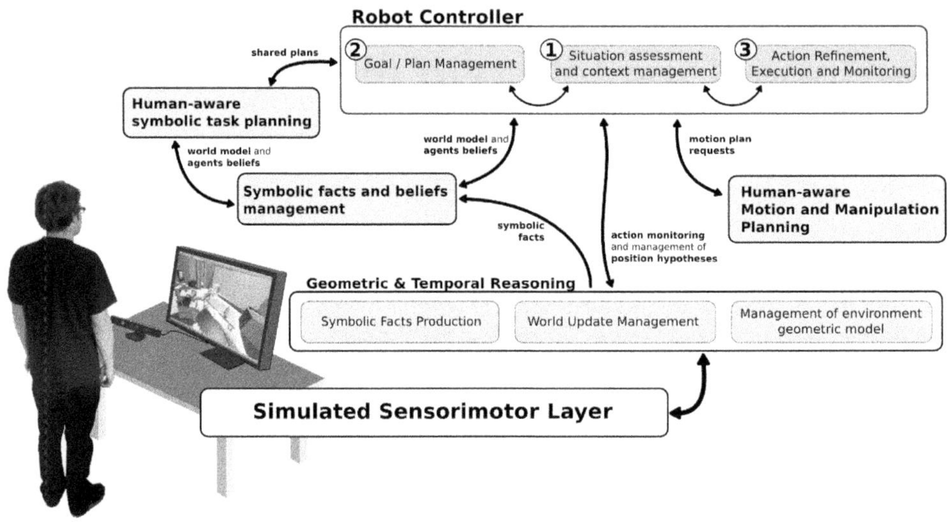

Figure 2: Example of integration in a software architecture: MORSE transparently replaces the real robot.

the avatar physically interacts (collision detection) with the surrounding furnitures. MORSE exports the position and posture of the avatar as a global joint state to be used by the real software components of the robot.

MORSE also offers a special sensor that exports abstracted informations of objects visible to the robot (called the *semantic camera*). This sensor typically exports the name, type (glass, table, bottle, etc.), color and location of objects. Since human-robot interaction often involves semantic-rich environments, this abstract sensor simplifies the experiments on such scenarii, by avoiding the added complexity of processing camera images to detect the objects of interest and exploiting the inherent knowledge of the simulated world.

3. CURRENT AND FUTURE EXPERIMENTS

Due to its nature, MORSE offers two main advantages compared to experiments on a real robot: light-weight deployment and repeatability. MORSE is already used for human-robot interaction in two laboratories, at the LAAS-CNRS, Toulouse, France, and at the Technical University of Munich, Germany.

Integration in the LAAS cognitive architecture: Figure 2 presents the current integration of MORSE in the LAAS' software architecture. In particular, both the human posture and the object features are integrated with SPARK [5], a module dedicated to geometric and temporal reasoning. This module is a key component providing a base of facts such as objects' relative placements, visibility and reachability by the agents present in the scene. It additionally provides a stable state of the world to the motion planners.

Human-Aware Motion Planning: The human avatar is being used to evaluate human-aware navigation planners at Munich TU. It enables developers to find and avoid errors before running on real robots. A simulated human tracker updates the position of the human and passes it to the robotic middleware. A video is available here: `http://vimeo.com/31257264`.

Interaction with Natural Language: Relying on the abstract sensors provided by MORSE, we plan to reproduce in simulation the results of a previous experiment [4] involving interactive verbal disambiguation. Natural language understanding relies on a dedicated language parsing and grounding module called DIALOGS, which is fed by a symbolic knowledge base build as a ontology server [3].

4. SUMMARY

We have presented the developments leading to an interactive experimental setup for Human-Robot Interaction. The development of particular functionalities of the robot architecture, especially those involving a quick response to changes in the environment induced by the human presence, will directly benefit from the HRI simulation capabilities provided by MORSE.

5. REFERENCES

[1] S. Chernova, N. DePalma, E. Morant, and C. Breazeal. Crowdsourcing human-robot interaction: Application from virtual to physical worlds. In *20th IEEE International Symposium in Robot and Human Interactive Communication*, 2011.

[2] G. Echeverria, N. Lassabe, A. Degroote, and S. Lemaignan. Modular openrobots simulation engine: MORSE. In *Proceedings of the IEEE ICRA*, 2011.

[3] S. Lemaignan, R. Ros, E. A. Sisbot, R. Alami, and M. Beetz. Grounding the interaction: Anchoring situated discourse in everyday human-robot interaction. *International Journal of Social Robotics*, pages 1–19, 2011.

[4] R. Ros, S. Lemaignan, E. A. Sisbot, R. Alami, J. Steinwender, K. Hamann, and F. Warneken. Which one? grounding the referent based on efficient human-robot interaction. In *19th IEEE International Symposium in Robot and Human Interactive Communication*, 2010.

[5] E. Sisbot, R. Ros, and R. Alami. Situation assessment for human-robot interaction. In *20th IEEE International Symposium in Robot and Human Interactive Communication*, 2011.

Vision-Based Attention Estimation and Selection for Social Robot to Perform Natural Interaction in the Open World

Liyuan Li, Xinguo Yu, Jun Li, Gang Wang, Ji-Yu Shi, Yeow Kee Tan, and Haizhou Li
Institute for Infocomm Research
1 Fusionopolis Way, Singapore, 138632
{lyli,xinguo,jli,gswang,jyshi,yktan,hli}@i2r.a-star.edu.sg

ABSTRACT

In this paper, a novel vision system is proposed to estimate attention of people from rich visual clues for social robot to perform natural interactions with multiple participants in public environments. The vision detection and recognition modules include multi-person detection and tracking, upper-body pose recognition, face and gaze detection, lip motion analysis for speaking recognition, and facial expression recognition. A computational approach is proposed to generate a quantitative estimation of human attention. The vision system is implemented on a robotic receptionist "EVE" and encouraging results have been obtained.

Categories and Subject Descriptors

I.4.8 [**IMAGE PROCESSING AND COMPUTER VISION**]: Scene Analysis:tracking, stereo, color

Keywords

Social robot, attention estimation, vision understanding

1. INTRODUCTION

The ability to be aware of people's attention and predict their intension in social interactions is an essential function for social robots. Non-verbal clues; such as motion trajectory, spatial proximity, body pose, gaze, head pose, hand gesture, and facial expressions; play an important role in social interactions involving multiple agents in open world environments. Human beings perceive such visual clues with less efforts for social interactions. However, it is still very challenging to enable a social robot to exploit such rich visual clues in interactions with multiple agents. In this paper, we explore this area by extracting and representing such rich visual clues for enhancing the ability of social robots to perform natural interactions in open world.

2. VISION DETECTION

Various state-of-the-art vision detection, tracking, and recognition modules, of which many were developed in our previous work, are employed in this system.

Our algorithm [2] for efficient and robust multi-person detection and tracking from a stereo camera for mobile service robots is implemented in this system. It provides the position and motion information of persons in the scene. We have proposed a novel model of Adaboost template for accurate recognition of upper-body pose from 2D depth images [3]. It recognizes the body's facing direction in real-world space. The Adaboost face detector from OpenCV is used for face detection from the head region of the tracked persons. Meanwhile, our recently developed novel gaze detection algorithm [4] is used to detect a pair of eyes looking at the robot. To be robust to variations across persons, poses, lightings, as well as head and body motion in natural social interactions, we propose a novel approach for speaking recognition from lip motion which consists of a mouth shape model based on Spectral Regression (SR), a mouth motion model based on spatial-temporal description, and a head motion descriptor. The features from the three models are combined and a SVM classifier is trained for robust recognition. We improved the LBP (Local Binary Pattern) based approach of facial expression recognition [1] by introducing different weights to original LBP Patterns so that the LBP patterns on the edges got high weights. The proposed feature is more suitable to describe the facial expression.

3. ATTENTION ESTIMATION

From the multi-person detection and tracking, we can obtain the 3D position (X_i, Z_i) of the ith person on the ground plane to the robot. Obviously, when there are several people around, the person closer to and directly facing the robot is the one who is most likely to talk to the robot. Let Z_c be the most comfortable distance for face-to-face human interaction. The visual clue of attention to the robot from his distance can be defined as

$$f_1(H_i) = \begin{cases} e^{-K_D(Z_i - Z_c)} & \text{if } Z_i \geq Z_c \\ e^{-(Z_i-Z_c)^2/\sigma_D^2} & \text{if } Z_i < Z_c \end{cases} \quad (1)$$

and the visual clue of the attention from the directional angle θ_i of the position from the robot is expressed as

$$f_2(H_i) = \cos(\theta_i) = Z_i/\sqrt{X_i^2 + Z_i^2}. \quad (2)$$

If a person wants to talk to the robot, he would keep standing. If he keeps moving across the scene, he might be a passerby. Let (X_i^{t-1}, Z_i^{t-1}) and (X_i^t, Z_i^t) be the positions of the ith person in the previous and current time steps. The visual clue of the attention from motion is computed as

$$f_3(H_i) = e^{-[(X_i^t - X_i^{t-1})^2 + (Z_i^t - Z_i^{t-1})^2]/D^2}, \quad (3)$$

HRI'12, March 5–8, 2012, Boston, Massachusetts, USA.
ACM 978-1-4503-1063-5/12/03.

Figure 1: Snapshot images from the example test sequence.

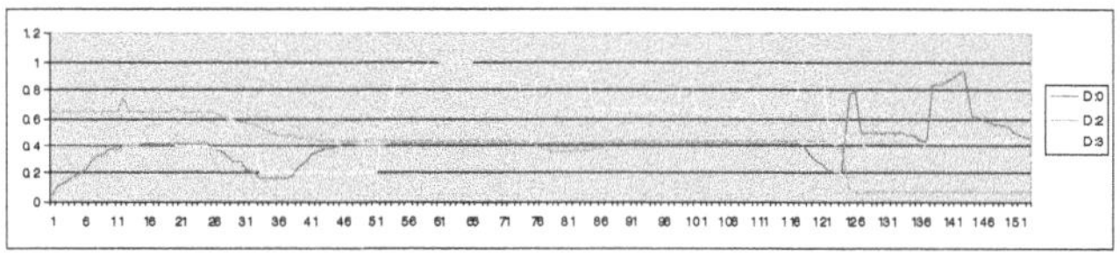

Figure 2: Plots of attention levels.

where the variance D is selected on human physical measurements. If the recognized upper-body pose of the ith person is α_i degree, the body pose to the robot is $\alpha_i - \theta_i$, where $\theta_i = \arctan(X_i/Z_i)$. The visual clue of the attention from the person's upper-body pose is defined as

$$f_4(H_i) = \cos(\alpha_i - \theta_i). \quad (4)$$

To be robust to the variations of detections, the evidence over a short time duration is extracted for visual clues from face, gaze, lip motion, and facial expression detections, ie.

$$f_5(H_i) = \frac{1}{T}\sum_{k=t-T}^{t} F(H_i^k) \quad f_6(H_i) = \frac{1}{T}\sum_{k=t-T}^{t} G(H_i^k)$$
$$f_7(H_i) = \frac{1}{T}\sum_{k=t-T}^{t} S(H_i^k) \quad f_8(H_i) = \frac{1}{T}\sum_{k=t-T}^{t} E(H_i^k) \quad (5)$$

where $F(H_i^k)$ and $G(H_i^k)$ are the outputs of face and gaze detections, $S(H_i^k)$ is the output of lip motion recognition, and $E(H_i^k)$ is the output of facial expression recognition. $E(H_i^k)$ equals '+1' for smile (happy), '0' for normal (neutral), and '-1' scowl (sad).

The visual clues are fused by using weighted sum to compute the attention level for the person H_i

$$AL(H_i) = \sum_{j=1}^{8} w_j f_j(H_i), \quad (6)$$

where the weights can be selected empirically.

4. EXPERIMENTS

We have conducted on-site experiments with the robot receptionist in various interaction scenarios. The results match with human judgements on the scenarios of interactions and dialogues. The videos of the testings are available online. The results on one test sequence is shown here. Four example images are displayed in Figure 1 and their attention level plots is shown in Figure 2. Visual features are displayed over human bodies within the bounding box. The circle over the left shoulder indicates detection of speaking from lip motion (red: speaking). The rectangle over the right shoulder indicates a facial expression captured (red: smiling; green: normal). The sequence demonstrates scenarios of three persons in the view who observe and test the robot, and discuss with one another in front of the robot. First, the two persons (ID:0-blue and ID:2-red) stand closely to the robot and observe how the robot performs. Then, the third person (ID:3-yellow) comes in. He is interested in the robot and try to test its performance. He tries to move to the central position among the people to catch the robot's attention, as shown in the first two images. As he is a later comer, his position is behind the two persons coming earlier. Later, the third person moves to the central position between the other two persons. He looks and smiles at the robot, as shown in the third image. The robot then noticed him, rotated to face him, and detected his smiling face and speaking lip motion for greeting. After having a dialogue with the robot, he is quite happy and turns his head to discuss with person ID:0 on his experience (after frame 121), so that one can observe the great drop of attention level.

5. CONCLUSIONS

We have presented a vision system for social robots to perceive the attention of people from rich visual clues in social interactions. To our knowledge, this is the first system to integrate so much visual information for attention perception. We present a computational way to quantitatively describe the attention from visual clues for social interaction in open world. Perception of attention in social interactions relates to social psychology, cognitive science, and communication theory. Future joint research could lead to a concrete understanding and better systems on quantitative evaluation of the attention in social interactions. This will enhance the ability of attention perception for social robots.

6. REFERENCES

[1] X. Feng, M. Pietikäinen, and A. Hadid. Facial expression recognition with local binary patterns and linear programming. In *7th International Conference on Pattern Recognition and Image Analysis: New Information Technologies (PRIA-7)*, 2004.

[2] L. Li, K. E. Hoe, S. Yan, and X. Yu. Ml-fusion based multi-model human detection and tracking for robust human-robot interfaces. In *IEEE Workshop on Applications of Computer Vision (WACV)*, 2009.

[3] L. Li, K. E. Hoe, X. Yu, L. Dong, and X. Chu. Human upper body pose recognition using adaboost template for natural human robot interaction. In *Seventh Canadian Conference on Computer and Robot Vision (CRV)*, 2010.

[4] X. Yu, W. Han, L. Li, J. Y. Shi, and G. Wang. An eye detection and localization system for natural human and robot interaction without face detection. In *12th Conference Towards Autonomous Robotic Systems (TAROS)*, 2011.

A Prototyping Environment for Interaction between a Human and a Robotic Multi-Agent System

Michael Lichtenstern, Martin Frassl, Bernhard Perun, Michael Angermann
Institute of Communications and Navigation
German Aerospace Center (DLR)
Oberpfaffenhofen, Germany
{m.lichtenstern, martin.frassl, bernhard.perun, michael.angermann}@dlr.de

ABSTRACT

In this paper we describe our prototyping environment to study concepts for empowering a single user to control robotic multi-agent systems. We investigate and validate these concepts by experiments with a fleet of hovering robots. Specifically, we report on a first experiment in which one robot is equipped with an RGB-D sensor through which the user is enabled to directly interact with a multi-agent system without the need to carry any device.

Categories and Subject Descriptors

H.5.2 [**User Interfaces**]: Input devices and strategies, Interaction styles, Prototyping, User-centered design

Keywords

human multi-robot interaction, heterogeneous multi-agent system, device-free, in situ multi-robot control

1. MOTIVATION

Robotic multi-agent systems (MAS) have, due to their high degree of parallelism and redundancy, huge potential in terms of task execution speed and robustness. Recent research has shown that such systems are well suited to carry out exploration or surveillance tasks. Moreover, these systems scale well if organized distributively and are robust against partial system failure if they realize a decentralized low-level artificial intelligence [3]. In particular, it is expected to be beneficial to combine low-level artificial intelligence for autonomy with human cognitive capabilities for high level supervision. In systems in which many robots form this low-level artificial intelligence and in which the level of autonomy that each robot can be granted varies over time, the higher-level intelligence needs to dynamically allocate its attention to where it is needed. We wish to enable a single user to simultaneously control a number of platforms in 3D-space, ideally without the need to glance at, type or touch a device with a graphical user interface. Instead the user shall be positioned within the environment with direct visual contact to the platforms, thereby facilitating uninterrupted awareness of the situation. Furthermore, a user shall be able to directly interact without bringing any device or requiring any interface components installed in the environment.

HRI'12, March 5–8, 2012, Boston, Massachusetts, USA.
ACM 978-1-4503-1063-5/12/03.

Figure 1: Snapshot of the "shuffle" experiment. The Pelican with the RGB-D sensor hovers in front of the user. The system is in the state "select", allowing the user to point to one of the three hovering Hummingbird in order to toggle its selection status

2. RELATED WORK

Olsen and Wood have introduced the *fan-out* of a human-robot team, i.e. the number of robots a human can control simultaneously [6]. They show that the achievable fan-out is determined by the ratio between the *activity time* and the *interaction time*, where activity time is the time the robot is actively effective after receiving commands from a user and the interaction time is the time that is takes for a human to interact with a given robot. Clare and Cummings have conducted experiments, in which significant numbers of participants performed missions involving multiple unmanned vehicles in a de-centralized simulation environment under increasing task load [1]. In our own previous work, we have shown how a single user can control a small fleet of micro aerial vehicles for the purpose of observing outdoor activities [4]. In contrast to the work presented here, this work employed GPS as a position sensor for the aerial vehicles and the user and required the user to drag desired positions on a small multi-touch display carried with him. Closely related to the work presented here, are experiments on selecting and commanding individual robots and groups of robots in a vision-based multi-robot system [2, 5]. In these experiment, a user is enabled to select robots from a set of ground-based robots (Roomba Create). Using cameras mounted on each robot, a robot can determine whether it is selected, either by detecting when the user looks at it [2], or when the human user makes a gesture that encircles a robot [5]. Colored LEDs are used to signal the selection status of each robot.

3. PROTOTYPING ENVIRONMENT

We have developed a prototyping environment in which the multi-agent system is comprised of four hovering aerial platforms (see Figure 1). All four platforms are quadrotors, three of them of type AscTec Hummingbird and one of them of type AscTec Pelican. The role of the Pelican is to provide an RGB-D sensor which is used to detect if a user is present and to recognize and process the user's input. The three Hummingbirds are the agents which are actually controlled by the user. A consumer-grade RGB-D sensor (Microsoft Kinect) is mounted on the Pelican and employed to detect the user's gestures and pointing. From the sensor's depth field, a pose estimation algorithm can derive a skeleton model of the user's body. To track the robots poses, we use a Vicon motion capture system which is capable to track a volume with a ground area of approximately 7.0 meters by 4.0 meters and a height of approximately 2.5 meters. As the user input is sensed by the RGB-D sensor and not by the motion capture system the user, in contrast to the robots, does not need to be equipped with markers or even be located inside the tracked volume. In our experiment "shuffle" the

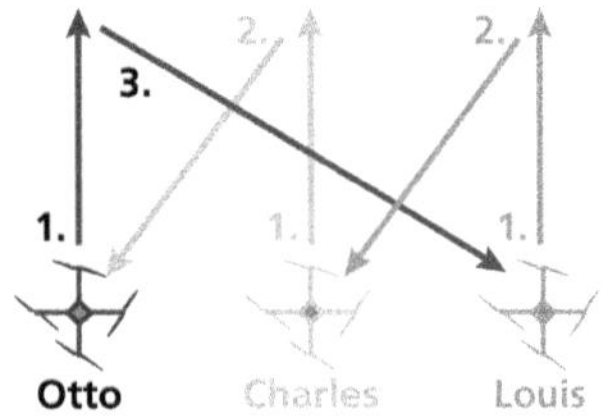

Figure 2: The illustration shows the sequence of movements to permutate three quadrotors.

task for the user is to rearrange three robots of type Hummingbird (see Figure 2). When the Pelican, equipped with the RGB-D sensor, detects a user entering the sensor's range, it is starting to hover at a distance of about 3.0 meters and at 1.0 meter height in front of the position of the user. The Hummingbirds are initially positioned on the ground behind the Pelican. The user selects all Hummingbirds by pointing at each of them with his right arm and confirming the selection by touching his right arm with his left hand. Afterwards he lifts his left arm above his shoulders to obtain control over the selected Hummingbirds. Now, relative position changes of the user's right hand are used to command relative position changes of the selected agents. Hence, when raising his right hand the quadrotors take off. The first task is to reposition all quadrotors about 1.5 meters forward, keeping the quadrotors hovering in about 1.0 meter height. One of the robots is then deselected, and the user repositions the remaining two quadrotors. In the next step, the selection is reversed, the single quadrotor is moved to it's target position and finally all quadrotors are landed.

4. RESULTS

We have carried out the experiment several times. All defined gestures are robustly detected. The RGB-D sensor reliably measures the pointing vector which is then dynamically transformed into the lab frame (see Figure 3). The resulting accuracy is sufficient to select robots and to control their motion. Coloured LEDs mounted on the quadrotors provide a good visual interface for the operator, who got instantaneous feedback on his actions. The experimental results show that it is feasible for a single user to control

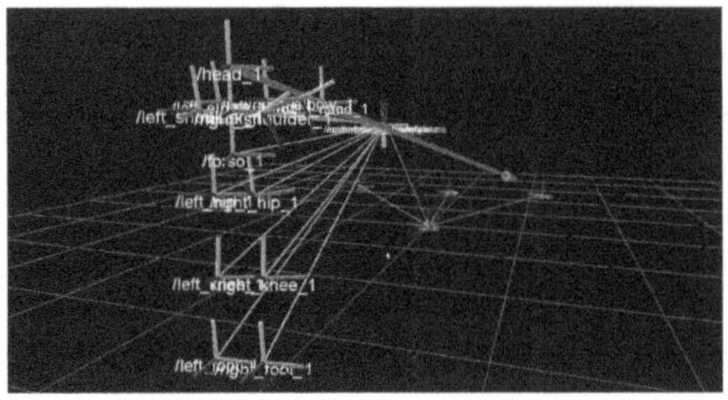

Figure 3: The positions of the RGB-D sensor, three quadrotors and the pose of the user. The red arrow represents the detected pointing vector.

a multi-agent system without carrying any form of personal device. To our knowledge, this is the first time that this has been achieved for a flying multi-agent system. The focus of the presented experiments was on the channel from the user towards the MAS ("input channel") by means of detecting gestures. However, we found that the channel from the MAS towards the user ("output channel") is of similar importance to signal acknowledgements or the current selection status. In our next steps we intend to investigate the capabilities and limits of this output channel and its influence on overall usability. While the experimental setup presented here involved the use of a motion capture system to measure the pose of the robots, we believe that similar results are obtainable outdoors, based on GPS positioning and attitude determination by means of inertial and magnetic sensors. The concept has been proven to work for the specific case of hovering aerial robots. The concept is fairly generic and should be applicable to other agents, such as ground based robots, underwater robots or heterogeneous multi-agent systems involving different robotic and/or human agents.

5. REFERENCES

[1] A. Clare and M. L. Cummings. Task-based interfaces for decentralized multiple unmanned vehicle control. In *Proceedings of AUVSI 2011: Unmanned Systems North America*, Washington D.C., USA, 2011.

[2] A. Couture-Beil, R. T. Vaughan, and G. Mori. Selecting and commanding individual robots in a vision-based multi-robot system. In *Proceeding of the 5th ACM/IEEE international conference on Human-robot interaction*, HRI '10, pages 355–356, New York, NY, USA, 2010. ACM.

[3] B. J. Julian, M. Angermann, M. Schwager, and D. Rus. A scalable information theoretic approach to distributed robot coordination. In *Proceedings of the IEEE/RSJ Conference on Intelligent Robots and Systems (IROS)*, San Francisco, California, USA, 2011.

[4] M. Lichtenstern, M. Angermann, and M. Frassl. Imu- and gnss-assisted single-user control of a mav-swarm for multiple perspective observation of outdoor activities. In *Proceedings of the ION ITM 2011*, San Diego, California, USA, 2011. The Institute of Navigation.

[5] B. Milligan, G. Mori, and R. T. Vaughan. Selecting and commanding groups in a multi-robot vision based system. In *Proceedings of the 6th international conference on Human-robot interaction*, HRI '11, pages 415–416, New York, NY, USA, 2011. ACM.

[6] D. R. Olsen and S. B. Wood. Fan-out: measuring human control of multiple robots. In *CHI'04*, pages 231–238, 2004.

Explaining Robot Actions

Meghann Lomas, Robert Chevalier, E. Vincent Cross II, R. Christopher Garrett, John Hoare, and Michael Kopack
Lockheed Martin Advanced Technology Laboratories
3 Executive Campus, Suite 600
Cherry Hill, NJ 08002
1 856.792.9681
{mlomas, rchevali, ecross, rgarrett, jhoare, mkopack}@atl.lmco.com

ABSTRACT
To increase human trust in robots, we have developed a system that provides insight into robotic behaviors by enabling a robot to answer questions people pose about its actions (e.g., Q: "Why did you turn left there?" A: "I detected a person at the end of the hallway."). Our focus is on generation of this explanation in human-understandable terms *despite the mathematical, robot-specific representation and planning system* used by the robot to make its decisions and execute its actions. We present our work to date on this topic, including system design and experiments, and discuss areas for future work.

Categories and Subject Descriptors
H.1.2 **[User/Machine Systems]**

General Terms
Design, Human Factors

Keywords
Robotic Actions, Natural Communications, Trust, Human-Robot Partnering, Explanations

1. INTRODUCTION
As robots become more frequently used in human-inhabited environments, it is increasingly important for people to maintain an appropriate level of trust in robots. Research has shown that no matter how capable an autonomous system is, if human operators do not trust the system, they will not use it [1].

Two key factors that contribute to human trust are predictability and a mechanism for social exchange, both of which are frequently reduced or lacking in robotic systems [2]. Without a mechanism for communicating information and intent, a robot may appear erratic and untrustworthy when, in fact, it is following a clear decision-making process. As in human partnerships, the ability to explain the information and logic behind decisions greatly increases trust by establishing an understanding of *why* a decision was made, and subsequently provides insight into future decisions.

The focus of our work is on enabling a robot to *explain its actions in human-understandable concepts and terms*, including what action it took, what information it had about the world at the time, *and* the logic behind its decision-making process. Because humans and robots use very different concepts when thinking about the world, robotic actions, information, and processes are rarely directly understandable by humans. Robots typically represent the world in coordinate-based terms (e.g., a grid representation of occupied areas, or a representation of detected color blobs in image coordinates); these representations do not align with human views of the world, which tend to be semantic representations of objects in space (e.g., the chair is near the desk). Robotic planning is done using mathematical formulas whose process and output are not readily expressed in semantic terminology. Additionally, the actions produced by the planner are influenced by many factors and based on continuous mathematical models, and so are not easily discretized.

To address these challenges, we have developed the Explaining Robot Actions (ERA) system, which includes a robotic world model capable of representing semantic and physically-based information, models of planning systems, and a query mechanism for obtaining the desired information from the robotic planning system to produce real-time, human-understandable answers to questions about a robot's behavior. For the purposes of this work, we focus on generating a semantic answer from robot-specific concepts and assume the use of a natural language system for parsing human questions.

2. RELATED WORK
To improve collaboration and co-existence between humans and robots, a considerable amount of work has gone into making robots expressive (e.g., in [3], gestures related to the robot's goals make the robot appear more intelligent and approachable). This supports increased engagement with people, but does not necessarily help clarify why or how a robot selects actions.

The idea of explaining behaviors is less developed, but has been explored in artificial intelligence (AI) research and in robotics. The area of explainable AI has addressed the challenge of explaining the state of agents in a simulation but focuses on after-action analysis to determine behavior models (e.g., [4]). For robotics—and most directly related to our work— robotic actions have been explained through visual timelines made up of action trees that describe the robot's recent actions [5]. Our approach differs from this work in that we enable verbal communication by enriching the robotic world representation and developing a mechanism for querying the representation and planning system for task-relevant information.

HRI '12, March 5-8, 2012, Boston, MA, USA.
ACM 978-1-4503-1063-5/12/03.

3. ERA SYSTEM

The ERA system (Figure 1) is designed to take in an already-parsed human query about the robot's behavior and output an answer in sentence form. The ERA system first determines what information is needed to answer the question. Based on the "subtext" of the question (i.e., the fundamental question being posed), ERA selects an answer template, which forms the framework for the robot's response. These templates are designed to be modular and combinable (e.g., "I am in *X* mode because *Y*" where *Y* is another template: "I detected object *A* at position *Z*."). To populate the template with the information specific to the question, ERA queries the world model and planning system, which have been integrated with and established as resources for the ERA component.

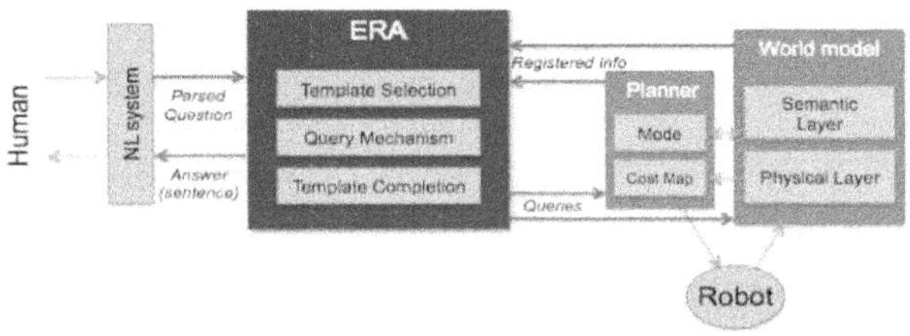

Figure 1: The ERA system determines what information is needed to answer the question and queries the two-layer world model and planner.

Because of their ubiquity and versatility, we assume the robot uses a cost map-based planner for low-level actions and a finite state machine high-level planner that selects the robot's mode based on world information (integration with other planners is left for future work). To represent the world, we use our two-layer world model, which incorporates a grid-based physical layer for planning and a semantic layer for a richer object representation and communication with people (see [6] for more details). This representation provides a translation that enables a semantic response understandable by people. Each resource registers the type of information it can provide (e.g., robot mode, object attributes) which allows ERA to select the information needed to complete the chosen template. Once completed, the sentence is output as an answer to the question.

4. PROOF-OF-CONCEPT EXPERIMENTS

To test and demonstrate the ERA system, we established a simulated search and rescue environment in Stage, and tasked a robot with searching the environment and escorting people to the nearest exit. "Operators" asked the robot questions about its behavior using a GUI interface. Two proof-of-concept experiments were performed. The first focused on retrieving information from the world model. For example, when queried "Tell me how you work," the robot described the type of planner it used, the grid types that combined to form the cost map, and the weights on each grid. These fields were populated with data registered by the world model and planning system. Similarly, when asked "Tell me what you've sensed," the robot responded with the information in the semantic layer of the world model.

In the second experiment, questions were asked that required the robot to intelligently query specific elements of the world model and planning system. Possible questions ranged from specific (e.g., "Why did you turn left going out of that room?") to general (e.g., "What are you doing?"). The robot used the ERA system and world model to respond semantically, e.g., "I am in 'searching for people' mode, heading to my goal, which is room 20 at [5.34, 20.11]." (N.B., in addition to supporting human teammates, this capability proved useful for developers producing and debugging the system and simulator.)

5. CONCLUSION AND FUTURE WORK

We have presented an approach for enabling a robot to explain its behavior in human-understandable concepts. Thus far, we have designed and developed a prototype system and algorithms that intelligently query semantic information from our layered world model and construct answers to a set of questions. While proof-of-concept experiments have shown the merits of this, future work includes:

1. Supporting more specific answers. Thus far, the responses from the ERA system have focused primarily on the robot's mode and world information. We have developed but not implemented algorithms that examine changes in the grids used to produce the cost map. These algorithms look for the most influential cells on the robot's route and determine what grid(s) correspond to the values of these cells. By understanding how those grids are constructed, we can determine what world information is most strongly affecting the robot's behavior.

2. Examining robot routes, not just modes or grids. We believe that examining multiple possible routes for the robot might provide a basis for explaining a route choice (e.g., "I took the left road because the right road is longer."). To do this, we plan to generate alternate paths and compare them.

3. Increasing the flexibility of the queries and responses. Humans tend to think of actions discretely (e.g., "left turn"), but the continuous nature of robotic motion means that these discrete actions are not easily associated with timestamps in the robot's operation. By integrating with a natural language understanding system, we plan to use context from the questions to pair a semantically described action with robotic motion.

6. REFERENCES

[1] Duez, P.P. and Zuliani, M.J. and Jamieson, G.A., "Trust by design: Information requirements for appropriate trust in automation." In *Proceedings of the 2006 Conference of the Center for Adv. Studies on Collaborative Research.* 2006.

[2] Lee, J.D. and See, K.A., "Trust in automation: Designing for appropriate reliance." *Human Factors*, Vol. 46, No. 1, Spring 2004, pp. 50-80.

[3] Takayama, L., Dooley, D., & Ju, W., "Expressing thought: Improving robot readability with animation principles." *Proceedings of the 6th International Conference on Human-Robot Interaction*, pp. 69–76. Lausanne, Switzerland. 2011.

[4] Core, M., Lane, H., Van Lent, M., Gomboc, D., Solomon, S., and Rosenberg, M., "Building explainable artificial intelligence systems." In *Proceedings of the National Conference on Artificial Intelligence*, volume 21. 2006.

[5] Brooks, D., Shultz, A., Desai, M., Kovac, P., and Yanco, H.A., "Towards state summarization for autonomous robots." In *Dialog with Robots: Papers from the AAAI Fall Symposium (FS-10-05).* 2010.

[6] Lomas, M., Cross, E., Darvill, J., Garrett, R., Kopack, M., and Whitebread, K., "A robotic world model framework designed to facilitate human-robot communication." *Proceedings of the SIGdial 2011 Conference*, Portland, OR. 2011.

Applying Politeness Maxims in Social Robotics Polite Dialogue

Looi Qin En
Hwa Chong Institution
661 Bukit Timah Road
Singapore 269734
+65 6468 3955
stuqel@i2r.a-star.edu.sg

See Swee Lan
Institute for Infocomm Research, A*STAR
1 Fusionopolis Way
Singapore 138632
+65 6408 2000
slsee@i2r.a-star.edu.sg

ABSTRACT
An important element of human-robot interaction, as with inter-human interaction, is conversation. Having previously suggested the Gricean maxims as suitable guidelines for social robotics dialogue, we discovered that a preferable alternative set of guidelines: politeness maxims. In this paper, we will introduce the politeness maxims and propose its enhanced applicability in human-robot interaction to create polite dialogue. Although no experimental results are available to support our proposition, the preliminary analysis of politeness maxims presents a promising future, as guidelines for the synthesis of polite robot dialogue. Through effective dialogue creation, the interaction with human and robots will be more pleasant, indicating a step forward in effective human-robot interaction.

Categories and Subject Descriptors
H.1.2 [**Models and Principles**]: User/Machine Systems – *human factors, human information processing, software psychology.*

General Terms
Human Factors, Languages

Keywords
Social Robotics, Polite Dialogue, Politeness Maxims, Social Context, Human-Robot Interaction, Human-Robot Communication

1. INTRODUCTION
As robots are more commonplace today than before, robots that are smaller, lighter and cheaper are able to handle complex tasks and exhibit varied behavior [1]. The move towards creating robots that respond to different contexts and different users has led to the earlier discussion of the applicability of Gricean maxims in social robotics dialogue [2].

However, upon further study, the Gricean maxims were found to be rigid and did not provide accurate guidelines which suited different contexts. In addition, previous studies have affirmed that Gricean maxims could not be universally applied because of culturally-determined social behavior [3]; conversation is an integral form of social behavior, and it is culturally determined. Gricean maxims do not account for these differences. It was demonstrated in [3] that the Malagasy people followed maxims which were contradictory to Gricean maxims during conversation.

Thus, we decided to study another set of conversational maxims, termed politeness maxims [4, 5], and their applicability in social robotics dialogue. Introduced as an alternative to Gricean maxims by Geoffrey Leech, the maxims are tend to be more applicable across different cultures. In a study of politeness in Chinese culture, Leech's maxims attempt to provide a missing link between the Gricean cooperative principle and the problem of how to relate sense to force [6].

In this paper, we will first discuss the politeness maxims Leech suggests and its pragmatics, followed by how dialog can be improved between humans and robots using these maxims.

2. POLITENESS MAXIMS
Politeness maxims were created by Geoffrey Leech, and they vary across cultures. Certain conversational behavior may be considered polite in a culture but queer or impolite in other cultures.

1. Tact Maxim: Avoid expressing detrimental beliefs to others, especially when it may cause harm. Instead, express beneficial beliefs to others. A demonstration of the maxim is given in [7], where the imposition on others is minimized and they feel less obliged, whilst others' interests, wants and needs are prioritized in the conversation.

2. Generosity Maxim: Beliefs that express or imply benefit to the individual himself should be minimized, whilst beliefs that express or imply benefit to others should be maximized. This maxim, unlike the Tact Maxim, is focused on the speaker and is a call for altruism.

3. Approbation Maxim: Avoid expressing negative criticism while generously praising others. This maxim serves a two-pronged purpose: firstly, to avoid disagreement and secondly, to flatter others through solidarity.

4. Modesty Maxim: Beliefs that express of imply self-praise should be minimized whilst beliefs that express dispraise of self should be emphasized.

5. Agreement Maxim: Avoid disagreements between the conversing individuals and try to strive for agreements.

6. Sympathy Maxim: Sympathy between the self and others should be achieved instead of antipathy.

HRI'12, March 5–8, 2012, Boston, Massachusetts, USA.
ACM 978-1-4503-1063-5/12/03.

Essentially, the politeness maxims convey a sense of modesty, and the speaker should be docile and submissive in order to be polite. The underlying principle of these maxims is the attitude the speaker has to the listeners'; if s/he is proud and arrogant, s/he will tend to prioritize his/her needs, wants and interests over the others and self-praise. Conversely, if the speaker is docile and modest, s/he will strive for agreements and avoid disagreements.

This is the most important difference between politeness maxims and Gricean maxims. The latter is focused on purpose of conversation; as suggested in [2], when the purpose of conversation is clear and attempts are made to achieve the purpose of the conversation, the Gricean maxims are followed. This results in a structured and less personal way of conversation that explain its limited application across cultures. Politeness maxims, however, address the subjective and personal influences on conversation; i.e. the attitude of the speaker. Thus, politeness maxims are arguably more effective guidelines, for it is more conceivable for Gricean maxims to be followed, yet impolite and rude conversation occurs, when the speaker is not altruistic.

3. IMPLICATIONS FOR HUMAN-ROBOT INTERACTION

As discussed earlier, politeness maxims is more thorough than Gricean maxims in understanding human conversation. Though the maxims were developed for inter-human interaction, they can be applied to human-robot interaction and more specifically, human-robot conversation.

We suggested for clear objectives in human-robot interaction in [2], but realized that this was already in effect and of little benefit towards enhancing human-robot conversation. A robot can achieve the goal of conversation with the human without engaging the human in polite dialogue. For example, a customer service robot in a fast-food restaurant (preliminary study of this scenario was conducted in [2]) can successfully take the customer's order in a direct conversation. According to Gricean maxims, the robot simply needs to start the conversation with "What do you want?" for such a question is not only direct but also achieves the objective of the conversation (to take the customer's order).

However, we find that the engagement could be thoroughly enhanced if the politeness maxims, specifically, the approbation maxim were applied to the situation. The robot could compliment the customer before taking the order: "Good Morning Sir. You look great today. Thank you for visiting (the fast food restaurant). May I know what would you like?" Evidently, the objective of the conversation is 'delayed' as the question suggesting the objective is said later, thus flouting the Gricean maxims, but it is arguable that the customer would feel more respected and well-treated.

We propose the reason for the enhanced effectiveness of politeness maxims in achieving engaging and polite social dialogue to be the ability to connect with humans emotionally. Human-robot dialogue is stereotyped to be monotonous and inhumane, thus if the politeness maxims can be successfully implemented in human-robot interaction, it is probably that engagement between human and robot will be improved.

4. CONCLUSIONS

Although the findings in this paper are preliminary and unsupported by experimental results, through the comparison of Gricean maxims with politeness maxims, we have discovered an important factor in the synthesis of polite social dialogue between humans and robots: the ability to connect with humans emotionally. Gricean maxims may be useful as general guidelines for creation of polite maxim dialogue, but politeness maxims are more useful when developing customized polite dialogue for specific instances.

Politeness maxims, though created for inter-human conversation, have great potential in the synthesis of polite robot dialogue with humans; if successfully applied, robots can engage in more personalized conversation with humans. As compared to Gricean maxims which drives conversation based on its objective and does not improve human-robot engagement, politeness maxims can significantly contribute to better human-robot interaction.

Future work will involve conducting experimental user studies on interaction between robots and humans to determine if politeness maxims do enhance human-robot interaction and appeal to humans better.

5. ACKNOWLEDGMENTS

The authors would like to thank Tristan Ang, Shawn Tan and Li Bing Qing for suggesting the research directions for this paper, as well as the staff of the Institute for Infocomm Research who has rendered invaluable assistance.

6. REFERENCES

[1] Arsenio, A. M. *Towards pervasive robotics*. Artificial Intelligence Lab—Massachusetts Institute of Technology. Available at http://groups.csail.mit.edu/lbr/hrg/2003/181.pdf

[2] Looi, Q. E. and See, S. L. 2011. The Applicability of Gricean Maxims in Social Robotics Polite Dialogue. In *Proceedings of the 6th International Conference on Human-robot interaction* (Lausanne, Switzerland). HRI'11. ACM, New York, NY, pp. 195-196. DOI= http://doi.acm.org/10.1145/1957656.1957728.

[3] Keenan, E. O..1976. *On the universality of conversational implicatures*. In Language and Society, 5. pp. 67-80.

[4] Leech, G. N. 1977. Language and tact. LAUT Series A Paper 46. University of Trier

[5] Leech, G. N. 1983. Principles of pragmatics. London/New York: Longman

[6] Gu, Y. 1990. Politeness phenomena in modern Chinese. In *Journal of Pragmatics*, Volume 15, Issue 2, pp. 237-257. Elsevier.

[7] Brown, P. and Levinson, S. 1987. Politeness: Some Universals in Language. Cambridge: Cambridge University Press.

Transfer from a Simulation Environment to a Live Robotic Environment: Are Certain Demographics Better?

Patricia L. McDermott, Alia Fisher, Thomas Carolan,
Mark R. Gronowski, Marc Gacy, and Michael Overstreet

Alion Science and Technology
4949 Pearl East Circle, Suite 200
011-303-442-6947
pmcdermott; afisher; tcarolan; agacy; mgronowski@alionscience.com; michael.overstreet@colorado.edu

ABSTRACT
The ability to remotely operate an unmanned vehicle while simultaneously looking for suspicious targets and then classifying those targets is not a trivial skill. This study looked at different training approaches to make better use of simulation as a first training step. When transferring to a live environment, the operators could be grouped into two categories according to whether they passed live training criteria or not. There were clear performance differences between these groups. The group that failed to pass criteria had poorer performance overall, more SA errors, and spent more time in training. Post-hoc analysis showed differences in the demographics between those who passed and those that did not. Male participants and younger participants were more likely to achieve criteria. There were no differences in gaming experience and perceived sense of direction.

Categories and Subject Descriptors
K.6.1 [**Management of Computing and Information Systems**]: Project and People Management – *training*.

General Terms
Measurement, Performance, Experimentation, Human Factors.

Keywords
Situation awareness, learning curves, demographics, sense of direction, teleoperation

1. INTRODUCTION
The purpose of this study was to investigate the usefulness of different training approaches for operating unmanned ground vehicles. Access to a live robot and testing environment can be scarce and therefore operators perform initial training via simulator. We focused on part-task training methods which would be useful in situations where the trainer does not have access to a single system that performs all aspects of the task to be learned.

This study investigated how well training in a simulated environment prepared operators for the live task with a real robot. This particular paper focuses on the differences between those who were able to achieve criterion in the live environment and those who failed to pass criteria.

HRI'12, March 5–8, 2012, Boston, Massachusetts, USA.
ACM 978-1-4503-1063-5/12/03.

2. METHODS
Training was a between-subjects variable with four levels. The levels specified whether the simulation and the live training consisted of part-task training, whole-task training or both. The conditions were: part task (PT) training via separate system simulations followed by PT and whole task (WT) training in the live environment; PT training via simulation and WT training in the live environment; PT and WT training via simulation and WT training in the live environment; and WT training in both the simulation and live environment. Post hoc, participants were separated into two groups: those that achieved the live criteria and those that failed to achieve criteria.

Thirty-nine volunteers between the ages of 18 and 39 (M = 27.13 SD = 5.57) completed the experiment, 24 males and 15 females. One participant that did not pass criteria *in simulator training* was excluded from the analysis.

The Soldier Machine Interface was used as the primary experimental testbed. Another software application presented part-task training for detection and identification of vehicles. For the live session, the unmanned vehicle was a Traxxas E-Maxx radio controlled car equipped with three network cameras. See McDermott and Fisher [3] for a full description of the robot and environment in a previous experiment. The Santa Barbara Sense of Direction Scale [2] was used to solicit self-ratings of spatial ability. The scale has been used in studies with of ground robotics operators in military reconnaissance missions [1] and the ability to learn routes with interference [4].

Participants completed training and practice scenarios in the simulation environment. All participants achieved criteria before attempting the live scenarios one week later. Participants remotely operated an unmanned ground vehicle along a predetermined route in an urban setting to look for enemy activity in the area. In order to pass, participants had to complete a scenario with zero wrong turns, zero collisions, 100% accuracy in detecting vehicles, 100% accuracy in identifying those vehicles (i.e., friendly, enemy, neutral), and within a time limit. After each scenario, participants completed a situation awareness (SA) record sheet in which they drew their route and marked the location of all vehicles encountered. The errors were weighted and combined to develop an overall SA score per scenario.

3. RESULTS
There were no differences among conditions in the number of participants who passed criteria. The ten participants that did not meet criterion were dispersed among the four training conditions.

Participants who failed to meet criteria performed poorer on the first live scenario than those who achieved criteria. A one-way analysis of variance on the differences between participants who met performance criteria in the live environment indicated a significant difference for trial 1 time, $F(1,37) = 4.80$, $p = 0.035$, Trial 1 collisions, $F(1,37) = 4.26$, $p = 0.047$, and trial 1 wrong turns $F(1,37) = 6.83$, $p = 0.013$. Those who failed to pass criteria also had more erratic learning patterns than those who did. For example, compare the mean learning curves for identification of vehicles (by training condition) of those who passed in Figure 1 to those in Figure 2 that did not pass.

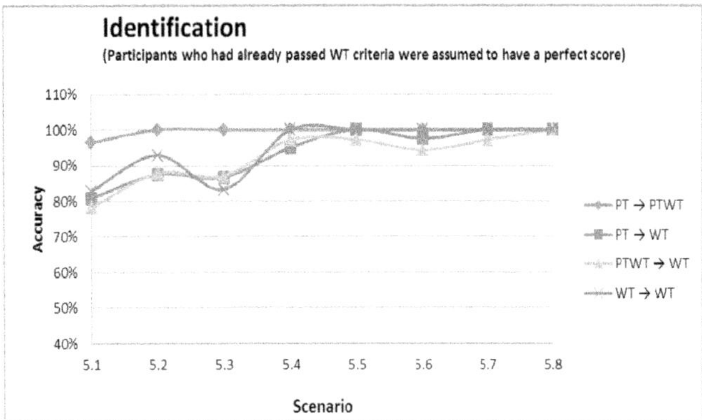

Figure 1. Learning curves for those who achieved criteria

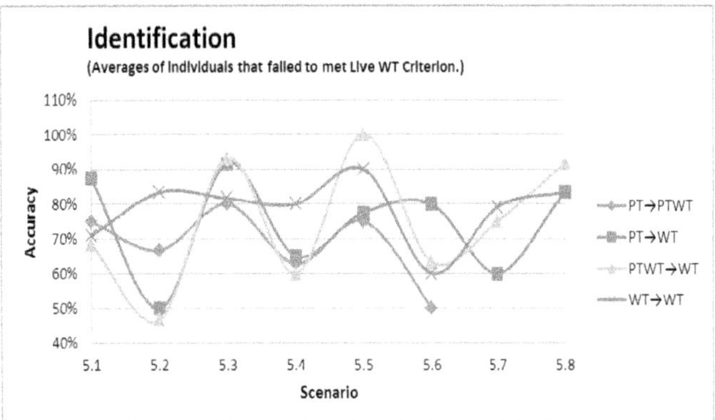

Figure 2. Learning curves for those who failed to achieve criteria

Those who failed to pass also had significantly more SA Errors, $F(1,37) = 14.10$, $p = 0.001$. Those who failed to pass had a mean of 6.83 SA errors and those who passed had a mean of 4.00 SA errors. As would be expected, those who failed to pass criteria spent a significantly longer time in live training, $F(1,37) = 29.01$, $p < 0.0001$. Those who failed to pass took almost twice as long (a mean of 29 minutes, 37 seconds versus 15 minutes, 40 seconds).

The demographics between the two groups were analyzed to determine if there were characteristics associated with those who failed to achieve criteria. There was a significant negative correlation between age and meeting criteria ($r = -0.338$, $p = 0.044$) and a significant positive correlation between gender and meeting criteria ($r = +0.608$, $p < 0.001$). This indicates that younger participants and male participants were more likely to meet performance criteria. There were no significant differences associated with video game experience or the Santa Barbara Sense of Direction Scale scores. Subsequent Analysis of Covariance did not yield significant differences due to training conditions.

4. DISCUSSION

There was a consistent pattern of poorer performance for those who failed to achieve criteria. It was evident in their first trial, learning curves, and overall time spent in training. It is interesting to note that the SA errors also followed the same pattern as the performance data. The same operators that struggled to achieve performance criteria were less able to maintain awareness of where the unmanned vehicle traveled and what vehicles were observed. It could be that they were so focused on performing the task accurately that they did not have mental capacity to devote to maintaining awareness of critical locations.

The participants who passed criteria were more likely to be male and to be younger. It is interesting to note which demographics did *not* differ between the two groups: video game experience and self-reported sense of direction. Future studies will include spatial ability tests in addition to the self report data.

A key point from the study is that poor task performance was associated with poor situation awareness, underscoring the need to measure both when assessing systems or operators. Another implication is that females and older operators may take longer to learn tasks to criteria, regardless of part-task training approach.

5. ACKNOWLEDGMENTS

This work is supported by the US Army Research Institute under Contract #: W91WAW-09-C-0081 titled Understanding the Impact of Training on Performance. The view, opinions, and/or findings contained in this paper are those of the author (s) and should not be construed as an official Department of the Army position, policy, or decision.

6. REFERENCES

[1] Chen, J.Y.C, (2010). Effects of operator spatial ability on uav-guided ground navigation. Proceeding of the 5th ACM/IEEE international conference on Human-robot interaction, 139-140. ACM, New York, NY.

[2] Hegarty, M., Richardson, A. E., Montello, D. R., Lovelace, K., & Subbiah, I. (2002). Development of a self-report measure of environmental spatial ability. *Intelligence, 30*, 425-447.

[3] McDermott, P. L. & Fisher, A. (2009). *The Tradeoff of Frame Rate and Resolution in a Route Clearing Task: Implications for Human-Robot Interaction*. Proceedings of the 53rd Annual Meeting of the Human Factors and Ergonomics Society, Santa Monica, CA.

[4] Wen, W., Ishikawa, T., & Sato, T. (2011). Working memory in spatial knowledge acquisition: Differences in encoding processes and sense of direction. Applied Cognitive Psychology, 25 (4), pp. 654-662.

A Probabilistic Framework for Autonomous Proxemic Control in Situated and Mobile Human-Robot Interaction

Ross Mead
University of Southern California
3710 McClintock Ave., RTH 423
Los Angeles, CA 90089-0781
(213)-740-6245
rossmead@usc.edu

Maja J Matarić
University of Southern California
3650 McClintock Ave., OHE 200
Los Angeles, CA 90089-1450
(213)-740-4520
mataric@usc.edu

ABSTRACT
In this paper, we draw upon insights gained in our previous work on human-human proxemic behavior analysis to develop a novel method for human-robot proxemic behavior production. A probabilistic framework for spatial interaction has been developed that considers the sensory experience of each agent (human or robot) in a co-present social encounter. In this preliminary work, a robot attempts to maintain a set of human body features in its camera field-of-view. This methodology addresses the functional aspects of proxemic behavior in human-robot interaction, and provides an elegant connection between previous approaches.

Categories and Subject Descriptors
I.2.9 [**Artificial Intelligence**]: Robotics – *operator interfaces.* H.1.2 [**Models and Principles**]: User/Machine Systems.

General Terms
Algorithms, Design, Experimentation, Human Factors, Theory.

Keywords
Human-robot interaction, proxemics, probabilistic models.

1. INTRODUCTION AND BACKGROUND
Proxemics is the study of the interpretation, manipulation, and dynamics of spatial behavior in co-present social encounters. Edward T. Hall coined this term, and proposed that psychophysical factors shaped by culture defined zones of intimate, personal, social, and public space [1]; these proxemic zones are characterized by the visual, auditory (voice loudness), olfactory, thermal, touch, and kinesthetic experiences of each interacting participant [2]. To facilitate situated and mobile human-robot interaction (HRI), we seek to develop functional and socially appropriate probabilistic models of proxemic behavior that can be implemented as controllers for sociable robots.

A number of rule-based spatial interaction controllers have been investigated in HRI [3, 4]. Interpersonal dynamic models, such as equilibrium theory [5], have also been evaluated with robots [6, 7]. Contemporary probabilistic modeling techniques have been applied to person-aware robot navigation in dynamic crowded environments [8], to goal-oriented navigation behavior with a human partner [9], and to calculating a robot approach trajectory to initiate interaction with a walking person [10].

HRI'12, March 5–8, 2012, Boston, Massachusetts, USA.
ACM 978-1-4503-1063-5/12/03.

In previous work, we took advantage of recent advancements in markerless motion capture (specifically, the Microsoft Kinect) to automatically extract low-level proxemics-relevant features based on metrics from the social sciences [11]. We used these features to recognize high-level spatiotemporal interaction behaviors, such as the initiation, acceptance, and termination of an interaction [12]. These investigations provided insights into the development of situated spatial interaction controllers for autonomous sociable robots; specifically, they (and related literature [2]) suggest an alternative approach to proxemic behavior that goes beyond simple distance and orientation by considering the sensory experience of each agent (human or robot) in a social encounter.

2. APPROACH
We propose a sort of mid-level probabilistic framework—between low-level spatial features [3, 4] and high-level interagent relationship features [6, 7]—that defines proxemic behavior based on how an agent will likely experience social stimuli (e.g., speech and gesture) in a co-present interaction. Previous approaches that addressed low-level spatial features are still relevant [3, 4], providing variables that are observable; specifically, we utilize interagent distance (d), the angle from the robot to the person (α), and the angle from the person to the robot (β) to generate priors—$p(d, \alpha, \beta)$.

Our probabilistic framework for proxemic behavior considers a set of desired (often latent) sensory features—$F = \{f_1, f_2, \ldots, f_n\}$; where f_i is the ith of n sensory features—to be experienced by an agent. These sensory features are conditioned over aforementioned spatial factors—$p(F \mid d, \alpha, \beta)$. We can then sample over possible agent poses (positions and orientations) in an environment, using simple Bayesian Inference to evaluate the posterior probability of the interagent distance and orientations conditioned on F (Equation 1). The pose with the maximum likelihood is selected as the location to conduct the interaction.

$$p(d,\alpha,\beta|F) = \eta \times p(F\ d,\alpha,\beta) \times p(d,\alpha,\beta)\ ;\ \eta = p(F)^{-1} \quad (1)$$

Based on this probabilistic framework, high-level interagent relationship features of proxemic behavior (e.g., amount of eye gaze [6, 7] or intimacy of topic [6]) can then define the desired sensory experience over social stimuli for each interacting agent.

3. IMPLEMENTATION
A proof-of-concept real-time proxemic controller has been developed that implements the proposed probabilistic framework. For this work, we utilized the PR2 robot from Willow Garage. The PR2 features a semi-holonomic mobile base, which affords it the ability to strafe, a valuable motion for proxemic control.

We first developed a simple rule-based dyadic (two-agent) proxemic controller that accepts the following input signals: desired and actual interagent distances (D and d, respectively); desired and actual angle from the robot to the person (A and α, respectively), and the desired and actual angle from the person to the robot (B and β, respectively, respectively). The controller attempts to minimize the error between each of the desired and actual distance and angle signals; the resulting linear velocities (v_x, v_y) and angular velocity (v_ω) are expressed as follows:

$$\begin{pmatrix} v_x \\ v_y \\ v_\omega \end{pmatrix} = \begin{pmatrix} \cos(\alpha) & -\sin(\alpha) & 0 \\ \sin(\alpha) & \cos(\alpha) & 0 \\ 0 & 0 & 1 \end{pmatrix} \begin{pmatrix} (d-D)\times\cos(\alpha-A) \\ \sin(\beta-B) \\ A-\alpha \end{pmatrix} \quad (2)$$

We then developed a higher-level spatial interaction controller based on our probabilistic framework of proxemic behavior. In this preliminary work, our feature set F contains the names of human joints (e.g., "head", "left shoulder", "right hand", "torso", etc.) to be detected by the Microsoft Kinect (Figure 1) [11]. This feature set was selected for its applicability in recognizing gesture parameters, such as *gesture locus* (the regions of the body involved by a gesture) [13]. A naïve model of $p(F \mid d, \alpha, \beta)$ has been established for a small group of users for a proof-of-concept; we are developing a larger study to better model these data.

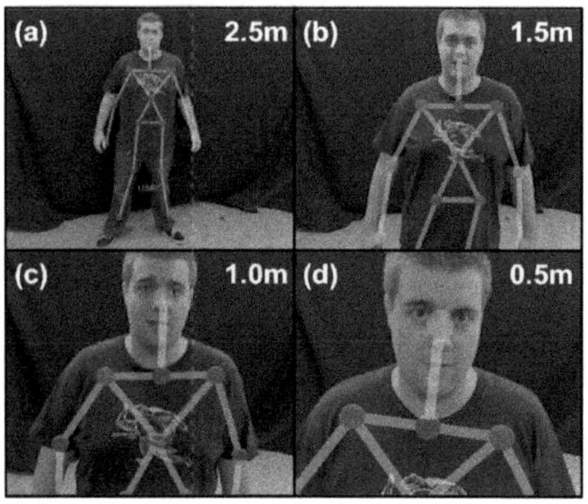

Figure 1. Body features that appear in the Kinect field-of-view at four distances: (a) 2.5m, (b) 1.5m, (c) 1.0m, and (d) 0.5m.

To sample robot poses, we discretize the world using a grid-based approach, where each grid cell serves as a pose (though, in future work, we will utilize a more efficient particle-based method). The pose that maximizes Equation 1 is used to determine the desired interagent distance and orientation parameters (D, A, B), which are sent to the rule-based proxemic controller (above) to generate the appropriate linear and angular velocities (Equation 2).

4. DISCUSSION AND FUTURE WORK

We have proposed and implemented a probabilistic framework for proxemic behavior production in HRI. This approach captures the functional aspects of spatial interaction by considering the sensory experience (in this case, human body features) of interacting agents, and sits as a mid-level representation between low-level spatial features [3, 4] and high-level interagent relationship features [6, 7]. A full data collection is underway to better model the desired features based on spatial parameters. In future work, we will consider using a coupled Dynamic Bayesian Network to improve goal pose estimation and to factor in the perspective of the human into robot control. All code will be part of the Social Behavior Library (SBL) in the USC ROS Packages Repository (http://sourceforge.net/projects/usc-ros-pkg/).

5. ACKNOWLEDGMENTS

This work is supported in part by an NSF Graduate Research Fellowship, as well as ONR MURI N00014-09-1-1031 and NSF CNS-0709296 grants, and the Willow Garage PR2 Beta Program.

6. REFERENCES

[1] Hall, E. T. 1966. *The Hidden Dimension*. Doubleday, Garden City, NY.

[2] Hall, E.T. 1963. A system for notation of proxemic behavior. *American Anthropologist*, 65, 1003-1026.

[3] Shi, C., Shimada, M., Kanda, T., Ishiguro, H., and Hagita, N. 2011. Spatial formation model for initiating conversation. *Proc. of Robotics: Science and Systems (RSS 2011)*. Los Angeles, CA.

[4] Kuzuoka, H., Suzuki, Y., Yamashita, J., and Yamazaki, K. 2010. Reconfiguring spatial formation arrangement by robot body orientation. *Proc. of the 5th ACM/IEEE Int'l. Conf. on Human-Robot Interaction (HRI 2010),* Osaka, Japan, 285-292.

[5] Argyle, M. and Dean, J. 1965. Eye-contact, distance and affiliation. *Sociometry*, 28, 3, American Sociological Association, 289-304.

[6] Mumm, J. and Mutlu, B. 2011. Human-robot proxemics: physical and psychological distancing in human-robot interaction. *Proc. of the 6th Int'l. Conf. on Human-Robot Interaction (HRI 2011)*, Lausanne, Switzerland, 331-338.

[7] Takayama, L. and Pantofaru, C. 2009. Influences on proxemic behaviors in human-robot interaction. *IEEE/RSJ Int'l. Conf. on Intelligent Robots and Systems (IROS 2009)*, St. Louis, MO, 5495-5502.

[8] Trautman, P. and Krause, A. 2010. Unfreezing the robot: navigation in dense, interacting crowds. *Proc. of the IEEE/RSJ Int'l. Conf. on Intelligent Robots and Systems (IROS 2010)*, Taipei, Taiwan, 797-803.

[9] Feil-Seifer, D. and Matarić, M. 2011. People-aware navigation for goal-oriented behavior involving a human partner. *Proc. of the IEEE Int'l. Conf. on Development and Learning (ICDL 2011)*, 2, Frankfurt Am Main, 1-6.

[10] Satake, S., Kanda, T., Glas, D. F., Imai, M., Ishiguro, H., and Hagita, N. 2009. How to approach humans?: strategies for social robots to initiate interaction. *Proc. of the 4th ACM/IEEE Int'l. Conf. on Human Robot Interaction (HRI 2009)*, San Diego, CA, 109-116.

[11] Mead, R., Atrash, A., and Matarić, M. J. 2011. Proxemic feature recognition for interactive robots: automating metrics from the social sciences. *Proc. of the 2011 Int'l. Conf. on Social Robotics (ICSR 2011)*, Amsterdam, Netherlands, 52-61.

[12] Mead, R., Atrash, A., and Matarić, M.J. 2011 Recognition of spatial dynamics for predicting social interaction. *Proc. of the 6th ACM/IEEE Int'l. Conf. on Human-Robot Interaction (HRI 2011)*, Lausanne, Switzerland, 201-202.

[13] Rossini, N. 2004. The analysis of gesture: establishing a set of parameters. *Gesture-based Communication in Human-Computer Interaction, Lecture Notes in Computer Science*, 2915, Springer Berlin/Heidelberg, 463-464.

Control of Human-Machine Interaction for Wide Area Search Munitions in the Presence of Target Uncertainty*

Pia E. K. Berg-Yuen[a]
Munitions Directorate
Air Force Research Laboratory
Eglin AFB, FL-32542, USA

Siddhartha S. Mehta[b†]
University of Florida - REEF
Shalimar, FL-32579, USA

Eduardo L. Pasiliao[c], Robert. A. Murphey[d]
Munitions Directorate
Air Force Research Laboratory
Eglin AFB, FL-32542, USA

ABSTRACT

In this report we describe the progress in developing a control architecture for human-in-the-loop wide area search munitions to reduce operator errors in the presence of unreliable automation and operator cognitive limitations. An optimal input tracking controller with adaptive automation uncertainty compensation and real-time workload assessment is developed to improve the system performance. Extensive simulations based on the experimental data involving 12 subjects demonstrate effectiveness of the presented controller.

Categories and Subject Descriptors

I.2.8 [**Artificial Intelligence**]: Problem Solving, Control Methods, and Search—*Control theory*; J.7 [**Computer Applications**]: Computers in Other Systems—*Military*

General Terms

Theory, Verification

Keywords

Control architecture, automation reliability

1. INTRODUCTION

The defense community is envisioning enhanced capabilities by introducing greater degree of munitions autonomy and the ability to strike multiple targets simultaneously from a single air platform in dynamic environments. One such complex system, the Wide Area Search Munition (WASM), combines the features of an unmanned aerial vehicle with those of traditional munition systems [7]. Bainbridge in the "Ironies of automation" [1] suggests that with increase in the complexity of a control system the need of a human operator would be more crucial and at the same time these complex systems can place a high demand on an operator's cognitive load. Automation reliability is also shown to have a widespread effect on operator attitudes, e.g., trust and behavior. It has been observed in [4] that unreliable automation along with high cognitive load can deteriorate the overall system performance significantly. Optimal system performance requires that strengths and limitations of both the human operator and the automation be taken into account when designing a control system. Recent solutions that include cognitive workload in the human-machine system design can be found in [3,5,6], where *a priori* known workload-performance or workload-task model is used to achieve the desired system performance by affecting the machine state.

*This research is supported in part by a grant from the US Air Force, Eglin AFB, grant FA8651-08-D-0108/025.
Email: [a]Pia.Berg-Yuen.ctr@eglin.af.mil, [b]siddhart@ufl.edu, [c]eduardo.pasiliao@eglin.af.mil, [d]robert.murphey@eglin.af.mil
[†]Corresponding author: (850)833-9350, Fax (850)833-9366

HRI'12, March 5–8, 2012, Boston, Massachusetts, USA.
ACM 978-1-4503-1063-5/12/03.

2. OBJECTIVE & METHOD

The objective of the presented work is to develop a control architecture for human-machine interaction (HMI) system that not only takes into account the *real-time operator cognitive workload* assessment but also the uncertainties associated with *automation reliability*. Specifically, we present a discrete-time Markovian process model and develop an event-based switching controller with adaptive target estimation to compensate for the automation errors.

2.1 System Model

The presented system consists of a set of WASMs (4, 8, or 16) traveling autonomously along the known desired trajectories. The autonomous operation of WASMs involves *sensing* the environment, *classifying* the perceived target as high-interest, low-interest, or false unit, and *communicating* the target information to a remotely situated operator. With respect to the received information, the operator performs a set of actions that can be broadly categorized as: 1) target labeling - marking a target as strike, no strike, or misidentification based on the operator's perception of target and classification accuracy, 2) target attack - requesting and assigning a weapon (WASM) to each target, and authorizing the attack, and 3) battle damage assessment - verifying the effect of attack on a target and poststrike reporting.

The model analyzed for the presented problem is a parallel Markov decision process (MDP) such that a new process is initialized for each target identified by a group of WASMs. At each state $s_i \in \mathcal{S} \subset \mathbb{R}$ there exists a finite set of actions $a_i \in \mathcal{A} \subset \mathbb{R}$ to choose from such that the transition $s_i \to s_i^+$ occurs when an action a_i is taken in s_i with the cost of action $c_i(s_i^+, s_i, a_i)$, where $i = 1, 2, \ldots, n$ denotes the number of MDPs (targets). Also, since the presented process is fully observable it can be modeled as a MDP.

2.1.1 *Target Uncertainty*

There exists an uncertainty in the classified target causing false alarms, missed targets, and elevating the risk of collateral damage. Target uncertainty is a result of limited efficiency of the pattern classification and image processing methods used by WASMs to classify a target based on the captured images. The uncertainty poses a challenge in determining an optimal action policy for a given MDP since the cost of action as well as the terminal state is function of the target type. We propose a novel sequential probability ratio test (SPRT) to adaptively estimate the target type using the actions taken by an operator with respect to the initial classification provided by an automated system. The estimates are obtained by sequential hypothesis testing where the null and alternate hypotheses are functions of the classification type I and type II errors. Thereby, the proposed method exploits superior human cognitive abilities to detect and rectify target uncertainty caused by the machine.

2.1.2 *Cognitive State Indicator (CSI)*

Real-time workload assessment is crucial to provide the operator adaptive-aiding when warranted by task demands. Cannon et al. [2] introduced an approach to detection of temporal changes in operator cognitive state based on monitoring the trends of CSI. CSI is a time-varying parameter obtained in real-time by projecting multidimensional EEG and EOG signals onto a continuous range by maximizing the Kullback-Leibler (KL) distance between distributions of the signals. CSI, denoted by $\mu(t) \in [0,1]$, indicates the degree to which the operator is engaged. A high cognitive load is identified when the rate of change of CSI exceeds an *a priori* determined threshold $\phi \in \mathbb{R}$, i.e., $\dot{\mu}(t) \geq \phi$.

3. CONTROLLER DEVELOPMENT

The control objective is to *assist* an operator in decision making to improve the performance of a closed-loop HMI system by minimizing the total cost of mission and the number of interactions or mission length. The control objective is achieved in two stages by 1) developing an optimal controller based on the above performance criterion and 2) tracking operator actions along the optimal control sequence obtained in stage 1 at each discrete event time k.

A receding horizon control problem is formulated to determine an optimal control sequence over the finite horizon of length N for an ith process in state s_i with terminal state s_i^t determined using SPRT-based target estimation. A new optimal control sequence is computed from each evolved state s_i^+ to s_i^t, thereby compensating for exogenous disturbances such as operator errors, high cognitive load etc. A cost functional J_i comprising of the cost of action $c_i(s_i^+, s_i, a_i)$ and number of interactions for each process is defined as

$$J_i \triangleq \alpha\lambda_i + \sum_{k=1}^{N} \beta c_i(s_i^+(k), s_i(k), a_i(k))$$

$$s_i(1) = s_i, s_i(N+1) = s_i^t, s_i^+(k), s_i(k) \in \mathcal{S}, a_i(k) \in \mathcal{A} \quad (1)$$

where λ_i is the number of interactions required to reach s_i^t and α, $\beta \in \mathbb{R}$ are positive weights. The optimal action at time k is obtained as the first step in the control sequence that minimizes the cost functional given by $u^*(k) = a_i^*(1)$.

The controller $a_i^*(1)$ provides an optimal action that minimizes J_i at each time k, however the operator action $a_i(k)$ may not be optimal. Therefore, we propose a tracking controller to maintain the operator action sequence along the desired optimal control sequence at time k. The tracking error is defined as $e(k) \triangleq a_i(k) - a_i^*(1)$ such that when $e(k) \neq 0$ the operator is notified with a corrective action, where the operator has the authority to accept or reject the corrective action. However in the presence of high cognitive workload, i.e., $\dot{\mu}(k) \geq \phi$, the control authority is switched to the machine and an optimal control $u^*(k)$ is applied until the operator cognitive load reduces such that $\dot{\mu}(k) < \phi$.

4. VALIDATION & CONCLUSIONS

Simulation results based on 36 experiments involving 12 different subjects are presented to verify the performance of the proposed controller. The psychophysiological data collected during each trial consisted of ECG, EEG (F7, Fz, Pz, T5, and O2), and EOG channels. Statistical model of operator behavior is obtained for different target types and the CSI is modeled as a complex Gaussian process using the available empirical data. The environment consisted of a group of 4 WASMs modeled using constant altitude unicycle dynamics and 6 stationary targets (3 high-interest, 2 low-interest, and 1 false unit) including a false unit misclassified as a high-interest target, i.e., unreliable automation. Average and standard deviation of the total mission cost for 1000 Monte Carlo trials is shown in Table 1 for 4 simulation scenarios - using the proposed method and when the controller, operator, or CSI measure were excluded from the simulation. Observations from 1 and 2: in the absence of the proposed controller the mission cost increases by 31%; from 1 and 3: unreliable autonomous systems incur more cost however, the system performance can be improved using human supervision to estimate the uncertainties; from 1 and 4: the presented CSI-based switching controller avoids performance degradation during high cognitive load conditions. Future efforts will focus on experimental validation using the WASM testbed at Wright-Patterson AFB, Ohio.

	Contr.	Oper.	CSI	Average cost	Cost std.dev.
1	✓	✓	✓	32.78	2.25
2	✗	✓	✓	42.97, 31% ↑	2.34
3	✓	✗	✓	37.06, 13% ↑	3.25
4	✓	✓	✗	34.22, 4.5% ↑	3.35

Table 1: Mission cost comparison for 4 simulations each based on 1000 random Monte Carlo trials. [Contr.:Controller, Oper.:Operator, ✓:included, ✗:excluded]

5. REFERENCES

[1] L. Bainbridge. Ironies of automation. *Automatica*, 19(6):775–779, Nov. 1983.

[2] J. A. Cannon, P. A. Krokhmal, R. V. Lenth, and R. Murphey. An algorithm for online detection of temporal changes in operator cognitive state using real-time psychophysiological data. *Biomed. Signal Proc. and Contr.*, 5(3):229–236, 2010.

[3] M. L. Cummings and C. E. Nehme. Modeling impact of workload in netwok centric supervisory control settings. In *Annual Sustaining Performance Under Stress Symp.*, College Park, MD, February 2009.

[4] S. Dixon, C. D. Wickens, and D. Chang. Unmanned aerial vehicle flight control: false alarms versus misses. In *Human Factors and Ergo. Soc. Annual Meet.*, 2004.

[5] D. J. Kim and A. Behal. Human-in-the-loop control of an assistive robotic arm in unstructured environments for spinal cord injured users. In *Proc. of ACM/IEEE Inter. Conf. on Human Rob. Interaction*, 2009.

[6] P. Rani, N. Sarkar, and J. Adam. Anxiety-based affective communication for implicit human-machine interaction. *Adv. Engg. Informatics*, 21:323–334, 2007.

[7] L. Warfield, T. R. Carretta, M. J. Patzek, J. R. Estepp, and J. K. O'Neal. Comparing manual and cooperative control mission management methods for wide area search munitions. Technical Report AFRL-RH-WP-TR- 2010-0063, Wright-Patterson AFB, OH, 2009.

Referent Identification Process in Human-Robot Multimodal Communication

Yuta Shibasaki[1] Takahiro Inaba[2] Yukiko I. Nakano[2]

[1]Graduate School of Science and Technology, Seikei University
3-3-1 Musashino-shi, Tokyo 180-8633, Japan
dm116220@cc.seikei.ac.jp

[2]Dept. of Computer and Information Science, Seikei University
us082013@cc.seikei.ac.jp
y.nakano@st.seikei.ac.jp

ABSTRACT

This paper presents a communication robot that can generate a referent identification conversation with human users. First, we conduct an experiment to collect face-to-face referent identification communication and investigate how the referent is identified by exchanging multiple speech turns between the participants. On the basis of the experimental observations, we implement a communication robot that can manage a referent identification conversation with a user by integrating the linguistic information obtained from speech recognition and the vision information obtained from a robot camera.

Categories and Subject Descriptors

H.1.2 [**Models and Principles**]: User/Machine Systems – *Human factors*. H.5.2 [**Information Interfaces and Presentation**]: User Interfaces – *Evaluation/methodology, User-Centered Design*.

General Terms: Design, Human Factors.

Keywords: Referent identification, Pointing, Multimodal conversation.

1. INTRODUCTION

Referent identification is essential in communication. In their experiment, Clark & Wilkes-Gibbs [1] demonstrated that the identification of a referent is a process of establishing mutual beliefs shared by a speaker and a listener. In face-to-face conversations, participants also frequently use nonverbal behaviors in referent identification, particularly pointing gestures. Speakers use verbal referential expressions accompanied by pointing gestures. Listeners, on the other hand, try to identify the referent by interpreting the speaker's multimodal presentation on the basis of the contextual information of the conversation. Therefore, referent identification is a co-constructive process between the conversational participants, and pointing is a communicative device used in the process.

While numerous studies have been conducted for recognizing pointing gestures in communication robots [2], this study focuses on the referent identification communication where a pointing gesture is a part of the communication. For identifying the referent while establishing joint attention, the integration of multimodal information such as verbal information and pointing gestures is indispensable [3, 4]. With the goal of establishing a model for a multimodal referent identification process, first, this study collects referent identification communication and analyzes the sequences of verbal and nonverbal behaviors in the corpus. Then, based on the analysis, we implement a communication robot that can identify a referent through verbal and nonverbal communication with the user.

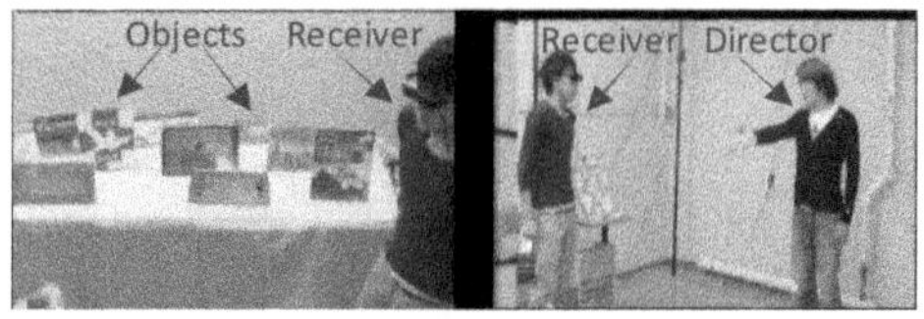

(a) Snapshot of experiment

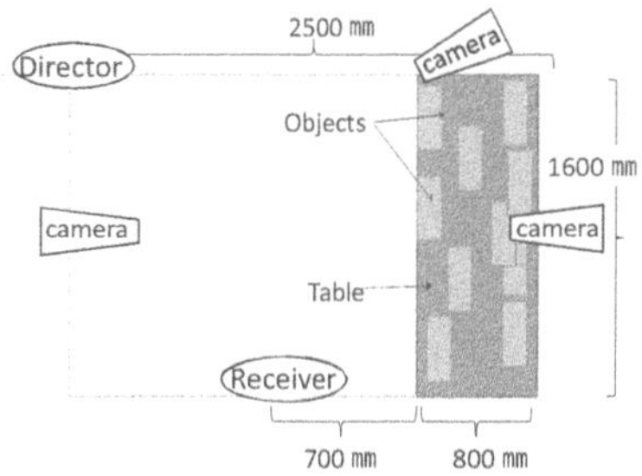

(b) Experimental settings

Figure 1. Snapshot of experiment and experimental settings

2. EXPERIMENT

2.1 Overview

First, we conducted an experiment to collect referent identification communication data. Figure 1 (a) shows a snapshot of the experiment. The labels of chocolates and candies were printed on paperboards and placed on a table. The subject role-played as a "director," and his/her task was to propose a better layout of the boards to the other person (an experimenter who role-played as a "receiver"). For example, "This one should be moved right next to the box that a girl is on." As shown in Figure 1 (b), the director stood 2.5 m away from the table, and the receiver stood close to the table. When the receiver redirected his body towards the director, she/he could not see the table, and vice versa. Each conversation was approximately 4 min long. Two female and three male subjects participated in the experiment as directors.

The subjects' upper body motions, including those of the arms, were recorded using a motion capture system (OptiTrack). In addition to the three video cameras shown in Figure 1 (b), the receiver wore a head-mounted camera that recorded the receiver's field of vision.

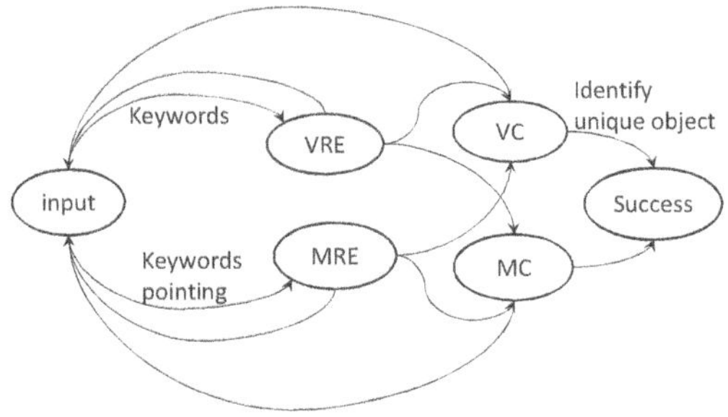

Figure 2. State transition model

2.2 Analysis of referent identification

We analyzed a sequence of utterances and pointing gestures performed by a director and the receiver in the referent identification communication, and found the following typical actions:

- Verbal referential expression (VRE): a target object is specified solely by verbal expressions.
- Multimodal referential expression (MRE): a target object is specified by combining verbal expressions and pointing gestures.

When VRE or MRE failed, the receiver asked clarification questions, which can be categorized as follows:

- Verbal clarification (VC): clarify the referent using only verbal expressions.
- Multimodal clarification (MC): clarify the referent using verbal expressions and pointing gestures simultaneously.

When the referent was not uniquely identified but a few candidates were found, the receiver used VC or MC to identify the target object. An example from the collected data is shown below:

[D1]: The <pointing> one at the back right corner. (MRE)
[R2]: At the back right corner. (VC)
[D3]: Yes.
[R4]: This <pointing> one? (MC)
[D5]: No. That <pointing> one. (MRE)
[R6]: That <pointing> one. (MC)
[D7]: Yes.

In [D1], the director referred to an object by mentioning the place of the object while pointing, and the receiver verbally clarified it. However, she/he failed to identify the object. Then, she/he clarified the object again by using a pointing gesture. Since the receiver pointed at a wrong object in [R5], the director corrected the error using a multimodal referential expression.

3. SYSTEM IMPLEMENTATION

On the basis of the observations mentioned in Section 2, we developed a state transition model for referent identification (Figure 2), and the system architecture (Figure 3) to implement a robot (Aldebaran's NAO) that can serve as a receiver.

In the speech recognition module (ASR), the system receives a speech input and recognizes keywords using a keyword spotting technique. If the speech sound is detected in the ASR module, the pointing recognition module perceives the user's pointing gestures from the motion capture data. It calculates angles for the direction from the user's right shoulder to the right wrist on the x-z plane (horizontal left-right direction) and on the x-y plane (vertical upward direction). If the shoulder-to-wrist vector crosses the region of the table, the system judges that the user is pointing

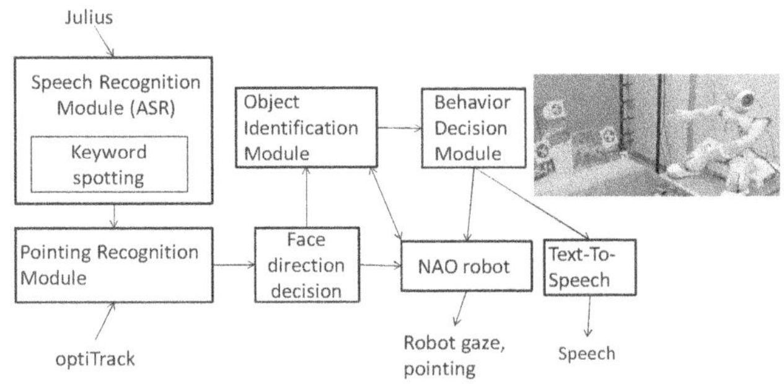

Figure 3. System architecture

at something. Once a pointing gesture is detected, the crossing point with the table region is assumed to be the pointing target, and the object identification (OI) module controls the robot head's direction when the robot searches the target object. We use graphic patterns as markers to identify the objects using the robot's camera. If the OI module finds a marker, it obtains information associated with the marker. Using this information, the OI module checks the consistency between the tagged information from the marker and the linguistic information (keywords) sent from the ASR module. If any pointing gesture is not recognized, the OI module changes the robot's face direction from left-to-right and right-to-left to find any marker whose tagged information matches the linguistic information.

In the behavior decision module, if the system identifies a unique object, the referent identification is successfully completed. If there are two candidates, this module generates clarification sequences, and the system shifts to either the "VC" or the "MC" state. In the current implementation, the system selects one of the candidates and generates a clarification question, such as "This one?" At this time, the system calculates the direction from the robot to the candidate, and based on this calculation, it calculates the angle of the robot's right arm to generate a pointing gesture. If more than two candidates are found, the system sends a speech command, "Which one?", and returns to the first step to start over (shift to state "input").

4. CONCLUSION

In this study, we collected and analyzed referent identification communication, and by integrating the visual and linguistic information, we implemented a robot that can identify a referent through a conversation with the user. As a future direction, we intend to improve the state transition network as well as the dialogue control mechanism to add more variation of the referent identification communication between the users and the robot.

5. REFERENCES

1. Clark, H. H. and D. Wilkes-Gibbs, *Referring as a Collaborative Process.* Cognition, 1986. **22**: pp. 1-39.
2. McGuire, P., et al. *Multi-Modal Human-Machine Communication for Instructing Robot Grasping Tasks.* in *International Conference on Intelligent Robots and Systems.* 2002.
3. Fransen, B., et al. *Using Vision, Acoustics, and Natural Language for Disambiguation.* in *HRI2007.* 2007.
4. Yu, C., M. Scheutz, and P. Schermerhorn. *Investigating Multimodal Real-Time Patterns of Joint Attention in an HRI Word Learning Task.* in *HRI2010.* 2010.

Listener Agent for Elderly People with Dementia

Yoichi Sakai[1] Yuuko Nonaka[2] Kiyoshi Yasuda[3] Yukiko I. Nakano[2]

[1]Graduate School of Science and Technology, Seikei University
3-3-1 Musashino-shi, Tokyo 180-8633, Japan
dm116217@cc.seikei.ac.jp

[2]Dept. of Computer and Information Science, Seikei University
us082103@cc.seikei.ac.jp
y.nakano@st.seikei.ac.jp

[3] Chiba Rosai Hospital
Ichihara-shi, Chiba 290-0003, Japan
fwkk5911@mb.infoweb.ne.jp

ABSTRACT

With the goal of developing a conversational humanoid that can serve as a companion for people with dementia, we propose an autonomous virtual agent that can generate backchannel feedback, such as head nods and verbal acknowledgement, on the basis of acoustic information in the user's speech. The system is also capable of speech recognition and language understanding functionalities, which are potentially useful for evaluating the cognitive status of elderly people on a daily basis.

Categories and Subject Descriptors

H.1.2 [**Models and Principles**]: User/Machine Systems – *Human factors*. H.5.2 [**Information Interfaces and Presentation**]: User Interfaces – *Evaluation/methodology, User-Centered Design.*

General Terms

Design, Human Factors

Keywords

Dementia, Conversational feedback.

1. INTRODUCTION

The number of people with dementia is increasing, and most of them are aged. It is difficult to find a sufficient number of caregivers for such patients. Besides providing physical assistance to those with physical impairments, it is important for assistive robots to provide communication functions [1]. Pollak [2] proposed three assistive functions for elderly people with cognitive impairment:

- Ensuring that the elder is safe and that he/she performs the necessary daily activities, and, if not, alerting a caregiver.

- Helping the elder to compensate for his/her impairment by assisting in the performance of daily activities.

- Assessing the elder's cognitive status.

In particular, we believe that a conversational humanoid can perform the third function stated above. The assessment should be natural, and it should not be unduly stressful for elderly people. If the agent system can assess the patient's cognitive status through conversations, that will be a more natural way of measuring the cognitive status of the patient. Moreover, interpersonal communication is one of the most preferable daily activities for elderly people. To accomplish this objective, we present a prototype of a listener agent, and a conversation log system that collects data for assessment.

(a) Snapshot of the agent

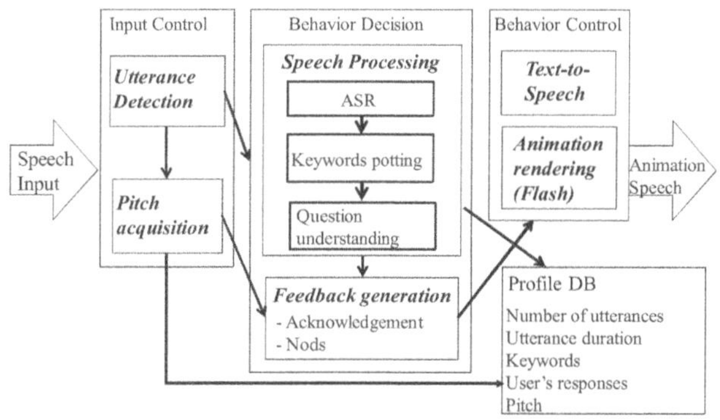

(b) System architecture

Figure 1. Agent snapshot and system architecture

2. LISTENER AGENT

An agent can effectively serve as a listener for people with Dementia if it is accepted as a companion by the patients. Previous studies on the acceptance of a robotic agent by elderly people reported that it is important for the agent to display social signals, such as smiling and head nods [3]; this enables the agent to gain the patient's trust and enhances intimacy [4]. Thus, initially, we implemented a simple virtual humanoid that only returns head nods and acknowledgement as conversational feedback.

2.1 System overview

Figure 1 (a) shows a snapshot of the proposed listener agent, and Figure 1 (b) shows the system architecture. To elicit responses from the user, the agent presented the user with a set of questions, one by one. We prepared the questions on the basis of those typically asked by doctors or nurses, such as inquiries regarding the patient's physical condition and meals, and more general topics, such as the patient's childhood memories and his/her locality.

Input Control Module: The speech recognition module detects the user's speech if the microphone input power exceeds a certain threshold. The threshold is adjusted according to the user. Then, the pitch acquisition module calculates the pitch information for a

Table 1. Feedback generation rules

Condition	Agent's behaviors
The current pitch *n* is more than 5% lower than the average pitch from *n-1* to *n-16*	nod
Following a pause of less than 2 s after the previous utterance, the speech input is detected,	acknowledgement ("Ee")
Pause of more than 2 s after the previous utterance or The pitch fall just before the utterance ends is less than 2% or The pitch just before the utterance ends is more than 10% higher than the average pitch of the current utterance.	acknowledgement ("Hai")

given speech input. These values are output to the behavior decision module.

Behavior Decision Module: Using the speech information obtained from the input control module, the behavior decision module updates the agent's internal state and determines the agent's actions. The agent enters the standby state when it asks the user a question. If the user's response is detected, the feedback generation rules (Table 1) are applied to the pitch information obtained from the input control module. Then, the behavior decision module determines the backchannel feedback and nod. The feedback generation rules were created on the basis of previous studies [5, 6]. If there is no response from the user after a certain amount of time, the agent asks the next question and awaits the user's response.

Behavior Control Module: This module executes the agent animations and produces synthesized speech. The 3D model of the agent was created using Poser 7. We also used Poser 7 to create key frames for the agent animations. Adobe Flash ver.10.0 was used for rendering the extracted key frame graphics to produce the agent animations on a web browser. The agent does not perform drastic motions; it simply moves its mouth while speaking, nods its head, and performs subtle idling behaviors.

2.2 Speech and language understanding

In addition to returning feedback to the user, the system needs to recognize and understand the user's verbal behaviors in order to store detailed data for the assessment of the patient's cognitive status. For this purpose, we add functions for speech recognition and natural language understanding, and we save the outputs from these mechanisms as a communication log in the profile DB. Sometimes, the patients may not enunciate their words; moreover, system robustness is of high importance. Hence, we employed simple but stable techniques for speech and language processing.

2.2.1 Speech recognition

We developed a dictionary and a language model from a transcription of patient speech data, and we applied it to Julius, a Japanese speech recognition engine. In addition to the speech recognition function, we used keyword spotting to extract important words from the patient's speech. Because a patient may not always enunciate his/her words, it is useful to recognize important words with higher reliability. We implemented a keyword spotting technique by calculating the similarity between a sequence of phonemes output from the speech recognition system and the phonemes for each keyword. The recognized speech information, such as keywords, pitch, utterance duration, and the number of utterances, is recorded in the profile DB.

2.2.2 Language understanding

First, we collected utterances from the patients in rehabilitation, and we created a list of possible utterances to be recognized. Then, by interviewing the nursing staff, we defined the response rules for each utterance. For language understanding, we employed a similarity-based method. After keyword spotting, a list of keywords is obtained. Then, the cosine distance between the word vector of the keyword list and that of possible utterances is calculated. The candidate with the highest similarity to the input keyword vector is selected as the result of language understanding. Once the user's input utterance is identified, the system response is selected according to the response rules. The response text is sent to the behavior control module, which produces speech sounds using a speech synthesis system, and agent animations are generated using Flash.

3. CONCLUSION

With the goal of creating an assistive conversational humanoid for people with dementia, we proposed an animated agent that interacts with the patient by asking questions. The agent not only produces conversational feedback during communication but also records speech and language information for assessing the cognitive status of the patient. We conducted an evaluation experiment to test whether such a conversational agent would be accepted by patients, and we found that the patients were willing to be engaged in conversation with the agent. We also found that the patients paid more attention to the agent when it returned backchannel feedback during the patient's speech. This suggests that the patients were more interested in communicating with conversational agents that produce social signals. In the future, in addition to collecting speech and language information, we need to analyze the patient's nonverbal behaviors, such as head motions, during the conversation. Moreover, in order to attract the user's attention, the system needs to automatically select topics of interest of the user on the basis of the user's profile and previous interactions.

4. REFERENCES

1. Bickmore, T., and Picard R. W. *Towards Caring Machines*. in *CHI 2004*. 2004.
2. Pollack, M., *Intelligent technology for an aging population: The use of AI to assist elders with cognitive impairments*, in *AI Magazine*. 2005. p. 9-24.
3. Heerink, M., et al., *Studying the acceptance of a robotic agent by elderly users.* International Journal of ARM, **7**(3), 2006.
4. Cassell, J., *Nudge, Nudge, Wink, Wink: Elements of Face-to-Face Conversation for Embodied Conversational Agents*, in *Embodied Conversational Agents*, J. Cassell, et al., Editors. 2000, MIT Press: Cambridge. p. 1-27.
5. Gratch, J., et al., *Virtual Rapport*, in *6th International Conference on Intelligent Virtual Agents*. 2006, Springer: Marina del Rey, CA.
6. Tsukahara, W. and N. Ward. *Responding to Subtle, Fleeting Changes in the User's Internal State.* in *CHI 2001*. 2001: ACM.

Can You Hold My Hand? Physical Warmth in Human-Robot Interaction

Jiaqi Nie[1], Michelle Park[1], Angie Lorena Marin[1], S. Shyam Sundar[1,2]

[1]Department of Interaction Science
Sungkyunkwan University
Seoul, Korea
+82-2-740-1889
nicol018nie@gmail.com

[2]Media Effects Research Laboratory
The Pennsylvania State University
University Park, PA, USA
+1-814-865-2173
sss12@psu.edu

ABSTRACT
This study investigates whether the temperature of a robot's hand can affect perceptions of the robot as a companion. Our research empirically analyzes the responses of 39 individuals randomly assigned to one of three conditions: (1) holding a warm robot hand or (2) holding a cold robot hand or (3) not holding a robot hand. The effects of this simulated 'human touch' on HRI were examined in the context of viewing a horror film clip. Results suggest that experiences of physical warmth and handholding increase feelings of friendship and trust toward the robot. However, the discrepancy between the expectation of an actual human touch and the mechanical appearance of a robot could result in negative effects.

Categories and Subject Descriptors
A.m [**Miscellaneous**]: Human-Robot Interaction; I.2.9 [**Artificial Intelligence**]: Robotics

General Terms
Experimentation, Human Factors

Keywords
Human Touch, Robot Handholding, Robot Warmth, HRI

1. INTRODUCTION
Touch, as a common communicative gesture, is an important aspect of social interaction among humans. It is therefore likely to be psychologically significant for human-robot interaction as well, given that we tend to anthropomorphize robots in every possible way, from morphology to interactivity. We approach this possibility by measuring the effect that robot handholding has on humans by studying participants' reactions to different body temperatures of a robot while experiencing a frightening situation.

Tactile Human-Robot Interaction (Tactile HRI) is a very broad field, which uses a variety of sensory approaches in order to approximate 'human touch' in robots. Prior Tactile HRI research has focused on such topics as sensor skins (i.e. soft skin vs. hard skin), interaction behavior (i.e., robot-centric vs. human-centric), and behavior development (i.e., learning algorithms to exploit human touch for new task-accomplishing behaviors)[1]. Although touch is recognized as being distinctly human, the warmth of the touch has not received much attention from researchers and designers. This is surprising because physical temperature is said to play an important role in sending interpersonal warmth (trust) information in human-human interactions. One recent study shows that even mere tactile experiences involving physical temperature serves to unconsciously affect one's impressions of -- and prosocial behavior toward--other people [2]. In the study, participants who were primed with temperature by briefly holding a cup of hot coffee perceived the personality of the target person as significantly warmer than those who had briefly held the iced coffee cup. They were also more likely to choose the gift for a friend than the gift for themselves. Accordingly, we suggest that 'human touch' with increased physical warmth during handholding would increase feelings of warmth in human robot interactions as well.

The overall objective of this study is to understand how human users respond to different levels of social contact with robots with varying degrees of physical warmth, with implications for tactile HRI. Based on the principles of CASA (Computers as Social Actors) theory [3], if the touch behavior of a robot, which was considered as a social actor by people, is positive, it will be judged as more human-like, and the perceived social presence and emotional attraction would be higher than those less human-like agents. Therefore, we hypothesize:

H1: 'Human touch' will increase perceived friendship, trust and human likeness of the robot.

H2: Perceived friendship, trust and human likeness of the robot will be enhanced with warm, as opposed to cold, 'human touch.'

H3: Perceived human likeness of the robot will mediate the relationship between touch and perceived friendship as well as trust toward the robot.

In interactions among humans, people use different kinds of social behaviors to not only mitigate the destructive effects of negative environmental events but also to reduce threat-related neural activation. This tendency is labeled "social regulation of emotion" by social psychologists [4]. The underlying mechanism of social regulation is articulated by social baseline theory[5], which states that individuals are hard-wired to assume close social proximity as a baseline affect regulation strategy. In other words, more satisfying relationships can be generated from an increased level of social contact (as in human touch), because it can have a greater effect on emotional regulation. Therefore, we propose:

H4: Human touch will attenuate perceived fright brought on by the horror film and increase enjoyment toward the movie.

H5: The attenuation of fright from the movie will be maximized in the condition of warm 'human touch.'

HRI'12, March 5–8, 2012, Boston, Massachusetts, USA.
ACM 978-1-4503-1063-5/12/03.

However, as Uncanny Valley theory notes [6], extreme human likeness from a social robot may yield a negative reaction due to the psychological gap between an immersive human touch and the mechanical appearance and behavior of the robot. Hence:

H6: Warm 'human touch' may induce the highest level of fright toward the robot than the other two conditions.

2. METHOD

2.1 Participants and Design

Thirty-nine college students were recruited to participate at a private university in Seoul, Korea. They were randomly assigned to one of three conditions: warm 'human touch', cold 'human touch' and no 'human touch.' The mean age of the participants was 26.1 years (SD=6.04).

2.2 Apparatus and Stimulus

The robot used in this experiment was RoboSapien RS-Media marketed by Wow Wee Company (see Figure 1). RoboSapien is a low-cost humanoid robot, 36 cm in height and 2.1 kg in weight. As the robot can be remote-controlled, the researcher hid the remote to make the participants believe that it was autonomous. RoboSapien was pre-programmed to say "hello" to participants in a male voice. The stimulus was a five-minute film clip from *Grudge*, an American horror movie, chosen because it included several horror scenes.

2.3 Procedure and Measurement

Participants were asked to complete a pre-questionnaire on a computer to assess their initial perception of robots and horror movies. Then, they were led to the experiment room, where they were greeted by RoboSapien (see Figure 1).

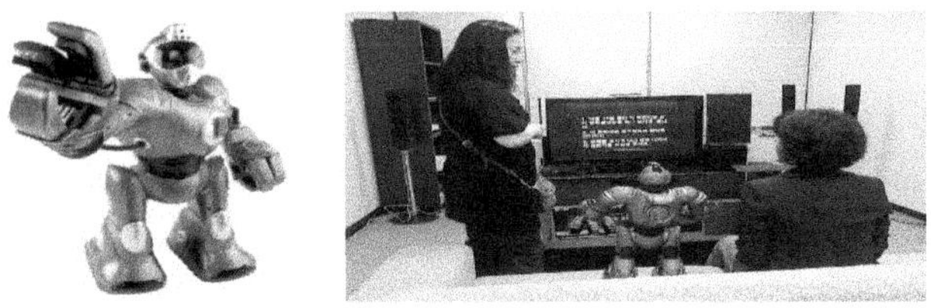

Figure 1. RoboSapien and the Experimental Setting

Participants were randomly assigned to one of the three experimental conditions. They were told to participate in a study of handholding with a robot while viewing the scary movie clip. In the no 'human touch' condition, participants watched the movie without handholding. In the cold 'human touch' condition, participants held the original mechanical hand of the robot during the movie-viewing. In the warm 'human touch' condition, a thin hot pad (15cm x 10cm) was pasted to the palm of the robot which can keep the robot's hand at a temperature of about 40 degrees Celsius. After viewing the clip, they were asked to complete a post-questionnaire on a computer, which assessed the levels of friendship and trust they experienced for the robot. Friendship was measured via 10 items: *companionship* (1 item), *help* (2 items), *intimacy* (2 items), *reliable alliance* (1 item) and *emotional security* (4 items). Trust was divided into 4 dimensions: *trust* (4 items), *reliability* (2 items), *immediacy* (4 items), and *credibility* (7 items). Human Likeness was measured with three scaled items. The degrees to which participants liked the movie and feared the robot were also measured. Prior attitudes toward horror movies and robots were measured as control variables. All items used a 7-point rating scale.

3. RESULT

An ANOVA was conducted to analyze the effects of 'human touch' in HRI. The results indicated that participants in the warm 'human touch' group reported a significantly higher degree of perceived friendship (M=2.64, SD=0.34) than those in the no 'human touch' condition [M=1.75, SD=0.48, $F(2, 35)=3.67$, $p< .05$]. Perceived trust levels in the warm condition (M=3.03, SD=0.78) were also significantly higher than levels in the cold 'human touch' (M=2.28, SD=0.75) and no 'human touch' conditions [M=2.10, SD=1.03, $F(2, 34)=3.97$, $p< .05$]. Participants in the warm condition also attributed greater human likeness to the robot (M=2.58, SD=1.07) than those in the cold (M=1.81, SD=1.17) and no touch conditions (M=1.62, SD=0.48), [$F(2, 33)=3.47$, $p< .05$]. Further, though there were no significant differences on perceived enjoyment and fright from the movie among the three conditions, a significant difference was found in perceived fright from the robot between the warm and no touch conditions. As we expected, people perceived higher level of fright from the robot in the warm 'human touch' condition (M=6.31, SD=1.03) than no human touch condition (M=4.71, SD=2.30). In sum, H2 was fully supported, H1 and H6 were partially supported, while H4 and H5 failed to receive support. Furthermore, the mediation effect of 'human likeness' on perceived friendship and trust toward the robot were found, using Baron and Kenny's method. Hence, H3 was supported.

4. DISCUSSION

It seems that even subtle cues like handholding and physical temperature can have a critical influence on human perception of robots. It appears that physical warmth translates into emotional warmth because of the human-likeness signified by tactile interaction. However, our data also indicate that touch can exacerbate the uncanny valley problem, calling for design solutions to arrive at an optimal level of 'human touch' for robots—one that matches human expectations.

5. ACKNOWLEDGMENT

This research was supported by World Class University program through the National Research Foundation of Korea funded by the Ministry of Education, Science and Technology (R31-2008-000-10062-0).

6. REFERENCES

[1] Argall, B. D. and Billard, A. G. 2010. A survey of tactile human-robot interactions. *Robotics and autonomous systems*, 1159-1176.

[2] Williams, L. E. and Bargh, J. A. 2008. Experiencing physical warmth promotes interpersonal warmth. *Science*, 322, 606-7.

[3] Reeves, B. and Nass, C. 1996. *How people treat computers, television, and new media like real people and places*. CSLI Publications and Cambridge university press.

[4] Berscheid, E. 2003. The human's greatest strength: Other humans.

[5] Coan, J. A. 2010. Adult attachment and the brain. *Journal of Social and Personal Relationships*, 27, 2, 210.

[6] Mori, M. 1970. The uncanny valley. *Energy*,7,4, 33-35.

Captain May I? Proxemics Study Examining Factors that Influence Distance between Humanoid Robots, Children, and Adults, during Human-Robot Interaction

Sandra Y. Okita
Columbia University
525 West 120th Street
New York, NY 10027
+1-212-678-4165
okita@tc.columbia.edu

Victor Ng-Thow-Hing
Honda Research Institute USA
425 National Avenue, Suite 100
Mountain View, CA 94043
+1-650-314-0428
vngthowhing@hra.com

Ravi Kiran Sarvadevabhatla
Honda Research Institute USA
425 National Avenue, Suite 100
Mountain View, CA 94043
+1-650-314-0434
RSarvadevabhatla@hra.com

ABSTRACT
This proxemics study examines whether the physical distance between robots and humans differ based on the following factors: 1) age: children vs. adults, 2) who initiates the approach: humans approaching the robot vs. robot approaching humans, 3) prompting: verbal invitation vs. non-verbal gesture (e.g., beckoning), and 4) informing: announcement vs. permission vs. nothing. Results showed that both verbal and non-verbal prompting had significant influence on physical distance. Physiological data is also used to detect the appropriate timing of approach for a more natural and comfortable interaction.

Categories and Subject Descriptors
H.1.2 [**Models and Principles**]: User Machine Systems—Software psychology; human factors.

General Terms
Human Factors, Design, Experimentation.

Keywords
Human Robot Interaction, Proxemics Study, Young Children.

1. INTRODUCTION
Current limitations in sensor technology make physical distance between robots and humans a crucial factor in human-robot interaction (HRI) to detect and respond to situational factors and user state. Closer proximity will allow better sensor range for facial expression detection, speech recognition, collision detection during collaborative tasks, and allow better choice of expressive communication (e.g., verbal, gesture). This study examines the factors influencing inter-agent distances, and ways to modulate physical distance for safe and effective HRI. The term *proxemics* first introduced by Edward Hall [1] identified proximity zones (e.g., social, personal, intimate), and studied the distances between people when interacting in shared social spaces (Figure 1). Hall found that human sensory perspectives influenced the kinds of communication modalities used. Walters et al. [2] found that people stay further away with synthetic voices opposed to real human voices, and likeability influenced distance. Most proxemics studies are with stationary robots [2, 3] or robots with wheels. If robot movement is more natural and familiar (e.g., walk like a human instead of vehicle), children may respond in ways that influence proxemics. To study proxemics with robots, various metrics and observable behaviors need to be recorded from different perspectives, such as individual pose information, behavioral cues, interpersonal distance and physiological metrics. Our study explores conditions that actively modulate the inter-agent distances by selecting different behavior strategies (verbal and non-verbal) that can influence personal space. Given these metrics, the robot can use these factors to adjust or predict the ideal distance between humans in relation to sensory input, comfort level, and method of communication. The knowledge can also be used to design appropriate verbal and non-verbal behaviors for robots. Given these metrics, the autonomous robot would be able to use these factors to adjust, or predict the ideal distance between humans in relation to sensory input, comfort level, and method of communication.

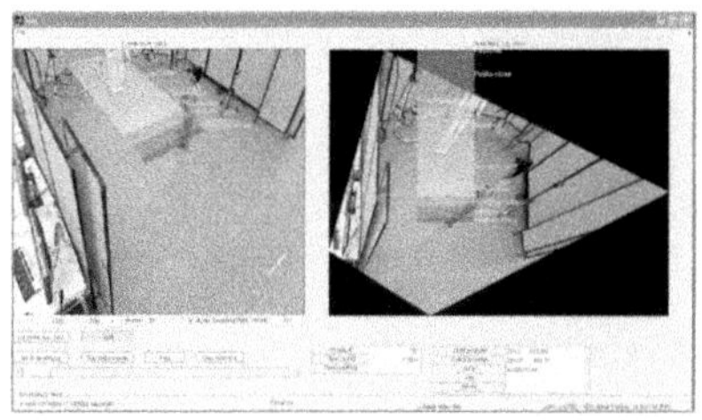

Figure 1. Semi-automatic tool that takes overhead video streams, and annotates the subject's angular orientations, distance from robot, and metric position (Hall's proxemics zones by color).

2. RESEARCH METHOD AND DESIGN
The study examines whether physical distance allowed by humans (adult or child) would differ under two specific situations, 1) *when human approaches robot*, and 2) *when robot approaches human*. Thirty children between the ages 5 to 7-years-old, and twelve adults participated in a one-to-one 20-minute session with Honda's humanoid robot. This study involved two parts. The first part "human approaches robot" was a 2x1 design where Age (adult vs. children) was the direct comparison. The second part "robot approaches human" was a 2x3 design that looked at Age (adults vs. children) and Type of Informing (Announce vs. Permission vs. Nothing). Through a WOZ environment [4] the robot and human engaged in conversation (e.g., hobbies). In part 1, *when human approaches robot*, we include the factor "prompting" to see if the robot's verbal (e.g., Feel free to come closer), or non-verbal prompting (e.g., beckoning) changes distance. In part 2, *when robot approaches human*, we include the factor "informing" like the child's game "Captain May I" to see if informing (e.g., robot makes announcement, asks permission, says nothing) changes distance.

3. RESULTS AND DISCUSSION

3.1 Part 1: Human Approaching Robot

There is a significant difference t(40)=2.2, p<0.05 between children and adults when initially approaching the robot. Without any prompting, adults show a 46% decrease in distance between themselves and the robot, while children only show a 25% decrease. Additional prompting (verbal and non-verbal) led to a decrease in physical distance for both children and adults where there were no significant difference between children (additional 14% decrease) and adults (additional 17% decrease). Non-verbal prompting (i.e., beckoning) led to another decrease by 12% for children, and %7 for adults. See figure 2 and 3. There was no significant difference at the end of the session between the two groups (Adults M=66% Children M=52%) t(40)=1.63, p=0.11.

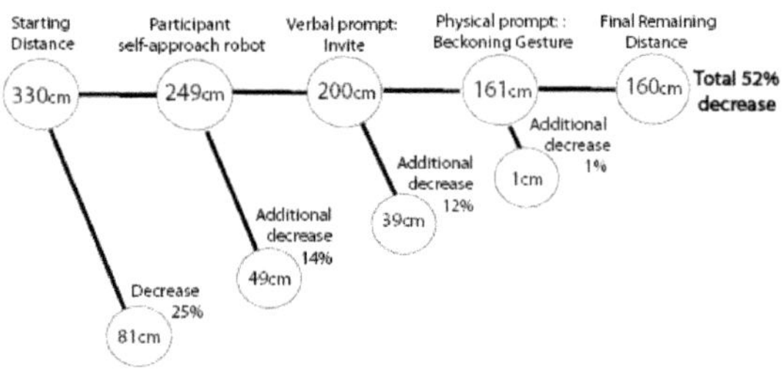

Fig.2. Avg. distance(cm) and %decrease in children

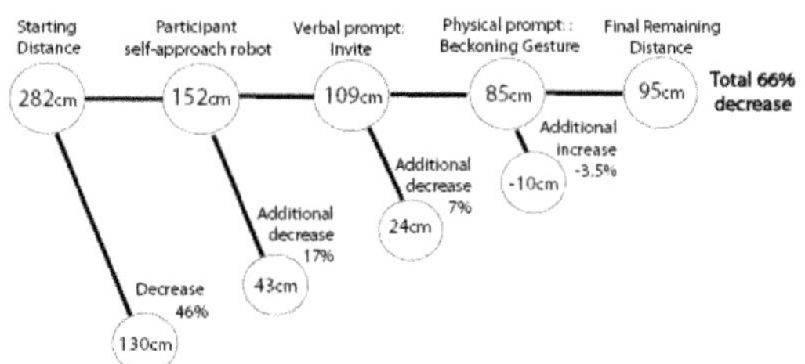

Fig.3. Avg. distance(cm) and %decrease in adults.

3.2 Part2: Robot Approaching Human

Type of Informing made a significant difference in % decrease for children at F (2, 27)=6.1, MSE=0.047, p<0.01. The simple act of "asking permission to take a step" significantly reduced the distance between robot and child compared to "making an announcement to take a step" or "saying nothing when taking a step". No difference was seen in adults (Figure 4). A Post-hoc Tukey's HSD tests showed that "permission" significantly decreased the distance compared to "announcement" (p<.01, Cohen's d=1.47, effect-size r=0.59), and approaching significance to "nothing (no warning)" style (p=.078, Cohen's d=1.16, effect-size r=0.50). See Figure 9. Another trend that we observed was that children seem to play safe by stopping the robot early, and then adjusting by stepping forward. This may have some design implications where the robot can approach a child but stop early to give the child a chance to adjust to give some control. Adults (see Figure 5), needed no adjustment and are quite consistent with the distance set. Skin conductance response (SCR) was used to measure physiological arousal levels in subjects to examine causal factors in participant behavior and physical distance (Figure 6). Video data and physiological data were synchronized to create potential data that can assist robots to predict and anticipate events (e.g., arousal level, robot movement). We are currently looking at SCR analysis in relation to conversations and events.

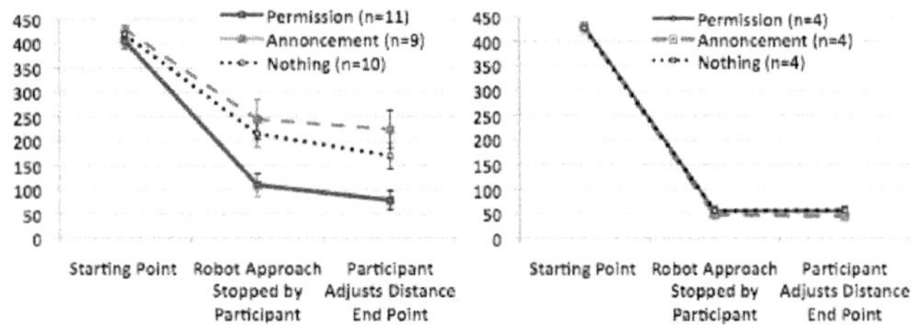

Figure 4. Inform effect on distance(cm) left-children, right-adult.

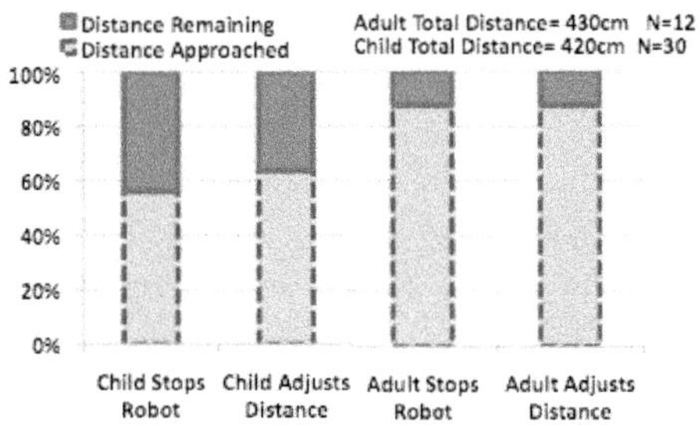

Figure 5. Remaining distance after stopping the robot, and adjustments made by child and adult.

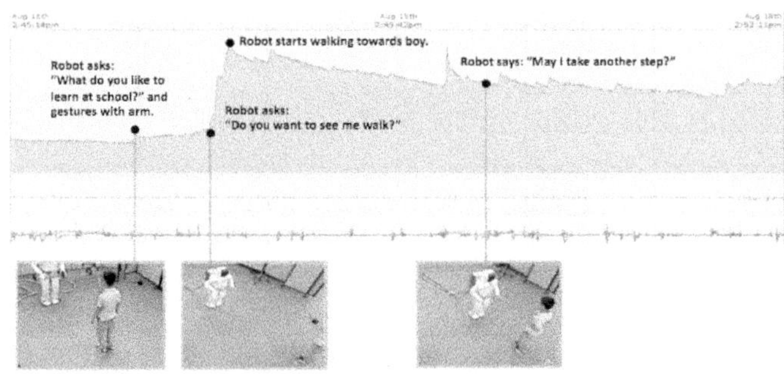

Figure 6. Examining physiological measures

4. SUMMARY AND FUTURE WORK

This human-robot proxemics study examined different factors that reduced distance. Findings employ a variety of strategies to increase the chances for successful detection and delivery of response. As HRI advances, delivery of appropriate behavior and dynamic adjustment in movement will be crucial in maintaining successful communication and collaboration.

5. REFERENCES

[1] Hall E. 1966. *The Hidden Dimension*, Anchor Books.

[2] Walters, M.L., D.S. Syrdal, K. L. Koay, K. Dautenhahn, R. te Boekhorst 2008. Human Approach Distances to a Mechanical-Looking Robot with Different Robot Voice Styles, in *Proc. of the 17th IEEE Int. Symposium on HRI Communication*, Germany.

[3] Takayama, L., Pantofaru, C. 2009. Influences on proxemic behaviors in human-robot interaction, in IEEE/RSJ *International Conference on Intelligent Robots and Systems.*

[4] Okita, S. Y., Ng-Thow-Hing, V, Sarvadevabhatla, R. K. (2011). Multimodal Approach to Affective Human-Robot Interaction Design with Children. ACM Transactions on Interactive Intelligent System (TiiS). 1,1, Article 5, 1-29.

Online Gaming with Robots vs. Computers as Allies vs. Opponents

Eunil Park[1], Ki Joon Kim[1], S. Shyam Sundar[1,2], Angel P. del Pobil[1,3]

[1]Department of Interaction Science
Sungkyunkwan University
Seoul, Korea
+82-2-740-1889
{pa1324, veritate}@skku.edu

[2]Media Effects Research Laboratory
The Pennsylvania State University
University Park, PA, USA
+1-814-865-2173
sss12@psu.edu

[3]Robotic Intelligence Laboratory
University Jaume-I
Castellon, Spain
+34-964-728-293
pobil@icc.uji.es

ABSTRACT

A 2 x 2 between-subjects experiment was conducted to examine the effects of the type of artificial agent (robot vs. computer) and the role of the agent (ally vs. enemy) on people's perceptions and evaluations of the agent when playing a video game. Participants perceived that playing the game with a robot was more enjoyable and easier than playing with a computer. Regardless of the agent type, participants reported that playing the game was more enjoyable when the agent played as an ally rather than as their opponent. Implications of notable findings are discussed.

Categories and Subject Descriptors

A.m. [**Miscellaneous**]: Human-Robot Interaction; I.2.9 [**Artificial Intelligence**]: Robotics

General Terms

Human Factors, Theory

Keywords

Artificial agent, online gaming, human-robot interaction

1. INTRODUCTION

1.1 Types of Artificial Agent

It is quite common for us to play games on our computers—not simply as a medium to compete with other gamers, but often playing against the computer itself. The latter situation raises the issue of "source orientation" [6]: Who or what are we orienting to when we play against a computer? Are we picturing the computer qua machine itself or are we imagining an unseen programmer in trying to beat the system? Studies in the CASA (computers as social actors) literature suggest that we make direct social attributions to the computer itself [5]. Given this direct source orientation to the machine, it stands to reason that a computer that is more social will serve to enhance users' responses to the system and thereby improve the quality of their user experience.

A good example of such as social computer would be the social robot [1]. Given their anthropomorphic features and socio-emotional interaction abilities, social robots can serve as credible proxies for human opponents in an online gaming context. On the other hand, they could also serve to heighten user expectations of humanness, thereby undermining the quality of user experience.

In order to investigate which of these alternatives is empirically valid, a direct comparison is necessary between the computer and the robot as one's opponent in an online game. Therefore, we posed the following question for study:

Research Question 1: When playing a video game online, what are the effects of the type of artificial agent (IV: robot vs. computer) on enjoyment of playing the game, intention to play the game again in future, perceived ease of playing the game, and perceived intelligence of the agent (DVs)?

1.2 Role of Artificial Agent

CASA studies have shown that when computers are labeled as team-mates, rather than as opponents, users perceive the computers to be more similar to themselves and friendler, think that the information provided by the computers is of higher quality and are more likely to cooperate with them [4]. In the domain of human-robot interaction, Hinds et al. [2] discovered that the role of robots does significantly affect users' perceptions; they found that the subordinate role assigned to a robot affects people's willingness to rely on and share responsibility with their robotic coworkers.

In light of these findings, an open question for our research is whether computers and robots will be treated differently based on their role in an online game context. Will the labeling of the computer/robot as "opponent" lead to more negative perceptions of the game experience compared to labeling it as a "team-mate"? Therefore, we posed the following research question for study:

Research Question 2: When playing a video game with an artificial agent (robot or computer), what are the effects of the role of the agent (IV: ally vs. enemy) on enjoyment of playing the game, intention to play the game again in future, perceived ease of playing the game, and perceived intelligence of the agent (DVs)?

2. METHOD

2.1 Design & Participants

The experiment was a between-subjects design with four conditions: 2 (types of artificial agent: computer vs. robot) x 2 (roles of the agent: ally vs. enemy). Participants were 60 female college students (average age=23.7, SD=3.9) with little experience in video games.

HRI'12, March 5–8, 2012, Boston, Massachusetts, USA.
ACM 978-1-4503-1063-5/12/03.

2.2 Procedure

Participants were assigned to one of the four conditions upon arrival at the laboratory. In the robot condition, participants were told that they were going to play a video game titled "Pro Evaluation Soccer 6" with a Nao humanoid robot. Although the robot did not actually play the game, its voice and gestures were programmed as if the robot was actively engaging in the game. The game was played by an experimenter in a room with a two-way mirror, where the experimenter could see and play the game without the participant's awareness.

In the computer condition, identical experimental procedure and setting were used, except that they were told that they were going to play the game with a computer. About half the participants in each condition were told that the computer/robot is their ally or team-mate while the other half were told that the computer/robot was their opponent in the game. Regardless of the experimental conditions and participants' gaming skills, the game results were manipulated in a way that participants always won the game.

After game-play, participants completed a post-experiment questionnaire, which included validated items from prior research relating to their assessments of enjoyment of playing the game (Cronbach's α=.81), intention to play the game again in future (α=.71), perceived ease of playing the game (α=.84), and perceived intelligence (α=.72) of the agent.

3. RESULTS

Results from a series of analyses of variance (ANOVAs) indicated that participants who played with the robot (M=4.70, SD=.66) experienced a higher level of enjoyment than those who played the game with the computer (M=3.77, SD=1.05), $F(1,56)=18.63$, $p<.001$. Participants who played with the robot (M=4.07, SD=.88) were also more likely to play the game again than those who played with the computer (M=3.11, SD=1.11), $F(1,56)=13.10$, $p<.01$ (Fig. 1).

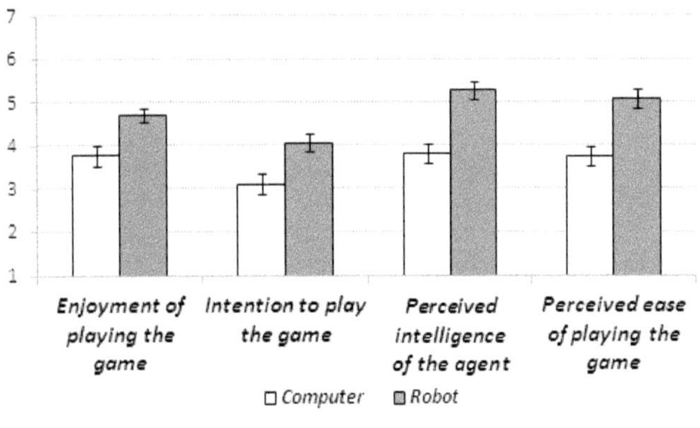

Figure 1. Effects of artificial agent type on DVs

In addition, participants who played with the robot (M=5.28, SD=.92) perceived that their gaming partner was more intelligent than participants whose gaming partner was the computer (M=3.81, SD=1.00), $F(1,56) =34.26$, $p<.01$. Participant perceived that playing the game with the robot (M=5.08, SD=.97) was easier than playing with the computer (M=3.76, SD=.97), $F(1,56)=27.16$, $p<.01$ (Fig. 1).

Lastly, the role of artificial agent had significant effects on enjoyment of playing the game. Regardless of the agent type, participants reported that playing the game was more enjoyable when their partner played as an ally (M=4.50, SD=1.14) rather than as their opponent (M=3.97, SD=.74), $F(1, 56)=5.89$, $p<.05$ (Fig. 2). However, no interaction effect was found between the type and role of the artificial agent.

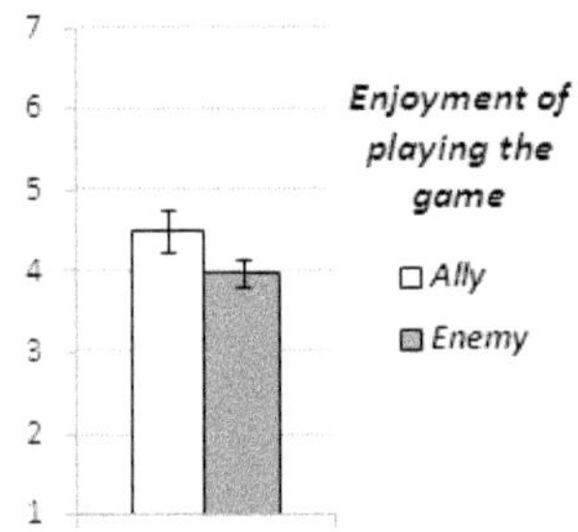

Figure 2. Effect of artificial agent's role on enjoyment

4. DISCUSSION

Our findings suggest that robot agents are generally more effective in providing greater enjoyment as well as more positive evaluations and experience than computer agents. This is perhaps because those who interacted with the robot experienced greater feelings of social presence. Social presence occurs when "technology users do not notice the artificiality of simulated nonhuman social actors" [3]. The humanoid robot's morphological features that resemble humans and other humanlike characteristics such as voice and hand gestures might have served to increase the authenticity of the gaming experience by imitating game-play with fellow humans, ultimately resulting in greater satisfaction with their experience. Future studies may explore the mediating effects of social presence in human-robot interaction as well as the effect of game outcome (e.g., winning vs. losing) on the perceived quality of the game experience.

5. ACKNOWLEDGMENT

This research was supported by World Class University program through the National Research Foundation of Korea funded by the Ministry of Education, Science and Technology (R31-2008-000-10062-0).

6. REFERENCES

[1] Breazeal, C. L. 2004. *Designing sociable robots.* Cambridge: The MIT Press.

[2] Hinds, P. J., Roberts, T. L., and Jones, H. 2004. Whose job is it anyway? A study of human-robot interaction in a collaborative task. *Human-Computer Interaction*, 19, 151-181.

[3] Lee, K. M. 2004. Presence, explicated. *Communication Theory*, 14, 27-50.

[4] Nass, C., Fogg, B. J., and Moon, Y. 1996. Can computers be teammates? *International Journal of Human Computer Studies*, 45, 669-678.

[5] Reeves, B. and Nass, C. 1996. *The media equation.* New York: Cambridge University Press.

[6] Sundar, S. S. and Nass, C. 2000. Source orientation in human-computer interaction: Programmer, networker, or independent social actor? *Communication Research*, 27, 6, 683-703.

The Effects of Immersive Tendency and Need to Belong on Human-Robot Interaction

Ki Joon Kim[1], Eunil Park[1], S. Shyam Sundar[1,2], Angel P. del Pobil[1,3]

[1]Department of Interaction Science
Sungkyunkwan University
Seoul, Korea
+82-2-740-1889
{veritate, pa1324}@skku.edu

[2]Media Effects Research Laboratory
The Pennsylvania State University
University Park, PA, USA
+1-814-865-2173
sss12@psu.edu

[3]Robotic Intelligence Laboratory
University Jaume-I
Castellon, Spain
+34-964-728-293
pobil@icc.uji.es

ABSTRACT

Do individual differences in dispositional behavioral tendencies, such as immersive tendency and need to belong, play a significant role in human-robot interaction? To answer this question, the present study conducted a 2 x 2 between-subjects experiment to examine the effects of immersive tendency (high vs. low) and need to belong (high vs. low) on individuals' perceptions of a social robot. Preliminary data analyses revealed that participants with a higher level of immersive tendency and need to belong showed greater attachment and trust towards the robot, and were more satisfied with their relationship with the robot than participants with a lower level of immersive tendency and need to belong. In addition, participants with a higher level of immersive tendency experienced greater feelings of social presence. Implications of notable findings are discussed.

Categories and Subject Descriptors

A.m. [**Miscellaneous**]: Human-Robot Interaction; I.2.9 [**Artificial Intelligence**]: Robotics

General Terms

Human Factors, Theory

Keywords

Immersive tendency, need to belong, social presence, attachment

1. INTRODUCTION

1.1 Immersive Tendency

Different individuals have different levels of immersive tendency, i.e., the degree to which one becomes "enveloped by, included in, and interacting with an environment that provides a continuous stream of experiences" [3]. Immersive tendency is considered a notable predictor of human-computer interaction because it largely determines the psychological state under which individuals focus on and interact with stimuli in an environment [9]. More importantly, this tendency is known to significantly affect technology users' feelings of social presence, which occurs when they do not notice the artificiality of simulated nonhuman social actors [6].

By extension, individuals with a high immersive tendency will be more likely to experience greater feelings of social presence, perceiving robots as genuine social actors with many humanlike attributes and deserving of social responses from human interactants. In contrast, those with a low immersive tendency will tend to perceive robots as mere artificial objects or machines. The present study, therefore, proposes the following hypothesis:

Hypothesis 1: Individuals with a higher level of immersive tendency will experience greater feelings of social presence when interacting with a robot than those with a lower level of immersive tendency.

1.2 Need to Belong

John Donne's widely quoted line "No man is an island" concisely captures the fundamental human desire to establish and maintain "lasting, positive, and significant interpersonal relationships" [1]. This innate human desire, referred to as the "need to belong" by Baumeister and Leary [1], is universally discovered in all humans across all cultures to some degree, and substantially influences our social interactions and emotional well-being. However, some individuals show greater need to belong than others.

Assuming that humans have a tendency to treat computers (and by extension, robots) as social actors [7] and apply rules of human-human interactions to human-robot interactions [6], we can expect that those individuals who have a high need to belong will be more likely to show greater attachment to robots because their innate desire to form social affiliations will motivate them to establish a strong bond with robots, whereas those with a low need to belong will be less motivated to form a close relationship with robots. The present study, therefore, proposes the following hypothesis:

Hypothesis 2: Individuals with a higher level of need to belong will show greater attachment and satisfaction when interacting with a robot than those with a lower level of need to belong.

2. METHOD

2.1 Design & Participants

A between-subjects experiment was conducted, with four conditions representing a fully-crossed factorial design featuring 2 levels of immersive tendency (high vs. low) and 2 levels of need to belong (high vs. low). Data were analyzed from 20 college students (8 males, 12 females). Participants' age ranged from 19 to 27, with a mean of 22.75 years (SD=2.08).

2.2 Procedure

Prior to the experiment, ten respondents reviewed three scripts prepared by the experimenters, and completed a questionnaire

HRI'12, March 5–8, 2012, Boston, Massachusetts, USA.
ACM 978-1-4503-1063-5/12/03.

assessing the neutrality of each script. Based on this procedure, the script rated as having the most neutral content was selected for the main experiment.

Upon arrival in the laboratory, participants completed a pre-experiment questionnaire measuring their levels of immersive tendency [9] and need to belong [5] (i.e., independent variables). Based on the distribution quantiles of participants' mean scores, 8 and 11 participants were identified as having a higher level of immersive tendency and need to belong respectively.

Participants were then introduced to Nao, a programmable autonomous humanoid robot, and engaged in casual conversations about their school life and responded to the robot's questions from the selected script. After the interaction, participants completed a post-experiment questionnaire on a 7-point Likert scale (1="not at all," 7="very much so"), which included items on their assessments of the following dependent variables.

2.3 Measures

Social presence of the robot was measured by an index composed of seven items (e.g., How much did you feel as if you were interacting with an intelligent being?; How much did you feel as if you were accompanied with an intelligent being?) [6].

Attachment to the robot was an index composed of nine items (e.g., I felt emotionally connected to this robot; I had a bond with this robot) [8].

Trust towards the robot was an index composed of four items (e.g., The robot was interested in my well-being, not just its own; The robot was truthful) [4].

Relationship satisfaction was an index composed of five items (e.g., Do you think you will be satisfied with your relationship with the robot?; Do you think the robot will meet your needs?) [2].

3. RESULTS

Hypothesis 1 was supported. Results from a series of 2 x 2 factorial analyses of variance (ANOVAs) revealed significant effects of immersive tendency on the dependent variables. Participants with a higher level of immersive tendency experienced greater feelings of social presence ($F(1,16)=6.61$, $p<.05$), showed greater attachment ($F(1,16)=7.58$, $p<.05$) and trust towards the robot ($F(1,16)=6.68$, $p<.05$), and were more satisfied with their relationship with the robot ($F(1,16)=8.75$, $p<.01$) than participants with a lower level of immersive tendency.

Hypothesis 2 was also supported. The ANOVAs revealed significant effects of need to belong on all dependent variables except social presence. Participants with a higher need to belong showed greater attachment ($F(1,16)=4.82$, $p<.05$) and trust towards the robot ($F(1,16)=4.59$, $p<.05$), and were more satisfied with their relationship with the robot ($F(1,16)=5.89$, $p<.05$) than participants with a lower need to belong.

However, no interaction effect was found between immersive tendency and need to belong.

4. DISCUSSION

These findings demonstrate that individual differences in immersive tendency and need to belong play a significant role in the quality of human-robot interaction. Theoretically, the tendency to mindlessly apply human attributes to robots is strengthened by one's willingness to relate to robots in social terms. Stronger immersive tendency suggests stronger CASA schema (of machines as social entities) in the minds of the user, thus leading to higher perceptions of social presence, and greater trust and attachment to the machine. Higher need to belong signals a gratification that is sought from the robot, and evidently fulfilled by simply interacting with it.

To the robot industry, these findings suggest that less emphasis be paid on engineering the robots to look and act like humans and more emphasis be put on understanding user psychology, with a view to segmenting the market. Certain populations (like gamers, who have high immersive tendencies, and retirees aging alone, with a pronounced need to belong) may be more receptive to social robots and therefore benefit more from them regardless of how the anthropomorphic the robot is in its morphology and actions.

5. ACKNOWLEDGMENT

This research was supported by World Class University program through the National Research Foundation of Korea funded by the Ministry of Education, Science and Technology (R31-2008-000-10062-0).

6. REFERENCES

[1] Baumeister, R. F. and Leary, M. R. 1995. The need to belong: Desire for interpersonal attachments as a fundamental human motivation. *Psychological Bulletin,* 117, 3, 497-529.

[2] Hendrick, S. S. and Hendrick C. 1995. Gender differences and similarities in sex and love. *Personal Relationships*, 2, 55-65.

[3] Jerome, C. and Witmer, B. 2002. Immersive tendency, feeling of presence, and simulator sickness: Formulation of a causal model. In *Proceedings of the Human Factors and Ergonomics Society 46th Annual Meeting,* 2197-2201.

[4] Johnson, D. and Grayson, K. 2005. Cognitive and affective trust in service relationships. *Journal of Business Research,* 58, 4, 500-507.

[5] Leary, M. R., Kelly, K. M., Cottrell, C. A., and Schreindorfer, L. S. 2007. *Individual differences in the need to belong.* Unpublished manuscript. Durham, NC: Duke University.

[6] Lee, K. M., Peng, W., Jin, S, and Yan, C. 2006. Can robots manifest personality?: An empirical test of personality recognition, social responses, and social presence in human-robot interaction. *Journal of Communication,* 56, 754-772.

[7] Reeves, B. and Nass, C. 1996. *The media equation.* New York: Cambridge University Press.

[8] Schifferstein, H. N. J. and Zwartkruis-Pelgrim, E. P. H. 2008. Consumer-product attachment: Measurement and design implications. *International Journal of Design,* 2, 3, 1-13.

[9] Witmer, B. and Singer, M. 1998. Measuring presence in virtual environments: A presence questionnaire. *Presence: Teleoperators and Virtual Environments,* 7, 3, 225-240.

Mechanical Model of Human Lower Arm

Borut Povse
R & D department for automation, robotics and electronic instrumentation
Vreskovo 68
Trbovlje, Slovenia
borut.dax@siol.net

Darko Koritnik
R & D department for automation, robotics and electronic instrumentation
Vreskovo 68
Trbovlje, Slovenia

Tadej Bajd, Marko Munih
Laboratory of Robotics, Faculty of Electrical Engineering
Trzaska 25
Ljubljana, Slovenia

ABSTRACT

The paper describes the development of a passive mechanical lower arm (PMLA) intended for physical human-robot interaction studies. Our research is focused on cooperation of a small industrial robot and human operator where collision is expected only between the robot end-effector and the lower arm of the human worker. A mathematical model of the passive human lower arm was built and adopted for the control of the PMLA. The mathematical model was optimized using the data from the experiments performed with human volunteers and implemented into the control scheme. The experiments with human volunteers were performed with safely low contact forces. The emulation system of the human lower arm was thoroughly evaluated in the robot impact experiments while using plane and line robot end-effector tools. During the experiment the impact force and the impact energy density were measured and compared to the measurements of the investigation with human volunteers. The PMLA proved to be a good emulation system of the passive human lower arm.

Keywords

Physical human-robot interaction, mechanical lower arm

1. INTRODUCTION

In the future we can imagine factories, where a great deal of work will be accomplished by robots. Work cells unsuitable for full automation will include a robot for help and assistance to the human worker. These robots will work in close cooperation with workers. When robots work near or with humans, safety is the primary concern. A real-world example from our industrial experiences, where the worker and the robot could collaborate, is a washing machine assembly line, where several phases could be united in order to save the place and shorten the production cycle time. Collision is predominantly expected between the robot end-effector and the lower arm, wrist, or hand of the human operator [1]. To study the effect of the impact between the robot and man, a passive mechanical lower arm (PMLA) was developed and equipped with inertial sensors (Fig. 1). The PMLA was built incorporating an advanced prosthetic material and a motorized elbow. The mathematical model of the human lower arm was developed and implemented into the closed-loop control of the PMLA motorized elbow. The structure of the model of the passive human lower arm is based on the mathematical model of human antagonistic pair of muscles and passive properties of the human elbow.

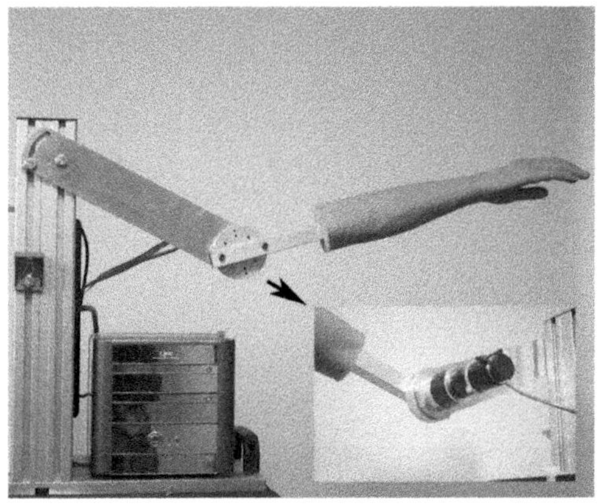

Figure 1: Passive mechanical lower arm (PMLA)

2. METHODOLOGY

The measuring system used in the investigation comprised the inertial sensors incorporating a set of two three-axis accelerometers ADXL203 and three gyroscopes ADXRS150 (Analog Devices, Inc.), three axis force and torque sensor (JR3, Inc.), and the optical measurement system Optotrak Certus (Northern Digital, Inc.). All data were logged by an xPC target computer. In experiments, the six-axis industrial robot collided with the unconstrained human lower arm or PMLA perpendicularly at a constant deceleration. The point of impact was positioned midway between the wrist and the elbow on the dorsal aspect of the lower arm. The line impact end-effector deceleration was changed incrementally from 1 m/s^2 to 5 m/s^2. The end-effector stop point was located inside the lower arm. The depth from the lower arm surface was changed from 10 mm to 30 mm by 10 mm steps. Altogether fifteen experiments were performed (1- low intensity, 15- high intensity). The injury intensities of the impact were evaluated by calculating the impact energy density [2]. The soft tissue injuries occur at an impact energy density higher than 2.52 J/cm^2, while hematoma or suffusions can occur below this value [3], [4]. Impact energy

HRI'12, March 5–8, 2012, Boston, Massachusetts, USA.
ACM 978-1-4503-1063-5/12/03.

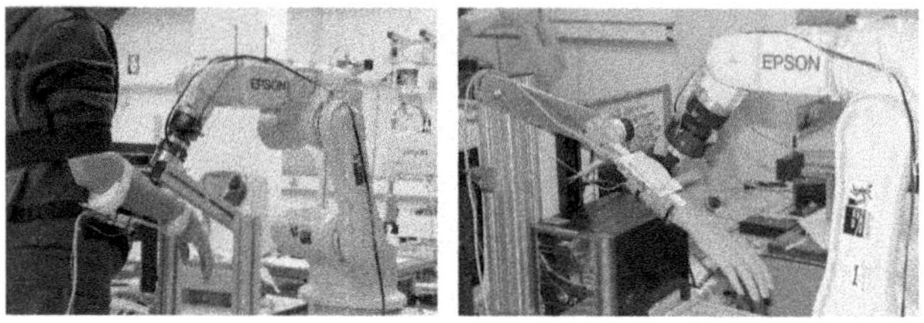

Figure 2: Experimental environment, robot and human volunteer(left), robot and PMLA(right)

density is calculated as

$$e_A = \frac{\int_{s_{\text{impact\_start}}}^{s_{\text{impact\_stop}}} F \cdot ds}{A_{\text{end-effector}}}. \quad (1)$$

3. RESULTS

The same experiments as with human volunteers were also performed with the PMLA. The quantities measured during the experiment, that provide the most information regarding the injury level caused by an impact, are the impact energy density and the maximal impact force. Relative deviation of the impact responses for six human volunteers compared to the impact responses of the PMLA are given in Figs. 3 and 4 for line impact robot end-effector. For a given experiment in the figure (e.g. experiment number 6, robot end-effector deceleration 2 m/s^2, stop point depth 30 mm) the relative deviation was calculated by subtracting the impact response of the human volunteer from the impact response of the PMLA. The result of the subtraction was divided by the impact response of the PMLA. Since there were six volunteers the procedure was repeated six times for each experiment. Results are for a given experiment and six volunteers presented using boxplot in terms of median values (bold solid lines), 25th and 75th percentile values (error boxes), and 5th and 95th percentile values (error bars). The red plus marks the data beyond the 5th and 95th percentile. The red rectangle defines the scope of impacts that cause higher impact energy densities and are the most important for evaluation.

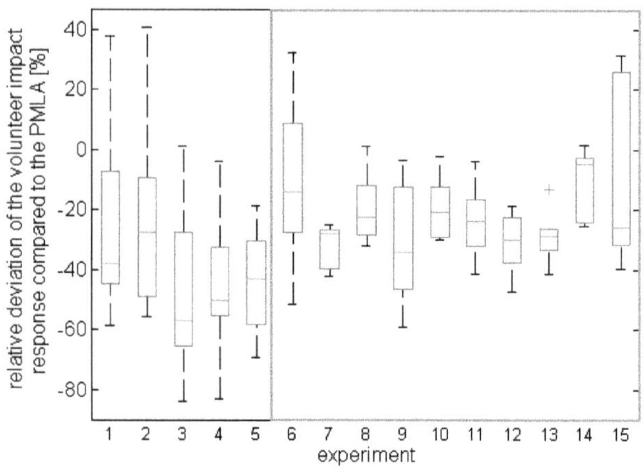

Figure 3: Relative deviation of the impact energy density of the volunteer impact response compared to the PMLA, line robot end-effector

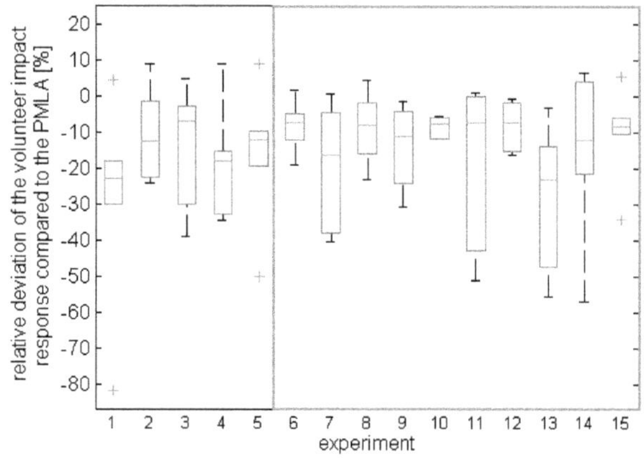

Figure 4: Relative deviation of the impact force of the volunteer impact response compared to the PMLA, line robot end-effector

4. CONCLUSIONS AND FUTURE WORKS

We developed a passive mechanical lower arm (PMLA). The results of the experiments indicate that the PMLA is a good emulation system of the passive human lower arm considering that the volunteers also had quite different lower arm properties (e.g. weight of the lower arm, soft tissue thickness). In our further investigations more dangerous experiments will be performed only with the PMLA. The robot end-effector will be displaced toward the point of impact along a straight line at a constant speed or constant acceleration.

5. ACKNOWLEDGMENTS

This work was supported in part by the European Union. "Operation part financed by the European Union, European Social Fund."

6. REFERENCES

[1] B. Povse, D. Koritnik, R. Kamnik, T. Bajd, and M. Munih, "Emulation system for assessment of human-robot collision," *Meccanica*, vol. 46, no. 6, pp. 1363–1371, 2011.

[2] S. Haddadin, A. Albu-Schaffer, A. D. Luca, and G. Hirzinger, "Evaluation of collision detection and reaction for a human-friendly robot on biological tissues," in *International Workshop on Technical Challenges and Dependable Robots in Human Environments (IARP 2008)*, (Pasadena, USA), 2008.

[3] S. Haddadin, A. Albu-Schaffer, and G. Hirzinger, "Safe physical human-robot interaction: Measurements, analysis & new insights," in *International Symposium on Robotics Research (ISRR2007)*, (Hiroshima, Japan), 2007.

[4] S. Haddadin, A. Albu-Schaffer, and G. Hirzinger, "The role of the robot mass and velocity in physical human-robot interaction - part 1: Unconstrained blunt impacts," in *IEEE International Conference on Robotics and Automation (ICRA2008)*, (Pasadena, USA), pp. 1331–1338, 2008.

Shared Gaze in Remote Spoken HRI During Distributed Military Operations

Zahar Prasov
U.S. Army Research Laboratory
Aberdeen Proving Ground, MD
zahar.prasov.ctr@us.army.mil

ABSTRACT

Collaboration between distributed human and robot partners during military operations is becoming more necessary. In order to enable efficient real-time communication, it is important to develop user interfaces that support robust spoken language understanding capabilities. As a step toward achieving this objective, this work examines the role of *shared gaze* between a human and robot during remote spoken collaboration engaged in a distributed military operations. Preliminary results have shown that an interface that supports shared gaze between a human and robot for a remote collaborative HRI search task has potential to improve automated language understanding as well as task efficiency.

Categories and Subject Descriptors

H.5 [**Information Interfaces and Presentation**]: *Miscellaneous*

General Terms

Algorithms, Experimentation

Keywords

Shared Gaze, Joint Attention, HRI, Remote Communication

1. INTRODUCTION

Military operations are increasingly depending on unmanned robotic technology, which can be used to enhance operational performance and soldier safety. In a typical scenario a soldier will interact with one or more semi-autonomous robots to perform a military mission. Ultimately, military operations will be performed by teams of human-robot peers. The challenge for enabling such peer-to-peer collaboration is to develop robots that support autonomous functionality as well as social interaction at appropriate times. Robot partners will need to have natural communication, information exchange, adaptation, learning, and training capabilities to optimally utilize the human-robot team dynamic. Moreover, robots will need to provide evidence of correct functionality to human partners in order to establish a certain level of trust. In this scenario, human partners will become responsible for understanding situational awareness, such as social and cultural context, changing tactics, meta-goals, etc [1].

To ensure safety of human team members during military operations, much of HRI needs to be remote and distributed, where human and robot team members are separated spatially or even temporally. Sophisticated human-robot paradigms will require multiple distributed humans to interact with robots and software agents to coordinate mission plans, human activities, and system resources. This involves designing efficient multimodal interfaces that facilitate communication between human and remote robot teammates. Recent work toward this goal has involved designing immersive visual, auditory, and tactile displays for HRI [3] as well as developing spoken language understanding in HRI [10, 11]. While there are many open research questions that must be addressed to enable such complex human-robot coordination, I am particularly interested in progressing human-robot communication for distributed military operations (e.g. search, patrol, surveilance) in cluttered urban environments. My work focuses on improving spatial language understanding during remote spoken HRI. As a part of this effort, I am currently investigating the role of *shared gaze* between a human and robot–that is, real-time awareness of each other's focus of attention–during remote spoken interaction.

2. SPATIAL LANGUAGE UNDERSTANDING

In situated human-machine dialogue, human users often need to make linguistic references to extralinguistic entities in the environment. Additionally, the most natural way to refer to an object may be via a spatial expression (e.g., "the truck to the right of the tree") rather than a more complex and descriptive noun phrase. Reliably resolving these references is critical for dialogue success and useful human-robot collaboration. However, reference resolution remains a challenging problem, partly due to limited speech and language processing capabilities caused by poor speech recognition (ASR), ambiguous language, and insufficient pragmatic knowledge [9]. Identifying and resolving spatial references provides additional challenges, namely ambiguities in the *frame of reference* [7, 5] (e.g. absolute, intrinsic, relative).

3. EYE GAZE DURING REMOTE COLLABORATION

As an attempt to address these challenges, my prior work has used eye gaze to disambiguate spoken language. Previous psycholinguistic findings have shown that eye gaze is tightly linked with language production and comprehension [2]. Additionally, it has been shown that speakers and listeners monitor each other's gaze shifts to perceivable objects in the environment and that this joint attention facilitates mutual rapport and comprehension [6]. Attending to an object that is being viewed by a conversational partner can provide valuable information about what is being talked about. For example, shared gaze between two human collaborators has been shown to improve task efficiency in a remote collaborative search tasks [8]. In this work two remotely located partners viewed a scene of a city and attempted to locate and reach consensus on a randomly appearing sniper target. The partners could communicate in one of the following conditions: shared speech, shared gaze (gaze cursors), or both. Shared gaze greatly increased the speed of consensus; that is location of the target by a participant whose partner had already found the target. Shared speech further improved task efficiency. Moreover, eye gaze has been shown to improve speech recognition, attention prediction, and reference resolution during human-machine dialogue [9]. Motivated by previous work, my hypothesis is that shared gaze plays an important role during remote spoken human-robot collaboration.

4. SHARED GAZE DURING REMOTE SPOKEN HRI

The objective of this work is to investigate the role of shared gaze between a human and robot during remote spoken collaboration engaged in distributed military operations. Shared gaze enables the partners to be aware of each other's real-time focus of attention as they collaborate. Such awareness enhances the ability of the partners to jointly attend to regions of interest as they speak about them. This, in turn, will allow the robot to disambiguate spoken language, including spatial expressions which may otherwise be difficult to interpret. Such awareness will also allow both the human and robot to detect a partner's inattention to important areas (e.g. potential military threats) and alert them to prompt a discussion. Preliminary results have shown that an interface that supports shared gaze between a human and robot for a remote collaborative HRI search task has potential benefits. For example, task efficiency is improved because spatial references produced by human participants tend to be short and simple expressions (e.g. "look here"). Such simple expression, rather than complex spatial expressions, are more likely to be correctly interpreted by the computational system.

Many open research questions persist regarding how to best use shared gaze in HRI and to what extent it can be beneficial. For example, it is unclear how to best represent "robot gaze" (i.e. focus of attention). A computational model of visual salience [4] can be used as a baseline, but it does not include discourse salience–which may also have an important impact on the focus of attention. Moreover, humans and robots may have different sensors which will affect their focus of attention. Additionally, it is unclear how to best incorporate shared gaze information for language interpretation and, in particular, what effect this has on interpretation of spatial references. My current work aims to address such research questions as a step toward remote spoken HRI during distributed military operations.

5. ACKNOWLEDGMENTS

This research was supported in part by an appointment to the U.S. Army Research Laboratory Postdoctoral Fellowship Program administered by the Oak Ridge Associated Universitites through a contract with the U.S. Army Research Laboratory.

6. REFERENCES

[1] J. Y. C. Chen, E. C. Haas, and M. J. Barnes. Human performance issues and user interface design for teleoperated robots. *IEEE Transactions on Systems, Man, and Cybernetics, Part C: Applications and Reviews*, 37(6):1231–1245, 2007.

[2] Z. M. Griffin and K. Bock. What the eyes say about speaking. In *Psychological Science*, volume 11, pages 274–279, 2000.

[3] E. C. Haas and J. B. F. van Erp. *Multimodal Research for Human Robot Interactions*, pages 271–292. Human-Robot Interaction in Future Military Operations. Farnham: Ashgate Publishing, 2010.

[4] L. Itti, C. Koch, and E. Niebur. A model of saliency-based visual attention for rapid scene analysis. *IEEE Transactions on Pattern Analysis and Machine Intelligence*, 20:1254–1259, 1998.

[5] C. Liu, J. Walker, and J. Chai. Ambiguities in spatial language understanding in situated human robot dialogue. In *AAAI 2010 Fall Symposium on Dialogue with Robots*, 2010.

[6] C. Moore and P. J. Dunham, editors. *Joint Attention: Its Origins and Role in Development*. Lawrence Erlbaum Associates, 1995.

[7] R. Moratz and T. Tenbrink. Spatial reference in linguistic human-robot interaction: Iterative, empirically supported development of a model of projective relations. *Spatial Cognition and Computation*, 6(1):63–107, 2006.

[8] M. Neider, X. Chen, C. Dickinson, S. Brennan, and G. Zelinsky. Coordinating spatial referencing using shared gaze. *Psychonomic bulletin review*, 17(5):718–724, 2010.

[9] Z. Prasov and J. Y. Chai. Fusing eye gaze with speech recognition hypotheses to resolve exophoric references in situated dialogue. In *Proceedings of the Conference on Empirical Methods in Natural Language Processing (EMNLP)*, 2010.

[10] M. Staudte and M. Crocker. Visual attention in spoken human-robot interaction. In *Proceedings of the 4th ACM/IEEE International Conference on Human Robot Interaction (HRI'09)*, pages 77–84, New York, NY, 2009. ACM.

[11] C. Yu, M. Scheutz, and P. Schermerhorn. Investigating multimodal real-time patterns of joint attention in an HRI word learning task. In *Proceedings of the 5th ACM/IEEE International Conference on Human-Robot Interaction (HRI'10)*, pages 309–316, New York, NY, 2010. ACM.

A Social Robot as an Aloud Reader: Putting together Recognition and Synthesis of Voice and Gestures for HRI Experimentation

Arnaud Ramey
RobotiscLab, Univ. Carlos III of Madrid
Leganés, Spain
arnaud.ramey@m4x.org

Javier F. Gorostiza
RobotiscLab, Univ. Carlos III of Madrid
Leganés, Spain
jgorosti@ing.uc3m.es

Miguel A. Salichs
RobotiscLab, Univ. Carlos III of Madrid
Leganés, Spain
salichs@ing.uc3m.es

ABSTRACT

Advances in voice recognition have made possible applications in robotics controlled by voice only. However, user input through gestures and robot output gestures both create a more vivid interaction experience. In this article, we present an aloud reading application offering all these interaction methods for the HRI-research robot Maggie. It gives us a testbed for user studies investigating the effect of these additional interaction methods.

Categories and Subject Descriptors

H.5.2 [**User interfaces**]: Voice I/O; I.2.9 [**Robotics**]: Operator interfaces; H.5.2 [**User interfaces**]: Theory and methods

General Terms

Design, Experimentation, Human Factors

Keywords

Gesture Recognition, ETTS, Multimodal Interaction

1. INTRODUCTION

Over the last few years, both voice recognition and synthesis have made great advances, which offers new possibilities in robotics, and especially makes possible creating voice-only controlled robots, such as the flower robot presented in [2]. In [4], it is made use of a robot with optional arms and eyes to test the attention of the users to the robot instructions. It underlines the gain of interest from the users when the robot performs, in addition to the voice, human-like gestures. It also stresses the difficulties of older users to use dialogue-only based interfaces, which suggests the use of additional interaction methods.

In this article, we present an application for the social robot Maggie converting it into an aloud reader: it can utter aloud some content such as books or news feeds. We couple both approaches: a voice-enabled interface with additional support for gesture-based input and output.

HRI'12, March 5–8, 2012, Boston, Massachusetts, USA.
ACM 978-1-4503-1063-5/12/03.

2. DESCRIPTION OF THE SYSTEM

The robot Maggie hardware and software architectures are presented in [1]. The latter is made of parallel modules synchronizing with one another thanks to short messages called *events*. A detailed description of the developed modules follows.

2.1 Voice recognition and synthesis

Speech recognition (*ASR*) in Maggie is based on the commercial software suite Loquendo ASR. For voice synthesis, the robot presents an Emotional Text-To-Speech (*ETTS*) architecture: it enables the conversion of a written text into sound, but with a voice depending on the emotional state of the robot. This affects a number of parameters, as the voice pitch, its speed, its volume, etc. It is based on the Loquendo ETTS software.

2.2 Gesture recognition

A gesture recognition system has been previously developed for Maggie [5]. It makes use of the Kinect time-of-flight camera. So-called *primitives*, which correspond to quick moves of the user's hand into a given direction, are detected. For instance, a sweeping move to the right (along the x axis) is a primitive. A full gesture is then represented as a sequence of these primitives. Each gesture is associated to an user-defined action, which is executed when the gesture is recognized.

2.3 Gesture synthesis

The robot is able to express her mood, if is paying attention, and other linguistic gestures as greetings, affirmation, agreement, pointing, etc.

Most of the gesture systems on Social Robotics are based on a *gestionary*, that is, a dictionary or a set of discrete gestures ([3], and many others). The system developed here also include the feature of generating gestures as "attitudes": they can be also derived on the fly as a dynamic movement based on sensory-motor feedback functions. For example, the robot can exaggerate a happiness attitude when the user is closer.

2.4 Core application: the aloud reader

The robot will read aloud some user-chosen content, while obeying user instructions and performing output gestures.

2.4.1 Input sources

The texts supplied as input can currently be of two types. Use can be made of English and Spanish literature books provided by the Gutenberg project, a project aiming at making available online books with an expired copyright. The robot can also subscribe to RSS feeds from important national and international newspapers, such as El Mundo.
The raw text can be extended thanks to metadata, which goal follows.

2.4.2 Application output

The main expected outputis the proper aloud reading, that is, is the vocal pronunciation of the wanted text. The text fetched as presented in 2.4.1, is cut into short chunks sent to the ETTS module.
Besides, [6] for instance showed that an ETTS system that does not take into account the emotional state of the content strongly lacks of expressivity. When encountered, the metadata tags, which come within the raw text, but are indicated via a simple markup language. affect the robot current emotion, resulting in changes in both its way of moving and speaking. as seen in 2.1 and 2.3. For instance, an excerpt, if it comes with a "sad" metadata tag, will be read with a sad voice and gestures showing a sad behavior.

2.4.3 Application control: input methods

The software architecture allows the application control with *events*, which can easily be emitted from any other process. Thus, it is easy to turn an existing program into a control module of the aloud reader. The current structure is visible in figure 1.
Voice orders can be also transmitted to the application thanks to the ASR system seen in 2.1. If the user utters the order *"go to next headline"*, the ASR module will convert that information into the corresponding event, which will make the robot perform the wanted action.
The gesture recognition system presented in 2.2 is used to give orders to the robot. In a similar way to the voice control, a sweeping gesture will be converted into the appropriate event to go to the next headline or next chapter.

3. RESULTS AND CONCLUSIONS

The aloud reader application has been implemented and tested in the robot Maggie. A sample picture of an end-user using it is visible in figure 2 The gesture recognizer lightweight workload allows it to run in the background with no delay on the recognition, as does the ASR and coupled with the agile handling of the queue of ETTS utterances, a fluent reading without pauses is generated.
We now aim at measuring how both input and output gestures change the user experience. A user study would allow us to quantify if they help maintaining the users' attention and how.

4. ACKNOWLEDGMENTS

The authors gratefully acknowledge the funds provided by the Spanish MICINN (Ministry of Science and Innovation) through the project *"A new Approach to Social Robotics (AROS)"*.

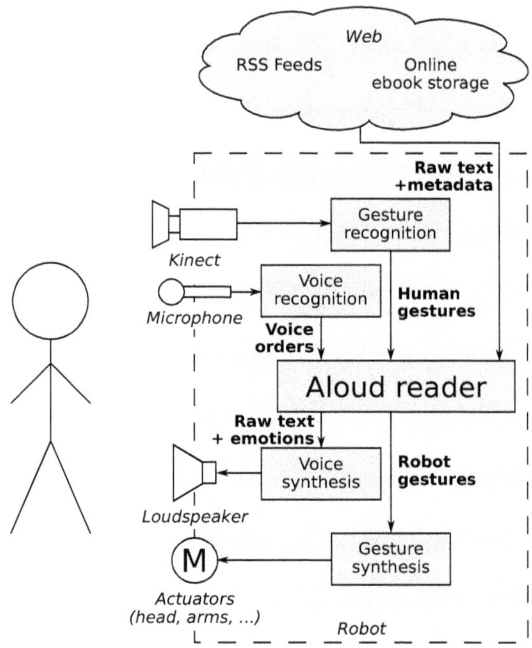

Figure 1: The current architecture of the aloud reader

Figure 2: An user listening to the robot Maggie as an aloud reader, which gestures show its current (happy) emotion.

5. REFERENCES

[1] R. Barber and M. Salichs. *A new human based architecture for intelligent autonomous robots*, pages 85–90. Elsevier, 2002.
[2] S. Chang, S. Ham, and D. Suh. Rohini: A robotic flower system for intuitive smart home interface. In *Control Automation and Systems (ICCAS), 2010 International Conference on*, pages 1773 –1776, oct. 2010.
[3] H. Ishiguro, T. Ono, M. Imai, T. Maeda, T. Kanda, and R. Nakatsu. Robovie: an interactive humanoid robot. *Industrial Robotics*, 28:498–503, 2001.
[4] H. Osawa, J. Orszulak, K. Godfrey, M. Imai, and J. Coughlin. Improving voice interaction for older people using an attachable gesture robot. In *RO-MAN, 2010 IEEE*, pages 179 –184, sept. 2010.
[5] A. Ramey, V. González-Pacheco, and M. A. Salichs. Integration of a low-cost rgb-d sensor in a social robot for gesture recognition. In *Proceedings of the 6th international conference on Human-robot interaction*, HRI '11, pages 229–230, New York, NY, USA, 2011. ACM.
[6] S. Roekhaut, J.-P. Goldman, and A. C. Simon. A model for varying speaking style in tts systems. *Proceedings of the 5th International Conference on Speech Prosody SP2010*, (Table 1):4–7, 2010.

Modified Social Force Model with Face Pose for Human Collision Avoidance

Photchara Ratsamee, Yasushi Mae, Kenichi Ohara,
Tomohito Takubo and Tatsuo Arai
Osaka University,1-3 Machikaneyama
Toyonaka, Osaka, Japan
ratsamee,mae,k-oohara,takubo,arai @sys.es.osaka-u.ac.jp

ABSTRACT

In order for robots to be a part of human society, their social accceptance is an important issue if smooth interaction with humans is to be achieved. We propose a modified social force model that allows robots to move naturally like humans, based on estimated human motion and face pose. We add to the previous model the effect of the force due to face pose, in order to predict human motion and compute the robot motion itself. Our approach was implemented and tested on a real humanoid robot in a situation in which a human is confronted with a robot in an indoor environment. Experimental results illustrate that the robot is able to perform human-like navigation by avoiding the human in a face-to-face confrontation. Our system provides accurate face pose tracking that allows a robot to have a more realistic behaviour compared to the original social force model.

Categories and Subject Descriptors

I.2.9 [**Computing Methodologies**]: Artificial Intelligence—*Robotics, Operator interfaces*; J.m [**Computer Applications**]: Miscellaneous

General Terms

Theory, Design

Keywords

Social Force Model, Face Pose, Human Tracking and Path Planning

1. INTRODUCTION

If robots are expected to share the same environment with humans, the ability to understand and act like humans is the main concentration for their social acceptance. We propose a modified social force model which is used for human motion prediction based on human face pose tracking. The modified social force model is also used in humanoid robot path planning, so that the humanoid robot is perceived as a human-being, due to the fact that its path is more natural and comfortable for humans in human-robot coexisting environments.

The social force model proposed by Helbing et al. [1] is a particle-based model where humans are considered as particles influenced by not only inner motivation to reach the goal, but also social and physical effects from other obstacles and other peoples. Many works use the social force model for realistic estimation of human motion. The techniques presented so far estimated the location and motion of people using laser range finder for a human leg-detection approach [2, 3]. However, these approaches still treated humans as just moving obstacles, since the social impact of the human position alone is not sufficient evidence during face-to-face confrontation. To supplement the social force model, we introduce face pose as a key social issue, which makes humans treat each other in a way that differs from the way they treat other moving obstacles when they plan to avoid or interact with them. our proposal is based on the fact that humans normally feel increasingly uncomfortable when they are observed and usually look in the direction they will take when avoiding another person, providing that person with information of the intended direction by their face pose.

Figure 1: Social interaction, i.e. a human and a robot passing by each other. Our robot estimates the intention of the human by analyzing the latter's face pose and behaves like a human, making the human feel safe.

In this paper, we extend human motion prediction by taking into consideration human face pose information too. The modified social force model integrates the force due to face pose for analysis of human motion through human-like robot motion. Instead of pursuing shortest collision-free trajectory, our robot is able to consider not only physical constraints, but also perceive social constraints by analyzing human face pose; it is also able to design a more natural path according to the modified version of the social force model. The overall idea is presented in Fig. 1.

HRI'12, March 5–8, 2012, Boston, Massachusetts, USA.
ACM 978-1-4503-1063-5/12/03.

Figure 2: Human position and face pose are estimated during walking process.

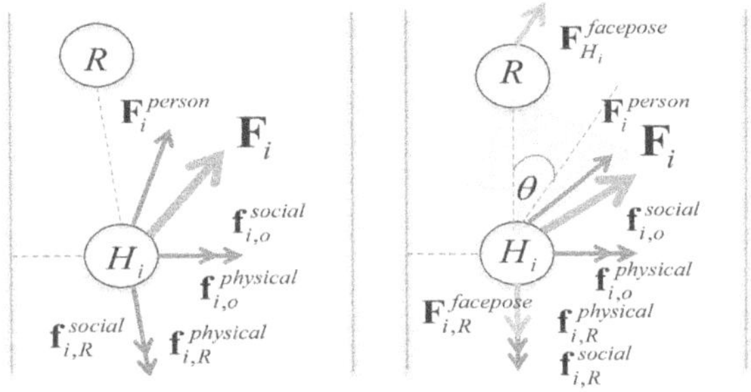

(a) (b)

Figure 3: The (a) original and (b) modified social force model based on face pose estimation.

2. THE MODIFIED SOCIAL FORCE MODEL WITH FACE POSE EFFECT

Basically, each individual human or robot has an intended goal and move according to their personal velocity. In the social force model, their motion is demonstrated by the superposition of 3 main forces: an attractive force that leads the human towards his/her intended goal, $\mathbf{F}^{goal}$, a repulsive force from other objects, $\mathbf{F}^{object}$, and a repulsive force from other humans, $\mathbf{F}^{human}$. Both $\mathbf{F}^{object}$ and $\mathbf{F}^{human}$ are result from a combination of social repulsive forces, $\mathbf{f}^{social}$ and physical repulsive forces, $\mathbf{f}^{physical}$. The resulting force, $\sum \mathbf{F}$, used for changing the motion of individual human H_i or robot R are defined as follows :

$$\sum \mathbf{F} = \mathbf{F}^{goal} + \mathbf{F}^{object} + \mathbf{F}^{human} \quad (1)$$

In face to face confrontation situation, subjects will look in the direction that they want to go to reduce uncomfortable feeling, and to show their own intentions. Hence, we propose a new force due to face pose which is modeled as :

$$\mathbf{F}^{facepose} = FSe^{\left(\frac{r_{i,R}-d_{i,R}}{s_R}\right)} \mathbf{v}_{i,R} \left(\lambda + (1+\lambda)\frac{1+\cos(\theta)}{2} \right) \quad (2)$$

FS specifies the strength of face pose effect. Its exponential growth of force is due to the range of the force, s_R, distance, $d_{i,R}$ and the sum of their radius, $r_{i,R}$. The term $\mathbf{v}_{i,R}$ is the face pose vector from H_i related to R which describes the direction of the force. The anisotropic factor is defined by λ and the different of face pose vector between them, θ. Therefore, our proposed resulting force is modeled as :

$$\sum \mathbf{F} = \mathbf{F}^{goal} + \mathbf{F}^{object} + \mathbf{F}^{human} + \mathbf{F}^{facepose} \quad (3)$$

Finally, $\mathbf{v}$ which describes the movements of human, H_i and path planning in robot, R, over time is derived from $\frac{d}{dt}\mathbf{v} = \frac{\sum \mathbf{F}}{m}$. An illustration of all forces is shown in Fig. 3.

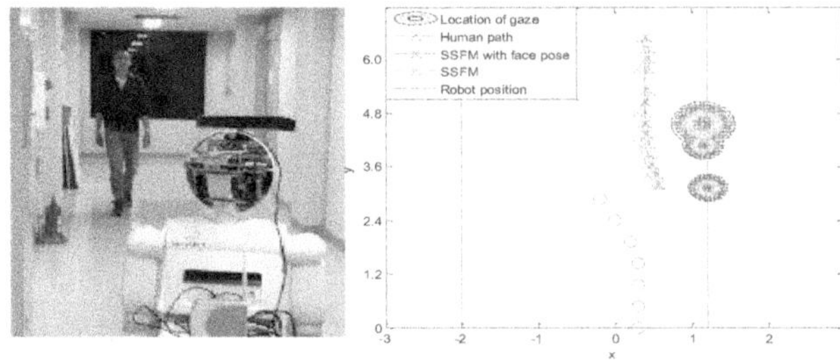

(a) (b)

Figure 4: (a) Experiment scene (b) The result of human tracking and robot path planning based on the social force model with face pose effect in the case of a robot avoiding the human in the corridor

3. PRELIMINARY RESULTS

Several experiments in a domestic environment where humans and robots are passing each other were performed. Estimating the goal from the observation of a tracked person includes the problem of intention or activity recognition, which can be understood by the face pose of the human as shown in Fig. 2. A face pose estimation method [4] to handle low resolution face images and scale change problem, especially the movement along the direction towards the camera was applied. The humanoid robot was able to track the human and his face pose and create a natural path using the proposed approach. Fig. 4 shows some examples of human trajectory prediction and robot path planning.

4. CONCLUSIONS

In this paper, we presented a modified social force model with an integrated face pose effect. Our approach extends human motion prediction and robot path planning by taking into account the influence of other people's face poses. This novel approach achieves precise human motion prediction and socially acceptable robot path planning.

5. REFERENCES

[1] Dirk Helbing and Peter Molnar. Social force model for pedestrian dynamics, Physical Review E, vol.51, no.5, pp.4282-4286, 1995.

[2] M. Luber, J. Stork, G. Tipaldi, and K. Arras. People tracking with human motion predictions from social forces, In *IEEE International Conference on Robotics and Automation (ICRA)*, pp. 464-469, 2010.

[3] Y. Tamura, T. Fukuzawa, and H. Asama. Smooth Collision human-robot coexisting environment, In *IEEE/RSJ International Conference on Intelligent Robots and Systems (IROS)*, pp. 3887-3892, 2010.

[4] P. Ratsamee, Y. Mae, T. Takubo, K. Ohara, and T. Arai. Face pose estimation in tracking person, In *Proc. Meeting on Image Recognition and Understanding (MIRU)*, Kanazawa, Japan, pp.660-667, 2011.

The Roomba Mood Ring: An Ambient-Display Robot

Daniel Rea
University of Manitoba
Department of Computer Science
Winnipeg, Manitoba R3T 2N2
umrea2@cc.umanitoba.ca

James E. Young
University of Manitoba
Department of Computer Science
Winnipeg, Manitoba R3T 2N2
young@cs.umanitoba.ca

Pourang Irani
University of Manitoba
Department of Computer Science
Winnipeg, Manitoba R3T 2N2
irani@cs.umanitoba.ca

ABSTRACT
We present a robot augmented with an ambient display that communicates using a multi-color halo. We use this robot in a public café-style setting where people vote on which colors the robot will display: we ask people to select a color which "best represents their mood." People can vote from a mobile device (e.g., smart phone or laptop) through a web interface. Thus, the robot's display is an abstract aggregate of the current mood of the room. Our research investigates how a robot with an ambient display may integrate into a space. For example, how will the robot alter how people use or perceive the environment, or how people will interact with the robot itself? In this paper we describe our initial prototype, an iRobot Roomba augmented with lights, and highlight the research questions driving our exploration, including initial study design.

Categories and Subject Descriptors
H.1.2 [**Models and principles**]: user/machine systems—*human factors*; H.5.3 [**Group and Organization Interfaces**]: user interfaces—*asynchronous interaction*

General Terms
Design, Experimentation, Human Factors

Keywords
Human-robot interaction, sociable robotics, group-robot interaction, ambient displays

1. INTRODUCTION
The use of peripheral ambient information for communication and interaction has garnered significant research interest (e.g., see [2]). Many have leveraged this communication channel to create *ambient displays*: 'glanceable' and 'unintrusive' electronic appliances that exploit our pre-attentive information processing abilities [2,3]. In this paper we posit that it can be useful to think of robots themselves as ambient displays, and to intentionally design them to use ambient cues for communication.

This is a timely consideration as Human-Robot Interaction research is increasingly targeting robots that enter people's everyday lives such as home utility or public service robots. These robots will often perform their tasks with high autonomy; even though robots and people will share spaces, people may only need to engage the robots for high-level direction or exceptional cases such as robot malfunction. Thus, the robot has an *ambient presence* in people's spaces, and people have *peripheral awareness* of the robot's actions and sounds, a presence which can be exploited by integrating ambient displays.

HRI'12, March 5–8, 2012, Boston, Massachusetts, USA.
ACM 978-1-4503-1063-5/12/03.

Figure 1. The modified Roomba on dark grey cement

One key difference between traditional ambient displays and our proposed robotic ones is that people tend to anthropomorphise robots more than other technologies. They attribute them with intentionality and animacy, and perhaps even assume robotic actions are signs of communication [1]. This is even the case when the robot is not designed as such, for example, people describe random quirks or technical attributes in terms of emotion or personality [4,5]. This raises the question of how the intrinsic anthropomorphic tendency will impact how people perceive the ambient display on the robot and the information it conveys.

In our work we will place an ambient-display robot into a public space to explore the question of how anthropomorphism relates to an ambient display. Our initial prototype and scenario is as follows: we place the robot in a café and enable people to report their mood to the robot by selecting a colour from a given palette (Figure 2). We mounted multi-color LEDs on an iRobot Roomba to broadcast ambient information in the form of a coloured halo (Figure 1). The robot aggregates all responses so that its colors abstractly and artistically represent the mood of the room. We do not do any explicit colour-to-emotion mapping or analysis, but rather directly use people's own selections.

Using this scenario, we conduct an exploratory study to observe how our ambient-display robot impacts the social café space. Our driving research questions are: 1) How is the ambient display robot received by people? 2) Do people tend to anthropomorphise the data from the ambient display, and if so, how is this manifested? 3) How does the ambient display integrate into the social space, e.g., does it impact how people use the space? Our investigation into these questions will serve as an important initial step toward understanding how people will interact with ambient-robot displays, and will provide groundwork for informing the design and development of robotic ambient appliances in general, for example, for other data sources, other robot types, and other display implementations.

2. PROTOTYPE

We designed our prototype to be used in a public space such as a café or lounge. We created a web interface (Figure 2) – accessible from a smart phone, laptop, or tablet – that people can use to interact with the robot. The web site presents the users with a palette and asks, "What colour do you feel right now?" Their choice (or "vote") is sent to the robot which will then broadcast that colour. The palette was constructed by selecting sample points along the hue, saturation, and lightness dimensions of the color space to create a palette of 36 colors [6].

We modified an iRobot Roomba with 16 bright, individually controllable RGB LEDs (BlinkM 24-bit lights). The lights are mounted to shine onto the floor around the robot in a halo (Figure 1). We use the Roomba's built-in cleaning mode.

We aggregate votes received and send real-time updates to the robot, displaying the results as a pie-graph style groupings of

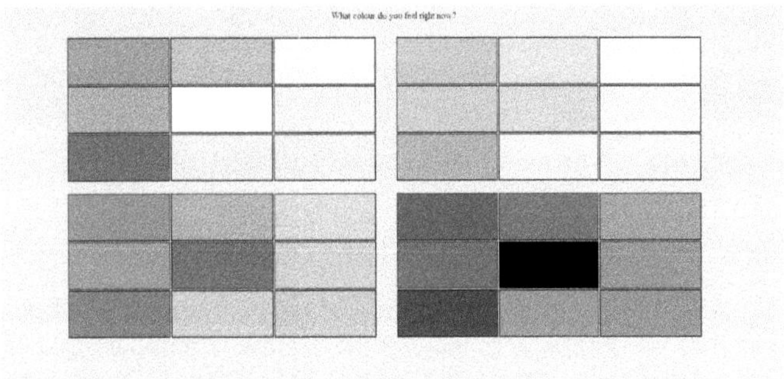

Figure 2: The web interface

color (Figure 1): we first sort votes by hue such that more votes create a larger colour region, and then calculate the saturation and lightness of the region using a weighted average over the votes in the region. While other mappings are possible, we resorted to a simple mapping in this first study.

3. EXPLORATORY STUDY DESIGN

We will perform an exploratory qualitative evaluation investigating how the robot is received by people, if anthropomorphism occurs (and how this impacts the reception), as well as how the social group is affected overall. The procedure is open ended: we will inform people of the robot's purpose, abilities, and its web interface with signs we will post throughout the room., We will passively solicit through these same signs for answers to an open-ended questionnaire about their impressions of our ambient robot display. In addition, through the web-interface tracking, we will analyze repeat usage patterns, both within a day and between multiple days, and the experimenters will take field notes of observations.

One expectation we have is that the robot will be anthropomorphized, and thus perhaps the colours will be interpreted as messages from, or emotions of the robot itself. At the same time, people will know that the information was reported by the group, which may change their interpretation. So we are curious to investigate how these robotic anthropomorphism and group communication aspects of the interface relate.

We also expect to see the robot affecting the social space. For example, if people perceive the colours as emotions, will the display affect abstract notions of the current *mood* of the environment? More specifically, what would happen if the robot shows many colours in the red spectrum and a person is not feeling "red"? They may choose to leave the environment early, or maybe they would leave feeling more "red". It may even make the user act more "red" to others in the space. The anthropomorphized robot's perceived emotions might evoke an empathic response, which may change how people perceive the physical space.

Another angle we will investigate is the sending and broadcasting of mood as a form of social media: people may use the robot as a bulletin board for their mood. It creates an intra-group communication situation, but with the robot as the medium instead of a traditional display.

Through these efforts we hope to gain insight into how ambient-display robots may be used in a social setting, and develop further targeted questions for continued investigation.

4. CONCLUSION

In this paper we present our initial exploration of *ambient display robotics*. We will investigate how such robots will be received and interpreted, how the ambient-display information may be anthropomorphized, and how such robots may impact environments they occupy.

We believe that a better understanding of how people interact with ambient-display robots will help in designing *group-robot interaction* that can enrich the way people interact with each other and their environments, and can enable people to actively shape their environments in new ways.

5. REFERENCES

[1] Sung, J.-Y, Guo, L., Grinter, R.E., and Christensen, H.I. 2006. "My Roomba Is Rambo": Intimate Home Appliances. *Proc. Ubicomp 2007*, pp. 145-162. ACM.

[2] Chang, A., Resner, B., Koerner B., Wang, X and Ishii, H., Lumitouch: An emotional communication device. *Proc. CHI EA 2001*, pp. 371-372. ACM.

[3] Zhao, A., and Stasko, J. What's Happening?: Promoting Community Awareness through Opportunistic, Peripheral Interfaces. *Proc. AVI 2002*, pp. 69-74. ACM.

[4] Forlizzi, J. and DiSalvo, C. Service Robots in the Domestic Environment: A Study of the Roomba Vacuum in the Home. *Proc. HRI 2006*, pp. 258-256

[5] Young, J. E.; Hawkins, R.; Sharlin, E. and Igarashi, T. Toward Acceptable Domestic Robots: Applying Insights from Social Psychology. *International Journal on Social Robotics*, 2009, 1(1);95-108

[6] Palmer, S.E, Schloss, K.B. An ecological valence theory of human color preference *PNAS 2010*, 107 (19):8877-8882

How to Use Non-Linguistic Utterances to Convey Emotion in Child-Robot Interaction

Robin Read
Centre for Robotics and Neural Systems
Plymouth University
Plymouth, United Kingdom
robin.read@plymouth.ac.uk

Tony Belpaeme
Centre for Robotics and Neural Systems
Plymouth University
Plymouth, United Kingdom
tony.belpaeme@plymouth.ac.uk

ABSTRACT

Vocal affective displays are vital for achieving engaging and effective Human-Robot Interaction. The same can be said for linguistic interaction also, however, while emphasis may be placed upon linguistic interaction, there are also inherent risks: users are bound to a single language, and breakdowns are frequent due to current technical limitations. This work explores the potential of non-linguistic utterances. A recent study is briefly outlined in which school children were asked to rate a variety of non-linguistic utterances on an affective level using a facial gesture tool. Results suggest, for example, that utterance rhythm may be an influential independent factor, whilst the pitch contour of an utterance may have little importance. Also evidence for categorical perception of emotions is presented, an issue that may impact important areas of HRI away from vocal displays of affect.

Categories and Subject Descriptors

I.2.9 [**Robotics**]: Operator Interfaces;
H.1.2 [**User/Machine Systems**]: Human factors

Keywords

Non-Linguistic Utterances, Child-Robot Interaction, Emotion, Affective Interpretation.

1. INTRODUCTION

Utterances made by robots need to be congruent with the user expectations. While it is well established that natural language plays a vital role in increasing interaction quality, breakdowns in natural language interaction are uncomfortable for users. In such cases, it is worth looking at alternative communication strategies. Solutions have been to constrain interactions to scripted scenarios, narrowing the scope of user responses, thus aiding their interpretation, or to switch to what has been termed "mother in law" mode, where the robot employs a set of general purpose utterances. These approaches have their limitations, as incorrect or repetitive linguistic responses are quickly identified by users. The work outlined here explores the potential of another alternative: non-linguistic utterances (NLUs).

NLUs have been used extensively and almost exclusively by the world of animation. R2D2, Chewbacca and WALL-E provide vivid examples of this. However, there is currently no principled understanding of how such utterances can be utilized in real social robots where there is great potential in the short term.

While having obvious shortcomings in comparison to natural spoken language, NLUs have a number of properties that make them particularly promising for HRI. Utterances are not bound to a particular language, an advantageous trait in multi-lingual and multi-cultural settings. There is no need to interpret semantic content from user utterances, thus settings that pose challenges for sensory equipment and technologies such as microphones and speech recognition are less problematic. Finally, as utterances are generally considered to contain less semantic content, the burden of interpretation lies with the human interactant, the *intelligent other*.

The majority of previous work has focused upon utilizing Text-To-Speech (TTS) technologies to generate and synthesize nonsense and gibberish *sentences* containing no or corrupted semantic content. These methods take inspiration from developments in affective Speech Synthesis [3]. Expression through sounds composed of beeps, squeaks and whirrs has received less attention. Existing work has generally focused on using simple, short single signal waves with raising or falling pitch frequencies as a means to express *positive* or *negative* states and intentions.

There are a number of commonalities in the previous work: most studies involved asking adult subjects to rate a small selection of stimuli within a laboratory environment frequently without the use of a real robot. This paper presents a brief overview of a recent study seeking to evaluate affective interpretations of young users rating a wide variety of NLUs more akin to sounds than spoken sentences, using a real robot in a real world environment. This approach allows assessing an important group of potential users whilst providing increased ecological validity in an environmental setting familiar to the subjects. The goal of this study was to gain insights as to how utterances may be coloured in order to steer affective user interpretations in a predictable manner.

2. EXPERIMENTAL SETUP

The study set out to probe a number of acoustic features of utterances that have been highlighted and queried in the related fields of speech synthesis and psychology, and their relation to NLUs. These features are the pitch base, range and contour of an acoustic signal, speech rate, the pause proportion to the spoken proportion, as well as the rhythm and tremolo.

HRI'12, March 5–8, 2012, Boston, Massachusetts, USA.
ACM 978-1-4503-1063-5/12/03.

Figure 1: The experimental setup: children listen to the Nao robot and record responses on a laptop. A supervisor is on hand during the experiments.

The study was conducted in a local primary school where children aged between 6 and 8 years were asked to evaluate non-linguistic utterances made by a robot[1] using a dynamic, animated affective measuring tool, the AffectButton [1] (Figure 1). The AffectButton dynamically interpolates between 9 prototypical facial expressions, encoding these expressions into a coordinate within a bi-normalized 3D affect space where the dimensions represent *Pleasure*, *Arousal* and *Dominance*. This novel approach to affective measurement has not been employed with NLUs before.

90 stimuli utterances were generated using a custom synthesis tool, NaoSqueak, composed of concatenated sound units consisting of single wave forms with modulated frequencies and amplitudes, as well as temporal modulations and presented to the subjects through the robot.

3. RESULTS

In total, 42 subjects took part in the study. Inspection of the data collected suggests categorical perception of affect (Figure 2) as there are densely populated regions of the affect space rather than a uniform distribution.

Our previous work has suggested that adults have promising levels of coherence in their affective interpretations of utterances [2], a finding that is to a certain extent echoed in the findings of this study with young children.

The results of this study suggest that the pitch contours of utterances do not play a pivotal role in colouring utterances. This is contrary to the hypotheses and findings of other studies from speech synthesis and psychology domains.

Further analysis also suggests that as the rhythm of an utterance becomes more pronounced, users tend to provide higher arousal ratings, while the other parameters yielded less clear results. We believe this is because affective colouring for such NLUs is a subtle and complex task where a variety of variables may be dependant upon each other.

4. CONCLUSIONS AND FUTURE WORK

The study found that children between 6 and 8 years old readily attribute emotion to a robot that produces non-linguistic utterances, such as clicks, whirrs and beeps. However, children do not appear to always be consistent in their interpretation of utterances, and it is expected that the interpretation can be coloured by contextual information and other modalities, such as gesture or facial expressions of the robot. It is striking how children show strong categorical

[1]An Aldebaran Nao robotic platform.

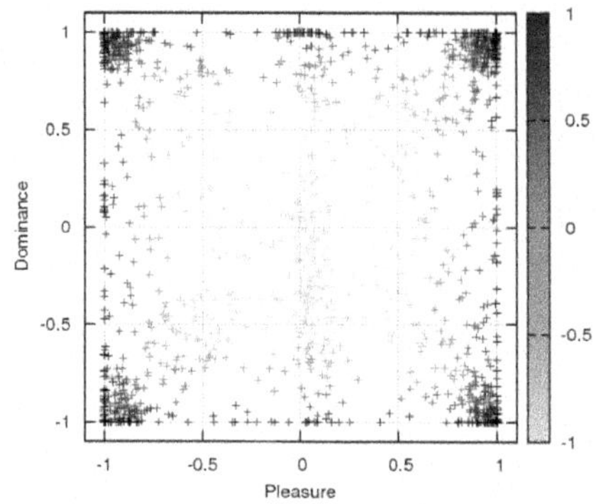

Figure 2: Plot of all data captured in the study - note the dense clusters of data points in the corners of the graph.

perception when interpreting the robot's utterances. There is no subtlety in their interpretation: the robot is –in their words– either sad, happy, angry, scared, surprised or tired, but they seldomly interpret utterances in more subtle emotions. We believe that upon closer inspection categorical perception will be observed in other modalities also, having an significant impact on the design of HRI for younger children: any effort to convey subtlety might be a lost effort.

The role of pitch contours in affective communication and colouring of vocalizations is subject to ongoing debate, having important implications for the generation of NLUs: it may be possible to generate utterances with *unique* contour shapes, in effect increasing the repetitor of utterances, whilst evoking similar affective responses in users.

The future direction of this work is to pursue how machine learning techniques may be adopted and employed to generate and synthesize utterances in an online manner. This avoids the need for databases of affectively labeled samples whilst still synthesising unique utterances that reduce the risk of the robot being perceived as a pre-programmed agent.

5. ACKNOWLEDGMENTS

This work is (partially) funded by the EU FP7 ALIZ-E project (grant 248116).

6. REFERENCES

[1] J. Broekens, A. Pronker, and M. Neuteboom. Real Time Labeling of Affect in Music Using the AffectButton. In *ACM Multimedia 2010 Workshop - Affective Interaction in Natural Environments (AFFINE)*, Firenze, Italy, 2010.

[2] R. Read and T. Belpaeme. Interpreting Non-Linguistic Utterances by Robots : Studying the Influence of Physical Appearance. In *ACM Multimedia 2010 Workshop - Affective Interaction in Natural Environments (AFFINE)*, Firenze, Italy, 2010.

[3] S. Yilmazyildiz, D. Henderickx, B. Vanderborght, W. Verhelst, E. Soetens, and D. Lefeber. EMOGIB : Emotional Gibberish Speech Database for Affective Human-Robot Interaction. In *Proceedings of the International Conference on Affective Computing and Intelligent Interaction (ACII2011)*, Memphis, Tennessee, 2011.

Ask, Inform, or Act: Communication with a Robotic Patient before Haptic Action

Timothy J. Martin, Allison P. Rzepczynski, and Laurel D. Riek
Computer Science and Engineering
University of Notre Dame
Notre Dame, IN
{tmartin6, arzepczy, lriek}@nd.edu

ABSTRACT

Currently in medical education, clinical students learn how to interact with real patients via simulated patients, which are inexpressive, teleoperated robot mannequins. We obtained five simulations that used such a robot to explore verbal communication between clinical students and the robot patient, specifically if the students sought approval before performing haptic-actions. We found that in our sample, student clinicians frequently acted without seeking approval or providing information to the robot patient. We hope to further our studies in order to identify if either current training of clinical students in communication is ineffective, or if the robot patients are too nonhuman-like and inexpressive to engender appropriate communication.

Categories and Subject Descriptors

I.2.9 [Artificial Intelligence]: Robotics; J.3 [Life and Medical Science]: Medical information systems; H.5.2 [User Interfaces]: Haptic I/O

Keywords

Robotics, patient simulation, clinical communication, haptics

1. INTRODUCTION

In the field of medical education, there has been a push toward allowing clinical students a chance to hone their procedural and communication skills in simulated environments. Students typically work on full-bodied simulated patients (i.e., teleoperated mannequins), which afford a safe, replicable learning environment before interacting with real ones [5].

The problem with the current state of the art of these mannequins is that they do not offer a lifelike experience for clinicians, because most are inexpressive – unable to show facial expressions, move their head, or gesture. We wondered how such inexpressive robots might affect clinical training, and performed an exploratory study to find out.

Last summer, we observed five clinical simulations at the Ysbyty Gwynedd Hospital in Bangor, Wales. These simulations were comprised of interdisciplinary teams of medical and nursing students undergoing training on a robotic patient mannequin. The Bangor robot has no facial expressivity, though its torso can breathe, its heart can beat, and it can experience physiological changes to medication. It can also "speak" via an internal speaker, thought its lips do not move. It is semi-autonomous, controlled by a technician and clinical instructors behind a one-way mirror.

HRI'12, March 5–8, 2012, Boston, Massachusetts, USA.
ACM 978-1-4503-1063-5/12/03.

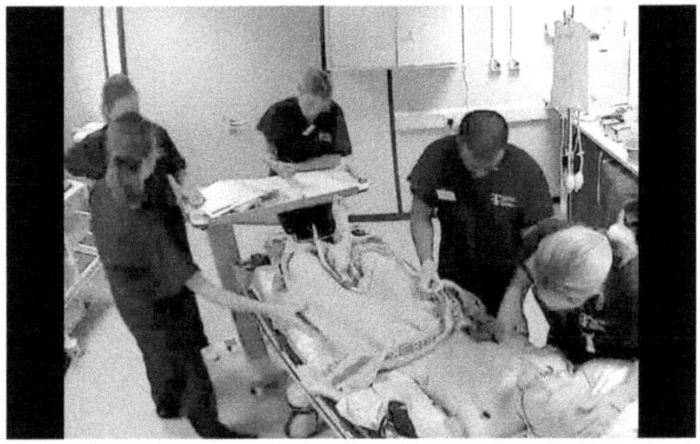

Figure 1: An example of a haptic action on the robotic patient.

According to American and Welsh medical standards, communication between a healthcare provider and a patient is an important aspect in doctor-patient relations [4, 6]. Patients who receive effective communication from their providers positively respond to their treatments and regimens [2]. Poor communication alienates patients and causes anxiety and confusion. Thus, doctors pledge to communicate honestly. When treatment actions are necessary, acute care providers seek approval from a patient before proceeding, unless the patients are unable to make conscious decisions [1].

We analyzed the Ysbyty Gwynedd Hospital simulations in order to explore how providers are being trained to communicate with patients. We were particularly interested to see if providers sought approval from the robotic patient before proceeding with any physical actions, such as interventions, diagnostics, taking vitals, and gown opening and closing.

To analyze this dataset, we first developed a haptic-action coding scheme, which is described in Section 2. Two independent coders then used the coding scheme on the dataset, which is described in Section 3. Section 4 describes our results, and Section 5 discusses the implications of them for both the robotics and the medical education communities.

2. HAPTIC-ACTION CODING SCHEME

We developed a haptic-action coding scheme to analyze the Bangor simulations. Despite surveying the literature we could find not find a scheme that encompassed our goals; instead, our coding scheme was based on typical actions that occur in acute care environments.

Each haptic-action was sorted according to type: interventions, gown, vitals, or diagnostics. *Interventions* included haptic-actions related to inserting an IV, taking blood, putting on an oxygen mask, spraying the throat, administering drugs, suctioning, and attaching ECG leads. *Gown haptic-actions* occurred whenever a student doctor or nurse opened or closed the patient's gown. *Vital haptic-actions* consisted of blood pressure and blood sugar readings, checking breathing with the stethoscope, attaching a finger monitor, and taking the patient's temperature. *Diagnostics haptic-actions* consisted checking the patient's pupils, their forehead for a fever, their pulse, their skin for clamminess or temperature changes, their airway, and their tongue.

For each haptic-action, we noted what verbal communication, if any, took place between the patient and the doctors or nurses. We determined that there were three types of verbal communication to the patient: *sought approval* (i.e., "Can I take your temperature?"), *gave information* but didn't seek approval (i.e., "I am going to take your temperature."), or *just acted* without saying anything to the patient.

3. METHODOLOGY

We used a dataset from Ysbyty Gwynedd Hospital's simulation day, where medical and nursing students interacted with a patient simulator robot. The dataset included eight fifth-year medical students from a six-year British medical school program and five final year nursing students from a three-year British nursing school program, along with one "behind the curtain" confederate nurse. Two anesthesiologist faculty members controlled the robot, as well as a technical operator who voiced for the patient according to faculty members' directions. All participants gave informed consent to participate in this research.

The students participated in five scenarios where they encountered patients having chest pain, left ventricular fluctuations, anaphylactic shock, septic shock, and an asthma attack. The students' task was to accurately identify the patient's condition based on their symptoms and then use an appropriate method to care for the patient. The confederate nurse would step in to guide the students when necessary.

Two independent coders analyzed the data using the haptic-action coding scheme. For this paper, we focused on seeking approval. As part of the American College of Emergency Physicians Code of Ethics, doctors should always communicate truthfully with their patients and obtain informed consent for treatment, unless the patient is in a dire situation [1]. The Welsh medical standards were found to be the same as American medical standards [6]. Thus, for this dataset, ideally clinicians would seek approval and/or give information before a doctor or nurse performed a haptic-action on the patient.

To test for inter-rater reliability, we calculated Krippendorff's α using SPSS. Then, to explore the data, we ran a MANOVA statistical analysis because we had multiple dependent variables - seeking approval, gave information, and just acted. These variables represented the degrees of communication that the provider shared with the patient before performing actions on the patient. Our fixed factor was the varying haptic-action types performed on the patient.

4. RESULTS

To determine inter-rater reliability between our coders, we calculated Krippendorff's α. Here, α was 0.88, which indicates high reliability.

Our MANOVA test indicated a significant effect of haptic action type on the form of verbal communication to the patient, $F(9, 108)$, $p < .001$. A discriminant analysis followed our MANOVA test to explore the relationships among the dependent variables (the levels of patient communication). For the *gown* haptic-actions and *diagnostic* haptic actions, we found a significant difference between *sought approval* and *just acting*. Figure 2 shows the average number of times each haptic action was performed across all five simulations. Our statistical analysis is ongoing, but from the graph it is clear that for opening and closing the gown and for taking diagnostics of the robot patient, it is more likely that clinicians just acted without first seeking approval or even telling the patient what they were doing.

5. DISCUSSION

Our results show that for our sample, student clinicians routinely acted without seeking approval or providing information to robot patients. Thus, as far as the medical education literature is concerned, these students weren't following the guidelines on informed consent.

It is unclear if this result is due to the training they received in patient interaction, or if it is due to the non-expressiveness of the robot. Further investigation is necessary to explore this question in depth. However, based on this and other work (c.f. [3]), we hypothesize a more expressive robot would lead to improved communication and patient empathy.

6. ACKNOWLEDGMENTS

The authors would like to thank Adam Ford and Einir Mowll at the Ysbyty Gwynedd Hospital in Bangor, Wales for allowing us to observe and record their clinical simulations.

7. REFERENCES

[1] American College of Emergency Physicians. (2011). Code of Ethics for Emergency Physicians. Last accessed Nov 2011, http://www.acep.org/content.aspx?id=29144

[2] Van Survellen, G. (2009). *Communication Skills for the Health Care Professional: Concepts, Practice, Evidence.* Jones and Bartlett.

[3] Riek, L.D., Rabinowitch, T., Chakrabarti, B., and Robinson, P. (2009). "Empathizing with robots: Fellow Feeling Along the Anthropomorphic Spectrum". *In Proc. of the 3rd Int'l Conference on Affective Computing and Intelligent Interaction (ACII).*

[4] Scheeres, H., Slade, D., Manidis, M., McGregor, J. and Matthiessen, C. (2008). Communicating in Hospital Emergency Departments. *Prospect Journal,* 23(2).

[5] Argani, C. H., Eichelberger, M., Deering, S., & Satin, A. J. (2011). The case for simulation as part of a comprehensive patient safety program. *American Journal of Obstetrics and Gynecology.*

[6] Healthcare Standards for Wales. (May 2005). Making the Connections Designed for Life. Last accessed Jan 2012, http://www.hiw.org.uk/ Documents/477/Healthcare Standards for Wales.pdf

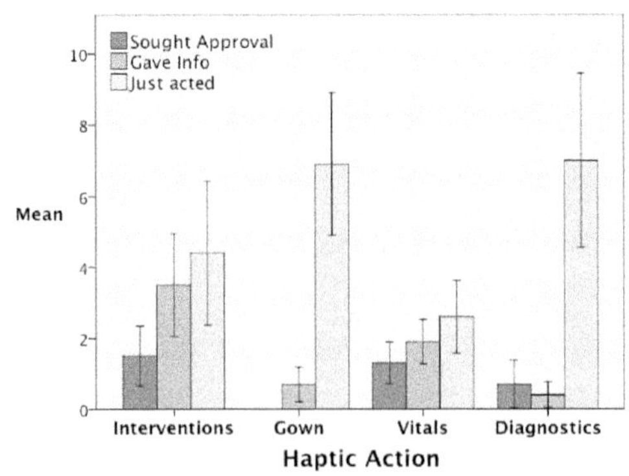

Figure 2: Mean number of haptic actions performed (across all simulations) with their degree of patient communication.

Creating Human-Robot Rapport with Mobile Sculpture

Tina Yue*, Alexandra E. Janiw*, Aaron Huus*, Salvador Aguiñaga*,
Megan Archer◆, Krista Hoefle◆, and Laurel D. Riek*

*Department of Computer Science and Engineering
University of Notre Dame
Notre Dame, IN 46556
{tyue, ajaniw, ahuus, saguinag, lriek}@nd.edu

◆Department of Art
St. Mary's College
Notre Dame, IN 46556
{marche01, khoefle}@saintmarys.edu

ABSTRACT

There is much discussion in the robotics community concerning the nature of people's impressions of robots. This pilot study employed the use of mobile robots coupled with artistic elements to create an environment conducive to human participation. PhotoBot took photos of participants (n = 16) in a gallery space and provided them with a physical copy of their image, while ProjectorBot displayed 3D Kinect imagery for participants to view. Participants completed a self-report measure of rapport (Bernieri's Rapport Criterion); the results of which suggest that they experienced a high degree of positive interaction with the robots.

Categories and Subject Descriptors

I.2.9 [**Artificial Intelligence**]: Robotics; J.5 [**Arts and Humanities**]: Arts, fine and performing.

Keywords: Mobile robots, human-robot interaction, kinect, rapport, sculpture

1. INTRODUCTION

To help make robotics more accessible (and acceptable) to the general public, there has been a trend in recent years toward "real world" human-robot interaction [3]. In addition to providing external validity from a methodological perspective, real-world deployments of robots afford researchers a chance to test out their sensing and motion algorithms in naturalistic settings. The recent release of the Kinect and Turtlebot robots, coupled with the ROS and PCL open-source communities, have made it very reasonable for members of our lab to design, build, and test ideas in real world settings.

We are particularly focused on non-verbal communication between people and robots, and for this work elected to focus on human-robot rapport. This is not a particularly easy concept to evaluate, particularly given the Turtlebot's limited expressivity. However, based on work by Walters et al. [8] and Byers et al. [4], we thought our robot displaying interesting proxemics behaviors while taking photographs might be a good starting point.

Inspired by efforts like Artbotics, where artists and computer scientists successfully collaborate to imagine new technological ideas [5], we formed a collaboration with the Art Department at St. Mary's College. Notre Dame computer science and engineering students worked with St. Mary's art students to conceptualize an idea they coined "technological nostalgia". High-tech robots could interact with participants by taking low-tech photographs (PhotoBot) and projecting 3D Kinect images (ProjectorBot).

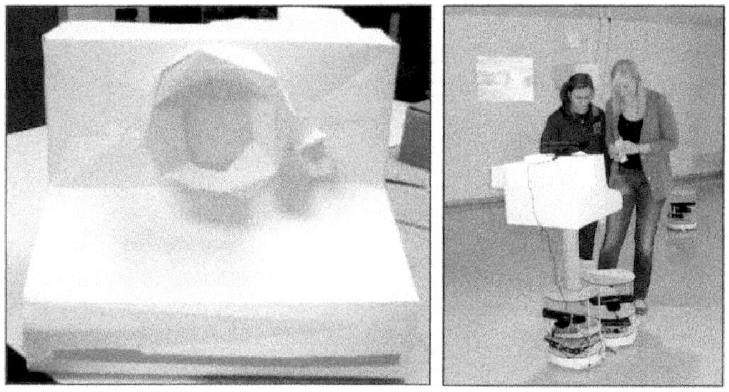

Figure 1. On left, the camera sculpture, on right, two participants interacting with PhotoBot (ProjectorBot is visible in the background).

In order to convey this notion, the artistic team developed sculptures that mounted atop the robots, and the robot team developed software to enable the robots to autonomously navigate and take and project photos.

We conducted our study at the Moreau Art Gallery at St. Mary's College, an open location with heavy foot traffic. After giving informed consent, participants were able to interact freely with the robots. They were encouraged to interact with PhotoBot and keep the photo taken of them, or ProjectorBot to see 3D images of themselves. They also could interact with two other Turtlebots (without sculptures) wandering around. Following the interaction, participants completed a self-report measure of rapport.

2. METHODOLOGY

2.1 Platform

Our main robots, PhotoBot and ProjectorBot, along with 2 additional sculpture-less robots, were built using the TurtleBot platform; a new mobile robot from Willow Garage that integrates an iRobot Create with a Microsoft Kinect sensor, allowing it to navigate its surroundings visually. The Create is a disc-like, wheeled platform, and the Kinect consists of an RGB camera and a depth sensor, which affords 3D motion capture and facial recognition.

PhotoBot and ProjectorBot each had a sculpture affixed to its top. PhotoBot's sculpture was a large Polaroid camera, and ProjectorBot's was a large overhead projector. (See Figure 1).

The interactivity between PhotoBot, the 2 sculpture-less robots, and the participants was direct. The robots autonomously traversed the gallery space using the open source *follower* program in ROS, following participants that moved within a certain distance of them. PhotoBot also ran a program that took photos and sent them to an attached portable printer every 25 seconds. Participants were encouraged to take these photos home with them.

In contrast, the interactivity between ProjectorBot and the participants was indirect. ProjectorBot had a Kinect sensor affixed to its back and a mobile projector on its front; projecting images on one of the gallery walls. Due to power constraints of the projector, the robot was immobile during the pilot.

Aesthetically, the design of both robots was intended to reference out-dated modes of technology, taking advantage of "technological nostalgia." By using forms that are familiar yet obsolete—a vintage Polaroid camera for PhotoBot and an overhead projector for ProjectorBot—we hoped to encourage a comfortable, interactive relationship between the robots and human participants.

2.2 Measures

Traditionally, human-robot rapport is not an easy concept to test [6]; even human-human rapport is not particularly easy to measure. In the literature, the most well-validated measure is the Rapport Criterion developed by Bernieri et al. [2]. This is a self-report, summed measure based on theoretical work by Tickle-Degnen and Rosenthal [7]. The measure tests three main constructs of rapport: positivity, mutual attention, and interpersonal coordination [2]. Its application in previous studies had a Cronbach's α of .94, indicating high reliability; it also has high test re-test reliability (c.f. Balaam et al. [1]).

The Rapport Criterion seemed well-aligned with the aims of our pilot study, and was also successfully used recently to measure interactional synchrony with an ambient display devoid of sound and facial affect [1].

We instructed participants to rate their interaction with PhotoBot based on the following characteristics: well-coordinated (+), boring (-), cooperative (+), harmonious (+), satisfying (+), comfortably paced (+), cold (-), awkward (-), engrossing (+), focused (+), involving (+), intense (-), friendly (+), active (+), positive (+), dull (-), worthwhile (+), and slow (-) [1]. This was an 18-item, 9-point Likert scale, where lower values indicated a lower level of rapport ("Not at all") and higher values indicated a greater level of rapport ("Extremely") [2]. Rapport scores for each participant were calculated by summing each rating after reversing the negatively worded items.

3. RESULTS

16 participants completed the Rapport Criterion measure after interacting with the various 4 robots. There was 1 male and 15 female participants with ages ranging from 16 to 70 years. (The high number of female participants can be attributed to the art gallery being located at women's college.) Eleven participants interacted with PhotoBot, and six of these had their picture taken by PhotoBot.

The mean Rapport score was 101.00 ± 19.91, with a range of 59 to 128. The maximum total possible score was 144 and the minimum possible score was 0. The participant that scored 59 did not interact with nor get their picture taken by PhotoBot. Overall, there was not a significant difference in Rapport score among the group who had their photo taken by PhotoBot and among those who did not.

It appeared that the general consensus of the participants was one of intrigue, enthusiasm, and amusement. The participants tended to personify the robots and treat them as thinking beings. Some participants commented on the novelty and the essence of the robots. For instance:

"I've never really interacted with robots in this manner."

"The robot looked pretty shaky... maybe it's shy!"

4. DISCUSSION

We aimed to create an environment that was conducive to human-robot rapport. We achieved this goal, though in future trials, we plan to choose a location that has increased foot traffic to increase our sample size. To better integrate the artistic elements of a base and sculpture for PhotoBot, enhanced weight distribution needs to be incorporated in future designs to improve the stability of the robot. Due to time constraints and technical obstacles, the extent to which we had hoped the participants would interact with the robots was not attained. Regardless, the Rapport Criterion scores indicate that the pilot was an overall success. An extension of this study could analyze if there is a significant difference amongst rapport scores of those who interacted with robots containing an artistic element and those who interacted with robots lacking one.

5. ACKNOWLEDGMENTS

The authors thank the Moreau Art Gallery at St. Mary's for use of their facilities. We would also like to thank Ziqi Zhang for her assistance with data collection.

6. REFERENCES

[1] Balaam, M., Fitzpatrick, G., Good, J., and Harris, E. Enhancing interactional synchrony with an ambient display. *Proc of the 2011 annual conference on Human factors in computing systems (CHI '11)*, ACM (2011).

[2] Bernieri, F.J., Gillis, J.S., Davis, J.M., and Grahe, J.E. Dyad rapport and the accuracy of its judgment across situations: A lens model analysis. *Journal of Personality and Social Psychology* 71, 1 (1996).

[3] Burke, J.L., Christensen, H.I., Menassa, R., Koeppe, R., Dyer, J., and Munich, M. HRI: The Real World. *Proceedings of the 6th international Conference on Human-Robot Interaction (HRI)*, ACM Press (2011).

[4] Byers, Z., Dixon, M., Goodier, K., Grimm, C.M., and Smart, W.D. An autonomous robot photographer. *Proceedings of the IEEE/RSJ International Conference on Intelligent Robots and Systems (IROS)*, IEEE (2003).

[5] Kim, H.J., Coluntino, D., Martin, F.G., Silka, L., and Yanco, H.A. Artbotics: community-based collaborative art and technology education. I*n ACM SIGGRAPH 2007 Educators Program, (2007).*

[6] Riek, L.D., Paul, P.C., and Robinson, P. When my robot smiles at me: Enabling human-robot rapport via real-time head gesture mimicry. *Journal on Multimodal User Interfaces* 3, 1 (2010).

[7] Tickle-Degnen, L. and Rosenthal, R. The nature of rapport and its nonverbal correlates. *Psychological Inquiry* 1, 4 (1990).

[8] Walters, M.L., Dautenhahn, K., te Boekhorst, R., Koay, K.L., Syrdal, D.S., and Nehani, C.L. An empirical framework for human- robot proxemics. Proc. of the AISB Symposium on New Frontiers in Human-Robot Interaction, (2009).

Unsupervised Clustering of People from 'Skeleton' Data

Adrian Ball, David Rye, Fabio Ramos and Mari Velonaki
Centre for Social Robotics / Australian Centre for Field Robotics
Rose Street Building J04, University of Sydney 2006 NSW, Australia
{a.ball I d.rye I f.ramos I m.velonaki}@acfr.usyd.edu.au

ABSTRACT

This paper investigates the possibility of recognising individual persons from their walking gait using three-dimensional 'skeleton' data from an inexpensive consumer-level sensor, the Microsoft 'Kinect'. In an experimental pilot study it is shown that the K-means algorithm—as a candidate unsupervised clustering algorithm—is able to cluster gait samples from four persons with a nett accuracy of 43.6%.

Categories and Subject Descriptors

I.5.3 [**Pattern Recognition**]: Clustering—*algorithms.*

Keywords

Gait analysis, unsupervised clustering, HRI

1. INTRODUCTION

Identification of individual people in human-robot interaction (HRI) can provide a more natural and intuitive experience for the person. From a human perspective it would be ideal if minimal changes to their behaviour or actions were required in order to interact with a social robot. Successful identification of individuals would allow a social robot to change its actions on the basis of any previous interaction with that person.

Gait recognition is a relatively new biometric technology that identifies people based on their walking gait. Physiological studies [3] have verified that gait characteristics can be used to identify people reliably. Gait recognition is an attractive method of identification due to the non-contact, non-invasive nature of data acquisition, and the possibility of recognition at a distance. Gait recognition is one of many techniques that can be used to distinguish between individuals. When integrated with information from other algorithms, gait information can substantially improve outcomes. For example, Zhou et al. [13] integrated gait and facial detection to identify people.

Successful recognition from gait requires that features be invariant to influences other than kinematic characteristics. In this paper we propose recognition by using a Microsoft 'Kinect' sensor, followed by K-means clustering using the angular information from the 'skeleton' extracted from the Kinect range image sequences. An unsupervised learning algorithm is proposed as this allows for the recognition of new people with no requirement for manual labelling. This characteristic is highly desirable for an extensible robotic system in HRI.

2. BACKGROUND

There are two main classes of methods used for gait recognition: model-based and model-free techniques [12]. Model-free gait recognition algorithms usually assess the silhouette of the human in the sagittal plane [6, 10] and generate gait energy images (GEIs). These GEIs are normalised grayscale images where the whiteness of each pixel depends on the frequency of occupation of that pixel by the person's silhouette during their walk cycle. Typically, principal component analysis (PCA) followed by image classification is performed on the GEI for person identification. Mori et al. [8] have also used silhouettes to extract gait phase information for recognition.

These model-free methods work quite well for classification [6, 11], however implicitly assume that the acquired data is not noisy. Su et al. [10] showed that when these methods are applied to real problems, noise invariably results in a significant reduction in classification accuracy.

Model-based methods [5, 12] often result in features suitable for signal processing techniques. This is because the feature generation algorithms exploit the oscillating movement of the lower body during a walk. Model-based methods are restricted to people walking along a linear path parallel to the image plane as other paths would provide variant data to the classifier.

The method proposed in this paper is a model-based technique for unsupervised clustering of gait samples. The human skeleton is approximated using an algorithm suggested by Shotton et al. [9]. The advantage of this method is that the walking path of the person is not restricted relative to the sensor viewpoint, as 3-D data are being used to extract gait information. This potentially overcomes problems with many vision-based gait identification techniques where recognition is sensitive to the camera view angle [7], or can fail if the walking motion is not parallel to the image plane. The algorithm used for generating the skeleton is also sufficiently fast to run in real time.

3. METHODOLOGY

A Kinect camera was placed in an office environment to capture images of people walking on arbitrary paths past

This work was supported by the Australian Centre for Field Robotics at the University of Sydney.

the camera. The skeletal information was extracted using Microsoft's Kinect SDK, and suitable walk half-cycles were subsequently extracted by hand. A walk cycle is defined here as the movement delimited by frames where the person's feet are a maximum distance apart, with the same foot in front. Due to the limited distance over which skeletons could successfully be extracted, walk half-cycles were used rather than full walk cycles as this choice yielded many more data sets.

Only the links of the lower body were considered for the purposes of clustering, as it is possible for the arms to be variously occupied while walking and this variation of arm data may lead to poor clustering. The features used for clustering were based on the lower limb joint angles. It was found that the limb length assigned by the Microsoft Kinect SDK skeletal algorithm changed significantly as the person walked across the camera field of view. This variation precludes distance-based features from being used for clustering, as the same person could walk past the camera twice and have two different skeletal representations. It was found that noise was also present in the link angle data. As the skeletal algorithm maintains link end-point connectedness for the duration of the walk, however, the relative change in joint position adds much less noise to angle-based features.

The features selected for clustering were the mean, standard deviation and maximum value of three angles for each of the left and right legs of the extracted skeleton. The angle of the upper leg relative to the vertical, the angle of the lower leg relative to the upper leg, and the angle of the foot relative to the horizontal were used, giving a total of 18 angular features.

The data were then "whitened" [2] to ensure that there was zero mean and unit covariance across all features. Borghese et al. [3] showed that while walking most body movement occurs in the sagittal plane, and working with angular-based features capitalises on that observation. As the number of features required for clustering was not large, there was no need to perform PCA, and the whitened data were clustered using K-means [4] with a Euclidean distance metric. The K-means algorithm should be regarded as only a candidate unsupervised clustering algorithm [1].

4. RESULTS AND DISCUSSION

The data set used for unsupervised clustering contained a total of 71 samples of walk half-cycles captured from four people. Table 1 shows the resulting confusion matrix. A combined clustering accuracy of 43.6% was obtained using K-means, significantly better than chance (25%).

Table 1: Clustering confusion matrix

	Label A	Label B	Label C	Label D
Person 1	**35%**	46%	8%	12%
Person 2	0%	**74%**	16%	11%
Person 3	21%	21%	**29%**	29%
Person 4	17%	17%	33%	**33%**

Table 1 shows that it is feasible to cluster individual people based on their gait using a straightforward unsupervised learning algorithm. Furthermore, this unsupervised clustering was done using a high-level three-dimensional representation of a human obtained from an inexpensive consumer-level sensor. Although direct comparisons to other work in the field cannot be made due to the different data that were used for clustering, our results are similar to those obtained by Su et al. [10] who recognised people from noisy data using a supervised classification algorithm.

It is interesting to note that the accuracy of clustering using the 18 features previously described ranges from 74% to 29%—which is not statistically better than chance. The features selected for this pilot study encode only lower body activation levels. It may be that the use of additional features—such as upper-to-lower-limb phase angles or energy ratios, or features that contain time or frequency information—may improve the discrimination between individuals.

5. REFERENCES

[1] A. Ball, D. Rye, F. Ramos, and M. Velonaki. A comparison of unsupervised learning algorithms for gesture clustering. In *Proc. 6th Int. Conf. on Human-Robot Interaction*, pages 111–112. ACM, 2011.

[2] C. Bishop. *Pattern Recognition and Machine Learning*. Springer, 2006.

[3] N. Borghese, L. Bianchi, and F. Lacquaniti. Kinematic determinants of human locomotion. *J. Physiology*, 494(3):863, 1996.

[4] R. Duda, P. Hart, and D. Stork. *Pattern Classification*, volume 2. Wiley, 2001.

[5] W. Isa, R. Sudirman, and S. Sh-Salleh. Angular features analysis for gait recognition. In *Proc. 1st Int. Conf. on Computers, Communications, & Signal Processing, 2005*, pages 236–238, 2005.

[6] H. Lee, S. Hong, and E. Kim. An efficient gait recognition based on a selective neural network ensemble. *Int. J. Imaging Systems and Technology*, 18(4):237–241, 2008.

[7] Y. Makihara, H. Mannami, and Y. Yagi. Gait analysis of gender and age using a large-scale multi-view gait database. *Computer Vision–ACCV 2010*, pages 440–451, 2011.

[8] A. Mori, Y. Makihara, and Y. Yagi. Gait recognition using period-based phase synchronization for low frame-rate videos. In *20th Int. Conf. on Pattern Recognition*, pages 2194 –2197, 2010.

[9] J. Shotton, A. Fitzgibbon, M. Cook, T. Sharp, M. Finocchio, R. Moore, A. Kipman, and A. Blake. Real-time human pose recognition in parts from single depth images. In *Proc. IEEE Conf. on Computer Vision and Pattern Recognition*, volume 2, pages 1297 –1304, 2011.

[10] H. Su, Z. Liao, and G. Chen. A gait recognition method using L1-PCA and LDA. In *Proc. 2009 Int. Conf. on Machine Learning and Cybernetics*, volume 6, pages 3198–3203, 2009.

[11] L. Wang, T. Tan, H. Ning, and W. Hu. Silhouette analysis-based gait recognition for human identification. *IEEE Trans. Pattern Analysis and Machine Intelligence*, 25(12):1505–1518, 2003.

[12] R. Zhang, C. Vogler, and D. Metaxas. Human gait recognition. In *2004 IEEE Comp. Soc. Conf. on Computer Vision and Pattern Recognition Workshops*, pages 18–18, 2004.

[13] X. Zhou and B. Bhanu. Integrating face and gait for human recognition. In *2004 IEEE Comp. Soc. Conf. on Computer Vision and Pattern Recognition Workshops*, pages 55–55, 2006.

Immersive Human-Robot Interaction

Anara Sandygulova[1], Abraham G. Campbell[2], Mauro Dragone[3], G.M.P. O'Hare[4]
CLARITY: Centre for Sensor Web Technologies, School of Computer Science and Informatics,
University College Dublin, Belfield, Dublin 4, Ireland
[1] anara.sandygulova@ucdconnect.ie [2] abey.campbell@gmail.com [3] mauro.dragone@ucd.ie
[4] gregory.ohare@ucd.ie

ABSTRACT
Networked robotic applications enable robots to operate in distant, hazardous, or otherwise inaccessible environments, such as search and rescue, surveillance, and exploration applications.
The most difficult challenge which persists for such systems is that of supporting effective human-robot interaction, as this usually demands managing dynamic views, changeable interaction modalities, and adaptive levels of robotic autonomy.
In contrast of sophisticated screen-based graphical user interfaces (GUIs), the solution proposed herein is to enable more natural human-robot interaction modalities through a networked immersive user interface. This paper describes the creation of one such shared space where to test such an approach, with both simulated and real robots.

Categories and Subject Descriptors
I.2.9 [**ARTIFICIAL INTELLIGENCE**]: Robotics – *Operator interface;* H.5.1 [**INFORMATION INTERFACES AND PRESENTATION**]: Multimedia Information Systems – *Artificial, Augmented, and Virtual Realities;* I.3.6 [**COMPUTER GRAPHICS**]: Methodology and Techniques – *Interaction Techniques.*

General Terms: Design, Experimentation, Human Factors.

Keywords: Human-Robot Interaction, Mixed Reality, User Interface.

1. INTRODUCTION
This work addresses the problems of achieving effective human-robot interaction by creating a shared, Immersive Human-Robot Interaction (IHRI) space, as a means by which to enable natural forms of interaction, e.g. based on gestures and dialogue, with remote mobile robots. The proposed system addresses the limitations of more traditional, GUI-based, networked solutions to remote robot monitoring and control, such as the cognitive fatigue caused by having to deal with unintuitive and cumbersome interfaces, especially when multiple robots are involved.
In our solution, the user is placed in a CAVE-like system called the LAIR [Denby et al. 2009], where a virtual environment mirroring the remote robot site is displayed around the user. Since the user's workplace can be purposefully engineered, for example through the additions of video cameras, microphones, and gesture & posture tracking interfaces, such an approach makes it easy to recognize both user's deictic gestures, such as pointing and gazing together with user's vocal instructions, before transmitting them to the remote robots. By mirroring the location of the real robots in the virtual environment, both robots and human users are given the illusion of sharing the same physical space. Instead of having to understand the robots' situation as depicted on a 2D screen, select on the graphical interface the particular robot to instruct, and click a target's destination, the user of such a system will be able to just look around, turn toward the robot, point to the target, and utter the instruction.
The remainder of this paper describes how we built upon these concepts to create a system where to test the effectiveness of natural forms of interaction with remote, networked robots.

2. RELATED WORK
Previous research has sought to address this problem of effective human-robot interaction with remote, networked robots. The Virtual Synergy [Tejada et al. 2003] interface is a means of communicating and monitoring a collective of mobile robots engaged in Urban Search and Rescue (USAR) operations. Robots and human operators in Virtual Synergy share a Collaborative Virtual Environment (CVE), which they "navigate" by controlling their avatar. Since the CVE mirrors the real, physical space where the robots operate, the position of the users' avatars can be relayed to the robots and used to help their search strategies. Tachi et al. [2001] developed an immersive telexistance cockpit where a human operator is provided with the information about the robot's site in the form of natural audio, video and force feedback gathered by the robot, de-facto assuming direct control of the robot's body. Odashima et al. [2003] exploited a virtual environment where an immersion type display captures human upper body motions to create virtual robot simulation and then presents the human with high presence of a virtual robot for greater perception of visual and auditory information.

Our IHRI approach differs from these works by utilising an immersive user interface. Additionally, our main focus is on testing effective human-robot co-operation attained through natural forms of interaction.

3. IMPLEMENTATION
The IHRI system is implemented on top of two pre-existing systems. The *Virtual Robotic Workbench* (VRW) [Dragone 2007] provided (i) the engine to control both physical and simulated robots, and (ii) the means to exchange telemetry and sensor information within UDP multicast groups. The IHRI system exploits the first feature to create self-autonomous robot control system that can be tasked through high-level goals, such *Explore*, representing the goal of wander in the environment trying to explore new sites while avoiding obstacles, and *MoveTo(?X,?Y),* representing the goal to be in a certain location (expressed in global, Cartesian coordinates). Once such an instruction is received by one of the robots, the robot is able to identify its location by using a pre-existing map of the environment where it operates, and also track its movements by analyzing the sensor

HRI'12, March 5–8, 2012, Boston, Massachusetts, USA.
ACM 978-1-4503-1063-5/12/03.

data gathered through its on-board laser and sonar range sensors. The robot's location, and the ID of the robot, is then multicasted to every entity logged to the workbench, including other robots (both real and simulated) and users operating user interfaces connected to the workbench.

The second system integrated in IHRI is the NeXuS Mixed Reality Framework [O'Hare et al. 2005], which provides the visualization and user tracking functionalities used in the CAVE. Specifically, a 3D replica of the robots' environment is projected around the user while an infrared tracking system and a track pad are used, respectively, to track the user's gaze, and enable the user to move within the virtual environment. As the user "travels" on the track pad, the whole scene is updated in a smooth motion and the position of the user is multicasted within the workbench. In addition, in order to experiment with gesture-driven control, a Nintento Wii Remote and the WiiGee [Schlömer et al. 2008] open-source gesture recognition library is used to recognize simple gestures, such as those used by the user to signal a robot to stop or to get close. Once these gestures are recognized, they are translated into instructions (such as *MoveTo(User)*), and transmitted to the robots. Finally, our setup allows the user to control (via voice commands) the opening and the closure of a 2D window showing the video feedback captured by the robot's camera, which is transmitted via real time video transmission implemented over the Java Media Framework (JMF).

4. EXPERIMENTS

To demonstrate the concept of IHRI and validate our system architecture, we run three experiments as depicted in Table 1:

Table 1. Interface Types Used in the Experiments

Experi-ment	Robot		Visualization & User Tracking	
	2D/Simulated	Real	**3D/Desktop**	**CAVE**
1	✓		✓	
2		✓	✓	
3		✓		✓

The first experiment (Figure 1) tested the system integration between the robot 2D simulation and the NeXuS framework running on a desktop in 3D.

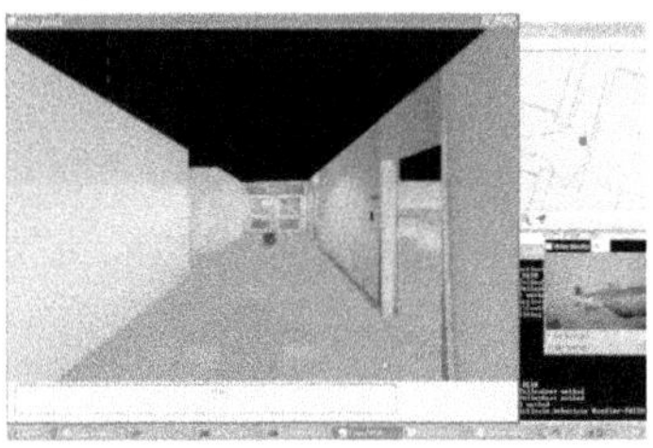

Figure 1. Simulated Robot & NeXuS on Desktop experiment, left: NeXuS snapshot, right: snapshot of 2D robot simulation

The second demonstration, shown in Figure 2, introduced the physical robots. Finally, we moved our tests to the CAVE, as shown in Figure 3. In order to exercise both robots' navigation skills and a simple set of gesture recognition modules, each experiment took the form of a simple Search-the-User game. Specifically, each robot was tasked to explore the environment and stop as soon as the user was in sight. At that point, the user was left to control the robot through the two gestures "Stop", and "Get Close", and also to open and close the video feedback window. The videos are available on the website http://ubirobot.ucd.ie/content/demonstration-files.

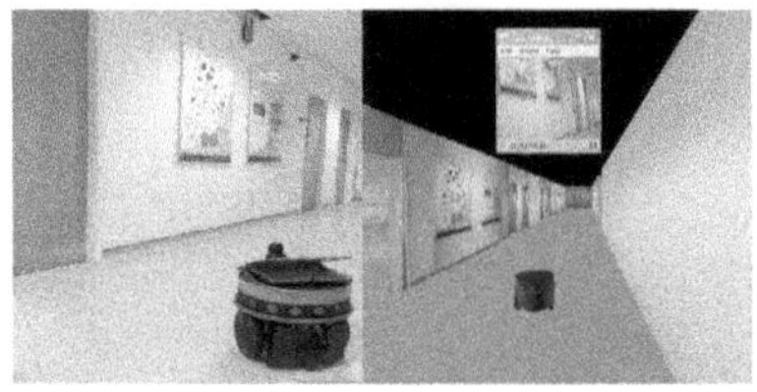

Figure 2. Physical Robot & NeXuS on Desktop experiment, left: robot in the office space, right: NeXuS snapshot

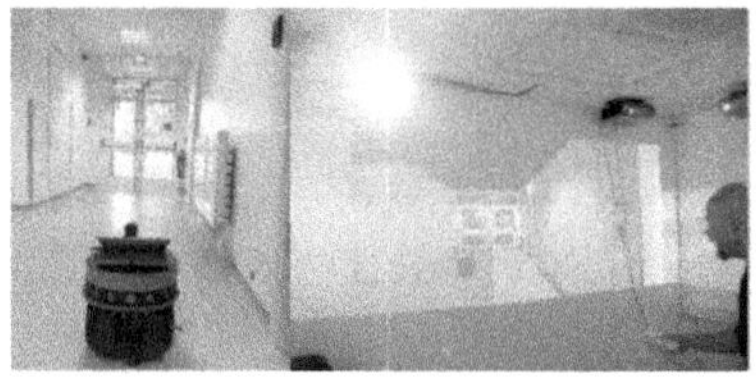

Figure 3. Physical Robot & NeXuS in CAVE experiment, left: robot in the office space, right: user in the CAVE

5. FUTURE WORK & CONCLUSION

The IHRI project addresses the questions of effective HRI by aiming to remove the dependency on traditional GUIs. In contrast, the user and the robots share the same space through Mixed Reality technology while the user can use natural forms of interaction to manage the remote robot.

Future work will build on the prototype and conduct evaluations of IHRI with the help of HRI metric techniques, including task-based analysis, experiments and usability studies.

6. ACKNOWLEDGEMENTS

This work is supported by Science Foundation Ireland under grant 07/CE/I1147 and IRCSET 'Embark Initiative' funding.

7. REFERENCES

[1] Denby, B., Campbell, A. G., Carr, H., and O'Hare, G.M.P. 2009. The LAIR: Lightweight Affordable Immersion Room. *Presence: Teleoper. Virtual Environ.* 409-411.

[2] Dragone, M. 2007. *SoSAA: An Agent-Based Robot Software Framework.* PhD thesis. University College Dublin.

[3] Odashima, T., Onishi, M., Luo, Z. W., and Hosoe, S. 2003. Immersion type virtual environment for human-robot interaction. In *Proc. of IEEE International Conference on Systems, Man and Cybernetics (SMCC).* 3651- 3656.

[4] O'Hare, G.M.P., Campbell, A.G. and Stafford, J.W. 2005. NeXuS: Delivering Behavioural Realism through Intentional Agents. In *the Proceedings of the AMT 2005.*

[5] Schlömer, T., Poppinga, B., Henze, N., and Boll, S. 2008. Gesture recognition with a Wii controller. In *Proceedings of the 2nd International Conference on Tangible and Embedded Interaction.*

[6] Tachi, S., Komoriya, K., Sawada, K., Nishiyama, K., Itoko, T., Kobayashi, M., and Inoue, K. 2001. Development of Telexistence Cockpit For Humanoid Robot Control. In *Proc. 32nd Int. Symp. Robotics (ISR2001).* 1483–1488.

[7] Tejada, S., Cristina, A., Goodwyne, P., Norm, E., and Tarapore, S. 2003. Virtual Synergy: A Human-Robot Interface for Urban Search and Rescue. In *Proc. of the AAAI 2003 Robot Competition.* (Acapulco, Mexico).

Don't Stand so Close to Me: Users' Attitudinal and Behavioral Responses to Personal Space Invasion by Robots

Aziez Sardar & Michiel Joosse
Institute of Informatics
University of Amsterdam
Science Park 508, 1098XH
Amsterdam, The Netherlands
{aziez}{mjoosse}@science.uva.nl

Astrid Weiss
ICT&S Center
University of Salzburg
Sigmund-Haffner-Gasse, A-5020
Salzburg, Austria
astrid.weiss@sbg.ac.at

Vanessa Evers
Human Media Interaction
University of Twente
Postbus 217, 7500 AE
Enschede, The Netherlands
v.evers@utwente.nl

ABSTRACT
When around humans, one might expect that a social robot would act according to the social norms people expect of each other. When someone does not adhere to a prevalent social norm, people usually feel threatened and uncomfortable. In comparison, insight is needed into what is perceived as socially normative behavior for robots. We conducted an experiment in which a robot approached a participant in order to determine the effect of personal space invasion. We manipulated the agent-type (human/robot) and the approach speed (slow/fast) of the agent towards the participant. Unexpectedly, our results show that the participants displayed more compensatory behavior toward the robot than toward the human confederate. Interestingly, we found that participants tended to trust the faster robot more compared and the faster human less. We consider these responses toward personal space invasion as an indication that people react differently to robots as they do to humans, and with more intensity.

Categories and Subject Descriptors
H.1.2 [**Models and Principles**]: User/Machine System – *human factors*. H.5.2 [**Information Interfaces and Presentation**]: User Interfaces – *evaluation/methodology, user-centered design*

General Terms
Measurement, Human Factors, Standardization.

Keywords
Human-Robot Interaction, immediacy, proxemics personal space invasion, social norms, avoidance behavior, approach speed.

1. INTRODUCTION
Humans interact with each other while adhering to an extensive set of social norms that offer guidance for when one is supposed to greet, eat, sleep and so on. When someone does not adhere to a prevalent social norm, people usually feel threatened and confused [1]. When in a human environment, one might assume that a social robot would act according to the social norms people expect of each other, so insight is needed into what is perceived as socially normative behavior for robots. The media equation [2] suggests that people will respond to robots as if these were social actors. However, research on Human-Robot Interaction indicates that people also respond differently to robots as they do to humans [3]. It is therefore unwise to presume that human social norms will carry unequivocally across to human robot interaction. The main question is: Do people expect robots to act according to the same social norms other people are held to? To investigate this question we conducted a first study (as starting point for a series of studies). The study presented in this paper is concerned with the socially normative behavior of maintaining personal space (proxemics) and builds upon the earlier work by Mumm and Mutlu [3].

2. THEORY & HYPOTHESES
Dosey & Meissels [4] tell us that our personal space is "to act (in part) as a buffer zone which serves as a protection against perceived threats to our emotional and physical well-being". Our personal space is therefore not only a physical, but also a psychological buffer zone. Personal space invasion occurs when either the close intimate (0.00m to 0.15m) or intimate zone (0.15m to 0.45m) is being invaded by someone who is not a close relation; this is experienced as disturbing and uncomfortable [1]. There are several factors that have been found to impact physical and psychological distance such as: eye-contact, gender, age, culture, physical attractiveness, culture, personality, and stress.

When the personal space of a human is invaded, it can cause several different (non-verbal) compensatory behaviors, such as physical proximity (step away or closer), facial expressions, body posture (leaning away or towards), and eye contact/gaze. However, these findings from social-psychology can only be generalized to HRI when a robot is perceived as a social actor or seen as somewhat similar to a human. Therefore, we expected that people would have stricter social norms for humans than for robots: Hypothesis 1 - People will display more compensatory behaviors when a human invades their personal space compared with when a robot does.

Moreover, we assumed that the approach speed will influence the perception of socially normative behavior perception. Butler et al. [5] identified an approach speed of 0.5 m/s as comfortable for study participants, an approach speed of 1.0 m/s as uncomfortable: Hypothesis 2 - A faster approach speed will lead to more compensatory behavior to personal space invasion.

3. METHOD & MEASURES
To investigate whether people expect robots to act according to human social norms, we conducted a 2x2 controlled between-participants laboratory experiment to determine the effects of personal space invasion by robots. We manipulated the *type of agent* (human confederate/robot) and the *speed of the approach* of the agent towards the participant (slow/fast). A total of 85 participants (50 males and 35 females) took part in the experiment; aged between 18 and 70 years (M = 25.14, SD = 9.82). The robot was operated from an adjacent room, in a

Wizard-of-Oz-setting. The speed with which the human confederate or robot approached the participant was either 'slow' (0.4 m/s) or 'fast' (1.0 m/s).

During the experiment the participant was put in front of a poster where s/he would perform a cognitive task (searching for something on the poster). After approximately one minute the agent (human or robot) would invade the participants personal space. This experimental scenario could be compared to a museum or airport situation in which someone is watching a picture or display and another observer comes too close to do the same (see figure 1). In order to measure the responses, the recorded *behavioral responses* were coded by two independent coders, focusing on immediacy cues, facial expressions, body posture and overall emotions. The *attitudinal measures* were based on several scales, such as perceived human-likeness of the robot, Negative Attitude toward Robot's, Source Credibility Scale and Interpersonal Attraction Scale (an overview on the measures can be found in [6]).

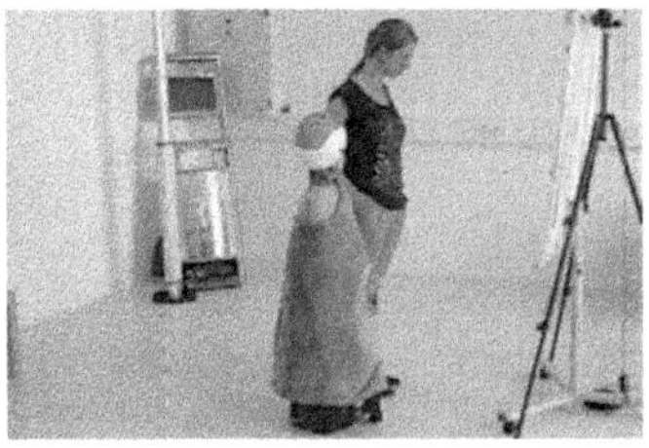

Figure 1. Study Setting.

4. RESULTS

The manipulation check revealed that our manipulation of the approach speed was successful. Comparing the slow and fast conditions, the 'slow' group indeed perceived the agent as slow ($U = 580.5$, $z=-2.839$, $p<0.01$, $M_{slow} = 4$, $M_{fast} = 3$). Similarly, the slow condition was perceived as more comfortable ($M = 2$) compared with the fast condition ($M = 4$) ($U = 396.5$, $z=-4.528$, $p<0.000$).

For the first hypothesis, we expected that when approached by a robot, participants would display less compensatory behaviors than when approached by a human confederate. Contrary to what we expected, the participants showed significantly more compensatory behavior in the robot condition ($M = 3.49$, $SD = .731$) than in the human condition ($M = 4.15$, $SD = .596$). We also expected that social norms between humans would be so strong that participants would expect the confederate to stop even if he moved towards them in the faster condition. For robots, however, we expected that the lack of established social norms would lead the participants to think that the robot may run into them.

For the second hypothesis, we predicted that the approach speed of the agent would intensify the compensatory behavior of the participant toward personal space invasion. A Kruskall-Wallis test revealed no significant differences. However, a trend was found that the approach speed influenced the participants' facial expressions ($U = 702.5$, $z = -1.792$, $p=0.073$, $r = -.19$). Participants had more pleasant facial expressions ($M = 4.15$, $SD = .596$) than unpleasant ($M = 4.15$, $SD = .596$) when the robot approached them with higher speed. This finding may suggest that participants felt more positive towards the fast robot compared with the slow robot. Interestingly, a significant interaction effect was found for the attitudinal measure *trust*. A Kruskall-Wallis test indicated that trust in the agent was significantly affected for agent type and approach speed ($H(3) = 8.59$, $p < .05$). The slower human confederate was perceived as more trustworthy. In contrast, the robot was perceived as more trustworthy when it approached with a faster speed. The interaction effect between agent type and agent speed for trust indicates that there may have been different social expectations for humans and robots. When the robot approached the participant with a slow speed, they trusted the robot less compared with the faster uncomfortable speed. For the human confederate, the opposite was true. A possible explanation is that the high concentration needed during the cognitive task may have led to the participant noticing the robot later and/or having less time to consider the robot's intention. In the faster condition, the speed of the robot and with that the increased sound of the wheels, may have made the participants more aware that a robot was approaching toward them so that personal space invasion was less surprising. Another explanation for the more positive evaluation of the faster robot might be that people found it more amusing to notice the robot approach and move in close.

5. CONCLUSIONS & OUTLOOK

Our experiment showed that participants displayed more compensatory behavior when a robot agent invaded their personal space compared with a human agent. Similarly, we conclude that people may not hold robots to human social norms but instead judge the robots functional intentions from the (social) behaviors the robot displays. We assume that the increased trust in the faster robot was due to the fact that it caused less of a surprise or amusement than the robot that slowly moved towards the participant. In future research, we are planning to replicate the experiment in a real museum setting to validate the results and to investigate other areas of interests in terms of social norms for robots, such as calling for attention and communication with strangers.

6. ACKNOWLEDGMENTS

We would like to thank Joris de Groot, who was a great help in every aspect of the research; the students who participated in the class Research Methodology and Techniques 2011; Bas Terwijn and Edwin Steffens for getting the Nomad Scout to work according to the requirements of the experiment.

7. REFERENCES

[1] Hall, E.T. The Hidden Dimension. DoubleDay, New York, 1966.

[2] Reeves, B. and Nass, C. The Media Equation: How People Treat Computers, Television, and New Media Like Real People and Places. Cambrigde University Press, 1996.

[3] Mumm, J and Mutlu, B. Human-Robot Proxemics: Physical and Psychological Distancing in Human-Robot Interaction. Proceedings of HRI 2006, 6 (2011), 331--338.

[4] Dosey, M.A. and Meissels, M. Personal Space and Self-Protection. Journal of Personality and Social Psychology, Vol 11, No. 2 (1969), 93-97.

[5] Butler, J.T. and Agah, A. Psychological Effects of Behavior Patterns of a Mobile Personal Robot. Autonomous Robots, 10 (2001), 185-202

[6] Joosse, M., Sardar, A.H., and Evers, V. BEHAVE: A Set of Meaures to Assess Users' Attitudinal and Physical Respones to Robot's Social Behaviors. Lecture Notes in Computer Science (2011)

Coupled Inverse-Forward Models for Action Execution Leading to Tool-Use in a Humanoid Robot

Guido Schillaci and Verena V. Hafner
Cognitive Robotics Group
Department of Computer Science
Humboldt-Universität zu Berlin
Germany
{guido.schillaci, hafner}
@informatik.hu-berlin.de

Bruno Lara
Cognitive Robotics Lab
Department of Computer Science
Universidad Autónoma del Estado de Morelos
Cuernavaca, Mexico
bruno.lara@uaem.mx

ABSTRACT

We propose a computational model based on inverse-forward model pairs for the simulation and execution of actions. The models are implemented on a humanoid robot and are used to control reaching actions with the arms. In the experimental setup a tool has been attached to the left arm of the robot extending its covered action space. The preliminary investigations carried out aim at studying how the use of tools modifies the body scheme of the robot. The system performs action simulations before the actual executions. For each of the arms, predicted end-effector positions are compared with the desired one and the internal pair presenting the lowest error is selected for action execution. This allows the robot to decide on performing an action either with its hand alone or with the one with the attached tool.

Categories and Subject Descriptors

I.2.6 [**Computing Methodologies**]: Artificial Intelligence—*Learning*

General Terms

Theory, Experimentation

Keywords

Internal models, inverse and forward models, action simulation, tool-use

1. INTRODUCTION

In this paper, we present preliminary results on a computational model for action execution. This model will be extended to implement action recognition. The Nao humanoid from Aldebaran has been adopted as robot platform. Internal models are used for predicting the end position of both arms. An extension tool is attached to the left hand of the robot to emphasize the power of predictor-controller pairs in selecting the best arm for action execution.

We understand the physical world according to our own experience. The faculties, capabilities and skills to dynamically interact with the world, which as adult humans we posses, emerge through a long process of tuning and rehearsing of *sensori-motor schemes*. A very important example is the acquisition of the sensori-motor schemes that code for the capabilities and reaches of our body. This set of schemes provides us with a notion which we exploit in order to perform tasks such as throwing, reaching and grasping, which have helped us through our development as a species. A proposed mechanism to code these schemes are internal models. In this work we investigate the capabilities of forward and inverse models.

Forward models were first proposed in the control literature as means to overcome problems such as the delay of feedback on standard control strategies and the presence of noise characteristic of natural systems [4]. A forward model is an internal model which incorporates knowledge about sensory changes produced by self-generated actions of an agent. Given a sensory situation S_t and a motor command M_t (intended or actual action) the forward model predicts the next sensory situation S_{t+1}. Much research has been done on computational forward models for action preparation and movement, with highly functional models that account, for example, for hand trajectory planning [3].

In cognitive robotics, an interesting implementation is presented by Dearden [2] where a robot learns a forward model that successfully imitates actions presented to its visual system, learning from a social context using a forward-inverse model pair.

While forward models (or predictors) present the causal relation between actions and their consequences, inverse models (or controllers) perform the opposite transformation providing a system with the necessary motor command (M_t) to go from a current sensory situation (S_t) to a desired one (S_{t+1}). Theories on how the human Central Nervous System (CNS) manages sensorimotor control suggest that a modular approach is implemented in the brain. Predictor-controller pairs, modelling different goal-directed actions, are run in parallel for assessing which one is more plausible in the given context. In [7], Wolpert suggested coupled inverse and forward models for action execution and recognition.

Tool use is an important skill that is acquired during early childhood in humans and requires several cognitive abilities related to sensorimotor interaction. Bril et al.[1] consider tool-use as an instance of a goal-directed action. Tool-use requires both an action and a cognition component. Maravita and Iriki [5] propose that we hold an adaptive body map comprising body posture and shape. They suggest that

HRI'12, March 5–8, 2012, Boston, Massachusetts, USA.
ACM 978-1-4503-1063-5/12/03.

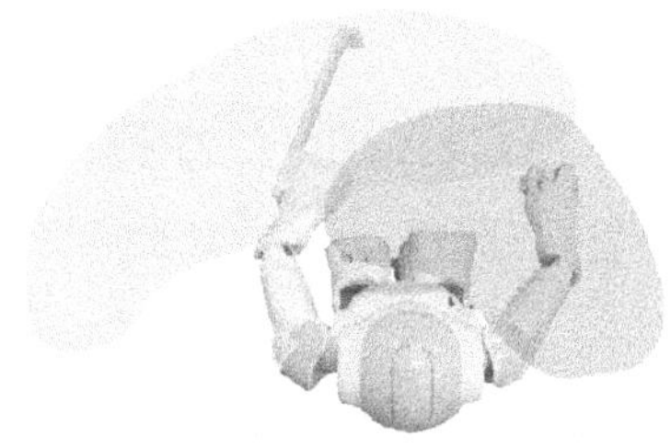

Figure 1: Top-down view of the Nao with an illustration of the reachability space for each end-effector.

the body schema is extended temporarily with the tools we are using ("as if our own effector (e.g. the hand) were elongated to the tip of the tool"). From this perspective, the extended arm experiment on the robot can be seen as the body of the robot being temporarily extended by a suitable tool for a specific task (namely reaching an object).

2. PRELIMINARY RESULTS

We implemented a system for generating reaching commands towards given points in the space. Figure 1 illustrates how different the action spaces of both arms can be, if the robot is provided with an extension tool on its left arm. While far-away positions can be reached with the left extended end-effector, other positions (like the ones close to the chest) are only reachable with the right arm. A forward-inverse model pair has been learned for each arm of the robot (see Figure 1) using a motor babbling mechanism previously implemented [6]. A k-Nearest Neighbours based algorithm has been adopted, together with the knowledge bases (one for each arm) collected during babbling, as the inference algorithm for both inverse and forward models.

Before executing the necessary motor command to reach a position, an internal simulation shown in Fig. 2 is run as follows: find the motor command M_t which given the sensory situation S_t, composed by the current coordinates of the arms, brings the system to the desired sensory situation S_{t+1}, composed by the coordinates of the goal position. This process is performed by a learned inverse model of each arm. Use the predicted motor command and the current sensory situation to find the next sensory situation. Again, this process is performed by a forward model of each arm. Predicted end-effector positions are compared with the desired one and the forward-inverse model pair with less error is selected for actual execution, namely moving the corresponding arm. Figure 3 shows the errors of both pairs when presented with different end positions estimated from a marker.

3. FUTURE WORK

These preliminary results make only use of one-step predictions, future work will include the use of the predicted sensory situation as input back to the system in what can be seen as long-term predictions. Internal models will also be used for goal-directed action recognition. Experiments on human-robot interaction will be performed for investigating whether a robot can direct user's action (like with pointing to an object) and how robot expectations on user reactions can be modelled as a prior distribution on the coupled inverse-forward models.

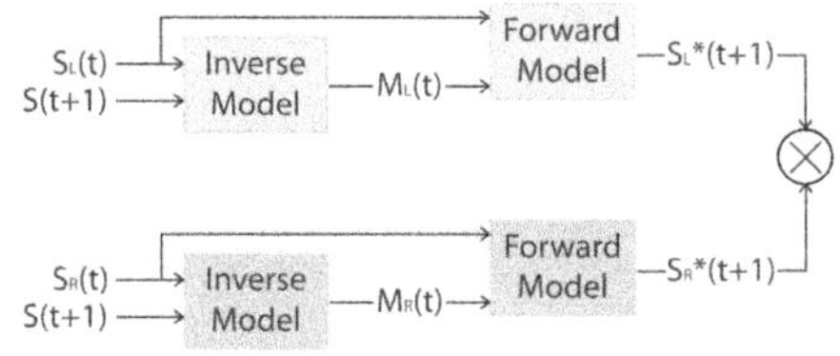

Figure 2: Inverse (controller) - forward (predictor) model pairs. $S_L(t)$ and $S_R(t)$ are the (x, y, z) coordinates of each end-effector at time t; $S(t+1)$ is the desired position to reach; $S^*_L(t+1)$ and $S^*_R(t+1)$ are the predicted final positions for each end-effector; $M_L(t)$ and $M_R(t)$ are the predictions of the motor configurations for reaching $S(t+1)$ from $S_L(t)$ and $S_R(t)$, respectively.

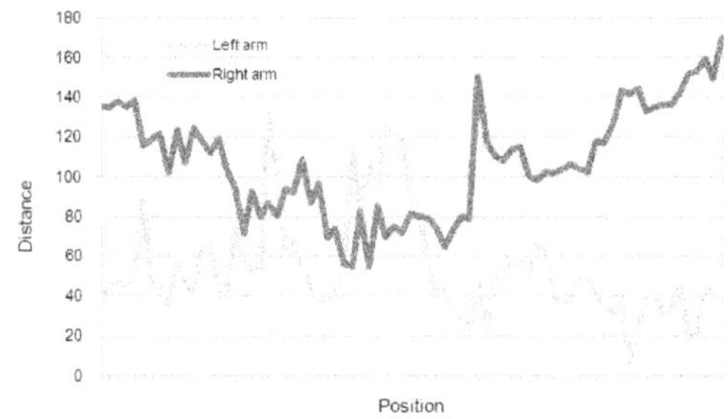

Figure 3: Distance between the predictions of end-effector position and the desired point.

Future work will also consider an active choice between several tools for a given task, and will include additional actions apart from reaching an object.

Acknowledgement

This work has been partially financed by the EU funded Initial Training Network (ITN) in the Marie-Curie People Programme (FP7): INTRO (INTeractive RObotics research network), grant agreement no.: 238486.

4. REFERENCES

[1] B. Bril, R. Rein, T. Nonaka, F. Wenban-Smith, and G. Dietrich. The role of expertise in tool use: Skill differences in functional action adaptations to task constraints. *Journal of Exp. Psyc.: Human Perception and Performance*, 36(4):825 – 839, 2010.

[2] A. Dearden. *Developmental learning of internal models for robotics*. PhD thesis, Imperial College London, 2008.

[3] M. Haruno, D. M. Wolpert, and M. Kawato. Mosaic model for sensorimotor learning and control. *Neural Computation*, 13:2201–2220, 2001.

[4] M. I. Jordan and D. E. Rumelhart. Forward models: Supervised learning with a distal teacher. *cognitive science*, 16:307–354, 1992.

[5] A. Maravita and A. Iriki. Tools for the body (schema). *Trends Cogn Sci*, 8(2):79–86, 02 2004.

[6] G. Schillaci and V. V. Hafner. Random Movement Strategies in Self-Exploration for a Humanoid Robot. In *Proc. of the Intern. Conf. on Human-Robot Interaction 2011*, pages 245–246, 2011.

[7] D. M. Wolpert, K. Doya, and M. Kawato. A unifying computational framework for motor control and social interaction. *Phil. Trans. of the Royal Soc. of London. Series B: Biol. Sc.*, 358(1431):593–602, March 2003.

Developing Guidelines for In-The-Field Control of a Team of Robots

Megha Sharma
University of Manitoba
Department of Computer Science
Winnipeg, Manitoba R3T 2N2
megha@cs.umanitoba.ca

James E. Young
University of Manitoba
Department of Computer Science
Winnipeg, Manitoba R3T 2N2
young@cs.umanitoba.ca

Rasit Eskicioglu
University of Manitoba
Department of Computer Science
Winnipeg, Manitoba R3T 2N2
rasit@cs.umanitoba.ca

ABSTRACT
In this work we explore the development of guidelines for creating "in-the-field" interfaces for enabling a single user to remotely control multiple robots. The problem of controlling a remote team of robots is complex, requiring a user to monitor and interpret robotic state and sensor information in real time, and to simultaneously communicate direction commands to the robots. The result is that a robot controller is often seated at a console; for many relevant applications such as search and rescue or firefighting this removes the user from the field of action, rendering them unable to directly participate in a task at hand.

Therefore, one challenge in HRI is to develop efficient interfaces that will enable a user to effectively control and monitor a team of robots in the field. In our project we explore various interface designs in terms of supporting this goal, taking the approach of involving a panel of professionals in the design process to direct exploration and development.

Categories and Subject Descriptors
I.2.9 [**Robotics**]: Operator Interfaces, H.5.2 [**User Interfaces**]: Interaction Styles, graphical user interfaces (GUI)

General Terms
Design, Experimentation, Human Factors

Keywords
Human robot interaction, Interface design, robot teams.

1. INTRODUCTION
Robots are emerging as important tools in professional contexts such as search and rescue, firefighting, or the military [1, 2]. In these situations robots are often even referred to as team members [3, 4], due to their complexity, capabilities, and semi-autonomous nature. However, this complexity means that it can be very challenging to both monitor the state of the robots, and their many environmental sensor readings, and to meaningfully give them instructions in real time. Thus an important goal is to create interfaces which effectively support interaction and collaboration between human and robotic team members.

One downside to many in-the-field robot control interfaces is that they often require a human team member to engage and immerse themself in complex laptop-like console interface, removing them from direct involvement in the task at hand (see, e.g., the iRobot Packbot control interface, and [5, 6, 7]).

HRI'12, March 5–8, 2012, Boston, Massachusetts, USA.
ACM 978-1-4503-1063-5/12/03.

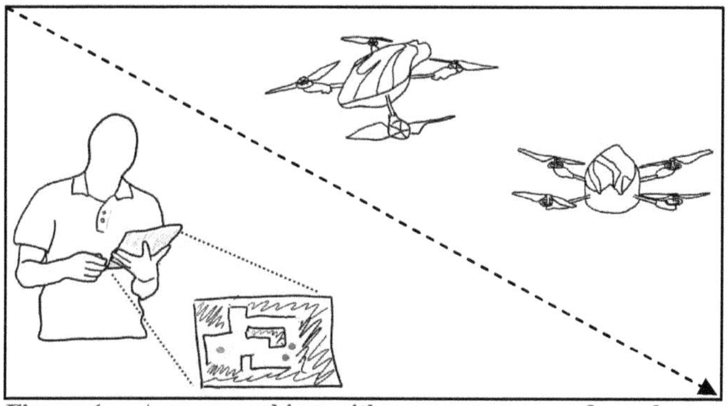

Figure 1 – A user working with a remote team of quadrotor robots for a mapping task.

Further, many existing systems attempt to minimize the complexity by having multiple human controllers per robot [8, 9]. In our work, we will investigate methods for enabling a single user in the field to interact with a team of remote robots while still being involved with the task at hand.

Our strategy for investigating this problem will be to create an expert panel of relevant professionals, for example, police officers, fire fighters and search and rescue workers, and to leverage their expertise for driving our interface design. This task and user-centered design approach keeps our investigations grounded in practical application constraints and the expert users provide a valuable user base for evaluation.

As an initial step we will focus solely on the task of using a team of robots to map an unknown space. We selected this scenario as it will be relevant to many professional contexts, it requires real-time monitoring and direction of robots to relevant areas as a task unfolds, and would be useful by a person in the field versus being removed from action at a console.

Specifically, we intend to explore the following research questions related to this task:

1. How do various device formats (e.g., cell phone, smart phone, tablet PC and UMPCs) integrate into real use for the tasks and scenario? How do they differ?

2. How do different kinds of input mechanisms lend themselves to use in the field, for example, single or multi touch gestures (single-handed or two-handed), portable keyboards (e.g., strapped to an arm or single-handed keyboards), or trackballs?

3. How can such portable interfaces support sliding-scale autonomy for robot control, that is, to enable the user to seamlessly go from directing an entire team at a high level to taking control and reading sensor data of an individual robot?

Through investigating these questions with an expert panel we intend to develop initial grounded guidelines for creating in-the-field interfaces for controlling a remote team of robots

2. EXPLORATION FOCUS

Here we briefly outline our methodology for exploring each of our research points.

We will explore the use of various device formats for in-the-field use. In particular we are interested in a range of touch screens, from small smart phones to large tablets, as they provide flexible input and rich display output in a very portable and compact package. We will investigate how these devices can be carried by a user, e.g., mounted on a forearm for direct use, easy-to-access belt for quick retrieval, etc. In particular, we are curious as to how these formats relate to particular professional tasks and to any existing equipment that must be used.

Touch input offers the advantage of enabling complex interactions using a single hand, either through multi-touch gestures or even a single point (e.g., using a thumb on a smart phone), but has the disadvantage of being hindered by gloves or dirty hands. Not only will we explore these high-level practical technology issues, but further, we intend to investigate which kinds of gestures and touch methods may be best suited to the given tasks and constraints. Further, we will compare how the displays can be coupled with non-touch input devices such as portable keyboards and joysticks.

We will investigate how interface designs can support sliding-scale autonomy, that is, enabling the user to seamlessly move between low-level robot interaction (individual view, battery levels, etc.) and high-level group interaction (macro-control, task-level state). What differentiates this work from other sliding-autonomy research is that we have the additional constraint of keeping people in the field, and so we will take a minimalist and "glanceable" approach to interface design and data presentation. We will explore existing methods and new approaches, grounding our explorations in our expert-panel input.

3. INITIAL PROTOTYPE

We will develop an initial prototype using many of our above-mentioned assumptions; this prototype will serve as an important exploratory platform and initial discussion point for our expert panel. A high-level overview of the prototype is shown in Figure 1: a user remotely controls two quad-rotor robots using a hand-held iPad touch interface to complete a mapping task.

We use AR-DRONE quadrotor flying robots as they provide precise and robust control, are easily remotely controllable, and have embedded cameras which can be used for camera-based SLAM (e.g., as available using the ROS toolkit) for mapping.

4. FUTURE WORK

Ultimately, we aim to develop grounded and directed guidelines for the design of in-the-field remote control robot interfaces. Such guidelines will be useful to help direct the continued development of such effective and easy to use interfaces for human-robot team collaboration.

The primary next step of our exploratory work is to narrow our focus to more targeted aspects of the larger problem. We believe this initial exploratory step is important to help select future questions which will be relevant to the target professional users.

One particular future work question which we will investigate is the design problem of displaying real-time robot sensor and state data, perhaps through meaningful aggregates of team data, in clear and effective ways to the user. For example, providing data overlap and certainty estimates, team-level summaries of such statistics as battery level, and so forth.

5. REFERENCES

[1] Davids, A. Urban search and rescue robots: From tragedy to technology. IEEE Intelligent Systems, 2002, 17(81--83).

[2] Garreau, J. Bots on the ground. Washington Post, www.washingtonpost.com/wpdyn/content/article/2007/05/05/AR2007050501009_pf.html, Visited April 9th, 2008, May 6, 2007.

[3] Hoffman, G., and Breazeal, C. L. Collaboration in Human-Robot Teams. In Proceedings of the American Institute of Aeronautics and Astronautics (AIAA) 1st Intelligent Systems Technical Conference, 2004. Chicago, IL, September 20–22, 2004.

[4] Wang, J., and Lewis, M. Human control for cooperating robot teams. In Proceedings of the 2nd ACM/IEEE Conference on Human-Robot Interaction, 2007. HRI '07, Washington, USA, March 10–12, 2007.

[5] Drury, J. L., Keyes, B., and Yanco, H. A. LASSOing HRI: Analyzing situation awareness in map-centric and video-centric interfaces. In Proceedings of the ACM/IEEE Conference on Human-Robot Interaction, 2007. HRI'07, Washington, USA, March 10–12, 2007.

[6] Drury, J. L., Riek, L., and Rackliffe, N. A decomposition of UAV-related situation awareness. In Proceedings of the 1st ACM SIGCHI/SIGART Conference on Human-Robot Interaction, 2006. HRI '06, Salt Lake City, USA, March 2–4, 2006.

[7] Kadous, W., Sheh, R. K.-M., and Sammut, C. Effective user interface design for rescue robotics. In Proceedings of the 1st ACM SIGCHI/SIGART Conference on Human-Robot Interaction, 2006. HRI '06, Salt Lake City, USA, March 2–4, 2006.

[8] Squire, P., Trafton, G., and Parasuraman, R. Human control of multiple unmanned vehicles: effects of interface type on execution and task switching times. In Proceedings of the 1st ACM SIGCHI/SIGART Conference on Human-Robot Interaction, 2006. HRI '06, Salt Lake City, USA, March 2–4, 2006.

[9] Yanco, H. A., and Drury, J. L. Classifying human-robot interaction: an updated taxonomy. In Proceedings of the IEEE International Conference on Systems, Man and Cybernetics, 2004. SMC '04, The Hague, Netherlands, October 10–13, 2004.

Is the Social Robot Probo an Added Value for Social Story Intervention for Children with ASD?

Ramona Simut, Johan Vanderfaeillie, Bram Vanderborght
Vrije Universiteit Brussel, Clinical and Life Span Psychology Department & Robotics Research Group, Brussels, Belgium +32487782543
ramona.simut@vub.ac.be

Cristina Pop, Sebastian Pintea, Alina Rusu, Daniel David
Babes-Bolyai University; Department of Clinical Psychology and Psychotherapy, Cluj-Napoca, Romania

Jelle Saldien
Howest University College, Industrial Design Center, Kortrijk, Belgium

ABSTRACT
In this paper, we describe the first results of using the robot Probo as a facilitator in Social Story Intervention for children with autism spectrum disorders (ASD). Four preschoolers diagnosed with ASD participated in this research. For each of them, a specific social skill deficit was identified, like Sharing toys, Saying Thank you and SayingHello, and an individualized Social Story was developed. The stories were told by both, the therapist and the robot,in different intervention phases. Afterwards, an experimental task was created where the child needed to exercise the ability targeted by the story. The results of this study showed that the participant needed a decreased level of prompt to perform the targeted behavior, when the story was told by the robot compared to the intervention with the human storyteller. Therefore, this preliminary study created great expectancies about the potential of Robot Assisted Therapy as an added value for ASD interventions

Categories and Subject Descriptors
J.4 Social and behavioral-Psychology

General Terms
Experimentation

Keywords
Robot Assisted Therapy, Autism Spectrum Disorders, Social skills Social Story Intervention, social robot.

1. INTRODUCTION
Children with Autism Spectrum Disorders (ASD) experience difficulties to engage in social interaction, and therefore lack learning opportunities in their classrooms and daily lives. Different therapies exist to attempt to lessen the deficits and family distress. Due to the recent advances in personal robots, the technology is becoming in the reach to be used as Robot Assisted Therapy (RAT). Hereby, the aim is to use robots as a facilitator [5] that mediate the interaction between the child and the human, and not to replace the human therapist. Emerging research shows, that autistic children proactively approach robots, that robots can act as a mediator between the child and the therapist, that robots can be used for play therapy and to elicit joint attention episodes between a child and an adult [6],[2]. The ability to express emotions is essential in social interaction. Therefore, different ASD therapies, like Social Stories developed by Carol Gray [4], focus on these aspects. Social Stories are short stories written or tailored to individuals with ASD to help them understand and behave appropriately in social situations. This paper discusses the first results of using the social robot Probo as Social Story telling agent for children with ASD. Probo's purpose is to serve as a multidisciplinary research platform for human-robot interaction focused on children. The robot is designed to act as a social interface by employing human-like social cues and communication modalities. With 20 motors in the head, the robot is able to express attention and emotions via its gaze and facial expressions [7]. To guarantee a safe physical child-robot interaction, compliant actuation systems and a layered structured with foam and fabric, are implemented.

HRI'12, March 5–8, 2012, Boston, Massachusetts, USA.
ACM 978-1-4503-1063-5/12/03.

2. PROCEDURE
Two boys and two girls participated in the study, aged between 4 to 9. The participants had an independent diagnosis of ASD which meets the criteria outlined in the DSM IV- TR [1]. For each of the four participants, a specific social skill deficit was identified (*sharing toys, saying thank you* and *saying hello*) and an individualized social story was created. The stories were developed using Gray's Social Story construction guidelines [4]. An ABAC/ACAB counterbalanced design was used. In the baseline phase (A), each child was observed during social interactions in order to measure the level of the abilities: to share toys, to say thank you or to say hello. The intervention phases (B and C) consisted in a Social Story intervention: during intervention phase (B), the story was read by the therapist, while in (C),the robot Probo told the story. After the Phase B / Phase C, the Social Story intervention was withdrawn to the baseline conditions and the participants were observed without receiving any intervention. Before the robotherapy phase (C), an habituation phase with the robot was conducted to let the child interact as much as he/she needed. The session of the intervention itself lasted approximately 15 minutes, one time per day, during the 6 days of intervention for each of the B and C phase. The Social Story reading was followed by a comprehensive check, which had to be answered with 100% accuracy. Afterwards, the child had to exercise the social ability that was targeted in the story. For example, if the ability targeted was to say *thank you*, the therapist created specific opportunities that were described in the story (someone is giving help/ a gift). All experiments were video recorded and two trained observers performed the analyses of video sequences. The dependent measurement was the level of prompt needed to provide the expected social response and was assessed using a 7-point Likert scale rating the amount of prompting necessary for a successful social interaction. Hereby, 6

indicated that the child did not respond to any prompt (gestural, physical or verbal) and a value of 0 was given if the participant independently and spontaneously engaged in an appropriate social interaction, without the need of any type of prompt. The scores between 0 and 6 were given for different types of prompts combinations (i.e. verbal and gestural). A more detailed description and analysis of the procedure and the outcomes of this study can be found in [8].

3. DATA ANALYSIS

For each intervention phase, the participants' performance measured by the level of prompt, is presented as box plots in Figure 1.

ANTONIA. Figure 1 for Antonia indicates that both interventions were associated with a decreased level of prompt needed to share the toys. However, the RAT Social Story intervention had a stronger effect on decreasing the level of prompt than the Social Story intervention without the robot. When Probo told the story, only gestural prompts were required to engage Antonia in the appropriate social interaction.

GEORGIA. Figure 1 for Georgia indicates a better performance of the target behavior during the two intervention phases, compared to the baseline phases. Similar to Antonia's case, compared to the Social Story intervention conducted by the human therapist, the RAT Social Story intervention appeared to have a stronger effect in decreasing the level of prompt Georgia needed to say thank you. When Probo was used as story telling agent, Georgia was mostly able to said thank you independently and spontaneously, without the need of any type of prompt.

MIHNEA. The graph analysis in Mihnea's case (see Figure 1) indicates a stronger effect in decreasing the level of prompt in the robot assisted intervention phase, as compared to the phase in which the story was told by the therapist. During Phase C, Mihnea was able to say hello spontaneously in half of the trials. Afterwards, in the second baseline phase A, Mihnea was able to say hello when he entered in a room where someone was present with mostly only gestural prompts.

NICU. As can be seen in Figure 1, Nicu, during the baseline phase, Nicu did not respond at all or only responded to the interaction after therapist's full prompt (gestural, physical and verbal). With the robot it was possible to decrease this average level by two points. Compared to the second baseline phase, the introduction of the Social Story intervention with the human therapist (phase B) was associated with a decrease of the level of prompt, but not as pronounced as in the case of robot assisted intervention (phase C).

4. CONCLUSION

The profound impairment in social interaction is considered to be the most important deficit that individuals with ASD experience across their life span. An emerging research field, Robot Assisted Therapy, consists of using social robots in therapeutical contexts for children with ASD in order to improve their social abilities. The social robot Probo was used as a facilitator in a Social Story intervention. The results suggest that enriched social environment (the presence of the robot) improved the effectiveness of a Social Story intervention, since the average prompt in Phase C is lower than in Phase B. In the robot assisted phase (C), in 40% of the interventions no prompts were necessary and a spontaneous engagement in the appropriate social interaction was obtained. When the Social Story intervention was conducted by a human therapist, no prompts were necessary only in 13% of the interventions from the B phase. Future studies are required for a more in depth analysis related to the ASD child- robot interaction and its implications.

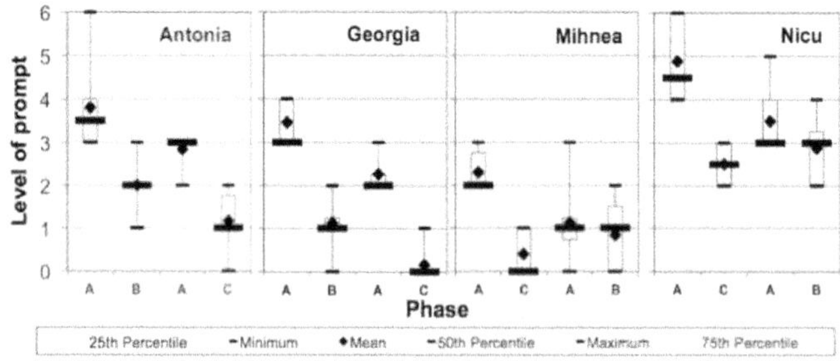

Figure 1: box plots of the data collected

5. ACKNOWLEDGMENTS

Our thanks to Autism Transylvania Association for allowing us to perform there the experiments, especially to the children that participated in the experiments and to their families. This work was supported by CNCSIS-Bucharest, Romania project PN-II-ID-PCE-2011-3-0484 and VUB-HOA16.

6. REFERENCES

[1] American Psychiatric Association. 2000.Task Force on DSM-IV.: *Diagnostic and statistical manual of mental disorders: DSM-IV-TR.* American Psychiatric Publishing, Inc.

[2] Dautenhahn, K., Werry, I., Rae, J., Dickerson, P., Stribling, P., Ogden, B. 2002. Robotic Playmates: *Analysing Interactive Competencies of Children with Autism Playing with a Mobile Robot.* Multiagent Systems, Artificial Societies, and Simulated Organizations 3.

[3] Goris, K., Saldien, J., Vanderborght, B., Lefeber, D. 2011. *How to achieve the huggable behavior of the social robot Probo? a reflection on the actuators.* Mechatronics 21(3), 490–500.

[4] Gray, C.2000. *The new social story book.* Future Horizons Inc.

[5] Pradel, G., Dansart, P., Puret, A., Barthelemy, C. 2010. *Generating interactions in autistic spectrum disorders by means of a mobile robot.* In: IECON 2010-36th Annual Conference on IEEE Industrial Electronics Society. pp. 1540–1545.

[6] Robins, B., Dickerson, P., Stribling, P., Dautenhahn, K. 2004. *Robot-mediated joint attention in children with autism: A case study in robot-human interaction.* Interaction Studies 5(2), 161–198.

[7] Saldien, J., Goris, K., Vanderborght, B., Vanderfaeilli, J., Lefeber, D. 2010. *Expressing emotions with the huggable robot Probo.* International Journal of Social Robotics, Special Issue on Social Acceptance in HRI 2(4), 377–389.

[8] Vanderborght, B., Simut, R., Saldien, J., Pop, C., Rusu, A.S.,Pintea, S., Lefeber, D., David, D.O.: *Using the social robot Probo as social story telling agent for children with ASD.* Interaction Studies (under review).

Animal-Inspired Human-Robot Interaction: A Robotic Tail for Communicating State

Ashish Singh
Department of Computer Science
University of Manitoba
Winnipeg, MB, Canada
ashish@cs.umanitoba.ca

James E. Young
Department of Computer Science
University of Manitoba
Winnipeg, MB, Canada
young@cs.umanitoba.ca

ABSTRACT
We present a robotic tail interface for enabling a robot to communicate its state to people. Our interface design follows an animal-inspired methodology where we map the robot's state to its tail output, leveraging people's existing knowledge of and experiences with animals for human-robot interaction. In this paper we detail our robotic-tail design approach and our prototype implementations, and outline our future steps.

Categories and Subject Descriptors
H.1.2 [**Models and principles**]: user/machine systems—*human factors*

General Terms
Design, Experimentation, Human Factors

Keywords
Human-robot interaction, animal-inspired interfaces.

1. INTRODUCTION
In this rapidly emerging field of HRI, many robots are designed and built as utility robots to help people in their homes and workplaces. Some argue that it will be important in these environments for robots to maintain a peripheral presence similar to that of Victorian servants [4,5]: that is, the robot should interrupt people as little as possible, but people should have peripheral awareness of the robot. This is important, for example, to identify when the robot requires assistance. One challenge with this goal is to enable the robot to communicate a range of states without being intrusive. In this paper we present a robotic tail interface as a passive means for a robot to broadcast its state.

The premise of our interface is that many people have intimate experience with tailed animals such as cats and dogs and understand much of their tail vocabulary [1], understanding which we can leverage for HRI. Although we concede that not everyone will understand a tail vocabulary, we note that most people at least have passing familiarity from the social stock of knowledge [6], for example, that dogs are happy when they wag their tail.

Several robots exist that use robotic tails, for example, the Sony AIBO, and Necoro the robotic cat [8,9]. In these cases the robotic tail was used to mimic the behavior and appearance of the actual animal, with less attention paid to the communication capabilities of the tail and how they can be used to communicate a utility robot's state [2]. In our work we extend this idea to focus on developing communication mappings and more capable tails to enable communication for even non-zoomorphic robots: for example, if an iRobot Roomba gets stuck, it could lower its tail to suggest a fearful state (Figure 1c), or could wag its tail to indicate that it found a particularly dirty spot that it was happy to clean.

HRI'12, March 5–8, 2012, Boston, Massachusetts, USA.
ACM 978-1-4503-1063-5/12/03..

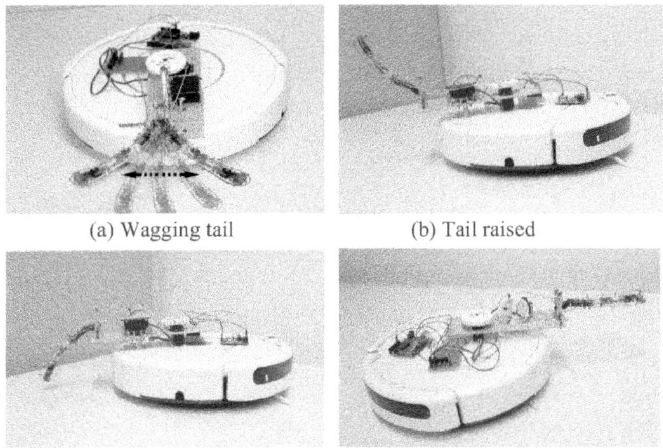

(a) Wagging tail (b) Tail raised (c) Tail lowered (d) Tail straight

Figure 1 – iRobot Roomba with a actuated-cable robotic tail.

Specifically, we modeled our initial prototype after a dog's tail: different animals such as dogs and cats have different tail vocabularies, and perhaps the dog is most commonly understood. Below we outline our approach to mapping robot state to tail state, and various prototype implementations and designs.

2. MAPPING ROBOT STATE TO TAIL
In order to leverage people's understanding of dog-tail vocabulary for HRI, we must first investigate how dogs themselves use tails to communicate, and, how people interpret this communication. For our initial prototype, and to provide a starting point for our engineering challenge, we informally explored dog tail language and behavior using pet-training resources. However, to develop meaningful robot state to tail mappings it will be important to investigate animal-behavior literature more deeply.

One challenge of adapting dog-tail language to a robot like the Roomba is that, with a real dog, the tail forms only one part of an overall expression. For example in Figure 1b, the robot could be angry or could be highly interested; for both a dog places its tail in this position to indicate high arousal, and uses other body language cues to specify the meaning more clearly. Thus we will investigate how relevant emotional models can be used to decompose an expression and explain which role the tail can play.

One such model which we will explore is the Pleasure-Arousal-Dominance Temperament Model [7], which decomposes an emotional state into the three dimensions. As with our dog example above, for example, both *anger* and *joy* would have high arousal but would be on opposite ends of the *pleasure* scale. In addition to providing a structure for tail vocabulary, this model also provides us with a mechanism for integrating the tail with other robot expression, for example, movement patterns or sounds.

3. PROTOTYPE DEVELOPMENT

We are currently exploring methods for implementing our robotic tail. Throughout this process we aim to procure smooth and fast movement over the tail range, enable continuous tail deformation to form a curve rather than piece-wise links, minimize noise from tail actuation, and minimize tail mechanism size for easy robot integration. Below we detail our current designs, a servo tail and actuated cable tail, and briefly discuss future prototype directions.

We implemented the servo-tail as a three-piece link of micro servo motors mounted on perpendicular axes to enable a wide range of tail movement. This prototype was unsuccessful as the servos were very noisy, moved too slowly for convincing behavior, and the piece-wise result did not resemble a tail.

Our current prototype is an actuated cable tail which we developed by loosely following animatronics hobbyist techniques for tails [10]: we modified Klixx toys and used servo-driven guide wires to enable rapid and smooth tail deformation (Figure 1, 2). With this setup the tail can move left-right and up-down, or both simultaneously to hit the corners, the tail curves when deformed into a smooth arc, and the servos leverage a wheel to achieve fast tail movements and direction changes.

Although this prototype gives convincing results, it unfortunately has noise problems, requires a large mounting mechanism, and the deformation shape breaks down at large deformations. Looking forward, we are investigating methods using, e.g., springs and guide wires, magnets and an outer skin, or nitinol wire (shape-memory alloy [3]), as means to improve movement range, deformation quality and smoothness, and reduce noise

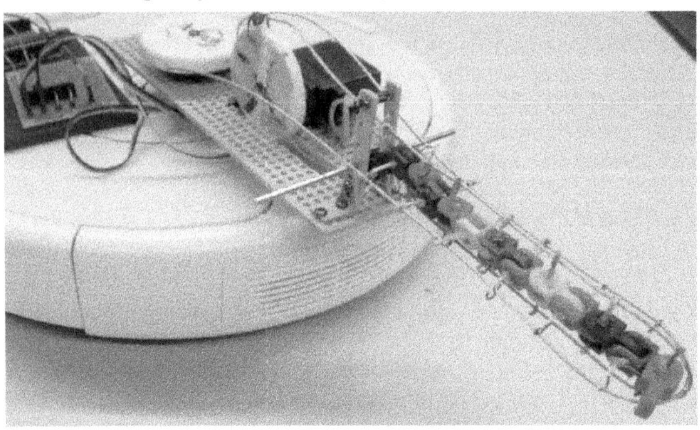

Figure 2 – actuated cable tail prototype

4. RESEARCH DIRECTIONS

We intend to conduct user evaluations to investigate questions regarding how the tail is received and how useful it can be as a communication mechanism. Questions for exploration include: if people will draw from previous experiences with animals to interpret a robot's tail movements, if we can develop effective robot state to tail mappings for domestic use-case scenarios, and the overall effect that the existence of the tail has on a person's perception of the robot. We will also investigate the differences based on user background, e.g., by classifying people as cat or dog people (or both or neither) based on their experiences, and investigating differences of expectations and reception. For people without prior animal experience, we may investigate what sort of "tail-vocabulary" training may be useful.

In addition, as we continue research into animal behavior and build targeted research questions about tail vocabularies, we intend to conduct studies to investigate, e.g., the effects of tail movement speed, pose locations and movement patterns. Finally, we intend to investigate the importance of particular physical tail properties such as length, thickness, color and appearance (e.g., robotic versus furry), or even different tail formats such as a lizard or elephant-style tail instead of a dog-style tail.

Overall, we believe that our animal-inspired HRI approach of having a robot use a tail for communicating state can provide an effective, easy-to-understand peripheral awareness channel for robots entering into people's everyday spaces.

5. REFERENCES

[1] Brown, S. E. (2004). The Human-Animal Bond and Self Psychology: Toward a New Understanding. *Society and Animals*, 12(1):67-86.

[2] Seewald, A. K. (2001). Entertainment Robots - Myth or Reality. *Proceedings of the Fourteenth International Florida Artificial Intelligence Research Society Conference* (pp. 504-508).

[3] Mills, J. W. (1992). Stiquito: A Small, Simple, Inexpensive Hexapod Robot. Technical Report 363a, Computer Science Department, Indiana University, Bloomington, IN 47405-4101.

[4] Young, J., Hawkins, R., Sharlin, E., and Igarashi, T. (2008). Toward Acceptable Domestic Robots: lying Insights from Social Psychology. *International Journal of Social Robotics*, 1(1):95-108.

[5] Hamill, L. & Harper, R. (2006). Talking Intelligence: a Historical and Conceptual Exploration of Speech-Based Human-Machine Interaction in Smart Homes. *Proceedings of the 1st International Symposium on Intelligent Environments* (pp. 121-128).

[6] Berger, Peter L., and Thomas Luckmann. (1967). The social construction of reality: a treatise in the sociology of knowledge. Garden City, N.Y.: Doubleday.

[7] Mehrabian, A. (1996). Pleasure-arousal-dominance: A general framework for describing and measuring individual differences in Temperament. *Current Psychology* 14(4):261-292.

[8] Weiss, A. a. (2009). "I Love This Dog"—Children's Emotional Attachment to the Robotic Dog AIBO. *International Journal of Social Robotics*, 1(3):243-248.

[9] Libin, A.V. and E. Libin. (2004). Person-Robot Interactions From the Robopsychologists' Point of View: The Robotic Psychology and Robotherapy Approach. *Proceedings of the IEEE*. 92(11):1-15.

[10] (n.d.). Retrieved September 20, 2011, from Wolftronix.com: http://www.wolftronix.com/tail/spine.

Spatial Language Experiments for a Robot Fetch Task

Marjorie Skubic[1], Laura Carlson[2], Jared Miller[2], Xiao Ou Li[2], & Zhiyu Huo[1]

[1]Electrical & Computer Engineering
University of Missouri
Columbia, MO 65211
1-573-882-7766
SkubicM@missouri.edu;
zhiyuhuo@mail.mizzou.edu

[2]Dept. of Psychology
University of Notre Dame
Notre Dame, IN 46556
1-574-631-8883
LCarlson@nd.edu;
jmille39@nd.edu; xiaoou_li@nd.edu

ABSTRACT
This paper outlines a new study that investigates spatial language for use in human-robot communication. The scenario studied is a home setting in which the elderly resident has misplaced an object, such as eyeglasses, and the robot will help the resident find the object. We present results from phase I of the study in which we investigate spatial language generated to a human addressee or a robot addressee in a virtual environment.

Categories and Subject Descriptors
H.5.2 [**Information Systems**]: User Interfaces – *Voice I/O.*

General Terms
Experimentation, Human Factors.

Keywords
Robots for elderly, Spatial language.

1. INTRODUCTION
Recent studies have shown that one of the top five tasks noted by seniors for assistive robots is help with fetching objects, for example, retrieving missing eyeglasses [1]. In addition, the most preferred domestic robot interface is natural language [2]. In this paper, we present the overview and initial results for a project designed to address the fetch task and study an appropriate language interface that allows the human users to communicate naturally and effectively with a robot.

When people communicate with each other about spatially oriented tasks, they typically use relative spatial references rather than precise quantitative terms, e.g., *the eyeglasses are in the livingroom on the table in front of the couch* [3]. Here, we explore the spatial language used in the context of the fetch task and investigate both language generation and understanding. A set of human subject experiments is planned, studying first college-age students and then adults over age 65 for comparison. Initial experiments are being conducted in a virtual environment (VE), which provides a controlled setting and is easier for capturing potentially subtle metrics between test conditions. In pilot work, the use of a VE was shown to have sufficient sensitivity to detect differences and also replicated key findings from work done in physical environments, e.g.,[4]. Later studies are planned with robots in the physical world.

HRI'12, March 5–8, 2012, Boston, Massachusetts, USA.
ACM 978-1-4503-1063-5/12/03.

2. EXPERIMENT PROTOCOL
The first round of human subject experiments has been completed with 128 college-age students. We investigated the type of spatial language used intuitively by participants when addressing either a human or a robot avatar. The VE included three rooms – a central hallway with a livingroom and a bedroom (Figure 1).

Each participant begins with a brief video illustrating the room layouts. At this point, candidate reference objects are shown but no target objects are included in the scene. The participant is then asked to explore the scene, to look for a specified target object which is now included in the VE. Eight target objects are used for the study: a book, cell phone, eyeglasses case, keys, letter, mug, notepad, and wallet; each participant has eight trials, one for each target object. After locating the target object, the participant is brought back to the hallway and is asked to give instructions to the avatar on the location of the target object. There are three test conditions: (1) the addressee, either human or robot, (2) the subject alignment with respect to the addressee, either aligned at 0 degrees or face to face, i.e., misaligned at 180 degrees, and (3) the instruction on the type of communications, either tell the addressee WHERE to find the target object or HOW to find it. The descriptions given by the participants were recorded, transcribed, and coded for analysis.

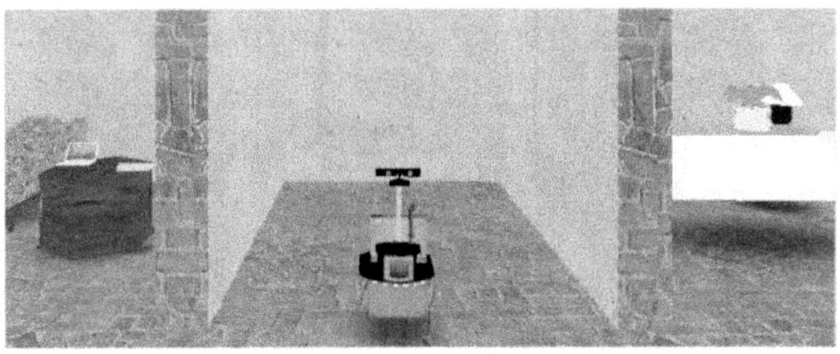

Figure 1. The virtual scene used for the experiments, showing the robot avatar in the hallway with the livingroom on the left and the bedroom on the right.

3. RESULTS
The more prominent results are summarized in Fig. 2-5. Participants had a preference for taking the addressee's perspective in all test conditions (Fig. 2), which agrees with prior HRI studies e.g., in [5]. Participants also had a strong preference for using descending descriptions (Fig. 3), that is, from outside to inside the room, e.g., in the bedroom on the table.

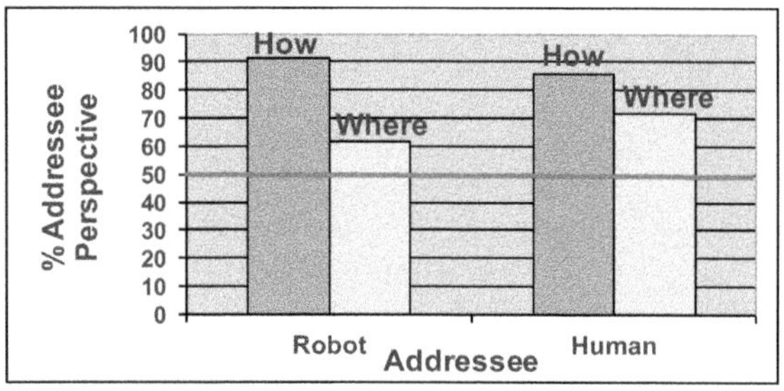

Figure 2. Percent of the time subjects used the addressee's perspective.

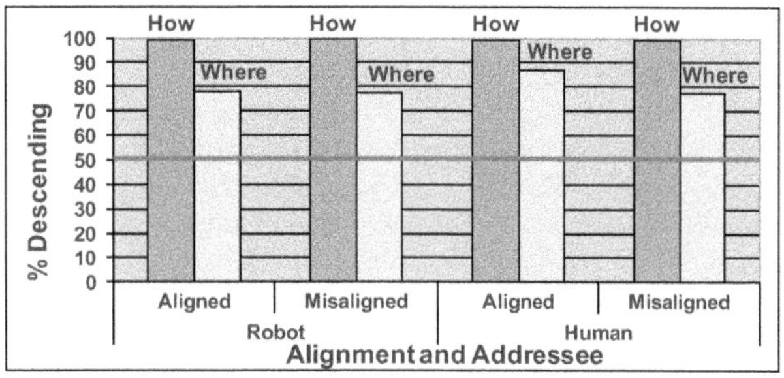

Figure 3. Percent of the time subjects used descending descriptions (outside to inside room).

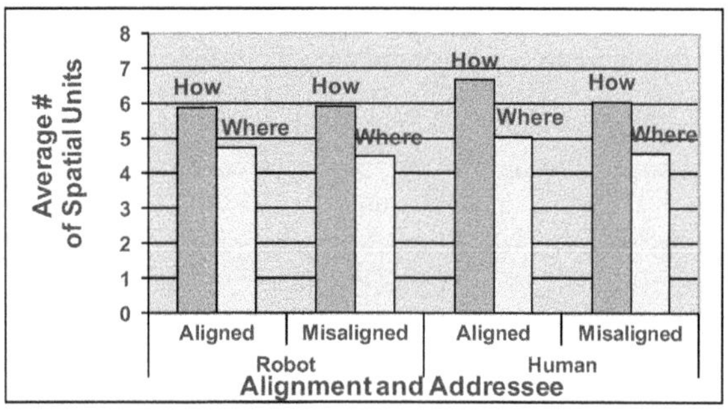

Figure 4. Average number of spatial units per description.

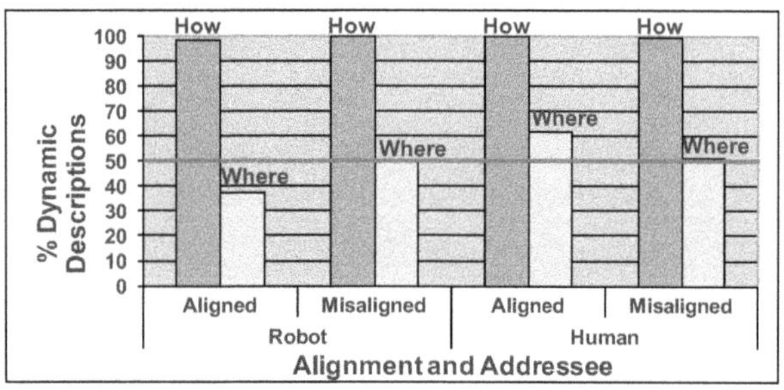

Figure 5. Percent of the descriptions that resemble step by step instructions to the target object, as opposed to static location descriptions.

We tested for differences between test conditions and found the only statistically significant differences were between the HOW and WHERE conditions. Fig. 2-5 all illustrate these differences. The HOW descriptions emphasized the preferences for addressee perspective and descending descriptions. In addition, participants used more spatial phrases on average for the HOW descriptions compared to the WHERE descriptions (Fig. 4). We also looked at whether subjects used static location descriptions (e.g., the book is in the bedroom on the table next to the bed) vs. dynamic descriptions that provide step by step instructions (e.g., go to the bedroom, move to the left side of the bed, look on the table and find the book). Not surprisingly, participants showed a very strong preference for these step by step instructions when asked to provide a HOW description (Fig. 5). It is perhaps more surprising that participants also used these step by step instructions about half the time for the WHERE descriptions.

There were no statistically significant differences between the robot and human addressees. The figures show some differences but none of these are statistically significant. It is possible that these college age subjects were viewing the human avatar in the VE as an anthropomorphic robot; we did not test for this. We anticipate that we may find differences with the older adults.

The most used spatial terms include on, to, right, and left. Top reference items include the furniture pieces (especially the table, bed, and desk) and house structure units (room, wall). Very few references were made to smaller objects lying on the table and desk next to the target objects.

3. CONCLUSIONS

The results of the experiments confirm the use of the addressee's perspective and show a preference for descending descriptions, which will be important in supporting a robot interface. They also show that the spatial descriptions depend on the task instructions given. The HOW descriptions convey step by step motion, and use many spatial units; the WHERE descriptions have no preference for conveying motion and use fewer spatial units. We are in the process of repeating the experiments with adults over 65 and will be especially looking for differences in the two age groups.

The results also highlight some of the challenges in implementing a robot fetch task in the physical world. We will need to recognize furniture items such as tables even when they are cluttered with objects, and support relative spatial terms with respect to the furniture. There are also significant challenges in speech recognition and natural language processing to translate the spatial descriptions into robot commands.

4. ACKNOWLEDGMENTS

This work is funded by the NSF under grant IIS-1017097.

5. REFERENCES

[1] Mitzner T., Smarr, C., Beer, J., Chen, T., Springman, J., Prakash, A., Kemp, C. and Rogers, W. 2011. Older Adults' Acceptance of Assistive Robots for the Home. Technical Report HFA-TR-1105. Georgia Tech Univ.

[2] Scopelliti, M., Giuliani, M., and Fornara, F. 2005. Robots in a domestic setting: a psychological approach. *Universal Access in the Information Society*, 4(2), 146-155.

[3] Carlson, L. A. and Hill, P. L. 2009. Formulating spatial descriptions across various dialogue contexts. In K. Coventry, T. Tenbrink, & J. Bateman (Eds.), *Spatial Language and Dialogue* (pp. 88-103). New York, NY: Oxford University Press Inc.

[4] Schober, M.F. 1995. Speakers, addressees and frames of reference: Whose effort is minimized in conversations about locations? *Discourse Processes*, 20, 219-247.

[5] Moratz, R., Fischer, K. and Tenbrink, T. 2001. Cognitive Modeling of Spatial Reference for Human-Robot Interaction, *Intl. J. on Artificial Intelligence Tools*, 10(4), 589-611.

Potential Measures for Detecting Trust Changes

Poornima Kaniarasu[1], Aaron Steinfeld[1], Munjal Desai[2], and Holly Yanco[2]

[1]Carnegie Mellon University, Pittsburgh, PA 15213
{kpoornima, steinfeld}@cmu.edu

[2]University of Massachusetts Lowell, Lowell, MA 01854
{mdesai, holly}@cs.uml.edu

ABSTRACT
It is challenging to quantitatively measure a user's trust in a robot system using traditional survey methods due to their invasiveness and tendency to disrupt the flow of operation. Therefore, we analyzed data from an existing experiment to identify measures which (1) have face validity for measuring trust and (2) align with the collected post-run trust measures. Two measures are promising as real-time indications of a drop in trust. The first is the time between the most recent warning and when the participant reduces the robot's autonomy level. The second is the number of warnings prior to the reduction of the autonomy level.

Categories and Subject Descriptors
H.1.2 [**User/Machine Systems**]: Human factors

General Terms
Experimentation, performance

Keywords
Trust, automation, experiments

1. INTRODUCTION
One of the key factors for the acceptance and safe deployment of robots is the degree to which a user trusts the robot. A main goal of our work is to model a person's current level of trust in a robot system so that it can be used to design robot interfaces and behaviors that foster appropriate levels of trust [2]. Mobile robots, especially those not designed for social interaction, are particularly interesting since they are likely to be task-oriented and therefore used for time-dependent activities, capable of damaging objects and hurting people, and unable to express their intent to bystanders (e.g., [6]).

There is an increasing amount of work exploring what factors are important for human trust in robots (e.g., [1-4]). In this work, we are explicitly examining two measures as potential real-time indicators that an operator has reduced their trust of a task-oriented robot. Specifically, we are focused on measures that can be internally observed by a robot without the use of traditional trust surveys. With these factors, on-line techniques will allow robots to recognize the human has lost trust and act accordingly.

2. METHOD
Our approach is to analyze data collected under an existing study. Details about the robot, interface, measures, experimental design, and results can be found in Desai et al. [2]. The study investigated how changing the robot's reliability influences people's use of robot autonomy and their trust in the robot system through experiments with participants operating a real robot through a slalom course. The experiment was designed to have a high workload so that the participants would need to use the autonomous capabilities of the robot in order to complete the task in time and to be able to complete the secondary task. We hypothesized that people would trust a robot system less when its reliability in autonomous mode decreased, switching to a shared mode. Therefore, the robot's reliability fluctuated between high and low levels during the experiment.

For this work, we examined the three conditions where users encountered reliability changes. These changes consisted of controlled dips in reliability based on pre-set locations (Figure 1). Audio messages indicating the cumulative number of wrong turns during the run and their penalties were issued as warnings to participants immediately after each navigation error occurred. These conditions were counterbalanced within the larger experiment. One of the measures collected at the end of each run was an autonomy trust survey developed by Jian, et al. [5].

Two potential measures were identified as possible indicators for a loss of trust. These were:

1. Time Between Events (*TBE*): The time between two events, whether a warning or a switch. This measure gives the time between two consecutive warnings (e.g., TBE = t_8-t_6 in Figure 2) or, in the case of a warned switch, the time between the most recent warning and a participant changing the autonomy level (TBE = t_3-t_2, in Figure 2).

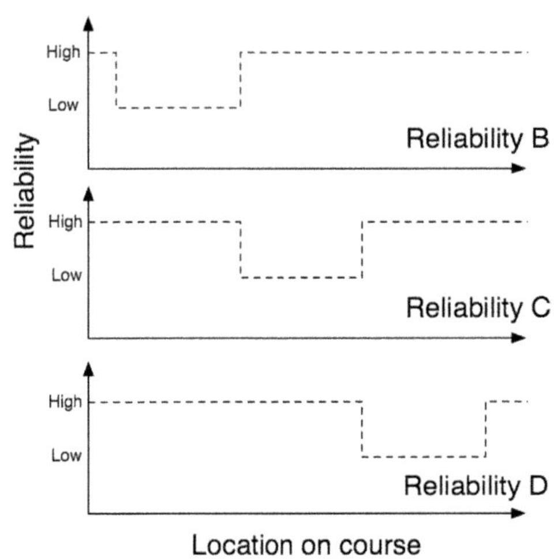

Figure 1. Reliability patterns for the analyzed conditions

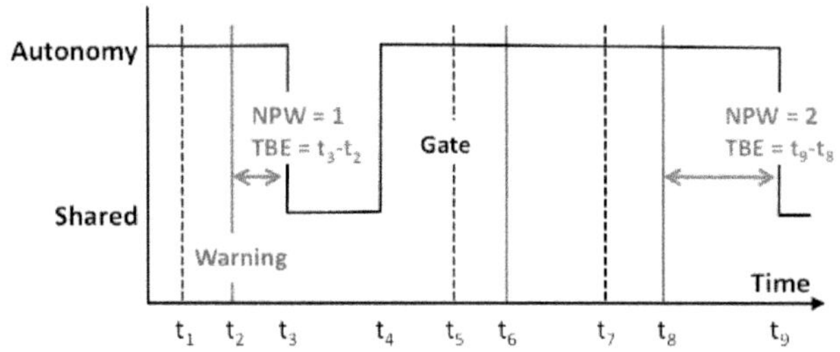

Figure 2. Timeline graph of the metrics TBE and NPW

2. Number of Prior Warnings (*NPW*): How many times the participant was warned prior to a switch. Every time a switch occurs, the NPW counter is reset to zero. If there is a warning but no switch, it accumulates (e.g., t_9).

3. RESULTS

3.1 Switching

We looked at how warnings impacted switching, which revealed that most of the switches were not due to a prior warning (Table 1). Also, the timing of when reliability dropped appeared to have an impact on switching behavior. This result aligns with other analyses by the team on switching behavior which showed that reliability drops in the middle of the run (C) led to increases in the number of mode switches. Additionally, switches away from autonomy were twice as slow as those seen for early drops in reliability (B) [2].

3.2 Warned switches

Next, we focused only on autonomy switches that were spurred by a warning. We also omitted switches where a user made a turn between the most recent warning and the switch since the warning would have been for the preceding turn. Participants could switch in two directions, either into or out of autonomy. ANOVA models were run for the Direction (Autonomy, Shared) and Reliability (B, C, D) on TBE and NPW. There was a significant interaction for TBE ($F(2, 26) = 4.3$, $p < 0.05$) where TBE was significantly higher for Auto-B than other combinations, showing that people took longer to recognize a reliability drop when it occurred near the start of the robot's use. There was also a marginal result for the main effect of Reliability ($F(2, 26) = 2.8$, $p < 0.1$) with reliability B being higher than the other two conditions. A quick power calculation revealed that this would likely become significant with a few more samples. This result aligned with a significant main effect for Reliability on NPW ($F(2, 26) = 5.7$, $p < 0.01$) where reliability B was higher than the other two conditions. No other effects were significant.

Correlation analyses were run for the two measures, the post-condition trust survey, and the percentage of time spent in autonomous mode during the condition. There were significant correlations for: NPW and TBE (0.475, $p < 0.05$), NPW and trust (-0.40, $p < 0.05$), and NPW and percent time using full autonomy (0.42, $p < 0.05$). In other words, for participants who had switched as a result of a warning, these results show that, (1) participants who accumulated more warnings responded slower than their peers to the most recent warning, (2) participants who accumulated more warnings had lower trust, and (3) participants who accumulated more warnings used more autonomy. While this last result seems to conflict with the second, we hypothesize that, when faced with a system that is repeatedly failing, users cannot distinguish between the robot's failures and their own, leading them to accept the easier control method. This result stresses the need to model a person's trust in a robot system, in order to find ways to encourage people to use the best autonomy mode for the current situation.

Table 1. Switching behavior

Condition	Warned	N (% of subtotal)
B	No	114
	Yes	11 (8.8%)
C	No	164
	Yes	17 (9.4%)
D	No	83
	Yes	11 (11.7%)

4. CONCLUSIONS

The impact of warnings on switching behavior seems to reinforce the earlier findings in Desai et al. [2] that early drops in reliability incur different behavior than middle or late reliability drops.

The correlation results have strong face validity. In the face of repeated alarms, a user will stop attending to them as quickly (NPW & TBE). Likewise, repeated alarms imply a system is untrustworthy. Therefore, Number of Prior Warnings and the user's response time to the warnings have the potential as real-time measures of a drop in trust.

5. ACKNOWLEDGMENTS

This work has been supported in part by the National Science Foundation (IIS-0905228 and IIS-0905148). Thanks to the larger team for all the hard work behind the design and execution of the source study.

6. REFERENCES

[1] I. Dassonville, D. Jolly, and A. Desodt. Trust between man and machine in a teleoperation system. *Reliability Engineering & Systems Safety*, 53(3):319–325, 1996.

[2] M. Desai, M. Medvedev, M. Vázquez, S. McSheehy, S. Gadea-Omelchenko, C. Bruggeman, A. Steinfeld, and H. Yanco. Effects of changing reliability on trust of robot systems. In *HRI '12: Proceedings of the 7th ACM/IEEE Int'l Conference on Human Robot Interaction*, 2012.

[3] A. Freedy, E. DeVisser, and G. Weltman. Measurement of trust in human-robot collaboration. *Collaborative Technologies and Systems*, Jan 2007.

[4] P. A. Hancock, D. R. Billings, K. E. Schaefer, J. Y. C. Chen, E. J. de Visser and R. Parasuraman. A meta-analysis of factors affecting trust in human-robot interaction *Human Factors: The Journal of the Human Factors and Ergonomics Society*, 53(5): 517-527, 2011.

[5] J. Jian, A. Bisantz, and C. Drury. Foundations for an empirically determined scale of trust in automated systems. *Int'l Journal of Cognitive Ergonomics*, 4(1):53–71, 2000.

[6] B. Mutlu and J. Forlizzi. Robots in organizations: the role of workflow, social, and environmental factors in human-robot interaction. In *HRI '08: Proceedings of the 3rd ACM/IEEE Int'l Conf.erence on Human Robot Interaction*, pages 287–294, 2008.

Affect Misattribution Procedure: An Implicit Technique to Measure User Experience in HRI

Ewald Strasser
ICT&S Center, University of Salzburg
+43.662.8044.4837
Ewald.Strasser@sbg.ac.at

Astrid Weiss
ICT&S Center, University of Salzburg
+43.662.8044.4838
Astrid.Weiss@sbg.ac.at

Manfred Tscheligi
ICT&S Center, University of Salzburg
+43.662.8044.4811
Manfred.Tscheligi@sbg.ac.at

ABSTRACT

This paper suggests new methodology for measuring User Experience in HRI. We suggest using implicit attitude to predict User Experience (Affect). Therefore we show a first validation study. The study uses short videos of a robot (IURO – Interactive Urban Robot) approaching a person and asking for the. IURO either approached a walking or a standing person. We measured people's implicit attitude towards the robot with the Affect Misattribution Procedure (AMP). The results show that a walking person being approached by the robot evolves an implicitly more negative attitude in the observing participant whereas corresponding questionnaire items showed no difference in attitude for the approach behaviour. We conclude from these results that measuring implicit attitude in HRI is valuable for the evaluation of the User Experience of a robot.

Categories and Subject Descriptors

D.1.2 [**Models and Principles**]: User/Machine Systems – Human factors, implicit attitude, User-Experience.

Keywords

Implicit attitude, user experience, approach behaviour, measureing

INTRODUCTION

In recent years we saw an increased desire in HRI research for a more sustainable and positive long-term experiences of robotic systems[1]. However, to systematically explore and measure user experience is a challenging endeavour. As people are often simply not able to put into words what they felt during the interaction on a reflective level [2], it is not enough to ask them about their feelings and perceptions of a robotic system. They find it difficult putting their finger on the exact feature what caused their emotions or, in the worst-case scenario, they distort their answer by trying to rationalise it.

Therefore we propose a more unobtrusive approach, the Affect Misattribution Procedure (AMP) [3] for the measurement of affection towards robots. Drawing on the USUS (Usability, Social Acceptance, User Experience, and Societal Impact) evaluation framework for HRI [4] we assume that emotions and affections are very central factors of User Experience.

The AMP [3], an accepted method in psychology for the measurement of implicit attitude, is easy to use and interpret. With the presented study we want to verify, if implicit attitude can be reasonably applied in HRI scenarios. The study was video-based with the goal being to investigate participants' attitudes towards a robot in different approach situations: a dynamic condition (in which the robot approached a walking person) and a static condition (in which the robot approached a standing person). According to Woods et al. [5] the use of video material of HRI instead of a live approach still produces valid results. Since it was our aim to investigate the applicability of the AMP in HRI the video approach had also the advantage that more standardised results could be obtained. Our goal was to investigate whether the AMP would be able to detect different attitudes towards the robot, compared to self-reporting questionnaires.

METHODOLOGY

The AMP was developed by Payne et al. [3] as a measurement for implicit attitudes. In the AMP participants are instructed to rate whether they perceive abstract patterns, such as Chinese characters, as pleasant or unpleasant. Prior to the Chinese character, another object appears. This is the object that we want to evaluate. The object is shown very short but in principle consciously visible which makes the AMP a priming procedure. The participants are instructed to disregard this object and value only the Chinese character. However, participants are more inclined to perceive the Chinese character as pleasant if the object before is favourable for them. The idea behind the AMP is to measure an affective-loaded concept, by transferring its affective context on a neutral sign such as the Chinese letter and make it therefore possible to gather an affective appraisal that would otherwise be hard to get (Figure 1 shows the timing of the AMP procedure).

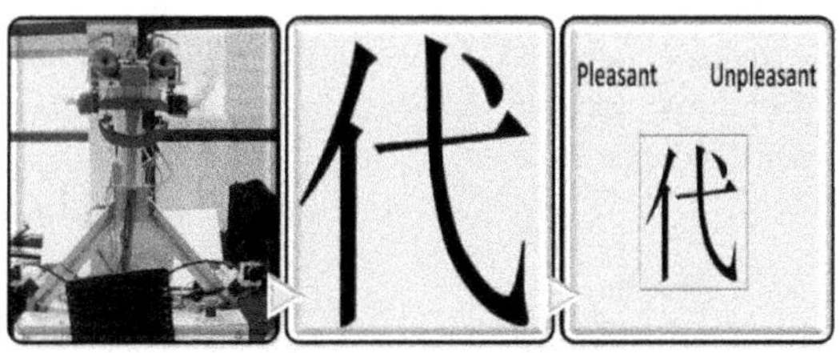

Figure 1: Affect Misattribution Procedure. From the left the pictures show: the prime, the Chinese letter (which should be evaluated) and the screen as displayed to the participants.

In the current study we "loaded" the Chinese letter with objects from the video of the human-robot interaction. It was either the human or the robot that were shown from the videos. The appraisal of the Chinese character provides us in this vein with a score for the bias towards the robot and the human.

HRI'12, March 5–8, 2012, Boston, Massachusetts, USA.
ACM 978-1-4503-1063-5/12/03.

PROCEDURE

In total, 30 participants were recruited for the experiment. Upon their arrival at the laboratory participants were instructed that their main task is to rate videos. They were told that the study should be self-explanatory and that they simply can follow the instructions on the screen, but that the researchers will stay in the room to answer potential questions. The whole experiment was programmed and presented in Inquisit [6] as this software enables combining the priming (needed for the AMP) and questionnaires (needed for the explicit scales).

First the participants had to fill in a demographic questionnaire. Then they were instructed how they were supposed to rate the videos. At that point they were shown three videos either showing the robot approaching a walking (straight, right, left) or a standing person (straight, right, left). After each video participants were asked to answer 3 different 9-point Likert scale questions which are shown later in the text. After all videos and all questions were presented in this manner, the AMP followed. The AMP as we used it was divided in an introduction (explaining that participants should rate Chinese letters in terms of being pleasant or unpleasant), a training block (to get used to the rating procedure) and a test block (the actual data for the later analysis).

RESULTS

For the analysis of the explicit attitude towards the robot we calculated a summative score for the three self-reporting questions of the conscious appraisal of the approach trajectory. Participants rated approach behaviour on a 9-point Likert scale.

- Whether they liked the approach trajectory of the robot (1- strongly disagree, 9- strongly agree)
- Whether they thought the speed of the robot was appropriate (1- too slow, 5- about right, 9- too fast)
- Whether they thought the stopping distance of the robot was appropriate (1- too close, 5- about right, 9- too far)

An initial analysis of the collected demographic data showed that we had tested many people who stated that they had previous experiences with robots. Since it is quite obvious that people with an established attitude towards robots would be significantly less vulnerable to our experimental condition (approaching a walking or standing person) we decided to exclude them from the analysis. The remaining sample consisted of 20 participants. An independent-sample t-test, for the AMP score, showed that participants preferred the static approach over the dynamic approach $t(20) = 2.11$, $p = .048$. In contrast, the self-reporting data of the questionnaires did not reveal comparable differences between the dynamic and the static condition [7].

DISCUSSION

The aim of this study was to verify the AMP can be used in HRI scenarios. For that reason we chose an HRI scenario where it is to some degree clear that our experimental conditions cause a different perception because when all other parameters are constant, two converging objects generate a higher speed compared to when only one object is moving. With that in mind we assume that these conditions generate a different impression in the observer. Results from the AMP confirmed this opinion. Further it showed that the less trivial but intuitive valid assumption that a robot should approach a standing person is true. However, had we taken only the responses from self-reports, we would have denied that such a difference in human perception exists for the observed HRI scenario. This observation is consistent with studies in psychology that implicit attitude has good incrementally validity with self-report measurements (e.g. [8]).

ACKNOWLEDGMENTS

This work is supported by the European Commission as part of the IURO project, see also **www.iuroproject.eu/project.eu**.

FUTURE WORK

As we showed the AMP adds a different perspective in User Experience research of robots. Despite this advantage the AMP has also a disadvantage; it can only evaluate one chosen (here pleasant/unpleasant) aspect of the implicit attitude towards a robot in every evaluation session. To be able to test different factors at once we designed another procedure which we call Reversed Affect Misattribution Procedure (R-AMP). R-AMP reveres the order of the AMP approach, which means that instead of priming the robot which should be evaluated the evaluation criteria is primed. As evaluation criteria we use the six basic emotions from Ekman [9] which are represented through emotional faces. The goal is to finally provide a standard measure for implicit Users Emotion in HRI.

REFERENCES

[1] Sung, J. Y., Christensen, H. I., and Grinter, R. E., Robots in the wild: understanding long-term use. (2009), 45-52.

[2] Bethel, C. L., and Murphy, R. R., Review of Human Studies Methods in HRI and Recommendations, International Journal of Social Robotics, vol. 2, (2010), 347-359.

[3] Payne, B. K., Cheng, C. M., Govorun, O., and Stewart, B. D., An inkblot for attitudes: Affect misattribution as implicit measurement, Journal of Personality and Social Psychology, vol. 89, no. 3, (2005), 277.

[4] Weiss, A., Bernhaupt, R., Lankes, M., and Tscheligi, M., The USUS evaluation framework for Human-Robot Interaction., in AISB2009: Proceedings of the Symposium on New Frontiers in Human-Robot Interaction, SSAISB, Edinburgh Scottland, (2009), 158-165.

[5] Woods, S. N., Walters, M. L., Koay, K. L., and Dautenhahn, K., Methodological issues in HRI: A comparison of live and video-based methods in robot to human approach direction trials. (2006), 51-58.

[6] Draine, S., Inquisit [Computer software], Seattle, WA: Millisecond Software, (1998).

[7] Gonsior, B., Landsiedel, C., Mirnig, N., Sosnowski S., Strasser, E., Złotowski., J., Buss, M., Kühnlenz, K., Tscheligi, M., Weiss, A., Wollherr, D., Impacts of Multimodal Feedback on Efficiency of Proactive Information Retrieval from Task-Related HRI., Journal of Advanced Computational Intelligence and Intelligent Informatics, (2012), Accepted for Publication.

[8] Gawronski, B., and Conrey, F. R., The Implicit Association Test as a measure of automatically activated associations: Range and limits, Psychologische Rundschau, vol. 55, no. 3, (2004), 118-126.

[9] Ekman, P., Universals and cultural differences in facial expressions of emotion: University of Nebraska Press Lincoln, 1971.

Policy Transformation for Learning from Demonstration

Halit Bener Suay
Sonia Chernova
Worcester Polytechnic Institute
100 Institute Dr., Worcester
MA, 01609
{benersuay, soniac}@wpi.edu

ABSTRACT

Many different robot learning from demonstration methods have been applied and tested in various environments recently. Representation of learned plans, tasks and policies often depends on the technique due to method-specific parameters. An agent that is able to switch between representations can apply its knowledge to different algorithms. This flexibility can be useful for a human teacher when training the agent. In this work we present a process to convert learned policies with two specific methods, Confidence-Based Autonomy (CBA) and Interactive Reinforcement Learning (Int-RL), to each other. Our finding suggests that it is possible for an agent to learn a policy with either CBA or Int-RL method and execute the task with the other with the benefit of previously learned knowledge.

Categories and Subject Descriptors

I.2 [**Artificial Intelligence**]: Robotics, Learning

General Terms

Algorithms

Keywords

Machine Learning, Learning from Demonstration, Robotics

1. INTRODUCTION

Existing robot learning from demonstration techniques can be grouped in three general categories: mapping world's state to decisions, making a plan to reach a goal, or modeling the system based on reward [1].

The difference between learning approaches naturally results in different representation of what the agent learns. Although it is possible to implement most learning methods in more than one way, approaches often have different characteristics such as learning only from rewards, or learning only from labeled examples, etc.

When handling real world problems, agents could take advantage of more than one approach to cope with various complicated situations. If an agent has a broader usage of a learned policy or plan, it can attack the problem it is trying to tackle from different angles. One way to do this is to convert learned policies with one technique to a representation or policy executable by another technique. By converting different types of policies to each other, an agent can learn a task using method A and can execute the same task using method B.

This can bring advantages for human-robot interaction as well. For instance, an agent that learns a task by asking questions to a human teacher, can execute the task to explore the world even in the absence of the teacher and receive its rewards from the world. Also, generalization over the knowledge obtained from different learning techniques can ultimately increase the efficiency of human-robot interaction.

In this work we present a process to convert learned policies with two different techniques to each other's representation. More specifically, conversion between the Interactive Reinforcement Learning policy (Int-RL) learned with Q-learning [3], and Confidence-Based Autonomy (CBA) policy [2] is explained. We first define the process, then introduce our preliminary results.

HRI'12, March 5–8, 2012, Boston, Massachusetts, USA.
ACM 978-1-4503-1063-5/12/03.

2. CONVERSION PROCESS

Int-RL and CBA agents are similar in that, neither of them keeps the temporal information in their policy. That is, the order of actions taken are independent of the world's state history. Int-RL agent tries to model human teacher's reward feedback signal H after randomly chosen actions, whereas the CBA agent directly asks the teacher which action to take if it is not "confident" enough.

From Int-RL to CBA

We define an Int-RL policy $\pi : S \mapsto A$, a learned policy with an estimated human reward signal $\hat{H}$. The agent chooses which action $a \in A$ to take in a state $s \in S$ using π. Instead of learning π directly, many reinforcement learning algorithms instead approximate the action-value function, $Q : S \times A \mapsto \mathbb{R}$, which maps state-action pairs to the expected real-valued return.

Here we assume that the task is previously learned and that we already have a $Q(s,a)$ function. We then define a label function L as:

$$L(s,Q) = \begin{cases} a & \text{if } \exists! \; max_a Q(s,a) \\ NULL & \text{otherwise} \end{cases} \quad (1)$$

where the label is equal to an action, if that action has the maximum Q-value and is the only one that does for a given state s. This equation filters out the states that the agent could not discover during the training (e.g. $\mathbf{q} =$

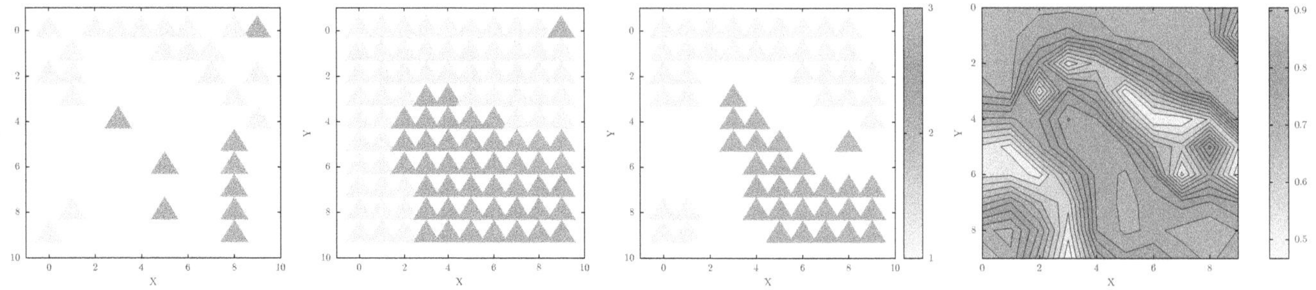

(a) Learned Int-RL Policy (b) CBA From Int-RL (c) Int-RL From CBA (d) $\mathcal{C}$ Confidence

Figure 1: Policy Representation. Color scale is the same for the first three plots. X and Y are the state variables.

$\langle 0.5, 0.5, 0.5 \rangle$) or the states where more than one action has the same maximum Q-value (e.g. $\mathbf{q} = \langle 0.7, 0.2, 0.7 \rangle$).

We obtain uniquely labeled states that we can feed to a classification or clustering function $\mathcal{C}(s, L(s, Q))$ that can classify multi-class input, and provide a class membership probability for an inquiry:

$$(a, c, db) \leftarrow \mathcal{C}(s, L(s, Q)) \tag{2}$$

where the returned action a is executed by the agent, if class membership probability (confidence) c is above a threshold τ_{conf}, which is determined by the decision boundary of the classifier, db [2]. Finally, upon execution of the task, $\mathcal{C}$ can be re-trained when $c \leq \tau_{\text{conf}}$ or the teacher wants to correct the agent's choice.

From CBA to Int-RL

Here we explain the reverse process, where we assume to have a trained classifier $\mathcal{C}$ (with the same properties mentioned previously) and our goal is to convert the state-action (input-output) pairs of $(s, \mathcal{C}(s))$ to a value function $Q(s, a)$ for an Int-RL agent to execute. We define:

$$V(s, \mathcal{C}(s)) = \begin{cases} \mathbf{V_d} & \text{if } d < \tau_d \textbf{ and } c > \tau_{\text{conf}} \\ \mathbf{V_{init}} & \text{otherwise} \end{cases} \tag{3}$$

$$\begin{aligned} \mathbf{V_d} &= \langle v_1, v_2, \ldots, v_n \rangle, \text{ where} \\ v_i &= \begin{cases} q_{\text{high}} & \text{if } i = a \\ q_{\text{init}} & \text{otherwise} \end{cases} \end{aligned} \tag{4}$$

$$\begin{aligned} \mathbf{V_{init}} &= \langle v_1, v_2, \ldots, v_n \rangle, \text{ where} \\ v_i &= q_{\text{init}}, \; \forall \; n \geq i \geq 1 \end{aligned} \tag{5}$$

In equations above, $\mathcal{C}(s)$ is the choice of action that the classifier returns for a given state s, $V(s, \mathcal{C}(s))$ is the conversion function that takes the classifier and the agent's state as input arguments, $\mathbf{V}$ is a vector of Q-values v_i, $n \in \mathbb{Z}^+$ is the size of the agent's action space, q_{init} is manually chosen initial Q-value, and $q_{\text{high}} > q_{\text{init}}$ is a manually chosen Q-value that is greater than initial Q-values, so that when the Int-RL agent is exploiting π, it is guaranteed to execute the action to which it is assigned. Details about CBA specific parameters d, τ_{d}, c and τ_{conf} are given in [2].

When applied over the state space S, Equation 3 returns the policy π necessary for an Int-RL agent to start executing the task. The agent can exploit π while the human teacher can shape it with the reward signal H.

3. RESULTS AND FUTURE WORK

To test the defined process, we designed a simple robot domain, where the state of the agent $s = (x, y) \in S$ is defined with two variables, $9 \geq x, y \geq 0$. In any given state agent has the choice of action $a \in A : \{1, 2, 3\}$. The agent randomly transitions between states and chooses a random action. We ran an experiment where a naïve human teacher trained a real humanoid robot, Aldebaran Nao, with Int-RL method. The teacher trained the agent for 10 minutes by giving a positive or negative reward after each action execution through a simple GUI.

Fig. 1a shows π that the Int-RL agent learned from the human teacher. Only action labels that are *not NULL* according to Equation 1 are shown. These (s, a) pairs, then are fed in an SVM classifier with Radial Basis Function Kernel, $\mathcal{C}$. Output of the classifier is shown in Fig. 1b. As can be seen, due to the inherent generalization abilities of the classifier, the agent has an idea about which action to take even for the states that were not discovered, or rewarded.

Fig. 1d shows the class membership probabilities for predicted actions by the classifier. The average confidence, $\tau_{\text{conf}} = 0.789$. With this information, we can then convert the policy to a Q-value function using Equation 3. The result is shown in Fig. 1c. Decision boundaries are low confidence zones (where $c < \tau_{\text{conf}}$). States in and around decision boundaries that are assigned to $\mathbf{V}_{\text{init}}$ are left blank.

Although our early results are promising, further research is necessary in machine learning and HRI directions. For machine learning, conversion with down/up-sampling the state-space, converting temporal planning policies to Markov DP representation and extending our process to continuous domains can be interesting. In terms of HRI, an important future work is to investigate the efficiency and comfort our process brings to interaction with naïve teachers in a real world domain. Particularly, evaluating the reflection of an agent's increased capability on its perceived intelligence by a human teacher can be useful for the field.

4. REFERENCES

[1] B. D. Argall, S. Chernova, M. Veloso, and B. Browning. A survey of robot learning from demonstration. *Robot. Auton. Syst.*, 57:469–483, May 2009.

[2] S. Chernova and M. Veloso. Interactive policy learning through confidence-based autonomy. *J. Artificial Intelligence Research*, pages 1–25, 2009.

[3] A. L. Thomaz and C. Breazeal. Adding guidance to interactive reinforcement learning. In *Proceedings of the Twentieth Conference on Artificial Intelligence (AAAI)*, 2006.

Exploration of Intention Expression for Robots

Ivan Shindev and Yu Sun
Computer Science and Engineering
University of South Florida
Tampa, FL 33620 USA
shindev@mail.usf.edu;
yusun@cse.usf.edu

Michael Coovert , Jenny Pavlova and Tiffany Lee
Department of Psychology
University of South Florida
Tampa, FL 33620 USA
coovert@usf.edu;
epavlova@mail.usf.edu;
ttl@mail.usf.edu

ABSTRACT

This paper presents a novel exploration on how to enable a robot to express its intention so that the humans and robot can form a synergic relationship. A systematic design approach is proposed to obtain a set of possible intentions for a given robot from three levels of intentions. A visual intention expression system approach is developed to visualize the intentions and implemented on a mobile robot and a manipulator to demonstrate the intention expression concept.

Categories and Subject Descriptors

H.5.2 [**User Interfaces**]: Interaction styles

General Terms

Human Factors

Keywords

Robot, intention expression, augmented reality

1. INTRODUCTION

When people meet a robot, they may be excited and curious, but those who are not experts in robotics usually have no idea what the robot can and will do. For example, when a mobile robot moves around in a human living environment, we act cautiously and nervously around it mostly because we do not know what it will do next and we need to prepare ourselves for its sudden and unexpected movements. We try to observe the robot carefully and use our cognition system to make sense of its motions and predict its intentions. Unfortunately, there typically is not a clear indication of the robot's next move, and the programming built into the robot is not observable by people who co-live with it. If, at all times, we could know the intention of the robot and its next move, we would feel much more comfortable and at ease around them.

A human not only has to understand the current motion of the robot but also must perceive the robot's intention in order to predict its next move(s) so as to actively collaborate with the robot and create synergy. To achieve this, one approach is to design robots that can express themselves as humans do - allow humans to identify the robot's intentions as we observe them in action (as humans do with other humans). A number of researchers in socially interactive robotics have suggested that in order to interact socially with humans, a software agent must be believable and life-like, must have behavioral consistency, and must have ways of expressing its internal states [1, 2]. However, it is unrealistic to require all the robots coexisting with humans to have human-like appearance, kinematics, and dynamics. A robot's motion, by nature, is far different from regular human motion we observe every day. The motions we consider to be unnatural are rooted in the physical properties and configurations of robots. A robot's material, actuators, and sensors are fundamentally different from humans, which yield rather different patterns of acceleration, deceleration, and the like. A robot is designed to amplify its strength and be efficient for certain tasks. The differences give it abilities that humans do not have, such as super speed and extreme precision.

To enable a robot to express its intention, we have developed an intention expression framework that allows the robot to derive its intention sets from its models and programmed tasks and select an optimal intention representation and expression with our study results on intention perception.

2. INTENTION EXPRESSION SYSTEMS

2.1 Mobile Robot

We have constructed an autonomous mobile system based on an Adept MobileRobots Pioneer 3-DX differential drive robot. A particle filter-based simultaneous localization and mapping (SLAM) algorithm and path planning algorithm were developed and integrated into the Robot Operating System (ROS). The software runs on an Acer Iconia duel-touchscreen laptop equipped with a Microsoft Kinect for 3D perception. The mobile robot can navigate fully autonomously with the Kinect sensor data, and it also can be tele-operated through wireless Internet. For intention expression, the system is equipped a Microvision SHOWWX+ 848 x 480 Scanning LASER projector for visual expression. The projector is rigidly mounted on the frame and pointing to the floor to render the robot's intention (Figure 1A).

After analyzing the tasks of the mobile robot, we can clearly derive its long-term intention, which is to reach a goal position defined in a task, while its mid-term intention can be an intermediate target such as avoiding an immediate obstacle. The long-term intention can be represented with

HRI'12, March 5–8, 2012, Boston, Massachusetts, USA.
ACM 978-1-4503-1063-5/12/03.

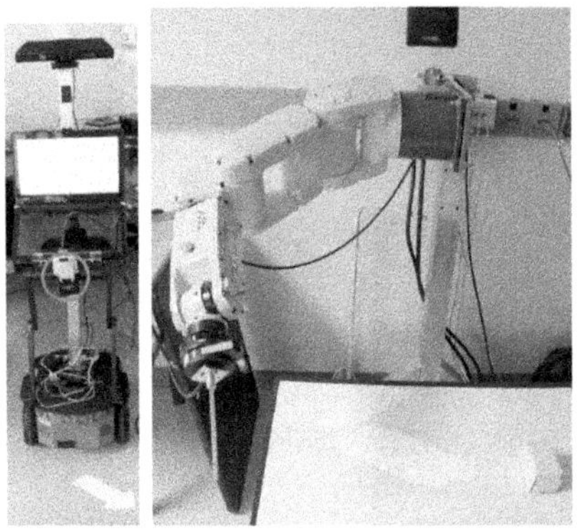

(A) (B)

Figure 1: Telemedicine mobile robot with intention displayed on the floor (A), manipulator with intention displayed with a LASER projector (B).

the trajectory generated by the robot's path planner, and the mid-term intention can be represented with a short and more-detailed path that the robot will take to get around the obstacle. The mobile robot's short-term intention could be its next motion, such as moving 30-degrees left with a 0.3 meter-per-second speed. It can be represented with a motion vector.

As the robot moves, the long-term and mid-term intentions can change if the robot computes a new path and trajectory with new perception data. For our studies, we displayed the intention on the ground and aligned the direction of the displayed map with a real-world environment to require less mental computation from human coworkers compared to displaying on a monitor (Figure 1A).

2.2 Robot Manipulator

We also have designed and implemented a visual expression system on our robotic manipulator - a 6-DOF Fanuc L200IC robotic arm. The same visual display element - a Microvision LASER projector - is attached to a stationary camera to form a camera-projector system (as shown in Figure 1B). The camera was originally setup for the manipulator as a part of a hand-eye system to automate manipulation tasks. One of the tasks is to flip a box with its stick. The location of the box is obtained from the camera, and then a sequence of motions is generated to carry out the flipping task. The intended pose is directly rendered on the workbench and superimposed onto the object. To accurately align the projection with the real-world objects the manipulator intends to handle, the poses and shapes of the objects are obtained from a camera, and the intrinsic and extrinsic parameters of the projector are calibrated relative to the robot coordinate system and the camera [3, 4, 5].

After a sequence of motions is planned, the projector displays the intended motion on the workbench and overlays the simulated intended target position with the box that should be flipped. The person supervising the task can easily judge if the planned motion is likely to be successful or not by simply glancing at the display. Figure 1B shows one example of the longer-term intention visual expression.

3. EVALUATION

A study was conducted with the mobile robot intention expression system to examine human understanding of directional information if the robot communicated information regarding its future direction by projecting. Participants in the study interacted with the robot in real time. The robot was moving and projecting arrows on the ground in front of it (Figure 1A). Two types of tests were performed: congruent (the projected arrows reflected the robot's movements) and non-congruent (the projected arrows were random and did not reflect the robot's movement). After each intention expression, the participants rated their confidence in the robot's upcoming movement on a scale of 1 (not at all confident) to 7 (completely confident). Each participant made a total of 10 ratings (4 congruent and 6 non-congruent). Data were collected from five young adults. The means for the congruent items ($M = 6.8$-7) greatly exceeded the means for non-congruent items ($M = 1$-1.2), $t(4) = 72.06$, $p<0.001$. The study showed that projecting arrows that represent the robot's current and future motion provide information with a high level of confidence to individuals approaching a robot, as opposed to not having the correct motion projected or not having a projection at all.

4. CONCLUSIONS

We have successfully implemented intention expression systems on two robotic systems - a mobile telemedicine robot and an industry robotic manipulator. The novel robot intention expression concept has the potential to allow people to perceive the robots' intention at all time when they are living with us, which will instill trust in regular people toward robots and significantly broaden the acceptance of robots in our lives. Equipped with a well-designed intention expression component, a robot will be viewed as reliable, truthful, and transparent, since people will be able to foresee and control its next motion at ease.

5. REFERENCES

[1] J. Bates. The role of emotion in believable characters. *Communications of the ACM*, 37:122–125, 1994.

[2] B. Blumberg. Old tricks, New Dogs: Ethology and Interactive Creatures. PhD thesis, MIT, 1996.

[3] M. Blackwell, C. Nikou, A. Digioia, T. Kanade. An image overlay system for medical data visualization. *Med Image Anal.* 4(1): 67-72, 2000.

[4] R. Raskar, J. Baar, P. Beardsley, T. Willwacher, S. Rao, C. Forlines. ilamps: Geometrically aware and self-configuring projectors. *ACM SIGGRAPH*, 809-818, 2003.

[5] A. Agrawal, Y. Sun, et al., Vision-guided robot system for picking objects by casting shadows, *International Journal of Robotics Research*, 29(2-3), 155-173, 2010.

Dimensions of People's Attitudes toward Robots

Daisuke Suzuki[1] Hiroyuki Umemuro[2]
Department of Industrial Engineering and Management, Tokyo Institute of Technology
2-12-1-W9-67 Ookayama, Meguro-ku, Tokyo 152-8552, Japan
+81-3-5734-2246
[1]suzuki.d.ad@m.titech.ac.jp, [2]umemuro.h.aa@m.titech.ac.jp

ABSTRACT
The purpose of this study was to investigate dimensions that construct people's attitudes toward robots. In the first phase, investigations using free description and group interview were conducted to extract potential elements for attitudes toward robots. In the second phase, a questionnaire battery was developed based on the elements extracted, and a survey investigation was conducted with the questionnaire. A factor analysis was conducted on the responses to the questionnaire, and nine factors were extracted as dimensions of people's attitudes toward robots.

Categories and Subject Descriptors
H.1.2 [**Models and Principals**]: User/Machine Systems – *human factors*. K.4.2 [**Computers and Society**]: Social Issues.

General Terms
Human Factors.

Keywords
Attitude, acceptance.

ACM 978-1-4503-1063-5/12/03.

1. INTRODUCTION
Robots are entering the domestic environment and can be helpful in our everyday lives. Older people may especially benefit from robots that support independent living in their own home of choice and community in a confident and comfortable manner.

On the other hand, robots can be perceived as too innovative a technology and may not be accepted, even though they are helpful. Therefore, it is essential to understand people's perceptions of robots. Research has been conducted to understand people's acceptance of domestic robots [1][2]. In particular, people's attitudes toward robots have been studied in relation to their acceptance [7][12]. However, individual differences in attitudes have not been fully studied.

As methods of measuring individual attitudes toward robots, the Negative Attitudes toward Robots Scale and Robot Anxiety Scale [10][11] have been proposed. However, these attitude scales measure only negative attitudes toward robots. Although negative attitudes are considered a good predictor of the success of human-robot interaction, people's attitudes may have more complex constructs. Studies on people's attitudes toward computers reveal multiple dimensions of computer attitudes. For example, Loyd and Gressard [8] reported three dimensions: liking, anxiety, and confidence. Jay and Willis [6] reported seven dimensions: comfort, efficacy, gender equality, control, dehumanization, interest, and utility. Hence, for better understanding of individual attitudes toward robots, multi-dimensional constructs of attitudes should be studied.

HRI'12, March 5–8, 2012, Boston, Massachusetts, USA.
ACM 978-1-4503-1063-5/12/03.

The purpose of this study was to investigate dimensions that construct people's attitudes toward robots. To include broad dimensions of attitudes, potential elements were extracted from people's free descriptions and interviews. A questionnaire-based investigation and a factor analysis were conducted based on these elements to extract attitude dimensions.

2. ELEMENTS CONSTRUCTING PEOPLE'S ATTITUDES TOWARD ROBOTS
In the first phase of the study, to extract the potential elements that construct people's attitudes toward robots, an investigation with a questionnaire and interview was conducted.

2.1 Methods
2.1.1 Participants
Participants were 64 Japanese older adults over age 60 and 19 Japanese undergraduate and graduate students.

2.1.2 Procedure and Stimuli
In the first part, an explanation of one sample robot at a time was given to participants. Then participants were asked to describe their own feelings and ideas freely on the questionnaire, in response to prompts such as "intention to use", "feelings when the robot comes to your home", and "points you like or dislike about the robot". This procedure was repeated for four sample robots: Roomba [5], a home-assistant robot [3], PaPeRo [9], and Paro [4].

In the second part, participants were asked to describe their thoughts about robots in general, such as "feelings and perceptions they might have when robots are in their home" and "relations with robots compared with relations with pets or friends".

After they completed their free descriptions, they were also interviewed as a group about feelings and thoughts they had had while answering the questionnaire.

2.2 Results
All responses to the questionnaire and interview were checked manually. Keywords that appeared related to attitudes toward robots were picked out, and duplications were eliminated. Finally, 125 keywords were extracted, hereafter referred to as attitude elements. These extracted elements could be categorized into 10 groups: "performance", "communication", "physical appearance", "social factors", "physical and mental factors of users", "adoption into current environment", "family member", "relation and chemistry", "living object vs. machine" and "miscellaneous".

3. DIMENSIONS OF ROBOT ATTITUDES
In the second phase of the study, a questionnaire-based investigation was conducted to extract dimensions constructing people's attitudes toward robots.

3.1 Method

3.1.1 Participants

Participants were 66 Japanese older adults aged 61 to 87 years (*M*=72.5, *SD*=6.2) and 135 Japanese younger adults in universities aged 19 to 38 years (*M*=22.2, *SD*=2.0). Among those, 47.0% of the older participants and 77.0% of the younger participants were males.

3.1.2 Procedure and Measurement

Participants were asked to complete the questionnaire at their own pace and return it to the investigator.

The questionnaire comprised two parts. The first part asked about participants' demographic information. The second part consisted of 125 question items corresponding to attitude elements extracted in the first phase. Participants were asked to indicate how each of the items matched their feelings and perceptions of domestic robots in general, by responding on seven-point Likert scales.

3.2 Results

A factor analysis was conducted on participants' responses. The number of factors was determined by decreasing degree of initial eigenvalues. The principal factor method with varimax rotation revealed a nine-factor structure. The cumulative contribution for the nine factors was 43.42%. These nine factors were considered dimensions of peoples' attitudes toward robots.

Variables that had high loadings on the first factor included "I think of robots as family members" and "I feel comfortable with robots". This factor was considered to represent familiarity with robots and thus was labeled "familiarity".

Variables that had high loadings on the second factor included "I want to train robots to fit my preferences" and "It is fun to use robots". This factor was considered to represent engagement with robots and therefore was labeled "involvement".

Variables such as "Using robots at home is distressing", "It's shameful to use robots", and "Robots will corrupt humans" had high negative loadings on the third factor. This factor was considered to represent a sense of self-respect and was labeled "self-esteem".

Variables that had high loadings on the fourth factor included "I have sufficient ability to use robots" and "I can learn how to use robots easily". This factor appeared to represent self-confidence in using robots and thus was labeled "self-efficacy".

Variables that had high factor loadings on the fifth factor included "Family or friends will help me use robots" and "I want to use robots if everybody uses them". This factor was considered to represent relations with other people and was thus labeled "social relations".

Variables that had high loadings on the sixth factor included "Robots are cute" and "Robots are beautiful". This factor appeared to be related to aesthetic appreciation. Thus, it was labeled "appealing appearance".

Variables such as "Robots are practical" and "Robots are easy to use" had high loadings on the seventh factor. This factor was considered to represent practical utility and was labeled "utility".

Variables that had high loadings on the eighth factor included "Robots with a variety of sounds should be developed" and "Robots with a variety of shapes should be developed". This factor was considered to represent appreciation of variations, and was thus labeled "variety".

Finally, on the ninth factor, variables such as "Robots are appropriate for sensitive jobs in places like hospitals" had high positive loadings. This factor appeared to represent the idea that robots should do assistive jobs in public but not private situations. Thus, this factor was labeled "helper".

4. CONCLUSION

This study investigated dimensions of people's attitudes toward robots. Nine dimensions representing robot attitudes were extracted. Based on these dimensions, multidimensional attitude scales can be developed in future research.

5. REFERENCES

[1] Broadbent, E., Stafford, R., and MacDonald, B. 2009. Acceptance of healthcare robots for the older population: Review and future directions. *International Journal of Social Robotics*, 1, 319-330.

[2] Ezer, N., Fisk, A. D., and Rogers, W. A. 2009. Attitudinal and intentional acceptance of domestic robots by younger and older adults. *Lecture Notes in Computer Science*, 5615, 39-48.

[3] Information and Robot Technology Research Initiative. 2008. Home-assistant robot. Retrieved November 18, 2011, from http://www.irt.i.u-tokyo.ac.jp/reform/081024/index.shtml

[4] Intelligent System Co., Ltd. 2005. Paro. Retrieved November, 18, 2011, from http://paro.jp/?page_id=241

[5] iRobot. 2011. Roomba. Retrieved November 18, 2011, from http://www.irobot-jp.com/product/index.html

[6] Jay, G. M., and Willis, S. L. 1992. Influence of direct computer experience on older adults' attitude toward computers. *Journal of Gerontology: Psychological Sciences, 47*, 250-257.

[7] Khan, Z. 1998. *Attitudes towards intelligent service robots* (TRITA-NA-P9821, IPLab-154). Sweden, Royal Institute of Technology, Numerical Analysis and Computing Science, Interaction and Presentation Laboratory.

[8] Loyd, B. H., and Gressard, C. 1984. Reliability and factorial validity of computer attitude scales. *Educational and Psychological Measurement, 44*, 501-505.

[9] NEC Corporation. 2005. PaPeRo. Retrieved November, 18, 2011, from http://www.nec.co.jp/products/robot/

[10] Nomura, T., Kanda, T., Suzuki, T., and Kato, K. 2006. Measurement of anxiety toward robots. *Proceedings the 15th IEEE International Symposium on Robot and Human Interactive Communication (RO-MAN 2006)*, 372-377.

[11] Nomura, T., Kanda, T., Suzuki, T., and Kato, K. 2008. Prediction of human behavior in human–robot interaction using psychological scales for anxiety and negative attitudes toward robots. *IEEE Transactions on Robotics*, 24(2), 442-451.

[12] Ray, C., Mondada, F., and Siegwart, R. 2008. What do people expect from robots?. *Proceedings 2008 IEEE/RSJ International Conference on Intelligent Robots and Systems, Nice, France,* 3816-3821.

Extending Chatterbot System into Multimodal Interaction Framework with Embodied Contextual Understanding

Jeffrey Too Chuan Tan
National Institute of Informatics
2-1-2 Hitotsubashi, Chiyoda-ku
Tokyo 101-8430, Japan
+81-3-4212-2518
jeffrey@nii.ac.jp

Tetsunari Inamura
National Institute of Informatics
2-1-2 Hitotsubashi, Chiyoda-ku
Tokyo 101-8430, Japan
+81-3-4212-2518
inamura@nii.ac.jp

ABSTRACT
This work aims to realize multimodal interaction with embodied contextual understanding based on the simple chatterbot system. A system framework is proposed to integrate the dialogue system into a 3D simulation platform, SIGVerse to attain multimodal interaction. The chatterbot AIML implementations are described in the achievement of the conversations with embodied contextual understanding in HRI simulations.

Categories and Subject Descriptors
H.5.2 **[Information Interfaces and Presentation (e.g., HCI)]**: User Interfaces – Interaction styles (e.g., commands, menus, forms, direct manipulation), Natural language; H.1.2 **[Models and Principles]**: User/Machine Systems – Human information processing

General Terms
Human Factors, Languages

Keywords
Chatterbot, Multimodal Interaction, Embodiment, Contextual Understanding

1. INTRODUCTION
In the development of human-robot interaction (HRI), often the research focus is concentrated on the building of robot's capability to *understand* the human's expressions verbally and nonverbally. However, can machines truly *understand* in the same sense as how humans comprehend in a conversation? It became an open question (refer to *Turing test* [1]), especially in the rise of Artificial Intelligence (AI) development.

Despite the fact that weak AI systems might demonstrate intelligence without actual reasoning ability like humans, the capability to deliver *appropriate* and *rational* responses may be useful in many robotics developments. One classical example is the development of chatterbot [1]. With a very fundamental pattern matching algorithm and stimulus-response database, chatterbot systems are constantly developed and widely applied in education, business service and e-commerce [1].

HRI'12, March 5–8, 2012, Boston, Massachusetts, USA.
ACM 978-1-4503-1063-5/12/03.

Taking the potential of chatterbot systems, this work proposed to extend the capability of chatterbot systems into multimodal interaction dimension to incorporate embodied contextual understanding for the applications in human-robot interaction. In contrast to many HRI developments [2], which involved complex modeling of robot intelligence, natural language processing and heavy computing, this work aims to adapt the lightweight concept of chatterbot systems to develop a multimodal interaction framework with embodied contextual understanding.

2. CHATTERBOT SYSTEMS
The first chatterbot, ELIZA was published in 1966 by Joseph Weizenbaum [1]. The main idea of the computer program is to recognize the cue words or phrases in the user's input and output the corresponding pre-programmed responses that are meaningful in the conversation. In the similar approach, the system was refined by many developments, and A.L.I.C.E. (Artificial Linguistic Internet Computer Entity) is one of the most successful developments that had been awarded the Loebner Prize [1].

The architecture of A.L.I.C.E. system is well developed, however, it is limited to verbal interaction (with speech recognition and synthesis). In HRI, nonverbal interaction with the concept of embodied contextual understanding plays an even more important role [2]. The physical body gestures, facial expressions and eye contacts, and the embodied contextual understanding of spatial information in the interaction environment are research themes of many developments in HRI. This work targets to propose an extended framework to realize multimodal human-robot interaction with embodied contextual understanding in a virtual environment.

3. SYSTEM FRAMEWORK
This work is developed based on a simulation platform named SIGVerse [3], to model the robot agent and the 3D virtual environment for embodied HRI. The entire system framework is divided into Server System, Client System and Service Providers (Fig. 1). On the Server System, SIGServer is the center of the simulation that runs core calculations. Agent Controller is executed on top of SIGServer as the agent's intelligence. Human user interfaces with the simulation via SIGViewer in the Client System. During the HRI simulation, the user controls his/her own virtual avatar through various multimodal User Interfaces: GUI, haptic devices, motion capture system (Kinect), voice commands, etc. Services Providers offer additional processing features (e.g. sensor processing, imagine processing) to the simulation. Particularly in this work, Dialogue Engine and Behavior

Recognition modules are developed, with Facial Expression and Speech Emotion modules under active development.

Dialogue Engine utilizes AIML (Artificial Intelligence Markup Language), an XML dialect used in A.L.I.C.E. to infer appropriate interaction responses based on the user's input. In the Behavior Recognition module, Mimesis Model method is applied to abstract input motion patterns with Hidden Markov models (HMMs). The distance between the HMMs represents the dissimilarity of the motion patterns for behavior recognition.

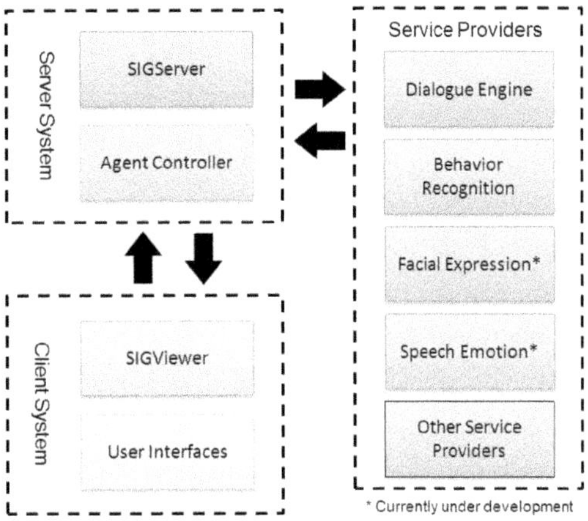

Figure 1. System Framework.

4. CONVERSATION WITH EMBODIED CONTEXTUAL UNDERSTANDING

This work develops extensions on top of AIML to integrate verbal and nonverbal inputs into Dialogue Engine to infer appropriate verbal and also nonverbal outputs as responses to realize multimodal interaction. The fundamental strategy is to deduce the nonverbal behaviors and represent them as additional parameters to be input along with the verbal expressions.

4.1 Body Behaviors

Body behaviors such as pointing gesture that involve directive reference in the interaction space, are commonly being studied in HRI. Using the Behaviors Recognition service provider, user's body gesture is analyzed to recognize possible meanings (in additional to the verbal expressions). If a pointing gesture is detected, the corresponding pointing target in the interaction environment is calculated. This information is then incorporated with the verbal expressions to be input into the Dialogue Engine for processing. The following illustrates an example of AIML model for pointing gesture interaction.

<set name="object">television</set>

<category>
<pattern>WHAT IS THAT</pattern>
<template>It is a <get name="object"/>.</template>
<category>

When a user questions the agent with the verbal expression, "What is that?" with a pointing gesture, the behavior is recognized and the pointing target (television) is set as the *object* in <set> tag. The agent will responses, "It is a television."

4.2 Emotional Expressions

Although the Facial Expression and Speech Emotion service providers are currently under development, simulated data is used for the multimodal interaction development. Facial expression and speech emotion are important elements to set the ambient mood of the interaction. Using the <topic> tag in AIML, the agent can response differently on the same question based on the user's emotion. Below is an example of AIML model for the agent to response greeting based on two different ambient moods (from simulated facial expression).

<topic name="happy">
<category>
<pattern>HELLO</pattern>
<template>Hello! I am glad to see that you are happy.</template>
<category>
</topic>

<topic name="sad">
<category>
<pattern>HELLO</pattern>
<template>Hello! Why you look so sad?</template>
<category>
</topic>

4.3 Output Actions

The agent's responses are not limited to just verbal expressions. By using the <system> tag, a shell command can be sent back to the agent controller to perform certain actions (e.g. execute a body gesture, set facial expression and speech emotion).

5. Conclusion and Future Work

The objective of this work is to realize multimodal interaction with embodied contextual understanding based on an extension of the simple Chatterbot dialogue system. A system framework is proposed to integrate the dialogue system into a 3D simulation platform, SIGVerse to achieve multimodal interaction. The AIML implementations are described to illustrate the conversations with embodied contextual understanding in HRI simulations. Comparison studies with conventional complex systems will be conducted in future to verify the effectiveness of this development.

6. REFERENCES

[1] Wallace, R.S. 2009. The Anatomy of A.L.I.C.E.. *Parsing The Turing Test*, Epstein, R., Roberts, G., and Beber, G., Ed. Springer Netherlands, 181-210. DOI= http://dx.doi.org/10.1007/978-1-4020-6710-5_13.

[2] Stiefelhagen, R., Ekenel, H.K., Fugen, C., Gieselmann, P., Holzapfel, H., Kraft, F., Nickel, K., Voit, M., and Waibel, A. 2007. Enabling Multimodal Human–Robot Interaction for the Karlsruhe Humanoid Robot. *IEEE Transactions on Robotics*. 23, 5 (Oct. 2007), 840-851. DOI= http://dx.doi.org/10.1109/TRO.2007.907484.

[3] Inamura, T., Shibata, T., Sena, H., Hashimoto, T., Kawai, N., Miyashita, T., Sakurai, Y., Shimizu, M., Otake, M., Hosoda, K., Umeda, S., Inui, K., and Yoshikawa, Y. 2010. Simulator platform that enables social interaction simulation – SIGVerse: SocioIntelliGenesis simulator. In *Proceedings of the IEEE/SICE Int. Symposium on System Integration* (Sendai, Japan, 21 – 22 December, 2010). SII '10. 212-217. DOI= http://dx.doi.org/10.1109/SII.2010.5708327.

Learning Verbs by Teaching a Care-Receiving Robot by Children: An Experimental Report

Fumihide Tanaka
Faculty of Engineering, Information and Systems, University of Tsukuba
1-1-1 Tennodai, Tsukuba, Ibaraki, Japan
fumihide@iit.tsukuba.ac.jp

Shizuko Matsuzoe
Graduate School of Systems and Information Engineering, University of Tsukuba
1-1-1 Tennodai, Tsukuba, Ibaraki, Japan
s1130184@u.tsukuba.ac.jp

ABSTRACT

We investigate the use of care-receiving robot (CRR) for the purpose of supporting childhood education. In contrast to the conventional teaching agents that are designed to play the role of human teachers or caregivers, the robot here receives cares from children. We hypothesize that by using this CRR, we may construct a new educational framework whose goal is to promote children's spontaneous *learning by teaching* through teaching the CRR. The paper describes an experiment for investigating whether a CRR can promote children's learning English verbs through teaching it.

Categories and Subject Descriptors

I.2.9 [**Artificial Intelligence**]: Robotics; K.3.1 [**Computers and Education**]: Computer Uses in Education

General Terms

Verification

Keywords

Care-receiving robot, CRR, learning by teaching, child-robot interaction, child education, learning support, learning reinforcement, direct teaching, robot ethics

1. INTRODUCTION

Most educational robots so far have been designed and developed to play the role of human teachers or caregivers. In other words, they were *caregiving* robots developed to teach or care for children. In contrast, we will deal with a reverse scenario here where the robots are *carereceiver* from the children. By introducing a *weaker* robot than the children and letting them teach the robot, we aim to promote children's spontaneous *learning by teaching* as a result.

The paper describes a study about the care-receiving robot (CRR), which originally started in 2009 [2, 3, 1]. Particularly, here we will reports the results of a field experiment investigating the effects on children's learning English verbs by teaching the CRR.

HRI'12, March 5–8, 2012, Boston, Massachusetts, USA.
ACM 978-1-4503-1063-5/12/03.

2. CARE-RECEIVING ROBOT (CRR)

A CRR is a robot that receives care from the people around it. The original concept of the CRR was first proposed in 2009 by the first author of the paper [2, 3]. Conceptually, the meaning of *care* is quite general and thus we can imagine many types of CRR. The most typical scenario using a CRR, we assume, is for learning support or learning reinforcement for children. In this scenario, teachers/parents decide on an educational topic and somehow ask the children to teach the topic to a robot. Then, children teach the robot, which initially is not good at solving the topic but shows improvement based on the children's instruction. It is expected that by introducing a CRR, children would, as a result, be highly motivated to complete the learning topic, thus providing indirect practice (learning by teaching) for the children.

3. A FIELD EXPERIMENT

We conducted a field experiment whose goal is to investigate whether the CRR that we introduce into the classroom can promote children's learning by teaching or not. We chose the venue of an English language school for Japanese children. Thanks to the cooperation of the Minerva Language Institute Co., Ltd. that manages 600 classrooms in Japan, we were fortunate to be able to conduct experiments in a classroom in Tsukuba. Our target subjects were children from 3 to 6 years of age who attend classes in this classroom. After we received approval from the Ethical Committee of the University of Tsukuba for this experiment, we started recruiting participants by explaining and advertising our research to the parents of the children. We used Aldebaran Robotics' Nao to implement the CRR. During the experiment, it was fully teleoperated (wizard-of-oz) using a teleoperation interface developed. Each session (one subject trial) lasted approximately 30 minutes. Figure 1 illustrates the flow chart of the experiment.

Based on repeated consultation with professional teachers at the Minerva Language Institute Co., Ltd., we decided to adopt a verb-learning game using cards. Experimental scenario follows the pattern of a regular lesson in the classroom: (1) Among the 4 word cards obtained during the pre-test, 2 cards are randomly chosen in advance and marked as "With CRR" and (2) From the 4 word cards, Experimenter #1 picks up one card randomly and asks the child its meaning, saying "Show us how to <the verb>." Since the verb is unknown to the subject at this moment, the subject cannot answer the question correctly. (3) Then, Experimenter #1 shows the correct answer by picking up the corresponding object and pronouncing the verb. If the card is marked "With CRR," the

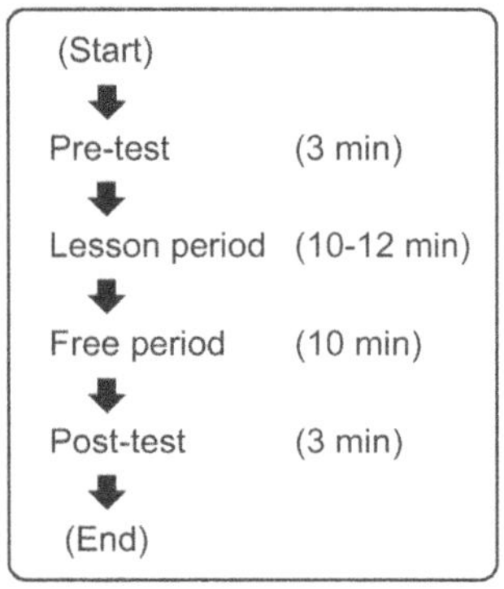

Figure 1: The flow chart of the experiment

game continues to step (4), otherwise they return to step (2), repeating the steps with another card. (4) Experimenter #1 turns to the CRR and asks it the same question. The CRR responds "Yes" and then says "Please pick up" while opening its right hand. Experimenter #1 hands the corresponding object to the CRR. The CRR then holds the object and makes a wrong movement. (5) Experimenter #1 says "No, that is incorrect." The CRR responds with "Please teach me." Then, Experimenter #1 takes the CRR by the hand and teaches it step by step (direct teaching). After the direct teaching is finished, the CRR plays back the movement correctly while pronouncing the verb. Then, Experimenter #1 says "Yes, that is correct." Experimenter #1 repeats the procedure for all 4 word cards to finish the first part of the lesson. The second part of the lesson is performed very similarly using the same 4 word cards. However, this time at step (5), Experimenter #1 asks the child to teach the CRR instead of Experimenter #1.

After the Lesson period is completed, the subject is told to play freely with the CRR by himself or herself for 10 minutes. Meanwhile, the CRR is teleoperated so as to follow steps of (4) and (5) as described above. At the end of each experimental session, a post-test is conducted to evaluate if the subject can answer the questions which he or she could not answer during the pre-test.

4. RESULTS

Figure 2 summarizes the results. Each bar represents the average percentage of post-test questions answered correctly by the 17 subjects. As we can see, the subjects picked up more correct graphic cards for verbs With CRR than without. This experiment was not designed to compare human teachers with the CRR, and, thus, the results should not be interpreted as showing the superiority of the CRR to human teaching. Nevertheless, the results reveal that the idea of the CRR can be implemented in the context of English verb-learning in a real classroom. It can promote and enhance children's learning by teaching; even verbs that are a challenge to learn with a minimum style of teaching (only about 20% were answered correctly at Without CRR).

5. CONCLUSIONS

The paper reported an experiment in which we introduced a CRR into a real classroom of English instruction for children to promote children's learning by teaching. The most important idea of this educational framework is to accelerate children's natural motivation to *caretaking* by introducing the

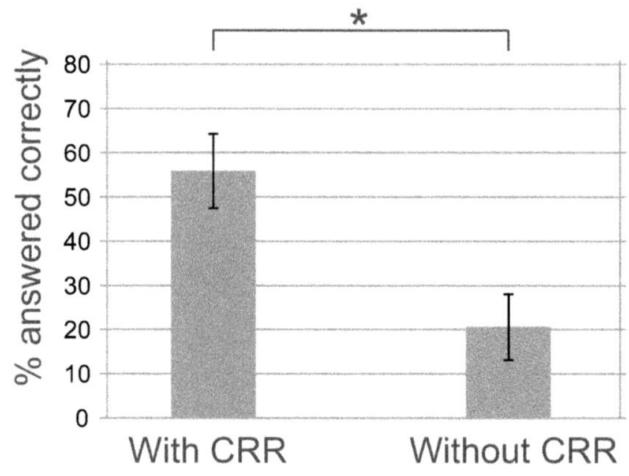

Figure 2: Post-test results: Each bar represents the average percentage of post-test questions answered correctly over 17 subjects. There was a significant difference ($P < 0.05$) in the average percentage answered correctly between verbs With CRR and Without CRR.

CRR. We believe this idea will provide a useful aspect for both the education and robotics research fields. We also realize that the basic idea of the CRR is very general; therefore, in the future, we will explore cases other than vocabulary learning, along with improving the requirements for designing better CRRs.

The most major deficit of the CRR used in the experiment would probably be the limitation of *its* learning capability. Since this was the first experiment of assessing the CRR for the purpose of children's learning by teaching, we did not put a learning factor into the robot, i.e., the CRR continued making a mistake. This surely made some of the subjects frustrated or irritated. If the CRR can keep learning as humans do, it may evoke a new motivation for the subjects. The appropriate learning dynamics of the CRR is our next important research target.

6. ACKNOWLEDGMENTS

We acknowledge the support from Minerva Language Institute Co., Ltd., the JST PRESTO program, and KAKENHI (23680020). We also thank the parents and children of the classroom for their support.

7. REFERENCES

[1] M. Ghosh and F. Tanaka. The impact of different competence levels of care-receiving robot on children. In *Proceedings of the IEEE/RSJ International Conference on Intelligent Robots and Systems (IROS 2011)*, pages 2409–2415, 2011.

[2] F. Tanaka and T. Kimura. The use of robots in early education: a scenario based on ethical consideration. In *Proceedings of the 18th IEEE International Symposium on Robot and Human Interactive Communication (RO-MAN 2009)*, pages 558–560, 2009.

[3] F. Tanaka and T. Kimura. Care-receiving robot as a tool of teachers in child education. *Interaction Studies*, 11(2):263–268, 2010.

A Tricycle-style Teleoperational Interface that Remotely Controls a Robot for Classroom Children

Fumihide Tanaka
Faculty of Engineering, Information and Systems, University of Tsukuba
1-1-1 Tennodai, Tsukuba, Ibaraki, Japan
fumihide@iit.tsukuba.ac.jp

Toshimitsu Takahashi
Graduate School of Systems and Information Engineering, University of Tsukuba
1-1-1 Tennodai, Tsukuba, Ibaraki, Japan
taka@bcl.esys.tsukuba.ac.jp

ABSTRACT

We consider the application of telepresence robots for supporting childhood education. One challenge here is to develop a teleoperational robot system that can be manipulated by children themselves. There are two requirements for realizing such a system. First, the system has to be sufficiently intuitive so that child users can control it without the need for detailed instructions. Second, the control of the system should have some amount of enjoyment so that child users do not get bored. To satisfy these requirements, we introduce a tricycle-style teleoperational interface that remotely controls a robot. We also report field tests that are currently being conducted at English learning schools for children in Japan.

Categories and Subject Descriptors

I.2.9 [**Artificial Intelligence**]: Robotics; H.4.3 [**Information Systems Applications**]: Communications Applications; K.3.1 [**Computers and Education**]: Computer Uses in Education

Keywords

Telerobotics, telepresence robot, distant communication, distant education, robotics for children, early childhood education, tricycle, teleoperational interface

1. INTRODUCTION

Most telepresence robots designed so far target adult operators [2, 1]. In contrast, we are currently developing a teleoperational robot system for children [3]. The goal is to remotely connect classrooms using the system, making it possible for children in one classroom to join activities in other classrooms in real-time.

There are two crucial requirements for the teleoperational robot system. First, since our target is younger children (3–6 years old), we cannot presuppose giving detailed instructions about how to use the system to them. Therefore, the system (especially its user interface) has to be sufficiently intuitive so that children can control it without the need for detailed instructions in advance. Second, since younger children have a short attention span, it is necessary to control of the system to have some degree of fun, so that they do not get bored using it.

To satisfy the two requirements, we are currently developing a tricycle-style teleoperational interface that remotely controls a robot. This paper reports the latest development of our research as well as field tests that are currently being conducted at English learning schools for Japanese children.

HRI'12, March 5–8, 2012, Boston, Massachusetts, USA.
ACM 978-1-4503-1063-5/12/03.

Figure 1: Tricycle-style teleoperational interface by which a child operator directly controls a remote slave robot (Figure 2). The operator can see a Skype monitor on a tablet PC and can also control the gripper hand of the robot.

2. TELEOPERATIONAL ROBOT SYSTEM

The teleoperational robot system that we are currently developing is composed of three components: a tricycle-style teleoperational interface, a server PC, and a slave robot. The operational signals sent from the tricycle-style teleoperational interface are transmitted over the internet to the server PC. Then, the signals are sent to the slave robot over the internet. All the signals are encrypted when they are transmitted over the internet (using SSH tunneling).

The tricycle-style teleoperational interface is equipped with encoders on its rear wheels (Figure 1) by which it can send the information pertaining to forward/backward/left/right movement to the slave robot through the server PC. The child operator rides the interface while wearing a data glove on his/her dominant hand. By clenching the hand, the child operator can remotely grasp an object through the slave robot. These operational signals are first sent via Bluetooth to the tablet PC installed on the handgrip, before being sent to the internet. The tablet PC is also connected via Skype to another tablet PC fixed on the slave robot by which the child operator can see a screen image and can communicate with people in the remote location.

The slave robot is built on a mobile robot platform, Pioneer P3-DX (Figure 2). Its movement is controlled so that it is synchronized with the movement of the tricycle-style

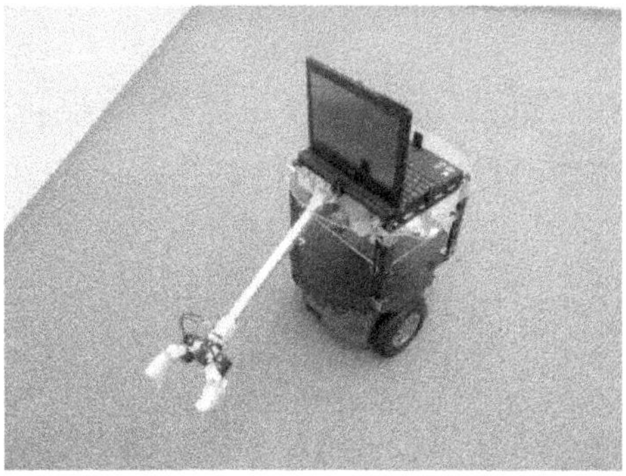

Figure 2: **The slave robot that is remotely controlled by the operator child. It has one DOF gripper hand and a Skype monitor PC. The mobile base is built on Pioneer P3-DX whose movement is controlled by the encoder signals sent from the tricycle-style teleoperational interface.**

teleoperational interface. The slave robot is also equipped with a gripper hand. At this moment, the child operator is able to control only its opening/closing hand (one DOF). A Dynamixel servomotor is used and the hand can grasp small objects in the classroom. The slave robot is about 80 cm tall, which is shorter than the average height of children 3–6 years old.

3. FIELD TESTS IN CLASSROOMS

To identify the current problem being faced by our system and to find ideal use cases for the system, we have been conducting field tests in real classroom environments. We are fortunate to have had the cooperation of the Minerva Language Institute Co., Ltd., which manages 600 English learning schools for children in Japan. With their help, we started field tests at a classroom in iias-Tsukuba, which is a big shopping mall located in Tsukuba city. In Japan, there is currently a huge demand for foreign language education, particularly for children. However, in reality, not many children there have had the opportunity to communicate with people outside Japan. It is therefore expected that our system will be welcomed if their children have the opportunity to join a classroom outside Japan in real-time.

By way of pilot trials that connected the classroom at Tsukuba with an international school in Yokohama, which is attended by children whose native language is English, we found that although still there existed some technical issues, children in both the classrooms welcomed the robot system with great enthusiasm. In particular, it was encouraging to see that Japanese children who were usually shy could communicate with the native English-speaking children through the system. Another interesting observation was that the children on the robot side were very motivated to help the slave robot. As a task, we asked the child operator to grasp alphabet sponges and build English words by putting them on a whiteboard. Owing to the limited performance of both its hardware and software, it was sometimes difficult for the operator to grasp the sponges with as much control as he/she desired. Nevertheless, the children surrounding the robot frequently helped it to grasp the sponges and build English words, which resulted in increased opportunities for the child operator in the remote location to communicate through the system.

Regarding the evaluation of the system, we are currently

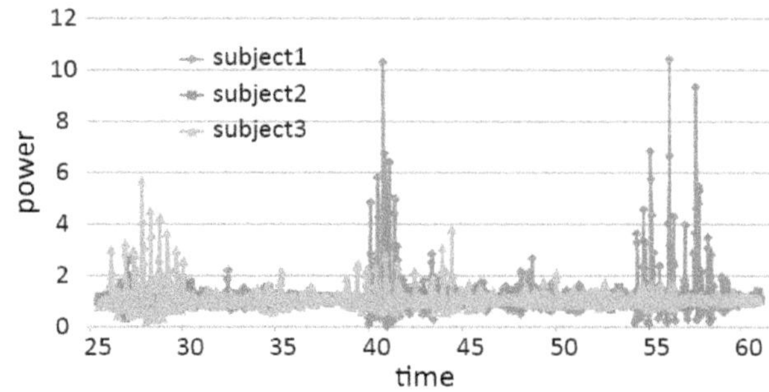

Figure 3: Sample data taken from the acceleration sensors used by children in classrooms.

investigating the use of acceleration sensors for the purpose of recognizing the movements of children. Children on both sides are asked to wear a small pouch in which is embedded a small acceleration sensor with a battery and a Bluetooth communication circuit. Figure 3 plots a test sample data obtained from one trial. We are currently trying to measure the level of engagement of children on both sides with the system on the basis of the analysis of their characteristic movements. By using videos taken during the trials, we first conduct the behavioral coding of the children. Then, by pairing the data obtained from the behavioral coding with the data from the acceleration sensors, we are trying to quantify the level of engagement.

4. CONCLUSIONS

The paper reported the development of a teleoperational robot system for children in a classroom. The key issue faced was in offering a controlling interface that satisfies the requirements of intuitive operability as well as in ensuring that it is fun for young users. We also described the introduction of acceleration sensors for the purpose of measuring the level of engagement of users with the system.

5. ACKNOWLEDGMENTS

We acknowledge the support from the Minerva Language Institute Co., Ltd., the JST PRESTO program, and the parents and children of the classroom.

6. REFERENCES

[1] S. O. Adalgeirsson and C. Breazeal. Mebot: A robotic platform for socially embodied telepresence. In *Proc. of the 5th ACM/IEEE International Conference on Human-Robot Interaction*, pages 15–22, 2010.

[2] S. Tachi. *Telecommunication, Teleimmersion and Telexistence.* IOS Press, 2003.

[3] F. Tanaka and T. Takahashi. Linking children by telerobotics: Experimental field and the first target. In *Proc. of the 6th ACM/IEEE International Conference on Human-Robot Interaction (HRI-2011), Late-Breaking Reports (LBR)*, pages 267–268, 2006.

Prosody-Driven Robot Arm Gestures Generation in Human-Robot Interaction

Amir Aly
Cognitive Robotics Lab/UEI
ENSTA-ParisTech
32 Blvd Victor, 75015, Paris, France
amir.aly@ensta-paristech.fr

Adriana Tapus
Cognitive Robotics Lab/UEI
ENSTA-ParisTech
32 Blvd Victor, 75015, Paris, France
adriana.tapus@ensta-paristech.fr

ABSTRACT

In multimodal human-robot interaction(HRI), the process of communication can be established through verbal, non-verbal, and/or para-verbal cues. The linguistic literature [3] shows that para-verbal and non-verbal communications are naturally synchronized. This research focuses on the relation between non-verbal and para-verbal communication by mapping prosody cues to the corresponding arm gestures. Our approach for synthesizing arm gestures uses the coupled hidden Markov models (CHMMs), which could be seen as a collection of HMMs modeling the segmented prosodic characteristics' stream and the segmented rotation characteristics' streams of the two arms' articulations [4] [1]. Nao robot was used for tests.

Categories and Subject Descriptors

I.2.9 [**Computing Methodologies**]: Artificial Intelligence—*Robotics*

General Terms

Experimentation

Keywords

HRI, CHMM, non-verbal and para-verbal mapping

1. INTRODUCTION

The important role of robots in our daily life shows the importance of making robots capable of generating intuitive multimodal communication with people under different interactional contexts. This research tries to understand the natural communication strategies of arm gestures that humans use in their daily interaction in order to transfer them to the robots. The database used in this research is the Stanford database available online [5] which is composed of many avatar videos. The data is acquired from human individuals using the PhaseSpace motion capture system and processed with the MotionBuilder system, which provides an approximation of the Euler rotations of the arms' joints.

HRI'12, March 5–8, 2012, Boston, Massachusetts, USA.
ACM 978-1-4503-1063-5/12/03.

Trajectory Class	Trajectory State
1	Pitch ↑ & Intensity ↑
2	Pitch ↑ & Intensity ↓
3	Pitch ↓ & Intensity ↑
4	Pitch ↓ & Intensity ↓
5	Unvoiced segment

Table 1: Voice Signal Segmentation Labels

2. TEMPORAL SEGMENTATION

2.1 Speech

Speech is analyzed in terms of the intensity and pitch curves of the voice signal. The inflection points (i.e., zeros crossing points of the rate of change of the curve) of the pitch and intensity curves, and the separating points between the voiced and the unvoiced parts of the voice signal are detected [1]. The trajectories between the detected points in the pitch and intensity curves are then labeled according to the behavior of these trajectories together (see Table 1). Speech signal is segmented into syllables constituting the states of the HMM of the voice stream. The authors in [2] defined the average duration of a syllable as 200ms and this duration can increase or decrease according to the nature of the syllable as being short or long. Practical tests proved that within a syllable of duration varying from 180ms to 220ms, the average number of trajectory classes in its corresponding pitch and intensity curves is around 5, therefore each 5 labels of the signal constitute a state in the HMM of the voice stream.

2.2 Arm Gestures

Arm gestures are characterized in terms of Euler angles of the six articulations (Elbow, Shoulder, and Wrist) of the two arms. Due to the mechanical limitations of the test platform (Nao robot), Euler rotations of the articulations are limited to be: Pitch and Roll - Shoulder; Yaw and Roll - Elbow; and Yaw - Wrist. Roll, Pitch, and Yaw rotations' data indicated in the database are segmented similarly to the voice signal by comparing the trajectories of the relevant Euler curves of each articulation and giving a label according to the behavior of these trajectories together (see Table 2 - the 5 states of the shoulder and elbow articulations represent the mechanical behavior of each articulation individually that is characterized by two Euler's curves at a time). However, for the wrist articulation, we used the following rule: if the rate of change of the specific trajectory of the yaw curve is increasing, it takes label 1; if it is decreasing, it takes labels

	Trajectory State	
Trajectory Class	Shoulder	Elbow
1	Pitch ↑ & Roll ↑	Yaw ↑ & Roll ↑
2	Pitch ↑ & Roll ↓	Yaw ↓ & Roll ↑
3	Pitch ↓ & Roll ↑	Yaw ↑ & Roll ↓
4	Pitch ↓ & Roll ↓	Yaw ↓ & Roll ↓
5	No Change	No Change

Table 2: Shoulder and Elbow Movements Segmentation Labels

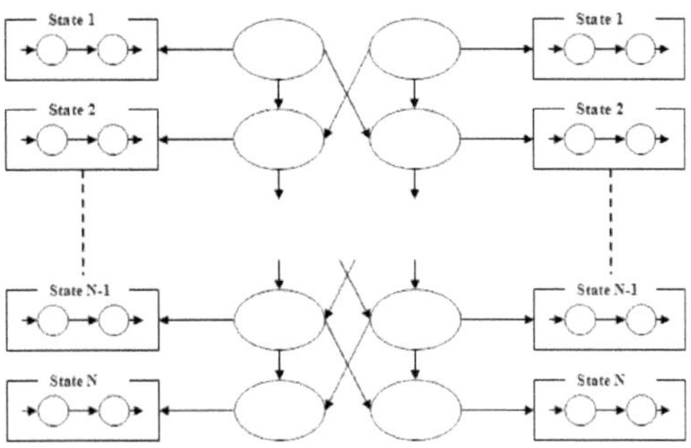

Figure 1: Coupled Hidden Markov Model CHMM mapping between speech to arm gestures' sequences

2; or it takes label 3 for no change. Arms articulations are modeled in terms of six independent HMMs, representing the mechanical rotations of the two arms (see Figure 1)[4]. Each state in the HMMs modeling arms' articulations represents a complete performed gesture, which is described by 4 labels of the obtained trajectory classes of each articulation.

3. SYNTHESIZED GESTURES

The synthesized Euler angles of each articulation are compared to the original Euler angles in terms of the similarity between the trajectory classes in terms of labels "1", "2", ..., and "5". (see Table 3). However, during human-human interaction, the generated arm gestures differ from one person to another in terms of the direction and the amplitude of the performed gesture. Therefore, the obtained scores of similarity between the original and synthesized trajectories could be considered as reasonable results (see Figure 2 and 3) because this research focuses on automatic robot arm gestures generation based only on human user prosody.

Articulation	Similarity Scores between Trajectory Classes
Left Shoulder	47%
Left Elbow	55%
Left Wrist	57%
Right Shoulder	52%
Right Elbow	59%
Right Wrist	61%

Table 3: Comparison between the original to synthesized gestures' trajectory classes

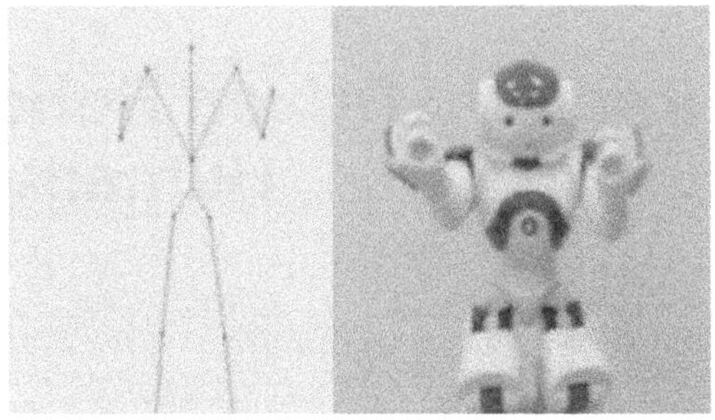

Figure 2: Typical view of the database test avatar and the Nao robot performing its own generated arm gestures

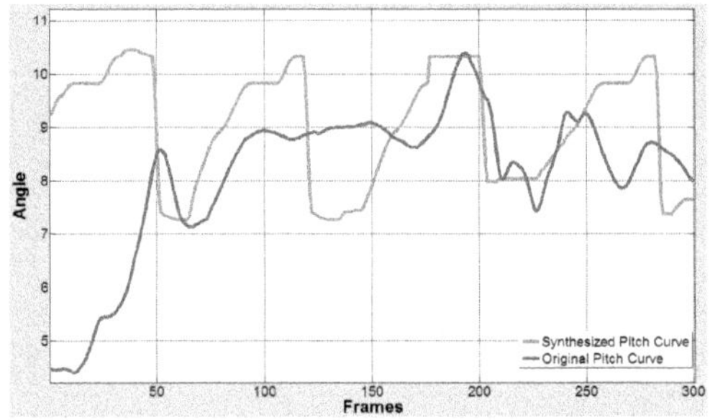

Figure 3: Original and Synthesized pitch curves for the left shoulder articulation

4. CONCLUSIONS

This research focuses on synthesizing robotic arm gestures based on human user speech characteristics (e.g., pitch and intensity of the signal). Our mapping system is based on the Coupled Hidden Markov Model (CHMM) that tries to find a coupling joint between the audio and gesture sequences. The obtained scores of similarity between the trajectories of the synthesized and the original Euler angles are in the range of 55%. Moreover, the synthesized gestures are similar to the real gestures and therefore still relevant to the interaction context and capable of conveying a similar message meaning to the message transmitted by the original gestures.

5. REFERENCES

[1] A. Aly and A. Tapus. Speech to head gesture mapping in multimodal human-robot interaction. In *In Proceedings of the European Conference on Mobile Robotics*, Orebro, Sweden, 2011.

[2] T. Arai and S. Greenberg. The temporal properties of spoken japanese are similar to those of english. In *Proceedings of Eurospeech*, pages 1011–1114, Rhodes, Greece, 1997.

[3] F. P. Eyereisen and J. D. D. Lannoy. *Gestures and Speech: Psychological Investigations.* Cambridge University Press, 1991.

[4] I.Rezek, P. Sykacek, and S. Roberts. Coupled hidden markov models for biosignal interaction modelling. In *Proceedings of the International Conference on Advances in Medical Signal and Information Processing (MEDSIP)*, 2000.

[5] S. Levine, P. Krahenbuhl, S. Thrun, and V. Koltun. Gesture controllers. In *ACM SIGGRAPH*, 2010.

Toward a Probabilistic Approach to Acquiring Information from Human Partners Using Language

Stefanie Tellex[a] Pratiksha Thaker[a] Robin Deits[b] Dimitar Simeonov[a] Thomas Kollar[c]
Nicholas Roy[a]
[a]Massachusetts Institute of Technology [b]Battelle [c]Carnegie Mellon University

Categories and Subject Descriptors

I.2.9 [**Computing Methodologies**]: Artificial Intelligence

Keywords

dialog, robotics, question-asking

1. INTRODUCTION

Our aim is to make robots that can naturally and flexibly interact with a human partner via natural language. Understanding natural language commands is a challenging problem because of the highly variable nature of unstructured linguistic input, and the use of ambiguous referring expressions that do not have a mapping to a unique object in the external world. When humans encounter this ambiguity in dialog with each other, a key strategy is to ask questions in order to better understand the situation. These questions are a crucial mechanism to clarify misunderstandings, explain problems, and gather more information. For example, if a person was asked to "Put the pallet on the truck," they might ask "Which one?" or "Do you mean the pallet on my right?" A response might be "Yes, the tire pallet" or "the red one near the truck."

In order for a robot to acquire information through dialog, it must have two capabilities: first, it must create a question that elicits useful information from the human partner, and second, it must exploit that information in order to infer better actions. These two problems are related, because what question to choose depends both on what the robot is uncertain about, and also on what types of answers the robot is likely to understand. This paper focuses on the second problem: how to understand answers to questions, leaving question generation to future work.

Understanding a human teammate's answers is difficult because of the large variety of responses the human could produce in response to the robot's questions. Previous approaches to robotic dialog understanding use a POMDP [Doshi and Roy, 2008, Rosenthal et al., 2011] with a fixed state/action space and a limited space of actions in response to the human's input. However, natural language is a powerful, compositional modality for expressing complex ideas that the robot may never have encountered during training. In our previous work [Tellex et al., 2011], we addressed this problem by introducing the G^3 framework, which converts a natural language command into a probabilistic graphical model or *grounding graph* that factors according to the compositional and hierarchical structure of language. Random variables in the model correspond to linguistic constituents in the command; inference in the model corresponds to finding *groundings*, or objects, places, paths, or events in the external world that map to linguistic constituents in the command.


```
Lift the tire pallet in the air, then proceed to deposit it to the right of the tire
pallet already on the table right in front of you.

Place a second pallet of tires on the trailer.
```

Figure 2: Sample commands from the corpus.

In this paper we describe an approach to asking questions that leverages the G^3 framework. Given a natural language command, a question, and an answer, the robot extracts grounding graphs for each linguistic construct. Then it finds linguistic constituents in the natural language command that refer to the same grounding in the external world, or that corefer. This problem, called *coreference resolution*, is a well-studied problem in computational linguistics [Jurafsky and Martin, 2008, section 18.1]. Next it merges variables in the graphs based on inferred coreference. Finally, it performs inference in the merged model, enabling it to infer the best set of groundings corresponding to the command, the question, and the answer.

Our approach is applicable not only to questions and answers, but also to understanding a sequence of commands that use coreference. For example, one command from our corpus is "Go to the second crate on the right. Pick it up and place it beside lonely crate." In order to pick up the correct pallet, the robot can use linguistic coreference to infer that "it" refers to "the second crate on the right," and then perform joint inference to find an action that corresponds to picking up that crate. An earlier version of this work appeared in Simeonov et al. [2011]; this work is an extension with evaluation results.

2. TECHNICAL APPROACH

When faced with a command, a question, and an answer, the system extracts grounding graphs from the spatial language input, finds coreferences, merges variables in the graphs according to the coreference information, and finally performs inference in the merged graph. To train and evaluate our system, we used the aligned parallel corpus of language paired with robot action described in Tellex et al. [2011]. This corpus consists of natural language commands paired with robot actions. Sample commands from the corpus appear in Figure 2.

Resolving linguistic coreferences involves identifying linguistic constituents that refer to the same entity in the external world. Although there are several existing software packages to address this problem, most are developed for large corpora of newspaper articles and generalize poorly to language in our corpus. Instead, we created a coreference system which is trained on language from our corpus. Following typical approaches to coreference [Stoyanov et al.,

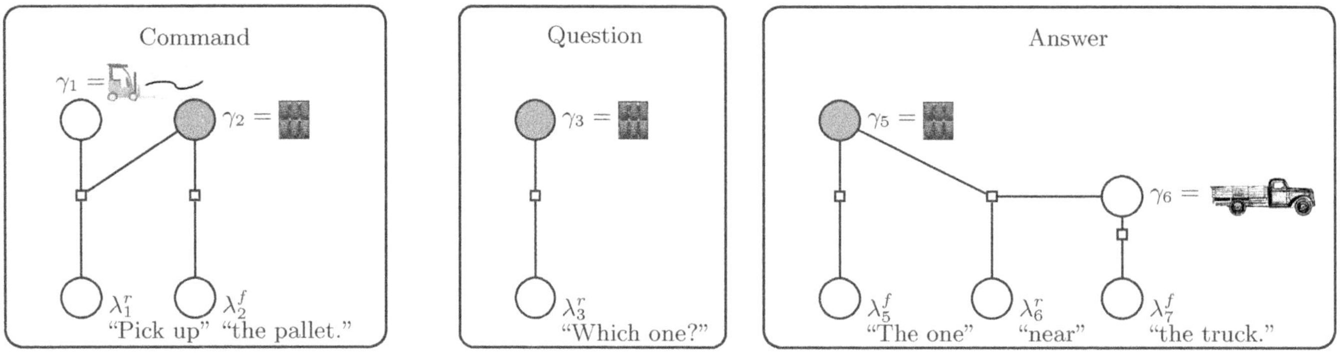

Figure 1: Grounding graphs for a three-turn dialog. The robot merges the three shaded variables.

2010], our system consists of a classifier to predict pairwise coreference, combined with a clustering algorithm that enforces transitivity. For the pair-wise classifier we used a log-linear model which is trained on the alignment annotations in our corpus: two constituents are coreferent if they are annotated with the same grounding in the external world. The model uses bag-of-words features. Since the pair-wise classifier does not guarantee that predicted coreferences are transitive, we post-process the classification results by taking the transitive closure of all links. On a corpus of commands, using ground-truth parses, our algorithm achieves an F-score of 0.889 on a held-out test set, and it achieves an F-score of 0.793 on a dialog corpus of commands paired with questions and answers.

Once coreferring variables have been identified, a merging algorithm creates a single unified grounding graph. Figure 1 shows a merged graph created from a command, a question, and an answer by our system. The λ variables correspond to language; the γ variables correspond to groundings in the external world, and the ϕ variables are *True* if the groundings correspond to the language, and *False* otherwise. For more information, see Tellex et al. [2011].

3. RESULTS

To evaluate the system, we used a subset of the corpus of natural language commands paired with robot actions described by Tellex et al. [2011], consisting of commands in which coreference is relevant. To focus the evaluation on the impact of the coreference algorithm, we used ground-truth parses. Within the test set, we see an increase in object grounding accuracy with merging. Inference using merging resulted in a total of 31 grounding variables having different inferred values, compared to not using merging. Of these, 13 were correct and resulted in changes from an incorrect grounding to a correct grounding. Many of these commands involved anaphoric expressions such as "Place it on the truck," in which coreference resolution enabled the system to find the correct grounding for the pronoun "it." Of the other nodes, many involved changes from one incorrect grounding to another and ambiguous noun phrases such as "the tire pallet" which could refer to more than one object, even while incorporating information from the merging algorithm. The robot could disambiguate these by asking questions. Others involved genuine errors, such as confusing "left" and "right," due to limitations in the system's semantic models.

We have also demonstrated the system end-to-end on an example dialog involving a question and an answer:

- **Person** "Pick up the pallet."
- **Robot** "Which one?"
- **Person** "The one that has boxes."

The system is able to parse the language, automatically extract coreference, and create a merged graph. By performing inference in the merged graph, it infers that the person was referring to a pallet of boxes, and not a tire pallet, which happened to be a slightly more likely grounding for the phrase "the pallet."

4. CONCLUSION

In this paper we presented preliminary results for a robot dialog understanding system that is able to ask the human user targeted questions and incorporate the inference into a probabilistic graphical model that factors according to the structure of language. We demonstrated that our framework is able to use information from commands, questions, and answers in order to infer more accurate actions from a corpus of realistic dialog collected on Amazon Mechanical Turk.

Our immediate next steps are to implement an algorithm for asking a question based on the entropy of variables in the grounding graph. By generating questions dynamically based on the robot's uncertainty about the mapping between the command and the external world, it can collect exactly the information it needs from the human user, maximizing the value of each interaction.

Our larger vision is to extend the framework to support question answering, enabling the robot answer questions about why it is acting in a certain way. Furthermore, this paper focuses on relatively short clarification dialogs. Supporting multi-turn dialogs in which complex activities are designed remains a challenging problem which requires more complicated discourse-level semantic structures, that represent the conversation at a more abstract level.

5. ACKNOWLEDGEMENTS

This work was sponsored under the Robotics Collaborative Technology Alliance by the U.S Army Research Laboratory, the Office of Naval Research under MURI N00014-07-1-0749, and Battelle. Their support is gratefully acknowledged.

References

F. Doshi and N. Roy. Spoken language interaction with model uncertainty: An adaptive human-robot interaction system. *Connection Science*, 20(4):299–319, 2008.

D. Jurafsky and J. H. Martin. *Speech and Language Processing*. Pearson Prentice Hall, 2 edition, May 2008. ISBN 0131873210.

S. Rosenthal, M. Veloso, and A. K. Dey. Learning accuracy and availability of humans who help mobile robots. In *Proc. AAAI*, 2011.

D. Simeonov, S. Tellex, T. Kollar, and N. Roy. Toward interpreting spatial language discourse with grounding graphs. In *2011 RSS Workshop on Grounding Human-Robot Dialog for Spatial Tasks*, 2011.

V. Stoyanov, C. Cardie, N. Gilbert, E. Riloff, D. Buttler, and D. Hysom. Reconcile: A coreference resolution research platform. Technical report, Cornell University, 2010.

S. Tellex, T. Kollar, S. Dickerson, M. Walter, A. Banerjee, S. Teller, and N. Roy. Understanding natural language commands for robotic navigation and mobile manipulation. In *Proc. AAAI*, 2011.

A Practical Comparison of Three Robot Learning from Demonstration Algorithms

Russell Toris
Worcester Polytechnic Institute
Worcester, MA
USA
rctoris@wpi.edu

Halit Bener Suay
Worcester Polytechnic Institute
Worcester, MA
USA
benersuay@wpi.edu

Sonia Chernova
Worcester Polytechnic Institute
Worcester, MA
USA
soniac@wpi.edu

ABSTRACT

Research on robot learning from demonstration has seen significant growth in recent years, but existing evaluations have focused exclusively on algorithmic performance and not on usability factors, especially with respect to naïve users. Here we present findings from a comparative user study in which we asked non-experts to evaluate three distinctively different robot learning from demonstration algorithms – Behavior Networks, Interactive Reinforcement Learning, and Confidence Based Autonomy. Participants in the study showed a preference for interfaces where they controlled the robot directly (teleoperation and guidance) instead of providing retroactive feedback for past actions (reward and correction). Our results show that the best policy performance in most metrics was achieved using the Confidence Based Autonomy algorithm.

Categories and Subject Descriptors

I.2 [**Artificial Intelligence**]: Robotics

General Terms

Human Factors

Keywords

Learning from Demonstration, Human-Robot Interaction

1. INTRODUCTION

Algorithms for robot *Learning from Demonstration* (LfD) seek to enable human users to expand the capabilities of robots through interactive teaching instead of explicit programming. Specifically, LfD methods enable new robot behaviors to be learned based on demonstrations of the desired actions performed by the teacher. The broad research goal is to develop an *intuitive* method of programming that is accessible to untrained, naïve, users [1]. Research in this area lacks a comparative study, especially among higher-level, symbolic, learning algorithms that focus on learning behavioral or task representations. Performance evaluations are typically focused on learning time compared against non-interactive methods.

This lack of a comparison leaves many critical questions unanswered. Without a comparison of existing methods, future researchers have no guidance with respect to what learning algorithms and interaction techniques are most effective, and what underlying algorithmic assumptions accurately reflect the real-world use case. These problems are further heightened by the fact that many LfD techniques have only been evaluated through use by expert roboticists.

To further investigate these issues, we present a comparison study of three established learning from demonstration algorithms, each representing a distinctively different variant of policy learning and utilize different demonstration techniques:

- **Interactive Reinforcement Learning (Int-RL):** A system model policy learning technique in which the reward signal is provided by the human user at runtime. The teacher interacts with the robot by providing reward and guidance input through an on-screen interface. [4]
- **Behavior Networks (BNets):** A planning-based policy learning technique. The teacher interacts with the robot through kinesthetic teaching, a type of *experienced demonstration*, in which the user physically guides the robot through the task. [3]
- **Confidence Based Autonomy (CBA):** A mapping approximation technique based on a classification algorithm. The teacher interacts with the robot by selecting the next action or correcting a past action choice through an on-screen interface. [2]

To compare the three algorithms, 31 participants were recruited to use each of the three methods to teach a robotic agent to accomplish a simple task. To accomplish this, we designed the *DBug domain* for the Aldebaran Nao humanoid robot, which required the robot to catch and remove small moving HEXBUG toys from a table.

2. THE DBUG EXPERIMENT

Within the DBug domain, the robot was given 3 preprogrammed actions: picking up a bug at a center pickup location (marked on the table with an 'X'), sweeping a bug inwards towards the pickup location, and waiting for 1 second (i.e. doing nothing). The wait action was an important action to include as it was useful in teaching the robot when it was not appropriate to either sweep or pickup. Using images from Nao's on-board camera and image processing with the OpenCV library, the robot was able to continuously track the x and y position of all bugs in its field of view (a 320x240 pixel image). The state space was defined

Table 1: Number of responses to "How well did the robot learn the task?"

	Not at All	Not Well	Well	Very Well
Int-RL	4	9	17	1
BNets	4	17	9	1
CBA	1	4	17	9

Table 2: Number of responses to "How quickly did the robot learn the task?"

	Very Slowly	Slowly	Quickly
Int-RL	11	16	4
BNets	8	11	12
CBA	3	9	19

to be the x and y position of the bug with the shortest Euclidean distance to the pickup zone. This state space was considered to be large enough such that the solution was not trivial, yet small enough that a user could make reasonable progress teaching the agent within a short period of time.

Upon arrival, the participant was given a broad overview of the task that was being trained as well as the capabilities of the Nao. Following the introduction, the participant used each algorithm for 10 minute intervals in a randomized order to train the robot. After each training session the participant was asked to fill out a brief questionnaire relating to the method that they had just used. At the conclusion of their session, a final general questionnaire was given which asked comparative questions about the three methods.

3. SUMMARIZED RESULTS

To evaluate the final learned policies for the three algorithms for each user, we reconstruct the final policy of each algorithm at the end of each training session using data collected during the experiments. These models were then queried for an action for every possible state. On average, our subjects taught the pick-up action and the sweep action using CBA with a higher accuracy as compared to the other methods followed by Int-RL. Subjects taught the wait action with higher accuracy using BNets. It is worth noting that the wait action was implicitly taught in BNets, that is, if the teacher doesn't demonstrate any action by moving the arms of the robot, then the agent learned to wait, for instance when there is no bug to pick up in sight. Based on the lower accuracy rate of the other two actions, we think that this implicitness is the reason why people had more success with BNets for this specific action.

In addition to assessing the user's performance with each algorithm, we are interested in looking at trends in the survey data. After each training session, the user filled out a algorithm-specific survey. In particular, two general questions appeared on each of the surveys: *How well did the robot learn the task?* and *How quickly did the robot learn the task?*. The total number of responses for each possible choice in each question are listed in Table 1 and Table 2.

When asked *Which method did you find easiest to learn?* 24 people answered CBA, 7 people answered BNets, and nobody said Int-RL. We understand from this that most people actually wanted to tell the robot what to do at each moment using a graphical interface and a mouse. When asked *If you were to teach a new task to a robot, which method would you prefer to use?*, 25 people answered CBA, 5 people answered BNets, and 1 person chose Int-RL. When we asked for the order of general preference of the methods, CBA was ranked first with 27 votes, Int-RL became second with 18 votes, and BNets ended-up being the last with 16 votes.

The general survey included space where participants were able to write free-form feedback about the algorithms in the study. Three common responses from the surveys include the need for additional feedback from the robot regarding what it was sensing, the need for better understanding of what the robot is "thinking", and the need of communicating with the robot more naturally (e.g. through speech). These three points, especially the importance of feedback, are surely not unprecedented findings; however, their importance is often underestimated in existing LfD methods. Our aim is to draw attention to these factors for future research.

4. CONCLUSION

Our results show that users were most proficient in and satisfied with teaching using the Confidence-Based Autonomy algorithm. Based on our participants' comments, this is mainly because naïve teachers like to see which actions are available to them and give commands directly to the robot. Although a lot of the participants liked the way they interacted with the robot for BNets (kinesthetic teaching), they were not happy with the performance of the result. In case of Int-RL, participants indicated that they could not see a satisfactory improvement in the action decision of the agent. This was mostly caused by the inherently slow learning rate of the algorithm; however, to the participants, this factor was not obvious. Our results tell us that it is important to keep the transparency of the robot's knowledge to its teacher during the whole process. Techniques for communicating the knowledge state of the robot in a effective way remains an important area for future work.

5. REFERENCES

[1] B. D. Argall, S. Chernova, M. Veloso, and B. Browning. A survey of robot learning from demonstration. *Robot. Auton. Syst.*, 57:469–483, May 2009.

[2] S. Chernova and M. Veloso. Interactive policy learning through confidence-based autonomy. *J. Artificial Intelligence Research*, pages 1–25, 2009.

[3] M. N. Nicolescu and M. J. Mataric. Natural methods for robot task learning: instructive demonstrations, generalization and practice. In *Proceedings of the second international joint conference on Autonomous agents and multiagent systems*, AAMAS '03, pages 241–248, New York, NY, USA, 2003. ACM.

[4] A. L. Thomaz and C. Breazeal. Adding guidance to interactive reinforcement learning. In *Proceedings of the Twentieth Conference on Artificial Intelligence (AAAI)*, 2006.

An Assistive Robot Contest: Designs and Interactions

Igor M. Verner
Department of Education in Technology and Science
Technion – Israel Institute of Technology
Haifa 32000, Israel
(972) 4829-2168
ttrigor@technion.ac.il

David J. Ahlgren
Department of Engineering
Trinity College
Hartford, CT 06106, U.S.A.
(860) 297-2588
David.Ahlgren@trincoll.edu

ABSTRACT
This paper considers engineering and educational challenges of the assistive robot contest RoboWaiter and interactions experienced by participants, both students and people with disabilities.

Categories and Subject Descriptors
K.3.1 [**Computers and education**]: Computer Uses in Education – *collaborative learning, computer-assisted instruction.*

General Terms
Design, Experimentation, Human Factors, Performance

Keywords
Assistive robotics, education, robot contest, design projects, HRI.

1. INTRODUCTION
One approach to raise awareness among engineering students about the needs of people with disabilities is to encourage them to solve design problems of assistive technology. There is a growing body of research focused on the creation of robots to assist people with reduced mobility is the daily need to move things around the house. A number of fetch-and-carry robots, addressing this need, have been recently developed and tested as laboratory prototypes for further research. Tapus, Mataric and Scassellati [1] suggested that establishing robot competitions directed to challenging problems of assistive robotics can have a stimulating effect on research and accelerate its practical implementation in the field. This suggestion is supported by the evidence that robot search and rescue competitions significantly promote robotics research, particularly of technical and human-centered aspects of HRI [2]. Sharing this view, we established the assistive robotics competition RoboWaiter in order to inspire students and young researchers in the subject and involve them in socially responsible education.

2. GENESIS OF THE COMPETITION
The inspiration for an assistive robotics competition came from the Connecticut Council on Developmental Disabilities which proposed to run an assistive robot competition jointly with the Trinity College Firefighting Home Robot Contest in Hartford, CT [3]. Persons with disabilities have been engaged in choosing the theme, developing the rules, and participating in the RoboWaiter events. Our collaboration for the development of the assistive robotics competition was developed through the following significant steps: opening the communication channel, clarifying mutual expectations, establishing a vision and developing a framework of the contest. Communication showed that people with disabilities are interested in robotics and that there are real needs for assistive robotic devices. Through discussion with roboticists, people with disabilities and caregivers a vision for the contest was created. The group chose a theme that represented an area of personal concern to Council members—the need for a person with disabilities to retrieve food from a refrigerator during an emergency, when a personal assistant is not present.

HRI'12, March 5–8, 2012, Boston, Massachusetts, USA.
ACM 978-1-4503-1063-5/12/03.

The educational robotics challenge that emerged from the contest planning consisted of two main elements:

Bringing people with disabilities in as clients. Following literature recommendations, we created a focus group in which people from the CTCDD could identify and describe the nature of the mobility-related problems that they encounter. The focus group put forward ideas for possible robotic solutions of the problems and their experimental design in the framework of the contest.

Integration of the RoboWaiter project in the robotics course. Since 2009, the assistive robotics project has been integrated in ENGR 120 (Introduction to Engineering Design—Mobile Robots), a first-year course at Trinity College. Student teams, each with three students, designed fetch-and-carry robots and participated in the RoboWaiter contest.

3. THE ROBOWAITER CHALLENGE
Challenging characteristics of robot operation in human environments include the following: presence of autonomous actors (people, pets, and robots), variability, combining manipulation and mobility, and use of specialized tools. These characteristics appear in varying degrees in the RoboWaiter contest task. The task simulates a situation where Grandpa, a person with a disability, wishes by means of an assistive robot to move a plate of food from a refrigerator to the table where he is sitting in a wheelchair. Grandma is standing in the kitchen during the robot operation. The action takes place in the arena that simulates a household kitchen. The "kitchen" is furnished by scaled miniature items including a table, a sink, a refrigerator, and a chair. When performing the task the robot, directed by a signal from the judge, should move from its home position to the refrigerator, pick up the plate, move to the table where Grandpa sits, place the plate on the tabletop, and return back to the initial position. This action must be fully autonomous. Robots must avoid collisions with the kitchen items and with the doll that represents Grandma. The judge examines the robot operation and measures the time from the start signal until the plate is placed on the tabletop.

The competition is run in two divisions. In the Standard Division all the objects are placed in the arena in known positions except for the chair, which is constrained in one direction only. A single shelf represents the refrigerator, and LEDs o serve as beacons that guide the robot to its destination. In the Advanced Division the positions of the Grandma doll and the chair change from run to run. The "intelligent" refrigerator model has a computer-controlled door opener that is actuated when a proximity sensor, embedded in the floor in front of the door, detects the presence of the robot. The two refrigerator shelves are equipped by LED plate-location beacons. The robot begins operation with receiving an audio signal of a certain frequency. If it detects the 3.5 kHz signal, it must pick up the plate from the bottom shelf; if 7.5 kHz, from the top shelf.

When modeling interactions with an assistive robot, we took human factors into account. Based on research observations, the RoboWaiter environment is organized so that the robot, while delivering food, will approach the doll representing Grandpa in the wheelchair from the right direction. Importance of human safety in assistive robotics is emphasized in the RoboWaiter contest rules so that the run is disqualified if the robot touches any doll or item.

Table I presents the RoboWaiter subsystems and requirements their design. Each subsystem presents a significant design challenge that requires research, detailed development, and careful testing.

Table 1. RoboWaiter subsystems

Subsystems & goals	Requirements
Robot Base: provide reliable locomotion.	Programmable, controllable, fast, capable of limited dead reckoning, well controlled turns.
Sensor System: enable navigation, object and plate detection.	Detect arena features (e.g. walls and obstacles); LED beacons on shelf & table; audio start signal; door and floor beacons.
Robot Arm: grab, hold and carry plate.	Grip plate; deployable/ retractable; robust; accommodate dimensional tolerances; 3D motion of gripper; good tactile behavior.
Software: verify, integrate, and tune all sub-systems.	Test and debug motor drives, sensors, arm/gripper; enable test of navigation, plate detection, obstacle avoidance, plate conveyance, return to home.

4. INVOLVEMENT OF CTCDD PEOPLE

Together with students, hobbyists, and engineers who design, demonstrate, and compete with their robots the RoboWaiter involves an additional category of participants – people connected to the Connecticut Council of Developmental Disabilities. For them RoboWaiter opens an opportunity of active social interaction in a public event where they can learn about modern assistive technology and personally engage in discussions with engineers on developing new assistive robots to answer their needs. The evolution and organization of RoboWaiter were in large part directed by persons with disabilities, who, as a result, have a high investment and sense of ownership and pride in the contest. CTCDD members joined the executive planning committee for RoboWaiter. Their evaluation activity including discussion of contest surveys, facilities, and accessibility issues, provides the basis for planning for the next event. Meetings also consider funding issues, and they serve as forums in which Council members discuss and make choices about such important contest-related questions as theme development, rules formulation, funding, public relations, participation, judging, and awards. The executive planning committee also chooses a keynote speaker, a leading expert in the field of assistive or rehabilitative robotics, who delivers a public lecture on the day of the competition. In addition to these planning responsibilities, members of the executive committee encourage attendance by persons with disabilities, recruit judges for the event, and oversee the scoring process on the day of the event.

5. Contest Surveys and Interviews

Reflections were collected by means of questionnaires and interviews with contestants (N=111), persons with disabilities and caregivers (N=34) during and after the 2009-2011 contests. Their findings are summarized below. The main reasons for participation in RoboWaiter were engineering challenge, curiosity in the assistive robotics contest, and humane challenge. Other reasons such as interest in assistive robotics job/scholarship opportunities, and social challenge were somewhat less important but still relevant. Evaluations of learning outcomes given by participants of the RoboWaiter were high and comparable with that given by participants of the Trinity College Fire-Fighting contest. Evaluations of RoboWaiter were higher in mechanics, robot manipulation and robot interaction with the environment while responses for the firefighting contest were stronger in the "take initiative", "make decisions", and "set priorities" areas. Respondents noted that participation in the contest motivated or even strongly motivated their interest in assistive technology, robotics, and in developing robots that can provide real assistance. The contest inspired them to think about a person with disability as a customer of assistive robotics who can help designers to make more realistic robots.

6. CONCLUSION

We characterize RoboWaiter as a system that has three interdependent and inseparable components: educational, social, and engineering. The educational component focuses on raising awareness of assistive robotics, exposing students to assistive robotics and associated design problems, and promoting creative thinking practices. The social component includes participation by persons with disabilities as planners, volunteers and judges, and engagement of these persons with contestants and supporters at the contest event. The engineering component is driven by the robot design problems defined by interpreting the RoboWaiter rules. Interactions promoted by the assistive competition helped the participants to recognize the social challenge of assistive robotics and to learn that engineering work can have importance beyond pure technical achievement.

7. REFERENCES

[1] Tapus, A., Mataric´, M. J., and Scassellati, B. 2007. The grand challenges in socially assistive robotics. *IEEE Robotics and Automation Magazine*. 14, 1, 35-42. DOI= http://www.cs.yale.edu/~scaz/papers/Tapus-RAM2007.pdf

[2] Yanco, H., and Drury J. 2007. Rescuing interfaces: A multi-year study of human-robot interaction at the AAAI Robot Rescue Competition. Autonomous Robots, 22, 4, 333-352. DOI=http://robotics.cs.uml.edu/fileadmin/content/publications/2007/yanco-drury-autonomous-robots-2007.pdf

[3] Verner, I., and Ahlgren, D. 2002. Fire-Fighting Robot Contest: Interdisciplinary Design Curricula in College and High School. *ASEE Journal of Engineering Education*. 91, 3, 355-359.

Interaction with Animated Robots in Science Museum Programs: How Children Learn?

Alex Polishuk
Israel National Museum of Science, Technology and Space (MadaTech)
Haifa 31448, Israel
972-48614438
alex.polishuk@madatech.org.il

Igor Verner
Department of Education in Technology and Science
Technion – Israel Institute of Technology
Haifa 32000, Israel
972-48292168
ttrigor@technion.ac.il

ABSTRACT
This paper examines learning through student interaction with animated robots, as implemented in robot theatre performances and workshops in the science museum.

Categories and Subject Descriptors
K.3.1 [**Computers and education**]: Computer Uses in Education – *collaborative learning, computer-assisted instruction.*

General Terms
Design, Experimentation, Human Factors, Performance

Keywords
Educational robotics; science museum; animated robot, learning through interaction, robot theatre.

1. INTRODUCTION
Educational robotics is a rapidly growing field which already significantly impacts education in schools, and recently has arrived on the doorstep of science museums. The literature discusses experiments of using robots in science museums as exhibits, remote facilitators, and entertaining guides. Science museums have started to offer learning activities using robot kits following the pioneer Computer Clubhouse program, started at the Boston Computer Museum [1]. Robotics workshops and other activities at the Gelfand Robotics Center of MadaTech have been conducted since 2000, with the participation of tens of thousands of school students annually. MadaTech in collaboration with Technion developed HRI learning activities using animated robots such as Nao, RoboThespian, Aibo and others. These activities have become popular, enjoying over 350,000 visitors in 2010. Here we consider implementation of interactions with animated robots in the museum programs and present a summary of our observations and student reflections.

2. LEARNING INTERACTIONS

2.1 Public Robot Theatre Performances
Robot theatre performances follow the long tradition of theatrical events which are known as an effective way for public communication of science and technology in museums. The robot theatre initiative of MadaTech is in line with the pilot theatrical programs that have been very recently introduced in other science museums [2]. We developed a series of theatrical pieces, in which a human actor demonstrates different robots: a vacuum cleaner Roomba, an Aibo robot dog, a humanoid Nao, and others. These performances introduce basic concepts of robotics, while demonstrating robot functionality and behavior repertoire. When presenting Roomba in action, the actor playfully explains the principles of its sensing and communication. Children interact with Roomba and test its reactions on stimuli such as spreading waste and colliding real obstacles and virtual walls.

The Aibo's presentation emphasizes its intelligence and emotion expression capabilities. Children interact with Aibo and through observing reactions on patting find the touch sensors on its back and head. They also test dog-like behaviors of the robot such as sitting down, playing with a pink ball or walking, in response to voice commands. The children reveal that the robot not only executes tasks but also expresses emotions by flashing red lights and emitted whining sounds (anger state) or green lights with joyful sounds (happiness state).

The culmination of the robot theater event is performance of Nao. It consists of three pieces: greetings, a Tai Chi exercise, and a dance in memory of Michael Jackson. In the greeting piece, the robot presents autonomous behaviors: standing-up, walking, gesturing and human-like speech. During the presentation, the actor introduces concepts such as stability, locomotion, and text-to-speech capability. In the Tai Chi exercise, the explanation is based on the use of two physics concepts: center of mass and equilibrium. Children are encouraged to practice Tai Chi along with the robot. They experience the difficulty of keeping balance during the exercise and discover that Nao performs better than some of the students! The Nao's dance demonstrates the robot's capability to perform various movements and synchronize motion with music (implemented in the pre-programmed mode). The accompanying explanation focuses on an analysis of the degrees of freedom of the robot used in the dance movements. Spectators were asked to compare Nao and humans in terms of degrees of freedom. Demonstrations are delivered in a specially-arranged 100-seat amphitheater. We evaluated children's attitudes towards the robot theatre performances by means of a survey. Answers were received from 63 respondents of ages 9-14, among them 35 middle school (MS) and 28 primary school (PS) students. The summary of their answers is given in Table 1.

2.2 Robotic animals workshop
The workshop was delivered to classes of 3rd and 4th grade students from nine primary schools (about 300 participants). The 18-hours curriculum was based on activities with the Lego Education We Do kit. The workshop topics included:

- Experimental testing of pre-programmed robot behaviors. Each pair of the students got a crocodile robot (equipped with an IR motion detector) and tested its responses to different "stimuli" movements.

HRI'12, March 5–8, 2012, Boston, Massachusetts, USA.
ACM 978-1-4503-1063-5/12/03.

Table 1. Robot theatre survey results

Narratives	Arguments	Proponents PS , MS
Science museum is a right place to present robots	The robot exhibits help to understand robotics. Robots are scientific products.	>90%, >90%
Demonstration of advanced robots add to students knowledge of robotics	I learned about degrees of freedom and how to count them. I learned that robots can speak and answer questions. Legged robots can perform complex movements and not fall down.	>80%, 58%
Nao's capabilities make it the "premier actor" of the robot theatre	Nao demonstrates capabilities similar and even higher than that of the human. It can dance, listen, answer questions, walk and seat and stay on one leg.	100%, 86%
The actor's explanations help to understand Nao's functions	The following explanations are mentioned: keeping equilibrium, visual perception, symbols recognition, locomotion.	>80%, >80%
Aibo's demonstrations are instructive and impressive	It is exciting, cool, impressive, cute, emotional, and behaves like a real dog. It shows how robots equipped by sensors function.	*Knowledge*: 27%, 62% *Impression*: 85%, 96%
Interaction with Roomba helps to understand how Nao and Aibo work	It demonstrated how robots equipped by sensors function.	85%, 38%
The theatre performance was impressive and enjoyable	I was impressed of seeing that robots behave so similar to humans.	100%, 100%

- Creating new behaviors of the given robot. The students programmed basic reactive robot behaviors using the graphical block language.
- Constructing, programming and testing a dog robot (with a tilt sensor) and its behaviors.
- Studying the robot motor and control systems by analogy between robotic and biological systems.
- Creating interactions between the crocodile and dog robots.

The workshop was followed up by an educational study aimed to explore characteristics of learning through interaction with robots. 48 students from two schools answered pre- and post-questionnaires; their interactions with robots were videotaped. Our findings indicate that through interactions with animal-like robots the young students gained understanding of the robotic system and the principles of its programming and operation. We observed numerous expressions of animistic thinking with regard to the robots. The students were closely involved in and excited by interaction with the robots. They collaborated to create robot-robot interactions. All the students highly evaluated the workshop. While some students were especially interested in building robots, the majority was attracted mainly by testing and creating robot behaviors.

2.3 Nao humanoid workshop

The workshop was conducted for 30 finalists of OlympiYeda 2011 - a nationwide program for 8th/9th graders (considered in our previous paper [3]). The program theme was "Sports science", and the workshop purpose was to learn mechanics of human locomotion through programming and operating motion behaviors of Nao. The students were divided into two groups, each used two Naos. The workshop stages:

- The movement and sensing capabilities of the robot Nao were introduced through analysis of analogies with features of the human body, such as: robotic cameras vs. human eyes, gyro and accelerometer vs. human vestibular system, robot engines and gears vs. human muscles, robot microphones and human ears. The ways in which the robot uses sensors data to move and interact with its environment were demonstrated.
- The hands-on session focused on programming Nao to imitate athletic movements using the Choregraphe language and simulator. The concepts of motors stiffness, parallel programming and error handling were explored.
- The students programmed Nao to mimic different typical sport movements. Keeping stability of the robot required them to analyze the center of gravity and the forces acting on the robot in different positions. For example, in order to mimic kicking a football the student had first to move the robot to lean over one leg and then perform the kick by the other leg, without falling down.

Students' reflections indicated that the workshop contributed to their knowledge in motion science. Part of the participants pointed that the most instructive activities were the interactive demonstrations, while others preferred the hands-on practices. As indicated by the observations, the most motivating activity was creating new movements of the humanoid robot. The students used their own bodies for designing realistic movements and for initial testing of robot's stability.

3. CONCLUSION

Our study of educational robotics programs for a mass audience of MadaTech provides indications of engaging school students in meaningful and joyful learning through interactions with animated robots. We found that a theatre performance of different modern robots can effectively introduce basic concepts of robotics. Workshops focused on playful analysis and creation of animal-like robot behaviors were a novel successful experience of robotics education for MadaTech and for our school system, for 3rd and 4th grade classes. The workshops have become in high demand for school visits to MadaTech. Regarding the Nao workshop for middle school students, modeling sport movements was an effective learning activity with the humanoid robot.

4. ACKNOWLEDGMENTS

Activities of the Gelfand Center for Modeling, Robotics and Communication are sponsored by Mark Gelfand, 2011 Honorary Doctor of the Technion.

5. REFERENCES

[1] Rusk, N., Resnick M. and Cooke, S.2009. Origins and guiding principles of the computer clubhouse. In Kafai, Y.et al. (Eds.), *The Computer Clubhouse: Constructionism and Creativity in Youth Communities*. NY: Teachers College Press, 17-25.

[2] Breazeal C. et al. 2003. Interactive robot theatre. *Communications of the ACM*, 46, 7, 76-85.

[3] Verner I. et al. 2011. Focusing a robotics education program on scientific and humane challenges: A museum case study. Proceedings of the IEEE EDUCON 2011.

A Survey on Robot Appearances

Astrid von der Pütten
University of Duisburg-Essen
Forsthausweg 2
47048 Duisburg
+49-203-379-1330
astrid.von-der-puetten@uni-due.de

Nicole Krämer
University of Duisburg-Essen
Forsthausweg 2
47048 Duisburg
+49-203-379-1330
nicole.kraemer@uni-due.de

ABSTRACT
Against the background of the uncanny valley hypothesis [1] and its conceptual shortcomings this study aims at identifying design characteristics which determine the evaluation of robots. We conducted a web-based survey with standardized pictures of 40 robots which were evaluated by 151 participants. A cluster analysis revealed six clusters of robots. The results are discussed with regard to implications for the uncanny valley hypothesis.

Categories and Subject Descriptors
K.4.m **[Computing Milieux]**: 4 COMPUTERS AND SOCIETY – Miscellaneous; J.4 **[Computer Applications]**: 4 SOCIAL AND BEHAVIORAL SCIENCES - Psychology

General Terms
Measurement, Design, Experimentation, Human Factors, Theory, Verification.

Keywords
Uncanny valley, human-robot interaction, robot appearances, design

1. INTRODUCTION
Since Masahiro Mori introduced his theory of the uncanny valley (UV) in 1970 [1] the UV is widely considered as a design obstacle to overcome. The theory states that increased human-likeness results in increased familiarity (and, thus, increased positive evaluation) until a certain peak is reached after which the system appears too humanlike and the positive evaluation reverses: the robot is perceived as creepy, causing eerie feelings in the user. However, Mori´s theory lacks precise definitions for its underlying concepts, i.e. familiarity or human-likeness. Previous studies tried to emulate the UV in very simple ways using only three items ("how humanlike/familiar/uncanny is this robot?"; see [2] for an overview). The items used in these experiments are often bi-polar adjective pairs (familiar-unfamiliar, humanlike-mechanical), but e.g. especially humanoid robots may be perceived as both: humanlike and mechanical. Moreover, these studies often use unstandardized material presenting the robots in front of different backgrounds, in different positions, performing different actions. The present study thus uses standardized stimulus material and a new methodological approach to investigate the influence of different robot appearances on evaluation. We aim at identifying certain design characteristics by the means of a cluster analysis.

HRI'12, March 5–8, 2012, Boston, Massachusetts, USA.
ACM 978-1-4503-1063-5/12/03.

2. METHOD
2.1 Stimulus Material
As stimuli we used standardized pictures of 40 robots. The pictures showed the full robot in front of a white background and the robot´s height was displayed above the picture. When possible, pictures were taken frontally in a still posture without movement (cf. figures 1-5 for pictures of the robots). Participants were randomly assigned to one of four questionnaire sets displaying the robots in different orders to avoid sequence effects. Participants rated each robot on a 5-point Likert Scale with 16 items (weak, intelligent, unfamiliar, likable, uncanny, pleasant, natural, attractive, dominant, threatening, competent, familiar, submissive, harmless, strange, eerie) from "I agree" to "I do not agree at all". We further asked how humanlike and how mechanical, respectively, participants perceived the robots to be on a 7-point Likert-Scale.

2.2 Sample
The sample comprises 144 complete questionnaires and seven data sets of participants who rated at least 20 of the 40 robots (N=151, 109 female, 42 male; age M=25.24; SD=8.27).

3. RESULTS
We first calculated a principal component analysis to reduce the number of variables for the following cluster analysis.

3.1 Principal Component Analysis
A principal component analysis was run on the 16 aforementioned items resulting in four factors: *1) Likability* (pleasant, likable, attractive, familiar, natural, intelligent, Cronbach´s α = .827, explains 29.83% of variance), *2) Threatening* (threatening, eerie, uncanny, dominant, harmless (reverse), Cronbach´s α = .886; explains 18.74% of variance), *3) Submissive* (incompetent, weak, submissive; Cronbach´s α = .653; explains 13.84% of variance) and *4) Unfamiliar* (strange, unfamiliar, Cronbach´s α = .674; explains 6.90% of variance).

3.2 Cluster analysis
A cluster analysis was run on 40 cases (40 robots). Six variables were entered into the cluster analysis: the four factors derived from the principal component analysis, namely pleasant, threatening, incompetence-submissiveness and strange-unfamiliar, and the two items humanlike and mechanical. A hierarchical cluster analysis using Ward's method produced six clusters, between which the variables were significantly different (for mean values refer to table 1).

Table 1. Mean values for factors included in cluster analysis

Cluster	Factor						N (robots included in cluster)
	Likability	Threatening	Submissiveness	Strange-Unfamiliar	Humanlike	Mechanical	
1	.44	-.52	.40	-.11	2.23	3.13	9
2	-.16	-.34	.67	.13	2.08	3.45	6
3	.06	.80	.09	-.64	4.56	2.05	2
4	.96	.31	-.35	-.82	4.65	2.21	2
5	-.17	.57	-.45	.07	2.76	4.08	11
6	-.07	-.04	.04	.45	1.94	3.78	10

The first cluster was characterized by small robots which were rated as likable, not threatening, submissive, rather not humanlike and medium mechanical (cf. figure 1). The robots in the second cluster were again rather small and rated as not threatening, submissive, not humanlike and medium mechanical, but they were less likable and more unfamiliar (cf. figure 2).

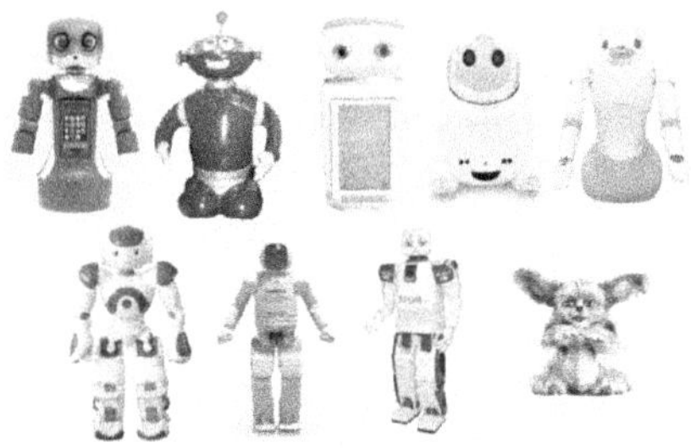

Figure 1. Cluster 1: Robovie MR2, Cosmobot, Autom, Papero, Riba, Nao, Asimo, Atom & Leonardo

Figure 2. Cluster 2: Asoy, ICat, Snackbot, Dynamoid, Ri-Man & Wakamaru

Figure 3. Cluster 3 (Geminoid HI-1 & Ibn Sina) and Cluster 4 (Geminoid DK & HRP-4c).

The cluster three and four are very similar. They both were essentially familiar, humanlike and not mechanical. Cluster three however (with Geminoid HI-1 and ibn Sina) was more threatening and less likable and cluster four (with Geminoid DK and HRP-4c) was highly likable and less threatening (cf. figure 3). Cluster five was rated as not likable, threatening, dominant, rather unfamiliar, rather not humanlike and very mechanical (cf. figure 4). And cluster six was characterized by not humanlike but mechanical robots rated as rather not likable, rather not threatening, submissive and very unfamiliar (cf. figure 5).

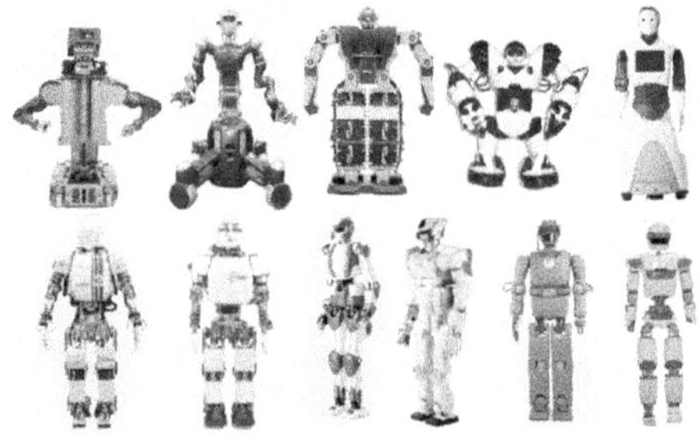

Figure 4. Cluster 5: PR2, Justin, Robonova, Robosapien, REEM-1390, Wabian, Kobian, HRP2, HRP3, REEM-1, Hwarang

Figure 5. Cluster 6: Mika, Luna, Olivia, Lucas, Kismet, Armar, Popo, Phobe, Robo-Head, Twendyone

4. DISCUSSION

We were able to identify different groups of robots which were rated significantly different on six dimensions. Interestingly, in two clusters robots were rated both low in human-likeness and low in "mechanical" indicating that in future research these adjectives should be used separately. The robots in those clusters with a high mean concerning "Threatening" are almost all rather tall and bi-pedal robots. In contrast to the humanoid robots (cluster 5), the android robots (cluster 3&4) were rated higher on "Familiarity", and on "Likability" although all three clusters were rated as "Threatening" which contrasts the UV hypothesis.

5. ACKNOWLEDGMENTS

We want to thank all the research groups and companies for their permission to use their pictures for this study. This work was partially supported by a doctoral fellowship of the German National Academic Foundation.

6. REFERENCES

[1] Mori, M. 1970. The uncanny valley. *Energy*, *7*(4), 33–35.

[2] MacDorman, K., and Ishiguro, H. 2006. The Uncanny Advantage of Using Androids in Cognitive and Social Science Research. *Interaction Studies*, *7*(3), 297–337.

Tele-operated Robot Control Using Attitude Aware Smartphones

Amber M. Walker
amber.m.walker@ou.edu

David P. Miller
dpmiller@ou.edu

University of Oklahoma
865 Asp Avenue
Norman, Oklahoma 73019

ABSTRACT

Smartphones have put video communications, computation, and proprioceptive sensing (e.g. accelerometers and gyros) into the hands of hundreds of millions of consumers. These small, microelectromechanical systems can be used in many applications, including remote control. This study proposes using smartphones with proprioception as handheld robot controllers and aims to determine feasibility of accelerometers as control inputs for tele-operation while defining heuristics for use. Initial results indicate accelerometers are suitable for tele-operation commands, but identify specific design characteristics meriting further investigation.

Categories and Subject Descriptors

A.0 [**General**]; H.5.2 [**User Interfaces**]: Input devices and strategies; I.2.9 [**Robotics**]: Operator interfaces

General Terms

Human Factors, Experimentation

Keywords

Smartphone, Tele-operation, Unmanned Ground Vehicles

1. INTRODUCTION

Hundreds of millions of smartphones have been sold since their introduction, possessing impressive computational power and a suite of microelectromechanical systems (MEMs) on board. Accelerometers, used to detect and measure motion, are one of these MEMs, and contribute to the user experience by switching orientations, self-orienting maps, and more [1]. Developers have used the commercial availability of these sensors to tailor them for use in handheld gaming platforms, pedometers, and remote control.

Aside from commercial uses for smartphone technology, government and industry are also taking note of smartphones' tailorability. The U.S. Army is examining the feasibility of equipping all soldiers with smartphones, as they are uniquely suited for dispersed and varying military missions due to their size, expanding capabilities, common platforms (iOS and Android), and agility [7]. Unmanned ground vehicles (UGVs) are one of the fastest growing segments of the U.S. military, and are almost always tele-operated, meaning "manually controlled by an operator at a distance that is too great for the operator to see what the robot is doing" [3]. In these situations, the operator can rely only upon the robot's sensors; in contrast, remote control occurs within line of sight of the robot and the operator can make his/her own judgments based on direct sensory input. Tele-operation tasks generally require more from an operator control unit (OCU) since tele-presence via video screen is necessary.

HRI'12, March 5–8, 2012, Boston, Massachusetts, USA.
ACM 978-1-4503-1063-5/12/03.

2. RELEVANT RESEARCH

Military labs have explored the feasibility of smartphones as UGV controllers, and so have a number of civilian institutions [4, 9]. Researchers at the University of South Florida demonstrated tele-operated smartphone control of the most popular military UGV, the iRobot PackBot. The controller uses a push button interface, not attitude control, but did prove video tele-presence via mobile device [5]. Commercially, the popular hobbyist quadcopter, AR.Drone (manufactured by the Parrot company), has done the same. Using an ad hoc wireless connection, the iPhone acts as both controller and video screen (for the on board camera); their interface supports both tilt inputs and push buttons to maneuver the quadcopter [11]. Given the available video feedback, it is possible that the AR.Drone could be tele-operated; however, it is almost always demonstrated where the operator maintains line of sight to the quadcopter and/or flies in large open areas where obstacles are limited.

While demonstrating various capabilities, these systems provide no real comparison of the advantages and/or disadvantages of using the OCU's attitude as a control input versus push buttons, joysticks and other traditional means. Tilt control has a number of perceived advantages over tactile or push-button interfaces: 1) it's more intuitive for users, especially in lieu of appropriate training, 2) it permits one-handed operation – important for multi-tasking users, and 3) it requires less attention i.e. time spent looking at the screen. Additionally, using tilt controls prevents unnecessary buttons from cluttering the small display, maximizing the area available for video feedback.

3. APPROACH

Touchscreen interfaces have been around long enough for basic heuristics to exist regarding button size, spacing, and menu display [10, 8]; however, few heuristics exist defining use of the device's MEMs e.g. accelerometers and gyroscopes. Apple provides a strong set of guidelines in their *iPhone Human Interface Guidelines*, but states only that developers should consider accelerometer values as they pertain to device orientation [2]. In tele-operated robot control, where accelerometer values are the main controller input, they should *not* be used to reorient the screen in response to device orientation. The research described below strives to identify heuristics, such as the one just defined, to inform future controller development using accelerometers in tele-operated robot control.

3.1 Preliminary Work

Preliminary work has focused on controller design and experiment set-up while the researchers await IRB approval. The test robot is a modified HPI Racing Crawler King, where a webcam has been added to provide the video feedback necessary for tele-operation (Figure 1(a)). The controller will be presented on an iPod Touch in landscape mode using the iPhly hardware to generate RF control signals from the application output [6]. Video is displayed in near full-screen, and a cross hair with floating bubble denotes the current tilt of the device when the dead man switch (touch anywhere on the screen) is active (Figure 1(b)).

Initial feedback from a small subset of users indicates that accelerometer control is feasible in tele-operated unmanned ground systems; however, there are some design issues that merit further consideration, e.g. accelerometer sensitivity. Individual operators have different expectations for how a "small" movement should translate to vehicle motion; therefore, formal testing should include a readily-available interface to adjust the sensitivity setting. Additional areas of focus for experimentation are described in the following section.

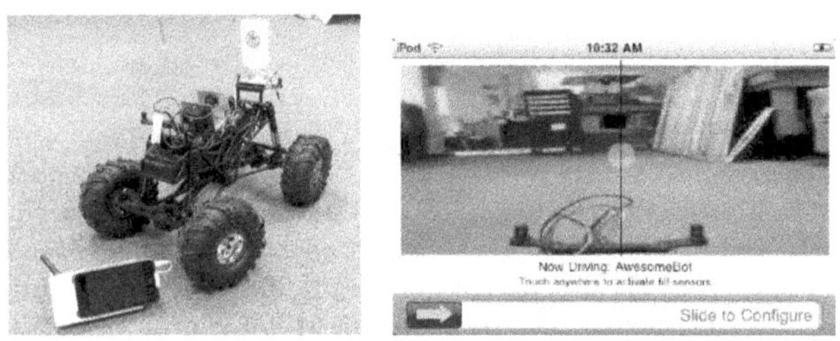

(a) HPI Racing Crawler King and iPhly Hardware (b) Controller Application

Figure 1: Robot System

3.2 Proposed Experiment

The authors propose to study the following major characteristics tied to the usability of an accelerometer equipped smartphone as a controller for a tele-operated vehicle.

1. Compare the efficiency and user feedback of the standard RC controller while viewing video feedback on the iPhone.
2. Conduct trials using the iPhone as both video screen and controller with a simple push button interface to control the robot.
3. Assess efficiency and user feedback when video is viewed on the iPhone and the robot is controlled via accelerometer input. A dead man switch will activate accelerometers, and operators will tilt forward, backward, and side to side to control the four-wheeled vehicle. A possible variation on this could test other methods for steering control e.g., rotation vs. tilt.
4. Compare/contrast the controller described in item 3 with two different "leveling" points. In one, the zero point[1] would be a defined point in space i.e. user must hold device at a 45 degree angle to activate the dead man switch and therefore the accelerometers; alternatively, the device could self-level meaning the the zero point is wherever the user activates the dead man switch. The results of this comparison have important usability implications, especially for users that require a more heads up controller environment. It will also help inform the researchers as to how users' mental models are built and how they perceive the controller *should* function and react.

Ideally, the results of these tests will help answer a number of questions. Aside from learning more about how users *expect* a controller to act, data collected will also measure general suitability (efficiency, precision of operation) of accelerometers in tele-operation. Do users prefer a tilt interface over other conventional interfaces, and if so how is accelerometer control best implemented? We hope to identify appropriate settings regarding the zero point, accelerometer sensitivity, and video display size/resolution.

4. REFERENCES

[1] One-Third of Mobile Phones to Use Accelerometers by 2010, Spurred by iPhone and Palm Pre. *M2PressWIRE*, 2009.
[2] Apple. iPhone Human Interface Guidelines, OCT 2011.
[3] J. Carlson and R. R. Murphy. How UGVs Physically Fail in the Field. *IEEE Transactions on Robotics*, 21(3):423 – 437, 2005.
[4] S. I. Erwin. Army under pressure to bring broadband to the battlefield. *National Defense Magazine*, September 2010.
[5] R. Gutierrez and J. Craighead. A Native iPhone Packbot OCU. In *2009 4th ACM/IEEE International Conference on Human-Robot Interaction*, 2009.
[6] A. Krupnik. Radio Control with iPhone, 2011.
[7] M. Milian. U.S. Army may soon equip troops with smartphones. *CNN*, July 2011.
[8] Y. Park and S. Han. Touch key design for one handed thumb interaction with a mobile phone: Effect of touch key size and touch key location. *International Journal of Industrial Ergonomics*, 40:68–76, 2010.
[9] R. A. Pettitt, E. S. Redden, and C. B. Carstens. Scalability of Robotic Controllers: An Evaluation of Controller Options. Technical Report ARL-TR-4457, Army Research Laboratory, May 2008.
[10] A. Smith. Pushing the right buttons: Design characteristics of touch screen buttons. *Usability News*, 11(2), 2009.
[11] G. Yarrish. Look, Ma! No Radio! *ModelAirplaneNews.com*, January 2011.

[1]The zero point is defined as the origin for the accelerometer output. Normally the iPhone is considered at 0,0,0 when resting flat on a table.

HRI Research: The Interdisciplinary Challenge or the Dawning of the Discipline?

Astrid Weiss
Institute of Informatics/ICT&S Center
University of Amsterdam/University of Salzburg
Science Park 904, 1098XH/ Sigmund-Haffner Gasse 18
Amsterdam, The Netherlands/ Salzburg, Austria
a.weiss@uva.nl/astrid.weiss@sbg.ac.at

ABSTRACT
The Human-Robot Interaction (HRI) research field has developed more and more over the past 10 to 20 years. It is still a relatively young community, which is in the process of developing its characteristics, such as being interdisciplinary, innovative, responsible, technical, and many others. Similarly, like in the development of the Human-Computer Interaction (HCI) community, being interdisciplinary is essential, but if we have a look on the current situation, HCI became more of an autonomous discipline nowadays. Where is the HRI community heading to in this respect? This paper should reflect in accordance to the "epistemic living spaces" concept on some stereotypical statements by researchers working in HRI. The reflection shows that in three phases of a researcher's career (orientation, positioning, and stabilizing and expanding) show a tendency towards the discipline and away from interdisciplinary work and that the forth phase (attachment) needs to be strengthened, independently of disciplinary or interdisciplinary approaches.

Categories and Subject Descriptors
A.m [MISCELLANEOUS]

General Terms
Human Factors.

Keywords
Interdisciplinary, disciplinary, HRI community, research field.

1. INTRODUCTION
HRI is a rather young research area existing approximately for 20 years. The annual IEEE RO-MAN conference series, originating in 1992 in Japan, has since then travelled across the world, reflecting this emerging new field. The growing interest for this research area can be determined by the inaugural annual ACM/IEEE International Conference on Human-Robot Interaction (HRI 2006). HRI is often considered to be a highly interdisciplinary research field, similar as HCI was coined in earlier times. A research field can be considered as research problems that need to be addressed by several disciplines in order to be solved. A discipline can be considered as a "thought collectives", with discipline-typical methods, approaches, and paradigms [4]. If we take a look on the development of HCI as a research field, it can be already coined to be a research discipline. There are HCI syllabuses, master and PhD programmes, you can get a Professor title for HCI, and there also text-books on HCI research methods. All these aspects are indicators for a disciplinary approach, besides the fact that HCI is not an institutionalized discipline at most universities, but most often part of computer science departments. But how will HRI develop, will it be the same development or will it stick closer to its technical origins related to engineering and Artificial Intelligence research?

This paper reflects on some of the often-mentioned comments made by researchers working on HRI topics and HRI projects. These statements will be interpreted in conjunction with the "epistemic living spaces" concept, developed by Felt et al. [1]. This concept is based on the coproductionist approach by Jasanoff, which stresses how sciences and society are intertwined [3]. Similarly, to Felt et al. "a person-centered view to draw the attention to the individual and/or collective perceptions and narrative re-constructions" [2] will be taken to understand the knowledge production in the research field of HRI. However, contrary to the work of Felt et al. [2] the interpretation is not based on actual interview or focus group data, but on often-mentioned statements by HRI researchers, which can be considered as stereotypical descriptions.

2. STEREOTYPICAL STATEMENTS
As mentioned above this paper presents stereotypical statements from researchers involved in HRI projects, which did not derive from specific interviews or focus groups. Rather they can be considered as often-heard informal statements in the community. These statements only serve as a starting point for reflection; therefore its "degree of reality" is not of utmost relevance. For each statement two viewpoints are presented: the potential point of view of a user-centered researcher (which subsumes disciplines such as psychology, social sciences, communication sciences, etc.) and the potential point of view of a robot-centered researcher (which subsumes engineers, computer scientists, software developers, etc.). This approach was chosen to depict the "creative tension" of disciplines in the field.

2.1 Coping with different knowledge
The first aspect that should be discussed is that in interdisciplinary research one has to cope with the fact that not everybody has the same background and knowledge on a common research topic/question. This can lead to the personal perception that one is missing skills to perform interdisciplinary research on his own (only a few people are interdisciplinary on their own, as little have finished two polarizing studies and even then discursive approaches are more fruitful for complex research fields, such as HRI, than single person projects). Stereotypical statements for this could be:

User-centered researcher: *"I am now taking this Stanford AI class, in order to know what roboticists are doing."*

Robot-centered researcher: *"I have read this social science book to know how to compile a questionnaire for my user study."*

HRI'12, March 5–8, 2012, Boston, Massachusetts, USA.
ACM 978-1-4503-1063-5/12/03.

Immersion with Robots in Large Virtual Environments

Xianshi Xie, Qiufeng Lin, Haojie Wu, Julie A. Adams, Bobby Bodenheimer
EECS department, Vanderbilt University, Nashville, TN, USA
[xianshi.xie,qiufeng.lin,haojie.wu,julie.a.adams,bobby.bodenheimer]@vanderbilt.edu

ABSTRACT

This paper presents a mixed reality system for combining real robots, humans, and virtual robots. The system tracks and controls physical robots in local physical space, and inserts them into a virtual environment (VE). The system allows a human to locomote in a VE larger than the physically tracked space of the laboratory through a form of redirected walking. An evaluation assessed the conditions under which subjects found the system to be the most immersive.

Categories and Subject Descriptors

H.5.1 [**Information Interfaces and Presentation**]: Multimedia Information Systems—*Artificial, augmented, and virtual realities*

Keywords

Mixed-reality, Human-Robot-Interaction, Immersion

1. INTRODUCTION

Humans and robots are increasingly operating together as teams in complex, demanding, and dangerous situations. It is necessary to empirically understand the scalability of human-robot teams in situations where the humans and robots are deployed in potentially large environments in order to develop future robotic systems. Virtual Environments (VEs) provide a good platform to meet such demands. Robots have been used in mixed-reality simulations, e.g., [2], but not for collaborative human-robot teams.

It is unknown how the inclusion of real robots will impact the perceived immersion of the VE and the virtual robots when the subjects are stationary or walking through the virtual space. In addition, the human and robot are physically collocated, but there is no requirement for virtual collocation. It is unknown how presence is affected when there is physical, but not virtual collocation.

To move humans through virtual environments larger than the tracked physical space, a form of redirected walking [3] was used. This method, called "resetting" [8, 9], permits free exploration of arbitrarily large spaces and preserves users' spatial awareness within the VE. Other methods, such as "walking in place" [7] may be possible, but suffer from various drawbacks from the perspective of interacting with robots. Methods employing only a joystick, for example, are not as good at preserving a user's spatial awareness [4].

2. EXPERIMENT

The physical tracking space is 5m x 5m. The Pioneer 3DX robot and human are tracked by eight Vicon cameras. The VE is viewed through a Nvis head mounted display (HMD). It is known that spatialized sound may increase the sense of presence in VEs [5]. This experiment used a Head-Related Transfer Function (HRTF) method. A recorded car engine sound, which is different from the real robot's high pitch sound, was used.

Twelve subjects, six males and six females, with ages ranging from 21 to 32 years old, participated in the experiment. The VE was a virtual town (about 750m x 750m), in which there were buildings, cars, scooters, etc. The experiment incorporated two locomotion conditions and three robot conditions. The locomotion conditions were Walking (W) and Standing Still (S). The robot conditions were With Real Robot (R), Without Real Robot and Sound (N), and Without Real Robot but including 3D Simulated Sound (Ss).

The walking locomotion condition required subjects to walk along a path, while a virtual Pioneer 3DX robot moved in a visible area simultaneously. The still locomotion condition required subjects to stand at some position, while a virtual robot moved along a straight path close to them. The real robot condition required the real robot to move the same distance (8-12 meters) in the physical space as the virtual robot traveled. The robot condition without the real robot or sound ensured that subjects heard nothing that sounded like the robot. The simulated sound robot condition provided a simulated robot sound and subjects were not told that the real robot was not moving.

Each subject completed three trials for the six conditions. The walking condition required subjects to walk a straight line and make two 90-degree-turns. There was one reset along the path. Subjects had 3-5 minutes to complete each trial block. After each block, subjects completed a questionnaire that included a modified version of the Slater-Usoh-Steed (SUS) questionnaire [1] measuring a user's sense of presence in the VE. The questionnaire was modified to ask similar questions about the robot to assess the impact of the real robot and sound on presence.

3. RESULTS AND DISCUSSION

The answers across the SUS questions were averaged [1] (Figure 1). A within-subject two-way ANOVA was performed. The dependent variable, subjects' average score, was measured across the three robot conditions (R, N and Ss) by the two locomotion conditions (W and S). The ANOVA results showed main effects for both locomotion ($F=9.5716$, $p=0.01$) and robot conditions ($F=7.0592$, $p<0.01$), but was not significant across conditions. Figure 1 indicates that walking increases subjects' sense of presence overall. A paired samples t-test for robot conditions was significant between N and Ss ($F=9.8268$, $p<0.01$), and between N and R ($F=6.1097$, $p=0.03$). This result may imply sound adds to the sense of presence, especially when there is no sound direction mismatch between a real robot and a virtual robot. A within-subject two-way ANOVA of the averaged response for the robot questions found a main effect of robot condition ($F=15.606$, $p<0.01$), but no main effect of locomotion and no interactions across conditions (see Figure 2). Given the main effect of the robot condition, a t-test for paired robot conditions found that the robot Ss condition is significantly different from both R ($p<0.01$) and N ($p<0.01$), thus implying subjects had a significantly better sense of presence when they heard the simulation sound. The robot questions demonstrate a similar pattern with regard to locomotion condition; however, the analysis is not significant.

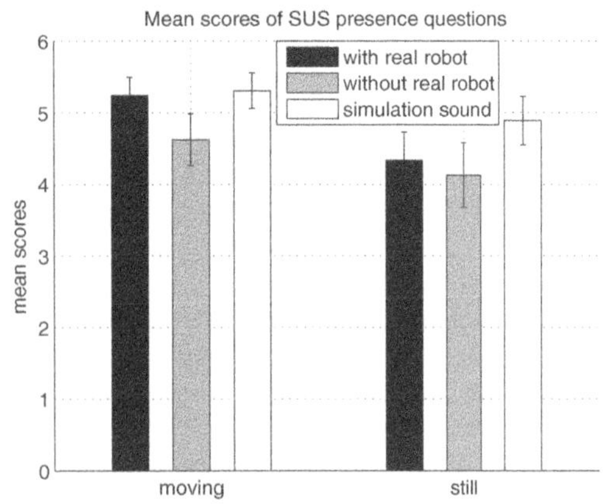

Figure 1: Mean SUS Score by condition. The error bar show one standard error of the mean.

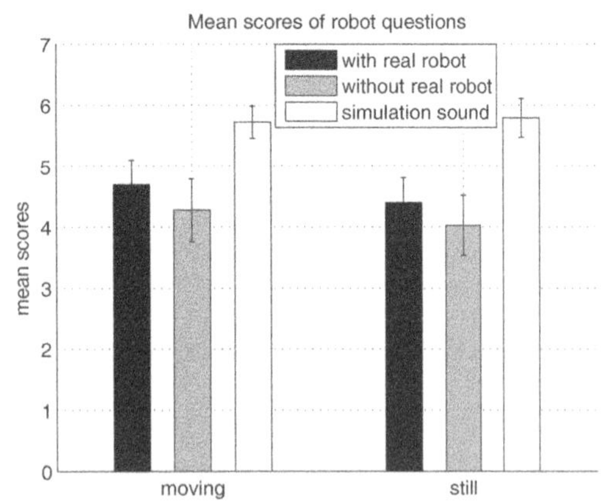

Figure 2: Mean robot score by condition.

4. CONCLUSIONS

The experimental results show that locomoting through the VE increased the user's sense of presence. Adding sound increased the user's perceived sense of immersion, and the simulated spatialized sound resulted in providing the highest perceived immersion. The results imply that a real robot is not necessary to provide a high sense of immersion if no physical interaction with the robot is required.

This experiment evaluated only a single virtual robot. Future work will include multiple virtual robots. Subjects' perceived immersion with multiple virtual robots may be substantially different. This experiment limited the locomotion and did not require subjects to explore the entire virtual town (750m × 750m), as was evaluated in [9]. Future work will assess how real physical interactions with the robot impact perceived immersion. Future work will also include a full-bodied self representation of the user, which has been shown to affect presence [6].

5. ACKNOWLEDGEMENTS

This material is based upon work supported by the National Science Foundation under Grants 0705863 and 0821640.

6. REFERENCES

[1] E. Catena, S. Arman, and M. Slater. Using presence questionnaires in reality. *Presence*, 9:497–503, 2000.

[2] I. Y.-H. Chen, B. MacDonald, and B. Wünsche. Mixed reality simulation for mobile robots. In *2009 IEEE Int. Conf. on Rob. and Auto.*, pages 922–927, 2009.

[3] S. Razzaque. *Redirected walking*. PhD thesis, Chapel Hill, NC, USA, 2005. Adviser-Brooks,Jr., Fredrick P.

[4] B. Riecke, B. Bodenheimer, T. McNamara, B. Williams, P. Peng, and D. Feuereissen. Do we need to walk for effective virtual reality navigation? physical rotations alone may suffice. In *Spatial Cognition VII*, volume 6222, pages 234–247. 2010.

[5] B. E. Riecke, A. Väljamäe, and J. Schulte-Pelkum. Moving sounds enhance the visually-induced self-motion illusion (circular vection) in virtual reality. *ACM Trans. Appl. Percept.*, 6:7:1–7:27, March 2009.

[6] M. Slater, B. Spanlang, and D. Corominas. Simulating virtual environments within virtual environments as the basis for a psychophysics of presence. *ACM Trans. Graph.*, 29:92:1–92:9, July 2010.

[7] J. N. Templeman, P. S. Denbrook, and L. E. Sibert. Virtual locomotion: Walking in place through virtual environments. *Presence*, pages 598–617, 1999.

[8] B. Williams, G. Narasimham, B. Rump, T. P. McNamara, T. H. Carr, J. Rieser, and B. Bodenheimer. Exploring large virtual environments with an hmd when physical space is limited. In *Applied perception in graphics and visualization*, pages 41–48, 2007.

[9] X. Xie, Q. Lin, H. Wu, G. Narasimham, T. P. McNamara, J. Rieser, and B. Bodenheimer. A system for exploring large virtual environments that combines scaled translational gain and interventions. In *Applied perception in graphics and visualization*, pages 65–72, 2010.

Effect of Scenario Media on Human-Robot Interaction Evaluation

Qianli Xu[1], Jamie Suat Ling Ng[1], Yian Ling Cheong[1], Odelia Yiling Tan[1]
[1]Institute for Infocomm Research, Agency for Science, Technology and Research (A*STAR)
1 Fusionopolis Way, #21-01 Connexis, Singapore
(65) 6408 2702
qxu, jamie, ylcheong, ytan@i2r.a-star.edu.sg

Ji Bin Wong[2], Benedict Tiong Chee Tay[2], Taezoon Park[2]
[2]School of Mechanical and Aerospace Engineering, Nanyang Technological University,
50 Nanyang Avenue, Singapore, 639798
(65) 6790 5894
z080221, bene0006, tzpark@ntu.edu.sg

ABSTRACT
Different media used to present the human-robot interaction (HRI) scenarios may affect users' perception of a robot in the user studies. We investigated how different scenario media (text, video, and live interaction) might influence user evaluation of social robots based on a controlled experiment. We found that multiple aspects of user acceptance were influenced by the scenario media. Moreover, more design problems and redesign proposals were elicited when users were exposed to media with higher fidelity. The results led to useful insights into choosing scenario media in HRI evaluation.

Categories and Subject Descriptors
H.1.2. [**Models and Principles**]: User/Machine Systems – *Human factors.* H.5.2. [**Information Interfaces and Presentation**]: User Interfaces – *Evaluation/methodology, User centered design.*

General Terms
Design, Human Factors.

Keywords
Social robot, Human-robot interaction, Scenario media.

1. INTRODUCTION
When evaluating human-robot interactions (HRIs), one should exploit the rich contextual information that can be represented as HRI scenarios [2,4]. In practice, researchers have used different media to deploy the scenarios, such as textual narratives, images or storyboards, videos, and physical prototypes [3]. These media convey HRI information with varying degrees of fidelity, thus they may have a different effect on users' perceptions of the product. However, little empirical evidence has been reported on the effect of scenario media on HRI evaluation.

In this study, we investigated how different media of HRI scenarios might influence user attitudes towards social robots. Based on the experimental results, we discussed how scenario media affect users' acceptance of social robots, and how effective the scenario media are in eliciting user feedback. These will help designers choose appropriate scenario media when carrying out HRI evaluations.

HRI'12, March 5–8, 2012, Boston, Massachusetts, USA.
ACM 978-1-4503-1063-5/12/03.

2. METHOD
We evaluated scenario media from two perspectives: (1) their effect on user acceptance, and (2) their effect on eliciting user feedback. User acceptance was evaluated using the Almere model [1]; and user feedback was collected in semi-structured interviews, which was transcribed so as to find out the number of design problems and redesign proposals suggested by users.

We built scenarios that described how a robot could help an elderly adult at home handle chores of daily life and engage in social life. The tasks of the robot included (1) fetching objects, (2) switching off electrical devices, (3) operating home appliances, (4) giving reminder of medication, (5) making emergency calls, and (6) chatting with the user.

The research compared a control condition, in which the subjects received no relevant HRI information (null scenario), to scenarios delivered in one of three formats: (1) a textual description illustrated with photographs (*text scenario*) (Figure 1), (2) a video showing the robot in operation with an actor (*video scenario*), and (3) a Wizard-of-Oz driven, live interaction between the subject and the robot (*live scenario*).

We recruited 60 participants (male: 17, female: 43; average age: 64.2, age range: 53-77). The participants were randomly assigned into four groups (15 in each group) to experience one of the four types of scenarios (i.e., null, text, video, and live). In the survey, participants answered the Almere questionnaire after going through the scenarios. Subsequently, we interviewed 15 subjects (5 each from the text, video, and live scenario groups), where participants were instructed to spot potential design problems and propose solutions based on what they experienced.

When you come to the dining table, Mika comes up to ask if you like coffee or tea. You ask for coffee; and Mika picks up a cup from the dining table and brings it to the coffee machine. After securing the cup under the nozzle, it sends a signal to the coffee machine to switch it on. When the coffee is ready, Mika holds the cup and brings it to the table…

Figure 1. A scenario of robot fetching a cup of coffee

3. RESULTS AND ANALYSIS
Effect of scenario media on user acceptance -We examined the user acceptance of the social robot with respect to 10 subscales in the Almere model (Figure 2). We conducted a one-way ANOVA to study the between-subject effect of scenario media. We found

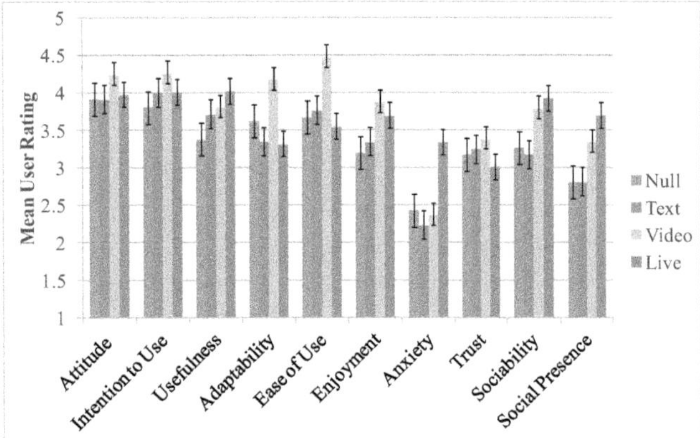

Figure 2. Mean rating of user acceptance based on Almere model (Horizontal axis: 10 subscales)

significant differences in user acceptance with respect to all 10subscales. This provided evidence that the user acceptance was different due to the exposure to different scenario media. Next, we scrutinized how scenario media influenced the user acceptance based on pair-wise comparisons between any two scenario media against individual subscales.

The text scenario had limited power in modulating user acceptance. Nevertheless, it did cause users' evaluation of *usefulness* of the robot and users' *intention to use* it to be different from the null scenario, as did the video and live scenarios. These two subscales were mainly related to the robot functionality and users' general attitude. Thus, the low-fidelity text scenario seemed to be as effective as its higher fidelity counterparts (i.e., video and live scenarios) in conveying such information.

The video scenario was most impactful in modulating user acceptance as it caused significant differences in all 10 subscales when comparing with the null scenario. Based on results of the pair-wise comparison, the video medium was the only one that caused a significant deviation from the null scenario in terms of robot's *ease of use* ($p<.01$) and *adaptability* ($p<.01$). Both subscales focused on how well the functions were fulfilled by the robot. It seemed that the video scenario was particularly effective in enhancing users' perceptions of the performance of robots on its intended functionality. The fluency with which the video scenario presented the interactive process between the robot and the human may have contributed to its capacity in this aspect.

The live scenario caused significant differences in 8 subscales when benchmarking with the null scenario. It was impactful in modulating the social aspects, such as *sociability* and *social presence*. It seemed that direct interaction with the robot was useful in evaluating the social aspects of the robot. Interestingly, the live scenario caused the highest *anxiety* level and lowest *trust* among all media. This can be explained by the constant mismatch between what people wanted the robot to do and what the robot offered during the interactions, whereas such a mismatch was not obvious in the other scenarios. As such, researcher should avoid premature deployment of live scenario because nonconformity of product capacity with user expectation may negatively influence user attitudes.

Effectiveness of scenario media in eliciting user feedback - User feedback was quantified by the number of design problems identified by users and constructive redesign proposals generated by them. The outcome was compared across different scenario media, whereas the null scenario was excluded from the comparison because the subjects did not have a meaningful reference in the evaluation. On average, a participant was able to identify 4.8, 7.7, and 11.2 design problems and put forth 2.6, 4.1, and 7.5 redesign proposals when exposed to text, video and live scenarios, respectively. The ANOVA test showed significant differences in the number of design problems identified ($F(2)=9.36$, $p=.002$) and the number of redesign proposals generated ($F(2)=5.37$, $p=.015$). Thus, the amount of user feedback was apparently dependent on the fidelity of scenario media - higher fidelity media led to more user feedback. This did not conform to findings in the evaluation of other types of products (e.g., website user interfaces) [3]. We suspected this was due to the unique characteristics of social robots, e.g. their proximity to human and their socio-capabilities. Thus, findings in those domains may not be applicable in HRI.

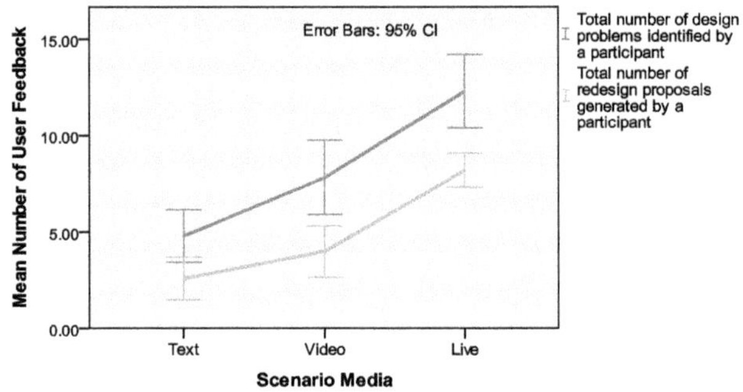

Figure 3. User feedback due to scenario media

4. SUMMARY

In HRI, prior work reported identical effect of video and live scenarios [4]. However, the results are limited to only one HRI mode. By planning multiple robot tasks, our results are applicable to more comprehensive HRI modes. We studied the effect of scenario media on user acceptance with refined granularity (10 subscales). To our best knowledge, our work is the first to systematically study the effect of scenario media on user acceptance at refined subscale levels. Based on the study, the contextual information conveyed by scenarios was apparently useful in modulating user aptitudes; and different media are effective in different aspects. Moreover, the capacity of scenarios in eliciting user feedback was dependent on the fidelity of the media; and live interaction was preferable in HRI evaluation for eliciting reflective user feedback.

5. REFERENCES

[1] Heerink, M., Kröse, B., Evers, V. and Wielinga, B. 2010. Assessing acceptance of assistive social agent technology by older adults: The almere model *Int J Soc Robot*, 2, 361-375.

[2] Mutlu, B. and Forlizzi, J. 2008. Robots in organizations: The role of workflow, social, and environmental factors in human-robot interaction. In *Proc. HRI'08* (Amsterdam, The Netherlands, March 12-15), 287-294.

[3] Walker, M., Takayama, L. and Landay, J. 2002. High-fidelity or low-fidelity, paper or computer? Choosing attributes when testing web prototypes. In *Proc. Human Factors and Ergonomics Society 46th Annual Meeting*, (Baltimore, USA, Sept 29-Oct 4), 661-665.

[4] Woods, S.N., Walters, M.L., Koay, K.L. and Dautenhahn, K. 2006. Methodological issues in HRI: A comparison of live and video- based methods in robot to human approach direction trials. In *Proc. RO-MAN 06*, (Hatfield, UK, September 6-8), 51-58.

Development of a Jenga Game Manipulator Having Multi-articulated Fingers

Tsuneo Yoshikawa
Ritsumeikan University
Kusatsu, Shiga 525-8577
Japan
yoshikawa@ci.ritsumei.ac.jp

Tatsuya Sugiura
Ritsumeikan University
Kusatsu, Shiga 525-8577
Japan
qwer.all.releave.140-691ru.ta28@gmail.com

Seiji Sugiyama
Ritsumeikan University
Kusatsu, Shiga 525-8577
Japan
seijisan@is.ritsumei.ac.jp

ABSTRACT

This paper describes current status of our effort to develop a robot system that can play Jenga game against human players. Unlike most of the previous Jenga robots that use grippers, this robot is equipped with a hand with two multiarticulated fingers covered by soft skin and does not have any major mechanical constraint in playing the game in a natural way as human players do. An experimental result of a game played between the robot and a human player is presented.

Categories and Subject Descriptors: I.2.9 [Robotics]: Manipulators

General Terms: Experimentation

Keywords: robot, manipulator, Jenga, multifingered hand

1. INTRODUCTION

Recently various game playing robots have been developed. Among them, there are several works on robots that play Jenga [1] [2] [3]. They all use either grippers that are custom-made for Jenga or simple grippers available on the market.

In this paper, we use a robot manipulator having a versatile hand with two multiarticulated fingers which we have been using for research of advancing grasping and manipulation skill of multifingered hands [4] [5] . Note that we have reported on a preliminary study of this system in a previous paper [6]. Since then several major improvements have been implemented: (1) Time needed for visual information processing was made shorter by replacing the omnidirectional camera by a simple monocular camera and by reducing the number of necessary images, (2) ordinary Jenga blocks commercially available are now being used for experiment due the the thinner fingertips we are using now (in [6] blocks made of two regular blocks bonded together were used, and (3) the fingers grasp a block from the side instead of grasping it from the top, making it possible to play with higher Jenga tower (In [6] playing with only three level towers were possible). Results of these improvements are presented in this paper.

HRI'12, March 5–8, 2012, Boston, Massachusetts, USA.
ACM 978-1-4503-1063-5/12/03.

2. TASKS NEEDED IN PLAYING JENGA

The tasks needed in playing the game can be divided into the following five basic tasks.

1. Recognition of the shape of the tower.
2. Selection of the block to be removed.
3. Performing the motion of removing.
4. Selection of the position for putting the removed block on the top level.
5. Performing the motion of placing the removed block back.

Tasks 2 and 4 correspond to the strategy and tasks 3 and 5 correspond to the skill of the game. Our approach for performing each of these tasks is explained below.

2.1 Recognition of Block Positions

The block positions in the tower is recognized using the images taken by a monocular camera attached to the base of the hand (camera in hand).

2.2 Selection of Block for Removal

Search for candidate blocks for removal is done based on the consideration that the tower should not fall after removing the block. The basic idea here is that a sufficient condition for the stability of the tower is that the vertical line going through the center of gravity of the set of all blocks from the top level to a certain level k (counted from the bottom) intersects with the support region defined by the smallest convex set including all the upper surfaces of the blocks in level $k-1$. If there are multiple candidate blocks, then we select one block based on a evaluation function representing the degree of danger. Currently the safest block is selected according to this degree of danger, although there can be other strategies that do not select the safest one.

2.3 Selection of Place for Putting Block

Selection of the place for putting back the removed block on the top level is also done in a similar way to that of removal block selection.

2.4 Removing Motion

One difficulty in playing Jenga game is that the size of the blocks slightly differ to one another. Because of this the difficulty of removing a block varies even if the geometrical condition of the block in the tower is the same. To cope with this problem, the following scheme is adopted. After grasping the block, the angle of the grasped block is slightly tilted. By this tilting, it is expected that the the contact point will move to the outside edge of the block and the

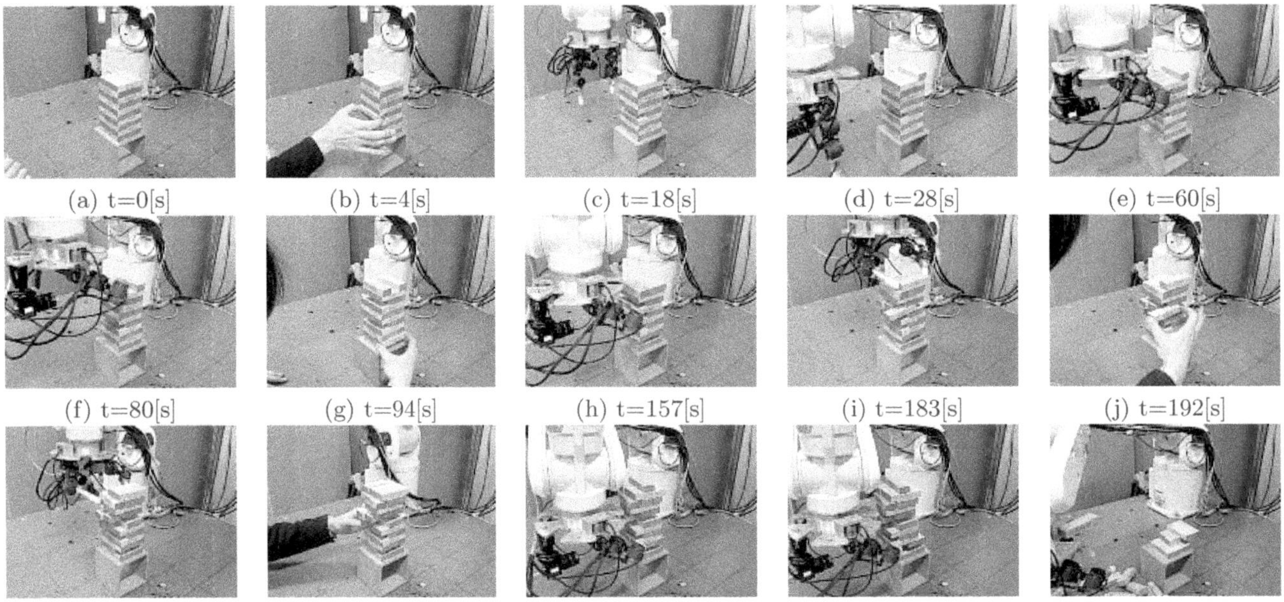

(a) t=0[s] (b) t=4[s] (c) t=18[s] (d) t=28[s] (e) t=60[s]
(f) t=80[s] (g) t=94[s] (h) t=157[s] (i) t=183[s] (j) t=192[s]
(k) t=240[s] (l) t=285[s] (m) t=339[s] (n) t=341[s] (o) t=351[s]

Figure 1: Experiment of match-up

distribution of the load among the blocks on the level of removal block will change, hopefully, in such a way that the load of the block of removal will decrease.

2.5 Putting Motion

When a human player performs the putting motion of a block on the top of the tower, it is done so that any large impact force may not be applied to the tower. One way for achieving this by the robot is to detect the contact between the block and the top surface of the top level of the tower by using the force sensors at the fingertips.

3. EXPERIMENT OF MATCH-UP

Experimental match-ups between the robot and a human player was performed using ten level tower consisting of 30 blocks. Fig.1 shows a case in which the robot performed very well. Fig.1(a) was the initial state. The human player took the turn first. The human player removed a block from level 5 (Fig.1(b)) and put it back to the top (Fig.1(c)). Then the human player signaled the completion of his turn. Note that the game proceeds automatically except for a signalling by the human player of the completion of his motion by clicking a key of a PC controller of the robot. Receiving this signal, the robot started to take two images of the tower from two neighbouring sides (Fig.1(c) and (d)) to estimate the positions of the blocks in the tower by using the camera. After repeating this process several times the robot, as its fourth move, tried to remove a block from level 2 (Fig.1(m) and (n)) but the tower collapsed (Fig.1(o)).

4. CONCLUSIONS

Although the present level of the robot system in performing the game is still very low, A variety of ideas could be tried in this system to improve the skill of the system. We believe that a robot system that has a hand similar to humans will be important for playing Jenga with human players.

5. REFERENCES

[1] T. Kröger, B. Finkemeyer, S. Winkelbach, L.-O. Eble, S. Molkenstruck and F. M. Wahl, "A Manipulator Plays Jenga," IEEE Robotics and Automation Magazine, vol.l5, no.3, pp.79-84, 2008.

[2] J. Wang, P. Rogers, L. Parker, D. Brooks, M. Stilman, "Robot Jenga: Autonomous and Strategic Block Extraction," In Proc. of The 2009 IEEE/RSJ Int. Conf. on Intelligent Robots and Systems, pp.5248-5253, 2009.

[3] S. Kimura, T. Watanabe, Y. Aiyama, "Force Based Manipulation of Jenga Blocks," In Proc. of The 2010 IEEE/RSJ Int. Conf. on Intelligent Robots and Systems, pp.4287-4292, 2010.

[4] Tsuneo Yoshikawa, "Multifingered Robot Hands: Control for Grasping and Manipulation," Annual Reviews in Control, vol.34, issue 2, pp.199-208, 2010.

[5] T. Yoshikawa, M. Koeda, and H. Fujimoto, "Shape Recognition and Optimal Grasping of Unknown Objects by Soft-Fingered Robotic Hands with Camera," Experimental Robotics: The 11th International Symposium, Springer Tracts in Advanced Robotics, pp. 537-546, Vol. 54, 2009.

[6] T. Yoshikawa, H. Shinoda, S. Sugiyama, and M. Koeda, "Jenga Game by a Manipulator with Multiarticulated Fingers," Proc. AIM2011, pp.960-965, 2011.

Robot Gesture and User Acceptance of Information in Human-Robot Interaction

Aelee Kim, Hyejin Kum, Ounjung Roh
Sangseok You
Department of Interaction Science
Sungkyunkwan University (SKKU)
Seoul, South Korea
+82 10 8640 4414
sangseok.you@gmail.com

Sukhan Lee
Department of Interaction Science
Intelligent Systems Research Center
Sungkyunkwan University (SKKU)
Gyeonggi-do, South Korea
+82 31 290 7150
lsh@ece.skku.ac.kr

ABSTRACT
This study explores how human users respond to coordinated and uncoordinated gestures of a robot as an information deliverer. A b etween-subject experiment was conducted using the Wizard of Oz method, with 63 participants randomly assigned to one of four con ditions (voice-only vs. no-gesture vs. coordinated gesture vs. unco ordinated gesture) taking an artwork class in a museum-like settin g. The robot was explaining the information of the artworks with modalities accordingly designed to each condition. Results showe d that the coordinated gesture was not aiding information delivery. However, there were notable relations between the coordinated ge sture and intimacy, homogeneity, and involvement. These results have theoretical implications for cognitive load of working memor y and practical implications for designing and deploying dynamic humanoid robots for museum tour guide.

Categories and Subject Descriptors
I.2.9 [**Artificial Intelligence**]: Robotics – *Commercial robots and applications, Kinematics and dynamics, Operator interfaces.*

General Terms
Experimentation, Theory

Keywords
human-robot interaction, robot gesture, information delivery

1. INTRODUCTION
Wheeled tour guide-robots have already been deployed in various museums worldwide. Most of the previous tour guide robots, however, focused more on the involved navigation task than on natural interaction with humans. A humanoid robot, on the other hand, offers a great potential for exploring intimate and multimodal interaction between humans and robots [1][2]. One of the most important motivations for many humanoid projects is that such robots could be capable of intuitive, multimodal interaction with people. This makes humanoid robots an ideal platform for museum guide projects, as the robots make use of multiple modalities such as speech, facial expressions and gestures.

Gesture is an important feature of social interaction, frequently used by human speakers to illustrate what they express in speech. Robot acting in a human-like way evokes social attributions that, in turn, make the human interaction partner believe that it has communicative intent. To induce such attributions, a communicative robot should generate social cues [3].

HRI'12, March 5–8, 2012, Boston, Massachusetts, USA.
ACM 978-1-4503-1063-5/12/03.

Our goal for this research was to stretch the field of inquiry by focusing especially on behavioral, cognitive, and social aspects of human-robot interaction and the social contexts surrounding human-robot interaction. The present study employed a humanoid robot Nao that we manipulated as a museum docent and that uses speech and gestures to interact with people. The main foci of the research are effectiveness of information delivery by coordinated gesture over uncoordinated gesture and an effect of robot presence on involvement in learning activity. The hypotheses of the research are as follows:

H1. Robot with coordinated gesture with the content will improve the effectiveness of the content delivery.

H2. Intimacy to the robot is highest in the coordinated gesture condition.

H3. Homogeneity with the robot is highest in the coordinate gesture condition.

H4. Involvement in the learning activity is highest in the robot with coordinated gesture condition.

Method

2.1 Participants and Independent Variables
Sixty-three college students were recruited to participate at a private university in Seoul, Korea. They were randomly assigned to one of four conditions: voice only without Nao, voice with Nao, voice with coordinated gesture, voice with uncoordinated gesture. The mean age was 23.95 years old.

2.2 Stimulus Material
The robot used in this experiment was Nao by Aldebaran Robotics, a medium-sized 57cm humanoid robot, fully programmable with embedded camera and speaker. In this research, we applied MP3 voice file for Nao to speak Korean recorded by professional voice actor.

2.3 Procedure
At the experiment room, the participants appreciate five artworks one at a time for approximately fifteen seconds per image. The works were displayed on the 70-inch LCD monitor. The artworks were selected from the modern art genre to rule out the bias from the prior knowledge and familiarity with them. In voice-only condition, the participants listened to the voice, without robot presence, explaining the artwork through the screen speaker, without Nao. Nao without gesture condition, the participants listened to the voice explaining the artwork from Nao but without gesture. Nao with coordinated gesture condition, Nao was explaining the artwork with both voice and gesture that was synchronized to the content. Nao with uncoordinated gesture condition, the participants listened to the voice from Nao that generate random motion in discordance with the content. Once the

appreciation of the works were completed, the participants left the experiment room and filled out a questionnaire, following which they were debriefed and dismissed.

Figure 1 Coordinated (left) and Uncoordinated gesture (right)

2.4 Measures

2.4.1 Recall test scores

Recall test was to measure effectiveness of information delivery by robot-supported information transfer. The participants were supposed to answer two questions in each of six artworks, questions that were about factual content from the robot's explanation of the artworks. Ten multiple-choice questions were coded as '1' if a participant's answer is correct, or as '0' if not correct.

2.4.2 Intimacy

Intimacy to the robot was measured via four items: likelihood to be friend, meaningful time with the robot, will to have personal relationship, and will to spend more time afterward [4]. They used a 7-point Likert scale, ranging from 'Not at all' to 'Very much' (Cronbach's α=0.90).

2.4.3 Homogeneity

These questions asked the degree to which participants agree to three notions: the robot is same as me, the robot acts like me, and the robot thinks like me. All questions were responded as 7-point scale, ranging from 'Strongly Disagree' to 'Strongly Agree' (Cronbach's α=0.93).

2.4.4. Involvement

This measure was identifying the degree to which participants were actively participated and concentrated on the overall learning activity. Involvement was measured via four items: emotional involvement, concentration, paying attention to the content, and perceived proactive reaction during the process. All questions had a 7-point scale, ranging from 'Not at all' to 'Very much' (Cronbach's α=0.88).

3. RESULTS

Overall, hypotheses were not supported by results. There were no differences in recall test score among all conditions. Means of 'Intimacy ($M=3.52$, $SD=1.27$, $F(3,58)=2.85$, $p<.05$)' and 'Involvement ($M=4.96$, $SD=0.98$, $F(3,56)=3.21$, $p<.05$)' were significantly higher in the uncoordinated gesture condition and commonly lowest in the coordinated gesture. Means of voice-only condition and uncoordinated gesture condition did not differ, and these two conditions marked the highest means in overall measures. 'Homogeneity ($M=2.89$, $SD=1.35$, $F(3,59)=1.26$, $p=0.29$)' showed no significant differences between among conditions, but demonstrated same directional tendency with other measures in terms of mean differences. In sum, the coordinated gesture indicated negative effect on overall measures. Thus, the results showed surprisingly reversed direction of all hypotheses, except for the recall test score.

To investigate why the coordinated gesture was in negative relationship and how these measures interact with each other, two mediation effects of coordinated were tested, using Baron and Kenny's method. Coordinated gesture had negatively affected intimacy, which led to decreased homogeneity and involvement.

4. DISCUSSION

It appears that human did not perceive robot's coordinated gesture positively in information transfer at the museum. The participants exhibited higher intimacy to the uncoordinated gesture, assuming that the participants felt more comfortable and natural when the robot tried not to imitate humanlike gesture. This unexpected phenomenon can be explained through multiple resource theory from psychology. That is, when two tasks compete for the same resource, one or both of the tasks degrades. In this research, the two tasks, the auditory information and the robot gestures might hinder the user's attention as the tasks were presented simultaneously, thus, they felt heavy load of working memory in allocating attention.

Even though this study failed to yield the positive effect of the coordinated gesture on information transfer, it still implies that there are notable relationships between coordinated gesture and intimacy, homogeneity, and involvement. If the robot failed to elicit high level of intimacy to users, it would hinder the formation of involvement of whole activity. Care should be taken to combine robot gestures with information transfer, as well as coordinated gesture itself. Users' involvement through natural gesture should be taken to designing robot-supported information transfer.

5. ACKNOWLEDGMENT

This research was supported by WCU (World Class University) program through the National Research Foundation of Korea funded by the Ministry of Education, Science and Technology (R31-2011-000-10062-0), by PRCP through NRF of Korea funded by MEST (2011-0018397), and by MKE, Korea under ITRCNIPA-2010-(C1090-1021-0008)(NTIS-2010-(1415109527))

6. REFERENCES

[1] S. Thrun, M. Beetz, M. Bennewitz, W. Burgard, A.B. Cremers, F. Dellaert, D. Fox, D. Haehnel, C. Rosenberg, J. Schulte, and D. Schulz. 2010. Probabilistic algorithms and the interactive museum tour-guide robot Minerva, *Int. Journal of Robotics Research, vol.19*, pp.972-999.

[2] F. Felix, B. Maren, G. Attila, G. Christoph, J. Dominik, S. Michael, B. Sven. 2009. The Humanoid Museum Tour Guide Robotinho, *Robot and Human Interactive Communication*, RO-MAN. The 18th IEEE International Symposium.

[3] C.Breazeal and B. Scassellati. 1999. How to Build Robots that Make Friends and Influence People. *In International Conference on Intelligent Robots and system.* IROS '99. Proceedings. IEEE/RSJ, volume 2. Pages 858-863.

[4] C. Cho and J. Leckenby. 1997. Internet-Related Programming Technology and Advertising. *Proceedings of the conference-american academy of advertising.* American academy of advertising. 1997. pages 69-79.

Establishment of Spatial Formation by a Mobile Guide Robot

Mohammad A. Yousuf, Yoshinori Kobayashi, Yoshinori Kuno, Keiichi Yamazaki*, Akiko Yamazaki**
Graduate School of Science and Engineering, Saitama University, *Department of Liberal Arts, Saitama University
255 Shimo-Okubo, Sakura-ku, Saitama 338-8570, Japan
**School of Media Science, Tokyo University of Technology
1404-1 Katakuramachi, Hachioji city, Tokyo 192-0982, Japan
{yousuf,yosinori,kuno}@cv.ics.saitama-u.ac.jp *BYI06561@nifty.com **ayamazaki@media.teu.ac.jp

ABSTRACT
A mobile museum guide robot is expected to establish a proper spatial formation with the visitors. After observing the videotaped scenes of human guide-visitors interaction at actual museum galleries, we have developed a mobile robot that can guide multiple visitors inside the gallery from one exhibit to another. The mobile guide robot is capable of establishing spatial formation known as "F-formation" at the beginning of explanation. It can also use a systematic procedure known as "pause and restart" depending on the situation through which a framework of mutual orientation between the speaker (robot) and visitors is achieved. The effectiveness of our method has been confirmed through experiments.

Categories and Subject Descriptors
H5.2. Information interfaces and presentation (e.g., HCI): User Interfaces – interaction style

General Terms
Performance, Design, Experimentation, Human Factors

Keywords
F-formation, human-robot interaction, mobile guide robot, pause and restart.

1. INTRODUCTION
Our research focus is to develop a museum guide robot that can naturally interact with the visitors at a museum. Kendon has shown that groups of people tend to form a set of systematic spatial arrangements (F-formation) in co-present face-to-face interaction, which allow all participants equal and direct access [1]. In human-human interaction, when people group themselves, they automatically form an F-formation. In human-robot interaction, if proper F-formation is not established during the initiation to talk, the robot should do some actions to establish F-formation. Goodwin showed some systematic ways (pause and restart) in which speakers obtain the attention of recipients [2]. There have been several studies about controlling the robot position during interaction with humans [3,4]. However, very few studies have revealed how a robot should behave to initiate the interaction with multiple visitors before starting explanation. In this paper, we focus on the issue of initiation to explanation in natural human-robot interaction. From the analyses about guide-visitor interaction in museums, we derive two interaction patterns for designing a mobile museum guide robot:

1) Robots should have the capability to formulate a proper spatial formation known as F-formation.
2) Robots should deploy "pause and restart" depending on the situation.

HRI'12, March 5–8, 2012, Boston, Massachusetts, USA.
ACM 978-1-4503-1063-9/12/03.

2. MOBILE GUIDE ROBOT SYSTEM
We have developed a museum guide robot system using humanoid robot Robovie-R Ver.3 (Vstone), (Overview is shown in Figure 1). Our system uses two general purpose PCs. For our vision system, we incorporate an omni-directional camera and two laser range sensors. The omni-directional camera is attached with the pole installed on the back of the robot. It can observe the target appearance. The two laser range sensors are attached with another long pole which is kept fixed just in front of the experimental area. We also attach an ellipsoid marker on the robot body, and put one of the two laser range sensors at the marker's height with the long pole. This laser range sensor detects the ellipsoidal marker to obtain the position and orientation of the robot while the other lager range sensor is used to track the position of the visitors.

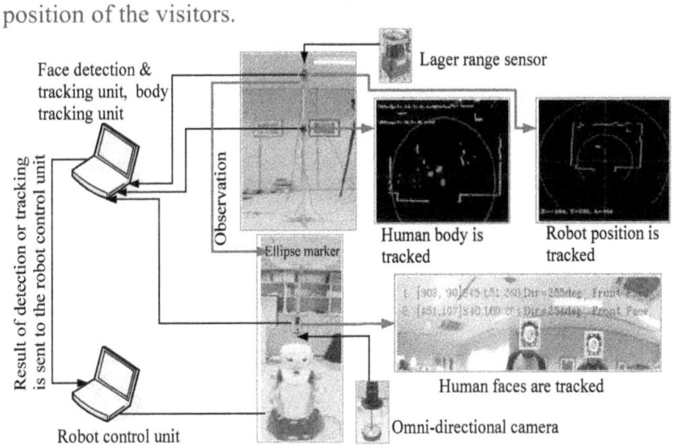

Figure 1. System overview of mobile guide robot

Our system consists of three software units: the face detection & tracking unit, the body tracking unit and the robot control unit. During the explanation of exhibits, the robot plays back predetermined bodily non-verbal actions, such as facing towards visitors, hand gestures, and pointing. We employ the particle filter [5] to track the position and orientation of visitor's body and head based on the position data and panorama image captured from the laser range sensor and the omni-directional camera respectively.

3. MODELING OF INTERACTION
For formulating a proper F-formation, the distance between the robot and the visitors (DRV), the distance between the robot and the exhibits (DRE) and FOV of the robot (RFOV) are considered. In our experiment, we consider:

- DRV: Ranges from 110 cm to 130 cm.
- DRE: Fixed as 110 cm in all cases of.
- RFOV: Set as a limit of 150^0.

Figure 2 shows the standing positions of the robot and the visitors during explanation to make an F-formation. After verifying F-formation, the robot observes the face direction of the visitors. If the visitors are not gazing at the robot at the beginning of the speaker's (robot) turn, the robot employs "pause and restart". The procedure to employ "pause and restart" is presented in Figure 3.

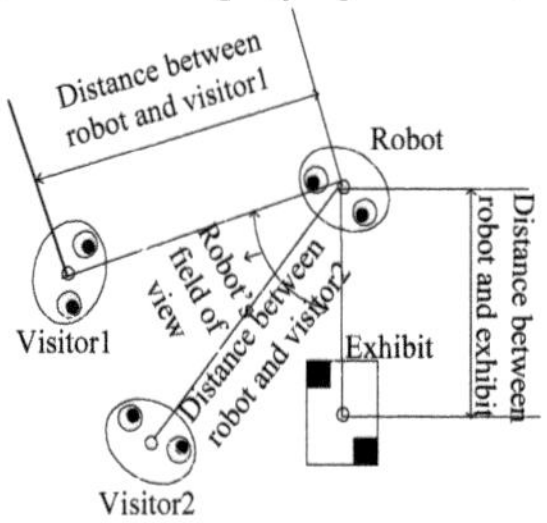

Figure 2. Position of robot and visitors to formulate F-formation

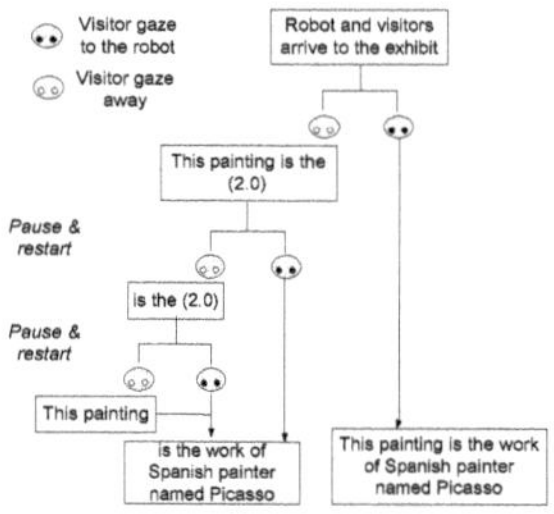

Figure 3. Procedure to implement "pause and restart

4. EXPERIMENT WITH GUIDE ROBOT

The robot based on the proposed model is compared with a robot that does not employ the proposed model.

a) Proposed Robot: The robot behaves based on the proposed model.
b) Conventional Robot: The robot does not care whether or not a proper F-formation is formed when starting explanation, and does not use 'pause and restart'.

A total of 16 people (8 groups of 2 people) participated in the experiment. Among the 8 groups, 4 groups participated in the experiment with the proposed robot and the rest 4 groups with the conventional robot. We did not inform the participants which one is our proposed robot system and which one is not. Before experiments, we gave them a brief explanation of F-formation and 'pause and restart.' The robots were programmed to explain two paintings to the visitors. Initially the robot stands in the middle of the experimental area. The proposed robot behaved as follows:

1) The robot guides them to the 1st exhibit.
2) The robot verifies whether F-formation is formed or not.
3) The robot verifies the face direction of the visitors. If face direction is not towards the robot, the robot starts the explanation with "pause and restart".
4) The robot and the visitors move to the next exhibit and repeat steps (2) and (3).
5) The robot returns to its initial position.

The conventional robot skipped steps (2) and (3).

Figure 4 shows an experimental scene using the proposed method. The robot guides the visitors to the 1st exhibit (Figure 4(a)). After verifying a proper F-formation and face direction of the visitors, the robot starts explanation (Figure 4(b). The red circle indicates that a proper F-formation is formed). After explaining, the robot moves to the 2nd exhibit. However, it finds that the visitors are still looking at the 1st painting. Then it turns its head towards the visitors (the red arrows in Figure 4(c)). Mutual gaze is established (the red arrows in Figure 4(d)). After arriving to the 2nd painting, the robot again verifies the F-formation (the red circle in Figure

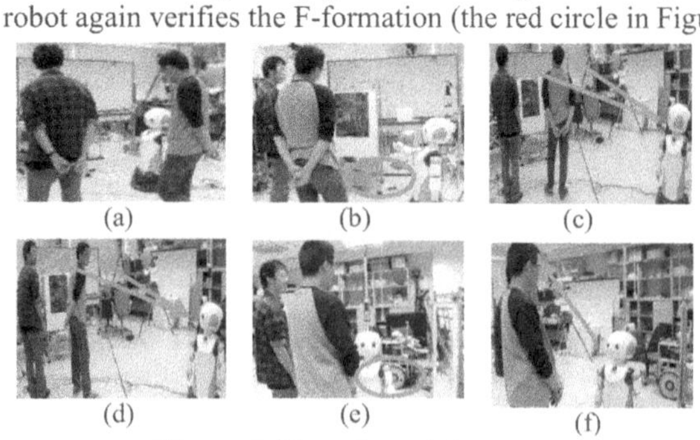

Figure 4. Experimental scene

4(e)) and face direction of the visitors (the red arrows in Figure 4(f)). In Figure 4(f), the robot employs "pause and restart" in this particular experiment as it notices that the visitors' face directions are not towards the robot.

5. RESULTS

After the experiments, the participants were asked to subjectively rate the robot's effectiveness on a seven-point Likert scale. The questionnaires items are as follows:

1) Evaluating the effectiveness of F-formation system.
2) Evaluating the effectiveness of "pause and restart".
3) Overall evaluation about the robot.

Overall evaluation: Figure 5 (a) shows the results of overall evaluation of the guide robots. The t-test shows a statistically significant difference between the two conditions ($t(14) = 3.15$, $p=0.0070$).

Effectiveness of F-formation: Figure 5(b) shows the result of effectiveness of F-formation. The t-test shows a statistically significant difference between the two conditions ($t(14) = 3.69$, $p=0.0024$).

Effectiveness of "pause and restart": Figure 5(c) shows the result of effectiveness of "pause and restart". The t-test shows a statistically significant difference between the two conditions ($t(14) = 6.27$, $p=0.0001$).

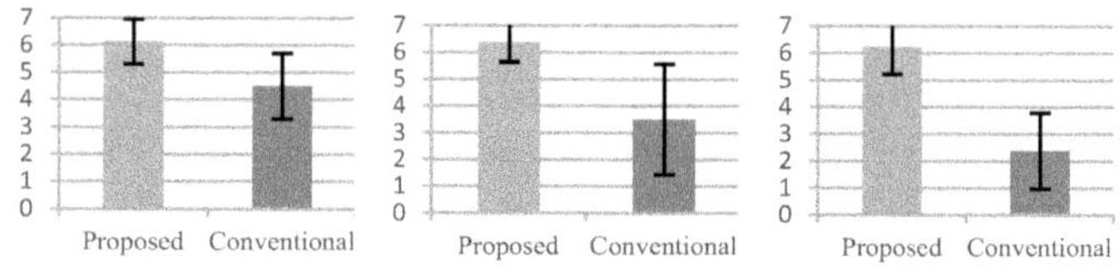

Figure 5. Experimental results. (a) Overall Evaluation (b) Effectiveness of F-formation (c) Effectiveness of "pause and restart"

6. CONCLUSIONS

We developed a museum guide robot system that is able to move from one exhibit to another in a museum gallery, to appropriately formulate an F-formation system, and also to employ "pause and restart" if necessary. We believe that the capabilities to naturally formulate a spatial formation and to use "pause and restart" are indispensable for social robots.

7. REFERENCES

[1] Kendon, A. 1990. *Conducting Interaction–Patterns of Behavior in Focused Encounters*. Cambridge University Press.

[2] Goodwin, C. 1980. Restarts, Pauses, and the Achievement of a State of Mutual Gaze at Turn Beginning, *Sociological Inquiry*, 50, 3-4, 272-302.

[3] Michalowski, M. P., Sabanovic, S., and Simmons, R.: A Spatial Model of Engagement for a Social Robot. 2006. In *IEEE International Workshop on Advanced Motion Control*, 762-767.

[4] Hüttenrauch, H., Eklundh, K. S., Green, A., and Topp, E. A. 2006. Investigating spatial relationships in human-robot interactions. *In IEEE/RSJ International Conference on Intelligent Robots and Systems (IROS2006)*, 5052-5059.

[5] Isard M. and Blake A.1998. Condensation – conditional density propagation for visual tracking, *International Journal of Computer Vision*, 29, 1, 5-28.

Navigating in Public Space: Participants' Evaluation of a Robot's Approach Behavior

Jakub Złotowski, Astrid Weiss, Manfred Tscheligi
ICT&S Center, University of Salzburg
Salzburg, Austria
firstname.lastname@sbg.ac.at

ABSTRACT

The results from an empirical study on the impact of a robot's approach trajectories on its social acceptance are presented. An online survey presenting short videos of a robot (IURO - Interactive Urban RObot) approaching a person in a public space and asking for help was shown to the users. IURO either approached a walking or standing person. The results show that walking participants preferred to be approached from the front left or front right direction rather than frontally. However, when they are standing all three approach directions were acceptable.

Categories and Subject Descriptors

A.m [**Miscellaneous**]: Social Robots

Keywords

social robot, public space, approach behaviour

1. INTRODUCTION

In the context of robots performing tasks in public space the group of potential inexperienced users is big, namely every bystander encountered on the way. In the IURO project we are exploring a scenario in which a robot without any map knowledge (no Internet or GPS access) has to navigate from one point in a city to another, using only information acquired from encountered strangers on the street. In that scenario it is necessary that the robot will be accepted in public space, as otherwise its chances of receiving help will decrease. In the future, where there will be multiple robots present in the environment and the novelty effect is no longer present, a robot requiring help from a stranger should be able to encourage the interaction and initiate it in a socially acceptable way. In the presented study we wanted to learn more about people's preferences for IURO's approach direction in public space.

2. RELATED WORK

In robotic research, people for long time were only perceived as obstacles that must be avoided by robots while navigating in populated areas. This has changed in the recent years with the development of social robots. It was shown that a seated person prefers to be approached by a robot in a fetch-and-carry task from the front left or right direction rather than frontally or from behind [1, 4]. Further research shown that there are other mediating factors, which can impact this preference, such as a person's experience with robots [2], gender [1] or in which part of the room she was standing or sitting [4]. This high context-dependence of approach behavior preferences emphasizes the need for validating these results in public space, which might differ from the previously explored domestic context. Satake et al. [3] proposed an approach model for robots which should initiate interaction in a shopping mall. However, while this model was successful in improving the robot's performance in initiating conversation, little is known whether people felt comfortable while being approached frontally.

A walking pedestrian approached frontally by an unknown robot might be afraid that it will not stop and collide with him. However, if a pedestrian is standing he could step aside in case of emergency. Therefore, we hypothesised that:

- H_1: People walking towards a robot will prefer to be approached either from the front left or front right rather than frontally.
- H_2: People standing in a public space will like frontal approach of a robot equally well as front left or front right trajectories.

3. METHOD

In order to have highly controlled conditions we have used a video-based method as proposed by Woods et al. [5], since it proved to provide reliable results for the hypotheses that we wanted to test. Short pre-recorded videos (approx. 15 sec) with audio of IURO's interaction with a human actor were used in the experiment (Figure 1). In each of these videos the robot approached the person from the (a) front left, (b) front right and (c) frontal direction and asked for the way to the assembly hall. After the actor showed the direction to IURO the video clip stopped. Furthermore, each of the above described approach trajectories was filmed in two conditions: (1) the approached person was walking and (2) the approached person was standing still. Thus, in total there were six videos. The actor was our colleague and he followed a prepared script.

4. PROCEDURE

The online survey with the videos was distributed via email on two scientific mailing lists, and to employees and students of a university via an online newsletter. It had a

HRI'12, March 5–8, 2012, Boston, Massachusetts, USA.
ACM 978-1-4503-1063-5/12/03.

Figure 1: napshot from the video showing the robot approaching a person.

one month run time (June 2011). The first part consisted of demographic questions, followed by 3 videos in which IURO approached either a walking (dynamic condition) or standing (static condition) person and asked him for the directions to the main assembly hall. The allocation was counterbalanced by chance. After each video which was displayed, the participants were asked to fill in a 5-point Likert scale (1 - strongly disagree, 5 - strongly agree) to state whether they liked the presented approach trajectory of IURO.

5. RESULTS

Overall, 182 participants answered the survey (94 participants saw the videos in the static and 88 participants in dynamic condition). There were significantly more male (n=118) than female (n=64) participants. The mean age of the participants was 33.42 years. Participants represented 41 countries. Over 60% of the participants have never been working in the field of HRI and never participated in a study in that field. However, 84% (n=153) of the participants were interested or very interested in technology in general and 73.1% (n=133) in robotics in particular.

In the dynamic condition, when the robot approached a walking person, Mauchly's test indicated that the assumption of sphericity had been violated [χ (2) = 16.05, p = .00] and the Greenhouse-Geisser correction has been applied for the ANOVA's degrees of freedom. There was a significant effect in the liking of the approach trajectories, F (1.71, 148.68) = 4.35, p = .02. Post-hoc comparisons showed that participants preferred when the robot approached a walking person from the front left side (M = 3.46, SD = .92) than frontally (M = 3.18, SD = .98) p = .049 and front right side (M = 3.51, SD = .8) than frontally p = .01. As we assumed in H_1, these results indicate that walking people should be approached from one of these sides, but not frontally.

In the static condition, Mauchly's test indicated that the assumption of sphericity had been met and no correction has been used. In contrast to the dynamic condition, when the robot approached a standing person, there was no significant effect in the preference of approach trajectories, which confirmed our H_2. In other words, people liked it the same when the robot approached them frontally or from a side.

The ANOVA revealed no significant relationship between the robot's approach behavior preferences and participants' age, experience with robots, gender or interest in technology.

6. CONCLUSIONS

The online survey presented in this paper gave insights on the approach behavior of the IURO robot. This study further emphasized that approaching a human in a socially acceptable manner in public space is not a trivial task in HRI. A robot that wants to initiate an interaction with a stranger can impact its chances of receiving help even before it asks for help.

Based on the presented results, we could derive implications for the IURO's approach behavior in public space. If a person is standing in a public space, a robot can approach from any of the directions. This result goes in line with previous findings that a person standing in the middle of a room finds it acceptable to be approached frontally [4]. However, in case a robot approaches a walking person, it should come either from the front left or front right direction.

In this study we have used a video based method and we are planning to evaluate the presented findings with a real robot in public space. Furthermore, there was only one robot and actor used in the study and the results cannot be generalized until they are replicated with other robotic systems. Finally, there are other aspects of approach behavior, such as proxemic behavior or approaching a group of people, which can impact robots acceptance in public space and should be addressed in future studies.

7. ACKNOWLEDGMENTS

This work was conducted within the IURO-project funded under the Seventh Framework Programme of the EU. The authors would like to thank Ewald Strasser for discussions on the study setup and Daniel Carton for help with the video material preparation.

8. REFERENCES

[1] K. Dautenhahn, M. Walters, S. Woods, K. Koay, C. Nehaniv, A. Sisbot, R. Alami, and T. Siméon. How may i serve you : a robot companion approaching a seated person in a helping context. In *Proceedings of the 1st ACM SIGCHI/SIGART conference on Human-robot interaction*, pages 172–179. ACM, 2006.

[2] K. Koay, E. Sisbot, D. Syrdal, M. Walters, K. Dautenhahn, and R. Alami. Exploratory study of a robot approaching a person in the context of handing over an object. *Proc. of AAAI-SS on Multi-disciplinary Collaboration for Socially Assistive Robotics*, pages 18–24, 2007.

[3] S. Satake, T. Kanda, D. F. Glas, M. Imai, H. Ishiguro, and N. Hagita. How to approach humans : strategies for social robots to initiate interaction. In *Proceedings of the 4th ACM/IEEE international conference on Human robot interaction*, HRI '09, pages 109–116, New York, NY, USA, 2009. ACM.

[4] M. L. Walters, K. Dautenhahn, S. N. Woods, and K. L. Koay. Robotic etiquette: results from user studies involving a fetch and carry task. In *Proceedings of the ACM/IEEE international conference on Human-robot interaction*, HRI '07, pages 317–324, New York, NY, USA, 2007. ACM.

[5] S. Woods, M. Walters, K. Koay, and K. Dautenhahn. Methodological issues in hri: A comparison of live and video-based methods in robot to human approach direction trials. In *Robot and Human Interactive Communication, 2006. ROMAN 2006. The 15th IEEE International Symposium on*, pages 51–58. IEEE, 2006.

Generation of Nodding, Head Tilting and Eye Gazing for Human-Robot Dialogue Interaction

Chaoran Liu
ATR Intelligent Robotics and Communication Labs
Kyoto, Japan
chaoran.liu@atr.jp

Carlos T. Ishi
ATR Intelligent Robotics and Communication Labs
Kyoto, Japan
carlos@atr.jp

Hiroshi Ishiguro
ATR Hiroshi Ishiguro Labs
Kyoto, Japan
ishiguro@sys.es.osaka-u.ac.jp

Norihiro Hagita
ATR Intelligent Robotics and Communication Labs
Kyoto, Japan
hagita@atr.jp

ABSTRACT

Head motion occurs naturally and in synchrony with speech during human dialogue communication, and may carry paralinguistic information, such as intentions, attitudes and emotions. Therefore, natural-looking head motion by a robot is important for smooth human-robot interaction. Based on rules inferred from analyses of the relationship between head motion and dialogue acts, this paper proposes a model for generating head tilting and nodding, and evaluates the model using three types of humanoid robot (a very human-like android, "Geminoid F", a typical humanoid robot with less facial degrees of freedom, "Robovie R2", and a robot with a 3-axis rotatable neck and movable lips, "Telenoid R2"). Analysis of subjective scores shows that the proposed model including head tilting and nodding can generate head motion with increased naturalness compared to nodding only or directly mapping people's original motions without gaze information. We also find that an upwards motion of a robot's face can be used by robots which do not have a mouth in order to provide the appearance that utterance is taking place. Finally, we conduct an experiment in which participants act as visitors to an information desk attended by robots. As a consequence, we verify that our generation model performs equally to directly mapping people's original motions with gaze information in terms of perceived naturalness.

Categories and Subject Descriptors

I.2.9 [**Robotics**]: Operator interfaces.

General Terms: Algorithms, Performance, Experimentation, Human Factors, Languages, Verification.

Keywords: Head motion; dialogue acts; eye gazing; motion generation.

1. INTRODUCTION

To allow smooth dialogue communication between humans and robots, besides the verbal (linguistic) information, nonverbal information such as head motion is also important for expressing paralinguistic information such as intentions, attitudes and emotions, and for increasing the robot's perceived lifelikeness.

Head motion naturally occurs during speech utterances, and can be either intentional or unconscious. In the former case, head motion may carry clear meanings in communication. For example, nods are frequently used for expressing agreement, while headshakes are used for expressing disagreement. Most of the time, however, head motion is unconsciously produced. Regardless of this difference, both types of motions are somehow synchronized with the speech utterances and transmit nonverbal information.

One of our motivations for the present work is to obtain a method for generating head motion from the speech signal in order to automatically control the head motion of a tele-operated humanoid robot (such as an android). Thus, the lifelikeness of a robot could be increased by imitating a human's natural head motion, and smooth human-robot communication can be expected.

In our previous work [1], we analyzed several sessions of free dialogue conversation between Japanese speakers, and found a strong relationship between head motion and dialogue acts (including turn taking functions). Nods occurred frequently during dialogue speech, not only for expressing dialogue acts such as agreement and affirmation, but also as indicative of syntactic or semantic units, appearing at the last syllable of the phrases, with strong phrase boundaries. At weak phrase boundaries where the speaker is thinking or indicates that he/she did not finish his/her speech, head tilts were frequently observed. Also, a rule-based nod generation model was proposed and evaluated.

In the present work, we extend our study of head motion generation by adding a head tilting generation model, based on human head motion analysis results. We then evaluate the proposed model in two types of humanoid robots: a female android, "Geminoid F", and a typical humanoid robot with less degrees of freedom, "Robovie R2".

The effects of an additional "face up" motion during utterances are also evaluated, with the goal of reducing perceived unnaturalness in robots that do not have a mouth (i.e., movable lips).

The rest of the paper is organized as follows. Section 2 describes past related works. In Section 3, a rule-based nodding/head tilting generation is described, and evaluated in two humanoid robots. In Section 4, a motion which can suggest utterance in robots which do not have movable lips is described and evaluated using a humanoid robot. Section 5 investigates the effects of gaze in our model and in mapping people's motions. Section 6 provides discussion and Section 7 concludes the paper.

2. RELATED WORK

Head motion analyses can focus on either of two problems: how to recognize a user's head motion and interpret its role in communication (e.g. [2-3]); or, how to generate a robot's head motions in synchrony with the robot's speech. The present work focuses on the latter problem of generating natural head motion while the robot is speaking.

Many studies have tried to find a correspondence between head motions and prosodic features, such as the fundamental frequency (F0) contours (which represents pitch movements), and energy

HRI'12, March 5–8, 2012, Boston, Massachusetts, USA.

contours, in several languages [4-9]. For example, emphasis of a word often goes along with head nodding, and a rise of the head can correspond with a rise in voice, in English [7]. For Swedish, words with focal accent are accompanied by a greater variation of the facial parameters (including head motions) than words in non-focal positions, in all expressive modes [8]. However, it is reported in [4] that, correlation among F0 and the 6DOF (rotation and translation) of the head motions was between 0.39 and 0.52 for English speakers, and between 0.22 and 0.30 for Japanese speakers. This shows that the use of F0 alone is not enough to generate head motions, and that the correspondence between F0 and head motions is language dependent. Also, most of these studies analyze read speech or acted emotional speech data. In [10], relations between head movements and the semantics of utterances are analyzed in Japanese spoken dialogue, by also considering speaking turn and speech functions. Regarding head motion generation, a system that uses corpus-based selection strategies to specify the head and eyebrow motion of an animated talking head is presented in [11]. The system considers syntactic, prosodic and pragmatic context for generating the motions. However, only data for one speaker was analyzed.

In [12], strong relationships between head motion and dialogue acts (including turn taking functions), and between dialogue acts and prosodic features, were reported. In the present work, we focus on the relationship between dialogue acts and head motion.

Dialogue act tags have been annotated for each phrase in a database of dialogue between several pair of speakers, according to the following set, based on the tags proposed in [13], taking into account dialogue acts such as affirmative or negative reaction, expression of emotions like surprise or unexpectedness, and turn-taking functions.

- **k** (keep): the speaker is keeping the turn; a short pause or a clear pitch reset is accompanied at strong phrase boundaries.
- **k2** (keep): weak phrase boundaries in the middle of an utterance (when no pause exists between phrases).
- **k3** (keep): the speaker lengthens the end of the phrase, usually when thinking, but keeping the turn (may or may not be followed by a pause).
- **f** (filler): the speaker is thinking or preparing the next utterance, e.g., "uuun", "eeee", "eettoo", "anoo" ("uhmmm").
- **f2** (conjunctions): can be considered as non-lengthened fillers, e.g., "dakara", "jaa", "dee" ("I mean", "so").
- **g** (give): the speaker finished talking and is giving the turn to the interlocutor.
- **q** (question): the speaker is making a question or asking for a confirmation to the interlocutor.
- **bc** (backchannels): the speaker is producing backchannels (agreeable responses) to the interlocutor, e.g., "un", "hai" ("uh-huh", "yes").
- **su** (admiration/surprise/unexpectedness): the speaker is producing an expressive reaction (admiration, surprise) to the interlocutor's utterances, e.g., "heee", "uso!", "ah!" ("wow", "really?")
- **dn** (denial, negation): E.g., "iie", and "uun" accompanied by a fall-rise pitch movement ("no", "uh-uh").

Among the analysis results in [1] for the relationship between head motion and dialogue acts based on dialogue between several pairs of speakers, it was shown that nodding is the head motion that most frequently occurs during dialogue speech, and that it frequently appears in backchannels (bc) and at the last syllable of strong phrase boundaries (k, g, q). At the weak phrase boundaries where the speaker is thinking, embarrassed, or indicates his/her speech utterance has not been concluded (f, k3), head tilting was frequently observed.

In the present work, we exploit these analysis results for creating a head motion generation model.

3. EVALUATION OF HEAD TILT GENERATION

Our rule-based head motion generation model for nodding was extended to include head tilting based on analysis results in [1], and the effects of head tilting control in two types of humanoid robot were evaluated.

3.1. Proposed method for head motion generation

In this proposed model, nods are generated in the center of the last syllable of utterances with strong phrase boundaries (k, g, q) and backchannels (bc), as in our previously proposed nod generation model [1], while head tilts are generated in the weak phrase boundaries where the speaker lengthens the end of the phrase or put a pause in the middle of a sentence due to disfluencies (f, k3). The utterance segmentation and the respective dialogue act tags were used for providing the timing of the nods.

Figure 1 shows examples of several shapes of nod and head tilt found in the database. The duration of the nods varied around 0.4 to 0.7 seconds. Note the presence of a slight upward motion, which often occurs before the characteristic down-up motion of nods. The duration of the head tilt samples varied in time from 0.8 to 1.5s, and was more dependent on the phrase length.

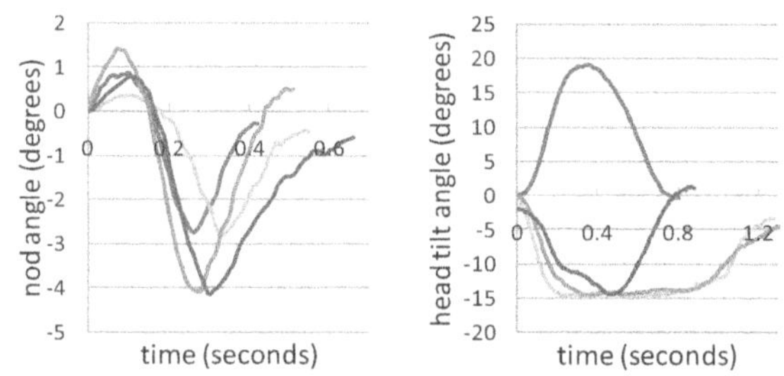

Figure 1. Examples of observed nod and head tilt shapes, extracted from the database.

Figure 2 shows representative nod and head tilt shapes, extracted from the database, which are used in the motion generation. The nodding generation model is the same used in our previous work [1].

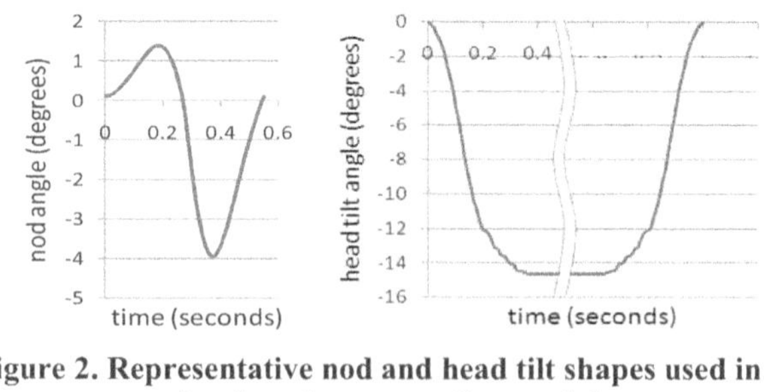

Figure 2. Representative nod and head tilt shapes used in the head motion generation model.

Fixed shapes for single nods and head tilts were used (i.e., the intensity and the duration of the nods, and the intensity of head tilts were kept the same), so that the effects of timing can be evaluated. Regarding head tilt, the tilt angle (of 15 degrees) was kept until the phrase finished, so that the length of the motion was decided according to the inter-phrase interval lengths.

3.2. Experimental Setup

Eleven conversation passages with durations between 10 to 20 seconds, including fillers and turn keeping functions (f and k3), were randomly selected from our database, and rotation angles (nod, shake and tilt angles) were extracted for each utterance. The duration of the conversation passages was limited to 10 to 20 seconds, since subjects have to compare a pair of motions for the same speech utterances, and such comparison would be difficult if each stimulus is too long. Also, as a dialogue act is attributed for each phrase unit, utterances of 10 to 20 seconds usually contain more than 10 phrases, i.e., some context information is still present in the dialogue passage.

Head rotation angles were computed by the head motion generation model described in the previous sub-section (NOD&TILT) for each conversation passage.

For comparison, we prepared two types of motion. One is a reproduction of the head rotations extracted from the original motion capture data (ORIGINAL). Another is the nod only motion proposed in [1] (NOD ONLY).

The motions were generated in two types of robots; one is a female android robot (Geminoid F) and the other is a humanoid robot (Robovie R2), as shown in Figure 3.

Figure 3. External appearance of Geminoid F and Robovie R2.

Robovie R2 has three degrees of freedom for its head, so that the rotation angles could be directly mapped to the actuator commands by a linear mapping. Geminoid F also has three actuators for the head (as shown in Figure 4), but their correspondence to the three rotation angles is not as straight-forward as for Robovie R2. A mapping function between the rotation angles and the android actuator commands proposed in our previous work was therefore applied [12].

Commands are sent each 20 ms for Geminoid F, and each 100 ms for Robovie R2, due to constraints in the current robot's hardware, but these rates are thought to be enough for head motion control purposes. Furthermore, for the android, the lip motion (jaw lowering motion) was reproduced from the distances between the nose and chin markers in the original motions.

Video clips were recorded for each stimulus, resulting in 33 stimuli (11 conversation passages and 3 motion types) for each robot type. For each trial, subjects saw a sequence of two videos with different head motion control types for the same conversation passage, so that they could compare the effects of changing the head motion control.

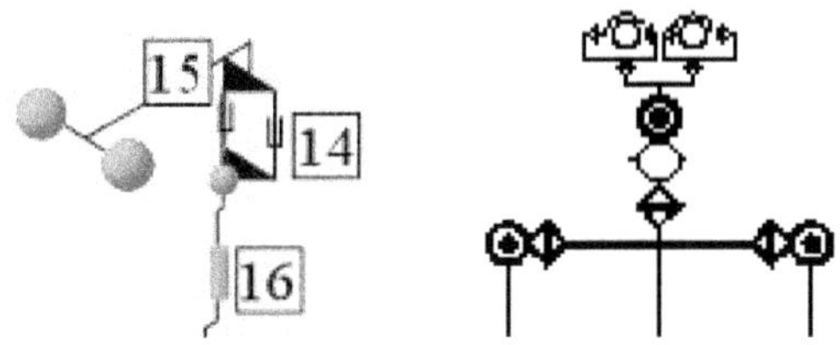

Figure 4. Head actuators for Geminoid F and Robovie R2.

Pairs of stimuli were presented to subjects in the following order:

- **NOD ONLY vs. NOD&TILT**
- **NOD&TILT vs. ORIGINAL**
- **NOD ONLY vs. ORIGINAL**

The video pairs were sorted in random order.

Subjects were asked to rate the naturalness of the motion for each stimuli, and preference scores for each pair of stimuli, according to the following questionnaire. Only one item was allowed to be chosen for each question.

- **Is the motion of the robot natural?**

clearly unnatural (1)
unnatural (2)
slightly unnatural (3)
difficult to decide (4)
slightly natural (5)
natural (6)
clearly natural (7)

- **Comparing the two stimuli, which one is more natural?**

the first is clearly more natural
the first is more natural
the first is slightly more natural
difficult to decide (both are natural or unnatural)
the second is slightly more natural
the second is more natural
the second is clearly more natural

38 paid subjects (18 male and 20 female) participated in the experiment. All subjects were not involved in robotics research, and ages ranged from 18 to 60 (average=34, stdev=13).

3.3. Experimental Results

Figure 5 shows average subjective naturalness scores for individual motion types, and for each robot type. Subjective naturalness was quantified using a scale of 1 to 7.

To understand the significance of these scores, a repeated measures analysis of variance (ANOVA) was conducted. A significant main effect was found ($F(2,36) = 26.152$, $p < 0.0005$ for Geminoid F and $F(2,23) = 19.109$, $p < 0.0005$ for Robovie R2).

Regarding the naturalness of individual motion, NOD&TILT was judged to be the most natural, while NOD ONLY and ORIGINAL obtained lower scores in both robot types. Possible reasons for these results are discussed in Section 3.4.

Comparing the two graphs in Figure 5, for all three motion types, scores for Geminoid F were higher than for Robovie R2. As quite similar motion was reproduced in the two types of robots, we believe the difference of external appearance causes this discrepancy. This is discussed in more detail in Section 3.4.

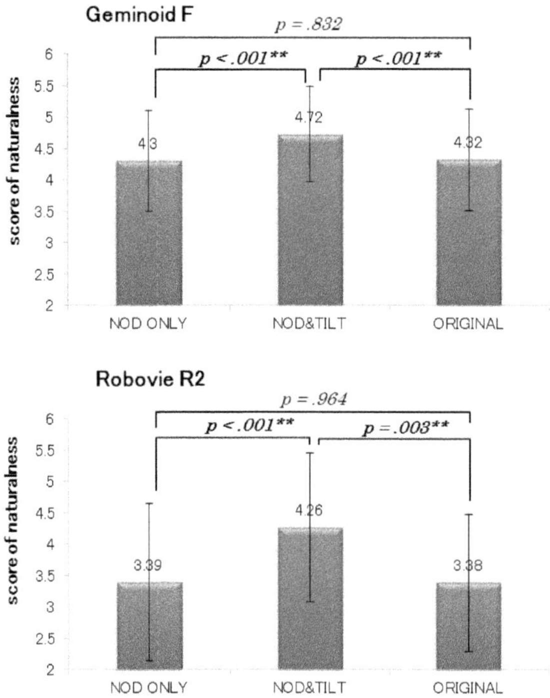

Figure 5. Subjective naturalness for each motion type (standardized value), for Geminoid F (top) and Robovie R2 (bottom). Error bars indicate Standard Deviation.

For Geminoid F, comparison between "NOD ONLY" and "NOD&TILT" shows that "NOD&TILT" (single nods plus head tilts with adequate timing) is significantly more natural than "NOD ONLY" ($p < 0.0005$), indicating the effectiveness of the head tilt motion generated in the specific weak phrase boundaries. Comparison between "NOD&TILT" and "ORIGINAL" showed that "NOD&TILT" model is more natural than "ORIGINAL" ($p < 0.0005$). For "NOD ONLY" and "ORIGINAL", it was expected that "ORIGINAL" would have higher preference scores. However, the results do not show a significant difference between these two cases. This means that simply trying to reproduce the original motion does not necessarily imply that natural motion will be generated.

For Robovie R2, we can observe that there were similar overall tendencies to the results for Geminoid F. "NOD&TILT" is evaluated significantly higher than "NOD ONLY" and "ORIGINAL" (*"NOD&TILT v.s. "NOD ONLY": $p < 0.0005$; v.s. "ORIGINAL": $p = 0.003$*).

We think a better control of the other rotation angles and the intensity of nodding may improve naturalness. However, the present results show that a "slightly natural" motion could be achieved even by using a very simple timing control model.

3.4. Further Analysis

Regarding the differences in the subjective naturalness scores between the two robot types, the results in Section 3.3 showed that when the same motion was used in both robots, the scores for Robovie R2 were lower score than for Geminoid F. One possible reason could be that Robovie R2 does not have movable lips. For the android, lip motion was reproduced from the distances between the nose and chin markers, so that the subjects can recognize the intervals of the speech utterance through both visual and auditory information. However, for Robovie R2, as head motion is not strongly related to phonetic features, it is difficult to feel that the speech is being uttered by the robot by only watching the robot's head motion. We think that this lack of visual information causes the lower scores in subjective naturalness for Robovie R2, as it does not have movable lips.

4. EVALUATION OF FACE UP MOTION FOR UTTERANCE SIGNATURE

To address the problem discussed in Section 3.4 of increasing perceived naturalness for robots without movable lips (such as Robovie R2) , we propose a slight "face up" motion (3 degrees) during speech utterance intervals, as an indication for utterance intervals. This upward motion is also often observed in natural speech, and should improve motion naturalness in a robot which does not have movable lips, such as Robovie R2.

4.1. Experimental setup

The head motion generation models "NOD ONLY" and "NOD&TILT" described in previous section 3.1 were combined with a face up motion during speech utterance intervals. We call these new generation models "NOD ONLY+" and "NOD&TILT+".

8 conversation passages were randomly selected from the previous experiment. For each conversation passage, six types of motion were generated, comprising "NOD ONLY+" and "NOD&TILT+" for Robovie R2 as well as "NOD ONLY" and "NOD&TILT" for both Robovie R2 and Geminoid F (the latter were used as baseline conditions for comparison).

Video clips were recorded for each speech segment accompanied by motion, resulting in 48 stimuli for Robovie R2 and Geminoid F. As in the previous experiment (in Section 3), subjects saw a sequence of two videos with different head motion control types for the same conversation passage, and were asked to rate the naturalness of the motion for each segment.

10 Japanese subjects (5 male and 5 female) participated in the experiment. All subjects were participants in their 20s (college students) not involved with robotics research.

4.2. Experimental Results

Figure 7 shows subjective naturalness scores for each motion type. As in the previous experiment, subjective naturalness was quantified using scale numbers of 1 to 7 where "1" was described as "clearly unnatural" and "7" was described "completely natural".

We compared the scores for motion types with upward motion (Robovie R2) against those without upward motion in both robots (the 2 baseline conditions). The middle bar in Figure 6 shows the subjective naturalness scores for the proposed "NOD&TILT+" model (with face up motion) for Robovie R2, while the left and right bars show the naturalness scores for "NOD&TILT" (without the face up motion) model for Robovie R2 and Geminoid F, respectively.

In place of lip movements synchronized to speech utterance, the face up motion during the speech utterance is expected to have a similar effect.

Analysis revealed a significant difference between "NOD ONLY+" and "NOD ONLY" for Robovie R2. However, a significant difference was not found between "NOD&TILT+" and "NOD&TILT" for the same robot. The reason for this is that the head tilt during the speech utterance served much the same purpose

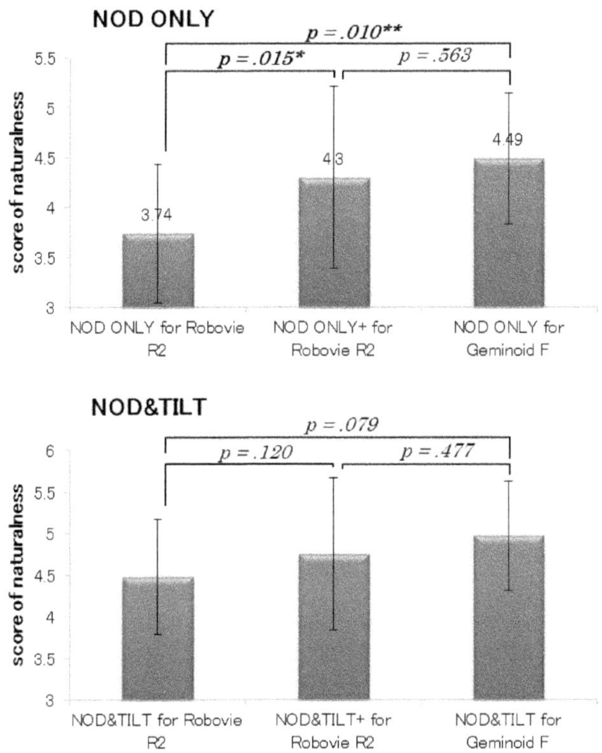

Figure 6. Subjective naturalness for NOD&TILT+, compared with NOD ONLY model and NOD&TILT model. Error bars indicate Standard Deviation.

as the face up motion by attracting the listener's attention before the final nod. It could be interesting in the future to further clarify the cause for these results and to investigate if there are other motions which can also be used in this way to improve perceived naturalness.

However, the face up motion was still not enough to make the subjective naturalness scores for Robovie R2 as high as for Geminoid F. As Geminoid F is a very human-like anthropomorphic robot, which is at first glance indistinguishable from real humans, external appearance could make the subjective naturalness results better. We plan to evaluate the effect of face up motion on other types of robots in the future.

5. EVALUATION OF HEAD MOTION AND EYE GAZING DURING HUMAN-ROBOT DIALOG INTERACTION

The results from the experiment in Section 3 left questions unanswered. First, how do subjects evaluate the models when they are interacting with a real robot and not merely watching video footage? And, why was the "ORIGINAL" motion, which we thought would be perceived as most natural, judged to be as unnatural as the "NOD ONLY" model and less natural than the proposed "NOD&TILT" model? For the latter question, the literature strongly points to the importance of gaze movements for a wide variety of interaction schemes, both in human science [15, 16] and for HRI [17, 18]. However, in the "ORIGINAL" motion, only head movements, with no gaze movement, were reproduced.

Therefore, we conducted an experiment with two purposes: to verify the goodness of our models when actual robots are used and to see if the addition of gaze movements would give the expected results for original motion.

5.1. Analysis of eye gazing

We first analyzed eye gazing in human-human dialogue conversations. Thirteen sessions of free conversations in our multimodal dialogue speech database were analyzed. The following tag set was used to annotate eye gazing based on the displayed video information from the speaker's viewpoint.

- **u**: upwards gaze;
- **d**: downwards gaze;
- **l**: to the left (from speaker's viewpoint);
- **r**: to the right;
- **ul**: up and to the left;
- **ur**: up and to the right;
- **dl**: down and to the left;
- **dr**: down and to the right;
- **no**: no gaze movement; (looking at conversation partner)

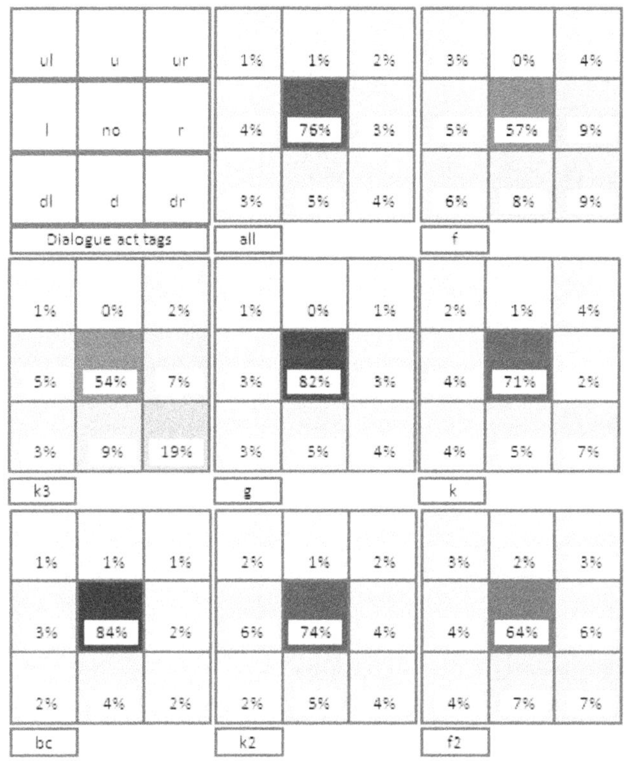

Figure 7. Distributions of the eye gaze tags for different dialogue acts: intensity for each square indicates how often speakers looked in that direction.

Figure 7 shows the distributions of the eye gaze tags for each dialogue act tag (described in Section 2), i.e., where conversation partners looked during the dialogue when speaking. The intensity of each square indicates how often speakers looked in that direction. The darker the background-color of the square the higher the frequency of that eye gaze act.

In Figure 7, it can be observed that in most cases (especially for "bc", "g", and "k"), the person speaking mostly looked directly toward the listener. However, the speaker looked away in about 50% of the cases when lengthening their previous utterance and thinking about what to say next ("f" and "k3"). In such cases, the speaker had a tendency to look downwards rather than upwards, when looking away. For "k3", speakers also looked frequently down and to the right (19% of the time).

The results in Figure 7 indicate that eye gazing has similarities with head tilting. In both cases, behavior reflects weak phrase boundaries when a speaker lengthens the end of the phrase or put a pause in the middle of a sentence due to disfluencies (f, k3). Figure 8 also shows a physical relationship (interdependency) between head tilting and eye gazing.

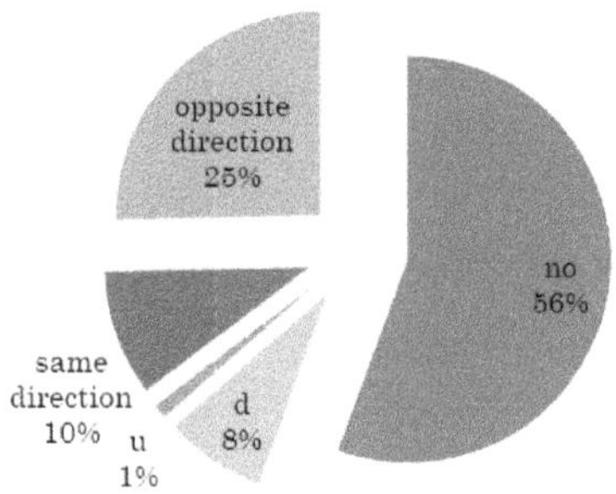

Figure 8. Gaze direction during speaker head tilting: "same direction" ("opposite direction") means gaze direction coincides (does not coincide) with the direction of the head tilt.

During head tilting, a speaker most often looked directly at the listener (56%), but also often looked in the direction opposite to the direction in which their head was tilted (25%). (E.g., if the speaker tilted their head left, they often looked to the right.) Sometimes the speaker also looked in the same direction as the head tilt (10%), or downwards (8%). The speaker rarely looked upward (1%).

The above results indicate that if a robot alters its gaze direction during a head tilt by looking in the opposite direction, this could be perceived as more natural. Based on this, we decided to also evaluate the control of gaze in addition to the head motion in the human-robot interaction experiment of the present section.

5.2. Experimental setup

In order to allow for comparison of different motion generation strategies in face-to-face human-robot interaction, we created a scenario in which subjects act as visitors to an information desk and robots play the role of receptionists. In this scenario, we can ask subjects to make the same questions to the robots, so that the effects of different motion generation strategies can be fairly compared, by using the same utterances with same lip motion and changing only the head motion and gazing strategies.

We first simulated a human-human interaction for the scenario above, and recorded audio and motion capture data for a female speaker playing the role of receptionist (collaborator). During the interaction, another speaker playing the role of the visitor asked several questions in order to induce the receptionist to naturally produce thinking behavior. The following items appeared in the interaction:

1. directions to IRC Lab.
2. what kind of research is done at IRC Lab.
3. how many staff work at IRC lab.
4. directions to the bathroom.
5. places to eat
6. Japanese restaurants
7. timetable for the bus
8. directions to Kobe
9. sightseeing spots in the vicinity
10. sightseeing spots in Nara

Two robots were used for the experiment: an android robot, Geminoid F, and a child-sized minimally-designed robot Telenoid R2 (see Figure 9). Telenoid R2, like Robovie R2, has a head rotatable about three independent axes and is less humanlike in appearance than Geminoid F and, but also has movable lips, which allows for clearer comparison of motion types.

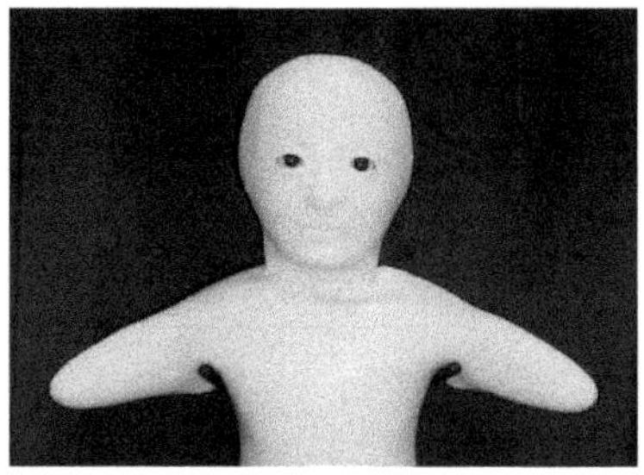

Figure 9. External appearance of Telenoid R2

At the start of the experiment, subjects were given instructions and told they would be playing the role of a visitor to an information desk where the two robots played the role of receptionists. The subjects then entered a room in which the two robots were positioned around a desk, and met and conversed with each robot one at a time. Specifically, subjects asked each robot the same questions in the list above, in a natural manner. After each question, the robot responded according to a remote control by the experimenters. This was conducted five times, for each of the five motion types described below in Section 5.4. After each session, subjects were asked to fill out a brief questionnaire. Answers for possible questions were prepared ahead of time by asking a collaborator to act out the scenario; audio, video and motion data of our collaborator were recorded and used as the original data to create the robot's answers.

Five motion control types were reproduced on the two robots.

- NOD ONLY
- NOD&TILT
- NOD&TILT with GAZING
- ORIGINAL
- ORIGINAL with GAZING

Motion control types "NOD ONLY", "NOD&TILT", and "ORIGINAL" are the same as described in Section 3 (the latter is generated by reproducing original motion from motion capture data).

During the analysis of eye gaze movements in Section 5.1, we found that when people tilt their heads, they often turn their gazes in the opposite direction. We applied this rule to generate eye gaze movements for the model 'NOD&TILT', thereby creating a new condition, "NOD&TILT with GAZING".

Eye gaze control for "ORIGINAL with GAZING" was based on annotated eye gazing tags corresponding to each moment.

For all five motion control types, lip motion was generated using a method proposed in [19], which is based on a rotation of the vowel space given by the first and second formants around the center vowel, and a mapping to the lip opening degrees.

Motion types were presented in a nearly random fashion, as we wished to have all subjects compare 3 types directly: "NOD&TILT", "NOD&TILT with GAZING", and "ORIGINAL with GAZING". Therefore, these 3 motion types were always presented together as a group either in this order, or reversed. The order of this group and the other two motion types was then determined randomly.

Subjects were asked to fill out a questionnaire for each motion control type to evaluate the naturalness of the robot's motion. The questionnaire had the following items:

- **Is the motion of the robot natural?**
- **Comparing with the previous one, which one is more natural?**

The options for the answers to these questions were the same as described in Section 3.2.

A total of 22 paid Japanese speakers (11 male and 11 female, aging from 18 to 50, average=29, stdev=11) participated in the experiment.

5.3. Experimental Results

Figure 10 shows subjective naturalness for individual motion types, and for each robot type. The results of subjective naturalness were quantified by using a 1 to 7 scale where "1" was described as "clearly unnatural" and "7" was described as "completely natural".

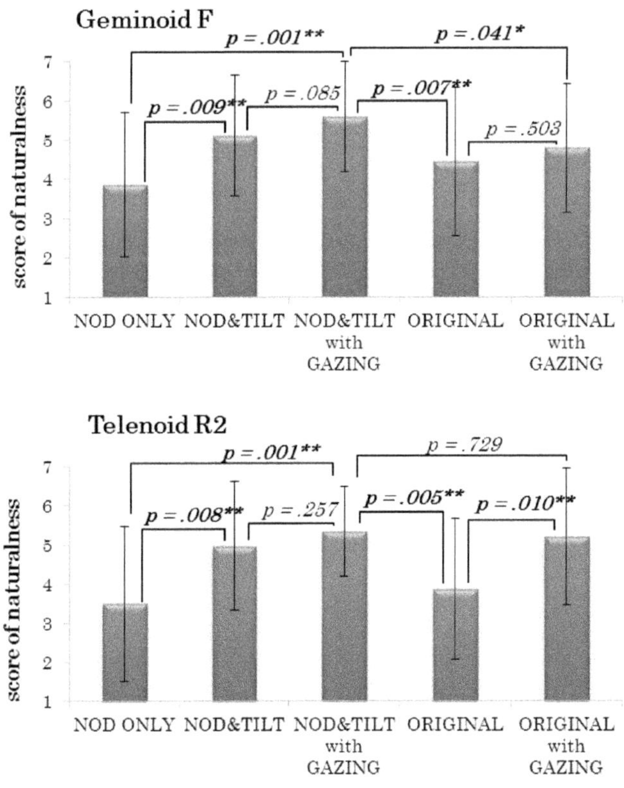

Figure 10. Subjective naturalness for each motion type and robot: Geminoid F (top) and Telenoid R2 (bottom). Error bars indicate Standard Deviation.

Comparing the performance of the motion control type using repeated measures ANOVA gave $F(4,18) = 4.451$, $p < 0.011$ for Geminoid F, and $F(4,15) = 4.558$, $p < 0.013$ for Telenoid R2.

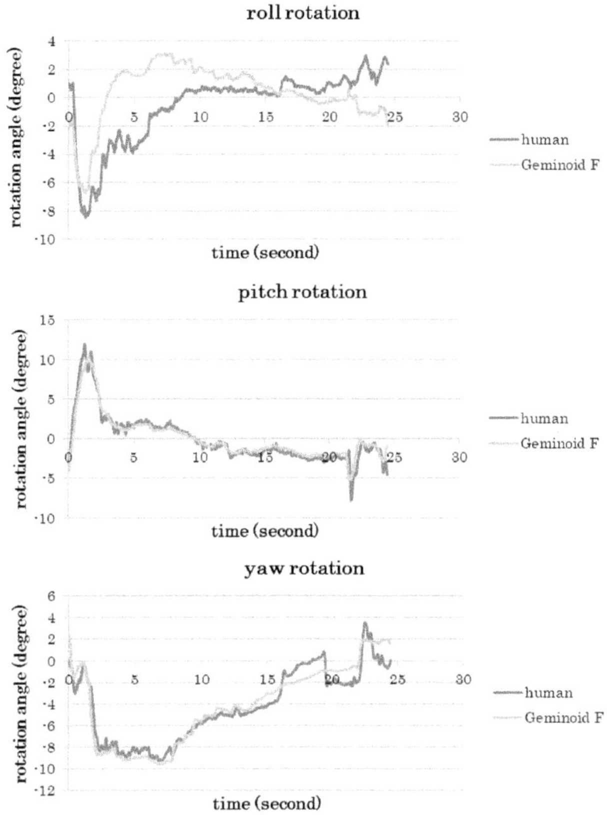

Figure 11. Motion data from human and Geminoid F

We can see that similar results were observed for motion control types "NOD ONLY", "NOD&TILT" and "ORIGINAL" as were observed previously in the video-based evaluation in Section 3.

Both for the Geminoid F and Telenoid R2, subjective scores for motion type "NOD&TILT with GAZING" are significantly higher than scores for motion types "NOD ONLY" and "ORIGINAL" (*"NOD&TILT with GAZING" v.s. "NOD ONLY": $p = 0.001$ for Geminoid F, $p = 0.001$ for Telenoid R2; "NOD&TILT with GAZING" v.s. "ORIGINAL": $p = 0.007$ for Geminoid F, $p = 0.005$ for Telenoid*), but comparison with "NOD&TILT" only shows non-significant or almost-significant differences (*$p = 0.085$ for Geminoid F, $p = 0.257$ for Telenoid R2*). This result is discussed in the next section..

For the Telenoid, "ORIGINAL with GAZING" was evaluated as highly as "NOD&TILT with GAZING" and significantly higher than "ORIGINAL" ($p = 0.010$) as expected. However, for the android, this was not the case; "NOD&TILT with GAZING" performed best. A comparisons between "ORIGINAL with GAZING" and "NOD&TILT with GAZING" shows significant difference ($p = 0.041$). We suspected that imperfect reproduction of the original motions was again the culprit. Specifically, Geminoid F cannot tilt her head about the roll axis without also rotating in the pitch axis due to the position of her head actuators (see Figure 4), causing her chin to move forwards. To confirm this, we recorded the robot's reproduced motion data using the motion capture system, and compared it with the original motion data which were captured from a human speaker. We found that pitch and yaw motions were reproduced well, but that roll motions were not accurately reproduced, causing a potential source for perceived unnaturalness (see Figure 11).

6. DISCUSSION

The effect observed from adding gaze to the developed generation models was not as great as we had anticipated. We speculate that two causes may exist. First, a person may communicate that they are thinking in different ways: a person may look to one side as if recalling something, tilt their head to one side, or do both. We think that doing both is not necessarily perceived as more natural, because it may convey only the same information (i.e., the person is thinking). Also, robots tend to be less complicated in terms of their communicative capabilities than people; due to this perception, it may be acceptable if some information (such as eye movements) which is normally present with humans is abstracted, or not present for robots. This highlights another interesting possibility; by removing extraneous, misleading or undesired communicative cues which can be observed in humans (such as involuntary twitching), robots could one day become capable of conveying information more clearly than actual humans; thereby, contributing much to service tasks in which such natural communication is an asset; e.g. teaching or theatre performances.

This study focused on exploring how generated nodding, head tilting, and gaze may contribute to the perceived naturalness of a communicating robot. Evaluation was performed both with video footage and with having participants converse face to face with the actual robots, during which three different humanoid robots were used.

However, the generality of the current work is necessarily limited by target group, choice of robot, cues for generating head motions, and the investigated motions. In this study, Japanese of various ages evaluated our robots for naturalness, but it remains to be shown if results will be exactly the same in different countries or for different cultures. Likewise, it would be interesting to investigate mechanisms for increasing perception of naturalness in non-humanoid communicative robots. Last, we can imagine that other motions during a conversation, such as posture shifts, or hand motions, could also have an interesting effect on naturalness.

7. CONCLUSIONS

A rule-based nodding and head tilting motion generation model was proposed based on dialogue act function tags, and evaluated in two humanoid robots, for utterances including thinking behaviors. Subjective scores showed that the proposed model including head tilting and nodding can generate head motion with increased naturalness compared to nodding only or directly mapping people's original motions.

Also, as an alternative for utterance motions, a "face up" motion was additionally proposed for robots without movable lips. It was shown that the inclusion of such a motion can improve the perceived naturalness of robots which do not have a mouth, such as the humanoid robot, Robovie R2.

An experiment conducted in which participants conversed with the robots face to face confirmed the validity of the previous video-based evaluation. As well, we found that where neck motions were properly reproduced in all three axes for the Telenoid that the original motions with gaze movements were perceived as most natural.

Remaining topics for future work include investigating the effects of eye gazing during head motions, and the development of methods for extracting linguistic and paralinguistic information from speech (such as phrase boundaries, disfluencies, and dialogue acts), in order to automate the generation of head motion commands.

8. Acknowledgment

This work was supported by JST CREST.

9. References

[1] C.T. Ishi, C. Liu, H. Ishiguro, N. Hagita. 2010. Head motion during dialogue speech and nod timing control in humanoid robots. Proc. *of IEEE/RSJ Human Robot Interaction* (HRI 2010), 293-300, 2010.

[2] C. Sidner, C. Lee, L.-P. Morency, C. Forlines. 2006. The effect of head-nod recognition in human-robot conversation. Proc. *of IEEE/RSJ Human Robot Interaction* (HRI 2006), pp. 290-296, 2006.

[3] L.-P. Morency, C. Sidner, C. Lee, and T. Darrell. 2007. Head gestures for perceptual interfaces: The role of context in improving recognition. Artificial Intelligence, 171(8-9): 568-585, June 2007.

[4] H.C. Yehia, T. Kuratate, E. Vatikiotis-Bateson. 2002. Linking facial animation, head motion and speech acoustics. *J. of Phonetics*, Vol. 30, pp. 555-568, 2002.

[5] M.E. Sargin, O. Aran, A. Karpov, F. Ofli, Y. Yasinnik, S. Wilson, E. Erzin, Y. Yemez, A.M. Tekalp. 2006. Combined Gesture-Speech Analysis and Speech Driven Gesture Synthesis. Proc *IEEE International Conference on Multimedia*, 2006.

[6] K.G. Munhall, J.A. Jones, D.E. Callan, T. Kuratate, E. Vatikiotis-Bateson. 2004. Visual prosody and speech intelligibility – Head movement improves auditory speech perception. *Psychological Science,* Vol. 15, No. 2, pp. 133-137, 2004.

[7] H.P. Graf., E. Cosatto, V. Strom, F.J. Huang. 2002. Visual prosody: Facial movements accompanying speech. Proc. *IEEE Int. Conf. on Automatic Face and Gesture Recognition* (FGR'02), 2002.

[8] J. Beskow, B. Granstrom, D. House. 2006. Visual correlates to prominence in several expressive modes. Proc. *Interspeech 2006 – ICSLP*, pp. 1272-1275, 2006.

[9] C. Busso, Z. Deng, M. Grimm, U. Neumann, S. Narayanan. 2007. Rigid Head Motion in Expressive Speech Animation: Analysis and Synthesis. *IEEE Trans. on Audio, Speech and Language Processing*, March 2007.

[10] Y. Iwano, S. Kageyama, E. Morikawa, S. Nakazato, K. Shirai. 1996. Analysis of head movements and its role in spoken dialogue. Proc. *International Conference on Spoken Language Processing* (ICSLP'96), pp. 2167-2170, 1996.

[11] M.E. Foster and J. Oberlander. 2007. Corpus-based generation of head and eyebrow motion for an embodied conversational agent. *Language Resources and Evaluation*, 41(3-4), 305-323, Dec. 2007.

[12] C.T Ishi, J. Haas, F.P. Wilbers, H. Ishiguro, and N. Hagita. 2007. Analysis of head motions and speech, and head motion control in an android. Proc. of *IEEE/RSJ International Conference on Intelligent Robots and Systems* (IROS 2007), 548-553, 2007.

[13] C.T. Ishi, H. Ishiguro, N. Hagita. 2006. Analysis of prosodic and linguistic cues of phrase finals for turn-taking and dialog acts. Proc. Interspeech'2006 - ICSLP, pp. 2006-2009, 2006.

[14] T. Minato, M. Shimada, H. Ishiguro, S. Itakura. 2004. Development of an android robot for studying human-robot interaction. Innovations in Applied Artificial Intelligence, Springer Verlag, pp. 424-434, 2004.

[15] M. DeBoer and A. M. Boxer. 1979. Signal functions of infant facial expression and gaze direction during mother-infant face-to-face play. Child Development, vol. 50, no. 4, pp. 1215–1218, 1979.

[16] S. R. H. Langton, R. J. Watt, and V. Bruce. 2000. Do the eyes have it? Cues to the direction of social attention. Trends in Cognitive Sciences, vol. 4, no. 2, pp. 50 – 59, 2000.

[17] F. Kaplan and V. Hafner. 2004. The challenges of joint attention. Interaction Studies, pp. 67–74, 2004. [Online]. Available: http://cogprints.org/4067/

[18] Y. Nagai, M. Asada, and K. Hosoda. 2006. Learning for joint attention helped by functional development. Advanced Robotics, vol. 20, pp. 1165–1181(17), October 2006.

[19] C. T. Ishi, C. Liu, H. Ishiguro, N. Hagita. 2011. Speech-driven lip motion generation for tele-operated humanoid robots. *International Conference on Auditory-Visual Speech Processing*, 2011

Designing Persuasive Robots: How Robots Might Persuade People Using Vocal and Nonverbal Cues

Vijay Chidambaram, Yueh-Hsuan Chiang, Bilge Mutlu
Department of Computer Sciences, University of Wisconsin–Madison
1210 West Dayton Street, Madison, WI 53706, USA
{vijayc, yhchiang, bilge}@cs.wisc.edu

ABSTRACT

Social robots have to potential to serve as personal, organizational, and public assistants as, for instance, diet coaches, teacher's aides, and emergency respondents. The success of these robots—whether in motivating users to adhere to a diet regimen or in encouraging them to follow evacuation procedures in the case of a fire—will rely largely on their ability to *persuade* people. Research in a range of areas from political communication to education suggest that the nonverbal behaviors of a human speaker play a key role in the persuasiveness of the speaker's message and the listeners' compliance with it. In this paper, we explore how a robot might effectively use these behaviors, particularly vocal and bodily cues, to persuade users. In an experiment with 32 participants, we evaluate how manipulations in a robot's use of nonverbal cues affected participants' perceptions of the robot's persuasiveness and their compliance with the robot's suggestions across four conditions: (1) no vocal or bodily cues, (2) vocal cues only, (3) bodily cues only, and (4) vocal and bodily cues. The results showed that participants complied with the robot's suggestions significantly more when it used nonverbal cues than they did when it did not use these cues and that bodily cues were more effective in persuading participants than vocal cues were. Our model of persuasive nonverbal cues and experimental results have direct implications for the design of persuasive behaviors for humanlike robots.

Categories and Subject Descriptors

H.1.2 [**Models and Principles**]: User/Machine Systems – *human factors, software psychology*; H.5.2 [**Information Interfaces and Presentation**]: User Interfaces – *evaluation/methodology, user-centered design*

General Terms

Design, Human Factors

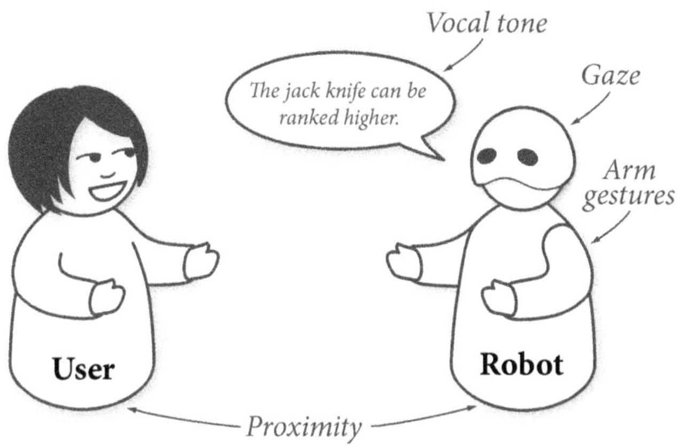

Figure 1: The vocal and bodily cues of persuasion in human-robot interaction.

Keywords

Persuasion, compliance, nonverbal immediacy, nonverbal cues, gaze, gestures, proximity, vocal tone

1. INTRODUCTION

Robots hold great promise as social actors that may positively affect and improve people's motivation and compliance in such areas as education [13], health [31], and well-being [29]. The success of these robots in motivating people will rely largely on their ability to *persuade*. But how could robots persuade people? And how can we design persuasive robots?

Research in human communication has identified a number of behavioral attributes that shape individuals' *nonverbal immediacy*—the degree of perceived bodily and psychological closeness between people—and suggested that individuals' nonverbal immediacy plays a key role in persuading others [35]. These attributes include primarily nonverbal behaviors, particularly *bodily* cues such as proximity, gaze, gestures, posture, facial expressions, touching and *vocal* cues such as vocal tone and expressions [43]. While a considerable amount of research in robotics has explored the role of nonverbal cues in human-robot interaction (e.g., [47, 5, 25]), the way in which these cues might shape the persuasive ability of a robot has not yet been studied.

A few studies in human-robot interaction have explored how robots might be designed as persuasive agents [48, 37]. While existing work highlights the importance of persuasion in human-robot interaction and provides some guidelines for robot design, how behavioral attributes of a robot might improve the robot's persuasiveness and to engender behavior changes in people has not yet ben studied. In this paper, we explore how manipulations in *bodily* cues, particularly proximity, gaze, and gestures, and *vocal* cues, specifically vocal tone and expressions, affect the persuasiveness of a humanlike robot. We designed a set of nonverbal behaviors for a humanlike robot and evaluated the effects of manipulations in these behaviors on people's perceptions of the robot's persuasiveness and their compliance with the robot in a human-robot interaction scenario (Figure 1).

In the next section, we review related work on persuasion and nonverbal behavior in human communication, related research on designing persuasive computer interfaces in human-computer interaction literature, and work to date in human-robot interaction on designing persuasive robots.

2. BACKGROUND

Persuasion, defined as "an attempt to shape, reinforce, or change behaviors, feelings, or thoughts about an issue, object, or action" [14, 15], is a key social process that enables cooperation, social influence, and attitude changes [42, 53]. An individual's persuasive ability is associated with a number of factors including the verbal and nonverbal behaviors of the individual, the dynamics of social interaction, and psychological and societal factors such as personality and social roles [46, 14, 15, 40].

Research in nonverbal communication has identified a number of nonverbal cues that shape *nonverbal immediacy*—the degree of perceived bodily and psychological closeness between people—and suggests that individuals' nonverbal immediacy is closely associated with their persuasive ability, likability, and attractiveness [35]. These cues include *bodily* behaviors such as proximity, gaze, gestures, posture, facial expressions, and touching as well as *vocal* behaviors such as vocal tone and expressions [43]. Table 1 lists the nonverbal cues that education researchers have identified as key behaviors displayed by teachers with high immediacy in the classroom.

Research results show that nonverbal immediacy and individual nonverbal cues such as gaze and proximity affect persuasion, particularly how much individuals comply with arguments and ideas presented by a speaker [8]. Mehrabian [35] argued that people's perceptions of an individual is shaped 7% by verbal cues (what the individual says), 38% by the tone of their voice, and 55% by their bodily cues, highlighting the role of nonverbal cues in how people evaluate others in social settings. In the context of public presentations, the amount of eye contact that the speaker maintains with the audience affects the persuasiveness of the speaker's message [9] and how much the audience complies with the speaker's message [6]. The proximity of a individual also affects people's compliance with requests made by the individual; closer proximity leads to higher compliance [16]. How much an individual mimics the nonverbal behaviors of others also affects their liking of the individual [10] and their perceptions of the individual's persuasiveness [50].

The persuasive effects of nonverbal immediacy have been studied extensively in research in educational psychology. These studies show that teacher nonverbal immediacy has a positive effect on the teacher's communicative effectiveness [1, 4] and perceived competence, caring, and trustworthiness [52]. Nonverbal immediacy also has positive effects on student motivation [11, 12], participation and attendance [44], and affective and cognitive learning [19]. These positive effects prevail across cultures [41], in large classes [36], and under high workload demands [39].

Behavioral cues	Cue affordances
Bodily cues	***Body***
Sits behind desk when teaching. (Reversed)	*Proximity*
Moves around the classroom when teaching.	*Proximity*
Sits on a desk or in a chair when teaching. (Reversed)	*Proximity*
Stands behind podium or desk when talking to the class. (Reversed)	*Proximity*
Gestures when talking to the class.	*Gestures*
Looks at the class when talking.	*Gaze*
Looks at board or notes when talking to the class. (Reversed)	*Gaze*
Has a very tense body position when talking to the class. (Reversed)	*Posture*
Has a very relaxed body position when talking to the class.	*Posture*
Smiles at individual students in the class.	*Facial Expressions*
Smiles at the class as a whole, not just individual students.	*Facial Expressions*
Touches students in the class.	*Touching*
Vocal cues	***Voice***
Uses monotone/dull voice when talking to the class. (Reversed)	*Vocal Tone*
Uses a variety of vocal expressions when talking to the class.	*Vocal Expressions*

Table 1: The bodily and vocal cues identified in educational settings to evaluate the effectiveness of teacher nonverbal behavior (adapted from [43]).

Research in human-computer interaction (HCI) has explored how principles of persuasion might be used to design computer interfaces that engender constructive changes in domains such as health, safety, education, and public policy [14, 15, 38]. Studies in this area have shown that computer interfaces that draw on the persuasive effects of nonverbal cues improve the likability [3] and persuasiveness of the interface [3, 2, 20, 38]. Users also perceive computer agents that display high behavioral realism to be more influential [20]. In immersive virtual environments (IVEs), avatars that maintain increased eye contact with people are perceived by users to be more persuasive [2].

Robotics researchers have developed a number of applications that draw on social behavior to deliver positive benefits in health management [31], weight loss [29], education [13], and energy conservation [37]. Recent studies have explored how robots might serve as persuasive agents, focusing particularly on the role of perceived robot gender in shaping the persuasiveness of a robot [48]. While this study highlights the importance of persuasion in human-robot interaction and provides some guidelines on how robot design might shape its persuasiveness, how behavioral attributes of the robot might shape the persuasiveness of a robot and affect user compliance are unknown.

In this paper, we explore how manipulations in *bodily* cues, particularly proximity, gaze, and gestures, and *vocal* cues, particularly vocal tone and expressions, affect the persuasiveness of a humanlike robot. The paragraphs below describe the hypotheses we have developed based on findings from research in human communication.

2.1 Hypotheses

Drawing on findings from human communication research, we developed the three hypotheses below on how nonverbal behavior, particularly the *bodily* and *vocal* cues of a robot, might affect the persuasiveness of a robot.

Hypothesis 1: Participants' perceptions of the persuasiveness of the robot and compliance with the robot's suggestions will be higher when the robot displays nonverbal cues (verbal and/or bodily cues) than when it does not display nonverbal cues to communicate with the participant. This hypothesis builds on research that suggests that nonverbal behavior facilitates persuasion in human interaction [40, 46].

Hypothesis 2: Participants' perceptions of the persuasiveness of the robot and compliance with the robot's suggestions will be higher when the robot displays only bodily cues than when it displays only vocal cues. This prediction follows argument that bodily cues might play a stronger role than vocal cues do in people's forming of impressions of others [35].

Hypothesis 3: Women's perceptions of the persuasiveness of the robot and compliance with the robot's suggestions will be higher than those of men in the presence of nonverbal cues. This hypothesis builds on the finding that women are more adept than men at reading nonverbal cues [24, 45].

3. METHOD

To test our hypotheses, we designed and conducted a laboratory experiment in which a humanlike robot sought to persuade participants performing the Desert Survival Task [30]. Below we describe our interaction design of the robot's nonverbal behaviors, the design of our experiment, the manipulations we created in the robot's behaviors, our measurements, and the participants of the experiment.

3.1 Design of the Robot's Behaviors

The design of the robot's nonverbal behaviors were guided by theories of human communication and empirical models derived from data collected from people in communication scenarios. We combined findings from research on nonverbal behavior (e.g., to identify what types of gestures should be designed into the robot) and data collected from a native English speaker performing the robot's utterances (e.g., to further identify the specific gestures that the robot should employ given the context of our experimental task) as bases for our design.

Gaze: [17] suggests that gaze cues communicate the social accessibility of an individual. Looking directly at a person indicates willingness to start a conversation with that person. Accordingly, we designed our robot's gaze behavior to look toward the participant when it spoke and to direct its gaze toward the computer screen when it referred to the Desert Survival items. According to Kendon [26], at the end of an utterance, speakers look toward their addressee, signaling that he/she has finished speaking and that the speaking floor is available for the other person. Accordingly, in our design, the robot looked toward the computer screen while referring to an item and moved its gaze back toward the participant at the end of its utterance.

Proximity: To identify the proximity cue that would contribute to achieving high nonverbal immediacy, we drew on the following classification of interpersonal space developed by Hall [23]:

* Intimate space – within *two* feet of the person.
* Personal space – between *two* and *four* feet of the person.
* Social space – between *four* and *twelve* feet of the person.
* Public space – between *twelve* and *twenty-five* feet of the person.

Research on the persuasive effects of proximity suggests that individuals' persuasive ability increases with closer proximity [16]. On the other hand, entering one's intimate space has the potential to evoke feelings of invasion and discomfort [49]. Therefore, in order to achieve the appropriate balance of increased persuasiveness and low discomfort, we designed our robot's proximity to vary across its partner's personal space boundary. The robot stood within the participant's personal space when it maintained a close proximity, while positioning itself beyond the personal space boundary and within the social space when it maintained a distant proximity.

Gestures: Research in nonverbal behavior suggests that people produce different types of gestures when they speak [34, 27, 28, 18] and that these gestures shape the persuasiveness of their speech [32]. While the literature in this area includes a large number of gesture categorizations and taxonomies, we have identified four types of gestures to be common across most such categorizations, (1) iconic, (2) metaphoric, (3) deictic, and (4) beat gestures, and integrated each type of gesture in the design of the robot's nonverbal behaviors. While the gesture literature helped us determine the types of gestures that the robot should employ, we identified the specific gestures that might accompany the robot's speech from an analysis of the data that we collected from the human performer. Figure 2 shows some of the gestures that the robot produced in alignment with its speech. Below we describe our implementation of each type of gesture into the robot in further detail.

Iconic Gestures: Iconic gestures are closely associated with the semantic content of speech, depicting a concrete event or object by creating a homology to aspects of the event or object. Our design of robot's iconic gestures involved identifying important gestures related to *quantifiers* and *action phrases* from the video data that we collected from the human performer and coding them into the robot. For action phrases, the robot depicted the actions in its speech using gestures. For instance, when the robot said "it would be best to drink the water as you become thirsty," it brought its right hand toward its face, depicting drinking water from a glass at the onset of the segment "drink the water." For quantifiers or comparative phrases, the robot illustrated the relative quantity of the event or object that he verbally described using gestures. For instance, when it said "the trench coat can be ranked higher" in the Desert Survival Task, it vertically raised its arms to above the level of its eyes to depict a "higher" quantity.

Metaphoric Gestures: Unlike iconic gestures, which represent concrete events or objects, metaphoric gestures depict abstract concepts such as knowledge or an idea. In human-to-human conversations, metaphoric gestures usually involve articulate fingers. A common example would be a ring hand, where the thumb and forefinger tips form a circle, to articulate PRECISE. In our design, the robot used one hand or both hands at the same time to form a gesture space containing the abstract ideas it described verbally. For instance, along with the utterance "Do you know that alcohol absorbs water?" the robot forms a gesture space containing the knowledge "alcohol absorbs water" by raising its hands in front of its chest with both hands facing to each other.

Deictic Gestures: Speakers use deictic gestures to direct attention toward concrete entities in the physical environment. These gestures are prototypically performed with

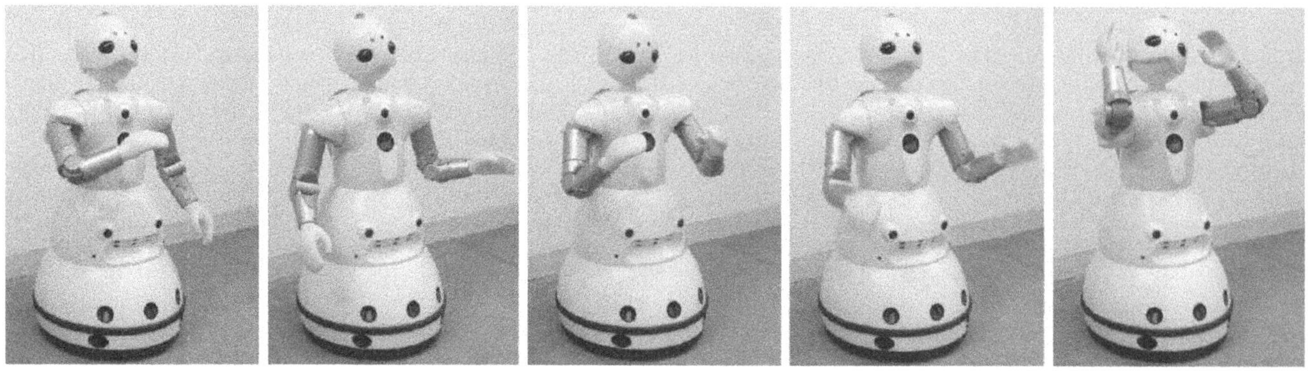

Figure 2: Examples of the gestures that we designed into the robot's nonverbal behavior: (a) *deictic gesture* while referring to itself, (b) *deictic gesture* while referring to the participant, (c) *metaphoric gesture* using both hands to form a gesture space containing an idea, (d) *beat gesture*, moving its arms rhythmically, and (e) *iconic gesture* depicting "higher."

the pointing finger, while pointing can also be involve an open palm or elbow. Our implementation of deictic gestures substituted finger pointing with hand pointing, because the design of our robotic platform does not involve articulate fingers. In our design, the robot used deictic gestures along with personal pronouns, pointing toward itself when it referred to itself, toward the participant when it referred to itself, and toward the environment when it referred to third parties. For example, it placed its right or left hand in front of its torso at the onset of the "I" in "I think the knife should be ranked higher" in the Task.

Beat Gestures: McNeill [34] describes beats as simple up-and-down movements that do not have discernible meanings, but are produced by speakers in conjunction with speech to maintain a rhythm and emphasize certain words or phrases. Our analysis of the video data showed that the speaker performed beat gestures in conjunction with explanations of complex or abstract ideas. In our design, the robot employed rhythmic up-and-down gestures when it explained complex ideas such as uttering "the body loses an enormous amount of water trying to process and remove the alcohol." However, it employed these gestures only it did not produce other forms of gestures.

Vocal Cues: Spoken language involves vocal cues that are used to distinguish lexical or grammatical meaning of words or phrases, add emphasis, and express emotion to verbal content. These cues are shaped by a number of parameters including vocal rate, pitch, and loudness that shape the expressivity of spoken language [7]. In our design, we manipulated vocal tone by varying the pitch of the robot's voice, as studies on the persuasive vocal tone is shown to affect compliance [8].

To implement the robot's speech and manipulate its vocal cues, we used the Festival text-to-speech system [51].

Condition	Vocal cues	Bodily cues	Proximity	Gaze	Gestures
1	No	No	*Outside personal space*	*Static*	*No*
2	No	Yes	*Within personal space*	*Dynamic*	*Yes*
3	Yes	No	*Outside personal space*	*Static*	*No*
4	Yes	Yes	*Within personal space*	*Dynamic*	*Yes*

Table 2: The four conditions of our experiment. We manipulated the robot's use of vocal and bodily cues.

We manipulated the vocal tone parameters of the synthesized speech, creating two versions of each utterance that are identical in verbal content but differ in intonation to be highly monotonic and highly expressive.

3.2 Experimental Design

We designed a two-by-two, between-participants study and created four conditions by manipulating the nonverbal cues that the robot employed: (1) no nonverbal cues, (2) vocal cues only, (3) bodily cues only, and (4) bodily and vocal cues. We measured how these manipulations affected participants' compliance with the robot's suggestions and their perceptions of the robot's persuasiveness.

The *vocal* cue manipulation involved changing the vocal tone and expressiveness that the robot used when it provided verbal suggestions to the participant. In the *no vocal cues* condition, the robot maintained a flat tone of voice for its entire speech, while the the *vocal cues* condition involved the robot varying its vocal tone at different points in its speech. These manipulations are described in more detail in the next subsection.

The second manipulation involved varying the robot's use of *bodily* cues, particularly proximity, gaze, and gestures. The proximity cue captured the bodily distance between the participant and the robot. We manipulated proximity to be either *within* or *outside* the personal space of the participant. We used a distance of two feet and four feet for the within and outside conditions respectively based on the classification developed by [22]. Gaze cues involved the direction in which the robot directed its head during speech and pauses. We manipulated the robot's gaze to be either *static* or *dynamic*. In the static condition, the robot looked straight ahead while talking to the participant, while in the dynamic gaze condition it divided its gaze between the participant and the computer screen, shifting its gaze at key points that were determined by its utterances. *Gestures* captured the robot's arm movements during speech. We manipulated whether gestures, particularly iconic, metaphoric, deictic, and beat gestures, accompanied the robot's speech, creating *gesture* and *no gesture* conditions. Table 2 lists the conditions in our experiment and the manipulation of the cues in each condition.

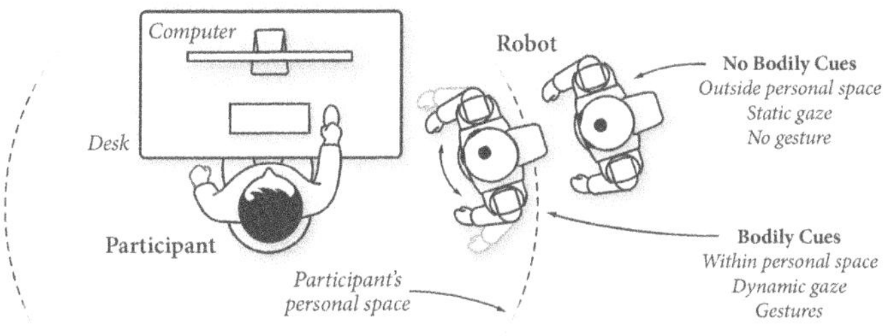

Figure 3: The spatial configuration of the robot and the participant. The robot gave suggestions while the participant was doing the task.

3.3 Experimental Task

We built on the Desert Survival Problem [30] to create a human-robot interaction scenario. In the prototypical task, participants imagine that they have crash-landed in the middle of a desert. They have items such as a torch, a jack knife, a bandage kit, a pair of sunglasses, and so on with them. Their task is to rank these items such that items that might increase their chances of survival receives a higher ranking. In the human-robot interaction scenario, the participants created an initial ranking of the items and the robot sought to persuade them to change their rankings of particular items. In each trial, the robot recommended that the participant rated the fourth-, sixth-, and ninth-ranked items higher, providing the rationale for each change. Participants were able to change their rankings once the robot provided its suggestions.

The robot's suggestions were identical across all conditions and the speaking time for the suggestions was approximately the same across items. The robot used the rankings that the participant provided through the computer interface to determine the appropriate suggestion and confirmations. The robot also followed common social interaction rituals such as greeting the participant at the beginning of the interaction and closing the interaction appropriately at the end of the experimental task. Figure 3 illustrates the spatial setup of the experimental task and the manipulations in the robot's bodily cues.

3.4 Experimental Procedure

Each trial of the experiment involved the following procedure. Participants were first given a brief description of the purpose and the procedure of the experiment. They were asked to review and sign a consent form after the brief introduction. Following their consent, they filled in a pre-experiment questionnaire that measured personality characteristics. The experimenter provided participants with more detail on the experimental task. Participants were brought into the experiment room and seated at a table with a computer. The robot greeted the participant and prompted the participant to start the experimental task. Following the end of the task, the participant left the room and filled in a post-experiment questionnaire on their perceptions of the robot, experience with the experiment, and basic demographic information.

The task and the entire experiment procedure in total took an average of 15 minutes and 25 minutes, respectively. The experiment was run in a dedicated space with controlled lighting and no outside distraction. Only the robot and the participant were present in the room during the experiment. The participants were paid $5 for their participation in the study.

3.5 Measurement

Our independent variables included whether the robot employed bodily cues, whether it employed vocal cues, and participant gender. The main dependent variables involved participants' perceptions of the robot's persuasiveness, their compliance with the robot's suggestions, and their perceptions of the robot's social and intellectual characteristics. We used both objective and subjective measurements to capture the dependent variables. Below we describe each measurement in detail.

Objective Measurements – Our measure of participants' compliance with the robot's suggestions involved the number of changes that the participants made in their rankings. This measure increased with changes in the direction of the robot's suggestions and decreased with changes against the robot's suggestions.

Subjective Measurements – We measured participants' perceptions of the robot and their experience with the task using a post-experiment questionnaire. The questionnaire included several items that measured their perceptions of the robot's persuasiveness, their perceptions of the robot's social and intellectual characteristics, their perceptions of the task, their affective state, and demographic information. Seven-point rating scales were used for all items. We conducted a factor analysis of all items that evaluated the robot to identify reliable scales and developed following three prominent factors that showed high reliability. The *Persuasiveness* scale involved three items: (1) persuasive, (2) helpful, and (3) relevant (Cronbach's $\alpha = 0.78$). *Intelligence* included two items: (1) knowledgable and (2) intelligent (Cronbach's $\alpha = 0.83$). The *Satisfaction* scale contained three items: (1) satisfied, (2) interesting, and (3) happy (Cronbach's $\alpha = 0.79$).

Research in persuasion suggests that personality traits affect persuasion; people with low levels of assertiveness are more easily persuaded [21]. In order to take the personality of the participant into account, we measured participants' assertiveness using a sub-scale of the International Personality Item Pool Interpersonal Circumplex (IPIP-IPC) [33]. The scale included four items: (1) "demand to be the center

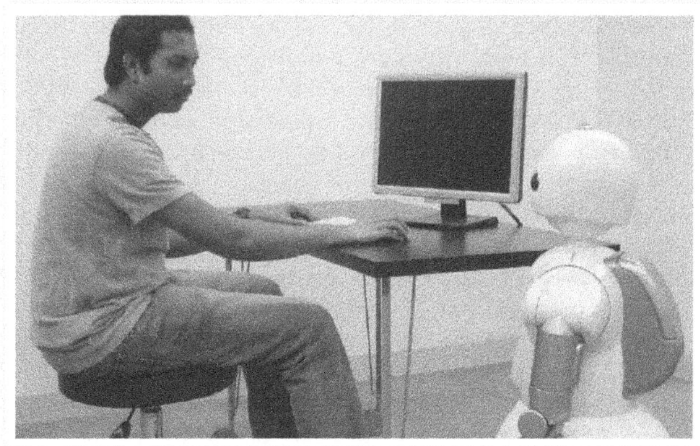

Figure 4: The experimenter demonstrating the experimental setup.

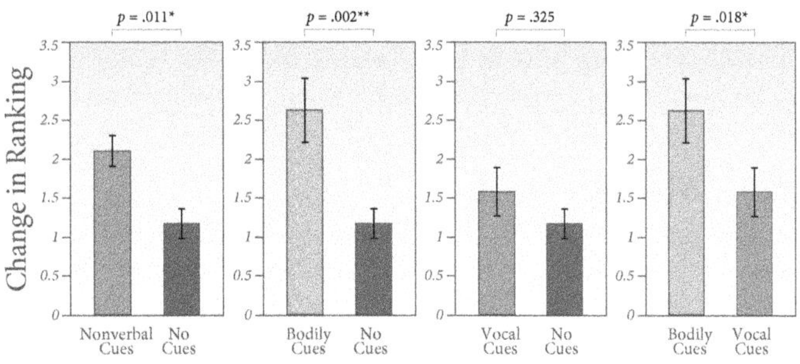

Figure 5: The results from our analysis of the objective data from the experiment.

of interest," (2) "do most of the talking," (3) "speak loudly," and (4) "demand attention" (Cronbach's $\alpha = 0.80$).

Manipulation Checks – The post-experiment questionnaire also included questions to confirm that our manipulations were successful in each condition. We asked participants to rate in a scale of one to seven how much they thought the robot was looking toward them, how much it moved its arms, and the expressiveness of its voice.

3.6 Participation

A total of 32 participants (16 males and 16 females) participated in the experiment. All participants were native-English speakers recruited from the University of Wisconsin–Madison campus. The ages of the participants varied between 19 and 49 with an average of 25.39. Participants represented a variety of university majors and occupations including sociology, biology, genetics, nursing, computer science, computer engineering, mechanical engineering, and finance. Participants were randomly assigned to the experimental conditions. We balanced the conditions for each gender, assigning four males and four females to each unique condition. Our demographic data showed that all participants rated their computer use 7 on a scale of one to seven ($M = 7, SD = 0$), while their familiarity with robots was moderate ($M = 3.67, SD = 1.71$).

4. RESULTS

Our analysis first checked whether the manipulations in the bodily and vocal cues were successful using analysis of variance (ANOVA). Participants rated the robot's use of bodily cues, particularly how much it moved its head and arms, significantly higher in the bodily cues condition than they did in the no bodily cues condition, $F(1,30) = 47.25$, $p < .001$, $\eta_p^2 = 0.612$ and $F(1,30) = 142.08$, $p < .001$, $\eta_p^2 = 0.826$ for head and arms, respectively, confirming that our bodily cues manipulation was successful. The results showed that our manipulation of the robot's vocal cues had a marginal effect on how expressive the participants rated the robot's voice, $F(1,30) = 3.77$, $p = .062$, $\eta_p^2 = 0.112$.

We used analysis of variance also to analyze how our manipulations affected our objective and subjective measures. Our first hypothesis predicted that participants' compliance with the robot's suggestions would be higher when the robot employed *nonverbal cues* (vocal only, bodily only, and vocal and bodily combined) than it would be when the robot did not employ any nonverbal cues. Our analysis confirmed our prediction; participants's compliance with the robot's suggestions were higher when the robot employed nonverbal cues ($M = 2.097$, $SD = 0.971$) than they did when it did not use any nonverbal cues ($M = 1.166$, $SD = 0.535$), $F(1,30) = 7.53$, $p = .011$, $\eta_p^2 = 0.212$.

Pairwise comparisons also showed that participants changed their ratings in a larger number of items based on the robot's suggestions when the robot employed *bodily cues* alone ($M = 2.626, SD = 1.160$) than when it did not use these cues ($M = 1.166$, $SD = 0.535$), $F(1,28) = 12.35$, $p = .002$, $\eta_p^2 = 0.306$. *Vocal cues* alone ($M = 1.583$, $SD = 0.868$), on the other hand, did not increase the participants' ratings compared with when the robots did not use these cues ($M = 1.166$, $SD = 0.535$), $F(1,28) = 1.00$, $p = .325$, $\eta_p^2 = 0.035$.

Our second hypothesis predicted that *bodily cues* alone would have a stronger effect on compliance than *vocal cues* alone would. Our analysis also supported this hypothesis; the change in participants' rankings of the items was higher when the robot used only bodily cues ($M = 2.626$, $SD = 1.160$) than it was when the robot employed only vocal cues ($M = 1.583$, $SD = 0.868$), $F(1,28) = 6.31$, $p = .018$, $\eta_p^2 = 0.184$. These results are illustrated in Figure 5.

Our last hypothesis predicted that female participants would comply with the robot's suggestions more than male participants would do when the robot employed nonverbal cues. Our analysis did not support this hypothesis; no interaction effect appeared between gender and cue factors, $F(1,24) = 0.122$, $p = .946$, $\eta_p^2 < 0.001$.

Our analysis showed that the effects of nonverbal cues, bodily cues alone, and vocal cues alone on the participants' compliance with the robot's suggestions did not appear in participants' subjective measures of the persuasiveness of the suggestions, though the participant's compliance with the robot's suggestions and their ratings of its persuasiveness were significantly correlated, $r(32) = .379$, $p = .032$. The analysis of the the intelligence measure showed a main effect of gender, as females rate the robot to be more intelligent, $F(1,28) = 4.95$, $p = .034$, $\eta_p^2 = 0.150$, and a marginal interaction effect between gender and the bodily cue manipulation, $F(1,28) = 3.50$, $p = .072$, $\eta_p^2 = 0.111$. Post-hoc tests showed that males evaluated the robot's intelligence significantly higher when it used bodily cues than they did when the robot did not use bodily cues ($p = .024$), while females evaluations of the robot's intelligence were not affected the robot's used of these cues ($p = .803$). The analysis of the satisfaction measure also showed a significant interaction effect between gender and the vocal cue manipulation, $F(1,28) = 4.41$, $p = .045$, $\eta_p^2 = 0.136$. Post-hoc tests showed that females were significantly more satisfied when the robot used vocal cues than they were when the robot did not use

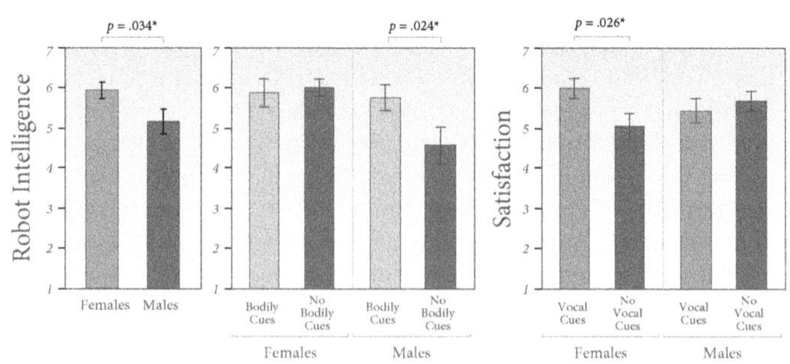

Figure 6: The results from our analysis of the subjective data from the experiment.

vocal cues ($p = .026$), while the robot's use of these cues did not affect males ($p = .537$). These results are illustrated in Figure 6.

Finally, we analyzed data from the personality scale to test whether participants' personality traits, particularly assertiveness, affected their compliance with the robot's suggestions and their ratings of the robot's persuasiveness. An analysis of covariance using assertiveness as a covariate showed that this personality trait was not a significant predictor of participants' compliance with the suggestions of the robot, $F(1,30) = 0.924$, $p = .344$, $\eta_p^2 = 0.030$, but was a significant predictor of their ratings of the robot's persuasiveness, $F(1,30) = 4.22$, $p = .089$, $\eta_p^2 = 0.123$, highly assertive participants rating the robot's suggestions low in persuasiveness.

5. DISCUSSION

The results from our experiment provided strong support for our first two hypotheses. We found that people complied with the robot's suggestions more when it employed nonverbal cues than they did when it did not employ nonverbal cues. Detailed analysis showed that the presence of bodily cues alone improved compliance, while the presence of vocal cues alone did not affect participants' compliance. The results also showed that bodily cues alone led to higher compliance in participants than vocal cues alone did. Our third hypothesis was not confirmed by our analysis, as participant gender did not affect participants' perceptions of the robot's persuasiveness and their compliance with the robot.

Limitations – The results presented here have a number of limitations. First, to create experimental conditions, we manipulated the "presence" of high-level categories of cues (e.g., bodily cues vs. no cues), instead of manipulating the presence of individual cues (e.g., gaze only, gestures only, etc.) or presenting these cues at different levels (e.g., more or less frequent). While a more finely granulated set of manipulations would have been more informative, the present study sought to confirm the link between nonverbal cues of a robot and its persuasiveness, only scratching the surface of how robots might persuade people through verbal and nonverbal cues. We plan to conduct follow-up studies to gain a deeper understanding of how individual cues might affect persuasion in human-robot interaction.

Second, our experimental task was based on the Desert Survival Task, which involved participants interacting with the robot in a hypothetical scenario. Whether the persuasive effects of the robot's nonverbal cues will affect decision making in real-world situations such as diet coaching, education, and public safety is unknown. Future work must investigate the generalizability of these results to real-world decision making and across a variety of situations.

Finally, the ethical implications of designing robots to persuade people require further consideration. We argue that persuasion is simply an effective form of communication and that persuasive robots should only be used in such domains as education, health, and wellbeing and only to support goals deemed appropriate by users.

6. CONCLUSION

Robots hold great promise as social actors helping to improve motivation and compliance in such as areas as education, health, and wellbeing. The success of these robots in being effective motivators will rely largely on their ability to *persuade*. In this paper, we explored how the nonverbal behaviors of a humanlike robot, particularly *bodily* cues such as proximity, gaze, and gestures, and *vocal* cues such as vocal tone and expressions, might affect the persuasiveness of a humanlike robot. We designed a set of nonverbal behaviors for a humanlike robot and evaluated, in a human-robot interaction study with 32 participants, the effects of manipulations in these behaviors on people's perceptions of the robot's persuasiveness and their compliance with the robot. Our results showed that the presence of nonverbal cues significantly improved people's compliance with the robot. Bodily cues alone improved compliance, while vocal cues alone did not. Bodily cues alone were also more effective in persuading participants than vocal cues alone were. Our model of persuasive nonverbal cues and experimental results have strong implications for the design of persuasive behaviors for humanlike robots. However, further work is required to understand how individual cues might affect persuasion, to what extent our results generalize to real-world situations, and to consider the ethical implications of designing persuasive technology.

7. REFERENCES

[1] J. Andersen. Teacher immediacy as a predictor of teaching effectiveness. *Communication yearbook*, 3(543-559), 1979.

[2] J. Bailenson, A. Beall, J. Loomis, J. Blascovich, and M. Turk. Transformed social interaction, augmented gaze, and social influence in immersive virtual environments. *Human Communication Research*, 31(4):511–537, 2005.

[3] J. Bailenson and N. Yee. Digital chameleons. *Psychological Science*, 16(10):814, 2005.

[4] J. Bradac, J. Bowers, and J. Courtright. Three language variables in communication research: Intensity, immediacy, and diversity. *Human Communication Research*, 5(3):257–269, 1979.

[5] C. Breazeal, C. Kidd, A. Thomaz, G. Hoffman, and M. Berlin. Effects of nonverbal communication on efficiency and robustness in human-robot teamwork. In *Intelligent Robots and Systems, 2005.(IROS 2005). 2005 IEEE/RSJ International Conference on*, pages 708–713. IEEE, 2005.

[6] R. Bull and E. Gibson-Robinson. The influences of eye-gaze, style of dress, and locality on the amounts of money donated to a charity. *Human Relations*, 34(10):895, 1981.

[7] D. Buller and R. Aune. The Effects of Vocalics and Nonverbal Sensitivity on Compliance A Speech Accommodation Theory Explanation. *Human Communication Research*, 14(3):301–332, 1988.

[8] D. Buller and J. Burgoon. The Effects of Vocalics and Nonverbal Sensitivity on Compliance A Replication and Extension. *Human Communication Research*, 13(1):126–144, 1986.

[9] J. Burgoon, N. Dunbar, and C. Segrin. Nonverbal influence. *The persuasion handbook: Developments in theory and practice*, pages 445–473, 2002.

[10] T. Chartrand and J. Bargh. The chameleon effect: The perception-behavior link and social interaction. *Journal of personality and social psychology*, 76:893–910, 1999.

[11] D. Christophel. The relationships among teacher immediacy behaviors, student motivation, and learning. *Communication Education*, 39(4):323–340, 1990.

[12] D. Christophel and J. Gorham. A test-retest analysis of student motivation, teacher immediacy, and perceived

sources of motivation and demotivation in college classes. *Communication Education*, 44(4):292–306, 1995.

[13] K. Dautenhahn. Robots as social actors: Aurora and the case of autism. In *Proc. CT99, The Third International Cognitive Technology Conference, August, San Francisco*, pages 359–374, 1999.

[14] B. Fogg. Persuasive computers: perspectives and research directions. In *Proceedings of the SIGCHI conference on Human factors in computing systems*, CHI '98, pages 225–232, New York, NY, USA, 1998. ACM Press/Addison-Wesley Publishing Co.

[15] B. Fogg. Persuasive technology: using computers to change what we think and do. *Ubiquity*, 2002(December):2, 2002.

[16] P. Glick, J. Demorest, and C. Hotze. Keeping Your Distance: Group Membership, Personal Space, and Requests for Small Favors. *Journal of Applied Social Psychology*, 18(4):315–330, 1988.

[17] E. Goffman. Behavior in public places, 1969.

[18] S. Goldin-Meadow. *Hearing gesture: How our hands help us think*. Belknap Press, 2005.

[19] J. Gorham. The relationship between verbal teacher immediacy behaviors and student learning. *Communication Education*, 37(1):40–53, 1988.

[20] R. Guadagno, J. Blascovich, J. Bailenson, and C. Mccall. Virtual humans and persuasion: The effects of agency and behavioral realism. *Media Psychology*, 10(1):1–22, 2007.

[21] G. Gudjonsson. Interrogative suggestibility: its relationship with assertiveness, social-evaluative anxiety, state anxiety and method of coping. *The British journal of clinical psychology/the British Psychological Society*, 27:159, 1988.

[22] E. Hall. A System for the Notation of Proxemic Behavior1. *American anthropologist*, 65(5):1003–1026, 1963.

[23] E. Hall. *The hidden dimension*, volume 6. Doubleday New York, 1966.

[24] M. Hoffman. Sex differences in empathy and related behaviors. *Psychological Bulletin*, 84(4):712–722, 1977.

[25] T. Kanda, M. Kamasima, M. Imai, T. Ono, D. Sakamoto, H. Ishiguro, and Y. Anzai. A humanoid robot that pretends to listen to route guidance from a human. *Autonomous Robots*, 22(1):87–100, 2007.

[26] A. Kendon. Some functions of gaze-direction in social interaction. *Acta psychologica*, 26(1):22, 1967.

[27] A. Kendon. An agenda for gesture studies. *Semiotic review of books*, 7(3):8–12, 1996.

[28] A. Kendon. *Gesture: Visible action as utterance*. Cambridge Univ Press, 2004.

[29] C. Kidd and C. Breazeal. A robotic weight loss coach. In *Proceedings of the 22nd national conference on Artificial intelligence - Volume 2*, pages 1985–1986. AAAI Press, 2007.

[30] J. Lafferty, P. Eady, and J. Elmers. The desert survival problem. *Experimental Learning Methods*, 1974.

[31] R. Looije, M. Neerincx, and F. Cnossen. Persuasive robotic assistant for health self-management of older adults: Design and evaluation of social behaviors. *International Journal of Human-Computer Studies*, 68(6):386–397, 2010.

[32] F. Maricchiolo, A. Gnisci, M. Bonaiuto, and G. Ficca. Effects of different types of hand gestures in persuasive speech on receiversÕ evaluations. *Language and Cognitive Processes*, 24(2):239–266, 2009.

[33] P. Markey and C. Markey. A Brief Assessment of the Interpersonal Circumplex. *Assessment*, 16(4):352, 2009.

[34] D. McNeill. *Hand and mind: What gestures reveal about thought*. University of Chicago Press, 1996.

[35] A. Mehrabian. *Silent messages*. Wadsworth Publishing Company, Inc., Belmont, CA., 1971.

[36] S. Messman and J. Jones-Corley. Effects of communication environment, immediacy, and communication apprehension on cognitive and affective learning. *Communication Monographs*, 68(2):184–200, 2001.

[37] C. Midden and J. Ham. Using negative and positive social feedback from a robotic agent to save energy. In *Proceedings of the 4th International Conference on Persuasive Technology*, Persuasive '09, pages 12:1–12:6, New York, NY, USA, 2009. ACM.

[38] G. Miller. The Art of Virtual Persuasion. *Science*, 317(5843):1343, 2007.

[39] T. Mottet, J. Parker-Raley, C. Cunningham, S. Beebe, and P. Raffeld. Testing the neutralizing effect of instructor immediacy on student course workload expectancy violations and tolerance for instructor unavailability. *Communication Education*, 55(2):147–166, 2006.

[40] P. Peters. Gaining Compliance Through Non-Verbal Communication. *Pepperdine Dispute Resolution Law Journal*, 7:87, 2007.

[41] L. Pogue and K. Ahyun. The effect of teacher nonverbal immediacy and credibility on student motivation and affective learning. *Communication Education*, 55(3):331–344, 2006.

[42] K. Reardon. *Persuasion in practice*. Sage Publications, Inc, 1991.

[43] V. Richmond, J. Gorham, and J. McCroskey. The relationship between selected immediacy behaviors and cognitive learning. *Communication yearbook*, 10(574-590), 1987.

[44] K. Rocca. College student attendance: Impact of instructor immediacy and verbal aggression. *Communication Education*, 53(2):185–195, 2004.

[45] J. Rosip and J. Hall. Knowledge of nonverbal cues, gender, and nonverbal decoding accuracy. *Journal of Nonverbal Behavior*, 28(4):267–286, 2004.

[46] C. Segrin. The effects of nonverbal behavior on outcomes of compliance gaining attempts. *Communication Studies*, 44(3):169–187, 1993.

[47] C. Sidner, C. Kidd, C. Lee, and N. Lesh. Where to look: a study of human-robot engagement. In *Proceedings of the 9th international conference on Intelligent user interfaces*, pages 78–84. ACM, 2004.

[48] M. Siegel, C. Breazeal, and M. Norton. Persuasive robotics: the influence of robot gender on human behavior. In *Proceedings of the 2009 IEEE/RSJ international conference on Intelligent robots and systems*, IROS'09, pages 2563–2568, Piscataway, NJ, 2009. IEEE Press.

[49] E. Sundstrom and I. Altman. Interpersonal relationships and personal space: Research review and theoretical model. *Human Ecology*, 4(1):47–67, 1976.

[50] R. Tanner and T. Chartrand. The Convincing Chameleon: Impact of Mimicry on Persuasion. *Journal of Consumer Research*, 34:754–766, 2008.

[51] P. Taylor, A. Black, and R. Caley. The architecture of the Festival speech synthesis system. In *The Third ESCA/COCOSDA Workshop (ETRW) on Speech Synthesis*, 1998.

[52] K. Thweatt. The impact of teacher immediacy, teacher affinity-seeking, and teacher misbehaviors on student-perceived teacher credibility. *National Communication Association, Chicago, IL*, 1999.

[53] P. Zimbardo and M. Leippe. *The psychology of attitude change and social influence*. McGraw-Hill New York, 1991.

How Do People Walk Side-By-Side? - Using A Computational Model Of Human Behavior For A Social Robot

Yoichi Morales, Satoru Satake, *Rajibul Huq, Dylan Glas, Takayuki Kanda, and Norihiro Hagita
ATR Intelligent Robotics and Communication Laboratories
2-2-2 Hikaridai, Seika-cho, Souraku-gun, Kyoto, Japan
{yoichims, satoru, dylan, kanda, hagita}@atr.jp *rajibul.huq@utoronto.ca

ABSTRACT
This paper presents a computational model for side-by-side walking for human-robot interaction (HRI). In this work we address the importance of future motion utility (motion anticipation) of the two walking partners.
Previous studies only considered a robot moving alongside a person without collisions with simple velocity-based predictions. In contrast, our proposed model includes two major considerations. First, it considers the current goal, modeling side-by-side walking, as a process of moving towards a goal while maintaining a relative position with the partner. Second, it takes the partner's utility into consideration; it models side-by-side walking as a phenomenon where two agents maximize mutual utilities rather than only considering a single agent utility. The model is constructed and validated with a set of trajectories from pairs of people recorded in side-by-side walking. Finally, our proposed model was tested in an autonomous robot walking side-by-side with participants and demonstrated to be effective.

Categories and Subject Descriptors
H.5.2 [Information Interfaces and Presentation]: User Interfaces - *Interaction styles*; I.2.9 [Artificial Intelligence]: Robotics

General Terms
Design, Experimentation, Human Factors.

Keywords
Human-robot interaction, side-by-side walking, path planning.

1. INTRODUCTION
Previous studies revealed a number of scenarios where mobile robots can serve walking people. For example, they have been used to guide people walking together while talking [7, 20]. Robots have also escorted groups of people as in [4]. When walking together, it is known that people tend to walk on a side-by-side formation [3], however, this knowledge was not used in the above studies. The importance of side-by-side walking for HRI is discussed in [19].

How can we enable a robot to walk with a person in a side-by-side formation? On the surface, it seems quite simple to produce a side-by-side walking situation; all that is needed is to let the robot go along the target person. But is this really so simple?

We found that it was not that simple. Figure 1 (b) shows a scene from one of our experiments illustrating the difficulty with such a simple assumption. In this situation, the person was stuck between the robot and an obstacle. This awkward situation happened because the robot simply tried to be aside her but was not aware of the situation that her front was blocked; from the person's perspective, she kept distance with the robot for safety or social reasons, and she did not tried to go close to the robot until her path was blocked. On the other hand, when the robot walks in a cooperative way, the robot is aware of the obstacle in front of the person and opens space while keeping distance (Figure 1 (a)), and they successfully walk in side-by-side formation.

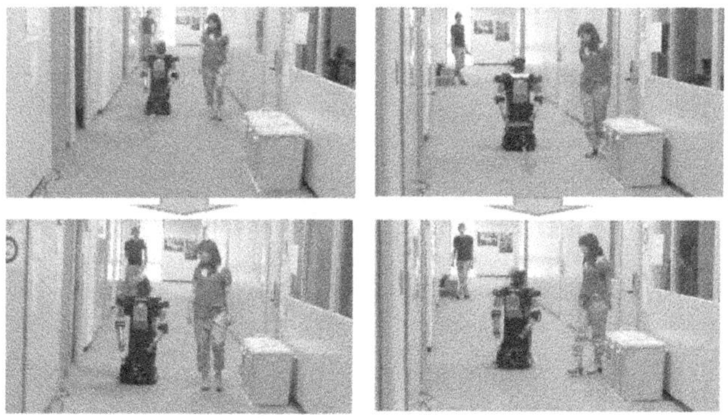

(a) Side-by-side walking model (b) A simple model to stay aside
Figure 1. **People walking side-by-side with a mobile robot.**

To create this model, we first observed how people walk with other people in a narrow environment that includes corners and obstacles. From these data we formulated a model for computing the utilities of future positions and used it to project the expected utility into the future for motion planning. The model was then implemented into a robot and experimentally evaluated.

2. RELATED WORKS
2.1 Human Science for Spatial Formation
Many studies have been conducted related to personal space and social distance. Hall considered the concept of proximity [9]. Spatial formation during conversation has been studied in [12]. In contrast, relatively only few studies have been conducted for spatial formation while walking. When there are only two people walking together, it is observed that people mostly walked side-by-side formation unless it is severely crowded; when three or more people are walking, they start to form more complex formations, such as "V" shapes and side-by-side formations [3].

In pedestrian modeling research, computational models of human walking behaviors have been developed. A social force model [10] simplifies people's computation as a combination of attractive force toward the goal and repulsive force toward nearby pedestrians. Moreover, recent studies have started to explore a model for group behavior, adding attraction force toward either the group's gravity center [17] or the positions of other members [28]. Predictions of future states are considered crucial for reproducing human walking behaviors [29].

2.2 Walking with a Robot

Spatial formation has been a focused topic in HRI for the last decade. Social distances with robots were observed and analyzed [8, 26] and reproduced in HRI [22]. Spatial formation during conversations have been replicated in HRI, too, including Kendon's concept of F-formation [14].

However, the above works mainly address static situations where people and robots interact while standing, and a few address human-robot spatial interaction for dynamic situations. Gockley et al. is one exception whose work describes a human-like way for a robot to follow a person [6]. There is a simulation study about a formation for shepherding a group of people [4]. Different from these engineering works, since side-by-side walk is very dynamic, we replicate a human computational model, which is the unique point of our study.

Kobayashi et al. took an engineering approach to implement side-by-side walking for a wheelchair robot. In addition to letting the robot go to the side position as a default, their robot wheelchair followed a caregiver when an obstacle was detected and made it stop when a caregiver stopped so that he can support tasks such as opening a door [13]. Prassler et. al. proposed an approach to coordinate the motion of a robotic wheelchair in a railway station moving side by side with the person [21] where they extrapolate the partner's velocity based on the past discreet trajectories. In these two works, contrary to our model, the prediction model only considers the linear extrapolation of velocity and does not use environmental characteristics.

2.3 Perspective Taking and Anticipation

The literature reported the importance of considering the partner's perspective for a robot that interacts with a human partner. A robot must understand the partner's perspective and act based on an estimate of the partner's view. For instance, in deictic interaction, a scene perceived by one agent does not necessarily resemble the partner's due to different viewpoints. Thus, implementing a perspective-taking capability in a robot improves efficiency[1, 25]. Further, beyond perception, anticipatory action was found to be effective in a joint work scenario [11]. We consider important to develop a robot that perceives situations in a similar way as humans and produces actions to improve the partner's future utility. In this work we study a side-by-side walking phenomenon built on the success of previous works that stressed the importance of perspective taking and anticipation.

3. SIDE BY SIDE WALKING MODEL

We propose a computational model that reproduces two-person side-by-side walking even when there are corners and obstacles in a narrow corridor. We took an analytical approach to find a computational model that explains human side-by-side walking. We collected trajectory data of people walking side-by-side and created a utility model that describes a walking motion that fits the data. Finally, we projected this utility model to predict people's walking behavior in the near future, which is a necessary step for planning robot motion. The proposed model uses a utility function to approximate the trajectory of a walking partner and outputs a trajectory similar to the one followed by a human when walking with a partner, we do not imply that humans use this exact model when walking together.

3.1 Data Collection

3.1.1 Tracking infrastructure

Human's trajectories while walking were measured using a laser-based human tracker. We used a network of 9 Hokuyo Top-URG laser range finders with a nominal detection range of 30 m. Each sensor was mounted at the top of a pole at a height of 83 cm. For this configuration, the sensors were set to scan an angular area of 180° at a resolution of 0.25°, at 40 Hz. The sensor poles were placed around the perimeter of the environment to track the positions of people. Range data was stored and analyzed offline to compute participant trajectories using the particle filter tracking algorithm presented in [5].

3.1.2 Procedure

We collected side-by-side walking trajectories of two people in a hallway environment. Fifteen pairs of participants without any knowledge of our research were paid for their participation. We asked each pair of participants to walk together from a start to a goal. For each pair, we collected 34 walking sessions by changing the obstacle positions (see Figure 2.). In total, 510 sets of side-by-side walking trajectories were collected. Each walking experiment lasted approximately 18 seconds. The participants walked in a side-by-side formation, except when avoiding obstacles or passing through narrow spaces. Since the trajectories were measured every 30 ms, we collected 238000 time frames of data. The first and last two seconds of each log were discarded to avoid frames where participants started or stopped walking.

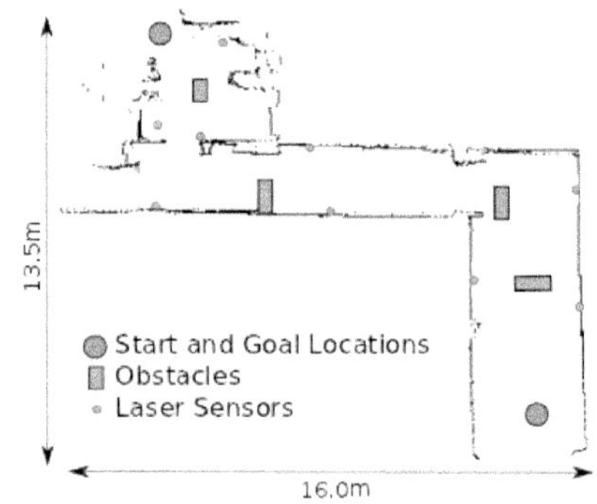

Figure 2. **Map of the hallway (16m x 13.5m) where two people walked together from a start to a goal locations.**

3.2 Modeling

3.2.1 Assumption

We believe that humans walk together while taking mutual decisions about their next motion. In this case, a person anticipates the partner's next motion and incorporates it in the planning of his own. Our assumption is that when people walk, they share information about their destination considering that subgoal-based path planning is common in human navigation [16]. Hence, we propose a model for side-by-side walking based on utility function and planning with anticipation.

3.2.2 Utility function for Walking Side by Side

We considered several parameters to compute the goodness of people's next motion. We included these parameters in a utility function and are briefly detailed below and illustrated in Figure 3.

Social relative distance (f_{R_d}): According to proxemics studies [9], people stay within a certain relative distance to the partner.

Relative angle (f_{R_a}): When walking together, people tend to walk in the same direction. This utility refers to the relative angle between the trajectories of the walking partners.

Relative velocity (f_{R_v}): Pairs of people coordinate their relative velocity, which is zero or close to zero. Hence, this utility contributes for having people walking at similar velocities.

Distance to obstacle (f_O): People prefer to walk away from obstacles. Positions far away from obstacles have more utility than positions close to obstacles (Eq. 1).

Sub-goal (f_S): While coordinating the position and motion toward the partner, the purpose of the walk is to arrive at the destination. Thus, one prefers to keep facing to the destination. This is computed as an angle between the next point and the subgoal.

Velocity (f_{M_v}): People have their own individual preference concerning walking velocity, referred to as their preferred velocity [10]. Thus, walking at a different speed than their preferred velocity is less comfortable. This utility differs from relative velocity in that it does not take into consideration the partner.

Angular velocity (f_{M_w}): People generally walk in straight lines or paths. This utility represents such a characteristic in which straight trajectories have more utility than non-straight trajectories.

Acceleration (f_{M_a}): When walking one minimizes energy consumption, thus, one minimizes the changes of speed.

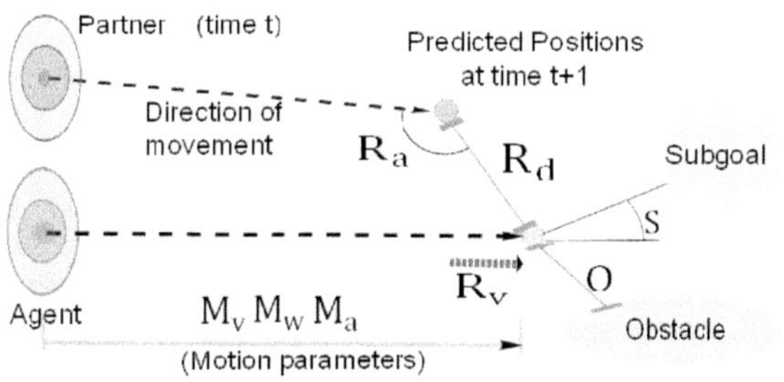

Figure 3. **Utility function parameters. Illustrated parameters correspond to the agent's point of view.**

Regarding the functions used in the utility computation, we modeled the **distance to obstacle** utility function as a step function (Eq. 1). People feel comfortable if they are far from obstacles (walls), but when they approach them, their desire to avoid obstacles rapidly increases:

$$f_o = -\left|\left(\frac{x}{a}\right)^{-2b}\right| \quad (1)$$

All other utility functions are modeled as a generalized bell function in the expression (Eq. 2), because their distribution has a clear peak and is roughly symmetrical. For example, Fig. 4 shows the histogram of the relative distance between the humans resulting from the trajectory analysis of the experiments in Section 3.1. This indicates that people's average distance during walking alongside was 0.75 m.

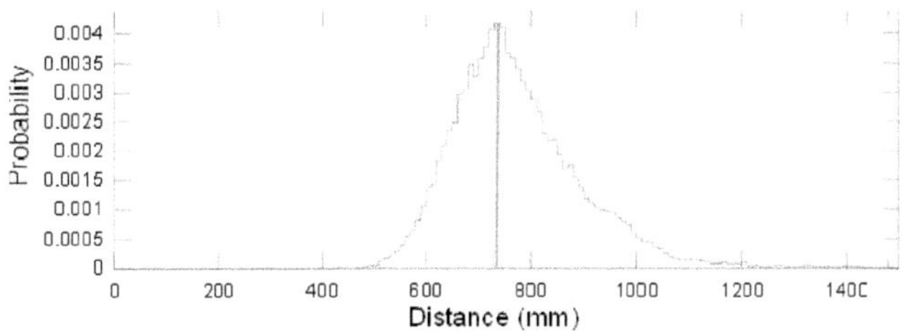

Figure 4. **Probability corresponding to the relative distance between center of mass of two people walking together.**

$$f_x = \frac{1}{1+\left|\left(\frac{x-c}{a}\right)^{2b}\right|} - 1 \quad (2)$$

This power function expression, which evaluates variable x, has three parameters. Parameter "a" is related to the tolerance of the function, parameter "b" corresponds to the curve steepness factor, and parameter "c" is the curve's center. The values of each variable are shown in Table 1.

In our model, we propose to compute the overall walking utility as a combination of individual utilities. We propose a utility function U_t for agent *i* positioned at point p_i toward its partner's position p_j as follows:

$$U(p^i, p^j) = k_O \cdot f_O + k_S \cdot f_S + k_{R_d} \cdot f_{R_d} + k_{R_a} \cdot f_{R_a} + k_{R_v} \cdot f_{R_v} + k_{M_a} \cdot f_{M_a} + k_{M_v} \cdot f_{M_v} + k_{M_w} \cdot f_{M_w} \quad (3)$$

In this expression variables f_x correspond to the utility functions of each variable x which are weighted by weighting constants k_x. Expression (3) outputs a higher utility when a walking agent and a partner are moving side by side towards a goal location.

3.2.3 Planning

We propose the following three models for the planning process:

Standard Prediction:

This is based on a commonly used method [13, 21] of side-by-side walking that uses linear extrapolation of the velocity to predict the position of the partner at the next moment and moves the robot to a position that is next to the partner (figure 5a). It predicts the partner's position at the next moment as a projection of the previous position with previous velocity, i.e.

$$\hat{p}^j_{t+1} = p^j_t + v^j_t \cdot \Delta t \quad (4)$$

where $\hat{p}^j_{t+1}$ is the predicted position of partner *j* at next time step *t+1* and p^j_t and v^j_t are the observed position and velocity at time step *t*. This model does not fully use the developed utilities (Eq. 3). Instead, it uses the following equation:

$$U_{St}(p_i, p_j) = k_O \cdot f_O + k_{R_v} \cdot f_{R_v} + k_{R_d} \cdot f_{R_d} + k_{R_a} \cdot f_{R_a} \quad (5)$$

Among all possible positions *P*, the next position is selected to maximize utility $U_{St}(p_i, p_j)$ in (Eq. 5):

$$\hat{p}^i_{t+1} = \operatorname{argmax}_{\{p^i|P^i\}} U_{St}(p^i, \hat{p}^j_{t+1}) \quad (6)$$

Self anticipation:

In this model an agent anticipates its future utility. Thus, it plans a motion to maximize its own utility (Eq. 3) in the next time step. It only anticipates its utility, not the partner's utility. Instead, it uses the linear extrapolation of velocity to predict the position of the partner at the next moment (Eq. 4). Fig. 5b illustrates how this model plans its next motion. This model's planning can be described with the following equation:

$$\hat{p}^i_{t+1} = \operatorname{argmax}_{\{p^i|P^i\}} U(p^i, \hat{p}^j_{t+1}) \quad (7)$$

The main difference in the implementation between standard prediction and self anticipation is the additional knowledge of subgoals in the environment together with the preferred linear velocity, angular velocity and acceleration.

Partner and self anticipation:

In this model an agent anticipates its future utility as well as its partner's future utility. It plans to output a motion to maximize its own utility as well as its partner's utility, both denoted with (Eq. 3), in the next time step. Fig. 5c illustrates how this model plans its next motion. This model's planning can be described with the following equation:

$$\hat{p}^i_{t+1} = \mathrm{argmax}_{\{p^i|P^i\},\{p^j|P^j\}}\{U(p^i,p^j)+U(p^j,p^i)\} \quad (8)$$

All models are implemented as a planner that searches for the next position among possible positions in the next step. The possible positions are implemented as a grid which we call the *anticipation grid* (P^i in Eqs. 6-8). The center of the grid is the linear extrapolation of the current velocity, i.e., $\hat{p}^i_{t+1}$. The grid is composed of a 7 x 7 matrix of cells of 0.20 m x 0.20 m in size. We set the time for prediction/anticipation (Δt) to be 1 second.

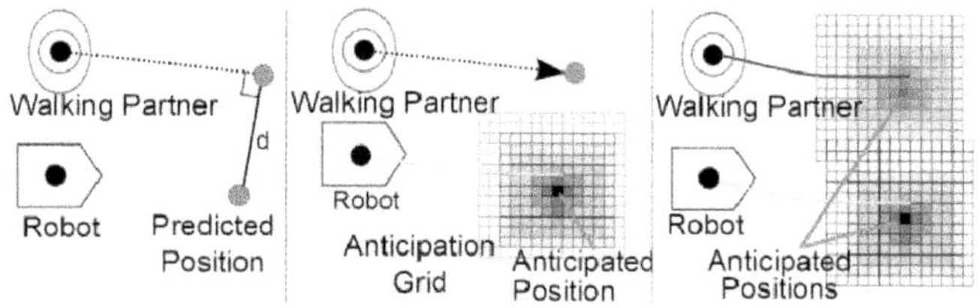

(a) Standard Prediction (b) Self Anticipation (c) Partner & Self Anticipation

Figure 5. **Different approaches for side by side walking.**

3.3 Analysis

3.3.1 Calibration of parameters

We manually calibrated the parameters for utility in Eqs. 3 and 5 from the observed data. The same set of data was used for calibration and testing. The same set of parameters is used for each of the three different planning methods: *standard prediction*, *self anticipation*, and *partner and self anticipation*. We took the following steps for parameter calibration:

1. Set initial parameters

   It is highly possible that the probability distribution (e.g., Fig. 4) for each characteristic is a consequence of each utility. Thus, we determined the initial parameters from the probability distribution. We set the utility function to output 0 when $x = x_{peak}$ and -1 when $x = x_{peak} \pm x_{sd}$, where x_{peak} is the peak and x_{sd} is the standard deviation for this probability distribution for variable x. We set c to be x_{peak}, a to be x_{sd}, and b to be 1. Each k was set to be 1.

2. Calibrate parameters for each utility function

   We used the human collected data to run a simulator and adjust the value of parameters a, b, and c. In this simulator an agent receives the initial position and the orientation of one of the persons, and then at step (p^i_t, p^j_t), the next position at $(\hat{p}^i_{t+1})$ is predicted, and prediction error $(|\hat{p}^i_{t+1} - p^i_t|)$ is computed. This process was done for different pairs of people and different obstacle configurations for each of the three planning methods.

3. Calibrate the integration parameter

   Finally, the best set of parameters k was adjusted. These constants determine the dominant factors in the utility functions.

The table 1 shows the calibrated parameters for *partner and self anticipation* method. The k parameter represents the balance among each utilities. It indicates that mainly the utility is affected by relative distance, angle, and distance to the object, all of which are also used in the equation (5) too. In addition, one fundamental difference is the considerable influence from the subgoal utility. The c parameter represents the peak of utilities.

3.3.2 Comparison of planning model

To evaluate our proposed utility function and planning method, we ran the simulator using the collected data. For each set of collected trajectories (we dropped non-useful trajectory sets which contained noise in the human trajectories, hence, 395 trajectories were processed in total), we used the utility function and planning method to simulate the person's motion (referred as *simulated agent* and *simulated trajectory*), while the partner's motion was read from actual collected trajectories. In the simulation, the initial position and orientation as well as the destination were given to the *simulated agent*, which were read from real trajectories. Then, for each sampling rate of 30ms, the *simulated agent* computes its next position from its previous position and the partner's real position. Finally, the goodness of fit is computed as the error in the distance of each time step of the *simulated trajectory* and its real trajectory.

The figure 6 shows the average and standard error of the error in distance from each planning method. A one-way repeated-measure analysis of variance (ANOVA) was conducted for the error. A significant main effect was revealed ($F(2, 788)=17.519$, $p<.001$, η_p^2 =043). A multiple comparison with Bonferroni method was conducted, which revealed that the error for *standard prediction* is significantly higher than that for *partner and self anticipation* and *self anticipation* ($p<.001$). No significance was found between *self anticipation* and *partner and self anticipation* ($p=1.0$). Thus, from this simulation, we confirmed that the anticipation is effective, but the effect of *partner and self anticipation* towards a simple *self anticipation* is not revealed. However, in the simulation, the partner agent moved based on the recorded trajectory and not reacted to the agent's motion. We wondered whether this would be due to a lack of reactivity in the simulation, and thus decided to test the methods with a real robot.

Table 1. Parameters determined for utility function

Parameter	a	b	c	k
f_{R_d}: Social relative distance (m)	0.45	2.00	0.75	0.25
f_{R_a}: Relative angle (rad)	0.08	3.00	$\pi/2$	0.32
f_{R_v}: Relative velocity (m/s)	0.20	1.20	0.0	0.01
f_O: Distance to Obstacles (m)	20.0	0.20	-----	0.11
f_S: Angle to Subgoal (rad)	0.75	1.00	0.0	0.20
f_{M_v}: Velocity (m/s)	0.90	1.6	1.10	0.05
f_{M_w}: Angular velocity (rad/s)	0.7	4.4	0.0	0.01
f_{M_a}: Acceleration (m/s^2)	1.50	1.0	0.0	0.01

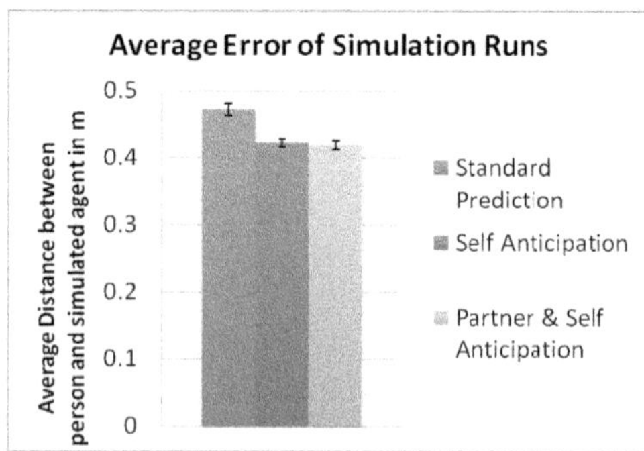

Figure 6. **Error distance estimation from the simulation results.**

4. IMPLEMENTATION

A robot walking together with a human requires that several functionalities be operating simultaneously. Fig. 7 shows the framework of our system. The environmental map and the positions of the robot and the human partner are fed to the path planner; the output of the system is the next position of the robot where the best utility is anticipated.

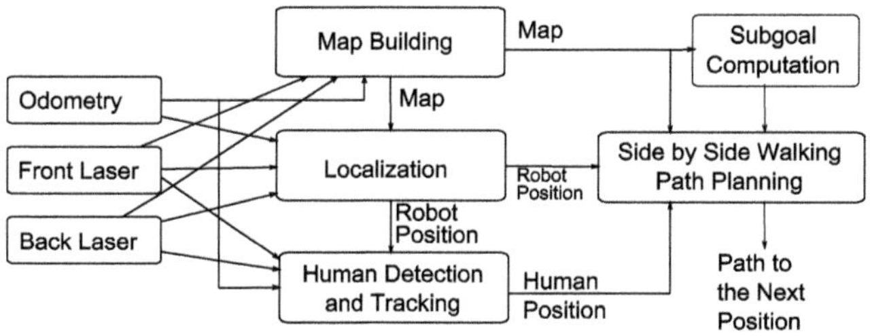

Figure 7. **System framework. The map building and subgoal computation are performed beforehand. Then the robot performs localization, human tracking, and Side-by-Side planning simultaneously. The output is the robot's next anticipated position.**

4.1 Hardware

We used Robovie, a communication robot characterized by its human-like physical expressions to interact with people through utterances and gestures. It is 1.20 m high and 0.40 m in diameter. For perception, it has motor encoders for proprioceptive sensing (Pioneer 3 mobile platform from Adept MobileRobots), and two 30 m range laser sensors (UTM-30LX from Hokuyo) covering 360° around the robot.

4.2 Human Detection and Tracking

For human detection and tracking we used range information from the robot's on-board laser sensors [15, 27]. The robot segments data into clusters to identify leg candidates and identify human candidates (leg pairs). The system initializes a human position at the center of two successive leg clusters and employs a particle filter (PF) per human for tracking the human position in successive cycles. The tracking is performed in an odometry based coordinate system and then transformed to global coordinates to avoid data jumps when the global position is corrected from the robot localization.

4.3 Map building and subgoal computation

We modeled the environment based on our previous study [16]. The model involves two tiers of representation. The first tier corresponds to a *survey view* of the environment. For this, we followed established procedures and state of the art algorithms for map building. To build a grid map, we applied ICP [2] for aligning consecutive laser scans . We used libraries provided in the slam6d framework [18, 23] for the final implementation in the map alignment. The resulting points were voted in a grid. We set the resolution of the grid as 0.05 x 0.05 m. The second tier of representation corresponds to a *route view*. Morales et al. proposed a path segmentation algorithm for retrieving a *route* [16]. In the algorithm, the Voronoi diagram of the grid map of the environment is computed, then, from the topological representation, redundant nodes are erased and remaining nodes are the *subgoals* humans walk in the environment. The obtained *subgoals* were used in the side-by-side walking path planner.

4.4 Robot Localization

We used a particle filter based method for localization with the grid map we built. Each particle contains a pose given by the state vector $\bar{\mathbf{x}} = \{x, y, \theta\}$ which is the position x and y and the orientation θ of the robot. The particle filter estimates the posterior using two laser sensors scanning the environment taking into consideration the measurement likelihood. The laser likelihood model used in this work computes $p(\mathbf{z}|\mathbf{x}, \mathbf{m})$ based on a laser scan $\mathbf{z}$ at a position $\mathbf{x}$ compared to the grid map $\mathbf{m}$ [24].

The map update depends on the state of the particle dispersion and the matching of the laser scans. The particles which are more likely to be correct after the map matching have a higher likelihood, then particle re-sampling is performed and the robot's pose is given by the average weight of the particles.

4.5 Side-by-side Walking Path Planner

The outputs from other modules is combined at this planner for side-by-side walking. The planner is implemented with the model described in Section 3. The position of the robot and the surrounding obstacles are fed from the localization module and the map. The position of the target human is fed from the human detection and tracking module. A moving average filter with a 1 second time window is applied for the positions when we compute velocities (for $f_{M_v}$and f_{M_w}) and acceleration (for f_{M_a}).

For planning, we used a 7x7 matrix of 0.20 m grids whose center is the linear extrapolation of the current velocity, i.e. $\hat{p}^i_{t+1}$. We set the time for prediction/anticipation (Δt) to be 1 second. With Eqs. 3 and 8, we computed the grid so that it can anticipate the best utility. The robot's forward and rotation velocities are computed to arrive at the center of the computed grid, which is sent to the mobile base.

Regarding the parameters for a robot, we found it necessary to adjust them. For instance, a person in the data collection might closely approach an obstacle, but this is not safe enough for a robot. Due to its locomotion constraints, when it approaches too close to a wall, it might have a collision. Similarly, due to the sensory noise in the human tracking module, the reading of the velocities is not as stable as the velocities measured with a human tracker with multiple laser-range finders used in the data collection. Thus, we manually adjusted the parameters to be feasible for real robots. For the comparison to be as fair as possible, in each planning method the same parameters were used.

5. Evaluation

5.1 Hypotheses and Prediction

While the anticipation model was found to be more effective than the standard prediction in the simulation, it does not necessarily ensure the usefulness of the proposed method with a real robot. There are two possible problems in the real world. First the dynamics of the real robot and second the behavior of real people towards the robot, both of which were not considered in the simulation. Further, in the simulation, the difference between the two anticipation methods was not revealed.

Hence, we conducted an evaluation experiment to confirm that the method works as designed and produces good side-by-side walking with real people. For that purpose, we primarily compared the proposed method (*partner and self anticipation* method) with the *standard prediction* method because the latter represents a traditional approach. Previous studies only proposed a method with linear extrapolation of velocity (which the *standard prediction method* performs) without proposing a way to anticipate either the robot or its partner's position. In addition, we expect to reproduce the improvement from *self anticipation* to *partner and self anticipation* in the evaluation.

With good side-by-side walking, a person's walking would probably be less hindered, so he/she could walk smoothly and consider the robot good. Based on this idea, we made this prediction:

> **Prediction**: since the robot with the *partner and self anticipation method* provides a **smoother** and **more natural** impression, it will be perceived as better in its **overall**

evaluation than either the one with the *standard prediction method* or with the *self anticipation method.*

5.2 Method

5.2.1 Participants

Fifteen Japanese people (nine males and six females whose average age was 21.9) were paid for their participation in the evaluation experiments.

5.2.2 Conditions

The experiment was a within-subject design with one factor, *control method.* Three conditions were prepared:

Standard prediction: the robot navigated with the *standard prediction method* (Eqs. 5 and 6) described in Section 3.2.

Self anticipation: the robot navigated with the *self anticipation method* (Eqs. 3 and 7) described in Section 3.2.

Partner and self anticipation: the robot navigated with the *partner and self anticipation method* (Eqs. 3 and 8) described in Section 3.2.

5.2.3 Procedure

Participants evaluated three methods of side-by-side walking. They walked in a side-by-side formation with the robot toward their destination along a given course. The side-by-side situation was explained as a situation where people talk during the walk and typically see their face each other.

Two courses were prepared: **with obstacles** (Fig. 8) and **without obstacles** (the same start positions and goal with the course shown in Fig. 8). This was to confirm the effect in situations with or without obstacles. Each participant walked each course three times with each condition presented in Section 5.2.2. The orders of the conditions were counterbalanced.

For each walk, participants stood at the start position (Fig. 8). After the experimenter confirmed that he/she was ready, the robot's program was initiated. The robot autonomously navigated based on the program reported in Section 4 until it arrived at the goal. Participants answered questionnaires after they arrived at the goal.

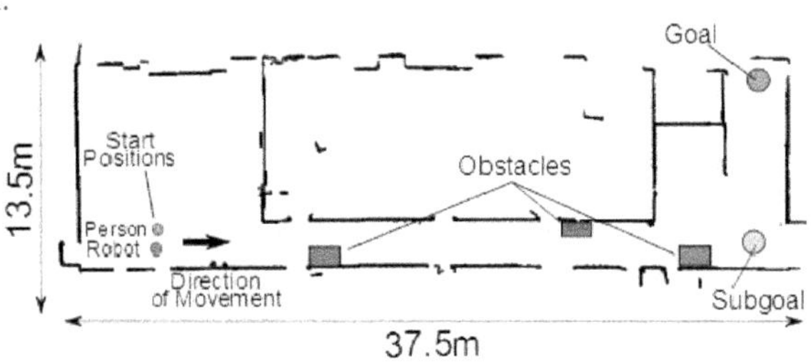

Figure 8. **Experimental Environment**

5.3 Measurement

After each session, participants filled out questionnaires with the following single item scales on a 1-to-7 point Likert scale for the following impressions:

- **Smoothness**
- **Naturalness**
- **Overall evaluation.**

5.4 Result

Fig. 9 shows the without-obstacle situation result. A one-way repeated-measure analysis of variance (ANOVA) was conducted. A significant main effect was revealed in *smoothness* (F(2, 28)=13.485, p<.000, η_p^2 =491), *naturalness* (F(2, 28)=13.000, p=.007, η_p^2 =.539), and *overall evaluation* (F(2, 28)=15.697, p<.001, η_p^2 =.529).

A multiple comparison with the Bonferroni method revealed that the ratings for *partner and self anticipation* are significantly higher than that for *standard prediction* (For *smoothness*: p=.001, *naturalness*: p=.005, and *overall evaluation*: p=.001). The difference between *standard prediction* and *self anticipation* are almost significant in all ratings (For *naturalness:* p=.065, *smoothness:* p=.084, *overall evaluation*: p=.069). The difference between *self anticipation* and *partner and self anticipation* are almost significant for *smoothness* (p=.051), not significant for *naturalness* (p=1.0), and significant in *overall evaluation* (p=.009).

Thus, in a situation without obstacles, participants clearly evaluated the *partner and self anticipation* method as smoother, more natural, and better overall than the *standard prediction* method. The difference between the *partner and self anticipation* and *self anticipation* methods was rather subtle, although overall participants preferred the former method.

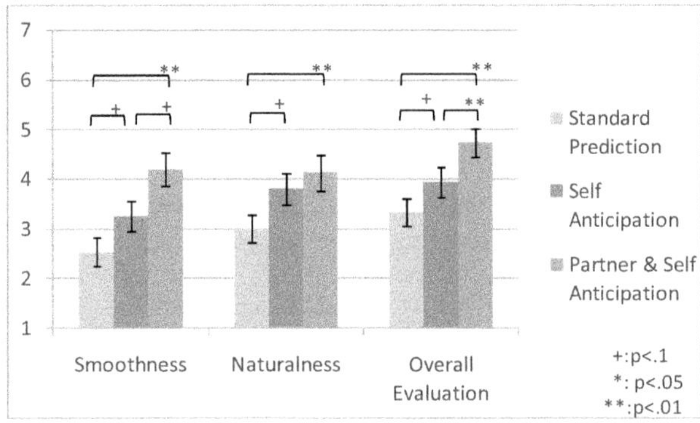

Figure 9. **Evaluation result (without obstacle)**

Fig. 10 shows the result for with-obstacle situation. A one-way repeated-measure analysis of variance (ANOVA) was conducted. A significant main effect was revealed in *smoothness* (F(2, 28)=17.982, p<.001, η_p^2 =.562), *naturalness* (F(2, 28)=8.788, p=.001, η_p^2 =.386), and *overall evaluation* (F(2, 28)=6.923, p=.004, η_p^2 =.331). A multiple comparison with Bonferroni method was conducted for each ratings.

The ratings for *partner and self anticipation* are significantly higher than that for *standard prediction* for all ratings (*smoothness*: p=.001, *naturalness*: p=.006, and *overall evaluation:* p=.010), and for *self anticipation* (*smoothness*: p=.001, *naturalness*: p=.034, and *overall evaluation:* p=.016). The difference between *standard prediction* and *self anticipation* was not significant (*smoothness:* p=.136, *naturalness*: p=.660, and *overall evaluation:* p=1.0).

Thus, in a situation with obstacles, participants evaluated the *partner and self anticipation* method as the best in all ratings. The difference between the *standard prediction* method and the *self anticipation* method was not significant. It seems that the presence of obstacles change the perception of the *self anticipation* method. Participants possibly evaluated it different from the *standard prediction* method in a without-obstacle situation (as, there are almost significant differences), but with obstacles it is evaluated rather similar to the *standard prediction* method.

Overall, the result follows our prediction. The robot *with partner and self anticipation method* provided more *smooth* and *natural* impression than other methods (in both situation for *standard prediction* method, and in with-obstacle situation for *self anticipation* method). The participants' *overall evaluation* for the *partner and self anticipation method* is better than their ratings for standard prediction method.

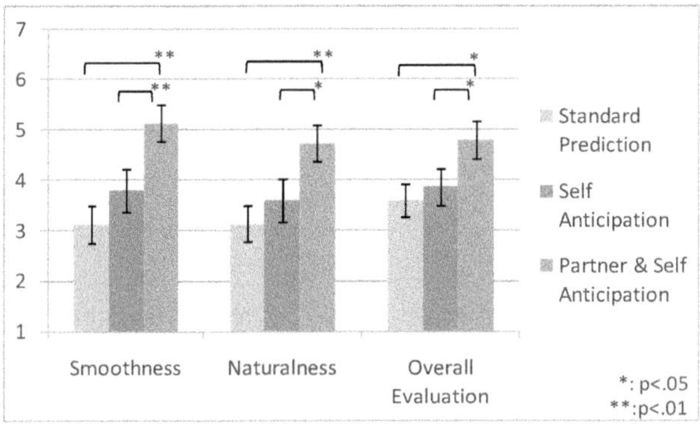

Figure 10. **Evaluation result (with obstacle)**

In the interview, we asked participants about perceived difference across the conditions. Participants mentioned that the robot kept a constant distance with which they felt comfortable while walking in all three methods; they also realized that the robot was adapting to the relative distance.

In relation with standard prediction, participants mentioned that the robot seemed to be nervous and sensitive to their movements, therefore they did not feel that the walking was smooth and natural. Some participants commented that the robot in the *self and partner anticipation* method made space for them, as shown in the figure 1 (a): "*I felt that when I was coming close to the trash can the robot moved away and gave me space to pass*". This suggests that the *self and partner anticipation* method successfully consider the partner's utility, and thus participants perceived better impression to it.

Figure 11 shows a scene when the robot opens space for the partner to walk while avoiding an obstacle with the *self and partner anticipation* method. The figure shows a transition of utility when the partner was approaching to the obstacle. The partner is represented as the circle at the top (red), and the bottom circle (blue) shows the current position of the robot. Their anticipated/planned position is shown to their right (yellow and blue circles) respectively. The anticipation grids are shown around the anticipated positions for both, the partner and the robot.

The notable moment is shown in the center of Figure 11 where the robot anticipated the person to go to the center of the corridor to avoid the obstacle, and planed its position towards the right side (bottom in the figure) of the corridor. Thus, in the right figure, when they were close to the obstacle, they are already in a course of jointly avoiding the obstacle still in a side-by-side formation.

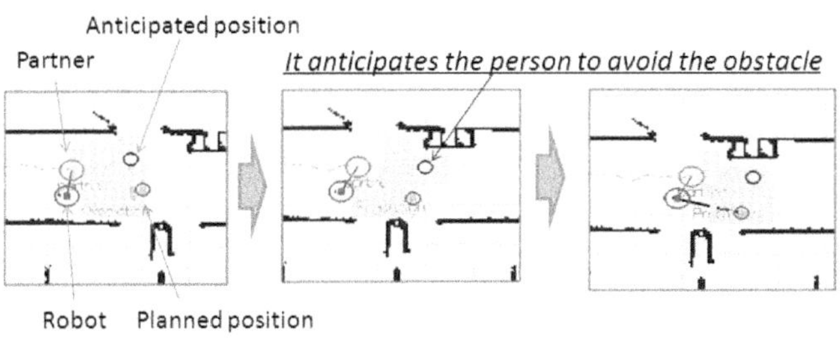

Figure 11. **Behavior of "self and partner anticipation" while avoiding an obstacle.**

Such motion is only possible with the partner and self anticipation method. In the case when *self anticipation method* faces the same situation, it does not open space for the partner. Figure 12 shows a scene of the *self anticipation* method experiment. It is the same situation as the figure 11 where they were approaching to the obstacle. At the moment where there is the obstacle in front of the partner, it failed to anticipate the partner's avoiding motion. It only noticed the partner avoiding motion after the partner started to move towards the center of the corridor.

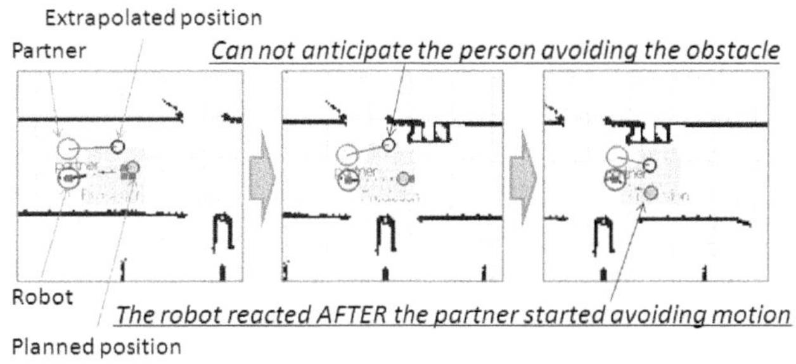

Figure 12. **Behavior of "self anticipation" while avoiding an obstacle.**

6. DISCUSSION

6.1 Contributions

The capability of side-by-side walking developed here promises to provide substantial pragmatic value. Many possible scenarios exist in daily interactions where a robot might walk side-by-side with a person. For example, consider a guidance application in a shopping mall, where a robot escorts a visitor to his destination. If they engage in conversation while walking, a side-by-side walking formation would be much more natural than a formation where the robot leads or follows.

6.2 Extending the Model

So far, the model is only for side-by-side situations with static obstacles, but we believe we can easily extend the method to more general situations. The basic idea of using utility is very close to the general idea in robotics to use potentials. The potential field method and its extension are widely used in robotics.

Our method requires subgoal information and assumes that when people walk, they share the information about their destination. At the same time, we consider it possible to extend our method to situations where a robot is not informed about the destination. From how people walk, it might be possible to infer the next subgoal of the person walking. For example, when a person is around a corner, a robot could have two hypotheses: going straight toward a subgoal or turning toward a subgoal in the branching direction. Based on how people walk, from the fitting with the assumed utility value, we consider it possible to predict the direction the person will take so that the robot can quickly update its hypothetical subgoal for moving. We consider this interesting future work.

6.3 Limitations

In this study, the parameters were not systematically calibrated, though it would be important to develop a method to calibrate them. Particularly a computational method to translate parameters used for human-human interaction to human-robot interaction would be beneficial. The reported parameters were calibrated for Japanese people, but it would require re-calibration for a different culture or context, as social distance alter along with context [9].

We did not specifically adjust parameters in favor of the proposed method for the experiment with the robot, though one could argue that the result would be due to the specific parameters we used. We believe that there are quantitative difference across the methods. For example, the *standard prediction method* and the *self anticipation method* lacks the capacity to quickly make a space for the partner (see Fig. 1 right and Fig. 12), as it does not predict the partner's motion constraints relate to obstacle (see Fig. 1 left and Fig. 11). The *standard prediction method* further lacks the capacity to move the robot toward the goal. Thus, we consider that the observed difference is due to the nature of the methods, but not due to the specific parameters.

The study only addressed peer-type relationship between people who walk in side-by-side, but leader-follower relationships would considerably change the situation. For example, as reported in [13, 21], in case a caregiver leads and a robotic wheelchair follows, it is clear who is leading, thus the *standard prediction model* would produce side-by-side walking as good as our proposed model. As observed, what is missing in *standard prediction method* is a capacity to take a lead and move toward the goal; the experimental result revealed that this is crucial in side-by-side walking with peer-type relationship.

7. CONCLUSION

We developed a computational model for side-by-side walking. A utility model describing how people prefer to move was built based on recorded trajectories of pairs of people walking side-by-side. Parameters for the model were chosen which best predicted the future walking motion of the people recorded in the trajectory data. This model was then used to generate appropriate walking behavior for an autonomous robot walking alongside a person. Our evaluations showed that the *partner and self anticipation* method, that is, projecting the future position of the partner based on the utility model and then planning an appropriate path for an autonomous mobile robot by using the same utility model to mimic human behavior, provided better results than simpler planning mechanisms.

8. ACKNOWLEDGMENTS

This research was supported by the Ministry of Internal Affairs and Communications of Japan. The authors thank the help and contribution of Thomas Kaczmarek and Trevor Condon.

9. REFERENCES

[1] Berlin, M., Gray, J., Thomaz, A. L. and Breazeal, C., 2006, Perspective Taking: An Organizing Principle for Learning in Human-Robot Interaction, *National Conf. on Artificial Intelligence (AAAI2006)*, pp. 1444-1450.

[2] Besl, P. and McKay, N., 1992, A method for Registration of 3-D Shapes, *IEEE Transactions on Pattern Analysis and Machine Intelligence*, vol. 14, pp. 239-256.

[3] Costa, M., 2010, Interpersonal Distances in Group Walking, *Journal of Nonverbal Behavior*, vol. 34, pp. 15-26.

[4] Garrell, A. ı. and Sanfeliu, A., 2010, Model Validation: Robot Behavior in People Guidance Mission using DTM model and Estimation of Human Motion Behavior., *IEEE/RSJ Int. Conf. on Intelligent Robots and Systems (IROS2010)*, pp. 5836-5841.

[5] Glas, D. F., Miyashita, T., Ishiguro, H. and Hagita, N., 2007, Laser Tracking of Human Body Motion Using Adaptive Shape Modeling, *IEEE/RSJ Int. Conf. on Intelligent Robots and Systems (IROS2008)*, pp. 602-608.

[6] Gockley, R., Forlizzi, J. and Simmons, R., 2007, Natural Person-Following Behavior for Social Robots, *ACM/IEEE Int. Conf. on Human-Robot Interaction (HRI2007)*, pp. 17-24.

[7] Gross, H.-M., Boehme, H.-J., Schroeter, C., Mueller, S., Koenig, A., Martin, C., Merten, M. and Bley, A., 2008, ShopBot: Progress in Developing an Interactive Mobile Shopping Assistant for Everyday Use, *IEEE Int. Conf. on Systems, Man, and Cybernetics (SMC2008)*, pp. 3471-3478.

[8] Hüttenrauch, H., Eklundh, K. S., Green, A. and Topp, E. A., 2006, Investigating spatial relationships in human-robot interactions, *IEEE/RSJ Int. Conf. on Intelligent Robots and Systems (IROS2006)*, pp. 5052-5059.

[9] Hall, E. T., 1966, *The Hidden Dimension: Man's Use of Space in Public and Private*, The Bodley Head Ltd.

[10] Helbing, D. and Molnár, P., 1995, Social force model for pedestrian dynamics, *Physical Review E*, vol. 51, pp. 4282-4286

[11] Hoffman, G. and Breazeal, C., 2007, Effects of Anticipatory Action on Human-Robot Teamwork, *ACM/IEEE Int. Conf. on Human-Robot Interaction (HRI2007)*, pp. 1-8.

[12] Kendon, A., 1990, Spatial Organization in Social Encounters: the F-formation System, in *Conducting Interaction: Patterns of Behavior in Focused Encounters*, A. Kendon ed., Cambridge University Press, pp. 209-238.

[13] Kobayashi, Y., Kinpara, Y., Takano, E., Kuno, Y., Yamazaki, K. and Yamazaki, A., 2011, Robotic Wheelchair Moving with Caregiver Collaboratively Depending on Circumstances, *extended abstracts on ACM Conference on Human Factors in Computing Systems (CHI2011)*. ,

[14] Kuzuoka, H., Suzuki, Y., Yamashita, J. and Yamazaki, K., 2010, Reconfiguring Spatial Formation Arrangement by Robot Body Orientation, *ACM/IEEE Int. Conf. on Human-Robot Interaction (HRI2010)*, pp. 285-292.

[15] Montemerlo, M., Thrun, S. and Whittaker, W., Conditional Particle Filters for Simultaneous Mobile Robot Localization and People-Tracking, *ICRA 2002*, pp. 695-701.

[16] Morales, Y., Satake, S., Kanda, T. and Hagita, N., 2011, Modeling Environments from a Route Perspective, *6th ACM/IEEE International Conference on Human Robot Interaction*,

[17] Moussaïd, M., Perozo, N., Garnier, S., Helbing, D. and Theraulaz, G., 2010, The walking behaviour of pedestrian social groups and its impact on crowd dynamics, *PloS one*, vol. 5,

[18] Nuchter, A., Lingemann, K., Hertzberg, J. and Surmann, H., 2007, 6d slam-3d mapping outdoor environments, *Journal of Field Robotics*, vol. 24, pp. 699–722.

[19] Peter H. Kahn, J., Freier, N. G., Kanda, T., Ishiguro, H., Ruckert, J. H., Severson, R. L. and Kane, S. K., 2008, Design Patterns for Sociality in Human-Robot Interaction, *ACM/IEEE Int. Conf. on Human-Robot Interaction (HRI2008)*, pp. 97-104.

[20] Pineau, J., Montemerlo, M., Pollack, M., Roy, N. and Thrun, S., 2003, Towards robotic assistants in nursing homes: challenges and results, *Robotics and Autonomous Systems*, vol. 42, pp. 271-281.

[21] Prassler, E., Bank, D. and Kluge, B., 2002, Key technologies in robot assistants: Motion coordination between a human and a mobile robot, *Transactions on Control, Automation and Systems Engineering*, vol. 4, pp. 56–61.

[22] Sisbot, E. A., Marin-Urias, L. F., Alami, R. and Simeon, T., 2007, A Human Aware Mobile Robot Motion Planner, *IEEE Transactions on Robotics*, vol. 23, pp. 874-883.

[23] Slam6d. *Slam6d - simultaneous localization and mapping with 6 dof*, Retrieved 2011, May, 20 http://www.openslam.org/slam6d.html.

[24] Thrun, S., Burgard, W. and Fox, D., 2005, *Probabilistic Robotics (Intelligent Robotics and Autonomous Agents)*, The MIT Press.

[25] Trafton, J. G., Cassimatis, N. L., Bugajska, M. D., Brock, D. P., Mintz, F. E. and Schultz, A. C., 2005, Enabling Effective Human-Robot Interaction Using Perspective-Taking in Robots, *IEEE Trans. on Systems, Man, and Cybernetics. Part A: Systems and Humans*, vol. 35, pp. 460- 470.

[26] Walters, M. L., Dautenhahn, K., Boekhorst, R. t., Koay, K. L., Kaouri, C., Woods, S., Nehaniv, C., Lee, D. and Werry, I., 2005, The Influence of Subjects' Personality Traits on Personal Spatial Zones in a Human-Robot Interaction Experiment, *IEEE Int. Workshop on Robot and Human Interactive Communication (RO-MAN2005)*, pp. 347-352.

[27] Wang, C.-C. and Thorpe, C., 2002, Simultaneous localization and mapping with detection and tracking of moving objects, *Proceedings of the IEEE International Conference on Robotics and Automation (ICRA 2002)*,

[28] Xu, S. and Duh, H. B.-L., 2010, A Simulation of Bonding Effects and Their Impacts on Pedestrian Dynamics, *IEEE Trans. on Intelligent Transportation Systems*, vol. 11, pp. 153-161.

[29] Zanlungo, F., Ikeda, T. and Kanda, T., 2011, Social Force Model with explicit collision prediction, *Europhysics Letters*, vol. 93,

A Techno-Sociological Solution for Designing a Museum Guide Robot: Regarding Choosing an Appropriate Visitor

Akiko Yamazaki
Tokyo University of Technology
1404-1 Katakuramachi, Hachioji City
Tokyo 192-0982, Japan
ayamazaki@media.teu.ac.jp

Keiichi Yamazaki
Saitama University
255 Shimo-Okubo, Sakura-ku
Saitama 338-8570, Japan
byi06561@nifty.com

Takaya Ohyama
Saitama University
255 Shimo-Okubo, Sakura-ku
Saitama 338-8570, Japan
tohyaman@cv.ics.saitamau.ac.jp

Yoshinori Kobayashi
Saitama University, JST, PRESTO
255 Shimo-Okubo, Sakura-ku
Saitama 338-8570, Japan
yosinori@cv.ics.saitamau.ac.jp

Yoshinori Kuno
Saitama University
255 Shimo-Okubo, Sakura-ku
Saitama 338-8570, Japan
kuno@cv.ics.saitamau.ac.jp

ABSTRACT
In this paper, we present our work designing a robot that explains an exhibit to multiple visitors in a museum setting, based on ethnographic analysis of interactions between expert human guides and visitors. During the ethnographic analysis, we discovered that expert human guides employ some identical strategies and practices in their explanations. In particular, one of these is to involve all visitors by posing a question to an appropriate visitor among them, which we call the "creating a puzzle" sequence. This is done in order to draw visitors' attention towards not only the exhibit and but also the guide's explanation. While creating a puzzle, the human guide can monitor visitors' responses and choose an "appropriate" visitor (i.e. one who is likely to provide an answer). Based on these findings, sociologists and engineers together developed a guide robot that coordinates verbal and non-verbal actions in posing a question or "a puzzle" that will draw visitors' attention, and then explain the exhibit for multiple visitors. During the explanation, the robot chooses an "appropriate" visitor. We tested the robot at an actual museum. The results show that our robot increases visitors' engagement and interaction with the guide, as well as interaction and engagement among visitors.

Categories and Subject Descriptors
H5.2. Information interfaces and presentation (e.g., HCI): User Interfaces – interaction styles.

General Terms
Performance, Design, Human Factors, Languages, Theory.

Keywords
Interaction analysis, museum, guide robot, human-robot interaction, computer vision.

1. INTRODUCTION
In studies of human-computer interaction, there has been an increasing recognition of the importance of human-robot interaction. This is due not only to improvements in technology, but also because the utility and usability of robots in providing explanations of things is becoming widely recognized. There are several recent advances in robot explanations to consider:

1. A robot can provide explanations in a real setting.
2. A robot can explain an object by coordinating verbal and non-verbal actions.
3. A robot can interact with multiple people at once, as well as interact with individual persons.

In terms of museums, explanations of exhibits are one of the most important services provided to visitors. Visitors experience real objects as exhibits in a museum, but they may require an explanation about them. Some visitors come alone, but more often they bring one or more companions such as a girl/boyfriend, family member, teacher, or friend. Consequently, this necessitates a way to provide explanations for visitors in multiparty settings in museums.

There are already several types of support used for such explanations in museums: labels, pamphlets, human guides, resource rooms, and digital technologies such as a PDA (personal digital assistant) utilizing headphones or earphones. The PDA is particularly useful and enhances the museum experience for visitors, who may wish to go from exhibit to exhibit at their own pace. The PDA tends to facilitate an individual experience, but it cannot enrich the experience of multiparty visitors as easily as when these visitors interact with an expert human guide [14]. Visitors to a museum often come accompanied by companions, and therefore represent a component of a social structure, so this social structure should be maintained and promoted.

There is growing interest among robotics researchers in facilitating the experience of multiparty visitors in museums. In order to support visitors' multiparty experience, museum guide robots should be able to draw the attention of visitors, and hold this attention throughout the course of the explanation as an expert human guide does. In recent studies, many researchers have been interested in applying to robots non-verbal behavior that humans employ in interaction, such as head-turning and nodding, body torque, and verbal pauses and restarts [12][13][7]. However, human guides coordinate their non-verbal and verbal actions precisely in order to draw and maintain the attention of multiparty visitors and involve them in the explanation. Thus far, while there has been work on robot gaze and footing during storytelling [8] [9], few studies have addressed robot verbal and non-verbal coordination.

In order to better understand how human guides coordinate their verbal and non-verbal actions, we conducted ethnographic studies, and based on their analysis of interaction between expert human guides and visitors they programmed their guide robot to turn its head towards the visitors at the ends of sentences [15]. We found that visitors are more likely to gaze and nod towards the robot, as displays of understanding, when the robot turns its head towards them than when it does not. We also found that appropriate coordination of verbal and non-verbal actions in the robot increases visitor engagement [17].

Since question-answer sequences were quite frequent in multiparty settings, we also studied how question-answer sequences facilitate guide-visitor interaction vis-à-vis the exhibit. We categorized questions asked by the guide into two types: (1) Type 1: questions that require the visitor to rely on his/her prior knowledge, and (2) Type 2: questions that require the visitor to examine and consider the exhibit. Expert guides tended to make frequent use of Type 2 questions [16].Their findings reveal that when the guide poses a question, he or she gazes towards the exhibit, and likewise the visitors gaze towards the exhibit while answering the question, nodding, and so forth. When asked Type 2 questions, visitors often displayed an increase in interest and engagement through verbal and non-verbal means. Based on these findings, we programmed their robot to pose Type 2 questions while gazing towards the exhibit. As expected, visitors answered the robot's questions while gazing towards the exhibit. Further, when the robot provided an answer visitors tended to nod and utter words while gazing towards the exhibit. This suggests that Type 2 questions increase visitor's displayed interest towards the exhibit in verbal and non-verbal ways.

In this study, in order to isolate how human guides and multiparty visitors interact and how they coordinate their verbal and non-verbal behavior, we conducted video ethnographic fieldwork in several museums and analyzed the data through an approach based on conversation/interactional analysis. In this ethnographic study, we focused on how expert human guides coordinate their verbal and non-verbal actions, because in the case of expert human guides their strategies and practices to increase multiple visitors' engagement at the same time while smoothly interacting with them make them quite superior to non-experienced human guides. One of the most effective strategies employed by expert human guides during question-answer sequences is how they choose an 'appropriate visitor' to ask a question. Based on our findings, in order to develop a robot that interacts with visitors more smoothly, we implemented verbal and non-verbal coordination in the robot. We tested this by having the robot explained the exhibit in real museum settings, whereupon we evaluated the results and proceeded to develop our robot further.

2. GUIDE-VISITOR INTERACTION IN MUSEUMS

We conducted video ethnographic fieldwork at several museums in order to find out how expert human guides conveyed their explanations to visitors. In order to clearly articulate the skills of the expert guides, we compared a situation where both expert and non-experienced guides explained the same exhibit to multi-party visitors at the Japanese Overseas Migration Museum. The Japanese Overseas Migration Museum houses valuable memorabilia of the migration of untold numbers of Japanese throughout history to many corners of the world. We videotaped the interaction of guides and visitors there on 31 January 2009 and 11 June 2011. In the first instance we used two video cameras to videotape an expert volunteer guide giving an explanation to three sets of multiparty visitors over the course of four hours. In the second instance we used three video cameras to videotape two non-experienced guides explaining the same exhibits to two sets of multiparty visitors over about four hours. We analyzed this data from the perspective of visitors' participation and the sequence of guides' questions, because guides frequently pose questions regarding exhibits to multiparty visitors, a strategy which we refer to as "creating a puzzle." We are especially interested in how visitors change their participation framework when the guide creates such a puzzle.

For instance, a guide may ask, "What does this express?" However, it is not the case that every guide uses the strategy of asking a question and receiving a response from multiparty visitors. Rather, this strategy was primarily used by "expert" guides as part of their repertoire of social practices and interactional strategies. Here, we attempt to employ these types of questions in our robot as indicative of an expert guide.

In his analysis of "dirty jokes," H. Sacks observed that speakers often create a puzzle in extended talk to multiple listeners [10]. When a speaker creates a puzzle, this invites multiple listeners to engage in the activity of providing an answer, thereby solving the puzzle. Sacks found that such a puzzle can package the information together with a solution. By creating a puzzle and a way to solve it, a speaker can transfer information to multiple people. Also, on occasion, the speaker can transfer the information towards a particular listener who offers a solution.

Casting a puzzle into the talk changes the participation of listeners from passive receptors to 'persons engaging in the activity for finding a solution.' Goodwin and Goodwin [3] examined participation frameworks that included not only speaker and listener but also a potential hearer, such as a bystander. In examining such participation frameworks, Goodwin and Goodwin made clear how participants change their orientation and engagement. M. Goodwin [2] analyzed boys' disputes, and revealed how one boy's storytelling about his opponent's cowardly behavior changed in the participation framework of boys who watched the dispute in question. At the beginning, some boys were audience members of the dispute, but became listeners of the story, and finally made an assessment of cowardice. In other words, the storyteller changed the framework of the other boys from passive to active listeners, and finally to speakers. When we address the issue of multi-party interaction, we have to consider not only the individual's understanding but also changes in the participation framework of all hearers who are co-present.

In this section, we will analyze how an expert guide gives an explanation and creates a puzzle in regards to an exhibit.

Fragments 1-3 are taken from the footage taken on 31 January, 2009. In Fragment 1 and Figure 1, an expert volunteer guide known by his skills explains a Vegetable Float to four visitors.

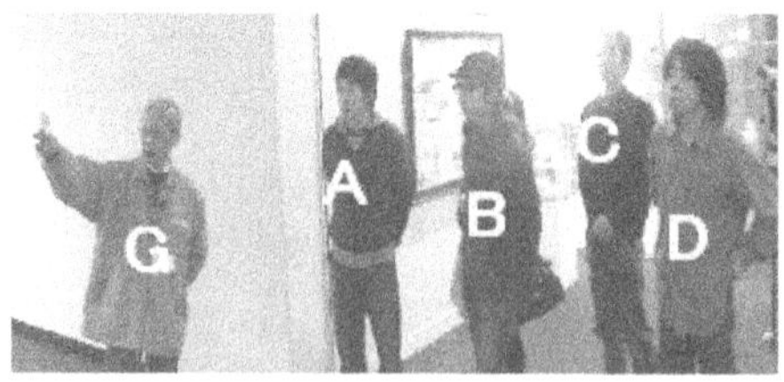

Figure 1. The guide explains a Vegetable Float

Fragment 1

G: Guide, Visitors: A, B, C, D

1 G: This: that:

((Pointing towards the roof of the float with his palm))

2 (1.0)

3 G: When you look at the roof you can see: (.) well:

4 (These) vegetables are not just pla: placed (on at random) right?

((G points towards the roof of the vegetable float))

((A gazes towards the upper side of the float))

5 (1.7)

6 A: ((nods))

7 G: What does it (=the vegetables) look like?

8 A: It is a flag, right?

9 G: Yes. It is the Stars and Stripes.

((The guide looks back towards all the visitors.))

In this fragment, the guide poses the first question that creates a puzzle by saying 'When you look at the roof you can see: (.) well: These vegetables are not just placed (on at random) right?' (lines 3 and 4). At the same time, this first puzzle becomes a clue to the second puzzle, 'What does it look like?'(line 7). Here, the guide draws the attention of all the visitors toward a specific point (the roof of the float), and then chooses visitor A, who responds by nodding. The guide then poses the second question directly to A, by asking 'What does it look like?'

Now we will examine the coordination of verbal and non-verbal actions between the guide and visitors in the sequence when the guide poses questions (creating a puzzle for the visitors to solve).

Fragment 2

Figure 2 Figure 3

3 G:When you look at the roof you can see: (.) well:

Figure 4 Figure 5

4 (These) vegetables are not just pla: placed (on it) right?

Figure 2. The guide points at the roof.

Figure 3. The guide turns around to the visitors.

Figure 4. The guide points at the roof again.

Figure 5. The guide poses a question while pointing at the roof.

In Fragment 2, when the guide gazes and points towards the roof while saying, 'When you look at the roof,' all four visitors gaze towards the roof (Figure 2). The guide stretches out the end of his sentence and turns back towards the visitors (Figure 3), then poses a question while gazing and pointing towards the roof of the float (Figure 4). At the end of this question, the guide looks at the visitors and asks another question (Figure 5). In this fragment, the guide monitors the movement of the visitors' gaze the first time he turns towards them. Then, by pointing towards the roof of the float and creating a puzzle, he monitors their gaze and response.

Fragment 3

5 (1.7)

6 A: ((nods)) (Figure 6)

7 G: What does it look like?

((G stops pointing and looks at A)) (Figure 7)

8 A: It is a flag, right?

9 G: Yes. It is the Stars and Stripes.

((The guide looks back towards all the visitors.)) (Figure 8)

Figure 6. The guide poses a question while pointing at the roof. A then nods.

Figure 7. The guide stops pointing and looks towards A. A then answers the question.

Figure 8. While giving the answer the guide turns toward the visitors. All the visitors then look at the exhibit.

In Fragment 3, the guide pauses 1.7 seconds after his question (line 5). During the pause, he monitors A's nodding as a positive response (Figure 6). The guide then poses a second question (a puzzle) towards the visitor who showed a positive response to the first question, rather than to all the visitors. The guide looks at visitor A (Figure 7) and asks, 'What does it look like?' (line 7). After hearing A's answer ('It is a flag, right?'), the guide turns back towards the visitors as a group, while saying 'Yes. It is the Stars and Stripes.' By gazing towards every visitor, the guide provides the answer (solution to the puzzle) not only to visitor A but also to the all co-present visitors (line 9, Figure 8). Through this question-answer (or puzzle-solution) sequence, the guide transfers the information package to the visitors, who then confirm the solution and information by examining the exhibit.

In this way, while coordinating his verbal and non-verbal actions, the guide first creates a puzzle, causing every visitor to gaze towards the roof. At the same time, the guide monitors the visitors' positive and negative responses. Then he chooses a particular visitor who displays a positive response such as nodding, and creates a second puzzle in the form of a question. When the guide provides the solution, he does so not only for the particular visitor who answered the question, but while also facing towards the other visitors, thereby prompting all of them to confirm the solution on their own. Through this sequence, the guide monitors the visitors' knowledge acquisition and understanding.

In Fragment 4 and Figure 9, conversely, a non-experienced temporary guide (actually a librarian) offers his explanation of the same exhibit (the vegetable float) to multiple visitors on 31 January, 2011.

Figure 9. A non-experienced guide explains the exhibit.

Fragment 4

G1: Guide, Visitors: A, B

[Figure 10] [Figure 11]

1 G1: When you look at the roof Well: you can see what kind of design?

2 (1.0)

[Figure 12] [Figure 13]

3 Can you (figure it out)? Well: it is the Stars and Stripes, the American flag. ↑

The guide points at the roof of the vegetable float while saying 'when you look at the roof' (Figure 10) then creates a puzzle by saying, 'Well: you can see what kind of design (has been made)?' He thus draws visitors' attention to the roof. After creating the puzzle, the guide turns his gazes towards visitor A and visitor B during a 1.0 second pause (Figure 11).

Figure 10. The guide creates puzzle by pointing.

Figure 11. The guide gazes towards visitors A and B.

However, visitors A and B are still looking at the roof, and therefore the guide's gaze does not meet with that of either visitor during this 1.0 second pause in line 2. The guide proceeds to create a puzzle by saying, "Can you (figure it out)?" in line 3. Both visitor A and visitor B keep looking at the roof, the guide's gaze failing to meet with either (Figure 12). Then the guide gives the solution to his puzzle by saying, "It is the Stars and Stripes, the American flag." During this utterance, both visitor A and visitor B continue looking at the roof (Figure 13). After the solution is given, at last visitor B comprehends the design of the roof of the vegetable float as shaped like the Stars and Stripes.

Figure 12. The guide gazes towards the visitors, although they keep looking at the roof.

Figure 13. The guide gazes towards the visitors while giving the solution, but they keep looking at the roof without watching the guide.

When we compare these fragments with the expert and non-expert guide creating the same puzzle in regard to the same exhibit, a significant difference is obvious. While the expert guide can choose a visitor who solves the puzzle immediately, the non-experienced guide could not choose an appropriate visitor to solve his puzzle, and had to solve it by himself. One of the strategies employed by expert guides is the duration of pauses. The expert guide employs longer pauses and during these he can monitor visitors' responses and choose a visitor to ask a question. When the guide solves their own puzzle on the other hand, they cannot draw the attention and involvement of multiparty visitors through question-answer sequences. We observed expert guides using the same strategy towards multiparty visitors at several different museums.

In designing a robot guide, we utilized the results gleaned from our analysis of expert guides interacting with multiple visitors. While all of the guides use practices and strategies related to creating a puzzle for the visitor, guides vary somewhat in the ways they deploy this practice. For instance, sometimes a guide provides the solution without first obtaining answers from visitors. We elected to choose the most successful, or "excellent," social practices and interactional strategies of expert guides from among the various museum guides and implement them in our robot.

3. ROBOT SYSTEM

3.1 Preliminary Experiment

We performed a preliminary experiment to examine how humans behave in response to a robot's questioning and how they feel depending on their knowledge of the answers of questions [5]. In that experiment, the robot turned its gaze towards multiple visitors. However, it chose an answerer at random. After the experiment, we asked the participants whether they had known the answers and whether they felt embarrassed when asked by the robot. The questionnaire results showed that the participants may have felt embarrassed when they did not know the answers. The video analysis revealed that the responses to the robot's gazing towards the participants could be classified into the following three types:

•Positive response: visitor produces a mutual gaze or nods during the last half of the robot's gaze distribution period, or looks straight at the robot during the robot's gaze distribution period without avoiding its gaze.

•Negative response: visitor avoids the robot's gaze without nodding during the last half of robot's gaze distribution period.

•Neutral response: visitor's actions are neither those in "positive" or "negative" above (does not nod or change head direction).

We found that participants who knew the answers often showed positive responses whereas those who did not may have displayed negative responses. Based on these findings, we developed a computer vision method of classifying human responses into the above three types by analyzing head movement (tracking approximate gaze movement). This method was applied to the stored video data of the preliminary experiment and was shown to be effective.

3.2 Robot System for Field Experiments

In the preliminary experiment, we utilized an omni-directional camera to include multiple visitors within a field of view, and a PC for image processing. Although the experiment using the stored video data suggested that the algorithm could work, the robot system required some improvements for field experiments. There were concerns that the image resolution might not have been sufficient to detect small head movements among multiple visitors, and that the system might not have worked in real time. Thus, we replaced the omni-directional camera with three pan-tilt-zoom (PTZ) cameras and used three PCs, each dedicated to processing images from one PTZ camera. Figure 14 presents an overview of the modified robot. In addition to the three PTZ cameras and three PCs for image processing, we used a laser range sensor (Hokuyo UTM-30LX) and another PC for processing the range sensor data and integrating all of the sensor processing results. The system detects and tracks the bodies of multiple visitors in the range sensor data. A PTZ camera is assigned to each detected body. The system is controlled to enable it to turn towards observed bodies. The processing of range sensor data is the same as that employed in the previous system. For detecting and tracking a face, and computing its direction, we used FaceAPI [11] in this system. The pan, tilt and zoom of each PTZ camera are automatically adjusted based on the face detection results, so that the face remains focused in the center of the image. The system can locate a human body with a margin of error of 6 cm for position and 6 degrees for orientation. It can measure 3D face direction within 3 degrees of error at 30 frames per second.

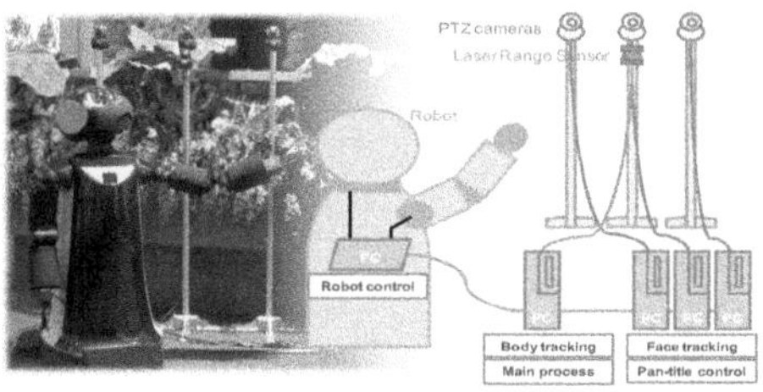

Figure 14. Overview of our guide robot system.

From the face direction results, the system can recognize the following behaviors:

Nodding: Recognized if the pitch movement of the head changes by more than +/- 5 degrees from its moving average observed in the past 10 frames (hereafter called "steady value").

Shaking: Recognized if the yaw movement of the head changes by more than +/- 5 degrees from its steady value.

Cocking: Recognized if the roll movement of the head changes by more than +/- 20 degrees from its steady value.

Gazing toward the robot: Recognized if the current head direction is towards the robot and has changed by more than 30 degrees from the steady value of the previous 10 frames.

Gazing away from the robot: Recognized if the current head direction is away from the robot and has changed by more than 30 degrees from the steady value of the previous 10 frames.

The above parameters and threshold values have been determined experimentally.

Figure 15 shows an example of the head gesture recognition process. The red lines indicate the horizontal (yaw) component of the head movement whereas the blue ones indicate the vertical (pitch) component. While asking questions, the robot distributes its gaze towards the visitors by turning its head, and recognizes their behavior. It chooses the answerer based on these behaviors recognition results. The selection method is described in the next section.

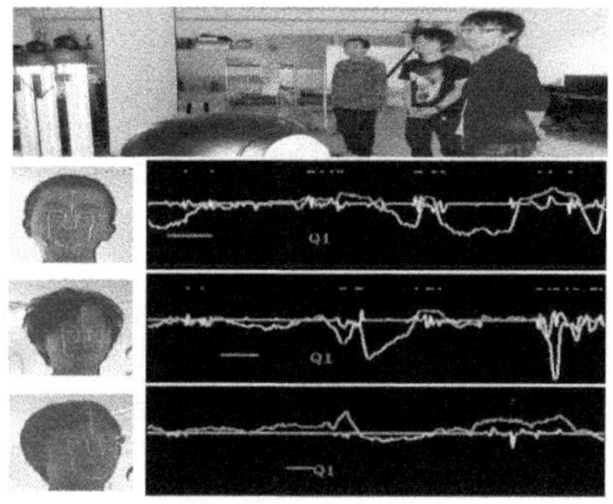

Figure 15. Example of human observations made using our sensing system.

4. FIELD EXPERIMENTS AT A MUSEUM

4.1 Experimental Setup

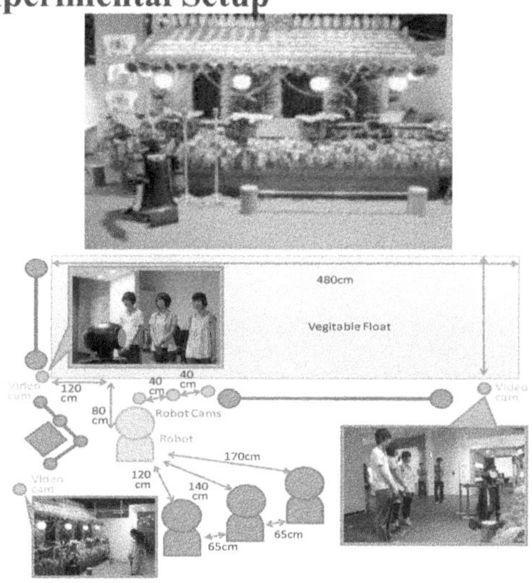

Figure 16. Experimental setup at an actual museum.

We performed experiments with the robot system at the Japanese Overseas Migration Museum. The robot performed the role of explaining the Vegetable Float exhibit, which was analyzed in the human-human interaction analysis discussed earlier. Figure 16 shows the experimental setup. The robot system can detect and track visitors. Thus human visitors can be at any position. We allowed them to be anywhere during the demonstration periods in order to test this. However, we specified approximate standing positions of three visitors in the experiments for human-robot interaction analysis in order to avoid any tracking errors that might happen when visitors move quickly and as a group. In fact, since human guides usually prompt visitors to move to appropriate positions, this setting may not be unrealistic (we are developing a robot that can initiate such a proper formation of visitors). It should be mentioned that the robot did not use its locomotive functions during these experiments.

We conducted our experiment on the third Sunday of August 2010. Since it was a Japanese summer vacation for schools, museum visitors mainly consisted of families, couples, friends, and parents and children. In this experiment, in which the robot explained the Vegetable Float for three visitors, we set a student dummy next to sets of two visitors who were couples and friends. The dummy kept its gaze held in one direction and did not react.

The participants were actual visitors who visited the museum on the day of the experiment. The effective number of participant sets was 31, while the number of valid participants was 71 (29 males and 42 females), with the total number being 93 if the dummy is included in each case. The robot conducted an explanation lasting 171 seconds, and we set 5-to-10 second pauses for awaiting visitors' answers. We programmed the robot's verbal and non-verbal actions while considering the following factors:

1. The coordination of verbal and non-verbal actions was programmed based on taking into consideration all the previous findings of our research group [5]. The robot displayed its change of orientation to visitors interactionally by turning its head and directing its gaze.
2. The robot created 1st puzzle for the visitors by posing questions and coordinating its verbal and non-verbal actions appropriately.
3. The second puzzle took the form of a pre-question in order to become a clue for solving 1st puzzle.
4. When the robot offered the second puzzle, it turned its head towards visitors and chose an appropriate visitor using the behavior recognition results in real time (the selection method is described below).
5. The robot gazed towards the selected visitor and asked a question (puzzle 3) that could be answered by referring to the object (i.e. exhibit).

The robot's utterance and bodily actions in the program are as follows. RT stands for the robot's verbal turn. The words inside the double parenthesis (( )) describe the robot's bodily actions. VT stands for a visitor's turn; however, in the program it consists of the robot pausing.

RT0: Hello. My name is Robovie. ((greeting))

RT1: This is a float made of vegetables.

((The robot turns its head towards the visitors at the end of the sentence.))

VT1: 6.0 second pause

RT2: In this float there is a "puzzle" hidden inside.

VT2: 4.7 second pause

RT3: Do you know what it is? (Puzzle 1)

((Robot points and gazes towards all visitors))

VT3: 7.0 second pause

RT4: When you look at the roof,

((Robot gazes and points towards the roof)) these vegetables are not just lined up there, right?

(Puzzle 2 and pre-question)

((The robot looks around at each visitor))

VT4 6.3 second pause

((The robot chooses a positive visitor during this pause))

RT5: What does it look like?

(Puzzle 3 =involvement question type 2)

((The robot gazes towards the positive visitor and asks a question))

VT5: (WOZ pause)

((Pause for response))

RT6: It is the Stars and Stripes.

((Solution to all three puzzles))

((The robot turns its head towards the roof, in order to prompt the visitors to confirm the solution by examining the roof).

4.2 Method of Selecting an Answerer

The robot selected an appropriate listener to answer based on the observed behaviors of the visitors that were recognized by the procedure described in the previous section. We had categorized visitors' demonstrated attitudes into Positive, Negative and Neutral. Nodding and gazing towards the robot were interpreted as a generally positive attitude, whereas shaking and gazing away from the robot were interpreted as a generally negative attitude. The basic strategy of answerer selection was thus to choose visitor who had demonstrated a positive attitude, if present, while avoiding those who had demonstrated negative attitudes.

For the second and third puzzles, the visitors' attitudes were categorizes as Positive if they nodded more than three times during the robot's question, and Negative if they shook their heads more than nine times; those who did neither were categorized as Neutral. We did not employ "gaze toward" and "gaze away" behaviors for this puzzle because visitors moved their gaze often and significantly since the target object (the roof) in question was large. Multiple visitors may demonstrate the same attitude, and consequently we assigned a score to each visitor via the following equations. The robot chose the visitor with the highest score as the answerer.

Positive: 100 + number of nods

Neutral: 50 + number of nods - number of head shakings

Negative: 0 - number of head shakings

4.3 Visitors' Responses to the Museum Experiment

Overall Tendencies

The overall tendencies of the visitors were as follows:

1. Visitors' positive response towards the robot using verbal and non-verbal behaviors was significant.
2. Interaction and conversation among visitors often occurred, especially among family members.

In the human guide-visitor interaction that we originally observed, conversation between the guide and visitors often occurred. However, conversation among visitors was not frequent.

One of the differences of interaction between the human guide-visitor and the robot guide-visitor dynamic is that the visitors tended to speak and offer their impressions on the exhibit to each other in front of the robot guide. While in our previous experiments using a museum guide robot (at the Science Museum in Tokyo and at the Ohara Museum of Art) we observed the same tendency, this tendency was more strongly marked in this experiment (Figure 17).

There are both advantages and disadvantages to the robot guide prompting an increase in interaction among visitors.

One advantage is that it may enhance overall 'enjoyment' of the museum experience, especially pleasure and affection among family members at the museum. Many visitors, especially families and couples, shared laughter warmly and happily during this experiment.

One disadvantage, on the other hand, is that visitors seemed to concentrate on the robot and its explanation, and were sometimes distracted from paying attention to the exhibit itself. However, to lessen this distraction, we programmed the robot to create puzzles like a human guide, posing Type 2 questions (i.e. ones that required the visitor to examine and consider the exhibit), as discussed earlier. Such puzzles seemed to invite the visitors to examine the exhibit and discuss the solution among themselves.

It should be mentioned that there is a methodological problem in evaluating human-robot interaction. When examining robot-visitor dyadic interaction, it is not difficult to code and quantify the responses of the visitor. However, in robot-visitor multiparty interaction, the responses of visitors depend on their activities at the moment, and so it is more difficult to code and quantify individual responses.

In relation to this, in multiparty interaction, we evaluated two aspects: the response of multiple visitors to the robot in general, and the response of the particular visitors chosen by the robot. When presenting puzzle 1, the robot made the query relevant to all the visitors by turning and gazing towards them. In assessing this movement, we focused not only on individual responses but also on how multiple visitors responded: one aspect is the response towards the robot simultaneously and coordinately, and a second aspect is the degree of discussion among the visitors. In presenting the results of this analysis, we want to clearly indicate how individual visitors responded towards the robot and how multiple visitors responded towards the robot as a group, through conversation, discussion, laughter, and gaze motion.

In assessing puzzles 2 and 3, we analyzed how the robot chose a particular visitor. After the robot provided the solution, we analyzed multiple visitors' responses as we did for puzzle 1. We then compared this with the visitors' responses observed when the robot asked a question related to prior knowledge.

Analysis of Visitors' Responses towards Puzzle1

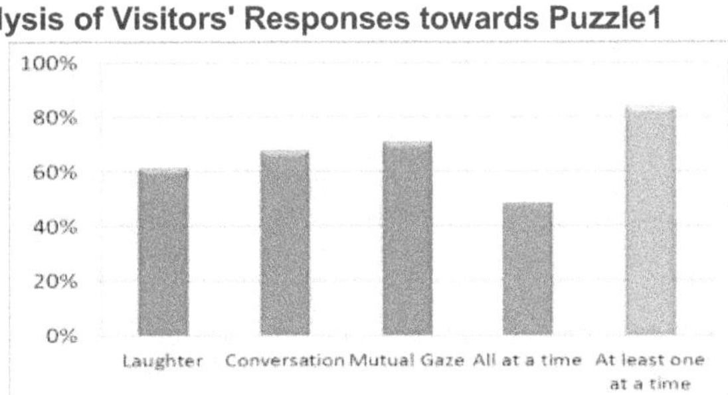

Figure 17. The interaction and engagement of visitors to the puzzle 1.

Figure 17 shows the interactions and engagement of visitors to puzzle 1. The robot creates its puzzle for the visitors, saying 'In this float there is a "puzzle" hidden inside. Do you know what it is?' while looking and pointing towards everyone (Figure 18). Here, 26 out of 31 cases showed interaction among visitors (sharing laughter, conversation, mutual gaze). Moreover, 19 out of the 26 cases showed sharing laughter, 21 cases showed conversation, and 22 cases showed mutual gaze. Therefore creating a puzzle seems to genuinely promote an increase in interaction among visitors. Every visitor looked at the exhibit, and some visitors examined it carefuly, some while putting on eye glasses. The robot clearly promoted the engagement of visitors.

Figure 18. The robot points towards each of the visitors.

Analysis of Puzzles 2 & 3

After creating puzzle 1, the robot started to say 'When you look at the roof' while pointing to the roof, and initiated puzzle 2 by saying 'When you look at the roof, these vegetables are not just lined up there, right?'. When the robot started to say 'When you look at the roof' while pointing, every visitor moved his or her gaze towards the roof. This suggests that such a puzzle increases the engagement of visitors with the exhibit. Then, the robot offered puzzle 2, saying 'these vegetables are not just lined up there, right?' while looking around at the visitors. We categorized responses towards this gaze into Positive, Neutral, and Negative responses by using our formula.

The number of positive responses was 36, neutral responses was 30, and negative responses was 5 (Figure 19) . The robot gazed towards the person demonstrating the most positive attitude among the three visitors and initiates puzzle 3 by posing another question, saying 'What does it look like?' The number of positive, neutral and negative responses to this puzzle were 26, 5, and 0 respectively (Figure 20).

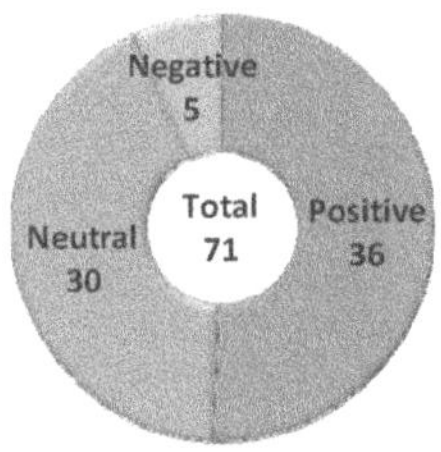

Figure 19. Judgements made by the robot.

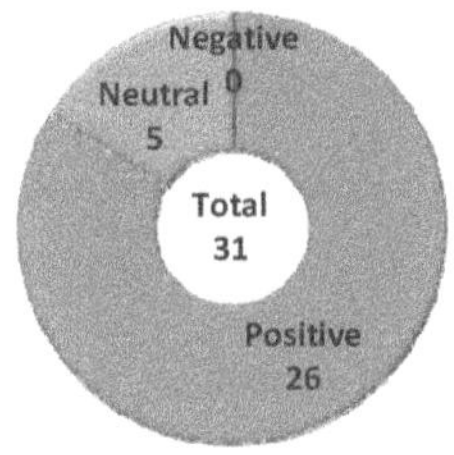

Figure 20. Ratio of responses of selected visitors.

According to the questionnaire, the number of visitors who knew the correct answer before the robot provided it was 14. Figure 21 shows the number of positive, neutral and negative responses of visitors who knew the correct answer (left), and that for visitors who did not (right). Since the number of visitors who knew the answer was too small, we cannot find significant differences although such visitors often showed positives response and never showed negative responses.

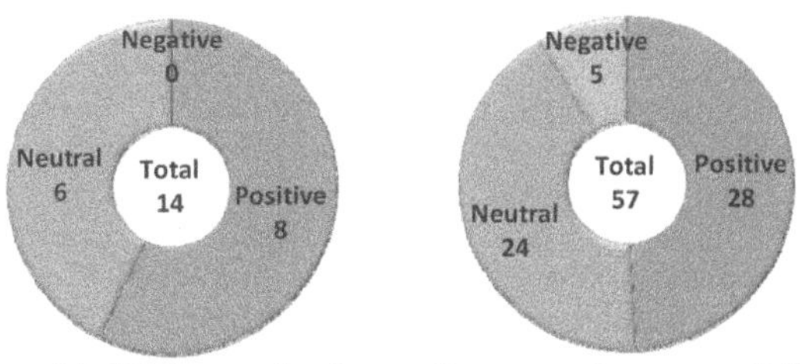

Figure 21. Visitors who knew the correct answer (left), and visitors who did not know beforehand (right) .

The number of cases that included at least one visitor who knew the correct answer was 8, while 14 visitors in these 8 cases knew the correct answer. The ratio of correct answers was 14/24 (58%), though the actual ratio of the robot choosing a visitor who knew the correct answer was 6/8 (75%). There were 2 cases in which the robot did not choose a person who knew the correct answer. Nevertheless, in 1 of these 2 cases, two visitors showed positive responses, and the robot chose another visitor.

Response of Multiple Visitors When the Robot Gave the Correct Answer

The robot gave the correct answer at the end, saying 'The Stars and Stripes' after puzzle 3 and looking around at the visitors. When the robot gave the correct answer, 61 out of 71 visitors gazed towards the roof of the vegetable float, and 41 visitors produced some utterances. 39 of these 41 showed a positive change in state of knowledge, such as by saying 'Oh,' 'Ah,' or repeating the answer [4].

In 38 out of the 41 cases, visitors uttered something and moved their gaze. In 17 of the 41 cases, the visitors collectively turned their gaze towards the roof.

Evaluation of Experiments

These experiments show that by coordinating verbal and non-verbal actions and creating a puzzle the guide robot prompted visitors' engagement with the exhibit and moreover encouraged interaction among the visitors.

In particular, when the robot presented puzzle 2 as an involvement question (type 2), visitors who knew the correct answer displayed positive responses (e.g. head nodding), and the robot recognized those responses. When the robot moved to puzzle 3, the visitors with positive responses uttered the correct answer. After the robot gave the solution, some visitors collectively moved their gaze and confirmed the answer. Moreover, they tended to utter "oh"-prefaced responses that indicated a change in their state of knowledge [4]. At the same time, visitors shared laughter among themselves. These behaviors show their affirmative and positive feelings towards each other vis-à-vis both the exhibit and the robot.

However, a few visitors knew the correct answer to our puzzles in this experiment but did not provide positive responses. One possible solution to this phenomenon, based on observations of expert guides, is that if there is no positive response offered we should add a clue as a second preliminary question to prompt a positive attitude.

5. CONCLUSION

In this paper, we discussed how our guide robot's use of puzzles for the visitors regarding an exhibit is an effective strategy to promote listening and reaction to the robot's explanation. We proposed a system in which the robot guide creates a puzzle and selects a visitor who shows positive responses towards it based on our ethnographic analysis of the interaction between human guides and visitors. We assessed this system's effectiveness by examining real visitors' behavior towards a robot in a museum. For designing a robot system that smoothly interacts and collaborates with humans, we must observe human social practices and interactional strategies, and program these effective behaviors into robot systems accordingly. In designing such robots, we have to consider these techno-sociological design issues and solutions, and to this end we can best tackle these issues through effective collaboration among engineers and social scientists.

REFERENCES

[1] Bennewitz, M., Faber, F., Joho, D., Schreiber, M. and Behnke, S. Towards a humanoid museum guide robot that interacts with multiple persons. In *Proc. HUMANOIDS 2005* (2005), 418-423.

[2] Goodwin, M. H. Tactical uses of stories: participation frameworks within boys' and girls' disputes. In *Tannen, D. ed., Gender and Conversational Interaction* (New York, Oxford University Press, 1993).

[3] Goodwin, C. and Goodwin, M. H. Participation. In *Duranti, A. ed., A Companion to Linguistic Anthropology* (Malden, MA: Blackwell, 2004), 222-244.

[4] Heritage, J. Oh-prefaced responses to inquiry. *Language in Society,* 27.3 (1998), 91-334.

[5] Kobayashi, Y., Shibata, T., Hoshi, Y., Kuno, Y., Okada, M. and Yamazaki, K. 'I will ask you': choosing answerers by observing gaze responses using integrated sensors for museum guide robots. In *Proc. RoMan 2010* (2010), 687-692.

[6] Kuno, Y., Sadazuka, K., Kawashima, M., Yamazaki, K., Yamazaki, A. and Kuzuoka, H. Museum guide robot based on sociological interaction analysis. In *Proc. CHI 2007* (ACM press, 2007), 1191-1194.

[7] Kuzuoka, H., Pitsch, K., Suzuki, Y., Kawaguchi, I., Yamazaki, K., Yamazaki, A., Kuno, Y., Luff, P. and Heath, C. Effect of pauses and restarts on achieving a state of mutual orientation between a human and a robot. In *Proc. CSCW 2008* (2008), 201-204.

[8] Mutlu, B., Hodgins, J. K. and Forlizzi, J. A. A storytelling robot: modeling and evaluation of human-like gaze behavior. In *Proc. HUMANOIDS 2006* (2006), 518-523.

[9] Mutlu, B., Shiwa, T., Kanda, T., Ishiguro, H. and Hagita, N. Footing in human-robot conversations: how robots might shape participant roles using gaze cues. In *HRI 2009* (ACM Press, 2009), 61-68.

[10] Sacks, H. Some technical considerations of a dirty joke. In *Schenkein, J. N. ed., Studies in the Organization of Conversational Interaction* (New York: Academic Press, 1978), 249-70.

[11] Seeing Machines Limited, FaceAPI. http://www.seeingmachines.com/product/faceapi/.

[12] Sidner, C. L., Lee, C., Kidd, C. D. and Rich, C. Explorations in engagement for humans and robots. *Artificial Intelligence,* 166.1 (2005), 140-164.

[13] Sidner, C. L., Lee, C., Morency, L-P. and Forlines, C. The effect of head-nod recognition in human-robot conversation. In *HRI 2006* (ACM Press, 2006), 290-296.

[14] Tallon, T. Introduction: mobile, digital, and personal. In *Tallon T. and Walker, K. eds. Digital Technologies and the Museum Experience: Handheld Guides and Other Media* (Alta Mira Press, 2008).

[15] Yamazaki, A., Yamazaki, K., Kuno, Y., Burdelski, M., Kawashima, M. and Kuzuoka, H. Precision timing in human-robot interaction: coordination of head movement and utterance. In *Proc. CHI 2008* (ACM Press, 2008), 131-140.

[16] Yamazaki, K., Yamazaki, A., Okada, M., Kuno, Y., Kobayashi, Y., Hoshi, Y., Pitsch, K., Luff, P., Lehn, D. V. and Heath, C. Revealing Gauguin: engaging visitors in robot guide's explanation in an art museum. In *Proc. CHI 2009* (ACM Press, 2009), 1437-1446.

[17] Yamazaki, A., Yamazaki, K., Burdelski, M., Kuno, Y. and Fukushima, M. Coordination of verbal and non-verbal actions in human–robot interaction at museums and exhibitions. *Journal of Pragmatics,* 42.9 (2010), 2398-241.

Panel

Robots in the Loop: Telepresence Robots in Everyday Life

Katherine M. Tsui
Department of Computer Science
University of Massachusetts Lowell
1 University Avenue, Lowell MA, USA
ktsui@cs.uml.edu

Stephen Von Rump
Giraff Technologies AB
Sintervägen 14
721 30 Västerås, Sweden
stephen.vonrump@giraff.org

Hiroshi Ishiguro
Osaka University, Japan

Leila Takayama
Willow Garage, USA

Peter Vicars
VGo Communications, USA

ABSTRACT
This year's human-robot interaction (HRI) conference focuses on "robots in the loop" and how robots are capable of enhancing the experiences of human users in everyday life and work. Telepresence robots allow operators the ability to participate in remote locations through their mobility and live bidirectional audio and video feeds. Using robotic telepresence, students with chronic illnesses are attending their regular classes, physicians are conducting virtual "home visits" for recovering patients, and remote teammates are having conversations beyond the office conference room.

This panel gathers experts from academia, business, and industry to discuss their experiences in developing robotic telepresences and "ah ha" moments reported from field use. Topics include how telepresence is defined, the practical use cases and application domains, the social and practical challenges encountered by operators and people physically present with the robots, and the implications for design of telepresence robots given these considerations.

Categories & Subject Descriptors: I.2.9. Artificial Intelligence: Robotics; H.5.m. Information interfaces and presentation: Miscellaneous.

General Terms: Human factors.

Keywords: Human-robot interaction, embodied video-mediated communication.

1. INTRODUCTION
The term telepresences emphasizes the idea of remotely "being there" in such a high fidelity manner it seems as though the experience was "in person" [2]. The Personal Roving Presence (PRoP) robots were the first internet controlled, untethered, terrestrial robotic telepresences, developed as research platforms in the late 1990s [3]; they were mobile robots with live two-way audio and video feeds and pointing devices. PRoPs enabled their users to wander around a remote space, converse with people, hang out, examine objects, read, and gesture, which are largely the goals of contemporary telepresence robots. In fact, many of the technical challenges (i.e., audio and video, navigation, safety, overall design) and social challenges (i.e., directed eyegaze, privacy) that were discussed in the early PRoP research still exist today.

HRI'12, March 5–8, 2012, Boston, Massachusetts, USA.
ACM 978-1-4503-1063-5/12/03.

Academic research projects push the boundaries of the leading edge technology and have long-term goals that can be realized in the commercial world given time. Fifteen years after PRoP, robotic telepresence is just becoming technologically viable. Recent advances in consumer technology and information services are poised to push robotic telepresence into the main stream. Smaller and faster computers, low-cost and high resolution webcams, high color LCD screens, relatively inexpensive range sensors (i.e., Hokuyo URG-04LX-UG01, Microsoft Kinect), and nearly ubiquitous wireless communication (i.e., 802.11 WiFi, 3G and 4G cellular services) have transformed robotic telepresence from being solely a research concept into commercially available products. In the last few years, we have begun to see how these robotic telepresences can actually be used in the real world.

2. PANEL DESCRIPTION
The target audience for this 1-hour 40-minute session is primarily professional or academic researchers interested in the domain space of robotic telepresence. We have invited panelists with experience in both industry and academia whose backgrounds range from business to communications to electrical engineering. The session is designed to encourage discussion and debate among the panelists and audience. A moderator will pose a series of questions including:

- What is "robotic telepresence" exactly? How is its definition influenced by characteristics of embodiment and overall function? How is a telepresence robot different than video conferencing?
- How can the robotic telepresence design accommodate both the needs of the operator and the people present with the robot? What functional capabilities are needed for specific use cases? Can these generalize to a common platform?
- How can human-human non-verbal conversational cues, such as eye gaze and weight shifting, be interpreted for robotic telepresence?
- What use cases, application domains, and markets are best suited for robotic telepresence? What application of robotic telepresence will rocket this technology into ubiquity?
- How can we change the stereo-typing of telepresence robots as an encroachment of people's privacy (e.g., the spying boss [1])?

3. INVITED PANELISTS

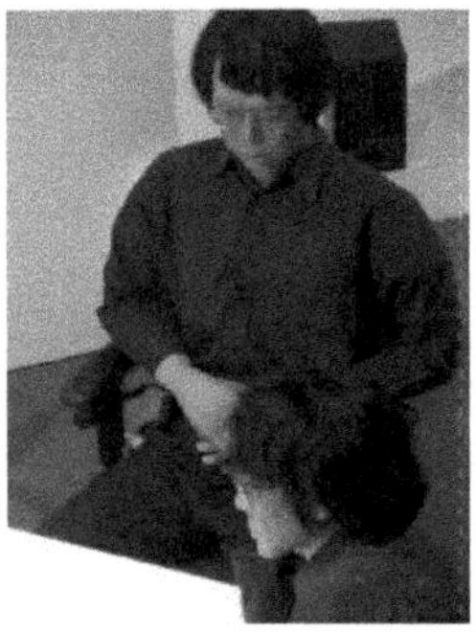

Hiroshi Ishiguro, *Professor, Osaka University, Japan*

Hiroshi Ishiguro received a D.Eng. in systems engineering from the Osaka University, Japan in 1991. He is currently Professor of Department of Systems Innovation in the Graduate School of Engineering Science at Osaka University and Director of Hiroshi Ishiguro Laboratory at the Advanced Telecommunications Research Institute. His research interests include distributed sensor systems, interactive robotics, and android science. He has published more than 300 papers in major journals and conferences, such as Robotics Research and IEEE PAMI. Hiroshi has developed many humanoids and androids, called Robovie, Repliee, Geminoid, Telenoid, and Elfoid; these robots have been featured in the media including Discovery Channel, BBC, and NHK. He has also received the best humanoid award four times at RoboCup. In 2007, Synectics Survey of Contemporary Genius 2007 selected him as one of the top 100 living geniuses.

Leila Takayama, *Research Scientist, Willow Garage, USA*

Leila Takayama is a Research Scientist in the area of human- robot interaction. With a background in Cognitive Science, Psychology, and Human-Computer Interaction, her current focus is understanding human encounters with robots in terms of how they perceive, understand, feel about, and interact with robots.

Leila completed her PhD in the Communication between Humans and Interactive Media (CHIMe) lab at Stanford University in 2008. She also holds a PhD minor in Psychology from Stanford, MA in Communication from Stanford, and BAs in Psychology and Cognitive Science from UC Berkeley (2003). During her graduate studies, she was a research assistant in the User Interface Research group at Palo Alto Research Center. Her thesis, titled Throwing Voices: Investigating the Psychological Effects of the Spatial Location of Projected Voices, won the Nathan Maccoby outstanding dissertation award. Prior to Willow Garage, she was a research scientist at Nokia Research Center.

Peter Vicars, *CEO, VGo Communications, USA*

Peter Vicars has served as the CEO of seven public and privately held technology companies over the past 25 years including Cayman Systems, Tekelec, and Chantry Networks. Peter began his career as a soft- ware developer, and progressed to company leadership through two stints running operations for high-volume hardware businesses. Since then, he has acquired deep domain experience in several areas of communications and software, and led businesses serving large and small enterprises, residential customers, and network operators. As CEO of Chantry Networks, he helped establish the company as a widely recognized technology leader in wireless networking and achieved strong returns for investors through the sale of the company to Siemens. Although mainly focused on business leadership, he has also retained his interest in the underlying technology. While working at Teloquent, he was awarded a patent for an innovative network architecture he developed. Peter is currently the CEO of VGo Communications in Nashua, NH.

Stephen Von Rump, *CEO, Giraff Technologies AB, Sweden*

Stephen Von Rump is the CEO and Founder of Giraff Technologies AB in Sweden. He received his MS and BS degrees in Electrical Engineering from Washington University. He has extensive startup experience as CEO of Metreos and Be Here Corporation, and public company experience as CEO of VTEL and GM of MCI's data and internet services business. He has also held various R&D assignments at MCI and AT&T Bell Laboratories, and been a Director of Quick Eagle Networks (formerly, Digital Link Corp) since November 1998. Stephen has spent much of his career in Internet communications, including the launch of numerous audio, videoconferencing and streaming products and services. He also has extensive international sales, marketing, and business development experience.

4. REFERENCES

[1] J. Markoff. The Boss Is Robotic, and Rolling Up Behind You. Available online at NYTimes.com, published September 4, 2010, accessed January 2012.

[2] M. Minsky. Telepresence. *Omni*, 2(9):45–52, 1980.

[3] E. Paulos and J. Canny. PRoP: Personal roving presence. In *Proceedings of the SIGCHI conference on Human factors in computing systems*, pages 296–303. ACM Press/Addison-Wesley Publishing Co., 1998.

Personalization in HRI: A Longitudinal Field Experiment

Min Kyung Lee[1], Jodi Forlizzi[1], Sara Kiesler[1], Paul Rybski[2], John Antanitis[1], Sarun Savetsila[2]
HCI Institute[1], Robotics Institute[2]
Carnegie Mellon University
5000 Forbes Ave
Pittsburgh, PA 15213
{mklee, kiesler, forlizzi, prybski}@cs.cmu.edu, {jantanit, ssavetsi}@andrew.cmu.edu

ABSTRACT
Creating and sustaining rapport between robots and people is critical for successful robotic services. As a first step towards this goal, we explored a personalization strategy with a snack delivery robot. We designed a social robotic snack delivery service, and, for half of the participants, personalized the service based on participants' service usage and interactions with the robot. The service ran for each participant for two months. We evaluated this strategy during a 4-month field experiment. The results show that, as compared with the social service alone, adding personalized service improved rapport, cooperation, and engagement with the robot during service encounters.

Categories and Subject Descriptors
I.2.9 [**Artificial Intelligence**]: Robotics – *Operator interfaces*

General Terms
Design, Experimentation, Documentation

Keywords
Personalization, social robot, human-robot interaction, service design, organization, mixed-method, field trial, HRI

1. INTRODUCTION
Robotic system services have strong potential for assisting people with everyday tasks in workplaces (e.g., [8]). Examples of current services include a hospital delivery robot, a rehabilitation coach, an assistive robot for the mobility impaired, and a shopping or museum guide. A robot that efficiently and correctly provides service is a prerequisite for success. For some services, however, it may be helpful if the robot is social and builds rapport with people.

Prior work in service and human-agent interaction research suggests rapport between people and a robot is critical when a positive service outcome depends on how well people trust and cooperate with the robot [2]. Stronger rapport between people and service providers was reported to increase people's satisfaction and willingness to cooperate with a service provider's recommendation and instructions [7]. Even in services that do not require high levels of cooperation on the part of the customer, social interaction and rapport can reinforce people's satisfaction with and loyalty to a service provider [7]. To create effective social robots, researchers have imbued robots with various social abilities. Some robots adhere to social norms [18], use relational languages [2], or have anthropomorphic or zoomorphic forms, so that people perceive them as an entity to which they can relate [12].

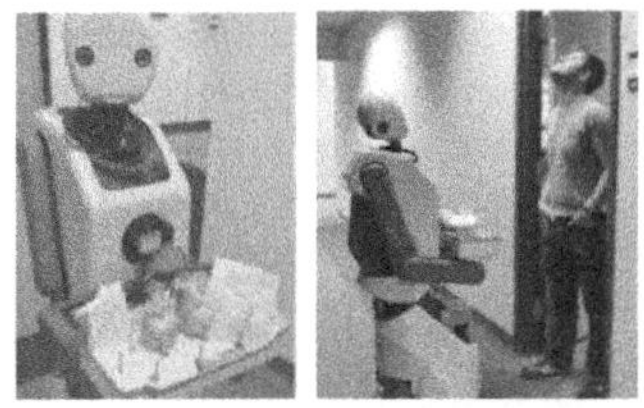

Figure 1. Snackbot carrying snack (left panel) and with a participant doing a neck stretch with the robot (right panel)

A research question that typically is not addressed in prior work is how to design meaningful social interaction and build rapport for repeated interactions. Styles of interaction successful for one time interaction may not be effective in building rapport over time. Existing work on social agents that used relational strategies over time suggests that people may lose interest in conversing socially with a robot once the novelty effect wears off [5]. Other work suggests that people's rapport with an agent may not increase after the first contact [2].

The work presented here asks whether personalizing a robotic service could help sustain people's interest in the robot, and takes a step toward designing and understanding personalization over time in the context of a service robot. We argue that for repeated usage, it will be helpful for the robot to be aware of its mutual experiences with users, and to use this information to personalize its interactions over time. In this manner, interactions with the robotic service become more relevant to individuals and groups who use the service over time, reinforce the rapport between people and robot, and sustain their engagement with the service.

We designed a personalized snack service for a workplace and evaluated it through a 4-month field experiment during which each customer interacted with the snack delivery robot, Snackbot [12] (Figure 1). Our results suggest that personalization positively affects how people relate to the robot and the service.

The contribution of this work is three-fold. First, we demonstrate the effects of personalization with memory in human-robot interaction. We also show changes in people's experiences with the robot over time through a longitudinal study, adding to a small but growing literature that investigates social HRI over repeated interactions [5]. Finally, we provide an example of applying a service design approach, and point to areas of opportunity and challenge, to help frame future work in this emerging area.

1.1 Personalization
Personalized service refers to any behavior occurring in a service interaction intended to individuate the customer and the service

experience [21]. Previous personalization research in human-computer interaction, HRI, and the service literature falls into three overlapping categories. Many social robot projects fall into a category in which researchers increase the friendliness or social presence of interactive systems, to make interaction feel social and personal. For instance, Pfeifer and Bickmore reported that users' exercise reports increased in accuracy when an interface included an anthropomorphic character [19]. A second category of work includes the design of interactive systems to fit users' preferences, and/or to allow users to customize these systems. Examples in HRI include the customization of a robot's appearance or personality [6], or task preferences [15]. Dautenhahn suggests a theoretical model for a personalized companion combining the first two approaches [4]. The third and smaller category of work consists of projects that personalize interactions over repeated encounters. In HRI, such robots might recognize when users return [5] or use the robot infrequently [9]. So far, these robots do not remember mutual interactions with users or maintain continuity in their interactions.

We advance research in the third category by using the history of a robot's repeated interactions with users to personalize its social interactions with users, rather than starting each interaction as if it were the first or second encounter. As a first step in this approach, we harness the snack delivery service robot's memory of users' snack choices, users' service usage patterns, and the robot's own behaviors when it was with users (particularly breakdowns). Prior encounters with each user are automatically logged in the system's memory and we use these data to create or update each new interaction with the robot. We hypothesized this personalization strategy would strengthen rapport between users and the robot, increase cooperation and engagement with the robot during service encounters, and increase service satisfaction. A robot that remembers and acknowledges its past interactions with users might give them the feeling of receiving special attention and personal recognition when they meet the robot again. The feeling of being treated as special is one of the reasons why customers build relationships with human service providers [7]. We suggest that personalized encounters will increase people's sense that the robot's dialogues are relevant to them, making interaction more engaging, and increasing the robot's persuasiveness.

There are some reasons to believe that personalization can backfire, however. Prior research shows that some people prefer self-service to interacting with a human service provider [17]. We thought this might be the case for a personalized social robot as well. Just as people sometimes feel more obligated to other people than they wish, a personalized robot might impose unnecessary social pressure and an unwanted feeling of obligation. A personalized robotic interaction also might increase people's concern for their privacy because the robot is tracking their behavior. These considerations led us to build a personalized robotic service and test it in a long-term field experiment.

2. ROBOTIC SERVICE DESIGN

Testing our personalization strategy in the workplace required us to design an end-to-end service that people would use. We designed a holistic service that comprised a website for customers to order snacks, desirable snack offerings, a semi-autonomous robot to locate offices and deliver snacks to customers, a database of snack deliveries and interactions [11]. We also used an out-of-sight operator to choose appropriate interactions from the pool of dialogue scripts and to fix unanticipated problems with the robot.

2.1 Components

The Snackbot service was comprised of a front end consisting of services that participants encountered directly, and a back end consisting of the underlying system that participants did not see.

2.1.1 Front end

Snack ordering website. Participants could order snacks using our snack ordering website [13]. They specified the snack type, delivery day, and their office number. Only those registered in the study could order snacks through the website.

Snacks. Snackbot delivered six different snacks—apples, bananas, oranges, Reese's peanut butter cups, Snickers candy bars, and chocolate chip cookies. We chose a mixture of snacks that were not always available in the workplace.

Robot. Snackbot [12], a 4.5-foot tall, anthropomorphic wheeled robot delivered the snacks. The robot can make head movements to each side, and up and down, and can animate its LED mouth display to smile, frown, or show a neutral expression. The robot uses its SICK LIDAR to navigate the office environment autonomously (with obstacle avoidance and path planning). In our study, because the website information was not linked to the robot, an operator manually specified the office delivery destinations. The robot used the Cepstral text to speech program with a male voice. The robot carried a web camera and a microphone on its chest to record interactions. Speech output was controlled remotely with a laptop connected to the robot through a wireless network. Despite all our efforts, the robot had significant limitations that were evident to participants. It followed pre-set scripts. There were frequent delays in the dialogue. Sometimes the system froze when there were wireless network communication problems. However, there were no differences in breakdown frequencies between the conditions of the study.

2.1.2 Back end

Robot control interface. Over the previous several years, we had developed a usable interface for operators. This interface allowed an operator to control the robot's navigation, nonverbal movements, and dialog system remotely. The interface showed the video feed from the robot, the robot's location on the building map, its head position, and a number of dialogue scripts. The operator could see a participant's actions through the video/audio feed on the interface.

Operator. An operator transformed the orders on the website to a delivery schedule, specifying a customer name, a snack name, and an office location to the robot control interface. The operator also loaded the snacks on the robot's tray, initialized the robot at the start of each delivery run, and localized it. The operator had three designated sitting locations in the workplace building not visible from participants' offices. The operator also opened any doors in the hallways to enable the robot to go through. According to the personalization condition and interaction timeline, the operator loaded an appropriate dialogue script and clicked each node based on what the human did. To know when problems occurred, operators used a robot control interface showing a video feed of participants interacting with the robot.

2.2 Interaction Design

The main interactions between the service and participants occurred through participants' website orders and interactions with the robot, the latter of which became a main focus of our design efforts. We constructed the interaction scripts before we launched the service, considering the events to take place and potential user choices and behaviors.

2.2.1 *Structure of interaction*

We created a prototypical interaction structure, informed by the interactions we observed between a hot dog vendor and his long-time customers. These interactions started with the vendor identifying the customer, greeting and engaging in small talk with the customer, engaging in the snack transaction, and then enacting social leave-taking. Below is one of the scripts that the robot operator could use in an early day in the trial.

[At the office door] Excuse me. I have an order for David. [Robot looks straight ahead.]
Hello, David. Nice to meet you [Robot looks up to make eye contact with David.]
{...social interaction...}
Please take your apple. [Robot looks down at the tray and then looks up at David.]
Thanks, David. Bye, I'm leaving now. [Robot looks straight]

The robot followed pre-set scripts, which did not allow for improvisations of the operator to maintain consistency across participant experiences. The robot's responses were constructed in a way that made sense regardless of the participants' response (e.g., "I see."), or had two alternative responses, each applied to a participant's yes or no answer. When the dialog scripts did not have appropriate responses to a participant's comment, the robot said, "I have no idea," or just laughed, "ha ha."

2.2.2 *Social interactions*

We created interaction dialogues that fit a workplace context, so the robot would be perceived as a member of the work organization (Table 1). The robot's responses also were designed to be agreeable, to emphasize similarity and honesty (e.g., admitting the inability to understand many topics).

Table 1. Social small talk topics

Categories	Topics	Examples
Temporal and seasonal	Days of the week, holidays (April Fool's Day, Memorial Day), seasons	"You've got something on your face! [pause] April Fool's!"
Organizational	Spring festival, mid-term and final exams, break	"Do you have any plans for carnival?"
Regional	Pittsburgh Pirates baseball team	"It is baseball season. Do you follow the Pirates?"
Task-oriented	Information or story related to snacks	"Bananas are a really good source of potassium and vitamin B6. Excellent choice."
Other	Joke, local weather	"It is a nice day today. I am glad to see you again and hope you are doing well."

2.2.3 *Personalized interactions*

For half of the participants, we built dialogues and planned interactions that used information from their prior interactions with the robot and snack deliver service (Table 2). We focused on users' snack choice patterns, service usage patterns, and the robot's prior behaviors. We did not personalize the interaction based on what participants said to the robot because it was not realistic with the current level of language technology. For the robot to personalize its interactions with participants, it had to be aware of its own prior behavior. One main way we accomplished that was to maintain a record of all breakdowns and mistakes in the service database so the robot could apologize for prior malfunctions. (In prior work, we have shown that apology can be helpful in rectifying mistakes [14].)

Table 2. Personalized topics

Categories	Topics	Examples
Snack choices	Users' favorite snacks; whether they stuck to healthy snacks; whether they seemed to like variety; group's snack consumption patterns	"By the way, it seems as though you really like [snack name]. This is the [nth] time you have ordered one. Are [snack name] your favorite snack?"
Service usage patterns	Whether they were regular weekly users; had they been in their office when the robot was there; times when they did not use the snack service	"I missed you during my snack deliveries [n] times so far. I am glad to finally see you again."
Robot's behaviors	Frequency of breakdowns and apology (no breakdowns to frequent breakdowns)	"I was thinking about my first month here. I realized that I broke down and made mistakes [n] times in front of you. Sorry for that, and thank you for being patient with me."

2.2.4 *Guiding interactions*

The current level of technology was not conducive to participant-initiated conversation. Therefore, the robot's interactions were designed to guide interaction. For example, instead of giving participants time to initiate conversations, the robot attempted to lead the conversation, for example, by asking questions. To address situations where the robot could not process human behaviors, the robot used dialogues to encourage participants or passersby to behave in a manner that it could process. For example, the robot sometimes said, "Can you please stand in front of me?" and "I have bad ears, so sometimes I cannot hear very well. Can you repeat, please?"

2.2.5 *Exceptional use cases*

Pretesting pointed to several situations other than snack transactions that the robot had to be prepared to address. For instance, some passersby took snacks from the tray without the robot's permission, or intentionally blocked the robot's path. In these cases, the robot made comments such as, "Please don't be rude. I am just a robot," and "please return the snack to a proper place. I have the campus police on my speed dial… Just kidding." Sometimes the robot broke down and stood in the hallway until it was debugged. In these situations, the robot communicated its status to people who approached, such as "I am not feeling well; my operators are fixing me."

3. METHOD

We conducted a field experiment from February to June, 2011 in a workplace to test the following hypotheses:

1. A personalized social robot will increase rapport and cooperation with the robot as compared with a sociable robot lacking personalization.

2. A personalized social robot will increase engagement during the service encounter as compared with a sociable robot lacking personalization.

3. A personalized social robot will increase satisfaction with a snack service as compared with a sociable robot lacking personalization.

3.1 Field Site

Our participants were distributed across 16 offices located in 10 hallways on one floor of an office building at Carnegie Mellon University. We randomized the assignment of conditions to hallways because participants within hallways could hear the interactions of the robot with their office mates or with those in adjacent offices. This adjustment assured non-contamination across conditions but did not allow for randomization at the individual level.

3.2 Experimental Design

The study was a two (Personalization vs. No Personalization) x two (Pre-personalization [Period 1] vs. Post-personalization [Period 2]) mixed factorial design (Table 3). We used interactions in Period 1 to collect baseline attitude scores and interaction behaviors. Baseline behaviors also were used to personalize the interactions in the Personalization condition. In general, Period 1 included each participant's first four interactions with the robot, and Period 2 included the rest of the interactions. However, for those who joined the service later (two in Personalization, three in No Personalization), we had to shorten their Period 1 to 2-3 interactions because at the end of June offices were being moved. Participants joining the later study who had 2-3 interactions in Period 1 were equally distributed across the conditions. In Period 2, participants interacted with the robot 5 times on average (Personalization M = 5.67 (SE = 0.59), No Personalization M = 4.33 (SE = 0.51), $F(1,20) = 2.96$, $p = .1$).

Table 3. Experimental design

Condition	Period 1 Robot Behavior	Period 2 Robot Behavior
Personalization	Social interaction	Social interaction + personalized interaction
No Personalization	Social interaction	Social interaction

3.3 Participants

We used flyers, postcards, and a snowball sampling method to recruit participants. The study required participants to have offices in our field site, and generally to be in their offices 2:30 p.m. to 4 p.m. at least one day a week. Thirty-two participants signed up; eight participants never placed an order, one participant left the organization, and two participants in the Personalization condition dropped out after two deliveries due to the inconvenient delivery schedule. We ended up with 21 participants, nine in the Personalization condition and 12 in the No Personalization condition. There were eight women ranging in age from 23–49 and 13 men ranging in age from 22-51. The participants included eleven graduate students, eight staff members, one post-doc, and one faculty member. All were members of a computer science school. Only one participant had prior exposure to the robot. Knowledge of programming did not statistically differ in the two conditions. Knowledge of robotics was a little higher in the No Personalization condition, but not statistically significant.

3.4 Procedure

The robot delivered snacks from 2:30–4pm Mondays, Wednesdays, and Fridays. We provided snacks for free to compensate users' participation in surveys and interviews. Participants could place an order anytime before noon on the day of snack delivery. If participants were not in their offices, their snack was placed in a paper bag and hung on their office door. Because we could not deliver snacks to all 21 participants in a day, those who joined the service early were retired from the study after two months of usage.

3.5 Data Sources

3.5.1 Interaction logs

The robot's camera and microphone recorded all interactions between the robot and participants. Except for one day when the robot's recording was turned off accidentally, and a few other cases when the camera was turned away from participants, 175 interactions were audio recorded and 161 interactions were video recorded when participants were in their offices.

3.5.2 Surveys

Participants completed a background survey after registering for the study, robot and service evaluation surveys at the end of Periods 1 and 2, and an exit survey. The background survey included questions about participants' demographic information, their snacking routines, and their orientations toward services, adapted from [14]. The evaluation survey included self-report measures of rapport development adopted from [1]. The exit survey measured participants' overall satisfaction with the service, and checks on the manipulation of personalization.

3.5.3 Interviews

The first author conducted 30–60 minute semi-structured interviews with the 21 participants at the end of the study. The interview protocol included questions about participants' positive and negative experiences with the robot and the service, their initial expectations and how their experiences with the robot changed over time, and how other people around them behaved. To avoid biasing the interview, the protocol did not include explicit questions about personalization.

3.6 Measures

3.6.1 Participants' service orientation

Our previous work showed that people's service orientations influenced their reactions to and satisfaction with a robotic service [14]. Therefore we used items (7-point Likert scales) from [14] to assess participants' food service orientation—relational vs. utilitarian. Using principle component analysis, we constructed a social orientation scale with three items (Cronbach's α =.78), and a utilitarian scale with six items (Cronbach's α =.52). Participants in the No Personalization condition (M = 5.31, SE = 0.34) had a higher social orientation than those in the Personalization condition (M = 4.07, SE = 0.40), $F(1, 20) = 5.55$, $p < .05$, so we included the social orientation scale as a control variable in our statistical analysis model.

3.6.2 Rapport

We measured rapport strength by using the constructs liking, closeness, and self-connection, suggested by the literature on relationship with brands [1] and politeness [3]. Subjective measures were included in the surveys, and behavioral measures were taken from participants' behavior during snack delivery. We first read all interaction transcripts, identifying behaviors that show participants liked the robot and felt close to it. We do not discuss behaviors equally exhibited in both conditions (e.g., greetings). Two coders coded for the following three behaviors.

Flattery and gift giving. These behaviors convey that people are cooperators, specifically, that the speaker wants to satisfy the hearer's wants [3]. We coded instances when participants complemented the robot (e.g., "you are inspirational to me," "I'm glad you came.") or gave a gift to the robot; Cohen's Kappa = .78).

Self-disclosure. Self disclosure indicates that two people feel close to each other [3]. We coded instances where participants shared information about themselves that was not solicited or goes beyond the typical response given to the robot (e.g., Snackbot: "Get ready for a new week." Participant B: "That's right. We'll see. We have a big presentation tomorrow. Hopefully we'll be okay."; Cohen's Kappa = .70).

Greeting using the robot's name. We coded instances when the participants greeted the robot using its name.

Closeness. The evaluation survey included two 7-point Likert items adopted from [1] (I have a personal relationship with the robot, I feel close to the robot; Cronbach's α = .76).

Self-connection. The evaluation survey included two 7-point Likert items adopted from [1] (Snackbot represents the personal service that I would want, The service fits my current lifestyle; Cronbach's α = .60).

3.6.3 *Cooperation*

Cooperation measures consisted of participants' responses to three requests the robot made in three visits towards the end of each participant's service experience. We invented cooperation tasks to meet the following criteria: participants would have to listen to the robot, comply with a request for a favor entailing a new behavior by the participant that would not happen without the request, and would be different in each case, to avoid learning or habit effects. We standardized the measures by transforming scores so that each distribution has a mean of zero and a standard deviation of 1.

Help request. The robot explained to participants that it needed to give visitors a tour of the building, and asked whether they could suggest good locations to add to the tour. We counted the number of locations that participants suggested.

Neck stretch. The robot explained to participants that taking a break has been shown to boost people's productivity. The robot said it knew how to do a neck stretching exercise that helps release the tension around a person's neck and shoulders. The robot asked participants whether they would like to try the exercise. We coded whether the participants completed the exercise with the robot or not (yes = 1, no = 0).

Mystery snack. The robot explained to participants that it was carrying a special "fresh and good" mystery snack. The robot asked whether participants would like to try the mystery snack instead of the snack that they ordered. The mystery snacks were baked goods such as a lemon bar or cupcake that had not been part of the service. We coded whether the participants took the mystery snack or not (yes = 1, no = 0).

3.6.4 *Engagement*

To measure engagement during service encounters, we coded participants' postures and facial expressions, which can indicate people's engagement in social interaction [10]. We do not discuss measures that did not differ between the two conditions (e.g., gaze, head nodding). We did not code proxemics because we could not reliably measure the distance between participant and robot from the recorded videos.

Facial expression. We coded for instances of smiling, laughter and general facial expression (positive, neutral, negative).

Standing posture. We coded whether participants were upright, leaning against the door, or leaning forward. The frequency of leaning forward did not vary by condition. Compared to leaning against the door, standing upright is a less relaxed behavior, and indicates a positive attitude, more attention to an addressee [10], and is exhibited when the addressee is of a higher status [16].

3.6.5 *Service satisfaction*

The exit survey included questions on participants' overall service satisfaction, their willingness to continue the service on a 7-point Likert scale, and how much they would be willing to pay per month to continue to use the service.

3.7 Analyses

3.7.1 *Analysis of interview data*

We transcribed the interviews and did thematic coding. Initial themes were used to create an affinity diagram. Based on these results, we chose to focus on unsolicited remarks that related to personalization (e.g., "the robot knew what I had chosen").

3.7.2 *Analysis of quantitative data*

We used a multi-level regression model to analyze the codes from the interaction log, comparing responses during Period 1 vs. Period 2. For the evaluation surveys, we used ordinary least squared regression analysis to measure rapport after Period 2, controlling for initial rapport after Period 1. For the exit survey, we used ordinary least squares ANOVA. We included the social orientation scale as a control variable in all the models, because, as noted above, social orientation differed between conditions.

4. RESULTS

Our results provide substantial evidence that personalization of the robot improved participants' service experience.

4.1 Overall Service Usage

There were 261 orders, on average 6 orders per day (SD = 4.53). On average, each participant ordered 12 snacks (SD = 3.96) throughout the study. The participants could order only one snack at a time. Excluding the times participants were not in their offices, they interacted with the robot 9 times on average (SD = 3.07). Each interaction averaged one minute and six seconds long (SD = 37 seconds), included 7 turns (SD = 2.28) by the participant and 8 turns (SD = 2.27) by the robot. The average number of words in participants' dialogues was 35.13 (SD = 23.08). The difference between conditions in interaction duration and number of turns was not statistically significant.

4.2 Manipulation Check

In the exit survey, we asked participants if the robot remembered their previous snack choices (Personalization M = 6.70 (SE = 0.56), No Personalization M = 4.31 (SE = 0.48), $F(2,19) = 9.38$, $p < .01$), other customers' snack choices (Personalization M = 6.63 (SE = 0.63), No Personalization M = 4.33 (SE = 0.50), $F(2,19) = 7.18$, $p = .02$), and how personal the service felt (Personalization M = 6.13 (SE = 0.44), No Personalization M = 4.90 (SE = 0.38), $F(2,19) = 4.01$, $p = .06$). These results show that the personalization manipulation was effective.

4.3 Rapport

As predicted in Hypothesis 1, recorded interactions show that participants exhibited social behaviors more frequently when the robot personalized its dialogues (see Figure 2).

Flattery and gifts. Participants in the Personalization condition were more likely to flatter the robot or to give it a gift during

Period 2 (M = 0.22, SE = .05) than during Period 1 (M = 0.07, SE = .05), F(1, 163.1) = 5.84, p < .05, and more than those in the No Personalization condition (M = 0.03, SE = .04), F(1, 34.7) = 9.16, p < .01; period x condition interaction, F = (1, 163.3) = 2.61, p = .1). Here is one example:

Participant E: (starts laughing). I have a snack for you.
Snackbot: Please take your orange.
Participant E: I have a snack for you Snackbot. It's a battery.
Snackbot: Thanks, [participant name]. Enjoy your snack.
Participant E: Bye Snackbot.
Snackbot: I hope you have a wonderful day. Goodbye.
Participant E: You too, enjoy your snack.

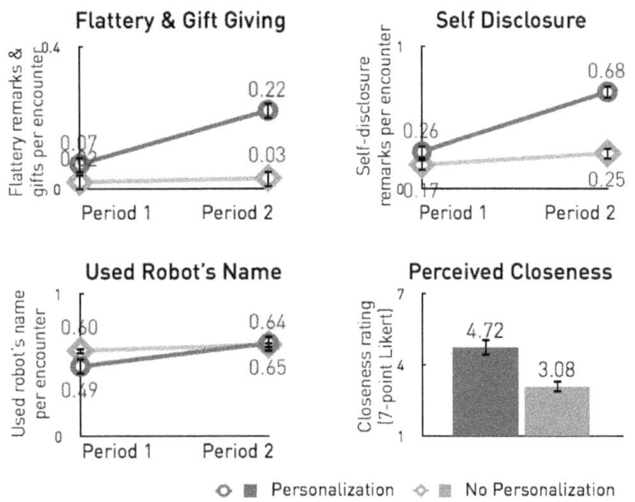

Figure 2. Measures of rapport

Self-disclosure. Participants also disclosed more about themselves in the Personalization condition during Period 2 (M = 0.68, SE = 0.10) than during Period 1(M = 0.26, SE = 0.11), *F(1, 162.4) = 14, p = .001*, and those in the No Personalization condition (M = 0.25, SE = 0.09), *F(1, 25.84) = 9.11, p < .01*; period x condition interaction, F(1, 159.5) = 4.92, p = 0.03).

Using the robot's name. Participants in the Personalization condition greeted the robot with the robot's name (i.e., "Hi, Snackbot") more frequently (M = 0.65, SE = 0.13) during Period 2 than Period 1 (M = 0.49, SE = 0.13), F(1, 143.7) = 5.23, p < .05. This result suggests a potential ceiling effect, but we could not think of reasons why there would be a ceiling in this rate.

Perceived closeness. Participants in the Personalization condition felt closer to the robot (M = 4.72, SE = 0.71) than those in the No Personalization condition (M = 3.08, SE = 0.52; F(3,16) = 3.05, p = .1) but the difference was only marginally significant. Perceived self-connection did not differ by condition.

4.4 Cooperation

Personalization increased participants' cooperation, as predicted in Hypothesis 2. We derived a summary measure of cooperation for each participant by standardizing scores on all three measures (see Figure 3) and calculating a mean for each person. The results showed people's willingness to cooperate with the robot was greater in the Personalization condition (M = 0.49, SE = .28) than in the No Personalization condition (M = -0.45, SE = .22), F(2,18) = 3.48, p = 0.02. We provide an example below.

Snackbot: I need to give a tour of [building] for visitors, I am still new to this building and I am not sure where to bring them. Could you suggest some interesting places in [building]?

Participant F (No Personalization condition): Snackbot, let's not be ridiculous, can I take my snack? Can I have my snack?

Participant L (Personalization condition): Let's see. You could visit the [exhibit name] on the first floor or the third floor. The second floor has a lot of cool other robotic stuff that you could check out or show people, I think they would like that [...].

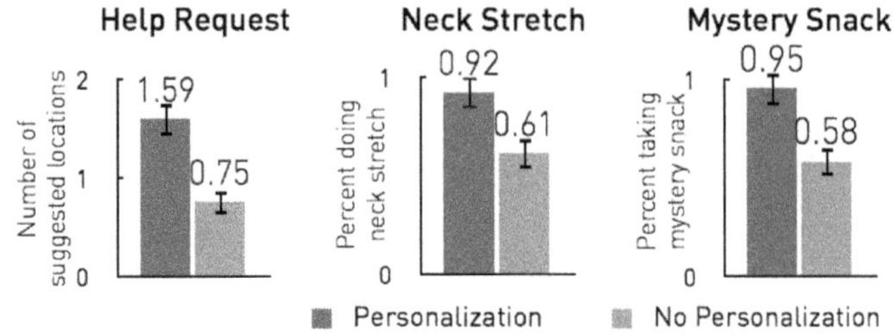

Figure 3. Measures of cooperation

4.5 Engagement

Participants' engagement with the robot appeared to be more sustained when the robot personalized its remarks (see Figure 4).

Laughing. Participants laughed more during personalized interactions during Period 2 (M = 1.53, SE = 0.36) than during Period 1 (M = 0.99, SE = 0.36), F(1,146.1) = 4.94, p < .05 and more than those in the No Participation condition (M = 0.70, SE = 0.32), F(1, 27.91) = 2.75, p = .10; period x condition interaction: F(1, 145.3) = 3.27, p = .07).

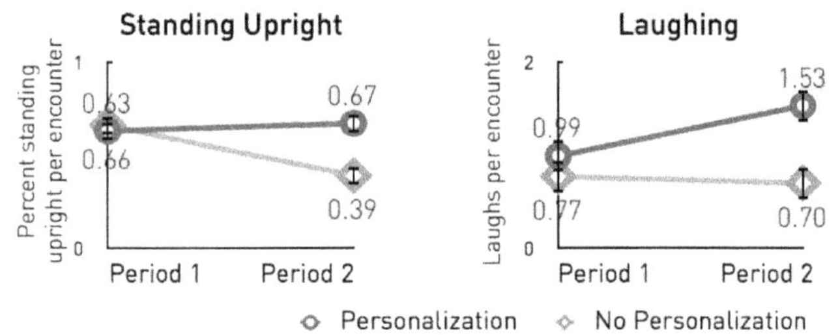

Figure 4. Measures of engagement

Standing posture. The percentage of the participants who sustained their upright standing posture did not change over time in the Personalization condition. In the No Personalization condition, the percentage of the participants who stood upright when interacting with the robot decreased from Period 1 (M = 0.66, SE = 0.1) to Period 2 (M = 0.39, SE = 0.1), F(1,140.2) = 11.25, p = .001. More participants in the No Personalization condition leaned against their office doors while interacting with the robot in Period 2, signaling higher status and/or less attention.

4.6 Service Satisfaction

The ratings of service satisfaction did not statistically differ by condition. Participants in both conditions were highly satisfied with the service (Personalization M = 6.05 (SE = 0.24); No Personalization M = 6.22 (SE = 0.21)), and were willing to continue the service (Personalization M = 6.40 (SE = 0.41); No Personalization M = 6.53 (SE = 0.35)). Participants in the Personalization condition said they would pay more to continue to use the service (M = $16.19, SE = 4.09) than those in the No Personalization condition (M = $12.4, SE = 3.48), but the difference was not statistically significant.

5. DISCUSSION

Our analyses suggest that personalizing the interactions with the robot reinforced participants' rapport, cooperation, and engagement. Our post-study interview results helped us understand how participants interpreted the personalization strategy. As noted above, in the interviews, we did not mention personalization, so the answers we received were unsolicited.

5.1 Receiving Personal Attention

Consistent with the literature on personalization, participants seemed to like personal attention from the robot. We designed Snackbot's personalization to build on real experiences between the robot and the person, creating an interaction that was unique to each participant. When the robot remembered even a small detail about a participant, for example, their favorite snack, it seemed to elicit feelings of closeness. For example, Participant N said:

*Surprisingly Snackbot knows that he never dies on me. (**Interviewer:** How did you feel about it?) So I feel good. I feel special that I communicate with Snackbot with no problem.*

By contrast, in the No Personalization condition, most participants expressed a desire to have more tailored interactions with the robot, as Participant U said:

But I felt like over time [...] if he shows up every week, Monday, Wednesday and Friday, you would hopefully learn [my] name or that the conversation would get to the point where it could be a little bit more personal.

The rapport created through personalization may have played a role in influencing people's willingness to cooperate or help the robot. Participant I in the No Personalization condition said during the interview that the robot's tour help question was one of his negative experiences with the robot:

I think it was mostly that you don't have enough of a rapport with it to answer that question. So if it was like someone—if it was like Justin or someone who works with me, I could be like "Oh we should show them the thing down in that lab where you work."

5.2 Sustaining Interest

According to the interviews, participants in the Personalization condition were more engaged with the robot over the course of the study. We surmise that the robot's interactions became more meaningful over time. For those in the No Personalization condition, interaction with the robot became less meaningful as participants realized that their conversation with the robot did not have any bearing on the robot's future behavior. This caused people to lose interest in conversing with the robot. By contrast, in the Personalization condition, the robot made comments based on its past performance or the participant's use of the service, building common ground and shared history. The robot's telling stories related to the participant each time caused excitement and expectation, as participants waited for new stories.

Participant L (Personalization condition): We even commented to each other a couple times; What do you think he's going to say today or do you think he's going to say something about carnival?

5.3 Disadvantages of Personalization

As in human interaction, personal conversation can create discomfort because people feel invested in the relationship. Some of Snackbot's personalized dialogues evoked negative responses, especially when participants felt uneasy about the behaviors that were the topics of conversation. The most sensitive topics pertained to participants' not being present when the robot arrived, and to their choice of snacks.

Participant M (Personalization condition): But then my most negative [feeling] was one time he said, "I notice that you always order Reese's Cups. You must really like Reese's Cups," and that was kind of awkward for me because it's like, "Oh, I'm the one ordering all the junk food, and eating junk food every day, and now he's pointing it out."

We were initially concerned that participants would have privacy concerns or feel more pressure to be social with the robot when the robot personalized its interactions. Participants mentioned that they did not have privacy concerns with the topics or events that the robot used to personalize. In both conditions, participants seemed to feel some pressure to be social and polite with the robot as the interactions took place in a social setting, the workplace, and others might hear these interactions.

6. IMPLICATIONS

In this section, we briefly address how to design successful personalization for repeated interactions.

When to use personalization. Human-robot interaction may benefit from personalized behaviors when it is important for the service to track and be aware of past service events. Customers know the business has a record of interactions and may expect a social robot to reflect these past interactions. For example, a snack delivery robot in a nursing home could be aware of what time meals were last served. Personalized behaviors may also be useful when the robot needs to be persuasive, for example, in choosing a healthy snack over an unhealthy snack, or when the robot needs help or input from customers [20]. Personalized behavior will be also useful in situations where the robot is assisting people doing boring and repetitive tasks since personalized behaviors over time can create surprise, joy and more engagement.

How to use personalization. We suggest personalization is best used to define a meaningful relationship between a robot and a person. As we learned in our study, the events that are selected to make meaning must be chosen carefully. For example, comments about liking a particular kind of candy were embarrassing rather than meaningful. Like human interaction, not all facts bear repeating. Consideration must be given to what critical moments in an interaction are and how they can be detected. For example, an assistive robot in a care facility might call out moments of independence and ability to complete activities of daily living rather than breakdowns or calls for assistance.

Challenges and opportunities. Individuals differ in their receptivity to personalization. It will be important to develop mechanisms to detect responses to specific strategies and ways for the robot to recover from mistakes. Personalization also offers new opportunities in services. One avenue for research will be to investigate personalization unique to robots; for examples, unlike humans, the robot has a perfect record of past interactions. In a setting where a human could not easily employ personalization techniques (e.g., a vendor in a big store), robots can personalize their interactions and change the dynamics of encounters. Another interesting avenue is self-aware robotic services. Compared to systems personalized to users' tracked behaviors, our attempt to use the robot's own tracked behaviors to personalize its interaction is relatively new. Our study suggests that it can be a promising area for the design of repeated interactions.

7. LIMITATIONS

Conducting a field experiment using a realistic service increased the ecological validity of our results but also entailed three notable constraints. First, we randomized conditions across the hallways to avoid contamination. Nonetheless, participants in the same hallway sometimes socialized during the Snackbot visit, and the existing culture of the hallway may have influenced the results reported in the paper. Second, we used different styles of personalization to elicit surprise and enjoyment. For this reason, we cannot distinguish among the effects of specific personalization tactics. We do not know whether our strategy would be as effective if only one of the personalization topics were used. Third, the robot took one or two more speaking turns in the Personalization condition than in the No Personalization condition. It could have been more effective simply because it spoke more.

Our study was also limited due to technical constraints. It was conducted on one floor of a computer science building, where the robot could operate reliably, with access to engineering help if it broke down. (Studying an organization's prototype within that organization is not uncommon for this reason [22].) None of our participants were part of the Snackbot development but some would have had a bias to like robots. Also, our study used a Wizard of Oz technique for the selection of nodes in the dialog script. When we asked participants if they believed the robot was autonomous, they wondered how much the robot was autonomous, but no one believed that they were communicating with an operator through the robot.

The snack service was operated as compensation for participating in the trial for at least two months. Free snacks may have contributed to high service satisfaction in both conditions. Also, we recorded all the interactions with participants' consent. Recording may have influenced participants' behaviors, as well. Finally, the robot was anthropomorphic. Generalizing the results to different service domains and robots will require further investigation.

8. CONCLUSION

Through a longitudinal study, we provide evidence that personalization with memory reinforces people's rapport, cooperation, and engagement with a robot. We also show changes in people's experiences with the robot over time. By presenting an example of a personalized robotic service, we offer insights on factors that other researchers can refer to when designing their systems. We hope this study inspires future research into how robots could be designed to engage people in a pleasurable and meaningful way over time.

9. ACKNOWLEDGMENTS

This work was supported by NSF grants 0624275 and 709077. We thank Alex McCluskey, Leonard Turnier, Katherine Cuti, Ahmad Shamsuddin, and Junior Baboolall for their assistance.

10. REFERENCES

[1] Aaker, J., Fournier, S., & Brasel, A. (2004). When good brands do bad. *Journal of Consumer research, 31*, 1-16.

[2] Bickmore, T. & Picard, R. (2005). Establishing and maintaining long-term human-computer relationships. *Transactions on Computer Human Interaction, 59(1),* 21-30.

[3] Brown, P., & Levinson, S. (1987). *Politeness: Some universals in language usage*. Cambridge University Press.

[4] Dautenhahn, K. (2004). Robots we like to live with: A developmental perspective on a personalized, life-long robot companion. *Proc. of ROMAN'04*, 17-22.

[5] Gockley, R., Bruce, A., Forlizzi, J., Michalowski, M., Mundell, A., Rosenthal, S., Sellner, B., Simmons, R., Snipes, K., Schultz, A.C., & Wang, J. (2005). Designing robots for long-term social interaction. *Proc. of IROS'05*, 2199-2204.

[6] Goetz, J., Kiesler, S., & Powers, A. (2003). Matching robot appearance and behavior to tasks to improve human-robot cooperation. *Proc. of ROMAN'03*, 55-60.

[7] Gwinner, K.P., Gremier, D.D., & Bitner, M.J. (1998). Relational benefits in services industries: The customer's perspective. *Journal of Academy of Marketing Science, 26(2),* 101-114.

[8] Hada, Y., Gakuhari, H., Takase, K., & Hemeldan, E.I. (2004). Delivery service robot using distributed acquisition, actuators and intelligence. *Proc. of IROS'04*, 2997-3002.

[9] Kidd, C. D., & Breazeal, C. (2008). Robots at home: Understanding long-term human-robot interaction. *Proc. of IROS'08*, 3230-3235.

[10] Knapp, M. L, & Hall, J. A. (2009). *Nonverbal communication in human interaction.* Cengage Learning.

[11] Lee, M.K., & Forlizzi, J. (2009). Designing adaptive robotic services. *Proc. of IASDR'09.*

[12] Lee, M.K., Forlizzi, J., Rybski, P.E., Crabbe, F., Chung, W., Finkle, J., Glaser, E., & Kiesler, S. (2009). The Snackbot: Documenting the design of a robot for long-term Human-Robot Interaction. *Proc. of HRI'09*, 7-14.

[13] Lee, M. K., Kiesler, S., & Forlizzi, J. (2011). Mining behavioral economics to design persuasive technology for healthy choices. *Proc. of CHI'11*, 325-334.

[14] Lee, M.K., Kiesler, S., Forlizzi, J., Srinivasa, S., & Rybski, P. (2010). Gracefully mitigating breakdowns in robotic services. *Proc. of HRI'10*, 203-210.

[15] Mason, M., & Lopes, M. (2011). Robot self-initiative and personalization by learning through repeated interactions. *Proc. of HRI'11*, 433-440.

[16] Mehrabian, A. (1970). A semantic space for nonverbal behavior. *Journal of Consulting and Clinical Psychology, 35(2),* 248-257.

[17] Meuter, M. L., Ostrom, A. L., Roundtree, R. I., & Bitner, M. J. (2000). Self-service technologies: Understanding customer satisfaction with technology-based service encounters. *Journal of Marketing, 64(3),* 50-64.

[18] Nakauchi, Y., & Simmons, R. (2002). A social robot that stands in line. *Autonomous Robots, 12,* 313-324.

[19] Pfeifer, L. M. & Bickmore, T. (2011). Is the media equation a flash in the pan? The durability and longevity of social responses to computers. *Proc. of CHI'10*, 777-780.

[20] Rosenthal, S., Veloso, M., & Dey, A. K. Is someone in this office available to help me? Proactively seeking help from spatially-situated humans. To appear in *Journal of Intelligent and Robotic Systems*.

[21] Surprenant, C.F., & Solomon, M.R. (1987). Predictability and personalization in the service encounter. *Journal of Marketing, 51(2)*, 86-96.

[22] Venolia, G., Tang, J., Cervantes, R., Bly, S., Robertson, G. G., Lee, B. & Inkpen, K. (2010). Embodied Social Proxy. *Proc. of CHI'10,* 1049-105.

Exploring the Role of Robots in Home Organization

Caroline Pantofaru, Leila Takayama, Tully Foote, Bianca Soto
Willow Garage, Inc.
68 Willow Road, Menlo Park, California, USA
{pantofaru,takayama,tfoote,sotob}@willowgarage.com

ABSTRACT

Technologists have long wanted to put robots in the home, making robots truly personal and present in every aspect of our lives. It has not been clear, however, exactly what these robots should do in the home. The difficulty of tasking robots with home chores comes not only from the significant technical challenges, but also from the strong emotions and expectations people have about their home lives. In this paper, we explore one possible set of tasks a robot could perform, home organization and storage tasks. Using the technique of *need finding*, we interviewed a group of people regarding the reality of organization in their home; the successes, failures, family dynamics and practicalities surrounding organization. These interviews are abstracted into a set of frameworks and design implications for home robotics, which we contribute to the community as inspiration and hypotheses for future robot prototypes to test.

Categories and Subject Descriptors

H.5.2. [**Information Interfaces and Presentation**]: User Interfaces—*User-centered Design*

Keywords

home robots, need finding, home organization, robot design

1. INTRODUCTION

In 1945, Vannevar Bush wrote of a future machine, the Memex, which would help to enhance human memory [4]. The Memex would store and organize records, books, and communications, all converted to a digital form. In many respects, this predicted future is now the present. Most music is now digital. Communication has also become digitized through email, social and news websites and academic publications. Books are increasingly being digitized. Interestingly, however, the elimination of physical books is being slowed not only by the screen readability, but also by human emotional attachment to the physical books themselves.

HRI'12, March 5–8, 2012, Boston, Massachusetts, USA.
Copyright 2012 ACM 978-1-4503-1063-5/12/03 ...$10.00.

Figure 1: Storage areas seen in informants' homes. These are the dark, cluttered and variable spaces that a robot tasked with organization will face.

The computerization of the organization of our social lives, work lives, and knowledge have been well studied (e.g. [21]). But what of our home lives, specifically the physical organization of objects? Technologists dream of putting robots in the home [10], however it is not clear what tasks these robots should perform. This paper explores the issues surrounding computers, in their physical incarnations as robots, helping to organize the physical *stuff* in our homes.

"Tidy the room." "Put stuff away." "Tell me where object X is." "Retrieve object X." "Help me throw stuff away." "Figure out how to fit all of my stuff into my home." These are all requests that might be made of a home robot. *Rosie* from *The Jetsons* television show, a human-sized robot maid, would have easily completed these tasks. Previous work has shown that our opinions of what robots should do are colored by popular culture [20, 13]. And yet, organization is a sensitive topic. Rosie was a fantastic machine, an AI, a member of the family, but even human family members become annoying when they move your things without permission.

Organization is such a complex task that entire industries have grown to serve the need, such as the National Association of Professional Organizers [15]. Unlike the tasks "wash the dishes" or "vacuum the floor", the task space of home organization is nuanced and emotional. To make matters more difficult, since personal robotics is still an emerging market, there are no established methods for exploring whether or how robots should perform a task in the home before the technology itself has been specified. To help direct research into robotics for home organization, this paper presents a set of frameworks for home organization and their related design implications discovered through the technique of *need finding* [2]. Need finding is a methodological tool borrowed from the product design community which rapidly explores a design space through a series of interviews and experiences. The interviews look for *needs*, the gaps between use, usability and meaning. The findings from the need finding process are meant to be inspirational and generative, allowing researchers to develop creative solutions. We hypothesize that applying need finding to robotics research will allow it to have broader impact and usability.

In the course of this exploration, we interviewed twelve people in their homes, discussing their approaches to organization and the emotions and practicalities around having help with organization. We then toured each home to discover first-hand the organizational methods (or lack thereof) in place. The set of interviews and tours were abstracted into frameworks and design implications intended to generate ideas for the role of robots within the home. We present these abstractions as guidance for others exploring the use of robots for organization within the home. These abstractions are not intended to be prescriptive nor universal, rather they should be tested with users once solutions to specific problems are prototyped.

2. NEED FINDING

We present *need finding* as a methodological tool for quickly learning about a user space, inspiring robotics research within that space, and grounding the resulting research.

Much (although certainly not all) of robotics research today is inspired by technology push - a technologist deciding to apply a technology to a problem. User-based research often does not start until after a prototype or system specification exists. This is a valuable method as researchers have spent years building intuition for their field. For robotics research, *need finding* can provide a complementary, user-driven source of inspiration and guidance, as well as refining technology push ideas to better fit an application space.

Need finding is a method that comes from the product design community [2]. The goal is to identify a set of fundamental user needs of the community a product aims to satisfy. The need finding process is summarized in Figure 2. Need finding begins with generating empathy for the user group through interviews, and sharing that empathy with other designers and researchers through conversations and media like videos. This is a concrete and analytic process. The results from the interviews are then abstracted into frameworks, often presented in graphical form such as the 2x2 charts in Figures 2 and 3. The lessons from the frameworks are then converted to design implications, which are meant to be generative, allowing many interesting solutions to evolve. This process can be iterated and interleaved as necessary. The process is expanded upon below, with a description of our own implementation for this paper.

2.1 Interviews and Observations

Need finding begins with identifying a community of potential product users. A very small team goes out to visit a sample of community members in the places where they do the activities that the product supports. Immersion in the interviewee's environment is the key to success and a distinguishing feature of need finding. This immersion inspires the participant to discuss details they might have otherwise forgotten, and allows the interviewer to quickly reconcile the interviewee's words with reality. It is important to note that

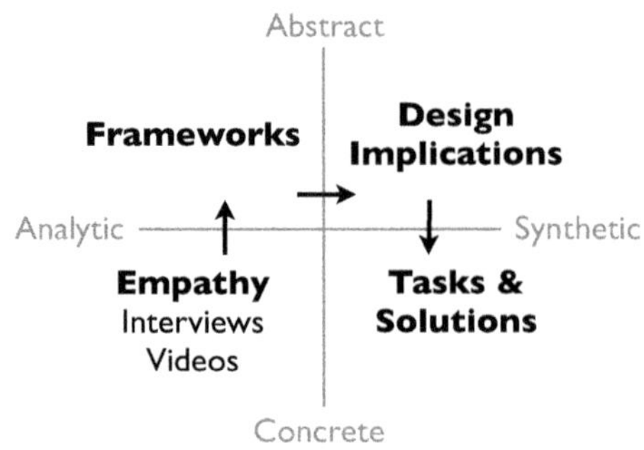

Figure 2: An overview of the *need finding* process.

relying on self-reported data alone is dangerous due to people's poor memories, as well as the social desirability bias (an inclination for respondents to say things they think will be received favorably.) There is even a standardized scale for measuring a person's inclination toward the social desirability bias [5]. These problems can be so serious that some usability experts suggest completely ignoring what users say, and instead watching what they do [16].

Along with getting a sense of the physical, cultural, and social environments, artifacts, and activities, the need finding team focuses on hearing good stories. The types of stories and observations that can be particularly interesting and fruitful for insight are contradictions (e.g., parents saying they want their children to exercise, but instead finding them in front of the television), spoken and unspoken social norms (e.g., not firing an employee via text message), and success or failure stories [7]. Interviewees' stories reflect situations in which reality did not match their mental model, thereby expressing an unfulfilled need.

The interview process also generates empathy towards the user group. This is an important step for technologists, especially those accustomed to technology push. Empathy allows technologists to be advocates for their users, designing for them more instinctively. Generating empathy is best done by meeting users and visiting their environment, however, it is not possible for each person to talk to each interviewee. Therefore, interviews should be presented to the entire team through videos, images, stories, or other media.

As compared to other types of user engagements, the sample size is relatively small; a sample size of eight to sixteen interviews is typical. Cultural Consensus Theory [18, 24] states that for a sufficiently culturally coherent group, a small number of interviews will yield responses which have a high probability of representing the entire group. So determining the group's level of cultural knowledge and agreement in turn determines how many interviews are required to obtain representative information.

Data gathered from the interviews forms the basis of the four components of the need finding process: generating empathy, creating frameworks, generating design implications and finally, creating solutions and prototypes.

2.2 Study Design

For this study, we interviewed people regarding the organization techniques, habits and difficulties within their homes. Each participant was interviewed in their own home, and the interviews were filmed. Each session began with a conversation in a comfortable room, encouraging participants to relax. Once a rapport was established, we asked to be taken on a tour of the home, focusing on organization issues. During the tour, we were able to reconcile the interviewees' stories with the realities of their homes. The interviewees also remembered extra details and told more stories during the tour. These visits lasted between 1.5 and 3 hours. An interesting difference versus traditional ethnography is that the interviewers may direct the conversation to keep on topic and make the process time-efficient. The goal is to extract useful information, not simply compile observations.

The interviews were structured but conversational, allowing for digression and topic reordering as necessary. That said, the basic interview guide was as follows:

Signing of the study agreement form.

Introduction: "We're interested in learning about your experiences with organizing your home. What's more important is that we'd like to get to know you as a person. So we're going to ask questions about your home, organizing, daily life, hobbies, chores, and much more." Robots and technology were not mentioned.

Interviewee general introduction: questions about the participant, their family, and their history in the home.

Description of a typical day

Description of the home: among the topics in this section were a description of rooms and their uses, the interviewee's favorite and least favorite spaces, any remodeling done, and any technology in the home.

Help in the home: topics included the amount of paid help, frequency of visits, assigned tasks, and the informant's feelings about having help in their home.

Organization: including whether the space was too big or small, storage solutions, approaches to organization, and family dynamics surrounding organization.

Home tour: a tour of the entire home, including areas such as closets.

Wrap-up: our interest in robots in the home.

The home tour was a vital component, allowing us to verify the interviewees' self-reported information with the reality of their homes. Examining private parts of a home, such as a bedroom closet, also accelerated the rapport-building process, allowing us to probe more deeply into previous comments. Entering interviewees' homes in this manner distinguishes this study from those conducted in a lab.

For this paper, 12 people were interviewed in 8 separate interviews. This group consisted of four married couples interviewed together, three married people interviewed without their spouses, and one single person. All of the interviewees were the principle figures in the home, organizing and directing activity. Ten of the twelve informants were found by a professional recruiting service, while the two professional organizers were found through professional contacts. All of the interviews centered around the topic of organization.

To increase the cultural coherence and competence of the informants, all of the interviewees lived in the San Francisco Bay Area, spoke English fluently, were not retired adults (ages 27-60) and were upper-middle class. Tightly specifying the user group should provide more reliable results given the group size [18, 24]. We hypothesize that this group is likely to be among early adopters of new technology and thus particularly relevant. The results described herein, however, may not generalize to other user groups.

Given the small number of interviews, it was important to extract useful information quickly and efficiently. One technique for doing so is to interview *extreme users*. Extreme users push the boundaries of a technique, technology or situation. They also communicate needs quickly because they feel them acutely. For example, someone with an average sized home may have some complaints about closet space, but those complaints are minimal. Because they do not encounter space constraints on a daily basis, it will take a long time for them to complain about closets in an interview. For someone with a very small space, however, the lack of storage is felt in daily attempts to dress, put groceries away, etc., so they will complain about the issue quickly and loudly.

The interviews presented herein include three types of extreme users. 1) People living in small spaces (under 600 square feet): spouses Sam and Mary, and single interviewee Tony. The small space constraint makes it difficult to store objects. 2) People living in large homes (above 4500 square feet), allowing them to accumulate possessions if desired: spouses Jeff and Beth, spouses Kate and Dave, and spouses Ann and Bob, and Jill (interviewed without her spouse). 3) Professional organizers, who help others organize their homes: Eva and Emma (interviewed without their spouses).

To develop the technologists' empathy for users, a key goal of this exercise, the interviews were conducted by a mixed team of need finding experts and technologists.

2.3 Frameworks, Implications and Solutions

Stories are the unit of analysis for need finding. Frameworks are extracted from the stories told by the informants relating "contradictions, spoken and unspoken norms, success and failure" [2]. The stories are used to identify the important dimensions (e.g. cognitive load or clutter level) of analysis and generate structured concepts such as timelines, hierarchies of needs, 2x2 diagrams (e.g. Figure 3), or any format that expresses an idea heard in multiple interviews. The frameworks are populated with specific examples from the interviews and are used to identify trends and gaps, which are then expressed as design implications.

Design implications are guidelines for a system design which will satisfy users. Even at this stage, it is preferable not to presuppose solutions. Consider the use-case example of an astronaut recording notes in space. A design implication might be that the astronaut must be able to record information while floating; another implication might be a need to recall the information within two seconds; finally an astronaut must always feel in control. None of these design implications presupposes that the astronaut will write the information, leaving open solutions such as voice recording.

With empathy for the user group, frameworks representing their needs and viewpoints, and design implications making these needs concrete, the process of generating solutions and prototypes is informed and responsive to the pull of user needs. The prototypes generated are hypotheses of what users want. Promising prototypes can be subjected to the traditional exercise of user testing to obtain more concrete feedback. Although this method seems linear, it is usually not. Most teams bounce back and forth between articulating stories, framework, insights, and solutions.

2.4 Related Approaches

Need finding is inspired by ethnographic research in cultural anthropology, conducting interviews and making observations. The methods differ beneath this superficial similarity, however. Ethnographic research aims to create a detailed, "thick description" of a culture [11]. Need finding, on the other hand, is concerned with rapidly extracting user needs and empathy for the culture being studied. The process also includes abstraction into frameworks describing these user needs, and the extraction of design implications to inspire new technologies. Finally, actual solutions to the needs are created, prototyped, and taken back to users.

A method similar to need finding is needs assessment [12]. The two methods have similar goals, both methods separate user needs from actual technological solutions. They differ, however, in that needs assessment aims to identify gaps between results and consequences, whereas need finding aims to identify gaps between use, usability, and meaning. The scope of needs assessment work is typically very broad, working at many levels of an organization (mega, macro, and micro). In contrast, the need finding work presented here is oriented toward driving research and development that will ultimately have impact for end-users.

3. FRAMEWORKS AND IMPLICATIONS

In this section, we present the findings from our need finding exercise. Each subsection describes a framework, the observations which lead to the framework abstraction, and the design implications of the framework.

The frameworks and design implications contained herein are hypotheses. As previously mentioned, need finding is an iterative process. To further validate these hypotheses, it will be necessary to design solutions inspired by the frameworks and encapsulating the design implications, and take those solutions back out to users for feedback and more traditional user studies.

3.1 Roles within the Home

Framework

Family members take on different roles within the home, taking responsibility for some tasks while delegating others.

In 5 of the 8 households we spoke to, one person was the *Organizer*. The Organizer determined storage methods and usually physically put objects away. More importantly, the Organizer kept the list of what-goes-where in their head. Family members would ask them when anything was needed. This hands-on approach was very important to the Organizers. They believed that they were more organized than their family, so it was necessary to keep control. In fact, even the suggestion of delegating the task made them nervous.

In homes with a dominant Organizer, the other family members were hands-off about organizing. They generally did not know where items were stored, except for the few things they used on a daily basis. They also did not care to expend more energy on organizing. It is interesting to note that none of the homes with an Organizer used lists of any sort to remember the locations of items or communicate those locations between family members. The Organizers reported being happy with the existing arrangement in which they were asked for items.

The Organizer versus family member dynamic was nuanced. In 4 of the 5 households with Organizers, the Organizer dominated the other family members (with respect to organizing), and felt like they ran the household. This was a source of tension within the family. Two of the Organizers' spouses teased the Organizers, rolling their eyes and using condescending language to describe their organizational tendencies. All of the Organizers called themselves "anal," "OCD," or other derogatory terms. The fifth home had a different dynamic, with the Organizer's spouse acting like a CEO (which he once was.) Jeff delegated the organization to his wife as a CEO would delegate to an assistant, retaining the feeling of control while remaining hands-off.

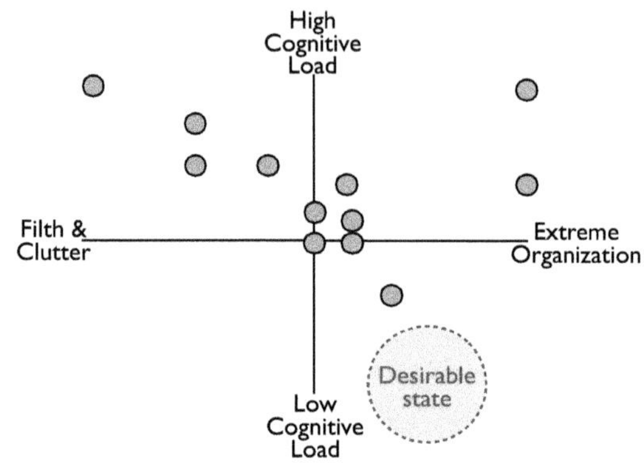

Figure 3: Framework: the informants' tradeoffs between cognitive load and organization in their homes. Each red dot represents a single informant. The green circle is a hypothesized desirable tradeoff.

In two of the other households, the organizational duties were spread more evenly among the family members. One person assumed a more managerial role, making sure tasks were completed, but overall the workload was shared. The final household was a single-member household.

Design implications

The family dynamics surrounding organization suggest multiple design implications. First and foremost, a robot must be designed for the entire family. Although it will often be 'owned' by one person, it will interact with everyone. The robot should allow people who are Organizers to feel like they have control over and are participating in the organizational process. Organizers want to save time, but they also need to know where everything is. An automation system which takes away this knowledge actually increases the Organizer's cognitive load as they are constantly wondering where things are. An organizing robot should take advantage of the stress-reducing power of human participation.

In contrast, the other family members should be able to interact with the robot in a hands-off manner, not exerting more effort than if a human were organizing. The other family members do not need to know where an object is in storage, but do need to access objects quickly.

A technology for organization should also not create tension between users by making them appear overly involved, or conversely overly lazy, to each other.

Finally, a robotic storage solution must avoid becoming a dependency. On a daily basis, the hands-off users can treat the technology as they currently treat the Organizer, allowing the robot to remember item whereabouts. But given that all technology fails eventually, there must be a non-automatic mode which allows users to find and retrieve their belongings. The worst case scenario was described in a paper about home automation in which the home automator passed away and no one else in the family could control the home [3]. If a robot fails, it must not leave a family unable to find their possessions; the robot cannot fail catastrophically.

3.2 Clutter and Cognitive Load

An interesting view of our interviewees emerges from considering the cognitive load they reported with respect to organizing, versus the actual order we saw in their homes. In this case, cognitive load was subjectively measured by how often the informants thought about organizing, regardless of whether that thought turned into action.

At one extreme, informant Tony did little to clean his small apartment and it was in a very disordered state. Tony did, however, report constantly think about cleaning, feeling guilty that he could not motivate himself into action.

Informant Ann also regularly thought about organizing, and did occasionally organize sections of her home. The large size of her home and her inability to throw things out, however, left some areas very disorganized even after cleaning. Clutter moved from one area to another, but never actually left the house. Ann repeatedly stated that the clutter in her home was very bothersome, and even suggested that her home was messy enough to be on a televised reality show. Although the reality we saw was not nearly drastic enough to be on television, there were certainly rooms that had accumulated a large amount of clutter over decades.

At the other end of the spectrum was Emma. Emma is a professional organizer and every item in her home was in its proper place, despite the disruptive presence of a baby. Much of her day, both professionally and at home, was spent organizing. Along with organizing communal objects and her personal possessions, Emma reported helping her husband organize his possessions, and returning objects to their "correct" locations after the housecleaning service finished.

Jill had also committed a large amount of time to organization, but focused instead on storage space. Having built her own home, she was able to include an impressive amount of storage cabinets, closets, etc. She had so much storage, in fact, that many of her cabinets were empty. Jill also came up with creative storage solutions, such as using large trash bins as table bottoms (hidden by a table skirt) to store infrequently used items such as Christmas decorations, as seen in the bottom-right of Figure 1. The result was an extremely organized home. Jill also reported being the only one in her household who knew where things were; when she was away, her husband would call her to ask about object whereabouts.

The examples above suggest that both too little and too much organization can be associated with high cognitive load. Is there an optimal tradeoff, a level of organization that can be achieved with less effort, but does not generate feelings of uneasiness for the homeowner? Informants Mary and Sam gave evidence of having a reasonable cognitive load versus organization tradeoff. Living in a very small apartment, Mary and Sam told of their organization challenges. By refraining from buying extraneous items for lack of storage, they kept their cognitive load as low as possible. However, they still had to play games like "Tetris" or "Jenga" when dealing with the objects in their closets. While they were managing day-to-day and reported satisfaction with their organization level, the possibility of being pushed into organizational chaos still weighed on their minds.

Kate, Dave and their seven children kept their cognitive load relatively low by distributing home organization among the family. Each person's load was low and the home was usually tidy. The possibility of chores being forgotten resulting in organizational chaos, however, was ever present.

Jeff had the lowest cognitive load among the informants. He had delegated all of the organization to his wife who was fastidious about cleanliness. However, Jeff had reserved for himself the task of organizing the items he needed on a daily basis, retaining control over his habits. Even this arrangement was not optimal, however, as Jeff was subjected to his wife's complaints about the need for more order and her cleaning efforts, thereby increasing his own cognitive load.

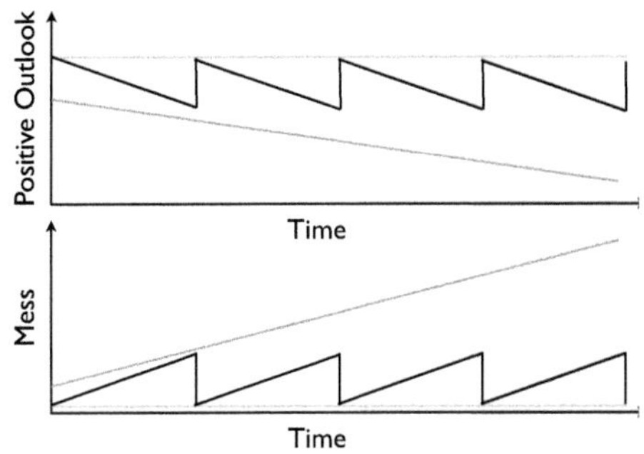

Figure 4: Framework: the correlation between a homeowner's positive outlook and home tidiness over time. Green: the desired but unachievable state of perpetual organization. Red: the degenerate case of unceasing increasing messiness and decreasing happiness. Black: the usual reality is a sawtooth pattern, worsening over time and then recovering in bursts of organization.

Framework

The tradeoffs between organization and cognitive load made by our informants are illustrated in Figure 3. We hypothesize that lowering the cognitive load requirement would be a more satisfying situation (the green circle in the diagram). In order to keep the cognitive load low, the resulting level of organization cannot be extreme (such as in Emma's house), but should be well beyond the tipping point into disorganization. This buffer of extra organization prevents concern that things could go wrong at any moment. Lower cognitive load can also be achieved by delegating organization, but there is a corresponding loss of control.

Note that none of the informants were in the lower-left quadrant of the diagram with both low organization and low cognitive load. This reflects the selection bias in our sample as we specifically recruited informants who self-reported organization challenges. This omission is not a hindrance for robot design, however. A problem which people do not care to solve is not a suitable task domain for which to invent new technology, there simply would not be a market.

Design implications

This framework implies that the current minimum for stress and cognitive load comes at below-maximum organization. Can robots push the amount of organization that can be achieved without increasing cognitive load? If not, robots should instead attempt to lower the cognitive load for creating the current buffer zone between organization and chaos. Increasing the cognitive load, even for a return of increased organization, is not valuable.

3.3 Organization Levels Fluctuate

"I get a bee in my bonnet," said interviewee Ann about organizing. Jill said, "You know, you just get into a mood to throw things out!" Kate and Dave said, "Sometimes we invite guests over to make us clean." In fact, all but one of our informants spoke about organizing and purging objects as discrete events, punctuating longer periods of declining orderliness. The level and length of decline varied, but the need to organize and purge came in strongly felt waves.

Purging events were explained in two ways. One was the impending arrival of guests. The second motivation was unknown, the informants simply said they felt compelled to tidy. The exact tipping point was unclear, but the feeling acute. Discomfort with the state of their home built up with the disorder, until finally something had to be done. All of the informants reported feeling happier and more satisfied with their homes after a bout of organization.

A counter-example to the cycle was Tony, who reported suffering from depression and was unemployed. He reported wanting to organize but being unable to find the energy for a cleaning binge. As a result, his apartment grew messier and Tony grew more upset over time.

Note that none of the informants maintained a constant state of order in their homes, the supposed ideal situation.

Framework

The organizing patterns and corresponding mental states of our informants are summarized in Figure 4. The ideal state of constant low messiness and highly positive outlook, which none of the informants attained, is given by the green line. The dismal reality of Tony's state is that the messiness continues over time while his outlook becomes increasingly more negative, as represented by the red line.

For most people, organization levels and mood are represented by the black sawtooth line. The mess in their homes increases, matched by a decrease in their satisfaction level. Then, guests visit or the urge to organize overtakes them, and order and happiness are restored. The cycle repeats.

Design implications

Robots as organizational aids have the potential to be tireless, organizing continuously. Working autonomously, robots could potentially minimize the overall mess in a home and prolong the positive outlook of the home's inhabitants.

It is likely, however, that robots will need human participation in the organizing process. Humans will facilitate the robot's actions, perhaps moving things from the robot's path as is done for a Roomba [8, 23]. Humans will certainly need to decide where to place items or when to discard them (as discussed in Section 3.6). In either case, robot algorithm design cannot expect humans to be consistently diligent about tending to the organizing task at all times. A robot should not continuously nag its owner.

A more successful strategy might be for a robot to work within the organizing patterns of its owner, grouping questions or tasks for when the human "has a bee in their bonnet" about organizing, all the while performing autonomous tasks in the background. The length of human-in-the-loop interventions will change with the owner, but it is clear that the person will tire of the organization task. The goal, however, does not need to be a spotless home, but instead to return the homeowner to a state of happiness with their home.

3.4 The Journey of Stuff

A commonly cited myth of organization is O.H.I.O, or Only Handle It Once. This mantra is often applied to mail or paperwork, and translates into the recipient completely dealing with the item (reading, addressing any requests, shredding or filing, etc.) at one time. Applied to objects within the home, O.H.I.O is interpreted as putting objects in their correct location as soon as they arrive, and immediately after each use. While this superficially seems like a good idea, the reality is quite different.

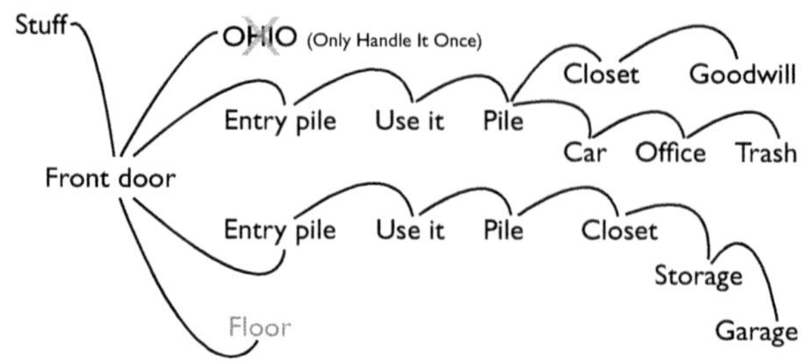

Figure 5: Framework: the journey of stuff over time. With time flowing from left to right, this figure shows examples of paths that objects take throughout their lifetime in the homes of our informants.

All of the informants save one told of the journey stuff takes through their homes and lives. Specifically, they all spoke about the occurrence of "piles". Mary and Sam, who live in a very small apartment, described an acceptable journey for their stuff as: it enters the home and is deposited in a pile near the entryway. They may use the object, and often deposit it back into the entry pile. At some point, the existence of the pile becomes a nuisance and the objects are put in their proper storage place, such as the closet. Closets are occasionally purged to make way for new things, depositing the old objects in the trash or giving them to a charity. Sam guiltily admitted that sometimes he is unable to organize his things and they end up stored in his car trunk, and then his office. Eventually these items are also stored or discarded.

Anther couple, Ann and Bob, were extremely upset about the seemingly unending journey of objects in their large home. They told of the entryway pile, but also of piles in other places. Piles of random objects in closets created minutes before guests arrived, then moved to the storage room when they had been selected for disposal, and eventually to the garage where they remained. Despite being marked as garbage, the objects were never disposed of; the piles formed temporary storage for transient objects.

Emma, a professional organizer with a meticulously organized home, admitted to the existence of temporary piles of mail and other small items on her kitchen counter. She fought the visual clutter by organizing the piles into attractive storage bins, but they were piles nonetheless.

Among our informants, only Tony actually only handled objects once. The result by Tony's own admission was a disorganized, messy, unsatisfying apartment. Tony would deposit objects wherever he used them, on the floor, on a chair, on his bed. Garbage and useful items alike were everywhere. Tony acutely felt failure about the disorder, but could not rectify the situation.

Framework

The framework that arises from this data is that stuff has a journey. Figure 5 illustrates some of the journeys discovered in our interviews. Objects move within a home between temporary storage, use, permanent storage, or disposal. Malone [14] wrote of a similar path for items on desktops, whose location on the desk and in piles aids human memory and productivity. A Steelcase Inc. office study discovered that piling patterns change with job responsibilities [22]. This journey of stuff can be satisfying or disappointing, but it is a reality. Only in degenerate cases is stuff only touched once.

Design implications

We hypothesize that O.H.I.O is an unattainable ideal that robots should not attempt to make a reality. Instead, robots should help to move stuff along its journey so that it does not stall. Piles of objects in a home are problematic when it seems like they will never dissipate. Acknowledged temporary piles, however, are necessary and acceptable. Robots should allow for the physical journey of objects and the accompanying mental journey of the home's occupants.

3.5 Organizing is Decision Making

When asked whether she would allow her housecleaning service to throw out unused items, Ann responded categorically, "No." All of the other interviewees had the same response, it was simply unacceptable for any other person, including their spouse, to decide what was garbage. Even Tony, incapable of summoning the strength to organize on his own, insisted on supervising his sister when she came to help; he would make the decisions.

Framework

Organizing is decision making, and decision making is hard. There is literature describing both phenomena, a full recital of which is beyond the scope of this paper. For example, Frost [9] links the inability to make decisions to hoarding. Baumeister et al. [1] and Danzigera et al. [6] describe decision fatigue, the declining ability to make decisions over time, necessitating rest between decision-making sessions. Schwartz [19] discusses how too many choices exhaust us and leave us unable to choose at all.

Interviewee Ann described the process of sorting through an overflowing closet as exhausting. She described beginning the process with good intentions, putting items in *keep* or *discard* piles. However, her willpower quickly drained, and the remaining unsorted items were simply deposited back into the closet. Even worse, having exhausted her stores of decision making energy, she was unable to actually throw out the items in the *discard* pile and instead deposited them in a storage room or the basement.

In addition to the difficulty of decision making, much of the inability to organize and discard now-useless items actually stems from the meaning of those items. Informant Eva, a professional organizer, categorized the stuff that clutters people's homes into two categorizes. The first category represents unfulfilled dreams, a past or potential reality the owner had to give up. The second category includes items kept just in case they might be needed one day.

We did in fact see many examples of the first category in our informants' homes. Eva herself kept scrubs from her previous career as a doctor. Ann kept baby items from her days as a young mother. Bob kept old technology, such as ten year old modems, from his days as an engineer. Tony kept his college textbooks in the hopes of a career which was slow to evolve. These items all reflected our informants' personalities. Keeping such items was not a decision based on utility, but rather on sentimentality mixed with denial.

Design implications

The task of home organization implicitly involves decision making. A robot will either have to make the decisions itself, or encourage people to decide. Our research suggests that there is only a small subset of items in the home for which a robot will be allowed to make decisions such as when to discard them, or even where to store them. For most objects, the robot will need to elicit human decisions.

To encourage decision making, the number of choices requested at one time should be minimal. Also, the robot should allow for temporary storage (such as piles) when decisions are simply too difficult. As previously discussed, allowing stuff to have a journey can mitigate the emotional turmoil associated with organizing. In the extreme, there should be allowances in robot behavior for objects with emotional significance which the robot may not touch at all.

3.6 My Way, Now

Framework

At work, there are compromises and inefficiencies, but we cope with them because we are paid to do so. This is not the case at home. At home, we expect things to go our way and be efficient, even the intrusion of family members on our intended path is frustrating.

Sam noted that when his small apartment became cluttered, every task took an unacceptable "five extra steps." Despite generally delegating organization to his wife, Jeff retained control over objects on his "critical path" through the day, allowing him to execute tasks efficiently and without compromise. Jill creatively organized her entire home, but did not touch her husband's work tools, which he needed on a daily basis. When organization implied extra work at task execution time for critical tasks (such as searching for keys on the way out the door), the informants withdrew the required items from the normal home organization scheme.

Design implications

A robot assisting with home organization should not make it more complicated for people to execute common tasks which they do efficiently in the absence of the robot. The robot should not complicate tasks in time, number of steps, or in the mental load required. Objects should never be placed where their owners cannot find them.

3.7 Peers, Parents, and the Work Ethic

Framework

Interestingly, despite our informants' abilities to pay for household help, they all had minimal to no hired help. The highest level of assistance was a housecleaning service which came once every two weeks. Nonetheless, many of the informants complained about the high cognitive load of keeping their homes organized. When asked why they did not hire help, the response was unanimously that they felt like organizing their home was a task they should do themselves. There was a bias toward a strong work ethic and personal responsibility with respect to the home.

This view of the home as the owner's responsibility also colored how the informants perceived family members' efforts to help with organization tasks. Tony recounted that his sister had come to help him organize his apartment. Her approach was to nag him into cleaning and she would make decisions without Tony's permission. Tony perceived his sister as parental and condescending, and it upset him.

During the interviews, when Ann and Beth recounted nagging their husbands to help organize, their respective husbands rolled their eyes. Both husbands viewed the nagging as a reminder that they were not completing a task they should be capable of executing.

On the other hand, conversations about organization could be conducted in a non-confrontational manner and without guilt if approached in a peer-to-peer manner. Tony told of a time when his brothers helped him clean. Instead of nagging him, they offered to help so that "I (Tony) could bring girls home." This offer of assistance was acceptable for Tony and did not hurt his pride. Both Mary and Sam and Kate and Dave worked as teams, sharing the organization load. This ownership over the task satisfied both partners.

Design implications

A robot which nags its owners into making decisions, implies failure or guilt in any way, or attempts to make important decisions for its owners, will read as a parent and should be avoided. Instead, a robot should help a person make decisions as a peer with shared goals. A robot should allow its owners to feel pride over accomplishing a task.

4. CONCLUSIONS

The study presented herein explored the needs and attitudes that people have with respect to organization in their homes. As a task for robots, home organization is difficult but potentially high impact. This study has discovered the deep needs, uncertainty, and family tensions which arise around organization. By relating these frameworks to design implications for robots, this paper aims to inspire and help direct future explorations into robot technologies.

In fact, the robotics community is already beginning to attempt basic organizational tasks such as putting away groceries [17]. Existing needs in the home like the ones discovered in this study, can inspire this research and direct it toward higher impact tasks and approaches. The design implications should also serve as hypotheses for user testing with new prototypes.

We have also presented the methodological tool of *need finding* for inspiring and grounding robotics research. This tool can be applied to any task domain, and is complementary to both other user-driven (needs-pull) and technology-push approaches to motivating research.

An interesting question for future studies on the topic of home organization revolves around generational issues. All of our interviewees were adults (ages 27 to 60), however, future generations may be different. Given the enormous amount and constant availability of information online, younger generations are learning to search for knowledge instead of memorizing. Will this search-based approach to knowledge extend to their homes? Will future homeowners be comfortable forgetting the whereabouts of objects and using technology to search for these objects as needed? The changes in how people relate to objects in their homes could prove to be fascinating.

5. ACKNOWLEDGMENTS

We wish to thank Anne Fletcher, Stu Glaser, Phoebe Kuo and Eitan Marder-Eppstein for helping to conduct and analyze the interviews.

6. REFERENCES

[1] R. Baumeister, E. Bratslavsky, M. Muraven, and D. Tice. Ego depletion: Is the active self a limited resource? *Journal of Personality and Social Psychology*, 74:1252–1265, 1998.

[2] S. Beckman and M. Barry. Innovation as a Learning Process: Embedded Design Thinking. *California Management Review*, 50(1):25–56, 2007.

[3] A. Brush, B. Lee, R. Mahajan, S. Agarwal, S. Saroiu, and C. Dixon. Home automation in the wild: Challenges and opportunities. In *CHI*, 2011.

[4] V. Bush. As we may think. *Atlantic Monthly*, 176(1):101–108, 1945.

[5] D. P. Crowne and D. Marlowe. A new scale of social desirability independent of psychopathology. *Journal of Consulting Psychology*, 24:349–354, 1960.

[6] S. Danzigera, J. Levavb, and L. Avnaim-Pessoa. Extraneous factors in judicial decisions. *Proc. of the National Academy of Sciences*, 108(17):6889–92, 2011.

[7] M. Fellman. Breaking Tradition. *Marketing Research*, 1:20–25, 1999.

[8] J. Forlizzi and C. DiSalvo. Service robots in the domestic environment: A study of the roomba vacuum in the home. In *Proc. HRI*, 2006.

[9] R. O. Frost and R. C. Gross. The hoarding of possessions. *Behaviour Research and Therapy*, 31(4):367–381, 1993.

[10] B. Gates. A robot in every home. *Scientific American*, pages 58–65, 2007.

[11] C. Geertz. From the Native's Point of View: On the Nature of Anthropological Understanding. *Culture Theory: Essays on Mind, Self, and Emotion*, 1984.

[12] R. Kaufman and F. W. English. *Needs Assessment: Concept and Application.* Educational Technology Publications, 1979.

[13] S. Kriz, T. Ferro, P. Damera, and J. Porter. Fictional Robots as a Data Source in HRI Research: Exploring the Link between Science Fiction and Interactional Expectations. In *ROMAN*, 2010.

[14] T. W. Malone. How do people organize their desks? implications for the design of office information systems. *ACM Transactions on Office Information Systems*, 1:99–112, 1983.

[15] NAPO. National Association of Professional Organizers. `www.napo.net`.

[16] J. Nielsen. First rule of usability: Don't listen to users. `www.useit.com/alertbox/20010805.html`, 2001.

[17] D. Pangercic and M. Beetz. TUM james goes shopping. `www.youtube.com/user/iasTUMUNICH#p/a/u/2/JS2zycc4AUE`, 2011.

[18] A. Romney, S. Weller, and W. Batchelder. Culture as consensus: A theory of culture and informant accuracy. *American Anthropologist*, 88(2), 1986.

[19] B. Schwartz. *The Paradox of Choice: Why More Is Less.* Ecco, 2004.

[20] M. Scopelliti, M. V. Giuliani, and F. Fornara. Robots in a domestic setting: A psychological approach. *Univ Access Inf Soc*, 4:146–155, 2005.

[21] A. Sellen and R. Harper. *The Myth of the Paperless Office.* MIT Press, 2002.

[22] Steelcase Inc. Six patterns of work. `http://www.steelcase.com/en/Resources/overview/Documents/six%20patterns%20of%20work.pdf`, 2006.

[23] J. Sung, R. E. Grinter, and H. I. Christensen. Domestic robot ecology. *International Journal of Social Robotics*, 2:417–429, 2010.

[24] S. C. Weller. Cultural consensus theory: Applications and frequently asked questions. *Field Methods*, 4:339–368, 2007.

The Domesticated Robot: Design Guidelines for Assisting Older Adults to Age in Place

Jenay M. Beer [1], Cory-Ann Smarr [1], Tiffany L. Chen[2], Akanksha Prakash[1], Tracy L. Mitzner[1], Charles C. Kemp[2], & Wendy A. Rogers[1]

[1]Georgia Institute of Technology
School of Psychology
654 Cherry Street
Atlanta, GA 30332
wendy.rogers@psych.gatech.edu

[2]Georgia Institute of Technology
Department of Biomedical Engineering
828 West Peachtree Street NW
Atlanta, GA 30308
charlie.kemp@hsi.gatech.edu

ABSTRACT
Many older adults wish to remain in their own homes as they age [16]. However, challenges in performing home upkeep tasks threaten an older adult's ability to age in place. Even healthy independently living older adults experience challenges in maintaining their home [13]. Challenges with home tasks can be compensated through technology, such as home robots. However, for home robots to be adopted by older adult users, they must be designed to meet older adults' needs for assistance and the older users must be amenable to robot assistance for those needs. We conducted a needs assessment to (1) assess older adults' openness to assistance from robots; and (2) understand older adults' opinions about using an assistive robot to help around the home. We administered questionnaires and conducted structured group interviews with 21 independently living older adults (ages 65-93). The questionnaire data suggest that older adults prefer robot assistance for cleaning and fetching/organizing tasks overall. However their assistance preferences discriminated between tasks. The interview data provided insight as to *why* they hold such preferences. Older adults reported benefits of robot assistance (e.g., the robot compensating for limitations, saving them time and effort, completing undesirable tasks, and performing tasks at a high level of performance). Participants also reported concerns such as the robot damaging the environment, being unreliable at or incapable of doing a task, doing tasks the older adult would rather do, or taking up too much space/storage. These data, along with specific comments from participant interviews, provide the basis for preliminary recommendations for designing mobile manipulator robots to support aging in place.

General Terms
Measurement, Design, Experimentation, Human Factors.

Keywords
Aging, Aging in Place, Assistive Robotics, Chores, Design Guidelines, Home, Home Environment, Mobile Manipulation, Needs Assessment, Older Adults, Service Robotics.

1. INTRODUCTION
1.1 The Challenges of Aging in Place
Maintaining one's independence is a primary goal of older adults and a key component to successful aging [1]. Older adults, defined as persons 65 years of age or older [11], encompass a growing segment of the world's population [27]. One way in which older adults may maintain independence is to age in place. Aging in place has been broadly defined in the literature as the ability for an individual to stay at home as he or she ages [e.g., 16, 20]. The definition of "home" may vary, comprising a variety of home-settings such as houses, apartments, retirement communities, or condominiums. Whatever the setting, the majority (80%) of older adults live independently in their own home, with a small percentage (5%) in a nursing home [18].

A growing area of research is investigating how to support aging in place, and retain or enhance one's capability to function later in life (i.e., successful aging). With age, there are general changes in cognitive, physical, and perceptual capability: working memory declines, physical strength decreases, and hearing and vision loss occur [15]. Age-related changes pose challenges for older adults to function independently and age in place.

Successful aging may be conceptualized by the selection, optimization and compensation model (SOC model) [2]. This psychological framework posits that successful aging is a process involving three components. *Selection* refers to development and commitment to personal goals (e.g., maintaining independence, aging in place). Throughout one's life there are many goals relevant to functioning. However, limited resources reduce the number of goals attainable. Therefore, older adults must either reduce their options by selecting a subset of goals, or develop new attainable goals. *Optimization* refers to the investment of time and energy to behaviors that support their chosen goals.

Compensation refers to the regulation of loss. That is, the use of compensatory or substitute mechanisms to prevent or balance age-related changes. According to Baltes and Baltes [2], compensation involves aspects of the "mind and technology". For example, the use of mnemonic strategies (i.e., memory aids) is an example of a psychological compensatory effort. Similarly, the use of technology and devices (e.g., hearing aids, wheelchairs, eyeglasses) are examples of technological compensatory efforts.

The selection, optimization, and compensation efforts of older adults to successfully age in place need to be considered in the context of the person *and* the environment. Aging in place requires the interaction between older adults and their management of environmental press, that is, the challenges in the environment that older adults need to cope with to maintain their independence. An example of environmental press is the challenge of maintaining one's home. Home upkeep tasks are a major component to aging in place. Aging in place can be threatened if older adults are experiencing difficulties in home upkeep tasks. This dynamic relationship between older adults and their environment was described in Lawton and Nahemow's ecological model of aging [21].

According to the ecological model of aging, a person can be characterized by their level of competence. A person's level of competence is dynamic across the lifespan, with age-related changes in physical, perceptual, cognitive, and social capabilities. Additionally, forces in the environment place physical, cognitive, and social demands on the individual. Lawton and Nahemow [21] stated that ideally an individual's competency should match the environmental press. If environmental press exceeds individual competence, maladaptive behavior and negative affect may result. To improve independence, persons must be considered within the context of the environment. With regard to aging in place, the environment is the home. The home places dynamic demands on older adults, requiring a variety of home upkeep tasks.

A number of studies have investigated the home maintenance needs of healthy older adults as they age in place. Human factors and needs assessment methodologies have shown that healthy older adults are able to maintain their homes, but not without difficulty. Doing laundry, cleaning, and making the bed were reported as difficult tasks [9, 13]. The data from Rogers and colleagues [24] suggested that 40% of the problems encountered were physical in nature, and 30% were cognitive in nature. Additionally, the previous needs assessment approaches suggested that many solutions to these difficulties include cessation (i.e., not doing the task or outsourcing), perseverance (i.e., still completing task even if it was more difficult or took longer), and compensation (i.e., behavior modification, changing the environment, or using tools or technologies) [14, 24].

As discussed in the SOC model [2], older adults may offset age-related changes and environmental press by using compensatory efforts, such as the use of technology. Robotics is a prospective technological effort that could compensate for environmental press. In particular, home robots have the potential to assist older adults in home upkeep, which is a major component to successfully aging in place, and therefore maintain independence.

1.2 The Potential for Home Robots

A variety of robots, both in development and commercialization, have been designed to assist with or perform home tasks. More are designed to help with physical aspects of housekeeping, ambulation, and social communication; compared to other activities (e.g., money management, grooming, laundry) [25].

It is important that robots are designed to meet older adults' needs and that older users are amenable to robot assistance. These considerations may increase the likelihood for home robots to be adopted by older adults. Acceptance of home robots may depend on the older adults' perception of their health status (physical and cognitive), where appreciation for the utility of a home robot increases as the older adults' health capabilities are at risk of declining [7]. There have been some efforts on developing robots for those with declined health capabilities, such as individuals with motor impairments [8, 19], dementia [28], or needing healthcare assistance [23].

However, it is crucial to also consider those healthy older adults who are independently living. Declines in cognitive, physical, and perceptual capabilities are a normal part of aging. Even independently living older adults encounter difficulty maintaining their home [13]. An effort to develop robots to assist independent older adults has the potential to promote aging in place, and avoid or postpone the need for assisted living or nursing home care.

Indeed, independently living older adults have expressed a willingness to use and accept home robots [12; 26]. More generally, older adults are willing to use home technological assistance, if the perceived benefits of using the technology are clear [22]. Furthermore, Giuliani and colleagues [17] have shown that older adults' solutions to difficulties they encounter in the home are highly task dependent, with technological assistance a viable solution to some problems (e.g., hearing aids were reported as a solution to hearing difficulties using the telephone). In terms of attitudes toward robot-specific assistance, within the healthcare domain, Broadbent and colleagues [4] have shown that older adults were more positive toward healthcare robots than family or staff. In particular, all participants reported the healthcare robot would be most useful in detecting falls and lifting heavy things.

In a large-scale questionnaire where participants were instructed to imagine a home-based robot, independent older adults indicated an interest in using robots in the home, particularly if the system is performance-directed (i.e., assists with critical infrequent home tasks) [12]. Conversely, Bugman and Copleston [5] conducted a questionnaire that indicated older adults preferred a robot to perform daily housework tasks. However, reasons why older adults held such preferences were not systematically evaluated.

Although the findings from the research outlined above indicate older adult acceptance of robots, their acceptance was task specific. Lacking is older adults' preferences for robot activities in the home coupled with reasons *why* they hold such preferences. The development of home robots as a technological compensatory effort requires a more detailed understanding of older adults' assistance preferences for using a robot for various home upkeep tasks. Research exploring independently living older adults' needs and preferences for robot assistance can guide developers to design robots more likely to aid older adults' aging in place. Moreover, robots developed with older adults' needs and preferences in mind may increase their acceptance.

2. GOAL OF CURRENT RESEARCH

In light of the SOC model [2], robots, as a compensatory technology, have potential to assist older adults to age in place. Successful integration of assistive robots into older adults' lives will require design of robots that meet their needs. It is not clear what tasks independent older adults would prefer a robot to assist with, the reasons why they hold such preferences, or their perceptions of robot assistance with maintaining their home. This understanding can inform the design of assistive robotics to support these tasks, and thereby aid older adults in aging-in-place. In this research study, we evaluated the potential of home robots to serve as a compensatory technology for older adults. Specifically, our research aimed to:

- Assess older adults' preference for assistance from robots or humans for home upkeep tasks (cleaning tasks, finding and fetching).
- Understand older adults' opinions of using a home robot.
- Consider the implications of findings for directing improvement efforts for the design of home assistive robots.

Our needs assessment approach used a combination of questionnaire and structured group interview methodology to provide a rich set of quantitative and qualitative data.

3. METHOD

3.1 Participants

We conducted five structured group interviews, with each group comprised of two to five independently living older adults (N = 21) between the ages of 65 and 93 (M=80.25 years, SD = 7.19). Participants were recruited from local senior centers and retirement homes. The structured group interviews were conducted at the recruitment locations.

We interviewed independently living older adults who reported themselves to be relatively healthy. Most participants lived in an independent residence in senior housing (66.7%) or in a house, apartment, or condominium (28.5%), whereas the remainder lived in a relative's home (4.8%). To ensure diversity the participants were recruited from two different senior centers, located in areas of metro Atlanta known to differ in socioeconomic status (SES). As such, the participants varied in their educational background (correlated with SES) such that 33.4% had less than formal college education, 38.1% had some college, and 28.5% had at least some post graduate training. Participants were also diverse with respect to race/ethnicity: 57.7% White/Caucasian and 42.9% Black/African American.

To assess participants' experience with robots, we administered a questionnaire measuring different aspects of robot familiarity as well as the frequency of using 13 different types of robots (e.g., manufacturing robots, entertainment/toy robots, personal robots, surgical robots). Participants reported some familiarity (i.e., heard about or seen robot) with various robots (e.g., surgical robots, manufacturing robots, entertainment/toy robots) but no to very little experience using any robots. Therefore, overall participants were somewhat familiar with, yet inexperienced using robots.

3.2 Personal Robot

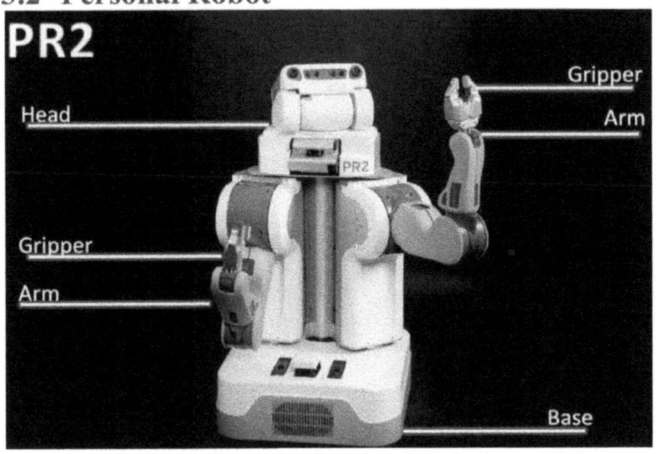

Figure 1. Video depicting layman descriptions of PR2.

The robot depicted in this study was the PR2 (Personal Robot 2), which was developed at Willow Garage, Inc. The PR2 is a commercially available mobile manipulator that is currently used by a number of robotics researchers. This robot has an omni-directional wheeled base with two 8 DOF arms/grippers, a telescoping spine, and a pan-tilt head that carries two stereo camera pairs and a LED texture projector. The PR2 is capable of autonomously navigating around a typical office environment, and manipulate some objects using its grippers.

Participants were introduced to the PR2 by viewing a narrated collage of video clips (see Table 1). The video introduced the robot (see Figure 1) and demonstrated its capabilities. The video clips were adapted, with permission, from Willow Garage's video library (see http://www.willowgarage.com/blog). The goal of the video was to show the robot's capabilities; however, we emphasized to the participants that the robot was not limited to what was shown in the video. The video served as a foundation for the older adults' discussion in the interview by ensuring the participants had similar expectations of the PR2's capabilities.

Table 1. Overview of video demonstrating PR2 capabilities.

	Overview of PR2 Video
Video chapter	*Description of what was shown*
Introduction	Layperson overview of robot's physical features (i.e., head, base, arms, grippers)
Capabilities	Navigating forward, backward, side to side; pushing a cart; grasping a variety of objects; and telescoping spine
Tasks	Opening a door; delivering drinks; delivering medication bottles; folding towels; plugging itself in; playing a game of billiards

3.3 Questionnaires

Prior to the structured group interview, demographic, health, and technology experience were collected using standardized materials developed by the Center for Research and Education on Aging and Technology Enhancement (CREATE) [10].

We developed an Assistance Preference Checklist to assess how assistance preferences vary (robot vs. human) as a function of task. We asked participants to imagine they needed assistance in everyday life and then indicate their preferences for human or robot assistance with 48 home-based tasks. We instructed participants to assume that the robot could perform the task to the level of a human. Participants rated their preference for assistance for each of the home-based tasks on a five point scale (1=only a human, 3=no preference, 5=only a robot).

3.4 Structured Group Interview Procedure

The interview script was designed following the research methodology outlined by Fisk et al., [15] and included a systematic development of script and questions, materials, selection of the interview environment, recruitment of participants, and training of moderators.

The interview sessions were stratified by gender to create a homogeneous group environment to encourage open discussions. On arrival to the structured group interview, participants provided written informed consent. Participants then completed questionnaires about robot opinions and experience. They were informed that the discussion would be digitally recorded and later transcribed for analysis. The moderator discussed the goals and topic of the structured group interview and rules for the conversation (e.g., not to interrupt other participants). The interview then followed a specific order, starting with the video depicting the PR2 and its capabilities, questions asking about the PR2 specifically, a brainstorm of tasks a robot may perform in the home, and finally a brainstorm of pros/cons to a robot performing specific tasks (i.e., medication management, finding and fetching items, and chores). Upon completion of the structured group interview, the participants completed questionnaires about their opinions about robots, and the Assistance Preference Checklist.

4. RESULTS

4.1 Assistance Preference Checklist Data

In this paper we focus on the data specific to cleaning tasks and fetching/organizing items, which are critical for aging in place. For a more detailed review of robot acceptance, see [25]. Collapsed across all tasks participants did not have a preference for human or robot assistance (*M*=2.99; *SD*=0.42). However, when each task was analyzed individually, histograms indicated preferences differed between tasks.

Task	*M*	*SD*	Histogram
Cleaning kitchen	3.71	0.78	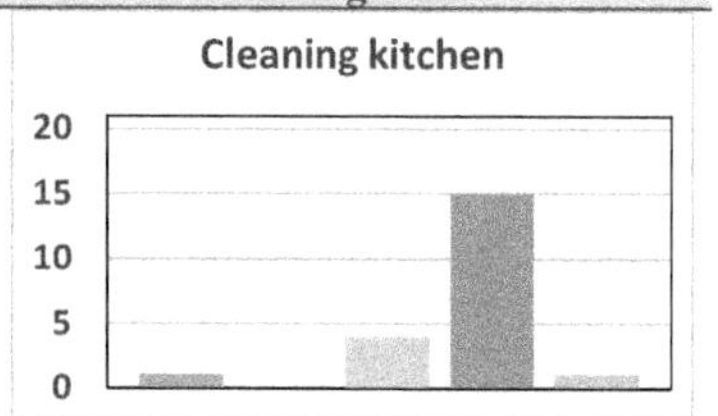
Sweeping/ scrubbing/ mopping	3.67	0.97	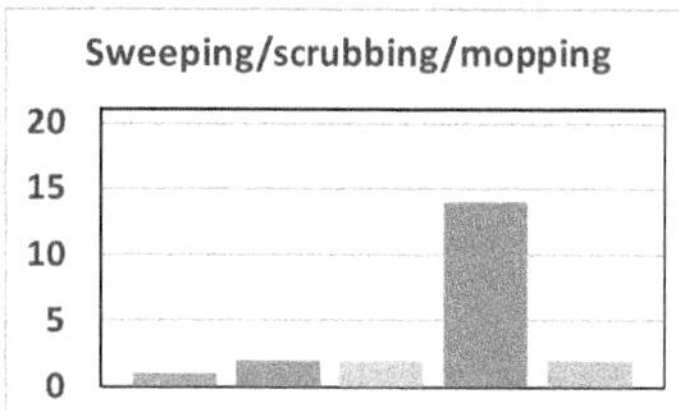
Cleaning bathrooms	3.57	0.93	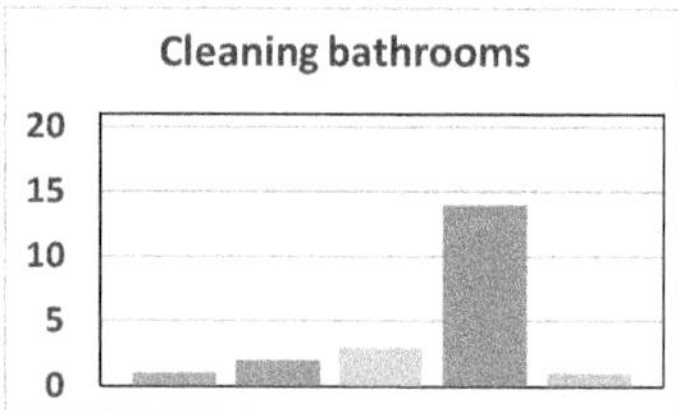
Cleaning windows	3.57	0.93	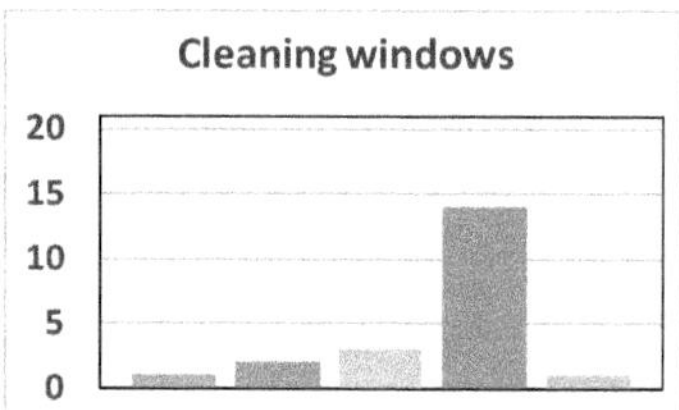
Making bed/ changing sheets	3.52	0.98	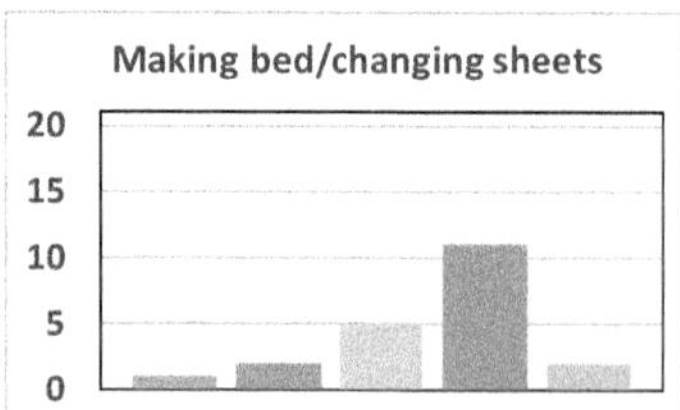

Only a human · Prefer a human · No preference · Prefer a robot · Only a robot

Figure 2. Cleaning Tasks – Prefer Robot.

Participants preferred assistance from a robot for 28 of 48 tasks ($M > 3.00$, where 3.00 = no preference), many of which were related to cleaning. Figure 2 depicts the top five reported cleaning tasks for which older adults indicated a preference for a robot performing (i.e., *M*=3.50 or higher).

For many cleaning tasks older adults indicated a preference for robot assistance. However, their preference did discriminate between tasks. For instance, Figure 3 suggests less clear preferences for certain tasks including sorting mail, doing laundry, and washing dishes by hand.

Task	*M*	*SD*	Histogram
Sorting, shredding, throwing away junk mail	3.29	1.19	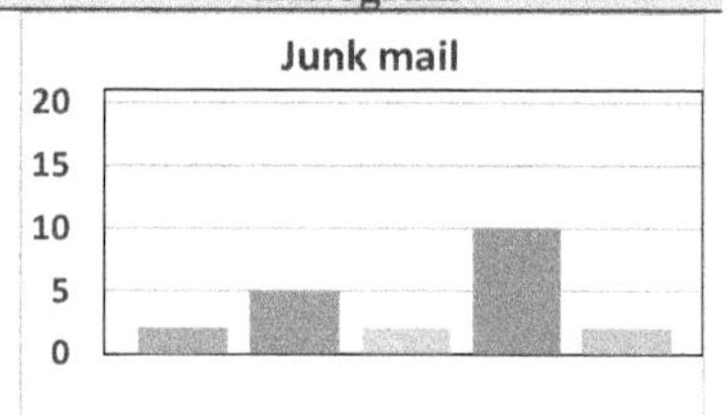
Doing laundry	3.29	0.96	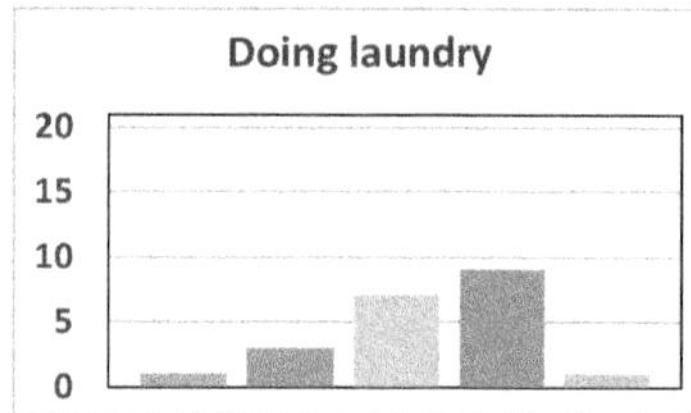
Washing dishes by hand	3.00	1.18	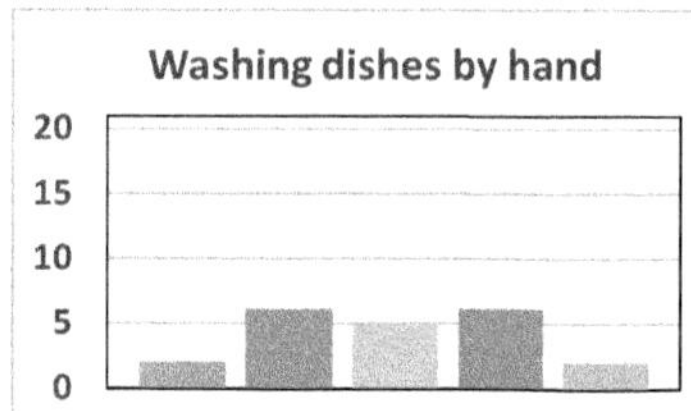

Only a human · Prefer a human · No preference · Prefer a robot · Only a robot

Figure 3. Cleaning Tasks – No Preferences.

The questionnaire contained tasks requiring fetching/organizing objects and included four aspects of manipulating objects: (1) fetching from floor/other rooms, (2) manipulating heavy objects, (3) reaching for objects (i.e., high or low), and (4) finding/delivering objects. As shown in Figure 4, these items indicated a clear preference (M = 3.5 or higher) for a robot to perform the task. In other words, none of the means regarding fetching/organizing home items averaged as "no preference" (M = 3.00) or "preferred to be performed by a human" ($M < 3.00$).

Although the Assistance Preference Checklist data above only indicate the types and nature of tasks that participants reported wanting (or not wanting) robot assistance with, the questionnaire does not indicate the reasons for such preferences. To gain insight about (or to provide an understanding as to) *why* older adults hold certain preferences for a robot's (particularly the PR2) assistance with home tasks, we analyzed the structured group interview data.

Task	*M*	*SD*	Histogram
Fetching objects from floor or other room	3.75	0.94	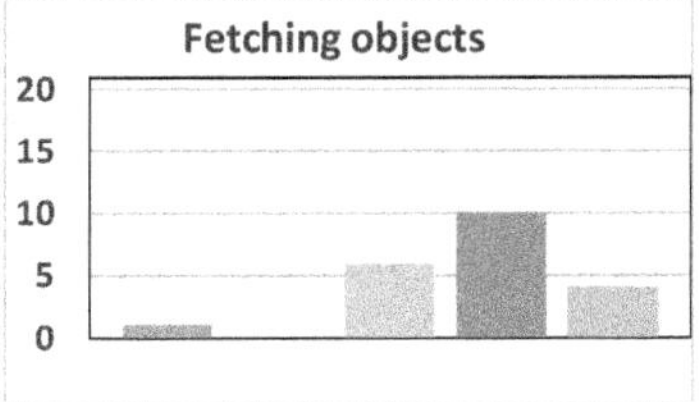
Picking up / moving heavy objects (e.g., furniture)	3.75	1.00	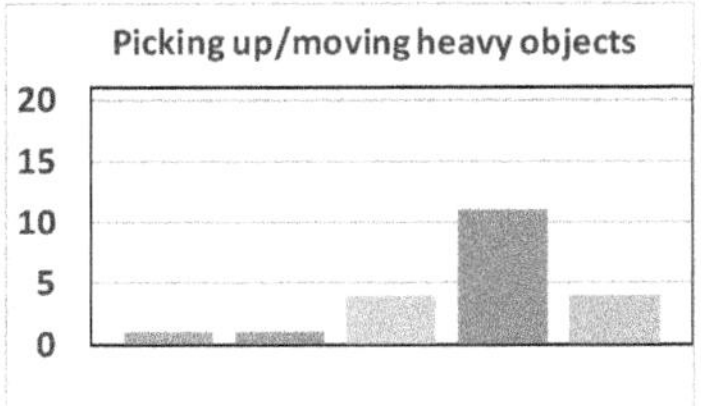
Reaching for objects	3.71	0.96	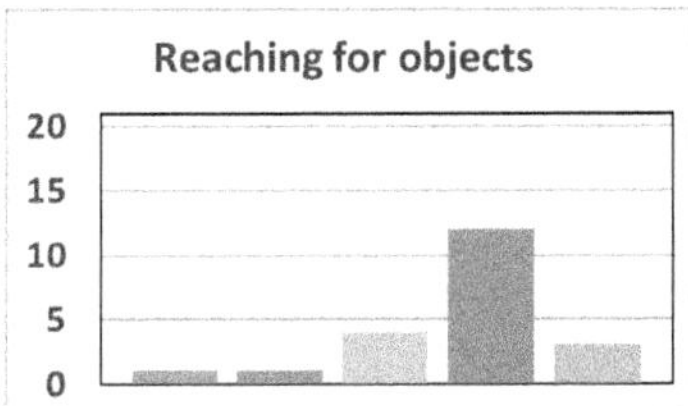
Finding / delivering items	3.62	0.92	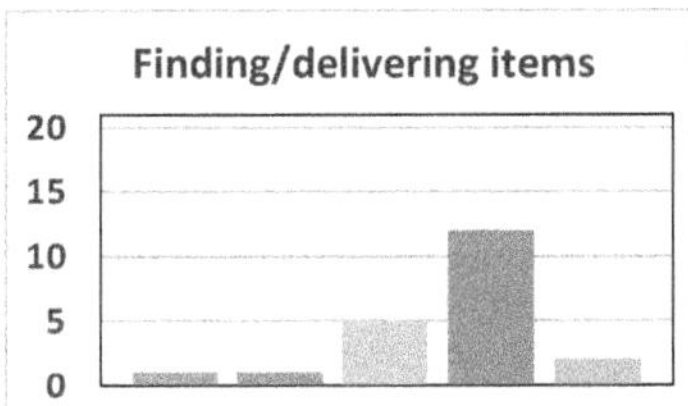

Only a human · Prefer a human · No preference · Prefer a robot · Only a robot

Figure 4. Fetching/Organizing Tasks – Prefer Robot.

4.2 Structured Group Interview Data

The following analysis focuses primarily on the structured group interview data pertaining to participants' comments in response to the prompt: "Now we would like you to think about a few specific tasks and discuss the pros and cons to a robot performing them." Below, we will report data on the tasks of home chores, and fetching and finding.

The structured group interview transcripts were analyzed according to a coding scheme to identify patterns and themes from the discussions. A coding scheme is an organized categorization of the information in the interviews. Each transcript was segmented by a primary researcher. Segments were based on what the participants explicitly stated and were determined as any utterance in which at least one pro or con was mentioned. Those segments were then coded by two independent researchers using qualitative data analysis software, MaxQDA10. Any discrepancies in the coding of the transcripts were addressed through discussion and refinement of the coding scheme by the two coders.

The goal of the coding scheme was to categorize high level pros and cons reported by the older adults. The coding scheme was based on both the literature and the nature of the participant comments. Tables 2 and 3 depict the categories of pros and cons, their definitions, and selected examples of participant quotes depicting each of these categories.

Table 2. Home Tasks – Pros of Robot.

Pro	Definition	Example Quotes
Compensation	Robot does difficult task, or task older adult can no longer perform.	"I lift all of that stuff out of the hamper, it's quite heavy so it would help with lifting." "If I find myself putting something away then... I forget where I put it. If the robot was there, I wouldn't worry about. It could always know where it was…"
Time Saving	The robot saves the older adult time.	"It will give me more time to go to my next activity *(chuckle)* which are bountiful." "Most of the time, I don't get all those things done on time, so a pro would be, it would enable me to get things done… it would allow you to do more things and do them on time."
Delegation of Undesirable Task	The robot performs a task that the older adult wishes not to do (i.e., older adult can still do task, just does not enjoy to do so).	"Do the things...it would be nice to have the robot do the thing that I don't like to do such as cleaning up" "I wouldn't have to get up and get it myself."
Effort Saving	The robot saves the older adult energy and reduces workload.	"No, just to do it. It saves time, energy, and I couldn't do it." "You might have more energy. You wouldn't get tired as quickly." …"Improve my health .
Optimization	The robot performs task at a high performance level.	"Robot should be perfect to a task." "The robot never gets tired."

Table 3. Home Tasks – Cons of Robot.

Con	Definition	Example Quotes
Damage to Environment	The robot may break something in the older adults' home (i.e., home itself or items in home).	"I'm thinking because it is a robot, I'm sure that at times it will break things as we break things. We drop dishes and we break them…" "I would think one of the cons would be, especially using a PR2… I guess it would have sensors to keep it from rolling over something… it may roll over and damage something..."
Dependency	The older adults would rather do task themselves or avoid over-relying on the robot.	"Well, con. If it makes you dependent." "It's actually probably good for us to do some of these things physically to keep, to keep moving"
Mental model	The older adults perception of the robots capabilities is limited (i.e., assume the robot is not capable of doing something).	"I keep thinking of it in terms of how it could help prepare my food but I don't know whether robots could cook." "I can see that if it does laundry, it needs to be able to sort by color. I can see that that would be a con and it couldn't do it."
Reliability of System	The robot either malfunctions or makes mistakes, therefore not performing the task correctly.	"You tell him to bring glasses, he brings you a pair of shoes." "And also if there is a malfunction on this robot. Some warning that let you know that something is wrong – a light or something..."
Storage and Space Limitations	The older adult has concerns about the robots size and/or storing it.	"Storage issues…" "[It] takes some space." "Well, I can't imagine having something big that we'd been shown that would help me"

5. DISCUSSION

5.1 Review of Findings

These results give insight into home upkeep tasks older adults would prefer a robot to assist with, and the reasons why they hold such preferences. In line with the SOC model [2], the data from the Assistance Preference Checklist questionnaire suggested that older adults are accepting of robots as a technological compensatory method. In fact, older adults expressed preference for robot assistance for 28 of 48 tasks. This analysis specifically focused on cleaning tasks and fetching/organizing objects. Many of these tasks were physical in nature (e.g., cleaning kitchen or bathroom, cleaning floors, finding and fetching), supporting previous research suggesting that older adults would benefit from assistance with physical home maintenance tasks [13]. One participant stated, "Well, if you have that arthritis for instance and you find it hard to get up and down in a chair, it [the robot] would fix things and it would certainly help you physically."

The interviews provided detailed information as to why older adults might prefer robot assistance. For an older adult to accept technology, such as robotics, the benefit has to be clear [6, 12]. Participants discussed the benefits of using a robot, e.g., "I think the robot for me is an amenity that enhances quality of life." In particular, older adults expressed interest in the robot assisting with difficult tasks, saving time, performing undesirable tasks, reducing effort, and performing tasks at a high performance level.

Tasks where older adults reported having no preference for a robot to perform included sorting mail, doing laundry, and washing dishes. Two of these tasks, laundry and washing dishes, would require the PR2 to come into contact with water. The older adults' concern with these tasks may be due to their fear of the robot getting wet. That is, throughout the transcripts participants mentioned the robot getting wet as a safety consideration according to their mental model (or knowledge) of the robot.

Further analysis of the structured group interview data indicated that older adults were concerned about becoming dependent on the robot. Malfunctions and limitations in the robot's capabilities were also discussed. We encouraged the participants to think of capabilities beyond what they viewed in the videos. However, their mental model of the PR2 was limited, and some questioned its capabilities. One said "Well, I'd say Hector [participant's name for the PR2], go find me the hat I had on yesterday and I mean I don't know how I'm limited to what it can fetch and find." Concerns also related to environmental damage (e.g., damaging items in the home) as well as the size/storage of the robot.

5.2 Design Recommendations

This needs assessment serves as a springboard for design guidelines relevant to the development of home assistive robots. The data provide insight into older adults' perceptions of robot assistance with home upkeep tasks, and thereby can inform the design of assistive robotics to support aging-in-place. By identifying not only the types of tasks older adults want help with, but also their perceived pros and cons to robot assistance, we can develop preliminary recommendations for a variety of design questions. We identified general preliminary recommendations (Table 4) for home robots, regarding the robot's physical capabilities, attributes, and means of interaction important in the design of assistive robots for the older adult population.

Table 4. Preliminary Design Recommendations.

Design Recommendation Parameters	
Parameter	*Description*
Customizability	Capability to tailor programs or behaviors to user preferences
Interaction / Collaboration	Ability to work as a robot-human team to meet a common goal
Manipulation	Level of dexterous manipulation (i.e., fine, gross)
Payload	Range of weight of expected objects for the robot to interact with
Range of motion	Large kinematic workspace to reach high/low/near/far
Storage / size	Physical attributes such as footprint, height, and mass

The older adult participants seemed to recognize the overall benefit of using a robot to assist them in the home. One participant stated, "It [the robot] would give that person a certain amount of independence in being in more control of their life." However, some older adults did express concern for depending or over-relying on the robot. An important aspect of aging-in-place is maintaining the older adults' independence [16].

A design aspect to consider is customizability (sometimes referred to as adaptability; [3]). The older adults discussed some tasks that they wanted performed in a specific manner to meet their individual preferences. For instance, one participant discussed the importance of her laundry to be folded and sorted in a certain way; another discussed his specific preference for meal preparation, "I like certain food cooked in certain way and like some veggie I don't like them too cooked, too dull..." There was some concern as to whether or not the robot could meet their individual preferences. One way to alleviate this concern is to explore ways in which the user could 'customize' the robot. For example, it may be important that the user have the option to teach the robot to perform certain tasks in a specific way, which could be achieved through robot learning.

For those older adults who may be fearful of depending on the robot, a design recommendation may be to facilitate human-robot interaction that encourages collaboration between the robot and human. For instance, the robot could lift heavy pots or pans from a low cabinet, but allow the older adult to cook with them. Or the robot could assist with making the bed (e.g., lifting the mattress, or assisting with the fitted sheet), but allow the older adult to complete the rest of the task. In this way, the older adult is still active and participating, however the robot is compensating the older adult's limitations by assisting with difficult aspects of the task. More research is needed, however, to understand the type and level of collaborative interaction appropriate for such tasks.

Regarding manipulation of objects, the older adults listed a variety of objects they would like the robot to interact with, including pots, pans, dishes, glasses, remote controls, clothing, and mattresses. Some concern was with regard to the robot's dexterity to manipulate breakable or fragile objects. The loss of fine motor control is a known age-related change [15]. Assistance with small objects also benefit for older adults. A participant said, "I'm getting slower and for example, I could spend fifteen minutes sometimes trying to latch a piece of jewelry and I'm so exhausted by the time I finish… I don't want to go out."

In maintaining one's home, previous research suggested that older adults experience physical difficulties [14]. Declines in physical strength that occur with normal aging make certain tasks, such as lifting heavy objects, difficult for older adults. Although the participants in this study mentioned a variety of objects for a robot to interact with, the questionnaire and structured group interview data suggested difficulty with heavy objects, such as moving furniture, or lifting a mattress to change linens: "I think making the bed is the most difficult thing for me […] I have a king sized bed and putting that fitted linen on is a job in the house." Thus a home assistive robot, capable of handing a high payload, would be of particular assistance to older adults.

Regarding the appropriate range of motion a home assistive robot would ideally be capable of, the Assistance Preference Checklist questionnaire suggested that picking up objects from the floor or other room was a task that older adults preferred a robot to perform (M=3.75, SD=0.94). Older adults' difficulty in bending and stooping has been suggested previously [14]. One participant from the current study stated that a benefit of robot assistance is that it "saves me from hurting my back when I bend down." Reaching for objects was mentioned, such as high shelves or into cabinets. It is recommended that a home assistive robot have a large kinematic workspace that includes being able to reach the ground, high shelves, and deep into drawers.

Finally, the older adults discussed concerns for storing the robot. One caveat to keep in mind is that this concern may be platform specific. The PR2 is a large robot, with a sizeable footprint (width and depth, 66.8 cm) and is relatively tall (ranges from about 1.3 to 1.67 meters). Many of the older adults interviewed in this study reported that they lived in a condo or apartment home, where storage was limited. When designing robots for the home environment, physical characteristics such as tucking arms, fitting into closets, and a small footprint/base should be considered.

In summary, there is much potential for home robots to support aging-in-place. The older adults we interviewed were quite open to the idea, for certain tasks. Their needs, opinions, and concerns provide guidance for designs that will be useful and acceptable. This study is a first step in better understanding robots for home tasks. The novelty of this work is the coupling of questionnaire data in addition to interview data to understand *why* they hold certain preferences. We did not want the participants to focus on pragmatic limitations of the current technology. Instead, we encouraged them to focus their discussion on the robot's potential, which may inform robot design. As such, this study provides useful design recommendations for development of home robots. The next step in this line of research would be to investigate older adults' opinions of seeing the robot in person, length of engagement, or robot assistance for older adults with impairments.

6. ACKNOWLEDGMENTS

This research was supported in part by a grant from the National Institutes of Health (National Institute on Aging) Grant P01 AG17211 under the auspices of the Center for Research and Education on Aging and Technology Enhancement (CREATE; www.create-center.org). This multidisciplinary effort between the HFA Lab (www.hfaging.org) and the HRL (www.healthcare-robotics.com) was inspired by collaboration with Willow Garage who selected the Georgia Tech as a beta PR2 site for research (www.willowgarage.com). Special thanks to Nachiketas Iyyengar for video development and Jennifer Megan Springman for assistance in data collection and creating experimental materials.

7. REFERENCES

[1] AARP. (2005, April). *Beyond 50.05 survey*. Retrieved February 13, 2009, from http://assets.aarp.org/rgcenter/il/beyond_50_05_survey.pdf

[2] Baltes, P. B., and Baltes, M. M. 1990. Psychological perspectives on successful aging: The model of selective optimization with compensation. In P. Baltes and M. Baltes (Eds.), *Successful aging: Perspectives from the Behavioral Sciences*, 1–34. New York: Cambridge University Press.

[3] Broadbent, E., Stafford, R., and MacDonald, B. 2009. Acceptance of Healthcare Robots for the Older Population: Review and Future Directions. *International Journal of Social Robotics,*1, 4 (2009), 319-330. DOI = http://dx.doi.org/10.1007/s12369-009-0030-6.

[4] Broadbent, E., Tamagawa, R., Patience, A., Knock, B., Kerse, N., Day, K., and MacDonald, B. A. 2011. Attitudes towards health-care robots in a retirement village. *Australas. J. Ageing*. DOI= http://dx.doi.org/10.1111/j.1741-6612.2011.00551.x.

[5] Bugmann, G., and Copleston, S. 2011. What Can a Personal Robot Do for You? In R. Groß, L. Alboul, C. Melhuish, M. Witkowski, T. Prescott & J. Penders (Eds.), *Towards Autonomous Robotic Systems*, 6856, 360-371. Berlin/Heidelberg: Springer.

[6] Caine, K.E., Fisk, A.D. and Rogers, W.A. 2007. Designing privacy conscious aware homes for older adults. In *Hum. Fac. Erg. Soc. P.* (Baltimore, MD, Oct 1-5, 2007). HFES'07.

[7] Cesta, A., Cortellessa, G., Giuliani, M.V., Pecora, F., Scopelliti, M. and Tiberio, L. 2007. Psychological implications of domestic assistive technology for the elderly. *PsychNology J.,* 5, 3.

[8] Choi, Y. S., Deyle, T., Chen, T., Glass, J. D. and Kemp, C. C. 2009. A List of Household Objects for Robotic Retrieval Prioritized by People with ALS. In *Proceedings of International Conference on Rehabilitation Robotics* (Kyoto, 23-26 June, 2009). ICRR '09. IEEE. DOI = http://dx.doi.org/10.1109/ICORR.2009.5209484.

[9] Clark, M. C., Czaja, S. J. and Weber, R. A. 1990. Older Adults and Daily Living Task Profiles. *Hum. Factors*, 32, 5(1990), 537-549.

[10] Czaja, S.J., Charness, N., Fisk, A.D., Hertzog, C., Nair, S.N., Rogers, W.A. and Sharit, J. 2006. Factors predicting the use of technology: Findings from the Center for Research and Education on Aging and Technology Enhancement (CREATE). *Psychol. Aging,* 21 (2006), 333-352. DOI = http://dx.doi.org/10.1037/0882-7974.21.2.333.

[11] Erber, J. T. (2005). *Aging and older adulthood.* Belmont, CA: Wadsworth.

[12] Ezer, N., Fisk, A.D. and Roger, W.A. 2009. More than a servant: Self-reported willingness of younger and older adults to having a robot perform interactive and critical tasks in the home. In *Hum. Fac. Erg. Soc. P.*, (San Antonio, TX, Oct 19-23, 2009). HFES'09, 136-150.

[13] Fausset, C. B., Kelly, A. J., Rogers, W. A. and Fisk, A. D. 2011. Challenges to aging in place: Understanding home maintenance difficulties. *Journal of Housing for the Elderly*, 25, 2 (May 2011), 125-141. DOI= http://dx.doi.org/10.1080/02763893.2011.571105.

[14] Fausset, C. B., Mayer, A. K., Rogers, W. A., and Fisk, A. D., 2009. Understanding Aging in Place for Older Adults: A Needs Analysis. In *Hum. Fac. Erg. Soc. P.*, (San Antonio, TX, Oct 19-23, 2009). HFES'09, 521-525.

[15] Fisk, A. D., Rogers, W., A., Charness, N., Czaja, S. J., and Sharit, J. 2009. *Designing for Older Adults: Principles and creative Human Factors Approaches* (2ed.). Boca Raton, FL: CRC Press.

[16] Gitlin, L. 2003. Conducting research on home environments: Lessons learned and new directions. *The Gerontologist,* 43, 5, 628-637. DOI = http://dx.doi.org/10.1093/geront/43.5.628

[17] Giuliani, M.V., Scopelliti, M., and Fornara, F. 2005. Elderly People at Home: Technological Help in Everyday Activities. *IEEE Workshop on Robots and Human Interactive Communication,* (2005), 365-370.

[18] Houser, A., Fox-Grage, W., and Gibson, M. J. 2006. Across the States 2006: Profiles of Long-Term Care and Independent Living. From http://assets.aarp.org/rgcenter/health/d18763_2006_ats.pdf.

[19] King, C., Chen, T., Fan, Z., Glass, J. D., and Kemp, C. C. 2011. Dusty: An assistive mobile manipulator that retrieves dropped objects for people with motor impairments. *Disabil. Rehabil.*, (Oct, 2011). DOI = http://dx.doi.org/10.3109/17483107.2011.615374.

[20] Lawton, M. P. 1990. Aging and performance of home tasks. *Hum. Factors,* 32, 5 (Oct 1990), 527-536.

[21] Lawton, M. P., and Nahemow, L. 1973. Ecology and the Aging Process. In C. Eisdorfer and M. Powell Lawton (eds). *The Psychology of Adult Development and Aging*, 619-674. Washington D.C.: APA.

[22] Melenhorst, A.-S., Rogers, W.A., and Bouwhuis, D.G. 2006. Older Adults' Motivated Choice for Technological Innovation: Evidence for Benefit-Driven Selectivity. *Psychol. Aging,* 21, 1 (Mar 2006), 190-195. DOI = http://dx.doi.org/10.1037/0882-7974.21.1.190.

[23] Pollack, M. E., Brown, L., Colbry, D., McCarthy, C. E., Orosz, C., Peintner, B., Ramakrishnan, S. and Tsamardinos, I. 2002. Autominder: an intelligent cognitive orthotic system for people with memory impairment. *Robot. Auton. Syst.*, 44 (2002), 273-282. DOI = http://dx.doi.org/10.1.1.12.4554.

[24] Rogers, W. A., Meyer, B., Walker, N., and Fisk, A. D. 1998. Functional limitations to daily living tasks in the aged: A focus group analysis. *Hum. Factors,* 40, 1 (1998), 111-125.

[25] Smarr, C.-A., Fausset, C. B., and Rogers, W. A. 2011. *Understanding the potential for robot assistance for older adults in the home environment* (HFA-TR-1102). Atlanta, GA: Georgia Tech, School of Psychology, Human Factors and Aging Laboratory. http://hdl.handle.net/1853/39670

[26] Smarr, C., Prakash A., Beer J., Mitzner T., Kemp, C. and Rogers W. In prep. Older Adults' Preferences for and Acceptance of Robot Assistance for ADLs.

[27] U.S. Census Bureau. 2011. *International data base. World population by age and sex.* From http://www.census.gov/ population/international/data/idb/worldpop.php].

[28] Wada, K., Shibata, T., Musha, T. and Kimura, S. 2008. Robot therapy for elders affected by dementia. *Engineering in Medicine and Biology Magazine*, 27, 4 (2008), 53-60. DOI = http://dx.doi.org/10.1109/MEMB.2008.919496.

The Effect of Monitoring by Cameras and Robots on the Privacy Enhancing Behaviors of Older Adults

Kelly Caine
Indiana University
School of Informatics and Computing
919 E. 10th Street, Bloomington IN
caine@indiana.edu

Selma Šabanović
Indiana University
School of Informatics and Computing
919 E. 10th Street, Bloomington IN
selmas@indiana.edu

Mary Carter
Indiana University
Psychological and Brain Sciences
1101 E. 10th Street, Bloomington IN
carterme@indiana.edu

ABSTRACT
This paper describes the results of an experimental study in which older adult participants interacted with three monitoring technologies designed to support their ability to age in place in their own home—a camera, a stationary robot, and a mobile robot. The aim of our study was to evaluate users' perceptions of privacy and their tendencies to engage in privacy enhancing behaviors (PEBs) by comparing the three conditions. We found that privacy concerns lead older adults to change their behavior in a home environment while being monitored by cameras or embodied robots. We expected participants to engage in more PEBs when they interacted with a mobile robot, which provided embodied cues of ongoing monitoring; surprisingly, we found the opposite to be true—the camera was the condition in which participants performed more PEBs. We describe the results of quantitative and qualitative analyses of our survey, interview, and observational data and discuss the implications of our study for human-robot interaction, the study of privacy and technology, and the design of assistive robots for monitoring older adults.

Categories and Subject Descriptors
H.1.2. [**Models and Principles**]: User/Machine Systems – Human Factors, Software Psychology; K.4.1 [**Computers and Society**]: Public Policy Issues – Privacy, I.2.9 [**Robotics**]: Operator Interfaces.

General Terms
Design, Experimentation, Security, Human Factors.

Keywords
Aging in place, home monitoring, home, older adult, privacy, human-robot interaction.

1. INTRODUCTION
The need to consider and protect user privacy is regularly cited as a benchmark for assistive robotic systems meant to be used in natural settings [12, 18, 20, 27]. Neither user perceptions of privacy nor their privacy behaviors, however, have been objects of empirical research within the domain of human-robot interaction. As autonomous and teleoperated robots that can collect data on people's behaviors and allow for remote operators to monitor others become more readily available to the public (e.g. Texas, QB, CareBot), the need to study the effects of robotic technologies on user privacy is becoming urgent. This is particularly true for rapidly growing application areas for monitoring and telecommunication robots, such as elder care [3].

Human-computer interaction researchers have studied privacy issues in relation to a variety of technologies including computers, the Internet, and memory aids [e.g. [7, 16], though relatively little work has examined privacy perceptions among older adults (see [5, 8, 9, 11, 23] for exceptions). Researchers suggest that robots, in comparison to previously studied information and communication technologies, may provide unprecedented protections and challenges to user privacy because of their embodiment, mobility, and novelty for users [3, 12, 23]. Much of this existing research suggests that people's privacy attitudes and behaviors are often out of synch [1, 17, 29]: while users might profess serious privacy concerns, they will disclose personal information freely in online and computer-based communication. One possible explanation for this discrepancy is that, because computer-based communication is radically different from the embodied, face-to-face interaction that people are accustomed to [6], this difference results in a lack of understanding of what information is disclosed, to whom, and what the consequences of these disclosures may be. Research in human-computer interaction, however, has not addressed the issue of embodied interaction as it pertains to privacy. Robots, which are not only embodied and mobile devices, but are often designed to resemble humans and/or perform human-like behaviors and functions, provide an opportunity for researchers to study whether an embodied interface will affect and possibly enhance user privacy protection in ways that other technologies do not.

In this paper, we study user privacy behaviors in the context of monitoring applications in the homes of older adults. Older adults comprise one of the most likely potential audiences for domestic assistive robotic technologies [12]; robots can enable older adults to age in place by providing telepresence and monitoring capabilities for caregivers (e.g. [3, 4, 22]), as well as various autonomous assistive services (e.g. [5, 26]. Our study focuses on teleoperation and monitoring as the most likely applications to be used in the near future. We compare two types of robots—mobile and stationary—to each other and to a wall-mounted camera to explore and understand the different privacy-related affordances of these devices from the perspective of older adult users. We also use our results to expand our understanding of the psychology of privacy with a particular focus on whether embodied and mobile devices make a difference in people's awareness of monitoring and their performance of privacy enhancing behaviors (PEBs; [6]). After providing an overview of assistive robotics for older adults and privacy as it relates to monitoring technologies in general and robots more specifically, we describe the results of our study and discuss their contributions to HRI and the study of technology and privacy.

HRI'12, March 5–8, 2012, Boston, MA, USA.

2. BACKGROUND

2.1 Assistive robots for older adults

The imminent use of robots in the daily aid and care of older adults brings up both the promise of helping people achieve their goals of aging in place [13] and concerns that robotic technologies might jeopardize the human rights of elders [27]. Based on an ethnographic study of the homes and everyday practices of older adults, [13] conclude that robotic technologies should conform to the local social ecology, support the values of independence and dignity, and also adapt to other actors in the environment to be appropriate for and helpful to older adults. Other researchers mention the possibility that using monitoring robots will deprive elders of much needed social interaction, infringe on their privacy, and cause loss of freedom and autonomy as robotic systems start making decisions and taking actions in the homes of elders [27]. [28] also suggest that the availability of robots will inevitably decrease older adults' access to the human touch, an often forgotten but necessary component of care. They also caution that robots as monitors and telepresence devices may have a negative impact on the relationship between caregivers and receivers as both parties become more reliant on technology. More research on specific robotic applications in different contexts of use is necessary for developing appropriate design guidelines and practices for assistive robots for older adults.

2.2 Privacy in human-computer interaction

While it is beyond the scope of this paper to review the literature related to privacy in HCI in its entirety (for a review see [15]), we want to highlight one particularly relevant issue we explore in this study: the finding that privacy attitudes often do not match privacy behaviors (e.g., [1]). For example, in the context of e-commerce, online disclosure of personal information while interacting with an e-commerce site did not match privacy preferences users stated previously [29].

[1] offers a number of possible explanations for this mismatch, including limited information, self-control problems, and bounded rationality. However, another possibility is simply *awareness* (see [6]). Because information and communication technologies do not offer the rich social experience provided by direct interaction with other humans, differences between privacy attitudes and behaviors may be explained by a lack of sufficient cues about the recipient, amount, and content of information being transmitted. Put simply, people's stated attitudes about privacy may differ from their actual privacy behaviors because they are not fully aware of what information they are revealing and to whom. While some recent work has been done to provide more cues to people as they make privacy decisions (e.g., [7]), this explanation remains underexplored. In relation to this study, this explanation suggests there is great potential for robots to represent feedback that increases user awareness more than other information and communication technologies because of their embodied nature.

2.3 Privacy in human-robot interaction

HRI scholars [12, 18] emphasize privacy as an important benchmark in the design of "increasingly humanlike" and assistive robots. [18] suggest that the presence of robots can affect a user's sense of privacy, particularly in relation to the robot's ability to monitor and collect information about the individual. In a user study with 12 older adults, [3] cite the concern for privacy as the second highest category mentioned by their research participants and suggest that features allowing users to refuse monitoring at will could address some of these concerns (see [8] for an independently derived conclusion and technical solution). [20] also suggest that privacy-oriented design is important for robots that operate in everyday human environments, pointing out that users may not be aware of the capabilities of robots and may like to be informed when they are being recorded or monitored.

Robotic embodiment has been identified and studied as an important factor in assistive applications for older adults [21], providing the user with a more naturalistic interface and encouraging the relationships with the technology. [12] propose that robots differ from other pervasive technologies because users can see when robots are observing them and send them away or avoid them, thus providing more possibilities for protecting user privacy (p. 430). Our study is a preliminary exploration of the effect of robots, as compared to more familiar technologies such as cameras, on privacy attitudes and behaviors of older adults.

2.4 Overview of the Study

We used an experimental approach to evaluate whether privacy attitudes and behaviors of older adults being monitored by a camera would differ from those of older adults interacting with embodied monitoring devices (i.e., robots). We were particularly interested in understanding the differences between the three monitoring conditions. To elicit privacy concerns, we had participants perform a task in which they imagined themselves planning a surprise party for a caregiver who was monitoring them. Further details of the experiment are described below.

3. METHODS

3.1 Participants

Participants were 18 older adults between the ages of 69 and 88 recruited from a local retirement community through posted advertisements and by word of mouth. There were 11 women and 7 men and all participants lived independently. Table 1 shows detailed demographic information.

Table 1. Demographic information for participants

	Overall (N=18)	Camera (n=6)	Stationary Robot (n=6)	Mobile Robot (n=6)
Age				
Mean	81	82	80	80
SD	(5.5)	(7.1)	(3.0)	(6.1)
Race				
White	100%	100%	100%	100%
Gender				
Female	61%	50%	50%	83%
Male	39%	50%	50%	17%
Income				
$30- 49k	6%	17%	0%	0%
$50- $74k	33%	0%	33%	67%
>$75k	56%	83%	50%	33%
Prefer not to answer	6%	0%	17%	0%
Education				
Some College	17%	0%	17%	33%
College	83%	100%	83%	67%
Living Situation				
Alone	56%	50%	67%	50%
W/Spouse	39%	50%	33%	33%
W/Family	6%	0%	0%	17%
Marital Status				
Married	39%	50%	33%	33%
Widowed	55%	50%	50%	67%
Never Married	6%	0%	17%	0%

3.2 Design

The study used a between subjects design. Participants were randomly assigned to one of three possible monitoring devices: camera, stationary robot, or mobile robot. We chose to use a camera because of its ubiquity as a monitoring technology and familiarity to users. The stationary robot was used as a control for the possibility that the robotic nature of the monitoring technology would influence participants. The mobile robot was the most embodied of our monitoring devices, as it followed the participant as they moved around the space during the experiment.[1]

The main objective of the study was *to evaluate users' perceptions of privacy of three types of monitoring technologies designed to support older adults' ability to age in place in their own home.* To accomplish this, we focused on the following research questions and hypotheses:

RQ1: Will older adults' level of comfort with household activities (CHA) change when they assume that they would be monitored?

- H_0: $|\Delta CHA| = 0$
- H_1: $|\Delta CHA| \neq 0$

RQ2: Which activities would be considered more comfortable to be performed while being monitored and which activities would be considered less comfortable?

- Exploratory/descriptive (no specific activities hypothesized)

RQ3: Will older adults' level of comfort with household activities (CHA) change differentially by device type?

- H_0: $|\Delta CHA|$ camera = $|\Delta CHA|$ stationary robot = $|\Delta CHA|$ mobile robot
- H_1: $|\Delta CHA|$ camera < $|\Delta CHA|$ stationary robot < $|\Delta CHA|$ mobile robot

RQ4: Will the proportion of older adults who engage in privacy enhancing behaviors (PEBs) differ by type of monitoring device?

- H_0: # of PEBs camera = stationary robot = mobile robot
- H_1: # of PEBs camera < stationary robot < mobile robot

RQ5: What types of PEBs do older adults utilize to maintain privacy while being monitored?

- Exploratory/descriptive (no specific behaviors hypothesized)

3.3 Materials

3.3.1 *Questionnaires*

Participants filled out four questionnaires at the beginning of the study: a demographics questionnaire, the Privacy Attitudes Questionnaire (PAQ; an extended version of the Westin segmentation index [17, 19]), the Negative Attitudes towards Robots (NARS [24]), and the modified comfort with household activities (CHA) questionnaire, which we developed using Frye and Dornisch's [14] disclosure topics as the basis of our questionnaire items. The demographics questionnaire enquired about participants' age, gender, household income, etc.; the PAQ enquired about participants' general privacy attitudes; the NARS measured people's existing attitudes towards robots; and the modified CHA assessed general levels of comfort performing activities within a home environment.

After participants had interacted with the monitoring technology, we administered two post-test questionnaires: The Collected Robot Scales and the modified CHA. The Collected Robot Scales questionnaire was used to assess the participants' perceptions of the home monitoring technologies they encountered; it combined the "Godspeed Scale" [2], which includes measures of animacy, anthropomorphism, likeability, perceived intelligence, and perceived safety, and [25]'s 6-level scale for "fear," "surprise," "disgust," and "unpleasantness." The modified CHA (as described above) was also used in the post-test, but this time participants were asked to answer considering the device they were saw during the study was being used in their home. Change scores (ΔCHA) were calculated for each question on the CHA by subtracting the post-CHA score from the pre-CHA score.

3.3.2 *Interview*

A verbal semi-structured interview was used to gather information about the participant's perceptions of and reactions to the monitoring technologies. The interview covered topics such as the participants' reasons for (or reasons for *not*) performing privacy enhancing behaviors, comfort with tasks, perceptions of the invasiveness of devices, and overall impressions of home monitoring technologies. One area of specific interest were PEBs. Each participant was asked if they engaged in any PEBs while performing the tasks assigned for the study. If the participant answered yes, the interviewer asked what type of behaviors the participant performed. If the participant answered no, the interviewer followed up to determine why no PEBs were employed. Researchers also asked each participant how they thought the monitoring technology they saw worked and whether they had any previous experience with monitoring technologies. Finally, participants were asked to describe situations in which the use of monitoring devices would be appropriate and inappropriate, as well as whether or not participants would use a monitoring technology such as the one they had interacted with in their home.

3.3.3 *Scenario and tasks*

To observe participants performing tasks appropriate for the home environment while giving them an opportunity to engage in PEBs, we developed a scenario in which participants imagined preparing a surprise birthday party for their caregiver. Participants received a task list to complete in planning the fictional surprise party; they were also asked to record completed tasks on the list. The tasks, in order of performance, included telephoning a guest to confirm their attendance, blowing a balloon, hanging a "Happy Birthday" banner, answering the door to receive a flower delivery, placing flowers in a vase, icing a cupcake, using sanitary wipes to clean their hands, putting on a tee shirt, and evaluating their appearance in a mirror. The tasks represented different levels of personal (e.g. putting on a t-shirt, cleaning hands) and informational disclosure (e.g. hanging banner, phoning guests) tasks.

3.3.4 *Monitoring Technologies*

Participants interacted with one of three types of monitoring technology participants during the study—a wall-mounted camera, a stationary robot, and a mobile robot. The wall-mounted camera was an Axis 215 PTZ camera. The stationary and mobile robots used the same robotic platform, an off-the-shelf Videre model mobile platform (See Figure 1), with a webcam mounted on top. The robot is approximately 40cm by 41cm by 15cm and is regularly used in university lab research with human subjects. The robot exhibited different behaviors in the stationary and mobile

[1] We use Dautenhahn's definition of embodiment as "the existence of perturbatory channels between the robot and its environment" [10], which in our case refers to its movement in response to the participant's actions.

robot conditions. In the stationary condition, the robot was placed in the room prior to the participant's entry and did not move from its spot throughout the study. In the mobile robot condition, a research assistant remotely operated the robot to follow the participant as they moved around the experimental space. The only interaction cues the robots provided to participants were physical presence in the case of the stationary robot and contingent movement in the case of the mobile robot.

Figure 1. (a) Robotic platform used in stationary and mobile conditions, (b) Wall-mounted pan-tilt-zoom camera.

3.3.5 Recording apparatus

We used a GoPro HD camera to visually capture participants' behavior throughout the study. Verbal data was collected using a digital voice recorder.

3.4 Procedure

The study took place at the R-House Living Lab at Indiana University, a five-room house used for research on human-robot interaction and the design of domestic interactive technologies. Once participants arrived at the R-House, researchers obtained informed consent. Participants were then asked to fill out the pre-test questionnaires and were randomly assigned to one of three monitoring device conditions: camera, stationary robot, or mobile robot. Next, participants were asked to give us the name of someone who could be their caregiver, should the need arise. Afterwards, participants were asked to imagine that the experimental space was a room in their home and that the person they had identified as a potential caregiver could monitor them using the technology they had been randomly assigned to (camera or stationary robot or mobile robot). The experimenter then introduced the monitoring technology, regardless of the condition, with the following script:

> *"The [TECHNOLOGY TYPE] you see here can collect images and sound like those from a video camera. Your caregiver can access this whenever they choose."*

After being introduced to the monitoring technology, participants were given a scenario that described a fictional surprise birthday party for the caregiver they imagined and asked to prepare for that surprise party by performing the tasks on a list provided for the purpose (see section 3.3.3 for a description of the tasks). The experimenter then left the room and observed the participants via camera until all tasks were completed or 30 minutes had passed. When the experimenter returned to the room, they conducted a semi-structured interview and administered the two post-test questionnaires. At the conclusion of the study, participants were debriefed as to the purpose of the study and remunerated $20 for their time. The entire study took approximately 1½ hours.

4. RESULTS

In the following section we discuss findings from each of the 5 research questions outlined in section 3.2.

4.1 Comfort with Household Activities

The first thing we were interested in was to understand how general comfort performing household activities would change in a non-monitored versus a monitored home. To determine this difference, we calculated the changes between the pre- and post-CHA score for each household activity (ΔCHA).

Overall 15 household activities changed by at least a quarter of a point (note that a positive change score indicates a *decrease* in comfort when in a home with a monitoring device).

Table 2. Activities where participant comfort decreased

	Participants (N=18) Mean ΔCHA (SD)
Engage in sexual activity with a partner	3.4 (2.6)
Engage in physical contact with an intimate partner	1.9 (3.3)
Take a shower	1.3 (2.1)
Engage in sexual activity by myself	1.2 (2.8)
Practice personal hygiene	0.8 (2.1)
Take a bath	0.8 (2.2)
Do personal finances (pay bills, write checks)	0.7 (1.6)
Have friends over	0.3 (1.2)
Blow my nose	0.3 (0.8)

There were 9 activities where comfort with performing decreased with the addition of a monitoring device in the home (see Table 2) and 6 activities where comfort increased (see Table 3).

Table 3. Activities where participant comfort increased

	Participants (N=18) Mean ΔCHA (SD)
Engage in a argument	-.72 (2.2)
Exercise	-.56 (1.6)
Dance around the house	-0.5 (2.0)
Drink alcohol	-.33 (1.6)
Express my political views	-.33 (1.3)
Watch any movie	-.39 (1.2)

In addition, we were interested in understanding whether level of comfort changed differently based on the type of monitoring device (camera, stationary robot, or mobile robot). A one-way ANOVA was used to test for differences in change scores by monitoring device type. The change score for two household activities—"engaging in sexual activity with a partner" and "drinking alcohol"—differed significantly by device type ($ps < .05$; see Table 4). None of the other change scores significantly differed by device type ($ps > .05$).

Table 4. ANOVA of ΔCHA

	Device Type				
	Camera *Mean* (SD)	Stationary Robot *Mean* (SD)	Mobile Robot *Mean* (SD)	F	*p*
Alcohol	0 (0)	.67 (1.37)	-1.67 (1.86)	4.88	.023
Sex	5 (1.55)	4.67 (1.37)	.50 (1.76)	15.36	.001

4.1.1 Alcohol

Bonferroni adjusted post-hoc comparisons of the three device types indicate that participants in the mobile robot group increased their comfort level with respect to consuming alcohol in a home containing the mobile robot whereas the Camera or Stationary Robot had very little change in their reported comfort with consuming alcohol, p = .025. There was no significant difference between the Camera and Stationary Robot (ps < .14).

4.1.2 Sex

Bonferroni adjusted post-hoc comparisons of the three device types indicate that participants in the mobile robot group changed their comfort level very little with respect to engaging in sexual activity with a partner in a home containing the mobile robot whereas participants in the Camera and Stationary Robot conditions reported significantly *more* concern, p = .001. Comparisons between the Camera and Stationary Robot were not statistically different (ps < .99).

4.2 Privacy Enhancing Behaviors (PEBs)

Privacy enhancing behaviors are behaviors that people engage in to avoid or alleviate privacy concerns [5]. We were interested whether PEBs differed by monitor type, what PEBs older adults would engage in, how frequently older adults engaged in PEBs, and to understand older adults PEB perceptions.

4.2.1 PEBs by Monitoring Type

The first thing we wanted to understand was whether or not there were differences in privacy enhancing behaviors by monitoring device type. Because our sample size was not large enough to use a chi-square test (requires minimum expected cell count of 5; our expected cell counts were 3), we used Fisher's exact test. Fisher's exact test is most appropriately used with a 2 x 2 contingency table, so we pooled participants into "robot" and "camera" conditions for this analysis (See Table 5 for the percentage of participants who engaged in PEBs by monitoring device type).

Applying Fisher's exact test, the proportion of participants in the camera condition who engaged in PEBs is significantly more than the proportion of participants who engaged in PEBs in the robot conditions (p = .057). While 67% of the participants in the camera condition engaged in PEBs only 17% of participants in the robot condition engaged in PEBs.

Table 5. Privacy enhancing behavior by monitoring type

PEB	Camera (n=6)	Stationary robot (n=6)	Mobile robot (n=6)
Yes	67%	17%	17%
No	33%	83%	83%

4.2.2 Quantity of PEBs

In addition to examining the presence of PEBs, we also wanted to understand, for participants who *did* engage in privacy behaviors, how often they engaged in PEBs over the course of the study. Overall, 6 of 18 participants engaged in PEBs during the session. However, there was a range in the *quantity* of PEBs per session. While most participants engaged in between 1 - 5 PEBs per session, 2 participants (one in the camera and one in the mobile robot condition) displayed 12 PEBs each.

4.2.3 Quality of PEBs

To understand the range of behaviors participants used to enhance privacy while being monitored we examined the video of each participant during the session. PEBs included:

- covering up camera with an object (e.g. painting, scotch tape)
- turning camera in opposite direction (on robot)
- self-censoring of speech during phone calls
- using an alternative location: setting the party up in a different room
- moving their bodies out of sight of camera (to another room, behind furniture)
- turning their back towards the camera and walking backwards
- obscuring objects with their body and/or other objects in the room (see Figure 2a for screenshot of participant hiding flowers behind her back)
- covering chest to hide "happy birthday" printed t-shirt (see Figure 2b for screenshot of participant hiding the contents of her t-shirt)

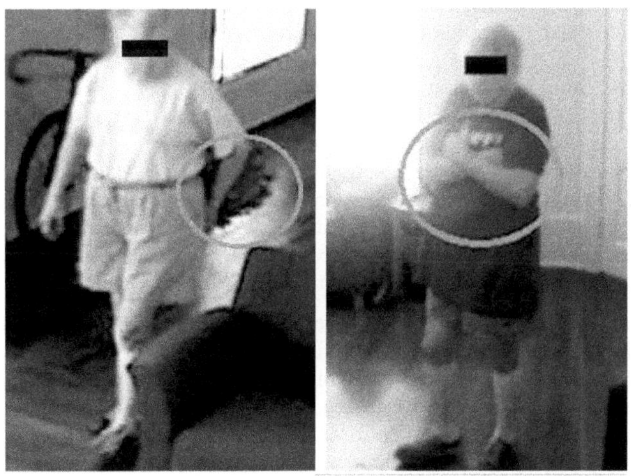

Figure 2. PEBs: (a) Participant hiding flowers behind their back, (b) Participant covering the "Happy Birthday" writing on their shirt with their arms

In addition to observable behaviors, we were also interested in participants' subjective self-report of their PEBs. When asked, many participants reported specific PEBs they had engaged in during the experiment. In the quotes below, participants who were in the camera condition are identified by 'cXX', those in the stationary robot condition by 'srXX', and those in the mobile robot condition by 'mrXX.'

Participants described how they tried to "hide" (mr025) from or "stay out of line" (c024) of the monitoring device. Similarly, others described how they occluded what they were doing using their body and objects in the room:

> *"I stood with my back to the camera so they couldn't see the number I was dialing.... The next task was to blow up the balloon so I got behind the chair... I left the balloon behind the chair until I had tape to put it under the camera."(c24)*

> *"[I] ... did everything [in the conference room] that I could, including hanging the banner. And when I came in here with the shirt I tried to keep my back faced this way." (c7)*

Three participants also reported censoring what they said on the phone to avoid letting the caregiver hear about the birthday party:

> *"I was pretty oblique... I avoided saying when it was or what it was about." (c24)*

> *"I just said she could come over... [instead of mentioning birthday plans]." (c20)*

> *"I figured she [my caregiver] wouldn't have any idea who it [party guest] was so it wouldn't make any difference because I would certainly be calling other people at other times... I didn't say anything... about the party."(c7)*

5. DISCUSSION

5.1 Comfort with Household Activities

5.1.1 Change in Level of Comfort

There were 9 activities where comfort with performing household activities decreased with the addition of a monitoring device in the home and 6 activities where comfort increased. Upon reflection, many of the activities where comfort decreased were activities that could be thought of as sensitive or personal. The 4 activities with the greatest decrease in comfort level were related to being nude and 3 of the 4 were related to intimate activities. On the other hand, many of the activities where comfort increased may be thought of as potentially dangerous activities during which older adults might get hurt. For example, engaging in an argument could raise one's blood pressure, while engaging in exercise or dancing around the house could result in a fall.

The evidence from the change in CHA scores indicates that participants understood the functionality of the monitoring systems, had privacy concerns about being watched while performing some activities (e.g., showering), but reported that they would be *more* comfortable performing some potentially dangerous activities in a home containing monitoring devices.

5.1.2 Change in Level of Comfort by Device Type

We were also interested in understanding whether level of comfort with household activities changed differently based on the type of monitoring device (camera, stationary robot, or mobile robot). We found that the change score for two household activities (consuming alcohol and engaging in sexual activity with a partner) differed significantly by device type.

Participants in the mobile robot condition reported increased comfort with respect to consuming alcohol while participants in the other two conditions changed their comfort level very little. One possible explanation for the increased comfort with the mobile robot rather than either the stationary robot or camera is that participants may have thought that the mobile robot could find them should they need assistance if they had too much to drink. Thus, the mobile robot could be thought of as helpful and therefore make older adults feel more comfortable.

As for the idea of engaging in sexual activity with a partner, participants comfort levels in the mobile robot condition changed very little from their initial comfort score to the score taken while considering engaging in the activity with the monitoring device present, whereas participants in the camera and stationary robot conditions reported significantly *more* concern when considering engaging in the activity with the monitoring device present. One possible explanation for this is that participants in the mobile robot condition may have thought they could "shoo" or tell the robot to leave the room should they decide to engage in such activity, or that perhaps the person controlling the robot would decide to leave of their own accord.

5.2 Privacy Enhancing Behaviors

Overall we found that older adults in each of the three monitoring conditions engaged in PEBs. The quantity of observed PEBs participants expressed ranged from 1 PEB to 12 PEBs over the course of the session. PEBs were varied and included covering up the camera with an object, censoring speech during phone calls, and covering up the "happy birthday" print on their t-shirts. In addition to these objective, observable behaviors, participants also described PEBs they exhibited in their own words. In general, they told us they changed their behavior to hide from the monitoring device so that their caregiver would not see them preparing for the party.

5.2.1 Older adults engage in PEBs

Taken together, this triangulated evidence from both objective and subjective sources suggests that *older adults will change their behavior in the home if they are concerned about their privacy due to in-home monitoring.* A number of researchers have suggested that monitoring devices, especially cameras, are associated with privacy concerns, however, the evidence reported here is the first behavioral study to demonstrate that privacy concerns lead older adults to change behavior in a home environment when they are monitored by a variety of monitoring devices, including embodied robots.

5.2.2 Incidence of PEBs Differs by Monitoring Type

In addition to examining whether older adults would change their behavior due to privacy concerns, we also examined how these behaviors differed by type of monitoring device. We expected participants in the mobile robot conditions to engage in *more* PEBs than participants in the stationary robot condition, and participants in the camera condition to engage in *fewer* PEBs than in either of the robot conditions. We expected this because the embodied presence of robots in the user's environment could make users more aware of being observed (e.g. [12]) and thus more likely to take actions to preserve their privacy. The results of this study, which show that more participants engaged in PEBs when being monitored by a camera, may be interpreted as evidence counter to the suggestion that users would be more aware of, or at least sensitive to, being observed. However, an alternative explanation relates to research on tele-operation and the elderly [3], which suggests that users may not be familiar with the capabilities of robots and therefore may not be aware of when they are being monitored. In contrast, people may already be quite familiar with cameras and may have already developed privacy practices in response to them. Admittedly, the robots in our study also had visibly attached cameras, but users' lack of familiarity with robots, or their interest in their novelty, may have trumped their recognition of the function of the mounted cameras.

To evaluate these explanations we examined the post-session interview data to understand how participants explained why they *did not* engage in PEBs. Users provided a variety of reasons for not engaging in PEBs. One reason, mentioned by four participants (sr12, sr10, sr12, mr21), was that they assumed the monitoring caregiver was in a position from which they would not be able to observe them at the time they were preparing for the party. In the words of one participant: *"I figured she would be on her way over and wouldn't be able to see it anyway"* (mr21). Another participant explained that the activity they engaged in would not be recognizable as preparation for a surprise party: "*I thought he might see me working. But I didn't think he would catch on;*" they also mentioned they were not bothered by the idea that their caregiver might see what they were doing, *"Anyway, I wouldn't be embarrassed if he saw me doing that" (sr17).*

Many participants (c9, sr19, c22) reported that it never occurred to them to engage in PEBs: "*No. I didn't think of it." (sr 19); "It never occurred to me" (c9).* One reason given for this was that participants were too engrossed in following the task-related directions to worry about being monitored: in c22's words "*I didn't know I was supposed to, I was just following directions."* The complexity of the task, which might have put the monitoring out of the participants' minds and led to fewer PEBs, is therefore a possible limitation of our study.

Some participants also questioned the technology's capabilities to monitor them: *"It wasn't at the right place, it didn't have the capability" (mr18).* One participant also mentioned not minding the robot *"It was moving around and for some reason or other it never bothered me. I paid no attention to it. As if it wasn't there" (mr013).* Such comments suggested participants perceived the robot as not being able to see what they were doing—it was relatively low to the ground, while the stationary camera was mounted in one corner of the room next to the ceiling affording a wider angle of view. This explanation is particularly reasonable for the stationary robot, which did not give any indication of monitoring the person. However, the mobile robot followed the participants around the space as they did the tasks, giving them an ostensibly clear indication that the monitoring function was indeed working. Another limitation of our study therefore relates to the differential effects of camera height on the participants' perception of the monitoring technology's effectiveness.

Our interpretation that participants may not be familiar with robots, and therefore less wary of their monitoring capabilities, was not born out by the interview data. When asked what they thought the monitoring technologies were doing, participants had reasonable explanations for all three conditions. For the camera, they mentioned that it could "*see me instantly like a TV, 24/7" (c08), "I'm assuming it just took a picture of me or could spot the motion" (c020), "It would probably pick up audio from anywhere in the room even if I was out of eyesight or camera sight" (c24).* The stationary robot was described similarly, as "*very much like Skype*" (*sr012*), or a *"remote TV and speaker type system" (sr011),* while participants describing the mobile robot focused on its responses to their movements: "*The motion attracts it every time" (mr06), "I felt like it was kind of following me around" (mr018).* More research is needed to figure out why participants did not respond by engaging in privacy enhancing behaviors while being monitored by the robot when they expressed understanding of its basic functions. One possibility is that participants ascribed more autonomous agency to the mobile robot than to the other two conditions, and were not as bothered by its ability to "see" and monitor them, as some of our participants suggested, as they would be if a person was doing the monitoring. This suggests future research to understand participants' differential perceptions of being monitored by a robot and a person.

An alternative explanation we will continue to explore is that participants in all conditions may not have understood that the caregiver they identified would be actively monitoring them while they were completing the tasks. For example, one participant told us they *"selected a time when [their caregiver] wouldn't be home*" (sr10), while another said that their caregiver would not be observing them because "*I know my daughter wouldn't be around all day"* (sr27). In future work, we will ensure that participants identify caregivers who they believe will be available during the study period. This alternative explanation also draws attention to the myriad and unexpected ways people protect their own privacy; in this case participants understood that although they *could* be monitored, they were not actively *being* monitored as their caregiver was otherwise engaged.

Other future work we are already pursuing will examine the effects of attitudes towards robots (as measured by the NARS) and privacy attitudes (as measured by the PAQ) on user perceptions of and reactions to the monitoring technologies. We plan to examine these, as well as other individual differences among participants, as possible differentiators for the varied reactions to the monitoring technologies.

6. CONCLUSIONS

This paper presented results from an experimental study of the privacy-related behaviors and perceptions of older adults interacting with three types of monitoring technology: a camera, a stationary robot (with camera), and a mobile robot (with camera). We were particularly interested in seeing how older adults reacted to the two robots in comparison to the camera, as there has been little empirical research on privacy behaviors in the context of human-robot interaction. While HCI researchers have investigated privacy with respect to many technologies (e.g., mobile, cameras, internet) our work is the first to consider the notion of embodied and interactive monitoring technologies, such as robots. The literature in HRI has so far not delved into empirical research on privacy behaviors around robots, but one of the expectations researchers have put forth is that robots might enable users to protect their privacy more effectively, since they are physically larger than cameras, make movements obvious to users, and can be asked to move out of the room and thus evaded when desired.

Our study specifically addressed this area at the intersection of HRI and HCI, looking at embodiment with respect to privacy. We hypothesized that an embodied, mobile monitoring technology would increase participants' use of PEBs, but we found the opposite to be true—fewer participants engaged in PEBs around robots. While we discussed potential explanations for this finding, more research is needed to evaluate these and other explanations. In the future, we propose to do more research to find out why this is the case, as we were only able to get a partial understanding from user comments during the final interviews.

7. ACKNOWLEDGMENTS

This research was supported in part by Indiana University through the Center for Law, Ethics, and Applied Research in Health Information, the School of Informatics and Computing (SOIC), and the SOIC Undergraduate Research Scholar Honorarium. We would like to thank B. Kechavarzi, M. Francisco, N. Michalik, M. Soladine, G. White, M. Kinkoph, and M. Degges for assistance in data collection, and our older adult participants.

8. REFERENCES

[1] Acquisti, A. and Grossklags, J. 2003. Losses, Gains, and Hyperbolic Discounting: An Experimental Approach to Information Security Attitudes and Behavior. In *Proceedings of 2nd Annual Workshop on "Economics and Information Security"*, UC Berkeley.

[2] Bartneck, C., Kulić, D., Croft, E. and Zoghbi, S. 2009. Measurement Instruments for the Anthropomorphism, Animacy, Likeability, Perceived Intelligence, and Perceived Safety of Robots. *International Journal of Social Robotics*, *1* (1), 71-81.

[3] Beer, J.M. and Takayama, L. 2011. Mobile remote presence systems for older adults: Acceptance, benefits, and concerns. In *Proceedings of 6th international conference on Human-robot interaction*, Lausanne, Switzerland, ACM, 19-26.

[4] Boissy, P., Corriveau, H., Michaud, H., Labonté, D. and Royer, M. 2007. A qualitative study of in-home robotic telepresence for home care of community-living elderly subjects. *Journal of Telemedicine and Telecare*, *13* (2), 79-84.

[5] Broadbent, E., Stafford, R. and MacDonald, B. 2009. Acceptance of Healthcare Robots for the Older Population: Review and Future Directions. *International Journal of Social Robotics*, *1* (4), 319-330.

[6] Caine, K. 2005. Understanding Privacy Behaviors and Misclosures *Unpublished Dissertation*, Georgia Institute of Technology, Atlanta, GA.

[7] Caine, K., Kisselburgh, L.G. and Lareau, L. 2011. Audience visualization influences disclosures in online social networks *Proceedings of the 2011 annual conference extended abstracts on Human factors in computing systems*, ACM, Vancouver, BC, Canada, 1663-1668.

[8] Caine, K., Zimmerman, C., Schall-Zimmerman, Z., Hazlewood, W., Jean Camp, L., Connelly, K., Huber, L. and Shankar, K. 2011. DigiSwitch: A Device to Allow Older Adults to Monitor and Direct the Collection and Transmission of Health Information Collected at Home. *Journal of Medical Systems*, *35* (5), 1181-1195.

[9] Caine, K.E., Fisk, A.D. and Rogers, W.A. 2006. Benefits and Privacy Concerns of a Home Equipped with a Visual Sensing System: A Perspective from Older Adults. *Proceedings of the Human Factors and Ergonomics Society Annual Meeting*, *50* (2), 180-184.

[10] Dautenhahn, K., Ogden, B. and Quick, T. 2002. From embodied to socially embedded agents-Implications for interaction-aware robots. *Cognitive Systems Research*, *3*, 397-428.

[11] Demiris, G., Oliver, D.P., Dickey, G., Skubic, M. and Rantz, M. 2008. Findings from a participatory evaluation of a smart home application for older adults. *Technology and Health Care*, *16* (2), 111-118.

[12] Feil-Seifer, D., Skinner, K. and Mataric, M.J. 2007. Benchmarks for evaluating socially assistive robotics. *Psychological Benchmarks of Human-Robot Interaction*, *8* (3), 423-429.

[13] Forlizzi, J., DiSalvo, C. and Gemperle, F. 2004. Assistive robotics and an ecology of elders living independently in their homes. *Hum.-Comput. Interact.*, *19* (1), 25-59.

[14] Frye, N.E. and Dornisch, M.M. 2010. When is trust not enough? The role of perceived privacy of communication tools in comfort with self-disclosure. *Computers in Human Behavior*, *26* (5), 1120-1127.

[15] Iachello, G. and Hong, J. 2007. End-user privacy in human-computer interaction. *Found. Trends Hum.-Comput. Interact.*, *1* (1), 1-137.

[16] Iachello, G., Truong, K.N., Abowd, G.D., Hayes, G.R. and Stevens, M. 2006. Prototyping and sampling experience to evaluate ubiquitous computing privacy in the real world *Proceedings of the SIGCHI conference on Human Factors in computing systems*, ACM, Montreal, Canada, 1009-1018.

[17] Jensen, C., Potts, C. and Jensen, C. 2005. Privacy practices of Internet users: Self-reports versus observed behavior. *International Journal of Human-Computer Studies*, *63* (1), 203-227.

[18] Kahn, P.H., Ishiguro, H., Friedman, B., Kanda, T., Freier, N.G., Severson, R.L. and Miller, J. 2007. What is a human? Toward psychological benchmarks in the field of humanrobot interaction. *Interaction Studies*, *8* (3), 363-390.

[19] Kumaraguru, P. and Cranor, L.F. 2005. Privacy indexes: A survey of Westin's studies. *ISRI Technical Report*.

[20] Lee, M.K., Tang, K.P., Forlizzi, J. and Kiesler, S. 2011. Understanding users' perception of privacy in human-robot interaction. In *Proceedings of 6th International Conference on Human-robot Interaction*, Lausanne, Switzerland, ACM, 181-182.

[21] Mataric, M.J. 2005. The Role of Embodiment in Assistive Interactive Robotics for the Elderly *Caring Machines: AI for the Elderly*, Arlington, VA.

[22] Michaud, F., Boissy, P., Corriveau, H., Grant, A., Lauria, M., Labonte, D., Cloutier, R., Roux, M.A. and Iannuzzi, D. 2007. Telepresence Robot for Home Care Assistance. In *Proceedings of AAAI Spring Symposium 2007 on Multidisciplinary Collaboration for Socially Assistive Robotics*, Stanford, CA.

[23] Mynatt, E.D., Melenhorst, A.-S., Fisk, A.D. and Rogers, W.A. 2004. Aware Technologies for aging in place: Understanding user needs and attitudes, IEEE Pervasive Computing, 36-41.

[24] Nomura, T., Kanda, T., Suzuki, T. and Kato, K. 2008. Prediction of Human Behavior in Human--Robot Interaction Using Psychological Scales for Anxiety and Negative Attitudes Toward Robots. *Robotics, IEEE Transactions on*, *24* (2), 442-451.

[25] Nonaka, S., Inoue, K., Arai, T. and Mae, Y. 2004. Evaluation of human sense of security for coexisting robots using virtual reality 1st report : Evaluation of pick and place motion of humanoid robots. In *Proceedings of IEEE International Conference on Robotics and Automation*, New Orleans, LA, 2770–2775.

[26] Pollack, M.E., Brown, L., Colbry, D., Orosz, C., Peintner, B., Ramakrishnan, S., Engberg, S., Matthews, J.T., Dunbar-Jacob, J., McCarthy, C.E., Thrun, S., Montemerlo, M., Pineau, J. and Roy, N. 2002. Pearl: A Mobile Robotic Assistant for the Elderly *AAAI Technical Report WS-02-02*.

[27] Sharkey, A. and Sharkey, N. 2011. Granny and the robots: Ethical issues in robot care for the elderly. *Ethics and Information Technology*, 1-14.

[28] Sparrow, R. and Sparrow, L. 2006. In the hands of machines? The future of aged care. *Minds and Machines*, *16* (2), 141-161.

[29] Spiekermann, S., Grossklags, J. and Berendt, B. 2001. E-privacy in 2nd generation e-commerce: Privacy preferences versus actual behavior. In *Proceedings of 3rd ACM Conference on Electronic Commerce*, Tampa, Florida, USA, ACM, 38-47.

Children Learning with a Social Robot

Takayuki Kanda Michihiro Shimada Satoshi Koizumi
ATR Intelligent Robotics and Communication Laboratory
2-2-2 Hikaridai, Keihanna Science City,
Kyoto, Japan
{kanda, m.shimada, satoshi}@atr.jp

ABSTRACT
We used a social robot as a teaching assistant in a class for children's collaborative learning. In the class, a group of 6th graders learned together using Lego Mindstorms. The class consisted of seven lessons with Robovie, a social robot, followed by one lesson to test their individual achievement. Robovie managed the class and explained how to use Lego Mindstorms. In addition to such basic management behaviors for the class, we prepared social behaviors for building relationships with the children and encouraging them. The result shows that the social behavior encouraged children to work more in the first two lessons, but did not affect them in later lessons. On the other hand, social behavior contributed to building relationships and attaining better social acceptance.

Categories and Subject Descriptors
H.5.2 **[Information Interfaces and Presentation]**: User Interfaces-Interaction styles

General Terms
Design, Experimentation, Human factors

Keywords
Communication robots, Robots for children, Field experiments

1. INTRODUCTION

1.1 Learning with Robots
Learning with robots is an active topic in recent research. Such small robot-kits as Lego Mindstorms are used as an visible example of math, physics, programming, and robotics in classrooms [13, 15]. Robots are also used as interactive exhibits to encourage children's interest in science and technology [3, 17]. Robots have also been successfully used for learning other subjects, such as foreign languages [7, 11]. Social behavior has been found to be effective in language education with a robot [16].

We extend the roles of robots in learning beyond such previous work. In contrast with previous studies, we used a robot as a teaching assistant in collaborative learning (Figure 1).

HRI'12, March 5–8, 2012, Boston, Massachusetts, USA.

Figure 1. Class with Robovie for learning Lego Mindstorms

1.2 Collaborative Learning
The importance of a *learner-centered* approach has been shown in learning science. In traditional *teacher-centered* approaches, teachers transfer as much information as possible to learners, often in lecture-style classes. On the contrary, in a learner-centered approach, the role of learners receives greater emphasis. Learners are expected to be self-motivated to gain knowledge and think for themselves. Teachers stimulate learners to be involved in such learning.

Collaborative learning, which is based on a learner-centered approach, is defined as "*a situation in which two or more people learn or attempt to learn something together*" [5]. People are motivated to explain, help, compete, and collaborate. For instance, when two people explain their own ideas, they try to convince one another, and through such conversations they typically expand their own ideas [14]. Thus, a collaborative learning situation can be created when we prepare an appropriate setting to cause interaction among people. Successful implementations include a class design with the Jigsaw method [18] as well as the use of computers [19], co-learner agents [10], and small robot-kits [4].

We believe that a social robot would be useful to facilitate both learner-centered and collaborative learning. It might be fluent enough to provide instruction and explain basic learning materials. We could also use a robot as a non-superior entity (e.g., one that does not know the answers) to help children relax and create spontaneous learning and problem-solving situations.

2. CLASS DESIGN

2.1 Design Principle
Our class design, which reflects collaborative learning, has two main features. First, to fulfill a learner-centered approach, instead of having the robot teach details to children, it only explains minimum information (i.e., using Lego Mindstorms) and lets children engage in problem-solving by themselves.

Second, based on the concept of collaborative learning [5], we created activities that cut across different levels: individual, within-group, and between-group. Children were asked to work on tasks; thus they first worked at the individual level. Here, we created a competitive situation; since four children worked at the same time on the same task, progress among groups could be compared. This is one activity at the within-group level. Moreover, children worked together (*discussion phase* as well as all the phases in the 6th and 7th lessons, explained below) to create activities at the within-group level. In addition, the class was designed to encourage children to compare their achievements with other groups during the 6th and 7th lessons. Such activities are at the between-group level.

2.2 Learning Goals

We have two learning goals that we expect children to achieve. First, we want the children to learn the basic knowledge about using Lego Mindstorms, including motors, touch sensors, and related programming. Second, we want them to understand how to apply this knowledge to a concrete problem. Thus, as in other educational activities [15], we created a mission that involved an *exploration robot* who explores a field to capture (touch and knock down) as many flags as possible in one minute. To make a robot for this mission, children must build a robot with two motors, a touch sensor, and a bumper and make a program to use these components (Figure 2).

2.3 Environments

The class used Lego Mindstorms NXT 2.0. In each lesson, depending on its contents, we only provided the relevant blocks to prevent excessive complexity. For example, we did not provide a touch sensor in the first lesson because the children didn't need it. Figure 3 shows the layout of the 7.4 x 5.6 meter room. There are desks in front of a display, where children watch videos and build robots. On the other side of the room, a 1.6 x 1.8 meter test field is located where children test their robots.

Figure 2. Exploration robot

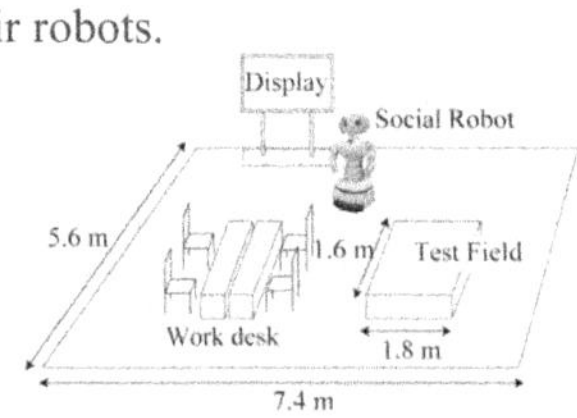

Figure 3. Layout of classroom

2.4 Details of Lessons

The class consisted of eight, two-hour lessons (Figure 4). The first five concentrated on imparting basic knowledge about Lego Mindstorms, specifically using one motor, followed by a touch sensor, two motors, programming, and combining them into an exploration robot. In each lesson, Robovie narrated and explained a video of a person manipulating the Lego blocks.

Each class consisted of a series of phases. The *basic task* was designed to be easy for everyone. Before the children started to work, Robovie showed a few minutes of a video about using Lego Mindstorms. For example, for the 1st lesson, it shows how to attach a motor, connect cables, and use the motor from a programming tool. After the explanation, the children worked freely on their task. In this phase, Robovie offered to show the video again. To ensure that all children learned the basic knowledge, Robovie asked them to bring their robots to the test field to check whether they correctly built the task.

After all children accomplished the basic task, the lesson moved to the next phase. Robovie showed a supplemental video explaining things that children need to know, such as the rules for the exploration robot (e.g., children should not touch the robot during the mission) and then moved to the *advanced task* phase. If they failed to finish the basic task in time, Robovie explained the supplemental video and moved to the *discussion phase*.

The *advanced task,* which is more difficult, requests children to explore things that are relevant to both the topics learned in that lesson's *basic task* and for the final goal of building an exploration robot. For example, in the 1st session, we asked the children to build a robot that runs fast.

At the end of the lesson, there are 15 minutes of *discussion phase* to provoke within-group activities. We asked the children to talk about what they learned and found to be useful for exploration robots. To elicit discussion, we asked them to write a summary their discussion for submission.

The 6th and 7th lessons were designed as within-group activities. In the beginning of the 6th session, they were asked to design plans for their exploration robots individually. Then they were asked to combine their plans within the group. After making a combination plan, Robovie asked them to decide roles, such as making the robot's body, the bumper (which is difficult), or the program. We also provided a score sheet to encourage children to evaluate their Lego robot in the test field. A list of the top scores of other groups was posted. Robovie explicitly encourages children to compete with these groups to encourage between-group activities. Note that these are scores of the virtual groups and the same scores were presented to all the experiment participants.

The 8th lesson tested children's individual learning achievement by measuring whether each child could build his/her robot alone. A human supervisor administered the test and did not allow children to communicate with each other. Children were allowed to use the test field, but only individually. Robovie was not present in the classroom.

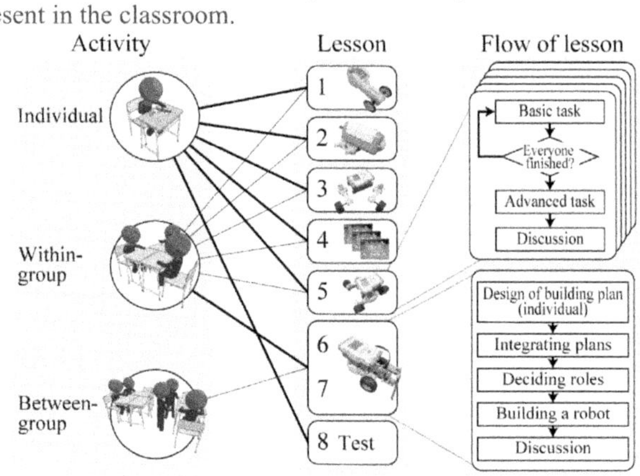

Figure 4. Configuration of eight lessons

3. SYSTEM

3.1 System Configuration

We used Robovie-R3 (Figure 1), which has two arms (4*2 DOF), a head (3 DOF), stand 110 cm tall, and is equipped with cameras, a microphone, and a speaker. Utterances are made with speech synthesis software. An operator chooses the utterances and video materials based on pre-determined rules using a teleoperation console. There are eight cameras attached to the room, which were used for the teleoperation as well as the later analysis.

3.2 Robovie's Behaviors

The study was conducted in Wizard of Oz style. We prepared the following behaviors (utterances and motions) that were controlled by the operator with the following rules.

3.2.1 Managing behavior

Phase control: Robovie speaks to the children to control phases. For example, it explains the *basic task* by saying, "Let me explain the task. Today, you are going to develop a robot that stops when a touch sensor is activated." Robovie reminds children about the tasks and time with such language as, "You only have five more minutes." Phase control behaviors are used when allocated time for each phase lasts. To judge the end of the *basic task* phase, Robovie checks whether a Lego robot build by a child is successful when he/she tests it in the test field and provides such feedback as "You did it." When all children are finished, it moves to the next phase.

Video explanation: Before the *basic* and *advanced tasks*, Robovie shows a video. For example, when it explains how to attach a motor, it shows a video of a scene where a person is attaching a motor (Figure 5). Robovie narrates the video: "Next, attach the motor to the body. See the three holes on the top of the orange part? Well, . . . " Such behavior was used during the *basic task* and *advanced task* phases. Robovie simply continues video explanations one by one.

Figure 5. Example of video material

Scolding: In our preliminary trial, children sometimes excessively deviated from tasks and caused conduct problems, such as bullying. To prevent them, when any interfering behavior was detected, Robovie scolded the child who was responsible. We define interference as cases where the tasks of others related to their Lego robots are obstructed with; for example, when a child was testing her robot in a test field, it would be considered interference if another child interrupted her task by either directly stopping her robot or making his robot collide with hers.

There are levels of scolding. First, Robovie simply asks the child to stop: "Taro (*name*), do not obstruct." While the majority of children don't go beyond this point, some had difficulty controlling themselves. When a child does not stop, Robovie uses harsher language. It next says, "Taro, stop interfering with others. Next time, I'll ask you to leave the room," and then, "Taro, stop that or leave the room." Finally, it orders the child to wait in the lobby and will keep asking until the child leaves the room.

3.2.2 Social behavior

These social behaviors are not necessarily for managing the class. A human teacher/tutor would use them for such social purposes as relation-building and encouragement.

Social: We prepared the following examples of social behaviors. First, while the children were working on their tasks, Robovie roamed and approached them in the test field and praised their achievements based on the behavior of their Lego robots. Robovie praised the first child to accomplish the given task: "You are the first to finish your robot. Great!" "Great, your robot stopped when it touched the wall!" If the child's work is a failure, it provides such encouragement as "Keep trying, I think you'll succeed soon. Good luck."

Second, it encourages competition in the *advanced task*. When two or more children simultaneously visit the test field, it approaches them and asks, "Whose robot can find the wall faster? Why don't you compare your robots?"

Third, it behaves sympathetically toward children who are struggling to solve the given problem. When a child does not bring successful work to the test field after more than 20 minutes, Robovie goes to the child's desk and asks about the progress without helping or proving answers.

Fourth, it stimulates children who seem bored with two rules. First, when a child stands in front of Robovie, it speaks to the child. Second, after 20 minutes have passed from the start of the basic or advanced task (typically this was when some children started getting bored), when Robovie finds a child who isn't working on the tasks (neither using Legos nor the computer), it asks: "How is your task coming?" If the child answers positively, it responds "Good, please keep up the good work." Otherwise, it responds, "Well, why don't you ask someone for help?"

Fifth, when a child enters the classroom at the start, Robovie chats with him/her. In cases where it praised the child in the previous lesson, it recalls that praise: "Taro, last time you were the first to finish making a robot that stopped with a touch sensor. Good luck again today." Otherwise, it chats with the child, e.g., "Do you have a lot of homework for summer vacation? You should do some everyday. I'm sure you'll do your best."

4. EXPERIMENT

4.1 Hypothesis and Prediction

In the literature, social behavior has successfully been used for relationship-building in a face-to-face learning context [16], in shopping assistance [9], and in encouraging exercise [2]. Since we believe that it would also be useful for relationship-building in children's collaborative learning, we made the following predictions:

> **Prediction 1**: A robot that exhibits social behavior will establish a stronger **relationship** with children than a robot that does not exhibit social behavior.

If the class works as designed, we should be able to identify a learning effect. This would be obvious if the class were managed by a human teacher; however, since the class is managed by a robot, we must confirm the effect. Further, it was reported that social behavior motivated people in teacher-oriented learning [16]. We hypothesize that social behavior will also be useful for encouraging children in collaborative learning. We thus made the following predictions:

> **Prediction 2-1:** Children will learn from the class and gain **basic knowledge** on how to use Lego Mindstorms for a given mission.

> **Prediction 2-2:** A robot that exhibits social behavior will motivate children to study and enable them to **achieve** higher learning than a robot that does not exhibit social behavior.

Previous research reported the effect of social behavior on enjoyment and social acceptance [9], even though it only involved short-term interaction. We hypothesize that this effect will be sustained for long-term interaction, so we made the following prediction:

Prediction 3: The children who learned with a robot that exhibits social behavior will perceive more **enjoyment** and greater **social acceptance** toward it than the children to whom it does not exhibit social behavior.

4.2 Condition

The study used a between-participants design and controlled one factor, *social behavior*. There were two conditions:

Non-social condition: Robovie only performed minimum behaviors for managing the class (explained in Section 3.2.1).

Social condition: In addition to the behaviors in the *non-social condition*, Robovie exhibited social behaviors (explained in Section 3.2.2) for social purposes.

In addition, for evaluating the learning effect, there is one within-participants factor: *prepost comparison*.

Note that Robovie's perception capability in both conditions was designed to be similar. When requested, Robovie explains the video materials again in both conditions, which shows a capacity to respond to certain spoken keywords. In the social condition, Robovie only responded to extra keywords when it initiated conversation. This policy resulted in Robovie ignoring many utterances from children, e.g., children asking about the remaining time but Robovie did not respond in either condition. In the social condition, Robovie praised children by observing the Lego robots in the test field, and in the non-social condition Robovie also showed the capacity to observe the Lego robots by judging the children's achievements in the basic tasks.

4.3 Participants

We targeted 6th-graders since they are mature enough to use Lego Mindstorms. 31 6th-graders (25 boys, 6 girls) participated in the experiment, all of whom responded to flyers distributed at elementary schools. The large gender disparity is also reported in other after-school activities with Lego Mindstorms, which might reflect societal expectations [13]. We formed applicants into groups of four children from different schools so that they didn't know each other from the beginning; due to the availability of participants, one group was formed with three children. Half of the groups were assigned to the *non-social condition* and the other half to the *social condition*. They were not paid.

4.4 Procedure

The children visited our research laboratory. The details of the lessons were described in Section 2. Each lesson was conducted on a different day during a 2-hour time slot. An adult assistant escorted the children to the door of the classroom without entering and instructed them to enter the classroom to study with the robot. During the class, before children start to work in either the *basic* or *advanced* tasks, Robovie announced that an adult could be summoned if a problem happened, e.g., computer malfunction for programming, or a lack of Lego parts; when children asked Robovie to call an adult, the assistant entered the room to fix the problem. The assistant did not provide any further support. Pretests were conducted before the 1st lesson, and posttests were conducted after the end of the 7th and 8th lessons. Short interviews were also conducted at the end of the 8th lesson. Due to tests and interviews, the children's learning time was shorter in the 1st, 7th, and 8th lessons.

4.5 Measurement

4.5.1 *Individual ability (pretest)*

The following measurements were taken before the 1st lesson.

Japanese and **math**: We chose a few items from publicly available tests. The scores were marked on a scale of 0-20.

Lego-building time: We measured the children's manipulation skills. We gave them a small Lego car and measured their time to build the same one. Typically they finished in 1-2 minutes.

4.5.2 *Perception of robot (posttest)*

The following measurements were taken by questionnaire using 1-to-7 point Likert scales. Since the children did not interact with Robovie during the 8th class, we administered the posttest after the end of the 7th session.

Relationship: We measured the strength of the relationship established with Robovie in terms of three scales: the degree of **perceived familiarization** (a single-item scale as used in [12]), degree of perceived **likeability,** and a scale for **desire to be liked** by Robovie. *Likeability* was measured with a five-item scale (like, friendly, kind, pleasant, and nice) reported in [1]. We developed a scale for *desire to be liked* that consisted of five items, such as "I wanted Robovie to like me," "I sometimes wondered whom Robovie liked the most," and "I don't care if Robovie doesn't like me."

Enjoyment: Five items [8] (enjoy talking, enjoy doing things with, enjoyable, fascinating, and boring) addressed to what extent children perceived Robovie as enjoyable.

Impressions: We asked their impressions of Robovie with two single-item scales: **teacher-like** and **friend-like**.

4.5.3 *Social acceptance (posttest)*

We used a scale developed by Heerink et al. [8] that consists of three items asking whether they would like to use the robot again on a 1-to-7 point scale.

Intention to use Robovie in Lego class: We asked the children whether they would like to participate in a Lego Mindstorms class using Robovie again.

Intention to use Robovie in other classes: We asked the children whether they would like to use Robovie again for different classes.

4.5.4 *Learning outcome*

We took the following measurements for the children's learning.

Basic knowledge (*pretest* and *posttest*): We asked the children to write how to build an exploration robot. We used a checklist with six items to code the answers. The test was conducted before the 1st session and after the 8th.

Basic achievement (*posttest*): To confirm what they had learned, we analyzed their Lego robots during the 8th lesson with a checklist of ten items, including whether motors were attached, connected with cables, and controlled by a program, and whether a touch sensor was attached, used in the program, and overall whether the program output meaningful behavior.

Overall achievement (*posttest*): The children built an exploration robot that captured as many flags as possible within a minute in three different fields. We measured the ratio of the captured flags. Achieving a high score required careful understanding of the problem, good ideas, and problem-solving by connecting what they know to the nature of the problem. For example, it is typically better to make a fast robot, since the time is limited, and a good bumper increases the robot's efficiency when exploring.

5. RESULT

5.1 General Trend

As we requested, most children attended all lessons. There were three absences in the *social* condition and six in the *non-social* condition due to such unavoidable reasons as colds. We requested the children who missed a class to ask others for help when facing something they had missed. No children dropped out.

Classes typically started quietly in the 1st and 2nd lessons since the children did not know each other yet. They often started to talk with each other after the 3rd lesson. In each lesson, during the basic task, their behavior was rather similar, trying to do a couple of things on their desk and bringing the robot to the test field when they thought they were finished. Their behavior during the advanced task showed more divergence. Some tried to improve their robots more, while some seemed to get bored in the middle. When a child found a good solution, others typically noticed and copied it. For example, in the 3rd lesson, children created a robot that could touch a wall and quickly return to the start point. They explored how to return the robot to the start position. Some groups found a strategy to count time during the turning motion to make an exact U-turn, and others discovered an alternative in which robots moved backward. Once one child made such a strategy, others often copied it.

In the group activities in the 6th and 7th lessons, they usually communicated in a friendly way about the design and allocation of their roles and worked in their decided roles. The discussion phase was less successful. While a few groups talked about the things that they learned, more than half had difficulty verbalizing what they learned and simply wrote comments one by one without talking.

During the seven lessons, Robovie used 1129.5 utterances (564.8 *phase control*, 556.5 *video explanation*, and 8.3 *scolding*) for each group in the *non-social* condition and 1542.5 utterances (523.8 *phase control,* 512.5 *video explanation*, 22.8 *scolding,* and 483.5 *social*) for each group in the *social* condition. There was some divergence in number because of the interaction with the children. For example, *video explanation* behaviors were used when children asked to replay the videos. The number of *scolding* behaviors was large in the *social* condition, most of which were directed to one specific child in one group in the *social* condition; the average in other groups was 5.7 *scolding* utterances, which is similar to the number in the *non-social* condition.

5.2 Verification of Predictions

5.2.1 Reliability of measurement

We computed the reliability of the scales. Cronbach's α was within a reliable range (enjoyment: .867, desire to be liked: .746, Intention to use Robovie in Lego class: .890, Intention to use Robovie in other classes: .924, and likeability: .885). For the measurement by coders, we computed Cohen's Kappa coefficient for each checklist item and took the average for each measurement. They show a reliable match between coders (basic knowledge: 0.892, basic achievement: 0.856).

5.2.2 Did social behavior improve relationships?

Figure 6 (a) shows the *relationship* results. The analyses were carried out using linear mixed-effects models having *social behavior* factor and *group* (which is nested within the *social behavior* factor) as fixed effect. There were significant main effects revealed in *social behavior factor* in *likeability* ($F(1,23)=9.969$, $p=.004$), *perceived familiarization* ($F(1, 23)=20.976$, $p<.001$), and *desire to be liked* ($F(1, 23)=6.034$, $p=.022$). *Group* factor was not significant in these analysis. This indicates that the children in the *social* condition show a stronger relationship with Robovie than the children in the *non-social* condition. Prediction 1 is supported.

We further analyzed the impression of Robovie (Figure 6(b)). An analysis with linear mixed-effects models revealed a significant main effect in *social behavior factor* in *friend-like* ($F(1, 23)=27.352$, $p<.001$). No significance was found in *teacher-like* ($F(1,23)=1.164$, $p=.292$). *Group* factor was not significant in both analysis. The children in the *social* condition perceived Robovie as more friend-like than the children in the *non-social* condition.

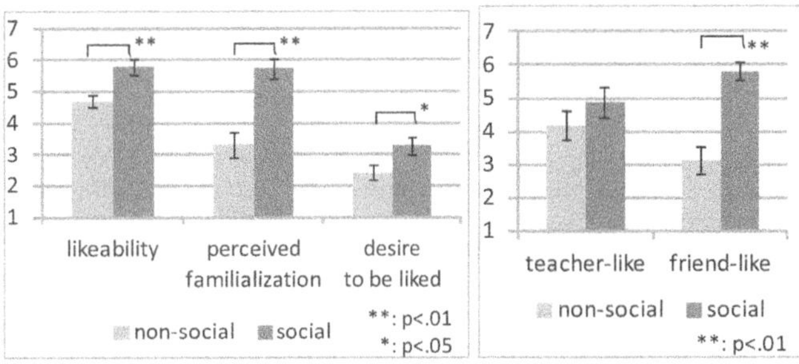

(a) relationship (b) impression

Figure 6. Children's perceived relationship with Robovie

5.2.3 Did social behavior improve learning?

Figure 7(a) shows the result for *basic knowledge*. We conducted a analysis with linear mixed-effects models having *prepost comparison (within-subject factor)*, *social behavior* factor, and *group* (which is nested within the *social behavior* factor) as fixed effect. A significant main effect was revealed in the *prepost* factor ($F(1,38.714)=229.089$, $p<.001$) and *group* ($F(6, 33.644)=2.777$, $p=.026$). No significance was found in *social behavior* ($F(1, 38.919)=.195$, $p=.661$) or in the interaction between social and *prepost* factors ($F(1, 38.714)=.004$, $p=.949$).

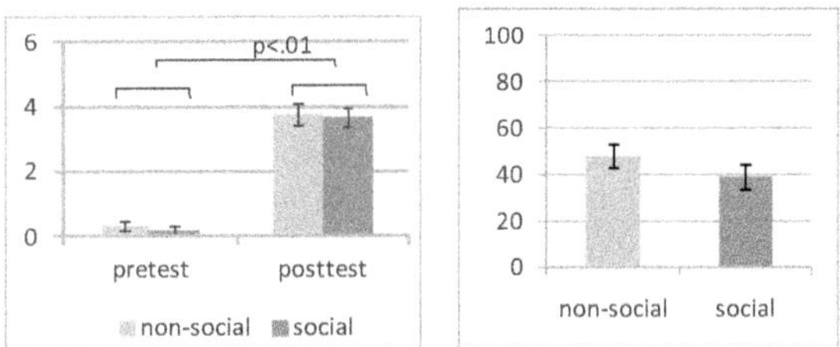

(a) knowledge (b) overall achievement

Figure 7. Learning effect

In addition, the *basic achievement* result (non-social: avg 9.94, sd 0.06, social: avg 9.93, sd 0.07) confirms the meaning of the learned *basic knowledge*. The children successfully built Lego robots that greatly involved the knowledge they were expected to have learned. Although we hope their scores would be perfect, in each condition, one child lost a point for failing to attach a bumper; this oversight made the robot's exploring less efficient, although the task could still be achieved. Overall, we consider that Prediction 2-1 is supported.

Figure 7(b) shows the *overall achievement* results measured in the 8th session. An analysis with linear mixed-effects models, having *social behavior* factor and *group* (which is nested within the *social behavior* factor) as fixed effect, reported no significance in *social behavior factor* ($F(1,23)=1.967$, $p=.174$), but reported significance in *group* factor ($F(6,23)=2.809$, $p=.034$). Although we thought that the children would be motivated by social behavior and thus it would contribute to improve the learning achievement, we did not find such an effect. Prediction 2-2 is not supported.

5.2.4 *Did social behavior improve social acceptance?*

Figure 8 shows the results related to Prediction 3. We conducted an analysis with linear mixed-effects models having *social behavior* factor and *group* (which is nested within the *social behavior* factor) as fixed effect. There were significant main effects in *social behavior factor* in *enjoyment* ($F(1, 23)=7.989$, $p=.010$) and *intention to use Robovie in other classes* ($F(1,23)=6.799$, $p=.016$). No significance was found in *intention to use Robovie in a Lego class* ($F(1,23)=1.031$, $p=.321$). *Group* factor was not significant in any of the analysis.

This result indicates that Robovie was well accepted by all children within the context of the Lego class, while children in the social condition showed a more general acceptance of Robovie, as indicated by the higher desire to use in another class. Prediction 3 is supported.

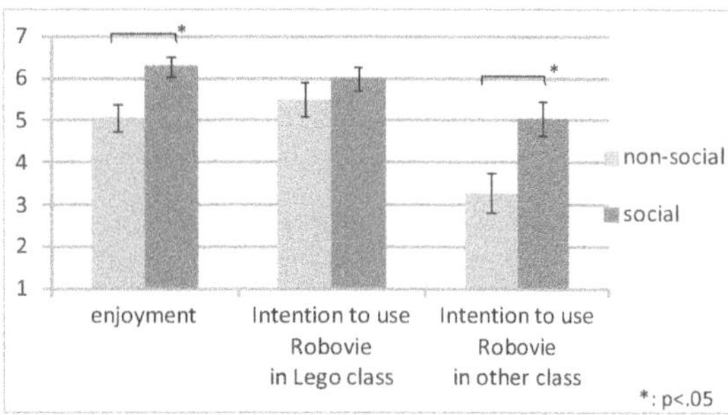

Figure 8. Social acceptance

5.3 Observations and Analysis

5.3.1 *Was a "learner-centered" class realized?*

We believe that we successfully made learner-centered classes. The children worked by themselves, learned, and achieved their tasks. They often explored innovations.

We interviewed them about their perceptions of the class with Robovie. Because they often faced difficulty articulating their impressions beyond simply saying that it was enjoyable, we asked them this question: "What do you think if there was an adult instead of Robovie in this Lego class". In Japan, schools most commonly have teacher-centered classes, and thus this question retrieves children's perceived differences due to having a learner-centered class with a robot instead of an adult.

Table 1 shows the result that was judged with two coders; Cohen's Kappa coefficient was 0.921. The children typically mentioned that they felt more relaxed, less pressured, and more enjoyment with Robovie, especially because adults get tense, exert pressure, and reduce enjoyment. But some children described Robovie as scary.

Another typical point is that human adults would provide more detailed explanations and instructions. One child made the following comment:

> *I'd learn more with an adult. Robovie didn't answer my questions.*

Another child commented on the same point from a different view:

> *A teacher gives me an answer when I ask. Robovie didn't answer, so I guess I need to think for myself. If there were a teacher, I would end up depending on the teacher. I could say, "I don't understand," and ask for the answer. Robovie didn't teach me. When I found an answer by myself, I felt happy.*

Overall, these suggest successful realization of a learner-centered class where children feel relaxed and can engage in problem-solving by themselves. Note that Chi-square test reported no significant difference across conditions in the result in Table 1.

Table 1. How children perceived the class with Robovie

	Social	Non-social
More relaxed, less pressured, and more enjoyment with Robovie	14/15	16/16
Robovie is scary	2/15	5/16
Human adults would provide more explanations and instructions	6/15	7/16

5.3.2 *How did the social behaviors affect learning?*

In the social condition, we observed that the children anticipated Robovie's praise and comments. Figure 9 shows a scene where the children waited for a moment when Robovie moved near the test field. One boy said, "Now is our chance to show Robovie," and they brought their robots to get Robovie's attention and comments. In Figure 10, Robovie praised another child's Lego robot, "Your robot works nicely. It detected the wall well"; then the praised boy proudly showed his robot to others. These scenes suggest that Robovie's social behavior motivated children to work more and to show their work to Robovie.

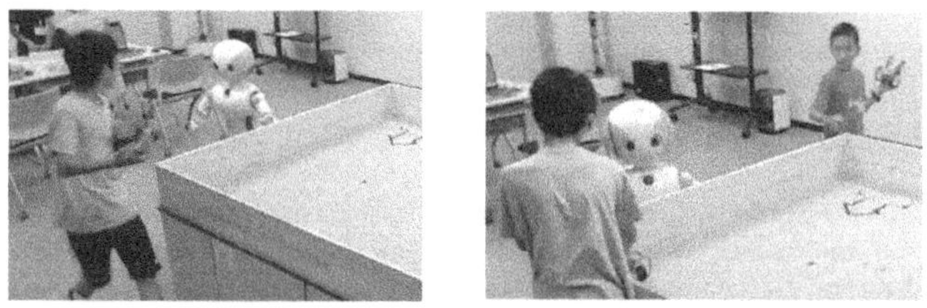

Figure 9. Children showing their Lego robots to Robovie

Figure 10. Boy showing his Lego robot to others

We analyzed whether social behavior significantly influenced the children. To quantify this motivational effect, we coded the **number of field tests** children conducted for their robot at the test field (Figure 11). Videos during the 1st to the 5th lessons were coded. The coders counted a field test as an instance when a child brought his/her robot to the test field and operated it. We also coded the **working time ratio** (Figure 12). A child is judged to be working when he/she was doing an activity with Lego blocks or a computer even when he/she was chatting with others; a child is judged to be not working when he/she was teasing Robovie, doing nothing, or doing apparently irrelevant things with the Lego blocks or computer. We observed a couple of teasing behaviors; some children put a cap on its head, and some stood in front of it to impede its locomotion. Note that the amount of teasing behavior each child conducted during the five lessons was 7.93 times (s.d. 11.09) in the *social* and 5.75 times (s.d. 5.40) in the *non-social* condition, which is not significantly different.

We conducted an analysis with linear mixed-effects models having *social behavior* factor and *group* (which is nested within the *social behavior* factor) as fixed effect. There were significant main effects revealed in *social behavior factor* in the *ratio* of the *working time* for the 1st and 2nd lessons (1st: $F(1,20)=90.583$, $p<.001$, 2nd: $F(1,23)=35.618$, $p<.001$) and the *number of field tests* for the 1st and 2nd lesson (1st: $F(1,20)=7.811$, $p=.011$, 2nd: $F(1,23)=4.487$, $p=.045$). No significance was found in *social behavior factor* in the later lessons. This indicates that social behavior motivated children to work more in the first two lessons, but this effect disappeared by the 3rd and subsequent lessons.

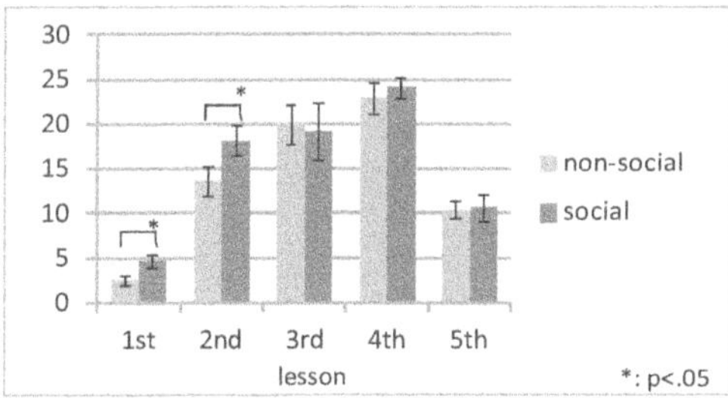

Figure 11. Field tests done by children

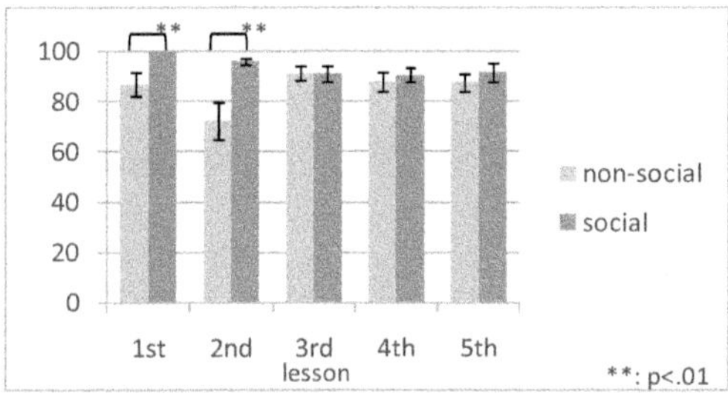

Figure 12. Ratio of working time

We also analyzed how children perceived the influence of Robovie's behavior by asking them to remember what Robovie said to them. We checked whether children mentioned three topics: *scolding*, *phase control*, and *social* (Mention of the *video explanation* was omitted because we first asked children to provide their comments for it). Since some children mentioned that *social behavior* was encouraging, we further classified whether they spontaneously mentioned the encouragement effect, based on whether a child reported positive feelings or improved motivation. Table 2 shows the result that was judged with two coders; the Cohen's Kappa coefficient was 0.836. The children in the social condition less frequently mentioned *scolding* (Chi-square test: $x2(1)=11.766$, $p<.01$) and more often mentioned *social behavior* ($x2(1)=15.746$, $p<.01$) than children in the non-social condition. No significant difference found in *phase control* behavior. The comments of six children showed this encouragement effect in social behavior:

I was happiest when I heard "I think you'll be successful soon."

When I was praised, I was happy. I felt like working more.

Meanwhile, one child claimed that the influence of social behavior was not encouraging:

Robovie said, "You are the first to finish the task." So I knew that I had done the task. It didn't feel like praise. I'd prefer to be praised by a human.

It is interesting that the children perceived the social behavior so differently.

Table 2. What the children remembered about what Robovie said

		Social	Non-social
Phase control		6/15	7/16
Scolding		4/15	14/16
Social		10/15	0/16
	Mentioned encouragement	6	0
	No mention of encouragement	4	0

6. DISCUSSION

6.1 Interpretation and Design Implication

First, our study demonstrated the successful use of a robot in a collaborative learning class. The effect of social behavior was not significant on learning achievement, suggesting that for immediate use, a configuration without social behavior would be economical. In contrast, better social acceptance was attained by the presence of social behavior, suggesting that it might be useful to motivate the children to use the robot if the subject of the class were less attractive.

Social behavior only showed an effect in the first two lessons. Such learning effects have been reported [16] in short-term interaction. Our result is consistent; both studies showed that social behavior is useful within a short-term range. This implies a "novelty effect."

On the other hand, the level of the class involvement might be another reason. Children were only able to use one sensor and one motor in the 2nd lesson. While this step was necessary for them to understand all the related knowledge, the class became less involving. They learned how to use two motors in the 3rd lesson; after that, the children had the potential to build exploration robots, gaining much freedom in the advanced tasks. In fact, the children in the non-social condition had a larger amount of time not working in the 1st and 2nd lessons than in the later lessons. Robots might more effectively motivate children when the class is less involving.

6.2 Limitations

We aim to reveal a way to use a social robot for collaborative learning. Even though our study unveiled certain aspects, we cannot generalize our findings only from this study. For example, we used self-selected participants, so perhaps their motivation for learning was high, although this is normal for after-school activities. If the children are less motivated, social behavior might motivate them to study more; this question remains open.

Regarding the comparison of social and non-social condition, we tested the differences in two pragmatic configurations. This necessitated changing a few factors across conditions. For example, to exhibit social behavior, Robovie roamed around the room; in the non-social condition, Robovie stood still in the corner while the children worked on their tasks. Consequently, the robot's gaze behavior was different across conditions. Nevertheless, we believe that our comparison provides a pragmatic design implication.

7. CONCLUSION

In the study, we created a class for Lego Mindstorm, in which a group of children learned with Robovie for 7 lessons. We consider that the study has two major contributions. First, it demonstrated a successful use of a social robot in a learner-centered class. Such usage is unique. In the class, the robot's role resembled a facilitator more than an interactive partner. Its lack of capacity (e.g., being unable to answer questions) is not necessarily a disadvantage, since it invites self-effort. Humans could certainly play the same role, although some children might be confused if a capable teacher refuses to help.

Second, the study revealed how social behavior affects children in long-term interaction. A few precedents exist, which primarily report to what extent interaction with a robot decreased or was sustained in a long-term fashion [6, 11, 20]. In contrast, our study reported that social behavior motivated children to work more in the class in the beginning and contributed to building perceived relationships with the robot. Note that such social behavior promoted children's intention to use the robot even after using it for seven lessons in the class.

8. ACKNOWEDGEMENTS

We thank Profs. Naomi Miyake, Jun Oshima, and Hajime Shirouzu for their helpful advice for the class design. We also thank Takahiro Miyasita and Hiroshi Ishiguro for their technical advices, and Kanako Tomita, Yuri Ichida, and Hiroyoshi Azuma for their help in conducting the experiment and the data analysis. This research was supported by a Grant-in Aid for Scientific Research, KAKENHI (21118008 and 21680022).

9. REFERENCES

[1] Bartneck, C., Kulic, D., Croft, E. and Zoghbi, S., 2009, Measurement Instruments for the Anthropomorphism, Animacy, Likeability, Perceived Intelligence, and Perceived Safety of Robots, *International Journal of Social Robotics*, vol. 1, pp. 71-81.

[2] Bickmore, T. W. and Picard, R. W., 2005, Establishing and Maintaining Long-Term Human-Computer Relationships, *ACM Transactions on Computer-Human Interaction (TOCHI)*, vol. 12, pp. 293-327.

[3] Burgard, W., Cremers, A. B., Fox, D., Hahnel, D., Lakemeyer, G., Schulz, D., Steiner, W. and Thrun, S., 1998, The Interactive Museum Tour-Guide Robot, *National Conf. on Artificial Intelligence (AAAI1998)*, pp. 11-18.

[4] Denis, B. and Hubert, S., 2001, Collaborative learning in an educational robotics environment, *Computers in Human Behavior*, vol. 17, pp. 465-480.

[5] Dillenbourg, P., 1999, What do you mean by 'collaborative learning'?, *Collaborative-learning: Cognitive and Computational Approaches*, pp. 1-19.

[6] Gockley, R., Bruce, A., Forlizzi, J., Michalowski, M., Mundell, A., Rosenthal, S., Sellner, B., Simmons, R., Snipes, K., Schultz, A. C. and Wang, J., 2005, Designing Robots for Long-Term Social Interaction, *IEEE/RSJ Int. Conf. on Intelligent Robots and Systems (IROS2005)*, pp. 1338-1343.

[7] Han, J., Jo, M., Park, S. and Kim, S., 2005, The Educational Use of Home Robots for Children, *IEEE Int. Workshop on Robot and Human Interactive Communication (RO-MAN2005)*, pp. 378-383.

[8] Heerink, M., Kröse, B., Evers, V. and Wielinga, B., 2010, Assessing Acceptance of Assistive Social Agent Technology by Older Adults: the Almere Model, *International Journal of Social Robotics*, vol. 2, pp. 361-375.

[9] Iwamura, Y., Shiomi, M., Kanda, T., Ishiguro, H. and Hagita, N., 2011, Do Elderly People Prefer a Conversational Humanoid as a Shopping Assistant Partner in Supermarkets?, *6th ACM/IEEE International Conference on Human-Robot Interaction (HRI2011)*,

[10] Ju, W., Nickell, S., Eng, K. and Nass, C., 2005, Influence of colearner agent gehavior on learner performance and attitudes, *ACM Conference on Human Factors in Computing Systems (CHI2005)*, pp. 1509-1512.

[11] Kanda, T., Hirano, T., Eaton, D. and Ishiguro, H., 2004, Interactive Robots as Social Partners and Peer Tutors for Children: A Field Trial, *Human-Computer Interaction*, vol. 19, pp. 61-84.

[12] Kanda, T., Shiomi, M., Miyashita, Z., Ishiguro, H. and Hagita, N., 2009, An Affective Guide Robot in a Shopping Mall, *ACM/IEEE Int. Conf. on Human-Robot Interaction (HRI2009)*, pp. 173-180.

[13] Lau, K. W., Tan, H. K., Erwin, B. T. and Petrovic, P., 1999, Creative Learning in School with LEGO@ Programmable Robotics Products, *ASEE/IEEE Frontiers in Educations Conference (FIE1999)*, pp. 12D4/26-12D4/31.

[14] Miyake, N., 1986, Constructive Interaction and the Iterative Process of Understanding, *Cognitive Science*, vol. 10, pp. 151-177.

[15] Mosley, P. and Kline, R., 2006, Engaging students: a framework using LEGO robotics to teach problem solving, *Information Technology, Learning, and Performance Journal*, vol. 24, pp. 39-45.

[16] Saerbeck, M., Schut, T., Bartneck, C. and Janse, M. D., 2010, Expressive Robots in Education: Varying the Degree of Social Supportive Behavior of a Robotic Tutor, *ACM Conference on Human Factors in Computing Systems (CHI2010)*, pp. 1613-1622.

[17] Shiomi, M., Kanda, T., Ishiguro, H. and Hagita, N., 2007, Interactive Humanoid Robots for a Science Museum, *IEEE Intelligent Systems*, vol. 22, pp. 25-32.

[18] Slavin, R. E., 1980, Cooperative Learning, *Review of Educational Research*, vol. 50, pp. 315-342.

[19] Stahl, G., Koschmann, T. and Suthers, D., 2006, Computer-supported collaborative learning: An historical perspective, in *Cambridge handbook of the learning sciences*, R. K. Sawyer ed., Cambridge University Press., pp. 409-426.

[20] Tanaka, F., Cicourel, A. and Movellan, J. R., 2007, Socialization between Toddlers and Robots at an Early Childhood Education Center, *Proceedings of the National Academy of Sciences of the USA (PNAS)*, pp. 17954-17958.

Blended Reality Characters

David Robert
Personal Robots Group
MIT Media Laboratory
20 Ames Street, E15-468, Cambridge, MA, USA
dr@media.mit.edu

Cynthia Breazeal
Personal Robots Group
MIT Media Laboratory
20 Ames Street, E15-468, Cambridge, MA, USA
cynthiab@media.mit.edu

ABSTRACT
We present the idea and formative design of a blended reality character, a new class of character able to maintain visual and kinetic continuity between the fully physical and fully virtual. The interactive character's embodiment fluidly transitions from an animated character on-screen to a small, alphabet block-shaped mobile robot designed as a platform for informal learning through play. We present the design and results of our study with thirty-four children aged three and a half to seven conducted using non-reactive, unobtrusive observational methods and a validated evaluation instrument. Our claim is that young children have accepted the idea, persistence and continuity of blended reality characters. Furthermore, we found that children are more deeply engaged with blended reality characters and are more fully immersed in blended reality play as co-protagonists in the experience, in comparison to interactions with strictly screen-based representations. As substantiated through the use of quantitative and qualitative analysis of drawings and verbal utterances, the study shows that young children produce longer, detailed and more imaginative descriptions of their experiences following blended reality play. The desire to continue engaging in blended reality play as expressed by children's verbal requests to revisit and extend their play time with the character positively affirms the potential for the development of an informal learning platform with sustained appeal to young children.

Categories and Subject Descriptors
H.5.m [Miscellaneous]: Human robot interaction; I.2.9 [Robotics]; I.3.6[Computer Graphics]; K.3.1 [Computer Uses in Education].

General Terms
Design, Experimentation, Human Factors, Theory.

Keywords
Human-robot interaction; Blended reality character; interreality portal; robot hutch; informal learning.

1. INTRODUCTION
In this paper, we explore the creation of a developmentally appropriate, technologically-mediated experience for young children between the ages of three and a half and seven.

To compliment the vivid and active pretend play, critical to children's development, this technological platform computationally models a *blended reality* context for a new type of play that takes place both on screen and in the real world as a fused and continuous space. This singular play context serves as the springboard for imaginative play activities, blurring the boundaries between screen-based and tangible, robotic media (see Figure 1).

Figure 1. Children play with Alphabot, a blended reality character as it migrates from the virtual to the physical

The Alphabot, a *blended reality character*, appears to seamlessly move on and off the screen, fluidly transitioning from a computer graphics character on screen, to a mobile robot in physical reality. The character's migration is enabled through a physical robot hutch enclosure, acting as a metaphorical portal between the real and the virtual. Passing through the interreality portal, the *blended reality character* maintains continuity and carries with it any changes that happen as a result of interactions with participants in the physical space.

This new context for play blends all of the affordances of the real, physical world in which children naturally develop, with the extensible space and potential of the digital world, in an intuitively accepted spatial arrangement as demonstrated in user studies.

In this paper, we will describe the concept and design of a blended reality character, and discuss how its unique interaction context supports young children's play in an appealing and fun environment. The design will reveal the system's foundation is rooted in a generalized approach to robotics as the design of "living characters," with specific extensions to the system to support children's participation in robotic gaming platforms for immersive learning and imaginative play.

This work attempts to understand what the impact of providing such an environment is on preschool-aged children's imaginative play. Using a system theories approach to frame the formal, experiential and cultural dimensions of blended reality, the scope of this paper confines itself to exploring the experiential domain of this framework, focusing on the playful interaction between young children and Alphabot, a blended-reality character (see Table 1).

Table 1. Blended reality framework

	Formal	Experiential	Cultural
Objects	physical blended virtual	human participants robot participants	medium
Attributes	object properties rules system properties rules	human interaction human-robot interaction state of the system	how, when, & why medium was created
Relationships	spatial behavioral (among objects)	social emotional educational playful	medium to culture
Environment	affects objects	context of play	culture

2. MOTIVATION

In our current, top-down media landscape, children often passively consume media produced by adult professionals. Early education specialist and founder of the influential Reggio Emilia approach to kindergarten, Loris Malaguzzi, claimed, *"each child has the right to be a protagonist."* [3] There is an urgent need to protect youth and empower them to shape their own media environments. The United Nations Convention on the Rights of the Child (CRC) adopted in 1989 confirmed this sentiment echoed by educators and media experts around the globe [21]. We hope the system presented in this paper and its evaluation can offer a perspective on what user-generated content by young children transcending cultural and linguistic boundaries might look like.

Alarming recent findings published in a 2009 report by Nielsen indicate the amount of screen time by kids aged two to five is on average more than 32 hours a week [16]. That's over an entire day a week that children are sitting sedentary in front of the television. Meanwhile, over the past three decades, childhood obesity rates in the United States have tripled. In 2010, First Lady Michelle Obama launched the Let's Move campaign stating, "the physical and emotional health of an entire generation and the economic health and security of our nation is at stake" [12]. This work seeks to address some of these issues by creating a novel context for imaginative play that transcends the limitations of current media and empowers children with the tools to physically engage with media both on and off the screen.

Nearly ¾ of American children play computer and video games [22]. Educational games offer a promising and untapped opportunity to leverage children's enthusiasm and help transform teaching and learning. Learning takes place best when children are engaged and enjoying themselves [21]. The literature on play is clear on the importance of creating opportunities for unstructured, imaginative play for preschool-aged children. Play is vital for the social, emotional, physical and cognitive development of young children [9]. If we want to create a future society of freethinking, tinkering problem-solvers we need to support our children's active creative exploration through playful, informal learning.

Inspired by the pioneering work of Joan Cooney, Gerald Lesser, Jim Henson and the Sesame Workshop folks who took charge and dedicated themselves to bringing their vision of accessible and fun education for all, it is our hope that this work plants the seed for an international effort to connect preschool aged children to each other through a playful, informal learning system built atop the foundations of a blended reality. The first step, and one of the core motivators of this work is to show that children have accepted blended reality as an extension of media, and are engaged with blended reality characters paving the way for fun and rewarding learning opportunities.

3. BACKGROUND AND RELATED WORK

3.1 Context for Robot Characters

The Personal Robots Group at the MIT Media Lab has long been driven by the vision to design sociable robots[1]. In an effort to create engaging robotic characters the group's researchers have drawn inspiration and critical insights from classic animation techniques [24]. If a robot were to be treated like a living character imbued with the illusion of life, and sustain an engaging interaction with a person, we playfully reasoned it would need a back-story, a context or world of its own. As one researcher jokingly asked: "Where do the robots *go* when the lights go out?" Based on the assumption that robotic characters, like animated characters, are perceived as more than the sum of their constituent parts, we imagined that providing them with their very own world might help people interacting with them move past constantly comparing robots to familiar life forms. Testing this fanciful hypothesis meant building a world for robots, a world that could blend into our own so that humans and robots could meet and play in a contextual middle-ground. We began by designing a home or robot hutch for our fluffy squash and stretch robot *Miso*. The idea of a robot hutch, however, dates back to 1950 when Grey Walter built a small home and recharging station for one of his famed analog Tortoise robots. He interchangeably referred to it as a kennel or robot hutch[7].

3.2 The Apple Yard

In [11], the authors advanced a definition of blended reality as the modeling of a "window" through which virtual objects enter the player's physical space. In the "Apple Yard" game prototype, a player used a wand to hit virtual apples metaphorically flying out of the screen. The unique features described included the ability for a player to interact directly with a virtual object in the physical world and the idea that the game's display screen is rendered as a "window" that connects the physical and virtual worlds. The potential benefits for health and fun were made apparent, due to the whole-body interaction capabilities of the single-player system designed.

Our own work also subscribes to a whole-body interaction design philosophy, but extends the definition of Blended Reality to include the user-perceived migration of phenomena from virtual to real, tangible objects able to move and interact with the user in physical space. Rather than designing experiences that orient the user towards the screen, our definition of blended reality also models a continuous environment, yet allows for further natural interaction that is not screen-dependent and is tactile and social. It incorporates the use of a metaphorical portal through which an embodied, appealing robotic character extends out of and onto the screen, providing a blended context for play that orients the multiple users towards each other and the mobile, robotic media in the physical interaction space.

3.3 Mixed reality robot gaming

With the goal of blurring the boundary between physical and virtual reality in order to provide a fused context for play, the authors of [19] implemented an interactive, mixed reality (MR) robot gaming platform. Procedurally animated, real-time computer graphics were synthesized live and displayed on a floor-mounted screen serving as a window into the robot character's 3D fantasy world as well as projected into the interaction space shared by both the human player and the robot. A virtual beach ball with the unique ability to transmediate between the floor space and the space in the screen seamlessly negotiated the interreality boundary and provided the main focus for a simple game of pong. Rather than control a character on-screen like in a traditional video game, the user's joystick tele-operated Miso, a tangible, physically embodied robot character as it played with its virtual companions. Special emphasis was placed on the importance of maintaining perceptual continuity by closely coupling the simulated world's physical laws to our material reality. This preliminary work set forth the technical underpinnings of modeling a singular, fused reality and documented the design considerations used in our current system. Whereas in [19] a virtual ball smoothly moved between both spaces, it was clear the next logical step would be to make a physical robot character appear to move across the physical/virtual divide.

4. BLENDED REALITY

Paul Milgram and Fumio Kishino defined Mixed Reality (MR) as "anywhere between the extrema of the *virtuality continuum*" [14]. In practice, the term refers to the merging of the real and virtual worlds to produce new environments where physical and digital objects co-exist and interact in real time.

We define blended reality as extending mixed reality, enabling the fluid movement of *blended reality characters* between the fully virtual and the fully physical.

This new, kinetically and visually continuous extension of media off the screen and into a mobile and interactive robotic character is made possible through the use of the *interreality portal.* Positioned at the boundary between the virtual and the physical, this enclosure or "robot hutch" opens and closes its servo-controlled doors concealing the robot character as it transits across the interreality boundary. This aids in maintaining the persistent illusion of life for the character as it continuously moves across what users perceive as a singular, fused context of play. Furthermore, it supports an immersive experience without requiring special equipment like head-mounted displays (typically used in mixed-reality robot scenarios) thereby encouraging whole body movement and collaborative face-to-face social play.

4.1 The physical (instrumented) space

The physical play space, measuring approximately 150 square feet provides ample room for up to three children to naturally and actively move through. The floor is cushioned by 42 white foam tiles creating a large floor screen. An aluminum truss system framing the space holds the projectors and Phasespace motion capture cameras used to track objects in the space [17]. Three large, sand-blasted acrylic panels make up the main rear-projection screen that measures 12 feet by 8 feet, providing an immersive display with an aspect ratio of 1.5:1. Three ultra short throw projectors mounted and aligned behind each one of the panels project a bright, stitched and cohesive image. Four short-throw projectors hung from the truss system and oriented downwards towards the white floor mats project the ground image. In addition, custom made wooden platforms attached to the truss hold four audio speakers that distribute sound through the space.

4.2 The digital space

The digital world is rendered in real-time on two dedicated graphics computers. The graphics system renders a total of approximately 7 million pixels (7,077,888) at interactive rates on two computers serving nine screens. One of the computers is allocated to rendering a 2304 x 1024 image stream for the main back wall screen while the other renders four 1024 x 768 image streams properly stitched and projected on to the floor. The remaining two additional outputs are used by the system operator to monitor and make live programming adjustments to the environment. The system makes extensive use of Graphics Processing Unit (GPU) accelerated methods to synthesize the resulting experience.

4.3 Unified coordinate space

Blended reality remaps user interactions in the physical subspace (recorded by the Phasespace motion capture system) into a unified coordinate space, computationally modeled as a superstructure including both the digital and physical spaces. The unified, blended reality coordinate space plots the Z or depth axis so that the zero-crossing matches the threshold point between physical and digital reality. This method clearly delineates the spaces yet permits an animator to smoothly interpolate an animated movement which begins in the digital space and ends in the physical space (and vice-versa) as one continuous movement. The system automatically detects the cross-over point and uses it to dynamically control either the digital, computer graphics character or the physical robot. This unified spatial definition further enables the computation of a superphysics model that takes into account the velocity and trajectory of the blended reality character, matching it's speed (incoming or outgoing through the portal) across the boundary in order to maintain kinetic continuity [6]. Techniques from [19] are extensively used and extended.

4.4 Engineering integrated experience

Blended reality play requires the careful orchestration of robotic and audio-visual media in response to user and environmental input. The core system programming is done in real-time with the results directly accessible and visible at all times. This coding *mise-en-scène* enables the designer to receive immediate feedback by tinkering with the "always on" world.

The system's pipeline is engineered in *Touch Designer*, a highly capable, advanced visual programming environment integral to the execution of this work. Synthesized graphics, event choreography, signal processing flow and inter-application communication run in a constantly evolving and experimental Touch project file. Our code provides the necessary animation for the blended reality character and conditions incoming sensor data, mapping it to fit various internal and external uses. Additionally, the software project hosts a scripted, internal logic responsive to event-based, environmental triggers and human interaction with the robot and the space itself.

A private, local network with static IP addresses enables communication between application modules running on separate computers. To ensure the robot can roam unfettered, the system communicates with the robot over a bi-directional XBEE series2 radio transceiver.

The Open Sound Control protocol was chosen as the main inter-application "glue" due to its widespread acceptance as a standard for connecting the most popular digital content creation applications. Both sound software packages (Ableton LIVE, Max/MSP) and the system's control hub, Touch Designer, communicate with each other over OSC.

5. BLENDED REALITY CHARACTER

A Blended reality character is designed to maintain visual and kinetic continuity between the fully virtual and the fully physical. The character is persistent in that it can only exist in one location (or subspace) at a time.

All appearance, movement, actions and attitudes must consistently build towards the character's style or personality. This essence must be maintained across the blended reality context of interaction in order to trigger a social response. Research shows a character doesn't have to look like a real person to give and receive real social responses [18]. Information about personality can come from anywhere. Inconsistencies in the presentation of characters, however, will diminish the purity of personality and thereby contribute to confusion and even dislike. The internal consistency of the character is doubly complicated by its dual representation in blended reality. A strong, consistent personality embodied in a simple form helps reduce complexity and delivers on expectations. By design, interactions with a blended reality character should be simple and causality must be clearly shown or it will fail. The approach used in designing this blended reality character took as its departure point the concept that children should be able to use what comes naturally as situated learners in the real world with real developmental needs and an insatiable appetite for play.

5.1 The Alphabot

Over three-hundred years ago, English philosopher John Locke in 1693 made one of the first references to alphabet nursery blocks "dice and playthings, with letters on them to teach children the alphabet *by playing*" (emphasis added) [13]. In the early 1700s, Friedrich Wilhelm August Froebel, the pioneer of the kindergarten movement introduced alphabet blocks and in 2003 the alphabet block was inducted into the National Toy Hall of Fame [2][15].

Alphabet blocks were one of the first educational toys for children. They are a mainstay of early learning and nearly every child has spent at least some time playing with alphabet blocks building critical social, creative, cognitive, motor and literacy skills. Traditionally, the tactile, tangible letter cut into the side of a block is a shape that can be traced by the finger of the child to form cross-sensory, multi-modal memories of the symbol.

Alphabot, an instance of a blended reality character, fashioned after a familiar wooden letter block was designed to be fun, safe and have a modular front face that could accept any symbol reacting to user input both on and off-screen.

Figure 2. (a)Alphabot and symbol tokens (b)virtual Alphabot

5.1.1 Physical Alphabot

The physical embodiment of the Alphabot blended reality character is a 12 inch cubed robot designed to resemble an alphabet block. The wooden robot is proportionally smaller than the youngest, standing child and moves predictably and slowly thus portraying a non-threatening demeanor. It enables interactions of many different kinds in the physical environment and can motivate specific play actions through immediate feedback. The top face is as an open tray, left experimentally ambiguous as white space in the design, inviting children's suggested use. Around the outside of the robot are four, small active-IR LEDs used in conjunction with the Phasespace motion capture system to localize the robot in space. The robot is tele-operated by an adult caretaker observing the child-robot interactions. The front velcro face of the robot, structurally supported by a thin sheet of clear acrylic, motivates specific action. Children can experiment with attaching and detaching wooden symbols with embedded RFID tags recognized by the robot and communicated wirelessly to the blended reality media system. Front and rear caster wheels supplement the robot's two DC motors actuating two wheels in a differential drive configuration.

5.1.2 Alphabot symbol token accessories

Alphabot's collection of tangible symbol tokens are laser-cut out of the same, high-quality Baltic Birch wood that the robot's body is made with. Serifa, a beautiful serifed font created by prominent Swiss typeface designer Adrian Frutiger in 1967, was chosen for its highly legible quality. Formative research by the Children's Television Workshop on *Ghostwriter,* pointed to the importance of resisting creative and unusual letter shapes and non-standard orientations when presenting text to children [5]. As Alphabot's symbols are intended to exist both in the real world as tangible letterforms as well as animated on-screen, research-validated best practices are employed to ensure the system is effective at clearly conveying information to children [4].

The system's symbols are a subset of numbers, shapes and letters including international characters. Following fabrication, each symbol is sanded down to smooth out the contours and coated with bright, non-toxic paint. Strips of adhesive velcro on the back provide an easy way to affix and detach the symbols from the robot's front face. Care was taken to design the symbols in a way that makes holding them a pleasure. The intention is to provide children with an opportunity to explore, manipulate and reflect upon the use of artifacts and their possible effects in blended reality play. Two blank square symbols coated with chalkboard paint invite customizations. Children of all ages have enjoyed using chalk to draw faces (or anything at all) for Alphabot.

5.1.2.1 RFID tags

To couple the symbols to the blended reality play experience, 16mm thumbnail-sized RFID button tags are inserted into the back face of each symbol. Each tag comes with a unique 32-bit ID code and is not reprogrammable. The carrier frequency of the tags is 125kHz which works well with the RFID reader (ID Innovations' ID-20) internally mounted inside the robot behind its front face. The range for the RFID reader to correctly identify the button tags is approximately two inches. The reader can read tags through various materials including wood. In this case, the RFID reader correctly identifies tags through the acrylic and velcro layers on the robot's front face.

Identified symbols immediately trigger the robot's LED to light up Alphabot's face. This lets the child know that the robot has recognized their input. The identified symbol is also sent wirelessly to the main environment control computer triggering visible and audible responses throughout the environment.

5.1.3 Digital Alphabot

Alphabot's virtual representation is designed to be as consistent as possible with the physical robot version of the character.

5.1.3.1 3D geometry

A simple geometric primitive box is all that's needed to represent Alphabot on the screen. This makes the geometry component of Alphabot's digital representation extremely lightweight and easy to render in real-time. In addition to leaving room for other important computations, keeping Alphabot geometrically simple creates opportunities for future migration of the blended reality character onto mobile platforms and other devices lacking 3D graphics prowess.

5.1.3.2 Surface appearance

Digital Alphabot's surface appearance leverages a common environment mapping technique known as cube mapping optimized for real-time rendering [8]. To match the appearance of the robot in physical reality, photographs of the physical character wearing each symbol are used as the source for the cube maps. A straight-forward, stock OpenGL Shading Language (GLSL) shader running on the graphics processing unit (GPU) permits the character's surface appearance to adapt to changing lighting conditions in the world, further integrating digital Alphabot into the final blended reality rendered scene [20].

In order to maintain character visual continuity throughout the blended reality play context the digital version of Alphabot's face displays the current symbol placed on the physical robot. To accomplish this, the robot's embedded software program continuously polls the RFID reader and wirelessly transmits a symbol ID to the main graphics and environment control computer rendering digital Alphabot. Upon receipt, the value is used to switch between all of the possible cube maps depicting the various symbols in the set. There is no noticeable latency as this process happens at faster than interactive rates.

5.1.4 Animating Blended Reality characters

Alphabot is animated using traditional keyframed animation as input into a procedural blending subsystem coupled with real-time procedural motion synthesis and performance animation techniques. Animation clips created in industry-standard, 3D animation content creation applications (e.g. Autodesk's Maya) are exported as FBX files, a platform-independent 3D data interchange format, and fed into a real-time procedural animation blending engine. This method allows for the integration of hand-crafted 3D animation clip playback, sequencing and event-triggered blending.

5.1.5 Limits of current system

The inexpensive hall-effector's sensors mounted to the back of the robot were unable to provide a stable quadrature signal in the current version of the robot. This restricted the possibility for making the robot autonomous. During the play test studies with children, however, the choice to have a research assistant tele-operate the robot helped acquiesce any parental concerns about the robot's safety around children. As the robot is currently being localized through the motion capture system, a location vector as well as an orientation (heading) quaternion are given, enabling future development of semi-autonomous behavior.

6. INTERACTION DESIGN GUIDELINES

As in traditional animation practice, everything starts with a story and a story often begins with a setting. We choose a setting of three hills on the outskirts of Alphabot city. The curved shape of the hills help enhance the dramatic effect of having Alphabot come out and play- moving from screen space representation to its mobile robot embodiment.

As part of the design philosophy we ensure that the core experience is designed for the appropriate number of child users. Leaving enough space for active, healthy children, we limit our interaction design and experience to a maximum of three children playing concurrently in the physical space.

Overall, safety is of utmost concern in all aspects of the work. The robot's motor speeds and thus movement are constrained and checked for safety at two different levels. Additionally, the physical interaction space is outfitted with soft, padded flooring.

Regarding visual style, when creating visual media for the system we ensure all contributions remain in cannon with the overall aesthetic design. By foregoing the use of a photorealistic style we opted for a naif, painterly world with plenty of white space for both the children's imaginations and the possible future inclusion of their own drawings and content.

7. EVALUATION

The study's goal was to evaluate the continuity of the blended reality character and its potential to engage young children in imaginative play scenarios. The design of the study was based on a classic comparison model.

7.1 Experimental conditions

Condition 1: A blended reality scenario in which the child plays with Alphabot (blended reality character) in its environment by exploring the causal effects of placing one of six tangible symbols on the robot and having it physically move into its hutch and watching its virtual representation continue up a hill on the screen in digital space. In this condition we tested 17 children: 11 boys and 6 girls between the ages of 3.5 to 7.

Condition 2: A video-game, virtual scenario of blended reality play with the Alphabot in which the child sits at a desktop computer and plays a symbol-placing game with the Alphabot (screen only) character using a mouse as input. In this condition we tested 17 children: 11 boys and 6 girls between the ages of 3.5 to 7.

The tasks in both conditions were analogous.

7.2 Observational methods

Unobtrusive audio/video recording observational methods were used to document each play test. A three-camera setup was used to record: 1. a wide-angle, rear shot, from the point-of-view of the child that framed the entire play scenario 2. a close-up, front shot of the child to enable observation of the child's face and 3. a master shot of the child during the post-play test interview.

7.3 Participants

We play tested 34 children three and a half years old to seven years of age. The development of imaginative play in children begins around two and reaches its peak between the ages of five and seven. We therefore designed the study's blended reality play scenario to be age-appropriate for children in that age range [21].

Seventeen children in each condition play tested the system. The age and gender distribution were evenly matched across both conditions. Children tested in one of the conditions were not permitted to play test the other condition.

7.4 Pre-play test protocol

Upon arrival of the child and their adult guardian or parent to the lab, a research assistant would greet them in the lobby and describe the play test to the parent. Parents were invited to observe the whole experiment but requested to pretend they were busy reading a magazine, in an effort to avoid having the child check in with them or seek their approval during the play test.

7.5 Play test scenario

Each play test lasted approximately ten minutes. During the play test a research assistant would begin by introducing the child to the Alphabot character on-screen (in digital space). The children were then invited to tickle the character by touching the character on the screen with their hands (Condition 1) or with the mouse-pointer (Condition 2). The researcher would model this interaction by tickling the on-screen character causing it to spin around or hop up and down. Following a verbal request for Alphabot to "come out and play," the character would appear to descend down the virtual hill in the on-screen, digital space and would promptly emerge into the physical space (Condition 1) or the virtual physical space (Condition 2), through the hutch or interreality portal. The character would then move around freely in the physical space stopping in front of the child and symbols spread out on the floor (Condition 1) or on-screen (Condition 2).

The researcher would proceed by demonstrating the process of changing the symbol on the character's face. In turn, the character would react by moving towards the hutch, the doors would open allowing the character to pass through and the character's virtual representation would be seen moving up the hill in digital space. Arriving at the top, the character's symbol would be displayed on its face where the child placed it, and express itself clearly, establishing causality. Throughout the main portion of the play test the child was given the autonomy to play freely and choose any symbol in any order or sequence, imagining the possible outcomes.

As the play test came to an end the sky in the digital space would darken simulating a sunset. The blended reality character would return to the hutch and move up to the hill and the child would be informed that Alphabot was "going to take a nap".

7.6 Interview protocol

Immediately following the end of the play test, a research assistant would ask the children if they wouldn't mind answering a couple questions. They would then escort the children out of the space to a separate area with a child-sized table and chairs.
The interview questions were structured to be brief, to the point and make use of age-appropriate language.

We also presented the children with a simplistic diagram of the blended reality play space depicting the screen, hutch and physical play space, including the framing of the screen and physical space. Children were asked to point to the location on the diagram in response to various questions. The interviewer coded the children's responses using a consistent nomenclature scheme across both conditions.

In addition, we asked each child if they wanted to give us any suggestions on how to improve Alphabot. This question also helped us determine their level of engagement and belief in the character.

Finally, we asked them to draw a picture for Alphabot and suggested that we could give it to him. We gave each child space to freely associate and tell us a story (reflecting on their experience).

7.7 Validated evaluation instrument

To help children understand what was expected of them, response options were presented to them in picture form using a Likert-type scale. For our study, we found that the Smiley Face Assessment Scale (SFAS) was a useful, validated evaluation instrument that removed confusion for the child. Explaining the 5-point continuum to the child by pointing at each face and describing them as "very happy, happy... sad," helped but was often not a requirement to get a clear answer. Hopkins and Stanley have shown that pictorial response scales are sometimes more effective for assessing attitudes, especially for children [10].

7.8 Results

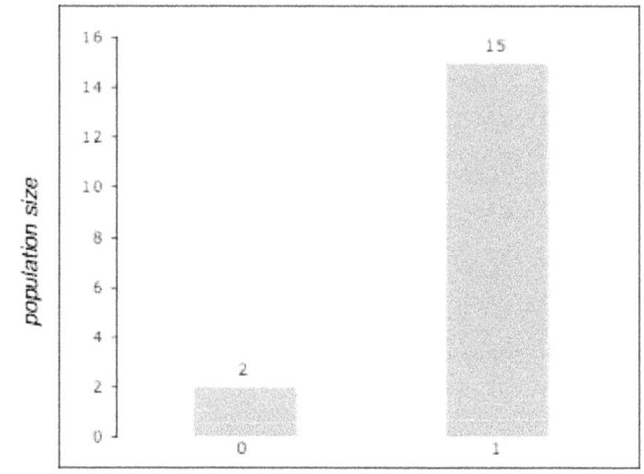

belief in blended reality character's persistence

7.8.1 Acceptance of Blended Reality character

Based on experimental results, our findings show that children aged three and a half to seven have accepted the idea of a blended reality character. To assess the character's continuity from physical space to digital space we asked children to identify the total number of Alphabot characters. Children who believed in the character's persistence across the blended reality play context were assigned a value of one, whereas those who identified the incorrect number of characters were assigned a zero value. Positive affirmation of the blended reality character's continuity across multiple forms of media (screen and robotic), also indicates belief in its world as the two are inextricably tied together.

7.8.2 Impact on character engagement and play

To uncover the core differences we studied the post-play test video interviews and tallied the number of children's responses to the question: "What would you want Alphabot to do?" This helped give us an indication of how deeply engaged they were with the blended-reality character.

Based on the large number of imaginative suggestions received from the children who participated in the blended reality play experience (condition 1) in comparison to the low number of answers in condition 2, we found that the blended reality play experience had a significant impact on children's post-play test verbal utterances and imaginative suggestions.

Following play tests in the first condition (blended reality play) 87% of the children tested made detailed suggestions on what they would want Alphabot to do, while 13% did not respond. In comparison, only 3 out of 17 children or 18% of the children who play tested condition 2 (virtual blended reality) replied with imaginative suggestions. Eighty-two percent of the children who did not experience blended reality play with Alphabot abstained from answering the question: "What would you want Alphabot to do?" showing a marked decrease in interest and engagement with the character when confined to the screen.

These results suggest that for the population tested, blended reality play experiences lead to a deeper engagement with a character able to fluidly migrate between a screen and physical reality in the form of a mobile robot in comparison to a strictly screen-based character. Furthermore, providing a blended reality play experience for children between the ages of three and a half and seven results in a notable increase in the number of post-play imaginative suggestions and creative ideation. Additionally, the study revealed a noticeable difference in the imaginative quality of the suggestions in both cases. In the control experiment (condition 2), children suggested that Alphabot should be able to dance and jump more. In condition 1 (blended reality) children also wished that Alphabot could dance and jump, as well as fly, play soccer, be a wind-up jack-in-the-box toy and go upside down. These qualitative differences also indicate deeper engagement with the blended reality character.

7.8.3 Situating the character

One of the compelling results the study uncovered in connection with the acceptance of the blended reality character was the children's views on where the blended reality character lived and played.

Asked to point to a spot in a diagram depicting the entire blended reality play context including the physical space, hutch and digital space, 65% of the children in condition 1 (blended reality) replied that the character lived in digital space and 35% replied that it lived in the hutch. None of the children replied that the character lived in physical space, despite playing with and touching the physical robot.

It is difficult to say whether the children conceived of the hutch as part of physical reality or as a distinct, liminal space between physical reality and the digital space on screen. Interestingly, the responses varied by gender, with the majority of boys (82%) asserting the character lived in digital space whereas the majority of girls (67%) replied that the blended reality character lived in the hutch.

Although almost two-thirds of the children in condition 1 thought the character lived in digital space, fifty percent replied that it played in physical space (with them). The second most common response (31% of the children) held that the character played in digital space. None of the children answered that it played in the hutch. Three of the girls did not answer making a gender comparison, in this case, difficult.

In the control experiment (condition 2: virtual blended reality video game), 65% of the children replied that the character played in digital space while 29% asserted that it played in virtual physical reality. Nine percent answered that it played in the hutch.

One of the boys in condition 2, got up from his chair and looked behind the flat-screen computer monitor when Alphabot went into the digital (screen space) in the game.

Results from this part of the study may prompt further investigation together with a deeper consideration of children's spatial reasoning abilities in light of their age and individual developmental stage. Given the unique spatial arrangement that blended reality affords, it may prove fruitful as a medium to explore what Dr. Howard Gardner, founder of the multiple intelligences theory, terms *spatial intelligence* as it emerges in young children [Gar83].

7.8.4 Interreality transit

The trend in providing richer detail and more imaginative responses in post-play test interviews of children who experienced blended reality play, in contrast to those who play tested the virtual version (condition 2), prevailed as evidenced by responses to the question: "What happens when Alphabot goes from "here" (pointing at the physical space in the diagram) to "there" (pointing at the digital space)?

In condition 2, children offered unvarnished, factual responses like, "It's triggered by a symbol" and "He changes himself". By contrast, blended reality play testers (in condition 1) came up with unexpected and imaginative explanations like, "He takes a train to get from here to there." Another explained that Alphabot had jumped through and yet another child simply answered, "Noise." Suffice it to say that making sense of these answers is challenging at best. What is apparent is the change in tonality between the more realistic answers given in condition 2 compared to the more inventive descriptions offered by children who engaged in blended reality play.

7.8.5 Symbol use

As an initial step towards a pedagogic use of Alphabot's symbol system to guide children's informal learning, we asked them to recall the effect of placing a symbol on the character in both conditions. Sixty-four percent of the interviewed children in both conditions verbally recalled a symbol. Some of the children that did not verbally recall a particular symbol during the interview, drew them when they were given time alone to reflect and draw freely. One child drew the Japanese symbol on Alphabot and added jet packs as well as two letter "P"s, a symbol not found in the play test set. Several children drew alphabots with hearts. Another child drew Alphabot with the number three and yet another with the letter "a". A four year old girl told us "I hope Alphabot gets to see my picture, I'm drawing alphabets."

7.8.6 Sustained appeal of blended reality play

Despite the relatively short (ten minute) duration of the blended reality scenario tested, several children expressed a desire to continue engaging in the experience. "I want to come back and play with Alphabot," one child mentioned. Another stated, "I want to play with Alphabot's friend in Japan". Children in the control experiment did not express similar wishes. They did not ask to replay the condition 2 video game of virtual blended reality. In comparison, a condition 1 play tester grew impatient with the interview process and asked, "Can we play with Alphabot now?" The long-term appeal of blended reality play is yet to be determined. Initial evidence, however, points towards children's desire to revisit and extend their play time with Alphabot in blended reality.

7.8.7 Children's drawings

Overall, children drew more pictures following their play experience in condition 1 (blended reality) than after condition 2 (virtual blended reality). In condition 1, children drew the blended reality environment often depicting themselves and the character together. In contrast, in condition 2 children drew the Alphabot character alone and did not draw themselves.

One of the interesting themes that emerged from children's drawings after experiencing condition 1 was the apparent switch or blending of spaces in their illustrations. Hills and flowers that they experienced existing strictly in the digital world were drawn in the representation of the physical space. In one case, a five year-old boy drew the whole blended reality context seemingly from the inside out.

These drawings do call for further analysis by an expert in the field. It is important to note, however, that these drawings should be respected for their own artistic merit and caution should be used when making interpretive assumptions.

8. CONCLUSION

The play test study's findings unequivocally demonstrate that young children (3.5 to 7 years old) have accepted blended reality characters and believe in their continuity and persistence across multiple forms of media.

As indicated through interviews and drawings, the play tests reveal significant qualitative and quantitative differences in children's engagement with blended reality characters over strictly screen-based characters. Deeper engagement is indicated by the length of verbal utterances, the more descriptive and imaginative qualities of children's responses to interview questions and the number of drawings produced.

In blended reality, children experience a deeper sense of immersion. The difference in the number of post-play test drawings in which children depicted themselves playing with the character in its blended reality world suggests that children see themselves as co-protagonists, immersed in blended reality play. Belief in the continuity and persistence of the blended reality character seems to be inextricably tied to belief in the persistence of the character's world.

The desire to continue engaging in blended reality play as expressed by children's verbal requests to revisit and extend their play time with the character shows the potential for development of an informal learning platform with sustained appeal to young children.

9. ACKNOWLEDGEMENTS

Thank you to all of the members of the Personal Robots Group for their support. Special thanks go to Adam Setapen for designing and fabricating the robot hutch and collaborating with Natalie Freed on the robot control software and embedded code. Additionally, Natalie Freed was instrumental in designing the robot's electronics. Many thanks to Professor Joe Blatt at HGSE and to the students of T-530 and HT-123. Thank you Fardad Faridi for providing beautiful artwork. Finally, thanks to the eval team, Julia Goldstein, Paulina Mustafa, Monica Bueno, parents and children for your trust in this work. The vision and community of the MIT Media Lab made this possible.

10. REFERENCES

[1] Breazeal, Cynthia. *Designing Sociable Robots.* MIT Press, Cambridge, MA, 2002.

[2] Brosterman, N., *Inventing Kindergarten.* Harry N. Abrams, New York, NY, 2002.

[3] Edwards, C. et al.,*The Hundred Languages of Children – The Reggio Emilia Approach – Advanced Reflections.* Ablex Publishing, Westport, 1998.

[4] Fisch, S., *Characteristics of effective materials for informal education: A cross-media comparison of television, magazines, and interactive media. of Effective Materials for Informal Education*. MediaKidz Research & Consulting, 2004.

[5] Ghostwriter Research, *Ghostwriter font study*, (Unpublished research report), Children's Television Workshop, New York, 1992.

[6] Ginatatus,V. et al., *Experimental evidence for mixed reality states in an interreality system*, Phys. Rev. E 75, 057201, 2007.

[7] Grey Walter, W., *An Imitation of Life,* Scientific American, 182(5): pp.42-45. 1950.

[8] Greene, N. *Environment mapping and other applications of world projections,* IEEE Computer Graphics Applications 6, 1986.

[9] Hirsh-Pasek et al., *A Mandate for Playful Learning in Preschool: Presenting the Evidence*, Oxford University Press, Oxford, UK, 2008.

[10] Hopkins, K. D., and Stanley, J. C., *Educational and psychological measurement and evaluation* (6th ed.). Englewood Cliffs: Prentice Hall Inc, 1981.

[11] Huynh, D. et al., *Exploring User Experience in "Blended Reality",*CHI, Montreal, Canada, 2006.

[12] Let's Move, www.letsmove.gov, 2011.

[13] Locke, J., *Some thoughts concerning education,* In Works of John Locke (Vol. 9)*,* Thomas Tegg, Sharpe and Son, London, 1823.

[14] Milgram, P. and Kishino, F., *A Taxonomy of Mixed Reality Visual Displays,* IEICE Transactions on Information Systems e 77-D, 1321-1329, Vol 12, 1994.

[15] *National Toy Hall of Fame.* http://www.toyhalloffame.org/toys/alphabet-blocks, 2011.

[16] Nielsen, A2/M2 *Three Screen Report: Television, Internet and Mobile Usage in the U.S.,* Vol. 5, 2nd Quarter, 2009.

[17] Phasespace, Inc. http://phasespace.com, 2011.

[18] Reeves, B. and Nass, C., *The Media Equation,* CSLI Publications, New York, NY, 1986.

[19] Robert, D. et al., *Exploring Mixed Reality Robot Gaming,* ACM TEI '11, Madeira, Portugal. 2011.

[20] Rost, Randi J., *OpenGL Shading Language,* 1st ed., Pearson Education, Inc., 2004.

[21] Singer, D. and Golinkoff, R. and Hirsh-Pasek, K., *Play = Learning*, Oxford University Press, New York, 2006.

[22] Thai, A.M. et al., *Game Changer: Investing in digital play to advance children's learning and health*, The Joan Ganz Cooney Center. www.joanganzcooneycenter.org/publications,2009.

[23] UN General Assembly, *Convention of the Rights of the Child*, United Nations, Treaty Series, Vol. 1577, p.3, 1989.

[24] Wistort, R., *TofuDraw: Choreographing Robot Behavior through Digital Painting,* MIT Media Lab Ms Thesis, 2010.

Modelling Empathic Behaviour in a Robotic Game Companion for Children: an Ethnographic Study in Real-World Settings

Iolanda Leite*, Ginevra Castellano†, André Pereira*, Carlos Martinho*, Ana Paiva*
*INESC-ID and Instituto Superior Técnico
Technical University of Lisbon, Portugal
{iolanda.leite,andre.a.pereira,carlos.martinho}@ist.utl.pt, ana.paiva@inesc-id.pt
†HCI Centre, School of Electronic, Electrical and Computer Engineering
University of Birmingham, United Kingdom
g.castellano@bham.ac.uk

ABSTRACT

The idea of autonomous social robots capable of assisting us in our daily lives is becoming more real every day. However, there are still many open issues regarding the social capabilities that those robots should have in order to make daily interactions with humans more natural. For example, the role of affective interactions is still unclear. This paper presents an ethnographic study conducted in an elementary school where 40 children interacted with a social robot capable of recognising and responding empathically to some of the children's affective states. The findings suggest that the robot's empathic behaviour affected positively how children perceived the robot. However, the empathic behaviours should be selected carefully, under the risk of having the opposite effect. The target application scenario and the particular preferences of children seem to influence the "degree of empathy" that social robots should be endowed with.

Categories and Subject Descriptors

I.2 [**Artificial Intelligence**]: Robotics—*Commercial robots and applications*; H.5 [**Information Interfaces and Presentation**]: User Interfaces—*User-centered design*

General Terms

Human Factors, Experimentation

Keywords

Social Robots, Affect Recognition, Empathy, Children.

1. INTRODUCTION

Robots are becoming closer and closer to humans. Yet, for them to truly be part of our routines and daily lives, they must be able to interact with us, humans, in a similar way as we interact with each other [4]. Demographic trends, the decreasing costs in technology and the emergence of commercial robots for domestic settings are some of the reasons why research in this area is becoming progressively more relevant. There are still many open challenges when developing social robots for real-world environments. For example, the role of affective interactions is still unclear. As social beings, relationships are very important to humans. It is often argued that empathy, defined as "an affective response more appropriate to someone else's situation than to one's own" [15, p. 4], facilitates the creation and development of social relationships, increasing similarity, fondness and affiliation [1]. Would social robots also benefit from being empathic In the field of HRI, there has been an increasing interest in studying different forms of empathy on user's attitudes towards robots [10, 23, 19]. However, most of these results were obtained in laboratory settings and the subjects were adults. When conducting studies with children, there are differences that should be carefully considered both in terms of methodology and data collection methods. For instance, with a few exceptions (e.g., [29]), most of the existing validated questionnaires are tailored for adults.

In this paper, we present a study that aims to understand how children perceive and interact with an empathic social robot in their school environment. To do so, we developed a robotic companion capable of recognising some of the children's affective states and react empathically to those states while playing a chess game. The presented study took place at a Portuguese elementary school where children interacted with one of the three versions of the robot: a neutral version, in which the empathic behaviours were not enabled, and two empathic versions that differed in the mechanism of selecting the robot's empathic behaviour. The results suggest that the empathic behaviours of the robot had a positive impact on children's perception of the robot.

The following section presents a state-of-the-art on HRI studies in real-world environments and empathic social robots. The scenario used for the study is then described. After that, we present the methodology used to collect and analyse the data, followed by the report of the most relevant findings of the study. Finally, we summarise the lessons learned and present some conclusions.

2. RELATED WORK

Our goal is to qualitatively evaluate children's reactions to an empathic robot in a real-world environment. As such, this section presents a brief literature review in two distinct but interconnected areas: HRI studies in real-world settings, with special emphasis on examples reporting children's interaction with robots in schools, and studies involving social robots with empathic capabilities.

Recently, there has been a growing interest in studying the interaction with robots in real-world settings such as homes [12, 28], workplaces [22], elderly-care facilities [24] or schools. In this last category, Kanda et al. [16] reported an exploratory long-term study with elementary school Japanese children and a social robot (Robovie) capable of calling children by their names, adapting its behaviour to each child using a pseudo-development mechanism (the more a child interacts with the robot, the more different behaviours the robot displays to the child), and also self-disclosure behaviours. Robovie was capable of engaging children over the two-month period that the experiment lasted, although with a slight decay from the second week. The children's main motivation for interacting with the robot was to become friends with the robot. Kozima et al. [17] undertook a similar study to investigate the interaction between toddlers and Keepon, a small robot designed to interact with children through non-verbal behaviours. The authors report that children's understanding of the robot changed over time from a mere "moving thing" to a "social agent".

The first empathy studies in HRI focused on mimicking the user's affective state, a particular aspect of empathy also designated as emotional contagion. For instance, Hegel et al. [14] reported a study with an anthropomorphic robot that recognises the user's emotional state through speech intonation and then mirrors the inferred state using facial expressions. The results suggest that users who interacted with this version of the robot found the robot's responses adequate both in terms of appropriateness to the social situation and timing, comparing with subjects who interacted with the robot without affective expressions. In another emotional mimicry study [23], it was found that most participants considered the interaction more satisfactory than participants who interacted with a version of the robot without mimicking capabilities. More recently, Cramer et al. [10] studied how empathy affects people's attitudes towards robots. In this study, there was a significant negative effect on user's trust towards a robot that displayed inaccurate empathic behaviour. Conversely, participants who observed the robot displaying accurate empathic behaviours perceived their relationship with the robot as closer. A relevant finding in empathy research is that empathy is correlated with social supportive behaviours [15]. Saerbeck and colleagues [25] investigated the effects of social supportive behaviours of an iCat robot on children's learning performance and motivation. The results indicate that simple manipulations in the robot's supportiveness, while maintaining the same learning content, increased student's motivation and scores on a language test.

Although significant research has been undertaken in these areas, there are still numerous issues that require further investigation. With this paper, we intend to add another contribution in this field by analysing children's interactions with an empathic social robot in a school environment.

3. SCENARIO

The scenario that we developed to study empathy in social robotic companions consists of an iCat robot that plays chess with children and provides affective feedback to the moves played on an electronic chessboard. The robot interacts with children by displaying facial expressions and verbal utterances. The interaction starts with the iCat waking up and inviting the child to play (e.g., by saying "let's play chess"). After every child's move, the robot reacts emotionally to that move by displaying a facial expression [21]. Then, it asks the user to play its move by saying the move in chess coordinates. After that, the robot waits for the children's next move, and so on. The robot's behaviour is totally autonomous, except for an initial parametrisation where the name of the child needs to be typed in (the iCat uses the child's name in some of the utterances).

Previous user studies using this scenario have suggested the need of integrating empathic behaviours in the robot [18]. Taking this into account, we developed an empathic model that contains two main parts, an affect detection system that recognises the valence of the user's feeling and an action selection mechanism that chooses the best empathic response to display at a certain moment, for a particular user.

3.1 Affect Detection

The empathic model includes an affect detector that allows the robot to infer the valence of the feeling experienced by the children. The affect detector, based on Support Vector Machines (SVMs) [7], processes in real-time behavioural and contextual features, providing as output a probability value for each of the following valence conditions: positive, neutral and negative. The system was trained with features extracted from the Inter-ACT corpus, an affective multimodal video corpus containing previous interactions of a group of children with the iCat robot [6]. Previous work in this scenario showed that children's non verbal behaviour can be used to predict their affective state [26]. Specifically, the behavioural and contextual features used as input for the system are:

1. *Probability of smile* – the average value of the probability of the user's smile over the previous six seconds of the interaction; the smile detector uses Support Vector Machines and processes geometric features derived from 2D and 3D facial landmarks extracted using the Seeing Machines faceAPI sofware.
2. *Eye gaze* – the user's eye gaze is estimated based on position and rotation values of the head, relative to the camera, tracked by faceAPI; it provides information on the amount of time the user has looked at the robot in the previous six seconds of the interaction.
3. *Game state* – a quantitative measure that describes who is in a position of advantage in the game [5].
4. *Game evolution* – an indicator of how the game is evolving for each player [5].

3.2 Empathic Strategies

When the affect detection system recognises that the user is not experiencing a positive feeling with a high probability (taking into account the features described in the previous subsection), in addition to the affective facial expression, the robot employs an empathic strategy. The empathic strategies implemented in the robot, inspired in the literature of empathy and prosocial behaviour [9], are the following:

1. *Encouraging comments* – include comments such as "don't be sad, I believe you can still recover your disadvantage".
2. *Scaffolding* – providing feedback on the user's last move and, if the move is not good, let the user play again.
3. *Offering help* – suggesting a good move for the user to play in his or her next turn. If the user accepts the help, he/she touches one of the robot's paws and the iCat tells the move.
4. *Intentionally playing a bad move* – for example, playing a move that allows the user to capture a piece.

Each one of these strategies might have a different impact in different users. To address this issue, the action selection mechanism uses a reinforcement learning algorithm [2]. This way, the robot can adapt its empathic behaviour by trial and error considering the previous reactions of a particular user to an empathic strategy (for more details on the adaptive algorithm see [20]). Thus, we are investigating not only whether the introduction of empathic behaviours improves the interaction, but also if the adaptation of the robot's empathic behaviour over time plays an important role as well. To evaluate these two aspects, for this study we developed three different parametrisations regarding the robot's behaviour:

1. *neutral* – the robot does not exhibit any empathic strategies. It simply comments the moves in a neutral way (e.g., "you played well", "bad move", etc.).
2. *random_empathic* – when the user is experiencing a negative feeling, the iCat selects one of the enumerated empathic strategies randomly.
3. *adaptive_empathic* – when the user is experiencing a negative feeling, the empathic strategies are chosen using an adaptive learning algorithm.

In addition to differences in selecting the empathic strategies, the robot's affective feedback is also different in the neutral condition. As empathy involves putting oneself in the other's perspective [15], in the two empathic conditions the robot's affective behaviour is user-oriented (i.e., it gets happy if the user plays good moves and sad if the user plays bad), while in the neutral condition the iCat will be sad if the user plays good moves and the other way around. This will be reflected in the facial expressions displayed after every user's move.

4. METHODOLOGY

To analyse how children respond to the empathic behaviours implemented in the robot, we conducted a field experiment with three different control groups (neutral, random_empathic and adaptive_empathic), as described in the previous section. With this ethnographic study, not only we intend to capture children's reactions to the robot in a real setting, but also to identify patterns of responses and explore new ways in which we can improve the design of empathic robots.

Figure 1: etting of the Experiment.

4.1 Experimental Setting

The study was conducted in a Portuguese elementary school where children have two hours per week of chess lessons. As illustrated in Figure 1, the experimental setting comprises the robot, an electronic chessboard, a computer where all the processing takes place, a webcam and a video camera. The webcam was used by the face tracking software to capture the children's expressions that served as input to the affect detection system, while the video camera was used to record all the interactions. There was a third video camera recording the interactions from a lateral perspective that does not appear in Figure 1. The setting was installed in the room where children have their weekly chess lessons. The goal was to integrate the robot in the natural environment where children usually play chess.

4.2 Participants

A total of 40 participants (19 male and 21 female), with ages between 8 and 10 years old, took part in the study. Children from this age group are sufficiently developed to participate in semi-structured interviews and are able to answer questions with some consistency [3]. Participants were randomised over the three control conditions. Since we followed a between groups design, our final sample consisted of 14 children in the adaptive_empathic condition, 13 in the random_empathic and 13 in the neutral condition. As the number of subjects in each condition is not equal, when appropriate, we present the results using percentages rather than frequencies.

One week before the study, parental consent forms were delivered to the parents by the school's chess instructor. The consent from contained a brief description of the project and the goal of the experiment, including a note mentioning that the sessions would be video recorded for future analysis. All parents authorised their children to participate in the experiment.

4.3 Data Collection

Participants were guided to the room where the setting was installed and were instructed to sit in front of the robot. After the calibration phase for the face tracking system to work properly (which lasted no more than one minute), children played a chess exercise with the iCat. The exercise was the same for all participants, and was suggested by the chess instructor so that the difficulty was appropriate for the children. Two experimenters were in the room with the child controlling the experiment and observing the interaction.

Table 1: tructure of the interview.

Question	Response Type
1. Do you believe that iCat recognised how you felt during the game?	yes-no
1.1. Why do you think so?	open
2. Did you experience any particular emotions during the game?	yes-no
2.1. Which emotions?	open
2.2. Do you think the iCat felt similar emotions?	yes-no
2.3. Why?	open
3. Do you think the chess game is more fun because of the iCat (than playing with a computer)?	yes-no
3.1. Why?	open
4. What are the main differences of playing against the iCat compared to playing against your colleagues?	open
5. How could we further improve the iCat?	open
6. Identify advantages and disadvantages of playing with the iCat.	open

Each child played, on average, 15 minutes with the robot. After that period, depending on the state of the game, the iCat either gave up (if it was in disadvantage) or proposed a draw (if the child was loosing or if none of the players had advantage), arguing that it had to play with another user. During this phase, in addition to direct observation from the two experimenters in the room, two video cameras recorded the interaction from frontal and lateral perspectives.

After playing with the robot, participants were oriented to another room where they were interviewed by another experimenter. The interviews were semi-structured, containing initial yes-or-no questions followed by open-ended questions that allowed children to justify and elaborate their answers. The structure of the interview is depicted in Table 1. Each interview lasted on average 10 minutes. The interviews were audio recorded and later on transcribed. The interview transcriptions were used as primary sources of data for content analysis.

4.4 Data Analysis

The interview transcriptions were coded according to the methodology proposed by Chi [8], using an iterative coding process. An initial coding scheme was obtained while reading and highlighting the main concepts in the text. Subsequent iterations allowed us to refine the initial coding scheme. After that, related codes were grouped into themes and categories. To ensure reliability during the coding process, two complete coding iterations were performed with an interval of one month by the same coder. Differences between these two coding steps were addressed by asking another coder to rate that segment. Finally, using the developed coding scheme, a comparative analysis between the three different control conditions was performed to understand how the robot's empathic behaviour affected children's perception of the interaction.

5. RESULTS

The results presented in this section are divided by the main themes in the open-ended questions of the interview. In the first two themes (questions 1 and 2), our goal was to understand whether children perceived the empathic behaviours expressed by the robot and how such behaviours had an impact on children's own emotions. After that, we looked for the main differences between playing with the iCat and playing with a computer or a colleague (questions 3 and 4). Finally, children were asked about the advantages, disadvantages and ways of improving the robot, with the aim of gaining a better understanding on how, in the view of children, would be the ideal interaction with a social robot (questions 5 and 6).

5.1 Perceived Empathy

In the first question of the interview, our goal was to investigate whether children believed that the robot "perceived" how they were feeling during the game, and see if they refer to the implemented empathic behaviours to justify their answer. In other words, we wanted to see if they considered that the robot was being empathic towards them. As depicted in Figure 2, when asked whether they believed that the iCat knew how they were feeling, almost all the children from the adaptive_empathic condition and nearly 77% in the random_empathic answered "yes", while in the neutral condition only about 50% of the children provided an affirmative answer. The iCat's comments and helpful behaviour seemed to play an important role for children who answered "yes", especially in the empathic conditions, as these two participants stated:

> P18: *"I think so, because when I played bad moves the iCat warned me about that. He said what I was thinking."*
> P21: *"Yes, because for example when I played a bad move, he [the iCat] told me that I should do better [...] and when I didn't know what to play sometimes he told me that he could help"*

While in the empathic conditions most participants used the robot's behaviour and facial expressions to justify their affirmative answers, most of the participants from the neutral condition considered that the robot knew how they were feeling due to the cameras and because the robot was placed in front of them. For example:

> P34: *"Yes, because he saw my face and managed to memorise it"*

This effect might have happened due to the initial calibration phase of the face tracking software, in which children were asked to move their heads to different sides while looking at the robot. This procedure was done in all the conditions, even when the affect detection system was not being used, i.e., for children in the neutral condition.

As for the participants who answered "no", most of them were not able to justify their answer. In the neutral group, some participants argued that the iCat was too focused on its own game to have time to understand their emotions.

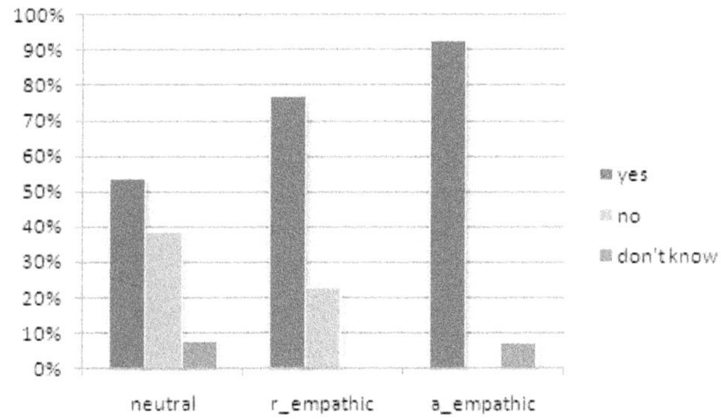

Figure : Children's answers to the question "Do you believe that the iCat recognised how you felt during the game?" in the three groups of users.

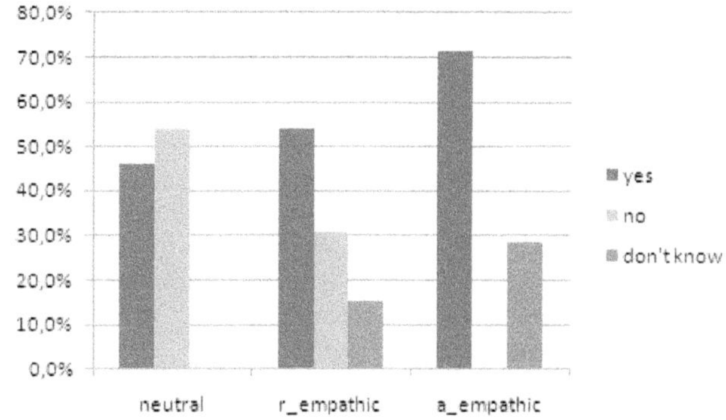

Figure 3: Children's answers to the question "Do you believe that the iCat felt emotions similar to the ones you felt?" in the three groups of users.

Table : Frequency of the reported affective and motivational states in the three conditions.

Category	neutral	r_empathic	a_empathic
happiness	3	1	7
nervous/anxious	1	3	0
enjoyed playing/fun	6	3	2
can't explain	1	2	2

Being a machine or a robot was another frequent reason given by participants who answered negatively:

> P25: *"No, because it's a robot and I don't think he feels the same things that people feel..."*

These results suggest that the implemented empathic behaviours influenced how children perceived the robot, as they frequently mentioned these behaviours to justify their affirmative answer. On the other hand, children who answered "no" often referred to the mechanical features of the robot. This leads us to believe that without the empathic behaviours, children perceive the robot more as a machine, and are less willing to suspend their disbelief during the interaction.

5.2 Experienced Emotions

The second group of questions was about children's experienced emotions while playing the game, and whether they believed that the iCat felt similar emotions. As in the previous group, we were looking for signs of perceived empathy: if children considered that the robot felt similar emotions, this would be an indicator of empathy. Most of the children in the three conditions answered affirmatively to the question "did you experience any particular emotions during the game " (respectively, 77% in the neutral condition, 85% in the random_empathic and 79% in the adaptive_empathic condition). When asked about particular emotions, we obtained a variety of affective and other motivational states, which in the final coding phase were grouped into the categories presented in Table 2. Children who interacted with the adaptive_empathic version of the robot reported more frequently "happiness" as one of the experienced emotions:

> P33: *"I was happy! I have waited so long to play with the iCat but it was worth it."*

Other participants reported that they felt anxious or nervous during the game. From the observations of the interaction, we noticed that the experimental apparatus (children being alone in the room with the experimenters, the presence of the cameras, etc.) may also have influenced children to feel this way. For example, one participant from the random_empathic condition described his affective state as follows:

> P38: *"I felt weird... I was playing with someone I've never played before [...] I was a bit nervous..."*

In addition to the reported affective states, a significant part of the children also described other positive states and reactions when asked about emotions, which can be explained by the fact that children's understanding of emotions is still under development at this age. As argued by Denham and colleagues [11], some children are more skilled than others in experiencing and using emotional vocabulary due to several factors such as gender, parent's emotional displays of affection and children's own past experience. Therefore, they used terms such as "enjoyed playing" or "I felt that I was having fun" to describe their emotional experience. Others even highlighted the fact that they were playing with a new robotic product, which is still not available to everyone. When asked about the emotions experienced during the game, one participant from the neutral condition answered:

> P34: *"I felt a big honour for playing with a robot that is not for sale yet!"*

As illustrated in Figure 3, more participants from the two empathic conditions considered that the iCat felt similar emotions to the ones they have reported, whereas in the neutral group this effect was not so substantial. Moreover, the affirmative answers in the neutral condition seem to be biased by an effect that we were unable to control: as children were very curious and expectant about the presence of the robot, the ones who played first spoke about their experience to the others. We noticed this because some of the children, at the end of the game, asked us questions such as "one of my colleagues told me that the iCat suggested him a good move... why didn't he do that with me ". Also, some children who played with the neutral version of the robot reported behaviours that do not happen in that condition (for example, the robot offering help). One of the participants even said that directly:

> P22: *"Yes, because I was told that he can do that."*

In sum, nearly all the children experienced emotions while playing chess with the iCat. However, the type of reported emotions changed slightly among conditions: while happiness appears in almost all the descriptions of children in the adaptive_empathic condition, children in the neutral and random_empathic conditions reported less positive emotions. An interesting finding was that negative affective states (e.g., anxious, nervous, etc.) were mentioned more often in the random_empathic condition, which may indicate that the introduction of empathic behaviours should be done carefully under the risk of confusing the users.

5.3 Comparing Robots with Computers

To gain a better understanding of how children perceive social robots and how the presence of the robot might influence their opinion of the chess game, we asked them what did they consider more fun, playing against the iCat or playing against a computer. All the participants except one (P18) agreed that playing the game against the iCat was more fun. The only participant who did not answered this way gave the following answer:

> P18: "For me it's the same, playing chess is always fun. But I had a lot of fun playing against the iCat".

As for the reasons why playing with the iCat was more fun than playing with a computer, there were some differences in the answers provided by participants from the different conditions. In both empathic conditions, children often referred to the helping behaviours of the robot as the main difference between playing with the robot when compared to playing with the computer. Also, some participants mentioned that playing with the iCat is more similar to playing with another person:

> P10: *"because the iCat is like a person but it's a robot."*
>
> P16: *"because he [the iCat] can say if I played well or bad, what he thinks about our moves... and also because we can talk to him."*
>
> P24: *"[the iCat] has emotions, and computers don't. He speaks and actually is a good friend and player."*

There were some aspects that were evenly used by participants in the three conditions. For example, the fact that they never played with a robot before, they need to play the iCat's moves while the computer plays automatically, the robot speaks, and also the robot's emotional and facial expressions.

5.4 Comparing Robots with Humans

After asking about the differences between robots and computers, we asked children what were the main differences between playing with the iCat or playing with their colleagues. There were differences between the answers provided by subjects from the adaptive_empathic condition and the two other conditions. Most children from the adaptive_empathic condition indicated as the main difference the fact that the iCat provides help and comments, or simply that playing with the robot it more fun than playing with their colleagues (for example, because the iCat allows them to take back moves). Conversely, many participants from the neutral and random_empathic conditions mentioned embodiment aspects of the robot as the main difference, and compared the robot's skills to the skills of a computer:

> P16: *"the iCat doesn't have arms and we need to play his moves, and I need to say yes when he proposes a draw. I need to play really well to beat him."*
>
> P38: *"the main difference is that [the iCat] is a robot, and my colleagues are regular people like me. Also we don't think the same way [...] the iCat thinks the way computers think."*

The expressivity and emotional behaviour of the robot was also frequently mentioned as one of the main differences, especially by participants from both empathic conditions:

> P26: *"[my colleagues] are not so fun, they don't smile when they are winning and don't get sad when they are loosing... I've never seen their feelings, but at least their faces... they never get very sad."*
>
> P17: *"[my colleagues] don't get so excited, they don't show so many emotions... and the iCat gives us advice, and people don't."*

Some children also highlighted the fact that the iCat is more focused in the game than their colleagues but, at the same time, does not try to rush the game:

> P33: *"my colleagues often say 'come on, play!' while I'm still thinking... and the iCat doesn't, he waits for my move, lets me think..."*
>
> P35: *"the iCat is very calm, he is always there focused in the game. [my colleagues] are always looking at something else and I am waiting, looking at the [chess] clock."*

5.5 Advantages, Disadvantages and Suggestions

The last part of the interview contained questions in which children were encouraged to provide suggestions on how we could improve the robot, and indicate advantages and disadvantages of the interaction. The answers to the question "how can we further improve the iCat " were coded into 4 different groups: suggestions regarding embodiment, behaviour and chess skills of the robot, and another category for the ones who considered that there is no need for improvement or were unable to name any suggestion. The category regarding embodiment included suggestions such as "the iCat should make less noise", "the robot should speak more slowly" or references to the addition of arms and other body parts, so that it could play its own moves. These suggestions regarding embodiment were more frequent in participants from the neutral condition. On the other hand, children from both empathic conditions provided more suggestions related to the chess skills of the robot. For example, some participants stated that to improve the iCat it should play chess more often, others that it should play an entire chess game instead of playing a chess exercise. As for the suggestions regarding the behaviour of the robot, while participants from the neutral condition proposed that the robot should be more helpful (or provide help when asked), children from the empathic conditions suggested improvements related to the expressivity and affective behaviour of the robot, for example:

> P34: *"he is joyful, and we get happy... it's like playing with real cats... it's very funny because he smiles and gets very emotional about our moves".*
>
> P33: "you could include a system for the iCat to say what he really feels about the moves I've done. I wouldn't be upset if sometimes the iCat say what he really feels..."

This last statement suggests that, even though the empathic behaviours of the robot improve the interaction in may ways, some users interpret these behaviours as less natural when compared to the behaviour of their chess mates. This result reinforces the idea that the robot should adapt its behaviour to the preferences of the child.

In terms of the advantages of playing with the robot, we did not found substantial differences in the advantages enumerated by children among the three conditions of the study, except for the fact that some children from the empathic conditions referred to the helping behaviour of the robot as one of the advantages. The most referred advantage indicated by the participants was that the iCat is a novel experience, as none of them had played with a robot before:

> P16: *"he is more funny because it's a robot, and we don't see that everyday."*
> P19: *"it is a novel experience, the iCat is a robot and I've never played with a robot before."*

Children also highlighted that they learn more when playing with the iCat due to the robot's comments during the game (when compared to playing with their colleagues or playing chess in the computer). Also, they considered that playing with the robot is more fun, which can be explained by the novelty effect of interacting with a social robot.

The majority of the participants said that there were no disadvantages or were unable to list any disadvantage. Only 6 children indicated disadvantages, 3 from the neutral condition, 1 from the random_empathic and 2 from the adaptive_empathic. Some children considered that playing the robot's moves is a disadvantage. However, others referred that as a positive aspect in earlier phases of the interview. Also, some children said that the robot speaks too fast, for example:

> P39: *"the disadvantage is that he is a robot, and does not have arms or legs. Legs aren't a problem, but the arms... sometimes I didn't understand what he said, it was a bit difficult to understand."*

We believe that the novelty associated with the interaction with a social robot contributed to the low number of disadvantages indicated by children. Also, we noticed that most children were not very comfortable in talking about disadvantages to the interviewer, which they considered one of the "builders" of the robot.

6. LESSONS LEARNED

The findings presented in the previous section reinforce the idea that empathy is an important mechanism so that social robots can interact naturally with people. Overall, children enjoyed the interaction and often mentioned the empathic behaviours of the robot as positive aspects. However, if not chosen appropriately, the robot's empathic behaviours might have a negative effect: while most of the children interacting with the adaptive_empathic version of the robot reported happiness as the most prominent emotion during the game, children who interacted with the random_empathic version of the robot reported more negative emotions such as "anxious" or "nervous". Also, some children felt that the robot was too nice for them, saying that they would not mind if the robot showed its "real feelings". These observations are important for deciding the "degree of empathy" in a social robot in two ways. First, the empathic behaviours (and the number of times they are employed) should be suitable to the application scenario where the robot is immersed. Second, as this might be a decision that depends a lot on the particular presences of users, the robot should be able to learn the best strategies for keeping that user in a positive affective state.

We also found that the amount of speech and comments of the robot seem to be appropriate for this type of scenario. None of the children seemed to notice that the robot does not have speech recognition capabilities, nor did they expressed any desire to interact with a more talkative robot. These results suggest that the perceptions of the game and the affective states of children are enough for generating believable verbal behaviour in social robots that act as game companions for children.

Our study also revealed some important lessons concerning the design of HRI studies with children. Collecting valuable data from 8 to 10-years-old children can be a challenging task. For example, some children were shy for being alone with the interviewer and provided short answers to the questions, or said that they did not know the answer. To address this issue, we suggest group interviews rather than individual interviews, since children tend to become less shy when they are in the presence of other children. At this age, some children still lack emotional vocabulary, or do not have a concrete idea of what an emotion is. Concerning this issue, we suggest the use of more visual methods to encourage children to talk about emotions (for example, by using a simplified version of the Geneva Emotion Wheel [27]).

Another observed effect was children's intrinsic tendency to please adults, also pointed out by Hanna and colleagues [13]. We noticed that very few participants were capable of enumerating disadvantages of the robot, but most of them provided many suggestions when asked about improvements for the iCat. As children are usually not very comfortable in discussing the negative aspects with adults (especially when they think that the adult owns the target of discussion), the interview should include constructive questions such as "how can we improve...", rather than directly asking for disadvantages.

7. CONCLUSIONS

In this paper, we presented a first evaluation of an autonomous social robot capable of performing an "affective loop", that is, recognising some of the user's affective states and displaying appropriate empathic behaviour. The study was conducted in a school setting, more precisely in the room where children usually have their chess lessons. The empathic behaviour of the robot was well understood by children, as they used it frequently to justify their answers to the empathy related questions in the interviews. The obtained results are in line with previous empathy studies with adults, in the sense that empathy facilitates the interaction and affects positively the perception of the robot. The findings also suggest that the selection of the empathic behaviours should be done carefully, under the risk of having the opposite effect. A similar result was discovered by Cramer et al. [10], where the presence of inaccurate empathic behaviours had a negative effect on adults' trust towards the robot.

Although more empirical work is necessary to validate the findings presented here, this study is an important contribution for the iterative process of designing an empathic robotic companion for children. Further research is also needed to verify if the results can be extrapolated for designing empathic social robots in different application domains and for other target users such as the elderly.

8. ACKNOWLEDGEMENTS

This research was supported by EU 7th Framework Program (FP7/2007-2013) under grant agreement no. 215554, FCT (INESC-ID multiannual funding) through the PIDDAC Program funds and a PhD scholarship (SFRHBD/ 41358/2007) granted by FCT.

9. REFERENCES

[1] C. Anderson and D. Keltner. The role of empathy in the formation and maintenance of social bonds. *Behavioural and Brain Sciences*, 25(1):21–22, 2002.

[2] P. Auer, N. Cesa-Bianchi, and P. Fischer. Finite-time analysis of the multiarmed bandit problem. *Machine learning*, 47(2):235–256, 2002.

[3] N. Borgers, E. de Leeuw, and J. Hox. Children as respondents in survey research: Cognitive development and response quality. *Bulletin de Methodologie Sociologique*, 66(1):60, 2000.

[4] C. Breazeal. Role of expressive behaviour for robots that learn from people. *Philosophical Transactions of the Royal Society B: Biological Sciences*, 364(1535):3527–3538, 2009.

[5] G. Castellano, I. Leite, A. Pereira, C. Martinho, A. Paiva, and P. McOwan. It's all in the game: Towards an affect sensitive and context aware game companion. In *Proceedings of ACII 2009*, pages 29–36. IEEE, 2009.

[6] G. Castellano, I. Leite, A. Pereira, C. Martinho, A. Paiva, and P. W. McOwan. Inter-ACT: An affective and contextually rich multimodal video corpus for studying interaction with robots. In *Proc. of the ACM International Conference on Multimedia*, pages 1031–1034. ACM, 2010.

[7] C.-C. Chang and C.-J. Lin. LIBSVM: A library for support vector machines. *ACM Transactions on Intelligent Systems and Technology*, 2:27:1–27:27, 2011. Software available at `http://www.csie.ntu.edu.tw/~cjlin/libsvm`.

[8] M. Chi. Quantifying qualitative analyses of verbal data: A practical guide. *Journal of the Learning sciences*, pages 271–315, 1997.

[9] B. Cooper, P. Brna, and A. Martins. Effective affective in intelligent systems: Building on evidence of empathy in teaching and learning. In A. Paiva, editor, *Affective Interactions*, volume 1814 of *Lecture Notes in Computer Science*, pages 21–34. Springer Berlin / Heidelberg, 2000.

[10] H. Cramer, J. Goddijn, B. Wielinga, and V. Evers. Effects of (in)accurate empathy and situational valence on attitudes towards robots. In *ACM/IEEE Int. Conf. on Human-Robot Interaction*, pages 141–142. ACM, 2010.

[11] S. A. Denham, J. Mitchell-Copeland, K. Strandberg, S. Auerbach, and K. Blair. Parental contributions to preschoolers' emotional competence: Direct and indirect effects. *Motivation and Emotion*, 21:65–86, 1997.

[12] J. Forlizzi. How robotic products become social products: an ethnographic study of cleaning in the home. In *ACM/IEEE Int. Conf. on Human-Robot Interaction*, pages 129–136. ACM, 2007.

[13] L. Hanna, K. Risden, and K. Alexander. Guidelines for usability testing with children. *Interactions*, 4(5):9–14, 1997.

[14] F. Hegel, T. Spexard, T. Vogt, G. Horstmann, and B. Wrede. Playing a different imitation game: Interaction with an Empathic Android Robot. In *Proc. 2006 IEEE-RAS International Conference on Humanoid Robots (Humanoids06)*, pages 56–61, 2006.

[15] M. Hoffman. *Empathy and moral development: Implications for caring and justice.* Cambridge Univ Press, 2001.

[16] T. Kanda, R. Sato, N. Saiwaki, and H. Ishiguro. A two-month field trial in an elementary school for long-term human–robot interaction. *Robotics, IEEE Transactions on*, 23(5):962–971, 2007.

[17] H. Kozima, M. Michalowski, and C. Nakagawa. A playful robot for research, therapy, and entertainment. *Int J Soc Robot*, 1:3–18, 2009.

[18] I. Leite, C. Martinho, A. Pereira, and A. Paiva. As time goes by: Long-term evaluation of social presence in robotic companions. In *Robot and Human Interactive Communication, 2009. RO-MAN 2009*, pages 669–674. IEEE, 2009.

[19] I. Leite, S. Mascarenhas, A. Pereira, C. Martinho, R. Prada, and A. Paiva. "why canŠt we be friends?" an empathic game companion for long-term interaction. In J. Allbeck, N. Badler, T. Bickmore, C. Pelachaud, and A. Safonova, editors, *Intelligent Virtual Agents*, volume 6356 of *Lecture Notes in Computer Science*, pages 315–321. Springer Berlin / Heidelberg, 2010.

[20] I. Leite, A. Pereira, G. Castellano, S. Mascarenhas, C. Martinho, and A. Paiva. Modelling empathy in social robotic companions. In L. Ardissono and T. Kuflik, editors, *UMAP 2011 Workshops*. Springer-Verlag Berlin Heidelberg, 2012.

[21] I. Leite, A. Pereira, C. Martinho, and A. Paiva. Are emotional robots more fun to play with? In *Robot and Human Interactive Communication, 2008*, pages 77 –82, aug. 2008.

[22] B. Mutlu and J. Forlizzi. Robots in organizations: the role of workflow, social, and environmental factors in human-robot interaction. In *ACM/IEEE Int. Conf. on Human-Robot Interaction*, pages 287–294. ACM, 2008.

[23] L. D. Riek, P. C. Paul, and P. Robinson. When my robot smiles at me: Enabling human-robot rapport via real-time head gesture mimicry. *Journal on Multimodal User Interfaces*, 3(1-2):99–108, 2010.

[24] A. Sabelli, T. Kanda, and N. Hagita. A conversational robot in an elderly care center: an ethnographic study. In *ACM/IEEE Int. Conf. on Human-Robot Interaction*, pages 37–44. ACM, 2011.

[25] M. Saerbeck, T. Schut, C. Bartneck, and M. Janse. Expressive robots in education: varying the degree of social supportive behavior of a robotic tutor. In *Proceedings of CHI 2010*, pages 1613–1622. ACM, 2010.

[26] J. Sanghvi, G. Castellano, I. Leite, A. Pereira, P. W. McOwan, and A. Paiva. Automatic analysis of affective postures and body motion to detect engagement with a game companion. In *ACM/IEEE Int. Conf. on Human-Robot Interaction*, pages 305–312. ACM, 2011.

[27] K. Scherer. What are emotions? and how can they be measured? *Social Science Information*, 44(4):695, 2005.

[28] J. Sung, H. Christensen, and R. Grinter. Robots in the wild: understanding long-term use. In *ACM/IEEE Int. Conf. on Human-Robot Interaction*, pages 45–52. ACM, 2009.

[29] D. Veenstra and V. Evers. The development of an online research tool to investigate children's social bonds with robots. *Human-Robot Personal Relationships*, pages 19–26, 2011.

Enhancing Interaction Through Exaggerated Motion Synthesis

Michael J. Gielniak
Georgia Institute of Technology
801 Atlantic Dr. NW
Atlanta, Georiga, USA
mgielniak3@gatech.edu

Andrea L. Thomaz
Georgia Institute of Technology
801 Atlantic Dr. NW
Atlanta, Georiga, USA
athomaz@cc.gatech.edu

ABSTRACT

Other than eye gaze and referential gestures (e.g. pointing), the relationship between robot motion and observer attention is not well understood. We explore this relationship to achieve social goals, such as influencing human partner behavior or directing attention. We present an algorithm that creates exaggerated variants of a motion in real-time. Through two experiments we confirm that exaggerated motion is perceptibly different than the input motion, provided that the motion is sufficiently exaggerated. We found that different levels of exaggeration correlate to human expectations of robot-like, human-like, and cartoon-like motion. We present empirical evidence that use of exaggerated motion in experiments enhances the interaction through the benefits of increased engagement and perceived entertainment value. Finally, we provide statistical evidence that exaggerated motion causes predictable human partner gaze direction and better retention of interaction details.

Categories and Subject Descriptors

I.2 [**Artificial Intelligence**]: Robotics—*Kinematics & dynamics, propelling mechanisms*; H.1 [**Models and Principles**]: [User/Machine Systems, Miscellaneous]

General Terms

Algorithms, Experimentation, Performance

Keywords

Exaggeration, cartoon-like motion, user study

1. INTRODUCTION

In human communication, the body is a high-bandwidth channel used for communicating spatial reference, disambiguating speech, inquiring for feedback, influencing others' behavior, identifying social goals or intentions, and directing attention. For a social humanoid robot to take advantage of its body like humans do, algorithms must autonomously modify its motion to include these social task elements.

Other than eye gaze and referential gestures (e.g. pointing), the relationship between robot motion and observer attention is not well understood. In addition to these explicit ways of directing attention, social robots need to communicate saliency and direct attention in all motions. Our research aims to do this with animation-inspired algorithms.

Exaggeration, a principle of animation [7], is defined in abstract terms as developing the *essence of an idea*, where a moderate version of something is replaced by a more extreme version. In theory, if a social robot needs to attract attention to a certain body part or region, that part should be exaggerated relative to the rest of its body. Exaggeration is a form of trajectory modulation used for emphasis.

We define a motion version to be exaggerated when it contains more contrast between two subspaces of motion (1) primary motion (i.e. task or intent of the motion) and (2) secondary motion (i.e. physics-based response to primary motion) as compared to the original motion that was used to create the exaggerated version. The open research question is how a social robot can autonomously generate exaggerated motion, rather than simply using such pre-designed motion.

In this paper, we present an algorithm to produce cartoon-like, exaggerated motion and present quantitative evidence from experiments that exaggerated motion used in social contexts can be used to control observer eye gaze. We show the benefits of exaggerated motion include: increased memory retention of interaction details, and improved interaction performance by keeping the human partner more engaged and more entertained throughout the interaction. We show that varying levels of exaggeration correlate to human expectations of robot-like, human-like, and cartoon-like motion.

2. RELATED WORK

Our literature survey uncovered no previous work adapting the animation principle of exaggeration to robots. However, the relationship between motion and subsequent observer attention direction has been studied in different contexts. E.g., unexpected changes in motion direction will attract attention [6]. And when watching videos, movement features, such as magnitude and vector, impact observer eye movement [2]. Humans attract the gaze of other humans by entering the space nearby; humans classify attention by variables such as head and body orientation [8]. In our work, we want to harness the relationships between motion and observer attention for use in social robots. Thus, we developed a motion synthesis algorithm, the output of which produces reliable, consistent predictability in observer gaze location.

Recently there has been an interest in applying animation principles to improve HRI. For example, surrounding action was designed to help communicate robot intent earlier [11]. In their related work, the robot motions were carefully designed by a Pixar animator. However, we are interested in making animation principles part of a robot's controller, so character-like motion can be generated autonomously, in real-time, during interaction with humans. Our own prior work examined autonomously generating other animation principles, namely secondary action [4] and anticipation [5].

In computer animation, there are many ways to create exaggeration. Small motion segments can be multiplied by a constant to create exaggeration [9]. Exaggeration of a motion signal can be created by changing filter amplitude in convolution [13]. Given an exaggerated and a moderate motion version, varying levels of exaggeration can be created by interpolation [12]. Motion can be transformed to the frequency domain, and exaggeration can be created by scaling only certain frequency bands [1]. Frequency domain tuning is difficult, non-intuitive and increases in complexity as the number of frequency bands increase. Often these techniques do not create noticable contrast between primary and secondary motion or they distort motion when all DOFs in motion are not systematically modulated together. Exaggeration in computer animation is an unfair comparison because breaking virtual world rules in the real-world can damage hardware (e.g., deforming bodies to exaggerate).

3. ALGORITHM

Exaggerated motion techniques developed for cartoon or virtual characters cannot be immediately applied to robots because fundamental differences exist in the real-world. The extent of exaggeration that can be produced safely on a robot is less than on cartoon or virtual character due to constraints such as torque or velocity limits of actual hardware. Our algorithm was designed to transform motion through exaggeration, while respecting real-world boundaries.

Exaggeration in joint-space can be produced by identifying the appropriate coordinates to modify for a given motion and then diminishing or amplifying these torque trajectories. Additional steps are necessary if the exaggerated motion must maintain certain features of the input motion (e.g. still be perceived as a representative exemplar of the input motion type). Exaggeration in other fields such as animation provide insight into how to appropriately adjust a trajectory for exaggerated motion effect. For example, important body parts must be amplified and less important body parts should be diminished so that gaze is directed away from regions of the motion which are less important.

We leverage actuation and timing information given in an input motion to appropriately exaggerate it, since the torques are arranged within this motion in magnitudes and directions relative to each other to produce a representative examplar. Logically then, relative DOF importance for exaggeration is measured by the amount of actuation in the degrees-of-freedom, which can be measured by torque.

Inspired by other research that examines motion decomposition by magnitude of actuation, we exploit the variance in actuation among all DOFs for a given motion to parameterize that motion according to a spectrum of actuation [14], [4], [3]. In doing so, relative importance of all DOFs remains constant in our algorithm, which ensures that as long as physical limitations (e.g. joint limits, torque limits) are not exceeded, the exaggerated version of the motion will also be an exemplar of the same type as the input motion.

Let q_j be the torque trajectory for DOF j of the original input motion with T equidistant time samples. For a robot with M degrees-of-freedom, the torque trajectories from the input motion are organized into an M x T column-stacked matrix $\tau = [q_0, q_1, ..., q_M]$. A singular value decomposition is performed on the covariance matrix in Equation 1 to obtain an M x M matrix of eigenvectors, denoted U, and an M x M matrix with eigenvalues along the diagonal, denoted Λ. The magnitude of the eigenvalues corresponds to a measure of the torque variance in the motion.

$$SVD((\tau - \mu)^T(\tau - \mu)) \tag{1}$$

where,
μ_i = mean torque of DOF i for the entire trajectory
μ = M x T stacked matrix, each column is $[\mu_1 ... \mu_M]^T$

The largest gap in the distribution of the eigenvalues defines a threshold, λ_{th}, which separates the corresponding eigenvectors into mostly actuated and near-unactuated eigenvectors. For reference, typically less than ten eigenvectors exist in the mostly actuated set. We define α_{algo} to quantify the amount of exaggeration for all DOFs in torque space. Larger values of α_{algo} equate to more exaggeration. Since exaggeration has a "polarizing" effect upon motion, pushing motion toward extremes (i.e. highly actuated increase in actuation, near-unactuated are diminished), all eigenvalues along the diagonal of Λ are modified according to the following four rules for DOF j, since $\alpha_{algo} \geq 1$:

1. If $\lambda_j < \lambda_{th}$ and $\lambda_j < 1$, then $\lambda_{j_{new}} = \lambda_j^{\alpha_{algo}}$.
2. If $\lambda_j < \lambda_{th}$ and $\lambda_j > 1$, then $\lambda_{j_{new}} = \lambda_j^{-\alpha_{algo}}$.
3. If $\lambda_j > \lambda_{th}$ and $\lambda_j < 1$, then $\lambda_{j_{new}} = \lambda_j^{-\alpha_{algo}}$.
4. If $\lambda_j > \lambda_{th}$ and $\lambda_j > 1$, then $\lambda_{j_{new}} = \lambda_j^{\alpha_{algo}}$.

These four rules are systematically designed to divide the two subspaces of a trajectory: primary and secondary motion. Rules one and two diminish the minimally existent actuation in near-unactuated torques in the original motion (i.e. reduce secondary motion); rules three and four exaggerate the torques in highly-actuated coordinates (i.e. amplifying primary motion). The composite effect of all four rules create the contrast necessary for exaggeration.

A new M x M diagonal matrix, denoted Λ_{new}, is formed from the new eigenvalues, and the original eigenvectors are used to determine the new torque matrix of exaggerated and diminished torques as shown in Equation 2. τ_{new} is the torque trajectory that is commanded to robot actuators to produce exaggerated motion.

$$\tau_{new} = U^T \Lambda_{new} U + \mu \tag{2}$$

The upper bound on α_{algo} is motion dependent and is a function of the maximum torque limits of the robot actuators. For safety, we place the upper bound on α_{algo} to be the minimum value that would cause any motor to exceed its torque limit.

Figure 1: SIMON: The hardware platform.

4. HARDWARE PLATFORM

The platform for this research is an upper-torso humanoid robot we call Simon (Figure 1). It has 16 controllable DOFs on the body and 4 per hand. Each arm has 7 DOFs (3 at the shoulder, 1 at the elbow, and 3 at the wrist) and the torso has 3 DOFs. Simon has 3 DOFs for the eyes, 2 per ear, and 4 for the neck.

5. HYPOTHESES

We have several hypotheses about the algorithm that we developed, the exaggerated motion it produces, and the benefits that it brings to HRI. These hypotheses are as follows:

- H1: Humans can perceive the difference between exaggerated and unexaggerated motion.
- H2: Exaggerated motion will appear to be more cartoon-like than unexaggerated motion.
- H3: Humans will feel more engaged and prefer interacting with a robot that displays exaggerated motion.
- H4: Exaggerated motion will enable human partners to remember more of the interaction details.
- H5: Exaggerated motion changes which body parts are salient for a given motion and directs attention to body parts that have higher kinetic energy.

6. EXPERIMENTAL DESIGN

To test these five hypotheses, we conducted two experiments using a storytelling task. Storytelling is suitable for our experimental goals since it minimizes the intellectual load of the human in the interaction, allowing humans to devote more attention to the robot. It engages the human enough to prevent overanalysis of motion. Furthermore, storytelling provides a context for the interaction and gives us a mechanism to test the effects of exaggerated motion on memory, to show instrumental HRI benefits.

Unlike other work that has focused on the benefits of *robot* gaze for storytelling [10], we manipulate robot motion to test the effects of *human* gaze without being responsive to the human. Robot gaze is irrelevant to our work because the robot is an actor in the story, not a storyteller or narrator.

In both experiments we use one of Aesop's fables adapted to include the robot as a main character. There are 13 motions that were designed to match the computer-synthesized story spoken in the background, so that the robot acted out what the narrator said. Context was completely omitted from the story. No objects were used, and no other characters in the story physically existed while the robot acted out the story. So as not to elicit attention, the robot made no intentional eye contact with participants.

Both experiments include two versions of 13 motions: original unexaggerated motions (UN), and exaggerated versions of each (EX), created using our algorithm (Section 3).

The 13 motions were shown in the same order to all participants because it was a story. The 13 motions (in story order) were: walk, look up, reach, look down, beckon, scan, shucks, grrr..., phooey, leave, point, hands open, and wave.

Four different experimental conditions were created for the story. They had identical story text, but differed in the motion types that accompanied the speech.

- AE: All 13 exaggerated motions in the story.
- EU: 7 EX followed by 6 UN.
- UE: 7 UN followed by 6 EX.
- AU: All 13 unexaggerated motions in the story.

6.1 Experiment 1: Testing Memory

Experiment 1 was designed to test whether exaggerated motion improves people's recollection of their interaction with the robot. 54 participants (36 male, 18 female; age 19-30) were recruited. Participants had no prior story knowledge. After participants watched one of the four story conditions, they rated the story motions according to 16 variables, each on a 7-level Likert scale. The 16 variables were: subtle, entertaining, realistic, exaggerated, expressive, stiff, accentuated, cartoon-like, life-like, emphatic, emphasized, natural, noticeable, engaging, smooth, and human-like.

After the Likert questionnaire, participants answered six fill-in-the-blank (FIB) questions and three short answer questions. The six FIB questions were verbatim from the narrator-spoken story, so that the objective was to fill in the exact word spoken from the story. Three were from each half of the story so that in conditions EU and UE, participants answered three FIB questions each from the UN and EX portions of the story. The answers were paired in both story halves so that these three answers were a location, an object, and an emotion. The correct answer pairs were synonyms in the story context. I.e., location: {high-above, vine}; object: {grapes, fruit}; emotion: {discouraged, unhappy}.

These pairings test the effect of exaggeration on memory substitution. In conditions where the same participant saw motions with and without exaggeration, we wanted to see if people more frequently substitute one answer for the other in the pair for EX or UN. A full analysis of this effect requires a high failure rate on correct responses to the FIB questions or a large number of participants so that enough data is collected to use appropriate statistical techniques.

Three short answer questions were designed to test H4:

- SA1: What was the title of the story that the robot just told you?
- SA2: What was the moral of the story?
- SA3: What was the color of the object that the robot was trying to reach?

Following the short answer questions, participants were asked to tell their favorite part of the story and their favorite motion from the story, and they were prompted for reasons why they selected these as their favorites.

6.2 Experiment 2: Testing Gaze Direction

The same exact story, motions, experimental layout, and four experimental conditions were repeated in Experiment 2. However, the faceLAB[1] system was used to track participant eye gaze direction. Left and right eye trajectories were captured for the duration of the entire story, from which we calculate the exact location on the robot's body that each eye of the participant was looking at during the story.

For Experiment 2, 68 participants (44 male, 24 female; age 19-30) were recruited. After the story finished, participants were asked to rate the robot according to each criteria using the same 16 variables on the 7-level Likert. Participants were then prompted for their favorite part of the story, their favorite motion from the story, and their reasons for both.

Then, participants were seated at a virtual model of the robot and given two controls: + (plus) and - (minus). These buttons indirectly controlled the value of α_{algo} for the exaggeration using a mapping function. Participants were asked to use the buttons to select the value of α_{exp} that produces the motions that they considered to be:

- HL: Most Human-like
- VP: Most Visually Pleasing
- CL: Most Cartoon-like
- RL: Most Robot-like
- LB: They Liked Best

Order of all the five values was randomized for each participant, but each participant saw the same order of the five values for all motions. Motion order was randomized for each participant. In order to intentionally bias our experiment away from the results we expected to achieve, the value of α_{exp} was always initially set at the value that corresponded to unexaggerated motion. The range on α_{exp} was limited between positive and negative one, which prevents exceeding the torque limits of any motor on the robot. This part of Experiment 2 was not done on hardware for safety.

7. RESULTS

The results from Experiments 1 and 2 are discussed together in support of the five hypotheses presented in Section 5 because data from both experiments is used to support the hypotheses. Discussion of the statistically significant variables from the Likert scales is distributed so that results are presented in support the appropriate hypotheses. The mean values from the Likert variables are shown in Table 1.

7.1 EX and UN are Perceptibly Different

First we test our algorithm to ensure it accomplished our goal of creating motion that is exaggerated. Although many of our Likert scales achieved statistical significance between participants in the AE and AU groups, three of the sixteen scales add evidence to support that our algorithm creates exaggerated motion: subtle, exaggerated, and accentuated.

First, ANOVAs were conducted on all the data from both experiments assuming that all four conditions (AE, EU, UE, AU) belong to the same distribution. F_{crit} for each variable in the Likert is 2.68. Respectively, the F-values achieved from these ANOVAs are 23.0 (subtle), 3.18 (exaggerated), and 4.85 (accentuated). Thus, for each measure at least one of the conditions is statistically different.

[1]faceLAB is a trademark of Seeing Machines.

Table 1: Average subjective responses to 16 Likert scale variables. 1 (not). 7 (variable).

Variable	AE	EU	UE	AU
Subtle	2.63	3.48	3.40	5.35
Entertaining	5.20	4.02	3.92	3.67
Realistic	2.87	3.84	3.59	4.21
Exaggerated	4.86	3.82	3.86	3.13
Expressive	4.59	4.08	4.22	3.39
Stiff	3.18	3.26	3.31	3.49
Accentuated	4.62	2.88	3.55	2.90
Cartoon-like	3.34	2.40	2.72	2.50
Life-like	3.24	3.85	3.70	3.62
Emphatic	4.36	3.86	3.77	3.05
Emphasized	4.47	3.76	3.57	2.87
Natural	3.90	3.52	3.50	3.70
Noticable	4.91	3.92	4.00	3.98
Engaging	5.24	3.84	3.67	3.53
Smooth	5.49	5.43	5.32	5.26
Human-like	4.12	3.56	3.72	3.67

Post-hoc pairwise t-tests were performed. For the subtle variable, three of the six pairings exhibit statistically significant results ($p<0.01$): (AE,AU), (EU,AU), and (UE,AU). And for both the accentuated and exaggerated Likert variable, three of the six pairings exhibit statistically significant results ($p<0.01$): (AE,EU), (AE,UE), and (AE,AU).

Participants who saw at least half of the motions modified by our algorithm indicated that the motions are less subtle than participants who saw only the original motions, and participants who saw only motions modified by our algorithm indicated that the motions are more accentuated and more exaggerated than participants who saw at least half of the original motions. Thus, we conclude that motions produced by our algorithm are less subtle, more accentuated, and more exaggerated than the input motions.

7.2 EX Appears More Cartoon-like Than UN

A good measure of success is whether the output of our algorithm maintains the same qualities and characteristics of its inspiration. H2 is a logical hypothesis, since the inspiration for exaggerated motion comes from animated and virtual characters. And by testing if EX is more cartoon-like than UN, we also are evaluating whether our algorithm accomplished one of its most fundamental goals. Two of the Likert variables provide evidence in support of H2, the hypothesis that exaggerated motions produced by our algorithm are more cartoon-like: realistic and cartoon-like.

ANOVAs conducted on all the data from both experiments assuming that all four conditions (AE, EU, UE, AU) belong to the same distribution yielded F-values of 3.84 and 4.26 for realistic and cartoon-like respectively. Both of these values exceed the F_{crit} of 2.68, indicating that for both measures at least one of the conditions is statistically different.

For the realistic Likert variable, three of the six pairings exhibit statistically significant results ($p<0.05$): (AE,EU), (AE,UE), and (AE,AU). These three results are the three pairings that compare participants who saw only motions produced by our algorithm compared with participants who saw at least half of the original motions. And for the cartoon-like Likert variable, one of the six pairings exhibits statistically significant results ($p<0.02$): (AE,AU). These results are the pair that compares participants who saw only mo-

Table 2: Average α_{exp} Participant Responses for Most Human-like (HL), Most Cartoon-like (CL), Most Robot-like (RL), Most Visually Pleasing (VP), and Liked Best (LB).

Motion	HL	CL	RL	VP	LB
Walk	0.131	0.910	0.056	0.877	0.844
Look Up	0.230	0.967	0.074	0.812	0.754
Reach	0.164	0.787	0.041	0.385	0.501
Look Down	0.607	0.959	0.434	0.8361	0.869
Beckon	0.230	0.956	0.066	0.771	0.574
Scan	0.517	0.869	0.246	0.689	0.877
Shucks	0.680	0.851	0.098	0.762	0.844
Grrr...	0.443	0.967	0.197	0.911	0.899
Phooey	0.639	0.926	0.080	0.836	0.756
Leave	0.541	0.754	0.221	0.803	0.639
Point	0.508	0.853	0.148	0.836	0.910
Hands Open	0.623	0.951	0.180	0.885	0.541
Wave	0.320	0.861	0.107	0.615	0.525
Average	0.433	0.893	0.150	0.771	0.733

tions produced by our algorithm compared with participants who only original motions. From these results, we conclude that motions exaggerated by our algorithm are less realistic but more cartoon-like than the input motions.

There is further quantitative evidence relevant to H2 from the data of experiment 2 where participants selected α_{exp} values that pertain to their subjective ratings of most human-like, most cartoon-like, most robot-like, most visually pleasing, and liked best (HL, CL, RL, VP, and LB, respectively).

The analysis was initially performed assuming that the qualities of HL, CL, RL, VP, and LB for α_{exp} are motion dependent. Thus 13 separate ANOVAs were performed, 1 per story motion, assuming all 5 groups belonged to the same distribution. 13 of 13 ANOVAs yielded F-values greater than F_{crit}, which means that for each motion, subjective settings for α_{exp} according to each of the 5 groups held at least one statistically independent pair of groups. 10 post-hoc pairings were performed for each motion, using all possible pairs of the five subjective qualities, yielding a total of 130 post-hoc pairings. 122 of the 130 pairings were statistically significant ($p<0.05$). The 8 pairings that failed statistical difference tests were all between VP and LB for the motions of reach, beckon, scan, shucks, phooey, leave, hands open, and wave. We conclude that by modulating α_{exp} (i.e. exaggeration) we can consistently create motions which are more robot-like, more human-like, and more cartoon-like, and the values of α_{exp} (and α_{algo}) are largely motion independent.

Average values across all motions for the α_{exp} setting from Experiment 2 are shown in the bottom row in Table 2. For consistency, prior to Experiment 2, all unexaggerated input motions to be used in the story were selected to have α_{exp} values between -0.2 and 0.0, and no α_{exp} for any exaggerated motion used in the story was less than 0.7. For reference, exaggerated motion values with an α_{exp} value of 1.0 correspond to exaggerated motion so that any more exaggeration would cause at least one motor to exceed its torque limit.

The average value of 0.89 for cartoon-like is consistent with the Likert results, which also found that the exaggerated motions were cartoon-like; it indicates that exaggerating motions by using our algorithm adds a cartoon-like quality to the motions. Robot-like motion, as defined by participants' expectations, tends to lack exaggeration, which may help explain the results regarding entertainment and engagement of exaggerated motion that we discuss in Section 7.4. Human-like motion tends to exhibit moderate levels of exaggeration, not near the torque limits for any motors, but also far from the values of robot-like exaggeration.

Table 3: P-values from post-hoc pairwise t-tests of the data in Table 2 for all 5 groups. Tests with p>0.05 are in gray. x = elsewhere in the table.

	HL	RL	VP	LB
CL	0.008	0.004	0.024	0.016
RL	0.023	x	0.009	0.012
VP	0.021	x	x	0.072
LB	0.027	x	x	x

The results from the motion-by-motion analysis for α_{exp} lead us to suspect that human-like, cartoon-like, and robot-like are motion independent. To show that these measures of HL, CL, and RL for exaggerated motion are truly independent of motion when equalized on the scale of α_{exp}, a singular ANOVA was performed on all the data in Table 2. The F-value of 59.71 from this analysis is greater than F_{crit} of 2.53, which means there is statistical difference between at least two of these measures independent of motion. 10 pairwise t-tests grouping data averages across all motions were performed to compare all possible pairings of HL, CL, RL, VP, and LB (p-values shown in Table 3).

Table 3 shows that across all motions, the qualities of human-like, cartoon-like, and robot-like are statistically different from all other variables in the study. Only statistical difference tests fail between visually pleasing and liked best (shaded gray in Table 3), which indicates that these two subjective measures for exaggerated motion may not come from independent distributions.

The final analysis that we performed in support of H2, was to evaluate 130 additional pairwise t-tests (10 per motion) for the α_{exp} values that correspond to the story motions and the distributions provided in Experiment 2 based upon HL, CL, RL, VP, and LB, to find out if the UN or the EX motions used in the story are statistically different from the human subjective measures. The ten pairings are all possible combinations of one member from each of the sets: {UN, EX} and {HL, CL, RL, VP, LB}. 65 of 65 pairwise t-tests for the UN motion pairs have $p<0.05$, which indicates that the choice of unexaggerated motion in the story did not coincide with our participants distributions of any of the five measures. The 26 motion pairs for EX with HL and RL have $p<0.05$, which indicates that our exaggerated motions used in the story were not robot-like or human-like, using participants' subjective responses as the measures of these two variables. However, for 13 of 13 (EX,CL) pairs there is no statistical difference ($p<0.05$) between exaggerated story motion and participants' expectations of cartoon-like motion.

7.3 EX Improves Interaction Performance

Thus far, the results have not established the benefits of exaggerated motion. In support of H3, HRI performance will be measured by how well participants can remember the story over a short period of time. Thus, evaluation of performance is based upon SA and FIB question answers.

Correct answers were verbatim from narrator speech to the 6 FIB questions (Table 4). Each of the 6 FIB answers were associated to one sentence and one motion in the story (either UN or EX). Thus, question answers can be grouped

Table 4: Percent of participants in each of four conditions that correctly answered the fill-in-the-blank question. EX and UN columns are percent of correct FIB question responses grouped according to motion associations cumulatively across all four conditions.

Answer	AE	EU	UE	AU	EX	UN
high above	71.4	84.6	30.8	42.9	77.8	37.0
grapes	92.9	76.9	38.5	64.3	85.2	51.9
discouraged	85.7	69.2	69.2	64.2	77.8	66.7
fruit	75.9	30.8	69.2	42.9	74.1	37.0
unhappy	85.7	61.5	84.6	71.4	85.2	66.7
vine	85.7	53.8	92.3	64.3	88.9	59.3

Table 5: Percent of incorrect answers for FIB question answer pairs substituted across contextual synonyms (1) location {high above, vine}, (2) object {grapes, fruit}, (3) emotion {discouraged, unhappy} in each of four conditions. EX and UN columns are percent of incorrect FIB question responses substituted across contextual synonyms, grouped according to the motion type executing when the substituted word was heard. The left column still defines the question by the intended correct response.

Answer	AE	EU	UE	AU	EX	UN
high above	75.0	50.0	66.7	12.5	71.8	31.3
grapes	0.0	66.7	75.0	20.0	37.5	43.3
discouraged	100.0	50.0	75.0	20.0	87.5	35.0
fruit	66.7	77.8	75.0	25.0	72.2	37.5
unhappy	0.0	80.0	50.0	0.0	40.0	25.0
vine	100.0	66.7	100.0	20.0	83.3	60.0

based upon associated motion type. The percent of correct FIB question responses grouped according to motion associations are shown cumulatively across all four conditions in the two columns furthest to the right in Table 4.

Comparing columns of (AE, EU, and UE) to (AU), more participants remember the story better when they watch some exaggerated motions. In the two columns on the right in Table 4 where the participant responses are grouped according to the associated motion type (EX or UN), the trend is stronger, which demonstrates the benefit that exaggerated motion helps people to remember the story better.

In Table 5, EX and UN are accumulated over the motion type that was executing when the substituted word was heard; this is different than in Table 4 because in Table 4, the UN and EX columns are tallied over the motion type associated with the text in the question.

As evident in Table 5, when participants saw an exaggerated motion, they remembered the words associated with the motion better than when they saw unexaggerated motion. When answers were incorrect, contextual synonyms heard during exaggerated motions more often get substituted for incorrect responses than from unexaggerated motions. This suggests that one benefit of exaggerated motion is to help retain interaction details in memory.

We see similar trends in the short answer questions. The moral of the story was explicitly mentioned during the 'hands open' gesture, which was the second to last motion in the story. The moral of the story was "Do not speak disparagingly of things that you cannot attain." Table 6 shows the percent of correct answers as a function of all four conditions and the two most common substitutions grouped according

Table 6: Percent of participants vs. SA2 (moral) answers. EX and UN columns are responses tallied across all four conditions, grouped by the motion type that was executing when the participant-provided answer was spoken by the narrator. #1 = "reach" substituted for "attain." #2 = "be discouraged by" substituted for "speak disparagingly of."

Answer	AE	EU	UE	AU	EX	UN
Verbatim Correct	35.7	7.7	30.8	0.0	33.2	3.8
Substitution 1	35.7	15.3	7.7	0.0	25.5	3.8
Substitution 2	7.1	30.7	23.1	7.1	19.0	15.1

Table 7: Percent of participants vs. SA3 (object color) answers. Not in Story = participant explicity wrote that the narrator did not say.

Answer	AE	EU	UE	AU
Not in Story	21.4	0.0	7.7	0.0
Blank; Knew	64.3	15.3	7.7	0.0
Blank; Didn't know	0.0	7.7	0.0	42.9
Purple	14.3	61.5	38.9	7.1
Green	0.0	15.3	38.4	0.0
Other	0.0	0.0	0.0	42.9

to the motion type seen when the participant-given answer was spoken by the narrator in the story.

Four instances of the word "reach" are heard in the story, and all occur during the first half of the story. One instance of "discouraged" occurs in the story and it occurs in the first half of the story. The two substitutions in Table 6 are "attain" replaced by "reach" and "speak disparagingly of" replaced by "be discouraged by." The short answer question about the story moral shows the same results as for the FIB questions. Exaggerated motions are associated to both (1) more correct responses and (2) a higher percentage of the answer substitions for incorrect responses made with contextual-synonyms heard during exaggerated motions.

The two short answer questions that asked participants to tell the story title and color of the object that the robot was trying to reach were not explicitly given in the story. Across all four conditions, the participants who saw only EX had the highest correct response rates in both Tables 7 and 8. Furthermore, feasible answers for color of the grapes such as green or purple for grapes are concentrated in the conditions that had at least half of the motions exaggerated. Blank answers were sorted during during the interview: *knew* (participant knew that the narrator did not say the answer) and *didn't know* (ambiguous or uncertain answers).

Based on the results presented, interaction task performance is improved with exaggerated motion because exaggerated motion leads to higher correct response rates for questions about the interactive experience with the robot. There is a distinct benefit of using exaggerated motion, when remembering details about the interaction is important.

7.4 Humans Prefer EX Over UN

In support of H4, two Likert variables provide evidence that humans prefer exaggerated motion: entertaining and engaging. ANOVAs conducted on all the data from both experiments assuming that all four conditions (AE, EU, UE, AU) belong to the same distribution yielded F-values of 3.56 and 3.17 for entertaining and engaging respectively. Both of

Table 8: Percent of participants vs. SA1 (story title) answers. Not in Story = participant explicitly wrote that the narrator did not say.

Answer	AE	EU	UE	AU
Not in Story	21.4	0.0	0.0	0.0
Blank; Knew	64.3	46.1	46.1	7.1
Blank; Didn't know	7.1	7.7	0.0	28.6
Adaptation of an Aesop Fable	7.1	46.1	46.1	14.3
The Robot & the [Sour] Grapes	0.0	0.0	7.7	42.8
Other Incorrect	0.0	0.0	0.0	7.1

these values exceed the F_{crit} of 2.68, which indicates that at least one of the four conditions is statistically different.

For the both the entertaining and engaging Likert variables, three of the six post-hoc t-test pairings exhibit statistically significant results ($p<0.05$): (AE,EU), (AE,UE), and (AE,AU). Participants who saw at least half of the story motions exaggerated through modification by our algorithm indicated that the motions are more entertaining and more engaging than participants who saw only original motions. Thus, motions produced by our algorithm are more entertaining and more engaging than the input motions.

Returning to the subjective α_{exp} data discussed in Section 7.2, in which participants were asked to find values of α_{exp} that are most human-like, most cartoon-like, most robot-like, most visually pleasing, and liked best, only 3 of 13 (EX,VP) and 5 of 13 (EX,LB) pairs showed statistical significance $p<0.05$; thus, the majority of motions showed no statistical difference between exaggeration and visually pleasing or exaggeration and liked best.

In regard to H4, participants also expressed a preference for exaggerated motion in favorite motion and story part. If all four conditions are included (i.e. AE, EU, UE, AU) from both experiments, then 62.2% of participants chose a favorite part of the story associated with an exaggerated motion and 64.8% of participants selected an exaggerated motion from one of the thirteen as their favorite motion in the story. However, in only two conditions did participants actually have a choice (i.e. for EU and UE). Excluding the other two conditions, these percentages increase to 75% and 80% respectively. Participants selected exaggerated motion or favorite story parts because these were described as "animated," "expressive," or "emotional."

7.5 EX Can Be Used To Direct Attention

The final set of results addresses whether exaggerated motion can be used to direct attention to salient body parts (H5). To draw meaningful conclusions from the eye gaze data collected with the faceLAB system, we need a measure of exaggeration that evaluates body part exaggeration, since the eye gaze data is a function of body part. α_{algo} is not sufficient because it corresponds to joint-space exaggeration.

Exaggeration in joint-space corresponds to an increase in energy for the motor moving the exaggerated DOF. Since the exaggeration for a particular body part is a function of all of the parent joints on the hierarchy relative to the body part, cumulative energy ratio (EX over UN) is used to represent exaggeration in Cartesian space.

The robot was discretized into 12 segments (i.e. non-overlapping body parts). Using the data from Experiment 2, the intersection of each of the two eye gaze trajectories with the 12 body part segments was determined and accumulated over all participants. Using these trajectories, percent of total trajectory time that each participant spent watching each body part could be determined for each motion in the story. Assuming both groups of motions (UN and EX), all body parts, and all motions belong to the same distribution, the ANOVA yielded an F-value of 67.4, which was greater than $F_{crit} = 3.9$; therefore post-hoc analysis is required.

Assuming each of the distributions are different according to body part and motion from the story, (UN, EX) data pairs exist. Each of these 156 pairs (12 body parts x 13 motions) represent distributions for the amount of time participants spent watching a particular body part for a particular motion. 151 of 156 pairwise t-tests were statistically significant ($p<0.05$), which indicates that participants watch the same body part in the same motion for different lengths of time. The five tests that failed to yield statistical difference were (scan, left hand), (scan, torso top), (scan, torso mid), (scan, head), and (wave, left hand). Scan and wave are both perfomed with the left hand; during gestures, humans focus on the symbol formed by the salient hand. Additionally, the head is near the left hand for a significant portion of the scan gesture, which could help explain these results.

The 156 pairwise t-tests were repeated using only data from EU and UE to create the distributions to determine if the same participants will change their behavior when they see exaggerated and unexaggerated motion. This time, 153 of 156 pairwise t-tests were statistically significant ($p<0.05$). The three pairings that were not statistically different with respect to percent of gaze time for a specific motion and body part were (scan, left hand), (scan, head), and (wave, left hand). Thus, the *same* participants will change their watching behavior when watching UN or EX.

To test whether attention is directed toward exaggerated body parts, we employ our Cartesian measure for exaggeration and plot the average percent time watching both exaggerated and unexaggerated motions against cumulative energy ratio (CNR) for the point trajectory that represents the body part centroid as a function of time (Equation 3).

$$CNR = \frac{\sum_{t=0}^{t=T_f} (v_x(t)_e^2 + v_y(t)_e^2 + v_z(t)_e^2)}{\sum_{t=0}^{t=T_f} (v_x(t)_u^2 + v_y(t)_u^2 + v_z(t)_u^2)} \tag{3}$$

where,
$v_i(t)$ = velocity of body part centroid in i-direction at time t, where i={x, y, or z}
T_f = final time sample for discrete motion trajectory
u = denotes unexaggerated motion trajectory
e = denotes exaggerated motion trajectory

The two plots of average percent time spent watching versus CNR for UN and EX are shown in Figures 2 and 3, respectively. Since trajectories for body part motions with higher CNR are more exaggerated, the horizontal axis is a measure of exaggeration. Each point on the plots represents the data for one body part from one motion. In Figure 3, on average participants spent more time watching body parts with more exaggeration. The absence of a trend in Figure 2 provides evidence that exaggeration produces this effect upon participant behavior. I.e. exaggerated motion is used to direct attention to body parts that are exaggerated more,

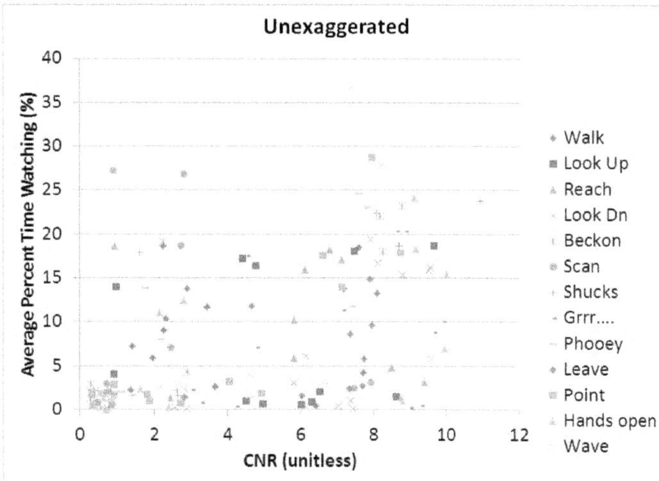

Figure 2: Average percent of time participants spent watching a specific body part in an unexaggerated trajectory vs. cumulative energy ratio for that body part centroid trajectory.

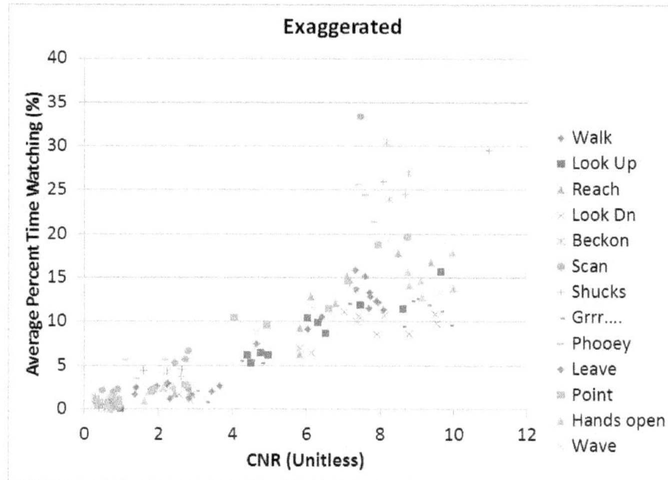

Figure 3: Average percent of time participants spent watching a specific body part in an exaggerated trajectory vs. cumulative energy ratio for that body part centroid trajectory.

and attention is directed away from body parts with diminished motion. We conclude that hypothesis H5 holds true.

8. DISCUSSION

Some might claim that the existence of a tunable parameter, α, is a disadvantage of our algorithm. Perhaps a systematic method for determing an "optimal" α for any motion is preferred. However, we believe that optimality is context-dependent (e.g. we provide optimal values that correspond to robot-like, human-like, and cartoon-like). Since we do not believe that one α value is optimal for all situations, α can be exploited to control the amount of exaggeration at any given time based on whether a robot needs to attract attention, increase engagement of human partners, or simply move in a more human-like manner.

9. CONCLUSIONS

We presented an algorithm that creates exaggerated variants of an input motion in real-time. Our experimental data confirmed that exaggerated motion is perceptibly different than the input motion, provided that the motion is sufficiently exaggerated. We found that various levels of exaggeration in motion correlate to human expectations of robot-like, human-like, and cartoon-like motion. Use of exaggerated motion enhances the interaction through the benefits of increased partner engagement and perceived entertainment value. We provided statistical evidence that exaggerated motion also increases retention of interaction details. Exaggerated motion changes the salient body parts and viewing durations when motion is executing so that observer gaze is directed toward exaggerated body parts for longer time periods and away from body parts with less energy.

10. ACKNOWLEDGMENTS

This research is supported by NSF grant IIS-0953181.

11. REFERENCES

[1] A. Bruderlin and L. Williams. Motion signal processing. In *Proceedings of the 22nd Annual Conference on Computer Graphics and Interactive Techniques*, pages 97–104, 1995.

[2] A. Bur et al. Dynamic visual attention: Motion direction versus motion magnitude. *Proceedings of the SPIE 20th Annual Symposium*, 6806, January 2008.

[3] M. Gielniak et al. Generalized motion through adaptation of velocity profiles. *ROMAN*, 2010.

[4] M. Gielniak et al. Secondary action in robot motion. *ROMAN*, 2010.

[5] M. Gielniak et al. Anticipation in robot motion. *ROMAN*, 2011.

[6] C. Howard and A. Holcombe. Unexpected changes in direction of motion attract attention. *Attention, Perception, and Psychophysics*, 72(8), 2010.

[7] O. Johnston and F. Thomas. *The illusion of life: Disney animation.* Disney, 1995.

[8] S. Lang et al. Providing the basis for human-robot interaction: A multi-modal attention system for a mobile robot. In *ICMI*, 2003.

[9] C. Liu et al. Motion magnification. In *SIGGRAPH*, 2005.

[10] B. Mutlu et al. A storytelling robot: Modeling and evaluation of human-like gaze behavior. In *Proc. Humanoids*, 2006.

[11] L. Takayama et al. Expressing thought: Improving robot readability with animation principles. *Proc. of HRI*, 2011.

[12] M. Unuma et al. Fourier principles for emotion-based human figure animation. In *Proceedings of the 22nd Annual Conference on Computer Graphics and Interactive Techniques*, pages 91–96, 1995.

[13] J. Wang et al. The cartoon animation filter. In *SIGGRAPH*, 2006.

[14] Y. Ye and C. Liu. Animating responsive characters with dynamic constraints in near-unactuated coordinates. *ACM Trans. Graph.*, 27(5), 2008.

The Illusion of Robotic Life

Principles and Practices of Animation for Robots

Tiago Ribeiro
INESC-ID, Instituto Superior Técnico,
Universidade Técnica de Lisboa, Portugal
Av. Prof. Dr. Cavaco Silva, 2744-016
Porto Salvo, Portugal
tiago.ribeiro@gaips.inesc-id.pt

Ana Paiva
INESC-ID, Instituto Superior Técnico,
Universidade Técnica de Lisboa, Portugal
Av. Prof. Dr. Cavaco Silva, 2744-016
Porto Salvo, Portugal
ana.paiva@inesc-id.pt

ABSTRACT

This paper describes our approach on the development of the expression of emotions on a robot with constrained facial expressions. We adapted principles and practices of animation from Disney™ and other animators for robots, and applied them on the development of emotional expressions for the EMYS robot. Our work shows that applying animation principles to robots is beneficial for human understanding of the robots' emotions.

Categories and Subject Descriptors

I.2.9 [**Artificial Intelligence**]: Robotics; H.5.1 [**Multimedia Information Systems**]: Animations, Evaluation/methodology; J.4 [**Social and Behavioral Sciences**]: Psychology; J.5 [**Arts and Humanities**]: Arts, fine and performing

General Terms

Design, Affective Computing

Keywords

Animation, Facial Expressions, Emotions

1. INTRODUCTION

Nearly one hundred years ago, the world witnessed the birth of animation movies, which has clearly marked the culture and art of the past century. Today technology has evolved, bringing us closer to having such animated features present in our real life. We are talking, of course, of animated robots. Robot use is scaling from the industrial production factories into our homes and into our lives, rising as a trend of becoming our own artificial companions and friends in the future[1]. One important aspect that has been gaining focus on social robots is non-verbal communication. This kind of affective behaviour plays an important role on the establishment and maintenance of long-term relationships, which is one of the goals for social robots [7]. Breazeal also showed that it is important for the user to understand the robot's behaviour, in order to maintain a social relationship [10]. Non-verbal communication can be achieved through several channels [6]. Our current work, however, focuses on the expression of emotions solely through animation [9, 8, 47, 2, 40, 19, 44]. We are especially interested in understanding how to develop a method of expressing emotions through animation in robots with constrained appearance and expression [10]. Van Breemen defines animation of robots as *the process of computing how the robot should act such that it is believable and interactive* [47]. We show that applying principles and practices of animation to robots portrays them with a higher degree of understandability when expressing their emotions.

We start by exploring the background that is necessary for our work, and by analysing related work from other authors. We then present our approach, and its two-phase development, with results for each. Finally we draw conclusions and outline our next steps.

2. BACKGROUND

We seem to know when to 'tap the heart'. Others have hit the intellect. We can hit them in an emotional way.

Walt Disney

Animation artists have struggled for creating believable emotional characters for many years now, and somehow have succeeded [5]. Several different emotional models have also been developed in the field of psychology and adapted for computation, so we must understand how these models can relate to and influence the expression of emotions.

2.1 Animation Principles and Practices

One approach for animating robots is making use of animation principles and practices. Most professional animators follow a set of twelve principles that result of more than 60 years of Disney™ productions. These were compiled in a book by Johnston and Thomas [46], the last two of the Nine Old Men[2], thus representing the bible of animation. Other animators have explored different perspectives than Disney™'s, such as Bob Clampett, Chuck

[1] LIving with Robots and intEractive Companions, www.lirec.eu

[2] A group of nine animators that worked closely with Walt Disney since the debut feature Snow White and the Seven Dwarfs (1937) and onto The Fox and The Hound (1981).

Jones and Tex Avery, with Warner Bros.™, who explored a deeper connection between emotion, exaggeration and character expressiveness. They have credit on short films featuring Bugs Bunny, Daffy Duck, Will E. Coyote, Porky Pig[3] and so on. William Hanna and Joseph Barbera are also mentioned as strong references from the Metro-Goldwyn-Mayer™Company, however, their focus was more story-driven than that of the Warner Bros.™productions, as we can find in the Scooby Doo and Tom & Jerry series[4]. Their animation is generally regarded as more simplistic, however still expressive as of the others [28]. Jim Henson's television puppet phow can also be viewed as interesting reference and inspiration for the animation of robots, because they deal with animating real physical characters that in most cases do not have much freedom of expressiveness [17].

2.1.1 *Disney™'s Twelve Principles of Animation applied to Robots*

Van Breemen [47] has already proposed to apply the Principles of Animation to robots, claiming *user-interface robots to have the same problem of early day's animations: they miss the illusion of life.* The Twelve Principles of Animation, described by Johnston and Thomas [46], should impact the creation of robot animated expressions. We present each principle below while relating them to robot animation.

Squash and Stretch
The movement and liquidness of an object reflects that the object is alive. One rule of thumb is that despite them changing their form, the objects should keep the same volume while squashing and stretching. This principle is hard to apply to robots, because robots are generally composed of rigid parts.

Anticipation
Anticipating movements and actions helps viewers and users to understand what a character is going to do. That anticipation helps the user to interpret the character or robot in a more natural and pleasing way.

Staging
Staging is related to the general set-up in which the character expresses itself. This principle is related to making sure that the expressive intention is clear to the viewer. Some ways of accomplishing this are by positioning lights, camera, music, characters and surrounding objects. In robots this suggests that we can use multi-modal expression with lights or sound.

Straight Ahead and Pose-to-Pose
Straight ahead action is more of a free method, when the animator knows what he wants to do, but hasn't completely foreseen it, so he starts on the first frame and goes on sequentially animating on to the last. Pose-to-Pose animation is used when the sequence is pre-planned, and is especially useful when making use of physics and synchronizing movements. We can regard this as having a robot to produce interactive, procedural animation (straight-ahead), or pre-designed animation, which can be synced and blended with other kinds of behaviours (Pose-to-Pose).

[3]These characters are ™of, and ©Warner Bros. Entertainment Inc.
[4]Scooby-Doo and Tom & Jerry are ™of, and ©Hanna-Barbera.

Follow-Through and Overlapping Action
This principle works as an opposite of Anticipation. When a character stops doing something, it shouldn't stop abruptly, for that causes an unnatural feeling. This follow-through animation is generally associated with inertia, but can also be used to emphasize the stop. A character that punches another one will first pull its body and arm back (anticipation), then punch (action), and slightly fall forward while trying to regain balance, give step or two or even fall down (follow-through/reaction). These principles have impact both on causing the impression that the robot is part of our natural world, and also to mark that an action has ended.

Slow In and Slow Out
This principle helps the motion of objects and characters to seem natural and pleasant to the viewers. Along with anticipation and follow-through, objects shouldn't be abrupt when they start or stop moving. A hand pulling back for a punch accelerates backwards, decelerates to a stop, and then accelerates forward to the punch. When a character smiles, its mouth doesn't suddenly turn to a smile, it slowly blends into it. In robot animation this principle is one of the major ones to apply, as it states that animations should softly blend one into another, or in and out of the character. Van Breemen [48] called this Merging Logic when he applied it in Philips's iCat[5] .

Arcs
This principle states that natural motions occur in arcs, so that should be taken into account when designing animation. When a person looks to the left and the right, it shouldn't just perform a horizontal movement, but also some vertical movement, so that the head will be pointing slightly upwards or downwards than it was while facing straight ahead. For robots this principle can be used both for pre-designed animations, and also for procedural ones.

Secondary Action
This kind of action does not directly contribute to the expression of the character, but aids in making it believable. When people speak to each other, they often scratch some part of their body, adjust their hair, or even look away from the person with who they are interacting. These actions are designated as secondary actions. One simple case of applying this to robots can be adding random blinks to the eyes, and adding a soft, slow sinusoidal motion to the body to simulate breathing (lat. *anima*).

Timing
Animation is, of course, all about timing, so it seems redundant to include timing as a principle. However, there is another purpose here. First, timing can help us to understand a motion to belong to a particular physical world. Timing of physical motion on Earth or on the Moon is very different. But timing can also be used as expression. A fast motion often suggests that a character is active and engaged on what it's doing. Saerbeck and Bartneck [40] have even recently

[5]©2004-2012 Koninklijke Philips Electronics N.V.

correlated acceleration of a movement with the perceived arousal of a robot. For robots the scalability factor of time is interesting to explore, as the same movement, pre-animated or procedural, can have different meanings depending on the timing used, which aids on reusing motions and animations with different emotions.

Exaggeration
This principle, along with Timing, is one of they key magic features in animation. Animated characters and objects do not need to follow the rules of our world and our physics. Exaggeration can be used to emphasize the robot's movements, expressions or actions, in order to make them more noticeable and convincing.

Solid Drawing
This is another principle that seems not to relate to robot animation, despite being able to be related to robot design. However, this principle also states some rules to follow while designing poses. A character should not stand stiff and still. We generally put more of our weight on only one of the legs, and often to not keep our hands at the same level. The main concept to get from this principle is asymmetry. Faces rarely exhibit the same expression on the left and right sides, and almost no poses are symmetrical, unless one wants to convey the feeling of stiffness.

Appeal
It is obvious that most people prefer beautiful, appealing characters than grotesque and ugly ones. However, this principle does not relate to "drawing pretty things", but to the fact that the viewers appeal should be taken into account when drawing and animating characters. If we want a viewer or user to love a character, then it should be beautiful and gentle. If we want them to hate a character it should be stiff and grotesque. Even if one wants to make viewers and users feel pity for a character (such as an anti-hero), then the character's motion and behaviour should generate that feeling, through clumsy and embarrassing behaviours. Motion and behaviours of robots should also be easy to understand, because if the users don't understand what they see, their appeal for the robot will fall.

2.1.2 Warner Bros™. and MGM™

The works by Warner Bros™. and MGM™pay special focus on one special principle that is Exaggeration.
Disney™'s principles already include this one, however, these other animators took this concept much further. Their exaggeration was especially physical, and generally included extreme distortions and breaking up a character or object in order to express emotions and key moments. Tex Avery was one of the greatest animators of all time, and created lots of concepts and gags that have remained not just as animation *clichés*, but also as expressive guidelines [11]. He was especially a master in exaggeration, being credited as the creator of the eyes-popping-out expression that is now sometimes called *the Tex Avery expression*, or just *a Tex Avery*. The practice of these artists is not as well documented as that of Disney™'s. However, while viewing their work one can understand not only a common line underneath them, which helps to define it as a style different than Disney™'s, but also notice that animation can do a lot more than what Disney™principles state.

2.1.3 Puppet Animation

If we are looking at artistic influences for animating robots, we must take a look at a genre that actually shares some resemblance with it. Puppet animation follows some of the guidelines that traditional animation follows. However, in practice, things work out very differently. Puppets are physical objects that are built in order to move and be expressive, and are subject to the laws of physics and our real world. If we replace the word *Puppets* with *Social Robots* in this last sentence, it would still be valid. Jim Henson created a puppet television show which has influenced several generations [17]. Their puppets are characterized by being very simple, and by not possessing many moving parts to be used in expression. Thus, animators had to develop their own non-verbal language suited for their puppets. One example is their puppet's head, which is generally characterized by an egg-like shape that is sliced somewhere in order to just open and close as a mouth. This mouth is frequently the only moving part of their head; even their eyes cannot gaze or shut. If we watch episodes of the series, we will find moments in which the characters cover their eyes with their hands in order to "close their eyes".
It is empirically clear that exhibiting emotions solely through facial expression is nearly impossible with a mouth that just opens and closes; we recall here that a simple smile is expressed especially with the corners of the mouth, and lower eyelids. So they generally associate whole body movement to the mouth in order to express emotions. For fear, one may find that the mouth trembles open, the hands grab the face, and the body assumes a posture of withdrawal. For happiness the mouth will open wide and the whole puppet will bounce around sort of crazy with its arms balancing freely to the movement. An angry expression is achieved by tilting the puppet (especially the head) against its object or character of hate, closing its mouth and pulling back its arms. As in most artistic inspirations, the best way to learn their principles and practices is by watching the episodes and eventually tag them with emotions and expressions, so that they may serve of reference.

2.2 Emotion Models

Before trying to express emotions with robots, it is important to analyse how emotions are being computed. We will not focus here on understanding which model or classification is better. Instead, we look to understand what kind of computational models might be used to trigger the expression of emotions, so that we can take into account than information.
One of the most referenced emotional models is FACS by Ekman and Friesen [16]. They establish that through facial expression, humans can universally recognize in other humans six basic emotions: anger, disgust, fear, joy, sadness and surprise. Albeit simple and universal, this set of emotions is somewhat limited, as humans can actually express and recognize much more. Another popular theory of emotions is that of Ortony, Clore and Collins [36], generally referred to as the OCC model, which defines 22 different emotional categories. However, this model is complex, and it isn't simple to understand how one can express each of those 22 emotions. Bartneck [3] has developed a model that

makes use of the OCC theory, and states that emotion processing is a five-phase process:

Classification What do I feel about what just happened?
Quantification How much do I feel about it?
Interaction How does this affect what I was already feeling?
Mapping What should I do to express this feeling?
Expression How should I do that?

For our work we are mostly interested in the Mapping and the Expression phases, however, it is important to understand what lies beneath them. Bartneck also suggests that the OCC emotions may somehow be mapped to the Ekman's expressions, but this mapping is not trivial. There are 11 positive emotional categories in the OCC model, but only one positive Ekman expression (smile). A workaround for this can be to contemplate the Interaction phase information for expression, so that, if possible, we can perform blending of expressions. Another popular model for emotions is the Pleasure-Arousal-Dominance (PAD), proposed by Mehrabian [31]. This is a three-dimensional model used to define an emotion in terms of three dimensions: Pleasure, Arousal and Dominance. This approach is scientific and adequate for computation, as emotions form a continuous space characterized by those three variables. One advantage of this model is that the transition from one emotion to another may be seamless, however, when transitioning from emotion A to emotion B, the transition may go through an emotion C, which can be invalid.

3. RELATED WORK

Several other authors have looked into developing a framework for expressive behaviour. Some have focused on describing languages for specification of hand and arm gestures, and on expressive dialogue acts [18, 12, 26, 25, 38, 13]. Badler [1] has presented an Expressive MOTion Engine (EMOTE) that implements LMA [21] using high-level parameters for human animation control; however, this solution is designed for anthropomorphic characters. Currently, the SAIBA framework [24] is becoming popular, but authors have used it mainly oriented at humanoid characters or robots [49, 34].

As to studying the expression of emotions with robots, various authors have done their own studies, and have therefore drawn conclusions on their specific robot [43, 44, 41, 10, 45, 52, 27, 19, 48, 23]. [51] have developed a robot that mimics human expressions by observation. Their robot recognizes expressions from videos, and maps the expressions' FACS features into the robot's own effectors.

Bethel [6] has strived into studying how to express emotions in robots that do not possess expressive capabilities. Her work focuses on robots that are mean to be functional, like search-and-rescue or military robots, and how to use multi-modal expression for the correct communication of the robot's emotional state and empathic behaviour. Multimodal expression of emotions with robots was also explored by Jung et al. [22] and Gorostiza et al. [19].

A mark on our research is the fact that we have found that most works [10, 15, 24, 34, 32] have integrated interaction with expression, and therefore have developed expressive behaviour for their specific interactive scenarios. We aim at studying how the definition of expressive behaviour may be dissociated from that interaction, while remaining useful for the interaction to trigger and produce such behaviour.

The need for an exhaustive and general comprehension on analysis, modelling and synthesis of facial expression in robots has been reported by Bartneck and Lyons [4]. Schröder et al. have proposed a general mark-up language that is not dependant of any emotion model or theory, thus marking a possible step for an abstract and broad specification of emotive behaviour [42]. However, their language provides only a structure to gather the description of an emotional expression, with no means on how to accomplish it. Moussa et al. have embarked on one approach for this, using MPEG-4 Facial Animation Parameters (FAP) [37] applied to a humanoid robotic face [33]. However, FAPs are designed for human faces and thus comprehend an extensive number of expressive features for the human face. Saerbeck and Bartneck have also attempted to correlate robotic motion with perceived emotions [40], and concluded that there exists a correlation between a robot's acceleration, and the perceived arousal.

3.1 Designing Emotion for Robots

Recently some authors have looked into design and communication concepts and practices so that they may be incorporated in the engineering of emotional expression. Hess [20] states that *the ability to well communicate emotions is relevant for both the encoder, who would like to be understood, and the decoder, who strives to understand.* Product design is an example of a field that has struggled between creating art/emotion and functional objects [35]. Regarding the design of believable social robots, Dautenhahn [14] divided the design process in two dimensions: the Universal dimension, in which the universal features of a behaviour or expression are abstracted; and the Abstract dimension, in which the designer of the behaviour or expression is free to be creative and develop a more artistically based result. Meerbeek et al. also follow the Universal vs. Abstract dimensions of design, stating that *since human expressions cannot be mapped one-to-one with expressions of the robot, we abstracted the human expressions first* [30]. The same authors have defended that the design of behaviour and expressions of robots should be a blend between an artistic approach and an iterative cycle to evaluate and refine the result, which follows the usual practice both in engineering and usability design. They also consider that using virtual 3D models for animating and visualizing the expressions of a robot is useful, especially if the virtual model is designed with resemblance to the real physical model and its behaviour [29]. [40] conclude that the perceived affective state of a robot is independent of the embodiment, suggesting that we can use motion design tools across different embodiments.

4. EXPRESSING EMOTIONS IN EMYS

Our EMYS robot has eleven degrees of freedom (Figure 1). The head can turn to the sides for panning and gazing, however, the eyes cannot look the other way for gazing or expression. One can notice that EMYS does not have any lower eyelids, which caused some difficulty in some of the expressions. Moreover, the only movement for the mouth (the Lower Plate), is to open and close. This and the lower eyelid absence made it difficult to empirically create a happy expression for Joy, because a smile comes especially from the

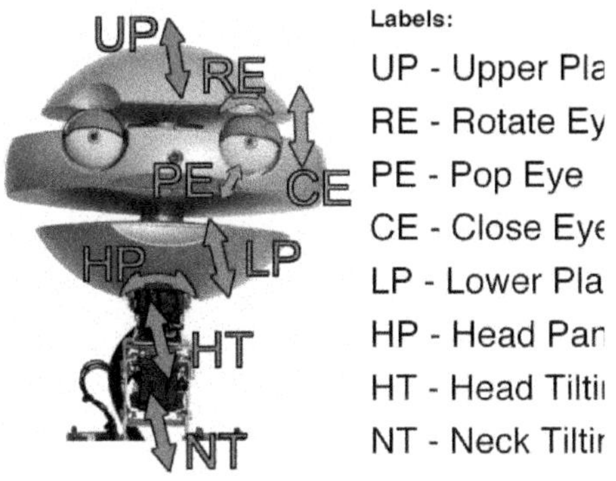

Figure 1: EMYS's facial expression features.

corners of the lips, and the lower eyelids, both of which are absent on EMYS's face. As to asymmetry, the only thing that we can do besides turning the whole face to one side, is closing the eyes and rotating the eyelids. The Upper Plate acts as middle eyebrows, by raising and lowering. Rotation of the eyelids is mostly used as lowering and raising of the outer eyebrows. Our approach for expressing emotions with the EMYS robot was inspired by the work and proposals of van Breemen [47, 48], Dautenhahn [14], Meerbeck et al. [30, 29] and Moussa et al. [33]. We started by creating a virtual interface for our robot using Autodesk 3ds max. The virtual model is the exact model used for the manufacturing of EMYS. Every time the virtual model is updated, a script communicates with the physical robot so that the latter is also updated and synchronized with the virtual one. All the degrees of freedom of the virtual model are also limited in order to provide the same freedom of the physical robot. This setup enabled us to use all the tools available for animation in 3ds max, thus providing us with nearly the same technological freedom as professional 3D animators. Another script provides us with export functionality, so that the animations created in 3ds max may be used by our own external .NET library, so that in the future we may use these animations in any other application that supports loading such type of library. We started to create expressions for EMYS in an iterative cycle, following [30, 29]. The first phase resulted in our initial expressions, which were then refined in the second phase.

5. PHASE 1: INITIAL EXPRESSIONS

For our initial expressions we analysed the work of Ekman and Friesen [16], thus deciding to create expressions for the six basic emotions that they claim to be universally recognized. They define these expressions in terms of FACS, which are features presents in the human face that can be used for expressing and recognizing emotions. MPEG-4 defines FAPs, another model of features that was especially defined for facial expression in virtual characters [37]. However, following [30], we abstracted the FACS and FAPs features into a simpler set, composed of only 9 valenced features that could more easily be mapped to a robotic face. Our approach here was therefore a more simplistic and cartoon-oriented version of [51]'s approach. We considered each feature to have 5 positive values, one neutral and other 5 negative values. For example, one feature is JA (Jaw), for which negative values means Jaw closed, neutral value means jaw relaxed, and positive values mean jaw open. Another example is LE, Lower-Eyelids, for which negative values correspond to closed eyelid (eyelid raised) and positive values to open eyelid (eyelid lowered).

From [16]'s descriptions, we defined the expression of each emotion in terms of ranges that could be set for each feature, so that each expression could have 5 different intensities (from 1 to 5, being that 0 would correspond to the default pose). For happiness, for example, an intensity of 1 (lowest) corresponded to having -1 on the Lower Eyelids (slightly closed), 0 on the Jaw (relaxed) and 1 on the Lip Corners (slightly raised). The intensity of 5 (highest) for happiness was defined as -5 for Lower Eyelids (totally pulled up), 5 for Jaw (mouth wide open as if laughing) and 5 for Lip-Corners (totally pulled up). This first part of our definition corresponds to Dautenhahn's Universal Dimension [14].

We created expressions for EMYS first using these definitions and then adding some creativity due to the fact that EMYS does not have, for example, Lower Eyelids, and that it supports Eyes-Popping. This second part corresponds to Dautenhahn's Abstract Dimension [14].

We followed by a preliminary evaluation, described in [39], to test if the human users were able to correctly identify our expressions. The results showed that Anger and Sadness were easily recognized, however Disgust was confused with Anger, and the other three we all confused between them.

6. PHASE 2: REFINED EXPRESSIONS

After the results from the first phase, we understood that due to the constrained appearance of EMYS's face, we needed to move away from a purely FACS approach and focus on some specific aspects. On this second phase we refined the initial expressions based on the application of the principles of animation just presented.

6.1 Development

Applying principles and practices of animation to robots is currently an artistic process. We believe that in the future we will be able to understand how to make these principles intrinsic to animation systems. Therefore we must first explore the ground before we can draw conclusions on how to generalize them, because each robotic embodiment will function and express itself in different ways. Sharing methods and discussing ideas with the scientific community is therefore, the first step we must take. Our method applied and worked specifically on the EMYS robot; However, it may inspire other researchers on how they may apply them on their own robotic embodiments. It also describes our line of thought on this subject, which can also inspire robot designers and creators, so that the design process may in fact consider these principles. Take exaggeration, for example: Instead of designing eyes that open and close normally, the designer may consider to create eyelids that can open wider than usual, or eyelids that can be extremely pressed up and together. Including controllable leds and displays on the robot also helps on staging, even if it does not seem necessary for the robot's functionality. EMYS has eyes that can pop out, providing us with a new feature to explore, even if functionally he doesn't need it. If he didn't physically support it however, we couldn't do it. We follow with the considerations we had on our robot.

For the distinguishing between Anger and Disgust, we made better use of Exaggeration, Timing and Solid Drawing on the latter one. We emphasized the asymmetry on the eyes (one closing more than the other), and increasing the horizontal panning. We also drove the head back into position in a more resistive way, and kept the eyes pressed during more time. To Anger we added some Exaggeration to the clenching of the eyes, by making the eyelids not only close, but tremble while closed. For both Surprise and Happiness we adjusted the Timing and Staging in order to better differentiate both expressions. Surprise was made more instant, and we also modified the eye-popping in order to make it more noticeable. For Happiness in particular, we drew inspiration from Jim Henson's work[17], in which the puppets bounce around when they are happy. Our robot could not bounce around, but could balance its head back and forth. Fear was also furnished with Exaggeration, also inspired by those puppets, by adding tremble to the chin, which really seemed to emphasize the fear factor, and also some Staging by adding a negative nod to the head, like if the character was pleading in despair. Sadness was kept nearly the same, because it already had good results. We also enhanced all of the expressions with respect to the principles of Arcs, Slow In/Slow Out and Follow Through animation.

6.2 Evaluation - Procedure

To test if human users were able to correctly identify our refinements we filmed the robot's expressions in order to run a video-based online survey. Several authors have supported that HRI studies could use videotaped scenarios as opposite to live interaction [2, 50]

Participants were first presented with information about the purpose and guidelines of the procedure. Evaluation consisted in visualizing a set of videos of EMYS performing an animation, and selecting, for each video, the emotion that was thought EMYS was expressing. Each emotion was expressed in three different intensities, totalling 18 videos, for the 6 basic emotions. We tested only intensities 1, 3 and 5 for each expression, so that the evaluation would not be too long. Each intensity of each expression was presented as an individual 4-second video, and the videos were all presented in a random order. Participants could play each video any number of times, and then choose an option from a forced-choice scale, regarding the perceived expression: *Anger*, *Disgust*, *Fear*, *Joy*, *Sadness*, *Surprise* or *Don't Know*. In the end of the survey we also collected some information about the cultural and artistic habits of the participants.

6.3 Evaluation - Participants

Data of 61 participants from 7 different countries was collected, with ages spanning 17 to 55 (mean age 27.3) and 49.2% being male. 42.6% were from Portugal, 18% from Hungary, 9.9% from Germany, United Kingdom and Poland, and the remaining 3.2% from The Netherlands and India. 73.8% of the participants appreciate Animation Movies, 31.1% of them appreciate Character-Driven Video Games and 23% appreciate Comic Books.

6.4 Evaluation - Results

Our new results showed a majority of correct answers for all of the expressions except for intensity 1 of *Disgust*.

Table 1: Results from our evaluation. *Abbreviations: Ang = Anger, Disg = Disgust, Sad = Sadness, Surp = Surprise, DK = Don't Know*

	Ang	Disg	Fear	Joy	Sad	Surp	DK
Anger1	83.6%	4.9%	1.6%		1.6%		8.2%
Anger3	88.5%	3.3%			1.6%		6.6%
Anger5	90.2%				3.3%		6.6%
Disgust1	1.6%	45.9%	1.6%	1.6%		1.6%	47.5%
Disgust3	19.7%	36.1%	1.6%	1.6%	1.6%	4.9%	34.4%
Disgust5	6.6%	55.7%		3.3%		4.9%	29.5%
Fear1		1.6%	72.1%	4.9%		11.5%	9.8%
Fear3	1.6%	4.9%	67.2%		1.6%		24.6%
Fear5	3.3%	4.9%	73.8%	1.6%		1.6%	14.8%
Joy1	4.9%	3.3%	3.3%	62.3%		13.1%	13.1%
Joy3	1.6%		1.6%	50.8%		16.4%	29.5%
Joy5	3.3%		3.3%	59.0%		21.3%	13.1%
Sadness1	1.6%	1.6%	3.3%	6.6%	50.8%	13.1%	23.0%
Sadness3		1.6%			95.1%		3.3%
Sadness5		4.9%			90.2%		4.9%
Surprise1	3.3%	6.6%		16.4%	1.6%	50.8%	21.3%
Surprise3	1.6%		11.5%	1.6%		67.2%	18%
Surprise5		3.3%	13.1%	4.9%		67.2%	11.5%

Table 2: Overall results from our online evaluation.

	None Correct	One Correct	Two Correct	All Correct
Anger	4.9%	3.3%	16.4%	75.4%
	8.2%		91.8%	
Digust	24.6%	31.1%	26.3%	18.0%
	55.7%		44.3%	
Fear	4.9%	19.7%	32.8%	42.6%
	8.2%		91.8%	
Joy	18.0%	21.3%	31.2%	29.5%
	39.3%		60.7%	
Sadness	3.3%	3.3%	47.5%	45.9%
	6.6%		93.4%	
Surprise	8.2%	18.0%	52.5%	21.3%
	26.2%		73.8%	

Table 1 shows the results. For each emotion that we expected to recognize, the table contains three lines, each corresponding to a different intensity for that emotion. The results for all the expressions yielded significant values. The columns contain the percentage of participants that, for each expression, selected each of the emotions displayed in the title of the column. In the first line, for example, *Anger1* shows that for intensity 1 of *Anger*, the majority of the respondents (83.6%) selected *Anger*, while 8.2% selected *Don't Know*, 4.9% selected *Disgust*, and the remaining 3.2% were equally divided between *Fear* and *Sadness*. Lines *Anger3* and *Anger5* correspond to intensities 3 and 5 of *Anger*.

Table 2 shows the overall results for each emotion, with the percentage of people that got all three intensities correct, only two intensities correct, only one of them, or none. For *AngerOverall*, for example, we see that most people (75.4%) selected the correct emotion for all three intensities that were shown, thus supporting that this expression was successfully

designed and expressed. In a summary, we consider that *Anger*, *Fear* and *Sadness* had very good results (over 90% got at least two of the expressions correct). *Anger* and *Sadness* were already expected to have good results, because of the first evaluation, however for *Fear* this was a great improvement. *Surprise* and *Joy* had fairly good results, as more than 60% of the participants identified correctly at least two of the three intensities shown. *Disgust* was our worst result. Intensity 1 of *Disgust* had slightly more participants selecting *Don't Know* (47.5%) than *Disgust* (45.9%). However for intensities 3 and 5, most participants identified the emotion correctly.
Nevertheless, Disgust was the only emotion for which less than 50% got at least two of the expressions correct.

7. CONCLUSIONS AND FUTURE WORK

We have given a step further into the design and development of non-verbal expression of emotions in social robots by making use of animation principles and practices. Our work derives from other authors who have given great contribute to this field of study, by reinforcing that not only psychologists, but also artists - especially animation artists - have an important role in supporting scientists on this quest. Expressing emotions in robots is thus equivalent to creating the illusion of life in robots: making people think and feel that the mechanical being they see in front of them actually has a *persona* and feelings. Our results were positive and support that we continue to develop our studies on this topic. One of the next steps that we plan is to study and integrate multi-modal expression in our model that is inspired both by art and psychology, using colors (through lights) and sound. We also plan on repeating our study on other robots, to try to find out what kind of general principles we can gather and apply in respect to animation and expression of robots.

To give the Coyote a look of anticipatory delight, I draw everything up: the eyes are up, the ears are up, and even the nose is up. When he is defeated, on the other hand, everything turns down. You can't do that as dramatically with human beings, although the emotions expressed are fully human.

Charles "Chuck" Jones

8. ACKNOWLEDGEMENTS

This work was supported by the EU FP7 ICT-215554 project LIREC (LIving with Robots and intEractive Companions). This research was supported by EU 7th Framework Program (FP7/2007-2013) under grant agreement no. 215554, FCT (INESC-ID multiannual funding) through the PIDDAC Program funds.

9. REFERENCES

[1] N. Badler, J. Allbeck, L. Zhao, and M. Byun. Representing and parameterizing agent behaviors. In *Proceedings of the Computer Animation*, Washington, DC, USA, 2002. IEEE Computer Society.

[2] C. Bartneck. Affective expressions of machines. In *CHI '01 extended abstracts on Human factors in computing systems*, NY, USA, 2001.

[3] C. Bartneck. Integrating the occ model of emotions in embodied characters, 2002.

[4] C. Bartneck and M. J. Lyons. Facial expression analysis, modeling and synthesis: Overcoming the limitations of artificial intelligence with the art of the soluble. In *Handbook of Research on Synthetic Emotions and Sociable Robotics*. IGI Global, 2009.

[5] J. Bates. The role of emotion in believable agents. *Commun. ACM*, 37:122–125, July 1994.

[6] C. L. Bethel. *Robots Without Faces: Non-Verbal Social Human-Robot Interaction*. PhD thesis, University of South Florida, June 2009.

[7] T. W. Bickmore and R. W. Picard. Establishing and maintaining long-term human-computer relationships. *ACM Trans. Comput.-Hum. Interact.*, June 2005.

[8] C. Breaseal and R. Brooks. Robot emotion: A functional perspective. *Who needs emotions: The brain meets the robot.*, Oxford University Press, 2004.

[9] C. Breazeal, , C. Breazeal, and P. Fitzpatrick. That certain look: Social amplification of animate vision. In *In Proceedings AAAI Fall Symposium, Socially Intelligent Agents - The Human in the Loop*, 2000.

[10] C. Breazeal. Emotion and sociable humanoid robots. *Int. J. Hum.-Comput. Stud.*, 59:119–155, July 2003.

[11] J. Canemaker. *Tex Avery: the MGM years, 1942-1955*. Turner Pub., 1996.

[12] J. Cassell, H. H. Vilhjálmsson, and T. Bickmore. Beat: the behavior expression animation toolkit. In *SIGGRAPH '01*, New York, NY, USA, 2001.

[13] N. E. Chafai, C. Pelachaud, and D. Pelé. Towards the specification of an eca with variants of gestures. In *IVA '07*, Berlin, 2007.

[14] K. Dautenhahn. Design spaces and niche spaces of believable social robots. *RO-MAN '02*, 2002.

[15] B. De Carolis, C. Pelachaud, I. Poggi, and M. Steedman. Apml, a markup language for believable behavior generation. Springer, 2004.

[16] P. Ekman and W. Friesen. *Unmasking the Face: A Guide to Recognizing Emotions From Facial Expressions*. Prentice Hall, 1975.

[17] C. Finch. *Jim Henson - The Works*. Random House, 1993.

[18] S. Gibet and T. Lebourque. High level specification and control of communication gestures: The gessyca system. In *Proceedings of the Computer Animation*, Washington, DC, 1999.

[19] J. F. Gorostiza et al. Multimodal human-robot interaction framework for a personal robot. In *RO-MAN '06*, 2006.

[20] U. Hess. The communication of emotion. *Emotions, Qualia, and Consciousness, Singapore*, 2001.

[21] J. Hodgson. *Mastering Movement: The Life and Work of Rudolf Laban*. Routledge, 2001.

[22] H. W. Jung, Y. H. Seo, Sahngwon, and H. S. Yang. Affective communication system with multimodality for a humanoid robot, ami. 2004.

[23] E. H. Kim, S. S. Kwak, K. H. Hyun, S.-H. Kim, and Y. K. Kwak. Design and development of an emotional interaction robot, mung. *Advanced Robotics*, 2009.

[24] S. Kopp, B. Krenn, S. Marsella, A. Marshall, C. Pelachaud, H. Pirker, K. Thórisson, and H. Vilhjálmsson. Towards a common framework for multimodal generation: The behavior markup language. *IVA '06*, 2006.

[25] S. Kopp and I. Wachsmuth. Synthesizing multimodal utterances for conversational agents: Research articles. *Comput. Animat. Virtual Worlds*, 2004.

[26] A. Kranstedt, S. Kopp, and I. Wachsmuth. Murml: A multimodal utterance representation markup language for conversational agents. In *Embodied Conversational Agents: Lets Specify and Compare them. WS at AAMAS 2002*, Bologna: Italy, 2002.

[27] B. J. A. Kröse, J. M. Porta, A. J. N. V. Breemen,

K. Crucq, M. Nuttin, and E. Demeester. Lino, the user-interface robot. *EUSAI 2003*.

[28] L. Maltin. *Of Mice and Magic: A History of American Animated Cartoons*. NAL Books, New York, 1987.

[29] B. Meerbeek, M. Saerbeck, and C. Bartneck. Iterative design process for robots with personality. In *AISB2009 Symposium on New Frontiers in Human-Robot Interaction*. SSAISB, 2009.

[30] B. Meerbeek, M. Saerbeck, and C. Bartneck. Towards a design method for expressive robots. In *HRI '09*, New York, 2009.

[31] A. Mehrabian. Pleasure-arousal-dominance: A general framework for describing and measuring individual differences in temperament. *Current Psychology*, 14(4):261–292, Dec. 1996.

[32] S. A. Moubayed, M. Baklouti, M. Chetouani, T. Dutoit, A. Mahdhaoui, J.-C. Martin, S. Ondas, C. Pelachaud, J. Urbain, and M. Yilmaz. Multimodal feedback from robots and agents in a storytelling experiment. 2008.

[33] M. B. Moussa, Z. Kasap, N. Magnenat-Thalmann, and D. Hanson. Mpeg-4 faps applied to humanoid robot head. *Immersive and Engaging Interaction with VH on Internet*, 2010.

[34] R. Niewiadomski, E. Bevacqua, M. Mancini, and C. Pelachaud. Greta: an interactive expressive eca system. *AAMAS '09*, 2009.

[35] D. A. Norman. *Emotional Design: Why We Love (or Hate) Everyday Things*. Basic Books, 2003.

[36] A. Ortony, G. L. Clore, and A. Collins. *The Cognitive Structure of Emotions*. 1988.

[37] I. Pandzic and R. Forchheimer. *MPEG-4 Facial Animation: The Standard, Implementation and Applications*. John Wiley & Sons, Inc., New York, 2003.

[38] H. Prendinger, S. Descamps, and M. Ishizuka. Mpml: a markup language for controlling the behavior of life-like characters. *J. Vis. Lang. Comput.*, 2004.

[39] T. Ribeiro, I. Leite, A. Paiva, J. Kedzierski, and A. Oleksy. Expressing emotions on robotic companions with limited facial expression capabilities. *IVA '11*, 2011.

[40] M. Saerbeck and C. Bartneck. Perception of affect elicited by robot motion. In *HRI '10*, New York, 2010.

[41] M. Scheeff, J. Pinto, K. Rahardja, S. Snibbe, and R. Tow. Experiences with sparky: A social robot. In *WS on Interactive Robot Entertainment*, 2000.

[42] M. Schröder, H. Pirker, and M. Lamolle. First suggestions for an emotion annotation and representation language. 2006.

[43] K. Suzuki et al. Intelligent agent system fßor human–robot interaction through artificial emotion. *Proceedings of the IEEE SMC*, 1998.

[44] T. I. T. Tojo, Y. Matsusaka and T. Kobayashi. A conversational robot utilizing facial and body expressions. *IEEE SMC '00*, 2000.

[45] F. Tanaka, K. Noda, T. Sawada, and M. Fujita. Associated emotion and its expression in an entertainment robot qrio. In *ICEC*, 2004.

[46] F. Thomas and O. Johnston. *The Illusion of Life: Disney Animation*. Disney Editions, 1995.

[47] A. van Breemen. Bringing robots to life: Applying principles of animation to robotics. In *CHI'04 WS on Shaping Human-Robot Interaction*, 2003.

[48] A. J. N. van Breemen. Animation engine for believable interactive user-interface robots. *IROS '04*, 2004.

[49] H. Vilhjálmsson, N. Cantelmo, J. Cassell, N. E. Chafai, M. Kipp, S. Kopp, M. Mancini, S. Marsella, A. N. Marshall, C. Pelachaud, Z. Ruttkay, K. R. Thórisson, H. Welbergen, and R. J. Werf. The behavior markup language: Recent developments and challenges. 2007.

[50] S. N. Woods, M. L. Walters, K. L. Koay, and K. Dautenhahn. Methodological issues in hri: A comparison of live and video-based methods in robot to human approach direction trials. In *Proceedings of the 15th IEEE International Symposium on Robot and Human Interactive Communication*, pages 51–58. IEEE, 2006.

[51] T. Wu, N. J. Butko, P. Ruvulo, M. S. Bartlett, and J. R. Movellan. Learning to make facial expressions. In *Proceedings of the ICDL*, Shanghai, China, 2009.

[52] M. Zecca, S. Roccella, M. C. Carrozza, H. Miwa, K. Itoh, G. Cappiello, J. J. Cabibihan, M. Matsumoto, H. Takanobu, P. Dario, and A. Takanishi. On the development of the emotion expression humanoid robot we-4rii with rch-1. *IEEE Humanoids '04*, 2005.

Trajectories and Keyframes for Kinesthetic Teaching: A Human-Robot Interaction Perspective

Baris Akgun[1], Maya Cakmak[1], Jae Wook Yoo[2], Andrea L. Thomaz[1]

[1]School of Interactive Computing
Georgia Inst. of Technology
801 Atlantic Dr., Atlanta, GA
{bakgun3,maya,athomaz}@cc.gatech.edu

[2]Computer Science Dept.
Texas A&M University
College Station, TX
jwookyoo@gmail.com

ABSTRACT

Kinesthetic teaching is an approach to providing demonstrations to a robot in Learning from Demonstration whereby a human physically guides a robot to perform a skill. In the common usage of kinesthetic teaching, the robot's trajectory during a demonstration is recorded from start to end. In this paper we consider an alternative, *keyframe demonstrations*, in which the human provides a sparse set of consecutive keyframes that can be connected to perform the skill. We present a user-study ($n = 34$) comparing the two approaches and highlighting their complementary nature. The study also tests and shows the potential benefits of iterative and adaptive versions of keyframe demonstrations. Finally, we introduce a hybrid method that combines trajectories and keyframes in a single demonstration.

Categories and Subject Descriptors

I.2.9 [**Artificial Intelligence**]: Robotics; H.1.2 [**Models and Principles**]: User/Machine Systems

General Terms

Experimentation, Design

Keywords

Learning from Demonstration

1. INTRODUCTION

The goal of Learning from Demonstration (LfD) is to enable humans to program robot skills by showing successful examples [4]. There are various ways that this demonstration can take place. In this work we focus on "kinesthetic teaching" whereby a human teacher physically guides the robot in performing the skill, as in Fig. 1.

Kinesthetic teaching has several advantages for LfD. Since the teacher is directly manipulating the robot there is no correspondence problem and demonstrations are restricted to the kinematic limits (e.g. workspace, joint limits) of the robot. Moreover, extra hardware/instrumentation, such as motion capture or teleoperation devices, is not necessary.

Figure 1: Simon interacting with a teacher.

While there has been much work on representations and learning algorithms, the usability of kinesthetic teaching has not been explored in depth. Kinesthetic teaching may be challenging for everyday users who do not have experience manipulating a robot arm with many degrees of freedom.

In many practical LfD applications (e.g., homes, schools, hospitals), the teacher will not be an expert in machine learning or robotics. Our research identifies challenges that everyday users face with common robot learning methods and investigate improvements to these methods to increase their effectiveness.

In the typical kinesthetic teaching interaction, and most LfD interactions, each demonstration is an entire state *trajectory*, which involves providing a continuous uninterrupted demonstration of the skill. In this paper, we explore the alternative of providing a sparse set of consecutive *keyframes* that achieve the skill when connected together. We present an experiment that compares these through quantitative measures, survey results and expert evaluations. We find that both are suitable to kinesthetic teaching from the user's perspective and both communicate different information. We also present two modified keyframe interactions and evaluate their utility. Based on our findings, we introduce a way to merge trajectory and keyframe demonstrations to take advantage of their complementary nature.

2. RELATED WORK

There are several methods for learning skills, which mostly fit into two groups: direct policy learning and cost/reward learning. Dynamical system approaches such as Stable Estimator of Dynamical Systems (SEDS) [13] and Dynamic Movement Primitives (DMP) [17] and mixture models (e.g. Gaussian Mixture Models as in [10]) fall into the former and inverse reinforcement learning (IRL) [1] or apprenticeship learning fall into the latter. These methods are designed for different skill types and all have their pros and cons. Apart from GMMs and DMPs, most methods require many training samples which is not suitable for a short-duration HRI study. However, DMPs and GMMs have either implicit or explicit time dependency. Most of the methods either cannot handle cyclic skills or they need to be reformulated. We chose GMMs as our skill learning algorithm for the following reasons: GMMs can be learned with a low-number of demonstrations, can be trained in interaction time, can handle cyclic skills as well as point to point skills, GMMs can be modified slightly to handle keyframe demonstrations (see Section 4.2.2), and the fact that we do not need temporal robustness (e.g. the robot will not be stopped in the middle of the motion) for our experiments.

In kinesthetic teaching, demonstrations are often represented as arm joint trajectories and/or end-effector path [9, 12]. Some also consider the position of the end-effector with respect to the target object of the skill [5, 11]. Typically start and end points of a demonstration are explicitly demarcated by the teacher. Most studies subsample the recorded data with a fixed rate [3, 5]. Demonstrations are often time warped such that a frame-by-frame correspondence can be established between multiple demonstrations [12].

Keyframes have been used extensively in the computer animation literature [16]. The animator creates important frames in a scene and the software fills in-between. In the LfD setting, an earlier work [15] utilizes via-points, which are similar to keyframes. These are extracted from continuous teacher demonstrations and updated to achieve the demonstrated skill. A recent approach is to only record keyframes and use them to learn a constraint manifold for the state space in a reinforcement learning setting [6]. In this paper we consider both trajectory and keyframe representations.

Human-robot interaction (HRI) has not been a focus of prior work on kinesthetic teaching, but there are a few examples. In [18], kinesthetic teaching is embedded within a dialog system that lets the user start/end demonstrations and trigger reproductions of the learned skill with verbal commands. A modification to the kinesthetic teaching interface is kinesthetic correction [7, 8], where the teacher corrects aspects of a learned skill in an incremental learning interaction by using a subset of joints in subsequent demonstrations. In another study [14], four types of force controllers that effect the response to users are evaluated for kinesthetic teaching. The study addressed human preferences on which controller was the most natural.

While various learning methods for kinesthetic teaching have been explored, there is a lack of studies with end-users testing the effectiveness of these techniques in terms of HRI. This is the focus of our work.

3. PLATFORM

Our robotic platform in this work is "Simon," an upper-torso humanoid social robot with two 7-DoF arms, two 4-DoF hands, and a socially expressive head and neck, including two 2-DoF ears with full RGB spectrum LEDs (Fig. 1). Simon is designed for face-to-face human-robot interaction. Simon's arms have compliant series-elastic actuators with stiffness that can be dynamically changed. We use *Microsoft Windows 7 Speech API* for speech recognition.

4. DEMONSTRATION METHODS

We explore three different ways for teachers to demonstrate skills: *trajectory demonstrations*, *keyframe demonstrations*, and *keyframe iterations*. In all three, users physically manipulate the robot's right arm, which is gravity compensated, to teach it skills (Fig.1). In this section we detail the implementation of each kinesthetic teaching mode.

4.1 Trajectory Demonstrations

4.1.1 Interaction

The teacher is informed that the robot will record all the movement they make with its right arm. The teacher initiates the demonstration by saying "New demonstration", moves the arm to make the robot perform the skill and finishes by saying "End of demonstration." This process is repeated to give as many demonstrations as the person desires. After a demonstration, the teacher can use the speech command "Can you perform the skill?" to have the robot perform the current state of the learned skill and adjust his/her demonstrations to attend to any errors.

4.1.2 Learning

Joint angle trajectories are recorded as the teacher moves the robot's arm to perform the skill. As mentioned previously, we use GMMs as our skill learning algorithm [10].

The data is subsampled in the time dimension to a constant length before being input to the learning algorithm. In our case, learning is done in an eight dimensional space (7 joint angles over time). First, we use the K-means algorithm with a constant k. The resulting clusters are used to calculate initial mean vectors and covariance matrices for the expectation-maximization (EM) algorithm. The EM algorithm is run to extract a Gaussian-Mixture Model (GMM) from the data. The resulting GMM has k sub-populations which is kept constant during our experiments. Gaussian-Mixture Regression (GMR) is used to generate a trajectory to perform the learned skill. The desired time dimension vector is given to GMR which in turn generates the joint positions. Note that the algorithm can learn from either a single or multiple demonstrations.

4.2 Keyframe Demonstrations

4.2.1 Interaction

The teacher is informed that the robot will only record the arm configuration when they say "Record frame", and it will not record any movements between these keyframes. The teacher can use the speech commands "New demonstration", "End of demonstration" and "Can you perform the skill?" in the same way as trajectory mode.

4.2.2 Learning

The resulting data from this interaction is a sparse trajectory of joint angles. If the teacher forgets to give keyframes for the start or the end position, these are added automatically. We generate time information for each keyframe using the inter-frame distance and a constant average velocity.

Learning is slightly different than the previous case, but the space is the same. Again K-means is the first step, but now the number k is chosen to be the maximum number of keyframes across all demonstrations provided for a skill. Then a GMM is learned in the same way as the trajectory version. To generate the skill, the GMM sub-population means are traversed by splining between them. We took such an approach since the GMM sub-population means obtained from the keyframe version will be of different nature than the ones obtained from the trajectory version. With keyframes, it is more likely to be a transition between two trajectory segments whereas with trajectories it is more likely to be a mid-point of a trajectory segment [10]. Thus, we need to control the velocity at each keyframe.

4.3 Keyframe Iterations

We implemented an augmented version of keyframe demonstrations, in which a new demonstration is an iteration of the current learned skill.

4.3.1 Interaction

In this mode, an initial demonstration is provided using keyframe demonstrations. Then the teacher can navigate through and edit the frames of this demonstration to create additional demonstrations. The teacher uses the speech command "Next frame", and "Previous frame" to navigate through the keyframe demonstration. At any keyframe, "Modify this frame" can be used to modify the arm configuration of the keyframe. "Add new frame" adds a new frame after the current frame. "Delete this frame" deletes the current frame. Also, the teacher can say "Play current demonstration" to see the modified demonstration before submitting it to the learning set. "Record this demonstration" submits the current demonstration to the learning set. As in the previous modes, the teacher can play the learned skill with "Can you perform the skill?"

4.3.2 Learning

Learning is the same as in keyframe demonstrations. We implemented iterations only for keyframes since implementing a similar extension for trajectory demonstrations is non-trivial. It is difficult to tie partial trajectories together. Moreover, speech commands might not give the necessary precision for navigating through a trajectory and modifying it. This is an interesting topic for future work.

5. EXPERIMENTS

Our experiments address three research questions: (Q1) When everyday people teach the robot, what are the effects of each demonstration type? (Q2) Does the teaching method have any effect on learning different types of skills? (Q3) Can simple extensions to keyframe demonstrations (iteration and adaptation) increase performance/preference?

5.1 Experimental Design

5.1.1 Skills

We differentiate between two types of skills. *Goal-oriented* skills are related with achieving a particular world state (e.g., finger tip on a point while avoiding obstacles.) *Means-oriented* skills, on the other hand, include a gesture or communicative intent. We use four skills of each type, see Fig.2.

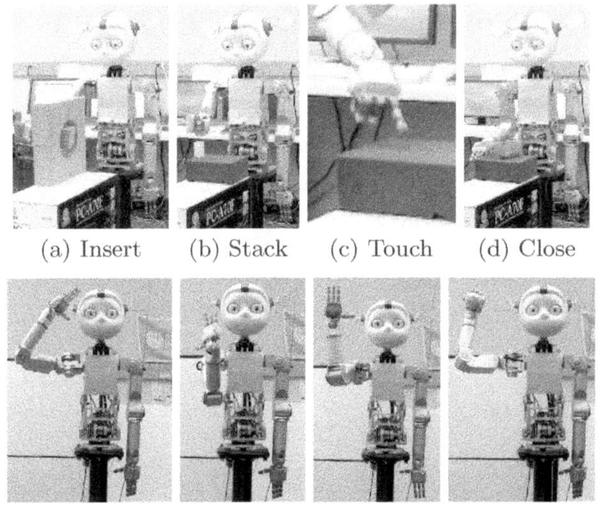

(a) Insert (b) Stack (c) Touch (d) Close
(e) Salute (f) Beckon (g) Raise (h) Throw

Figure 2: Goal-oriented (a-d) and means-oriented (e-h) skills.

The goal-oriented skills are as follows. (Fig. 2(a-d)). **Insert**: insert the block in hand through the hole without touching other blocks. **Stack**: stack the block in hand on top of another block on the table. **Touch**: touch a certain point with the finger tip. **Close**: close the lid of a box without moving it.

There is a single goal position per skill for the ease of the experiment. Multiple goal positions would prolong the experiment and our aim is not to analyze the generalization properties of the methods but rather to analyze the utility of the keyframe demonstrations from a user's perspective.

The means-oriented skills are as follows. (Fig. 2(e-h)). **Salute**: perform a soldier's salute. **Beckon**: perform a gesture asking someone to come closer. **Raise-hand**: raise the robot's hand as if it is asking for permission. **Throw**: perform a throwing gesture with a ball (without actually releasing the ball).

5.1.2 Conditions

Our experiment has four conditions, and we use a within-subject design. Three conditions correspond to the teaching methods in Sec.4. In addition, a fourth condition tests the effect of the initial demonstration in keyframe iterations.

- *Trajectory Demonstrations (TD):* Participants give one or more trajectory demonstrations for each skill.
- *Keyframe Demonstrations (KD):* Participants give one or more keyframe demonstrations for each skill.
- *Keyframe Iterations (KI):* Participants give keyframe iterations to teach the skills.
- *Keyframe Adaptation (KA):* Participants start with a predefined, slightly failing skill (e.g. touch is off by a few centimeters), instead of giving her/his own initial demonstration. They use the KI interaction to navigate and edit the frames to improve this skill.

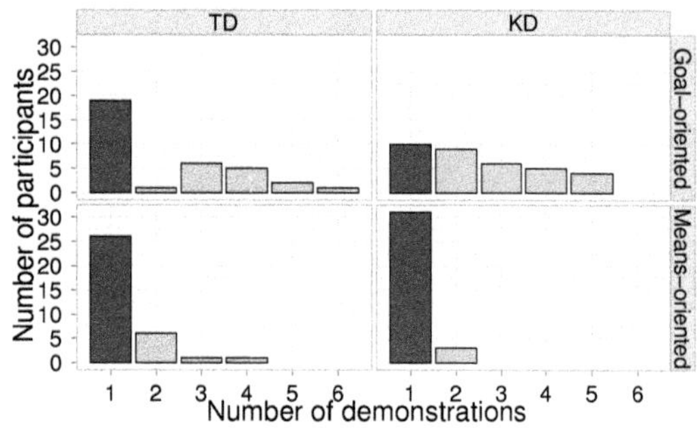

Figure 3: Histogram of number of demonstrations provided by participants in KD and TD conditions.

For Q1, effect of demonstration type, we compare TD and KD conditions. For Q2, learning of different skill types, we compare, *goal-oriented* and *means-oriented* skills in TD and KD conditions. For Q3, effect of extensions to keyframes, we first compare KD and KI, then compare KI and KA.

5.2 Experimental Protocol

The participants first teach the robot in the TD and KD conditions. The order of these two is counterbalanced. After these two conditions, KI and KA conditions followed. Thus the order was $(TD \mid KD) \rightarrow KI \rightarrow KA$. This ordering is to prevent biasing a participant in the KI condition with the predefined skill we use in the KA condition. We varied the type of skill taught to the robot across the TD and KD conditions, each participant taught one *means-oriented* skill, and one *goal-oriented* skill in these modes. Only *goal-oriented* skills were used in KI and KA conditions, to reduce experiment duration.

At the beginning of each condition, participants taught a *pointing* skill to the robot for familiarization with the condition. Participants were also allowed to move the robot's arm to practice before recording demonstrations.

5.3 Measures

We asked 7-point Likert-scale questions, administered after each condition, about *Feel*, *Naturalness*, *Ease*, and *Enjoyability*. We also asked open-ended questions after the first two conditions and at the end of the experiment.

Two different methods were used to measure the quality of the different types of learned skills. We evaluated the *goal-oriented* skills with three levels of success criteria, and the *means-oriented* skills with expert ratings.

The performance of goal-oriented skills were scored separately by the two of the authors, using three levels of success criteria Success-PartialSuccess-Fail. The scoring was based both on the recoded videos of the experiment and on the skill performances recreated on the robot. In the few cases where there was disagreement, the two coders revisited the example and reached a consensus on the scoring.

Unlike the goal-oriented skills, success for *means-oriented* skills is subjective. Therefore, we used expert ratings of the recreated movements to evaluate the performance. The experts, whose specialities are in computer animation, are asked to answer three 7-point Likert-scale questions for all means-oriented skills taught by all participants. The questions are about *appropriate emphasis*, *communicating intent*, and *closeness to perfection*.

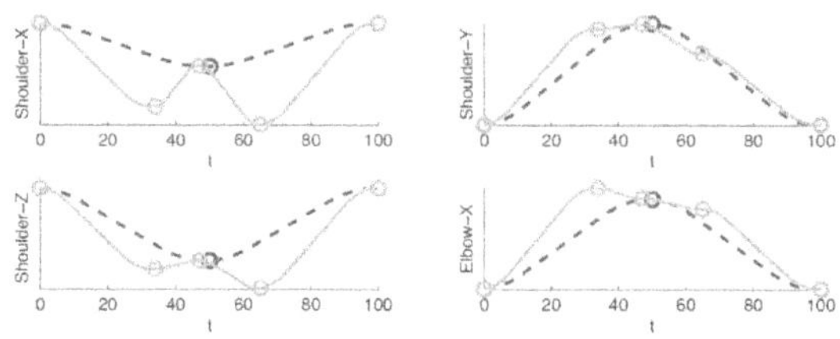

Figure 4: An example, in the KD condition, of forgetting obstacle avoidance keyframes in a first demonstrations (dashed line), and providing them in a second (solid line) while teaching the *Touch* skill.

We also measured the number of demonstrations, the number of keyframes, the time stamps for every event, and all trajectories of the joint movement during demonstrations.

6. RESULTS

We conducted a study with 34 participants (6 females, 28 males between the ages of 19-47), who were undergraduate and graduate Georgia Institute of Technology students with no previous machine learning and robotics experience. 22 of the participants taught Simon in all four conditions, while 12 only performed the first two conditions. The means-oriented skills were rated by two animation experts.

6.1 Trajectory vs Keyframe Demonstrations

First we compare the TD and KD experimental conditions, making five observations. [1]

6.1.1 Single demonstrations are common

Users were able to see what the robot has learned after each demonstration and either decide to move on or give another demonstration. Fig. 3 shows the number of demonstrations provided by participants in TD and KD. We see that teaching with a single demonstration was common in both modes. For goal-oriented skills, a larger portion of the participants provided a single demonstration in the TD condition than in the KD condition (19 versus 10). It was common in the KD condition to forget to provide keyframes that allow the robot to avoid obstacles while trying to achieve the goal. These frames were provided by participants in subsequent demonstrations after observing the performed skill colliding with obstacles (*e.g.* see Fig. 4). For means-oriented skills, teaching with a single demonstration was more common in the KD condition than in TD (31 versus 26).

6.1.2 Trajectory demonstrations may be better at teaching goal skills in a single demonstration.

Table 1 provides the distribution of participants according to the success of the goal-oriented skills they taught[2]. More participants achieved success in TD as opposed to KD (15 versus 5) when they taught with a single demonstration.

The large number of single demonstration instances is an artifact of our experimental design. The skills used in our experiments were chosen to be fairly easy to achieve, there was only a single goal location and participants were allowed to practice a particular skill before providing an actual

[1]We note that all of our observations reported in this section did not vary across particular skills.

[2]We treat success levels of skills as ordinal data.

Table 1: Number of participants who achieved different levels of success for goal-oriented skills.

Cond.	# of demo.	Success	Partial Success	Fail
TD	Single	15	4	1
	Multiple	1	5	8
	Total (%)	**16 (46)**	**9 (27)**	**9 (27)**
KD	Single	5	5	1
	Multiple	4	9	10
	Total (%)	**9 (27)**	**14 (41)**	**11 (32)**
KI	Single	4	0	2
	Multiple	6	4	6
	Total (%)	**10 (46)**	**4 (18)**	**8 (36)**
KA	Single	6	2	1
	Multiple	8	4	1
	Total (%)	**14 (64)**	**6 (27)**	**2 (9)**

demonstration of the skill. We observed that this practice opportunity was used more in the TD condition, where people often practiced enough to be able to teach the skill in a single demonstration. To quantify this observation we analyzed the total movement of the arm during the practice sessions measured in the 7DOF joints space. We designate "minimal practice" to mean that movement in the practice session is less than 10% of average movement of all practice sessions. We observed that 17 practice sessions in the KD condition can be classified as minimal under this definition, while only 4 of the TD sessions are minimal. This supports our anecdotal observation that practice is more likely to be skipped in the KD condition.

Secondly, as mentioned earlier, participants often do not think of providing keyframes for obstacle avoidance in their first demonstrations. In some cases this does not effect skill success in terms of achieving the goal (*i.e.* partial success) and participants could be satisfied by this since they were not explicitly told to avoid collisions. A large portion of the participants who provided a single demonstration in the KD condition at least achieved partial success.

6.1.3 *Keyframe demonstrations may result in preferable means-oriented skills*

Table 2 summarizes the expert ratings for the means-oriented skills taught by participants. Both experts rated the means-oriented skills learned in KD condition higher in all three dimensions on average. The difference was only significant for *closeness to perfection*, and the difference is marginally significant when the three scales are averaged (Z=2740, p=0.06 on Wilcoxon signed rank test). This distinction is partly related to the difficulty of moving a 7-DOF arm smoothly in the TD condition.

6.1.4 *Participants like both trajectory and keyframe*

Analyzing participant's Likert responses, we found all ratings were biased towards higher values, and none of the measures showed a statistical difference between TD and KD (based on paired Wilcoxon signed rank tests). We observe that participants' ratings are correlated with their success in teaching the goal-oriented skills (r=.31, p<.001 in Spearman's rank correlation test, assuming Fail:1, Partial:2 and Success:3). As a result, when the participants are grouped

Table 2: Expert ratings of means-oriented skills: Median and Coefficient of Dispersion

Cond.	Expert	*Emphasis*	*Intent*	*Perfection*
TD	1	5.5 (0.27)	5 (0.33)	5 (0.35)
	2	3 (0.29)	3.5 (0.38)	4 (0.3)
KD	1	6 (0.21)	6 (0.17)	6 (0.2)
	2	4 (0.21)	4 (0.24)	5 (0.22)
TD versus KD (Wilcox s.r. test)		Z=2679, p=0.10	Z=2677, p=0.11	Z=2796, p=0.03

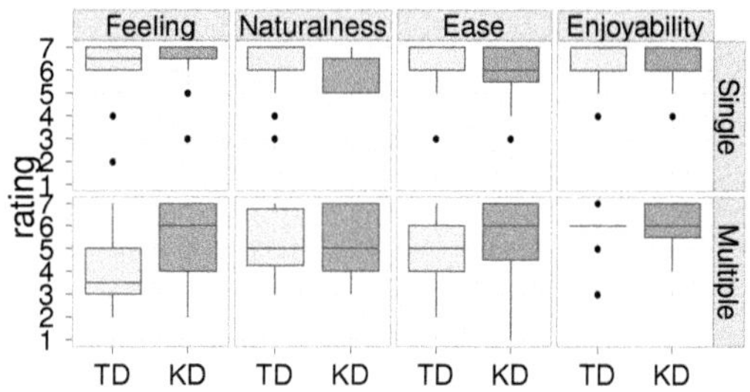

Figure 5: Subjective ratings of TD and KD conditions for goal-oriented skills separated by the number of demonstrations provided by the participant.

into ones that provide a single demonstration and ones that provide multiple demonstrations, we find that participants who provided multiple demonstrations felt more comfortable with keyframe demonstrations (V=98, $p < 0.05$ in unpaired Wilcoxon signed-rank test). This difference is not seen in participants who provided single demonstrations.

6.1.5 *Trajectory demonstrations require less time*

Providing one demonstration in the TD condition took participants on average 19.34*sec* (SD=7.65), while it took 34.37*sec* (SD=12.79) in the KD condition. There are two reasons for this difference. First one is that participants could freely move the arm before providing a keyframe in the KD condition. Thus, they used more time during the demonstration to think about the next keyframe that they wanted to provide and adjust the arm for it. This is supported by the comparison of all arm movements in the KD condition between keyframes (which was recorded for reference) and arm movements in the trajectory demonstrations provided in the TD condition. We find that the average arm movement per demonstration per person in the TD condition is about 82% of that of the KD condition, although this difference is not statistically significant ($t(112)=1.54$, p=.13 on *t*-test). The second reason is that the overhead of the speech commands to record keyframes in the KD condition. Given that the average number of keyframes is 6.71 (see section 6.3) and assuming giving a record keyframe command takes a second, both reasons seem to be valid.

In the TD condition, since all movements are recorded, participants must constantly progress and cannot pause or adjust as in the KD condition. As mentioned earlier, one manifestation of this was a more thorough practice session prior to TD, as compared to KD. Additionally, while not shown in this experiment, we expect that TD has a higher mental workload than KD.

6.2 Goal- vs. Means-oriented Skills

6.2.1 *Different objective functions for each skill type*

As seen in Fig 3, a much larger fraction of participants provide a single demonstration for teaching means-oriented skills, in both TD and KD. Across both conditions, the average number of demonstrations provided for goal-oriented skills (2.37, SD=1.45) is significantly larger than the number of demonstrations provided for means-oriented skills (1.22, SD=0.53) ($t(84)=6.18$, $p<0.001$ on *t*-test). This highlights a fundamental difference between the skill types: while goal-oriented skills have a well defined objective function, means-oriented skills are subjective and under-specified. Means-oriented skills can vary a lot and were often satisfactory for the participants after a single demonstration.

Open ended questions in our survey reveal more about the difference in the objective functions for the two types of skills. We asked participants to indicate their criteria of success for each skill that they taught. We observe that 15 participants mentioned achieving the goal as their criteria for goal-oriented skills (*e.g.* "The action would most accurately meet its end goal", "Performing the task correctly", "Touching the point perfectly") while 11 participants mentioned at least one style-related criteria for means-oriented skills. 4 participants mentioned *naturalness* (*e.g.* "more fluid and natural performance", "how naturally Simon emulate the demonstration"), 4 participants mentioned *human-likeness* (*e.g.* "with human characteristics", "seeming less robot-like") and 6 participants mentioned *smoothness* (*e.g.* "how smooth and liquid the motion of the arm is", "more fluid motion", "no choppy movements").

6.2.2 *Characteristics of provided keyframes are different for each skill type*

The average distance between keyframes in the 7DOF joint space for goal-oriented skills is much smaller (around 47%) than the average distance for means-oriented skills ($t(38)=-3.94$, $p<.001$ on *t*-test). We hypothesize that participants are providing different types of keyframes within a single demonstration. For goal-oriented skills we see a distinction between keyframes that are instrumental to the goal of the skill, and the keyframes that lets the robot avoid obstacles. Similarly in means-oriented skills we see a distinction between keyframes that actually give the skill its meaning and make it recognizable and keyframes that are waypoints. Participants provide a large number of frames that are close to one another around the goal of goal-oriented skills. For means-oriented skills, they provide less frames that are separated by a larger distance. For both types of skills the waypoint keyframes or obstacle avoidance keyframes tend to be further apart. We do not observe a statistical difference in the average number of keyframes for goal-oriented skills (6.75, SD=1.89) and means-oriented skills (6.21, SD=2.17) ($t(65)=1.11$, $p=.27$ on *t*-test).

6.3 Keyframe demonstrations vs. Iterations

A larger fraction of the participants achieve success in the KI condition as compared to the KD condition (Table 1). Since we did not counter-balance the order of these two conditions, this difference partially involves the improvement that comes with more experience in teaching. However these results show that the iteration process was effectively used by the participants, despite the increased number of commands and the complex interaction cycle. Note that 6 out of the 22 participants in the KI condition did not use the iteration process, *i.e.* they were satisfied the skill performance after the initial demonstrations, which is provided in exactly the same way as in the KD condition (Fig. 6). From the participants who used the iteration process, 10 participants provided a single iteration, however the iteration often involved several editing commands.

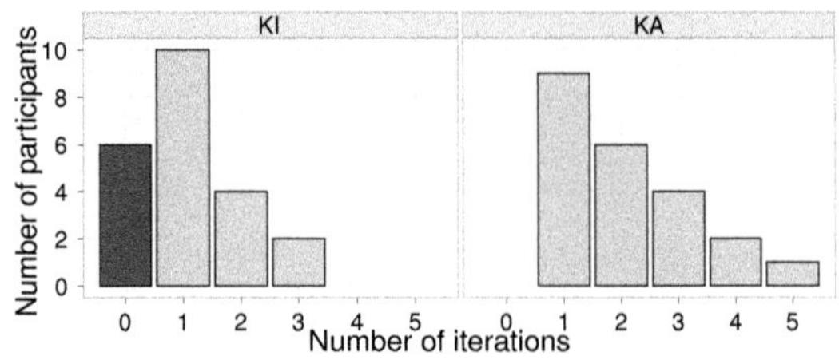

Figure 6: Histogram of number of iterations provided by participants in KI and KA conditions. For the KI condition "0" indicates the participants who only provided an initial demonstration and did not provide any iterations.

An interesting observation is that the number of keyframes in the demonstrations given by a participant varies less in the KI condition. The average number of keyframes provided within a demonstration for each participant is not very different in the KI (7.62, SD=1.48) and KD (6.71, SD=1.91) conditions. However the standard deviation in the number of provided keyframes across demonstrations of a participant seems larger in the KD condition (1.26, SD=0.94) as compared to the KI condition (0.46, SD=0.58). By starting from the previous learned skill, the iteration process limits the number of keyframes in the provided demonstrations which can be an advantage while learning from keyframe demonstrations if the initial demonstration is relatively good.

6.4 Effects of the Starting Skill for Iteration

The fraction of participants who achieve success is largest for the KA condition with 64% among all the others (TD: 46%, KD: 27%, KI: 46%, Table 1). As in the KI condition, experience in teaching the robot might be contributing to this improvement. However, this result indicates that an iterative process starting from a rough, often failing skill is potentially the best option in terms of achieving successful goal-oriented skills.

Our survey involved a question asking whether having a rough skill to start from in the KA condition made it easier or harder, or whether it did not matter in comparison with providing the initial demonstration themselves as in the KI condition. 12 participants responded that it made it easier, while 5 said "harder" and 5 said "did not matter".

7. DISCUSSION

7.1 Benefits of Demonstration Methods

The results of our experiment show that both trajectory and keyframe demonstrations are viable methods of interaction for kinesthetic teaching in LfD. Each has advantages and users seem positive towards both of them.

Trajectory demonstrations are clearly intuitive for a naïve user, and there is the benefit that many existing LfD methods are designed for learning skills from trajectory data. Trajectories allow complicated skills to be taught, and are particularly appropriate when speed information is a key component. However, it might be hard for users to manipulate a high-degree of freedom robot or sustain smooth trajectories over the course of a demonstration. In our experiment, this resulted in longer practice sessions for trajectory demonstrations, as well as means-oriented skills that achieved lower expert ratings.

Keyframe demonstrations are robust to these noisy and unintended motions during a demonstration. Their sparse nature result in a modular representation, which may be useful in generalizing a skill to new situations. For example, existing motion planning methods can easily be used to navigate between keyframes to execute salient aspects of the skill while avoiding obstacles. Additionally, it may be easier to deal with time alignment between multiple demonstrations.

A drawback of keyframes is the lack of timing information. We observed that some participants tried to achieve slower movements or stops by providing a large number of very close or overlapping keyframes. Several participants mentioned wanting speed related commands.

7.2 Skill Types and Demonstration Methods

Our experiment reveals the different nature of goal-oriented and means-oriented skills. The former is defined by *success* while the latter by *style*. Moreover, in the goal-oriented skills, only a portion of the skill's motion contributes to success whereas in means-oriented skills the entire motion contributes to the style.

People gave more demonstrations for goal-oriented skills. Since means-oriented skills can vary a lot, they were often satisfactory after a single demonstration. This was particularly true for keyframe demonstrations, since some users had a hard time manipulating the robot's arm, especially during the start of a skill. In goal-oriented skills, this usually did not impact task success (e.g. initial motion of the arm was not that important for pointing). However, this does have an impact when style is the objective. Thus, users often needed to correct the style by giving multiple demonstrations for the means-oriented skills in trajectory mode.

For goal-oriented skills, participants often gave multiple demonstrations due to the lack of fine control with keyframes, most notable being the timing (velocity) information. The robot often did not perform a skill as intended after the first demonstration in keyframe interactions, prompting users to improve the skill with more demonstrations.

We also observed that skill types have an effect of types of keyframes that are provided by the user. Waypoint keyframes are common in both of the skill types. Goal keyframes (keyframes that are closer together near a goal) and style keyframes (keyframes that are placed strategically to do the gesture, more apart) can clearly be seen respectively for goal-oriented and means-oriented skills.

7.3 Designing Keyframe Interactions

We explored different interaction mechanisms for a keyframe approach. Since keyframes temporally segment the demonstration, it is easy to apply an iterative interaction mechanism, and our experiment showed that people were able to use this to achieve greater skill success. We also saw that in an iterative interaction, people do not stray too far from their initial demonstration, thus emphasizing the importance of the starting skill. Our experiment showed that people were even able to use the iterative process to adapt a starting skill that was not their own, and many said that this made the teaching process easier.

As mentioned above, all keyframes are not equal, people think about them in different ways (e.g., goal frames, via points, etc.). A future extension would be to devise interaction mechanisms that are specific to each keyframe type. The distinction between these types of keyframes is important information for the underlying learning algorithm that human partners can easily provide.

7.4 A Hybrid Mode of Interaction

From this study, we found kinesthetic teaching is a viable method for teaching a humanoid robot, and that both keyframe and trajectory demonstrations have their advantages.

We think that the ability to provide both keyframe and trajectory information in the context of a single demonstration will be useful and intuitive for a variety of skills and even combination of skills (e.g. scooping and then serving). Hence, we are developing a new interface for LfD which merges trajectory and keyframe demonstrations in a single interaction. This *hybrid* interaction scheme allows the teacher to give both keyframes and trajectory segments in their demonstration (see in Figure 7). During a demonstration, the teacher can provide a keyframe by moving the arm to a desired position and saying "Go Here". At any point, the user can say "Like this" to initiate a trajectory demonstration and "That's it" to finish the segment. The teacher can combine these in any order resulting in four different kinds of demonstration: pure keyframe, single trajectory, segmented trajectory, and hybrid demonstrations.

We recently demonstrated this approach at the AAAI 2011 LfD challenge [2], on the PR2 robot, where anecdotal evidence[3] shows that this hybrid-mode was intuitive for conference goers. We have preliminary results with hybrid mode from a pilot study where three users tested the method in a scoop and pour task. These users gave positive comments about the interaction and their skills were successful.

We think the hybrid approach will give users more tools at their disposal to program robots in ways they find intuitive. Our future work will focus on the experimental validation of the hybrid mode of interaction, both with naïve and expert users and for LfD methods other than kinesthetic teaching.

7.5 Other Considerations

As mentioned previously, our aim was not to assess the generalization properties of the methods. One way to achieve generalization is to represent state as end-effector coordinates with respect to the goal. A more elaborate method is presented in [10]. This can be applied to trajectories, keyframes and hybrid demonstrations.

The presented tasks were simple but these interaction approaches may be even more beneficial for more complex tasks. For example in the case of a bi-manual LfD task, the user can position the arms individually and proceed with keyframes, which is usually impossible for trajectories.

[3] `http://www.youtube.com/watch?v=Ng7SYetzkXI`

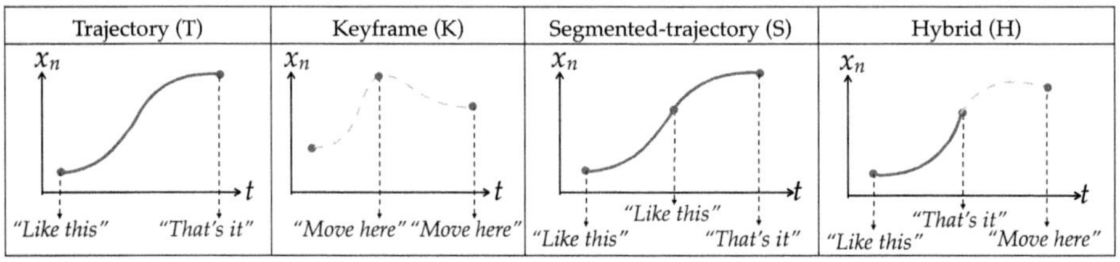

Figure 7: The possible interaction flows of the hybrid mode. The dots correspond to start/end points or keyframes, the solid lines to user demonstrated trajectories and the dashed lines to splines between keyframes.

8. CONCLUSIONS

We compared different methods of interaction for kinesthetic teaching in LfD with everyday people. Our study focused on the effects of different demonstrations, and we showed that trajectory and keyframe demonstrations have their relative advantages. We also explored different interaction schemes that a keyframe representation makes possible (iterations and adaption) and showed their success with human teachers. Finally, based on these observations, we introduced a hybrid mode of interaction in which the user can chain together keyframe and trajectory segments.

Acknowledgement

We acknowledge Sung Hyun Park and Karen Feigh for their contributions to the pilot version of this study. This research is supported by NSF CAREER grant IIS-0953181.

9. REFERENCES

[1] P. Abbeel and A. Y. Ng. Apprenticeship learning via inverse reinforcement learning. In *In Proceedings of the Twenty-first International Conference on Machine Learning*. ACM Press, 2004.

[2] B. Akgun, K. Jiang, M. Cakmak, and A. Thomaz. Learning tasks and skills together from a human teacher. In *Proceedings of the Twenty-Fifth AAAI Conference on Artificial Intelligence*, pages 1868–1869. The AAAI Press, 2011.

[3] H. B. Amor, E. Berger, D. Vogt, and B. Jun. Kinesthetic bootstrapping: Teaching motor skills to humanoid robots through physical interaction. *Lecture Notes in Computer Science: Advances in Artificial Intelligence*, 58(3):492–499, 2009.

[4] B. Argall, S. Chernova, B. Browning, and M. Veloso. A survey of robot learning from demonstration. *Robotics and Autonomous Systems*, 57(5):469–483, 2009.

[5] A. Billard, S. Calinon, and F. Guenter. Discriminative and adaptive imitation in uni-manual and bi-manual tasks. *Robotics and Autonomous System*, 54(5):370–384, 2006.

[6] S. Bitzer, M. Howard, and S. Vijayakumar. Using dimensionality reduction to exploit constraints in reinforcement learning. In *Intelligent Robots and Systems (IROS), 2010 IEEE/RSJ International Conference on*, pages 3219–3225, 2010.

[7] S. Calinon and A. Billard. Incremental learning of gestures by imitation in a humanoid robot. In *In Proc. of the ACM/IEEE International Conference on Human-Robot Interaction*, pages 255–262, 2007.

[8] S. Calinon and A. Billard. What is the teacher's role in robot programming by demonstration? - Toward benchmarks for improved learning. *Interaction Studies. Special Issue on Psychological Benchmarks in Human-Robot Interaction*, 8(3), 2007.

[9] S. Calinon and A. Billard. Statistical learning by imitation of competing constraints in joint space and task space. *Advanced Robotics*, 23(15):2059–2076, 2009.

[10] S. Calinon, F. Guenter, and A. Billard. On learning, representing and generalizing a task in a humanoid robot. *IEEE Transactions on Systems, Man and Cybernetics, Part B. Special issue on robot learning by observation, demonstration and imitation*, 37(2):286–298, 2007.

[11] E. Gribovskaya and A. Billard. Learning nonlinear multi-variate motion dynamics for real- time position and orientation control of robotic manipulators. In *Proceedings of IEEE-RAS International Conference on Humanoid Robots*, 2009.

[12] M. Hersch, F. Guenter, S. Calinon, and A. Billard. Dynamical system modulation for robot learning via kinesthetic demonstrations. *IEEE Transactions on Robotics*, 24(6):1463–1467, 2008.

[13] S. M. Khansari-Zadeh and A. Billard. Learning Stable Non-Linear Dynamical Systems with Gaussian Mixture Models. *IEEE Transaction on Robotics*, 27(5):943–957, 2011.

[14] M. Lopez Infante and V. Kyrki. Usability of force-based controllers in physical human-robot interaction. In *Proceedings of the 6th international conference on Human-robot interaction*, HRI '11, pages 355–362, 2011.

[15] H. Miyamoto, S. Schaal, F. Gandolfo, H. Gomi, Y. Koike, R. Osu, E. Nakano, Y. Wada, and M. Kawato. A kendama learning robot based on bi-directional theory. *Neural Netw.*, 9:1281–1302, November 1996.

[16] R. Parent. *Computer animation: algorithms and techniques.* Morgan Kaufmann series in computer graphics and geometric modeling. Morgan Kaufmann, 2002.

[17] P. Pastor, H. Hoffmann, T. Asfour, and S. Schaal. Learning and generalization of motor skills by learning from demonstration. In *IEEE Intl. Conference on Robotics and Automation*, 2009.

[18] A. Weiss, J. Igelsboeck, S. Calinon, A. Billard, and M. Tscheligi. Teaching a humanoid: A user study on learning by demonstration with hoap-3. In *Proceedings of the IEEE Symposium on Robot and Human Interactive Communication (RO-MAN)*, pages 147–152, 2009.

SDT: A "KonKon" Interface to Buildup the Connotation Interactions

Yu Arita
arita@icd.cs.tut.ac.jp

Hirota Shinya
hirota@icd.cs.tut.ac.jp

Yuta Yoshiike
yoshiike@icd.cs.tut.ac.jp

P. Ravindra S De Silva, Michio Okada
ravi@tut.jp, okada@tut.jp
Interactions and Communication Design Lab, Toyohashi University of Technology, Toyohashi 441-8580, JAPAN

ABSTRACT

The advantage of the proposed approach is that users can invent their own communication protocol (based on the knock patterns) to communicate with the creature. This approach is a novel concept to establish future human-robot communication protocols within many contexts or human centric applications. However, through symbolic communication (knock-based communication) a human is able to convey to the robot to adapt and communicate with their personalized communication protocol.

Categories and Subject Descriptors

H.1.2 [**Models and Principles**]: User/Machine Systems-Human Factors; H.5.2 [**Information Interfaces and Presentation**]: User Interfaces—*Evaluation/methodology, User-Centered Design*

General Terms

Design, Human Factors

Keywords

Interpersonal coordination, visually-mediated connectedness, COLUMN

1. INTRODUCTION

The video shows an SDT containing two robotic creatures (a pot and a dish) to interact with users (Figure 1). A user can convey their intentions through a KonKon interface which can be considered as the minimal social cues required for communicating with the creatures. The creatures use a "number of knocking" state (within a specific period) with a Temporal Difference (TD) learning model to learn and adapt to the user's intentions. In particular, the creatures are capable of moving different directions on the table (right, left, straight, and back) based on the user's knock. Through a delimitation of the creature's adaptation (movement), the user can convey his/her intentions by a number of knocking sounds while looking at the creature readjustment. The creature is capable of performing mutually adaptive learning through the TD model in real time to satisfy the user's goal or intentions.

Figure 1: Users interact with DT

2. ACKNOWLEDGMENTS

This research has been supported by Grant-in-Aid for scientific research of KIBAN-B (21300083) from the Japan Society for the Promotion of science (JSPS).

Human-Swarm Interaction through Distributed Cooperative Gesture Recognition

Alessandro Giusti
Dalle Molle Institute for Artificial Intelligence (IDSIA)
Lugano, Switzerland
alessandrog@idsia.ch

Jawad Nagi
Dalle Molle Institute for Artificial Intelligence (IDSIA)
Lugano, Switzerland
jawad@idsia.ch

Luca M. Gambardella
Dalle Molle Institute for Artificial Intelligence (IDSIA)
Lugano, Switzerland
luca@idsia.ch

Stéphane Bonardi
Biorobotics Laboratory (BioRob)
École Polytechnique Fédérale de Lausanne (EPFL)
Lausanne, Switzerland
stephane.bonardi@epfl.ch

Gianni A. Di Caro
Dalle Molle Institute for Artificial Intelligence (IDSIA)
Lugano, Switzerland
gianni@idsia.ch

Categories and Subject Descriptors

I.2.9 [**Robotics**]; I.2.11 [**Distributed Artificial Intelligence**]: Coherence and coordination; C.2.4 [**Computer Communication Networks**]: Distributed applications

General Terms

Algorithms

Keywords

Gesture recognition, Distributed consensus, Swarm robotics

1. VIDEO OVERVIEW

The video presents the first results of a Swiss-funded project focusing on *symbiotic peer-to-peer interaction and cooperation between humans and robot swarms*. As a first step, we considered human-swarm interaction, and selected the use of hand gestures to let a human communicate with a swarm of relatively simple mobile robots. In our scenario, a hand gesture encodes a command, that the swarm will execute. The robots that we used are the *foot-bots*, developed in the *Swarmanoid* project [1].

Hand gestures are a powerful and intuitive way to communicate, and do not require the use of additional devices. However, real-time *vision-based* recognition of hand gestures is a challenging task for the single foot-bot, due to its limited processing power and field of view. We investigated how to exploit *robot mobility*, *swarm spatial distribution*, and *multi-hop wireless communications*, to let the robots in the swarm: (i) implement a *distributed and cooperative sensing* of hand gestures, and (ii) robustly reach a *consensus* about a gesture.

The first step was to use 13 foot-bots to collect 70,000 hand gestures images representing finger counts (from 0 to 5) and five furniture-like shapes. With this data set we trained a *Support Vector Machine*, which is used by the robots for individual gesture classification and generation of an *opinion vector*, assigning a probability to each known gesture [3].

Following glove detection, the robots *move* to adapt their viewpoint for better sensing and maximize gesture information mutually gathered by the swarm as a whole. Using *multi-hop message relaying*, robots spread their opinions throughout the swarm. A *distributed consensus protocol* let the robots integrate and weight the different opinions.

When a robot has gathered enough evidence, it takes a *decision* for the hand gesture, and sends it into the swarm. Different decisions compete with each other, with the one assessed with the *highest confidence* that eventually wins.

The video shows two examples of command *execution*. For the furniture-like shapes, robots send their decisions to the *Roombot* system, where modular robots 'build' the furniture [2]. For the finger count, after receiving a 'two', the swarm splits in two groups moving in opposite ways.

2. ACKNOWLEDGMENTS

This research was supported by the Swiss National Science Foundation (SNSF) through the National Centre of Competence in Research (NCCR) Robotics (`www.nccr-robotics.ch`).

3. REFERENCES

[1] *Swarmanoid: Towards humanoid robotic swarms.* FET-OPEN project funded by the European Commission, `http://www.swarmanoid.org`.
[2] *Roombots: Modular robotics for adaptive and self-organizing furniture.* `http://biorob.epfl.ch/roombots`.
[3] J. Nagi et al. Max-pooling convolutional neural networks for vision-based hand gesture recognition. In *Proc. of IEEE ICSIPA*, Kuala Lumpur, Malaysia, 2011.

Field Trials of the Block-Shaped Edutainment Robot HangulBot

Sonya S. Kwak[1], Eun Ho Kim[2], Jimyung Kim[3], Youngbin Son[4], Inveom Kwak[4], Jun-Shin Park[5], Eun Wook Lee[6]

[1]Div. of Humanities & Soc. Sciences, POSTECH, Pohang, Korea
sonakwak@postech.ac.kr

[2]Culture Mechatronics Technology Center, KITECH, Cheonan, Korea
kimeunho@kitech.re.kr

[3]Miraemulsan
Seoul, Korea
miraemulsan@gmail.com

[4]Dept. of Computer Science & Eng., POSTECH, Pohang, Korea
{sonyoungbin, klauski}@postech.ac.kr

[5]Dept. of Mech. Eng., POSTECH
Pohang, Korea
lonelywing@postech.ac.kr

[6]Dept. of Chem. Eng., POSTECH
Pohang, Korea
hossun1@postech.ac.kr

ABSTRACT

The objective of this study was to develop an edutainment robot which provides multi-sensory learning experiences to improve users' space perception and creativity. In particular, we focused on developing educational content for the study of the Korean alphabet (Hangul) using the robot. On the basis of the phonemic and modular nature of Hangul, we devised a block-shaped edutainment robot for the study of Hangul. The robot known as "HangulBot" is composed of a consonant block and a vowel block. By rotating and rearranging those blocks, a user can create different characters. To enable the robot to perceive the arrangement of the blocks and the distance between a consonant block and a vowel block, IR LEDs and photo transistors were used. The eight IR LEDs in the consonant block generate different radiation signals, and the vowel block perceives the arrangement of the blocks by receiving the signals. The distance between the two blocks is estimated by measuring the thresholding, and the corresponding sound of each arrangement is then played through a speaker installed in the vowel block. We executed two short-term field trials with a twenty-seven month old child in June of 2011 and November of 2011 to ascertain children's initial reaction to HangulBot and how their reaction would change over time. While the results are preliminary, we noted several interesting findings. First, after several trials by the mother, the child felt comfortable with HangulBot. Second, the child intuitively followed the corresponding speech sounds which were generated by HangulBot according to the arrangement of the blocks. That is to say, the sound generated after the arranging the block intuitively induced the child to follow the sound. Third, the child's initial reaction to HangulBot was mostly block play, but after five months later, her reaction to the robot included not only block play but also active learning of the Korean alphabet. This result indicates that HangulBot could be an effective edutainment tool which improves space perception and creativity as well as linguistic abilities by stimulating both sides of the brain.

Categories and Subject Descriptors

I.2.9 **[Artificial Intelligence]**: Robotics - *commercial robots and applications*; K.3.1 **[Computers and Education]**: Computer Uses in Education

General Terms

Design, Human Factors

Keywords

Human-robot interaction, Edutainment robot, Korean alphabet

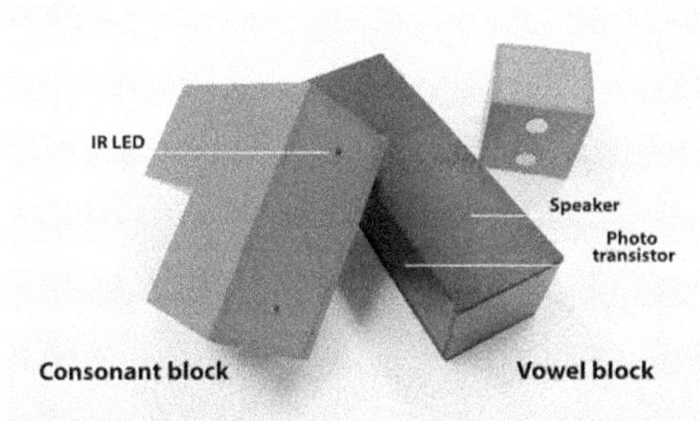

Figure 1. A prototype of HangulBot.

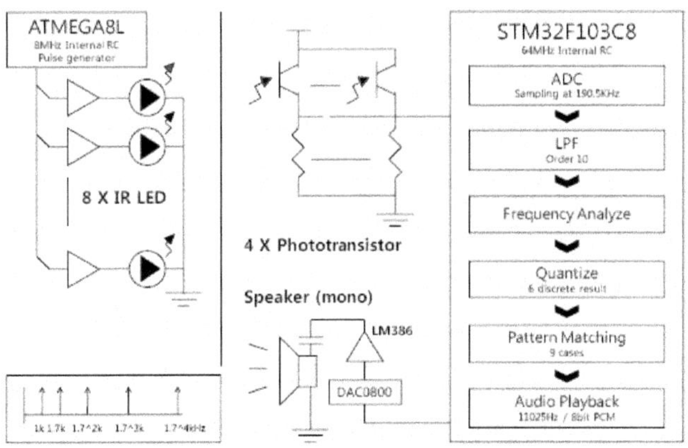

Figure 2. Architecture of HangulBot.

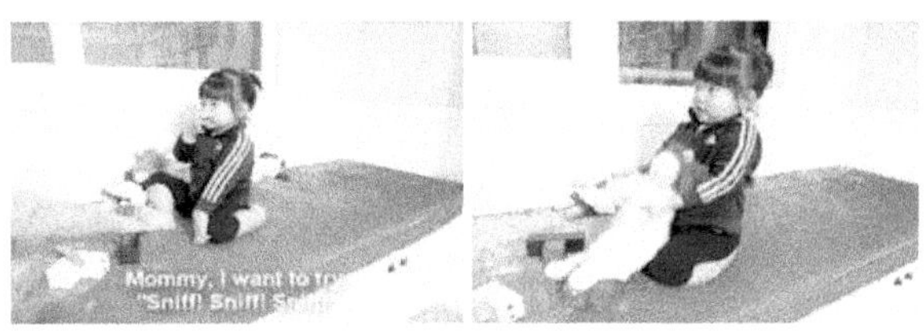

Figure 3. Field trials of HangulBot.

Human-Human vs. Human-Robot Teamed Investigation

Caroline E. Harriott[x] Glenna L. Buford[x] Tao Zhang* Julie A. Adams[x]

[x] Department of Elec. Eng. and Comp. Sci.
Vanderbilt University
Nashville, TN, USA
+1 (615) 322 - 8481

* Department of Library Sciences
Purdue University
West Lafayette, IN, USA

caroline.e.harriott@vanderbilt.edu, glenna.l.buford@vanderbilt.edu, zhangtao2000@gmail.com
julie.a.adams@vanderbilt.edu

ABSTRACT
Clips from an evaluation where participants, each paired with either a human or robot partner, were deployed to search a hallway for suspicious items, in a manner similar to tactics used by first responders handling bomb threats are presented. The teams used natural, verbal communication to collaborate, determine where hazards were located, and which items were suspicious. The video demonstrates that the investigations in both conditions played out in a similar manner and participants were able to complete the investigations successfully with a robot partner; however, sometimes the participants were uncertain how to interact with the robot.

Categories and Subject Descriptors
H.1.2. [**Information Systems**]: User/Machine Systems – *Human Factors;* I.2.9. [**Artificial Intelligence**]: Robotics.

General Terms
Performance, Experimentation, Human Factors

Keywords
human-robot peer-based teams

1. INTRODUCTION
The presented evaluation paired a participant with either a human or robot partner. The teams were deployed to search a hallway for suspicious items in a manner similar to tactics used by first responders handling bomb threats. Each team member had specific responsibilities; the participants investigated trash cans, recycling bins and bulletin boards, while their partner monitored air quality and checked fire extinguishers. When an item was deemed suspicious, both team members took photographs. The teams used natural, verbal communication to collaborate, determine where hazards were located, and which items were suspicious.

Throughout the evaluation, participants wore a point-of-view camera which recorded the investigation, and they took photographs using a digital point-and-shoot camera. The video presents segments from six participants, two in human-human teams and four in human-robot teams. Ultimately, this evaluation was completed in order to investigate differences in human performance between the two conditions.

A previous evaluation indicated that the participants had similar task completion success, yet experience lower overall workload when working with the robot partner [1]. Preliminary results indicate a similar trend. The video demonstrates that the investigations in both conditions played out in a similar manner and participants were able to complete the investigations successfully with a robot partner. The video also demonstrates some interaction uncertainties encountered by the participants in the human-robot condition.

2. HUMAN-HUMAN TEAMS
The human-human teams were comprised of an experimenter introduced as a trained first responder and a participant. The experimenter held a Fujitsu slate tablet PC with a conversational script. The experimenter informed the participant that this computer was an air quality monitor, camera and communication device for transferring information to the incident commander.

3. HUMAN-ROBOT TEAMS
The human-robot teams were comprised of a participant and a Pioneer-3DX robot. The robot autonomously navigated the hallway using a laser range finder, RFID tag reader, and the Player [2] software system. The robot's speech was based on a pre-programmed script, with room for the experimenter to adapt it to the actual conversation. An experimenter monitored the robot's movement and initiated speech utterances from a remote location, unbeknownst to the participant. The experimenter had an interface indicating the robot's expected location on a map and hear the conversation through a microphone on the participant's headset.

4. ACKNOWLEDGEMENTS
The authors thank Eli Hooten, Sean Hayes, Travis Service and Corey Wollaeger. This research is supported by AFOSR award FA9550-09-1-0108, National Science Foundation Grant IIS-0643100, and an Office of Naval Research MURI Program award N000140710749.

5. REFERENCES
[1] Harriott, C.E., Zhang, T. and Adams, J.A. 2011. Evaluating the applicability of current models of workload to peer-based human-robot teams. In *Proceedings of the 6th International Conference on Human-Robot Interaction* (Lausanne, Switzerland, March 6-9, 2011). 45-53.

[2] Vaughan, R.T., Gerkey, B.P., and Howard, A. 2003. On device abstractions for portable, reusable robot code. In *Proc. of the IEEE/RSJ Intl. Conf. on Intelligent Robots and Systems*, (Las Vegas, Nevada, October 2003). 2121-2427.

Acting Lesson with Robot: Emotional Gestures

Heather Knight
Carnegie Mellon University, Robotics Institute
5000 Forbes Avenue
Pittsburgh, PA
heatherbot@cmu.edu

Matthew Gray
Northeastern University, Department of Theatre
360 Huntington Avenue
Boston, MA
ma.gray@neu.edu

ABSTRACT

In this video, real-life acting professor Matthew Gray tutors Data the Robot (a Nao model) to improve his expression of emotion via Chekhov's Psychological Gestures. Though the video narrative is fictional and the robot actions pre-programmed, the aim of the dramatization is to introduce an acting methodology that social robots could use to leverage full body affect expressions.

The video begins with Gray leading Nao in traditional human actor warm-up exercises. Next, Gray shows Data a video of his students practicing Chekhov psychological gestures [4][11]. Finally, Data tries out some 'push' gestures himself. By pairing the 'push' gesture with text, the viewer is intended to unconsciously associate the words with an outpouring of emotion. Finally, Data's programmer, Knight, arrives to pick up the robot from his lesson, "until next time."

This video playfully introduces full-body emotional gestures. The benefit of such movement-based full-body expressions is that they do not necessarily require a robot to have human-like facial expressions nor humanoid form to be effective (though the interplay of psychological gesture with multi-modal expressions could provide fertile terrain for future research). Instead, these full-body motions are translations of an actor's motive/intent that suffuse the whole form (e.g. expansion, sluggishness, lightness).

HRI'12, March 5–8, 2012, Boston, Massachusetts, USA.
ACM 978-1-4503-1063-5/12/03.

We note that there are various schools of physical theater dedicated to understanding movement [5]. Related investigations in the robotics world that have applied acting method or practice to social robot design or architecture also include [2][3][6][7][8][9][10].

As Blaire writes about in her text on acting and neuroscience [1], the discovery of mirror neurons in our brain have led some dramaturges to theorize that audience members simulate the gestures of the performers through their own neural circuitry for interpretation. If so, full body gestures may be able to tap into our emotional experience in a uniquely human way. We hope this will be the first of several spirited demonstration videos that explore intersections wherein human acting methodologies might benefit the development of robot non-verbal expressions.

Categories and Subject Descriptors

J.5 [**Arts and Humanities**]: Arts, fine and performing**

Keywords

Acting Methodology, Entertainment Robots, Gesture, Robot Theater, Social Robots

REFERENCES

[1] Blair, R.: Actor, Image, and Action: Acting and Cognitive Neuroscience. Routledge, New York (2008)

[2] Breazeal, C., et al.: Interactive robot theatre. IROS (2003)

[3] Bruce, A., et al.: Robot Improv: Using drama to create believable agents. AAAI (2000)

[4] Chekhov, Michael: On the Technique of Acting, HarperPerennial, New York(1995)

[5] Callery, D.: Through the Body: A practical guide to Physical Theatre. London: Nick Hern Books. (2001)

[6] Demers, L.: Machine Performers: Neither Agentic nor Automatic. HRI Workshop on Collaboration w Arts (2010)

[7] Hoffman, G., Kubat, R.: A Hybrid Control System for Puppeteering a Live Robotic Stage Actor. RoMAN, (2008)

[8] Knight, H.: A Savvy Robot Standup Comic: Online Learning through Audience Tracking. Workshop paper, ACM TEI (2010)

[9] Knight, H.: Eight Lessons Learned about Non-Verbal Interaction through Investigations in Robot. ICSR (2011)

[10] Lu, D., Smart, W.: Human Robot Interaction as Theatre. Ro-MAN (2011)

[11] Rumohr, F.: Michael Chekhov, Psychological Gesture, and the Thinking Heart. Chapter in Movement for Actors, ed. Potter, N. Allworth Press, New York (2002)

MARIOBOT: Marionette Robot that Interact with an Audience

Woong-Ji Kim
IIPS Lab.
INHA University
Incheon, Korea
clrblue84@gmail.com

Sun-Wook Choi
IIPS Lab.
INHA University
Incheon, Korea
swchoi@inhaian.net

Chong Ho Lee
IIPS Lab.
INHA University
Incheon, Korea
chlee@inha.ac.kr

ABSTRACT

A marionette is a puppet operated by human manipulator using strings and control bar. It is an ancient and universal form of performing art. Marionettes have been used for entertainment for centuries because people are fascinated with unique and funny movements of marionettes.

In this video, we introduce a new type of entertainment robot for theatrical performances. Our MARIOBOT (marionette robot) is driven by a robotic controller that consists of eight motors and their coupling components. Each motor pulls the string connected to corresponding marionette's joint. This robot freely moves along the stage hung by a mobile base platform.

In order to show the capabilities of our robot actor on the stage, this video with scripts demonstrates the performance of the robot. We also show a few audience-robot interaction methods, as from the perspective of the modern performances on the stage, the feature of interaction with the audience draws much of interests and appeals to the audience [1]. The audience loses attention quickly during a passive performance. Therefore, we need to consider appropriate HRI techniques to keep the audience's attention lively. However typical stage environment in the theater is not well-conditioned to accommodate the vision based HRI methods, because of: luminance of theater, distance between robot and the audience, etc. To overcome these problems, we present effective methods for interaction with the audience and the situations related to these methods on the video together with the system overview.

Categories and Subject Descriptors

J.m [**Computer Applications**]: Miscellaneous – *image processing, robotics, control.*

General Terms

Algorithms, Performance, Design, Human Factors.

Keywords

Marionette, robot, performance, audience-robot interaction.

Acknowledgments

This research was financially supported by the Ministry of Knowledge Economy (MKE), Korea Institute for Advancement of Technology (KIAT) through the Inter-ER Cooperation Projects.

HRI '12, March 5–8, 2012, Boston, Massachusetts, USA.
ACM 978-1-4503-1063-5/12/03.

1. REFERENCES

[1] Davenport, G. 1994. Seeking Dynamic, Adaptive Story Environments. *IEEE Multimedia*, vol.1, no.3.

Introducing Students Grades 6-12 to Expressive Robotics

David V. Lu
Dept. of Computer Science and Engineering
Washington University in St. Louis
davidlu@wustl.edu

Ross Mead
Computer Science Department
University of Southern California
rossmead@usc.edu

Every year, tens of thousands of middle school and high school students participate in robotics competitions, such as Botball, FIRST, and VEX. This provides them with an excellent introduction to the ins and outs of building robots and programming them to autonomously accomplish specific tasks. However, the rules of many of these competitions often limit or prohibit human interaction with the robots. As a result, students are not exposed to and are, thus, not encouraged to think about human-robot interaction (HRI) and its potential impacts on society.

To start getting a new generation of students thinking about HRI, we held a workshop entitled "Expressive Robotics: Motion and Emotion" at the 2011 Global Conference on Educational Robotics. The focus of the workshop was to teach the students how to program robots to express emotion and intent just through the robot's physical actions. The robot used by each participating group was not unlike the ones they used in competitions: an iRobot Create base provided a mobile platform, upon which stood a three degree-of-freedom arm, serving as an articulated spine and head.

Each group was first asked to program the robot in a way that expressed one of Eckman's six basic emotions. They were then tasked with implementing a simple keyframe animation system to control the robot's limited degrees of freedom, which they could then modify to illustrate certain principles from theatre and animation. They used their animation system to tell a story purely through robot motion and interactions with props, as seen in Figure 1. Students were encouraged to focus on making the robot's emotion and intent apparent to a human audience.

Ultimately, the workshop proved to be a great success; it was very well-received by all involved, who encouraged us to make this an annual event at the conference. We felt that this type of workshop served as a fine proof-of-concept for introducing students grades 6-12 to human factors in robotics, and could be extended throughout the entire K-12 education pipeline. Furthermore, we believe that exposure to sociable robotics has the potential to increase interest and self-efficacy of underrepresented student populations, particularly girls, in STEM-related activities.[1]

[1]The video accompanying this abstract can be viewed on YouTube at `http://tinyurl.com/hri2012vid101`. To view all `http://tinyurl.com/expressiverobotics`.

HRI'12, March 5–8, 2012, Boston, Massachusetts, USA.
ACM 978-1-4503-1063-5/12/03.

Figure 1: Frame from the student film 'Stranded'

Table 1: Workshop Schedule

9:00	Lecture: Introduction and Emotion
9:30	Activity 1: Emotion Motion
10:00	Activity 2: Evaluating Emotion
10:15	Lecture: Keyframing
10:45	Activity 3: Keyframe Animator
11:15	Lecture: Telling a Story and Storyboarding
11:30	Activity 4: Lunch and Brainstorming
12:30	Activity 5: Storyboarding
13:00	Activity 6: Take 1
14:00	Lecture: Principles of Animation
14:30	Activity 7: Polishing and Filming
16:30	Activity 8: Robots Go To Hollywood
17:00	Lecture: Closing Remarks

Categories and Subject Descriptors

I.2.9 [**Artificial Intelligence**]: Robotics

Keywords

expressive robotics; HRI; educational robotics

Multi-user Multi-touch Command and Control of Multiple Simulated Robots

Eric J. McCann, Sean McSheehy, and Holly A. Yanco
University of Massachusetts Lowell
Department of Computer Science
One University Avenue, Lowell, MA
{emccann, smcsheeh, holly}@cs.uml.edu

ABSTRACT

This video demonstrates three users sharing control of eight simulated robots with a Microsoft Surface and two Apple iPads using our Multi-user Multi-touch Multi-robot Command and Control Interface.

The command and control interfaces are all capable of moving their world camera through space, tasking one or more robots with a series of waypoints, and assuming manual control of a single robot for inspection of its sensors and teleoperation. They display full-screen images sent from their user's world camera, overlaid with icons that show the position and selection state of each robot in the camera's field of view, dots that indicate each robot's current destination, and rectangles that correspond to each other user's field of view.

One multi-touch interface runs on a Microsoft Surface, and the others on Apple iPads; they all have the same functional capabilities, other than a few differences due to the form factor and touch sensing method used by the platforms. The Surface interface is able to interpret gestures that include more than just finger tips, such as placing both fists on the screen to make all robots stop and wait for new commands. As iPads sense touch capacitively, they do not support detection of such gestures. The Surface interface allows its user to move their world camera while simultaneously teleoperating one of the robots with our Dynamically Resizing Ergonomic and Multi-touch Controller (DREAM Controller) [1, 2]. On the iPads, however, the command and control mode and teleoperation mode are mutually exclusive.

The robots are simulated in Microsoft Robotics Developer Studio. Each user's world camera has similar movement capabilities to a quad-copter. The UDP communications between users and robots are all handled by a single server that routes messages to the appropriate targets, allowing scalability of both the number of robots and users.

Categories and Subject Descriptors

H.5.2 [**User Interfaces**]: Graphical user interfaces, Input devices and strategies, Interaction styles, Screen design, User-centered design; H.5.3 [**Group and Organization Interfaces**]: Collaborative computing, Synchronous interaction

General Terms

Design

Keywords

Human-Robot Interaction, Multi-touch

1. ACKNOWLEDGEMENTS

This work has been supported in part by the Army Research Office (W911NF-07-1-0216) and Microsoft.

2. REFERENCES

[1] M. Micire, M. Desai, J. Drury, E. McCann, A. Norton, K. Tsui, and H. Yanco. Design and Validation of Two-Handed Multi-Touch Tabletop Controllers for Robot Teleoperation. In *Proc. of the Intl. Conf. on Intelligent User Interfaces*, 2011.

[2] M. J. Micire. *Multi-Touch Interaction for Robot Command and Control.* PhD thesis, Computer Science Department, University of Massachusetts Lowell, December 2010.

HRI'12, March 5–8, 2012, Boston, Massachusetts, USA.
ACM 978-1-4503-1063-5/12/03.

Encounters: From Talking Heads to Swarming Heads

Damith C. Herath
MARCS Auditory Laboratories
University of Western Sydney
Locked Bag 1797
Penrith NSW 2751, Australia
d.herath@uws.edu.au

Christian Kroos
MARCS Auditory Laboratories
University of Western Sydney
Locked Bag 1797
Penrith NSW 2751, Australia
c.kroos@uws.edu.au

Stelarc
MARCS Auditory Laboratories
University of Western Sydney
Locked Bag 1797
Penrith NSW 2751, Australia
stelarc@stelarc.org

ABSTRACT

Robots at home and work has been a key theme in science fiction since the genre began. It is only now that we see this come in to realization, albeit in very basic forms such as the robot vacuum cleaners and various entertainment robotic platforms. In this video we highlight a number of projects woven around the iRobot Create research robot platform and an embodied conversational agent called the Prosthetic Head - an installation work by Stelarc. We start the visual journey by taking a satirical look at some of the parallels between a commercial communication product and the Prosthetic Head. The journey then moves through telepresence robotics, gesture based robot human interaction. The robots featured in the video are driven by an attention and behavioral system. Finally, the video concludes with a preview of the "Swarming Heads" - an interactive installation.

General Terms

Design

Keywords

HRI, ECA, tele-presence, gesture-based control, mobile robots, distributed embodiment

Acknowledgments

We would like to thank three anonymous reviewers for their insightful comments and we wish to acknowledge the support of NHMRC/ARC grant TS0669874. We are also highly grateful for contributions made by the Professional Experience Students at the School of Computing and Mathematics, UWS.

HRI'12, March 5–8, 2012, Boston, Massachusetts, USA.
ACM 978-1-4503-1063-5/12/03.

Johnny-0, a Compliant, Force-Controlled and Interactive Humanoid Autonomous Robot

François Ferland
francois.ferland@usherbrooke.ca

Arnaud Aumont
arnaud.aumont@usherbrooke.ca

Dominic Létourneau
dominic.letourneau@usherbrooke.ca

Marc-Antoine Legault
marc-antoine.legault@usherbrooke.ca

François Michaud
francois.michaud@usherbrooke.ca

Université de Sherbrooke
Sherbrooke, Québec, Canada

Figure 1: Johnny 0

Categories and Subject Descriptors

I.2.9 [**Robotics**]: Kinematics and dynamics

General Terms

Algorithms, Experimentation

Keywords

Humanoid robot, Force-control

ABSTRACT

Johnny-0 [2], shown in Figure 1, is our new humanoid robot which integrates an expressive face on an orientable head, two arms with 4 degrees of freedom (DOF) each and grippers, mounted on an omnidirectional, non-holonomic mobile platform. Our underlying goal with Johnny-0 is to design a platform capable of natural reciprocal interaction (motion, language, touch, affect) with humans, to address integration issues associated with advanced motion, interaction and cognition capabilities on the same platform, and their use in unconstrained real world conditions. To do so, compliance is a necessity to provide natural and safe interactions.

One distinctive element of Johnny-0 is that it uses force-controlled actuators (called Differential Elastic Actuators - DEA for active steering of its mobile platform, and for interactive control of its 4-DOF arms. Compliance at the mobile platform level allows a person to physically guide the robot without having to push it from a specific location on the platform [1]. Motion can also be constrained to avoid obstacles and collisions, providing natural physical interaction with the robot. Impedance control of each joint enables infinite combination of arm behaviors, from zero impedance for free movement with gravity compensation, to high stiffness constraining the arms to precise positions or ranges of movement. Stiffness can be configured to create virtual constraints in cartesian space, providing force feedback to the user about movement's limitations of the arms. For instance, stiffening the arms in certain poses could indicate to the user that the arms are restrained to move into a specific volume. Beyond these limits, any pushing or pulling force can be perceived by the mobile base, and can be interpreted as an intention to move the robot around.

Combining compliance to other sensors (e.g., Kinect motion sensor) and a robot head capable of facial expression allows Johnny-0 to detect incoming people and adjust the impedance of its actuators accordingly (e.g., extend its gripper to greet them), and express its state based on how people physically interact with it (e.g., displaying surprise when the user move the arms beyond specific limits).

HRI'12, March 5–8, 2012, Boston, Massachusetts, USA.
ACM 978-1-4503-1063-5/12/03.

Acknowledgments

This work was supported by the Natural Sciences and Engineering Research Council of Canada (NSERC), the Canada Research Chair and the Fonds Québécois de la Recherche sur la Nature et les Technologies (FQRNT).

1. REFERENCES

[1] J. Frémy, F. Ferland, L. Clavien, D. Létourneau, F. Michaud, and M. Lauria. Force-controlled motion of a mobile platform. In *IEEE/RSJ International Conference on Robots and Systems*, pages 2517–2518, 2010.

[2] F. Michaud, F. Ferland, D. Létourneau, M. Legault, and M. Lauria. Toward autonomous, compliant, omnidirectional humanoid robots for natural interaction in real-life settings. *Paladyn Behavioral Robotic Journal*, 1(1):57–65, 2010.

Situation Understanding Bot Through Language and Environment

Daniel J. Brooks[1] Constantine Lignos[2] Mikhail S. Medvedev[1] Ian Perera[2]
Cameron Finucane[3] Vasumathi Raman[3] Abraham Shultz[1] Sean McSheehy[1]
Adam Norton[1] Hadas Kress-Gazit[3] Mitch Marcus[2] Holly A. Yanco[1]

[1]University of Massachusetts Lowell [2]University of Pennsylvania [3]Cornell University
{dbrooks, mmedvede, ashultz, smcsheeh, anorton, holly}@cs.uml.edu
{lignos, mitch}@cis.upenn.edu {cpf37, hadaskg}@cornell.edu
iperera@cs.rochester.edu vraman@cs.cornell.edu

ABSTRACT

This video shows a demonstration of a fully autonomous robot, an iRobot ATRV-JR, which can be given commands using natural language. Users type commands to the robot on a tablet computer, which are then parsed and processed using semantic analysis. This information is used to build a plan representing the high level autonomous behaviors the robot should perform [2] [1]. The robot can be given commands to be executed immediately (e.g., "Search the floor for hostages.") as well as standing orders for use over the entire run (e.g., "Let me know if you see any bombs.").

In the scenario shown in the video, the robot is asked to identify and defuse bombs, as well as to report if it finds any hostages or bad guys. Users can also query the robot through this interface. The robot conveys information to the user through text and a graphical interface on a tablet computer. The system can add icons to the map displayed and highlight areas of the map to convey concepts such as "I am here."

The video contains segments taken from a continuous 20 minute long run, shown at 4x speed. This work is a demonstration of a larger project called Situation Understanding Bot Through Language and Environment (SUBTLE). For more information, see `www.subtlebot.org`.

Categories and Subject Descriptors

I.2.9 [**Artificial Intelligence**]: Robotics—*Autonomous vehicles, Operator interfaces*; I.2.7 [**Artificial Intelligence**]: Natural Language Processing—*Language parsing and understanding*

Keywords

Natural Language Interface

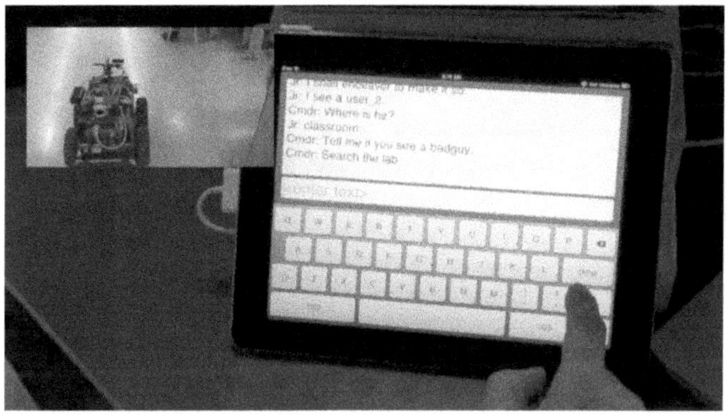

Figure 1: Autonomous vehicle being controlled by natural language interface

1. ACKNOWLEDGMENTS

This work was funded by Army Research Office MURI (W911NF-07-1-0216). We would like to acknowledge Munjal Desai, Eric McCann, and Kate Tsui of UMass Lowell for their help with the project. Video background music "Sweet Idleness" by Pocketpod; `http://pocketpod.newgrounds.com`.

2. REFERENCES

[1] J. Allbeck and H. Kress-Gazit. Constraints-Based Complex Behavior in Rich Environments. *Proceedings of the 10th International Conference on Intelligent Virtual Agents*, Sept. 2010.

[2] H. Kress-Gazit, G. Fainekos, and G. Pappas. Temporal Logic-based Reactive Mission and Motion Planning. *IEEE Transactions on Robotics*, 25(6):1370–1381, 2009.

HRI'12, March 5–8, 2012, Boston, Massachusetts, USA.
ACM 978-1-4503-1063-5/12/03

How to Sustain Long-Term Interaction between Children and ROBOSEM in English Class

Jeonghye Han
Department of Computer Education
Cheongju Nat'l Univ. of Education
Sugok 135, Cheongju, Chungbuk,
South Korea, 361-712
hanjh@ cje.ac.kr

Bokhyun Kang, Seongju Park
R&D center, Yujin Robot Co.,Ltd.
#601, Namsung Plaza, 345-30,
Gasan-dong, Seoul, Korea
{bhkang,sampark}@yujinrobot.com

Seongwook Hong
Grafizix Co., Ltd., DMC R&D Center
1649 Sangam-dong, Mapo-gu, Seoul,
121-270, Korea
boss@grarfizix.com

ABSTRACT
According to studies confirming that robot assisted learning (RAL) can positively contribute to improving learners' motivation and achievement for language learning [1,4], RAL is facing its diffusion through the demand of parents and government. However, the biggest obstacle to the long-term interaction between humans and robots is the robots' low-success rate of visual and voice recognition, as well as the limitation of artificial intelligence for the daily-life HRI [3]. This study demonstrated ROBOSEM's ability to sustain long term interaction between children and a robot in an elementary English class from the pilot studies with IROBIQ, called Langbot [1].

Figure 1. Scene of English class assisted by ROBOSEM

Five factors are of concern in sustaining the long-term interaction between children and ROBOSEM, as shown in Figure 2: (1) enhancing the recognition ability of ROBOSEM with class materials, such as marker hats, bracelet watches embodied RFID tags, Wiimocon, etc; (2) sharing the birth story of ROBOSEM, which works to increase children's tolerance toward weak recognition from the result of [2]; (3) making a favorable impression, such as by flashing children's faces on the screen; (4) telling the history of a child's personal learning activities; and (5) tele-operation by the intelligence of a human being.

Categories and Subject Descriptors
H.4.3 Communications Applications: Computer conferencing, teleconferencing, and videoconferencing

Keywords
Robot, English, Human Robot Interaction, RAL (Robot Assisted Learning), Long-term Interaction

HRI'12, March 5–8, 2012, Boston, Massachusetts, USA.
ACM 978-1-4503-1063-5/12/03.

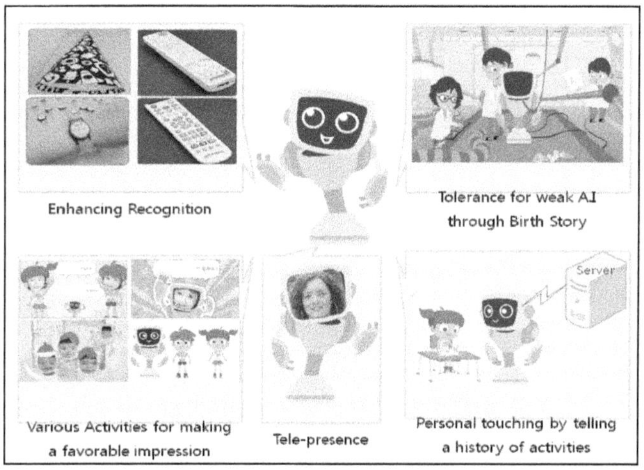

Figure 2. Five factors for long-term HRI

ACKNOWLEDGMENTS
This research was supported by the diffusion and dissemination program for intelligent robots of KIRIA (Korea Institute for Robot Industry Advancement Ministry of Knowledge and Economics) and MKE (Ministry of Knowledge Economy).

REFERENCES
[1] Han, J., Lee, S., Kang, B., Park, S., Kim, J., Kim, M., & Kim, M. "A Trial English Class with a Teaching Assistant Robot in Elementary School", *Proceeding of the 5th ACM/IEEE ACM/IEEE Human Robot Interaction*, ISBN:978-1-60558-404-1, Osaka, Japan, March 2-5, 2010, pp.335.

[2] Hur, Y. & Han J., "Analysis on Children's Generosity toward a Teaching Assistant Robot Storytelling on a Innate Weak Recognition", *Proceedings of the 8th Application and Processing Information System*, Japan, 2009

[3] Kanda, T., Sato, R., Saiwaki, N. & Ishiguro, H., "A Two-Month Field Trial in an Elementary School for Long-Term Human–Robot Interaction", Robotics, IEEE Transactions on, 23(5), 2007, pp. 962-971.

[4] Park, S., Han, J., Kang, B. & Shin, K., "Teaching assistant robot, ROBOSEM, in English class and practical issues for its diffusion", *Proceedings of the IEEE ARSO(A workshop on advanced Robotics and its SOcial impacts) conference*, 2011, Oct. 2-4, Half-moon bay, California, USA

Implementing Human Questioning Strategies into Quizzing-Robot

Takaya Ohyama[1]
tohyaman@cv.ics.saitama-u.ac.jp

Yasutomo Maeda[1]
maeda@cv.ics.saitama-u.ac.jp

Chiaki Mori[1]
mori@cv.ics.saitama-u.ac.jp

Yoshinori Kobayashi[1, 2]
yosinori@cv.ics.saitama-u.ac.jp

Yoshinori Kuno[1]
kuno@cv.ics.saitama-u.ac.jp

Rio Fujita[1]
s08LL137@mail.saitama-u.ac.jp

Keiichi Yamazaki[1]
byi06561@nifty.com

Shun Miyazawa[2]
homageofhj@gmail.com

Akiko Yamazaki[2]
ayamazaki@media.teu.ac.jp

Keiko Ikeda[3]
osakakeiko@hotmail.com

[1]Human-Robot Interaction Center, Saitama University
255 Shimo-Okubo, Sakura-ku, Saitama 338-8570, Japan

[2]PRESTO, Japan Science Technology Agency
4-1-8 Honcho Kawaguchi, Saitama 332-0012, Japan

[3]Tokyo University of Technology
1404-1 Katakuramachi, Hachioji City, Tokyo 192-0982, Japan

[4]Kansai University
3-3-35 Yamate-cho, Suita City, Osaka 564-8680, Japan

ABSTRACT
From our ethnographic studies on various kinds of museums, we discovered that guides routinely propose questions to visitors in order to draw their attention towards both his/her explanation and the exhibit. The guides' question sequences tend to begin with a pre-question which serves to not only monitor visitors' behavior and responses, but to also alert visitors that a primary question would follow. We implemented this questioning-strategy with our robot system and investigated whether this strategy would also work in human-robot interaction. We developed a vision system that enables the robot to choose an appropriate visitor by monitoring a visitor's response from the initiation of a pre-question to the following pause. Results indicate that this questioning-strategy works effectively in human-robot interaction. In this experiment, the robot asked visitors about a photograph. At the pre-question, the robot delivered a rather easy question followed by a more challenging question (Figure 1). More participants turned their head away from the exhibition when they were not sure about their answer to the question. They either faced away from the robot, or smiled wryly at the robot or at each other. These types of behaviors index participants' states of knowledge, which we could utilize to develop a system by which the robot could choose an appropriate candidate by computational recognition.

HRI '12, March 5-8, 2012, Boston, Massachusetts, USA.
ACM 978-1-4503-1063-5/12/03.

ACM Classification Keywords
H5.2. Information interfaces and presentation (e.g., HCI): User Interfaces – interaction styles.

General Terms
Design, Experimentation, Human Factors.

Keywords
Interaction analysis, human-robot interaction, computer vision.

Figure 1. The robot asks an easy pre-question followed by a challenging question.

ACKNOWLEDGMENTS
This work was partly supported by KAKENHI (21300316, 22243037, 23252001).

Whole-Body Imitation of Human Motions with a Nao Humanoid

[Video Abstract]

Jonas Koenemann
Humanoid Robots Lab
Department of Computer Science
University of Freiburg, Germany
jonas.koenemann@yahoo.de

Maren Bennewitz
Humanoid Robots Lab
Department of Computer Science
University of Freiburg, Germany
maren@informatik.uni-freiburg.de

ABSTRACT

We present a system that enables a humanoid robot to imitate complex whole-body motions of humans in real time. For recording the human motions, any sensor system capable of inferring the joint angle trajectories can be used. In our work, we capture the human data with an Xsens MVN motion capture system consisting of inertial sensors attached to the body. Our framework converts the human joint angles to the robot's joint angles in real time. Here, we use a mapping between the human's and the robot's joints to ensure feasibility of the motion. The focus of our system lies in ensuring static stability when the motions are executed which is a challenging task, depending on the complexity of the movements. To avoid falls of the robot that might occur when using direct imitation of the joint angle trajectories due to the different weight distribution, we developed an approach that actively balances the center of mass over the support polygon of the robot's feet. At every point in time, our approach ensures that the robot is in a statically stable configuration, i.e., that the ground projection of the center of mass lies within the convex hull of the foot contact points. To achieve this, we apply inverse kinematics given valid foot positions that satisfy the stability criterion and generate the corresponding leg joint angles. In more detail, our system first finds valid positions for the robot's feet by determining a target plane and its orientation, so that the feet can be placed planar and the robot's center of mass is over the support polygon. The new positions of the feet are chosen as the projection on the target plane. Afterwards, the corresponding leg joint angles are calculated via inverse kinematics. To determine whether the configuration is in the double support modus, and if not, which foot is the stance foot, we evaluate the position of the center of mass relative to the feet.

As can be seen in the experiments with a Nao humanoid, our approach leads to a highly stable imitation of challenging human movements (see also Fig. 1). In contrast to recent approaches that capture human data using a Kinect-like sensor and only imitate arm movements while keeping the body static, our system can deal with complex, whole-body motions. Note that our approach does not require a prior learning phase but computes stable configurations online and almost in real time as can be seen in the accompanying video.

We are currently working on imitating motions to learn complex navigation actions such as climbing up staircases or walking down ramps. Our system can also be used for teleoperated tasks that include whole-body movements where stability needs to be guaranteed in order to successfully fulfill the mission.

Figure 1: Nao imitating complex whole-body motions. The robot actively balances its center of mass to achieve static stability.

Categories and Subject Descriptors

J.m [**Computer Applications**]: Miscellaneous

General Terms

Experimentation

Keywords

Humanoid Robots, Imitation

HRI'12, March 5–8, 2012, Boston, Massachusetts, USA.
ACM 978-1-4503-1063-5/12/03.

Roboscopie

A Theatre Performance for a Human and a Robot [Video Abstract]

Séverin Lemaignan, Mamoun Gharbi, Jim Mainprice, Matthieu Herrb, Rachid Alami
CNRS-LAAS, 7 av. du Colonel Roche, F-31077 Toulouse, France
Université de Toulouse, UPS, INSA, INP, ISAE, LAAS, F-31077 Toulouse, France
surname.name@laas.fr

Categories and Subject Descriptors

I.2.9 [**Computing Methodologies**]: Artificial Intelligence—*Robotics*

General Terms

Experimentation, Human Factors

Keywords

Theater Performance, Human-Robot Interaction

1. THE PERFORMANCE

Theatre with robotic actors is an emerging field, with a few previous published results [1, 2, 3].

On the 14th of October 2011, we performed for a general public audience (over 300 persons) a 18 min long live theatre play, acted by professional actor Xavier Brossard and the LAAS/CNRS PR2 robot. The play was created and directed by Nicolas Darrot, a mixed-media artist from Paris.

The PR2 has been programmed in a 2-months course, reusing several software components developed at LAAS/CNRS, including the 3D environment for situation assessment SPARK, the ontology-based knowledge base ORO and the natural-language processor DIALOGS.

This video abstract presents the storyline of the play, and underlines some of the significant outcome for the human-robot interaction community.

The full-length version of the video, along with downloads of the open-source components, is available from `www.laas.fr/roboscopie`.

2. STORYLINE

The play discusses how humans and robots can find a common ground for understanding each other, by living in a kind of frontier world, where real objects are replaced by abstract, disembodied counterparts.

Xavier and PR2 share a white, almost empty, stage. To get the robot to see his world, Xavier must keep being recognized by the human tracking module that lies on the wall, and must stick everywhere 2D barcodes, instead of real objects. The robot can read and identify these barcodes, and while the stage gets covered by the tags, the robot constructs for itself a 3D world with the right-looking objects: a phone, a lamp, a hanger...

While Xavier is drawing by hand more and more of these 2D tags, the robot tries to offer its help. It brings first a bottle of water, then a fan... which blows away all Xavier's code. Angry, Xavier leaves, and PR2 remains alone.

The night comes, and the robot decides to explore the stage, scattered with those barcodes on the ground. On the next morning, Xavier discovers that the robot's 3D model is a mess, full of random objects: an elephant, a boat, a van... Xavier resets the robot model and starts to tidy up the place. The robot decides to help with a trash bin, but suddenly gives up and a new program starts: a home-training session. Xavier starts the exercises, but as the program goes along, the robot looks more and more menacing, up to the point that Xavier shouts "Stop!".

Xavier eventually shows one after the other the objects to the robot, explaining they are all fake, and like the robot, we realize that everything was just an experiment.

3. SIGNIFICANCE FOR HRI

A first noteworthy achievement of this project from the human-robot interaction point of view is the use and display of the set of research tools developed at LAAS in front of a general audience: while the show had been precisely scripted and rehearsed, the robot was running the same software components we use on a daily basis in the laboratory.

By building the performance storyline on the current, actual state of robotic research, the play also put light on three key questions of today's human-robot interaction: how the human and the robot can understand each others (the robot tries to help but remains intrusive)? how to share and coexist in a common living space? how roles build up between the human and the robot (who dominates)?

4. REFERENCES

[1] C. Breazeal, A. Brooks, J. Gray, M. Hancher, J. McBean, D. Stiehl, and J. Strickon. Interactive robot theatre. *Communications of the ACM*, 46(7):76–85, 2003.

[2] C. Lin, C. Tseng, W. Teng, W. Lee, C. Kuo, H. Gu, K. Chung, and C. Fahn. The realization of robot theater: Humanoid robots and theatric performance. In *ICAR 2009*, pages 1–6. IEEE, 2009.

[3] N. Mavridis and D. Hanson. The IbnSina center: An augmented reality theater with intelligent robotic and virtual characters. In *RO-MAN 2009*, pages 681–686. IEEE, 2009.

HRI'12, March 5–8, 2012, Boston, Massachusetts, USA.
ACM 978-1-4503-1063-5/12/03.

Demonstrating Maori Haka with Kinect and Nao Robots

Thammathip Piumsomboon, Rory Clifford, Christoph Bartneck
HIT Lab NZ, University of Canterbury
Private Bag 4800, 8140 Christchurch, NZ
thammathip.piumsomboon@pg.canterbury.ac.nz

ABSTRACT

In this video, "Nao Haka", four robots and a haka leader perform a traditional Maori Haka. The Haka leader, who performs the main actions is supported by Aldebaran Nao Robots, which are controlled by an external performer, using a Microsoft Kinect as the input device. This device allows for full-body user tracking. This Video was made as a supportive gesture towards the All Blacks Rugby World Cup Campaign 2011.

Categories and Subject Descriptors: I.2.9 Robotics: Operator Interfaces

General Terms: Performance, Design, Experimentation, Human Factors.

Keywords: Robot Interaction, Robotic Puppeteering..

1. INTRODUCTION

Traditionally, Maori Haka is a war dance performed by Maori warriors before engaging combat with rival tribes or more notoriously known in this age on New Zealand Rugby fields. Haka war dances are also regularly practiced around New Zealand by Kapa Haka groups and in many schoolyards across the country, which helps promote New Zealand heritage and culture. The Particular Haka for this performance was chosen to honor the sensitive cultural aspects of Maori History. The more notorious and controversial Haka, Kamate which was adopted by the All Blacks, was decided to be detrimental to local cultural values if performed.

To support our national team, the mighty All Blacks in the 2011 Rugby World Cup final, a traditional haka was performed by robots, all of which actions are being controlled in a puppeteer style fashion by a solo performer. The Group is also valorously lead by a human performer due to system limitations. It should be noted that the robots are manufactured in France, who's rugby team played against the All Blacks in the 2011 Rugby World Cup. which has raised some controversy. The French lost to the All Blacks 7-6, so we like to think that our performance provided some good spirit and energy (mana) to the All Blacks.

A capable Performer was requisitioned who could correctly perform the war dance as the group leader. A second person who was also capable of performing the dance had control over the four Nao Robots from Aldebaran Robotics where the Microsoft Xbox 360 Kinect Camera system was used for body pose tracking [1].

HRI'12, March 5–8, 2012, Boston, Massachusetts, USA.
ACM 978-1-4503-1063-5/12/03 .

An open-source library, *nao$_o$penni* [`http://www.ros.org/wiki/nao_openni`], is used to recognize the humanÕs poses and map them on to Nao, providing gesture control without the need to attach any sensors to the user. The OpenNI [`http://openni.org/`] or Open Natural Interaction software library provides the functionality to estimate the full body poses of the user, with the exception of the hands, feet and the head. The user performs a calibration pose, called the Psi pose, to initiate tracking, and must remain within the KinectÕs field of view throughout the tracking session.

To interface the Nao robot to the Kinect camera, three libraries were used, called stacks, provided by Robot Operating System (ROS) [`http://www.ros.org/wiki/`]. The 'nao' stack, developed at the Humanoid Robots Lab at the Albert-Ludwigs-Unversitaet, provides joystick teleoperation, odometry, joint state, and a basic robot model for Nao devices. The 'openni' stack, integrated into ROS by Wim Meeussen, provides critical functions for processing data from the Kinect camera. Lastly, the *'nao$_o$penni'* stack, developed by Halit Bener Suay at Worcester Polytechnic Institute, provides the necessary mapping of body pose between the KinectÕs user pose and Nao.

The power and speed of execution in a Haka is critical to the effect that it has in devastating enemy moral. It can be seen that the Robots performance was quite noticeably poor, which lead them to become the supportive dancers. The update rate of the Nao pose was set to provide adequate performance without putting too much load onto the Nao robots CPU. Unfortunately at this rate there is around a one to two second delay between the user input and Nao's response. We also found that the Haka performance involved a significant amount of hip rotation, which proved difficult to map to the Nao. For future improvements, we want to reduce the Nao's response time and impose a constraint to prevent Nao's arms from hitting itself. Furthermore, allowing Nao to imitate userÕs leg and hip movement, while keeping its balance, is also desirable. We hope to discover a greater range of map-able movements for better control of the Nao Robots.

2. REFERENCES

1. T. Leyvand, C. Meekhof, W. Yi-Chen, S. Jian, and G. Baining,"Kinect Identity: Technology and Experience" Computer, vol. 44, pp. 94-6, 2011.

The Cocktail Party Robot: Sound Source Separation and Localisation with an Active Binaural Head *

Antoine Deleforge
INRIA Grenoble Rhône-Alpes
655, Av. Europe, 38330 Montbonnot, France
Antoine.Deleforge@inria.fr

Radu Horaud
INRIA Grenoble Rhône-Alpes
655, Av. Europe, 38330 Montbonnot, France
Radu Horaud@inria.fr

ABSTRACT

Human-robot communication is often faced with the difficult problem of interpreting ambiguous auditory data. For example, the acoustic signals perceived by a humanoid with its on-board microphones contain a mix of sounds such as speech, music, electronic devices, all in the presence of attenuation and reverberations. In this paper we propose a novel method, based on a generative probabilistic model and on active binaural hearing, allowing a robot to robustly perform sound-source separation and localization. We show how interaural spectral cues can be used within a constrained mixture model specifically designed to capture the richness of the data gathered with two microphones mounted onto a human-like artificial head. We describe in detail a novel EM algorithm, we analyse its initialization, speed of convergence and complexity, and we assess its performance with both simulated and real data.

Categories and Subject Descriptors

G.3 [**Mathematics of Computing**]: Probability and Statistics—*Multivariate statistics, Probabilistic algorithms*; I.2.7 [**Natural Language Processing**]: Speech Recognition and Synthesis

General Terms

Algorithms, Theory, Experimentation

Keywords

Blind source separation, computational auditory scene analysis, EM algorithm, learning

1. INTRODUCTION

There is an increasing interest in robots able to communicate with people in the most natural way, e.g., Fig. 1.

*This work was supported by the European project HUMAVIPS, under EU grant FP7-ICT-2009-247525.

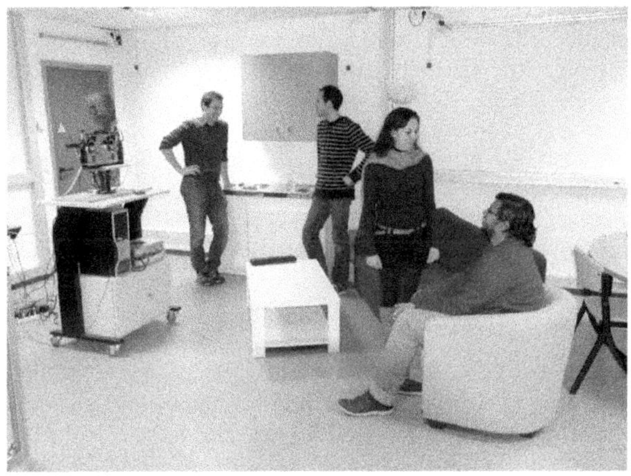

Figure 1: A *cocktail party robot* should be able to communicate with people in the most natural way. One fundamental task that such a robot should be able to accomplish is to localize the speakers in a room and to separate their emitted speech signals, all in the presence of music, background noise, and reverberations.

Such robots must be endowed with the ability to reliably process and understand sensory inputs, e.g, visual or auditory data, gathered in unconstrained physical situations. Within the field of computational auditory scene analysis (CASA) tremendous progress was made in speech recognition. Nevertheless, current approaches work well with a single sound source often recorded with a close-range microphone. A more natural setup, e.g., humanoids with their own on-board microphones, implies to deal with much more complex and challenging acoustic inputs involving auditory data of various kinds (speech, prosody, music, electronic devices, etc.) and originating from sparsely located multiple sound sources, all in the presence of noise, attenuation and reverberations.

A classical example illustrating the difficulty of modeling such situations is the well known *cocktail party problem* (CPP) [9]. While human listeners solve this problem routinely and effortlessly, we note that it has not been properly addressed from the perspective of *robot audition*. Two key aspects of CPP are localization and separation of several sound sources. We believe that principled solutions to these problems are some of the prerequisites for addressing higher-level tasks in HRI such as speech and music recognition, verbal communication, dialog handling, etc.

In this paper we propose a new method for solving for sound-source separation and localization based on a generative probabilistic model and on *active binaural hearing.* More precisely, we promote a novel robot audition paradigm based on interaural spectral features, namely the *interaural level difference* (ILD) and the *interaural phase difference* (IPD) and on a constrained mixture model specifically designed to capture the richness of the binaural data recorded with a robot head endowed with a human-like *head related transfer function* (HRTF). We formally derive an EM procedure that alternates between separation (E-step) and localization (M-step). We show how our system can be fully automatically and efficiently trained using an audiomotor map. We analyse our algorithm in detail and we assess its performance with both simulated and real data.

There is behavioral and physiological evidence that humans use binaural cues in order to infer the direction of a sound. Two such cues seem to play an essential role, namely the ILD (already mentioned) and the *interaural time difference* (ITD). A number of computational models have been recently developed for robust sound localization and sound tracking based on ITD and/or ILD [19, 24]. However, it is well known that the spatial information provided by interaural-difference cues within a restricted band of frequency is spatially ambiguous, particularly along a roughly vertical and front/back dimension [15]. To avoid these ambiguities, more accurate sound localization models incorporate the HRTF, e.g. [12]. These approaches are based on the fact that the particular shape of the head, pinna and torso act as a filter depending on the emitting 3D location of the sound source (distance, azimuth, and elevation). However, HRTF databases are subject-specific and room-dependent (noise, reverberations, room geometry, etc.), and only a handful of HRTF databases are available in practice [1, 21]. This makes them hardly applicable to a real robotic application. To overcome these issues, the HRTF of a specific robot can be automatically learnt using *audio-motor maps* [8, 10]. Such maps are built by recording a static full-spectrum sound source from different motor states of the robot.

The problem of sound source separation has been thoroughly studied in the last decades and several interesting approaches were proposed. For example, [4, 20] and many others achieve separation with a single microphone, based on known acoustic properties of speech signals, and are therefore limited to a specific type of input. Other techniques such as independent component analysis (ICA) [6] or multi-microphone techniques require as many microphones as the number of sources. Several other methods use binaural localization cues for source separation [11, 14, 23]. In [11] acoustic inputs at different frequency channels are clustered over time by means of some assumptions on the emitted signals, and an HRTF data look-up table is used to find their corresponding positions in space. Once exact locations are known, up to two sources can be separated using the HRTF at each frequency channel.

Our method is based both on clustering (localization) and on spectral masking (separation). Spectral masking, also called binary masking, allows the separation of an arbitrary number of sources from a mixed signal, with the only assumption that a single source is active at every frequency-time point (f, t). This is referred to as the *W-disjoint orthogonality* assumption [25] and it has been shown to hold, in general, for simultaneous speech signals; It is particularly well suited for binaural recordings in realistic environments. Recently, [14] proposed a probabilistic model for multiple sound source separation based on interaural spatial cues. For each sound source, a binary mask and a discrete distribution over interaural time delays is provided. This can be used to approximate the azimuth angle of the source with a front-back ambiguity, if the distance between the microphones is known.

It appears that there are very few methods that formally combine 2D localization and separation. The first originality of our approach is to formally cast the localization and separation tasks into a generative probabilistic model that is solved very efficiently with a novel EM algorithm. Full details of this algorithm and its initialization as well as favorable comparisons with recent binaural-based separation methods are presented in a companion paper [7]. The second originality of our approach is to use an active robotic head in order to automatically learn *audio-motor maps* for a very large set of sound-source directions located in the far field. Such a training phase incorporates an implicit model of the HRTF and leads to a high precision both in terms of source separation and of localization. A thorough evaluation of the method with simulated and real data is provided, and puts forwards this approach as a promising future tool for auditory human-robot interactions.

The remainder of this paper is organized as follows: section 2 describes a binaural sound representation, section 3 presents in detail the formal model as well as its associated EM algorithm, section 4 describes the data acquisition and recording technique, section 5 presents our validation method, experiments and results. Concluding remarks and directions for future work are discussed in section 6.

2. BINAURAL SOUND REPRESENTATION

Both sound source localization and separation require a proper representation of the perceived data. For localization, one needs a content-independent representation that contains as much spatial information as possible. For separation, one needs a representation preserving all the richness of the original signals. In this paper we put forward two *binaural* representations, namely ILD and IPD spectrograms.

As already mentioned in section 1, our sound source separation method is based on binary masking. This technique consists in "filtering" the original signal by weighting all frequency-time points corresponding to the target source with 1 and all the other points with 0. Consequently, the perceived signal needs to be described within a time-varying spectral representation, a *spectrogram.* Spectrograms associated with each one of the two microphones are computed using a short-term fast Fourier transform (FFT) algorithm [2]. We use a time window of 64ms with 8ms overlap between two consecutive windows, thus yielding $T = 126$ time-windows for a signal lasting 1s. Since sounds are recorded at a sample rate of 24,000Hz, each time-window contains 1,536 samples, multiplied by a Hann window padded with zeros, for a total length of 2,048 samples. Each window is then transformed via FFT to obtain complex coefficients of $F = 1,024$ positive frequency channels between 0 and 12,000Hz. We denote with $s_{f,t}^{(k)} \in \mathbb{C}$ the spectrogram value at frequency-time point (f, t) of a signal emitted by a sound source k and with $s_{f,t}^{(\mathrm{L})}$ and $s_{f,t}^{(\mathrm{R})}$ the spectrogram values perceived by the left- and right-microphone respectively. The spectrogram of the *emitted acoustic level* is defined by:

$$a_{f,t}^{(k)} = 10 \log(|s_{f,t}^{(k)}|^2) \tag{1}$$

while the spectrogram of the *perceived acoustic level* is defined by

$$a_{f,t}^{(\mathrm{LR})} = 10 \log(|s_{f,t}^{(\mathrm{L})}|^2 + |s_{f,t}^{(\mathrm{R})}|^2) \quad (2)$$

Fig. 2-(a) and (b) show an example of spectrograms of emitted and perceived acoustic levels.

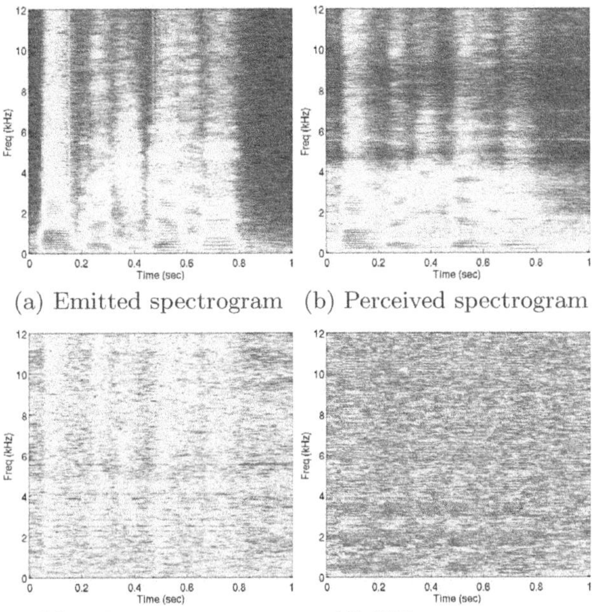

(a) Emitted spectrogram (b) Perceived spectrogram

(c) ILD spectrogram (d) IPD spectrogram

Figure 2: Spectrograms corresponding to a male utterance emitted from the left-hand side of the binaural head.

The W-disjoint orthogonality assumption implies that a single sound source emits at a given frequency-time point (f,t). Let k be that source, the HRTF model provides a relationship between emitted and perceived spectrogram points at (f,t):

$$s_{f,t}^{(\mathrm{L})} = h^{(\mathrm{L})}(\boldsymbol{x}_k, f) s_{f,t}^{(k)} \text{ and } s_{f,t}^{(\mathrm{R})} = h^{(\mathrm{R})}(\boldsymbol{x}_k, f) s_{f,t}^{(k)} \quad (3)$$

where $\boldsymbol{x}_k \in \mathbb{R}^3$ denotes the 3D position of sound source k in a robot-centered coordinate frame (the origin of this frame is the midpoint between the two microphones and the x, y, and z axes pointing towards the left microphone, the head-top and in front of the head) and $h^{(\mathrm{L})}$ and $h^{(\mathrm{R})}$ denote the left and right HRTF. The latter functions depend on both the emitter's position and the frequency and they act as linear filters on the emitted signal. The HRTFs are mainly determined by the shape of the head, pinna and torso of the listener, e.g, the robot-mounted dummy-head in our case. The *interaural transfer function* (ITF) $\hat{I}$ is defined by the ratio between the left- and right-HRTF:

$$I(\boldsymbol{x}_k, f) = \frac{h^{(\mathrm{R})}(\boldsymbol{x}_k, f)}{h^{(\mathrm{L})}(\boldsymbol{x}_k, f)} \in \mathbb{C} \quad (4)$$

We can now define the interaural spectrogram as the ratio between the left- and right-microphone spectrograms:

$$\hat{I}_{f,t} = \frac{s_{f,t}^{(\mathrm{R})}}{s_{f,t}^{(\mathrm{L})}} \quad (5)$$

From (3), (4) and (5) one may notice that if the source k emitting at time-frequency point (f,t) is located at $\boldsymbol{x}_k$ then $\hat{I}_{f,t} \approx I(\boldsymbol{x}_k, f)$. The equality is only approximate due to the sensor noise and FFT errors. Therefore, at a given frequency-time point, the interaural spectrogram value $I_{f,t}$ does not depend on the emitted spectrogram value $s_{f,t}^{(k)}$ but only on the source position $\boldsymbol{x}_k$. We finally define the *ILD spectrogram* α and the *IPD spectrogram* ϕ as the log-amplitude and phase of the complex interaural spectrogram $\hat{I}_{f,t}$, e.g., Fig. 2-(c) and (d):

$$\alpha_{f,t} = 20 \log |\hat{I}_{f,t}| \in \mathbb{R} \text{ and } \phi_{f,t} = \arg(\hat{I}_{f,t}) \in]-\pi, \pi] \quad (6)$$

3. SEPARATION AND LOCALIZATION

This section starts by describing how a training set of interaural parameters can be built out of a sound dataset annotated with sources positions. After formally stating the problem, we then introduce our novel probabilistic model for interaural parameters in a mixture of point sound sources. We finally detail a new version of the EM algorithm for this model, allowing to achieve simultaneous sound sources separation and localization based on learned positions.

3.1 Building a Training Set

We first explain how a training set of interaural parameters can be built out of a sound dataset annotated with sources positions. Let $\mathcal{X} = \{\boldsymbol{x}_n\}_{n=1}^N$ be a set of 3D coordinates in the robot centered frame that we already described above. If a single sound source n emits white noise from $\boldsymbol{x}_n \in \mathcal{X}$, let $\{\alpha_{f,t}^n\}_{f=1,t=1}^{F,T}$ and $\{\phi_{f,t}^n\}_{f=1,t=1}^{F,T}$ be the perceived ILD and IPD spectrograms. The *mean ILD vector* $\boldsymbol{\mu}(\boldsymbol{x}_n) = (\mu_1^n \dots \mu_f^n \dots \mu_F^n)^\top \in \mathbb{R}^F$ associated with $\boldsymbol{x}_n$ is defined by taking the temporal mean of α at each frequency channel:

$$\mu_f(\boldsymbol{x}_n) = \frac{1}{T} \sum_{t=1}^{T} \alpha_{f,t}^n \quad (7)$$

Similarly, the *mean IPD vector* $\boldsymbol{\xi}(\boldsymbol{x}_n) = (\xi_1^n \dots \xi_f^n \dots \xi_F^n)$ is the temporal mean of the IPD spectrogram ϕ at each frequency channel:

$$\xi_f(\boldsymbol{x}_n) = \arg \frac{1}{T} \sum_{t=1}^{T} \exp(j\phi_{f,t}^n) \in]-\pi, \pi] \quad (8)$$

The components of the mean IPD vector are estimated in the complex domain in order to avoid problems due to phase circularity, as suggested in [16]. White noise is used because it contains equal power within a fixed bandwidth at any center frequency: the source n is therefore the only source emitting at each frequency-time point (f,t), and μ_f and ξ_f are thus approximating the log-amplitude and phase of $\hat{I}(.,f)$. The set $\mathcal{X}$ of 3D sound-source positions as well as the mappings $\boldsymbol{\mu}$ and $\boldsymbol{\xi}$ will be referred to as the training data to be used in conjunction with the sound-source separation and localization algorithm described below.

3.2 Problem Formulation

We suppose that there are K sound sources randomly located in the robot's environment and that these K sources emit simultaneously sounds with unknown spectrograms from the unknown locations $\{\boldsymbol{x}_k\}_{k=1}^K \subset \mathcal{X}$. From the acoustic inputs perceived by the robot's microphone pair, one can build the ILD and IPD spectrograms $\{\alpha_{f,t}\}_{f=1,t=1}^{F,T}$ and $\{\phi_{f,t}\}_{f=1,t=1}^{F,T}$ as already described. The goal of the sound-source separation and localization algorithm is to associate each perceived frequency-time point (f,t) with one of the sound sources and to estimate their 3D locations.

The observed $\alpha_{f,t}$ and $\phi_{f,t}$ are useful only if there is a significant signal at (f,t). To identify such *significant observations* we estimate the perceived acoustic level $a^{\text{LR}}_{f,t}$ with (2) and we retain only those frequency-time points for which the acoustic level is above a threshold, $a^{\text{LR}}_{f,t} > \epsilon_f$. One empirical way to choose the thresholds (one at each frequency) is to measure the highest perceived acoustic level at each f in the absence of emitting sound sources. These thresholds are typically very low compared to perceived acoustic levels due to natural sounds, and allow to filter out frequency-time points corresponding to the "room silence". We call *level mask* the associated spectral mask. We denote by M_f the number of significant observation at f and by $\alpha_{f,m}$ and $\phi_{f,m}$ the m-th significant ILD and IPD observations at f. The *observed data* will be denoted by $\mathbf{A} = \{\alpha_{f,m}\}_{f=1,m=1}^{F,M_f}$ and $\mathbf{\Phi} = \{\phi_{f,m}\}_{f=1,m=1}^{F,M_f}$.

We also introduce the *missing data*, i.e., a set of *unobserved variables* $\boldsymbol{z}_{f,m} \in \{0,1\}^K$, such that $z_{f,m,k} = 1$ if observations $\alpha_{f,m}$ and $\phi_{f,m}$ were generated by source k, and $z_{f,m,k} = 0$ otherwise. The W-disjoint orthogonality constraint can therefore be written as:

$$\sum_{k=1}^{K} z_{f,m,k} = 1 \quad \forall(f,m) \tag{9}$$

Variables $\boldsymbol{z}_{f,m}$ are also called *data-to-source assignments* at (f,m), and $\mathcal{M}_k = \{z_{f,m,k}\}_{f=1,m=1}^{F,M_f}$ corresponds to the binary spectral mask associated with sound-source k. Finally, we denote with $\mathbf{Z} = \{\boldsymbol{z}_{f,m}\}_{f=1,m=1}^{F,M_f}$ the set of all missing data. The problem of simultaneous localization and separation amounts to estimate the sound-source locations $\{\boldsymbol{y}_k\}_{k=1}^K$ and the masking variables $\mathbf{Z}$, given the number of sound sources K and the observed data $\mathbf{A}$ and $\mathbf{\Phi}$.

3.3 A Constrained Mixture Model

We assume that both the ILD and IPD observed data are drawn from normal distributions. The conditional likelihood of the observation $\alpha_{f,m}$ given its assignment to sound source k, i.e., $z_{f,m,k} = 1$ and located at $\boldsymbol{x}_k$ is therefore drawn from a 1D Gaussian distribution centered at $\mu_f(\boldsymbol{x}_k)$ and with variance $\sigma^2_{f,k}$:

$$\begin{aligned} P(\alpha_{f,m}|z_{f,m,k},\boldsymbol{x}_k,\sigma_{f,k}) &= \mathcal{N}(\alpha_{f,m}|\mu_f(\boldsymbol{x}_k),\sigma^2_{f,k}) \\ &= \frac{1}{(2\pi)^{1/2}\sigma_{f,k}} \exp\left(-\frac{(\alpha_{f,m}-\mu_f(\boldsymbol{x}_k))^2}{2\sigma^2_{f,k}}\right) \end{aligned} \tag{10}$$

For simplicity, the expression $z_{f,m,k} = 1$ is replaced by $z_{f,m,k}$ in probabilities. Since the IPD data lie on the circle $]-\pi,\pi]$, they should be modeled by a circular normal distribution. As proposed in [14], we approximate the wrapped normal distribution with a 1D Gaussian. The conditional likelihood of the observation $\phi_{f,m}$ given its assignment to sound source k located at $\boldsymbol{x}_k$ is given by:

$$\begin{aligned} P(\phi_{f,m}|z_{f,m,k},\boldsymbol{x}_k,\rho_{f,k}) &= \angle\mathcal{N}(\phi_{f,m}|\xi_f(\boldsymbol{x}_k),\rho^2_{f,k}) \\ &= \frac{1}{(2\pi)^{1/2}\rho_{f,k}} \exp\left(-\frac{\Delta(\phi_{f,m},\xi_f(\boldsymbol{x}_k))^2}{2\rho^2_{f,k}}\right) \end{aligned} \tag{11}$$

where $\rho^2_{f,k}$ is the IPD variance associated with source k at frequency f and the Δ function is defined by

$$\Delta(x,y) = \arg(e^{j(x-y)}) \in]-\pi,\pi] \tag{12}$$

The distribution $\angle\mathcal{N}(\xi,\rho^2)$ approximates the normal distribution on the circle $]-\pi,\pi]$ when ρ is small relative to 2π. Preliminary experiments on IPD spectrograms of white noise showed that this assumption holds in the general case.

Note that in our model, the source positions act as latent constraints on Gaussian means and thus *tight* the observations at different frequency channels. Such an approach can also be used to tight auditory and visual observations, as recently done in [13].

As emphasized in [14], the well known correlation between ILD and IPD does not contradict the assumption that Gaussian noises corrupting the observations are independent. Under this assumption, the conditional likelihood of the observed data $(\alpha_{f,m},\phi_{f,m})$ given its assignment to source k located at $\boldsymbol{x}_k$ and with variances $\sigma^2_{f,k}$ and $\rho^2_{f,k}$ is

$$\begin{aligned} &P(\alpha_{f,m},\phi_{f,m}|\boldsymbol{z}_{f,m},\boldsymbol{x}_k,\sigma_{f,k},\rho_{f,k}) = \\ &\mathcal{N}(\alpha_{f,m}|\mu_f(\boldsymbol{x}_k),\sigma^2_{f,k}) \times \angle\mathcal{N}(\phi_{f,m}|\xi_f(\boldsymbol{x}_k),\rho^2_{f,k}) \end{aligned} \tag{13}$$

The prior probability of assigning an observation to a sound source writes:

$$P(z_{f,m,k}) = \pi_{f,k} \tag{14}$$

Finally, we denote by Θ the set of all the parameters of our binaural mixture model:

$$\Theta = \left\{\{\boldsymbol{x}_k\};\ \{\pi_{f,k}\};\ \{\sigma^2_{f,k}\};\ \{\rho^2_{f,k}\}\right\}_{f=1,k=1}^{F,K} \tag{15}$$

3.4 The Separation-Localization Algorithm

The problem of both source separation and source localization can now be expressed as an optimal parameter estimation problem, namely the maximization of the observed-data log-likelihood over the parameters Θ:

$$\widetilde{\Theta} = \arg\max_{\Theta} \mathcal{L}(\mathbf{A},\mathbf{\Phi};\Theta) \tag{16}$$

In order to keep the model as general as possible, there is no assumption on the emitted sounds as well as the way their spectra are spread across the frequency-time points. Therefore, we assume that all the observations are independent:

$$\begin{aligned} \mathcal{L}(\mathbf{A},\mathbf{\Phi};\Theta) &= \log P(\mathbf{A},\mathbf{\Phi}|\Theta) \\ &= \sum_{f=1}^{F}\sum_{m=1}^{M_f} \log P(\alpha_{f,m},\phi_{f,m}|\Theta) \end{aligned} \tag{17}$$

It is well established that the direct optimization of (17) is difficult because of the presence of many local maxima. Therefore we recast the problem within the framework of *maximum likelihood with missing data*, i.e., $\mathbf{Z}$ in (13), which is traditionally solved via expectation-maximization (EM). The EM algorithm alternates between estimating the *expected complete-data log-likelihood* (E-step) using the current model parameters and maximizing this likelihood over the model parameters given the current posterior probabilities (M-step). The posterior probability $r_{f,m,k}$ of $z_{f,m,k}$ is defined by:

$$r_{f,m,k} = P(z_{f,m,k}|\alpha_{f,m},\phi_{f,m};\Theta) \tag{18}$$

As it will be detailed below, the E-step of the proposed algorithm computes the posterior probabilities of assigning each spectrogram point to a sound source k (*separation-step*), while the M-step maximizes the expected complete-data log-likelihood with respect to the model parameters Θ, namely the priors, the variances and, most notably, the source locations $\{\boldsymbol{x}_k\}_{k=1}^K$ (*localization-step*).

One of the most interesting properties of EM algorithms is that the log-likelihood $\mathcal{L}$ in (16) is increased at each EM iteration and hence it converges to a local maximum. An outline of the proposed EM method is provided in Algorithm 1. The finally estimated posterior probabilities allow to compute the binary spectral masks $\mathcal{M}_k$ associated with each source k while the final parameters $\boldsymbol{x}_k$ provide estimates for the source locations. $v^{(p)}$ denotes the value of variable v at iteration p, and $v^{(\mathrm{P})}$ denotes its final value. The E-step, M-step, initialization and convergence check are detailed below.

Algorithm 1 Separation-Localization EM

```
Input: A, Φ, {μ(x_n)}_{n=1}^N, {ξ(x_n)}_{n=1}^N, K
Output: {x_k^(P)}_{k=1}^K, {M_k}_{k=1}^K
Θ^(0) := initialize
p := 0
while !converged do
    p := p + 1
    {r_{f,m,k}^(p)} := E-step(Θ^(p-1))
    Θ^(p) := M-step({r_{f,m,k}^(p)}_{f,m,k})
end while
M_k := (k == argmax_{k'} r_{f,m,k'}^(P))_{f=1,m=1}^{F,M_f}
```

Based on Bayes' formula and on marginalization rules, the E-step computes the current posterior probabilities conditioned by the previously estimated parameters, i.e., (13):

$$r_{f,m,k}^{(p)} := \frac{\pi_{f,k} P(\alpha_{f,m}, \phi_{f,m} | \boldsymbol{z}_{f,m}; \Theta^{(p-1)})}{\sum_{i=1}^{K} \pi_{f,i} P(\alpha_{f,m}, \phi_{f,m} | \boldsymbol{z}_{f,m}; \Theta^{(p-1)})} \tag{19}$$

The expected complete-data log-likelihood can now be written as:

$$\begin{aligned} Q(\Theta|\Theta^{(p-1)}) &= E_{(\mathbf{Z}|\mathbf{A},\boldsymbol{\Phi};\Theta)}[\log P(\mathbf{A}, \boldsymbol{\Phi}, \mathbf{Z}|\Theta)] \\ &= \sum_{f=1}^{F} \sum_{m=1}^{M_f} \sum_{k=1}^{K} r_{f,m,k}^{(p)} \log \pi_{f,k} P(\alpha_{f,m}, \phi_{f,m} | \boldsymbol{z}_{f,m}; \Theta) \end{aligned} \tag{20}$$

The M-step maximizes (20) with respect to Θ:

$$\Theta^{(p)} := \widetilde{\Theta} = \operatorname*{arg\,max}_{\Theta} Q(\Theta|\Theta^{(p-1)}) \tag{21}$$

By combining (13) with (20) the problem becomes equivalent to minimizing:

$$\sum_{f=1}^{F} \sum_{m=1}^{M_f} r_{f,m,k}^{(p)} \left(\log\left(\frac{\sigma_{f,k}^2}{\pi_{f,k}}\right) + \log\left(\frac{\rho_{f,k}^2}{\pi_{f,k}}\right) + \frac{(x_{f,m} - \mu_f(\boldsymbol{x}_k))^2}{\sigma_{f,k}^2} + \frac{\Delta(\phi_{f,m}, \xi_f(\boldsymbol{x}_k))^2}{\rho_{f,k}^2} \right) \tag{22}$$

which can be easily differentiated with respect to $\{\pi_{f,k}\}_f$, $\{\sigma_{f,k}\}_f$ and $\{\rho_{f,k}\}_f$ to obtain closed-form expression, conditioned by $\boldsymbol{x}_k$, for the optimal values of these parameters. These expressions are then subtituted in (22), which is evaluated for all $\boldsymbol{x}_k \in \mathcal{X}$ to find the optimal position $\hat{\boldsymbol{x}}_k$, and deduce all the other optimal parameters. Interested readers can find a detailed solution in [7].

As already mentioned, EM converges to a local maximum of the observed data log-likelihood function $\mathcal{L}$. However, the non-injectivity of the interaural functions μ_f and ξ_f leads to a very large number of these maxima, especially when the set of learned positions $\mathcal{X}$, i.e., section 3.1, is large. This makes the algorithm to be very sensitive to initialization. A common way to avoid being trapped in local maxima may be to initialize the parameters at random, but such a strategy cannot be directly applied here: First, because the cardinality of the parameter set Θ is very large and second, because there is no straightforward way to initialize the variances $\sigma_{f,k}^2$ and $\rho_{f,k}^2$. Another possibility may be to randomly initialize the source assignment variables $\mathbf{Z}$ and then proceed with the M-step, but extensive experiments with simulated data revealed that the algorithm very rarely converged to a global maximum (in less than 0.1% of the cases). We therefore decided to adopt a method that combines these two initialization strategies by randomly perturbating both the source locations and the source assignments.

This lead us to develop a *stochastic initialization* procedure similar to the stochastic EM (SEM) algorithm [5]. The idea of exploiting stochasticity to escape from local maxima is a commonly used principle in global optimization [26]. The SEM algorithm includes a stochastic step (S) in between the E and the M steps, during which random samples $R_{f,m,k} \in \{0, 1\}$ are drawn from the posteriors $r_{f,m,k}$. These samples are then used instead of the posterior probabilities during the M-step. To initialize our algorithm, we first set all the posterior probabilities to $1/K$ and then proceed through the following step sequence: S M* E S M, where the M*-step is a variation of the M-step in which the sources' positions are drawn randomly from $\mathcal{X}$ instead of computing $\{\hat{\boldsymbol{x}}_k\}_k$. Experiments with simulated data showed that this technique converged to a global optimal solution in over 10% of the cases. More advanced techniques may also be used to improve the convergence rate, and are detailed in [7].

In practice, twenty stochastic initializations are used in order to increase the chances of correct convergence, and only the one providing the best log-likelihood after two iterations is eventually iterated until the convergence criteria is satisfied. The algorithm stops either when the log-likelihood gain is less than 1%, or after $p_{\max} = 20$ iterations. Indeed, from two hundred simulated experiments (see section 5.2), we concluded that, on an average, the algorithm converges in eleven iterations.

3.5 Algorithm Complexity

Both the E- and S-steps are linear in the total number of significant observations, $\sum_f M_f$ and in the number of sound sources, K. The M-step is linear in the number of frequency channels F and in the number N of source locations available in the training set $\mathcal{X}$. In the case of simulated data using one-second long signals composed of two sources, the values of these parameters are: $K = 2$, $F = 1024$, $\sum_{f}, M_f = 70,000$, and $N = 10,800$. With a Matlab implementation executed on a 2.53GHz Intel-Xeon processor, we obtained the following average running times: 49ms (E), 1030ms (S), 2520ms (M) and 24ms (M*). The most time-consuming part of the algorithm is its initialization (20 stochastic initializations iterated 2 times) which takes 195.9 seconds, while each subsequent EM iteration takes 2.569 seconds, amongst which 98% corresponds to selecting all optimal $\hat{x}_k$ in $\mathcal{X}$. Careful algorithm and software optimization will allow, however, to obtain realistic execution times needed for human-robot interactions scenarios, e.g., one to two seconds.

4. DATA ACQUISITION

In order to build the training set introduced in section 3.1, we developed a technique to learn a large number of sound source locations in an entirely unsupervised and automated way using the motor system of a binaural robot head. This technique is initially inspired from the sensorimotor theories of early development in psychology, suggesting that experiencing the sensory consequences of voluntary motor actions was necessary for an organism to learn the perception of space [17]. In particular [3] argued that naive organisms such as humans and echo-locating bats could learn sound localization based solely on acoustic inputs and their relation to motor states. This idea was experimentally validated using a robot system in [8].

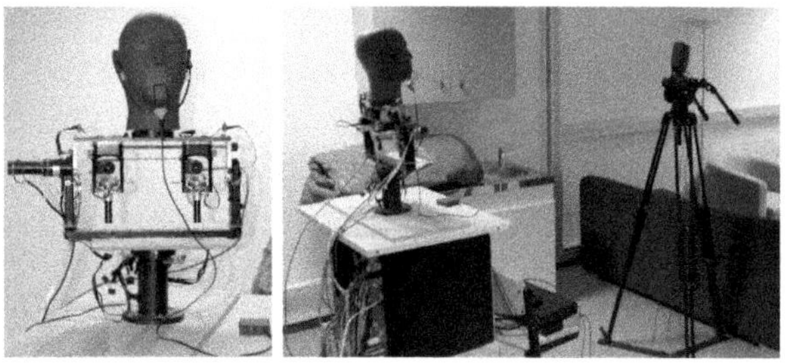

Figure 3: A binaural head is placed onto an agile device that can perform precise and reproducible pan and tilt motions (left). The emitter (a loud-speaker) is placed in front of the robot head at approximately 2.7 meters (right).

Sound acquisition is performed with a Sennheiser MKE 2002 acoustic dummy-head linked to a computer via a Behringer ADA8000 Ultragain Pro-8 digital external sound card. The head is mounted onto a robotic system with two rotational degrees of freedom: a pan motion and a tilt motion (see Fig. 3). This device was specifically designed to achieve precise and reproducible movements. The emitter – a loud-speaker – is placed at approximately 2.7 meters ahead of the robot, as showed on Fig. 3. The loud-speaker's input and the microphones' outputs were handled by two synchronized sound cards in order to simultaneously record and play. All the experiments were carried out in real-world conditions, i.e., a room with natural reverberations and background noise due to computer fans. All the recordings are publicly available[1]. We believe that such a large audio-motor data set has no equivalent today, and is therefore a contribution in its own right.

Rather than placing the emitter at known 3D locations around the robot, it was kept in a fixed reference position while the robot recorded emitted sounds from different motor states. Consequently, a sound source direction is directly associated to a pan-tilt motor state (ψ, θ) rather than a 3D point in space. A robot trained in such a way would therefore be able to perform the head movement pointing toward an emitting sound source in an entirely unsupervised way, without needing the inverse kinematics, the distance between microphones, or any other parameters.

Recordings were made from 10,800 uniformly spread motor states: 180 pan rotations ψ in the range $[-180°, 180°]$ (left-right) and 60 tilt rotations θ in the range $[-60°, 60°]$ (top-down). The associated set of 3D source positions $\mathcal{X} \subset \mathbb{R}^3$ could be deduced using the direct kinematic model of the robot given in [8]. However, the speaker was located in the far field of the head during experiments (> 1.8 meters), and [18] showed that HRTFs mainly depend on the sound source direction while the distance has fewer impact in that case. That is why sound source locations will be expressed with angles in the rest of the paper.

At each motor state, five binaural recordings of one second each were made while the speaker emitted different sounds. Sound 1 corresponds to white noise, and was used to build the training set (section 3.1). Sounds 2, 3 and 4 form the test set (see section 5.3), and correspond respectively to a woman pronouncing *"Bonjour!"*, a man pronouncing *"Un petit café?"*, and a flute melody. Sound 0 corresponds to "room silence" and was used to determine the perceived acoustic level threshold ϵ_f during tests (see section 3.2).

5. EXPERIMENTS AND RESULTS

5.1 Performance Evaluation

Algorithm 1 was tested and validated using the database described in the previous section. In all presented results, a source is considered as *correctly localized* if the algorithm could locate it within 2° of absolute angular error in both azimuth and elevation. For real mixtures (section 5.2), the separation was evaluated using standard metrics, namely Source-to-Distortion Ratio (SDR) and Source-to-Interferences Ratio (SIR), proposed in [22]. These metrics are based on the decomposition of the estimated signal into the target signal, the error term coming from other interfering signals, and the error term coming from unexplained artifacts. Another term can be added to evaluate background noise reduction, but this is outside the scope of this work. Once the decomposition is achieved, the SDR is defined by the ratio of energy between the target and the error terms, while the SIR is the ratio of energy between the target and the interference term only. Both these metrics are expressed in decibels. As they are computed from 1D signals, the masked left and right microphone spectrograms were converted back to temporal signals using the inverse FFT, and concatenated to evaluate the quality of binaural-based separation.

SDR and SIR scores obtained with our algorithm are given together with upper and lower bounds. The upper bound corresponds to the SDR of the ground truth mask or *oracle mask* [25], which is set to 1 at every spectrogram point in which the target signal is at least as loud as the combined other signals and 0 everywhere else. The lower bound corresponds to the SDR and SIR in the original mixture. The level mask described in section 3.2 was applied to all signals so that only separation of significant observations could be evaluated.

For tests made on simulated spectrograms, since there is no 1D signal to compare with, the separation was evaluated by a pointwise $\neq$ operation between estimated and oracle binary masks: we define the *mask error* by the ratio of points in the estimated mask that differs from the oracle mask.

[1] `http://perception.inrialpes.fr/~Deleforge/CAMIL_Dataset/`

5.2 Simulated Data

To validate our model we first tested the algorithm on simulated data. We simulated ILD and IPD spectrograms with $F = 1024$ frequency channels from 0 to 12,000Hz, and 10 to 126 significant observations per channel (M_f randomly drawn for each f). First, K source positions $\boldsymbol{x}_1 \dots \boldsymbol{x}_K$ are randomly drawn from $\mathcal{X}$. Then, each spectrogram point (f, m) is randomly assigned to one of the K sources, and we generate ILD and IPD observations $\alpha_{f,m}$ and $\phi_{f,m}$ using the distributions $\mathcal{N}(\mu_f(\boldsymbol{x}_k), \sigma_f^2(\boldsymbol{x}_k))$ and $\angle\mathcal{N}(\xi_f(\boldsymbol{x}_k), \rho_f^2(\boldsymbol{x}_k))$, where $\sigma_f^2(\boldsymbol{x}_k)$ and $\rho_f^2(\boldsymbol{x}_k)$ correspond to ILD and IPD variances of white noise emitted from $\boldsymbol{x}_k$, and are estimated from the dataset. A total of 200 mixtures composed of 2 to 5 sources were generated, and K was set to the correct number of sources for each test. Table 1 shows the percentage of correctly localized sources, the mean mask error obtained over all sources, and the mean mask error obtained with randomly generated binary masks.

Table 1: Mean localization and separation results for 200 simulated mixtures of 2 to 5 sources.

K	Correctly Localized(%)	Estimated Mask Error(%)	Random Mask Error(%)
2	99.2	20.7	50.0
3	91.2	25.4	55.6
4	54.1	28.9	62.5
5	14.4	29.3	68.0

These results show that our model is very well suited for both separation and localization tasks. A very high localization accuracy is achieved for mixtures composed of 2 and 3 sources. When the number of sources is higher, the number of observations per source decreases while the number of local maxima of the log-likelihood function increases, leading to poorer results.

5.3 Real Mixtures from Learned Positions

Three sounds were used to test the algorithm with real data: male speech, female speech, and flute melody (see section 4). Mixtures were obtained by summing raw binaural signals from the dataset, corresponding to randomly drawn positions in $\mathcal{X}$. ILD and IPD spectrograms were computed from these mixtures, and only significant observations corresponding to points with a higher acoustic level than the corresponding background mixtures were kept, as described in section 2.

First, 200 mixtures for each possible pair of test sounds were generated, resulting in 1,200 localization-separation tasks. Mean results obtained with our method for all these tasks as well as mean results for correctly localized sources only are shown in Table 2. SDR and SIR scores are compared to those of the original mixture and oracle mask.

A very high localization rate is obtained, with 84.6% of the 1,200 test sound sources localized with less that 2° error in both azimuth and elevation. In addition, significant improvements of SDR and SIR ratios over the original mixtures are achieved. This is particularly true for SIR ratios that almost reached oracle ratios in some tests. This shows that although our algorithm generates some artifacts in the target sound signal during the spectral masking process, it is able to considerably reduce the volume of the interferer and thus significantly improve the perceived quality. Finally, we note that SDR and SIR ratios are always better when the algorithm could correctly localize the sound source, which demonstrates the importance of localization cues for sound sources separation.

Table 2: Mean angle error, SDR (Source-to-Distortion Ratio) and SIR (Source-to-Interferences Ratio) for real mixtures of 2 sound sources.

	Ang Err(°)	SDR(dB)	SIR(dB)
Original	-	0.05	0.05
Us (All sources)	**11.5**	**4.11**	**14.3**
Us (Loc: 84.6%)	**0.18**	**5.17**	**15.2**
Oracle	-	17.9	39.6

A second experiment was made with 200 mixtures of the three test sounds emitting altogether from random positions in $\mathcal{X}$, resulting in 600 localization-separation tasks. Results are displayed in a similar fashion in Table 3.

Table 3: Mean angle error, SDR and SIR for real mixtures of 3 sound sources.

	Ang Err(°)	SDR(dB)	SIR(dB)
Original	-	-3.61	-3.61
Us (All sources)	**35.5**	**-4.03**	**3.21**
Us (Loc: 45.2%)	**0.18**	**0.09**	**7.29**
Oracle	-	14.5	32.7

Although performances significantly decrease for this very challenging task (a 1s mixture of 3 equally loud sounds is hard to decipher even for humans), we note that the algorithm could still accurately localize almost half of the 600 individual sources, while significantly improving their SDR and SIR with respect to the original mixture.

5.4 Real Mixtures from Unlearned Positions

The last experiment is also the most challenging one. Indeed, we tested the ability of our algorithm to separate sound sources emitting from positions outside the training set. More precisely, we made several test recordings in which two loud speakers were simultaneously emitting different sounds while the robot stayed in its reference position (0°, 0°). One of the loud speaker was placed in a *frontal* position corresponding to the training dataset, while the second one was manually placed in 21 *side* positions around the robot, with a 10° to 90° azimuth distance and 0° to 30° elevation distance. These experiments were repeated for the six possible pairs of test sounds, resulting in 252 separation-localization tasks. The frontal loud speaker was correctly localized at (0°, 0°) in 95.2% of the mixtures tested. Localization performances for the side loud speaker were not evaluated, as no ground truth was available. Table 4 shows the mean SDR and SIR scores of each source obtained with our approach, the original mixture, and the oracle mask.

Despite a decrease of performances with respect to sound mixtures from learned positions, the significant improvement of SIR obtained in realistic cocktail-party like scenarios puts forward our approach as a promising tool for auditory human-robot interactions.

Table 4: Mean SDR and SIR of frontal and side speakers

	Frontal SDR(dB)	Frontal SIR(dB)	Side SDR(dB)	Side SIR(dB)
Original	-0.28	-0.28	0.32	0.32
Us	**-1.09**	**3.01**	**-3.30**	**5.69**
Oracle	18.29	40.64	18.63	40.13

6. CONCLUSION AND FUTURE WORK

Traditionally, computational auditory scene analysis was addressed with either a close-range or an array of microphones and using simulated or anechoic audio data. We propose to bridge the gap between constrained and unconstrained audio analysis and to apply CASA to HRI. We propose a system integrating the audio-motor abilities of a robot within a unified framework, and performing sound-source separation and localization in a realistic cocktail-party like scenario. By providing a genuine audio-motor database and presenting encouraging results obtained from these data, we presented a benchmark for the unexplored field of sensorimotor learning for robot audition.

One of the most interesting and promising directions will be to extend our model to a continuous space of sound source positions. This could be done using the manifold structure of interaural parameters studied in detail in [8]. By approximating this manifold by local tangent spaces, the size of the training set could be considerably reduced, thus speeding up the M-step, while improving the localization of sound sources from unknown places. Dynamic models incorporating moving sound sources and head movements could also be included based on this idea. Finally, a "garbage" source class could be added to our model in order to better deal with background and non-point sources. We believe that these ideas combined with careful algorithm and software optimization could lead to a novel *robot hearing* paradigm within the emerging field of human-robot interaction.

References

[1] R. V. Algazi, R. O. Duda, D. M. Thompson, and C. Avendano. The CIPIC HRTF Database. *IEEE ASSP Workshop on Applications of Signal Processing to Audio and Acoustics*, pages 92–102, Oct. 2001.

[2] J. Allen. Short-term spectral analysis, synthesis, and modification by discrete fourier transform. *IEEE Trans. Acous., Speech and Signal Process.*, 25(3):235–238, 1977.

[3] M. Aytekin, C. F. Moss, and J. Z. Simon. A sensorimotor approach to sound localization. *Neural Computation*, 20(3):603–635, 2008.

[4] S. Bensaid, A. Schutz, and D. T. M. Slock. Single microphone blind audio source separation using EM-Kalman filter and short+long term AR modeling. In *Latent Variable Analysis and Signal Separation*, pages 106–113, 2010.

[5] G. Celeux and G. Govaert. A classification EM algorithm for clustering and two stochastic versions. *Computational Statistics and Data Analysis*, 14(3):315–332, 1992.

[6] P. Comon and C. Jutten. *Handbook of Blind Source Separation, Independent Component Analysis and Applications.* Academic Press (Elsevier), Feb. 2010.

[7] A. Deleforge and R. Horaud. A latently constrained mixture model for audio source separation and localization. In *Latent Variable Analysis and Signal Separation*, Tel Aviv, Israel, March 2012.

[8] A. Deleforge and R. P. Horaud. Learning the direction of a sound source using head motions and spectral features. Technical Report RR-7529, INRIA, Feb. 2011.

[9] S. Haykin and Z. Chen. The cocktail party problem. *Neural Computation*, 17:1875–1902, 2005.

[10] J. Hörnstein, M. Lopes, J. Santos-Victor, and F. Lacerda. Sound localization for humanoid robots – building audio-motor maps based on the HRTF. In *Proc. of IEEE/RSJ IROS*, pages 1170–1176, 2006.

[11] F. Keyrouz, W. Maier, and K. Diepold. Robotic localization and separation of concurrent sound sources using self-splitting competitive learning. In *Proc. of IEEE CIISP*, pages 340–345, Hawaii, Apr. 2007.

[12] F. Keyrouz, Y. Naous, and K. Diepold. A new method for binaural 3D localization based on HRTFs. In *Proc. of IEEE ICASSP*, volume 5, May 2006.

[13] V. Khalidov, F. Forbes, and R. P. Horaud. Conjugate mixture models for clustering multimodal data. *Neural Computation*, 23(2):517–557, Feb. 2011.

[14] M. I. Mandel, R. J. Weiss, and D. P. W. Ellis. Model-based expectation-maximization source separation and localization. *IEEE Trans. on Audio, Speech and Lang. Proc.*, 18:382–394, Feb. 2010.

[15] J. C. Middlebrooks and D. M. Green. Sound localization by human listeners. *Annual Review of Psychology*, 42:135–159, January 1991.

[16] J. Nix and V. Hohmann. Sound source localization in real sound fields based on empirical statistics of interaural parameters. *Journal of the Acoustical Society of America*, 119(1):463–479, 2006.

[17] J. K. O'Regan and A. Noe. A sensorimotor account of vision and visual consciousness. *Behavioral and Brain Sciences*, 24:939–1031, 2001.

[18] M. Otani, T. Hirahara, and S. Ise. Numerical study on source-distance dependency of head-related transfer functions. *Journal of the Acoustical Society of America*, 125(5):3253–61, 2009.

[19] N. Roman and D. Wang. Binaural tracking of multiple moving sources. *IEEE Trans. on Acoust., Speech and Signal Process.*, 16(4):728–739, 2008.

[20] S. T. Roweis. One microphone source separation. In *Advances in Neural Information Processing Systems*, volume 13, pages 793–799. MIT Press, 2000.

[21] B. Shinn-Cunningham, N. Kopco, and T. J. Martin. Localizing nearby sound sources in a classroom: Binaural room impulse responses. *Journal of the Acoustical Society of America*, 117(5):3100–3115, 2005.

[22] E. Vincent, R. Gribonval, and C. Févotte. Performance measurement in blind audio source separation. *IEEE Transactions on Audio, Speech & Language Processing*, 14(4):1462–1469, 2006.

[23] H. Viste and G. Evangelista. On the use of spatial cues to improve binaural source separation. In *Proc. Int. Conf. on Digital Audio Effects*, pages 209–213, 2003.

[24] V. Willert, J. Eggert, J. Adamy, R. Stahl, and E. Koerner. A probabilistic model for binaural sound localization. *IEEE Transactions on Systems, Man, and Cybernetics–Part B*, 36(5):982–994, 2006.

[25] O. Yılmaz and S. Rickard. Blind separation of speech mixtures via time-frequency masking. *IEEE Transactions on Signal Processing*, 52:1830–1847, 2004.

[26] A. Zhigljavsky and A. Žilinskas. *Stochastic Global Optimization.* Springer, 2008.

Multi-party Human-Robot Interaction with Distant-Talking Speech Recognition

Randy Gomez
Academic Center for Computing and Media Studies, Kyoto University
Sakyo-ku Kyoto, JAPAN
randy-g@media.kyoto-u.ac.jp

Tatsuya Kawahara
Academic Center for Computing and Media Studies,Kyoto University
Sakyo-ku Kyoto, JAPAN
kawahara@media.kyoto-u.ac.jp

Keisuke Nakamura
Honda Research Institute
Wako-shi, JAPAN
nakamura@hri.jp

Kazuhiro Nakadai
Honda Research Institute
Wako-shi, JAPAN
nakadai@hri.jp

ABSTRACT

Speech is one of the most natural medium for human communication, which makes it vital to human-robot interaction. In real environments where robots are deployed, distant-talking speech recognition is difficult to realize due to the effects of reverberation. This leads to the degradation of speech recognition and understanding, and hinders a seamless human-robot interaction. To minimize this problem, traditional speech enhancement techniques optimized for human perception are adopted to achieve robustness in human-robot interaction. However, human and machine perceive speech differently: An improvement in speech recognition performance may not automatically translate to an improvement in human-robot interaction experience (as perceived by the users). In this paper, we propose a method in optimizing speech enhancement techniques specifically to improve automatic speech recognition (ASR) with emphasis on the human-robot interaction experience. Experimental results using real reverberant data in a multi-party conversation, show that the proposed method improved human-robot interaction experience in severe reverberant conditions compared to the traditional techniques.

Categories and Subject Descriptors

H.5.2 [**Information Interfaces and Presentation**]: User Interfaces—*Voice I/O*

General Terms

Algorithms, Design, Human Factors, Performance

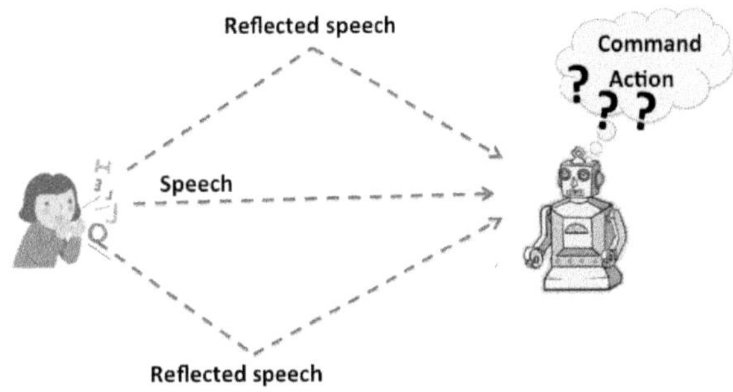

Figure 1: Real environment scenario.

Keywords

Robot Audition, Multi-party Interaction, Dereverberation, Robustness in Speech Recognition

1. INTRODUCTION

The notion of robots being a fixture in a manufacturing assembly line is a thing in the past. Recent developments in the use of professional service robots is gaining popularity and is expected to grow dramatically in the next decade. Thus, a potential need for autonomous robots equipped with complex mechanical functions in performing different tasks to work side by side with humans. An example of this is a companion robot in healthcare facilities, hospitals and even at homes. The robot may assist patients performing simple tasks (e.g in taking pills on schedule, pick up news papers etc.). It can also provide help in accomplishing strenuous activities such as lifting and moving the patient from one place to another. As robot applications are geared toward collaboration with humans in day to day activities, it is safe to assume that its functionality is not just merely mechanical in nature, but it is becoming more personal as well. Given this scenario, it is desirable to have a seamless human-robot interaction where one can communicate with each other at ease. Since speech communication is one of the most natural interface, its use is paramount towards a more effective human-robot interaction. Surveys show that humans are more comfortable in using voice commands when interact-

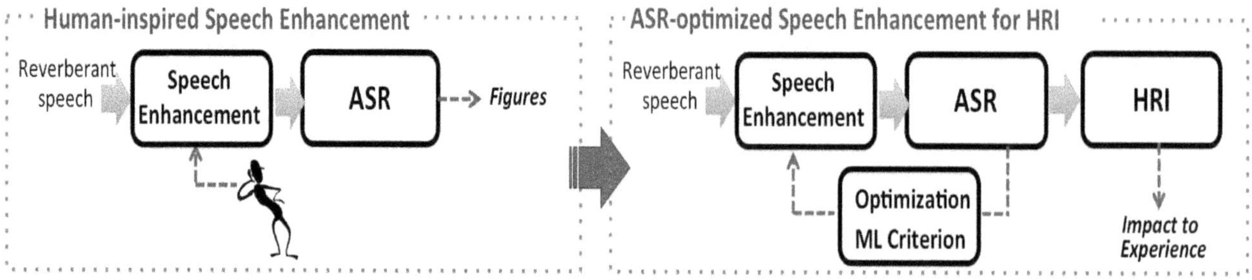

Figure 2: General concept of a speech enhancement technique optimized in conjunction with the ASR.

ing with robots, rather than using a keyboard. In fact, it was found out that ordinary users tend to speak freely when using speech-based systems. User interest is high when the robot recognizes and understands correctly the spoken commands, otherwise, the tendency of becoming dissatisfied is imminent when they have to repeat the spoken command again and again. The latter is most likely to happen in real environments where robots are deployed. Distant-talking speech recognition is a very difficult task due to the effects of reverberation as depicted in Fig. 1. This is a phenomenon attributed by the reflections of the speech signal within the room enclosure. Due to the delays in the time of arrival of the reflected speech signal, the observed signal at the microphone is a smeared version of the actual speech. In a reverberant scenario, the late reflection is considered to be the most detrimental to the ASR. Depending on the speaker-robot distance and the room characteristics, the effects of reverberation may drastically degrade ASR performance. The degradation may be severe that would render the system to be useless for human-robot interaction application. Thus, it is important to enhance the corrupted speech signal prior to speech recognition and understanding. Conventional methods address this problem to improve intelligibility in human perception. This technique is later adopted to achieve robustness in human-robot communication application. Although this works well, the enhancement process is carried-out independently from the automatic speech recognition (ASR) system. We note that computers recognize speech differently from humans.

In this paper we will show a method in optimizing human-inspired speech enhancement techniques to be closely linked to the ASR system for improved human-robot interaction experience as depicted in Fig. 2. The objective of the proposed method is to look at the overall impact of the speech enhancement as far as human-robot interaction experience is concerned. We tailor-fit the speech enhancement technique with the ASR system using maximum likelihood (ML) criterion and evaluate its impact to actual human-robot interaction experience. Although existing methods have been proposed regarding optimization of speech enhancement techniques together with the ASR [2][10][13], these works only tackle the *Speech Enhancement->ASR* scenario and stops right there. However, there is more to speech enhancement than just recognition performance (ASR figures). We need to investigate whether the speech enhancement translates into improving human-robot interaction experience. Thus, in this paper we consider the complete scenario: *Speech Enhancement->ASR->human-robot interaction.* Moreover, we evaluate the effectiveness of the proposed approach using real reverberant data in a multi-party conversation scenario which is not addressed in [2][10][13](single-party with no interaction). We used the Honda Research Institute robot platform called "Hearbo" shown in Fig. 3 to interact with humans and answer queries.

The organization of the paper is as follows; in Section 2, we show the background of achieving robustness in a speech-based human-robot interaction system. In Section 3, we present the traditional speech enhancement techniques, followed by the proposed optimization method, linking it with the ASR in Section 4. In Section 5, we discuss the experimental set-up, followed by the results and discussion in Section 6. We will conclude this paper in Section 7.

Figure 3: The Hearbo robot platform used for human-robot interaction experiments at Honda Research Institute, Japan.

2. BACKGROUND

A block diagram of a robust speech-based human robot interaction system is shown in Fig. 4. In this figure, a human-robot interaction is initiated from the human side through speech communication. Due to channel effects, the observed signal is smeared with the reflected speech (i.e. late reflection). The two main components of Fig. 4 are summarized as follows:

2.1 Speech Enhancement

Signal processing technique is employed to suppress the effects of late reflection, recovering the original speech. During the processing of the observed speech, artifacts are consequently introduced to the enhanced speech. While, humans tend to be more robust to the presence of artifacts

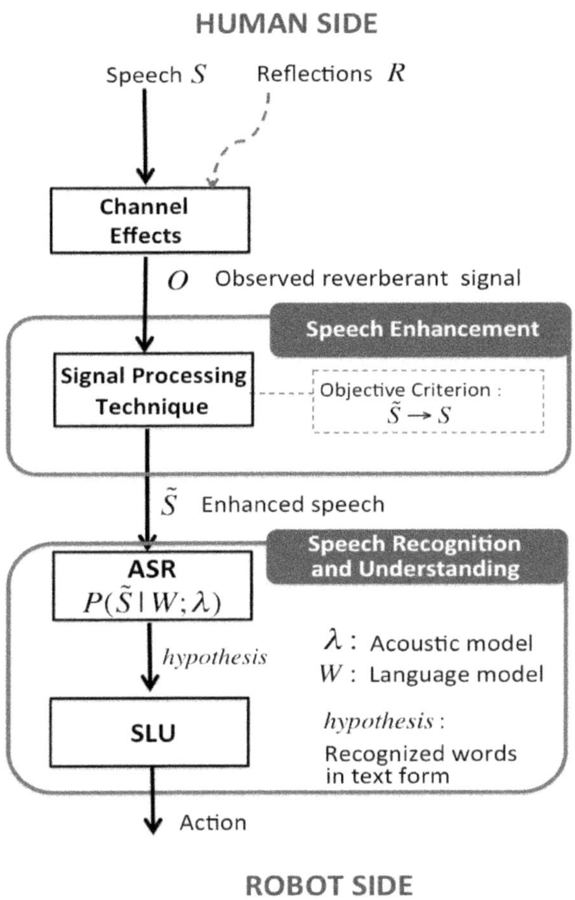

Figure 4: Block diagram of basic robust human-robot interaction.

[1], the ASR is very susceptible. Since speech enhancement was originally used for human perception, this technique operates mainly on improving sound perception. Hence, the loose criterion $\widetilde{S} \rightarrow S$.

2.2 Speech Recognition and Understanding

The enhanced speech signal is then recognized by evaluating the likelihood $P(\widetilde{S}|W;\lambda)$ by the ASR. W is the language model trained by using text database, and λ is the acoustic model trained using speech database. A mismatch between the training condition of λ and the observed reverberant data O degrades the ASR performance especially when the latter is trained with clean speech data. Thus, speech enhancement is employed to process O resulting to $\widetilde{S}$. This minimizes the effect of mismatch. The ASR system outputs a *hypothesis* (in text format) reflecting the recognized speech. This is then processed using conditional random field (CRF) to pick-up the slots that convey meaning for spoken language understanding. In this part here, the unnecessary entries in the sentence are removed, selecting only the slot that conveys meaning to our application. The selected slot is then searched in the database.

3. TRADITIONAL SPEECH ENHANCEMENT

In this paper, we consider three different speech enhancement methods designed for human perception. We adopt the reverberant model in [2] that treats the additive effect of late reflection to the clean speech

$$O \approx S + R \tag{1}$$

where O, S and R are the observed reverberant speech, clean speech and the late reflection respectively. Enhancement is defined as processing O and obtaining the estimate $\widetilde{S}$.

3.1 Wavelet Thresholding

A technique that operates in the wavelet domain (w), where the observed speech is enhanced based on thresholding [7]. The enhanced speech estimate is expressed as

$$\widetilde{S}(w) = \begin{cases} 0 \quad if \mid O(w) \mid \leq \; thr \\ sign(O(w))(\mid O(w) \mid - thr) \\ \quad if \mid O(w) \mid > \; thr. \end{cases} \tag{2}$$

The threshold thr, is derived using the wavelet coefficients of the late reflection. Any value that goes lesser than this, belongs to the subspace occupied by the late reflection. Thus, based on thr, Eq. (2) can be interpreted as setting the subspace occupied by late reflection to zero, and implementing a magnitude subtraction in the observed signal subspace. The threshold that defines the late reflection subspace can be calculated [7] as

$$thr = \sigma\sqrt{2\,\log(L)}, \tag{3}$$

where L is the length of the late reflection signal $R(w)$ with variance σ^2.

3.2 Spectral Subtraction

This is a classic speech enhancement technique which operates in the frequency domain (f) originally proposed for denoising in [3] [4]and later expanded to address reverberation in [5][6]. The short-term Fourier transform estimate of the enhanced speech is given as

$$\widetilde{S}(f) = \begin{cases} \sqrt{|S(f)|^2} \;\; e^{j\phi(f)} & \text{if } |S(f)|^2 > 0 \\ \beta |O(f)|^2 \;\; e^{j\phi(f)} & \text{otherwise,} \end{cases} \tag{4}$$

where $|S(f)|^2$ and $|O(f)|^2$ are the short-term power spectra of the clean speech and the observed reverberant signal respectively. β is the constant flooring coefficient derived experimentally. The phase $\phi(f)$ is derived from the observed signal. Since the clean speech is not available in real scenario, we can approximate its power spectrum by subtracting the power of the observed signal with the power of the late reflection. We note that the latter can be directly estimated as described in [5][6]. Thus,

$$|S(f)|^2 \approx |O(f)|^2 - |R(f)|^2. \tag{5}$$

Here, $|O(f)|^2$ and $|R(f)|^2$ are the short-term power spectra of the observed reverberant signal and late reflection, respectively.

3.3 Wiener Filtering

This technique is based on the Wiener Kinchine theorem, where the observed reverberant signal is filtered with a

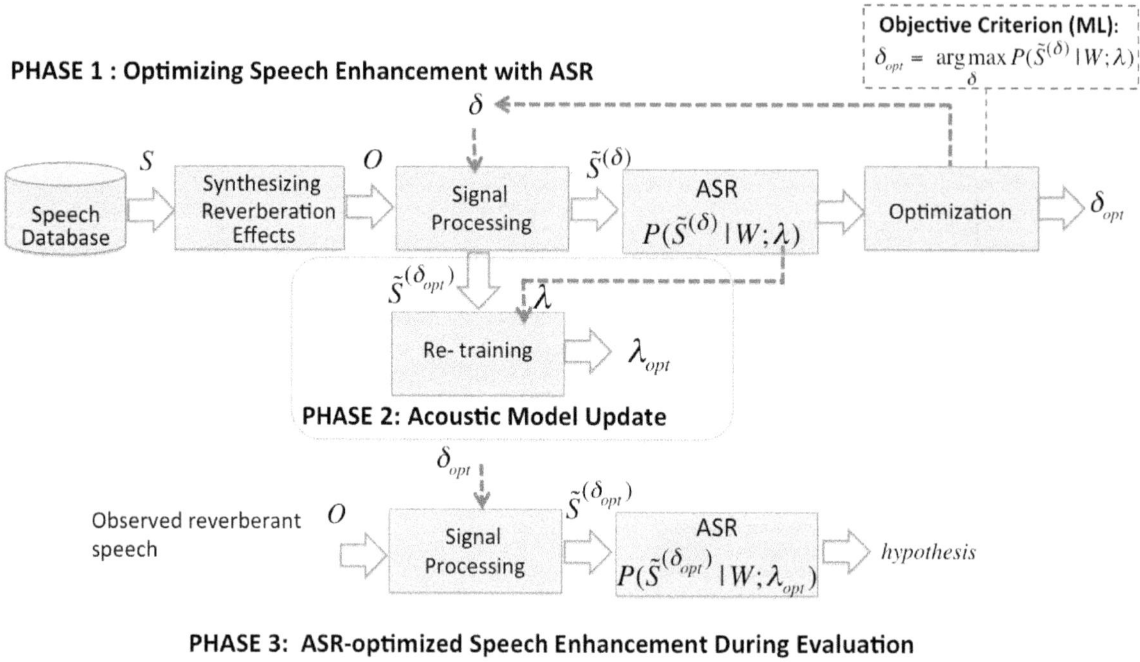

Figure 5: Method of optimizing speech enhancement with the ASR.

Wiener gain for enhancement [11]. Originally developed for denoising, and later expanded to address reverberant condition [10]. The general expression of the Wiener gain [10] is given as

$$\kappa = \frac{S(w)^2}{S(w)^2 + R(w)^2}, \tag{6}$$

where $S(w)^2$ and $R(w)^2$ are the short-term power in the wavelet domain (w) of the clean speech and the late reflection respectively. The latter can be directly estimated as described in [5][6], while the former is estimated by subtracting the observed signal with the late reflection power similar to the one described in spectral subtraction. Wiener Filtering in the wavelet domain is carried out by weighting the observed reverberant wavelet coefficients $O(w)$ with the Wiener gain as,

$$\widetilde{S}(w) = O(w) \ . \ \kappa. \tag{7}$$

In Eq. (7), the Wiener weighting κ dictates the degree of suppression of the late reflection to the observed signal. If the late reflection power estimate is greater than the estimate of the speech power, then κ may be set to zero or a small value, attenuating the effect of the late reflection and enhancing observed reverberant signal to $\widetilde{S}(w)$.

4. PROPOSED ASR-OPTIMIZED SPEECH ENHANCEMENT

Although the traditional speech enhancement techniques used for ASR applications are performing well, robustness can be further achieved when optimizing directly with the ASR. When the reverberant speech is processed, artifacts are introduced to the enhanced speech. The human auditory system together with the brain may be robust to these artifacts but not the ASR. Since the ASR processes speech differently from that with humans, it is logical to closely link the speech enhancement technique with the ASR. In Fig. 5 we show a method of optimizing the speech enhancement technique jointly with the ASR. The details are as follows,

4.1 Optimization Process

- PHASE 1
  The reverberant signal O is synthesized using clean speech S database and the room impulse response [8] in a method described in [9]. This approach is a common practice, since gathering real reverberant data for training is difficult. Then, δ parameter is introduced in the signal processing resulting to the enhanced signal $\widetilde{S}^{(\delta)}$. This parameter is used to control the amount of suppression of the late reflection, and in effect, controls the amount of artifacts introduced during the enhancement process. Consequently, δ is optimized iteratively with the ASR using the ML criterion, in which δ_{opt} that maximizes the model likelihood of the ASR is selected. The use of optimal δ_{opt} ensures the ASR to handle the artifacts without performance degradation. The corresponding δ_{opt} when used in enhancing the signal guarantees an improved in ASR performance compared to any other δ values.

- PHASE 2
  Model is updated with data $\widetilde{S}^{(\delta_{opt})}$ that are processed using ASR-optimized enhancement technique. This results to a new model λ_{opt} which guarantees a better performance than using the initial λ for the ASR.

- PHASE 3
  In the actual evaluation, the incoming real reverberant data observed in the microphone is enhanced using

Figure 6: Different views of the actual human-robot interaction experiment (Referred as Room 5)

δ_{opt}. Subsequently, we used the optimized model λ_{opt} for the ASR.

4.2 Expansion of the Enhancement Techniques

Specifically, when the δ parameter is introduced to the enhancement techniques discussed in Sec. 3, Eq'ns (3), (5) and (6) become

$$thr = \delta \ \sigma\sqrt{2\log(L)}, \tag{8}$$

$$|S(f)|^2 \approx |O(f)|^2 - \delta|R(f)|^2 \tag{9}$$

$$\kappa = \frac{S(w)^2}{S(w)^2 + \delta R(w)^2}, \tag{10}$$

for wavelet enhancement thresholding (Sec. 3.1), spectral subtraction (Sec. 3.2) and Wiener filtering in the wavelet domain (Sec. 3.3), respectively. The δ parameter, together with the optimization method shown in Fig. 5 adjust the degree in suppressing the late reflection. Consequently controlling the artifacts (by-product of the enhancement process), in conjunction with the ASR (for minimum impact).

5. EXPERIMENTAL SET-UP

The multi-party interaction setting is shown in Fig. 6, taking place at a very reverberant room referred to as Room 5(reverberation time of 940 ms). The direct distance between the robot and each speaker is approximately 4 meters. Other room configurations used in our experiment are shown in Fig. 7. We used Hearbo as our robot platform. This is equipped with microphone array processing technology [12].

The topic of the multi-party conversation is about different fish varieties used in preparing sushi and sashimi (Japanese traditional dishes). There are two participants in the conversation, Japanese and English language speakers. Although the conversation is in English, they only know the name of the fish in their language (i.e. Japanese or English). While the two are engaged in the conversation process, the robot is in listening mode. As the conversation progresses, either one of the two speakers stumble upon a fish name, and the other party tries to explain further. Upon realizing that there is no way the other party would identify the fish, the speaker would ask Hearbo the Japanese/English equivalent name of the particular fish. While in listening mode, Hearbo identifies if a question is being asked and automatically distinguishes whether the Japanese or the English speaker asked the question. This is done by checking the acoustic likelihood of the utterance using both Japanese and

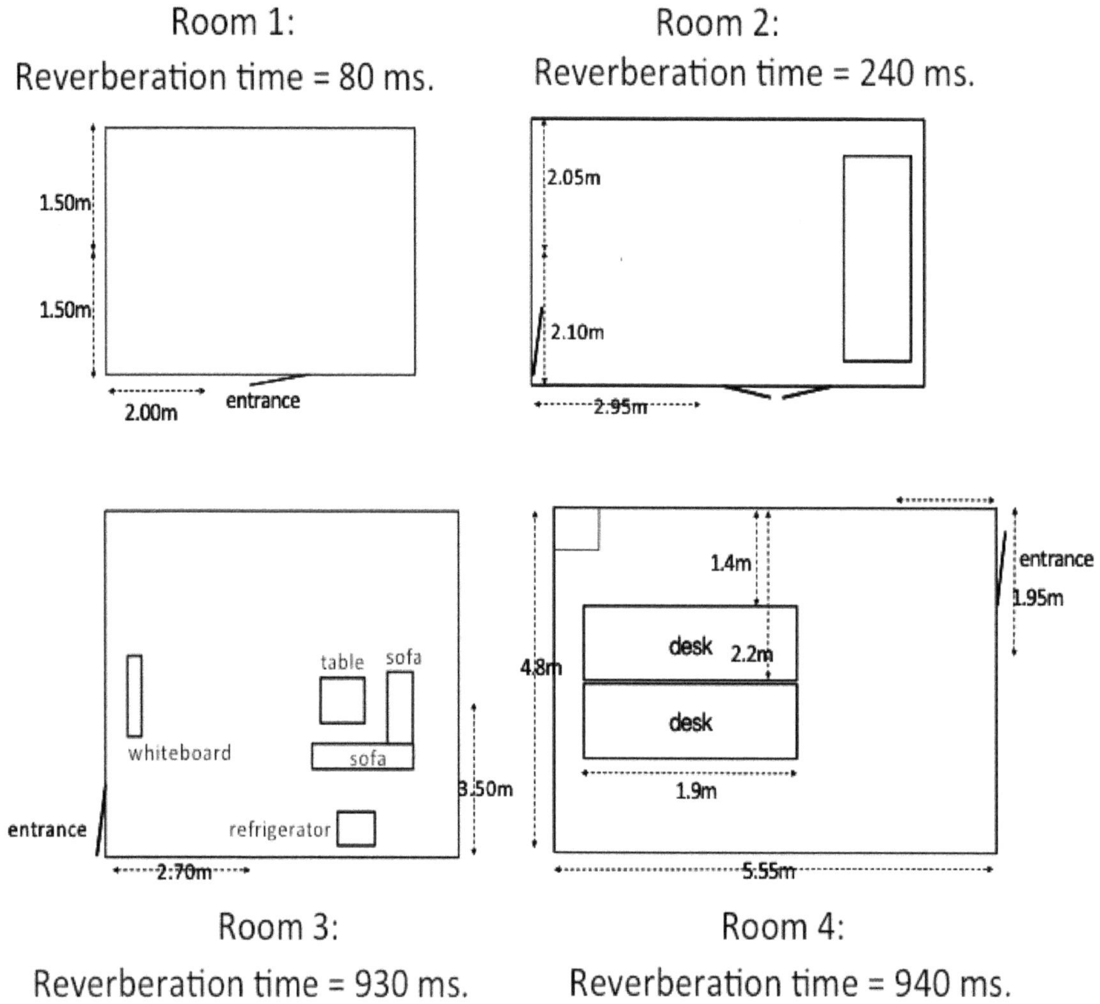

Figure 7: Other room configurations used in the experiment.

English acoustic models. We note that the English speaker asks Hearbo in English while the Japanese speaker asks in Japanese. Then, Hearbo moves its head towards the appropriate speaker and gives the equivalent fish name. Thus, if the question is in Japanese, Hearbo answers the English name and vice versa. Consequently, the two speakers continue to engage in the conversation until one of them asks Hearbo again. The sample of the multi-party conversation is shown in Fig. 8. The actual conversation may be longer than that shown in Fig. 8.

A total of 10 speakers participated in the multi-party human-robot interaction experiment (5 speakers for each language). Each speaker asks 10 fish names to Hearbo. We used both Japanese and English (triphone HMMs) acoustic models trained with the Japanese Newspaper Article Sentence (JNAS) with 20K vocabulary and the Wall Street Journal, respectively. We use a simple dialog system for evaluation, training only a single slot for fish name. We experimented in the condition of reverberation time: Room 1=80 ms , Room 2=240 ms, Room 3=930 ms, Room 4=940 ms, and Room 5 = 940 ms.

English Speaker : I went to a sushi bar yesterday and ordered a Sweetfish. Have you tried it ?

Japanese Speaker: Hmmm... Not really. Can you explain further what a Sweetfish is ?

English Speaker: This is an edible fish, mostly consumed in East Asia. Its flesh has a distinctive, sweet flavor with "melon and cucumber aromas".

Japanese Speaker: Not a clue.

English Speaker : It is usually cooked like barbecue on a stick.

Japanese Speaker: I give up.

English Speaker: Hearbo, can you tell us what Sweetfish is in Japanese ?

Hearbo: Sweetfish is commonly called as Ayu in Japanese.

Figure 8: Multi-party human-robot interaction.

Table 1: Automatic Speech Recognition (ASR): Performance in % Word Accuracy (Japanese/English)

Methods	Room 1	Room 2	Room 3	Room 4	Room 5
No enhancement					
Reverberant data	75.9 / 72.1	64.3 / 60.2	23.4 / 20.5	20.2 / 17.4	17.1 / 14.2
Wavelet Thresholding					
Traditional (Sec. 3)	77.5 / 74.0	67.1 / 63.4	34.8 / 31.8	31.9 / 29.7	30.5 / 29.8
ASR-Optimized (Sec. 4)	81.6 / 78.7	73.0 / 70.3	46.3 / 43.6	43.4 / 40.8	42.0 / 40.1
Spectral Subtraction					
Traditional (Sec. 3)	78.2 / 75.6	68.7 / 66.0	36.6 / 33.4	34.0 / 31.5	33.2 / 30.7
ASR-Optimized (Sec. 4)	82.4 / 79.2	74.5 / 71.3	48.9 / 46.1	46.3 / 43.6	45.2 / 43.0
Wiener Filtering					
Traditional (Sec. 3)	79.0 / 76.2	70.2 / 67.5	39.4 / 36.3	36.0 / 33.4	35.1 / 32.7
ASR-Optimized (Sec. 4)	83.7 / 80.9	76.7 / 73.6	51.7 / 48.9	48.5 / 46.5	47.7 / 46.0

Table 2: Human-machine interaction: Hearbo responding with a correct fish name % Hits (Japanese/English)

Methods	Room 1	Room 2	Room 3	Room 4	Room 5
No enhancement					
Reverberant data	90.0 / 86.0	82.0 / 78.0	38.0 / 34.0	34.0 / 30.0	34.0 / 30.0
Wavelet Thresholding					
Traditional (Sec. 3)	90.0 / 86.0	84.0 / 82.0	44.0 / 42.0	44.0 / 42.0	44.0 / 42.0
ASR-Optimized (Sec. 4)	92.0 / 88.0	86.0 / 84.0	64.0 / 62.0	64.0 / 60.0	64.0 / 60.0
Spectral Subtraction					
Traditional (Sec. 3)	90.0 / 86.0	84.0 / 82.0	44.0 / 42.0	44.0 / 42.0	44.0 / 42.0
ASR-Optimized (Sec. 4)	92.0 / 88.0	86.0 / 84.0	64.0 / 62.0	64.0 / 60.0	64.0 / 60.0
Wiener Filtering					
Traditional (Sec. 3)	92.0 / 88.0	86.0 / 84.0	44.0 / 42.0	44.0 / 42.0	44.0 / 42.0
ASR-Optimized (Sec. 4)	92.0 / 88.0	88.0 / 86.0	64.0 / 64.0	64.0 / 60.0	64.0 / 60.0

6. RESULTS AND DISCUSSION

For the quantitative evaluation, we investigated the ASR performance and the actual human-robot interaction experience defined as the percentage in which Hearbo correctly answers the 10 fish names being asked. Moreover, a simple qualitative assessment was conducted by asking the participants' opinion regarding their overall impression of the the proposed method (i.e. satisfied or not satisfied).

6.1 Quantitative Results

Table 1 shows the ASR results in detail for both Japanese and English (referred as "*hypothesis*" in Fig. 4). This result takes into consideration the deletion and insertion errors caused by the ASR in recognizing the utterance. We show the speech recognition performance when no enhancement is employed using a clean acoustic model, together with the results when using traditional speech enhancement techniques. Consequently, the results of both the traditional and the ASR-optimized speech enhancement techniques are compared. A summary is shown in Fig. 9, where the average results for the traditional and ASR-optimized enhancement techniques are presented. It is apparent that all traditional speech enhancement techniques improved speech recognition performance compared to the method that employed no enhancement. Moreover, when these enhancement techniques are optimized in conjunction with the ASR, speech recognition performance is further improved. The improvement of the ASR-optmized enhancement techniques is more evident in extremely reverberant conditions (i.e. Rooms 3, 4 and 5).

In Table 2, we show the detailed results of the actual human-robot interaction experience in which Hearbo correctly answers the fish name (referred as "ACTION" in Fig. 4). This is the result after extracting the slot from the output of the ASR (*hypothesis*). Table 2 is summarized in Fig. 10. This results show how the speech enhancement is translated to improving human-robot interaction experience. The proposed method has minimal impact in Room 1 (80 ms) and Room 2 (240 ms). However, almost 100% improvement is achieved in Room 3 (930 ms), Room 4 (940 ms) and Room 5 (940 ms) when compared to the method with no enhancement. The proposed ASR-optimization works well at higher reverberation time. This means that the perception of improved human-robot interaction experience is more evident under severe reverberant conditions. Although ASR recognition performance is insightful, the figures are more dynamic than the actual human-robot interaction. For example, improvement in ASR performance for Room 1 and Room 2 (see Fig. 9) does not necessarily translate to an improvement in human-robot interaction experience as depicted in Fig. 10. Thus, human-robot interaction experiment is preferred when infering improvement in human-robot interaction experience.

6.2 Qualitative Results

As a result of the survey, the participants were satisfied with the human robot interaction set-up. The perception of satisfaction can be explained further in Fig. 11. This figure shows the percent improvement in human-robot interaction experiment of both the traditional and the proposed ASR-optimized speech enhancement methods as a function of reverberation time. This result impacts user satisfaction experience. In this particular experiment, we synthetically generate data of different reverberation time [9][2] since it is difficult to gather real reverberant data. To synthetically

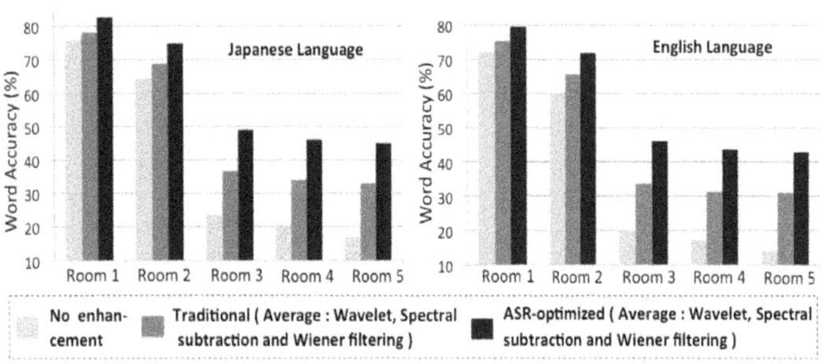

Figure 9: Average % Word Accuracy (ASR performance).

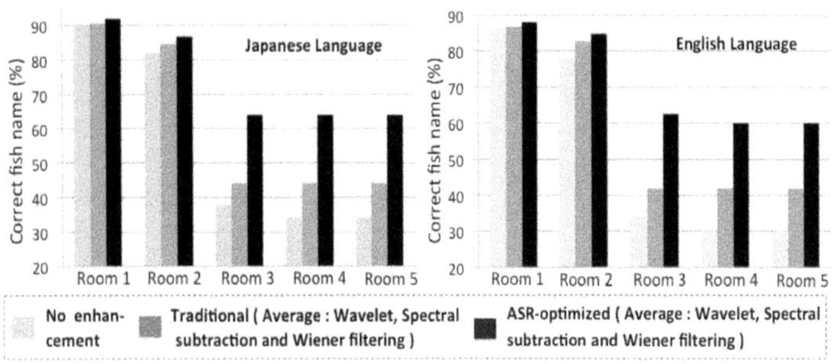

Figure 10: Average Hits (Hearbo responding correctly).

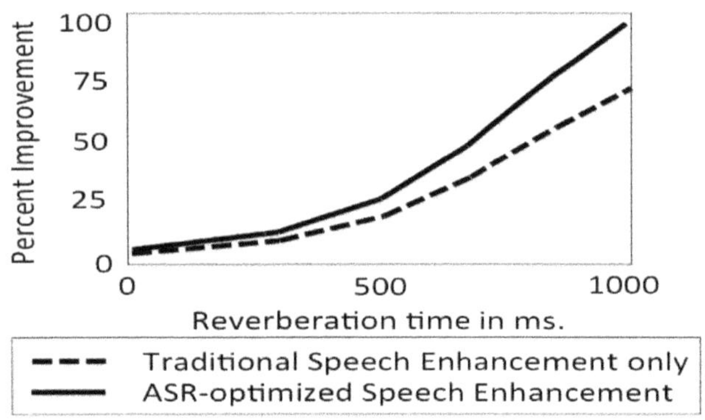

Figure 11: Percent improvement in human-robot experience when using traditional and ASR-optimized speech enhancement as a function of reverberation time.

generate a reverberant data, we need an approximation of the room characteristics in the form of room impulse response (RIR). This can be measured physically or modelled mathematically [9]. Then, the clean speech is convolved with the RIR to generate the synthetic reverberant data. We enhanced the data with both the traditional and the proposed method. Then, we measure the percentage of improvement in terms of Hearbo's correct response. This figure shows significant improvement under severe reverberant conditions. Further, this validates both user satisfaction and is consistent with our results in Table 2.

7. CONCLUSION

We have presented a method that integrates the traditional speech enhancement techniques in relation with the back-end ASR system for improved human-robot interaction experience. In this paper, we have expanded three independent speech enhancement techniques that operate in different domains and optimized based on the ASR. All of these, show consistent improvement in human-robot experience especially under severe reverberant conditions. In this paper we have dealt more on the effect of seamless human-robot interaction with optimized speech enhancement, rather than solely focusing on ASR results. Currently, we are only dealing with the effects of reverberation. In our future works, we will consider the contribution of noise jointly with reverberation.

8. REFERENCES

[1] S. Vaseghi "Advanced Signal processing and Digital Noise reduction", Wiley and Teubner, 1996.

[2] R. Gomez and T. Kawahara, "Robust Speech Recognition based on Dereverberation Parameter Optimization using Acoustic Model Likelihood" *In Proceedings IEEE Transactions Speech and Acoustics Processing*, 2010

[3] S.F. Boll, "Suppression of acoustic noise in speech using spectral subtraction" *In Proceedings IEEE Int. Conf. Acoust., Speech, Signal Proc. ICASSP* pp 208-211, Apr. 1979.

[4] S. Kamath and P. Loizou,"A Multi-Band Spectral Subtraction Method for Enhancing Speech Corrupted by Colored Noise" *In Proceedings IEEE Int. Conf. Acoust., Speech, Signal Proc. ICASSP* 2002.

[5] R. Gomez, J. Even, H. Saruwatari, and K. Shikano, "Fast Dereverberation for Hands-Free Speech Recognition" *IEEE Workshop HSCMA*, 2008

[6] R. Gomez, J. Even, H. Saruwatari, and K. Shikano, "Distant-talking Robust Speech Recognition Using Late Reflection Components of Room Impulse Response" *In Proceedings IEEE Int. Conf. Acoust., Speech, Signal Proc. ICASSP*, 2008

[7] D.L. Donoho, "Denoising by soft thresholding", *IEEE Trans. Info. Theory* 1995.

[8] Y. Suzuki, F. Asano, H.-Y. Kim, and T. Sone, "An optimum computer-generated pulse signal suitable for the measurement of very long impulse responses"

[9] H. Kuttruff, "Room Acoustics" *Spon Press*, 2000

[10] R. Gomez and T. Kawahara , "Optimizing Spectral Subtraction and Wiener Filtering for Robust Speech Recognition in Reverberant and Noisy Conditions" *ICASSP*, 2010.

[11] E. Ambikairajah and G. Tattersall, "Wavelet Transform-based Speech Enhancement" *In Proceedings International Conference on Speech and Language Processing ICSLP*, 1998

[12] H. Nakajima, K. Nakadai, Y. Hasegawa and H. Tsujino, "Adaptive Step-size Parameter Control for real World Blind Source Separation" *In Proceedings IEEE Int. Conf. Acoust., Speech, Signal Proc. ICASSP*, 2008.

[13] R. Gomez, T. Kawahara, "Optimization of Dereverberation Parameters based on Likelihood of Speech Recognizer" *In Proceedings of Interspeech*, 2009.

Do You Remember that Shop? - Computational Model of Spatial Memory for Shopping Companion Robots -

Takahiro Matsumoto Satoru Satake Takayuki Kanda Michita Imai Norihiro Hagita
ATR Intelligent Robotics and Communication Laboratory
2-2-2 Hikaridai, Keihanna Science City,
Kyoto, Japan
{matsugen, satoru, kanda, michita, hagita}@atr.jp

ABSTRACT
We aim to develop a shopping companion robot that can share experience with users. In this study, we focused on the shared memory acquired when a robot walks together with a user. We developed a computational model of memory recall of visited locations in a shopping mall. The model was developed with data collection from 30 participants. We found that shop size, color intensity of facade, relative visibility, and time elapsed are the influencing features for recall. The model was used in a scenario of a shopping companion robot. The robot, Robovie, autonomously follows a user while inferring the user's memory recall of shops in the visited route. When the user asks the location of other shops, Robovie replied with *destination description*, referring to the known locations inferred with the model of the user's memory recall. With this scenario, we verified the effectiveness of the developed computational model of memory recall. The evaluation experiment revealed that the model outputs shops that the participants are likely to recall, and makes the directions given easier to understand.

Categories and Subject Descriptors
H.5.2 **[Information Interfaces and Presentation]**: User Interfaces- Interaction styles

General Terms
Algorithms, Design, Experimentation, and Human Factors.

Keywords
Environment Modeling, Computational model of memory, Social human-robot interaction

1. INTRODUCTION
Robots are expected to work as human companions in daily life, such as in an office [20] or a shopping mall [12]. Robots are expected to build relationships with people through repeated interactions [11, 13]. Getting to know each other more, people would be expected to obtain "common grounding" about robots [14]. If a robot acquires knowledge about users, a robot is able to engage in efficient conversation with common grounding [25]. The question is, how to acquire common grounding? Does a robot obtain it explicitly from a user's input?

Humans often share experience while being together. Below is an example of conversation built on shared experience by two friends window-shopping (Figure 1).

A: I'm a little tired. Want to get a coffee?

B: There's a cafe next to *the book store*.

A: Sounds good!

In this conversation, the destination "a cafe" is explained with the nearby landmark "the bookstore". In a study of direction giving, it is reported that people use "*destination description*", in which they refer to a known landmark to explain *where* the destination is [7] (see section 2.1 for more details). Such "*destination description*" can be used when people are familiar with an environment.

Figure 1: Sample interaction between two friends window-shopping based on shared experience

How to achieve this interaction based on shared experience? The above conversation is successful when both share the experience of seeing *the bookstore*, and thus they are able to recall it. A robot would have a permanent memory, and thus be able to recall all the details of the scenes it experienced. Is such an accurate memory the capability that enables the interaction based on shared experience? However, people do not have such an accurate memory, and thus do not remember all the things they have seen. We consider that a robot would need to simulate people's perception and cognition capability.

This paper reports our study in which we developed a computational model of spatial memory. The paper addresses an example use of the model for a direction-giving scenario with a shopping companion robot.

2. RELATED WORKS
2.1 Spatial cognition
Studies in spatial cognition have reported how people remember environments. The fundamental structure is considered to be a "cognitive map", in which people build a layered representation of an environment consisting of *landmark*, *route*, and *survey* knowledge [10]. When people explore the environment by themselves, they typically form a representation with *route perspective,* that is, a mental tour of the environment [28]. This

process was modeled in a robot [17]. However, while the *route perspective* addresses a static representation of the environment, these studies did not reveal the dynamic aspect of memory, i.e. which shop/location is well recalled.

A few studies address the dynamic aspect relevant to recall. Nothegger et al. have modeled salient features in an environment for a comprehensive route direction. They found that a shop with a larger facade area, visibility, and color intensity would be associated with larger visual attraction, and thus coulde be a good landmark [18]. Schroder et. al., covered diverse features, including such as relative location, age, emotions [22]. While these studies inform us about potential features to be considered, it did not model memory and thus does not provide a computational model for memory recall.

In direction giving, it is known that people use turn-by-turn expression, describing actions to be taken at each turn from the current location, such as "to go to the cafe, please go straight down this corridor, and turn right at the book store, ..." [5]. In this way, people explain *how* to arrive at the destination. In contrast, Dethlefs et al. revealed that when people are familiar with an environment they use *destination description* for route direction, which refers to a known landmark such as "the destination is *near* the baseball stadium" [6]. In this way, people explain *where* the destination is. While it involves the concept of a "known" place thus implying the model of memory recall, in their study the salience of an individual location is not computed from the observed feature, but from an online corpus; while such an approach would be useful for a well-known place, it would be difficult to apply their method to a companion robot that works in non-well-known place.

2.2 Perspective-taking and shared experience

In recent HRI studies, researchers have started to develop a computational model of human-like perception and cognition. Joint-attention capability was replicated by imitating humans' gaze mechanism [21]. Perspective-taking is also a pioneering example of a computational model of human-like cognition, simulating what a partner can see in deictic interaction [1, 26]. A computational model was also developed for route perspective [17]. Further, building on similarity in perception, human-like gazing is considered to be the key feature for engagement [23].

The importance of memory was reported in relation-building for long-term interaction. Bickmore et al, included memory of interaction for enabling a long-term relationship with an agent [3]. Gockley et al. implemented story-telling interaction [11]. These studies report the importance of memory in long-term interaction.

However, none of the previous studies addressed how to build shared memory in spatial interaction.

3. COMPUTATIONAL MODEL OF RECALL OF LOCATIONS

We developed a computational model of humans' recall of locations after they walked along a route. The model computes a likelihood of recall from relevant features, such as the visibility of locations and time passed from the person passing the location. The model was built through data collection of humans' recall, and analysis of relationships between the recall and environmental features.

3.1 Data Collection

The purpose of the data collection is to gather the *recall rate of locations,* which is defined as the probability of a person recalling a shop. The data collection was conducted at a shopping mall. Figure 2 shows a map of the building. There are 36 shops in the shopping mall. Thirty Japanese native speakers (14 males and 16 females, 20.2 years old on average) participated to the data collection. They were paid for the participation.

We consider that it is difficult to directly observe the probability of recall from a single person. Instead, we decided to collect data from many people for the same situation, and computed the ratio of people who recall a location. The participants were asked to walk the same route. Three routes were prepared (see Figure 2). Routes were chosen to include main streets where customers typically go for window-shopping, and to not overlap with each other so as to include different shops. Each route was traversed in both a forward and backward directions. This compensated for the effect of different viewpoints, as well as various elapsed times. Thus, in total, there were six route patterns. Each participant walked three routes out of six .

Each participant was instructed to assume that they were window-shopping. They were not asked to remember the shops; instead, they were asked to behave as they would when window-shopping. After each walk, the experimenter asked whether the participant recalled the visible shop on the route. To prevent participants from pretending to recall, the experimenter first asked whether a participant recalled a shop with a specific feature, e.g., "Are there any accessory shops?" When he/she said there was such a shop, the experimenter asked him/her to describe the shop in detail. Only when he/she could provide some details did we judge that he/she recalled the shop.

Figure 2: *Traversed routes* in data collection

3.2 Data Analysis

To build the model, we conducted an analysis to reveal environmental features on the route that are useful for predicting recall rate of locations.

3.2.1 Features

We collected a couple of possible features that would correlate with humans' *recall rates* of locations. Mainly we referred two studies Nothegger et. al., [18] and Schroder et. al., [22] and retrieved the features that can be used by a robot.

1) Preparation

First, to compute features, we developed a computer program to simulate users' views related to visibility of shops (Figure 3). The program builds 3D polygon data of the environment. It uses perspective transformation and z-buffering, well known 3D-graphics algorithms, to convert 3D polygon data to 2D pixel data for the given viewpoint. It outputs *perceived size of a shop* as an amount of 2D pixels, denoted as $visibility_i(vp)$ for shop i where vp represents the current view point (x, y, θ) (Fig. 4). The height of view (h) was set at 150 cm. For each *traversed route* (denoted as r), the program computed vp for every 125 ms.

Figure 3: Simulated view of a person

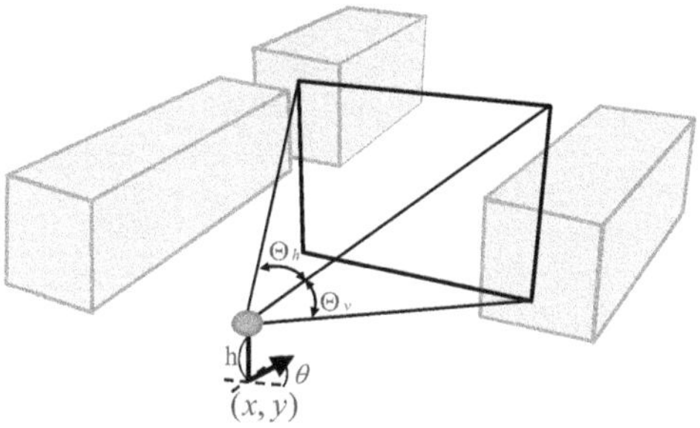

Figure 4: Definition of Viewpoint

Regarding the width of view, it would be not a trivial issue to decide the simulated horizontal view because there are a couple of possibilities. Perhaps, people might focus on a narrow view while walking, as they would concentrate on the frontal direction. However, people might use a wider view, or even turn their heads, as they would search for directions to go in. For this problem, there is a successful modeling of pedestrian' decision-making behavior, in which a simulated agent was successful with a horizontal view angle (Θ_h) of 80 degrees [27]. We used this value for the horizontal, and we experimentally decided on a vertical view angle (Θ_v) of 60 degrees to be similar enough to the view of people walking.

2) Visual features

We included the following features (Fig. 5):

- **Facade size (F)**: We computed the facade size of a shop that is visible from the travelled route.

- **Shop size (S)**: As the study [18] was in a city, facade size typically represented the size of a shop; however, in a shopping mall, some shops are in rectangular in shape with a small facade on the route people pass.

- **Color of facade (C)**: Following the procedure used in [18], we computed means of RGB value within the facade area from a pre-recorded picture of each shop in advance. While the average of R-G-B values was used in [18], we analyzed not only the average of R-G-B but also each R, G and B value. We finally used the G value because the model fits well with it most (described later in single factor mode of Sec. 3.2.2).

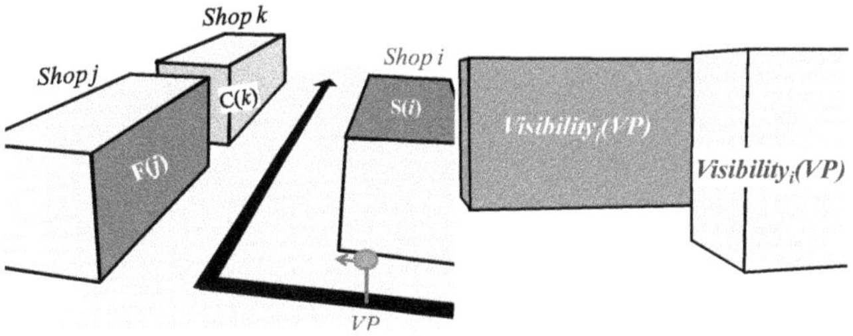

(a) Static features (b) features from VP

Figure 5: Visual Features

Visibility changes along with viewpoint. Here, there are two possible ideas. First, we speculated that people would recall shops that were more exposed, not only in terms of perceived size, but also in terms of time of exposure. The relationship between duration of exposure and memory recall has been found in other contexts, e.g. memory of digits displayed [16]. To reflect this idea, we included the following features.

-**Integral of visibility (IV)**: This is computed as the sum of visibility of the shop, defined as follows.

$$IV_i(r) = \sum_{vp \in r} visibility_i(vp) \tag{1}$$

where i represents the ID of the shop, r is the *traversed route*, and vp is the view point.

Alternatively, we speculate that if a shop is too small, even though it is seen for a long time, we would not pay attention to it. Instead, an intense scene would be well recalled. Based on this idea, we speculated that the scene that has the largest visibility would explain the rate of recall.

Max of visibility (MV): This is computed as the max of visibility of the shop, defined as follows:

$$MV_i(r) = \max_{vp \in r}(visibility_i(vp)) \tag{2}$$

We further speculated that people would only pay attention to meaningful structures (e.g. shops), but not other structures such as the floor or ceiling of a corridor. For example, if there is only one shop in a person's sight, one would likely to pay attention to the shop, even though the shop is small (Figure 6, left). On the other hand, if there are many shops in a person's view, he/she would pay attention to a bigger shop, but not so much to a smaller shop (Figure 6, right). To reflect this idea, we computed the ratio of the visibility of the shop to all other visible shops, defined as follows:

$$RV_i(vp) = \frac{visibility_i(vp)}{\sum_j visibility_j(vp)} \tag{3}$$

Similar to the features for simple visibility, we prepared two versions of integration of relative visibility over time.

- Integral of relative visibility:

$$IRV_i(r) = \sum_{vp \in r} RV_i(vp) \tag{4}$$

- Max of relative visibility:

$$MRV_i(r) = \max_{vp \in r}(RV_i(vp)) \tag{5}$$

Figure 6: Relative visibility
(Shop marked with a white polygon is in the same size in pixels)

Table I: Regression with single factor model

	Best model	R^2
Facade size	log(x)	.285
Shop size	log(x)	.287
Color of façade	x^2	.138
Max of visibility	x^2	.180
Max of relative visibility	x^2	.241
Integral of visibility	log(x)	.283
Integral of relative visibility	log(x)	.308
Time elapsed	x^2	.129
Number of shops observed	x^2	.068
Around corner	x	.043
Around the end of corridor	x	.177
Around salient structure	x	.015

3) Temporal features

A recent event is often recalled more than one in the past [15]. We speculated that this would be the case for the recall of shops. For example, we often remember a shop that we just passed even if the shop is small. We included the following feature:

- **Time elapsed (T)**: In our data, the recall rate was not only associated with distance, but also with time. Instead, the time elapsed was approximated from the time needed to travel the *traveled distance* with average speed, which was 1.2 m/sec in our data. The *traveled distance* was computed from the most recent point on the route where the target shop was visible.

Alternatively, one would speculate that memory capacity limits the recall of shops. If it is influential, people would be less likely to remember a shop when he/she see more shops after seeing the shop. Thus, we included the following feature.

- **Number of shops observed**: We compute the number of shops observed after the target shop was seen.

4) Structural features

In [22], structural features are included, such as "near the corner". We included binary variables about whether each shop is **around corner**, **around the end of corridor**, and **around salient structure** (elevator, escalator, stair, and exit). For each feature, we used distance threshold 3 m to judge whether it is "around" corner, end of corridor, and a salient structure.

3.2.2 Analysis

Single Factor Model for each feature

We run the regression analysis for each feature with a linear model as well as log, exponential, and square model. Table I shows the result. For each feature, the model in the table shows the best model that produced the highest coefficient of determination (R^2). For example, the *log model* was the best for *shop size* and *integral of visibility*, and the *square model* fits well for most other features.

The color of façade has four candidates of the calculation: R, G, B and average of them. We select the fittest calculation from R^2 and finally the G is selected ($R^2 = .138$).

Regression model for all features

Next, we included all the features in the analysis. For each feature, we used the best model found in the first analysis. With the step-wise method [7], we run a regression analysis. We set the threshold of inclusion to be .05 and for exclusion to be .1. It provides the solution having features of *maximum of relative visibility, size of shop, color of facade* and *time elapsed,* to predict *recall rate.* The obtained model is shown in the equation,

$$\hat{p}_i(r) = \beta_{MRV}(\max_{vp \in r} RV_i(vp))^2 + \beta_S \log_{10}(S_i) + \beta_C C_i^2 + \beta_T T_i(r)^2 + \alpha \quad (6)$$

where $\hat{p}_i$ is the predicted *recall rate* of the shop *i*, and α and β are coefficients specified in Table II .

The model's coefficient of determination (R^2) is 0.53 and RMSE is 15.75. As there is considerable complexity in humans' cognitive processes, we believe that this result shows a reasonably good prediction of the recall rate. Figure 7 shows the plot of predicted and actual *recall rate.* It seems that the model successfully predicts the *recall rate* particularly when actual *recall rate* is higher than 0.8.

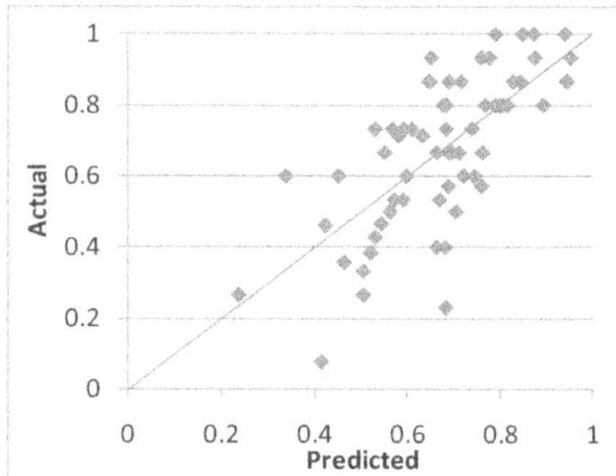

Figure 7: Result of regression analysis

Table II: Obtained regression model

Feature	Symbol	Coefficient	Std. coefficient
Max of relative visibility (*MRV*)	β_{MRV}	.256	.401
Shop size (*S*)	β_S	.120	.314
Color of facade (*C*)	β_C	$-.766*10^{-5}$	-.222
Time elapsed (*T*)	β_T	$-.202*10^{-5}$	-.235
Residual	α	.400	-

In Table II, two coefficients are shown. The standardized coefficient can be used when we interpret the contribution of each feature. As the *max of relative visibility* has the highest value, it is the most influential feature among all included features. It is well fit to the actual *recall rate.* For example, dry-cleaning store (MRV=0.35) has the lowest value (0.77), but the book store (MRV=0.96) has the highest value (1.0). The (non-standardized) coefficient is used to compute the predicted *recall rate.* For instance, after 100 seconds of elapsed time, the recall rate is predicted to drop 2.02% ($100 \times 100 \times -.202 \times 10^{-5}=0.0202$), and after 200 seconds it would drop 8.08% ($200 \times 200 \times -.202 \times 10^{-5}=0.0808$). A shop with 100 m^2 size would gain 12% better recall than a shop with 10 m^2 size ($\log_{10}(100) \times .120=.24$ and $\log_{10}(10) \times .120=.12$). The coefficient for color is negative, meaning that people recall better when the color of a shop is darker. This would be counter-intuitive, but in the shopping mall, as floor and ceilings are colored with very bright colors, a shop with darker color is rather perceived as highlighted.

4. Shopping Companion Robot

We developed a shopping companion robot for a scenario in which it escorts a visitor, carries his/her basket/bag, and provides a route direction. It has a module that simulates a user's spatial memory, particularly about the recall of visited locations.

4.1 Hardware Architecture

Robovie-II is used. It is characterized by its human-like physical expressions to interact with people through utterances and gestures. It is 120 cm high and 40 cm in diameter. The robot has motor encoders for odometry computation (P3 mobile platform from Pioneer) and two 30 m range laser sensors (UTM-30LX from Hokuyo) attached at a height of 8 cm. We used it with maximum speed of 650 mm/sec.

4.2 Software Architecture

Our system (Fig. 8) consists of three functions: **navigation**, **spatial memory**, and **conversation**. To **navigate** the robot, the system has three modules: *localization, human detection,* and *human tracking*. Regarding the **spatial memory,** we extended our previous study on the modeling environment with *route-based representation* [16], as well as using the computational model for *simulation of recall*. It holds a traversed route based on the route-perspective representation, and the recall module updates the list of locations that the user is likely to recall. Two modules are involved for **conversation**, *keyword detector* and *direction giving*. There was a human operator present, who corrected errors in *localization* and *human detection* modules, and operated the *keyword detector* (the modules that an operator was involved in are colored gray in the figure).

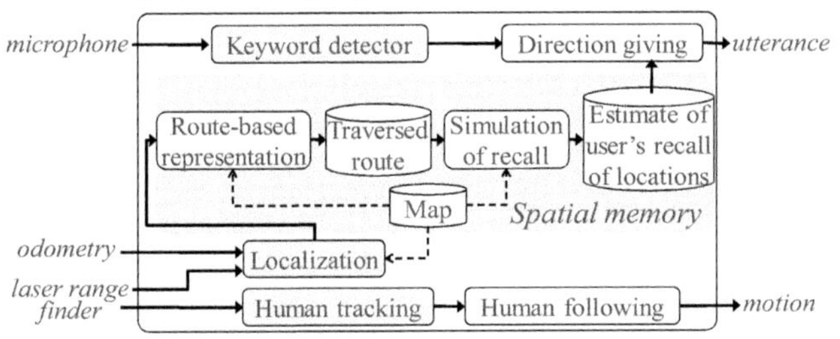

Figure 8: Architecture of Companion Robot

4.3 Navigation

The purpose of navigation modules is to keep track of the position of the robot and the target person, and follow him/her.

4.3.1 Localization

For robot localization, we use a particle filter with the ray tracing approach with a grid map [4, 24]. The grid map is built from odometry and laser scanner data. To build a grid map, we applied ICP [2] for aligning laser scan data, and the global relaxation technique for increasing the consistency of the odometry data. This module is called every 10 msec and updates the robot's position. To assure the safety of the robot, a human operator is requested when the system detected potential error in localization. This warning rarely happened; we experienced it only twice during the experiment with 19 participants.

4.3.2 Human tracking

This module consists of two steps: 1) detecting human legs from laser sensors, and 2) tracking human position from the detected legs. For the first step, we assume the size of human legs to be within 100 mm. We apply a segmentation algorithm to detect human legs. For the second step, we apply a particle filter, which models the relation between human motion and motions of his/her legs. It estimates human's position for every 25 msec.When this module fails to track the target person, it requests a human operator for help.

4.3.3 Human following

To generate a following motion, the system computes velocity of the robot to move toward the location of the person. This module draws a line to the followed user and applies the line-following algorithm for every 30 msec. In addition, we prepared a simple safety mechanism to prevent collisions with dynamic obstacles, such as shopping carts and other customers. It uses the idea of a dynamic window approach [8] to limits the forward velocity to be safe enough to stop before colliding with the nearest object.

4.4 Spatial Memory

There are two processing steps. First, output from the localization module is stored as "*traversed route*", which is a series of segments on a topological map. The *recall module* updates an estimate of the user's memory based on the current *traversed route* information.

4.4.1 Route-based representation

We used the *route-perspective* representation [17] as the base of spatial memory. First, to abstract the history of navigation, trajectory acquired through the localization module is converted to the series of segments on a topological map, which is built by applying the Voronoi method to the grid map. When new input of a position (x,y) is obtained from the localization module, given that *traversed_node_list* is initialized with the first node as the start position (x_0, y_0), the following algorithm is applied:

1) *insertion*: If the current location is close enough to a node (n_i) on a topological map (Fig. 9) (within three meters, which is the half-width of the largest route in the mall), and if the most recent node in the *traversed_node_list* is not n_i, it puts the node n_i into the *traversed_node_list.traversed_node_list.*

2) *segmentation*: When *insertion* occurs, a subgoal-action analysis [17] is applied to remove n_{j-1} if it is redundant.

3) *update*: Otherwise, it remembers the *current position* (x, y) as the *latest_position.*

It outputs the *traversed route* as the list of nodes stored in the *traversed_node_list.*

$$r = n_1, n_2, \ldots, n_j, n_c \qquad (7)$$

where n_i (i=1,...j) represents a node in the *traversed_node_list*, and n_c is the *latest_position.*

4.4.2 Simulation of recall

This is the module that simulates a user's recall. It uses the *traversed route* to compute the *likelihood of recall* with the model described in eqn. (6) for all shops that were visible on the *traversed route*. The computation result is stored to the *estimate of user's recall of locations*. The *estimate of user's recall of locations* preserves the list of shops that the user is likely to recall. A shop i is listed in the *estimate of user's recall of locations* if $\hat{p}_i(r)$ is larger than a threshold. We adjust the threshold to maximize the usefulness of direction giving. With a higher value, fewer shops are in *memory of locations*. Thus a robot might use a distant shop and provide less comprehensive descriptions. A lower value results in higher risk where a user doesn't recall the landmark. With the chosen threshold, the robot provides reasonably good direction giving. We empirically set the threshold value to be 0.8.

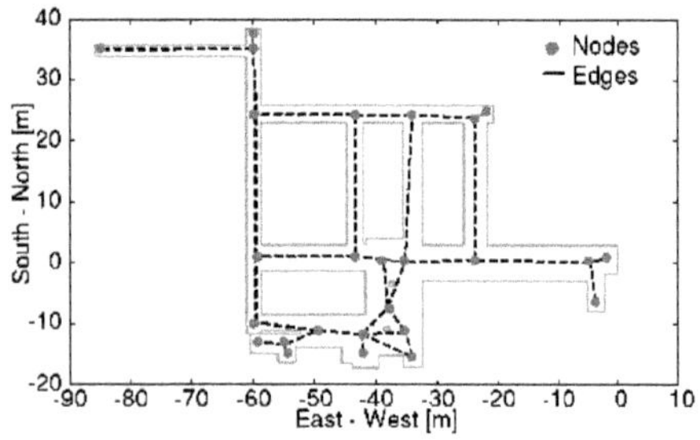

Figure 9: Topological Map of Shopping Mall

4.5 Conversation

4.5.1 Keyword detection

This module is operated by a human operator. In the scenario, we assume the user asks either "where is *X*? (name of shop)" or "where is *Y* (name of products/goods) sold?". Imitating a response of speech recognition software, an operator selects one of the keywords registered in advance to the system when a user spoke to the robot. There are 17 names of shops (such as "*Academia*" (the name of a book store)) and 17 names for attribution of shops (such as "eating facility", "glasses shop", "a shop for children's clothing") registered for our scenario.

4.5.2 Direction giving

It outputs the *destination description*. To make the description acceptable to the user, it refers to the *estimate of user's recall of locations*. It consists of the following two steps:

1) Finding a shop

The system has its own database about the shops in the shopping mall. Each entry of a shop has an attribution of *name*, *location*, and *a list of keywords of products* the shop sells. A simple word matching is conducted with the keyword detected by *keyword detection*. When multiple shops are matched, for simplicity, the system chooses the closest one from the current location.

2) Generating a route direction

Given that the shop s_d is chosen in the step of *finding a shop*, the system uses the following rule set (from rule 1 to rule 6) to generate an utterance for a route direction:

1) If s_d is contained in the *memory of locations*, the robot says "it's sold in *the shop named* N_d *(the name of the shop* s_d*)*".

2) If a shop next to s_d, defined as s_{next}, is contained in the *memory of locations*, the robot says "it's sold in *the shop named* N_d, *which is next to the shop* N_{next} ".

3) If a shop opposite s_d, defined as $s_{opposite}$, is contained in the *memory of locations*, the robot says "it's sold in *the shop named* N_d, *which is opposite the shop* $N_{opposite}$ ".

4) If a shop near s_d, defined as s_{nearby}, is contained in the *memory of locations*, the robot says "it's sold in *the shop named* N_d, *which is the near the shop* N_{nearby} ". We used a distance threshold of 10 m to select s_{nearby}.

5) In case where conditions (a) - (d) below are satisfied, the shop s_d faces the route where a known shop s_{on_route} faces, and s_d is located on this route after a corner near the shop s_{on_route}. This means that the user could reach to the shop s_d if he/she did not turn the corner near the shop s_{on_route} and instead went straight. Thus, the robot says "it's sold in *the shop named* N_d, *which is on the route after you see the shop* N_{on_route} ".

a) The shop s_{on_route} is contained in the *memory of locations*

b) The shop s_{on_route} faces a route segment $n_i n_{i+1}$ (the segment between the node n_i and n_{i+1}) in the *traversed_node_list*

c) The shop s_{on_route} is close to the corner n_{i+1} (within 6m)

d) The shop s_d faces a (straight) route segment $n_i n_k$, the segment $n_i n_k$ includes the segment $n_i n_{i+1}$, and s_d does not face the segment $n_i n_{i+1}$.

6) Otherwise, it is not possible to use *destination description*. The system could generate a turn-by-turn description. Due to the focus of this study, the system gives up providing the route direction, and instead says "Sorry, such shop is not nearby. May I guide you somewhere else?"

In all utterances, the robot points and looks at the direction of the shop when it refers.

5. EVALUATION

We conducted an evaluation experiment for the developed model of recall of locations. The robot was used in a window-shopping scenario in a shopping mall. Participants walked with the robot while asking the robot to provide route directions.

5.1 Hypotheses and Prediction

In the *destination description*, it is vital that locations referred to are known to the listener. We consider that if the computational model and our implementation is appropriate enough, the system should be able to provide a useful route direction with *destination description*. We compared the system with a system that does not have a recall model. We made the following prediction:

Prediction: The route direction provided by the robot with the *spatial memory model* includes locations that people **recall more**, and thus is perceived as **easier to understand** than the one provided by a robot *without the spatial memory model*.

Note that since *destination description* [6] is found to be more effective when appropriate, we did not compare with a turn-by-turn direction-giving and focused on the evaluation of the effectiveness of the memory model on *destination description*. The effects of turn-by-turn direction-giving were demonstrated in previous studies [19].

5.2 Method

5.2.1 Participants

Nineteen native Japanese speakers (12 males and 7 females whose average age was 23.4 years old) participated in the experiments and were paid for their participation.

5.2.2 Conditions

The experiment was a within-subject design having one factor, *spatial memory*. Two conditions are prepared:

With model: The system has the *recall module* as described in section 4; thus the developed recall model is used.

Without model: Instead of the *recall module*, it assumes that all the locations visible from the *traversed route* are recalled. It results in providing *destination description* using any shops seen in the traversed route.

5.2.3 Procedure

Participants were instructed to go window-shopping with a robot. The robot was introduced as a shopping companion robot that carries their bags. Participants were allowed to walk freely, but not allowed to go back nor visit the same intersection again to have them explore different locations.

Participants were told that the robot is able to provide route direction. During walking, they were instructed to ask about a location by an experimenter through a transceiver twice; for each location, they were instructed to ask about a location where a specific product is sold (e.g. "Japanese fan", and "Japanese food"). The name of the product was provided by an experimenter at the moment each participant asked. An experimenter who did not know the purpose of this experiment randomly choose a location that can be taught with the *destination description*. This procedure was to limit the direction-giving to the type of *destination description* so that the developed function was evaluated in the experiment.

Participants participated in both conditions, and the orders of conditions were counterbalanced.

5.3 Measurement

We took the following measurements after each session:

Ease of recall (landmark): We asked about "ease of recall of locations used as the landmark in route direction" with a single item scale.

Ease of understanding (destination): We asked about "ease of understanding the route direction" with a single item scale.

All questionnaire items were with a 1-to-7 point Likert scale. A brief interview was conducted at the end of the experiment.

5.4 Result

Figure 10 shows the result. A one-way repeated-measure analysis of variance (ANOVA) was conducted. A significant main effect was revealed in *ease of recall (landmark)* ($F(1, 18)=6.881$, $p<.017$, $\eta_p^2=.277$), and *ease of understanding (destination)* ($F(1, 18)=7.245$, $p<.015$, $\eta_p^2=.287$). Participants found the route directions provided by the robot with the recall model included landmarks that were easier to recall and easier to understand than the ones provided by the robot without the recall model. Thus, the prediction is supported.

In the interview, we specifically asked the users about the difference of the two methods. For five participants, difference between two methods was not obvious. This would be due to the chance that *without model* condition could also provide direction with recalled landmark, as it randomly chose landmarks. There are 12 participants out of 19 reported that direction giving is useful, and thus they would choose to use the robot (with the proposed method) than the other if they used a robot for shopping.

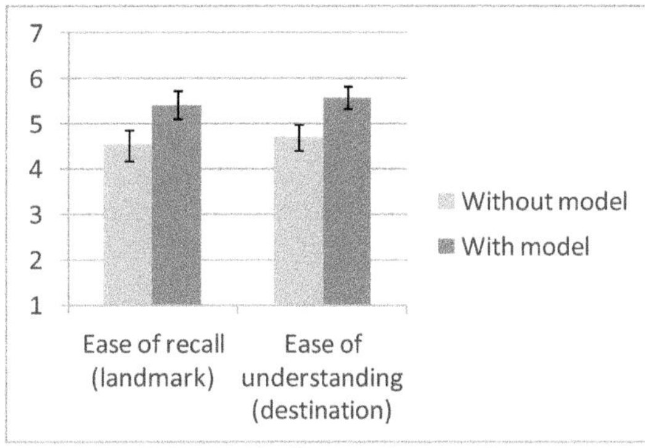

Figure 10: Evaluation result

6. DISCUSSION

6.1 Design Implications

We chose a direction-giving scenario as one example. There would be a couple of usages of the memory recall model. The model would enable a number of dialogs related to locations. In our daily life, we frequently refer to known locations. For instance, there would be used in a clarification dialog,

A: Shall we rest at *the cafe*?

B: Which one? The one in front of *the bookstore*, or the one next to *the drug store*?

Moreover, shared experience would be a fundamental capability in our daily interaction among people, and the capability of being aware of salient features in an environment would enhance cognitive similarity of a robot, and possibly make the robot a more acceptable companion/partner. It would be nice if we could have a dialog with a robot such as the one below.

User: I found a nice bag somewhere...

Robot: Was it in *the clothing shop* next to *the glasses store*?

6.2 Extending Model Beyond First-time Visit

One of the limitations of our current method is that it only addresses the recall within a short duration because it is modeled from a few minutes of data just after people walked. Although it was enough for our scenario, this would limit the further use of the model. For instance, predicted recall rate decreases in a function of square of the elapsed time in our model; thus, according to the model in its specification, no one will remember anything after a while. This would be a possible approximation for people during walking; as they keep gaining new information, shops they have seen in the beginning would be barely recalled if they travel a long distance. However, as some models in memory study use logarithmic decay function [9] to explain long-lasting features of the decay process, if we measure the recall after they finished walking, the shape of the curve could be modeled in a different way.

Related to the above discussion, the more important feature to be included in future work is the fact that people repeatedly reconstruct memory [9]. We expect that repeated visits to the same location would cause a reconstruction, and thus we can recall shops that we visit often. Similarly, in our scenario, the robot refers to shops in its dialog to provide route directions. Such conversation would make the relevant memory more recalled. We consider that these reconstructions occurred when being with a robot could be modeled with a simple extension of the model, e.g. by adding terms related to the reinforcement of recall with either re-visiting or use in conversation.

6.3 Limitations

Our model involves a couple of parameters such as coefficients in the regression model. It is difficult to prepare general parameters with a single study; therefore, we modeled the parameters for a particular indoor environment. While it enables us to develop a proof of concept, it is probably necessary to adjust the parameters to use the model in other environments.

Although the recall was reasonably approximated to the extent we can use it in the direction giving scenario, the computational model would not reflect a true characteristic of the cognitive process humans would perform. For instance, high-order attention is not considered in the model; one would be interested in a particular type of shop (e.g. clothing) and thus recall more. To include such high-order attention in the computational model, further study is necessarily.

7. CONCLUSION

We developed a recall model for locations in a window-shopping scenario. With data collection, we built the model. It approximates that a shop that is recently seen, visually large, large in terms of floor size, and dark in facade color is recalled well. The robot uses the model to estimate a user's perspective on recallable locations in a direction-giving interaction. The evaluation experiment revealed the effectiveness of the model for direction-giving with *destination description,* which requires a known location as a landmark.

We believe the study demonstrates a case where a companion robot shares experience with a user implicitly through its observation, and their interaction is successfully built on their shared experience. This case was only for a few minutes of "being together"; as we expect a companion robot to share much longer time with users, we consider it our future work to explore what constitutes shared experience on a longer time scale.

8. ACKNOWLEDGMENTS

This research was supported by the Ministry of Internal Affairs and Communications of Japan.

9. REFERENCES

[1] Berlin, M., Gray, J., Thomaz, A. L. and Breazeal, C., 2006, Perspective Taking: An Organizing Principle for Learning in Human-Robot Interaction, *National Conf. on Artificial Intelligence (AAAI2006)*, pp. 1444-1450.

[2] Besl, P. and McKay, N., 1992, A method for Registration of 3-D Shapes, *IEEE Transactions on Pattern Analysis and Machine Intelligence*, vol. 14, pp. 239-256.

[3] Bickmore, T. W., Caruso, L., Clough-Gorr, K. and Heeren, T., 2005, 'It's just like you talk to a friend' relational agents for older adults, *Interacting with Computers*, vol. 17, pp. 711-735.

[4] D. Fox, W. B., and S. Thrun, 1999, Markov localization for mobile robots in dynamic environments, *Journal of Artificial Intelligence Research*, vol. 11, pp. 391-427.

[5] Daniel, M.-P., Tom, A., Manghi, E. and Denis, M., 2003, Testing the Value of Route Directions Through Navigational Performance, *Spatial Cognition & Computation*, vol. 3, pp. 269-289.

[6] Dethlefs, N., Wu, Y., Kazerani, A. and Winter, S., 2011, Generation of Adaptive Route Descriptions in Urban Environments *Spatial Cognition & Computation: An Interdisciplinary Journal* vol. 11, pp. 153-177.

[7] Draper,N.R. and Smith,H Applied Regression Analysis, *Wiley-Interscience, Hoboken, New Jersey,* pp.307-312, 1998

[8] Fox, D., Burgard, W. and Thrun, S., 1997, The Dynamic Window Approach to Collision Avoidance, *IEEE Robotics & Automation Magazine*, vol. 4, pp. 23-33.

[9] Friedman, W. J., 1993, Memory for the time of past events, *Psychological bulletin*, vol. 113, pp. 44-66.

[10] Garling, T. and Evans, G. W., 1991, *Environment, Cognition, and Action: An Integrated Approach*, New York : Oxford University Press.

[11] Gockley, R., Bruce, A., Forlizzi, J., Michalowski, M., Mundell, A., Rosenthal, S., Sellner, B., Simmons, R., Snipes, K., Schultz, A. C. and Wang, J., 2005, Designing Robots for Long-Term Social Interaction, *IEEE/RSJ Int. Conf. on Intelligent Robots and Systems (IROS2005)*, pp. 1338-1343.

[12] Gross, H.-M., Boehme, H., Schroeter, C., Mueller, S., Koenig, A., Einhorn, E., Martin, C., Merten, M. and Bley, A., 2009, TOOMAS: Interactive Shopping Guide Robots in Everyday Use - Final Implementation and Experiences from Long-term Field Trials, *IEEE/RSJ Int. Conf. on Intelligent Robots and Systems (IROS2009)*, pp. 2005-2012.

[13] Kanda, T., Shiomi, M., Miyashita, Z., Ishiguro, H. and Hagita, N., 2010, A Communication Robot in a Shopping Mall, *IEEE Transactions on Robotics*, vol. 26, pp. 897-913.

[14] Kiesler, S., 2005, Fostering Common Ground in Human-Robot Interaction, *IEEE Int. Workshop on Robot and Human Interactive Communication (RO-MAN2005)*, pp. 729-734.

[15] Koriat, A., Goldsmith, M. and Pansky, A., 2000, Toward a psychology of memory accuracy, *Annual Review of Psychology*, vol. 51, pp. 481-537.

[16] Mackworth, J. F., 1962, The visual image and the memory trace, *Canadian Journal of Psychology*, vol. 16, pp. 55-59.

[17] Morales, Y., Satake, S., Kanda, T. and Hagita, N., 2011, Modeling Environments from a Route Perspective, *6th ACM/IEEE International Conference on Human Robot Interaction*,

[18] Nothegger, C., Winter, S. and Raubal, M., 2004, Selection of Salient Features for Route Directions, *Spatial Cognition & Computation*, vol. 4, pp. 113-136.

[19] Okuno, Y., Kanda, T., Imai, M., Ishiguro, H. and Hagita, N., 2009, Providing Route Directions: Design of Robot's Utterance, Gesture, and Timing, *ACM/IEEE Int. Conf. on Human-Robot Interaction (HRI2009)*, pp. 53-60.

[20] Rosenthal, S., Biswas, J. and Veloso, M., 2010, An effective personal mobile robot agent through symbiotic human-robot interaction, *International Conference on Autonomous Agents and Multiagent Systems (AAMAS 2010)*, pp. 915-922.

[21] Scassellati, B., 2002, Theory of Mind for a Humanoid Robot, *Autonomous Robots*, vol. 12, pp. 13-24.

[22] Schroder, C. J., Mackaness, W. A. and Gittings, B. M., 2011, Giving the 'Right' Route Directions: The Requirements for Pedestrian Navigation Systems, *Transaction in GIS*, vol. 15, pp. 419-438.

[23] Sidner, C. L., Lee, C. and Lesh, N., 2003, Engagement By Looking: Behaviors for Robots When Collaborating with People, ed.

[24] Thrun, S., Burgard, W. and Fox, D., 2005, *Probabilistic Robotics (Intelligent Robotics and Autonomous Agents)*, The MIT Press.

[25] Torrey, C., Powers, A., Marge, M., Fussell, S. R. and Kiesler, S., 2006, Effects of Adaptive Robot Dialogue on Information Exchange and Social Relations, *ACM/IEEE Int. Conf. on Human-Robot Interaction (HRI2006)*, pp. 126-133.

[26] Trafton, J. G., Cassimatis, N. L., Bugajska, M. D., Brock, D. P., Mintz, F. E. and Schultz, A. C., 2005, Enabling Effective Human–Robot Interaction Using Perspective-Taking in Robots, *IEEE Trans. on Systems, Man, and Cybernetics. Part A: Systems and Humans*, vol. 35, pp. 460- 470.

[27] Turner, A. and Penn, A., 2007, Evolving Direct Perception Models of Human Behavior in Building Systems?, *Pedestrian and Evacuation Dynamics 2005*, pp. 411-422.

[28] Tversky, B., 1993, Cognitive maps, cognitive collages, and spatial mental models, *Spatial Information Theory A Theoretical Basis for GIS*, pp. 14-24.

Color Anomaly Detection and Suggestion for Wilderness Search and Rescue

Bryan S. Morse
Brigham Young University
3361 TMCB
Provo, Utah 84602
morse@byu.edu

Daniel Thornton
Brigham Young University
3361 TMCB
Provo, Utah 84602
danielthornton@byu.edu

Michael A. Goodrich
Brigham Young University
3361 TMCB
Provo, Utah 84602
mike@cs.byu.edu

ABSTRACT

In wilderness search and rescue, objects not native or typical to a scene may provide clues that indicate the recent presence of the missing person. This paper presents the results of augmenting an aerial wilderness search-and-rescue system with an automated spectral anomaly detector for identifying unusually colored objects. The detector dynamically builds a model of the natural coloring in the scene and identifies outlier pixels, which are then filtered both spatially and temporally to find unusually colored objects. These objects are then highlighted in the search video as suggestions for the user, thus shifting a portion of the user's task from scanning the video to verifying the suggestions. This paper empirically evaluates multiple potential detectors then incorporates the best-performing detector into a suggestion system. User study results demonstrate that even with an imperfect detector users' detection increased significantly. Results further indicate that users' false positive rates did not increase, though performance in a secondary task did decrease. Furthermore, users subjectively reported that the use of detector-based suggestions made the overall task easier. These results suggest that such suggestion-based systems for search can increase overall searcher performance but that additional external tasks should be limited.

Categories and Subject Descriptors

I.2.9 [**Artificial Intelligence**]: Robotics—*Operator Interfaces*; I.4.9 [**Image Processing and Computer Vision**]: Applications

Keywords

Wilderness Search and Rescue, Search and Detection, Unmanned Aerial Vehicles, Anomaly Detection, User Study

1. INTRODUCTION

Wilderness search and rescue (WiSAR) is the task of finding missing persons in wilderness areas. This is highly time-sensitive, not only because of increasing danger to the search subject (missing person) but also because the search area increases over time. An efficient way to augment search is to use semi-autonomous unmanned aerial vehicles (UAVs) with cameras that transmit live video and telemetry data to a ground-based team [15]. With sufficient video resolution, searchers can identify traces of the subject from the air.

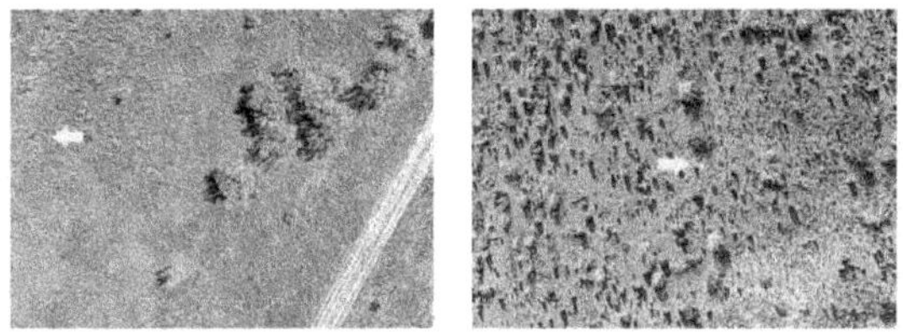

(a) Grassy Location (b) Desert Location

Figure 1: Examples of video frames from a search scenario. Targets are marked with yellow arrows: (a) blue blanket and (b) white shirt.

Even with the aid of aerial video, it can be difficult to identify signs of a search subject. Figure 1 shows examples of typical search video frames. Image resolution is limited both by the camera (constrained by the payload capacity of inexpensive UAVs) and by the need to cover as much ground as possible. In addition, the UAV and hence the video can move quickly, disorienting searchers and giving them little time to detect targets. This can be compensated for somewhat by enhancing the spatiotemporal presentation of the video [19] or visually enhancing the hue and saturation of color values [24], but detecting objects of interest remains a challenging problem. Consequently, searchers may miss signs of the search subject, even when seen by the camera.

In practice, the task of observing search video consists of two elements: *detection* and *analysis*. An object of potential interest must first attract the user's attention as worthy of inspection, then further analysis must determine whether it is indeed of interest. Objects of interest may include the missing person, abandoned clothing, camping or hiking gear, or other personal items. Such objects may also vary in difficulty of detection; a blanket (Figure 1(a)) may be easier to detect than a shirt (Figure 1(b)).

Detection is the task of quickly identifying possible signs of the search subject. It might be something as simple as seeing something that "looks out of place", especially when searching wilderness areas with primarily naturally-occurring content. When performed by a person, detection may be responded to by a reflexive action, such as a keystroke, accompanying the appearance of an unusual or significant object. The person then tries to determine, through inspection of the imagery, whether the object is likely to be a positive sign. Since detection is less dependent on domain knowledge it is a good candidate for automation, while analysis is best left to the human operator.

Anomaly detection clearly has strong similarities to automated target recognition (ATR) systems, which have also been used to assist human operation of semi-autonomous UAVs (e.g., [8, 11, 12]). However, such ATR systems usually have a notion of a specific target (in our case, the missing person) rather than the more ill-posed problem of finding clues that indicate the recent presence of the search subject.

This work primarily relies on leveraging color information in video of the search area to detect signs of a missing person, a technique referred to as *spectral anomaly detection.* For the best chances of success in applying spectral anomaly detection to this domain, multiple detection methods were implemented and compared empirically. Using the temporal and spatial information in the video stream, unusually colored pixels are aggregated into larger objects.

Because of the sensitive nature of search and rescue, and likely imperfections in automated anomaly detection, the detection task is best treated as a collaborative effort between human and autonomy. Indeed, the very question is ill-posed: find anything that "looks out of place". Therefore, the goal of this work is to use automated detection as an aid for visual search rather than a replacement for it.

The addition of automation into human tasks is not without potential issues. Research has shown that automation can affect users' behavior in various ways, including misuse of the automation [23]. In particular, the results of using automation aids for human-robot interaction can vary greatly depending on the task [16] or even across users [7]. In addition to empirical evaluation of potential detector methods, this paper also contributes the results of a user study designed to evaluate the effects of detector-based suggestions.

2. EFFECTS OF AUTOMATION

Automated detection systems have been studied for many years, including their impact on users' behavior in the presence of imperfect detection. Although such systems can increase user performance in many situations, there are numerous examples of ways in which these systems can be misused or disused [23].

Introducing automation can result in inattention either through a false sense of *complacency* or through *automation bias*, in which the saliency and authoritative nature of the recommendations from the automation affect user behavior and compliance [22]. Because anomaly detectors will inevitably miss potential targets of interest this can result in *errors of omission*, in which users similarly miss potential targets by focusing primarily on the automated suggestions. Because the detector will also inevitably make make incorrect suggestions, false alarms can serve both as a source of distraction and as a source of *commission errors* (more false alarms similarly raised by the user).

The effects of false alarms can be even more pronounced for detection tasks with rare targets or events [23]. If targets appear relatively infrequently, as is the case with search, even a detector with a low false-alarm rate will result in many more false alarms than correct detections (hits). The effects of these false alarms is an important question to address when evaluating the use of detector-based suggestions.

Some studies have shown that imperfect automation has the potential to detract rather than enhance users' performance [11, 12], especially when performing multiple tasks simultaneously. An abundance of either false alarms or missed detections can lead to disuse of the automation. Similarly, frequent misses by the detector may lead to users not making use of it (and thus not gaining the benefit of the suggestions).

Other studies have shown that for UAVs with ATR systems, "human-robot teams can benefit from imperfect [ATR] automation even under high workload conditions" [8, 9]. This suggests that incorporation of imperfect (and even ill-posed) anomaly detectors may behave similarly.

3. COLOR ANOMALY DETECTION

The literature on detection in video is full of various methods, but many of these are not suitable for the particular needs of wilderness search and rescue. Background subtraction [18, 28] can be used to find objects not normally found in a scene, but only if the background has been previously observed. This is not the case when searching a new area, though it does have potential for repeated sweeps. Detectors designed for specific objects (e.g., [21, 27]) also do not apply since the nature of the objects of potential interest is typically not known ahead of time. Detectors based on other domain knowledge [4, 14] might be of use—for example, the method in [4] is useful for finding large man-made objects in natural scenes by using texture distributions for large natural regions versus large man-made regions. However, this reliance on texture means that this method works only for objects significantly larger than those in WiSAR search video, which may only be several pixels in size.

Of course, the one known target is the missing person, so it is tempting to use methods for detecting humans [20, 21, 31]. However, the body of work in this area has focused almost exclusively on detection of humans that are moving and/or upright relative to the camera. Since aerial video sees the person only from above (or perhaps at a slightly oblique angle), these methods cannot be used here unless by chance the person is lying on the ground. Similarly, the speed at which the aircraft passes over the area makes detection and analysis of much-slower human motion difficult.

Finally, while the term *anomaly detection* occurs in the video-processing literature (e.g., [17]), it is commonly used to describe *behavioral* anomalies (such as might be associated with a security threat) rather than simple visual anomalies.

For these reasons, we choose to focus here on anomalous color detection: the finding of unusually colored objects in a scene. This is a subclass of the broader problem of spectral anomaly detection, most of the work for which has focused on hyperspectral images. We adapt here some of these methods for use with RGB video images.

3.1 Spectral Anomaly Detection

A common approach to hyperspectral anomaly detection is to model the statistical distribution of spectral signatures with one or more multivariate normal distributions [3, 25, 26]. This model is then used to identify pixels whose spectral signatures are statistical outliers [5]. The normal distribution is most often used for its simplicity. Once the mean vector and covariance matrix have been calculated, outliers can be identified using a threshold on the Mahalanobis distance [13]. In a multivariate normal distribution, the Mahalanobis distances are distributed according to the chi-square distribution with cumulative distribution function

$$F(d_M; k) = P(k/2, d_M/2) \quad (1)$$

where k is the dimensionality of the multivariate normal and P is the regularized Gamma function. The distance threshold can therefore be chosen to encompass a desired probability. In the case of one-dimensional data, this method yields the well-known bell-curve confidence intervals. When data points are RGB triples, $k = 3$.

A multivariate normal distribution rarely characterizes all of the colors in a natural scene, though. This means any effective spectral anomaly detector must perform some transformation on or clustering of the data for it to fit the assumption of normality. Such procedures are referred to here as *normalization*. Once the values have been normalized, the mean vector and covariance matrix are estimated in order to calculate the Mahalanobis distance of each pixel.

3.2 The RX Algorithm

Perhaps the simplest normalization method is the RX algorithm [25]. It assumes that each pixel is drawn from a multivariate normal distribution but that the mean and variance of the distribution change across the image. The variance is generally assumed to change more slowly than the mean, so much so that it is common to use the same variance estimate for the entire image but to calculate this using a spatially-varying mean [6]. The mean is usually calculated within a window near, but not including, the immediate neighborhood of the pixel. Once this local mean has been subtracted, it is straightforward to calculate the covariance matrix and thereby the Mahalanobis distance of each pixel. Apart from the Mahalanobis distance threshold, the only parameters to this algorithm are the radius of the outer included neighborhood, R, and the radius of the inner excluded neighborhood, r, of the local neighborhood.

These steps can be thought of as an unsharp masking operation, resulting in a residual error image. Next, a color transformation is applied to the error image with the inverse of its color covariance matrix. Finally, the dot product of each transformed error vector with the corresponding untransformed error vector is computed and stored, resulting in a gray-scale image of Mahalanobis distances, to which is applied a threshold chosen with Equation 1.

Adaptations of the RX algorithm include exchanging the covariance matrix for the correlation matrix [6] or combining local parameter estimation with clustering [1].

3.3 Clustering Methods

A common normalization method is to divide the image pixels into clusters using methods such as vector quantization and k-means [1, 3, 26]. In none of these examples do the authors explicitly state that their clusters are normal in shape, but all use a Mahalanobis distance threshold, which implies an assumption of normality. The BACON algorithm [2, 26] explicitly chooses the distance threshold using the chi-square distribution, as discussed previously.

Gaussian Mixture Modeling (GMM) can be considered a form of fuzzy clustering. It assumes that the true distribution can be modeled by a mixture of normal (Gaussian) distribution components. The GMM for a set of data is usually found using Expectation-Maximization (EM) [10]. If properly estimated, each mixture component may be considered a fuzzy cluster. Unlike clusters produced by k-means or vector quantization, these clusters are designed to be normally-distributed, but this is a much more costly process.

3.4 Robust Methods

Most clustering approaches, such as CBAD [3] and GMM, simply use the sample mean and covariance matrix of each cluster, but a more robust approach to outlier detection is the BACON algorithm [2, 26]. BACON aggregates sample points within a cluster into an inlier set by gradually increasing the threshold on Mahalanobis distance, re-estimating the mean vector and covariance matrix at each step. This iterative estimation is more robust to outliers, thus ensuring that the outliers can be correctly identified. It is also more costly than simply calculating the sample mean and covariance of the entire image since it requires multiple iterations with sample sizes approaching the full set.

4. DETECTOR EVALUATION

To evaluate which detection methods work best for this domain, we implemented the following four methods:

1. The RX algorithm [25]
2. Vector quantization (as used in CBAD [3])
3. K-means clustering
4. The EM algorithm [10], initialized using k-means

The BACON algorithm [2] for robust outlier nomination was also implemented to see if it could improve the results of the best spectral detector.

4.1 Data Collection

In order to evaluate the different detectors, a set of test images was collected. These images are of natural scenes containing a few foreign man-made objects. A ground-truth labeling of the objects within each image was created for fast and repeatable testing.

To best control the content of the images, the scenes were set up carefully and deliberately. Two natural scenes were used: a grassy location and a desert location, each typical of wilderness search environments. These two locations were carefully chosen to minimize the likelihood of man-made objects in the scene. A small number of man-made objects were then placed at each location for use as visual targets. These targets ranged in size from a t-shirt to a small blanket. Each target consisted of one or two solid-colored objects. Six targets were placed in the first scene and five targets were placed in the second scene. Thus, each scene contained mostly naturally-occurring objects, with only a few foreign man-made objects.

A professional aerial photographer captured aerial imagery of each scene using a digital camera mounted on a small, remote-controlled plane. The photographer then flew the camera over the area, capturing both high-resolution still images and standard-resolution digital video.

The video and images of each scene were then reviewed carefully by visual inspection and with the aid of temporally local mosaics [19]. In addition to the target objects placed in the scene, a number of other objects that could reasonably be considered foreign to the environment were seen. For the grassy location, these included the pilot and two other people, two vehicles, and multiple nearby buildings (video only). For the desert location, these included the pilot and a vehicle (video only), a plastic grocery bag (photographs only), a white box, and a bright orange object.

All anomalies, including the accidental objects listed, were manually labeled in the digital stills on a per-pixel basis. These label maps were then used to tune and compare the different spectral anomaly detection methods.

4.2 Spectral Detector Evaluation

An automated test suite was built for fast and repeatable evaluation. The test suite calculates a Receiver Operating Characteristic (ROC) curve for each anomaly detection method by varying the detector's threshold and plotting the true positive rate (TPR) against the false positive rate (FPR). The comparison metric for the different methods is the area under the ROC curve.

At least one method (the RX algorithm) is sensitive to the size of objects in the image. Therefore, the full-size stills as well as the corresponding label images were subsampled to get object sizes similar to those seen in the video but not so small as to hinder visual detection, and the images were subdivided to produce stills comparable to the video frames. Each of the 278 still images produced 24 video-

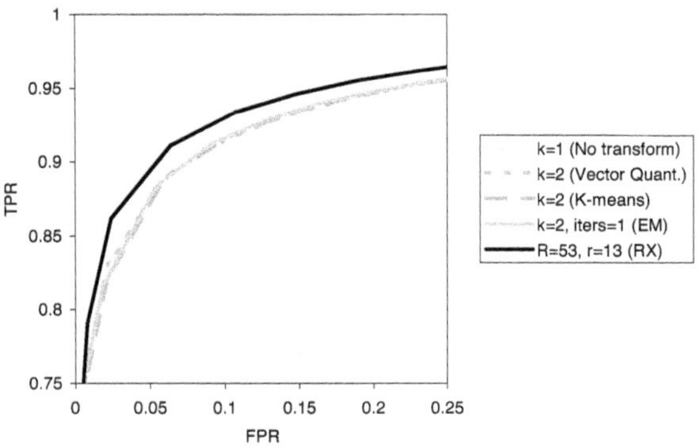

Figure 2: ROC curves for each method

frame-sized subimages, for the equivalent of about 3.7 minutes of manually-labeled high-quality video.

In order to cover as near as possible the full range of false positive values, the target false positive rate was varied from 0% to 100%. This target value was used, in connection with Equation 1, to determine the Mahalanobis distance threshold for each test. Each ROC curve was comprised of 40 tests.

4.3 Detector Evaluation Results

The detector that performed the best overall was the RX algorithm with $R = 53$ and $r = 13$ (Figure 2). While there were significantly worse settings for RX, comparable results were found in a fairly broad range of the parameter space.[1]

The second best detector was the degenerate clustering case of $k = 1$. This case is the same for all clustering methods as it performs no clustering or normalization of the data. Comparable results were found for each clustering method with $k = 2$, but larger values of k showed a decrease in performance overall, even though more clusters sometimes performed better for selected images.

4.3.1 Why Clustering Approaches Struggle

In further analyzing the performance of the various cluster-based approaches, values of k ranging from two to four would often outperform the degenerate $k = 1$ case on many images. In fact, these would often outperform the RX method as well. These results show that clustering can work very well in many cases with appropriate parameter tuning, but the degenerate $k = 1$ case still performs best overall for the clustering approaches, and the RX method in general outperformed that. The problem with using clustering in this domain is that it is sensitive to the content of the scene [3]. The content of the scene can change frequently as the plane flies over different areas. If only one type of ground cover is present, k should be very low. For more types of ground cover, it should be higher to correctly model the background. The correct number of clusters to use will then change as more or fewer types of ground cover are in view.

In contrast, the best window size for RX is primarily determined by the *projected sizes* of the targets and other objects [3], which is a function of true object size, viewing distance, and camera resolution. In aerial search, altitude is controlled to keep targets large enough for detection while maximizing ground coverage [15]. (For our system's camera resolution, this range is 60–100 meters for targets the size of a person.) Thus, the projected target size should easily fall within a predictable range. Since target size is less variable and easier to predict in this domain than scene content, RX should be preferable to a clustering approach.

Although there are numerous methods in the literature for dynamically adjusting the number of clusters (e.g., [29]), these methods were not explored further because dynamic adjustment of the number of clusters on a constant basis was considered too slow for processing of live search video.

4.3.2 BACON

The best performing normalization method, RX, was combined with a robust outlier detection method, BACON [2], to try to improve performance. The ROC curve area with BACON (97.17%) was slightly higher than with RX alone (96.93%), but this increase is less significant than the one between no normalization (96.46%) and RX.

Although using BACON produced minimally better results, it was also much slower (for reasons given in Section 3.4). It was also harder to control in terms of tuning the desired false alarm rate than RX alone. Therefore, BACON was not further used for detection in this work.

5. SYSTEM AND USER STUDY

To evaluate the effectiveness of suggestions made by an automated detector for unusually colored objects, we implemented a simple system with detector-based suggestions and conducted a user study to compare user performance both with and without the aid of such suggestions. In particular, we designed the study to investigate potential effects identified previously for automated systems (Section 2):

- How does the introduction of detector-based suggestions affect the user's overall detection sensitivity?
- Do false positives from the detector increase the user's false positive rate?
- Do these false positives distract in such as way as to cause the user to miss potential targets of interest?
- How do these false positives affect a user's performance on a secondary task?

5.1 Presentation and Interface

There are numerous ways that the search imagery and detector's suggestions could be presented to a searcher, and the choice of presentation will certainly have an effect on the searcher's performance. In order to keep the implementation and analysis tractable, one simple user interface was implemented and evaluated for this study.

Each participant was asked to view a series of eight aerial video clips and mark foreign or man-made objects. Participants placed marks (displayed as red circles) on the video with a single mouse-click and could similarly remove marks. To reduce effects of hand-eye coordination, participants were given the option of freezing the video display to examine or mark objects, after which display would resume with the "live" video search. Each time a participant marked an object in the video, the location and time were recorded. Unless a mark was removed by the participant, it was also logged in a final list of markings for that participant.

All aerial videos were presented using temporally-local mosaics, a presentation method previously developed in order to expand the spatiotemporal window of observation opportunity for the user [15]. The presentation order was counterbalanced and the order of the videos was randomized. For each participant, four of the eight video clips were randomly selected and marked with suggestions from the detector.

In addition to the primary task of target detection, participants were also given a secondary task in order to assess

[1]A more complete reporting of the performance of each method for various parameter settings can be found in [30].

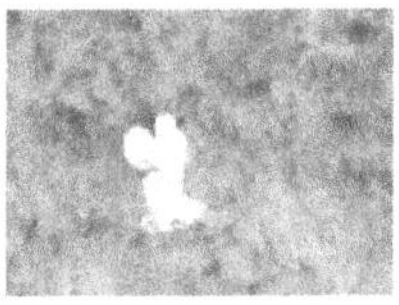

(a) Tan shirt w/ shorts

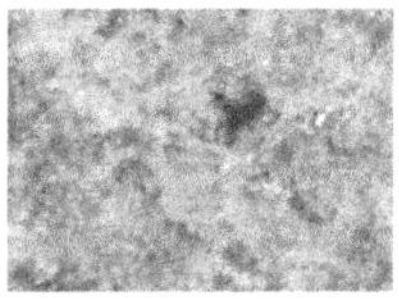

(b) Blue blanket

(c) Red shirt

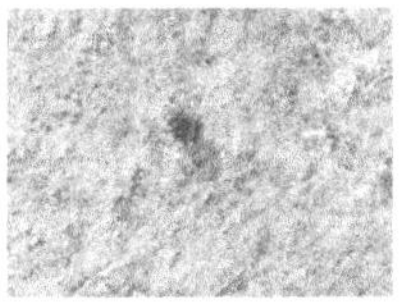

(d) Black shirt

(e) Green bag

Figure 3: Target objects at the grassy location

the effect on this of primary task automation. For the secondary task, users were asked to count discrete tones played during each video clip. Some clips contained a series of only low-pitched tones, while other clips contained tones of two different pitches. After each exercise, the participant was asked to report the number of low tones and (if present) the number of high tones played during the exercise. We feel this to be an ecologically valid secondary task, as search-and-rescue personnel are often required to monitor audible communications while performing their primary tasks.

These options resulted in four presentation/task combinations: suggestions with only low tones only, suggestions with both high and low tones, no suggestions with low tones only, and no suggestions with both high and low tones.

After completing a brief demographic survey, each user walked through a set of on-screen instructions. This consisted of a number of explanatory example images followed by two practice video clips before beginning the exercises. In both practice clips, the participant viewed the same video sequence but with different presentation methods. The presentation method for the first practice clip was generated randomly, with the second being the complement of the first.

5.2 Video Clips Used

As explained in Section 4.1, aerial video was taken at both a grassy and a desert location, with many of the same target objects being used at both locations. For each location, four one-minute video clips were selected for the user study. The number of targets visible in each clip ranges from zero to seven. The complete set of eight video clips included two appearances each of 12 target objects for a total of 24 targets. See Figures 3 and 4 for higher-resolution images of each target, and Figure 1 for examples of their placement and size in full video frames.

Each of the 12 target objects consisted of one or two man-made objects. Each man-made object is on the order of a person in size and consists of a single color: red, blue, green, orange, tan, black, or white. Some of these physical objects were used at both locations. For example, the same blue blanket was laid out on the ground in the grassy location (Figures 3(b) and 1(a)) and draped over sage brush at the desert location (Figure 4(a)). The same red shirt is laid out

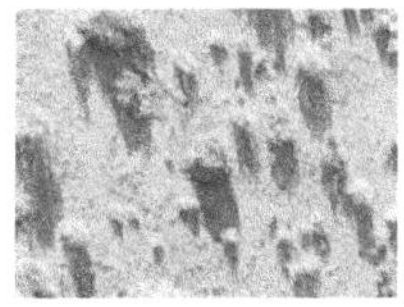

(a) Blue blanket

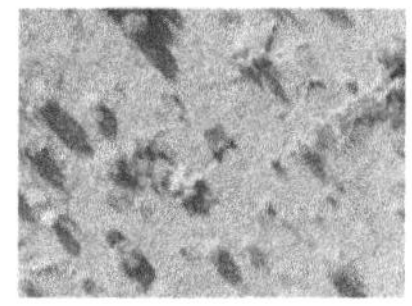

(b) Black shirt

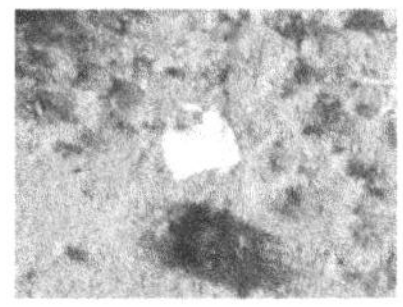

(c) White shirt

(d) Red shirt w/ shorts

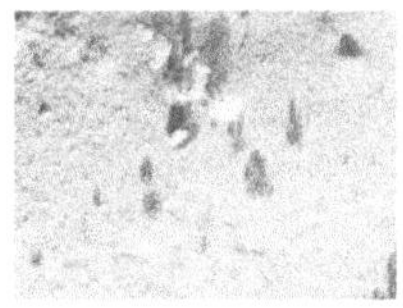

(e) Green bag

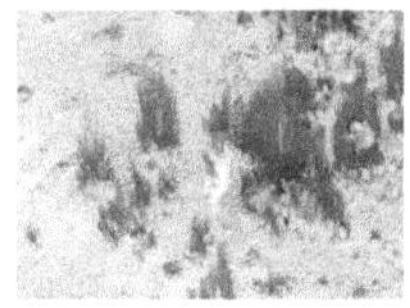

(f) Orange object

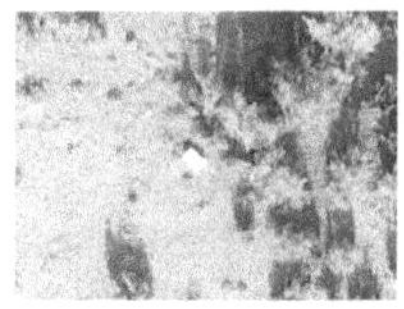

(g) White box

Figure 4: Target objects at the desert location

by itself in the grass (Figure 3(c)) but paired with the tan shorts in the desert (Figure 3(c)).

In addition to intentionally-placed targets, several other man-made objects were discovered at each location, and video clips were carefully chosen to exclude most of these objects. Two unintentional objects were included as targets because they were difficult to exclude, were of the right size, and consisted of solid colors (Figures 4(g) and 4(f)). (The detector does not require solid-colored objects, but we used them here for consistency.)

Because some targets appear twice and several physical components are reused between targets, a training effect is possible. In a pilot version of the study, participants were asked if they noticed any repeated objects and all of them answered in the negative, suggesting that different appearances of the same object were sufficiently unique and that the randomized order of presentation was sufficient.

Four of the eight video clips presented to each user included target suggestions, and the other four did not. Suggestions were presented as light blue circles (Figure 5) with one circle for each anomaly found by the detector. The size and location of each circle was made to encompass a region twice the size of the anomaly's bounding box.

For the pixel-wise detection step, an RX detector was used. The inner radius for the RX convolution kernel should be large enough to exclude most of the target object, while the outer radius needs to be just large enough to accurately sample the surrounding region. While the optimal settings discovered in ground-truth evaluation were 13 and 53, respectively, the inner radius was increased here to 26 to produce better results with some of the larger targets. The false-positive rate for the pixel-wise detector was adjusted to 1 in 10 million pixels.

To compensate for noisy imagery (either from acquisition

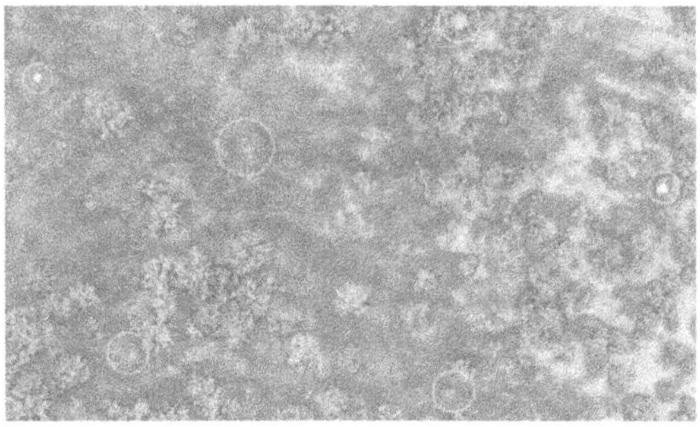

Figure 5: Suggestions as blue circles

or transmission), we also implemented a spatiotemporal filtering step in order to present to the user only object-level anomalies rather than stray pixels. Potential anomalies were ignored if they appeared for less than 3 frames or in less than 91% of their known temporal extent. Objects of potential interest were also restricted to those that had contained at least 43 anomalous pixels in at least one frame and touched the border of their first or last frame. The best parameters for the object filtering step were determined empirically using the ground-truth and the object lists from this aggregation step. The settings were chosen by varying each parameter, observing the resulting object list in the user study interface, and subjectively choosing a good trade-off point between the number of true positives and the number of false positives. Of the 24 target objects, 11 overlapped with suggestions for a 45.8% true-positive rate.

With eight video clips and four presentation methods, there were 32 unique exercises to choose from. The presented sequences were generated so that each participant would view each of the eight video clips once and each of the four presentation methods twice. If this constraint resulted in multiple choices, exercises were then chosen to ensure equal frequency among the 32 exercises across all participants. Other constraints on the selection included preventing any presentation method, exercise, or two-clip subsequence from occurring more frequently than the others.

5.3 Data Gathering Methods

The final set of results consisted of data logs from 35 users. Seven clips were shown to each of the participants, with one clip accidentally skipped by one participant. Three presentation methods were shown 70 times, with one shown 69 times. Nine of the exercises were shown 8 times, with 23 shown 9 times.

Ground truth markings were created by hand using the user study interface. One ground truth marking was made for each of the 24 target objects. Once all of the participant data had been gathered, all user markings, suggestions, and ground truth markings were grouped into clusters. Each marking was put in the same cluster as any other markings whose centers lay within its radius, resulting in a total of 535 clusters. Each of the 24 ground truth markings belongs to a unique cluster. Of these ground truth clusters, 11 include one or more suggestions and 21 include one or more participant markings. Out of all 535 clusters, only 132 included one or more suggestion markings. Markings removed by the participant were included in the clustering step but ignored in all other considerations.

In reviewing the data for the secondary task, two types of input errors became apparent: so-called "fat-finger" errors (pressing two adjacent number keys when only one was intended), and skipping a tone-count question (which was erroneously recorded as zero). These errors appeared to be random, and we attribute them to the interface for gathering the users' responses rather than the task itself. Data exhibiting such obvious errors (e.g., a response of "87" when the correct answer was "7") were removed manually prior to evaluating the users' performance.

	No Suggestions	Suggestions
Detection Rate	52.57%	61.14%
False Positive Rate	2.88%	2.44%
Sensitivity (d')	1.96	2.25
Response Bias (c)	0.92	0.84

Table 1: User performance on the primary task both with and without automated suggestions.

5.4 Results and Statistical Analysis

An analysis of variance using least-squares means showed that the effect of the suggestions on the primary task performance was significant ($F(1,33) = 5.69, p = 0.02$). Users' detection rates increased by 16.30% when aided by suggestions from the anomaly detector ($M = 61.14\%, SE = 2.77$) as compared to normal search without suggestions ($M = 52.57\%, SE = 2.57$). The difficulty of the secondary task did not produce a significant effect on the detection rate.

No statistically significant difference was found for either the false positive rate (FPR) or positive predictive value (PPV) for the different presentation methods or level of secondary-task difficulty. Of the four combinations, the lowest estimated FPR was 2.44% for suggestions and only low tones. The highest was an FPR of 2.88% for no suggestions with both high and low tones. Estimates for PPV ranged from 49.88% for suggestions and only low tones to 50.84% for no suggestions with both high and low tones. None of the differences in FPR or PPV were statistically significant.

Secondary task performance was estimated using the log of the mean squared error in reported tone counts. No significant difference in log MSE was found between exercises with low tones and those with both high and low tones, suggesting that this aspect did not affect their performance. However, the log MSE increases from 0.5269 to 0.8843 when suggestions are added ($p = 0.0047$), suggesting that the presence of suggestions did have an effect on user attention.

5.5 Discussion and Findings

The likely reason that users detected more targets is that the anomaly detector suggested ones that they otherwise would have missed. But it is also worth exploring the effect of automated suggestions on targets not identified by the anomaly detector, specifically whether the presence of other suggestions distracted the users from these targets or caused them to exhibit complacency. This could potentially decrease the user detection rate for targets that the anomaly detector missed and thus trade increased detection of suggested targets for decreased detection of non-suggested ones.

To see if this effect existed, we broke down user performance further by whether the target was suggested by the detector (11 targets) or was not (13 targets), the results of which can be found in Table 2. For targets the automated detector missed, there was no significant difference in user detection whether the detector was otherwise making suggestions (38.0%) or not (37.3%). This suggests that the presence of other suggestions did not impair the detection of non-suggested targets.

For targets found by the automated detector, users exhibited a significant increase in detection with suggestions

	No Suggestions	Suggestions
Not Marked	37.2%	38.0%
Marked	68.1%	87.9%

Table 2: User detection rates by whether the automated detector identified the target.

(87.9%) compared to without suggestions (68.1%). This suggests that for targets suggested by the detector the contribution of the suggestions is even stronger (29.1%).

It is interesting to note that users still missed some targets even if were correctly identified by the automated detector. This suggests that, even if automatically detected, these targets were still difficult for human users to verify in such a live-flight scenario.

These results plus those from Section 5.4 suggest the following regarding the questions identified previously:

- User detection rates increased by 16.3% from 52.6% to 61.1% when using detector-based suggestions. We would not expect user detection to increase for objects not identified and suggested by the detector, and this was indeed the case (consistently approximately 38%). For objects suggested by the detector, user detection increased 29.1% from 68.1% to 87.9%.
- Although the anomaly detector made frequent false suggestions, we did not detect a significant change in the users' false positive rates when using these suggestions. This suggests that, given modest operational limits, users were able to effectively and quickly filter out incorrect suggestions.
- It did not appear that the presence of false suggestions served as a distractor and caused the users to miss targets of interest. As shown in Table 2, their performance on non-suggested targets appeared to be unaffected by the presence of false suggestions.
- While the use of suggestions increased performance on the primary detection task, it also decreased performance on the secondary tone-counting task. This suggests that filtering false suggestions might require increased user attention. However, it might also suggest that the suggestions shifted more attention to the primary task and away from the secondary task. While the latter is a good thing in terms of the search, it is a factor that should be considered when considering other roles the video searcher might be performing. It should be further noted that 60% of the users subjectively reported that they found the overall combination of tasks easier with suggestions than without.

5.6 Study Limitations

These specific results are for a single simple interface only, and clearly many other factors might be considered. For example, the form in which the suggestions appeared in the interface—their size, coloring, or contrast—would likely affect user performance. Indeed the interface used to present the video itself would also affect the performance.

The parameters of the detector, especially the false positive rate, would also affect the amount to which suggestions aid the searcher. If the false positive rate is too high, it is likely that users would be overwhelmed with processing suggestions, resulting in both decreased detection and increased false positives. In practice, the searcher would also likely be given the ability to adjust this to suit their preferences and comfort level, but this was not included in this study.

The study was limited to fairly consistent (mid-day outdoor) illumination, though this should not generally affect the proposed methods. As long as the variation in lighting is not so extreme as to cause the object's color to no longer appear different from the surrounding environment, the anomaly detector can still function correctly. Changes in daytime lighting also affect shadows, which we found frequently appear as near-black regions of the image and comprise a sufficiently large enough fraction of the image so as to be considered part of the normal environment.

Finally, this study was limited to two search environments typical for Wilderness Search and Rescue and did not encompass a wider range of potential search environments. We have found that some types of search environments generate more potential anomalies than others or are simply more difficult to search. The findings here should generalize to search environments where the targets of interest can be distinguished from the surroundings by their coloring. The environment may consist of a variety of colors, which need not be known ahead of time. Similarly, the targets can consist of multiple colors, which also do not need to be known, as long as one or more of them is atypical of the environment. The methods here would struggle when the target was distinguished by other characteristics (shape, texture, etc.). They would also generate frequent false positives for environments that routinely contain non-target anomalies–i.e., colors that are natural to the scene but do not comprise large parts of it. For these situations, it might be beneficial to allow users to mark these irrelevant anomalies and to explicitly mask these colors in the detector.

6. CONCLUSION

This paper has presented a method for aiding users performing UAV-assisted wilderness search and rescue by incorporating a spectral anomaly detector to suggest unusually colored objects. The contributions include an empirical evaluation of candidate spectral anomaly detectors to evaluate how well each works within this domain, a method for filtering out stray anomalously colored pixels and grouping pixels into potential targets, and a simple method for presenting these potential targets as suggestions to the user.

The empirical evaluation of the various detectors shows that while all of the methods compared here can achieve nearly comparable performance with appropriate parameter tuning, the detector with RX normalization had slightly better overall performance and was much more robust to parameter settings. This robustness relative to clustering-based approaches is likely due to the dependence of these more on the size of potential objects (which stays fairly consistent under search conditions) than on the variation in ground cover (which does not and affects clustering).

The results of the accompanying user study suggest that the use of automated detectors to aid human observers of aerial search video can increase user performance in WiSAR tasks. Even with a partially effective detector (which is perhaps the best that can be hoped for in many situations) that has a much higher false-positive rate than hit rate, user detection can increase without corresponding increase in false positives. User detection of targets suggested by the anomaly detector increased while their detection of non-suggested targets was unaffected. However, the filtering of incorrect suggestions (or a potential shift of attention caused by the suggestions) might require limiting secondary tasks placed on the user.

These findings should transfer to similar search domains where objects of potential interest might be identified by unusual coloring relative to the normal search environment. Actual performance will of course vary depending on the task, the relative clarity of the targets, the detector used, and the interface used to present the suggestions, but incorporation of suggestions from an automated detector should be considered when designing interfaces for searching video.

7. ACKNOWLEDGMENTS

The work was partially supported by the National Science Foundation under grant numbers 0534736 and 0812653. Any opinions, findings, and conclusions or recommendations expressed are those of the authors and do not necessarily reflect the views of the National Science Foundation.

We would like to thank Dr. Dennis Eggett of BYU's Department of Statistics for his help with the design and statistical analysis of the user study. We would also like to thank Jim Walker, Wallace Barrus, and Craig Randall for volunteering their time and expertise with collecting the images and video used for the empirical and user studies.

8. REFERENCES

[1] E. Ashton. Detection of subpixel anomalies in multispectral infrared imagery using an adaptive Bayesian classifier. *IEEE Trans. Geoscience and Remote Sensing*, 36(2):506–517, Mar 1998.

[2] N. Billor, A. S. Hadi, and P. F. Velleman. BACON: blocked adaptive computationally efficient outlier nominators. *Computational Statistics & Data Analysis*, 34(3):279 – 298, 2000.

[3] M. Carlotto. A cluster-based approach for detecting man-made objects and changes in imagery. *IEEE Trans. Geoscience and Remote Sensing*, 43(2):374–387, 2005.

[4] Y. Caron, P. Makris, and N. Vincent. A method for detecting artificial objects in natural environments. In *16th International Conference on Pattern Recognition*, volume 1, pages 600–603, 2002.

[5] V. Chandola, A. Banerjee, and V. Kumar. Anomaly detection: A survey. *ACM Comput. Surv.*, 41(3):1–58, 2009.

[6] C.-I. Chang and S.-S. Chiang. Anomaly detection and classification for hyperspectral imagery. *IEEE Trans. Geosci. and Remote Sensing*, 40(6):1314–1325, 2002.

[7] J. Y. C. Chen. Individual differences in human-robot interaction in a military multitasking environment. *Journal of Cognitive Engineering and Decision Making*, 5(1):83–105, 2011.

[8] E. de Visser and R. Parasuraman. Adaptive aiding of human-robot teaming: Effects of imperfect automation on performance, trust, and workload. *Journal of Cognitive Engineering and Decision Making*, 5(2):209–231, 2011.

[9] E. J. de Visser and R. Parasuraman. Effects of imperfect automation and task load on human supervision of multiple uninhabited vehicles. *Human Factors and Ergonomics Society Annual Meeting Proceedings*, 51(18):1081–1085, 2007.

[10] A. P. Dempster, N. M. Laird, and D. B. Rubin. Maximum likelihood from incomplete data via the EM algorithm. *Journal of the Royal Statistical Society. Series B (Methodological)*, 39(1):1–38, 1977.

[11] S. R. Dixon and C. D. Wickens. Automation reliability in unmanned aerial vehicle control: A reliance-compliance model of automation dependence in high workload. *Human Factors*, 48(3):474–486, 2006.

[12] S. R. Dixon, C. D. Wickens, and D. Chang. Unmanned aerial vehicle flight control: False alarms versus misses. In *Proceedings of the Human Factors and Ergonomics Society 48th Annual Meeting*, 2004.

[13] R. O. Duda, P. E. Hart, and D. G. Stork. *Pattern Classification (2nd Edition)*. Wiley-Interscience, 2000.

[14] L. M. Fletcher-Heath, L. O. Hall, D. B. Goldgof, and F. R. Murtagh. Automatic segmentation of non-enhancing brain tumors in magnetic resonance images. In *Artificial Intelligence in Medicine*, pages 43–63, 2001.

[15] M. A. Goodrich, B. S. Morse, D. Gerhardt, J. L. Cooper, M. Quigley, J. A. Adams, and C. Humphrey. Supporting wilderness search and rescue using a camera-equipped mini UAV. *Journal of Field Robotics*, 25(1-2):89–110, 2008.

[16] R. C. Johnson, K. N. Saboe, M. S. Prewett, M. D. Coovert, and L. R. Elliott. Autonomy and automation reliability in human-robot interaction: A qualitative review. *Human Factors and Ergonomics Society Annual Meeting Proceedings*, 53(18):1398–1402, 2009.

[17] A. Mecocci, M. Pannozzo, and A. Fumarola. Automatic detection of anomalous behavioural events for advanced real-time video surveillance. In *International Symposium on Computational Intelligence for Measurement Systems and Applications*, pages 187–192, Jul 2003.

[18] A. Mittal and D. Huttenlocher. Scene modeling for wide area surveillance and image synthesis. In *IEEE Computer Society Conference on Computer Vision and Pattern Recognition*, pages 160–167, 2000.

[19] B. Morse, D. Gerhardt, C. Engh, M. Goodrich, N. Rasmussen, D. Thornton, and D. Eggett. Application and evaluation of spatiotemporal enhancement of live aerial video using temporally local mosaics. In *IEEE Conference on Computer Vision and Pattern Recognition*, pages 1–8, Jun 2008.

[20] H. Nanda and L. Davis. Probabilistic template based pedestrian detection in infrared videos. In *IEEE Intelligent Vehicle Symposium*, pages 15–20, 2002.

[21] C. Papageorgiou, M. Oren, and T. Poggio. A general framework for object detection. In *International Conference on Computer Vision*, pages 555–562, 1998.

[22] R. Parasuraman and D. H. Manzey. Complacency and bias in human use of automation: An attentional integration. *Human Factors*, 52(3):381–410, 2010.

[23] R. Parasuraman and V. Riley. Humans and automation: Use, misuse, disuse, abuse. *Human Factors*, 39(2):230–253, 1997.

[24] N. Rasmussen, D. Thornton, and B. Morse. Enhancement of unusual color in aerial video sequences for assisting wilderness search and rescue. In *15th IEEE International Conference on Image Processing*, pages 1356–1359, Oct 2008.

[25] I. Reed and X. Yu. Adaptive multiple-band CFAR detection of an optical pattern with unknown spectral distribution. *IEEE Trans. Acoustics, Speech and Signal Processing*, 38(10):1760–1770, Oct 1990.

[26] T. Smetek and K. Bauer. Finding hyperspectral anomalies using multivariate outlier detection. In *IEEE Aerospace Conference*, pages 1–24, Mar 2007.

[27] J. Solka, D. Marchette, B. Wallet, V. Irwin, and G. Rogers. Identification of man-made regions in unmanned aerial vehicle imagery and videos. *IEEE Trans. PAMI*, 20(8):852–857, Aug 1998.

[28] C. Stauffer and W. Grimson. Adaptive background mixture models for real-time tracking. In *IEEE Computer Society Conference on Computer Vision and Pattern Recognition*, 1999.

[29] C. A. Sugar and G. M. James. Finding the number of clusters in a dataset: An information-theoretic approach. *Journal of the American Statistical Association*, 98:750–763, January 2003.

[30] D. Thornton. Unusual-object detection in color video for wilderness search and rescue. Master's thesis, Brigham Young University, December 2010.

[31] L. Zhao and C. Thorpe. Stereo- and neural network-based pedestrian detection. *IEEE Trans. Intelligent Transportation Sys.*, 1(3):148–154, 2000.

Levels of Embodiment: Linguistic Analyses of Factors Influencing HRI

Kerstin Fischer
University of Southern Denmark
IFKI, Alsion 2
DK-6400 Sonderborg
0045-6550-1220
kerstin@sitkom.sdu.dk

Katrin Lohan
Bielefeld University
Universitätsstr. 23-25
D-33615 Bielefeld
klohan@uni-bielefeld.de

Kilian Foth
University of Hamburg
Department of Informatics
Vogt-Koelln-Str. 30
D-22527 Hamburg
foth@informatik.uni-hamburg.de

ABSTRACT
In this paper, we investigate the role of physical embodiment of a robot and its degrees of freedom in HRI. Both factors have been suggested to be relevant in definitions of embodiment, and so far we do not understand their effects on the way people interact with robots very well. Linguistic analyses of verbal interactions with robots differing with respect to physical embodiment and degrees of freedom provide a useful methodology to investigate factors conditioning human-robot interaction. Results show that both physical embodiment and degrees of freedom influence interaction, and that the effect of physical embodiment is located in the interpersonal domain, concerning in how far the robot is perceived as an interaction partner, whereas degrees of freedom influence the way users project the suitability of the robot for the current task.

Categories and Subject Descriptors
J.4 [**Computer Applications**]: Social and Behavioral Sciences

General Terms
Human Factors

Keywords
Embodiment, robot simulations, degrees of freedom, linguistic analysis, verbal human-robot interaction

1. INTRODUCTION
Previous work has identified considerable problems in defining the notion of embodiment. Ziemke [33], for instance, suggests six different dimensions along which the notion can be defined: Embodiment as structural coupling (between the organism and its environment); historical embodiment (comprising also the history of previous structural couplings); physical embodiment, accounting for the common sense notion of having a physical body; organismoid embodiment, having at least to some extent a similar form and sensorimotor equipment as living organisms; and organismic embodiment, i.e. being a living organism; and finally, social embodiment (concerning the functions of embodiment in social interaction). These dimensions reflect the different criteria discussed from a broad range of perspectives and in different disciplines, for instance, cognitive science, biology, robotics and cognitive linguistics. In contrast, other authors focus on a minimal definition of embodiment; for instance, Dautenhahn et al. [3] aim at identifying criteria that define the core of the concept. The authors suggest perturbatory bandwidth and structural variability to be central criteria; they define perturbatory bandwidth in terms of the sensitivity of the system to react to, and to act upon, its environment, i.e. in the amount of sensors and the way the information from the sensors is made use of. More sensors and more degrees of freedom thus allow a robot to interact more with its environment and hence the robot will be embodied to a greater degree. Structural variability, on the other hand, refers to the system's ability to adapt to, and change through, the interaction with its environment. Their definition allows the quantification of degrees of embodiment: $DOM_{S,E} = f(x,y,t)$, where the degree of embodiment (DOM) of a system S is defined with respect to an environment E as a function of the vectors x and y and time t. The two vectors represent perturbatory bandwidth and structural variability respectively.

Concerning the effects of these different levels of embodiment, previous work in HRI has mostly focused either on the role of anthropomorphic properties of artificial agents on the one hand, or on the role of physical embodiment on the other; regarding the former, numerous studies have shown that anthropomorphic features of robots trigger social responses from participants (e.g. Sproull et al. [29], Koda & Maes [17], Nass [26]); regarding the latter, most studies suggest that users prefer real robots over simulations and present robots over telepresent ones; thus, physical embodiment seems to play a crucial role. Other kinds and degrees of embodiment have hardly been investigated, and so far it is unclear what exactly the different findings are caused by. Consequently, we still understand little of the effects of different types and degrees of embodiment on the processes involved in human-robot interaction (cf. Kiesler et al. [16]). In this study, we approach aspects of embodiment using a relatively new methodology that has several advantages for the investigation of the effects of certain design decisions on HRI; in particular, we address the role of physical embodiment on the one hand and of different degrees of freedom of the robot on the other. We will show how users' perceptions of systems with different embodiment differ using linguistic analysis which will provide additional information on the influences of certain degrees and kinds of embodiment.

HRI'12, March 5–8, 2012, Boston, Massachusetts, USA.

2. RELATED WORK

Related work concerns, for instance, the evaluation of the robot, for instance, with respect to its enjoyability, attributions of intentions and judgments concerning its social presence. These effects are usually assessed using post-experimental questionnaires. In addition, most studies also address one or more behavioral measures which concern, for example, the ability of the robot to persuade the user to do something, to speed up the users' performance, or to facilitate learning.

The results of the different studies do not allow unanimous conclusions, however. While several studies demonstrate an effect for physical embodiment, others do not. For instance, Lohan et al. [20] investigate the effect of physical embodiment on participants' behavior in interaction and show that participants interact more with a real robot than with a simulated one in terms of eye gaze. However, studies focusing on other variables do not find consistent effects of physical embodiment. For instance, Komatsu & Abe [18] compare the effects of a simulated and a physically embodied robot in a setting in which the robots distracted the participants from their current activity. Dependent variables were whether participants followed the robots' invitations, where participants looked and their success rate in the game. There was a tendency for more participants to switch the activity with the physically embodied robot than with the simulated robot, yet there were no effects on looking times or task completion. In a follow-up study, Komatsu & Kuki [19] find manipulations of speaker expectations to influence participants' willingness to follow the simulated and the physical robot' invitations to the same degree. In a further study, Komatsu & Kuki [20] find that expectation manipulation (by means of an text stressing the life-like interactive capabilities of the robot) makes up for the initial disadvantage of the simulation. Similarly, Hoffmann & Krämer [13] find pleasantness of physical versus simulated robots to depend on the task to be solved by the user: In task-based problem solving, the physical robot was evaluated better, whereas in a conversational setting, the screen-based robot was rated higher. The authors furthermore find no differences with respect to social presence between the two conditions, yet their participants felt more control in the scenario with the agent displayed on a screen. Moreover, consistent with the other studies, there were no significant differences in participants' performance in the two tasks depending on the condition. Finally, contrary to the results by Komatsu and colleagues, in their experiment physical embodiment had no effect on the persuasiveness of the respective robot. To sum up, there seem to be effects for physical embodiment, yet they seem to be mediated by other factors.

Bainbridge et al. [1] investigate the role of the physical presence of a robot using a real and a tele-present robot in their experiments. They find in post-experimental questionnaires that participants give generally higher scores to the physically present than to the video-displayed robot, though very few differences in questionnaire items individually reached significance. Regarding the behavioral measures, they find no significant differences between conditions regarding the reciprocation of the robot's greeting, yet participants followed the robot's unusual request to trash some expensive books significantly more often in the situation in which the robot was physically present. Finally, they find more participants to respect the robot's personal space in the physical presence condition than in the video condition. These results are in contrast to the results by Kidd & Breazeal [15] who find no differences between a present robot and one whose image is displayed on the screen. Instead, they report considerable differences in the perception of physically embodied robots in comparison with an animated character. Thus, they find presence to have no impact, whereas the reality of the agent makes a difference.

Wainer et al. [31] aim to disentangle the effects of physical embodiment and presence in a task-based scenario. Their participants (mostly computer scientists and even roboticists) find the embodied robot to be more appealing and perceptive of the world than the virtual robot, and in a post-experimental questionnaire, they judge the present physically embodied robot to be more enjoyable and helpful than both the virtual and the tele-present robot. There are no significant differences between the simulated and the tele-present robot, and the present, embodied robot scores highest in all comparisons. Likewise, Kiesler et al. [16] compare four different conditions, in which the robot is a) present, b) remote and projected, c) simulated and present and d) simulated, remote and projected. They find physical embodiment to matter such that users spend more time with the physically embodied robot, disclose less and attribute more human behaviors and intentions to it, while presence has no impact on these outcomes. In contrast, presence made a difference on users' behavior (healthy eating) after the experiment and on their judgments of the robots' enjoyability.

Regarding degrees of freedom, there are no conclusive results indicating that the amount of flexibility of a robot should influence the way people interact with the respective system, either. For instance, Lusk & Atkinson [23] investigate degrees of embodiment in pedagogical agents; the dimension they address concerns the amount of movement of the animated figure. The authors compare students' learning behavior in three conditions: a multimedia environment either equipped with a locomoting agent in form of a parrot which uses gaze and gesture to direct attention, the same agent which was however minimally embodied such that it remained static on the screen, and a voice only condition with no animated character. The authors find a higher level of understanding for the static agent compared to the no agent condition and significantly better transfer of the material learned in the animated agent condition than in the no agent condition. The effects are however small, and the behavior of the agent seems to play a rather neglectable role, at least with respect to learning.

To sum up, previous work shows some tendencies regarding possible effects of physical embodiment, realism, and degrees of freedom, yet there is no conclusive picture emerging. Thus, a qualitative methodology targeting users' cognitive representations of their artificial communication partner may be useful to investigate the role of different degrees and kinds of embodiment.

3. METHODOLOGY

Since the methodology proposed is rather unusual in HRI research, we describe it in some detail.

The key principle exploited in this method is the notion of recipient design (Sacks et al. [28]). People have been found to choose the linguistic features of their utterances on the basis of their understanding of the needs of the current communication partner; for instance, in child-directed speech, caregivers adjust their utterance length to the child's knowledge of the words within that utterance, thus taking considerable knowledge about their communication partner into account (Roy et al. [27]). Similarly, caregivers adjust syntactic, semantic and pragmatic features of their utterances to their children's respective receptive capabilities (Cross et al. [2]). Concerning speech to robots, recent studies also show that users' linguistic choices in interaction are correlated with their understandings of these robots (Fischer [7], Fischer et al. [8]). Given that speakers design their utterances so that they are well suited for the particular communication partner in the current situation, investigating the linguistic choices speakers make can inform us about what users judge the communication partner to have problems with and what they consider them to be good at. Thus, the association of particular linguistic features with their functions in interaction provides us with qualitative data on speakers' mental models of their artificial communication partners.

The method proposed has a number of advantages:

- it relies on objectively observable behavior and avoids the pitfalls of self-reports;
- it is comprehensive, broad and well suited both for discovery and for hypothesis testing;
- it measures users' behavior in the moment; Fussel et al. [10], for instance, show that ratings of anthropomorphism may differ considerably depending on whether they are elicited during interactions or afterwards (see also Takayama [30]);
- and it is online and can thus be exploited for adapting to the respective user once the relationship between linguistic features and user preferences has been established.

The methodology therefore allows us not only to identify different behaviors as effects of different degrees and types of embodiment, but it provides us also with information on the functional effects of different degrees and types of robot embodiment.

The methodology employed consists of three parts: 1) data elicitation in controlled experimental human-robot interaction scenarios; 2) quantitative analysis of the linguistic features occurring; 3) qualitative analysis of the functions the linguistic choices users make fulfill in the respective data set.

Data elicitation takes place in experimental settings that are controlled concerning all external factors and differ only with respect to the variable investigated. Within the data elicitation sessions, participants are not restricted in what they say or how they behave. They are asked to interact verbally with the robot, irrespective of whether or not the robot produces utterances (yet) itself (in the current scenario this is particularly credible since the robots employed were designed to resemble young children).

The data analysis concerns the linguistic properties of participants' utterances in the interaction with the robot. These properties serve as dependent variables and provide the analyst with evidence for particular ways of understanding the affordances of the respective robot, tasks and situation. Thus, the linguistic choices participants make reveal the mental models they build up of the robot as a communication partner during the interaction.

The data are recorded and subsequently transcribed, that is, an orthographic version of the spoken utterances is created. This work is typically done by assistants who are ignorant about the goals of the study, and transcriptions are usually crosschecked in order to guarantee a trustworthy rendering of the spoken data.

The quantitative data analysis is done half-automatically using computational tools. The qualitative analysis is based on linguistic research on the functions of the linguistic features in question.

4. EMPIRICAL STUDY

The study comprises the elicitation of human-robot interactions in three different conditions that differ concerning the robots involved. In the current study, the independent variable concerns the kind of robot and the degrees of freedom of the robots employed. The different data sets were thus elicited with the same tasks the participants had to fulfill, namely to explain the use of some household objects (e.g. a lamp, a salt carrier, a bell) and some toys (e.g. cups, blocks and a bag with rings) to the respective robot. The dependent variables investigated are users' spontaneous linguistic behaviors in these tasks.

4.1 Robots

The first robot (henceforth Akachan) is simulated such that it is presented on a screen. However, it interacts with its environment by means of eye-gaze, which is driven by a saliency model (Nagai & Rohlfing [25]). This visual attention module detects salient locations in a scene which stand out in terms of color, intensity, changes in brightness and motion. The calculation of the salient point follows the model by Itti et al. [14]. The robot, originally called 'Babyface', was designed to resemble an infant, yet in order not to influence participants, we used the Japanese translation 'Akachan' to refer to it in the experiments. In addition to the saliency-based eye-gaze, the robot blinks randomly with its eye lids and opens its mouth in a random fashion.

The second robot (henceforth iCub I) is the humanoid robot iCub (Metta et al. [24]), which is physically embodied with 53 degrees of freedom, yet to make it comparable to Akachan in Condition 1, it was enabled only to move its eyes. Its gaze behavior was controlled by the same mechanism as Akachan in the first condition.

Figure 1: left: Babyface/Akachan; mid: iCub; right: a participant explaining how to stack cups

The third robot (henceforth iCub II) is again iCub, but here the robot is allowed to move its head together with its eyes. Thus, of the 53 degrees of freedom available to the robot, only those controlling the eyes, the head and the neck were enabled. In this way, the robot exhibits exactly the same functionality as the other two robots – it gazes at a salient object or movement, yet it does so by moving both head and eyes. The saliency module was the same as in the other two conditions. The coordination between eye and head movements was modeled based on Lopes et al. [22].

4.2 Participants

In the first condition, 30 participants interacted with Akachan. In the second condition, eight participants interacted with iCub I (which was only moving its eyes), and six participants interacted with iCub II (which was moving its eyes and head). All participants were students or staff from Bielefeld University from various different disciplines. Participants were between 17 and 63 years old and recruited by word-of-mouth.

4.3 Data Elicitation

Data elicitation proceeded in three different conditions: In condition 1, participants were seated at a table across a screen showing the simulated robot Akachan. The experimenter placed objects on the table in front of the participant, which the participant then had to explain to the robot. The robot followed the instructions with its eyes.

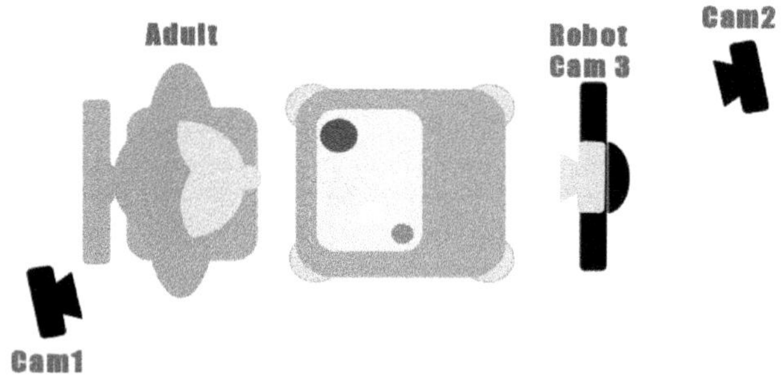

Figure 2: Experimental set-up for Condition 1 using the simulated robot Akachan

In the second condition, participants were seated across the table of the humanoid robot iCub. Again, objects were placed in front of the participant with the request to explain these objects to the robot. As in Condition 1, the robot follows the instructions by means of eye gaze.

The set-up for the third condition is identical to the second condition with the only difference that the robot (iCub II) in this condition moves eyes and head together.

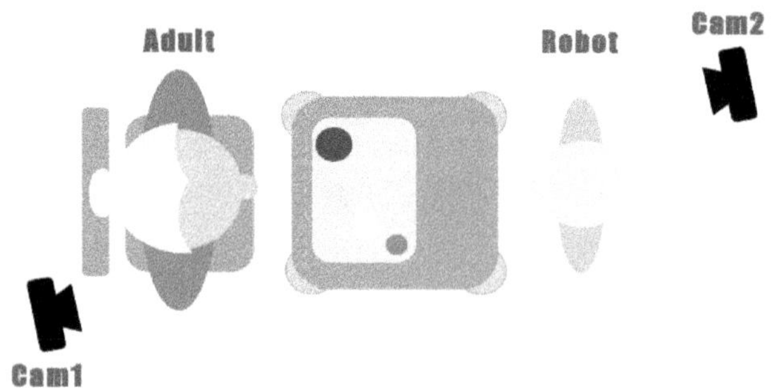

Figure 3: Experimental set-up for Conditions 2 and 3 using the humanoid robot iCub

4.4 Data Analysis

The utterances produced by the participants were manually transcribed and syntactically analysed. The linguistic analysis was carried out using the constraint-based parser described in Foth et al. [9]. This system performs morphological classification and syntactic and referential dependency analysis on the word level and assigns every dependency to one of 35 syntactic classes. The output format allows the quick computation of basic frequency counts such as mean length of utterance (MLU) or category distribution, but also supports searches among inflected words for their stems, or for the syntactic roles of words. The label set employed allows distinctions such as those between subjects, direct objects, and indirect objects, or between active and passive voice, to be retrieved easily. To rule out distortions of the results due to any systematic imperfections of the parsing accuracy, all analyses were fully verified for correctness manually. The linguistic analysis concerns three different factors: verbosity, complexity and interactivity.

Linguistic verbosity

The first general property investigated concerns the amount of speech presented to a communication partner, i.e. linguistic verbosity. The verbosity measures tell us about how much effort speakers spend on each task and how much information they consider suitable or necessary for their communication partner to understand, thus providing indirect information about speakers' recipient design for their respective communication partners. Moreover, the number of different words tells us about the suspected competence level of the communication partner. Thus, to begin with, for each corpus we counted the total number of words for further analyses and the number of different words per speaker in each of the six tasks (diversity) as well as number of utterances per task.

Complexity of utterances

The second measure concerns the complexity of utterances; a very common measure of sentence complexity is the **MLU**, the mean length of utterance. To calculate the MLU, we simply divided the number of words per speaker by the number of utterances by the same speaker. By utterance we understand all turn-constructional units, that is, units consisting of clause complexes, of single clauses, but also smaller units, such as noun, verb or prepositional phrases, answer particles and feedback signals that occur independently (Sacks et al. [28]).

Another measure of complexity, and at the same time a feature revealing the suspected competence of the communication partner, concerns the **concreteness** versus **abstractness** of terms used, for instance, *cup*, *bowl*, or *block* versus *object, container* or *obstacle* (cf. Fischer [5]).

Furthermore, some structures are more complex than others. The **passive**, for instance, is a structure that introduces a perspective in which the patient or undergoer of an action is foregrounded and the agent is backgrounded. The construction is also formally quite complex and thus a useful indicator for assumed competence.

Sentence complexity is also reflected in the number and type of objects used. In particular, we distinguish between **direct objects, indirect objects** and **object complement clauses**, for instance, *she hit it, she gave him the ball*, and *she said that it is sad*, respectively. As, for instance, Hawkins [12] shows, these three types of objects exhibit increasing degrees of complexity.

Relative clauses, such as *the man who walks on the other side of the street is my uncle*, have been found to be good indicators of suspected partner competence and linguistic proficiency; thus, in human-robot interaction speakers only use **relative clauses** if they are certain to be understood or if their partner uses them as well (Fischer [5]). We therefore take uses of relative clauses as an indicator for complexity here.

Embedding is a composite feature, combining all structures that can be embedded in the main sentence structure, such as relative clauses, object complement clauses, dependent main clauses, subclauses, appositions, infinitival complements, and subject clauses. In particular, we use the following definitions: Subclauses are subordinate clauses like whenever he goes to school, *he feels sick*. Appositions are added elements, such as *see the button, the red one*. An example for an infinitival complement is *she wants to go* and for a subject clause *what she really wants is love*.

Interpersonal function

A third property concerns the amount of social information used and the degree with which speakers involve their communication partner. One such feature concerns the sentence type, in particular, imperative, declarative, interrogative or infinitive mood. The declarative is generally used to make assertions. Furthermore, instructions by means of **declarative** sentences are very common, thus avoiding that the speaker directly imposes his or her wishes onto the communication partner, as it is the case with a simple **imperative**, such as, for instance, *move*! In German, imperatives are however often toned down by means of **modal particles**, sentence medial particles that serve politeness and grounding functions (cf. Fischer [6]). In the current data sets, the down-toned imperative occurs frequently in expressions with attention getting functions, such as *guck mal* (*look*).

In situations without a concrete addressee (cf. Halliday & Matthiessen [11]), or with a highly unfamiliar addressee, such as a computer or robot (cf. Fischer [5]), instructions and explications using the **infinitive** are very common, for instance: *den blauen nehmen*; this corresponds roughly to the English use of the gerund, as in, for example, *no smoking*.

Moreover, speakers can ask **questions** to involve their addressees, or they can use understanding **checks**, such as tag questions like *doesn't it* or *don't you* in English and *ne?* in German.

Also, personal pronouns are useful indicators of the relationship between speakers in a communicative situation. For instance, speakers may avoid addressing the partner, using the impersonal form *man* (*one*). Alternatively, speakers can address their partner using *du* (*you*), or they can refer to themselves with or without including the partner, using either *ich* (*I*) or *wir* (*we*). Similarly revealing regarding the degree with which the communication partner is involved is the use of the **vocative**, for instance, the partner's first name.

The absolute occurrences of these features, besides the verbosity features, were counted per person in these conditions and divided by the number of utterances used by this person. The numbers underlying the statistical comparison are thus the numbers occurring per number of speakers' utterances.

5. RESULTS

The analyses reveal that many linguistic features remain the same in all conditions. Especially regarding the amount of speaking, i.e. the verbosity measures, there are no differences, and also most complexity measures yield very similar results across conditions. There are, however, significant differences between all three conditions, as well as differences that only hold between two of the three conditions, pointing to different effects of physical embodiment on the one hand and degrees of freedom on the other.

Table 1: ANOVA Results for all Linguistic Features; * = p< .05, ** = p< .001; ᵗ = p< .20

	Akachan	iCub I	iCub II	F
diversity	7.9 *(0.2)*	8.9 *(0.1)*	6.7 *(0.1)*	1.30383
MLU	8.4 *(2.7)*	8.2 *(4.0)*	6.0 *(2.3)*	1.740385 ᵗ
concrete	0.75 *(0.34)*	1.0 *(0.6)*	0.9 *(0.4)*	0.27753
abstract	0.05 *(0.06)*	0.06 *(0.09)*	0.02 *(0.02)*	0.80275
passive	0.41 *(0.08)*	0.42 *(0.04)*	0.008 *(0.01)*	0.61220
direct obj	0.03 *(0.03)*	0.46 *(0.2)*	0.25 *(0.1)*	49.97369**
ind.object	0.03 *(0.05)*	0.03 *(0.07)*	0.07 *(0.02)*	0.39219
embedding	0.19 *(0.09)*	0.22 *(0.17)*	0.11 *(0.10)*	1.42739
subclauses	0.11 *(0.07)*	0.13 *(0.10)*	0.07 *(0.05)*	1.18885
rel.clause	0.03 *(0.03)*	0.03 *(0.02)*	0.005 *(0.01)*	1.34224
copula	0.04 *(0.04)*	0.19 *(0.16)*	0.19 *(0.1)*	17.04208**
declarative	0.92 *(0.3)*	1.11 *(0.6)*	0.76 *(0.3)*	1.93717 ᵗ
infinitive	0.06 *(0.05)*	0.08 *(0.1)*	0.09 *(0.09)*	0.61691
imperative	0.02 *(0.03)*	0.17 *(0.07)*	0.08 *(0.07)*	2.24545 ᵗ
mod.particle	0.10 *(0.07)*	0.21 *(0.15)*	0.19 *(0.12)*	4.71764*
question	0.04 *(0.05)*	0.03 *(0.05)*	0.01 *(0.02)*	0.91876
check	0.001 *(0.0)*	0.002 *(0.0)*	0.02 *(0.04)*	3.29369*
vocative	0.01 *(0.03)*	0.04 *(0.05)*	0.16 *(0.14)*	16.84286**
du (*you*)	0.03 *(0.09)*	0.06 *(0.07)*	0.02 *(0.04)*	0.51796
man	0.20 *(0.2)*	0.10 *(0.2)*	0.15 *(0.2)*	1.13165
ich (*I*)	0.15 *(0.15)*	0.17 *(0.17)*	0.12 *(0.08)*	1.12149
wir (*we*)	0.04 *(0.06)*	0.12 *(0.15)*	0.04 *(0.07)*	2.87466

It turns out that the vocative, i.e. the use of the robot's name, differs significantly in all conditions (see Figure 4). The mean in Condition 1 is M= 0.01 (sd= 0.03), in Condition 2 M= 0.04 (sd=0.05) and in Condition 3 M=0.16 (sd=0.13), t (Condition 1 vs. 2) = -2.29, p < .03; t (Condition 1 vs. 3) = -5.69, p < .001; t (Condition 2 vs. 3) = -2.26, p < .05. The results show that the number of instances of the robot's name increases with increasing physical embodiment of the robot and with increasing degrees of freedom.

We turn now to paired comparisons between conditions. Figures 4 to 6 illustrate the different possible distributions between the linguistic features investigated.

5.1 Simulated versus physical robot (eyes only)

The first comparison concerns the role of physical embodiment, i.e. Condition 1 (Akachan) versus Condition 2 (iCub I). Several features can be found that are influenced by the robot's physical embodiment; these are the number of instances of the personal pronoun 'we' (Condition 1 M=0.04, sd=0.064; Condition 2: M=0.12, sd=0.15, t=-2.29, p< .03), the number of modal particles

(Condition 1 M=0.10, sd=0.07; Condition 2 M=0.21, sd=0.15, t=-2.73, p< .01), the number of direct objects (Condition 1 M=0.03, sd=0.03; Condition 2 M=0.46, sd=0.24, t=-9.72, p< .001), and the number of utterances containing a copula, i.e. a form of 'to be' as the main verb of the sentence (Condition 1 M=0.04, sd=0.04; Condition 2 M=0.21, sd=0.16, t=-4.99, p< .001, see also Figure 5). Furthermore, there is a tendency (t=1.96, p= .058) for more imperatives in Condition 2 (M=0.05, sd=0.07) compared to Condition 1 (M=0.02, sd=0.02).

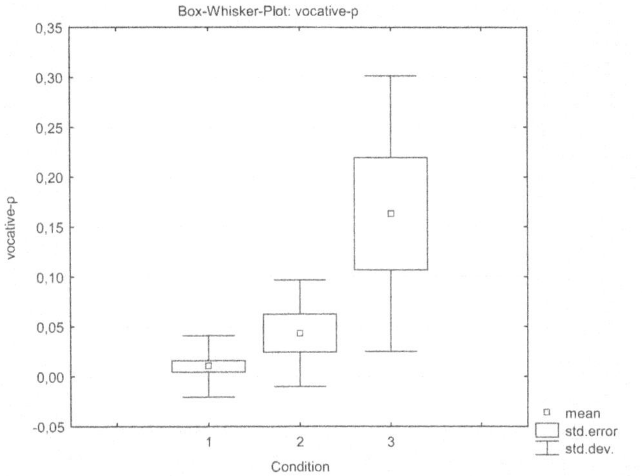

Figure 4: Results for the vocative in the three conditions

5.1 Simulated versus physical robot (eyes only)

The first comparison concerns the role of physical embodiment, i.e. Condition 1 (Akachan) versus Condition 2 (iCub I). Several features can be found that are influenced by the robot's physical embodiment; these are the number of instances of the personal pronoun 'we' (Condition 1 M=0.04, sd=0.064; Condition 2: M=0.12, sd=0.15, t=-2.29, p< .03), the number of modal particles (Condition 1 M=0.10, sd=0.07; Condition 2 M=0.21, sd=0.15, t=-2.73, p< .01), the number of direct objects (Condition 1 M=0.03, sd=0.03; Condition 2 M=0.46, sd=0.24, t=-9.72, p< .001), and the number of utterances containing a copula, i.e. a form of 'to be' as the main verb of the sentence (Condition 1 M=0.04, sd=0.04; Condition 2 M=0.21, sd=0.16, t=-4.99, p< .001, see also Figure 5). Furthermore, there is a tendency (t=1.96, p= .058) for more imperatives in Condition 2 (M=0.05, sd=0.07) compared to Condition 1 (M=0.02, sd=0.02).

5.2 Physical robot (eyes only) versus physical robot (eyes and head)

The comparison between Conditions 2 and 3, i.e. between iCub I and iCub II, reveals the impact of the robot's head movements in Condition 3. Significant differences that point to an influence of the amount of degrees of freedom of the robot concern, besides the vocative, also the amount of passive constructions employed (Condition 2 M=0.04, sd=0.03; Condition 3: M=0.08, sd=0.01, t=2.19, p< .05). Thus, there are many more instances of the passive construction in Condition 2, the condition with iCub I, which is only using its eyes. Furthermore, there is a statistical tendency towards more direct objects (Condition 2 M=0.46, sd=0.24; Condition 3: M=0.26, sd=0.12, t=1.87, p< .09).

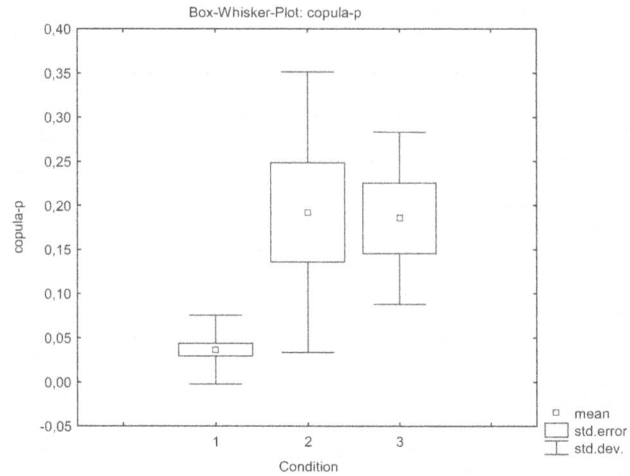

Figure 5: Results for the copula in the three conditions

5.3 Simulated versus physical robot (eyes and head)

Finally, we compare Condition 1, interactions with Akachan, with Condition 3, interactions with iCub II, who is moving both eyes and head. Significant differences concern, besides the vocative, also the amount of understanding checks employed (Condition 1 M=0.001, sd=0.004; Condition 3: M=0.017, sd=0.04, t=-2.35, p< .03). Furthermore, there are significantly more direct objects in Condition 3 than in Condition 1 (Condition 1 M=0.03, sd=0.03; Condition 3: M=0.25, sd=0.12, t=-9.35, p< .001). There is moreover a tendency for increased uses of the imperative in Condition 3 (Condition 1 M=0.016, sd=0.03; Condition 3: M=0.045, sd=0.07, t=-1.80, p= .08). There are also significantly more uses of the copula in Condition 3 than in Condition 1 (Condition 1 M=0.037, sd=0.04; Condition 3: M=0.19, sd=0.05, t=-6.43, p< .001). In addition, there are significantly more modal particles in Condition 3 than in Condition 1 (Condition 1 M=0.10, sd=0.07; Condition 3: M=0.19, sd=0.12, t=-2.44, p< .02). Finally, the evaluation also shows that users' mean length of utterance is significantly shorter in the third condition than in the other two conditions (Condition 1: M= 8.4, sd 2.7; Condition 2: M= 8.2, sd= 4.0; Condition 3: M= 6.0, sd= 2.3; t= 0.167, p< .05; see also Figure 6).

6. DISCUSSION

The distribution of the results is readily interpreted in the light of the functions the respective linguistic features fulfill in interaction. Notable is first that with the exception of the number of direct objects and the MLU, none of the linguistic features concerning linguistic complexity were found to differ significantly between the conditions. Instead, the differences concern interactional features of language: the use of the robot's name to direct its attention, the use of the inclusive personal pronoun 'we', the use of modal particles which indicate common ground between speaker and hearer (Diewald [5], Fischer [7]), the use of understanding checks and the tendency to use imperatives, a grammatical form that encodes the addressee directly. These features all concern interpersonal relationships, and the different amounts with which they are being produced indicate increasing amounts of interactivity over the three conditions, with the more embodied robot iCub II being taken as a more serious communication partner than the simulated robot.

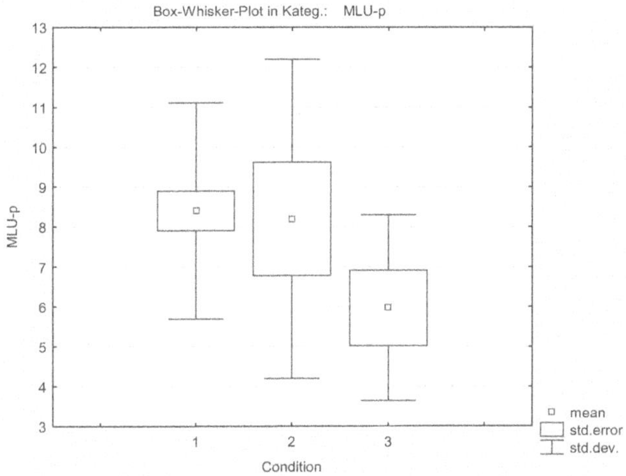

Figure 6: Results for Mean Length of Utterance (MLU) in the three conditions

The use of the copula and the direct object do not seem to fit in this pattern; however, in the current dialogs, the copula is used to introduce an object before something is asserted about it. For instance, participants say something like "iCub, this is a lamp, and to switch it on, you pull the string." The use of the copula 'is' is thus an indicator for the use of a tutoring strategy by means of which a task is decomposed into smaller steps. The same holds for the direct objects: Sentences with a single agent and a direct object are very straightforward and easy to understand and they constitute the most common utterances containing a verb in child-directed speech (Zeschel [32]).

The passive however is often an indicator for problems concerning the interpersonal relationship because it focuses on an event without stating the agent, for instance: "a lamp is switched on by pulling the string." In this utterance using the passive, a direct address of the communication partner is avoided (in contrast to, for instance, "you pull the string" or even "pull the string"). Increased use of the passive thus indicates that participants are uncomfortable with their communication partner.

The functional analysis of the choices made by the participants thus points to differences in the interpersonal relationships participants build up with the robots. This finding corresponds to previous findings concerning the 'enjoyability' or 'pleasantness' of physically embodied robots, which have been found in post-experimental questionnaires. However, while post-experimental questionnaire studies do not allow any conclusions with respect to why participants find the interactions more enjoyable, the current results show that these differences are likely to be due to the interpersonal relationship with the communication partner such that the robot is accepted as of the same kind ('we'), that tutor and robot share common ground (modal particles) and that the robot is an independent entity with a name who can be summoned in order to draw its attention (vocatives).

Furthermore, the results show that not only physical embodiment matters but also the robot's degrees of freedom; the linguistic features differing between the simulation and the iCub with head movement condition show that a physically embodied robot that uses its body at least to some degree is considered to be more likely to profit from the tutoring than a simulated robot or a robot that does not make use of its body in a way that suggests task fulfillment. Thus, linguistic differences between the iCub with head condition and the other two conditions concern, besides interpersonal differences, also task decomposition and presentation (mean length of utterance, direct objects, understanding checks). Thus, while the degrees of freedom of the agent in the study by Lusk & Atkinson [23] did not contribute to task completion and hence had no impact on the interaction, whether the robot can move its body is crucial concerning the relevance of teaching it to stack cups onto each other. Thus, the suggestion made here is that the number of degrees of freedom matters to participants in this study because the robot's movements are indicative of the robot's capabilities and thus its credibility as a partner for the task at hand.

7. CONCLUSION

In this study, we have employed a new methodology to address the effect of different degrees and types of embodiment of a given system on human-robot interaction. The method employed allowed us to identify significantly different behaviors in interaction with the three systems investigated. Moreover, the different linguistic behaviors observed could be related to different ways of understanding the robotic agents and thus to the factors determining the influence of degrees of a robot's embodiment on human-robot interaction. In particular, it was found that the two factors studied influence both the interpersonal relationship between human user and the robot and the amount of tutoring the robot received.

8. DESIGN IMPLICATIONS

The current investigation has shown that not only the robot's physical embodiment, but also its degrees of freedom influence human-robot interaction. The robot behaviors displayed were found to be relevant especially in the light of the tasks it is meant to fulfill. Thus, the robot should use its degrees of freedom in a way that is in accordance with its capabilities.

ACKNOWLEDGMENTS

This research was partially funded by the European Union in the framework of the ITALK project under grant number 214668.

REFERENCES

[1] Bainbridge, W.A., Hart, J.W., Kim, E.S. and Scasselati, B. 2010. The Benefit of Interactions with Physically Present Robots over Video-Displayed Agents. *International Journal of Social Robotics* 1-2.

[2] Cross, T., Johnson-Morris, J.E. and T.G. Nienhuys. 1980. Linguistic Feedback and Maternal Speech: Comparisons of Mothers Addressing Hearing and Hearing-impaired Children. *First Language* 1, 3:163-189.

[3] Dautenhahn, K., B. Ogden and T. Quick. 2002. From Embodied to Socially Embedded Agents – Implications for Interaction-aware Robots. *Cognitive Systems Research* 3: 397-428.

[4] Diewald, G. 2006. Discourse Particles and Modal Particles as Grammatical Elements. In Fischer, K. (ed.): *Approaches to Discourse Particles*. Amsterdam: Elsevier.

[5] Fischer, K. 2006. *What Computer Talk is and Isn't: Human-Computer Conversation as Intercultural Communication.* AQ, Saarbrücken.

[6] Fischer, K. 2007. Grounding and common ground: Modal particles and their translation equivalents. In A. Fetzer and K. Fischer (eds.), *Lexical Markers of Common Grounds*. Studies in Pragmatics 3. Amsterdam: Elsevier.

[7] Fischer, K. 2011. Interpersonal Variation in Understanding Robots as Social Actors. *HRI'11*, Lausanne, Switzerland.

[8] Fischer, K., Foth, K., Rohlfing, K. and B. Wrede,. 2011. Mindful Tutors: Linguistic Choice and Action Demonstration in Speech to Infants and a Simulated Robot. *Interaction Studies* 12, 1: 134-161.

[9] Foth, K., Menzel, W. and Schröder, I. 2000. A Transformation-based Parsing Technique with Anytime Properties. *4th Int. Workshop on Parsing Technologies,* IWPT-2000, pp. 89-100.

[10] Fussell, S.R., Kiesler, S., Setlock, L.D. and Yew, V. 2008. How people anthropomorphize robots. *Proceedings of Human-Robot Interaction 2008* (pp. 145-152). NY: ACM Press.

[11] Halliday, M.A.K. and Matthiessen, C. 2004. *Introduction to Systemic Functional Grammar*. London: Arnold.

[12] Hawkins, J.A. 1994. *A Performance Theory of Order and Constituency*. Cambridge University Press.

[13] Hoffmann, L. and Krämer, N.C. 2011. How Should an Artificial Entity be Embodied? Comparing the Effects of a Physically Present Robot and its Virtual Representation. *Proceedings of the HRI 2011 Workshop on Social Robotic Telepresence*, Lausanne, Switzerland, pp. 14-20.

[14] Itti, L., Koch, C. Niebur, L. et al. 1998. A model of saliency-based visual attention for rapid scene analysis. *IEEE Transactions on pattern analysis and machine intelligence*, vol. 20, no. 11, pp. 1254–1259.

[15] Kidd, C. and Breazeal, C. 2004. Effect of a Robot on Uer Perceptions. *IEEE/RSJ International Conference on Intelligent Robots nd Systems*, pp. 3559-3564.

[16] Kiesler, S., Powers, A., Fussell, S. and Torrey, C. 2008. Anthropomorphic Interactions with a Robot and Robot-like Agent. *Social Cognition* 26, 2: 169-181.

[17] Koda, T. and P. Maes. 1996. Agents with Faces: The Effect of Personification. *Fifth IEEE International Workshop on Robot and Human Communication.*

[18] Komatsu, T. and Abe, Y. 2008. Comparing an On-Screen Agent with a Robotic Agent in Non-Face-to-Face Interactions. In Prendinger, H., Lester, J. and Ishizuka, M. (eds.), *IVA 2008*. Berlin, Heidelberg: Springer, pp. 498-504.

[19] Komatsu, T. and Kuki, N. 2009a. Can Users React toward an On-Screen Agent as if they are Reacting toward a Robotic Agent? *HRI'09*, La Jolla, CA, pp. 217-218.

[20] Komatsu, T. and Kuki, N. 2009b. Investigating the Contributing Factors to Make Users React toward an On-Screen Agent as if they are Reacting toward a Robotic Agent. *18th IEEE International Symposium on Robot and Human Interactive Communication*, Toyama, Japan.

[21] Lohan, K., Gieselmann, S., Vollmer, A.L., Rohlfing, K. and B. Wrede. 2010. Does Embodiment Affect Tutoring Behavior? *Proceedings of ICDL'10.*

[22] Lopes, M., Bernardino, A., Santos-Victor, K. Rosander, and C. von Hofsten,. 2009. Biomimetic Eye-Neck Coordination. *ICDL 2009.*

[23] Lusk, M.M. and R.K. Atkinson. 2007. Animated Pedagogical Agents: Does their Degree of Embodiment Impact Learning from Static or Animated Worked Examples? *Applied Cognitive Psychology* 21: 747-764.

[24] Metta, G. Sandini, G. Vernon, D. Natale,, L. and F. Nori,. 2008. The iCub humanoid robot: an open platform for research in embodied cognition. *PerMIS: Performance Metrics for Intelligent Systems Workshop.* Aug, 2008, pp. 19–21.

[25] Nagai, Y. and Rohlfing, K.J. 2009. Computational analysis of Motionese toward scaffolding robot action learning. *IEEE Transactions on Autonomous Mental Development* 1: 44-54.

[26] Nass, C. 2004. Etiquette Equality: Exhibitions and Expectations of Computer Politeness. *Communications of the ACM* 47, 4: 35-37.

[27] Roy, B., Frank, M.C. and Roy, D. 2009. Exploring word learning in a high-density longitudinal corpus. *Proceedings of the 31st Annual Meeting of the Cognitive Science Society.* Amsterdam, Netherlands.

[28] Sacks, H., Schegloff, E.A. and Jefferson, G. 1974. A Simplest Systematics for the Organization of Turn Taking for Conversation. *Language* 50: 696–735.

[29] Sproull, L., Subramani, R., Kiesler, S., Walker, J., and Waters, K. 1996. When the interface is a face. *Human-Computer Interaction* 11, 97-124.

[30] Takayama, L. 2009. Making Sense of Agentic Objects and Teleoperation: In-the-moment and Reflective Perspectives. *Late Breaking Results of Human Robot Interaction* (HRI), San Diego, CA, pp.239–240.

[31] Wainer, J., Feil-Seigfer, D.J., Shell, D.A. and Mataric, M.J. 2007. Embodiment and Human-Robot Interaction: A Task-Based Perspective. *16th IEEE International Conference on Robot and Human Interactive Communication*, Jeju, Korea.

[32] Zeschel, A. forthcoming. Basic Argument Structure Constructions in Child-Directed Speech. *Journal of Child Language.*

[33] Ziemke, T. 2003. What's that Thing Called Embodiment? In: Alterman and Kirsh (eds.) *Proceedings of the 25th Annual Conference of the Cognitive Science Society* (pp. 1134-1139). Mahwah, NJ: Lawrence Erlbaum.

Tell me when and why to do it!

Run-time Planner Model Updates Via Natural Language Instruction

Rehj Cantrell
Cognitive Science Program
Indiana University
Bloomington, IN
rcantrel@indiana.edu

Kartik Talamadupula
Dept. of Computer Science
Arizona State University
Tempe, AZ
krt@asu.edu

Paul Schermerhorn
Cognitive Science Program
Indiana University
Bloomington, IN
pscherme@indiana.edu

J. Benton
Dept. of Computer Science
Arizona State University
Tempe, AZ
j.benton@asu.edu

Subbarao Kambhampati
Dept. of Computer Science
Arizona State University
Tempe, AZ
rao@asu.edu

Matthias Scheutz
HRI Laboratory
Tufts University
Boston, MA
mscheutz@cs.tufts.edu

ABSTRACT

Robots are currently being used in and developed for critical HRI applications such as search and rescue. In these scenarios, humans operating under changeable and high-stress conditions must communicate effectively with autonomous agents, necessitating that such agents be able to respond quickly and effectively to rapidly-changing conditions and expectations. We demonstrate a robot planner that is able to utilize new information, specifically information originating in spoken input produced by human operators.

We show that the robot is able to learn the pre- and postconditions of previously-unknown action sequences from *natural* language constructions, and immediately update (1) its knowledge of the current state of the environment, and (2) its underlying world model, in order to produce new and updated plans that are consistent with this new information. While we demonstrate in detail the robot's successful operation with a specific example, we also discuss the dialogue module's inherent scalability, and investigate how well the robot is able to respond to natural language commands from untrained users.

Categories and Subject Descriptors

I.2.7 [**Artificial Intelligence**]: Natural Language Processing—*Language Parsing and Understanding*; I.2.8 [**Artificial Intelligence**]: Problem Solving, Control Methods, and Search—*Plan execution, formation, and generation*

General Terms

Design

Keywords

HRI Communication, Language processing, Learning about the environment, Learning from interaction, Enabling technologies

1. INTRODUCTION

Robots are currently being used in and developed for critical HRI applications (e.g., search and rescue) in which they must smoothly adapt to new information, and communicate quickly and effectively with human operators despite serious challenges imposed by the task and environment. These scenarios tend to induce high stress levels in human operators, while rapidly-changing conditions and expectations combine to increase operators' cognitive loads. If robots are be truly helpful in such scenarios, they must be able to communicate naturally and react flexibly to dynamic conditions by, for example, updating previous goals and accepting new goals according to changing information about the world. In this paper, we describe a robot architecture capable of understanding new action sequences presented via natural language and integrating the new knowledge to allow the robot to reason about the new abilities (e.g., in problem-solving).

The architecture described here features a planner that utilizes new information, such as state updates from sensory processing, at run time (e.g., to generate a plan to enter and explore a room when informed of a newly-detected open doorway). We outline extensions to the planner that allow us to update its list of available actions (e.g., to include a new capability to enter rooms by pushing closed doors). We then demonstrate that the robot is able to learn the preconditions for and the postconditions of previously-unknown action sequences from a natural language description (e.g., "if you are at a closed door and you push it one meter, you will be in the room"), and immediately apply this knowledge to generate and update plans.

In previous work [3, 2] we described a natural language human-robot interaction system that uses a trainable parser to operate on spoken natural language input, producing incrementally and in parallel (1) a semantic representation, and (2) identification of and information about new word forms. In this paper, we expand on this in two key ways. First, the robot's learning capabilities are expanded to include learning pre- and postconditions of an action. For example, given a robot's limited conceptual knowledge, it may not be immediately clear that, when one is in the hallway and goes through a door, one is then in a room. Our method allows the robot to accept information given verbally, which a planner can then combine with the robot's existing knowledge in order to produce logical insights. Second, our method of referent grounding is expanded to allow the robot to infer the existence of entities that have not been explicitly mentioned. For example, when the robot knows that doors are connected to rooms, mention of a door implies that there is also a room.

The structure of the paper is as follows. First we summarize relevant previous work and provide an overview of the system in Section 2. Next we describe the relevant modules in detail: in Section 3, we detail the dialogue module using an example, and in Section 4, we discuss the robot's planner. This is followed by, in Section 5, a brief evaluation of the informal application of the system to human language elicited from naive subjects. Finally, we discuss our conclusions and future work in Section 6.

2. BACKGROUND

Planners have improved significantly in terms of scalability and speed (c.f. [12]), but this has come at the cost of standardizing assumptions about the nature of the input. In particular, modern planners typically assume full up-front knowledge of the world as well as of the underlying model that the robot is subject to—the *closed world and model* assumption (see section 4.2 for a detailed discussion of this). However, this assumption is unsuited for dynamic scenarios like search and rescue, where the world is constantly changing with new objects and conditions continually being discovered. Moreover, even the model is subject to change (e.g., when new capabilities are learned at run time). A number of systems have considered execution monitoring and plan repair or replanning upon the discovery of an inconsistent execution state, notably [11] and [7] among others. There has also been some work on systems that control service robots, the most relevant being [1]. Yet, no robotic architecture has the ability to integrate new discoveries at all levels—including run-time model updates in the planner—in order to take best advantage of the new information.

Similarly, many natural language processing systems require that all necessary knowledge be provided at startup. However, a number of projects have begun to investigate robotic architectures that can learn actions and action sequences. For example, [14, 18, 13] learn action sequences and a single associated precondition: the environment in which the sequence is learned is assumed to be the only environment in which the sequence is valid. While this assumption may be sufficient for certain tasks and in certain domains (e.g., kitchen tasks tend to be performed only in the kitchen) it limits the usefulness of the learned sequence (e.g., if the robot must be explicitly taught a sweeping task in each room of the house).

Similarly, [9, 8] does not allow postconditions to be specified, though preconditions are verbally stated. The input language is limited to a structured fragment of English. An example of the required input is, "Environment with initial conditions $\emptyset$. You start in r_1 [a location] with initial conditions $\emptyset$. Go to $\{r_1, ..., r_2\}$ infinitely often. Call Medic iff Person is found. Call Fireman iff Gas is detected."

[10] allow operators to specify an action, such as **grasp the doll**, and its stated effect, **be grasping the doll**. This corresponds to the following instruction: until agent x senses that it is grasping the doll, agent x should continue to perform a grasping action on the doll. However, the instruction has to be presented with a specific input structure: "Until you are grasping D; Holds you should grasp object D," which is then translated into a formal procedural description that allows the specification of preconditions but not postconditions. The formal language into which the commands are translated does list effects, but these effects are stopping conditions, rather than, as in our method, propositions that become true upon performing the action sequence.

Each of these projects addresses important aspects of action learning. However, a more comprehensive capability to, at run time, integrate knowledge learned via natural language at all levels throughout the architecture is necessary before robots will be able to, for example, use natural language input to update the planner's knowledge and even underlying assumptions. Robotic architectures must be developed that allow for flexible actions and plans, based on new knowledge derived from flexible natural language input. This necessitates the ability to expand concept knowledge both within the planner (e.g., new action concepts that are made available) and within the robot's dialogue module. In this paper, we describe a method of explicitly defining new actions, including pre- and postconditions, using natural language in a way that allows the robot's reasoner to be updated, making new capabilities available at run time.

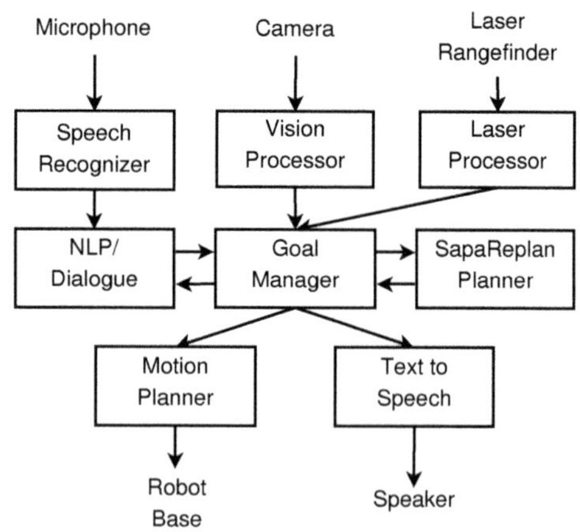

Figure 1: Architecture overview.

2.1 System Overview

Figure 1 provides an overview of the DIARC architecture used below. In particular, it shows the tight integration of the dialogue module and the planner with the goal manager. The goal manager maintains a database of procedural knowledge (i.e., actions that the architecture is capable of performing, along with their pre- and postconditions and information about the types of entities that could be encountered; [15]). The goal manager attempts to ensure that progress is made toward the achievement of all of the agent's goals by collecting information from the sensor servers and dispatching commands to the effector servers as required. Actions for multiple goals can execute in parallel, however, when resource conflicts arise, they are mediated in favor of the goal with the highest priority.

When the dialogue module determines that speech input consists of a command, a new goal is generated to carry out the command and submitted to the goal manager, which provides updates on the state of that goal as needed. Similarly, when a new action definition is distilled from natural language input, the dialogue module submits the new action to the goal manager, making it available to the goal manager from that point on. Natural language processing is described in greater detail in Section 3.

The planner's domain is initialized at system start-up by the goal manager, based on the contents of its knowledge base, ensuring that the planner has the most up-to-date information about the robot's capabilities. Moreover, because the planner can accept run-time updates, the domain can be augmented by newly-learned actions that can be used to improve subsequently-generated plans. Similarly, as the robot interacts with the world, the planner receives notification of state updates (e.g., external world states obtained from sensors, internal states such as sub-goal completion) from the goal manager; if an update triggers replanning, the new plan is returned to the goal manager in the form of an action script that can be directly executed to achieve the associated goals. The planning component is described in greater detail in Section 4.

This architecture is designed to enable the augmentation of a robot's capabilities at run time in response to operator input. For example, consider a search and rescue scenario in which a robot is searching a building that is unsafe for human exploration. Instead, the robot communicates with a human operator via a wireless link-up. At the beginning of the exploration task, the robot understands that it should enter any rooms it encounters as it traverses a hallway, and its procedural knowledge base includes an action sequence that will enter a room through an open door. Furthermore, its opportunistic planner, upon receiving a state update indicating a newly-discovered open doorway, produces a new plan to enter and search the room before continuing to explore the hallway. However, state updates regarding closed doors do not result in plans to enter and search rooms, as there is no known action sequence that will lead to the robot being in the room given the "closed door" precondition.

What if, during the search operation, the human operator is informed that the building's doors are all designed to unlatch when the fire alarm is triggered? In that case, the robot should be able to push the doors open, allowing it to search rooms behind closed doors. Without the ability to learn new actions via natural language and propagate that knowledge through the architecture, the operator is faced with a choice. The robot could be allowed to continue searching only rooms with open doors—but those could be the very rooms that are most likely to have been evacuated already. Or the architecture could be shut down while the operator adds that information and any associated actions to the robot's procedural knowledge base—but that would cost valuable time, and also assumes that the operator has the skills necessary to reconfigure the robot in this way.

Our solution, in contrast, allows the operator to transmit the information to the running system via natural language, associating appropriate pre- and postconditions with a previously-known action (initially applicable only in other contexts): "If you are at a closed door and you push it one meter, you will be in the room." The new information can be given by an operator with limited knowledge of the details of configuring the architecture and integrated with little or no delay, allowing the robot to search the rooms where it is most likely to find victims.

This motivating example will be used throughout the remainder of the paper, as we describe the unique features of the architecture that make it possible: sophisticated natural language processing capabilities, and an opportunistic planner that can accept world model updates, in addition to state updates, at run time.

3. NATURAL LANGUAGE PROCESSING

In this section, we describe the robot's dialogue module. For descriptive ease, it is described in a modular, sequential manner; in fact, however, all facets of processing occur incrementally and in parallel. At any time the module can be queried for a completely-processed semantic representation of the utterance as it currently stands.

In particular, this means that NLP and dialogue management are done in a tightly-integrated fashion. As utterances are recognized, mentions of entities are tracked, allowing incremental referent grounding. Other discourse information that is tracked includes, for example, points of confusion for the system, such as unknown words and utterances that do not fully describe the desired action.

NLP utilizes a number of initial knowledge sources, in particular, a dictionary containing primitives that can be combined into complex forms; an annotated corpus to train the syntactic parser; and optionally a database containing contextual knowledge (e.g., about interactions between entities). Entries in this database are represented in the same way as are semantic representations produced by the dialogue module (and perceptual modules such as vision) so that the database could in principle be built entirely from things the robot is told or is able to learn.

To demonstrate key steps of the process, we describe here how the dialogue manger handles the example sentence, "if you are at a closed door and you push it one meter, you will be in the room". The goal of semantic processing for this sentence is to identify the action definition, preconditions, and postconditions. Specifically, we must correctly segment the natural language input as follows:

preconditions you are at a closed door
action definition you push it one meter
postconditions you will be in the room

We will first describe reference and anaphora resolution, and then parsing, which produces a literal semantic representation. The final subsection will describe semantic manipulation in accordance with real-world principles known to the robot. After that, we will describe how the new action is given to the goal manager.

3.1 Reference Resolution

Each time a discourse entity (currently identified by the existence of a noun or a pronoun) is encountered during sentence processing, an object is created to hold all information about that entity. Each object may represent an independent entity, or may be reference to another entity, indicating that the two entities corefer. For example, when an indefinite phrase "a closed door" is encountered, an object is created that holds the information that it represents an entity of type **door** which exhibits the property **closed**. According to the determiner "a" this object represents an independent entity, which is first asserted to exist in this statement (i.e., it must not refer to a previous discourse entity).

By contrast, when a definite phrase such as "the room" is encountered, an object is created that holds the information that it represents an entity of type **room**; however, based on the determiner "the", the entity must be a reference. This alerts the robot that it must locate some referent (e.g., among previous discourse entities or by using its perceptual modules). Anaphors such as pronouns are treated similarly.

3.2 Parsing

As implied above, the goal manager requires a partial parse, ideally with breaks between distinct phrases, in order to integrate the new information into its knowledge base. Such a parse is assumed, and is realistic only in the current domain-limited context. In section 5.1, we briefly touch on this issue when trying alternate phrasings of our instruction, but future work must explore ways of compensating for a less-than-idea partial parse.

Given such a parse, the preconditions, action definition, and postconditions may be derived from a variety of verbal input frameworks such as: "if [preconditions are true] if [action definition is performed] then [postconditions will be true]", "if [preconditions are true] and [action definition is performed] then [postconditions will be true]", or "if you [action definition] when [preconditions are true] then [postconditions will be true]".

For this reason, a dependency-based parser that uses lambda calculus to produce semantic representations [3] is trained on a set of phrases labeled as a precondition, an action, or a postcondition. The phrases were extracted from the CReST corpus [6] and manually modified. Each phrase is returned as a pair comprising a semantic representation and one of these labels. For example, the input "if you are at a closed door and you push it one meter you will be in the room" results in the list of labeled phrases:

entities $door(x) \wedge agent(q) \wedge Rs.object(s) \wedge Rr.room(r)$[1]
pre $closed(x) \wedge outside(q, x)$
action $pushed(q, s, 1)$
post $in(q, r)$

In other words, the robot has segmented the natural language input into several distinct pieces of information. (1) the entities under discussion include a door, an agent, an ungrounded object s (which must be grounded in terms of previous discourse entities), and an ungrounded room represented (which must be grounded in terms of previous discourse entities or the environment). (2) This action is relevant only when the agent is outside the closed door (3) from which position, if the agent pushes s 1 meter, (4) the agent will then be in the room. The literal meaning of the input has now been established, but some important information is still missing. In particular, what object must be pushed, and what room is the robot entering? The semantic manipulation step provides the missing pieces.

[1]The operator R indicates that its argument is a reference which must be grounded to a discourse entity, a real-world entity, or a **theoretical entity**, an entity that the robot infers must exist (e.g., if a door is mentioned a room must exist because doors occur only in relation to rooms). An entity of this type will be introduced in the next subsection.

3.3 Semantic Manipulation

Assumptions and gaps in the input sentence are addressed by consulting the background knowledge accessible to the dialogue module, which includes the following: $\forall x[door(x) \rightarrow \exists y, z.room(y) \wedge location(z) \wedge doorconnected(x, z, y)]$ or "for each door, there exists a room y and a space z that is connected to that room by the door". To be "outside a door" is to be in z, giving us the following: $\forall x[door(x) \rightarrow \exists y, z.room(y) \wedge location(z) \wedge doorconnected(x, z, y) \wedge (outside(q, x) \leftrightarrow at(q, z))]$. Logically combining this with the explicit semantics, we arrive at:

entities $door(x) \wedge agent(q) \wedge Rs.object(s) \wedge Rr.room(r) \wedge room(y) \wedge location(z)$
pre $closed(x) \wedge outside(q, x) \wedge doorconnected(x, z, y) \wedge at(q, z)$
action $pushed(q, s, 1)$
post $in(q, r)$

Now the robot knows about the particular room y that is connected to a location z outside the door, which allows an agent at z to enter y. However, entities s and r remain unresolved. In this case, the sentence is a theoretical description of an action and its consequences, so all entities need only be grounded within the discourse, *not* in the real world. One consequence of this is that entities that contain assertions of existence such as "a door" need not be grounded. However, if the sentence were an instruction/statement pair such as "push a door 1 meter and then you will be in a room," the robot would need to ground "a door" to a single, perceivable real-world entity in order to follow the instruction.

As described above, resolution is done by comparing what is known about such a reference with what is known about each independent entity and identifying those that are semantically compatible. If this leaves several candidates, a selection is made based on heuristics such as recency of mention and parallel grammatical function. In this utterance, there are two entities flagged for resolution. The first is the pronoun "it". In this case, only one object has yet been mentioned, the door. Thus "it" can be resolved right away, resulting in the following semantic representation:

entities $door(x) \wedge agent(q) \wedge Rr.room(r) \wedge room(y) \wedge location(z)$
pre $closed(x) \wedge outside(q, x) \wedge doorconnected(x, z, y) \wedge at(q, z)$
action $pushed(q, x, 1)$
post $in(q, r)$

The second reference, the noun phrase "the room", refers to an entity not found in the *explicit* content of the sentence. However, as we saw above, the entity was implied. This implication was made explicit through the inclusion of the robot's background knowledge of the world. Thus we resolve r and arrive at:

entities $door(x) \wedge agent(q) \wedge room(y) \wedge location(z)$
pre $closed(x) \wedge outside(q, x) \wedge doorconnected(x, z, y) \wedge at(q, z)$
action $pushed(q, x, 1)$

post $in(q, y)$

These parts can now be assembled and sent to the goal manager as described next.

3.4 Submitting the New Action

The new capability is introduced into the goal manager via a method call: `associateMeaning(action definition, preconditions, postconditions)`. This tells the goal manager that it needs to `associate` a particular action sequence (the `action definition`) with the given `preconditions` and `postconditions`. Note that in this case, as in many others, there are other postconditions, or side effects (e.g., the door will be open), but for the current example, if we have entered the room, the fact that, out of all potential ways the room could be entered, it happened to be through the mechanism of opening a closed door, is not explicitly treated.

Once the goal manager has processed the new capability, it can be used in response to subsequent requests by the operator, and also presented to the planner. We describe how the planner utilizes this information next.

4. PLANNING CAPABILITIES

An indispensable requirement in supporting a flexible dialogue module on a robot is a robust planner that can handle changes to the world model and the problem at hand, and re-plan taking these modifications into account. If the human specifies (during execution) that the agent must push the door to a room in order to enter that room, the robot must be in a position to understand the implications of that directive. If there are goals that can only be achieved by entering that room, the robot must update its understanding of the world and infer that the new capability now allows it to achieve those goals. It is these tasks that are performed by the planner: (1) the task of updating the robot's model of the world and understanding the implications of those changes, and (2) processing changes to the facts and goals in the robot's knowledge.

We employ the `Sapa Replan` [4] planner, an extension of the metric temporal planner `Sapa` [5]. `Sapa Replan` is a state-of-the-art planner that can handle actions with costs and durations, partial satisfaction of goals and changes to the world and model via replanning. Of these, the most relevant to the problem of dynamic natural language input is the ability to model and use changes to the world to the robot's advantage. The planner is mainly concerned with two such changes: changes to the world facts and the problem at hand currently, and modifications to the overall model of the world that the robot is using.

4.1 Changes to the Model

The planning model is an abstraction of the dynamics of the world and is used by the planner to guide the robot in its pursuit of the various goals assigned to it. The constituent components of such a model are usually the types of objects in the agent's world, the predicates that describe the relationships between these objects, and the actions available to the agent. An initial model is made available to the planner (and hence the robot) along with a description of the world (described in section 4.2). Changes to this model are generally harder to process than changes that are restricted to specific facts, since they involve the modification of several facts that are distributed throughout the planner's knowledge base. Additionally, verifying the consistency and correctness of an executing plan—a non-trivial task at best when facts about the world are changing—is rendered even more difficult by the extensive role that the model plays in the process of plan synthesis.

Unlike changes to facts in the world that can either be discovered by the robot's sensors or specified by a human in the loop, changes to the model are more likely to be entirely specified by human operators using natural language. One reason for this is the fact that such updates are much more complex and often invoke the robot's capabilities and the dynamics of the world's evolution. Examples of utterances that force updates to the model include, for example, statements that give new information about how entities in the world are interrelated and how the robot should interact with real-world entities, such as the directives produced by naive human users in Section 5. These are additional capabilities that are being specified to the robot during execution, and they may result in a change in achievability as far as the current goals to be fulfilled are concerned.

Currently, such directives are supported in the planner by modifying the planner's knowledge base. A reference to the concerned action[2] is given and the new capability is added to the effects of that action. The planner process is then restarted and the search for a new plan begins, since the added capability may have altered the reachability of some goals. We demonstrate in the following section that, in the example scenario, this method is sufficient to achieve our current goals. When considering the problem of plan validity under modifications to the planning model in general, though, the system may either replan from scratch, or update and re-use the current plan.

Replan from Scratch: Given a new version of the domain model (with updates), the planner runs again in order to come up with a plan that completely replaces the currently executing plan. This is the approach that we employ currently.

Plan Re-use: The planner analyzes the current plan with respect to the updates received and adds, removes, or updates an action.

The addition of an action to the domain does not affect the validity of the current plan. Other metrics associated with the problem may change, since a new plan may now be available, but no change is necessitated if a sufficient plan is already executing. Likewise, the removal of a non-participating action (i.e., an action that does not participate in the currently executing plan), no change is necessary.

However, removal an action that does participate in the currently executing plan is a more complex case, and requires a detailed analysis of the sub-goals (and top-level goals) that the action supports, and the nature in which its removal will affect those goals. Similarly, when parts of an action are updated (addition or deletion), a more complex analysis of the various commitments entailed by the currently executing plan is required.

While a full analysis of these methods followed by a selective implementation of the best approach is work for the future, the effectiveness of the current approach is demonstrated in the next section.

[2]For this work, we restrict our attention to directives that expand on the robot's capabilities, and that are associated with a particular action.

4.2 Changes to the World

Dynamic environments like search and rescue also have a stream of constantly updating information about world facts associated with them. This information is quite distinct from changes to the overall model, since it is specific to that particular instance and the state of the world at the current instant. Any agent that seeks to effectively achieve a given set of goals in such conditions must be in a position to process and take advantage of this information. Agents that are unable to do so run the risk of executing plans that are no longer valid for the given circumstances. Knowledge about these changes can come from two main sources, as mentioned previously: the first is through the robot's own exploration of the world, and the feedback received via its sensors; the other is through natural language (i.e., human-issued knowledge and directives). The planner must take these changes into account and ensure that the plan that is currently being executed remains valid and effective. If this ceases to be the case, the robot must replan, taking into account the new knowledge and any changes to objectives, and pass a new plan for execution on to the agent.

The `Sapa Replan` planner that directs the robotic agent is comprised of two main components – the execution monitor and the planner – working in a tight loop. The planner receives the initial state information and goals and dispatches a plan to the monitor that is sent to the robot for execution. The monitor then listens in for updates from the world, which can be generated from either of the two sources mentioned previously. In particular, updates can include new objects, timed events, goal additions or modifications, and a new time point from which execution is expected to resume. The update syntax is as follows:

```
(:update
 :objects
    red3 - room
 :events
    (at 125.0 (not (at red2)))
    (at red3)
    (visited red3)
 :goal (visited red4) [500] - hard
 :now 207.0)
```

These problem updates cause the monitor to interrupt the planner and restart it after updating the internal problem representation. However, handling the arrival of new goals that are contingent upon objects and facts initially unknown to the agent is a challenge for most modern automated planners. This is due to the fact that planners assume a closed world and model at the outset—that is, they expect full knowledge of the initial state and the model of the world, and up-front specification of all goals. Information that is left unspecified initially is assumed to be false in order to gain closure over the knowledge about the world. Instead of this, what is required is a way to exclude certain classes of objects from this closure requirement, and a way to specify additional information about such objects (such as goals associated with them). To this end, we extended `Sapa Replan` with the concept of *Open World Quantified Goals* (OWQG) [17] which are constructs that provide a compact way of specifying conditional reward opportunities over an open set of objects.

For example, a human commander may assign the robot a soft goal of reporting the location of all injured humans, with the additional information that these injured humans may be found inside rooms. Such a directive would be translated to an OWQG using the following syntax:

Figure 2: A Videre ERA equipped with a USB web-cam, a Hokuyo Urg laser range-finder, a microphone, and a quad-core Linux PC for onboard processing exercising its newly-acquired capability of opening closed doors to look for red boxes.

```
(:open
  (forall ?r - room
    (sense ?hu - human
     (looked_for ?hu ?r)
      (and (has_property ?hu injured)
           (in ?hu ?r))
  (:goal (reported ?hu injured ?r)
           [100] - soft))))
```

While space does not allow a full exposition of the syntax and the manner in which OWQGs are handled by the planner, additional information may be found in [16].

5. DISCUSSION

The natural language processing and planning capabilities described in Sections 3 and 4 were implemented and tested on the robot shown in Figure 2 in a scenario similar to the one described in Section 2 (see the video at The video at `http://www.youtube.com/watch?v=NEhBZ205kzc`). The search target in the demonstration is a red box and a microphone is used instead of radio communication, but otherwise the demonstration is faithful to the original motivating example. Sphinx 4 was used for speech recognition.[3].

Although this example demonstrates that the architecture is able to learn the action and use it for problem-solving, it is clear that to be of general use, the robot must be able to learn more than one action. Furthermore, examples produced by researchers intimately familiar with the details of their own systems, while instructive for demonstrating the robot's operation, are necessarily informed by detailed knowledge of the system, so it is unclear how widely successful even this single example would be given an uninitiated user. Therefore, in this section we first discuss the system's

[3]`http://www.cmusphinx.org`

inherent general scalability (i.e., to different action descriptions), then investigate how successfully the robot is able to handle the types of natural language instructions produced by naive users for this particular example using the results of a preliminary pilot eliciting natural language input from a small sample of untrained users.

5.1 System Scalability

So far, only one example has been presented. However, several facets combine to contribute to the system's scalability to other scenarios. In particular, the dialogue module is characterized by the ability to scale up to larger domains due to three key features. First, the parser is trainable from annotated data. Second, given an initial set of primitives, new terms describing complex combinations of these primitives can be learned and added to the dictionary. Third, parses often fail around certain predictable points such as conjunctions. These points can indicate, among other things, phrase breaks, allowing a parser's weak point to be used to produce a more flexible set of of processable input types. These characteristics combine to contribute to the dialogue module's extensibility, producing the ability to successfully analyze the semantics of new natural language input.

Moreover, the interface via which a new capability is submitted to the goal manager is flexible enough to accept a wide variety of definitions, requiring only the basic elements described in Section 3: the action description, a (possibly empty) list of preconditions, a list of postconditions (or effects) of the given action sequence, and (optionally) a name. Those elements are sufficient to make the new capability available for use by the goal manager, and to allow it to forward the information to the planner.

Finally, `Sapa Replan`'s flexible mechanisms for defining new domain elements make it possible to update the world model at run time with any construct that would be permissible at system start-up. This, combined with a design that allows it to respond to changes to the world, ensure that any newly-learned capability can be injected into the running planner and used to improve subsequent plans.

To test this scalability, we tried a variety of different phrasings on the same instruction. Simple changes such as alternating conjunctions (e.g., "if" replacing "and", "then" replacing null) and reordering information (e.g., "if you push a closed door one meter when you are at it then you will be in the room") worked easily. A resolution error arose with "to go into a room when you are at a closed door push it one meter" — "it" was incorrectly found to be coreferent with "room". Similar errors will be corrected with more fine-grained semantic knowledge of object types.

Next, an assortment of disfluencies were tried. "Uh" insertion in a variety of locations (e.g., between prepositional phrases and their head verb, between object phrases and their head verb) posed little problem. One case resulted in a failure to attach "closed" to "door" (i.e., the system would not understand this instruction is needed only when the door is not already open). Repetition of non-content words (e.g., "to uh to go") was also unproblematic, with the repetitions left unattached. Repetition of content phrases was slightly more problematic. "Push it uh push it one meter" resulted in two sequential steps, "push it" and "push it one meter" being added to the instructions.

However, even though these examples were constructed after the system was designed (i.e., they had no effect on the design of the system), they likely are not representative of the kind of language naive users would produce. In the next section, we elicit language from naive users.

Table 1: Naive users instructions to Gogol.

continue forward	✓
detect a closed door push on the door and enter	✓
open the door	✓
go straight into the door and keep pushing	
go up to a wall and push on it	✓
um you need to push on the door to make it open	
push the door open	✓
if you detect a closed door push on it	✓
when you think you've hit a wall push on it	✓
push on every wall you find	

5.2 Naive Subject Evaluation

Because several aspects of any NLP system are heavily reliant on phrasing, it is important to understand what type of natural language *naive* users produce if we want to evaluate the general applicability of the system.

As a pilot study, a small sample of naive users (10) were shown diagrams depicting Gogol, a non-humanoid robot, as it navigates a hallway searching for blocks. The users were told that Gogol is able to sense both open and closed doors. To maintain consistency with the demonstration run, "closed" doors are actually depicted slightly open and subjects were told that Gogol considers them "closed". They were then told, "Unbeknownst to him, if he detects a closed door, he can push on it. Besides moving him into the room, this action moves the door into a position that Gogol considers 'open'." Subjects were then instructed to "tell Gogol how to go into rooms with closed doors". The instructions did not explicitly state that Gogol did not know the meaning of the verb "open", only that he did not know **how** to perform the action. "Open" was not used as a verb anywhere in the description of the scenario or the instruction.

Users responses are shown in Table 1. The dialogue module correctly parsed seven of the ten responses. Most of the resulting instructions were syntactically simple, with only two users making use of more than simple imperatives. While all instructions included an action or series of actions only half included something identifiable as a pre- or post-condition. All instructions assumed the action "push" did not require a distance or any explicit stopping point (i.e., no one told Gogol how far or long to push or when to stop). Most problematically, some responses suggested that their speakers held incorrect assumptions about the robot, in particular a lack of trust in the robot's ability to detect closed doors. In the case where the robot was instructed to push on "every wall [it] finds". the robot may find that it follows instructions perfectly, but still is unable to complete its goals in a timely manner.

In short, it is clear that in its current state, the system is unlikely to work with completely naive users, as they don't understand the key elements that a robot needs to integrate and use a new command. For scenarios such as the one envisioned here, this is probably not a fatal shortcoming, as search and rescue robot operators can reasonably be expected to receive training that affords them a level of familiarity sufficient to understand the relevant issues without, requiring that they have a developer's view of the architecture. On the other hand, the capabilities described here could also be of great general utility as robots become increasingly common in our society. The results of this small user study suggest that this will require substantial improvements to the architecture.

6. CONCLUSIONS AND FUTURE WORK

We presented a robotic architecture featuring a planner that uses discovered information to produce new and updated plans. The robot can learn action sequences with defined pre- and postconditions from natural language descriptions, and immediately apply this knowledge to improve planning. We demonstrated a successful real-world implementation on a physical robot and showed that the system is scalable to a large set of possible inputs, but found that it is unable to handle input from naive users with limited understanding of the robot's capabilities. In addition, we noted limitations in the planner's update process.

As the initial evaluation revealed, naive users are not necessarily effective at providing sufficient information to fully specify the learned action sequence; additional mechanisms will likely be required to successfully deploy the architecture for use with the general public. We are considering two approaches to this problem: detecting missing information in operator descriptions during the natural language processing phase so the robot can request clarification, and performing a "mental" simulation with the new capability to determine whether it seems likely to be effective in the real world.

When the planner's world knowledge and/or model knowledge is updated, the current plan is replanned from scratch, which as previously described is not necessarily the most effective method. Future work will include evaluating the possible ways of deciding whether and how to reevaluate a plan and implementing the most successful procedure.

7. ACKNOWLEDGMENTS

This work was in part funded by ONR MURI grant #N00014-07-1-1049.

8. REFERENCES

[1] Sugato Bagchi, Guatam Biswas, and Kazuhiko Kawamura. Interactive task planning under uncertainty and goal changes. *Robotics and Autonomous Systems*, 18(1-2):157–167, 1996.

[2] Rehj Cantrell, Paul Schermerhorn, and Matthias Scheutz. Learning actions from human-robot dialogues. In *20th IEEE International Symposium on Robot and Human Interactive Communication*, 2011.

[3] Rehj Cantrell, Matthias Scheutz, Paul Schermerhorn, and Xuan Wu. Robust spoken instruction understanding for HRI. In *Proceedings of the 2010 Human-Robot Interaction Conference*, March 2010.

[4] William Cushing, J. Benton, and Subbarao Kambhampati. Replanning as a Deliberative Re-selection of Objectives. *Arizona State University CSE Department TR*, 2008.

[5] Minh Binh Do and Subbarao Kambhampati. Sapa: A multi-objective metric temporal planner. *Journal of Artificial Intelligence Research*, 20(1):155–194, 2003.

[6] Kathleen Eberhard, Hannele Nicholson, Sandra Kübler, Susan Gunderson, and Matthias Scheutz. The Indiana "Cooperative Remote Search Task" (CReST) Corpus. In *Proceedings of LREC-2010*, Valetta, Malta, 2010.

[7] Russell Knight, Gregg Rabideau, Steve Chien, Barbara Engelhardt, and Rob Sherwood. Casper: Space exploration through continuous planning. *IEEE Intelligent Systems*, pages 70–75, 2001.

[8] Hadas Kress-Gazit, Georgios Fainekos, and George J Pappas. Translating structured english to robot controllers. *Advanced Robotics*, 22(12):1343–1359, 2008.

[9] Hadas Kress-Gazit, Georgios E. Fainekos, and George J. Pappas. From structured english to robot motion. In *Proceedings of the 2007 IEEE/RSJ International Conference on Intelligent Robots and Systems*, San Diego, 2007.

[10] Pat Langley, Nishant Trivedi, and Matt Banister. A command language for taskable virtual agents. In G. Michael Youngblood and Vadim Bulitko, editors, *AIIDE*. The AAAI Press, 2010.

[11] Solange Lemai and Félix Ingrand. Interleaving temporal planning and execution: IxTeT-eXeC. In *Proceedings of the ICAPS Workshop on Plan Execution*. Citeseer, 2003.

[12] Silvia Richter and Matthias Westphal. The lama planner: Guiding cost-based anytime planning with landmarks. *Journal of Artificial Intelligence Research*, 39(1):127–177, 2010.

[13] Paul E. Rybski, Jeremy Stolarz, Kevin Yoon, and Manuela Veloso. Using dialog and human observations to dictate tasks to a learning robot assistant. *Journal of Intelligent Service Robots, Special Issue on Multidisciplinary Collaboration for Socially Assistive Robotics*, 1(2):159–167, April 2008.

[14] Paul E. Rybski, Kevin Yoon, Jeremy Stolarz, and Manuela M Veloso. Interactive robot task training through dialog and demonstration. In *HRI '07: Proceedings of the ACM/IEEE international conference on Human-robot interaction*, pages 49–56, New York, NY, USA, 2007. ACM.

[15] Matthias Scheutz, Paul Schermerhorn, James Kramer, and David Anderson. First steps toward natural human-like HRI. *Autonomous Robots*, 22(4):411–423, May 2007.

[16] Kartik Talamadupula, J. Benton, Rao Kambhampati, Paul Schermerhorn, and Matthias Scheutz. Planning for human-robot teaming in open worlds. *ACM Transactions on Intelligent Systems and Technology*, 1(2), 2010.

[17] Kartik Talamadupula, J. Benton, Paul Schermerhorn, Rao Kambhampati, and Matthias Scheutz. Integrating a closed world planner with an open world robot: A case study. In *Proceedings of the Twenty-Fourth AAAI Conference on Artificial Intelligence (AAAI '10)*, Atlanta, GA, July 2010.

[18] Kevin Yoon and Paul E. Rybski. Teaching procedural flow through dialog and demonstration. In *Proceedings of IROS'07, the 2007 International Conference on Intelligent Robots and Systems*, pages 807–814, San Diego, CA, Oct–Nov 2007.

Talking with Robots about Objects: A System-Level Evaluation in HRI

Julia Peltason, Nina Riether, Britta Wrede and Ingo Lütkebohle
Applied Informatics Group
Bielefeld University, Germany
{jpeltaso|nriether|bwrede|iluetkeb}@techfak.uni-bielefeld.de

ABSTRACT

We present the design process, realization and evaluation of a robot system for interactive object learning. The system-oriented evaluation, in particular, addresses an open problem for the evaluation of systems, where overall user satisfaction depends not only on the performance of the parts, but also on their combination, and on user behavior. Based on the PARADISE method known from spoken dialog systems, we have defined and applied internal and external metrics for fine-grained and largely automatable identification of such relationships. Through evaluation with n=28 subjects, indicator functions explaining up to 55% of variation in several satisfaction metrics were found. Furthermore, we demonstrate that the system's interaction style reduces the need for instruction and successfully recovers partial failures.

Categories and Subject Descriptors

C.4 [**Performance of Systems**]: Measurement techniques; I.2.9 [**Artificial Intelligence**]: Robotics; I.2.7 [**Artificial Intelligence**]: Natural Language Processing

General Terms

Measurement, Performance, Human Factors

Keywords

Human-robot interaction, object learning, social robotics

1. INTRODUCTION

Interaction provides many benefits for (semi-)autonomous robotic systems, e.g., to set goals, to efficiently enlist human help, and to explain the system's capabilities and limitations, for better human-robot-cooperation [2, 4]. It is well accepted that a combination of modalities and capabilities improves such interactions. However, understanding the effects of dynamic, interactive combinations of capabilities remains very challenging.

One way to evaluate this is to systematically vary each capability, but while this is certainly a powerful approach, it is infeasible to conduct it for large numbers of capabilities. Furthermore, it is difficult in such an approach to determine the *relative* importance of each capability.

Therefore, in this contribution, we will, firstly, present the design and realization of a full robotic system, and, secondly, its *system-level evaluation*. The system realizes an interactive, social object learning scenario, where object labels are learned during spoken language interaction.

As a concrete example of the benefit of such evaluations, consider the following: For object learning, object recognition performance depends both on the algorithm and on training image quality. The latter, then, depends on both the moment when the training image is taken, as well as demonstration behavior (i.e., hands in front of the object or similar), both of which are influenced by the dialog strategy. So, which of these is most important in the end?

Similar *explorative* questions arise in many larger systems, and we believe that this constitutes one of the major issues for robotics research today. One possible way to address it is through regression analysis, which relates user's subjective judgements with metrics that describe system behavior. Specifically, we have chosen a method widely used for speech-based dialog systems called PARADISE [15].

An important prerequisite to use PARADISE, however, is the definition, and efficient capture, of insightful metrics. In particular, automatic capture of metrics would greatly increase practicality. Therefore, we propose to analyze system message logs in combination with two general action models, one for interaction episodes (Interaction Patterns [12]) and one for system activities (the Task-State-Pattern [10]).

Related Work. Despite structural similarities to spoken dialog systems, the PARADISE approach is currently little used for HRI. One of the few studies is by Foster et al. [5], who explored the predictive power of interaction effort measures. These are found to be significant, but only explain a relatively small amount of variation (between 4 and 12%). In comparison, typical dialog system evaluations find values between 30 and 70% [5], and we found between 17 and 55%.

A later study by Dethlefs et al. [3] compares a binary task success metric with a scoring scheme that takes difficulties into account. They find that the binary metric is not significantly related to user satisfaction, while the more detailed scores are. In combination with an interaction-turn measure, they account for 62% of variation. This suggests that the process is of crucial importance for user satisfaction.

Both of these studies take an external view, with relatively few (three to five) metrics that are manually annotated. In contrast, our approach looks *inside* the system, by directly analyzing message logs. This inside view allows us to determine not only how the system performs, but also why.

2. SYSTEM DESCRIPTION

In our interactive learning scenario, a number of previously unknown objects are present on a table (see figure 1(b)) and participants can show them to the robot, label them, and verify what has been learned. Moreover, the robot can also take initiative and ask for object labels itself, in a mixed-initiative setting. Most parts of the system are autonomous, which includes real object and speech recognition to create realistic challenges. Two operator-assisted (Wizard-of-Oz) components are used for reference resolution and region selection during robot initative (cf. section 2.3). In the remainder of this section, we present the interaction strategy design, architecture and capabilities, as well as the dialog principles.

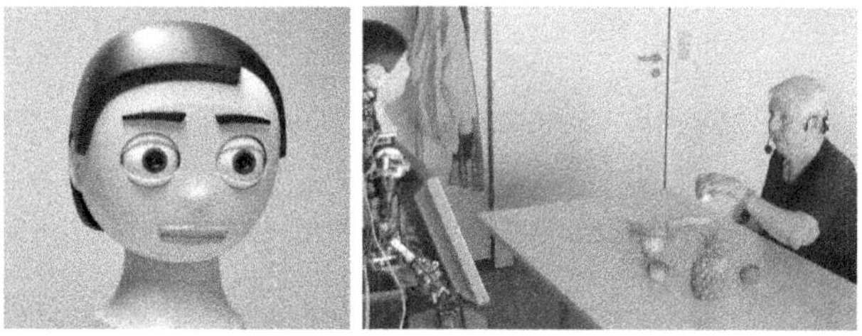

(a) The Flobi head [7] (b) Demonstrating a fruit

Figure 1: Scenario overview

2.1 Iterative Interaction Design

For the design of dialog systems, it is important to know which capabilities users expect in a given situation. While the variability of human behavior means that not everything will (or has to) be supported, missing important aspects in this stage could lead to errors that overshadow other evaluation results. We handled this using pre-studies which we'll describe in the following. In addition to these, we also conducted an earlier study on the robot's initiative, to validate the mixed-initiative setting [8].

Initially, we focused on speech recognition, as it has often been reported to be a major contributor to user satisfaction (e.g., [14]). In an explorative user study with 10 participants, from a total of 392 user utterances 27% were not understood correctly. Among these, out-of-capability utterances (i.e., the utterance is beyond the system capabilities) represented the by far largest portion (54%, i.e., 15% of all utterances). More than half of these were caused by the user attempting to reverse the roles after the initial steps, proceeding to demonstrate objects on their own, which was not possible in that system. As a consequence, this capability has now been added.

Based on re-examination of an earlier Wizard-of-Oz (WOz) study on object teaching by our colleagues, Lang et al. [6] (targeting facial expressions), with 11 participants, we determined the dialog strategies used. All interactions had a very similar structure, consisting of an opening part, a task-related part and a closing part. 36% of the interactions additionally feature transitional phrases that introduce the task-related part. In the opening phase, introducing each other (82%) and exchanging pleasantries (18%) are frequent. Aside from object demonstrations, the task-related phase consists of checking learned objects (45%) and transitional phrases between the objects (36%). Praising the robot for correctly learned objects turned out to be universal (100%). The task-related part may include closing remarks (36%).

Based on this information, we designed the interaction strategy of the present scenario. In general, most of the observed strategies could be realized. Some particularly rich descriptions went beyond the current system capabilities and were ignored, however. For example, the object descriptions sometimes included functional description of objects that are not yet supported. Table 1 lists the resulting interaction capabilities of the system.

Beside the out-of-capability utterances, out-of-vocabulary utterances (i.e., the meaning is within capabilities, but the particular wording is not in the configured set) constitute a significant portion of non-understandings (12%). Thus, enhancing the speech recognition grammar became another focus of the current iteration. Here, a balance had to be found between broad grammar coverage and acceptable recognition performance. We opted for an inclusive grammar, so some errors will inevitably remain. In order to cope with these, robust key-word matching techniques were applied for interpretation of recognition results.

Purpose	Example interaction
Greeting	User: Hello, Flobi Flobi: Hi, hello.
Introducing	User: My name is Anneliese. Flobi: Hello, Anneliese. Nice to meet you.
Exchanging pleasantries	User: How are you? Flobi: Fine, thank you!
Task transition	User: I am here to show you some objects. Flobi: Alright, let's start!
Attracting attention	User: Flobi, look here! Flobi: Yes, I'm looking.
Object demonstration	User: This is a yellow lemon. Flobi: Alright, a yellow melon. Is that correct? User: No, it's a lemon. Flobi: Alright, a lemon. Is that correct? User: Yes, that's correct. Flobi: Okay, I am going to learn the lemon. Flobi: I have learned the lemon.
Object query	Flobi: What is that? User: This is an apple. (remainder similar to object demonstration)
Listing learned objects	User: What have you learned so far? Flobi: I have learned a lemon and an apple.
Checking	User: Flobi, do you know this one? Flobi: I will have a look at it. Flobi: This is an apple.
Praising	User: Well done, Flobi. Flobi: Thank you.
Restart	User: Restart. Flobi: Okay, let's start over again.
Transitional phrases	User: Would you like to learn another object? Flobi: Yes, I want to learn them all.
Closing task	User: Okay, that's it. Flobi: Are you leaving already?
Parting	User: Good bye, Flobi. Take care! Flobi: Bye, see you later.

Table 1: Interaction capabilities of the system. The blocks are: i) Opening, ii) task-related, iii) closing.

2.2 Architectural Concepts

Our system uses a component-based, layered architecture with event-based, asynchronous communication, whose main ideas have previously been reported in [8, 9]. The components of the system are shown in in figure 2.

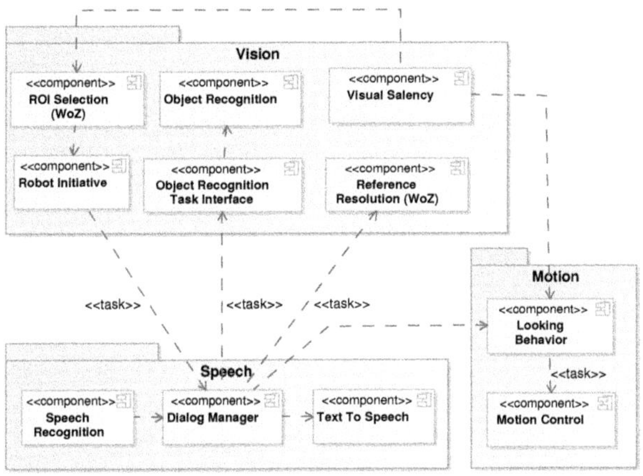

Figure 2: System components. Dashed lines indicate general event notifications, while lines labelled with *task* indicate task communication.

A central concept, with particular relevance for this paper, is the Task-State-Pattern model for representing system activities [10]. It represents all actions of the system as a pair of i) state and ii) goal specification. The set of possible states is the same for all tasks of the system. The goal specification, on the other hand, is specific to to the task.

Communication in the system occurs through event notifications on every transition. Example transitions include *initiated* for starting it, *completed* for success and *failed* for failure. There are also transitions for updating the goal or aborting tasks. Through the corresponding states, we can track all tasks in a generic manner, which forms the basis for the task-success metrics (cf. section 4.3).

Moreover, the dialog manager uses this general model to trigger activities and inform the user about their progress in a consistent fashion (e.g., when object recognition or learning begins and ends). This detailed reporting gives the user clear information on the system's state. Lastly, observation of ongoing tasks is also used to coordinate the gaze feedback module with current dialog activity.

2.3 System Capabilities

The hardware used includes the Flobi anthropomorphic robot head (cf. figure 1(a)), a close-talking microphone, and a flat-screen display, used to show the selected object in robot-initiative (but not with user-initiative).

Apart from basic social talking skills (greeting, complimenting, etc.), the system has three main capabilities: i) Having the user teach an object, ii) asking for an object on its own initiative, and iii) answering requests about object labels and known objects. All of these can be triggered and answered through a variety of verbal utterances (cf. table 1). Furthermore, the system also orients its gaze at the current most (visually) salient point or the user, as appropriate for the task state.

Compared to the previous scenario [8], an important difference is the operator-assisted reference resolution and region of interest selection. Despite our general goal of using only autonomous behavior, we found through the WOz study on object teaching that referencing behavior varies considerably, and this observation was confirmed by the present user study, where 8 different referencing strategies could be identified, as listed in table 2. Not all of these could be acceptably automated for now. Hence, the system detects references on its own, but for the moment resolving is done by an operator selecting the appropriate object region. The robot will then gaze at the selected region. Requests may also be rejected, e.g., when no object is visible in the robot's cameras.

Referencing strategy	% of total	% of user references
Robot	28%	n/a
Lifting	19%	30%
Isolated object	14%	22%
Pointing	13%	22%
False positive	11%	n/a
Touching	10%	16%
Non-visual	5%	7%
Isolated-object + pointing	1%	1%
Isolated-object + touching	1%	1%

Table 2: Ratios for referencing strategies (rounded).

For the region of interest selection, we previously used a saliency model [8], which was found to be not predictable enough for repeatable study results, however. Therefore, in the present system, a human operator chose one of the objects visible when the system had initiative.

The remainder of the system is autonomous, however, and for all of the functions, vision and speech interact in a tightly integrated manner, with each step modeled as a task, and integration based on event notifications.

2.4 Interaction Patterns

The mixed-initiative interaction strategy was realized with the PaMini (for Pattern-Based Mixed-Initiative HRI) dialog management approach that we have suggested recently [12]. This approach relies on generic Interaction Patterns that model recurring conversational structures, such as making a suggestion or negotiating information. Interaction Patterns are defined at an abstract level, but can be tailored with an application-specific configuration. They can be initiated either by the human (e.g., Human Object Test) or by the robot (e.g., Robot Information Request).

Within a pattern, the above system tasks may be initiated, and in turn the patterns react to task state changes. As an example, figure 3 depicts the Human Object Test pattern, which was used to model the user's object recognition requests ("Flobi, what is that?"). Within this pattern, two sequential tasks are initiated (first a *reference resolution* task, then a *lookup* task), and the pattern combines system task states (such as *task failed*) with appropriate robot dialog acts (such as *R.apologize*), thus relating the domain level with the conversation level. Similarly, the pattern used for the user's object demonstration ("This is an apple") initiates a *resolve reference* task first, then a *learn object* task.

For each interaction purpose in table 1, a corresponding Interaction Pattern was configured, plus several additional patterns as variations. Overall, the dialog consisted of 20 Interaction Patterns, each of which is capable to respond to numerous different wordings for a specific dialog act. At run-time, multiple patterns can be active at the same time, and can be interleaved to achieve a more flexible interaction style. The information about the interaction patterns used during an interaction, and the sequence of states they took, forms the basis for the dialog efficiency and quality metrics (see section 3).

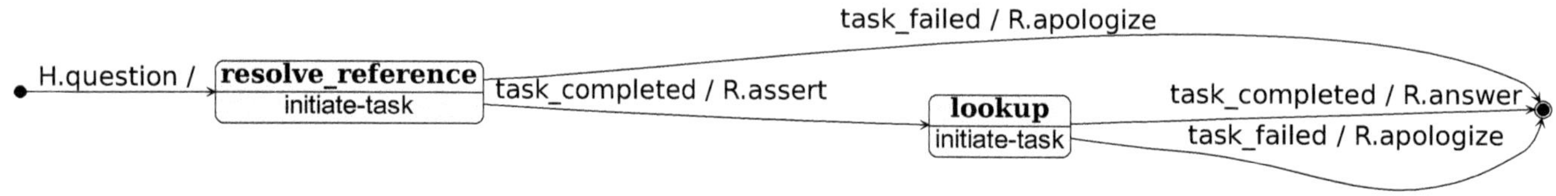

Figure 3: The *Human Object Test* pattern, which is used to model the user's object recognition requests.

3. PARADISE EVALUATION FOR HRI

The PARADISE evaluation approach has originally been proposed for spoken dialog systems [15], where it has become a de-facto standard. Its goal is to explore the relative importance for, and contribution of, the functions of a system for user satisfaction. The system is characterized using objective metrics, and user satisfaction is determined through subjective judgments. The relationship between these is then established using multiple linear regression (cf. section 4.5). A critical aspect to achieve a meaningful PARADISE analysis is the choice of metrics. Ideally metrics should be standardized, but this is not yet the case. We hope to make one step in this direction with the description of our procedure.

Choosing Subjective Measures: Subjective measures are usually determined through a questionnaire and here, a number of good proposals for HRI and dialog are available. We recommend using a combination of these to capture various aspects. For general impression of the system, we have used Bartneck et al.'s "godspeed" proposal [1], aggregated by category. For interaction-related metrics, we have adapted a questionnaire for evaluation of the COMIC system [16]. See section 4.4 for details on the metrics used.

Choosing Objective Measures: The objective measures need to capture what actually happened, i.e., what the system and the user did. They are assigned to three categories: task success, dialog quality and dialog efficiency.

Measures are based on system logs, but need to reflect system performance, so we cannot simply count messages. Instead, we have based our metrics on two generalizations: Firstly, user and system utterances are assigned to Interaction Patterns (cf. 2.4), which describe a single interaction episode. From this, we can compute the length and the number of steps for each episode. Secondly, messages about system activities are assigned to the task they pertain to, using the Task-State-Pattern (cf. 2.2), which describes the outcome (success/failure) of all tasks, grouped by type. Thus, we exploit the Interaction Patterns to gain information about interactional aspects of system performance (i.e., about dialog quality and efficiency), whereas the Task State Pattern provides information at task level (i.e., about the task success).

Interaction Patterns and Tasks are a form of system model which we also use at runtime. However, this is not a prerequisite for defining metrics, it just simplifies computation.

4. USER STUDY

The robot system was evaluated via a user study in which participants interacted with the system. It focuses on the relationships between the various objective, system-based measures contributing to dialog efficiency, dialog quality and task success (see 4.3), and the subjective variables (see 4.4). Thus, our evaluation was aimed to detect general connections among system based measures and subjective reactions to the system (see 5.3).

4.1 Participants

From 32 participants who took part, we used 28 recordings (14 male, 14 female), with the remaining four excluded due to technical problems. Most of the participants had been recruited at a university event for the general public and thus represented a wide age range, with mean age at 33.5 years, minimum 21 and maximum 79. On a scale from 1 (none) to 6 (lots of), the average rating for knowledge of computers was at 5.07, of speech systems at 2.52, of robot systems at 1.96 and of programming experience at 2.26. They were compensated for their participation in the experiment.

4.2 Instruction

In order to study natural demonstration behavior, participants received as little instruction as possible. They received written instructions, specifying that they were to engage in interaction with the robot Flobi, and that Flobi was supposed to learn object labels during interaction. They were also advised to check that the robot had actually learned the labels. It was, however, not specified how they should present and check objects. They were told that they should interact with the robot as long as they wish, with 5-10 minutes recommended as a guideline. Also, they were informed that they could begin the interaction by greeting the robot, and end the interaction by saying goodbye. In addition, participants were advised not to be discouraged by speech recognition problems, and that they could repeat or rephrase their utterance in such cases. Lastly, an emergency phrase ("Restart") was provided. The interactions were in German.

4.3 Objective Measures

A wide range of objective measures has been collected, most of which were derived from system message logs. For each component, the relevant messages were logged, such as speech recognition results, text-to-speech output, Interaction Pattern state changes as well as object recognition and reference resolution tasks. With these log data, a detailed reconstruction of the interaction can be achieved. A few measures have been annotated manually based on the video material, such as inappropriate robot utterances, or the correctness of the robot's answer on a test question. In total, we used 22 of these for evaluation (see table 3).

The **dialog efficiency** measures capture the rapidity of the interaction and include for example the duration of interaction, the number of user and robot utterances within a certain time unit, the mean length of user utterances, or the number of objects learning episodes within a certain time unit. The **dialog quality** measures address the smoothness of the interaction. We considered for example gaps, overlaps, repairs and label corrections. The **task success** measures concentrate on the outcome of the interaction with respect to object learning. Among others, we measured the proportion of successful reference resolution and object learning tasks, the proportion of correct robot answers on test questions, and the user's out-of-capability utterances.

4.4 Subjective Measures

In addition to the objective measures described above, we collected subjective measures based on a questionnaire the participants were asked to complete after the interaction with Flobi. The questionnaire consisted of 50 items, that we aggregated into seven category measures. The first four categories, **dialog efficiency**, **robustness**, **cooperativeness** and **usability**, refer to the interaction itself. These interaction-oriented items are roughly based on the evaluation of the COMIC dialog system [16], which we adapted for our specific scenario. The remaining three categories, **likeability**, **perceived intelligence** and **animacy** address the participants' impression of the robot. They were adopted from the standardized Godspeed questionnaire[1] [1]. In addition, the questionnaire included five single (summarizing) Likert-scale questions, targeting the overall impression of **ease**, **efficiency**, **clarity**, **pleasantness** and **understandability** of the interaction.

4.5 Computing Performance Functions

As PARADISE is not yet widely used in HRI, we will shortly summarize the procedure for performing it here, and place it in the appropriate context. For more background, see [15].

The model uses stepwise multiple linear regression to make predictions about subjective measures, like user satisfaction, based on several objective performance dimensions, like task success, dialogue quality, and dialogue efficiency. Multiple linear regression analysis is widely used to assess the degree to which an outcome or dependent variable is related to a set of predictor or independent variables. It specifies a model in which a dependent variable Y represents a linear combination of several parameters, including multiple independent variables, of the form

$$Y_i = (b_0 + b_1 x_{1i} + b_2 x_{2i} + \ldots + b_n x_{ni}) + \epsilon_i$$

The b_0 term represents a constant, whereas b_1 is the coefficient, or weight, of predictor x_1 and ϵ_i stands for the error term, or residual, of the equation.

In a stepwise approach, the best-fit predictors from a number of possible predictors are selected, by excluding predictors that do not contribute significantly to the outcome variable (Y). Taken together, the linear combination of the selected predictor variables and their computed coefficients can then be applied to make predictions of the dependent variable [13]. This technique also allows assessment of the relative contribution of each of the predictors towards the outcome variable (the β-value).

The multiple linear regression model can be referred to as general linear model with only one outcome or dependent variable. Along with path analysis, factor analysis and canonical correlation, it is a special case or subform of the more complex structural equation modeling [13]. Different from generalized linear models (not to be confused with general linear models), residuals in multiple linear regression are required to be normally distributed, whereas in generalized linear models they may belong to distributions from the exponential family [11].

[1] However, we skipped the categories **anthropomorphism** and **perceived safety**, as we considered them irrelevant for the scenario at hand.

5. RESULTS

In this section we present results of the objective and subjective measures, and on the relationship between them, described by several performance functions.

5.1 Objective Measures

Table 3 shows a sample of the results of the objective measures. On average, interactions lasted 11.07 minutes, during which 9.43 objects were taught successfully (0.91 per minute). In the mean, there were 3.55 user utterances and 10.91 robot utterances per minute. The mean value of gaps between utterances was 1.5 seconds, whereas utterances overlapped during 20% of the speaking time. 14% of the robot's utterances were repair utterances, and 6% were inappropriate, i.e. the robot misunderstood a "valid" user utterance. When participants demonstrated objects to the robot, reference resolution succeeded in 79% of all cases. From these, 78% could be learned, i.e. 62% of the overall demonstrated objects. The robot could answer 53% of the test questions, 55% of them correctly, i.e. 29% of the overall test questions. It must be admitted that while the participants were generally very accepting, this is not the final task performance we had expected from a state-of-the-art recognition algorithm as used here. While it is obviously the combination of all errors, it is still puzzling. We surmise that very variable lighting conditions may play a role, as well as a sub-optimal choice of training images at a time when the user's hand is still visible. We will have to investigate this further.

That said, we were pleased to see that the out-of-capability utterances could be reduced considerably in comparison with the previous iteration of the scenario (from 15% to 6%). This means that with our preparatory analysis of the WOz study and the resulting extensions of the system, the system capabilities were successfully adapted to the users' expectations. Table 4 breaks down the remaining out-of-capability utterances into subcategories. Attempts to correct the robot's answer on object check questions were a common error cause (which points out a deficiency in the interaction strategy), followed by meta comments on object organization ("let me put this one here"). Trying to demonstrate more than one object at a time was also common. Also, a significant portion of out-of-capability utterances referred to the robot's gaze direction ("where are you looking?"). Other categories include canceling an on-going object learning action, teaching object categories ("these are fruits"), asking the robot to repeat its utterances, asking it to proceed to the next label, or teaching colors.

Finally, we'd like to note the remarkably high standard deviation, indicating significant differences between the single interactions.

5.2 Subjective Measures

Table 5 shows the results for the subjective measures. All aspects were generally rated positively, with ratings of more than 4 (on a scale from 1 to 6) for ease of use, clarity, pleas-

Abbreviation	Measure	Min	Max	Mean	Stdev
Time	Duration of interaction (in minutes)	5.10	21.14	11.07	4.05
UU/min	User utterances per minute	3.16	7.90	3.55	2.20
RU/min	Robot utterances per minute	5.44	15.19	10.91	2.50
Obj	# Objects learned	2	19	9.43	3.92
Obj_{corr}	# Correctly learned objects	1	17	8.79	3.85
ObjDemo	# Object demonstrations	2	25	10.07	5.97
Obj/min	Objects learned per minute	0.2	1.6	0.91	0.39
$Steps_{ObjDemo}$	Average interaction steps for object demo	4.23	6.80	5.80	0.60
$Steps_{ObjReq}$	Average interaction steps for object request	0.00	4.00	3.41	1.21
Gaps	Length of global pauses (in seconds)	0.53	3.32	1.50	0.62
Overlaps	% of time UU and RU overlapping	0%	49%	20%	14%
Repair	% of RU dealing with repair	4%	38%	14%	8%
RU_{inapp}	% of inappropriate RU	1%	26%	6%	6%
NPattern	# Completed interaction patterns	18	69	43.50	13.65
Ref	Success rate of reference resolution	36%	100%	79%	15%
Learn	Success rate of object learning	45|35.6%	100|79%	78|62%	20|15%
Check	Success rate of object check	0%	100%	53%	29%
$Check_{correct}$	Correctness rate of object check	0|0%	100|53%	55|29%	38|20%
UU_{ooc}	Out-of-capability UU	0	28	7.64	6.80
UU_{ooc} %	Percentage of out-of-capability UU	0%	21%	6%	5%

Table 3: Objective measures. Blocks are: i) dialog efficiency, ii) dialog quality, and iii) task success. Percentages are rounded to full integers.

Category	Percentage
Correct object check	17%
Meta object-organization	17%
More than one object	13%
Gaze	12%
Cancel learning	5%
Categories	5%
Repeat	3%
Next	3%
Color	3%
Others (task-related)	20%
Others (non task-related)	2%

Table 4: Ratios for out-of-capability (rounded).

Measure	Min	Max	Mean	Stdev
Ease of use	1	6	4.25	1.43
Efficiency	1	6	3.57	1.45
Clarity	1	6	4.29	1.12
Pleasantness	1	6	4.04	1.43
Understandability	1	6	4.50	1.43
Efficiency	1.20	5.20	3.30	0.96
Usability	1.60	5.00	3.52	0.87
Robustness	1.60	6.00	4.23	1.07
Cooperativeness	2.33	6.00	4.05	0.88
Likeability	3.00	6.00	4.80	0.93
Perceived Intelligence	2.50	6.00	3.83	0.85
Animacy	1.86	6.00	3.55	0.93

Table 5: Results of subjective measures. Blocks are: i) Single-item measures, ii) aggregated measures.

antness, understandability, robustness, cooperativeness and likeability and more than three for efficiency (both as single-item and as aggregated measure), usability, perceived intelligence and animacy.

5.3 Performance Functions

As outlined before (cf. section 4.5), we computed predictor functions for the various subjective measures. The objective measures from section 4.3 were included as potential predictors.

5.3.1 Presentation

Table 6 presents the resulting performance functions for our subjective outcome variables. The obtained results were cross-validated using data splitting. The values of R^2 describe the amount of variance of the outcome variable that is accounted for by the model. These scores range from .171 (perceived intelligence) to .553 (usability) for our data, meaning for instance that 55.3% of the variance of the subjective measure "usability" is explained by the respective predictor function.

The factors in the predictor functions show which objective measures affected the subjective ratings, and their weights specify how much. However, the absolute numbers for the weights are not easily interpretable, because they depend on the range and unit of the factor in question. Therefore, the next-column ("β values") gives normalized weights, which represent the standardized b-coefficient for each predictor, which is independent from the range and unit of measurement of that specific objective predictor. Hence, they are directly comparable to each other, giving a better insight into the "true" weight or importance of the respective predictor. Positive coefficients indicate a positive relationship between predictor and outcome, negative coefficients stand for a negative relationship.

The significance column states whether each predictor of the particular function made a significant contribution to the subjective measure. Note that for two subjective measures no predictor function could be obtained, meaning that no single predictor or combination of multiple predictors could make a significant contribution to the outcome variable.

5.3.2 Interpretation

Most of the performance functions resulting from the PARADISE evaluation (shown in table 6) appear plausible, but some of them point out unexpected relationships between measures that help to identify deficiencies of the system. The performance functions generally exhibit high R^2 values, indicating that they do explain much of the variance in the data.

For instance, the number of objects learned per minute is a strong predictor for the **ease of use**, together with the overall number of demonstrated objects and the duration of interaction, suggesting that users who found interaction easy and were successful in teaching objects tended to have longer interactions with the system.

In contrast, **usability** as an aggregated measure from

Measure	Function	R^2		β	Significance
Ease of use	-0.30 + 4.33(Obj/min) + 1.15(ObjDemo) + 0.12(Time)	.471	Obj	.792	$p<.001$
			ObjDmo	.466	$p<.05$
			Time	.399	$p=.059$
Efficiency	No significant model				
Clarity	4.63 - 6.92(Repair) + 1.39(Obj/min)	.367	Repair	-.467	$p<.01$
			Obj	.336	$p<.05$
Pleasantness	3.58 + 0.18(Learn)	.184	Learn	.465	$p<.05$
Understandability	3.78 + 0.14(Ref)	.196	Ref	.443	$p<.05$
Efficiency	1.42 - 5.09(Repair) + 0.61($\text{Steps}_{\text{ObjDemo}}$) - 0.06($\text{UU}_{\text{ooc}}$) - 0.10($\text{Steps}_{\text{ObjReq}}$)	.441	Repair	-.403	$p<.05$
			$\text{Steps}_{\text{ObjDemo}}$	.394	$p<.05$
			UU_{ooc}	-.404	$p<.05$
			$\text{Steps}_{\text{ObjReq}}$	-.316	$p=.081$
Usability	4.03 - 0.22($\text{Steps}_{\text{ObjDemo}}$) + 2.39(Obj/min) - 0.21(UU/min) + 0.03(NPattern)	.553	$\text{Steps}_{\text{ObjDemo}}$	-.747	$p<.001$
			Obj	.736	$p<.001$
			UU/min	-.521	$p<.01$
			NPattern	.391	$p<.05$
Robustness	5.07 - 6.08(Repair)	.183	Repair	-.428	$p<.05$
Likeability	No significant model				
Cooperativeness	1.89 + 1.79(Obj/min) + 0.001(Gaps)	.446	Obj_{corr}	.698	$p<.001$
			Gaps	.418	$p<.05$
Perceived Intelligence	3.02 + 0.001(Gaps)	.171	Gaps	.171	$p<.05$
Animacy	3.67 + 1.09(Obj/min) - 0.118(UU/min)	.281	Obj	.443	$p<.05$
			UU/min	-.332	$p=0.074$

Table 6: Performance functions. Blocks are: i) Single-item measures, ii) aggregated measures. See section 5.3.1 for an explanation of the terms.

questions on control, predictability, concentration, clarity of when to speak is explained by a different function. Here, not only the number of objects learned per minute has an impact, but also how many interaction steps were required for learning, and the number of completed interaction patterns (which can be interpreted as a measure for the general interaction success, taking into account not only the task-related but also the social aspects). Somewhat surprisingly, the user utterances per minute had a negative impact on usability, and also on the **perceived intelligence**. Qualitative analysis revealed that for users who tend to talk rather fast and keep talking even during the robot's utterances, there is a risk of cumulating delays in the robot responses, as the user utterances are simply queued and processed one after another. This indicates the need for a more flexible turn-taking behavior that enables the robot to suspend or abort its own utterances, or to ignore user utterances if there is a newer one present.

Asking for **efficiency** directly did not yield a significant model. However, estimating **efficiency** as an aggregated measure (from questions on time required for object learning, clarity of referred objects, and general functionality of the system) suggests as predictors the number of repair utterances, interaction steps required for object demonstrations and object requests, as well as the out-of-capability utterances. The interaction steps for object demonstrations contribute positively, which may seem surprising at first sight. Looking at the interaction pattern used for object demonstrations, we realized that a high number of steps indicates multiple object corrections, while a low number of steps indicates that already the reference resolution fails. Thus, an explanation for the positive contribution of interaction steps might be that failures of reference resolution have a stronger negative effect on efficiency than correcting a misunderstood label, and are even more frustrating.

The factors that influence **clarity** of interaction are the number of repair utterances and the number of objects learned per minute, with the former being a major measure for interaction success and the latter for task success. The **pleasantness** is predicted (not very strongly, though) only by the overall number of objects learned, but our measures possibly miss some factors that are relevant for pleasantness.

Interestingly, we note a significant impact of reference resolution failures on the **understandability** score. We attribute this to the fact that for such failures, the robot only reports that it could not determine what the user referred to, not why. This provides very little information towards good error recovery, leaving users guessing. Qualitative analysis confirms this interpretation: several users explicitly asked the robot why it was not able to determine the referred object. This result points out again the frustrating effect of reference resolution failures, and the importance of more informative feedback in error conditions.

The **robustness** of the system as an aggregated measure (from questions on reliability and robustness of the system) is predicted by the proportion of repair utterances, but not by measures that specifically refer to object learning and recognition. This demonstrates that a robust interaction can alleviate deficiencies at task level.

Cooperativeness (aggregated from questions on the robot's readiness to interact, interest, attentiveness and autonomy) is affected not only by objects learned per minute, but also by gaps between utterances. While this may seem less obvious at first sight, it might be another evidence for the crucial role of turn-taking: longer gaps between utterances indicate a smooth turn-taking, and thus better speech recognition performance. It is probably for the same reason that gaps contribute to **perceived intelligence** as well.

6. DISCUSSION

We presented the design process and implementation of an integrated robot system for object learning within intuitive natural interaction. Further, we described its evaluation based on user study in which unexperienced users interact with the system and then rate different aspects of user satisfaction. To determine the relative importance of the various system functions for the subjective judgements, we related them to objective measures of the interaction through a PARADISE-style regression analysis.

A novel approach for such an evaluation has been the automatic definition and computation of objective system metrics based on, firstly, Interaction Patterns, which describe self-contained interaction episodes, and secondly, Task-State-Machines, which describe all system activities. This automated approach has allowed us to compute over 20 metrics, to perform fine-grained analysis. The resulting performance functions can significantly explain up to 55% (avg 33%) of variation in user satisfaction, which demonstrates that the metrics are meaningful.

Apart from providing information about the relative importance of the system's functionality, these functions also revealed unexpected relations between objective system measures and certain aspects of user satisfaction. For example, turn-taking was related to usability and perceived intelligence in unexpected ways. Moreover, it helped us to focus additional qualitative analysis on the most important of the many possible aspects. For example, it is a general tenet of error reporting to provide enough information for error recovery. However, this is not always achieved. In this case, the error reporting for reference resolution failures was apparently not helpful enough, which was clearly identified by the PARADISE analysis.

We also described our iterative strategy for designing the dialog strategy, and the user study confirmed intuitiveness of interaction, as indicated e.g., by a considerable decrease of out-of-capability utterances compared to previous iterations, despite little instruction. The PARADISE analysis also showed that error recovery in the dialog is generally seen as positive, if it leads to a successful completion. The remaining out-of-capability utterances, however, point out issues for improvement, e.g., the somewhat unclear gaze direction of the robot.

One aspect that we'd like to focus more on in future work are the large individual differences between users, both in task and in dialog success. There is no clear explanation in the data analyzed, but this certainly requires further investigation to assist users with (initial) difficulties. Similarly, while for some subjective metrics the performance functions explain about half the variation, which we consider a very good value, there are some metrics where only 20% of variation is explained, which is less helpful. A first step here could be to determine participants' initial expectations and predispositions and include them in the analysis.

7. ACKNOWLEDGMENTS

We gratefully acknowledge partial support by the German Aerospace Center (DLR) with funds from the Federal Ministry of Economics and Technology (BMBF) due to resolution 50RA1023 of the German Bundestag. We'd also like to thank Annika Rothert for assistance with preparing and executing the user studies. Last, but not least, we'd like to thank the anonymous reviewers for their exceptionally helpful commentary.

8. REFERENCES

[1] C. Bartneck, E. Croft, and D. Kulic. Measuring the anthropomorphism, animacy, likeability, perceived intelligence and perceived safety of robots. In *Metrics for Human-Robot Interaction Workshop*, 2008.

[2] P. R. Cohen and S. L. Oviatt. The role of voice input for human-machine communication. *National Academy of Sciences*, 92(22):9921–9927, 1995.

[3] N. Dethlefs, H. Cuayáhuitl, K.-F. Richter, E. Andonova, and J. Bateman. Evaluating Task Success in a Dialogue System for Indoor Navigation. In P. Lupkowski and M. Purve, editors, *Semantics and Pragmatics of Dialogue 14*, pages 143–146. Polish Society for Cognitive Science, 2010.

[4] T. Fong, C. Kunz, L. M. Hiatt, and M. Bugajska. The human-robot interaction operating system. In *1st ACM SIGCHI/SIGART conference on Human-robot interaction*, HRI'06, 2006.

[5] M. E. Foster, M. Giuliani, and A. Knoll. Comparing objective and subjective measures of usability in a human-robot dialogue system. In *Joint 47th Annual Meeting of the ACL and 4th International Conference on Natural Language Processing*, 2009.

[6] C. Lang, M. Hanheide, M. Lohse, H. Wersing, and G. Sagerer. Feedback interpretation based on facial expressions in human–robot interaction. In *International Symposium on Robot and Human Interactive Communication (RO-MAN)*. IEEE, 2009.

[7] I. Lütkebohle, F. Hegel, S. Schulz, M. Hackel, B. Wrede, S. Wachsmuth, and G. Sagerer. The Bielefeld Anthropomorphic Robot Head "Flobi". In *IEEE International Conference on Robotics and Automation*, 2010.

[8] I. Lütkebohle, J. Peltason, L. Schillingmann, C. Elbrechter, B. Wrede, S. Wachsmuth, and R. Haschke. The curious robot - structuring interactive robot learning. In *IEEE International Conference on Robotics and Automation*, ICRA '09, 2009.

[9] I. Lütkebohle, J. Peltason, L. Schillingmann, C. Elbrechter, S. Wachsmuth, B. Wrede, and R. Haschke. A mixed-initiative approach to interactive robot tutoring. In *Towards Service Robots for Everyday Environments*, STAR-76. Springer, 2011.

[10] I. Lütkebohle, R. Philippsen, V. Pradeep, E. Marder-Eppstein, and S. Wachsmuth. Generic middleware support for coordinating robot software components: The Task-State-Pattern. *Journal of Software Engineering for Robotics*, 1(2), 2011.

[11] J. A. Nelder and R. W. M. Wedderburn. Generalized Linear Models. *Journal of the Royal Statistical Society*, 135(3):370–384, 1972.

[12] J. Peltason and B. Wrede. Modeling human-robot interaction based on generic interaction patterns. In *AAAI Fall Symposium: Dialog with Robots*, 2010.

[13] B. G. Tabachnik and L. S. Fidell. *Using multivariate statistics*. Allyn & Bacon, Boston, fifth edition, 2007.

[14] M. Walker, C. Kamm, and D. Litman. Towards developing general models of usability with paradise. *Natural Language Engineering*, 6:363–377, 2000.

[15] M. A. Walker, D. J. Litman, C. A. Kamm, and A. Abella. PARADISE: a framework for evaluating spoken dialogue agents. In *EACL 8*, 1997.

[16] M. White, M. E. Foster, J. Oberl, and A. Brown. Using facial feedback to enhance turn-taking in a multimodal dialogue system. In *HCI International*, 2005.

Human-Agent-Robot Teamwork

Jeffrey M. Bradshaw, Ph.D.
Florida Institute for Human and Machine Cognition (IHMC)
Pensacola, FL USA 32502
+1 (850) 202-4462
jbradshaw@ihmc.us

Virginia Dignum, Ph.D.
Catholijn M. Jonker, Ph.D.
Delft University of Technology
Delft, The Netherlands
M.V.Dignum@tudelft.nl
C.M.jonker@tudelft.nl

Maarten Sierhuis, Ph.D.
Agent iSolutions
Moffett Field, CA
sierhuis@ejenta.com

Categories and Subject Descriptors
D.m [**MISCELLANEOUS**]: Software psychology

Keywords
HART, Human-Agent-Robot Teamwork

CONTRIBUTION AND BENEFIT
Teamwork has become a widely accepted metaphor for describing the nature of multi-robot and multi-agent cooperation. By virtue of teamwork models, team members attempt to manage general responsibilities and commitments to each other in a coherent fashion that both enhances performance and facilitates recovery when unanticipated problems arise. Whereas early research on teamwork focused mainly on interaction within groups of autonomous agents or robots, there is a growing interest in leveraging human participation effectively. Unlike autonomous systems designed primarily to take humans out of the loop, many important applications require people, agents, and robots to work together in close and relatively continuous interaction. For software agents and robots to participate in teamwork alongside people in carrying out complex real-world tasks, they must have some of the capabilities that enable natural and effective teamwork among groups of people. Just as important, developers of such systems need tools and methodologies to assure that such systems will work together reliably and safely, even when they have been designed independently.

The purpose of the HART workshop is to explore theories, methods, and tools in support of humans, agents and robots working together in teams. Position papers that combine findings from fields such as computer science, artificial intelligence, cognitive science, anthropology, social and organizational psychology, human-computer interaction to address the problem of HART are strongly encouraged. The workshop will formulate perspectives on the current state-of-the-art, identify key challenges and opportunities for future studies, and promote community-building among researchers and practitioners.

WORKSHOP OVERVIEW
The workshop will be structured around four two-hour sessions on themes relevant to HART. Each session will consist of presentations and questions on selected position papers, followed by a whole-group discussion of the current state-of-the-art and the key challenges and research opportunities relevant to the theme. During the final hour, the workshop organizers will facilitate a discussion to determine next steps. The workshop will be deemed a success when collaborative scientific projects for the coming year are defined, and publication venues are explored. For example, results from the most recent HART workshop (Lorentz Center, Leiden, The Netherlands, December 2010) will be reflected in a special issue of *IEEE Intelligent Systems* on HART that is slated to appear in January/February 2012.

HRI'12, March 5–8, 2012, Boston, Massachusetts, USA.
ACM 978-1-4503-1063-5/12/03.

Advances in Tactile Sensing and Touch based Human-Robot Interaction

Fulvio Mastrogiovanni, Giorgio Cannata
Department of Communication, Computers and System Sciences,
University of Genova, Via Opera Pia 13, 16145, Genova, Italy

Lorenzo Natale, Giorgio Metta
Italian Institute of Technology, via Morego 30, 16163, Genova, Italy

ABSTRACT
The problem of "providing robots with the sense of touch" is fundamental in order to develop the next generations of robots capable of interacting with humans in different contexts: in daily housekeeping activities, as working partners or as caregivers, just to name a few.

In a low-level perspective, through tactile sensing it is possible to measure or estimate physical properties of manipulated or touched objects, whereas feedback from tactile sensors may enable the detection and safe control of the interaction between the robot and objects or humans. In a high-level perspective, touch-based cognitive processes can be entailed by developing robot body self-awareness capabilities and by differentiating the "self" from the "external space", thereby opening new relevant problems in Robotics.

The objective of this Workshop is to present and discuss the most recent achievements in the area of tactile sensing starting from the technological aspects, up to the application problems where tactile feedback plays a fundamental role.

The Workshop will cover, but will not be limited, to the following three areas:

1. Technological aspects of robot artificial skin design and implementation including advanced transduction devices, large-scale sensing technologies, embedded electronics, system level solutions, etc.
2. Software and algorithmic aspects related to tactile data processing: software engineering, robot control, touch-based reactive behaviors, touch classification, object recognition, etc.
3. Cognitive issues related, but not limited, to skin-based behaviors and task level control, including: human-robot interaction, learning and assistive technologies, etc.

Categories and Subject Descriptors
[I.2.9] Robotics, Sensors.

General Terms
Measurement, Performance, Design, Experimentation.

HRI'12, March 5–8, 2012, Boston, Massachusetts, USA.
ACM 978-1-4503-1063-5/12/03.

Keywords
Tactile Sensing, Robotics, Human-Robot Interaction, Algorithms, Large-scale, Skin, Touch, Robot Control.

1. WORKSHOP OVERVIEW
The Workshop is full day.

Three sessions are (tentatively) planned addressing: (i) robot skin technologies, (ii) touch-based recognition and classification and (iii) touch-based human-robot interaction tasks.

Two invited talks are planned for each session and 8 to 12 contributed presentations are planned.

Talks will last 40 minutes for invited speakers, and 20 minutes (depending on the number of contributed papers) for contributed presentations.

A final round table discussion involving the participants to the Workshop is planned to highlight the future evolution of touch-based technologies and applications.

The Workshop is planned to start at 9:00am and to close at 6:00pm, with breaks at 10:20, 12:40 (lunch), 15:20.

2. TARGET AUDIENCE
The Workshop is expected to be a forum for researchers working in the areas of tactile sensing, robot grasping and interaction control, haptics, human-robot interaction, cognitive robotics etc.

3. POSITION PAPER REQUIREMENTS
Researchers presenting their results at the Workshop are requested to submit a 2 to 4 pages extended abstract, including figures, pictures, graphs and references. The extended abstracts will form the Workshop Proceedings.

4. ACKNOWLEDGMENTS
The Workshop organization has received funding from the European Community's Seventh Framework Programme (FP7/2007-2013) under grant agreement no. 231500 (ROBOSKIN).

Gaze in HRI: From Modeling to Communication

Frank Broz
University of Hertfordshire
Dept. of Computer Science
Hatfield, UK
f.broz@herts.ac.uk

Hagen Lehmann
University of Hertfordshire
Dept. of Computer Science
Hatfield, UK
h.lehmann@herts.ac.uk

Yukiko Nakano
Seikei University
Dept. of Computer and Information Science
Tokyo, Japan
y.nakano@st.seikei.ac.jp

Bilge Mutlu
University of Wisconsin-Madison
Dept. of Computer Science
Madison, WI, USA
bilge@cs.wisc.edu

ABSTRACT

The purpose of this half-day workshop is to explore the role of social gaze in human-robot interaction, both how to measure social gaze behavior by humans and how to implement it in robots that interact with them. Gaze directed at an interaction partner has become a subject of increased attention in human-robot interaction research. While traditional robotics research has focused work on robot gaze solely on the identification and manipulation of objects, researchers in HRI have come to recognize that gaze is a social behavior in addition to a way of sensing the world. This workshop will approach the problem of understanding the role of social gaze in human-robot interaction from the dual perspectives of investigating human-human gaze for design principles to apply to robots and of experimentally evaluating human-robot gaze interaction in order to assess how humans engage in gaze behavior with robots.

Computational modeling of human gaze behavior is useful for human-robot interaction in a number of different ways. Such models can enable a robot to perceive information about the state of the human in the interaction and adjust its behavior accordingly. Additionally, more human-like gaze behavior may make a person more comfortable and engaged during an interaction. It is known the gaze pattern of a social interaction partner has a huge impact on one's own interaction behavior. Therefore, the experimental verification of robot gaze policies is extremely important. Appropriate gaze behaviour is critical for establishing joint attention, which enables humans to engage in collaborative activities and gives structure to social interactions. There is still much to be learned about which properties of human-human gaze should be transferred to human-robot gaze and how to model human-robot gaze for autonomous robots. The goal of the workshop is to exchange ideas and develop and improve methodologies for this growing area of research.

Categories and Subject Descriptors

H.1.2 [**Information Systems**]: User/Machine Systems

HRI'12, March 5–8, 2012, Boston, Massachusetts, USA.
ACM 978-1-4503-1063-5/12/03.

Keywords

gaze, social behavior, human-robot interaction

Dr. Frank Broz is a post-doctoral researcher in the Adaptive Systems Research Group at the University of Hertfordshire. He currently studies mutual gaze in human conversational pairs to inform the development of robot gaze controllers as part of the ITALK project (Integration and Transfer of Action and Language Knowledge in Robots). He received his B.S. in computer science from Carnegie Mellon University and his Ph.D. in robotics from the CMU Robotics Institute. His dissertation topic was, "Planning for Human-Robot Interaction: Representing Time And Human Intention." He organized the AAAI spring symposium "It's All in the Timing: Representing and Reasoning About Time in Interactive Behavior" in 2010 and the ICDL-EpiRob special session on "Social Gaze: From Human-Human to Human-Robot Interaction" in 2011.

Dr. Hagen Lehmann is a post-doctoral researcher in the Department of Computer Science in the Adaptive Systems Research Group at the University of Hertfordshire, UK. He is currently involved in the iTalk and the AURORA project. Within the AURORA project he works on the development of a therapeutic robot to help children with autism to enhance their social skills. Within the iTALK project he works on mutual gaze in human conversational pairs to inform the development of robot gaze controllers and on child like language acquisition through the interaction with humans in the iCUB robot. He received his M.S. in Psychology from the Technical University Dresden and a Ph.D. in Computer Science from the University of Bath. He worked in different interdisciplinary research fields examining possible reasons for the evolution of social structure in non-human primates and the role of social dominance in this process. While studying at the Max-Planck Institute for Evolutionary Anthropology in Leipzig he worked with different primate species and human infants. The topic of this research was social gaze behaviour and its role in human social evolution. He co-organized the ICDL-EpiRob special session on "Social Gaze: From Human-Human to Human-Robot Interaction" in 2011.

Yukiko Nakano is an associate professor in the Department of Computer and Information Science at Seikei University, Japan, and leading the Intelligent User Interface Laboratory. She received her M.S. in Media Arts and Sciences from Massachusetts Institute of Technology and Ph.D. in Information Science and Technology from the University of Tokyo. In the research field of Conversational Agents, she has been working on establishing models of multimodal grounding and conversational engagement based on attentional information, and applying the models to communication between user and conversational agents. She organized "the 2nd Workshop on Eye Gaze in Intelligent Human Machine Interaction", at IUI 2011, and an AISB symposium on "Conversational Informatics for Supporting Social Intelligence & Interaction: Situational and environmental information enforcing involvement in conversation" in 2005.

Dr. Bilge Mutlu is an Assistant Professor of computer science, psychology, and industrial engineering at the University of Wisconsin-Madison where he directs a research program on designing social robots and their applications in education, health and wellbeing, and collaborative work. He received his Ph.D. from Carnegie Mellon University's Human-Computer Interaction Institute in 2009.

URL

http://homepages.stca.herts.ac.uk/~hri/

TIMES

Start: 1pm
End: 5pm

ROS and Rosbridge

Roboticists out of the loop

Christopher Crick
Brown University
Providence RI
chriscrick@cs.brown.edu

Graylin Jay
Brown University
Providence RI
tjay@cs.brown.edu

Sarah Osentoski
Bosch Research and
Technology Center
Palo Alto CA
sarah.osentoski@us.bosch.com

Odest Chadwicke Jenkins
Brown University
Providence RI
cjenkins@cs.brown.edu

ABSTRACT

The advent of ROS, the Robot Operating System, has finally made it possible to implement and use state-of-the-art navigation and manipulation algorithms on widely-available, inexpensive standard robot platforms. With the addition of the Rosbridge application programming interface, interface designers and applications programmers can create robot interfaces and behaviors without venturing into the specialized world of robotics engineers. This tutorial introduces ROS and Rosbridge, and shows how quickly and easily these tools can be used to design and conduct large-scale online HRI experiments, access algorithms for autonomous robot behavior, and leverage the huge ecosystem of general-purpose web-based and application-oriented software engineering for robotics and HRI research. Tutorial attendees will learn the basics of autonomous and teleoperated navigation and manipulation, as well as interface design for online interaction with robots. During the tutorial they will design and write their own remote presence application, as well as develop strategies for incorporating autonomy and dealing with data collection.

Categories and Subject Descriptors

I.2.9 [**Computing Methodologies**]: Artificial Intelligence—*Robotics*

General Terms

Design, Human Factors

Keywords

Robot middleware, protocols, HRI

1. SPEAKERS

- **Christopher Crick** is a postdoctoral research associate at Brown University. He has contributed ROS drivers for visual recognition and the AR Drone flying robot, as well as Rosbridge support for multi-robot systems. He has also developed and published Rosbridge-based HRI studies in human perception and robot learning.
- **Graylin Jay** is a researcher at Brown University and developer specializing in language tools. He is the lead developer for the base Rosbridge framework.
- **Sarah Osentoski** is a Research Engineer at the Bosch Research and Technology Center. She contributes to the PR2 Remote Lab project and the remote_lab stack that enables web visualization of ROS robots.
- **Odest Chadwicke Jenkins** is Associate Professor of Computer Science at Brown University. His recent work focuses on web-scale robot learning and casting robots as web services to broaden public accessibility to state-of-the-art robots.

2. MOTIVATION

Since its inception, robotics has been plagued by the fact that every lab had its own robot, unlike any other, which required the care and feeding of a platoon of graduate students well-versed in the intricacies and idiosyncracies of that machine's low-level control systems. HRI only intensifies this problem, owing to the fact that successful HRI research requires not only extensive expertise in robot programming and robotics algorithms, but in interface design and experimental procedures as well. This has made HRI an intensely collaborative field, with progress dependent upon close cooperation among people who contribute different skillsets.

It has lately, at last, become possible to decouple some of the elements of the long and intricate development cycle which has always been the rule in HRI, resulting in multiple points of failure or delay and difficulty in generalizing results beyond singleton experimental studies. ROS, Rosbridge and rosjs make it much easier to port studies from one robot to another, to design and instantly test robot interfaces, to leverage algorithms developed elsewhere, to engage large numbers of experimental subjects over standard web interfaces, and to apply arbitrary programming languages, tools and libraries to problems in robotics and HRI.

HRI'12, March 5–8, 2012, Boston, Massachusetts, USA.
ACM 978-1-4503-1063-5/12/03.

This tutorial will demonstrate how much more easily and quickly HRI experiments can be designed, tested and fielded on a large scale by using the tools we are developing.

3. AUDIENCE

This tutorial is designed to appeal to researchers in HRI who occupy several different niches. For the roboticists who program robot interactions and behaviors, ROS provides a huge array of pre-existing code implementing most common algorithms in robotics, and running on a large variety of platforms. Rosbridge makes it possible to pull that functionality into a user's existing software and systems, seamlessly and easily. However, the tutorial is also explicitly intended for HRI researchers who are not themselves roboticists. Rosbridge and rosjs make it easy to design interfaces and implement robot behaviors without bringing robotics experts into the loop.

A passing familiarity with web technologies (HTML and Javascript) and a desire to work with robots are the only prerequisites for the tutorial. The program will be of interest to the following:

- Anyone looking for an introduction to ROS
- Roboticists interested in using Rosbridge to apply ROS's algorithms and tools to their own preexisting software, robots, and development methodologies
- HRI researchers looking to shorten the development cycle between experimental design, implementation and subject testing
- Researchers interested in collecting large-scale HRI data over the internet
- Designers interested in producing functional robotic interfaces quickly and without requiring constant input from roboticist programmers

4. OVERVIEW

1. Introduction to robot middleware (approx 45 minutes)
    (a) Message protocols
    (b) ROS messages and services
    (c) Using and understanding messages
2. What ROS offers (approx 100 minutes, plus break)
    (a) Platform independence
    (b) Tools and algorithms
        i. Navigation
            A. Localization, path planning and AMCL
            B. Mapping and SLAM
        ii. Manipulation
            A. Object recognition and the Point Cloud Library
            B. Action servers
        iii. Control and state machines
    (c) Visualization and RVIZ
3. Afternoon: Rosbridge (approx 45 minutes) *Note: attendees who are already familiar with ROS can attend only the afternoon session, starting here*
    (a) Messaging protocol and JSON
    (b) Connecting to ROS from anywhere
    (c) Managing a robot through a Rosbridge connection
    (d) Authentication
    (e) Data collection
4. Building online interaction tools (approx 90 minutes, plus break)
    (a) Javascript, rosjs and wviz
    (b) Serving video with mjpeg_server
    (c) Rosjs and wviz features
        i. WebGL visualization
        ii. Teleoperation
        iii. Pick and place
        iv. Interactive markers
    (d) Teleoperation in fifteen lines
5. Designing large-scale experiments in HRI (approx 30 minutes)

5. REFERENCES AND LINKS

[4] lays out the design philosophy behind ROS. A huge amount of information, including tutorials, installation tools for various robots, and an exhaustive listing of available tools can be found at http://www.ros.org. Rosjs is first described in [3] and is described at http://www.ros.org/wiki/rosjs. Rosbridge and the philosophy behind its creation are found in [1]. Both Rosbridge and rosjs are available for download at http://code.google.com/p/brown-ros-pkg, and can also be used as part of a standard ROS installation. [2] describes HRI experiments conducted using the infrastructure and methodologies to be covered in the tutorial.

6. REFERENCES

[1] C. Crick, G. Jay, S. Osentoski, B. Pitzer, and O. C. Jenkins. Rosbridge: Ros for non-ros users. In *Proceedings of the 15th International Symposium on Robotics Research*, 2011.

[2] C. Crick, S. Osentoski, G. Jay, and O. C. Jenkins. Human and robot perception in large-scale learning from demonstration. In *Proceedings of the 6th ACM/IEEE Conference on Human-Robot Interaction*, 2011.

[3] S. Osentoski, G. Jay, C. Crick, B. Pitzer, C. DuHadaway, and O. C. Jenkins. Robots as web services: reproducible experimentation and application development using rosjs. In *Proceedings of the 2011 IEEE International Conference on Robotics and Automation*, 2011.

[4] M. Quigley, B. Gerkey, K. Conley, J. Faust, T. Foote, J. Leibs, E. Berger, R. Wheeler, and A. Ng. Ros: an open-source robot operating system. In *Proceedings of the Open-Source Software Workshop of the International Conference on Robotics and Automation*, 2009.

Cognitive Science and Socio-Cognitive Theory for the HRI Practitioner

Jeffrey M. Bradshaw, Ph.D.
Florida Institute for Human and Machine Cognition (IHMC)
40 South Alcaniz Street
Pensacola, FL USA 32502
+1 (850) 202-4462
jbradshaw@ihmc.us

J. Chris Forsythe, Ph.D.
Sandia National Laboratories
Albuquerque, NM USA

+1 (505) 844-5720
jcforsy@sandia.gov

ABSTRACT

This tutorial provides a synopsis of key findings and theoretical advances from cognitive science and socio-cognitive theory, with examples of how the results of this research can be applied to the design of human-robotic systems. Topics covered will run the gamut from basic cognitive science (e.g., perception, attention, learning and memory, information processing, multi-tasking, conscious awareness, individual differences) to socio-cognitive issues (e.g., theories of social interaction, dynamic functional allocation, mixed-initiative interaction, human-agent-robot teamwork, coactive design, theory of organizations). Additionally, the tutorial will address new technologies that attempt to leverage the current state of theory (e.g., neuroergonomics, brain-machine interfaces, detection of cognitive states, robotic prostheses and orthotics, cognitive and sensory prostheses). Throughout the tutorial, the presenters will give descriptions and demonstrations of working systems that exemplify the principles being taught. Separately, the presenters have given highly-successful tutorials on relevant subjects at workshops and conferences such as CHI and HCI International, as well as in a variety of industrial and government settings. In this tutorial, they propose to bring together their experience to bear on issues of specific interest to the HRI community.

Categories and Subject Descriptors

D.m [**Miscellaneous**]: Software psychology

Keywords

Cognitive Science, Human-Agent-Robot Teamwork, Human-Robotic Interaction, Socio-Cognitive Theory

Contribution and Benefit Statement

One of the strengths of the HRI community is that its practitioners come from a wide variety of professional disciplines. One consequence of this fact, however, is that many more practitioners are adept in the use of existing HRI techniques in their applications than are proficient in the principles of human cognition and social behavior that allow them to develop and rigorously evaluate new innovations. Moreover, even those who are well-schooled in these foundational principles may find the gap between theory and practice so great as to make it difficult to connect their knowledge to their research programmes. This tutorial provides a synopsis of key findings and theoretical advances from cognitive science and socio-cognitive theory, with examples of how the results of this research can be applied to the design of human-robotic systems.

HRI'12, March 5–8, 2012, Boston, Massachusetts, USA.
ACM 978-1-4503-1063-5/12/03.

Biographical Sketches

Jeffrey M. Bradshaw (Ph.D., Cognitive Science, University of Washington) is a Senior Research Scientist at the Florida Institute for Human and Machine Cognition (IHMC). Formerly, he led research groups at The Boeing Company and the Fred Hutchinson Cancer Research Center. Jeff has been a Fulbright Senior Scholar at EURISCO in Toulouse, France; an Honorary Visiting Researcher at the University of Edinburgh; a visiting professor at the Institut Cognitique at the University of Bordeaux; is former chair of ACM SIGART; and former chair of the RIACS Science Council for NASA Ames Research Center. He served as a member of the National Research Council (NRC) Committee on Emerging Cognitive Neuroscience Research in the Next Two Decades, as a scientific advisor to the Japanese NEC Technology Paradigm Shifts initiative, and as an advisor to the HCI and Visualization program at the German National AI Research Center (DFKI). He currently serves as a member of the Board on Global Science and Technology for the National Academies and as an external advisory board member of the Cognitive Science and Technology Program at Sandia National Laboratories. Jeff served for over a decade on the Board of Directors of the International Foundation for Autonomous Agents and Multiagent Systems. In 2011, he received the Web Intelligence Consortium Outstanding Contributions Award. He has many years of editorial board experience on professional journals and in the organization of international conferences and workshops. He serves as co-editor of the Human-Centered Computing Department for *IEEE Intelligent Systems.* Among many other publications, he edited the books *Knowledge Acquisition as a Modeling Activity* (with Ken Ford, Wiley, 1993) and *Software Agents* (AAAI Press/The MIT Press, 1997). Jeff has participated in NASA Blue Sky Study Groups for the "Human-Centered Vision of Mars Exploration" and for the "Small Pressurized Rover." From 2002-2006, his KAoS framework was used as part of a NASA series of annual two-week field tests of human-robot teams performing simulated planetary surface exploration at the Mars Desert Research Station. Jeff was sponsored by DHS to undertake detailed simulation studies of the use of human-robot teams to secure facilities at Port Everglades. He has also led the ONR-sponsored NAIMT and Coordinated Operations projects where a team of humans and heterogeneous robots performed field exercises aimed at port reconnaissance, and robot-assisted detection and apprehension of intruders.

Chris Forsythe (Ph.D., Experimental Psychology; M.S., Biopsychology, University of Memphis) is a Distinguished Member of the Technical Staff for Sandia's Cognitive Science and Applications Department. Chris helped found Sandia National Laboratories' cognitive systems program in 1999, including playing a pivotal role in activities leading to Sandia establishing a

Grand Challenge in cognition and establishing Cognitive Science and Technology as a lab-wide Focus Area. Chris is co-inventor and patent holder for Sandia's computational framework for modeling human cognition and initiated the development of automated knowledge capture technologies. Chris currently manages a portfolio of projects that encompasses research addressing individual differences in the neurophysiology of human performance, advanced training technologies development, and human-machine systems integration,. Chris serves as Sandia interface to government sponsors that include Office of Naval Research, Naval Air Systems Command and DARPA. Chris has served on government advisory panels for DARPA, Office of Naval Research and DTRA. In 2006, he received the DARPA IPTO Outstanding Service award and in 2009 was given the Federal Laboratories Consortium Mid-Continent Notable Technology Development Award. Chris has over forty peer-reviewed publications and has edited two books in the fields of cognitive psychology, human factors and intelligent systems. He has won three best paper awards from the Human-Computer Interaction International Conference (2009, 2011).

References

[1] Bradshaw, J. M., Dignum, V., Jonker, C., and Sierhuis, M. (eds.) Special issue on Human-Agent-Robot Teamwork. *IEEE Intelligent Systems*, Jan/Feb 2012, in preparation.

[2] Bradshaw, J. M., Paul Feltovich, and Matthew Johnson. Human-Agent Interaction. In *Handbook of Human-Machine Interaction*, edited by Guy Boy. Ashgate, 2011, pp. 283–302.

[3] Bradshaw, Jeffrey M., ed. *Software Agents*. Cambridge, MA: The AAAI Press/The MIT Press, 1997.

[4] Bradshaw, Jeffrey M., Paul Feltovich, Hyuckchul Jung, Shri Kulkarni, William Taysom, and Andrzej Uszok. Dimensions of adjustable autonomy and mixed-initiative interaction. In *Agents and Computational Autonomy: Potential, Risks, and Solutions*. edited by Matthias Nickles, Michael Rovatsos and Gerhard Weiss, 17-39. Berlin, Germany: Springer-Verlag, 2004.

[5] Dixon, K., Hagermann, K., Basilico, J., Forsythe, C. Siegfried, S. Schrauf, M. & Kincses, W. (2009). Improved team performance using EEG- and context-based cognitive-state classifications for a vehicle crew. *Proceedings of the International Conference on Human-Computer Interaction,* San Diego, CA.

[6] Ford, K.M., and J.M. Bradshaw, eds. *Knowledge Acquisition as Modeling*. New York City, NY: John Wiley, 1993.

[7] Forsythe, C. (2005). Cognitive modeling for augmented cognition. In D.D. Schmorrow (ed.) *Foundations of Augmented Cognition*, Lawrence Earlbaum, Mahwah NJ, 546-547.

[8] Forsythe, C. & Giordano, J. (2011). The need for a U.S. strategy addressing brain science and its ramifications for national security. *Synesis*.

[9] Forsythe, C., Bernard, M., Xavier, P., Abbott, R., Spped, A. & Brannon, N. (2005). Using psychologically plausible operator models to enhance operator performance. *Proceedings of the Human Factors and Ergonomics Society 51st Annual Meeting*, Denver, CO.

[10] Forsythe, C., Goldsmith, T.E. & Bernard, M.L. (2005). *Cognitive Systems: Cognitive Models in Systems Design*, Lawrence Erlbaum, Mahwah, NJ.

[11] Giordano, J., Forsythe, C. & Olds, J. (2010). Response to Marks: Neuroscience, Neurotechnology and National Security: Toward an Ethic of Responsible Action and Need for Precautionary Process, *American Journal of Bioethics*.

[12] Green, C., Griffin, D., Blascovich, J., Bradshaw, *et al*. *Emerging Cognitive Neuroscience Technologies and Applications*. National Research Council. Washington, DC: The National Academies Press, 2008.

[13] Johnson, Matthew, Jeffrey M. Bradshaw, Paul J. Feltovich, Robert R. Hoffman, Catholijn Jonker, Birna van Riemsdijk, and Maarten Sierhuis. Beyond Cooperative Robotics: The Central Role of Interdependence in Coactive Design. *IEEE Intelligent Systems*, May/June 2011 (vol. 26 iss. 3), pp. 81-88.

[14] Magliano, J.P., Skowronski, J.J., Britt, M.A., Guss, D. & Forsythe, C. (2008). What do you know: How perceivers use goals to make inferences about others. *Cognition*, 106, 594-632.

[15] Trumbo, M, Stevens, S., Hendrickson, S., Abbott, R. Haass, M. & Forsythe, C. (2011). Individual differences and the science of human performance. *Proceedings of the International Conference on Human-Computer Interaction, Orlando, FL.*

Author Index

www.ingramcontent.com/pod-product-compliance
Ingram Content Group UK Ltd.
Pitfield, Milton Keynes, MK11 3LW, UK
UKHW051415290926
15471UKWH00009B/1613